4th International Conference on Natural Resources and Technology

IOP Conference Series: Earth and Environmental Science
Volume 1115

Sumatera Utara, Indonesia
29-30 August 2022

ISBN: 978-1-7138-6454-7
ISSN: 1755-1307

Printed from e-media with permission by:

Curran Associates, Inc.
57 Morehouse Lane
Red Hook, NY 12571

Some format issues inherent in the e-media version may also appear in this print version.

Institute of Physics
Dirac House, Temple Back
Bristol BS1 6BE UK

Phone: 44 1 17 929 7481
Fax: 44 1 17 920 0979

techtracking@iop.org

Additional copies of this publication are available from:

Curran Associates, Inc.
57 Morehouse Lane
Red Hook, NY 12571 USA
Phone: 845-758-0400
Fax: 845-758-2633
Email: curran@proceedings.com
Web: www.proceedings.com

TABLE OF CONTENTS

PREFACE

The 4th International Conference on Natural Resources and Technology (ICONART) 2022 was held on August 29-30, 2022 with a theme of *Natural Resources and Technology for Achieving Sustainable Development Goal through Academic, Industry*, and *Community* and a subtheme of *Resilience and Innovation Research on Sustainable Natural Resources and Technology Post-Covid 19*. The present conference was initiated by Center of Excellence for Mangrove (*Pusat Unggulan Iptek Mangrove*) was collaborated with Faculty of Forestry, Universitas Sumatera Utara, JSPS Core to Core Program, e-Asia Joint Research Program which was fully supported by Center of Excellence for Mangrove, Universitas Sumatera Utara, Medan, Indonesia and Ministry of Education, Culture, Research, and Technology, Republic of Indonesia.

The conference covered six major topics in the related fields such as (i) Natural science and natural product, (ii) Natural resource technology, (iii) Information systems of tropical resources, (iv) Tropical biodiversity, (v) Food science and food technology, and (vi) Ethnobotany and ethnozoology. Due to the increasing cases of Covid-19 in Indonesia and other countries which pose a significant impact on our social gathering, work culture and travel restrictions, the committee has agreed on conducting this current conference was hybrid (conducted at Hotel Grandhika, Medan) and a Zoom meeting. Opening ceremony, speech of committee was done offline in hotel. All Parallel Presentation 1 in the morning session was carried out offline at hotel with six rooms. Parallel presentation 2 (afternoon session) was performed virtually through Zoom meeting with six room breakouts. Two panel session of key note speakers in morning and afternoon were performed hybrid (Dr Sasmito and Dr Nuryawan came to present in hotel).

Notablen keynote speakers were invited to present the current issue and challenge in their respective field of studies. Special thanks to Prof. Dr. Tadashi Kajita, University of the Ryukyus, Japan; Dr. Sigit D. Sasmito from National Universiy of Singapore, Singapore; Prof. Petar Antove from University of Forestry, Sofia-Bulgaria; Lubos Kristak, Ph.D, University of Zloven, Slovakia; and Arif Nuryawan, Ph.D from Universitas Sumatera Utara, Indonesia.

We received 141 papers in total through the online submission system for abstract system (https://ocs.usu.ac.id/ICONART/iconart2022) and Morressier as Conference submission management system for full papers from several countries in which some of them were later accepted or declined for presentation. Selected papers (129 articles) will be published in the next volume or issue in IOP Conference Series: Earth and Environmental Sciences by IOP Publishing (Bristol, UK). We are hoping that the publication will benefit our participants to access their subject of interest and also for the advances of publication by

UniversitasSumatera Utara as an academic institute.

Finally, the organizing committee would like to express our most profound appreciation to Directorate of Research, Technology and Community Service, Ministry Education, Culture, Research, and Technology, Republic of Indonesia, Rector of Universitas Sumatera Utara, Vice-Rectors, Dean and Vice Deans of Faculty of Forestry, Head of the Center of Excellence for Mangrove, Universitas Sumatera Utara, Toba Pulp Lestari, Medan, North Sumatra, Indoneia for their support and cooperation during the conference. Special gratitude was also directed to all organizing committee of the 4th ICONART 2022 for their active contributions, hard work, and cooperative teamwork in the preparation of the conference. Until then, we are looking forward to your participation again in the next event of 5th ICONART 2023.

Mohammad Basyuni
Chairman of Organizing Committee
4th ICONART 2022

ORGANIZING COMMITTEE

1. Honorary Chair	
USU Rector	: Dr. Muryanto Amin, S.Sos, M.Si
Vice Rector 3 of USU	: Prof. Dr Popy Anjelisa Zaitun Hasibuan, S.Si, M.Si, Apt
Dean of Faculty of Forestry	: Dr. Rudi Hartono, S.Hut., M.Si
Vice Dean 2 of Faculty of Forestry	: Pindi Patana, S.Hut, M.Sc
Vice Dean 3 of Faculty of Forestry	: Dr. Agus Purwoko, S.Hut, M.Si
Head of Center of Excellence for Mangrove	: Prof. Mohammad Basyuni, S.Hut, M.Si, Ph.D
2. General Chair	
Chair	: Prof. Mohammad Basyuni, S.Hut, M.Si, Ph.D
Co-Chair	: Dr. Apri Heri Iswanto, S.Hut, M.Si
3. Program/Technical	
Chair	: Dr. Arif Nuryawan, S.Hut, M.Si
Co-Chair	: Dr. Iwan Risnasari, M.Si, M.Si
Members	: 1. Dr. Masitta, S.Si, M.Si 2. Desrita, S.Pi, M.Si 3. Julia Syahriani, S.Pi, M.Si
4. Liaison	
Chair	: Dr. Arida Susilowati, S.Hut, M.Si
Co-Chair	: Dr. Ir. Iriany, M.Si
Members	: 1. Dr.Bejo Slamet, S.Hut, M.Si 2. Dr. Ir. Yunasfi, M.Si 3. Astrid Fauzia Dewinta, S.St.Pi, M.Si
5. Local Committee	
Chief	: Dr. Nurdin Sulistiyono, S.Hut, M.Si
Co-Chief	: Dr. Jumirah, Apt., M.Kes
Secretary	: 1. Era Yusraini, STP, M.Si 2. Ahmad Baiquni Rangkuti, S.Hut,M.Si
Treasurer	: Dr.Erni Jumilawaty, S.Si, M.Si

Co-Treasurer	: Dr. Deni Elfiati, SP, MP
Publicity	: 1. Dr. Etti Sartina Siregar, S.SI, M.Si 2. Dr. Taslim 3. Ir. Netti Herlina, MT 4. Khairunnisa S.Pi, M.Si
6. Members	: Ridwanti Batubara, S.Hut, MP Ipanna Enggar Susetya, S.Pi, M.Si Syafriandy, ST Rafika Suryani, SH Ari Wahyu Risma Putri, A.Md Anggia Murni, SS, M.Si Jefri Fauzul Mikzan, S.Kom

Peer Review Statement

All papers published in this volume have been reviewed through processes administered by the Editors. Reviews were conducted by expert referees to the professional and scientific standards expected of a proceedings journal published by IOP Publishing Publishing.

- **Type of peer review:** Double Anonymous
- **Conference submission management system:** Morressier
- **Number of submissions received:** 141
- **Number of submissions sent for review:** 129
- **Number of submissions accepted:** 104
- **Acceptance Rate (Submissions Accepted / Submissions Received × 100):** 73.8
- **Average number of reviews per paper:** 2
- **Total number of reviewers involved:** 22
- **Contact person for queries:**
  Name: Prof. Mohammad Basyuni, Ph.D
  Email: m.basyuni@usu.ac.id
  Affiliation: Department of Forestry, Faculty of Forestry, Universitas Sumatera Utara, Medan, North Sumatera, Indonesia

Shorebird community and diversity in Bagan Serdang Beach, North Sumatra

E Jumilawaty[1,2*], A L A Nasution[1] and S Siregar[1]

[1]Department of Biology, Faculty of Mathematics and Natural Sciences, Universitas Sumatera Utara, Medan, North Sumatra 20155, Indonesia
[2]Center of Excellence for Mangrove, Universitas Sumatera Utara, Medan 20155, Indonesia

[*]E-mail: erni1@usu.ac.id

Abstract. The occurrence of shorebirds is strongly influenced by environmental factors, especially food availability. Shorebirds are closely linked to their habitat conditions, so they seek out feeding sites according to their life cycle. Coastal conditions that have changed due to human activities may affect the assemblage of shorebird community. The aim of this study is to determine the diversity of shorebird species and the factors that influence them. Bird observations were made for three days using the point count method while the number of individuals was estimated by using "block method". The results revealed 17 shorebird species from the families, Charadriidae, Scolopacidae and Sternidae. The dominant species were *Charadrius mongolus*, *Pluvialis fulva* and *Sterna hirundo*. The number of species in the morning observations was higher than in the afternoon. The diversity index for migratory birds ranges from 1.16 to 1.96, which is relatively low. This study showed that mudflats in Bagan Serdang Beach, Pantai Labu District is occupied by shorebirds in this region.

1. Introduction

Pantai Labu is a district located in Deli Serdang Regency, North Sumatra. The coastal area of Pantai Labu is an area that has experienced human exploitation for various activities, namely: coastal tourism, aquaculture, settlements and intensive sand mining [1]. The existence of these activities might have a detrimental effect by polluting the coastline. The mix of land and marine life found in mangrove ecosystems gives them distinctive traits and a plethora of potential biological resources. Mangrove forest habitats are distinctive because they are covered in a layer of mud and fine silt, which causes the soil to be highly salinized and experience tidal inundation cycles. Mangrove roots protect the coastline and serve to settle the muds and sands [2].

A group of aquatic birds called the shorebird population reside in the coastal environments, which include several migratory species. Shorebirds migrate from the northern to the southern hemisphere in order to escape the threat of winter, in which during the session, there is a limited supply of food in their home range [3]. Large populations of individual shorebird dwell together in groups, and when they venture out in search of food, they form groups to defend themselves.

Shorebirds occupy the mangrove forest as a resting spot, low tide mud flats as a feeding location, and wetland regions as breeding grounds [4]. The assembly of birds during meals also serves to harass prey that is buried in the mudflats [5].

Since the community primarily uses the Pantai Labu region as a tourist destination, it may indirectly or directly contribute to degradation in habitat quality, damage to mangroves, contaminated shorelines owing to inconsiderate waste disposal, and impact to the ground utilized by shorebirds for foraging. The present study deals with the inventory of shorebird in Bagan Serdang Beach, as a part of Pantai Labu to assess their population and diversity in the region.

2. Methodology

This study was conducted in February 2022 on the eastern coastline of North Sumatra with a study site located at Bagan Serdang Beach (98°50'53" E, 3°42'08" N), Pantai Labu District, Deli Serdang Regency.

Figure 1. Map of research location

Observations were carried out by survey on three consecutive days (07:00–11:00 AM, 03:00–06:00 PM) using point-count and block method to count the bird individuals and visual identification of shorebird species with the aid of monoculars and binoculars. The identification process followed some list of Indonesian shorebirds and a scientific guidebook [6,7]. The species abundance, Shannon's diversity index (H), and Shannon's equitability index (E_H) of shorebirds are calculated as follow [8]:

$$Species\ abundance\ (Pi) = ni/N$$

$$Shannon's\ diversity\ index\ (H) = -\sum\left[\left(\frac{ni}{N}\right) ln\left(\frac{ni}{N}\right)\right]$$

$$Shannon's\ equitability\ index\ (E_H) = H/ln\ S$$

Where, ni is the number of individual of i-species. N is the total number of all individuals. ln is the natural log. S is the total number of unique species.

3. Results and Discussion

This study recorded 17 shorebird species from Bagan Serdang Beach within the order of Charadriiformes and three families (Charadriidae, Scolopacidae, Sternidae) as presented in Table 1. Table 1 shows that there were no notably different species present during the course of the three observation days. The first day recorded 15 species or its peak, followed by 14 on the second and 13 on the third. The weather (wind, temperature and rainfall) and tides have an impact on changes in the presence of species and the quantity of species. The findings revealed that the family Scolopacidae has

47% species than Charadriidae, (29 %) and Sternidae (24 %) (Figure 1). The duration of the tidal period, which creates a stretch of mud for foraging and other activities, has a significant impact on the number of species in this family of shorebirds. In addition , the dominance of Scolopacidae in the region is confirmed as revealed from a study in Percut Sei Tuan [9] .

The presence or daily fluctuation of the number of species and individual birds in an area is influenced by environmental factors such as wind , rain , tides , food availability and disturbances by both visitors and fishermen . The change in tidal time is the most decisive thing for the presence of The difference in tides on the day of shorebirds , this is closely related to the time of observation .. observation greatly affects observations in terms of transportation in water areas that can only be reached at high tides [10]. Water level, water level fluctuations, vegetation, salinity, terrain, food type, ease of acquiring food, wetland size, and wetland connectivity are some of the habitat factors that have an impact on the existence of shorebirds in a particular area [11].

The dominant species obtained from observation data for three days at Bagan Serdang Beach are *Charadrius mongolus* and *Pluvialis fulva* (Table 1) where these species are always present and their numbers always dominate in the morning and evening compared to other birds. This is because these birds always make communities or groups when looking for food, where if *C. mongolus* flies then the

P. fulva group will follow.

Table 1. List of shorebirds in Bagan Serdang Beach, Pantai Labu District.

Family	No	Scientific names	Local names	Observation (days)			IUCN Status
				1	2	3	
Charadriidae	1.	*Charadrius dubius*	*Cerek Kalung-Kecil*	8	1	3	LC
	2.	*C. leschenaultii*	*Cerek Pasir-Besar*	1	-	-	LC
	3.	*C. mongolus*	*Cerek Pasir-Mongolia*	452	116	152	LC
	4.	*Pluvialis fulva*	*Cerek Kemrnyut*	75	44	42	LC
	5.	*Vanellus cinereus*	*Trulek Kelabu*	-	-	4	LC
Scolopacidae	6.	*Actitis hypoleucos*	*Trinil Pantai*	6	4	7	LC
	7.	*Limnodromus semipalmatus*	*Trinil Lumpur-Asia*	5	33	28	NT
	8.	*Limosa limosa*	*Biru-Laut Ekor-hitam*	-	20	-	NT
	9.	*Numenius arquata*	*Gajahan Besar*	2	1	8	NT
	10.	*N. madagascariencis*	*Gajahan Timur*	5	5	5	EN
	11.	*N. phaeopus*	*Gajahan Penggala*	1	2	5	LC
	12.	*Tringa cinereus*	*Trinil Bedaran*	6	-	28	LC
	13.	*T. stagnatilis*	*Trinil Rawa*	2	4	9	LC
Sternidae	14.	*Chlidonias hybrida*	*Dara-Laut Kumis*	26	-	21	LC
	15.	*Sterna nilotica*	*Dara-Laut Tiram*	4	3	7	LC
	16.	*S. bengalensis*	*Dara-Laut Bengala*	7	5	-	LC
	17.	*S. hirundo*	*Dara-Laut Biasa*	45	34	67	LC
Total				645	272	386	

Note: (-) = Absent, IUCN = The International Union for Conservation of Nature's Red List of Threatened Species, LC = Least concern, NT = Near threatened, EN = Endangered.

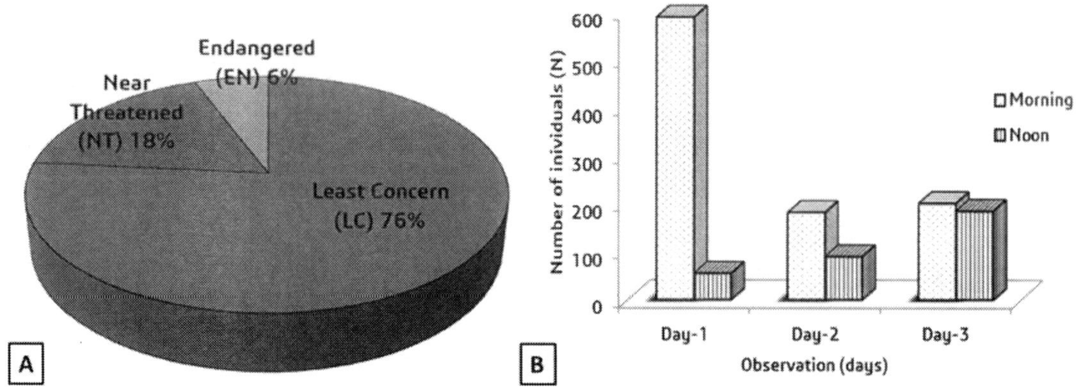

Figure 2. Percentage of shorebird population based on family (A) and number of individuals during morning and noon observations of shorebirds (B) in Bagan Serdang Beach.

Even if the area of mudflat is increasing broader due to the impact of the decline of mangroves, birds, one of the high-level consumers, may struggle to locate food. This will have an impact on the variety of shorebird species. The number of species and individual birds will be impacted by land conversion (land conversion), which decreases the extent of mangrove forests. If a community has a large number of different species, it is considered to have high species diversity [12].

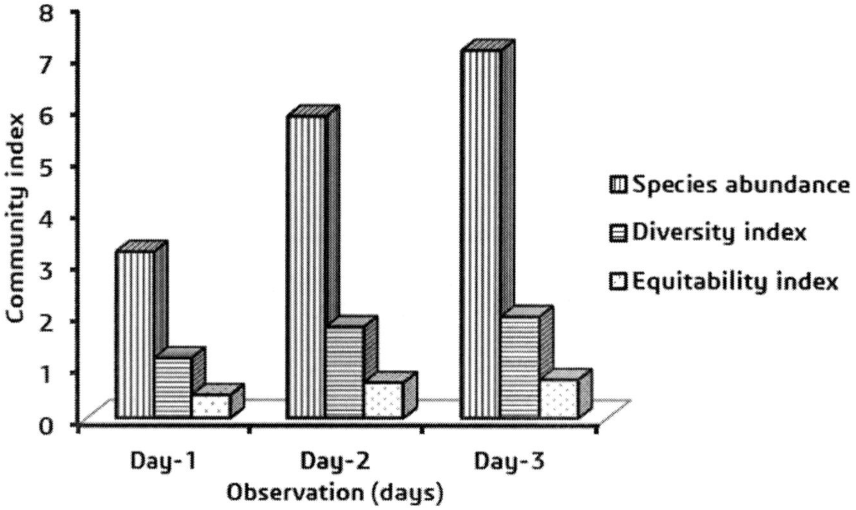

Figure 3. Species abundance, Shannon's diversity index (*H*), and Shannon's equitability index (E_H) of shorebirds in Bagan Serdang Beach.

Some bird species are not significantly impacted by vegetation conditions. For instance, certain bird species, such as *Actitis hypoleucos* and *Numenius phaeopus* are reliant on wetlands. Although indirectly, many sorts of birds also need vegetation as a location to nest, the presence of various birds is significantly determined by the availability of food. Therefore, the mangrove is the most beneficial environment for birds in terms of the supply of food and nesting sites [13]. According to IUCN 3.1, there are three categories of threat for waterbirds in Bagan Serdang: Endangered (EN) for 1 species (*N. madagascariencis*), Near-threatened (NT) for 3 species, and Least concern (LC) for 13 species (Table 1). According to this information, even if this area is a part of the Important Bird Areas (*Daerah Penting bagi Burung*/ DPB) for the Sumatran region, it is still important to protect the mudflat area for the existence of shorebirds.

Along with the increasing observation time (day-3), the ecological index of shorebirds in this research area also increases. The highest abundance of shorebirds was obtained at 7.11 ind/total, while the Shannon diversity index was 1.96 and the Shannon uniformity index was 0.74 (Figure 2). Based on these results, it can be seen that it is still possible to increase the observation or survey period even though the initial data obtained is still poor in the context of biodiversity. The diversity index value is strongly influenced by species richness and individual abundance. External factors that greatly affect species richness and individual abundance in Pantai Labu District including dispersal, emigration and immigration of waterbirds every month, area size, landscape and distance between habitats and disturbances. Species richness is a characteristic of diversity which is influenced by external factors (immigration, migration and dispersal) and local factors (competition, predation) [14].

If there are enough food resources, shorebirds will migrate to the region to feed and reside there to supply their requirements. Variations in habitat conditions, plant species variety, feed availability (source), diversity and composition of vegetation, security and comfort conditions, and many other variables affect the quantity and species of avifauna species including human disturbances. Fish, shrimps, crabs, and other aquatic organisms that can be found in wetlands, are beneficial food sources for shorebirds [15]. These subjects should be well investigated in the future to draw a correlation factor to the existence of shorebirds in the coastline area of Bagan Serdang Beach.

4. Conclusion

Community assemblage of migratory bird in Bagan Serdang Beach, Pantai Labu District consisted of 17 species classified into 3 families namely Scolopacidae (8 species), Charadriidae (5 species) and Sternidae (4 species). The results of 3-day observation revealed the higher bird count in the morning than in the noon. Based on the Shannon's diversity index (*H*), the biodiversity of shorebird in the area is still regarded as poor while further investigations are needed to record the possibility of increasing species and individuals during the migratory period in the future.

References

[1] Doloksaribu D C N, Barus T A and Sebayang K 2020 *IOP Conf Ser Earth Environ Sci* **454** 012086

[2] Piersma T and Lindstrom A 2004 *Ibis* **146** 61

[3] Bakker W, Ens B J, Dokter A, van der Kolk H J, Rappoldt K, van de Pol M, Troost K, van der Veer H W, Bijleveld A I, van der Meer J, Oosterbeek K, Jongejans E and Allen A M 2021 *Estuar Coast Shelf Sci* **259** 107458

[4] Putra C A, Farajallah D P and Mulyani Y A 2017 *Hayati J Biosci* **24** 16

[5] Bocher P, Robin F, Kojadinovic J, Delaporte P, Rousseau P, Dupuy C and Bustamante P 2014 *J Sea Res* **92** 115

[6] Sukmantoro W, Irham M, Novarino W, Hasudungan F, Kemp N and Muchtar M 2007 *Daftar Burung Indonesia No. 2* (Bogor: Indonesian Ornithologist's Union)

[7] Van Balen B, Mackinnon J, Phillipps K, Rahardjaningtrah W, Adikerana A, Martodihardjo P, Supardiyono E K, Sumadipura S and Kartikasari A 1992 *Burung-burung di Sumatera, Jawa, Bali, dan Kalimantan* (*Termasuk Sabah, Sarawak, dan Brunei Darussalam*) (Bogor: Puslitbang Biologi LIPI)

[8] Krebs C J 1972 *Ecology: The Experimental Analysis of Distribution and Abundance* (New York: Harper & Row)
[9] Jumilawaty E 2021 *IOP Conf Ser Earth Environ Sci* **713** 012041
[10] Jumilawaty E, Mardiastuti A, Prasetyo L B and Mulyani Y A 2016 *Media Konservasi* **16** 108
[11] Ma Z, Cai Y, Li B and Chen J 2010 *Wetlands* **30** 15
[12] Huang S C, Shih S S, Ho Y S, Chen C P and Hsieh H L 2012 *Restor Ecol* **20** 76
[13] Hamzati N S and Aunurohim 2013 *Jurnal Sains dan Seni Pomits* **2** 2337
[14] Lisovski S, Gosbell K, Minton C and Klaassen M 2021 *J Animal Ecol* **90** 2005
[15] Quammen M L 1984 *Ecology* **65** 529

Identification of the *PLAG1* polymorphisms in Peranakan Ongole cattle population

M R Akmalputra[1], S Sukaryo[1], Y Yanti[1], J Riyanto[1], T A Bramastya[1], S D Volkandari[2], P Sudrajad[3] and M Cahyadi[1*]

[1]Department of Animal Science, Faculty of Agriculture, Universitas Sebelas Maret, Surakarta 57126, Indonesia
[2]Research Center for Food Technology and Processing, Research Organization for Agriculture and Food, National Research and Innovation Agency, Gunung Kidul 55861, Indonesia
[3]Research Center for Animal Husbandry, Research Organization for Agriculture and Food, National Research and Innovation Agency, Bogor 16915, Indonesia

[*]E-mail: mcahyadi@staff.uns.ac.id

Abstract. Genetic factor plays a very important role in determining of cattle performance. A gene which is reported to be controlling cattle growth and body size is pleomorphic adenoma gene 1 (*PLAG1*). Therefore, the goal of this study was to identify *PLAG1* polymorphisms in PO cattle population. A total of 43 unproductive PO cows from the slaughterhouse at Kebumen were used in this study. Blood samples were collected individually when the cattle were slaughtered. Furthermore, DNA genome was harvested from blood samples and used as DNA template for polymerase chain reaction (PCR) targeting g.48308 C>T, g.32212 (19 bp indel) and g.45233 T>C polymorphisms. g.48308 C>T and g.45233 T>C SNPs were genotyped using PCR-restriction fragment length polymorphism (PCR-RFLP) technique and a 19 bp indel of *PLAG1* was directly genotyped by observing PCR products. Both allele and genotype frequencies were calculated to test Hardy-Weinberg equilibrium (HWE) status. The results showed that g.48308 C>T SNP and 19 bp indel varied. Three genotypes of the g.48308 C>T SNP namely CC, CT and TT were found in this study and two genotypes (DD and DI) of 19 bp indel were also identified. Moreover, heterozygous CT genotype frequency of g.48308 C>T SNP was 0.51 and T allele frequency was 0.60. In term of 19 bp indel, DD genotype frequency was 0.88 and D allele frequency was 0.94. In conclusion, two polymorphisms of the *PLAG1* were identified in PO cattle population.

1. Introduction

Indonesia is a tropical country which is rich in biodiversity including ruminant livestock such as cattle. Cattle breeds which are widely developed in Indonesia mostly come from *Bos javanicus* (Bali cattle) and *Bos indicus* such as Peranakan Ongole (PO) cattle breed. It is a beef and draught cattle breed which is favorable to be raised by smallholder farmers in Indonesia due to its advantage traits such as adaptive to tropical condition. It is very popular in some parts of Java Island due to its superior performances [1]. The

PO cattle is historically a cross breed between Indonesian local cattle (Jawa cattle) and Ongole cattle imported from India since 1904. It is furthermore developed from generation to generation by farmers in Central and East Java [2].

Phenotypically, PO cattle has white to grey coat color, black hair at the end of the tail and around the eyes, large body size, a hump above the shoulder and dewlap, long horns and prick-ears [3]. The PO and Brahman cattle breeds are sometimes very difficult to be differentiated due to similar phenotypic appearances. However, PO cattle has bigger body measurements than Brahman cattle in terms of withers and waist heights [4]. Mature PO bulls and cows can reach 600 and 450 kg, respectively. Moreover, body weight gain of PO cattle is ranged from 0.4 to 0.8 kg per day [5]. Generally, factors affecting cattle growth including PO are environment, genetics and their interactions. Genetic factor plays a key role in determining cattle performance and a gene associated with cattle growth and body measurement traits is pleomorphic adenoma gene 1 (*PLAG1*).

PLAG1 is a sub family members of pleomorphic adenoma gene encoding a group of zinc-finger proteins along with *PLAGL1* and *PLAGL2* [6]. These three genes have homolog structure but different functions [7]. Recently, Cahyadi *et al* [8] identified a single nucleotide polymorphism (SNP) of the *PLAG1* and its significant association with body length of Bali cattle. The same SNP, g.48308 C>T SNP, is also associated with body height and chest girth of five cattle breeds in China [9]. In addition, a 19 bp indel of PLAG1 is also significantly associated with growth-related traits in various cattle breeds in China [10]. Previous study reported polymorphism of *PLAG1* in PO cattle population is very limited, especially PO cattle from Kebumen which is known as the source of superior PO calves. Therefore, the study goal was to find out polymorphisms of the *PLAG1* fragment in PO cattle originated from Kebumen.

2. Materials and methods

2.1 Ethical approval

Ethical Clearance Commission of National Research and Innovation Agency (BRIN) approved all animal procedures related to samples of PO cattle in this study with decree No. 34/Klirens/III/2021.

2.2 Collection of blood samples and DNA extraction

This study used 43 blood samples of unproductive PO cows taken from slaughterhouse around Kebumen, Jawa Tengah. Their average age was 5.2 years old and in good health. Blood samples were collected from the jugular vein when slaughtered as much as 3 mL using an 18G vacutainer needle containing the anticoagulant ethylene diamine tetra acetic acid (EDTA). The genomic DNA was extracted from blood samples using the High Salt Method according to Montgomery and Sise [11]. The quality of genomic DNA was checked using 0.8% agarose gels under the UV light.

2.3 Amplification of PLAG1 fragments

Amplification of *PLAG1* fragments were conducted to identify three polymorphisms of the *PLAG1*, namely g.48308 C>T, 19 bp indel and g.45233 T>C. A total volume of 25 μL PCR reaction consisting 12.5 μL My Taq HS Red Mix 2X (Bioline, UK), 9.5 μL nuclease-free water, 1 μL each primer (Table 1) and 1 μL DNA templates was performed using a MiniAmp Thermal cycler machine (Thermo Fisher Scientific, Singapore). The PCR program was started by pre-denaturing at 95°C for 5 minutes; and followed by denaturation at 95°C for 30 seconds, annealing for 30 seconds and extension at 72°C for 30 seconds for 35 cycles. The reaction was completed by final extension at 72°C for 10 minutes. Annealing temperature for each primer pair is presented in Table 1. Moreover, PCR products were examined with 2% agarose gels stained by ethidium bromide (EtBr) at a voltage of 110 volts for 35 minutes using submarine electrophoresis system and visualized with Glite UV Gel Doc System (Pacific Image Electronic Co., Ltd., New Taipei City, Taiwan).

Table 1. The list of primers to amplify *PLAG1* fragments

Polymorphism	Primer (5' to 3')	Ta (°C)	Amplicon (bp)	RE	Ref.
g.48308 C>T	F: gcgcgtatcagtcaaggacat R: cctttgcctgttgctttccc	58/45	628	*Sac*II	[9]
19 bp indel	F: tccgaacaacaggtgagggagaaat R: ccacttcaggggtgctctaggtttg	60/30	142/123	-	[10]
g.45233 T>C	F: gcgtgaaggagaagaagcac R: gatcgggttatagggagggc	56/45	767	*Bcl*I	[8]

Ta is annealing temperature; RE is restriction enzyme; Ref. is reference

2.4 Genotyping of the PLAG1

Two polymorphisms, g.48308 C>T and g.45233 T>C SNPs, were genotyped using RFLP. To detect g.48308 C>T SNP, 628 bp of PCR product was digested by *Sac*II. The reaction contained 20 µL PCR product, 1 µL *Sac*II and 2 µL green buffer and then incubated at 37°C for an hour. In term of g.45233 T>C SNP, *Bcl*I was used to digest 767 bp of PCR products. The same reaction set up with the first SNP was applied. Furthermore, the results of genotyping were checked using 2% agarose gel stained by EtBr under the UV light. A 100 bp marker ladder was used as standard size of DNA bands. On the other hand, 19 bp indel variation was checked by observing PCR product sizes.

2.5 Allele and genotype frequencies

Allele and genotypes frequencies of PLAG1 polymorphisms were analyzed using formula as follows:

$$H = p^2 + 2pq + q^2 \quad (1)$$

where H is expected value, p is 1st allele frequency and q is 2nd allele frequency [12]. The Chi- square test (X^2) was used to test whether the *PLAG1* alleles and genotypes were following Hardy-Weinberg equilibrium law (HWE). The formula is as follows:

$$X^2 = \sum(O - E)^2/E \quad (2)$$

The X^2 value is Chi-square, O is observed genotype value and E is expected value. The alpha was set at 5%. Allele and genotype frequencies are in HWE if the probability value is less than 3.84.

3. Results and discussion

3.1 DNA extraction

Genomic DNA isolated from blood samples of PO cattle are qualitatively presented in Figure 1. Genomic DNA has been successfully harvested. The quality of genomic DNA in this study was also good which is indicated by the existence of clear and bright DNA bands. The level of clarity of DNA bands is one of the determining factors for the success of DNA extraction. Puspitaningrum *et al.* [13] stated that the quality of extracted DNA can be proven by the clear and bright visualization of DNA bands under the UV light. Good quality genomic DNA is necessary as a key factor in PCR process. In addition, due to large size of genomic DNA, their DNA bands in agarose should be in the top of the agarose gel, on the same line with the largest

IOP Conf. Series: Earth and Environmental Science 1115 (2022) 012002 doi:10.1088/1755-1315/1115/1/012002

size of lambda marker. In accordance to opinion of Prayitno and Nuryadani [14], smaller DNA fragments move faster than larger DNA fragments and vice versa.

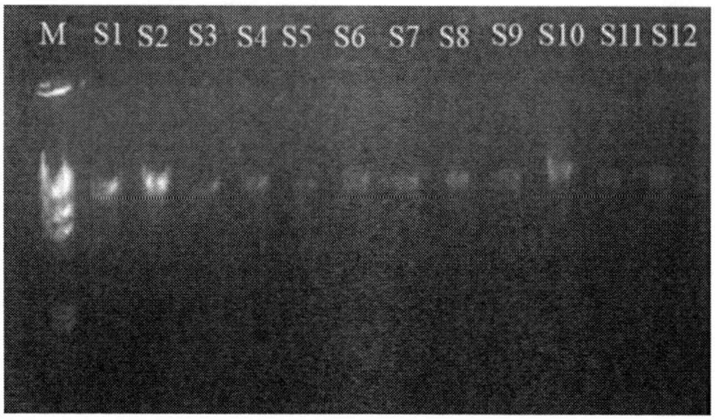

Figure 1. Genomic DNA of PO cattle. M is lambda marker ladder, S1-S12 are genomic DNA

3.2 Amplification of the PLAG1 fragments

Three fragments of *PLAG1* have been amplified in this study (Figure 2). They were 628 bp to identify g.48308 C>T SNP (Figure 2a), 123/142 to identify 19 bp indel (Figure 2b) and 767 bp to identify g.45233 T>C SNP (Figure 2c), respectively.

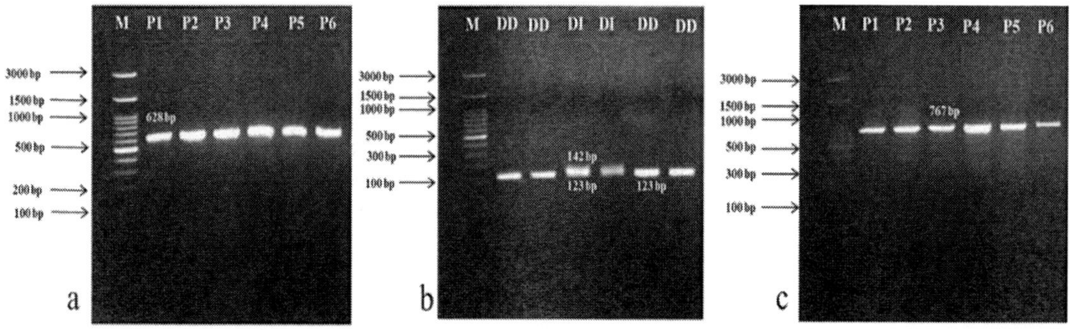

Figure 2. The PCR products of three fragments of the *PLAG1*. a. PCR product for identification of g.48308 C>T SNP; b. PCR product for identification of 19 bp indel; c. PCR product for identification of g.45233 T>C SNP.

The PCR products of this study were specific which is indicated by the single DNA band, especially for g.48308 C>T and g.45233 T>C SNPs. In term of 19 bp indel located in intron 3, there were two specific DNA bands sized 123 and or 142 bp as reported previously by Xu *et al* [10]. The g.48308 C>T SNP is a polymorphism located in 3' untranslated region (3'UTR) and the size of PCR products to identify this SNP is in accordance to previous study reported by Zhong *et al.* [9]. Additionally, g.45233 T>C SNP located in exon 4 of *PLAG1* was successfully amplified and the PCR product size was similar with Cahyadi *et al* [8]. Specific DNA bands is produced only when all reaction components working simultaneously and in optimum condition [15].

3.3 Genotyping of the PLAG1 polymorphisms

Three polymorphisms have been analyzed in this study. The g.48308 C>T SNP and 19 bp indel of *PLAG1* were polymorphic. Three genotypes of g.48308 C>T SNP, CC, CT and TT, have been identified. The homozygous CC genotype is indicated by two DNA bands, 522 and 106 bp; heterozygous CT genotype is indicated by 628, 522 and 106 bp; and homozygous TT is indicated by 628 bp of DNA bands (Figure 3a). In term of 19 bp indel, two genotypes, homozygous DD and heterozygous DI, have been identified (Figure 2b). The DD genotype is indicated by 123 bp and DI genotype is indicated by 142 and 123 bp. The DI genotype was not clearly showed because 19 bp differences could not be separated perfectly using agarose gels. On the other hand, g.45233 T>C SNP of *PLAG1* was monomorphic (Figure 3b). The RFLP method is an effective approach to genotype individual sample which is cheap and still powerful to be used in wide-range of research area. The restriction enzyme is able to cut nucleotide sequence when it recognizes restriction site within the DNA fragment [16]. Different pattern of DNA bands showed different individual genotype.

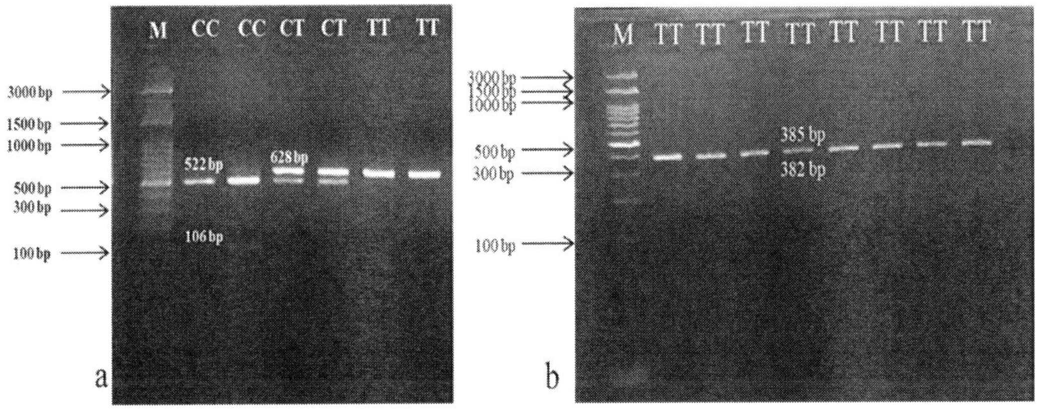

Figure 3. Genotyping of *PLAG1*. a. Genotyping of g.48308 C>T SNP using *Sac*II restriction enzyme; b. Genotyping of g.45233 T>C SNP using *Bcl*I

Table 2. Genotype and allele frequencies of the *PLAG1* polymorphisms

Polymorphism	Genotype frequency			Allel Frequency		X^2	*P-value*
g.48308 C>T	TT	TC	CC	T	C		
	0.35	0.51	0.14	0.60	0.40	0,211	0,9
19 bp Indel	DD	DI	II	D	I		
	0.88	0.12	0.00	0.94	0.06	0,164	0,686
g.45233 T>C	TT	TC	CC	T	C		
	1.00	0	0	1.00	0	-	-

3.4 Genotype and allele frequencies

The heterozygous CT genotype was 0.51 which is found as the highest genotype frequency of g.48308 C>T SNP with C allele as a minor allele frequency was 0.4 (Table 2). Additionally, homozygous DD genotype of 19 bp indel was the most genotype found in PO cattle population with minor-allele frequency (I allele) was 0.06. The HWE calculation showed that the P-value more than 5% which means the genotype distributions for g.48308 C>T SNP and 19 bp indel in PO cattle population was in equilibrium. This study used PO cattle from small-holder farmers located around Kebumen Regency. Therefore, the possibilities of non-random mating, selection and significant immigration of population ware rarely occurred due to traditional raising management. Nussbaum [17] explain that a population follows HWE law when the population used in the study is large, random mating, non-significant migration, no new mutation and no selection against any particular genotype.

4. Conclusion

Two of three polymorphisms in *PLAG1* observed in this study, namely g.48308 C>T SNP and 19 bp indel, were polymorphic. On the other hand, g.45233 T>C SNP did not vary (monomorphic). Further study should be conducted to evaluate association between these polymorphisms with growth traits in PO cattle.

Acknowledgment

This study was partially funded by the Institute for Research and Community Service of Universitas Sebelas Maret through research group scheme with contract number: 254/UN27.22/PT.01.03/2022.

References

[1] Direktorat Jenderal Peternakan dan Kesehatan Hewan 2013 *Statistik Peternakan dan Kesehatan Hewan Livestock and Animal Health Statistics 2013* (Jakarta: Direktorat Jenderal Peternakan dan Kesehatan Hewan Kementrian Pertanian)
[2] Febri H I I 2018 *Karakteristik Peforman Sapi Peranakan Ongole (PO) Kecamatan Palang dan Tuban Kabupaten Tuban* (Malang: Universitas Brawijaya)
[3] Standar Nasional Indonesia 2015 *Bibit Sapi Potong-Bagian 7: Pernakan Ongole* (Jakarta: Badan Standarisasi Nasional)
[4] Bramastya T A, Sukaryo S, Dhiaurridho M I, Riyanto J, Volkandari S D, Sudrajad P and Cahyadi M 2022 *IOP: Conf. Ser. Earth Environ. Sci.* **1001** 012015
[5] Santosa K and Andoko A 2012 *Bisnis Penggemukan Sapi* (Indonesia: Agromedia)
[6] Hou J, Qu K, Jia P, Hanif Q, Zhang J, Chen N, Huang B and Lei C 2020 *Anim. Genet.* **51** 87
[7] He P, Zheng L, Wang J and Zhang Y 2017 *Appl. Biol.* G17121736
[8] Cahyadi M, Sukaryo S, Dhiaurridho M I, Bramastya T A, Yanti Y, Riyanto J and Sudrajad P 2022 *Vet. World* **15** 782
[9] Zhong J L, W J, Xu, Wang J, Wen Y F, Niu H, Zheng L, He H, Peng K, He P, Shi S Y, Huang Y Q, Lei C Z, Dang R H, Lan X Y, Qi X L, Chen H and Huang Y Z 2019 *Gene* 689
[10] Xu W, He H, Zheng L, Xu J W, Lei C Z, Zhang G M, Dang R H, Niu H and Qi X L 2018 *Gene* 675
[11] Montgomery G W and Sise J 1990 *J. Agric. Res.* **33** 441
[12] Nei M and Kumar S 2000 *Molecular Evaluation and Phylogenetic* (New York: Oxford University Press Inc)
[13] Puspitaningrum R, Adhiyanto C and Solihin 2018 *Genetika Molekuler dan Aplikasinya* (Yogyakarta: Penerbit DeePublish)
[14] Prayitno E and Nuryandani E 2011 *Nusantara Bioscience* 3 1
[15] Cahyadi M, Wibowo T, Pramono A and Abdurrahman Z H 2020 *Food Sci. Anim. Resour.* **40** 628
[16] Gupta S and Yel L 2020 *Middleton's Allergy: Principles and Practice: Molecular Biology and Genetic Engineering* (Amsterdam: Elsevier)
[17] Nussbaum R L 2016 *Thompson & Thompson Genetics in Medicine: Genetic Variation in Populations* (Philadelphia: Elsevier)

Exploration of high-quality wood-producing trees on the Universitas Sumatera Utara (USU) Medan campus's green space

A H Iswanto[1,4,5*], I M Ginting[2], H H Rachmat[3], L Hakim[1,4], T Sucipto[1,4]

[1]Department of Forest Product, Faculty of Forestry, Universitas Sumatera Utara. Kampus USU Padang Bulan, Medan 20155, Indonesia
[2]Department of Silviculture, Faculty of Forestry, Universitas Sumatera Utara. Kampus USU Padang Bulan, Medan 20155, Indonesia
[3]Forest Research and Development Center, Ministry of Environtment and Forestry. Jl Raya Gunung Batu Bogor. 16118. Indonesia
[4]JATI-Sumatran Forestry Analysis Study Center, Universitas Sumatera Utara. Kampus USU Padang Bulan, Medan 20155, Indonesia
[5]Center of Excellence for Mangrove, Universitas Sumatera Utara, Medan 20155, Indonesia

E-mail: *apri@usu.ac.id

Abstract. Urban green space has various benefits, including as a conservation area for different plant species. As a green open space in Medan, the Universitas Sumatera Utara (USU) campus plays an important role. It performs a dual function as a means of conserving various types of trees that have multiple benefits, including as a producer of wood. Until now, information about the diversity of wood-producing tree species in the green space of the USU campus has not been found. This research was conducted to fill in the gaps in information regarding the diversity of wood-producing species in the green space of the USU campus, Medan. This research was conducted using a comprehensive tree inventory method by investigating the types, uses, and levels of wood strength found in various growing trees. The results showed that 19 tree species had durable class I to II (very durable and durable), 83 tree species had strength class I-III, 38 did not have durable class information, and 36 did not have strength class information. Two tree species that have an endangered status are *Shorea parvifolia* and *Pterocarpus indicus*.

1. Introduction

Green space from two words, namely: space and green. According to Warren [1], green space is defined as land covered with some form of vegetation. Green space is a piece of vegetated land allocated for recreational purposes or a park surrounding an area for residential or industrial use. One of the essential benefits of urban forests for human health is that they can be a solution for reducing air pollution and interception [2-3]. Trees can remove a large amount of air pollution. A study in the United States shows that trees in urban areas remove 711,000 metric tons of air pollution every year [4]. In addition, trees are also helpful as carbon stores where carbon is a significant contributor to climate change [5]. The more mature the tree, the more carbon it will store in its biomass [6].

Reducing the amount of carbon released into the atmosphere is one of the main ways to limit the impact of climate change [7].

In addition to storing carbon, urban forests can also help supply affordable energy for people who need it [8]. The energy in question can be in the form of wood produced by trees that grow as plant communities in green open spaces. Wood is an environmentally friendly material and is a renewable resource, so it plays a significant role in the world [9]. As a construction material that can benefit economically [10], wood has a high strength-to-weight ratio and is the most widely used and efficient structure. Parameters that determine the requirements of wood for structural use include the level of strength, density, stiffness, and wood density [11]. These parameters are commonly used worldwide to determine the strength value of wood [12].

Maintaining and enhancing biodiversity is necessary for sustainable urban sustainability [8]. There is a growing concern that if biodiversity loss continues, ecosystems will inevitably lose much of their ability to function effectively [13]. As a green open space in the city of Medan, the USU campus plays an important role and performs a dual function of conserving various types of trees with multiple benefits. Previous research found that there are 121 native and exotic tree species [14] that have multiple functions, both as food producers [15] and as producers of medicines for the surrounding community [15]. However, until now, information regarding wood-producing tree species in the green space of the USU campus has not been found. Therefore, this research was conducted to fill in the gaps in information related to the diversity of wood producing tree species in the green space of the USU campus, Medan.

2. Research Methods

This study was carried out at the USU campus in Padang Bulan Village, Medan Tuntungan District, Medan City, North Sumatra Province. The USU campus is at an altitude of ± 56 meters above sea level and has an average temperature of 27.0 – 28.9 °C and humidity of 78.97 - 85.57% [16].

This research was conducted using a comprehensive tree inventory method by investigating the types, uses, and levels of wood strength found in various growing trees. Visual observations were also carried out to determine the physical characteristics of the observed wood. A total of 121 tree species [15] were identified to obtain data in the form of individual numbers, conservation status, general use, and wood strength level, which were obtained based on secondary data from various related publications. Usability information is typically obtained through the use of the surrounding community and the community.

3. Result and Discussion

3.1. Number of species

Trees are one of the largest sources of wood commonly used to produce relatively large amounts of wood. According to [17], the need for wood has increased along with the increase in population. Wood is used for many purposes because it has a unique structural material for various purposes ranging from pulp and paper, construction, building, acoustics, medicine, and so on [18]. Based on this, wood sources can be obtained from various trees with different characteristics.

Based on the research results, it is known that the number of individual trees of each type found in the green space of the USU campus tends to vary. The highest number of individuals was found in *Swiethenia macrophylla*, totaling 1347. The smallest number of individuals, each as much as one, is found in several species including: *Annona squamosa*, *Stelechocarpus burahol*, *Gymnostroma sumatranum*, *Shorea parvifolia*, *Shorea laevis*, *Shorea leprosula*, *Elaeocarpus grandiflora*, *Pithecelobium dulce* variegate, *Peltophorum pterocarpum*, *Guazulma ulmifolia*, *Ficus hispida*, *Moringa oleifera*, *Psidium guajava* Red. Malaysia, *Baccaurea deflexa*, *Premna corymbosa*, and *Chrysophyllum cainito*.

The Tri Dharma educational forest within the USU campus area has influenced the high number of *Swiethenia macrophylla* individuals, with *Swiethenia macrophylla* accounting for nearly 90% of the species composition in the artificial forest. In addition, *Swiethenia macrophylla* is considered a suitable species to be planted in various green spaces and yards because it has strong stems, lush canopy, and adaptability. This also increases the number of *Swiethenia macrophylla* individuals in the green space of the USU campus as a whole. Meanwhile, some tree species with the smallest number are thought to be caused by random distribution and the lack of seed dispersers. Individual trees that grow accidentally cause random distribution, which in certain locations are used to increase development as an effort to expand buildings or locations with other designations. This causes trees that grow in these locations vulnerable to be destroyed in the development process which aims to clear the land. In addition, the limited number of animals in the study area caused the distribution of seeds to be disrupted. The low number of animals is caused by a crowded environment and tends to produce noise which makes animals reluctant to stop or stay. This is consistent with statements [19-20] that state that human presence is considered dangerous to animals and areas visited by humans may be avoided by seed dispersers.

Figure 1. Tri dharma educational forest

3.2. *Wood quality*

Over the past 20 years, wood construction has become necessary because of its unique intrinsic properties as a building material and its sustainability [21]. Wood is a biological material susceptible to various biotic agents such as termite attacks and fungal degradation. These biotic agents are a significant threat to the service life of wood [22]. Termites are insects that most often eat dead plant material from wood, leaf litter, animal waste, soil, and cellulose. Wood with various types has a different resistance to termite attack, influenced by factors such as moisture content, hardness, lignin content, and resin [23]. The factor of durability and strength of wood is a characteristic that is an essential requirement in its purchase and use [24]. Besides being important in its use, the durability of wood is also important for global carbon balance because wood biodegradation is essential for carbon and nutrient cycling in forest ecosystems [25].

According to the research findings, only 19 of the 121 tree species are classified as having wood of quality I to II (very durable and durable). Meanwhile, 43 species are known to have wood with moderate to very low durability (III to V), and 38 other species have not had test results related to the level of durability of their wood (Table 1). Various factors influence the durability of wood. Environmental factors in the tropics are thought to encourage different tropical species to adapt to be more resistant to multiple destructive pathogens. According to [26], woody species under tropical conditions generally deteriorate more rapidly than in areas with a temperate or boreal climate. Based on this, it is important to understand the broader resilience of tropical species. There are various types of wood-destroying pathogens, however, fungi cause the most damage [27]. Based on these circumstances, different tree species are suspected of producing extractive substances. According to [28-29], extractive substances are very important for trees as a defense mechanism against microbes. Furthermore, physicochemical properties such as location, growth rate, age, wood portion, and microenvironment affect wood durability [30-31], resulting in different durability classes for wood.

Table 1. Index of tree species on the USU campus, along with their strength and durability classes

No	Species	Family	Total	Utility	Durable class[1]	Strength class[2]	Conservation status	Source (1.2)
1	*Mangifera odorata*	Anacardiaceae	2	Light construction, building construction, furniture, crafts, and other services are available.	V	II	Not on the Redlist	[24]
2	*Mangifera foetida*	Anacardiaceae	2	Light construction, building construction, furniture, crafts, and other services are available.	II-III	II	Not on the Redlist	[24]
3	*Mangifera indica*	Anacardiaceae	272	Light construction, building construction, furniture, crafts, and other services are available.	V	III	Not on the Redlist	[32]
4	*Anacardium occidentale*	Anacardiaceae	2	Wood for carpentry, plywood, firewood, and charcoal [33]	-	-	Not on the Redlist	
5	*Bouea macrophylla*	Annonaceae	3	Light construction, building construction, furniture, crafts, and other services are available.	V	II	Not on the Redlist	[24]
6	*Polyalthia longifolia*	Annonaceae	625	Firewood and light construction	-	-	Not on the Redlist	
7	*Annona muricata*	Annonaceae	29	Firewood and light construction	V	IV	Least concern (LC)	[24]
8	*Annona squamosal*	Annonaceae	1	Firewood and light construction	-	-	Not on the Redlist	
9	*Annona montana*	Annonaceae	2	Firewood and light construction	-	-	Least concern (LC)	
10	*Stelechocarpus burahol*	Annonaceae	1	Firewood and light construction	-	-	Not on the Redlist	
11	*Alstonia scholaris*	Apocynaceae	67	Chests, matches, concrete molds, crafts, puppet shows, masks, and other items are available.	V	IV-V	Least concern (LC)	[33]
12	*Plumeria acuminata*	Apocynaceae	8	Light construction, building construction, furniture, crafts, and other services are available.	IV-V	II	Not on the Redlist	[24]

13	*Plumeria alba*	Apocynaceae	13	Firewood and light construction	-	-	Not on the Redlist	
14	*Plumeria rubra*	Apocynaceae	12	Light construction, building construction, furniture, crafts, and other services are available.	IV-V	II	Least concern (LC)	[24]
15	*Cerbera manghas*	Apocynaceae	2	Firewood and light construction	V	III-IV	Least concern (LC)	[24]
16	*Araucaria heterophylla*	Araucariaceae	77	Firewood and light construction	-	-	Vulnerable (VU)	
17	*Agathis dammara*	Araucariaceae	60	Furniture, packing crates, veneer and plywood, pulp, and other materials.	IV	III	Vulnerable (VU)	[33]
18	*Handroanthus chrysotrichus*	Bignoniaceae	7	Firewood and light construction	-	-	Not on the Redlist	
19	*Spathodea campanulata*	Bignoniaceae	29	Firewood and light construction	V	IV	Least concern (LC)	[24]
20	*Ceiba pentandra*	Bombacaceae	2	Firewood and light construction	V	V	Least concern (LC)	[24]
21	*Calophyllum inophylum*	Callophyllanacea e	5	Buildings, crates, under-roof light construction, power poles, carts, canoes, bearings, barrels, golf bat heads, and more.	II-IV	II-III	Least concern (LC)	[33]
22	*Casuarina junghuhniana*	Casuarinaceae	167	Light construction, building construction, furniture, crafts, and other services are available.	II	I	Not on the Redlist	[24]
23	*Casuarina equisetifolia*	Casuarinaceae	248	Light construction, building construction, furniture, crafts, and other services are available.	II-III	I-II	Least concern (LC)	[24]
24	*Gymnostoma sumatranum*	Casuarinaceae	1	Firewood and light construction	-	-	Not on the Redlist	
25	*Garcinia atroviridis*	Clusiaceae	14	Firewood and light construction	-	-	Not on the Redlist	
26	*Garcinia dulcis*	Clusiaceae	1	Light construction, building construction, furniture, crafts, and other services are	IV	II	Least concern (LC)	[24]

				available.				
27	*Garcinia mangostana*	Clusiaceae	37	Light construction, building construction, furniture, crafts, and other services are available.	II	I	Not on the Redlist	[24]
28	*Terminalia cattapa*	Combretaceae	147	Light construction, building construction, furniture, crafts, and other services are available.	IV	II-III	Least concern (LC)	[24]
29	*Terminalia mantally*	Combretaceae	44	Firewood and light construction	-	-	Least concern (LC)	
30	*Platycladus orientalis*	Cupressaceae	77	Firewood and light construction	-	-	Near threatened (NT)	
31	*Shorea parvifolia*	Dipterocapaceae	1	Plywood, veneers, building frames, beams, walls, doors, windows, floors, boats, bridges, crates, musical instruments, and other materials	III-V	III-IV	Endangered (EN)	[33]
32	*Shorea laevis*	Dipterocarpaceae	1	Light construction, building construction, furniture, crafts, and other services are available.	I-II	I-II	Vulnerable (VU)	[33]
33	*Shorea leprosula*	Dipterocarpaceae	1	Plywood, veneers, building frames, beams, walls, doors, windows, floors, boats, bridges, crates, musical instruments, and other materials	III-V	III -IV	Near threatened (NT)	[33]
34	*Shorea macrophylla*	Dipterocarpaceae	69	Plywood, veneers, building frames, beams, walls, doors, windows, floors, boats, bridges, crates, musical instruments, and other materials	III-V	III-IV	Least concern (LC)	[33]
35	*Diospyros discolor / D. durionoides*	Ebenaceae	4	Light construction, building construction, furniture, crafts, and other services are available.	III	I	Not on the Redlist	[24]

36	*Diospyros celebica*	Ebenaceae	2	Fancy furniture, sculptures, carvings, lathes, fans, decorative tools, veneer brushes, musical instruments, and other items are available.	I	I	Vulnerable (VU)	[33]
37	*Muntingia calabura*	Elaocarpaceae	38	Firewood and light construction	-	-	Not on the Redlist	
38	*Elaeocarpus grandiflora*	Elaocarpaceae	1	Firewood and light construction	-	-	Not on the Redlist	
39	*Aleurites moluccana*	Euphorbiaceae	22	Firewood and light construction	IV-V	IV-V	Least concern (LC)	[32]
40	*Antidesma bunius*	Euphorbiaceae	28	Light construction, building construction, furniture, crafts, and other services are available.	IV	II	Not on the Redlist	[24]
41	*Hevea brasiliensis*	Euphorbiaceae	59	Light construction, building construction, furniture, crafts, and other services are available.	IV-V	II-III	Least concern (LC)	[34]
42	*Hura crepitans*	Euphorbiaceae	5	Firewood and light construction	-	-	Not on the Redlist	
43	*Acacia mangium*	Fabaceae	18	Light construction, building construction, furniture, crafts, and other services are available.	III	II-(III)	Least concern (LC)	[24]
44	*Adenanthera pavonina*	Fabaceae	249	Light construction, building construction, furniture, crafts, and other services are available.	II-I	I	Least concern (LC)	[24]
45	*Bauhinia purpurea*	Fabaceae	26	Firewood and light construction	-	-	Least concern (LC)	
46	*Caesalpinia pullcherrima*	Fabaceae	2	Firewood and light construction	-	-	Least concern (LC)	
47	*Cynometra cauliflora*	Fabaceae	2	Light construction, building construction, furniture, crafts, and other services are available.	III	I	Not on the Redlist	[24]
48	*Pithecelobium dulce variegata*	Fabaceae	1	Firewood and light construction	-	-	Not on the Redlist	

49	*Delonix regia*	Fabaceae	49	Firewood and light construction	-	-	Least concern (LC)	
50	*Peltophorum pterocarpum*	Fabaceae	1	Ships and train carriages, decorative panels, veneer, plywood, energy wood, pulp and others.	III-IV	II	Not on the Redlist	
51	*Lauchaena leucochepala*	Fabaceae	19	Firewood and light construction	V	-	Not on the Redlist	[35]
52	*Falcataria moluccana*	Fabaceae	12	Firewood and light construction	IV-V	IV-V	Least concern (LC)	[35]
53	*Parkia timoriana*	Fabaceae	2	Temporary building, veneer, plywood, carpentry, furniture, cabinet, molding, crates, matches, chopsticks, fishing net floats, pulp and paper, wood for energy. [38]	III	-	Least concern (LC)	[36]
54	*Pterocarpus indicus*	Fabaceae	289	Furniture, poles, planks, bridges, boats, tobacco pipes and decorative veneers.	II-(I-IV)	II-(I-IV)	Endangered (EN)	[33]
55	*Samanea saman*	Fabaceae	138	Light construction, building construction, furniture, crafts, and other services are available.	IV	III	Least concern (LC)	[24]
56	*Tamarindus indica*	Fabaceae	305	Light construction, building construction, furniture, crafts, and other services are available.	I	I	Least concern (LC)	[24]
57	*Pithecellobium dulce*	Fabaceae	4	Firewood and light construction	-	-	Least concern (LC)	
58	*Senna siamea*	Fabaceae	5	Firewood and light construction	I	-	Least concern (LC)	[37]
59	*Instia bijuga*	Fabaceae	2	Beams, pylons, planks, houses, bridges, bearings, boats, floors, furniture and lathes.	I-II	II-I	Vulnerable (VU)	[32]
60	*Erythrina crista*	Fabaceae	72	Firewood and light construction	-	-	Least concern (LC)	
61	*Limahnalia crenullata*	Gentinaceae	2	Firewood and light construction	-	-	Not on the Redlist	
62	*Gnetum gnemon*	Gentinaceae	29	Light	V	II	Least concern	[24]

				construction, building construction, furniture, crafts, and other services are available.			(LC)	
63	*Pheronema canescens*	Lamiaceae	5	Roof frames, house poles, bridges, furniture, veneers, cabinets and others.	III	I-III	Least concern (LC)	[33]
64	*Persea americana*	Lauraceae	17	Firewood and light construction	IV	IV/V	Least concern (LC)	[38] [39]
65	*Cinnamomum verum*	Lauraceae	7	Interior fittings, panels, cabinet furniture, wardrobes, chests, and other items are available.	IV-V	III-II	Not on the Redlist	[40]
66	*Cinnamomum burmanii*	Lauraceae	2	Light construction, building construction, furniture, crafts, and other services are available.	IV	III-II	Not on the Redlist	[24]
67	*Baringtonia asiatica*	Lecytidaceae	30	Light construction, building construction, furniture, crafts, and other services are available.	V	III	Least concern (LC)	[24]
68	*Lagerstromia speciosa*	Lythraceae	9	Building houses, planks, beams, poles, door frames, windows, bridges, bearings, boats, wheels, planks, walls, floors, pestle, barrels and lathes [24]	II-III	II-III	Not on the Redlist	[32]
69	*Durio zibenthinus*	Malvaceae	66	Light construction, crates, cigar boxes, plywood, clogs and cheap furniture.	IV-V	II-III	Not on the Redlist	[33]
70	*Theobroma cacao*	Malvaceae	5	Firewood and light construction	-	-	Not on the Redlist	
71	*Hibiscus tiliaceous*	Malvaceae	9	Light construction, building construction, furniture, crafts, and other services are available.	III	III-II	Least concern (LC)	[24]
72	*Guazuma ulmifolia*	Malvaceae	1	Light construction, building	IV-V	III	Least concern (LC)	[24]

IOP Conf. Series: Earth and Environmental Science 1115 (2022) 012003 doi:10.1088/1755-1315/1115/1/012003

				construction, furniture, crafts, and other services are available.				
73	*Sterculia foetida*	Malvaceae	7	Light construction, building construction, furniture, crafts, and other services are available.	III	II-III	Not on the Redlist	[24]
74	*Magnolia champaca*	Mangnoliaceae	4	Light construction, building construction, furniture, crafts, and other services are available.	II	III	Least concern (LC)	[24]
75	*lansium domesticum*	Meliaceae	13	Light construction, building construction, furniture, crafts, and other services are available.	III	II	Not on the Redlist	[24]
76	*Swiethenia macrophylla*	Meliaceae	1347	Decorative veneer, plywood, furniture, boats, houses, sculptures, carvings, and so on are examples of products made from wood.	III	I-III	Vulnerable (VU)	[33]
77	*Swiethenia mahagoni*	Meliaceae	214	Decorative veneer, plywood, furniture, boats, houses, sculptures, carvings, and so on are examples of products made from wood.	III	I-III	Near threatened (NT)	[33]
78	*Melia azedarch*	Meliaceae	2	Chests, boards, veneers, plywood and furniture.	IV-V	III-II	Least concern (LC)	[32]
79	*Sandorium koetjape*	Meliaceae	9	Firewood and light construction	IV-V	III-IV	Least concern (LC)	[24]
80	*Ficus benjamina*	Moraceae	69	Light construction, building construction, furniture, crafts, and other services are available.	V	III	Least concern (LC)	[24]
81	*Ficus microcarpa*	Moraceae	7	Firewood and light construction	-	-	Least concern (LC)	
82	*Ficus variegata*	Moraceae	2	Firewood and	V	V	Least concern	[41]

				light construction			(LC)	
83	*Ficus elastica*	Moraceae	17	Light construction, building construction, furniture, crafts, and other services are available.	V	II	Not on the Redlist	[24]
84	*Ficus racemosa*	Moraceae	18	Light construction, building construction, furniture, crafts, and other services are available.	V	III	Least concern (LC)	[24]
85	*Ficus ampelas*	Moraceae	3	Firewood and light construction	V	III-IV	Least concern (LC)	[24]
86	*Ficus hispida*	Moraceae	1	Light construction, building construction, furniture, crafts, and other services are available.	V	III	Least concern (LC)	[24]
87	*Ficus lyrata*	Moraceae	7	Firewood and light construction	-	-	Not on the Redlist	
88	*Artocarpus integer*	Moraceae	2	Light construction, building construction, furniture, crafts, and other services are available.	II-(I)	II	Not on the Redlist	[24]
89	*Artocarpus heterophylla*	Moraceae	120	Light construction, building construction, furniture, crafts, and other services are available.	II-III	II-III	Not on the Redlist	[24]
90	*Artocarpus altilis*	Moraceae	31	Light construction, building construction, furniture, crafts, and other services are available.	-	II	Not on the Redlist	[42]
91	*Moringa oleifera*	Moringaceae	1	Firewood and light construction	-	-	Not on the Redlist	
92	*Syzygium aqueum*	Myrtaceae	56	Plywood, light construction materials, floors, furniture, telephone poles, bridges, rail sleepers, veneer, fiberboard, firewood, charcoal, and pulp are all	III	II	Not on the Redlist	[37]

				examples of products made from wood. [38]				
93	*Syzygium malaccense*	Myrtaceae	28	Firewood and light construction	-	-	Least concern (LC)	
94	*Syzygium cumini*	Myrtaceae	24	Firewood and light construction	-	-	Least concern (LC)	
95	*Syzygium paniculatum*	Myrtaceae	334	Firewood and light construction	-	-	Not on the Redlist	
96	*Syzygium polyantum*	Myrtaceae	9	Light construction, building construction, furniture, crafts, and other services are available.	-	III	Not on the Redlist	[43]
97	*Psidium guajava*	Myrtaceae	65	Light construction, building construction, furniture, crafts, and other services are available.	III	II	Least concern (LC)	[24]
98	*Psidium guajava Red. Malaysia*	Myrtaceae	1	Firewood and light construction	-	-	Not on the Redlist	
99	*Myristica fragrans / M. Phillipinensis*	Myristicaceae	6	Light construction, building construction, furniture, crafts, and other services are available.	V	III	Data deficiens (DD)	[24]
100	*Averrhoa carambola*	Phyllantaceae	12	Light construction, building construction, furniture, crafts, and other services are available.	IV	II	Not on the Redlist	[24]
101	*Averrhoa bilimbi*	Phyllantaceae	13	Light construction, building construction, furniture, crafts, and other services are available.	V	III	Not on the Redlist	[24]
102	*Baccaurea deflexa*	Phyllantaceae	1	Light construction, building construction, furniture, crafts, and other services are available.	III	II	Not on the Redlist	[24]
103	*Pinus merkusii*	Pinaceae	8	Residential buildings, flooring, furniture, matchboxes and sticks, pulp, power poles,	IV	III	Vulnerable (VU)	[32]

				wood wool boards and plywood.				
104	*Podocarpus polystachyus*	Podocarpaceae	2	Light construction, building construction, furniture, crafts, and other services are available.	IV	II-III	Vulnerable (VU)	[24]
105	*Neolamarckia cadamba*	Rubiaceae	2	Matches, wrapping crates, concrete moulds, children's toys, pulp, clogs and light construction.	V	III-IV	Not on the Redlist	[24]
106	*Morinda citrifolia*	Rubiaceae	11	Light construction, building construction, furniture, crafts, and other services are available.	IV	II	Not on the Redlist	[24]
107	*Flacourtia rukam*	Salicaceae	6	Light construction, building construction, furniture, crafts, and other services are available.	III	I	Not on the Redlist	[24]
108	*Salix babylonica*	Salicaceae	3	Firewood and light construction	-	-	Not on the Redlist	
109	*Dimocarpus longan*	Sapindaceae	44	Firewood and light construction	-	-	Near Threatened (NT)	
110	*Nephelium lapaceum*	Sapindaceae	126	Light construction, building construction, furniture, crafts, and other services are available.	III	I-II	Least concern (LC)	[24]
111	*Filicium decipiens*	Sapindaceae	243	Firewood and light construction	-	-	Not on the Redlist	
112	*Schleichera oleosa*	Sapindaceae	17	Light construction, building construction, furniture, crafts, and other services are available.	III	I	Least concern (LC)	[24]
113	*Pometia pinnata*	Sapindaceae	40	Bridges, furniture, flooring, boats, handlebars and sports equipment.	V	II (I-III)	Least concern (LC)	[33]
114	*Manilkara zapotta*	Sapindaceae	7	Firewood and light construction	-	-	Not on the Redlist	
115	*Manilkara kauki*	sapindaceae	3	Light construction.	I	I	Not on the Redlist	[24]

				building construction, furniture, crafts, and other services are available.				
116	*Mimusops elengi*	sapindaceae	707	Floors, furniture, tools, sculptures, carvings, bridges, ships and bearings.	I-II	I	Least concern (LC)	[32]
117	*Chrysophyllum cainito*	Sapotaceae	1	Firewood and light construction	-	-	Not on the Redlist	
118	*Aquilaria malaccensis*	Thymeliaceae	4	Firewood and light construction	-	-	Critical endengered (CR)	
119	*Gmelina arborea*	Verbenaceae	25	Firewood and light construction	IV-V	III-IV	Least concern (LC)	[44]
120	*Tectona grandis*	Verbenaceae	94	Construction, pylons, beams, bridges, roof trusses, sills, dams, furniture, ships, flooring, rail equipment and medicine.	II	II	Not on the Redlist	[33]
121	*Premna corymbosa*	Verbenaceae	1	Light construction, building construction, furniture, crafts, and other services are available.	III	II	Not on the Redlist	[24]

In addition to durability, wood strength is an important consideration, as it is one of the factors that determine wood quality, according to [24]. The results showed that as many as 83 tree species in the green space of the USU campus were classified as moderate to very strong (III-I). There are more species with an excellent strength class compared to their durability. This indicates that strong wood does not always have durable properties. Some species with good strength but poor durability are *Mangifera odorata*, *Garcinia dulcis*, *Cinnamomum verum*, *Morinda citrifolia*. Meanwhile, it has low strength and durability for carpentry wood species such as *Melia azedarach*. This is presumably due to extractive substances produced during its growth, as revealed by [45,46].

In this study, several tree species were found that did not have information related to their durable class and their strength class. Some species that do not have information related to wood quality which includes durable and strength classes, are *Anacardium occidentale*, *Annona squamosa*, *Annona montana*, *Stelechocarpus burahol*, *Plumeria alba*, *Handroanthus chrysotrichus*, *Garcinia artrovidis*, *Terminalia mantally*, *Hura crepitans*, and several other species. The lack of information on the quality of wood from several types is thought to be because the focus on utilizing these trees is not on the wood. The analysis results show that some of these species are generally used as fruit producers and ornamental plants. The most common use of this type is used as firewood which incidentally does not require particular information about the quality of the wood owned.

The high use of wood for various purposes causes the demand for good quality wood to increase. This impacts the population of different wood-producing species, decreasing until they are threatened with extinction. The results showed that of the 121 tree species in the green space of the USU campus, 2 (two) of them had the status of Endangered (EN) or were threatened with extinction. The two types are *Shorea parvifolia* and *Pterocarpus indicus*. Both of these species are recorded to have superior quality wood, which is classified as strength in class II and III, and durability, which is classified in class II to III. This indicates that the protection of wood species is very important, and substitution of wood use must be carried out immediately to protect these species from the threat of extinction.

4. Conclusion

There are 19 tree species with durable class I to II (very durable and durable), 83 tree species with strength class I-III, 38 species with no information on durable class, and 36 species with no information on strength class. *Shorea parvifolia* and *Pterocarpus indicus* are two critically endangered tree species.

References

[1] Warren JL 1973 Green space for air pollution control *Technical report (50) Schoolof Forest Resources* (North Carolina State University: 118)
[2] Nowak D J, Hirabayashi S, Doyle M, McGovern M and Pasher J 2018 *Urban For Urban Gree* **29**
[3] Nowak D J, Crane D E and Stevens J C 2006 Air pollution removal by urban trees and shrubs in the United States *Urban For Urban Gree* **4**
[4] Nowak D J 1993 Atmospheric carbon reduction by urban trees *J of Environ Manage* **37** 3
[5] Schwab J 2009 Planning the urban forest: Ecology, economy, and community development *Chicago, IL: American Planning Association*
[6] Skoff J B T and Cavender 2019 *Plants, People, Planet* **1**
[7] Food and Agricultur Organization of the United Nations (FAO) 2016 *Food And Agriculture Organization Of The United Nations* Rome
[8] Rocha M F V, Veiga T R L A, Soares B C D, deAraujo A C C, Carvalho A M M and Hein P R G 2019 *Floresta e Ambiente* **26** 3
[9] Ramagea M H, Burridgeb H, Busse-Wicherc M, Feredaya G, Reynoldsa T, Shaha D U, Wud G, Yuc L, Fleminga P, Densley-Tingleye D, Allwoode J, Dupreec P, Lindenb P F and Schermane O 2017 *Renew Sustain Energy Rev* **68**
[10] Bacher M and Krosek S 2014 *For Wood Tech* **88**
[11] Pretzsch H and Rais A 2016 *Wood Sci Technol* **50**
[12] Cardinale B J, Duffy J E, Gonzalez A, Hooper D U, Perrings C and Venail P 2012 *Nature* **486**
[13] Susilowati A, Rangkuti A B, Rachmat H H, Iswanto A H, Harahap M M, Elfiati D, Slamet B and Ginting I M 2021 *Biodiversitas* **22** 5
[14] Rangkuti A B, Susilowati A, Harahap M M and Iswanto A H 2021 *IOP Conf Series: Earth and Environtmental Science* **918** 012022
[15] Susilowati A, Ginting I M, Rachmat H H, Dwiyanti F G, Iswanto A H and Sucipto T 2022 *IOP Conf Series : Earth and Environmental Science* **977** 012016
[16] BPS City of Medan 2022 *Medan Municipality in Figures* 2022 *BPS Kota Medan* (Medan)
[17] Nasution N S, Ulfah D and Arryati H 2019 *Jurnal Sylva Scienteae* **2** 6
[18] David S N and Aleru K K 2016 *Int'l Journal Of Agric And Rural Dev* **19** 2
[19] Bötsch Y, Tablado Z and Jenni 2017 *Proceedings of the Royal Society of London B: Biological Sciences* **284** 20170846
[20] Haigh A, Butler F, O'Riordan R and Palme R 2017 *Biological Conservation* **211**
[21] Hansel A, Sandak J, Sandak A, Mai J and Niemz P 2022 *A Review Wood Material Science and Engineering* **17** 3
[22] Knapic S, Santos J, Santos J and Pereiral H 2018 *Ciencia y tecnología* **20** 3
[23] Cruz H, Jones D and Nunes L 2015 *Springer International Publishing Switzerland*
[24] Oey D S 1990 *Secific gravity of Indonesia wood and its significance for practical use* (Forest products research and development centre: Bogor)

[25] Ulyshen M D, Horn S, Brownie C, Strickland M S, Wurzburger N and Zanne A 2020 *Biol Invasions*
[26] Wong A H H, Morsing N, Henriksen K and Salmiah U 2004 *Above ground microbial decay test of biocide treated and untreated wood exposed to Danish and humid tropical climates Int Res Group* (Wood Preserv Document No IRG/WP: 04-20306)
[27] Franc,a F J N, Franca T S F A, Arango R A, Woodward B M and Vidaurre G B 2017 *For Prod J* **67** 3
[28] Silva A, Gartner B and Morrell J 2007 ASTM J Test Eval **35** 3
[29] Valette N, Perrot T, Sormani R, Gelhaye E and *Morel-Rouhier M 2017 Fungal Biol Rev 31 3*
[30] *Gonzalez J M and Morrell J J 2012* Wood Fiber Sci **44** 4
[31] Van-Acker J, De Windt I, Li W, *Van-den-Bulcke J 2014 Critical parameters on moisture dynamics in relation to time of wetness as factor in service life prediction In* (Proceedings IRG annual meeting IRG/WP 1-20555)
[32] Martawijaya A, Kartasujana, Kadir K, Mandang Y I, Prawira S A 1989 *Atlas Kayu Indonesia jilid II* (Badan Penelitian dan Pengembangan Kehutanan Bogor: Indonesia)
[33] Catarino L, Menezes Y and Sardinha R 2015 *Sci Agric* **72** 5
[34] Suheryanto D 2010 *Pengaruh Konsentrasi Cupri Sulfat terhadap Keawetan Kayu Karet* (Seminar Rekayasa Kimia Dan Proses, E-06 Semarang Jurusan Teknik Kimia, FakultasTeknik, Universitas Diponegoro)
[35] Jasni 2016 *Jurmal Penelitian Hasil Hutan* **34** 3
[36] Lessy I, Ohorella S and Karepesina S 2018 *Jurnal Agrohut Fakultas Pertanian Universitas Darussalam Ambon* **9** 1
[37] Sutisna U, Wardani M, Kalima T, Mandang Y I, Hadjib N, Pari G, Sumarni G, Abdurrohim S, Barly, Iskandar M I, Rachman O, Basri E, Lisnawati Y, Iskak and Doom F 2004 Atlas Kayu Indonesia Jilid III *Departemen Kehutanan, Badan Penelitian Dan Pengembangan Kehutanan, Pusat Penelitian Dan Pengembangan Teknologi Hasil Hutan*, Bogor
[38] Hendrati R L and Hidayati N 2014 Budidaya Johar (Cassia seamea) untuk antisipasi Kondisi Kering *IPB Press* Bogor
[39] CEN-European Committee for Standardization 1994 Standard EN 350-1/2 Durability of wood and wood-based products - Natural durability of solid wood Part 41 Guide to the principles of testing and classification of the natural durability of wood Part 2: *Guide to natural durability and treatability of selected wood species of importance in Europe*
[40] Fuentes-Talavera F J, Silva-Guzmán J A, Rodríguez-Anda R, Lomelí-Ramírez M G, Sanjuán-Dueñas R and Richter H G 2011 *Madera y Bosques* **17** 1
[41] Muslich M and Sumarni G 2008 *Jurnal Penelitian Hasil Hutan* **26** 1
[42] Samuel A O 2021 *Wood Properties And Natural Durability of Artocarpus Altilis (Parkinson Ex FA Zorn) Fosberg* (Thesis University of Ibadan, Ibadan)
[43] [PKKI] Peraturan Konstruksi Kayu Indonesia 1961 PKKI NI – 5 1961 *Direktorat Jenderal Cipta Karya Departemen Pekerjaan Umum* Bandung
[44] Alrasjid H and Widiarti A 1992 Teknik Penanaman dan Pemungutan Gmelina arborea (Yamane) Informasi Teknis No 36 *Pusat Penelitian dan Pengembangan Hutan* Bogor
[45] Alañón M E, Pérez-Coello M S, Díaz-Maroto I J, Martín-Alvarez P J, Vila-Lameiro P and Díaz-Maroto M C 2011 *Forest Ecology and Management* **262** 2
[46] Zanuncio A J V, Colodette J L, Gomes F J B, Carneiro A C O and Vital B R 2013 *Ciência Florestal* **23** 4

The diversity of the Polypodiaceae in University green space

A Susilowati[1,3,4*], I M Ginting[1], H H Rachmat[2], D Elfiati[1,3,4], T Sucipto[1,3] and H Nadeak[1]

[1]Faculty of Forestry, Universitas Sumatera Utara. Jl. Tridharma Ujung No.1, USU Campus-Medan 20155, Indonesia
[2]Research Center for Ecology and Ethnobiology, National Research and Innovation Agency, Jl. Raya Jakarta-Bogor Km 46, Cibinong-Bogor 16911
[3]JATI-Sumatran Forestry Analysis Study Center, Universitas Sumatera Utara. Jl. Tridharma Ujung No.1, Kampus USU Medan 20155, Indonesia
[4]Center of Excellence for Mangrove, Universitas Sumatera Utara, Medan 20155, Indonesia

*E-mail: arida.susilowati@usu.ac.id

Abstract. Polypodiaceae is one of the fern families by the highest number species and plays an important role in the ecosystem. The polypodiaceae family is also known for its pharmacological properties. The University of Sumatera Utara campus, as one of green spaces in Medan, has a various variety of plants from undergrowth to higher plants. However, no information on the diversity of Polypodiaceae epiphytic species on the USU campus has not been recorded. The aim of the research is to identify and evaluate the diversity of Polypodiaceae species on the USU campus. For those purpose an inventory method in all green spaces on the USU campus. Species identification was accomplished by morphological identification of founded epiphytes. The relative density and dominance values have been used to calculate species diversity (H'). The study revealed the presence of 6 (six) Polypodiaceae epiphytic species on the USU campus, namely *Microsorum diversifolium*, *Microsorum pustulatum*, *Pyrrosia piloselloides*, *Phymatosorus scolopendria*, *Pyrrosia eleagnifolia* and *Pyrrosia lanceolate*. The highest diversity index was found in *Pyrossia piloselloides* with an H' value of 0.345 and the lowest diversity was found in *Microsorum pustulatum* with an H' value of 0.001.

1. Introduction

Pteridophytes are a group of vascular plant often found in tropical and temperate forests [1]. Pteridophytes, a group of primitive vascular plants, has at least 13,600 species with geographical distribution all around the world [2]. This plant group has a wide and varied habitat, ranging from land, epiphytes and aquatic habitats in ecosystems ranging from coastal ecosystems to high mountain forests [3]. Pteridophytes have megaphylls (larve leaves), midribs and microfils (small leaves). Pterodophyte with megafil has a distinctive rhizome, which is the main stem of the pteridophyte. Pterodophyte with megafil has a characteristic in the form of the rhizome, which is the main stem of the pteridophyte. Rhizomes are frequently found on the surface of the substrate, creeping, erecting, or forming stems [4]. Ferns have morphological characters that are used as determinants of ferns in terms of size, shape, type and location of spores [5,6]. Polypodiaceae, Cyatheaceae, Hymenophyllaceae, Thelypteridaceae and Pteridaceae are the five families with the largest percentage in the pteridophyta group. These five families account for almost half 50% of total diversity [7].

The Polypodiaceae family has the most species diversity of any fern family [8]. The majority of members of this tribe are epiphytes and one of the most abundant groups of pantropical vascular epiphytes in subtropical and tropical forests [9,10]. Polypodiaceae contributes to the ecosystem as an epiphyte by serving as a nesting site for small insect [11], producing litter and nutrient cycles [12], nitrogen and water fixation [13] and providing a suitable microhabitat for various organisms living above the canopy [14]. Several Polypodiaceae species are also potential sources of pharmacological with anti-diabetic [15], anti-cancer [16] and anti-inflammatory properties [17]. According to [18], the Polypodiaceae family contains neoflavonoids, which are biogenetic and structurally related to flavonoids and isoflavonoids. Polyphenols, as flavonoids, could be the basic source for cancer prevention and treatment [19-21].

Universitas Sumatera Utara (USU) is one of the largest campuses in North Sumatra which has an area of about 120 ha. Besides functioning as an educational facility, the USU campus also severals role as one of the green spaces in the city of Medan. Previous research found that there are at least 121 tree species on the USU campus [22]. The high diversity of tree species is not supported by information on the existence of epiphytes, especially those with high potential for utilization, such as Polypodiaceae. In fact, this information is crucial for optimizing utilization and conservation strategies in the area. Therefore, the research aims to identify and evaluate the diversity of Polypodiaceae species on the USU campus.

2. Research Methods

2.1. Research location

This research was conducted in USU campus, Padang Bulan-Medan with total area of 120 Ha. The USU campus is located at an altitude of 25 to 37.5 meters above sea level with temperatures ranging from 23.2°C to 33.2°C. The average relative humidity was 85%, while the lowest humidity was recorded at 65% [23]. Since the campus's founding in 1952, a variety of native and exotic trees have been iregullar planted. The variety of tree species will produce porophytes for a wide range of epiphytes.

2.2. Data Collection and Analysis

Data were collected using the inventory method with morphological observations of all individual epiphytic. As primary data, morphological characteristics of the epiphytes and the host tree were recorded. For further analysis, data on the number of species, individuals, conservation status and species utilization were gathered. Species diversity was determined using data analysis of relative density (RD), species richness (D), diversity and evenness. The formula below is used to calculate the relative density.

$$\text{Relative density of species (RD)} = \frac{\text{Number of individual species}}{\text{Total number of trees}} \text{x } 100 \quad (1)$$

$$\text{Relative abundance of species (Pi)} = \frac{\text{Relative density of species}}{100} \quad (2)$$

The obtained values were then classified into five categories: high (RD > 5.00), frequently (4.00 RD 4.99), rarely (3.00 RD 3.99), very rare (1.00 RD 2.99) and threatened if the RD value was less than 1.0 or 0 [24]. The diversity index is also calculated to determine the level of species richness using the formula according to [25].

$$d = (S - 1)/\ln N \quad (3)$$

Where d = index of species richness, S = number of species, ln = natural logarithm and N = number of all species in the research location. The species richness index can be classified into three categories, namely low, medium and high with a value limit of d < 2.5, 2.5 < d < 4.0 and d > 4.0. Meanwhile, the level of species diversity d was calculated using the Shannon Wienner index (H') according to [26].

$$H' = -\sum Pi \ \ln Pi \tag{4}$$

Where Pi is an individual of species i and individuals of all species contained in a community that is observed with ln as the natural logarithm.

3. Result and Discussion

3.1. Species density of Polypodiaceae in the green open space of USU campus

Epiphytes have a close relationship with trees, particularly the function of attaching as a host tree. According to [27], epiphytic orchids depend on their hosts' availability and are strongly influenced by the microclimate. This illustrates that the existence of trees and epiphytes is closely linked. The findings revealed that there were six species of Polypodiaceae in the USU campus (Table 1). Despite belonging to the same family, these six species have distinguishable morphologies (Figure 1).

Table 1. Diversity of Polypodiaceae species at the USU campus, Medan.

No	Polypodiaceae species	number of individual	Percentage	RD	Pi	H'	Utilization potential	number of host trees
1	*Microsorum diversifolium*	13	0,078%	0,077542	0,000775	0,005	Potential as an ornamental plant	1
2	*Microsorum pustulatum*	3	0,018%	0,017894	0,000179	0,001	Potential as anti-tumor [29]	1
3	*Pyrrosia piloselloides*	4150	24,754%	24,75395	0,24754	0,345	As an antioxidant and anticancer [30]	51
4	*Phymatosorus scolopendria*	549	3,275%	3,274679	0,032747	0,111	Potential as a medicinal plant [31]	11
5	*Pyrrosia eleagnifolia*	4039	24,092%	24,09186	0,240919	0,342	Potential as an ornamental plant	39
6	*Pyrrosia lanceolata*	45	0,268%	0,268416	0,002684	0,015	Potential as medicinal plants [32]	3

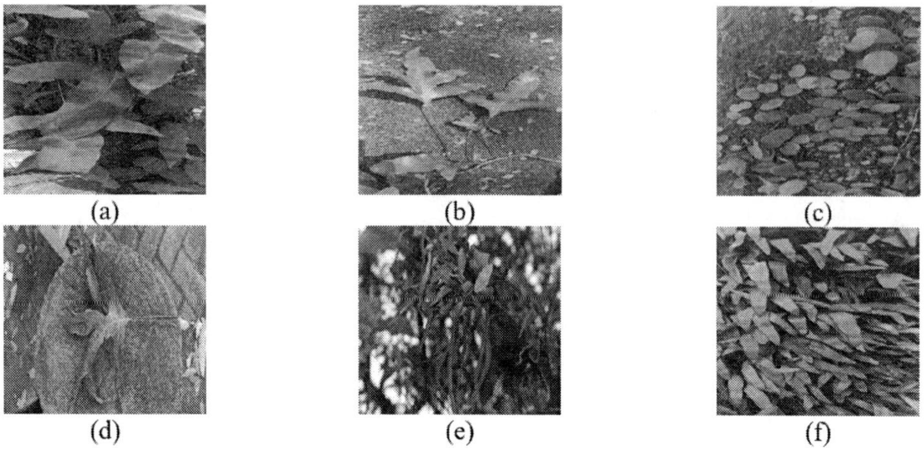

Figure 1. Polypodiaceae species in a green open space on the USU campus, Medan : (a) *M. diversifolium*, (b) *M. pustulatum*, (c) *P. piloselloides*, (d) *P. scolopendria*, (e) *P. lanceolata*, (f) *P. eleagnifolia*.

According to this study, *P. piloselloides* is the most common Polypodiaceae epiphytic species. A total of 4,150 individual *P. piloselloides* were found with a diversity value of 0.345 and classified as low. This value is highest compared to other species, this. Another Pyrrosia member, *P. eleagnifolia*, has a diversity value of 0.342 (lower). The genus Pyrrosia's high value of diversity is supported by high adaptability and morphological forms that are well adapted to environmental conditions. *P. eleagnifolia* is a harsh, adaptable fern that can survive in dry conditions due to its fleshy midrib, dense hair cover and reduced lamina area, according to [28]. The high diversity of these two species is thought to be due to their good adaptability.

These two species, in addition to having a high diversity value, occupy more diverse host trees than other species. This is indicated by the high number of host tree species occupied by these two types of epiphytes. *P. piloselloides* was found in 51 host tree species, while *P. eleagnifolia* was found in 39 different host tree species. This implies that the large number of individuals is directly proportional to the number of host trees occupied. Previous research discovered at least 121 tree species in the green open spaces of the USU campus [22], suggesting that this genus accounts for nearly 50% the tree species on the USU campus.

In conversely to Pyrrosia, *M. pustulatum* and *M. diversifolium* had the lowest diversity values. Both of these species belonged to the genus Microsorum and their H' values were 0.001 and 0.005, respectively (Table 1). Only one species of host tree was found to be host to these two species. Microsorum is presumed to have a more specific habitat, requiring special requirements for host trees to grow. M. pustulatum is an epiphytic fern that grows to climb on the shady branches of large trees, according to [28]. This also contributes to the low value of diversity obtained.

3.2. Strategies for the use and conservation of Polypodiaceae in green open spaces of the USU campus

Based on observations, the presence of Polypodiaceae epiphytes in USU green spaces is still under utilized. Until now, it has been assumed that Polypodiaceae epiphytes are only used as ornamental plants on campus. In fact, four of the six species were discovered to have medicinal plant potential. *M. pustulatum* contains phytochemicals that are anti-tumor [29], *P. piloselloides* has anti-cancer and antioxidant properties [30], as well as *P. scolopendria* [31] and *P. lanceolate* [32] have medicinal properties. The low utilization of epiphytes is due to a lack of information obtained by the surrounding community.

The presence of epiphytes attached to host trees suggests the existence of host trees plays an important role in the abundance of Polypodiaceae epiphytes on the USU campus. The relationship formed by Polypodiaceae epiphytes with host tree species suggests that the destruction of host trees may have a negative impact on Polypodiaceae epiphytic populations in USU campus. The campus development program which usually includes the removal of various types of trees, will contribute to a reducing in the population of these epiphytes. Therefore, it is necessary to increase and enrich the host tree species in order to save this species from extinction.

4. Conclusion

A total of 6 (six) Polypodiaceae epiphytic species were found in the USU campus, namely *M. diversifolium, M. pustulatum, P. piloselloides, P. scolopendria, P. eleagnifolia* and *P. lanceolata*. The highest diversity index found in *Pyrossia piloselloides* with an H' value of 0.345 and the lowest diversity value found in *Microsorum pustulatum* (H':0.001). The enriching porophyte species is highly recommended as an effort to conserve epiphytes and increase the diversity of epiphytes in the USU Medan campus.

References

[1] Dixit R D 2000 *Indian Fern J* **17** 77
[2] Watkins J E J, Catherine C, Robert K C and Robbin C 2006 *Am J Bot* **93** pp 73
[3] Mildawati, Sobir, Sulistijorini and Chikmawati T 2020 *Biodiversitas* **21** 3200
[4] Christenhusz M J M and Chase M W 2014 *Ann Bot* **113** 571
[5] Apriyanti N, Jaya S D and Madang K 2017 *Jurnal Pembelajaran Biologi* **5** 113
[6] Suryadi, Muhibuddin, Hasanuddin, Samingan and Nurmaliah C 2020 *Journal of Physics: Conference Series*, 1460
[7] Tjitrosoepomo G 1993 *General Taxonomy (Basic Plant Taxonomy)* (Yogyakarta: Gadjah mada University Press)
[8] PPG I A 2016 *J Syst Evol* **54** 563
[9] Gentry A H and Dodson C 1987 *Ann Mo Bot Gard* **74** 205
[10] Benzing D H 2004 Forest canopies. San Diego: Elsevier 175
[11] Gotsch S G, Nadkarni N and Amici A 2016 *J Trop Ecol* **32** 455
[12] Putz F E and Mooney H A 1991 *The Biology of Vines* (Cambridge University, Cambridge, UK)
[13] Stanton D E, Chávez J H, Villegas L, Villasante F, Armesto J and Hedin LO 2014 *Funct Ecol* **28** 1274
[14] Scheffers B R, Phillips B L and Shoof L P 2014 *Global Ecol Conserv* **2** 37
[15] Chai T T, Elamparuthi S, Yong A L, Quah Y, Ong H C and Wong F C 2013 *Bot Stud* **54** 55
[16] Gonzalez S, Gilaberte Y and Philips N 2010 *Photochem Photobiol Sci* **9** 559
[17] Choudhry S Z, Bhatia N, Ceilley R, Hougeir F, Lieberman R and Hamzavi I 2014 *J Drugs Dermatol* **13** 148
[18] Christensen LP 2018 *Elsevier* 147
[19] Asensi M, Ortega A, Mena S, Feddi F and Estrela J M 2011 *Crit Rev Clin Lab Sci* **48** 197
[20] Shukla S, Meeran S M and Katiyar S K 2014 *Cancer Lett* **355** 9
[21] Alvarado-Sansininea J J, Sánchez-Sánchez L, López-Muñoz H, Escobar M L, Flores-Guzmán F and Tavera-Hernández R 2018 *Molecules* **23**
[22] Susilowati A, Rangkuti A B, Rachmat H H, Iswanto A H, Harahap M M, Elfiati D, Slamet B and Ginting I M 2021 *Biodiversitas* **22** 2839
[23] BPS Statistic of Medan Municipality 2020 *Medan Municipality in Figure 2020 BPS-Statistic of Medan Municipality, Medan*
[24] Ogwu M C, Osawaru M E, Obayuwana O K 2016 *Appl Trop Agric* **21** 46
[25] Margalef R 1958 Information Theory in Ecology General System 3 56
[26] Shannon C E and Weaver W 1949 *The Mathematical Theory of Communication* (The University of Illinois Press: Urbana. IL. USA)

[27] Mondragon D, Valverde T and Hernandez-Apolinar M 2015 *Trop Ecol* 56 1
[28] Brownsey P J and Perrie L R 2014 *Manaaki Whenua Press* Lincoln
[29] Neneng L, Pujirahadju A, Ferayani F and Widayat T 2020 *Biosaintifika* **12** 422
[30] Wulandari E T, Elya B, Hanani E and Pawitan J 2013 *International Journal of PharmTech Research* **5** 119
[31] Sujatha S, Ramya R I, Gayathri M and Catharina S S 2018 *Int Res J Pharm* **9** 168
[32] Gnana D R R and Catharin S S 2014 *International Reasearch Jornal of Pharmacy* **5** 879

IOP Conf. Series: Earth and Environmental Science 1115 (2022) 012005 doi:10.1088/1755-1315/1115/1/012005

Assessing the impacts of local intention on community-based wildlife ecotourism in Tangkahan, Gunung Leuser National Park (GLNP) and the development strategy in the post-COVID-19 outbreak

R O P Situmorang[1*] and M Hussain[2]
[1]Population Research Center, National Research and Innovation Agency (BRIN), Gd. Widya Graha Lt. X, Jl. Jend. Gatot Subroto No. 10, Central Jakarta 12710, Indonesia
[2]Department of Forestry, National Chung Hsing University, 145 Xingda Rd., Taichung City 402, Taiwan (R.O.C)

*Email: pita_80s@yahoo.com

Abstract. Theory of Planned Behavior (TPB) has been implemented to evaluate the structure of behavioral change of the local communities in the forest and wildlife conservation program in Tangkahan, the eastern part of GLNP. This study has the purpose to reveal the determinant factors of local community behavior to support forest and wildlife conservation through ecotourism activities and to formulate a strategy to recover wildlife ecotourism in the post-COVID-19 outbreak. The study was conducted in Tangkahan through a survey method to collect the data. The data were analyzed using linear regression analysis. This study found that the determinant factors of behavioral intention are the conservation attitude and perceived behavioral control and behavioral intention significantly affects the behavior to participate in conservation activities. To improve the visiting number and the ecotourism service in Tangkahan in the post-COVID 19 outbreak, the community should transform and develop the management system from a conventional to virtual/digital method, develop and follow the COVID-19 protocol of ecotourism, re-arrange the ticket fee and the tour packages to cope with domestic visitors and improve the access to the area by constructing the roads and provision of transportation services.

1. Introduction

Protected areas are globally considered a key strategy for the conservation of natural environments and species. Among the various types of protected areas, the IUCN categorized national parks as protected areas in category II which holds objectives for biodiversity conservation, tourism and recreation, research and education sites and cultural heritage [1-5]. *Gunung Leuser* National Park (GLNP) or Mount Leuser National Park located in Sumatra Island, Indonesia, is a protected forest that has the main function to preserve biodiversity particularly endangered and unique species [6,7]. As a tropical rainforest, the forest covers an area of 830,268.95 ha. It contains abundant biodiversity where 536 plant species and around 3,500 animal species have been identified so far in the GLNP [8]. In more detail, Jay [9] noted that the Leuser ecosystem is home to over 100 mammals, 380 birds, over 95 reptiles and amphibians and rafflesia (the tallest and the largest flowers in the world). The key species of GLNP are Rafflesia (*Rafflesia arnoldii*), orang-utan (*Pongo* spp), Sumatran elephant (*Elephas maximus* sumatrensis), tiger (*Panthera tigris* sumatrae) and rhino (*Dicerorhinus sumatrensis*) with critical endanger status by IUCN [9].

Although having the status of a protected area, GLNP is still under the threat of deforestation [10]. The GLNP Agency [11] reported that in 2000, a total of 18,089 ha of the forest area has been felled. In 2015, TROPENBOS-UNESCO reported that the total deforested area has reached 42,488 ha with a deforestation rate of 972.4 ha/ year during 1990-2014 [12,13]. Jay [9] stated that illegal logging is still a significant cause of deforestation in Leuser. The national park, which is located in two provinces, 11 districts and adjacence a hundred villages, has different problems in many spots that must be handled specifically. The limitation of the government's personnel and facilities to protect the forest has weakened the control over forest and law enforcement leading to illegal logging and encroachment in GLNP [14,15]. Besides that, the ignoring of local community participation [16], a poverty problem [17,18] and the available demands and marketing chains of illegal logs [19], have increased deforestation. To control deforestation, state-initiated conservation programs should continue to work closely with a wide variety of actors in the field, particularly communities that reside in and around forests [20,21].

Community-based ecotourism is a form of community-based resource management (CRBM) that carries the potential to solve many problems in protected forests, such as illegal logging, forest encroachment and human-wildlife conflicts [22,23]. The concept here is the devolution of rights and responsibility of forest management from government to community organization [24,25]. Community-based wildlife ecotourism management (CBWEM) has been promoted to conserve wildlife and protect the wildlife's natural habitat from human disturbances, improve local people's livelihood and provide environmental education to both tourists and local people [23,26-28].

Wildlife ecotourism has been implemented in some parts of GNLP, such as in Bukit Lawang to conserve Sumatran orangutans (*Pongo abelii*) since 1973 [10,29] and in Tangkahan to conserve Sumatran Elephant (*E. maximus sumatranus*) since 2002 [30,31]. It has significantly reduced the disturbances to the forest by empowering local communities. The long history of the co-existence of local people and the forest is interesting to be studied, to understand the community's needs and motivation to support the sustainability of the forest.

To understand the deterministic likelihood of performing specific behavior, such as in forest conservation, the theory of planned behavior (TPB) has been widely applied in assessing pro-environmental behavior. This theory explains that the intention to perform a behavior is influenced by attitude, subjective norms and perceived behavioral control. Attitude is influenced by individuals' beliefs about the outcomes (positive or negative outcomes) or attributes of performing behaviors. Subjective norm is a social pressure to perform or not to perform any activity that influences the intention and behavior [32].

Many studies have been conducted in GLNP, however, research on the assessment of social construction of community motivation in forest and wildlife conservation in GLNP is still limited. This study aimed to reveal the determinant factors of local community behavior to support forest and wildlife conservation through ecotourism activities. The study area is in Tangkahan which performs jungle and elephant ecotourism. This study also has the purpose to describe the current management of Tangkahan wildlife ecotourism and to formulate a strategy for the new normal period of the COVID-19 Pandemic. As we know, the COVID-19 pandemic has changed social life and rapidly improved communication technology. This has yielded high competition among service providers and trading industries.

2. Methodology

The research was conducted in the Tangkahan wildlife-ecotourism area which is the utilization zone of the south-eastern part of GLNP and Namo Sialang Village (Figure 1).

IOP Conf. Series: Earth and Environmental Science 1115 (2022) 012005 doi:10.1088/1755-1315/1115/1/012005

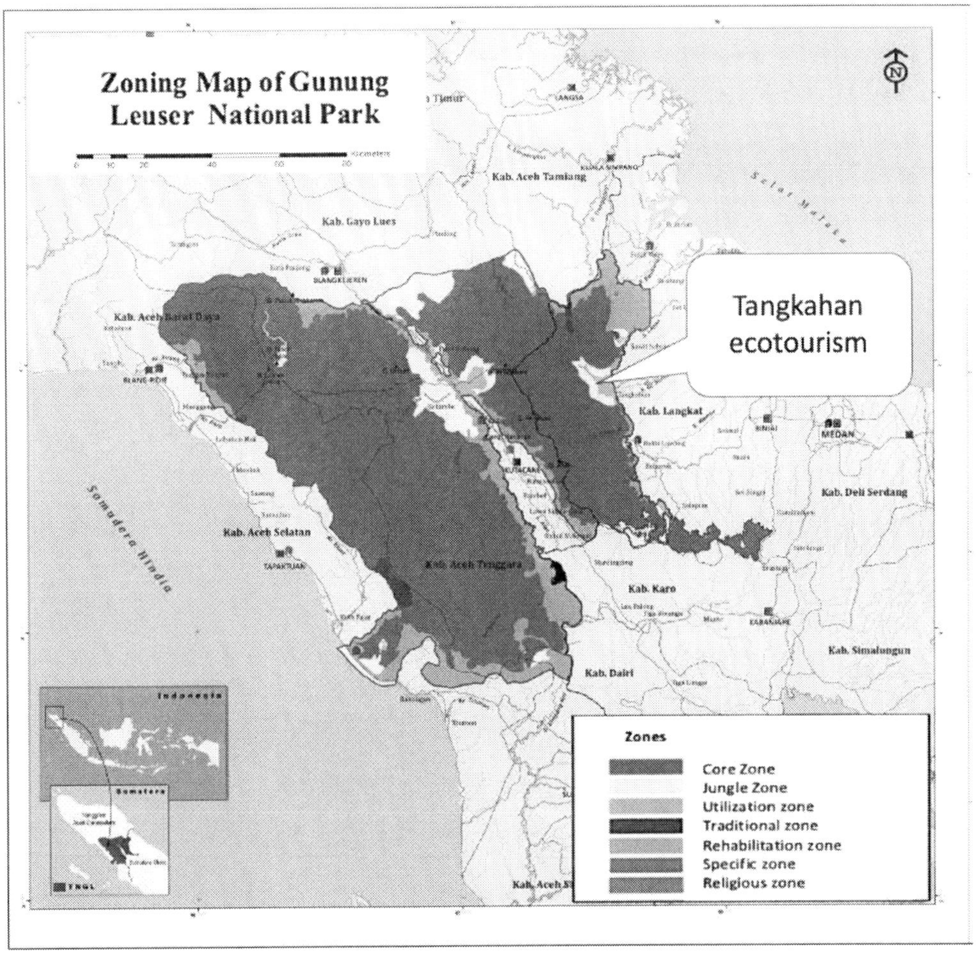

Figure 1. The zoning map of GLNP based on SK Dirjen PHKA No. 35/IV-SET/2014

The data was collected from April to June 2017 and the observation of the current management was continued in December 2021. The total 110 responders included the local people (formers loggers, entrepreneurs, ecotourism staff and residents), the village head and Conservation Response Unit (CRU) management. CRU is a center for the conservation of translocated elephants from conflict areas. Data collection was done by the survey method. An in-depth interview was also performed with the key informants selected by their experience, role and knowledge of the history and the management of the Tangkahan ecotourism. The key informants consisted of former loggers (5 people), initiators of ecotourism or environmentalists (3 people), government officers of GLNP Office Affair (3 people) and ecotourism staff (3 people).

The questionnaire consisted of seven sections, i.e.: (1) the perception of the outcomes/impacts of forest and wildlife conservation programs on the local community; (2) attitude towards wildlife conservation; (3) subjective norm on the community participation in forest and wildlife conservation programs; (4) perceived control; (5) behavioral intention; (6) behavior; and (7) the impact to forest security. The answers were scored following a 5-level Likert scale. Descriptive statistical analysis was first conducted to evaluate the category of ecotourism impacts on forest protection and security using interval scoring. For this analysis, the interval of categories is determined by dividing the gap between the highest and lowest scores (in this research, they are 5 and 1 respectively) by the number of expected categories. This research used five categories. Therefore, the range between categories is 0.80 and the interval of categories is presented in Table 1.

Table 1: Interval class

Interval	Category
1.00 - 1.80	Very low
1.81 - 2.60	Low
2.61 - 3.40	Medium
3.41 - 4.20	High
4.21 - 5.00	Very high

To reveal the determinant factors of community intention in the development of wildlife ecotourism management in Tangkahan, this study employed a study framework presented in Figure 2. The model was evaluated using simple linear regression for one independent variable (BEN—> ATT; BI —>B; B–>IMPCT) and multiple linear regression for more than one independent variable (SN, ATT, PBC—>B).

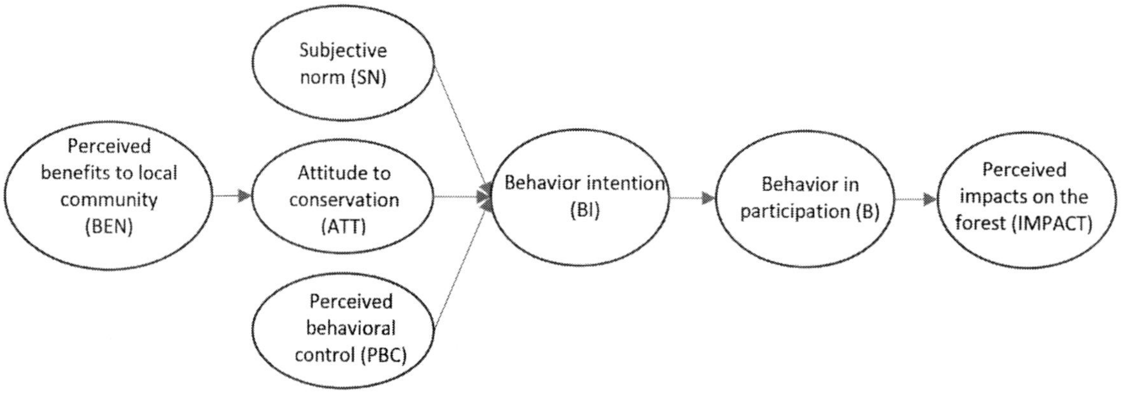

Figure 2. Research framework

3. Result and Discussion

3.1. Establishment of forest and wildlife conservation through ecotourism in Tangkahan

Tangkahan ecotourism in Namo Sialang Village was established in 2002 through signing memorandum of understanding (MoU) between Tangkahan Ecotourism Community (TEC, originally named *Lembaga Pariwisata Tangkahan*) with the GLNP Agency. The TEC was established to reduce illegal logging in GLNP in the 90s and to overcome conflicts with neighboring villagers (Bukit Lawang community) from 1998 to 2001 who first implemented ecotourism in GLNP. Illegal logging activities had steadily contacted the jungle tracking tour area managed by the Bukit Lawang community and increased gradually. Eventually, many cases of illegal logging went to court for the hearing and imprisoned several Namo Sialang loggers. It encouraged the government and environmentalists to find solutions for those environmental and socio-economic threats in Namo Sialang.

Intensive meetings between Namo Sialang villagers and NGOs such as Flora and Fauna International (FFI) and Indonesian Ecotourism Network (Indecon) had yielded a conclusion to form ecotourism in Tangkahan because this area has a huge potential for ecotourism. In 2001, the Tangkahan Simalem Ranger was established and started to promote jungle tourism and river-based tourism. The community soon proposed the legal management of Tangkahan ecotourism to the government and they both signed the MoU in 2002. In that MoU, they collaborated to sustainably manage 17,000 ha of GLNP through ecotourism and agreed that the entry fee should be shared for state revenue. After the operation, the tourist object was improved through the development of the CRU to conserve captive Sumatran elephants in Tangkahan. Since then, the Tangkahan is more popular as the elephant conservation and tourism center and it has attracted international and domestic visitors.

Through the presence of captive elephants in Namo Sialang, the community has integrated forest tourism and wildlife conservation in Tangkahan. Until now, the trained local people maintain 10 captive elephants synergized with the conservation program of Sumatran elephants as a critically endangered species by IUCN in 2012. Visitors can interact with the trained elephants such as feeding, bathing, taking pictures and raiding through rivers and forests. The area surrounded by GNLP has an astonishing rainforest that contains huge biodiversity and wild animals including orangutans (*Pongo* spp), one of the key species in Sumatra. The forest area also has other interesting tourist attractions such as waterfalls, caves, hot springs and rivers that can be explored while doing jungle tacking in GNLP [33].

3.2 The impacts of wildlife ecotourism on forest protection and forest security

The impacts of the development of wildlife-based ecotourism are presented in Table 2.

Table 2. Community's perception of the impact of the development of wildlife-based ecotourism on forest protection and forest security in Tangkahan

Impacts of ecotourism	Score	Categories
Forest Security	4.40	Very high
Forest cover improvement	4.12	High
Wildlife conservation	3.98	High
Biodiversity conservation	3.79	High

Based on the results in Table 2, community-based wildlife ecotourism improves forest protection and conservation, which is categorized as very high and high. The increasing security of the forest is the highest impact on ecotourism development in Namo Sialang. Based on the local people's description, the condition of the forest in Tangkahan was very poor before 2002 due to illegal logging. Since the operation of ecotourism, the local community has been willing to protect the forest from disturbances and let the trees grow naturally for promoting ecotourism purposes that have improved the forest cover. The elephant ecotourism program in CRU Tangkahan also practices an elephant flying squad patrol program that has significantly reduced human and wild elephant conflicts (HEC) and helped to conserve wildlife. The routine patrol of the elephant squads (the team consists of trained elephants and *mahouts* or elephant shepherds) has driven wild elephants away from human areas to the core zone that can avoid the trapping and poisoning of elephants. This program is also implemented in other elephant conservation and training centers in Sumatra [23]. The last, the maintenance of forest for ecotourism purposes also improve biodiversity conservation in GLNP. The Tangkahan ecotourism promotes jungle tracking and camping to educate visitors about forest biodiversity. This also allows them to have a magnificent experience to see key species in GLNP, such as the rafflesia flower and orangutans, or to find the prints of wild animals (tigers, rhinos), such as nests, dung and footprints.

3.3 Evaluating the model

The reliability test was first conducted to evaluate the reliability of the data. The result is presented in Appendix 1. The result shows that Cronbach's alpha of all the items is above 0.30, which means that the data are a good fit and reliable to be continued for further analysis.

Furthermore, the determinant factors of local behavior intention to wildlife ecotourism were assessed using regression analysis and the result is presented in Table 3.

Table 3. The relationship of the evaluated variables

Relationships	Beta Coef.	*p*-Value	R_{adj}^2
BEN —>ATT	0.732	0.000**	0.532
SN —> BI	0.095	0.392	0.617
ATT—> BI	0.328	0.002**	
PBC —> BI	0.572	0.000**	
BI —> B	0.691	0.000**	0.649
B —> IMPACT	0.378	0.001**	0.109

Note: * significant at $p= 0.05$, ** significant at $p= 0.01$

Table 3 explains that perceived benefits of ecotourism significantly influence the community's attitude toward conservation activities. Perceived behavioral control significantly influences behavioral intention to conserve forest and its resources, while subjective norm does not significantly affect the behavioral intention. This result explains that the increasing attitude on conservation and the perceived behavioral control (it refers to the community's capability to run the ecotourism) increases the behavioral intention to participate in forest conservation in the form of ecotourism. Likewise, behavioral intention influences community behavior to participate in wildlife ecotourism. The last, behavior significantly influences the perceived outcomes of ecotourism in protecting the forest.

Community perception on benefits has a significant and positive relationship to attitude. It means that when the community has a high perception of the benefits of ecotourism or when they have gained high benefits from ecotourism, their attitude towards conservation activities in Tangkahan is also high and vice versa. As mentioned by scholars, socio-economic motives can influence community attitudes to join a community-based conservation program [34-36]. The positive attitude to conservation programs can be affected by economic benefits gained from forest resources, the need for social cohesiveness, the need for accessing land/land rights, developing cultural identity and developing organizational capability. In Tangkahan Ecotourism, wildlife ecotourism provides socio-economic rewards to communities by increasing job opportunities and improving income levels [30,31,35]. The research of Purwoko et al [31] calculated the economic value of Tangkahan Ecotourism based on travel cost was 72,708,168,000/year (equivalent to the US $505,514.6). Likewise, from the social aspect, ecotourism has driven a better social life compared to the social life before the operation of ecotourism in Tangkahan. The respondents described that before 2002 most children in Tangkahan were not going to school. The condition of the houses was very poor and no toilets. Additionally, access to Namo Sialang as a remote area was very difficult. After the ecotourism operation commencement, many environmentalists and researchers visited this area and accompanied the communities to improve their life (in education, skills and entrepreneurship), which gradually developed the village into an industrialized village.

Regarding the relationship between subjective norm, attitude and perceived behavioral control towards behavioral intention, previous studies about conservation programs are consistent with this study. Maleksaeidi and Keshavarz [37] found a significant relationship between attitude and BI, an insignificant relationship between SN and BI [38]; and a significant relationship between PBC and BI [38,39]. Behavioral intention is a key factor of individual participation/actions in conservation activities that can be affected by internal factors (individual aspects) and other drivers (other actors and information media). Attitude influences the behavioral intention of Namo Sialang villagers because they believe that jungle and wildlife ecotourism can improve their livelihood and harmonize the community needs and the authority's will. The subjective norm on the other hand is not a significant determinant factor of behavioral intention. It indicates that the community is not influenced by other drivers/actors (such as neighboring villagers, government and NGOs) when they are willing to participate in forest and wildlife conservation programs in the present time. It is reasonable because Tangkahan ecotourism has existed for a long time (since 2002) that has formed personal motives, rather than been influenced by external actors.

Intention to participate in conservation activities significantly affects behavior. It depicts that the people that have more will also realize it in real actions, such as joining the community activities and spending more time/money/personal resources on collective actions. As the outputs/impacts, the behavior to participate in wildlife ecotourism and conservation has created the responsibility of the community that improves the quality of the forest and protects biodiversity.

3.4 Strategies to adopt for maintaining wildlife ecotourism during Covid 19's time

The outbreak of COVID-19 has crippled many republics, especially those whose economy relies on tourism. Many countries have faced severe economic crises amid this pandemic which has created circumstances that require us to reshape our existing tourism policies for post COVID era. The necessary strategies should be adopted to cope with a possible future pandemic outbreak, and it is essential to anticipate the post-COVID crisis situation and to make necessary plans [40].

Tourism, as one of the key elements of Indonesia's economy, has been hit by the pandemic, including ecotourism in Tangkahan, GLNP. The local community that depends on their livelihood from community-based ecotourism has suffered during the Covid pandemic. The post-outbreak known as the new normal era allows reviewing and revising the existing tourism policies to recover ecotourism. However, the management of ecotourism in the post-COVID outbreak needs to be adjusted to the new social life and should follow the government's regulations or policies. Therefore, strategies both short-term and long-term dealing with a possible future pandemic need to be formulated in Tangkahan. In this regard, the following measures should be taken on a short-term and a long-term basis:

- To consider the management transformation from a conventional method to a digital method through improving technology (computer and internet-based) to promote virtual interaction for booking and other services. For instance, developing a complete, clear and updated information on the website and providing tour packages (including transportation services), scheduling, ticketing and payment on the website can help the visitors to be convenient to select and order the trip. Some attractive tourist objects and attractions, such as interactions with captive elephants, river cruising, river tubing, hot spring and waterfalls need to be promoted on the website by providing complete information and interesting presentation (video or pictures) [41]. To support this, it requires strengthening rural communities in their locality to adopt communication technology thus, to promote socio-economic sustainability.
- On a short and long-term basis, in the zones of ecotourism, the new construction sites and the architectural designs should be modernized and should follow the protocol of the COVID outbreak. For instance, the arrangement of catering, hotel facilities and public facilities on the site helps to reduce physical contact between the service providers and visitors.
- In the past, wildlife ecotourism in Tangkahan used to solely depend on foreign visitors. Therefore, the ticket for jungle tracking and elephant tourism is very expensive and is not affordable to domestic visitors. The present situation has changed. There are many travel restrictions both in the departure and destination countries that have caused the reluctance of international travelers to travel abroad. In this regard, the community needs to re-arrange the tour packages in Tangkahan to attract more domestic visitors, for example, reducing the entry fee for the jungle tracking and observing the elephants, but keep paying attention to conservation and animal welfare rules.
- The main problem visitor face while visiting Tangkahan is the bad road access to the area. This area is remote (about 100 KM from Capital City/Medan) and limited in public transportation. The community should propose to authority for the construction of better roads and provide transportation services for a short-term strategy. The community can improve promotion and collaboration with travel agencies that provide more transportation facilities to their customers.

4. Conclusion

TPB has been implemented to evaluate the structure of behavioral change in the local communities in the conservation program in Tangkahan. The behavioral intention for conservation activities is the key to behavioral change, where when the individuals have a high intention, their participation in the conservation activities is also high. The determinant factors of behavioral intention are the attitude towards conservation and perceived behavioral control. Economic and social rewards gained by the community for being active in wildlife ecotourism management significantly influence the community's attitude to conservation programs. The history of shifting of community livelihood from direct services of forest to environmental services through ecotourism depicts a high dependency of the community on the forest and its resources. The management of ecotourism needs to be reviewed and revised to solve problems during the COVID-19 pandemic and post-outbreak. The optional strategies are the transformation of management from a conventional to virtual/digital method, developing and following the new COVID-19 protocol of ecotourism, re-arranging the ticket fee and the tour packages to increase domestic visitor number and improving the access to the area by constructing the roads and provision of transportation services.

Acknowledgment

We acknowledge the Aek Nauli Standardizing Institute (Formely named BP2LHK), for funding the research, the member of Kelti Sosek (Johansen Silalahi, Salman, Firman) for their help to collect the data and all parties who have supported the manuscript.

References

[1] Wang P, Li N, Ye H and Ye H 2008 *Land* **11** 1566
[2] Kijazi A J H 2014 *Int J Protected Areas and Conserv* **20** 27
[3] Kaffashi S, Radam A, Shamsudin M N, Yacob M R and Nordin N H 2015 *Forests* **6** 2345
[4] De Castro E V, Souza T B and Thapa P 2015 *Int J Protected Areas and Conserve.* **21** 51-62
[5] Muhumuza M and Balkwil K 2013 *Int J Biodiversity* **2013** 1
[6] Djufri 2015 *Proc of Nat Seminar on Indonesia people and biodiversity* **1** 1543
[7] YOSL-OIC 2009 *Guide Book to The Gunung Leuser National Park.* (Medan, Indonesia: Yayasan Orang Utan Sumatera Lestari-Orangutan Information Centre YOSL-OIC)
[8] Ministry of Forestry of the Republic of Indonesia (MOFRI) 2003 *Submission for nomination of tropical rainforest heritage of Sumatra* (Jakarta, Indonesia: Directorate General of Forest Protection and Nature Conservation Ministry of Forestry)
[9] Jay D 2015 *Leuser at a crossroads* (Medan, Indonesia: Leuser Ecosystem Alliance)
[10] Purwoko A, Kuswanda W, Situmorang R O P, Hutapea F J, Saputra M H and Pasaribu P H 2022 *Sustainability* **14** 11328
[11] GLNP Agency 2007 *Jejak Leuser Bulletin* **3** 5
[12] Purwanto E 2015 *Tropenbos Infosheet* **3**
[13] Purwanto E 2015 *Anti-encroachment strategy in the tropical forest heritage area of Sumatra: Towards a new paradigm* (Jakarta: TROPENBOS International Indonesia Programme-UNESCO)
[14] Hoare A 2015 *Illegal logging and related trade: The Response in Malaysia* (The Royal Institute of International Affairs Chatham House, London)
[15] Qian J, Xu B, Yang H and Nie Y 2016 *Open J For* **6** 39
[16] Ginting Y, Dharmawan A H and Sekarjakrarini S 2010 *Jurnal Transdisplin Sosiologi, Komunikasi, dan Ekologi Manusia* **4** 39
[17] Adinugroho W C 2009 *Illegal Logging, A Disaster for the Indonesian Forestry World Which Is Not Finished* (Bogor, Indonesia: Bogor Agriculture Institute)
[18] Obasi F A, Agbo F U and Onyenekwe C S 2015 *African J Agri Research* **10** 1233
[19] Elias P 2012 *Logging and the Law: How the U.S. lacey act helps reduce illegal logging in the tropics* (Cambridge: Union of concerned scientists)

[20] McShane T O and Wells M P 2004 *Integrated conservation and development? Getting biodiversity projects to work towards better conservation and development* (New York, USA: Columbia University Press) 3
[21] Rajpoudel N, Fuwa N and Otsuka K 2014 *The impacts of a community forestry program on forest conditions, management intensity and revenue generation in the Dang district of Nepal* (Tokyo, Japan: National Graduate Institute for Policy Studies)
[22] Bandara T W M T W 2020 *Int J Sci Research Publications* **10** 437
[23] Kuswanda W, Garsetiasih R, Gunawan H, Situmorang R O P, Hutapea F J, Kwatrina R T, Karlina E, Atmoko T, Zahrah M, Takandjandji M and Gunaryadi D 2022 *Diversity* **14** 420
[24] Nelson A, Bidwell P and Sillero-Zubiri C 2003 *A review of human-elephant conflict management strategies* (People and Wildlife Initiative; Conservation Research Unit of Oxford University)
[25] Situmorang R O P and Yen T M 2021 *Quarterly Journal of Forest Research* **43** 81
[26] Situmorang R O P and Barus S P 2015 *Proc The 3rd INAFOR Conference 21-22 October 2015 Bogor Indonesia*
[27] Situmorang R O P and Kuswanda W 2019 *Biodiversitas J Biol Diversity* **20** 2858
[28] Situmorang R O, Silalahi J and Kuswanda W 2020 *Jurnal Penelitian Sosial dan Ekonomi Kehutanan* **17** 83
[29] Rijksen H D 1974 *Biol Conserve* **6** 20
[30] Ni'am L, Koot S and Jongerden J 2021 *Geoforum* **127** 162
[31] Purwoko A, Nurrochmat D R, Ekayani M, Rijal S and Garura H L 2022 *Sustainability* **14** 8272
[32] Montano D E and Kasprzyk D 2015 *Health behavior: Theory, research and practice* **70** 231
[33] Tangkahan Ecotourism 2019 About Tangkahan https://tangkahan.id/en/about/
[34] Ruiz-Mallén I, Schunko C, Corbera E, Rös M and Reyes-García V 2015 *Ecol Society* **20** 33
[35] Nilsson D, Gramotnev G, Baxter G, Butler J R, Wich S A and McAlpine C A 2016 *Conserv Biol* **30** 816
[36] Soe K T and Yeo-Chang Y O U N 2019 *For Pol Econ* **100** 129
[37] Maleksaeidi H and Keshavarz M 2019 *Glob Ecol Conserv* **20** e00698
[38] Apipoonyanon C, Kuwornu J K M, Szabo S and Shrestha R P 2020 *J Sustain For* **31** 184
[39] Primmer E and Karppinen H 2010 *For Pol Econ* **12** 136
[40] Cherkaoui S, Boukherouk M, Lakhal T, Aghzar A and El Youssfi L 2020 *E3S Web of Conferences* **183** 01003
[41] Purwoko A and Thoha A S 2021 *IOP Conf Ser Earth and Environ Sci* **912** 012065

IOP Conf. Series: Earth and Environmental Science 1115 (2022) 012005 doi:10.1088/1755-1315/1115/1/012005

Appendix

Appendix 1. Reliability test of the variables

Code	Description		Cronbach's Alpha
BEN1	I have seen a local industry growth in my village	0.838	0.85
BEN2	I have gotten job opportunities to my family and the community	0.823	
BEN3	My family income is improved from the ecotourism activities	0.817	
BEN4	The physical development of my village is improved	0.83	
BEN5	Ecotourism promotes the culture to visitors	0.85	
BEN6	Ecotourism has improved education life in my village	0.837	
BEN7	Ecotourism improves my cognition and skills in conservation	0.822	
BEN8	Ecotourism improves my capacity in business / entrepreneurship	0.834	
BEN9	Ecotourism improves my capacity in organization matters	0.849	
SN1	Joining the forest conservation is influenced by other local residents (peers)	0.747	0.731
SN2	My family often discuss the conservation activities that affects my participation in joining the forest conservation program	0.583	
SN3	The accompaniment of the government/NGOs affects me to join the conservation activities	0.574	
SN4	Personally, I am obligated to help improve the forest conservation around me	0.723	
ATT1	I believe that wildlife ecotourism can improve people's cognition and awareness of environmental issues	0.858	0.893
ATT2	I think that wildlife ecotourism improves forest security	0.862	
ATT3	I think that forest quality should be improved	0.878	
ATT4	Biodiversity and wild animal conservation can be improved through community support	0.875	
ATT5	Forest destructions diminish the quality of the environment and create disaster	0.902	
ATT6	I believe that ecotourism is a good approach to community development	0.874	
ATT7	I think that my contribution is helpful to support forest protection and conservation	0.888	
PBC1	I have a capacity to work in the management of CRU (elephant conservation center)	0.879	0.907
PBC2	I have an ability to maintain the management of CRU	0.866	
PBC3	I have a capability to guide the public in wildlife ecotourism	0.883	
PBC4	I have skills to arrange the wildlife conservation program in Tangkahan	0.895	
PBC5	I can obtain external resources (NGO/government) to develop the conservation activities in Tangkahan	0.911	
BI1	I spent time in conservation activities	0.913	0.932
BI2	I spent money/resources to support the forest and wildlife conservation program in Tangkahan	0.906	
BI3	I use my skills to improve the success of the forest and wildlife conservation program	0.919	

BI4	I am willing to promote the forest and wildlife conservation program	0.922	
BI5	I am willing to persuade/fight for the destroyers of the forest	0.922	
B1	I have attended the community meetings regularly	0.885	0.872
B2	I have actively given ideas to solve the problems of the forest and wildlife conservation program in my village	0.841	
B3	I spent my time to community activities/to improve my skills	0.836	
B4	I participate in financial to support the forest and wildlife conservation program	0.85	
B5	I developed the network to increase the forest and wildlife conservation program	0.847	
B6	I motivate other members to join CBEM or to improve their participation	0.852	
B7	I actively promote the forest and wildlife conservation program through Tangkahan ecotourism	0.863	
IMPACT1	Ecotourism has improved the forest security in Tangkahan/GLNP	0.818	0.864
IMPACT2	Ecotourism has improved the forest covering around my village	0.794	
IMPACT3	Ecotourism has improved wildlife conservation (particularly elephants)	0.85	
IMPACT4	Ecotourism has improved biodiversity protection and conservation in GLNP	0.84	

IOP Conf. Series: Earth and Environmental Science 1115 (2022) 012006 doi:10.1088/1755-1315/1115/1/012006

Evaluation on the growth pattern and biological conditions of Sembilang (*Plotosus canius*) in East Aceh

A P A Samad[1*], M Jamil[2], C Gustiana[2], S Anzitha[2], T Mahyuddin[2] and B Baihaqi[3]

[1]Department of Aquaculture, Faculty of Agriculture Universitas Samudra, Aceh 24416, Indonesia
[2]Department of Agribusiness, Faculty of Agriculture Universitas Samudra, Aceh 24416, Indonesia
[3]Department of English Language Education, Faculty of Teacher Training and Education Universitas Samudra, Aceh 24416, Indonesia

[*]Email : agus.putra.samad@gmail.com

Abstract. Sembilang (*Plotosus canius*) is a delicious and nutritious fish that belongs to the family of Plotosidae and has great demand in coastal communities. Therefore, this research aimed to examine the data on growth patterns, morphometric and meristic measurements, and feeding habits of sembilang in the East Aceh mangrove area. The data collection included growth patterns, fish and fin length, and body width, and then the meristic measurements were carried out by counting the number of fin rays. The results showed that the length and weight of the fish ranged from 22-43 cm and 86-296 gr, respectively. Furthermore, the correlation coefficient r is 0.986 and the regression equation is $W = 0.024L^{2.410}$. Its main diet is shrimp, with an IP rate of 47%, and they also consume feeds such as worms, small fish, crabs, and pieces of leaves. In conclusion, the growth pattern of sembilang (*Plotosus canius*) in the East Aceh mangrove area was negative allometric.

1. Introduction
Mangrove areas are brackish waters where the sea and freshwater meet [1], resulting in various aquatic biota [2,3], including finfish. In plain view, Sembilang has similarities with catfish, but what distinguishes it is its habitat [4,5]. It lives in or around estuary waters directly opposite the sea, having the soil and mud as substrates at a depth of 0-10 m [6]. Furthermore, it belongs to the carnivorous fish that prey on small fish, gastropods, mollusks, and crustaceans [7,8].

Sembilang is a fishery commodity with mass cultivation and sustainable potential because of its meat quality and high economic value [9,10]. However, efforts to manage natural resources have not been taken seriously until now, and there is increasing consumer interest in sembilang, resulting in high fishing activities in the wild. This condition can lead to a decline in the population of this fish [7].

For sustainable management, an overview of the biological aspects and eating habits of sembilang is required, thereby the data can be used as initial information for implementing aquaculture systems and preventing overfishing [9]. Therefore, this research aims to analyze the growth patterns such as the relationship between length, morphometric and meristic measurements, condition factors, and eating habits. It was also conducted to collect the data that can be used as a source of information for the community and policymakers in preserving the commodity in the waters of East Aceh.

2. Materials and Methods

2.1. *Time and place*
This research was conducted in the East Aceh mangrove area for 95 days from March to June 2021. The sample used was 100 sembilangs, and sampling was conducted directly in the field. In contrast, morphometric and meristic data measurements and observations of feed consumption were carried out at the Aquaculture Laboratory of Samudra University.

2.2. *Research methods*
During measurement, all the fish were stunned using ice cubes to prevent the release of poisonous pectoral and dorsal fins. The morphometric measurements were then carried out using a ruler, and the parameters measured were total, standard, head, and fin length, as well as body, eye, and mouth opening width. On the other hand, the meristic measurement uses a magnifying glass to calculate the total number of fin rays of the fish.

2.3. *Data analysis*
Descriptive and quantitative analyses were carried out on the research data, then presented in tables and graphs. The parameter data displayed includes all length-weight, morphometric, and meristic relationship data, condition factors, eating habits, and water quality conditions obtained during the research. The formula as below:

$$W = aL^b \qquad (1)$$

whereas: W = fish weigtht (gr), L = fish length (cm), a and b = constanta.

$$Kn = \frac{W}{aL^b} \qquad (2)$$

Whereas: Kn = condition factor, W = average fish weight (gr), L = average fish length (cm), a dan b = regression constanta

3. Results and Discussions

3.1. *Morphometric measurement*
The data of the sembilang (*P. canius*) morphometric measurements are shown in Table 1.

Table 1. Morphometric measurement of *P. canius* in East Aceh mangrove area

Morphometric Parameter	Value (cm)
Total length	32.89 ± 4.35
Standard length	29.57 ± 5.33
Head length	6.32 ± 0.66
Anal fin length	16.54 ± 1.02
Caudal fin length	2.55 ± 0.23
Pectoral fin length	3.43 ± 0.35
Ventral fin length	1.25 ± 0.23
Dorsal fin length	2.85 ± 0.32
Body width	3.51 ± 0.22
Eye width	0.65 ± 0.08
Mouth opening	4.34 ± 0.43

Table 1 shows that the average range of total and standard length in the morphometric measurement is 32.89 and 29.57 cm, respectively. In addition, the head, anal, and tail lengths were 6.32, 16.54, and

2.55 cm, respectively. Furthermore, the average body width was 3.51 cm, with a mouth opening width of 4.34 cm. These research results are consistent with the previous data, which showed that sembilang in the Tanjung Pinang estuary waters had a total length range of 29 -39 cm [11].

3.2. Meristic Measurement

The meristic measurement results of sembilang showed four fin rays on the dorsal fin, consisting of 1 hard fin and 3 soft rays. This value is different from the number of anal, pectoral, and ventral fin rays, ranging from 10-11, respectively, while the tail fin ranges from 5-12. The results of meristic measurements are shown in Table 2.

Table 2. Meristic measurement of *P. canius* in East Aceh mangrove area

Meristic Parameter	Number of Fin Rays
Dorsal fin rays	4.73 ± 0.13
Anal fin rays	10.04 ± 0.04
Ventral fin rays	10.02 ± 0.82
Pectoral fin rays	10.13 ± 0.92
Caudal fin rays	7.72 ± 0.57

According to Suharti et al [12], some fish body parts can be measured or counted, such as the number of scales and fin rays. In this research, the meristic measurements showed that the average number of dorsal and anal fin rays was 4.73 and 10.04, respectively. Furthermore, the number of ventral and pectoral fin rays is 10.02 and 10.13, respectively, and the average number of caudal fin rays was 7.72.

3.3. Length-weight Relationship

The length and weight relationship data are shown in Table 3.

Tabel 3. Length and weight relationship of *P. canius*

Parameters	Value
Sample (fish)	100
Total length (cm)	22-43
Body weight (gr)	86-296
Regression coefficient (a)	0.035
Regression coefficient (b)	2.410
Determination coefficient (R^2)	0.963
Correlation coefficient (r)	0.986
Regression equation (power)	$W=0.024L^{2.410}$
T-test of b value	T value > t table
Growth pattern	Negatif allometric

The figure below shows the relationship between the length and weight of sembilang during the research:

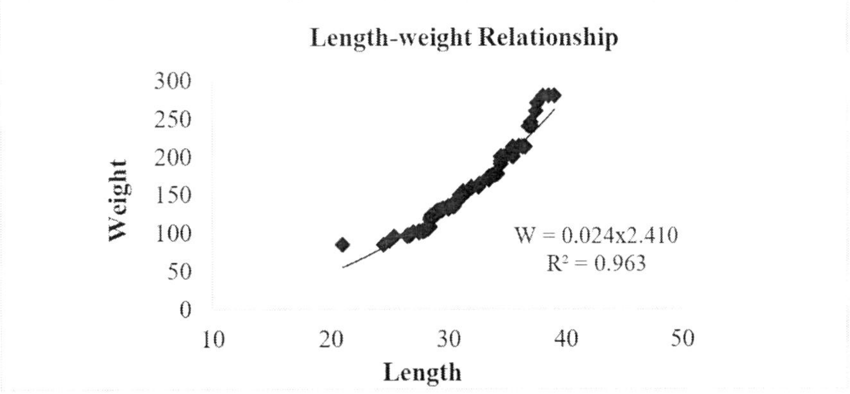

Figure 1. Relationship of the length and weight of *P. canius*

The information about the relationship between the length and weight of a fish species is necessary for estimating its fishery potential [3] because this activity can aid in determining the composition of organism availability in nature, the age of gonad maturity, mortality, life cycle, growth, and reproduction of a living organism. According to Samad et al [13-14] factors such as water quality conditions, availability, and continuity of feed, size of seed stocking, sex, and the number of other fish that utilize the same resource are involved in the fish growth process.

This research result showed that the total length and weight of sembilang ranged from 22-43 and 86-296 gr, respectively. Furthermore, the relationship analysis between the length and weight showed the regression equation $W = 0.024L^{2.410}$ with a coefficient of determination R^2 and b value of 0.963 and 2.410, respectively. The data showed that the sembilang growth in this region was negative allometric. This result is consistent with [7], stating that when the value of $b < 3$, then the organism's growth is negative allometric. However, when $b > 3$, then it is positive allometric. [15] added that the swimming behavior of fish also influenced the b value. In this case, an active swimming fish had a higher b value than the passive. Furthermore, [16] explained that the difference in the value of b is related to the amount of energy used in movement activities and gonadal development.

3.4. Condition Factor

The results of the sembilang condition factor analysis are shown in table 4.

Tabel 4. Condition factor of *P. canius*

Weight (gr)	Condition Factor
86-112	0.918 – 1.531
113-139	0.919 – 1.102
140-166	0.941 – 1.141
167-193	0.924 – 1.024
193-220	0.960 – 1.083
221-247	0.936 – 1.063
248-274	1.027 – 1.057
275-296	1.057 – 1.126

Table 4 showed that the lowest condition factor was found in the weight group of 86 – 112 gr, which was in the range of 0.918 – 1.531, while the highest is 275 – 296 gr ranging from 1.057 – 1.126 is probably due to the different distribution of fish. Furthermore, the high and low value of the condition factor is caused by the high difference in the population density of organisms in a place [16].

Furthermore, [6,17] reported that fish living in high-density waters tend to have low condition factor values due to the narrow space for movement, thereby affecting growth and food struggle.

3.5. Index of preponderance

Based on the observations, crabs, shrimp, worms, fish pieces, and leaves were found in the intestines of the sampled fish, indicating that they are carnivores because they feed on animals. The analysis results using the index of preponderance are shown in Figure 2.

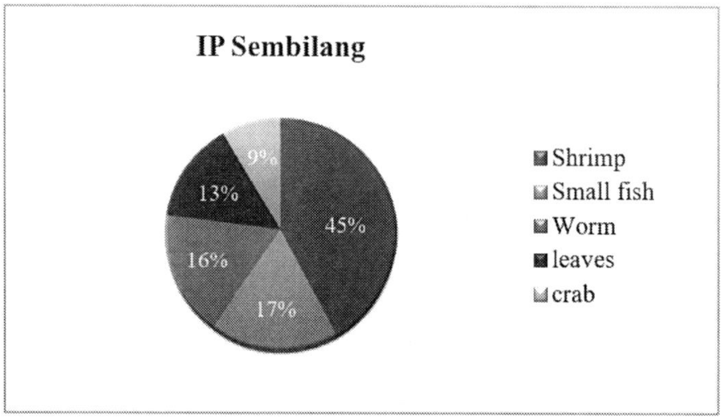

Figure 2. Index of preponderance of *P. canius*

According to Angraini et al [7], the grouping based on the Index of preponderance (IP) values for main, supplementary, and complementary food was IP > 40%, IP 4-40%, and IP < 4%, respectively. Based on these groupings, the main food of fish was shrimp, with a value of 45%, followed by complementary, namely pieces of fish, worm, leaves, and crabs, with a value of 17%, 16%, 13%, and 9%, respectively. According to [11], crustaceans are the most common, and both shrimp and crabs belong to this group. Therefore, the main food of *P. canius* was crustaceans, with a value range of 34.4 -50.2%. The results of [10] also showed that the most common organisms found in *P. canius* were crustaceans with a value range of 58-65%. According to [18], crustacean organisms are the main food with a value of 79%. However, the number of organisms eaten by fish varies due to the distribution in each region [19]. In general, several factors influence fish-eating habits, namely habitat, fish preference for certain types of food, size, color, season, and age.

3.6. Water quality

The results of water quality measurement are as shown in table 5:

Tabel 5. Water quality parameter

Parameter	Value
pH	6.5 – 8.1
Temperature	28.5 – 30.2 °C
Dissolved Oxygen	5.5 – 5.8 mg/l
Salinity	27.0 – 32.2 ppt

The temperature range at the research site was 28.5-30.2 °C, and the pH value was 6.5-8.1, which is considered optimum for the life of the fish [7]. The same results of water quality measurements around the mangrove forests of the east coast of Aceh were reported by Samad et al [20]. Furthermore, the DO value was 5.5 – 5.8 mg/L, which is consistent with water quality in the Sumbawa mangrove waters [19]. The salinity in this research, which ranges from 27.2 – to 30.4, is in accordance with the results of [7].

The growth pattern of sembilang (*Plotosus canius*) is negative allometric. However, the length and weight relationship show a very close correlation, with r = 0.963, indicating that the longer the fish body, the heavier it is. In addition, the length and weight of each were 22 – 43 cm and 86 – 296 gr, respectively. These results indicated that the mangrove area was relatively narrow for these species; hence, it affected their growth.

Acknowledgment
Authors wishing to thanks to all the staffs of the Aquaculture Laboratory who have facilitated the provision of space and equipments during this research.

References
[1] Hasanah F T 2020 *J Geogr* **20** 13
[2] Syahrial S, Saleky D, Samad A P A and Tasabaramo I A 2020 *J. Sumberd Akuatik Indopasifik* **4** 1
[3] Syahrial S, Anggraini R, Samad A P A, Ikhsan N, Saleky D and Hasidu 2020 *J ENGGANO* **5** 2
[4] Irawan H 2019 *Intek Akuakultur* **3** 1
[5] Sari C K, Dewita and Sumarto 2018 *J Online Mhs Fak Perikan dan Ilmu Kelaut Univ Riau* **5** 1
[6] Fatah K and Asyari A 2017 *BAWAL Widya Ris Perikan Tangkap* **3** 4
[7] Angraini R M, Desmelati, and Sumarto 2019 *Berk Perikan Terubuk* **47** 1
[8] Fatah K and Asyari 2011 *BAWAL* **1** 1
[9] Yulianto T, Atmadja W K, Zulpikar Z, Ariska R and Suryanti A 2020 *Depik* **9** 3
[10] Muharam D, Sulistiono and Riani E 2021 *J Teknol Perikan dan Kelaut* **11** 2
[11] Safitri D, Susiana S and Suryanti A 2021 *J Akuatik lestari* **4** 2
[12] Suharti R, Saktiawan K Y, Rachmad B, Triyono H and Zulkifli D 2018 *J Kelaut dan Perikan Terap* **1** 1
[13] Samad A P A, Santoso U, Lee M C and Nan F H 2014 *Fish Shellfish Immunol* **36** 2
[14] Samad A P A, Amin M, Baihaqi B, Hatta M and Ayuzar E 2021 *Depik* **10** 3
[15] Muchlisin Z A, Nurfadillah N, Arisa I I, Rahmah A, Putra D F and Nazir M 2017 *Biodiversitas J Biol Divers* **18** 2
[16] Hasan U and Afriani D T 2021 *J Aquac Indones* **1** 1
[17] Aufar T F Z, Kunarso K, Maslukah L, Ismunarti D H and Wirasatriya A 2021 *Indones J Oceanogr* **3** 2
[18] Triyulianto T 2018 *Intek Akuakultur* **2** 1
[19] Saraswati N L G R A, Rustam A Y, Salim H L, Heriati A and Mustikasari E 2017 *J Segara* **13** 1
[20] Samad A P A, Agustina P and Herri M 2020 *J Ekon dan Pembang* **11** 1

Growth and exploitation rate of Dwarf Whipray (*Brevitrygon heterura*) landed from Tanjung Beringin, Serdang Bedagai Regency, North Sumatera Province

V R Manurung*, M Nababan, S F P Nasution, Desrita, J S Hasibuan and H Husaini

Aquatic Resources Management Program Study, Faculty of Agriculture, Universitas Sumatera Utara, Medan, North Sumatera 20155, Indonesia.

*Email: vindyrilani.m@usu.ac.id

Abstract. Stingrays are a group of fish that have high economic potential landed at TPI Tanjung Beringin. The dominant type of stingray landed at TPI Tanjung Beringin is the Dwarf Whipray (*Brevitrygon heterura*) or commonly called Tuka-Tuka by the local community. This research wa carried out during July and September of 2021 at TPI Tanjung Beringin, survey method with a census technique, namely a method by seeking various information and various facts regarding the biological aspects. This study aims to determine the growth and exploitation rate of the dwarf whipray. There were 82 dwarf whipray, which were then measured for length and weight. The results of the study concluded that dwarf whipray (*Brevitrygon heterura*) which landed in TPI Tanjung Beringin had a negative allometric growth pattern, namely length growth was greater than weight growth. The exploitation status of the dwarf whipray (*Brevitrygon heterura*) in the waters of the Malacca Strait is classified as overfishing.

1. Introduction

Serdang Bedagai Regency has considerable potential in the field of fisheries, such as capture fisheries, aquaculture and public waters [1]. Tanjung Beringin is one of the sub-districts in Serdang Bedagai Regency. This sub-district borders the Malacca Strait and is a fishery production center in Serdang Bedagai Regency, where there is a Fish Auction Place (TPI) to support fishery production in the area.

One group of fish that has high economic potential that landed in TPI Tanjung Beringin is a member of the Dasyatidae or stingrays. Stingray production in the province of North Sumatera is high. Statistical data information from Ministry of Maritime Affairs and Fisheries, the production of stingray catches in 2018 reached 22,868.47 tons [2]. The dominant type of stingray landed at Tanjung Beringin TPI is the dwarf whipray (*Brevitrygon heterura*) or commonly called the tuka-tuka by the local community. The dwarf whipray is a demersal fish that live the bottom of the water. Dwarf whipray live in shallow waters and continental shelves to a depth of 50 m, sometimes found in coastal areas of bay waters. The dwarf whipray is benthic in coastal habitats, and occurs over soft substrates, often near coral reefs from the surface to a depth of 50 m [3].

The dwarf whipray is one of the fish that is in demand by the people in the area. The demand for this fish which is always high has resulted in fishermen continuously catching fish to fulfil market demand. It is feared that the continuous activity of catching dwarf whipray will cause a decrease in the stock of dwarf whipray in the waters. Therefore, it is necessary to study the growth and exploitation rate of the dwarf whipray that landed at Tanjung Beringin, North Sumatera province in order to obtain information that forms the basis for the management of the dwarf whipray.

2. Methodology

2.1. Research site

This research wa carried out during July and September of 2021. *Brevitrygon heterura* was caught in Malacca Strait and landed at Tanjung Beringin, Serdang Bedagai Regency, North Sumatera Province. Research location can be seen in Figure 1.

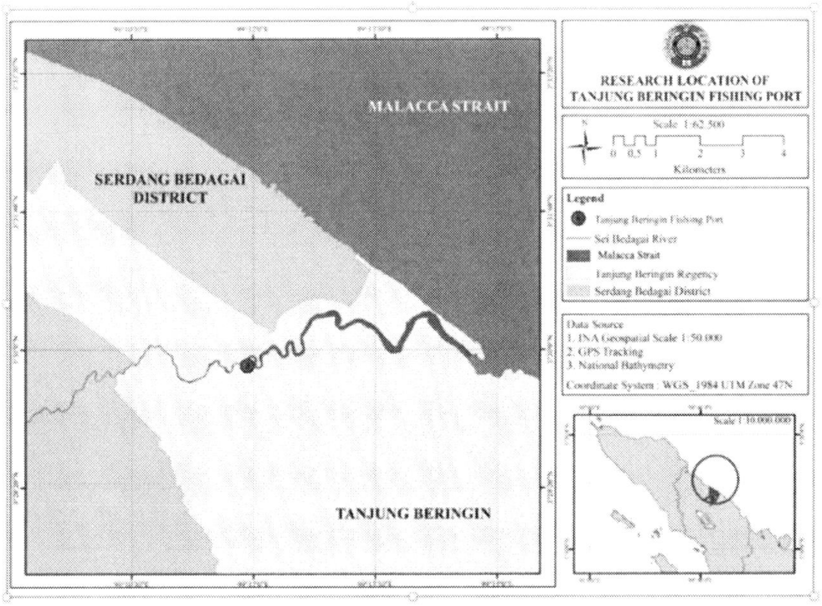

Figure 1. Research location

2.2. Tools and materials

The tools used in primary data collection include writing instruments, roll meter with an accuracy of 1 mm, a digital camera, scales, Microsoft Excel 2010 and the FISAT II program. The material used in this study was the dwarf whipray (*Brevitrygon heterura*) as the object of research.

2.3. Research procedure

This research methodology combines a survey method with a census strategy, namely a method by seeking various information and various facts regarding the biological aspects of fish species [4]. Sampling of stingrays was carried out at TPI Tanjung Beringin.In this study, both primary and secondary data were collected. Primary data was obtained by direct sampling of fish at TPI Tanjung Beringin as a result of fishing catches carried out once a month. After being cought, the fish were measured for thei overall length and wet weight. While secondary data is the form of data on the average surface temperature of the water.

2.4. Data analysis

Sex Ratio. The sex ratio is the ratio between the sexes of the fish in the waters. In statistics, the concept of the ratio is the proportion of a certain population to the total population with the sex ratio formula is [5]:

$$P = \frac{A}{B} x\ 100\% \quad (1)$$

Information :
P = Sex Ratio (%).
A = Number of male / female fish.
B = Total individual male and female fish (tail).

Furthermore, to test the balance of the sex ratio, the following formula is used [6]:

$$X^2 = \sum_{i=1}^{n} \frac{(oi-ei)^2}{ei} \quad (2)$$

Information :
X^2 = Chi Square (the value of the random variable X^2 whose number of samples is close to Chi squared).
Oi = Frequency of observed male or female fish k – i.
Ei = The number of expected frequencies of male and female fish.

2.4.1 Relationship length weight. Weight can be thought of as a function of length. The relationship between length and weight can be determined by the formula [7]:

$$W = a\ L^b \quad (3)$$

Information:
W = Weight (gram)
L = Length (mm)
a and b = Constant

The growth pattern is determined from the value of the constant b (slope) obtained from the calculation of length and weight through the following hypothesis:
H0 : value b=3, growth pattern is isometric
H1 : value of b≠3, growth pattern is allometric, namely:
a) if the value of b>3, positive allometric
b) if the value of b<3, allometric is negative

In order to accept or reject the hypothesis put forth in order to ascertain the value of b, a T test is conducted. After that, the value of tcount is compared with the value of 2table so that the decisions that can be taken are as follows:
Tcount> Ttable, then reject H0
Tcount< Ttable, then fail to reject H0

2.4.2 Condition factor. The condition factor is determined by the formula:

$$FK = \frac{W}{aL^b} \quad (4)$$

Information:
FK = condition factor
L = total length of fish (mm)
W = Weight of fish (grams)
a and b = Constant

2.4.3 Mortality and exploitation rate. The total mortality rate (Z) was estimated using the Jones and Van Zalinge method packaged in the FISAT II program. Mortality of natural rate (M) is estimated by using the following formula [8]:

$$\text{Log } M = -0.0066 - 0.279 * \text{Log }(L\infty) + 0.6543 * \text{Log }(K) + 0.4634 * \text{Log }(T) \quad (5)$$

Information :
M = Mortality of natural
L = Asymptotic length of Von Bertalanffy
K = Growth coefficient of Von Bertalanffy
T = Average water surface temperature (^{0}C)

The fishing mortality rate (F) can be determined using the following formula:

$$F = Z - M \tag{6}$$

The total mortality rate (Z) was estimated using the Jones and Van Zalinge method which was packaged in the FISAT II program. The exploitation rate (E) is determined by comparing fishing mortality (F) to total mortality (Z) [9]:

$$E = \frac{F}{F+M} = \frac{F}{Z} \tag{7}$$

The fishing mortality rate (F) or the optimum exploitation rate is as follows [8]:

$$\text{Foptimum} = \text{M and Eoptimum} = 0.5 \tag{8}$$

The optimum Exploitation value is 0.5. So if the exploitation value is more than 0.5 then it can be said to be an indication of overfishing conditions, especially due to catching.

3. Results and Discussion

3.1. *Morphology of dwarf whipray (Brevitrygon heterura)*

The dwarf whipray (Brevitrygon heterura) which landed in TPI Tanjung Beringin has morphological characteristics that sized small and oval in shape, has a pointed snout. This is according to [3] which states that the dwarf whipray is a very small type of stingray with an oval-shaped disc. The dwarf whipray have a body width of only 24 cm with a total length of 46 cm.

Figure 2. The Dwarf Whipray (*Brevitrygon heterura*)

The dorsal surface is pale greenish brown while the ventral surface is white with dusky disc margins. The dwarf whipray have a tail that is not too long where the tip of the tail is rounded in adult females. There are small spines at the base of the tail and there are 1 or 2 caudal stings in the middle of the tail. This is consistent with [3] which states that the dwarf whipray has a short to slightly elongated tail, which is 1-1.7 times the width of the body, where the tip of the tail is rounded in adult female fish. There are 1 or 2 lance-shaped stinging spines in the middle of the tail.

3.2. Sex ratio

There were 82 dwarf whipray (*Brevitrygon heterura*) obtained during the study, consisting of 40 females and 42 males (Table 1).

Table 1. Sex Ratio of *Brevitrygon heterura*

Month	Gender		Sex Ratio
	Female	Male	
July	8	8	1:1
August	18	16	1.12:1
September	14	18	1:1,27
Total	40	42	1:1.04

This it can be obtained sex ratio between female and male is 1: 1.04. Thus, the number of male and female fish is said to be unequal. This refers to [10] which says that male and female fish in the waters are said to be balanced if they have a ratio of 1:1, where male and female fish have the same number. According to [11], the difference in the number of female and male fish in a waters is caused by fish activity in the waters, adaptability, genetic factors, food, fishing factors, migration, changes in new fish species in an existing population, and growth patterns.

3.3. Length frequency distribution

The following is a graph of the frequency distribution of the length of the dwarf whipray that landed at the TPI Tanjung Beringin can be seen in Figure 3. The most caught dwarf whipray were in the length class 343-375 mm, which was 23 fish. While the smallest size group is in the 442-474 mm class, which is as many as 2 fish, where the fish with the largest size is in this class.

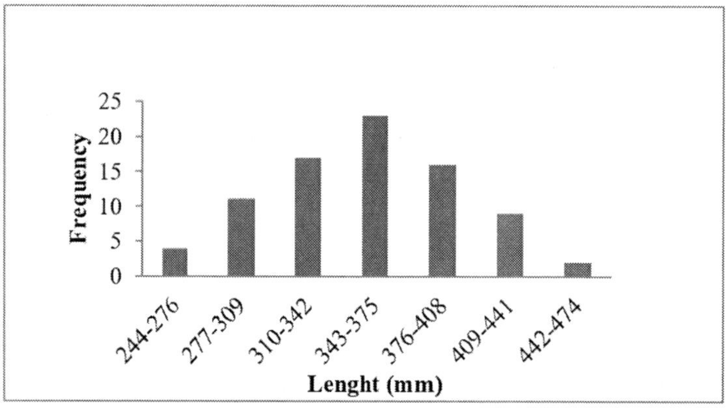

Figure 3. Length frequency distribution graph

Meanwhile, the fish with the smallest size were in the class 244-276 mm as many as 4 fish. According to [2] the dwarf whipray have a body width of only 24 cm with a total length of 46 cm. The body width of an adult male stingray is 16-18 cm and that of a female stingray is 17 cm.

3.4. Length-weight relationship

The study of all data from the relationship between length and weight reveals that the relationship between length and weight has the following equation $W = 0.0073L^{1.7374}$ with a determination value (R^2) = 0.3783 and a correlation coefficient (r) = 0.6151 in figure 4. Meanwhile, for female fish it has the equation $W = 0.000002L^{2.7557}$ with r = 0.6522 in figure 5 and male fish is $W = 0.00151L.^{9754}$ with r = 0.8161 in Figure 6.

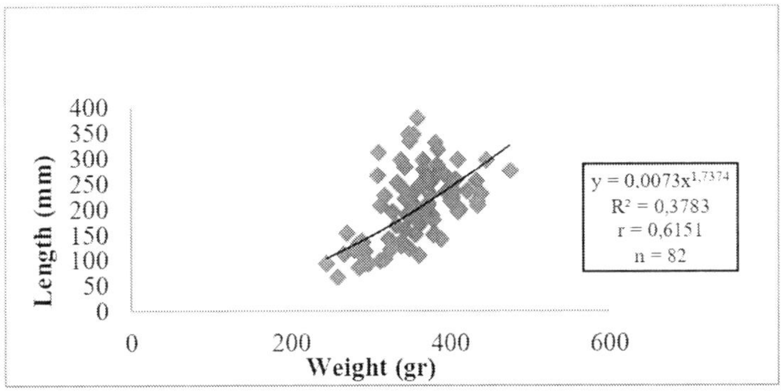

Figure 4. Length-Weight Relationship of Female and Male Brevitrygon heterura

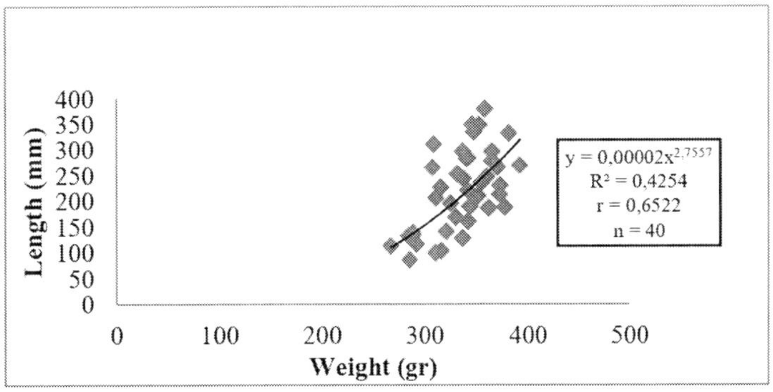

Figure 5. Length-Weight Relationship of Female Brevitrygon heterura

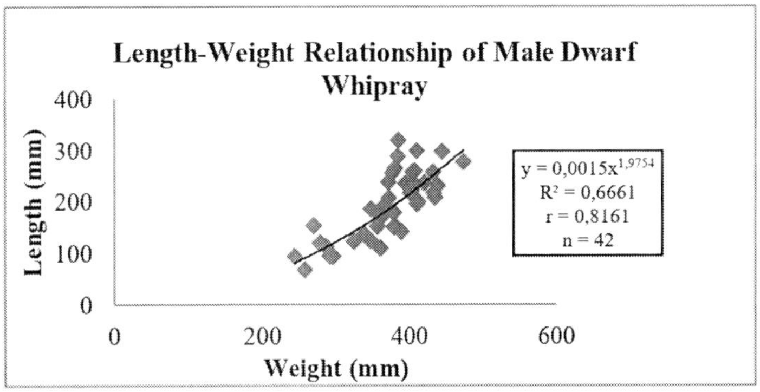

Figure 6. Length-Weight Relationship of Male Brevitrygon heterura

Based on the above equation, it is known that the value of b for all fish is 1.7374 while for female fish b = 2.7557 and male fish b = 1.9754. This shows that the dwarf whipray has a negative allometric growth pattern, namely the growth in length is greater than the growth in weight. According to [12], the growth pattern is divided into two, namely isometric and allometric. It is said to be isometric if the value of b = 3 where the weight growth pattern is proportional to the length growth pattern. Meanwhile, when the value of $b \neq 3$ shows an allometric growth pattern where the weight growth pattern is not proportional to the length growth pattern.

The correlation coefficient (r) of female and male the dwarf whipray are 0.6522 and 0.8161, respectively. The value of (r) in female dwarf whipray is lower than that of male whose r value is close

IOP Conf. Series: Earth and Environmental Science 1115 (2022) 012007 doi:10.1088/1755-1315/1115/1/012007

to 1 which describes a close relationship between the two. According to [6] The correlation coefficient (r), which can be calculated using the formula R^2: where R^2 is the coefficient of determination, demonstrates the close link between the length and weight of the fish. A value that is near one indicates a close relationship between the two, and a value that is far from one indicates a distance between the two.

3.5. Condition factor

The results of condition factors (K) for female and male *Brevitrygon heterura* at Tanjung Beringin are based on negative allometric growth patterns (length growth is greater than weight) as can be seen in Table 2.

Table 2. Condition factors of *Brevitrygon heterura*

Gender	Amount	Condition Factor (K)	Information
Male	42	0.654-1.664	Less flat
Female	40	0.666-2.139	Less flat
Total	82	0.545-2.011	Less flat

The condition factor value of all *Brevitrygon heterura* is 0.545-2.011, where condition factor male ranged from 0.654-1.664 and female 0.545-2.011. This shows that both male and female *Brevitrygon heterura* have less flat bodies. According to [13] says that if the K value of a type fish 1-3, then the condition of fish is flat. The condition factor values for fish with slightly flat bodies (thin). Variations in condition factors in fish are strongly influenced by body size, age, sex, gonad maturity and behavior before and after spawning.

3.6. Exploitation rate and mortality

The calculated mortality is the total mortality rate (Z), the natural mortality rate (M) and also the fishing mortality rate (F). To estimate the natural mortality rate is Pauly's formula used with the average surface temperature of the waters of the Malacca Strait 30.8°C. The results of the analysis of the estimated mortality rate and exploitation rate of *Brevitrygon heterura* can be seen in Table 3.

Table 3. Mortality and exploitation rate of *brevitrygon heterura*

Parameter	Observation Month (July-September)
Fishing Mortality (F)	1.170
Natural Mortality (M)	0.924
Exploitation Rate (E)	0.559
Total Mortality (Z)	2.094

The total mortality rate (Z) of the dwarf whipray is 2.094 per year, the natural mortality rate (M) is 0.924 and the fishing mortality rate (F) is 1.07. This shows that the fishing mortality rate is higher than the natural mortality rate. According to [14], the high rate of fishing mortality and the decreasing natural mortality rate can also indicate the alleged occurrence of growth overfishing conditions, namely the small number of old fish because young fish do not have time to grow due to being caught so that fishing pressure on these stocks should be reduced to reach the optimum condition, namely the mortality rate, catch equals the natural mortality rate predators, illness, and age all have an impact on natural mortality, but environmental factors like maximum length, growth rate and average also have an impact [15].

The exploitation rate of the dwarf whipray landed at Tanjung Beringin is 0.559, meaning that 55.9% of the deaths of the dwarf whipray are caused by catching. According to [9] the higher the fishing mortality rate (F), the higher the exploitation rate (E). The optimal exploitation value is 0.5. So if the exploitation value is more than 0.5, it can be said that the exploitation conditions of the dwarf whipray in the waters of the Malacca Strait have experienced overfishing or excessive fishing activities. This assumption is supported by [16] it is suspected that the *Brevitrygon heterura* has undergone a 30–49% reduction in population over the last three generation lengths (26 years) and it is assessed as Vulnerable.

4. Conclusion

The *Brevitrygon heterura* that was observed at Tanjung Beringin has a negative allometric growth pattern, meaning that the growth in length is increasing faster than the growth its weight. The exploitation status of *Brevitrygon heterura* in the waters of the Malacca Strait is classified as overfishing.

Acknowledgment

This research was supported by TALENTA Lembaga Penelitian Universitas Sumatera Utara 2021 No. 6789/UN5.1.R/PPM/2021. The author also thanks Universitas Sumatera Utara for support as well.

References

[1] Nainggolan H L, Tampubolon J and Ginting A 2018 *Indonesian Journal of Fisheries Science and Technology* **15** 139
[2] Statistik KKP https://statistik.kkp.go.id
[3] Last P R, White W T, de Carvalho M R, Seret B, Stehmann M F W and Naylor G J P 2016 *Rays of The World* (USA : Cornell University Press) **1** 533
[4] Pais M P and Cabral H N 2018 *PeerJ* **6** 1
[5] Codrijah U and Faizah R 2021 *IOP Conf. Ser Earth Environ Sci* **890** 012054
[6] Walpole R E 1992 *Pengantar Statistik* (Jakarta : Gramedia Pustaka Utama)
[7] Effendie M I 1979 *Metode Biologi Perikanan* (Bogor : Yayasan Dewi Sri) **3** 112
[8] Sparre P and Venema S C 1999 *Introduksi Pengkajian Stok Ikan Tropis* (Jakarta: Badan Penelitian dan Pengembangan Pertanian) 438
[9] Pauly D 1984 *International Center for Living Aquatic Resources Management ICLARM Studies and Reviews* **8** 325
[10] Effendie I 2002 *Biologi Perikanan* (Bogor: Yayasan Pustaka Nusatama) **163**
[11] Hart P J B and Reynolds J D 1997 *Handbook of Fish Biology and Fisheries* (UK: Blackwell Publishing) **1** 225
[12] Bintoro G, Lelono T D, Setyohadi D, Maharani H and Rihmi M K 2021 *IOP Conf. Ser Earth Environ Sci* **919** 012018
[13] Susetya I E, Syahara L, Soemaryono, Fadhilah A and Yusni E 2020 *IOP Conf. Ser Earth Environ Sci* **454** 012134
[14] Barnett L A K, Branch T A, Ranasinghe R A and Essington T E 2017 *Old Growth Fishes Become Scarce Under Fishing Current Biology* **27** 5
[15] Manurung V R, Nasution S F P, Nababan M, Desrita and Hasibuan J S 2022 *IOP Conf Ser Earth Environ. Sci* **977** 012114
[16] The IUCN Red List of Threatened Species https://www.iucnredlist.org

IOP Conf. Series: Earth and Environmental Science 1115 (2022) 012008 doi:10.1088/1755-1315/1115/1/012008

Welfare management of Sumatran Tiger (*Panthera tigris sumatrae*) in Medan Zoo and Siantar Zoo

M Zahrah[1*], M Ulfa[1] and R P Saragih[2]

[1]Forestry Faculty of Universitas Sumatera Utara, Medan, North Sumatera, Indonesia
[2]Forestry Undergraduate Program Students, Universitas Sumatera Utara, Medan, North Sumatera, Indonesia

[*]Email: marifatin@usu.ac.id

Abstract. *Panthera tigris sumatrae* is one of endemic species of Sumatera Island included in the endangered species red list by the IUCN. The efforts to preserve the sumatran tiger is with conservation activities both in-situ and ex-situ. One form of ex-situ conservation is to keep animals in the zoos. This study aims to assess the welfare level of Sumatran tigers are in Medan Zoo and Siantar Zoo. Aspects of tiger welfare are measured through 5 variables basedon five animal freedom, namely the free of hunger and thirst, the free of environmental discomfort, the free of pain, wounds and diseases, the free of natural behavior and the free of fear and suffering. Data collection through direct observation of animal management and assessment of minimum animal welfare standards. Interviews with veterinarians and animal keepers about human resources and management activities carried out in supporting animal welfare. The assessment is carried out by the management, namely veterinarians, animal keepers, researchers and visitors to obtain objective results. The implementation of sumatran tiger welfare management in Medan Zoo has an average of 76.9 with the category is good and for Siantar Zoo has an average of 95 with the category is very good. Recommendation that needs to be considered from the results of this study is that Medan Zoo needs to add enrichment in the cages that tigers can behave naturally.

1. Introduction

Sumatran tigers (*Panthera tigris sumatrae*) live in isolated populations on Sumatera Island resulting in habitat loss, illegal trade and reduced prey animals in nature [1]. According to IUCN records, in 2021 the sumatran tiger population is less than 400 individuals [2].

Tigers (*Panthera tigris*) are a wildlife that can be found in various types of habitats ranging from tropical forests, swamps and mangrove forests, to grasslands at the foot of the Himalayan mountains so tigers can be said to be very adaptive wildlife [3]. The Sumatran tiger (*Panthera tigris sumatrae*) is one of the six subspecies on earth and the only subspecies that can still be found in Indonesia after the extinction of the Bali tiger (*Panthera tigris balica*) and the Javan tiger (*Panthera tigris sondaica*) around the 1930s and 1970s [4].

Forest conversion, habitat degradation, tiger and prey hunting and tiger conflict with humans are the main causes of the decline in Sumatran tiger populations. One of the efforts to preserve sumatran tigers is through conservation activities; this can be done in-situ and ex-situ. In-situ conservation is a preservation effort carried out in its natural habitat, and ex-situ conservation is a preservation effort carried out outside its natural habitat [5]. One form of ex situ conservation is through conservation institutions, among others, is a Wildlife Park or Zoo that has a goal for conservation, education, research and recreation.

Animal welfare is the physical condition and psychology of animals to adapt to their environment. Concern for animal welfare has been a concern for the past 15-20 years. Today, zoos andwildlife sanctuaries around the world are deeply committed to implementing animal welfare in their management [6].

This study aims to analyze the management system and measure the level of welfare of Sumatran tigers at Medan Zoo and Siantar Zoo. This research is expected to be a source of information and reference materials for future research and become a consideration for the managers of Medan Zoo and Siantar Zoo in optimizing efforts to improve animal welfare.

2. Method

This research was conducted at Medan Zoo and Siantar Zoo in August – September 2021. The tools used in this study are stationary, camera, mobile phone and meter. The material used in this study is a stuffing sheet of animal welfare factors based on the Regulation of the Director General of Forest Protection and Conservation of Nature (called in Indonesian is Direktur Jenderal Perlindungan Hutan dan Konservasi Alam (Dirjen PHKA)) number: P.6/IV-SET/2011, while the object of the study is the Sumatran Tiger (*Panthera tigris sumatrae*). Data was collected through direct observation of the object studied in the form of animal management and assessment of minimum standards of animal welfare. Interviews with veterinarians and animal keepers about human resources and management activities carried out in supporting animal welfare.

The method used in analyzing animal welfare management in both conservation agencies is to providevalue to each variable. The meaning of the value 1 = bad, 2 = less, 3 = enough, 4 = good and 5 = satisfactory. The assessment is aimed at several predefined components based on five principles of animal welfare. The assessment is carried out by the management namely veterinarians and animal keepers, researchers and visitors to obtain objective results. The assessment was carried out based on the Regulation of the Director General of PHKA No. 6/2011; while the scoring score is as seen in Table 1.To calculate animal welfare values using the following formula:

$$score\ of\ assessment = \frac{\Sigma\ weight\ value}{5}$$

Table 1. Classification of animal assessment

No	Classification of assessment	Score
1	Very Good	80,00 – 100
2	Good	70,00 – 79,00
3	Enough	60,00 – 69,00
4	Less	< 60

Source: Regulation of the Director General of PHKA No. 6 of 2011

3. Results and Discussion

3.1. Condition of Sumatran tigers

Sumatran tigers located in Medan Zoo were 6 individuals, consisting of 4 males and 2 females are the result of breeding carried out by the managers of a pair of Sumatran tigers from Tanjung Balai; while Sumatran tigers located in Siantar Zoo was just 1 individual born in Aceh natural forest in 2004 and translocation to Siantar Zoo in 2008.

3.2. Management of feed components

Sumatran tigers are Carnivora and act as a predator for herbivorous animals such as deer (*Cervus unicolor*), wild boar (*Sus scrofa*), kijang (*Muntiacus muntjak*), beruk (*Macaca nemestrina*) and sometimes even prey on honey bears (*Helarctos malayanus*); but in captivity these animals cannot be given natural feed because their prey animals are impossible to sacrifice as tiger feed, considering some of these prey are protected animals.

The types of Sumatran tiger feed given in Medan Zoo are fresh chicken, beef and pork meat. The feeding schedule is on Mondays, Wednesdays and Saturdays given 5 kg of chicken meat feed. On Tuesdays and Thursdays are given 1 kg of beef and 2 kg of chicken meat. On Sundays are given a feed of 1 kg of wild boar and 3 kg of chicken meat, while friday sumatran tigers in Medan Zoo fast to maintain the tiger's diet so as not to be overweight. During the pandemic, Sumatran tigers were only fed chicken meat of 4 kg.

Feeds of the Sumatran tiger located in Siantar Zoo are fresh meat, consisting of 67% chicken meat and 33% beef. The type of feed given every day is 2 kg of beef and 4 kg of fresh chicken meat. Feeding is carried out 3 times a day, namely: morning 1 kg of chicken meat, noon 2 kg of beef and afternoon 3 kgof chicken meat. Based on the quantity of feed, tigers in Medan Zoo and in Siantar Zoo do not differ much from the quantity of tiger feed in Ragunan Wildlife Park and Bandung Wildlife Park; That is approximately 10% of the tiger's body weight per day. The average body weight of adult tigers is 60 kg so sumatran tiger feed is given 3-6 kg per day [7].

The basic needs that must be available in nature for Sumatran tigers to survive in their natural habitat are sufficient prey animals, water sources and vegetation cover [8]. Some things that must be considered in the provision of feed for Sumatran tigers in the captivity are a sufficient amount of feed,good and fresh feed quality and also the types of feed given. Water is one of the most important components for animals, because it is used for drinking and soaking. Water is provided in each Sumatran tiger cage both in the exhibit cage, drying cage and inhibit cage. In addition to good water quality, the drinking place is also cleaned every day so that the water is always clean. The availability of drinking water is also always there at all times and sufficient for sumatran tigers.

3.3. Management of components of the cage and environmental comfort

Cages are one of the most important things in captivity, because the life of animals are fully anchored in the cage. Some things to note are the use of cages (such as for demonstrations, rest, pregnancy and childbirth) as well as the shape and size of the cage, cage facilities and cage cleanliness. Medan Zoo has 4 cages are inhabited by 6 individual Sumatran tigers; whose placement is as follows: 2 cages are each inhabited by 2 female individuals and 2 cages are each inhabited by 1 individual male. Unlike thecase with Siantar Zoo which only has 1 cage inhabited by of male Sumatran tiger. Generally, in each cage is divided into 2 parts, namely the exhibit cage and the inhibit cage. The exhibit cage is used for tigers to move during the day. Siantar Zoo has one additional cage, namely a drying cage for all tiger species kept there. Sumatran tigers which are in the exhibit cage usually only consist of one individual just like other wildlife parks such as Ragunan Zoo Jakarta, but in the mating season can reach 2 individuals consisting of a male and a female. Tigers that are in the exhibit cage are usually male tigers because they are considered more aggressive and move more than tigresses [9].

The exhibit cage in Medan Zoo is sized 26.80×9.20 m^2 and is designed to be open (not roofed), while the exhibit cage in Siantar Zoo is sized 11.30×11.90 m^2. The size of the cage is by the needs of tigers to be able to move freely in the exhibit cage. [10] report that the size of the exhibit cagein Surabaya Zoo is sized 9×8 m2 and the inhibit cage is sized 2.5×2.5 m2.

The capacity of tigers in the exhibit cage is one Sumatran tiger per cage. The exhibit cage at Medan Zoo is designed like the natural habitat of sumatran tigers such as made shade for tigers sheltered from the heat of the sun and bad weather, there is a small pond for tiger drinking and there is a large pond asa barrier with a fence so that tigers cannot jump out of the fence. The pool in the exhibit cage has the benefit of being a place for tigers to bathe and swim when the weather is hot.

Just like the exhibit cage in Medan Zoo, Siantar Zoo also has a shelter for tigers. The tree where the tiger shelters and climbs is installed wire so that the tiger does not climb too high which can endanger the tiger and the tiger fears is detached from the cage. There is also a pool for tigers to swim and soak in when the weather is hot. In addition, there are also wooden blocks as a means for tigers to scratch so that tiger hooves are maintained and not injured. To keep training tigers with natural behavior jumping and catching prey provided rails at tiger food places that are usually done once a week. The exhibit cage at Siantar Zoo is made of glass and is bounded by a 1-meter-high fence for the safety of visitors and also the comfort of tigers. An overview of the exhibit cage of the Medan zoo and the SiantarZoo as shown in Figure 1.

(a) (b)

Figure 1. The Exhibit Cage in (a) Medan Zoo, (b) Siantar Zoo

The inner cages in Medan Zoo have a size of 3.15×1.8 m2 and are designed to be closed and have a part foreach tiger. In inner cages there is a bed made of wood which serves not only as a bed, but also as a means for tigers to sharpen nails. In addition, in the inner cage there is also a small drinking place. Theinner cages are also where tigers are given feed to make it easier for nurses to monitor tiger feed consumption. Health monitoring, vaccine injections and sumatran tiger check-ups are also carried out in inner cages. In addition, the inner cage also serves as a place for tiger isolation if anyone experiences pain or stress. The cage inhibits in Medan Zoo and Siantar Zoo as shown in Figure 2.

(a) (b)

Figure 2. The Inhibit Cage in (a) Medan Zoo, (b) Siantar Zoo

The inner cage in Siantar Zoo measuring 3×3.9 m^2, is a resting place for Sumatran tigers and a place of activity from 05.00 p.m. to 07.00 a.m. The inner cage is also where the tiger eats and performs health checks. The facilities in these cages are beds made of planks and wooden blocks to sharpen claws for tigersand outside the cage there is a trench for water disposal when cleaning the cage. The inner cages located in Medan Zoo and Siantar Zoo are not designed according to the natural habitat of Sumatran tigers. The floor of the inner cage is made of cement and there is no vegetation in this cage. For Siantar Zoo there is a clamp cage that is in the inner cage, which serves to check the health of the tigerand also when the tiger is in surgery. When the tiger will do the examination the tiger will be put in the clamp cage then clamped but will not hurt the tiger. The purpose of the clamp cage is to make it easierfor veterinarians to check tigers and so that tigers are not often sedated when health checks will be carried out.

Looking at the size of the cages in the two Conservation Institutions, it can be said that the exhibit andinhibit cages in Medan Zoo and Siantar Zoo are included well. This refers to the report from [9] that the area of the Sumatran tiger cage in Ragunan Wildlife Park Jakarta has a size of 50×20 m2 and also 20×16 m2; while the size of the tiger cage in Taman Safari Indonesia is 10.5×5 cm2 and the second cage measures 10.5×7.5 m2.

Siantar Zoo has a drying cage for tigers, measuring 4.9×3 m2. This drying cage is designed to be moreopen than the inner cage. The drying cage is made for the sunbathing tiger when it is finished swimmingand when the cage is finished cleaning. The facilities in the drying cage are not much different from thefacilities in an inner cage; what distinguishes it is that the size of the drying cage is wider than the inner cage and the location of the drying cage is more open and exposed to sunlight, especially in the morning.

Cleaning the cage is done every day because it will greatly affect the health of sumatran tigers and avoid disease. Care for the cleanliness of the cage at Siantar Zoo, including the place to drink and the environment around the cage is carried out every day. Cleaning the cage is done by cleaning the dirt and food waste of the tiger, spraying water and by brushing the floor thoroughly. In each cage there is wastewater damming channel. In addition to cleaning the cage also rejuvenate the grass and trees in the exhibit cage once a week. The pool and drinking place in the exhibit cage are also cleaned every day and also brushed so as not to moss. Different conditions at the exhibit cage in Medan Zoo is poorly maintained during the study. Weeds in the exhibit cage were never trimmed, and the drinking place lookeddirty and mossy, but unlike the exhibit cage, the inner cage and also the sewer around the tiger cage arealways cleaned every morning even though there is still garbage in some places.

Based on the assessment of cage comfort and environment, both Medan Zoo and Siantar Zoo belong to the good category because the condition of the cage has sufficient ventilation and lighting. According to [11] ventilation is a requirement so that the air can be exchanged so that it is not easy to transmit disease. The exhibit cage is given a shaded place for tigers to shelter from the heat and bad weather. Tiger cages are also designed in such a way that they will not harm tigers. The trees in and around the tiger cage have been ensured to be safe not to harm the tiger, and also make sure that electrical equipment is not in the cage so that it is not possible to harm the tiger.

Siantar Zoo has a special room for pregnant, giving birth and lactating tigers. Tiger that will give birthis separated into one closed cage, the cage is covered with a tarp so that it will not be visible from the outside with the aim that tigers do not feel disturbed by the presence of others by the provisions of the [12] that tigers that will give birth must be given access to a quiet and safe room frominterference from other animals, others person and noise. In this area should be installed a video monitorthat presents a remote view so that the tiger can be closely monitored.

The cages are designed to allow the tiger to move freely and interact with its environment. Siantar Zooprovides enough space for tigers to move freely and behave naturally. Tigers are given enrichment every one time two weeks the tiger will be given feed in the form of live chickens to train the tiger to hunt and pounce on prey. Once a week the tiger will be trained to pounce on prey by giving feed that is hung on the rails that are inside the exhibit cage. This is by the AZA statement [12] that the development of enrichment initiatives should focus on providing behavioral enrichment by the natural behavior of tigers, such as: stalking, hunting, playing or finding prey. Management of aspects free to behave naturally can be seen from the adequacy of space and cages, security and enrichment facilities. Enrichment programs are very important to prepare animals to havegood abilities to adapt, compete and survive when released into the wild. Enrichment can also be said as an effort to reduce the possibility of tigers getting involved again in conflict with humans [13].

[14] said that allowing tigers to behave naturally as in their habitat is difficult to do in cages. However, [15] explain that animal keepers can do creativity and take innovative approaches suchas stalking, pouncing, running, chasing, climbing and recognizing odors. Sumatran tiger in Siantar Zoowas

supported by good facilities, so that tigers can freely move and behave naturally. In the cage some trees can train tigers to climb, wooden blocks to sharpen tiger claws and ponds for tigers to swim and soak in when the weather is hot.

Sumatran tigers in Medan Zoo spend a lot of time to resting, because of the lack of enrichment contained in the cage so the tiger moves slightly and spends a lot of time to rest. This is very different from the natural behavior of tigers in their natural habitat so it's feared that it can change their natural behavior.

3.4. Management of the animal health components

The health aspect of animals must receive serious attention because it concerns the welfare of the animals being kept. The management of aspects free of wounds, pain and disease can be seen from several factors as follows: health of animals in captivity, frequency of health checks, health records, availability of health workers, health facilities and handling of sick animals [13]. In addition, [16] states that animal health facilities include hospital buildings, clinics, laboratory rooms, quarantine rooms, isolation rooms, veterinarians and medical personnel.

The medical treatment facilities at Siantar Zoo are complete and also well maintained, the treatment room is always in clean condition. The animal clinic also provides protective clothing that is stored well in a sterile place. All medicines, vaccines and other medicinal products are stored safely and shouldonly be given to Sumatran tigers with the permission of a veterinarian. Waste medicines such as syringes, needles, chemical waste and expired drugs are disposed of in special trash cans and under thesupervision of health workers. Somewhat different from the conditions in the Medan Zoo, the sumatrantiger health management in the Medan Zoo is not good enough, there are no important facilities such aslaboratory rooms and operating rooms, while the quarantine room is there but it is poorly maintained. The management of chemical waste and medicines in Medan Zoo is the same as in Siantar Zoo, which is disposed of in a special place with the supervision of health workers. An overview of the animal clinics in both conservation institutions is shown in Figure 3.

(b)

(b)

Figure 3. The Veterinary Clinic in (a) Medan Zoo, (b) Siantar Zoo

Sumatran tigers health checks at Medan Zoo and Siantar Zoo are carried out by veterinarians. During the period of the study no symptoms of tigers with signs of illness were found. Health checks are carried out by checking the feces and leftovers from the tiger. Examination of tiger feces is carried out once every 3 months. The veterinarian will also conduct a medical examination based on the animal keeper's report.

Disease prevention can be done with good and clean feed management. Another thing that supports tiger health is the administration of vaccines, vitamins and deworming. Sumatran tigers that are in Medan Zoo are given vitamins every 1 month in the form of renovit, calcium and iron. Deworming is carried out once every 6 months. Sumatran tiger feces examination is carried out every 3 months to find out whether the tiger being demonstrated is in a healthy physical condition. Examination of feces and administration of deworming Sumatran tigers in Siantar Zoo is carried out once in three months. The reason for administering deworming every three months is because within a period of three months the tiger has shown signs of helmesiation (deworming). In addition, vaccine administration is also carried out once every 1 year. According to [11] physical examination should be carried out annually, and stool examination should be carried out once every three months to determine if there are parasites in the body of Sumatran tigers.

The management of tiger health components in Siantar Zoo is satisfactory. Sumatran tigers that are demonstrated in good health, are always carried out health checks and timely administration of vaccines by a veterinarian. [11] states that animal captives must have one or more veterinarians who are specifically familiar with medical programs. The veterinarian should be available 7 days a week and 24 hours a day. Veterinarians always keep well-recorded and complete animal medical record data by the provisions of the [12] that the collection of medical and pathological data recordingdata is essential for any medical program. This record contains information about vaccinations, surgicaldiseases, anesthetic episodes, parasitic problems, weight gain, drugs received, blood and other data onclinical pathologies.

The health component of animals in captivity that must be considered is not only their physical health but also psychological health is a well-being factor that is no less important. [17] said that stress greatly affects the welfare of animals according to the magnitude of the biological damage caused by the stress. Although there are several ways stress can be avoided or cured, if it cannot be avoided, stress can cause death in animals. Therefore, animals in captivity must be strictly concerned about their comfort and welfare of these animals. Cages can be a stressful environment for tigers if they feel threatened. This can be minimized by providing hiding space in the cage [18]. [19] said that habitat factors such as the density of canopy cover and bush abundance can minimize stress levels in tigers. Habitats with the opposite conditions and lack of availability of prey will trigger stress levels for tigers. Animal captivitymust provide comfort for the animals in it because the animals kept in captivity are for conservation, education, research and recreation purposes [20].

For this reason, it is necessary to carry out management related to aspects of being free from fear and suffering. The results of interviews with animal keepers are known that to avoid stress on animals, efforts are made, among others, by making warning boards so as not to disturb animals, and making high fences so that visitors cannot make physical contact that can interfere with the comfort of animals;as [21] said that animals will suffer stress when animals experience emotional and physical disturbances.

The condition of the Sumatran tigers in the two conservation institutions studied did not experience stress, fear or suffering. This can be seen with a normal appetite. Even the Sumatran Tigers in Medan Zoo have adapted well, it is proven that they already have tiger cubs. Sumatran tiger at Siantar Zoo was not disturbed by the presence of visitors. Tigers do not show stress traits such as frequent pacing and biting their tails.

Sumatran tigers that are in Medan Zoo every 2 years are mated. This mating process begins with the introduction of males and females to avoid fighting tigers. The process of introducing tigers is not in a short time, namely by placing tigers in adjacent cages. Then the tiger will be put together when the female has shown signs of lust. During the mating process, the veterinarians and animal keepers will monitor them to anticipate any quarrel between the two tigers. Female tigers in captivity are more sensitive to the presence of other tigers, especially male tigers; while male tigers are not [22].

3.5. The assessment results of the welfare factors of the Sumatran tiger

In general, the achievement of implementation at the Medan Zoo is a good category with an average score of 76.9. The best component value was the aspect of being free from fear and suffering; The reasons were the number of animal keepers was sufficient, the performance of animal keepers was considered quite good, there was special handling for newly arrived tigers and efforts in handling animals experiencing stress. According to [21] these components are very important to support the health of animals in captive. On the other hand the assessment of the free to behave naturally still has a low score due to the lack of cage enrichment. The management of the free aspects of natural behavior can be seen from the adequacy of space and cages as well as the security and enrichment facilities [13]. The Siantar Zoo belongs to the excellent category, since all assessment components have an excellent value and an average value of 95. Based on the results of the assessment, it shows that tiger management in the two zoos meets animal welfare requirements. The assessment results can be seen in Table 2.

Table 2. The achievement of Sumatran tiger welfare implementation in Medan Zoo and Siantar Zoo

No	Components	MZ PSZ	MZ	PSZ	MZ	PSZ	MZ	PSZ
		Weight	Total Score		Scoring		Score Weighted	
1	Free from hunger and thirst	30	58	74	3.8	4.6	114	138
2	Free from environmental discomfort	20	39	45	3.9	4.5	78	90
3	Free from pain, injury and disease	20	65	65	4.3	5	86	100
4	Free to behave naturally	15	29	29	2.5	4.8	37.5	72
5	Free from fear and suffering	15	37	40	4.6	5	69	75
	Average						76.9	95
	Value						B	A

information: MZ (Medan Zoo), PSZ (Siantar Zoo)

4. Conclusion

Tiger management system both in Medan Zoo and Siantar Zoo has supported animal welfare. The level of tiger welfare in Medan Zoo is included in the good category, while in Siantar Zoo is included in the excellent category.

Acknowledgments

The author would like to thank the Research Institute of the Universitas Sumatera Utara for the funding provided for the research. We also give appreciation to the management and staff of the Medan Zoo and the Siantar Zoo, who have helped so that this research can be completed.

References

[1] Linkie M, Haidir I A, Nugroho A and Dinata Y 2008 *Biological Conservation* **141** 2410
[2] IUCN International Union For Conservation of Nature and Natural Resources 2015 *The IUCN Red List of Threatened Species* www.iucnredlist.org
[3] Sanderson E W, Forrest J and Loucks C 2006 *Setting Priorities for Conservation and Recovery of Wild Tigers: 2005-2015: The Technical Assessment. WCS, WWF, Smithsonian and NFWS-STF* (New York and Washington DC)
[4] Hadadi O H, Hartono and Haryono E 2015 *Majalah Geografi Indonesia* **29** 40
[5] Peraturan Menteri Kehutanan Republik Indonesia 2012 *Lembaga Konservasi* (UU P.31/ Menhut/-ii/2012)
[6] Mellor D J and Beausoleil N J 2015 *Animal Welfare* **24** 241
[7] Yolanda, Rusdi and Supiyani A 2017 *Jurnal Bioma* **13** 100
[8] Dinata Y and Sugardjito J 2008 *Jurnal Biodiversitas* **9** 222
[9] Yultisman, Azizah M and Wardoyo S E 2019 *Jurnal Sains Natural Universitas Nusa Bangsa* **9** 29
[10] Ganesha A and Aunurohim 2012 *Jurnal Sains dan Seni ITS* **1** 48
[11] [GFAS] Global Federation of Animal Sanctuaries 2018 *Standard of Felid Sanctuaries*, (Washington DC)
[12] [AZA] Association of Zoos and Aquariums, Tiger Species Survival Plan 2016 *Tiger Care Manual* (Association of Zoos and Aquariums, Siver Spring, MD)
[13] Saab R, Siregar P G and Setia T M 2019 *Journal Biodiversitas* **20** 2790

[14] Shepherdson D J, Mellen D J and Hutchins M 1998 *Captive Environments For Small Felids Second Nature, Evironment Enrichment For Captive Animals Smithsonian Institution* (Wahington and London)
[15] Shepherdson D J, Carlstread K, Mellen J D and Seidensticker J 1993 *Zoo Biology* **12** 203
[16] [PKBSI] Perhimpunan Kebun Binatang Se-Indonesia 2000 *Pengelolaan Taman Satwa di Indonesia* (Jakarta: PKBSI)
[17] Moberg G P 2000 *Biological Response to Stress: Implication For Animal Welfare* (Wallingford oxon: CABI Publishing)
[18] Clubb R and Vickery S 2006 *Streotypic Animal Behaviour Fundamentals and Applications to Welfare* (CABI. ISBN 1-84593-042-8 E-Book: 58)
[19] Malviya M, Kumar V, Mandal D, Sarkas M S, Nigam P, Gopal R, Sankar K, Umapathy G and Ramesh K 2018 *Hystrix the Italian Journal of Mammalogy* **29** 195
[20] Pitsko L E 2003 *Wild Tiger In Capacity A Study of The Effects of The Captive Environment on Tiger Behavior* (Virginia Polytechtic Institute And State University)
[21] Jordan B 2005 *Wild and Captive Animals* **24** 515
[22] Rouck M D, Kitchener A C, Law G and Nellisen M 2005 *Journal Animal Welfare* **14** 229

Stock Assessment of Indian Mackerel (*Rastrelliger kanagurta*) Landed at Tanjung Beringin's Auction Place Serdang Bedagai, Sumatera Utara

J S Hasibuan[1,2*], M A P Sidabutar[1], J Sinaga[1], Desrita[1] and V R Manurung[1]

[1]Department of Management Aquatic Resources, Faculty of Agriculture, University Sumatra Utara, Medan, North Sumatera, Indonesia
[2]Center of Excellence for Mangrove, Universitas Sumatera Utara, Medan 20155, Indonesia

*Email: juliasyahriani.h@usu.ac.id

Abstract. Indian Mackerel (*Rastrelliger kanagurta*) is one of the main catch fishery at Tanjung Beringin's Auction Place, Serdang Bedagai, Sumatera Utara. The purpose of this research is to analyze the stock status of indian mackerel landed at Tanjung Beringin. This research was conduted in July – September 2021. The method of research used purposive sampling method. The total sample fish of indian mackerel were 272 individuals. The research showed Indian mackerel's sex ratio was 1: 1.060. Indian mackerel's growth pattern was negative allometric. Value estimation of male and female's $L\infty$ was 268.80 mm and 295.05 mm. Value K of male mackerel was bigger than female. But value of L_c and L_m male mackerel were smaller than female. The total rate of mortality was 2.466/year, natural mortality was 0.684/year and fishing mortality (F) was 1.782/year. The rate of exploitation was 0,723 classified as high exploitation level (overfishing).

1. Introduction

Serdang Bedagai Regency has a large diversity of natural resources, especially in the field of fisheries. One of the catches in the waters of Serdang Bedagai is mackerel. Indian mackerel (*Rastrelliger kanagurta*) is the catch of fisherman that has important economic value, has a dense meat texture, and contains a high nutritional value [1].

Indian mackerel catches land at the Fish Auction Place. Fish Auction Place (TPI) is a fishery activity infrastructure that aims to promote and improve the business and welfare of fishermen [2]. Tanjung Beringin's Auction Place production of Indian mackerel in 2020 was 676224 kg. The exploitation and utilization rate indian mackerel is high in the waters of the Malacca Strait, causing the condition of Indian mackerel stocks to decline due to overfishing and poor management policies.

The purpose of the research was analyzing the stock status of Indian mackerel (*R kanagurta*). The benefit of this research is that it can become a basic recommendation of indian mackerel's management plan so that it can be carried sustainable natural resources of Malacca strait.

2. Materials Method

2.1. *Research site*

This study was carried out in July – September 2021 in Serdang Bedagai, Sumatera Utara (Figure 1).

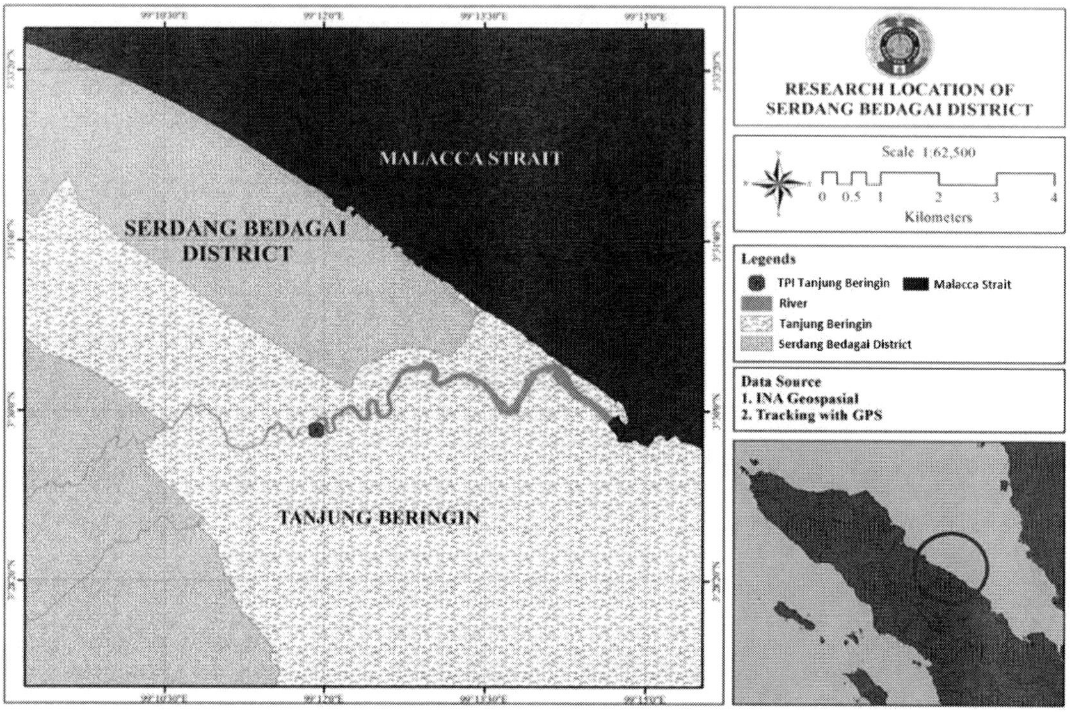

Figure 1. Tanjung Beringin's Auction Place

2.2. *Procedures*

The tools used in this study were stationery, millimeter blocks, cellphone cameras, analytical scales, surgical tools such as knives, scissors, gloves, paper labels, safety pins, trays, thermometers, coolbox, and GPS (*Global Positioning System*). The tools to process the data are Microsoft Excel 2019 and the FISAT II program. The materials used in this study were male mackerel, dry and wet wipes, and disinfectant.

Fish samples were taken 3 times, and carried out for 3 months with an interval of 1 month and 1 time from June to September 2021. There were 272 male mackerel fish observed. The fish were brought to the laboratory for measurement of total length (mm), fish body height (mm), fish weight measurement (g), and measurement of fish meristic characteristics. Furthermore, the fish were dissected to determine the morphology of the gonads and their development and sex based on their primary characteristics. Other primary data were obtained through fishermen's debriefing.

2.3. *Analysis of data*

2.3.1. Sex ratio. According to [3], knowing the balance of male and female in sex ratio's formula by:

$$X = \frac{M}{F} \quad (1)$$

Where :
x = sex ratio
M = total of mackerel's male
F = total of mackerel's female

2.3.2. Length-weight relationship. Length weight of relationship can be known by the formula [4] :

$$W = a\,L^b \tag{2}$$

Where :
W = Mackerel's weight (gram)
L = Mackerel's length (mm)
a and b = contant

2.3.3. Condition factor. Length and weight can be estimate of condition's indian mackerel with formula [4]:

Isometic can calculated if b = 3, with formula:

$$\mathrm{FK} = \frac{10^5 \mathrm{W}}{\mathrm{L}^3} \tag{3}$$

Allometric can be calculated if $b \neq 3$ used the formula:

$$\mathrm{FK} = \frac{\mathrm{W}}{\mathrm{aL^b}} \tag{4}$$

Where :
FK = condition factor
W = mackerel's weight (gram)
L = mackerel's total length (mm)
a and b = constant

2.3.4. Distributed in the length-frequency. The estimated fish groups were analyzed using length measurements which were distributed in the length-frequency class so that the length class of *i* had a frequency (*fi*) using Microsoft Excel. Determination of the normal distribution of the size group was estimated by FISAT II.

2.3.5. The first size of catch fish (Lc). Estimate the first size of Indian markarel with method [5]:

$$S_L = \frac{1}{1+exp(S_i - S_2 * L)} \tag{5}$$

$$\ln\left[\frac{1}{SL} - 1\right] = a - bL \qquad L_c = -\frac{a}{b} \tag{6}$$

Where:
SL = estimated value
L = the mean of length class (mm)
a dan b = constant on the logistic curve formula
L_c = the length of the first fish caught (mm)

2.3.6. The first size of ripe gonads (Lm). Estimate the first size average of gonad mackerel maternity with the Spearman Karber method [6] :

$$m = \left[x_k + \left(\frac{x}{2}\right)\right] - \left(x \sum P_i\right) \tag{7}$$

The 95% confidence interval for log Lm is defined by:

$$Antilog\left(\mathrm{m} \pm 1.96\sqrt{X^2 \sum \frac{p_i \times q_i}{ni-1}}\right) \tag{8}$$

Where:

m = the log of fish length at first gonadal maturity
x_k = the mean of last class of mature gonad's mackerel
x = the log of increase in mean of length
p_i = propotion of mature gonad's mackerel in the length class of *i*
ni = total of mackerel's length class of i
qi = 1-p
Lm = the length of first mackerel's matured gonad

2.3.7. Mortality and exploitation rate. The natural mortality rate (M) is estimated with formula [5]:

$$\mathbf{Log\,M = -0.0066 - 0.2790 \times Log\,(L_\infty) + 0.6543 \times Log(K) + 0.4634 \times Log(T)} \quad (9)$$

Where:
M = natural mortality
L_∞ = von bertalanffy equation's asymptotic length
K = von bertalanffy equation's growth coefficient
T = average water surface temperature (^{0}C)

Comparing the fishing mortality to the total mortality can be determined the exploitation rate with formula [7]:

$$\mathbf{E = \frac{F}{F+Z} = \frac{F}{Z}} \quad (10)$$

2.3.8. Von Bertalanffy's growth. Indian mackerel's growth can be determined using model [6] :

$$\mathbf{L_t = L_\infty \left[1 - e^{-K(t-t_0)}\right]} \quad (11)$$

Where :
Lt = length of mackerel's age t
L∞ = theoretical maximum length of mackerel
K = growth coefficient
t0 = theoretical age at mackerel's length is zero

Meanwhile, in estimating the value t_0 (theoretical age at mackerel's length is zero). The equation [7] is :

$$\mathrm{Log}\left(-t_0\right) = -0.3922 - 0.2752\ \mathrm{Log}\ L_\infty - 1.038\ \mathrm{Log}\ K \quad (12)$$

3. Result and Discussion

3.1. Sex ratio

There were 272 male Indian mackerel fish observed in this study, consisting of 132 mackerel male and 140 mackerel female. Sex ratio of indian mackerel can be described with balance of male and female mackerel as 1: 1.060. The factor that causes the sex ratio conditions in waters to vary, such as differences in behavior, movement, reproduction, environmental conditions, and fishing activities [8].

Table 1. Indian mackerel's sex ratio at TPI Tanjung Beringin

Months	Frequency		Ratio	Description
	Male	Female	Male: Female	
July	73	50	1 : 0.684	Not balanced
August	23	39	1 : 1.696	Not balanced
September	36	51	1 : 1.417	Balanced
Total	132	140	1 : 1.060	Balanced

3.2. *Length-weight relationship*

The analysis of the length weight relationship showed that the mackerel's growth pattern was negative allometric with male b-value = 2.679 and female b-value = 2.380. The value of b < 3 indicates that the mackerel male's length growth is more dominant than growth of mackerel's weight, with a coefficient of determination is 91% and 85%, which means the mackerel's length can explain body weight. So, male mackerel's growth pattern has a negative allometric or a b value less than 3 which means the fish's body condition is average [1] is presented in Figure 2.

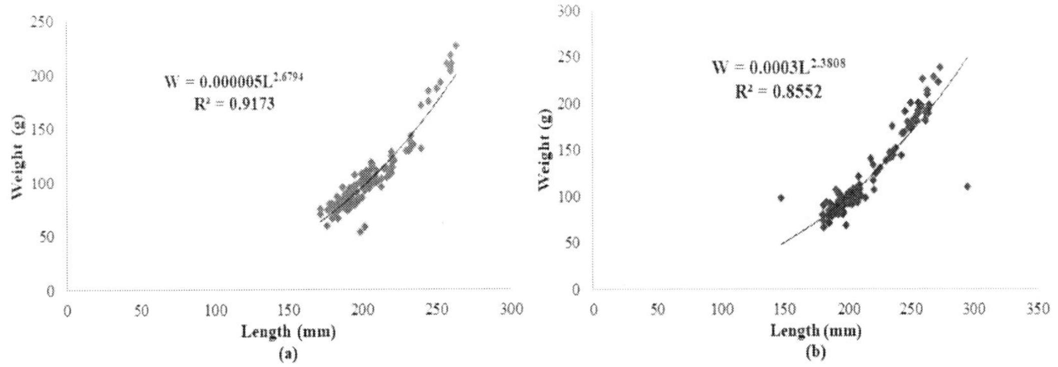

Figure 2. The growth pattern of Indian mackerel (*R. kanagurta*); (a) male, (b) female

3.3. *Condition factor*

The range of the male mackerel condition factor was 0.611 – 1.297 where the highest condition factor was in July, while the female condition factor ranged from 0.481 – 2.210 where the highest condition factor was in September condition factor values of Indian mackerel presented in Table 2.

Table 2. Indian mackerel's condition factors at TPI Tanjung Beringin

Sex	Average	Range	Description
Male	1.084	0.612 – 1.297	Rather flat
Femal	1.098	0.481 – 2.210	Rather flat

The condition factor value of male mackerel is 0.611 – 1.297. If the condition factor value of 1-2 indicates that the body is rather flat. [5] stated that the value of condition factors in fish with slightly flat bodies ranging from 1 to 2, while in fish with less flat bodies it ranged from 2 to 4.

3.4. *Distribution of length-frequently*

The cohort line is an indication of mackerel's growth which is indicated by a shift right. Each month has 1 size group (cohort) where the cohort is a group of individuals of the same biota and comes from the same spawning place [9]. The size group from July to September is presented in Figure 3.

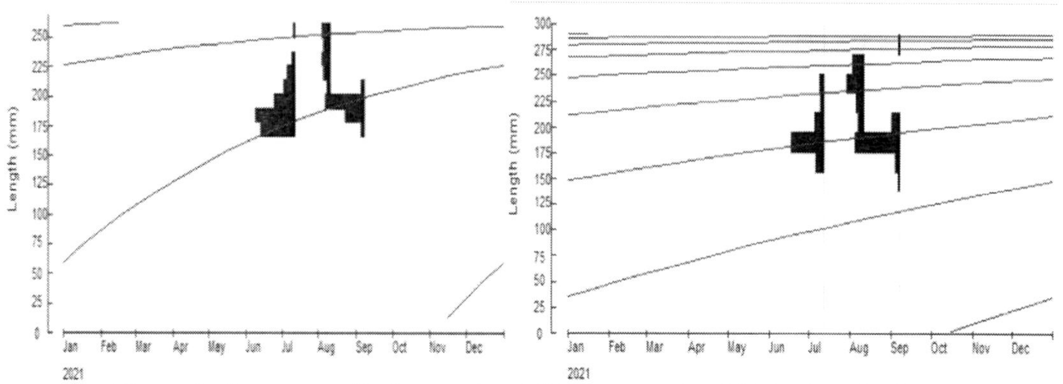

Figure 3. Size group of indian mackerel (*R. kanagurta*) at TPI Tanjung Beringin

3.5. *Von Bertalanffy's growth*

Method of ELEFAN-1 can be described value of growth model with the growth coefficient (K), asymptotic length value (L∞), the estimated theoretical age value when theoretical age at mackerel's length is zero (t0) is presented in Table 3.

Tabel 3. Estimation of Indian mackerel growth parameters at TPI Tanjung Beringin

Parameters of growth	Male	Female
L∞ (mm)	268.80	295.05
K (mm/moth)	1	0.570
t_0 (bulan)	-0.529	-0.288
Lt (mm)	$268.80\,(1\text{-EXP}^{(-1(t+0.529))})$	$295.05(1\text{-EXP}^{(-0.57(t+0.288))})$

The results of growth can be estimated values L∞ of 268.80 mm male and 295.05 mm female. The growth coefficient values of male and female mackerel of 1.00 and 0.579 mm/month are included in the high category. that the higher the K value, the faster the fish reach their asymptotic length and the faster the fish will die [10]. The growth graph based on the von Bertalanffy equation is presented in Figure 5. The male mackerel's von Bertalanffy equation shows an increase in fish length in reaching its asymptotic length over time.

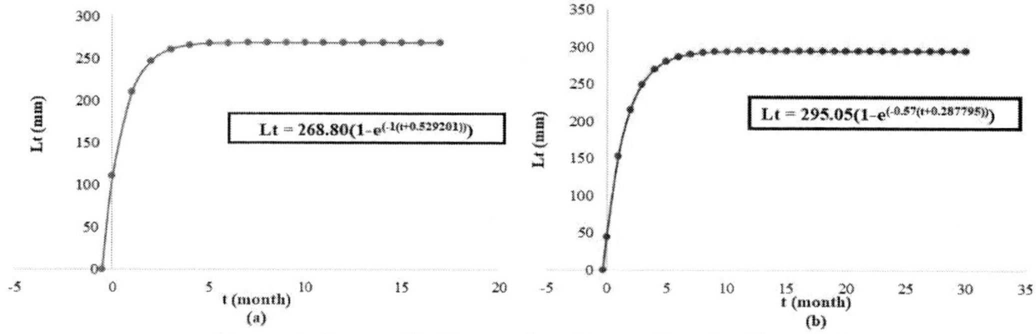

Figure 4. Curve of Indian mackerel's von Bertalanffy growth

3.6. *The size of the first captured and the size of the first maturity of the gonads*

In the research, the first size of male mackerel caught 197.255 mm and 210,273 mm for female fish, while the first gonad length were 217,079 mm for male and 222,446 mm for female. The value of the size of the fish caught for the first time in male and female fish was smaller than the size of the maturity of the first gonad, so this condition indicated overfishing growth conditions. Catching mackerel whose gonads are immature is caused by fishing gear with inappropriate or smaller mesh sizes [11]. The graphs of the length at first maturity of the gonads and the size at first caught by male and female mackerel are presented in Figure 6.

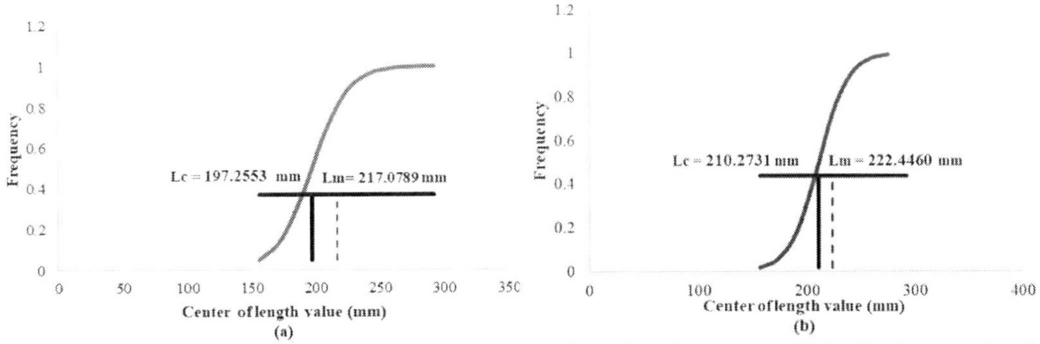

Figure 5. The first size of gonadal maturity (L_m) and the first time caught (L_c) of Indian mackerel

3.7. *Mortality and exploitation rate*

The Indian mackerel's capture mortality was 1.782/year, which is bigger than the natural mortality of 0.684/year. This shows that the mortality factor of Indian mackerel landed in TPI Tanjung Beringin is more dominantly caused by fishing activities. The mortality value and fish exploitation rate are presented in Table 4.

Table 4. Mortality and exploitation rate of Indian mackerel (*R. kanagurta*) at TPI Tanjung Beringin

Parameters	Total (per year)	
	This research (2022)	Other research [13] (2017)
Total of mortality (Z)	2.466	2.837
Natural mortality (M)	0.684	0.891
Capture mortality (F)	1.782	1.946
Exploitation (E)	0.723	0.685

The exploitation level of Indian mackerel was 0.723. This value exploitation of Indian mackerel is more than 0.5 indicating that the Indian mackerel has been overexploited. This shows that fishing activities are very influential on changes in fish populations in waters [13].

4. Conclusions

The study of the stock of Indian mackerel (*Rastrelliger kanagurta*) landed at the Tanjung Beringin's Auction Place showed the occurrence of overfishing based on several parameters such as mortality values, and generally caught at immature gonads (*growth overfishing*).

Acknowledgments

The authors want to thank to Septi, Jihan, Rifna, Rizky, and Eben as partner in this research.

References

[1] Harteman E 2015 *Journal of Tropical Animal Science*) **4** 6
[2] Agustiari A M, Saputra S W and Solichin A. 2017 *Journal of Maquares* **6** 33
[3] Effendie M I 1997 *Biologi Perikanan [Fisheries Biology]* (Yogyakarta: Yayasan Pustaka Nusantara)
[4] Effendie M I 2002 *Biologi Perikanan* (Yogyakarta:Yayasan Pustaka Nusatama) 163
[5] Sparre P and Venema S C 1998 *Introduksi Pengkajian Stok Ikan Tropis*. Buku 1: Manual (FAO: Puslitbang Perikanan Balitbang Pertanian Indonesia)

[6] Udupa, K S 1986 *Fishbyte* **4** 8
[7] Pauly D 1984 *Fish Population Dynamics in Tropical Waters: a Manual for Use Programmable Calculators* (Manila: International Center for Living Aquatic Resources Management ICLARM)
[8] Sadewi S P, Mashar A and Boer M 2018 *Journal of Tropical Fisheries Management* **2** 45
[9] Dewi K T, Barus A and Desrita 2012 *Analisis Pertumbuhan dan Laju Eksploitasi Ikan Tongkol (Auxis Thazard) yang Didaratkan di KUD Gabion Pelabuhan Perikanan Samudera Belawan Sumatera Utara* [Skripsi] (Medan : Universitas Sumatera Utara)
[10] Permatachani A, Boer M, Kamal MM 2016 *Jurnal Teknologi Perikanan dan Kelautan* **7** 107
[11] Dahlan M A, Omar S B A, Tresnati J, Umar M T, Nur M 2015.*Jurnal Torani* **25** 25
[12] Wulandari Y U L I, Utomo B, Desrita D 2017 *Aquacoastmarine* **5** 1
[13] Masrikat 2012 *Journal of Coastal Development* **15** 276

Feed management and health condition of Sumatran Elephant (*Elephas maximus sumatranus*) in Conservation Response Unit (CRU), Tangkahan, Langkat, North Sumatra

K Berliani[1*], P Patana[2] and W Azmi[3]

[1]Department of Biology, Faculty of Mathematics and Natural Sciences, Universitas Sumatera Utara, Medan, North Sumatra 20155, Indonesia.
[2]Department of Forestry, Faculty of Forestry, Universitas Sumatera Utara, Medan, North Sumatra 20155, Indonesia.
[3]Conservation Response Unit (CRU) Tangkahan, Namu Sialang Village, Batang Serangan, Langkat, North Sumatra 20852, Indonesia.

[*]Email: kaniwa@usu.ac.id

Abstract. The present study investigates the feeding management and health condition of elephants through a descriptive and field observation approach by interviewing the mahouts and clinical workers in Tangkahan CRU, Langkat Regency, North Sumatra. The ex-situ conservation unit promoted two feeding strategies namely the natural or grazing activity in the forest and additional feed (food stocks) collected or purchased by the workers in the form of coconut fronds, elephant grasses, bananas, watermelons, and etc. In addition to the collected fruits or crops, the CRU workers also prepared concentrates or pudding made from a mixture of palatable crops including rice, sticky rice, green beans, sugars to increase the food supply for the elephants. The clinical workers in the form of Vesswic is a team of veterinarians, paramedics and administrative personnel who focus on animal health including the Sumatran elephants. The Tangkahan CRU was then proven to promote the welfare of elephants through a series of effective feeding and health management.

1. Introduction

One of the important components in the management of wild animals in ex situ is the management of feed and animal health. Animal life has its limits, and if the quality and quantity of the feed are inadequate, this is frequently the principal limiting factor for the survival of animal life. Feed management is the control and administration of feed that satisfies the biological and technological requirements in accordance with the demands of the animal; from an economic standpoint, it is seen as inexpensive and is readily attainable on an ongoing basis. Therefore, it is very important to know the level of food management and health of Sumatran elephants (*Elephas maximus-sumatranus*) at the Tangkahan Conservation Response Unit (CRU) as an ex-situ habitat that supports the survival of elephants. This condition of feed and health management will determine the level of welfare of the elephants in the ex-situ habitat. Sumatran elephant feed comprises of all plant types that elephants may consume, including both organic and inorganic substances, which can be partially or totally digested and do not harm the elephants. One of the key elements in determining the effectiveness of ex-situ conservation is feed. In light of the fact that food management and elephant health are factors that promote the body growth and development while in captivity, the better feed management, the better the ex-situ conservation outputs of elephants.

2. Methodology

This research was conducted from October to December 2020. The study area is located in the Tangkahan Conservation Response Unit (CRU) area, Namo Sialang Village, Batang Serangan District, Langkat Regency, North Sumatra, Indonesia. Tangkahan is a buffer zone area of Gunung Leuser National Park (TNGL). A qualitative, descriptive approach was used to explore the feed management and health condition of elephants in CRU. A convenience sample of mahout (elephant keeper) were interviewed followed with direct observation in the field during the feeding, feed storage, distribution, and interaction between mahouts and elephants. The data is analyzed descriptively to point out the general view of information regarding our context of study.

3. Results and Discussion

Based on our observation, there are two programs in feeding management of elephants through the supply of natural feeding (grazing) and additional feeding (food stocks). The grazing activity was managed to anticipate the lack of variety of feed provided by CRU management for elephants while in captivitiy. This variety of feed is considered important to meet the nutritional deficiencies and mineral salts in the body. Feed has an important role for animals, both for the growth of young elephants and for maintaining life and energy and reproducing for adult elephants. Another function of feed is to maintain endurance and health. The type of feed given must be of good quality and in sufficient quantities. According to Siregar [1], a high-quality and sufficient amount of nutrients, including energy, protein, carbs, fat, vitamins, and minerals—all of which are required in the proper and balanced ratios—are found in a high-quality feed. The feeding rate of elephants may be influenced by their body size, age, and sex [2] depending on the area, weather and ecosystem [3]. Elephants have a predisposition for certain plant species, and they are picky eaters that will consume multiple taxa from diverse plants [3]. By employing the best feed-finding procedures, energy demands will be met [4]. The satisfaction of the mineral salts that elephants consume is also a part of this ideal technique.

In the beginning of feed session, elephants will bathe in the river before and after participating in the elephant herding program (Figure 1A). This bath activity is to cleanse the body of germs and ectoparasites attached. After bathing, the elephant will be directed into the forest of Gunung Leuser National Park (TNGL). The shepherding program is carried out every day from 10:00 AM to 03:00 PM (Figure 1B). The purpose of this program is for elephants to get a variety of feeds in their habitat that provides abundant natural food and natural mineral salts in salt licks.

Figure 1. Documentation of elephants bathing session (A) and shepherding (B) in Tangkahan CRU

White et al [5] reported that elephant habitat naturally provides a wide variety of feed in the form of grass, bark, wood, roots and fruits. Elephants are likely to consume as soon as they locate food, and as foraging distance increases, their feeding activity will likely decrease [6]. In addition, Sumatran elephants (*Elephas maximus-sumatranus*) are browser animals (bush eaters), folivores (leaf-eaters), frugivores (fruit eaters), seeds eaters and other parts of plants to meet their mineral needs. For adults weighing between 3,000 and 4,000 kg, elephants require roughly 250 kg of feed daily [7,8]. Depending on the locale, weather, and ecology, this megaherbivore eats more than 400 diverse and highly diverse plant species, including leaves, fruits, grasses, midribs, young stems, flowers, bark, and lianas [9-11]. Elephants are, however, also highly fussy eaters and are impacted by the type of vegetation [12]. There are some feed components that are chosen over others due to the selective character of the available feed and the feeling of the feed ingredients before and during consumption [13]. Elephants use their selective feeding habits as one strategy to get the nutrients they need to construct their own diets. This demonstrates that elephant preference for palatable food is a naturally occurring phenomenon.

Figure 2. Documentation of feed stocks for elephants in Tangkahan CRU: Coconut fronds (A) Elephant grasses (B), Bananas (C), and Watermelons (D)

In the process of consuming natural food, elephants must choose from a change in the physical structure that differs across plant species including phenological level, structure type, chemical composition, relative or absolute abundance and distribution patterns of plants. A plant may be ingested often or in big amounts, but this does not always indicate that the elephant like the plant. The choice of this feed is based on the supply of feed. In a landscape with a variety of shrubs, grasses, and trees, it is challenging to evaluate the relative abundance of plant species, making it impossible to identify the elephant's preferred meal. Thus, the level of preference of a particular plant or plant part is known when elephant feeding activities are faced with the simultaneous availability of several different species of plants [9].

In addition, elephants also show a phenomenon of preference and palatability in choosing their food. Palatability is a performance characteristic of feed ingredients as a result of the physical and chemical conditions possessed by feed ingredients which are reflected by their organoleptic properties such as appearance, smell, taste (bland, salty, sweet, sour, bitter), texture and temperature which are the main strengths. attract and stimulate animals to consume them [14]. Palatability is a very important factor to determine the level of feed consumption. Therefore, despite the fact that all feed types offer varying amounts of protein, carbs, fat, minerals, and vitamins, elephants avoid consuming feed types that contain chemical compounds that are bad for their digestion. In addition, the measure of palatability response is based on the results of the central nervous system integrating taste and internal state signals and then combined with cues arising from previous associations [15].

Figure 3. Distribution of collected feeds (elephant grass) by workers (A) across the river (B).

The supplementary feeding program for elephants aims to meet the nutritional needs of elephants that are not obtained in their natural habitat. The feeds were coconut fronds (*Cocos nucifera*), king or elephant grass (*Pennisetum purpureum*), bananas, pumpkins, watermelons and pineapples (Figure 2). Every day after the elephants return from herding, this feed is given to them. The weight of the grass and coconut fronds will be taken before feeding the elephants. At the Tangkahan CRU, each elephant receives 20 kg of coconut midrib and 70 kg of king grass per day. In the meantime, the different fruit varieties are solely distributed based on the stock in the storage facility. Next, the feed is distributed across the river to the elephant enclosure using a pulley (Figure 3). The grass was given to the elephant as soon as they crossed the river to the other side. The Tangkahan community garden provided access to feed materials such coconut fronds, king grass, bananas, pumpkins, watermelons, and pineapples. King or elephant grass is particularly managed or intentionally planted. To provide for the need for elephant feed, this is done. This king grass was chosen because it is a productive plant with attributes that help enhance soil quality [16]. The CRU management formed a farmer group in the Tangkahan community to plant king grass. Furthermore, the farmers take turns selling this type of feed to the CRU management. The need for additional feed of this type has provided economic benefits to the Tangkahan community. The management will purchase the feed from outside Tangkahan if the sorts are not found in the hamlets of Kuala Buluh and Kuala Gemoh, ensuring that the feed is always accessible.

Elephants are provided with additional nutrition in the form of edible plants with high water, carbohydrate, and fiber contents. Berliani et al [17] reported that rice (*Oryza sativa*) with an average of 22.57 kg/hour/individual to 23.79 kg/hour/individual, was the most consumed crop plant by elephants in the study area. Furthermore, bananas (*Musa* sp) reached an average of 18.93 kg/hour/individual to 21.14 kg/hour/individual. The selection of feed types is also sufficient for other nutritional values, namely organic substances including nitrogen compounds (protein and non-protein), fats (lipids), carbohydrates, and vitamins; and inorganic substances (minerals) consisting of essential minerals (Ca,

P, Mg, Na, K, Cl, S, Fe, Cu, I, Zn, Cr, etc) and non-essential minerals. Therefore, the CRU management also provides additional feed in the form of concentrate as a supplement.

The feed concentrates are composed of rice, sticky rice, bran, green beans, sugar and salt (Figure 4A-4D). These ingredients are cooked into pudding or porridge-like consistency with a composition of 7.5 kg of rice, sticky rice, and green beans, 3 kg of sugar, 1 kg of salt, and bran. In order to prevent the pudding from becoming overly soft after cooking, the bran is not included in the pudding recipe excessively. Pudding is prepared by combining the ingredients in a bowl and adding water thereafter. The pudding is cooked and ready to chill after being stirred for the following two hours. Each week, the elephants receive this pudding twice (Figure 4E).

Figure 4. Formulation of concentrate as additional food (A) for elephants prepared from rice, sticky rice, green beans, etc; Addition of sugar into formulation (B), Mixing and cooking the food (C), Physical appearance of concentrate (D), and feeding the elephants (E)

As a result, elephants are fed a balanced diet. This is a result of the requirements that feed given to animals must be balanced (containing all the nutrients needed by animals in the right amount to fulfill all physiological functions of the body), high in nutritional value, sufficient (meeting the amount according to the needs of the animal (age, sex, production status, season), palatable (according to the animal preferences and habits), continuous (available constantly throughout the animal's lifetime), and does not interfere with health. Feeding forage and pudding will not have a significant effect on high body weight gain in a short time. Therefore, supplementary feeding must be given continuously with good quality and quantity to maintain body condition. The CRU management also follows the elephant feeding strategy by taking into account factors related to the feed niche of each type of elephant being managed, namely body size and its relationship to metabolic rate, the anatomical and physiological composition of the digestive tract, the ratio between the volume of the digestive tract and body size, and special feeding structure.

The health of the Sumatran elephant in the Tangkahan ex-situ conservation is influenced by environmental conditions, feed, management, the presence of disease and others. CRU management is assisted by the Veterinary Society for Sumatran Wildlife Conservation (Vesswic) in handling elephant health, both for health maintenance, disease prevention and treatment of Sumatran elephants in Tangkahan. Vesswic is a team of veterinarians, paramedics and administrative personnel who focus on animal health including the Sumatran elephant. Vesswic conducts examinations, provides medicines and vitamins to maintain health, stimulate growth and reproduction of Sumatran elephants. Clinical examination of elephants is also carried out through inspection, palpation, percussion, and laboratory tests to support or confirm the diagnosis. Every day, the behavior, hunger, and exterior physical appearance of the elephant are observed to perform routine health checks. The mahout will alert the veterinarian for additional care if there are any changes in behavior, hunger, stool shape, urine color, or external sores on the elephant. Using the tools and medications at their disposal, veterinarians conduct examinations and administer treatments. Every three months, worm medication is administered in an effort to prevent worms, whereas every six months, immunizations are administered in an effort to prevent tetanus. Medicines are administered orally or intravenously. The drugs were embedded within fruits when being fed to the elephants (Figure 5). If an elephant is sick with a serious illness it will be isolated, thereby reducing interaction with other individuals.

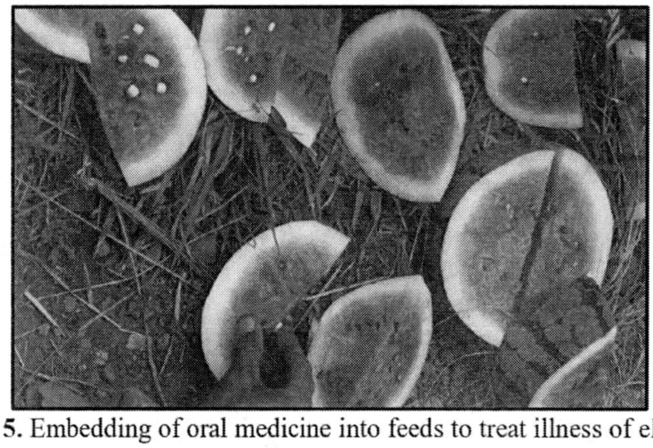

Figure 5. Embedding of oral medicine into feeds to treat illness of elephants

On the other hand, the CRU management tries to maintain elephant health by looking at the sanitation and hygiene management aspects of the elephant enclosure. Sanitation of the enclosure environment must be maintained because it will affect the quality of the enclosure and the health of the elephant. One of them to avoid the cause of parasitic disease or harmful bacteria. The mahouts take turns cleaning the feed and burning it. The success of ex-situ habitat management is determined by many factors, including health. Furthermore, elephant health is influenced by environmental conditions, feed, management, diseases and disorders. Thus, the management of feed and elephant health at the Tangkahan CRU is considered to be good, because it takes these factors into account. In this condition, it is hoped that the existence of the Sumatran elephant in the Tangkahan CRU can be maintained properly according to the meaning of conservation, namely protecting, preserving and utilizing.

CRU is an ex-situ elephant conservation program initiated by the Aceh Natural Resources Conservation Center (BKSDA). These elephants are involved in the conservation program of the TNGL area. Elephants are used as a means of patrolling the forest from illegal logging and monitoring areas of human-wildlife conflict. Furthermore, Tangkahan is also directly adjacent to TNGL which consists of several balanced biotic and abiotic ecosystems, so that it has a high diversity of flora and fauna. Thus, the TNGL area provides a variety of forage plants and mineral salts to meet the body's needs of wild animals. These mineral salts are also obtained at certain locations that are sources of mineral salts. Therefore, the salt location in Tangkahan is an important area for wildlife including elephants in maintaining their health and survival. This is supported by Riba'i et al [18] that the availability of salt locations in elephant habitat will determine the level of welfare of these animals in carrying out their daily activities.

4. Conclusion

The CRU Tangkahan elephant feed management aims two programs, namely a grazing program and an extra-feeding program, to promote the animal's growth and development and improve ex-situ conservation. In order to maintain elephant health, daily assessments of behavior, appetite, and external physical appearance are monitored.

Acknowledgments

The authors gratefully acknowledged that this research was supported by *Deputi Penguatan Riset dan Pengembangan Kementerian Riset dan Teknologi/ Badan Riset dan Inovasi Nasional* year 2020 with a Contract Number: 11/AMD/E1/KP.PTNBH/2020.

References

[1] Siregar S B 1994 *Ransum Ternak Ruminansia* (Jakarta: Penerbit Swadaya)
[2] Poole J 1996 *Coming of Ages with Elephants* (New York: Hyperion Press)
[3] Fowler M E and Mikota S K 2006 *Biology, Medicine, and Surgery of Elephants* (Iowa: Blackwell Publishing Professional)
[4] Sukumar R 1990 *J Tropical Ecol* **6** 33
[5] White L J T, Tutin C E G and Fernandez M 1993 *Afr J Ecol* **31** 181
[6] Blake S 2002 Ecology of Forest Elephant Distribution and its Implications for Conservation (Thesis: University of Edinburgh, UK)
[7] Lekagul B and McNeely J A 1975 *The Mammals of Thailand* (Bangkok: The Association for the Conservation of Wildlife)
[8] Eltringham S K 1982 *Elephants* (USA: Blandford Press)
[9] Sukumar R 2003 The Living Elephants: Evolutionary Ecology, Behavior, and Conservation (New York: Oxford University Press)
[10] Santosa Y, Supartono and Thohari M 2011 *Media Konservasi* **16** 149
[11] Abdullah, Supriadi, Syafrianti D, Khairil, Daud A M and Zulfikar 2021 *IOP Conf Ser Earth Environ Sci* **922** 012055
[12] Tchamba M 1996 Elephants and their Interactions with People and Vegetation in the Waza-Logone Region, Cameroon (Dissertation: University of Utrech, Netherlands)
[13] Owen-Smith N and Chafota J 2012 *J Mammal* **93** 698
[14] Foguekem D, Tchambal M N, Gonwouo L N, Ngassam P and Loomis M 2011 *Sci Res Essays* **6** 3577
[15] Rasmussen L E L and Munger B L 1996 *Anat Rec* **246** 127
[16] Gonggo B M, Hermawan B and Anggraeni D 2005 *Jurnal Ilmu-Ilmu Pertanian Indonesia* **7** 44

[17] Berliani K, Alikodra H S, Masy'ud B and Kusrini M D 2018 *J Phys Conf Ser* **1116** 052015
[18] Riba'i, Setiawan A and Darmawan A 2012 *Tengkawang Jurnal Ilmu Kehutanan* **2**

Composition and diversity of tropical forest vegetation at Sibolangit tourism park, North Sumatra, Indonesia

M Ulfa[1,2]*, M M Harahap[1,2], R Rambey[1,2], Y E Ritonga[3], A Zahirah[1], R R Silalahi[1] and A R Sitinjak[1]

[1]Faculty of Forestry, Universitas Sumatera Utara, Kampus USU 2 Bekala, Deli Serdang 20353, Indonesia
[2]JATI-Sumatran Forestry Analysis Study Center, Deli Serdang 20353, Indonesia
[3]BIOTA- Biologi Pencinta Alam of North Sumatra, Medan, Indonesia

*Email: mariahulfa@usu.ac.id

Abstract. Vegetation is an important component of preventing erosion and drought, regulating water and hydrology, improving soil physical, chemical, and biological properties, balancing the carbon dioxide and oxygen in the air, conserving biodiversity, effecting other biota in ecosystem, and supporting the ecotourism potential. The effect varies depending on the structure and composition of the plants that comprise the area's vegetation formation. Sibolangit tourism park, with an area of 24.85 Ha, was established in 1980 by Minister of Agriculture Decree No. 636/Kpts/Um/1980. This study aims to examine the importance value index and diversity index of species in the Sibolangit tourism park area. The data was sampled using a simple random sampling method with a 5% sampling intensity. The observation plots were randomly placed to cover all the study sites. Species, diameter, height, and the number of each individual were recorded during field observation. The study recorded a total of 67 species in Sibolangit Tourism Park. The dominant species at the tree stage was *Pterocarpus indicus* with an IVI of 28%. *Mitrephora obtusa* was the dominant species at the pole and sapling stages with the IVI of 52% and 35%, respectively. At the seedling stage, *Strombosia ceylanica* was the dominant with an IVI of 62%. The vegetation diversity at the study site was in the moderate range with the 2.71 of Shannon-Wiener diversity index.

1. Introduction

Vegetation is an important component of preventing erosion and drought, regulating water and hydrology [1,2], improving soil physical, chemical, and biological properties [3], regulating the balance of carbon dioxide and oxygen in the air, conserving biodiversity, particularly Wallacea flora and fauna [4,5], effecting other biota in ecosystem [6,7], and supporting the ecotourism potential [8]. The effect varies depending on the structure and composition of the plants that comprise the area's vegetation formation [9].

Vegetation analysis can be used to describe the composition of an area. According to PP No. 28 of 2011, Natural Tourism Park (TWA) is a conservation area whose use is centered on nature tourism and recreation. Aside from that, it can be used in a variety of other activities such as research, education, and raising awareness about nature conservation and germplasm sources. Sibolangit tourism park, with an area of 24.85 Ha, was established in 1980 by Minister of Agriculture Decree No. 636/Kpts/Um/1980. Because of its high biodiversity, BBKSDA of North Sumatra suggested this area is ideal for use as a natural laboratory in the development of education and conservation. Sibolangit tourism park represents unspoiled ecosystems and has unique, rare, and beautiful natural communities as well as landscapes with natural tourism potential [10].

The integrity of the vegetation that composes the tourism park is a determining factor in the area's ability to provide optimal and long-term benefits. Vegetation analysis should be performed regularly to ensure that all changes in the composition, structure, and diversity of vegetation are properly recorded. This is the fundamental basis for both sustainable forest management and species conservation plans [11]. Previous research has reported important index values in several conservation areas, including Punti Kayu Tourism Park [12], Sungai Dumai Tourism Park [13], Camplong Tourism Park and Baumata Tourism Park [11]. This study aims to examine the importance value index and diversity index of species in the Sibolangit tourism park area.

2. Materials and Methods

This research was undertaken at Sibolangit Tourism Park. The study site was located in Sibolangit, Deli Serdang Regency, Sumatera Utara Province. The data was sampled using a simple random sampling method with a 5% sampling intensity [12]. The plot size for each vegetation stage was 20x20 m^2 (tree), 10x10 m^2 (pole), 5x5 m^2 (sapling), and 2x2 m^2 (seedling). The observation plots were randomly placed to cover all the study sites. Species, diameter, height, and the number of each individual were recorded during field observation.

The data were then analyzed to measure the Importance Value Index (IVI). IVI of tree pole stages is the sum of relative density, relative frequency, and relative dominance [9]. The relative dominance was not included for IVI of sapling and seedling stages [9]. The species diversity of each vegetation stage was calculated using the Shannon-Wienner Diversity Index [14].

3. Result and Discussion

3.1. Vegetation composition and important value index

Vegetation is a combination of various species of plants that grow together in one location to form a unity and interact closely with one another. The study recorded a total of 67 species in Sibolangit Tourism Park, i.e. *Acacia caffra, Actinodaphne* sp. *Aglaia macrocarpa, Alangium rotundifolium, Alstonia angustiloba, Antidesma montanum, Aquilaria malaccensis, Arthocarpus asiatica, Arthocarpus elasticus, Arthrophyllum diversifolium, Artocarpus dadah, Artocarpus elasticus, Baccaurea motleyana, Bellucia pentamera, Bischofia javanica, Calophyllum teysmannii, Canarium* sp., *Carallia lucida, Ceiba paletonia, Ceiba veletonii, Chukrasia tabularis, Cinnamomum parthenoxylon, Claoxylon longifolium, Dendrocnide stimulans, Diospyros buxifolia, Durio zibethinus, Eugenia densiflora, Ficus ribes, Gymnacranthera paniculate, Hibiscus similis, Horsfieldia* sp., *Knema laurina, Lansium domesticum, Leea aculeata, Lithocarpus bennettii, Lithocarpus elegans, Lithocarpus gracilis, Litsea accedens, Litsea* sp., *Macaranga gigantea, Macaranga tanarius, Macropanax maingayi, Mallotus peltatus, Mangifera foetida, Mastixia trichotoma, Microcos tomentosa, Mitrephora obtusa, Myristica conferta, Myristica inners, Neonauclea* sp., *Oroxylum indicum, Peronema canescent, Pinus merkusii, Psychotria viridis, Pterocarpus indicus, Pterocymbium tinctorium, Samanea saman, Shorea leprosula, Sterculia campanulata, Sterculia macrophylla, Strombosia ceylanica, Swietenia mahagoni, Syzygium cumini, Syzygium polyanthum, Tabernaemontana macrocarpa, Terminalia copelandii,* and *Toxicodendron radicans*. The study site had a higher number of species than the other tourism park from the previous study at Punti Kayu Tourism Park (24 species) [12].

Table 1. The ten species with highest IVI at tree stage

No	Species	Relative Density	Relative Frequency	Relative Dominance	IVI
1	*Pterocarpus indicus*	15%	11%	3%	28%
2	*Pinus merkusii*	6%	2%	16%	24%
3	*Litsea sp*	2%	4%	17%	23%
4	*Mitrephora obtusa*	11%	11%	0%	22%
5	*Strombosia ceylanica*	9%	10%	3%	22%
6	*Arthrophyllum diversifolium*	9%	5%	5%	18%
7	*Pterocymbium tinctorium*	3%	2%	8%	13%
8	*Macropanax maingayi*	6%	4%	1%	10%
9	*Bischofia javanica*	1%	2%	6%	9%
10	*Toxicodendron radicans*	4%	3%	1%	7%

Species composition is the arrangement of vegetation from each stage of growth in a habitat. 49 species of tree were documented in Sibolangit Tourism Park and the ten species with the highest IVI were *Pterocarpus indicus, Pinus merkusii, Litsea* sp., *Mitrephora obtusa, Strombosia ceylanica, Arthrophyllum diversifolium, Pterocymbium tinctorium, Macropanax maingayi, Bischofia javanica,* and *Toxicodendron radicans*, respectively (Table 1).

According to their ecological value, species are ranked using the Important Value Index (IVI). This indicator is used to identify the dominating species in a certain community at each stage of growth [15]. *Pterocarpus indicus* was the dominant species in this location at the tree stage with an IVI of 28%. *Pterocarpus indicus* species have a high percentage of life indicating the ability of plants to adapt to their environment [16].

Table 2. The ten species with the highest IVI at pole stage

No	Species	Relative Density	Relative Frequency	Relative Dominance	IVI
1	*Mitrephora obtusa*	18%	14%	20%	52%
2	*Arthrophyllum diversifolium*	16%	14%	17%	46%
3	*Macropanax maingayi*	10%	9%	10%	29%
4	*Strombosia ceylanica*	10%	11%	8%	29%
5	*Toxicodendron radicans*	9%	6%	6%	20%
6	*Pterocarpus indicus*	6%	6%	5%	17%
7	*Sterculia campanulata*	5%	5%	7%	17%
8	*Baccaurea motleyana*	3%	5%	4%	12%
9	*Litsea accenden*	2%	3%	4%	9%
10	*Macaranga tanarius*	3%	3%	2%	8%

The pole phase usually starts to have better adaptability and stable growth. The vegetation can access more sunlight to support its growth and life at this stage [17]. A total of 23 species were found at this site at the pole stage. *Mitrephora obtusa* was the dominant species with the highest IVI of 52% (Table 2). This species was also dominant at the tree stage (22% of IVI). At the tree and pole stage, stem diameter affects the value of the important index along with the density and frequency of a species [18].

A diversifolium, M maingayi, and S ceylanica were also dominant in the study location. A previous study stated that *A diversifolium* was also one of the dominant species at Mount Bulusaraung [18]. The Importance Value Index of a species illustrates that the existence of that species is more stable or has the opportunity to be able to maintain the growth and sustainability of the species [19].

Table 3. The ten species with the highest IVI at sapling stage

No	Species	Relative Density	Relative Frequency	IVI
1	*Mitrephora obtusa*	22%	13%	35%
2	*Strombosia ceylanica*	10%	10%	20%
3	*Arthrophyllum diversifolium*	9%	10%	19%
4	*Eugenia densiflora*	5%	5%	10%
5	*Mastixia trichotoma*	3%	5%	8%
6	*Pterocarpus indicus*	3%	5%	8%
7	*Antidesma montanum*	3%	3%	7%
8	*Macropanax maingayi*	3%	3%	7%
9	*Aglaia macrocarpa*	2%	3%	6%
10	*Litsea sp*	2%	3%	6%

The high IVI of a species at the sapling and seedling stage is caused by the density and frequency that are highly and evenly distributed throughout the area [18]. 34 species were recorded at sapling stage with *Mitrephora obtusa* as the dominance species (35% of IVI) (Table 3).

The various IVI values indicate the influence of the environment condition, temperature, humidity, and the ability of species to compete in the struggle for nutrients, light, sun, and growth space with other species, all of which have a significant impact on the growth of the diameter of vegetation and human activities in direct contact with the forest.

Table 4. The ten species with highest IVI at seedling stage

No	Species	Relative Density	Relative Frequency	IVI
1	*Strombosia ceylanica*	36%	26%	62%
2	*Mitrephora obtusa*	17%	15%	32%
3	*Baccaurea motleyana*	24%	6%	30%
4	*Arthrophyllum diversifolium*	7%	6%	12%
5	*Pterocarpus indicus*	4%	6%	10%
6	*Litsea accendes*	2%	3%	5%
7	*Macropanax maingayi*	2%	3%	5%
8	*Antidesma montanum*	1%	3%	4%
9	*Eugenia densiflora*	1%	3%	4%
10	*Leea aculeata*	1%	3%	4%

The high IVI at the seedling stage indicates the regeneration ability of a species ([17]. Of the 67 species documented at Sibolangit tourism park, only 19 species were found at the seedling stage. [11] also found 19 species of seedling at Camplong tourism park. The number of species was higher than that was recorded at Baumata tourism park, which was 5 species of seedling.

Strombosia ceylanica was the dominant species (IVI of 62%) followed by *M. obtusa, B. motleyana, A. diversifolium*, and *P. indicus* (Table 4). *S. ceylanica, M. obtusa*, *A. diversifolium*, and *P. indicus* were species found in all stages of growth in the study site, however, it cannot be generalized to every species [19].

3.2. Vegetation diversity at Sibolangit tourism park

The Shannon-Wiener diversity index has been widely used in environmental research to measure species richness and diversity because of its simplicity of estimation and interpretation [20]. This index provides prior information about the tree community found in a location [22]. The highest diversity was found in the growth rate of trees and saplings with values of 3.23 and 3.03, respectively.

On the other hand, seedlings and poles have low values (Figure 2). Therefore, the vegetation diversity at the study site was in the moderate range with the 2.71 of diversity index [26].

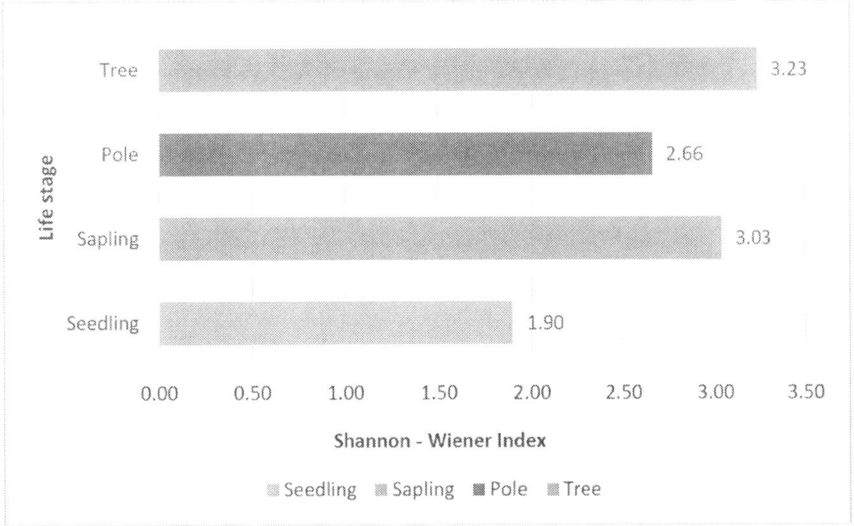

Figure 1. Shannon – Wiener diversity index in each vegetation life stage

The research location is a degraded secondary natural forest due to the increase in community needs for forest resources [23,24]. A previous study estimated that between 2009 and 2014 there has been a decrease in the total forest cover area of around 30 ha in the Sibolangit District [25].

4. Conclusions

This study recorded a total of 67 species in Sibolangit Tourism Park. The dominant species at the tree stage was *Pterocarpus indicus* with an IVI of 28%. *Mitrephora obtusa* was the dominant species at the pole and sapling stages with the IVI of 52% and 35%, respectively. At the seedling stage, *Strombosia ceylanica* was dominant with an IVI of 62%. The vegetation diversity at the study site was in the moderate range with the 2.71 of Shannon-Wiener diversity index.

Acknowledgments

This research was fully funded by University of Sumatera Utara through TALENTA research grant with the contract number: 136/UN5.2.3.1/PPM/SPP-TALENTA USU/2021. We also thank BBKSDA Sumatera Utara and the management of Sibolangit Tourism Park who supported us during the field research.

References

[1] Naharuddin 2017 *Jurnal Hutan Tropis* **5** 134
[2] Rambey R, Susilowati A, Rangkuti A B, Onrizal O, Desrita, Ardi R and Hartanto A 2021 *Biodiversitas* **22** 3250
[3] Masnang A, Sinukaban N, Sudarsono and Gintings N 2014 *Jurnal Agroteknos* **4** 32
[4] Pitopang R 2012 *Natural Science* **1** 85
[5] Naharuddin, Bratawinata A, Hardwinarto S and Pitopang R 2016 *Jurnal Warta Rimba* **4** 28
[6] Ulfa M, Ikejima K, Poedjirahajoe E, Faida L R W and Harahap M M 2018 *Reg Stud Mar Sci* **24** 296
[7] Zhao S, Li Z and Duo L 2021 *Bulletin of Entomological Research* **111** 553
[8] Ulfa M and Harahap M M 2022 *IOP Conf Ser Earth Environ Sci* **977** 012094
[9] Indriyanto 2006 *Ekologi Hutan* (Jakarta: PT Bumi Aksara)

[10] Ginting I A, Patana P and Rahmawaty 2013 *Peronema Forestry Science Journal* **2** 74
[11] Kristinawati I, Almulqu A A, Adrin and Renoat E 2022 *Florea: Jurnal Biologi dan Pembelajarannya* **9** 66
[12] Rulianto A, Bintoro A, Winarno G D and Safe'i R 2019 *Jurnal Biologi Eksperimen dan Keanekaragaman Hayati* **5** 7
[13] Ikhsani H 2021 *Wahana Forestra: Jurnal Kehutanan* **16** 25
[14] Magurran A E 1988 *Ecological Diversity and Its Measurement* (Princeton: Princeton University Press).
[15] Susilowati A, Elfiati D, Rachmat H H, Yulita K S, Hadi A N, Kusuma Y S and Batu S A L 2020 *Biodiversitas* **21** 3025
[16] Fitriani D E, Hatta G M and Effendy M M 2019 *Jurnal Sylva Scienteae* **2** 834
[17] Hartini K S and Silitonga I L 2019 *IOP Conf Ser Earth Environ Sci* **374** 012049
[18] Asrianny, Paweka C B, Achmad A, Oka N P and Achmad N S 2019 *Jurnal Perennial* **15** 32
[19] Dendang B and Handayani W 2015 *Pros Sem Nas Masy Biodiv Indon* **1** 691
[20] G Barrantes and L Sandoval 2009 *Revista de biología tropical* **57** 3
[21] Susilowati A, Rangkuti A B, Rachmat H H A, Iswanto A H, Harahap M M, Elfiati D, Slamet B, and Ginting I M 2021 *Biodiversitas* **22** 5
[22] Harahap M M, Poedjirahajoe E, Santosa S H M B and Ulfa M 2021 *J. Sylva Indones* **4** 02
[23] Harahap M M, Rahmawaty, Kurniawan H, Rauf A and Ulfa M 2021 *IOP Conf. Ser. Earth Environ. Sci* **912** 1
[24] Rahmawaty, Harahap M M, Rauf A and Kurniawan H 2021 *IOP Conf. Ser. Earth Environ. Sci* **752** 01
[25] Barbour M G, Burk J H and Pitts W D 1987 *Terrestrial Plant Ecology* (Menlo Park, CA: Benjamin/Cummings Publishing Co)

Assessment of the tree damage level in Medan Beringin Park

M M Harahap[1,2,3,*], A G Ahmad[1,3], J Simatupang1, M Ulfa[1,2,3] and P Y A P Wirabuana[4]

[1]Faculty of Forestry, Universitas Sumatera Utara, Medan, North Sumatra, Indonesia
[2]JATI – Sumatran Forestry Analysis Study Center, Jl. Tri Dharma Ujung No. 1, Medan, North Sumatra 20155, Indonesia
[3]KEHATI – Link, Center for the Study of Biodiversity and Environmental Services, Universitas Sumatera Utara, Medan, North Sumatra, Indonesia
[4]Department of Forest Management, Universitas Gadjah Mada, Sleman, Yogyakarta 55281, Indonesia

*Email: mhr_mrghy@usu.ac.id

Abstract. Medan is the capital city of North Sumatra Province and the biggest city on the island of Sumatra. Big cities in Indonesia generally have problems with environmental degradation. One of the efforts to preserve the environment in urban areas is to provide public green spaces dominated by trees. In the process of its life, the tree experiences many disturbances and interventions from the surrounding environment. It can make it grow abnormally or become sick. One of the negative impacts of this is when extreme weather occurs and causes trees to break easily, fall, and fall on whatever is underneath. It can cause various material losses and even fatalities. The research has conducted at Beringin Park in Medan. It is one of the favorite places for people to gather and enjoy the freshness of clean air. Research on tree health surveys and inventories in public green spaces is important to do as one of the efforts to mitigate the occurrence of fallen tree disasters. This research is aimed to take an inventory of tree health in Beringin Park as an mitigation effort of the fallen tree disaster in Medan. Vegetation parameters were collected by using the census method at the pole and tree stage. The Shannon – Wienner diversity index has used to assess the biodiversity level of the research location. Then damage level index was analyzed by using Forest Health Monitoring Method. The results showed that the tree vegetation in Beringin Park was in the healthy category based on the Tree Damage Index Value which ranged from 1.20 to 4.10.

1. Introduction

Medan is the capital city of North Sumatra Province as well as the largest city on the island of Sumatra. Based on its geographical position, Medan is located at 3° 30' 00" to 3° 43' 00" North Latitude and 98° 35' 00" to 98° 44' 00" East Longitude. Big cities in Indonesia generally have problems with environmental degradation. Some of city issues are dense population and developed area [1]. One of the efforts to preserve the environment in urban areas is to provide public green spaces which are dominated by trees. As it is well known that trees that dominate a land will be able to provide various environmental services, including; the availability of clean and fresh air [2], soothing shade, absorbing waste particles scattered in the air, absorbing noise pollution, providing a microclimate as well as soil and water conservation services etc.

The existence of trees in the Medan, especially in public green spaces, is inseparable from the lives of residents. However, the growing tree is not always in a healthy condition. In the process of its life, the tree experiences many disturbances and interventions from the surrounding environment which can make it grow abnormally or become sick. When a tree becomes sick and grows abnormally, it will also have a negative impact on the surrounding environment. One example is when there is extreme weather that causes trees to break easily, fall and fall on whatever is under them. This can cause various material losses and even fatalities.

Research on tree health surveys and inventories in several public green spaces is important to do as one of the efforts to mitigate the occurrence of fallen tree disasters. This is because the data and the availability of tree health information are not yet available and can be accessed in general. Thus, the results of this study are expected to provide up-to-date data and information for those in need and as input for the City Government in carrying out tree maintenance [2] and developing public green spaces in Medan.

2. Materials and Methods

The research was conducted at the Beringin Park, Medan, North Sumatra. The research was carried out from April 2021 - October 2021. Various tools were used in the field, including; research location map, GPS, digital camera, hagameter, phiband, thallysheet and stationery. The material used in this research is tree vegetation in Medan Beringin Park. Vegetation parameters were collected by using census method at the pole and tree stage. The Shannon – Wiener diversity index was used to assess the biodiversity level of the research location. Then damage level index was analized by using Forest Health Monitoring Method.

3. Results and Discussion

3.1. Biodiversity

The Shannon-Wiener biodiversity index is often used to describe the level of diversity in an ecosystem [4]. Based on the results of the analysis of this location has a diversity index value of 1.99 (Figure 1). This value is in the medium category. This value also describes normal or stable environmental conditions without any significant disturbance. Environmental interactions both sourced from inside and outside the ecosystem can affect biodiversity. The results of these interactions can be observed by paying attention to growth, form of development, emergence and loss of species [5] as well as the movement of fauna.

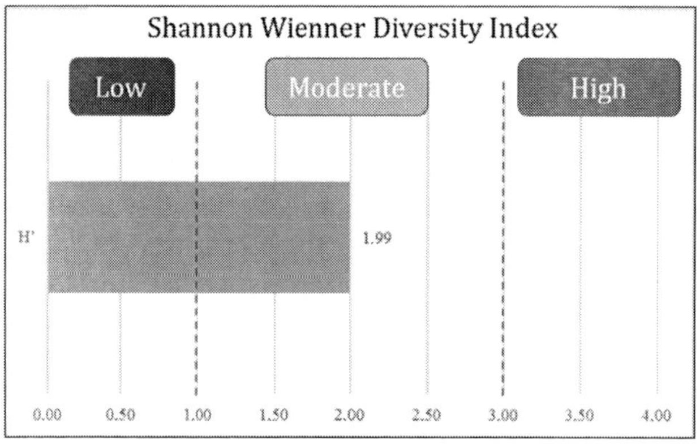

Figure 1. Diversity of trees in Medan Beringin Park

3.2. Damage of trees

The Forest Health Monitoring method used in this study examines tree damage based on location, type and severity (location, type and severity). These three parameters will determine the value of the tree damage level index of the resulting tree. Therefore, proper assessment of these three parameters will determine how severe the damage is being experienced by tree vegetation in each location. The most common location of damage found at this location was the upper stem at 29% (Figure 2). Furthermore, the most damage found is the rootstock and the top. Damage to the stem is the most dominant thing found in the Beringin Park. Damage to this location causes disturbances in the process of transporting nutrients in plant body parts [6]. As it is known that the stem is the support for the top of the canopy. Therefore, stem damage needs to be anticipated as early as possible so as not to pose a potential hazard to visitors who come. Beringin Park has 10 types of damage found in the tree vegetation (Figure 3). The damage can cause the decreasing of environmental sevices of trees, such as carbon sink and biomass [7]. Beringin Park has a diverse level of severity. The value of this parameter ranges from 0% - 39%. The severity level most frequently found is in the range of 20% - 29% (Figure 4). Several types of tree damage found at the Beringin Park location are presented in Figure 5.

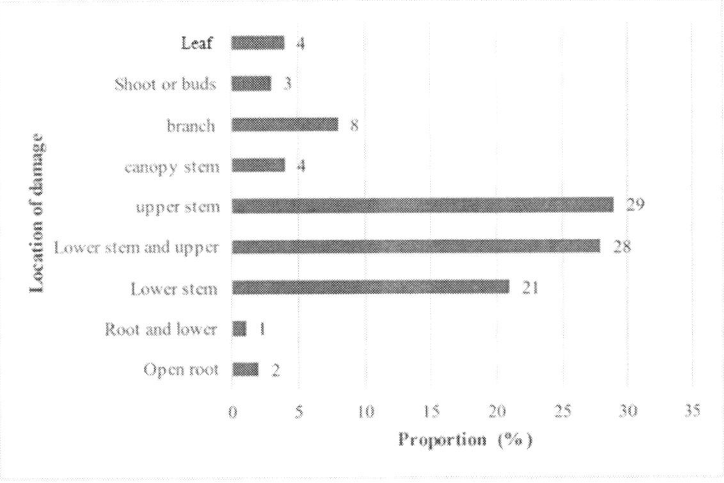

Figure 2. Location of damage to tree vegetation in Beringin Park

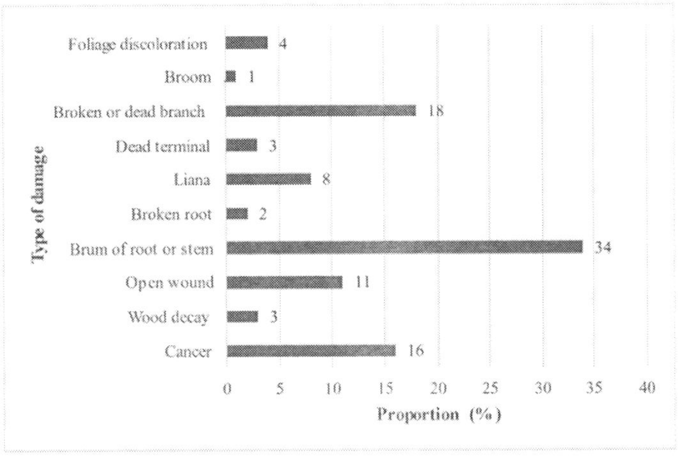

Figure 3. Types of damage to tree vegetation in Beringin Park

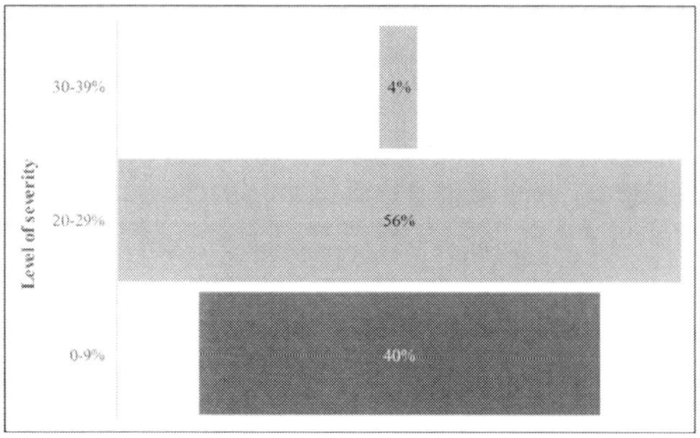

Figure 4. Severity of the tree vegetation in Beringin Park

Figure 5. Some of tree damage in the Beringin Park (a and d are dead branch, b is open wound and c is cancer)

The results of the assessment using the tree damage index in Taman Beringin have a score of 1.20–4.10. Therefore, the tree vegetation in this location is included in the healthy category. However, various tree damage has been found at these two locations. Therefore, routine and regular maintenance activities are still needed [8,9] so that the existing vegetation can provide optimal environmental service functions to visitors and avoid the potential for a fallen tree disaster.

4. Conclusions

The Shannon Wiener biodiversity index value found at the Beringin Park location is only 1.99. This means that the condition of diversity at this location is moderate with a fairly stable environment. The tree vegetation in the Beringin Park is categorized as healthy based on the Tree Damage Index Value which ranges from 1.20 to 4.10.

Acknowledgments

Our research was funded by The Universitas Sumatera Utara TALENTA Fund 2021 (Contract Number: 138/UN5.2.3.1/PPM/SPP-TALENTA USU/2021). We would like to thank the anonymous reviewers and colleagues who have provided constructive and valuable advices. We also thank ICONART 4 committee.

References

[1] Zaitunah A, Samsuri, Anggie Y, Mandalahi, and Syaufina S 2022 *J. Sylva Indonesiana* **5** 68
[2] Rahmawaty, Sembiring I E P, Batubara R and Patana P 2018 *IOP Conference Series: Earth and Environmental Science* **166** 012020
[3] Harahap M M, E Poedjirahajoe, S H M B Santosa and M Ulfa 2021 *J. Sylva Indonesiana* **4** 87
[4] Salsabila R, Hariyadi and N Santoso 2021 *Jurnal Sylva Lestari* **9** 86
[5] Safe'i R, Erly H, Wulandari C and Kaskoyo H 2018 *Perennial* **14** 32
[6] Safe'i R and Tsani M K 2016 *Kesehatan Hutan: Penilaian Kesehatan Hutan Menggunakan Teknik Forest Health Monitoring* (Plantaxia: Yogyakarta)
[7] Rahmawaty, Sitorus N A and Rauf A 2017 *Indonesia Malaysian Forester* **80**
[8] Susilowati A, Ahmad A G, Silalahi V and Rachmat H H 2018 *IOP Conf. Series: Earth and Environmental Science* **203** 012028
[9] Susilowati A, Rangkuti A B, Rachmat H H, Iswanto A H, Harahap M M, Elfiati D, Slamet B and Ginting I M 2021 *Biodiversitas Journal of Biological Diversity*, **22** 2839

Tree diversity in Medan Ahmad Yani Park

M M Harahap[1,2,3,*], A G Ahmad[1,3], J Simatupang1, M Ulfa[1,2,3] and P Y A P Wirabuana[4]

[1]Faculty of Forestry, Universitas Sumatera Utara, Medan, North Sumatra, Indonesia
[2]JATI – Sumatran Forestry Analysis Study Center, Jl. Tri Dharma Ujung No. 1, Medan, North Sumatra 20155, Indonesia
[3]KEHATI – Link, Center for the Study of Biodiversity and Environmental Services, Universitas Sumatera Utara, Medan, North Sumatra, Indonesia
[4]Department of Forest Management, Universitas Gadjah Mada, Sleman, Yogyakarta 55281, Indonesia

*Email: mhr_mrghy@usu.ac.id

Abstract. Medan is the capital city of North Sumatra Province. It is one of the biggest cities on Sumatra Island. Big cities in Indonesia generally have problems with environmental degradation. One of the efforts to preserve the environment in urban areas is to provide public green spaces. Ahmad Yani Park is a green space dominated by trees that grow compactly and tightly in 1.3 ha of area. The research has been conducted at Ahmad Yani Park, one of the favorite places for people to gather and enjoy the freshness of clean air. Research on tree diversity in several public green spaces is needed to conduct as one of the efforts to provide biodiversity data and information in urban. This research is aimed to analyze the tree structure, composition, and diversity in Ahmad Yani Park to support sustainable development goals, especially in an urban area. The parameters of vegetation collecting by using census methods. The condition of vegetation and environment were assessed by using relative density, Shannon Wienner Diversity Index, and Evenness Index. The Ahmad Yani Park contained 240 individuals of 32 species It dominated by 62 individuals of Mahogany (*S. macrophylla*). The dominant vegetation is in 20 - 150 cm diameter class (59%). It is a mature phase of tree. While the juvenile and seedling stage have 26% and 15%. The Ahmad Yani Park has 2.75 score of Shannon Wiener Diversity index. It means this park is moderate diverse with a fairly stable environment.

1. Introduction

Big cities in Indonesia generally have problems with environmental degradation. There are many activities in city can generate the pollutant. Providing public green spaces which are dominated by trees is one of the efforts to preserve the environment in urban areas. It is well known that domination of trees in a land will be able to provide various environmental services, such as providing clean and fresh air [1], soothing shade [2], absorbing waste particles scattered in the air, absorbing noise pollution, providing a microclimate etc.

Ahmad Yani Park is a green space in Medan, North Sumatra, Indonesia. It is dominated by growth trees compactly and tightly within 1.3 ha of area. Trees through the metabolic process, namely photosynthesis, can absorb carbon dioxide (CO_2) in the atmosphere and turn it into organic compounds [3]. This study examines the diversity of tree in the Ahmad Yani Park. Thus, the results of this study will be able to provide up-to-date data and information on tree diversity in public green spaces. Furthermore, by knowing that data and information through scientific studies will be used in the context of tree maintenance and the development of green spaces [1] by the government and the community.

2. Materials and Methods

The research was conducted at the Ahmad Yani Park, Medan, North Sumatra, Indonesia (Figure 1). The research was carried out from September 2021 - October 2021. Various tools were used in the field, including; research location map, GPS, digital camera, hagameter, phiband, thallysheet and stationery. The material used in this research is tree vegetation in Ahmad Yani Park. The parameters of vegetation collecting by using census methods. The condition of vegetation and environment were assessed by using relative density, Shannon Wienner Diversity Index and Evenness Index.

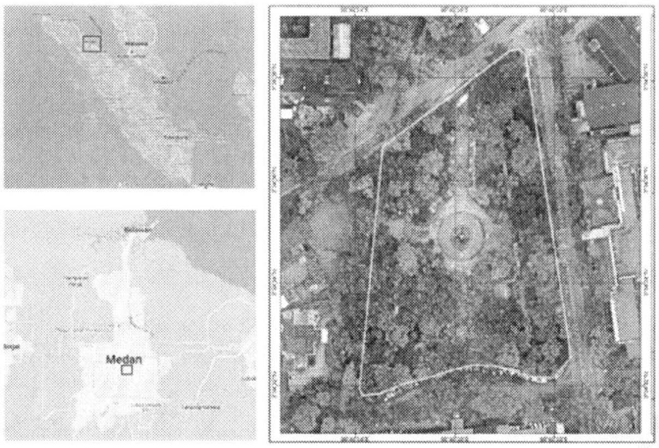

Figure 1. Research location in Ahmad Yani Park

3. Results and Discussions

3.1. Tree structure and composition in Ahmad Yani Park

The results of research on the composition and structure of trees at the seedling, sapling, pole, and tree levels in Ahmad Yani Park, Medan, found 30 species with a total number of 240 individuals. A total of 62 individuals who dominate are the type of Mahogany (*Swietenia macrophylla*). Other types of vegetation that were commonly found were Butterfly (*Bauhinia blackheana*) with 23 individuals and Pulai (*Alstonia scholaris*) (Figure 2c) with 20 individuals. Meanwhile, the types of vegetation with the least number of individuals were found in Bungur (*Lagerstroemia speciosa*), Noni (*Morinda citrifolia*), Kepuh (*Sterucila feotida*), Jackfruit (*Artocarpus heterophyllus*), Suren (*Toona sureni*), and Sawo (*Manilkara zapota*) amounted to 1 individual for each species. The top three dominated species were founded in Ahmad Yani Park can be seen in Figure 2.

The results of research on the composition and structure of trees at the seedling level in Ahmad Yani Park, Medan were not found. This is because the garden floor is planted with grass and partly installed with paving blocks by the manager so that there is no growing space for seedlings to grow naturally. For vegetation at the sapling and pole levels, there are 14 species with a total of 98 individuals. The dominant type of stake and pole level is Jambu bol (*Syzygium malaccense*) totaling 18 individuals.

The data obtained for the tree level contained 20 tree species planted with the highest number of trees being Mahogany (*S. macrophylla*) as many as 55 individuals with an average diameter of 58.7 cm. Mahogany (*S. macrophylla*) has reached a very old age that can be seen from the average size of the diameter. The abundance of Mahogany (*S. macrophylla*) due to its use as a street shade tree, can be seen from the planting of Mahogany evenly on the outskirts of Ahmad Yani Park which borders the highway.

a b c

Figure 2. Several dominant species (a) Mahogany (*S. macrophylla*), (b) Butterfly (*B. blackheana*), and (c) Pulai (*A. scholaris*)

In accordance with the statement of [4], which stated that the Mahogany is a type of road protection tree that is widely planted in several cities in Indonesia because of its tall tree architecture and round and shady canopy. Meanwhile, for the tree level, the lowest population is Breadfruit (*Arthocarpus comunis*) and Tabebuya (*Handroanthus chrysotrichus*) with 1 individual tree each. The tree diameter class by life stage data in Ahmad Yani Park can be seen in Figure 3. Based on the structure of the vegetation, the dominant vegetation is in 20 - 150 cm diameter class (59%). It is a mature phase of tree. While the juvenile and seedling stage have 26% and 15%. It indicates the stands dominated by the mature stage. It is indicates the vegetation are maintained on trees or mature phase.

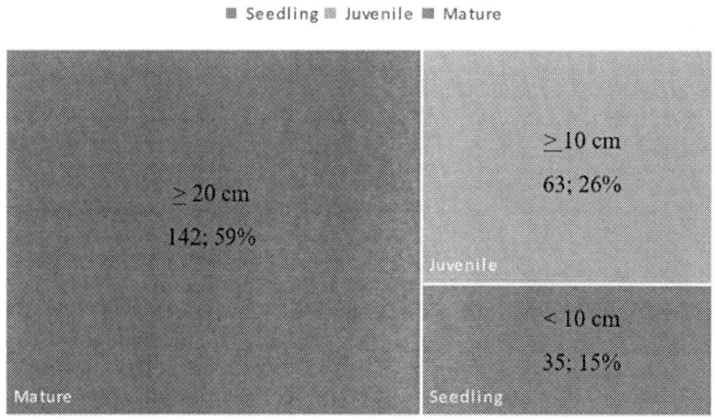

Figure 3. Diameter class of trees by life stage in Ahmad Yani Park

The tree species with the largest diameter was Pulai (*A. scholaris*). It had a diameter of 150 cm, followed by the Mahogany (*S. macrophylla*) which had a diameter of 130 cm, and the Roda tree (*Hura crepitans*) with a diameter of 111 cm. Based on research conducted by Silalahi [5], Pulai as a shade tree that is commonly found in Indonesia. The highest size of this species can reach 60 meters and a diameter can reach 130 cm.

Tree is a type of vegetation can absorp carbon dioxide in larger amount. The planted vegetation in Ahmad Yani Park is dominated by forestry plants. Vegetation with tree habitus dominates The Ahmad Yani park. It can provide the function of green space as an absorber of pollutants and forming a microclimate is quite optimal. In accordance with previous research [6], in addition to natural forests and plantation forests which were mostly developed for industrial timber needs, the presence of trees in urban areas had an important role in carbon sequestration. This adds to the benefits of the function of trees in urban areas, other than as a spatial conditioner, oxygen producer, animal habitat, and water catchment area.

3.2. Relative density, diversity and evenness index

Information on relative density, species diversity index, and species evenness index is needed for planning and efforts to conserve biodiversity. Monitoring of vegetation composition will also be necessary to maintain biodiversity and its ecosystem [7]. For tree species conservation, it is important to examine the current status of species diversity, composition and abundance as it will provide guidance in their management and valuable reference for assessment, as well as increase our knowledge in identifying ecologically useful species [8]. The relative density of each species founded in all life stages can be seen in Table 1.

Table 1. The Relative density of various trees species in all life stage in Ahmad Yani Park

No	Local Name	Scientific Name	Relatif Density (%)
1	Angsana	*Pterocarpus indicus*	4,17
2	Asam glugur	*Garcinia atroviridis*	0,42
3	Beringin	*Ficus benjamina*	0,83
4	Buah Roda	*Hura crepitans*	1,25
5	Bungur	*Lagerstroemia speciosa*	0,42
6	Buni	*Antidesma bunius*	2,08
7	Cemara Laut	*Casuarina equisetifolia*	1,25
8	Dadap	*Erythrina variegata*	0,83
9	Durian	*Durio zibethinus*	1,25
10	Glodokan	*Polyathia longifolia*	4,58
11	Jambu bol	*Syzygium malaccense*	7,50
12	Kamboja	*Plumeria obtusa*	1,25
13	Kepuh	*Sterucila feotida*	3,75
14	Kepundung	*Baccaurea racemosa*	0,42
15	Kerai payung	*Filicium decipiens*	1,25
16	Ketapang Biola	*Ficus lyrata*	1,67
17	Kupu-kupu	*Bauhinia blackheana*	9,58
18	Mahoni	*Swietenia macrophylla*	25,83
19	Mangga	*Mangifera indica*	5,83
20	Manggis	*Garcinia mangostana*	0,42
21	Matoa	*Pometia pinnata*	4,17
22	Melinjo	*Gnetum gnemon*	1,25
23	Mengkudu	*Morinda citrifolia*	0,42
24	Nangka	*Artocarpus heterophyllus*	0,42
25	Pulai	*Alstonia scholaris*	8,33
26	Rambutan	*Nephelium lappaceum*	1,67
27	Sawo	*Manilkara zapota*	0,42
28	Sentul	*Sandoricum koetjape*	0,83
29	Sukun	*Arthocarpus comunis*	0,83

No	Local Name	Scientific Name	Relatif Density (%)
30	Suren	*Toona sureni*	0,42
31	Tabebuya	*Handroanthus chrysotrichus*	0,83
32	Tanjung	*Mimusops elengi*	5,83
	Total		100

The highest relative density value of 27.07% was found in the type of Mahogany (*S. macrophylla*). The highest relative density value for it was included in the abundant category with 62 individuals. The high relative density value was caused by the planting of this species which was evenly distributed on each edge of the garden. According to [9] It is able to grow for decades, is resistant to physical disturbances, is not easily affected by pests and diseases, does not fall easily with a strong wooden structure, grows straight up with a high crown. The lowest relative density was found in several tree species, including Noni (*Morinda citrifolia*), Bungur (*Lagerstroemia speciosa*), Glugur Acid (*Garcinia atroviridis*), Kepuh (*S. feotida*), Mangosteen (*Garcinia mangostana*), Jackfruit (*Artocarpus heterophyllus*), Sawo (*Manilkara zapota*), and Suren (*Toona sureni*) by 0.42%. The lowest relative density value is included in the very rare category. This is because the management places more emphasis on planting shade trees than on fruit trees.

The species diversity index (H') is an index that expresses community structure and ecosystem stability. The higher the value of the species diversity index, the better the stability of the ecosystem. The value of species diversity in the ecosystem in Ahmad Yani Park has a value of 2.75. This value indicates that the diversity index in the Ahmad Yani Park ecosystem is included in the medium diversity category. In the calculation of the value of the evenness of species, the total value of the evenness of tree species in Ahmad Yani Park is 0.50. It indicates there are several species dominate other species in the community. The evenness value of tree species of 1 indicates that there is one or several species that dominate other species in the community. The value of the species evenness index can be used as an indicator of the type that dominates among each species in the community [10]. The species evenness index value of 1 indicates that in this habitat type there is no dominance of one species over another. The evenness index of this species is the level of similarity between the number of one species and the number of other species.

4. Conclusions

The Ahmad Yani Park contained 240 individuals of 32 species It dominated by 62 individuals of Mahogany (*S. macrophylla*). The dominant vegetation is in 20 - 150 cm diameter class (59%). It is a mature phase of tree. While the juvenile and seedling stage have 26% and 15%. The Ahmad Yani Park has 2.75 score of Shannon Wiener Diversity index. It means this park is moderate diverse with a fairly stable environment.

Acknowledgments

Our research was funded by The Universitas Sumatera Utara TALENTA Fund 2021 (Contract Number: 138/UN5.2.3.1/PPM/SPP-TALENTA USU/2021). We would like to thank the anonymous reviewers and colleagues who have provided constructive and valuable advices. We also thank ICONART 4 committee.

References

[1] Rahmawaty, Sembiring I E P, Batubara R and Patana P 2018 *IOP Conference Series: Earth and Environmental Science* **166** 012020
[2] Harahap M M, Poedjirahajoe E, Santosa S H M B and M Ulfa. 2021 *J. Sylva Indonesiana.* **4** 87
[3] Junaedi A 2008 *Info Hutan* **5** 1
[4] Mansur M and Pratama B A 2014 *Jurnal Biologi Indonesia* **10** 149
[5] Silalahi M 2019 *Jurnal Pro Life* **6** 136

[6] Samsoedin I and Wibowo A 2012 *Jurnal Penelitian Sosial dan Ekonomi Kehutanan* **9** 43
[7] Susilowati A, Rangkuti A B, Rachmat H H, Iswanto A H, Harahap M M, Elfiati D and Ginting I M 2021 *Biodiversitas Journal of Biological Diversity* **22** 2839
[8] Ogwu M C, Osawaru M E and Obayuwana O K 2016 *Applied Tropical Agriculture* **21** 46
[9] Manik E, Latifah S and Patana P 2016 *Peronema Foresty Science Journal* **5** 1
[10] Santosa Y, Ramadhan E P and Rahman D A 2008 *Media Konservasi* **13** 1

IOP Conf. Series: Earth and Environmental Science 1115 (2022) 012014 doi:10.1088/1755-1315/1115/1/012014

Mapping and Population Structure of *Shorea platyclados in* Martelu Purba Nature Reserve

KS Hartini[1], AB Rangkuti[1], BP Manalu[1], SW Andini[1], A Insani [1].

[1]Faculty of Forestry, Universitas Sumatera Utara.
Kampus USU 2 Bekala, Pancurbatu, Deli Serdang, Indonesia. 20355

Email: kansih_sri_hartini@usu.ac.id

Abstract. In its natural habitat, *S. platyclados* is abundant in Martelu Purba Nature Reserve, North Sumatra. However, like the Dipterocarpaceae family, this species has limitations in regeneration related to the uncertain flowering phase and the seeds' recalcitrant nature. This condition can be a long-term threat to the sustainability of *S. platyclados*. Efforts should be made to select mother plants as a source of short-term quality seeds, to support the breeding program and the conservation of *S. platyclados*. Monitoring of mother plants will be more effective if mapping data is available. Therefore, it is important to map the mother plant of *S. platyclados*. In addition, the population structure needs to be identified and related to trends in its sustainability. The goal of this study was to map the mother plant of *S. platyclados* in Martelu Purba Nature Reserve and to examine its population dynamics.. This research was conducted in Martelu Purba Nature Reserve, Simalungun Regency, North Sumatra. There are 195 hectares of designated population there. Combining the path method and the plotline, the sampling methodology was applied. Fifty (50) plots were observed across the sample area of 2 hectares (1% Sampling Intensity), which was separated into three lanes. Observations were made on a 20 m x 20 m plot to measure the vegetation at the tree level (diameter 20 cm), a 10 m x 10 m observation sub-plot to measure the vegetation at the pole level (diameter 10cm - 20m), a 5 m x 5 m observation sub-plot to measure the vegetation at the sapling level (diameter 10 cm), and a 2 m × 2 m observation sub-plot to measure the vegetation at the seedling level. The mother plants were recorded as many as 309 individuals in 50 observation plots, or 155 individuals/ha spread evenly in the relatively sloping Nature Reserve area. The distribution structure of the height and diameter of the mother plants indicates a good adaptation, resembles a standard curve. The population structure of *S. platyclados* is curved like an inverted J, indicates that this population is quite stable and sustainable. However, monitoring at the pole level is crucial based on a gap in the presence of this regeneration compared to others.

1. Introduction

Indonesia's biodiversity is very high. This high biodiversity is related to the condition of Indonesia, which is in a tropical climate, with always warm temperatures, high humidity, and high rainfall. This condition causes various plants to grow and adapt well. Around 400 species, or 70% of the world's Dipterocarpaceae, are found in Indonesia [1]. One is *Shorea platyclados*, located in hills or mountains between 700 to 1,300 meters above sea level [2].

S. platyclados, known as Meranti Merah, is an economically important wood tree species traded internationally. Additionally, the bark of this tree includes chemical elements that can be used as antifungal, antibiotic, cancer-fighting, and other medicines. [3,4]. However, due to widespread overlogging of natural forests [5,] limited natural regeneration, and sluggish recovery rates, the demand for its wood products has led to a catastrophic decline in their habitats [6, 7]. *S. platyclados* is listed as Critically Endangered (Endangered A1cd ver 2.3) on the Red List of Threatened Species maintained by the International Union for Conservation of Nature (IUCN), which indicates a population loss of at least

50% over the last ten years [8]. Extinction may occur if population decline outpaces conservation efforts.

The abundance of *S. platyclados* in its natural environment, the Martelu Purba Nature Reserve in North Sumatra, runs somewhat counter to the IUCN's assertion [9]. However, due to the unpredictable flowering phase and the resistive character of the seeds, this species experiences regeneration restrictions similar to those experienced by the Dipterocarpaceae family. The sustainability of *S. platyclados* may be in risk due to this problem in the long run. To promote the breeding program and the conservation of *S. platyclados*, efforts should be made to choose mother plants as a source of short-term quality seeds. If mapping data is provided, monitoring of mother plants will be more successful. Therefore, mapping the *S. platyclados* mother plant is crucial. In addition, trends in sustainability must be identified and tied to the population structure. Populations have a life cycle similar to how animals grow, diversify, and maintain themselves. To determine group characteristics like birth and death rates, age ratios, and genetic compatibility, just the population may be employed [10]. Populations are controlled by physical elements such as weather, water currents, chemical factors that limit pollution, and other physical factors in ecosystems with low variety that are prone to unpredictably exogenous disruptions. In contrast, biology and natural selection control population growth in extremely diversified ecosystems. Young people typically make up the majority of a population that is growing quickly[11]. The goal of this study was to map the mother plant of *S. platyclados* in the Martelu Purba Nature Reserve and to examine its population dynamics.

2. Method

2.1. Plotting and Sampling Intensity

North Sumatra's Simalungun Regency's Martelu Purba Nature Reserve served as the site of this study. There are 195 hectares of designated population there. The narrative and path were merged in the sampling technique [12, 13]. Three contour-cutting observation lines made up each sampling unit. Fifty (50) plots were observed across the sample area of 2 ha (1% SI), which would be divided into three lanes (Figure 1.). A 20 x 20 m plot was used for measurements of vegetation at the tree level (diameter 20 cm), a 10 x 10 m plot for measurements of vegetation at the pole level (diameter 10 cm to 20 m), a 5 x 5 m subplot for measurements of vegetation at the sapling level (diameter 10 cm), and a 2 x 2 m subplot for measurements of vegetation at the seedling level.

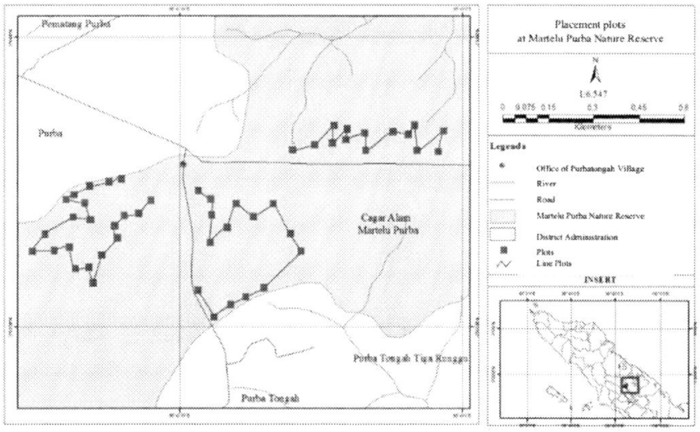

Figure 1. Map of location research

2.2. Mapping mother plant of S. platyclados

Create coordinate points using the Global Positioning System (GPS) of each *S. platyclados* tree found on the plots that have been made in the field. Record the tally sheet's diameter, height, and merchantable height [14]. Place a map of the Martelu Purba Nature Reserve over the coordinates..

2.3. Population Structure

The formula [10] was used to analyze the plant population pattern based on plant density at each stage:

Density is calculated as (total number of people)/(area (ha)). The following standards apply to seedlings, saplings, poles, and trees [15]:

1. A seedling that has grown from a germination point to 1.5 meters in height.
2. Saplings that are 1.5 meters tall and up to young trees with a diameter of less than 10 cm.

3. Pole (pole), young trees with a diameter of 10 to 20 cm.
4. Mature trees 20 cm and larger in diameter.

3. Results and Discussions

3.1. Abundance of mother plants in Martelu Purba Nature Reserve

The mother plant is identified by referring to the tree's growth rate, namely regeneration with a diameter of 20 cm up. The resurrection is assumed to have entered the generative phase at this growth rate, so flower primordia have begun to form. This study found 309 individual trees spread over 50 observation plots. Stand structure based on diameter and height is shown in Figures 1 and 2.

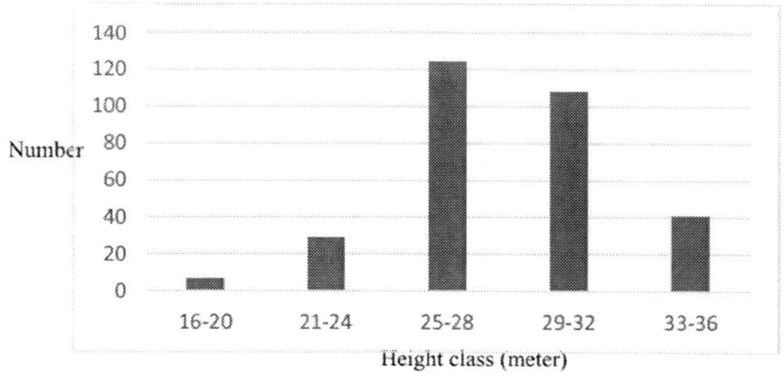

Figure 2. Stand structure of mother plants based on height distribution

Figure 2 demonstrates that the population of S platyclados inhabits a highstratum of 25 m and above in the Martelu Purba Nature Reserve. This indicates that this species belongs to strata B and A in its position in Martelu Purba Nature Reserve. This most elevated position suggests that the growth of *S platyclados* is a competitor of other species in obtaining sunlight. This is also a widely reported phenomenon of naturaltropical forest conditions dominated by the Dipterocarpaceae family.

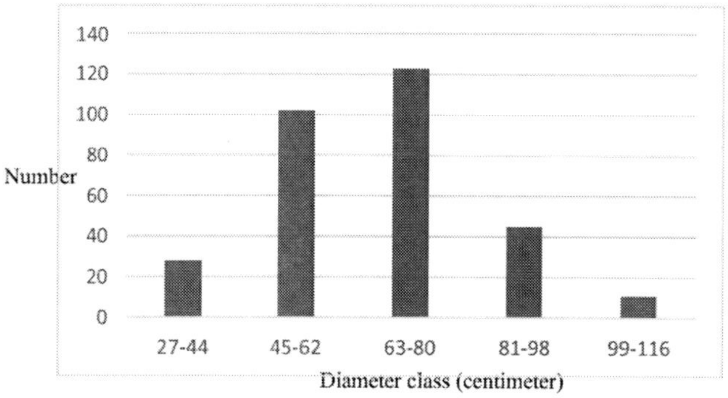

Figure 3. Stand structure of mother plant based on diameter distribution

Based on its diameter distribution, Figure 3 demonstrates that the population of S platyclados in the Martelu Purba Nature Reserve resembles a standard curve, which is typically a vertical structure in plantation forests. According to the history of Martelu Purba Nature Reserve, this location is indeed the location of a reforestation and land rehabilitation project using the genus Shorea that was carried out in the 1940s. The growth of *S. platyclados* indicates that this species can adapt well to Martelu Purba Nature Reserve.The decrease in the number of trees with large diameters (81cm up) suggests that this species may be felling.

3.2. Map of Mother Plant

All mother plants were observed in the plot, and their coordinates were taken and then plotted on a map of the Martelu Purba Nature Reserve location. Results of the map plotting are displayed in Figure 4.

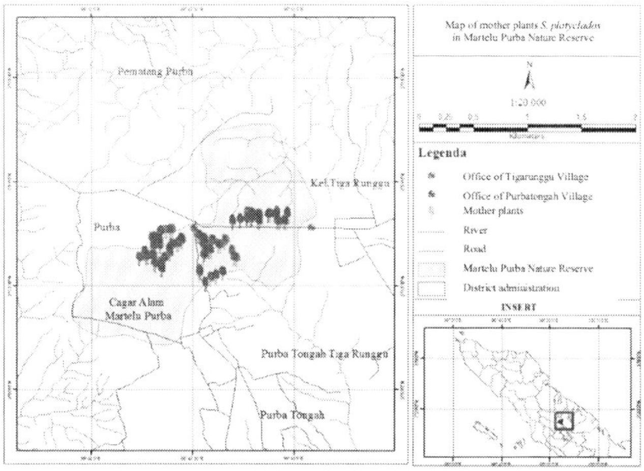

Figure 4. Mother plant *S. platyclados*' location on a map of the Martelu Purba Nature Reserve

This mother plant distribution map will be handy for monitoring the flowering phase of each individual as it is known that the flowering stage of this species tends to be irregular, which can inhibit the regeneration process. For this reason, continuous monitoring is necessary. With these coordinates and distribution points, monitoring becomes more organised, and administration becomes more organised, and this is very important considering the long life cycle of Shorea. This mother plant can be utilized as a source of vegetative seeds, such as cuttings, grafts, and others, in addition to monitoring flowering as a source of generative seeds, although this still needs more investigation.

3.3. Population Dynamics

The number of *S. platyclados* individuals distributed according to growth stage served as the basis for determining the population structure. Table 1 and Figure 5 depict the population structure and pattern.

Table 1. *S. platyclados* population structure in Martelu Purba Nature Reserve

No.	Growth Level	Numbe r	Density (individual/ha)
1	Seedling	64	3,200
2	Sapling	129	1,032
3	Pole	2	4
4	Tree	309	155

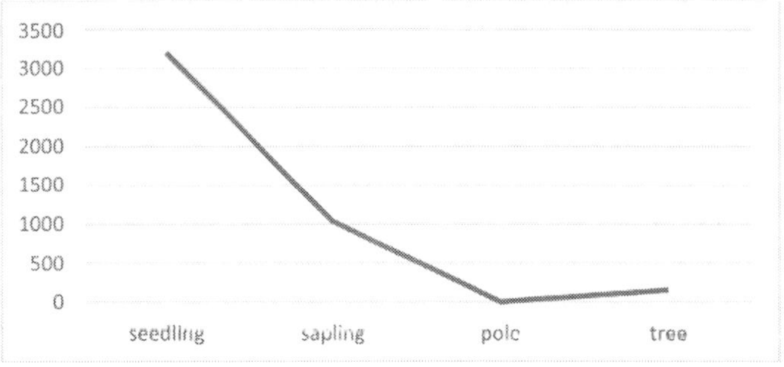

Figure 5. The population structure curve of *S. platyclados*

The population structure curve is curved like an inverted J. As evidenced by the vast number of seedlings, these data indicate that *S. platyclados* can regenerate effectively. In natural forests, there is a tendency to reduce the number of individuals/ha during the growth stage known as inverted J. Although Martelu Purba Nature Reserve was formerly the reforestation and land rehabilitation project site and was only converted into a conservation area in 1993, it can be seen from the various existing growth rates. This population is slowly approaching the population structure in typical natural forests [16]. In standard type forests, the availability of stands will produce a very high inverted J curve, ensuring the long-term viability of perspectives [17]. The regeneration of the pole level in this population shows the lowest number; it can be assumed that there is a disturbance at the regeneration level. Therefore, intensive security measures are needed by the Martelu Purba Nature Reserve manager, which should also involvethe local community. The findings of this study differ from those of earlier investigations [9]. In 2021, most of the Dipterocarpaceae families in Southeast Asia had abundant flowering and fruiting, including *S. platyclados* in the Martelu Purba Nature Reserve. This mass flowering and fruiting caused the regeneration of seedlings and saplings to be found when this research was carried out.

The horizontal structure of forest vegetation, which exhibits an inverted J curve with advancing time, will reveal a secondary succession process [18]. A decline in tree density can be caused by a variety of factors, including the frequency of tree regeneration, which produces new juvenile stages (seedlings/saplings), and the tree collapsing because it is old and dying. The disturbance could be caused by encroachment, burning, the introduction of exotic species, and other human activities. The minimal thickness of a plant population varies with location and species [19]. For instance, in the tropics, a plant population needs at least 25 mature individuals to live in a 1 km2 area; yet, measurements at the research site indicate that *S. platyclados* has a tree density of 155 trees/ha, or 15,500 trees/km2. Furthermore, [20] asserted that a serious situation exists when there are less than 50 mature trees in a 100 km2 area. The results indicate that this population can be considered sustainable.

4. Conclusions

To locate the mother plant's exact location, it is essential to chart the location of the mother trees of S. platyclados in the Martelu Purba Nature Reserve. This map will help collect data on the flowering phase and other data related to regeneration efforts. The mother plants were recorded as many as 309 individuals in 50 observation plots, or 155 individuals/ha spread evenly in the relatively sloping CA area. The distribution structure of the height and diameter of the parent tree indicates a good adaptation.

The population structure of *S. platyclados* demonstrates that this population is relatively stable and sustainable. However, monitoring at the pole level is urgent; based on data, a gap in the presence of this regeneration is found compared to others.

5. Acknowledgements
This study is a component of the USU TALENTA-funded study "Growth of Wilding Meranti Batu (Shorea platyclados) at Martelu Purba Natural Reserve," with grant number 135/UN5.2.3.1/PPM/SPP-TALENTA USU/2021. Our appreciation also extends to the students who helped with data collecting at the North Sumatra Center for Natural Resources Conservation, as well as the Head of Resort and Staff at Martelu Purba Nature Reserve.

References

[1] Kusmana C, Hikmat A. 2015. Keanekaragaman Hayati Flora di Indonesia. *Jurnal Pengelolaan Sumberdaya Alam dan Lingkungan*, 5 (2) : 187-198.
[2] Cao CP, Gailing O, Iskandar Z, Siregar, Ulfah J, Finkeldey R. 2009. Genetic Variation in Nine Shorea Species (Dipterocarpaceae) in Indonesia revealed By AFLPs. *Tree Genetics and Genomes*, 5: 408-412
[3] Djarwanto, Damayanti R, Balfas J, Basri E, Jasni, Sulistianingsih I M, Andianto, Martono D, Pari G, Sopandi A, et al. 2017 Pengelompokan Jenis Kayu Perdagangan Indonesia (Bogor: FORDA Press)
[4] Saroyobudiyono H and Aisyah S 2006 A trimer resveratrol compound from *S. platyclados* Sloot trunk leather (DIPTEROCARPACEAE) *Jurnal Penelitian Sains & Teknologi* 17 11-16
[5] Wardani M and Susilo A 2017 Evaluation of the Existence of *Shorea platyclados* Slooten ex Endert in the Bukit Daun Protected Forest. *Widyariset* 3 151-160
[6] Pulan D E and Buot Jr I E 2014 Leaf architecture of Philippine Shorea species (Dipterocarpaceae) *Int.Res.J. Biological Sci.* 3 19-26
[7] Rosdayanti H, Siregar U J and Siregar I Z 2019 Karakteristik penciri morfologi daun meranti (*Shorea* sp) pada area ex-situ KHDTK Haurbentes. *Media Konservasi* 24 207-215
[8] [IUCN] International Union for The Conservation of Nature 2008 IUCN Red List of Threatened Species
[9] Rangkuti AB, KS Hartini, A Susilowati, R Rambey, H Arinah, and L Irmayanti. 2022. Morphological characteristics and abundance of red meranti (*Shorea platyclados*) in Martelu Purba natural reserves. *IOP Conf. Ser.: Earth Environ. Sci.* 959 012016. doi:10.1088/1755-1315/959/1/012016.
[10] Odum E P 1998 Basics of Ecology Third Edition Translation Tjahyono Samingan (Yogyakarta: Gadjah Mada University Press)
[11] Hardiansyah 2010 Introduction to Plant Ecology (Banjarmasin: Faculty of Teacher Training and Education UNLAM)
[12] Ludwig JA and JF Reynolds. 1988. Statistical Ecology. Simultaneously. Canada
[13] Mueller-Dombois D and H Ellenberg. 1974. Aims and Methods of Vegetation Ecology. John Wiley and Sons Inc. New York
[14] KS Hartini and B Ompusunggu. 2019. Choosing plus tree of Salagundi (*Roudholia teysmanii*) at Salib Kasih Tourism Area. IOP Conf. Ser.: Earth Environ. Sci. 260 012162. doi:10.1088/1755-1315/260/1/012162
[15] Soerianegara I and Indrawan A 1998 Indonesian Forest Ecology (Bogor: Faculty of ForestryBogor Agricultural University)
[16] KS Hartini, A Susilowati and M Rizky. 2021. Distribution patterns and population structure of salagundi (*Rhodoleia teysmanii*) From North Tapanuli-NorthSumatra. *IOP Conf. Ser.: Earth Environ. Sci.* 912 012078. doi:10.1088/1755-1315/912/1/012078
[17] Suwardi A B 2013 Journal of Biology 12 168
[18] Widiyanti P and Kusmana C 2014 *Journal of Tropical Silviculture* 5 69

[19] Tati S 1998 Paper Capita Selecta Seminar Post Graduate Biology ITB (Bandung: ITB Bandung Unpublished)
[20] Mace G M and Lande R 1991 Conservation Biology 5 148

IOP Conf. Series: Earth and Environmental Science 1115 (2022) 012015 doi:10.1088/1755-1315/1115/1/012015

Manifestation of endoparasitic helminths in Sumba Horse (*Equus caballus*) in Citra Pesona Ladangku Animal Park

L Sinaga and M Tanjung*

Department of Biology, Faculty of Mathematics and Natural Sciences, Universitas Sumatera Utara, Medan, North Sumatra 20155, Indonesia

*E-mail: masitta@usu.ac.id

Abstract. Gastrointestinal parasitic infections can cause horses to experience diarrhea, weight loss, weakened physical condition and even death. This horse disease is a great danger that can reduce horse performance.The research on the description of gastrointestinal endoparasites in Sumba horses (*Equus caballus*) at Citra Pesona Ladangku animal park, Pancur Batu, Deli Serdang, North Sumatra. The aim of the study was to determine the type of endoparasites, prevalence and the intensity of parasite attacks. The sample testing was carried out at the Parasitology Laboratory, Medan Veterinary Center using the EPG Mc. Masters method. The results showed that from 18 Sumba horses examined, 6 types of endoparasites were found, namely *Strongylus* sp., *Trichonema* sp., *Dictyocaulus arnfieldi*, *Strongyloides* sp., *Triodontophorus* sp. and *Ascaris* sp. The prevalence of *Strongylus* sp. was 86% in the usual category and the intensity attack was 1.530 in the Super infection category, *Dictyocaulus arnfieldi* was 53% in the frequent category and the intensity attack was 1.161 in the Super infection category, *Trichonema* sp. was 38% in the common category, and the intensity attack 300 categorized in the very severe category, *Strongyloides* sp. was 22% in the often category and the intensity was 430 in the very severe category, *Triodontophorus* sp. was 16% in the often category and the intensity was 500 in the very severe category and *Ascaris* sp. had prevalence 14% in the often category and the intensity was 100 in the severe category. The gastrointestinal parasitic infections in Sumba horses need special attention.

1. Introduction

Horses are rare pets in Indonesia which are experiencing a decline in population. According to data from the Central Statistics Agency, the horse population in Indonesia in 2018 reached 42.977 individuals that experienced a decline in 2020 to 36.351 individuals. The fact that horses are not properly bred and are therefore more susceptible to a variety of illnesses is one of the factors contributing to the diminishing horse population [1]. Disease control in horse rearing is very important. Poor sanitation can reduce environmental quality and can trigger infectious endoparasitic diseases of the gastrointestinal tract in horses. The decline in horse health and immunity can be influenced by a climate suitable for worm breeding, poor environmental conditions, feed contaminated with infective larvae, stress, hormone levels and pregnancy in mares [2]. In addition, immune system issues, fighting, dietary deficits, and changes in environment all contribute to the incidence of parasite diseases [3].

Gastrointestinal (GI) helminths that often infect the digestive tract in horses are *Trichostrongylus axei, Oxyuris equi, Triodontophorus serratus, Strongyloides westeri, Strongylus edentatus, Habronema muscae, Parascaris equorum, Probstmayria vivipara, Strongylus vulgaris, Gasterophilus nasalis, Anoplocephala magna* and *Anoplocephala perfoliata* [4]. The larvae and eggs of parasitic helminths that enter the body and develop in the digestive tract are thought to be the cause of parasitic infection. The prevalence of gastrointestinal helminths in horses was 87% which probably due to the condition of the horses that lack nutrition, feed and drinking water that have been contaminated with infective larvae, while the intensity of gastrointestinal parasites obtained by *Ascaris equorum, Paramphistomum* sp., *Haemonchus* sp., and *Strongyloides westeri* [2,4]. Strongyle infection is at higher risk in young cattle than in older horses [5,6].

In North Sumatra, the Citra Pesona Ladangku Animal Park is one of the breeding facilities with a population of horses. Although gastrointestinal tract infections in horses may not result in abrupt death, they can be severe in adult animals, which can lower output and work capacity, while in young animals, they can delay growth and development [7]. The objective of this study is to detect the gastrointestinal helminths in horse population at Citra Pesona Ladangku and to assess the severity of parasitic infestation as a baseline data for welfare management.

2. Materials and Method

2.1. Fecal sampling

Feces were taken directly in the morning at 9 am from 18 individuals at Citra Pesona Ladangku Animal Park, Pancur Batu, Deli Serdang Regency, North Sumatra. Approximately 5 to 10 g of fresh fecal samples were taken from each Sumba horse, put in a plastic bag, tied with rubber and labeled with a code. The samples were stored in cold condition prior GI helminth identification at the Laboratory of Parasitology, Regional I Veterinary Investigation and Testing Center Medan.

2.2. Sample preparation

The preparation of fecal samples followed the protocol of E.P.G (Egg Per Gram) Mc. Master or a flotation technique to yield a significant amount of helminth eggs using a saturated solution. Approximately 2 g of fecal sample were suspended into 58 mL of saturated salt solution, mixed well and filtered (100-mesh). The filtrates were pipetted and checked under compound microscope for parasite identification and count.

2.3. Data analysis

The data were subjected to descriptive analysis with the parameters of parasitic GI helminths, species prevalence and intensity (infestation) [8].

3. Results and Discussion

Based on the microscopical investigation on 18 samples of Sumba horse feces at the Citra Pesona Ladangku Animal Park, we documented six species of parasitic helminths as presented in Table 1. The investigation of the horse feces from Sumba revealed endoparasite infection from the nematode class, but no larvae of cestodes and trematodes were discovered.

Table 1. Endoparasites of Sumba Horse at Citra Pesona Ladangku Animal Park

No	Class	Order	Family	Species
1	Nematoda	Ascaridida	Ascarididae	*Ascaris* sp.
2		Strongylida	Dictyocaulidae	*Dictyocaulus arnfieldi*
3		Rhabditida	Strongyloididae	*Strongyloides* sp.
4		Strongylida	Strongylidae	*Strongylus* sp.
5				*Trichonema* sp.
6				*Triodontophorus* sp.

Environmental conditions including the location where Sumba horses are released that has been polluted by nematode worms can cause samples of horses to become infected by nematodes. Infection occurs through forage feed and drinking water taken by Sumba horses. The life cycle of nematode larva do not need an intermediary host, allowing its eggs to hatch in environments suitable for their stage of development and infecting horses through larvae that are consumed through tainted feed and water [9].

Since the life cycle of this class is more complicated and cestode parasites have an indirect life cycle, no larvae species from this class were discovered. Cestode worms could not be detected because they were young and immature when they were inspected, or because there were few eggs that either did not coalesce in the feces or were still in the larval stage. Trematodes require intermediary host, which makes their transmission and reproduction were discovered in digestive tract. Proglottids are infrequently identified in horse dung, and stool testing on horses has a rather poor sensitivity. The prevalence of endoparasites in Sumba Horse is presented in Table 2.

Table 2. Prevalence of endoparasites in Sumba Horse at Citra Pesona Ladangku Animal Park

No	Endoparasites	Prevalence (%)	Category
1.	*Strongylus* sp.	86	Usual
2.	*Trichonema* sp.	38	Common
3.	*Dictyocaulus arnfieldi*	53	Frequent
4.	*Strongyloides* sp.	22	Often
5.	*Triodontophorus* sp.	16	Often
6.	*Ascaris* sp.	14	Often

The highest prevalence of endoparasitic helminths was recorded on *Strongylus* sp. followed by *Trichonema* sp., *Dictyocaulus arnfeldi, Strongyloides* sp., *Triodontophorus* sp., and *Ascaris* sp. The source of pasture used by Sumba horses at the research location may be contaminated with infectious larvae and it could be the reason for the high incidence of *Strongylus*. It takes around 3 days for the eggs of the Strongylida family to develop into infectious larvae. By ingesting grass that has been polluted with infectious larvae, the larvae can infect into animal bodies. When animals ingest the parasites, the larvae can enter through the skin or fecal-contaminated areas, travel to the heart organ, and eventually mature into adult worms in the digestive tract. More than 80–90% of horses worldwide are affected with strongylosis, one of the most significant internal parasite illnesses in horses [10,11]. The second most prevalent parasite in this study was *D. arnfeldi* which is characterized by its adult form which are slender and medium-sized worm with a whitish-grey color. This parasite has a digestive system and a nervous system but does not have an excretory system [12]. *Dictyocaulus arnfieldi* is a lungworm species that infects donkeys, horses, ponies, and zebras, and is found worldwide [13]. In addition, the intensity of endoparasitic attacks on Sumba Horse is presented in Table 3.

Table 3. Intensity of endoparasite infection in Sumba Horse at Citra Pesona Ladangku Animal Park

No.	Endoparasites	Intensity	Category
1.	*Strongylus* sp.	1.530	Super infection
2.	*Trichonema* sp.	300	Very severe
3.	*Dictyocaulus arnfieldi*	1.161	Super infection
4.	*Strongyloides* sp.	430	Very severe
5.	*Triodontophorus* sp.	500	Very severe
6.	*Ascaris* sp.	100	Severe

Based on the results, the endoparasites infect all horses with a range from 100 (*Ascaris* sp.) or *severe* to 1.531 (*Strongylus* sp.) ind/horse or *super infection*. The growth of *Strongylus* sp. may be facilitated by the concentrate feed at Citra Pesona Ladangku Animal Park in the form of tofu waste, which may account for the high intensity of the endoparasite in Sumba horses [14]. In addition to feeding factors, *Strongylus* spp. are photophobic, which means the larvae will rise to the grass surface in the morning and fall back to the grass bottom during the day to avoid high temperatures [15]. Sumba horses that grazed in the morning may expose themselves to the contaminated green, however this issue may be solved by providing wilted grass as horse feed to reduce the population of viable *Strongylus* infective larvae.

The survival and development of parasitic worms has influencing factors, including: a humid environment, climatic conditions and rainfall. Humid conditions in captivity can also support the development and life cycle of nematode worms. The poor sanitation conditions of the cages lead to high nematode worm infections in Taman Citra Pesona Ladangku and facilitate the life cycle of nematode worms. When opposed to open cages, closed cages are supposed to have moderate temperatures but high humidity levels. Unfavorable circumstances for worm growth and development might be found in areas with high humidity.

4. Conclusion

Six species of endoparasitic helminths were recovered from the fecal samples of Sumba Horse in Citra Pesona Ladangku Animal Park i.e *Strongylus* sp., *Trichonema* sp., *Dictyocaulus arnfieldi, Strongyloides* sp., *Triodontophorus* sp., and *Ascaris* sp. *Strongylus* sp. was documented as the most important endoparasite in our study as indicated from its high prevalence and intensity. Controlled maintenance management is required to lower the prevalence of endoparasite infection in Sumba horses at the study site.

References

[1] Campbell M L H and Sandøe P 2015 *Vet Rec* **176** 436
[2] Apriliawati E, Mufasirin, Wurliana, Hastutiek P, Suwanti L T and Tehupuring B C 2019 *J Parasite Sci* **3** 83
[3] Khamesipour F, Taktaz-Hafshejani T, Tebit K E, Razavi S M and Hosseini S R 2020 *Vet Med Sci* **7** 25
[4] Teixeira W F P, Felippelli G, Cruz B C, Maciel W G, Favero F C, Gomes L V C, Buzzulini C, Prando L, Bichuette M A, Lopes W D Z, de Oliveira G P and da Costa A J 2014 *Braz J Vet Parasitol Jaboticabal* **23** 534
[5] Mezgebu T, Tafess K and Tamiru F 2013 *Open J Vet Med* **3** 267
[6] Enigidaw S, Assefa A, Mekonnen N and Belete S 2015 *Am-Eurasian J Sci Res* **10** 228
[7] Hinney B, Wirtherle NC, Kyule M, Miethe N, Zessin K and Clausen P 2011 *Parasitol Res* **108** 1083
[8] Margolis L, Esch G W, Holmes J C, Kuris A M and Schad G A 1982 *J Parasitol* **68** 131
[9] Soulsby E J L 1982 *Helminths, Arthropods, and Protozoa of Domesticated Animals* (London: Baillière Tindall)
[10] Khan M A, Roohi N and Rana M A A 2015 *J Animal Plant Sci* **25** 1
[11] Nielsen M K, Monrad J and Olsen S N 2006 *Vet Parasitol* **135** 47
[12] Bowman D D 2014 *Georgis' Parasitology for Veterinarians* (St. Louis, Missouri: Elsevier)
[13] Smith B P 2009 *Large Animal Internal Medicine* (USA: Mosby)
[14] David K H 2012 *Controlling Internal Parasites of Horse* (New York: Saunders)
[15] Tydén E, Enemark H L, Franko M A, Hoglund J and Ostermann-Lind E 2019 *Vet Parasitol* **276** 100013

Identification of gastrointestinal endoparasites in sheep (*Ovis* sp.) at various ages in Arjuna Farm, North Sumatra

S M S Napitupulu and M Tanjung*

Department of Biology, Faculty of Mathematics and Natural Sciences, Universitas Sumatera Utara, Medan, North Sumatra 20155, Indonesia.

*E-mail: masitta@usu.ac.id

Abstract. Gastrointestinal endoparasite is an infection that causes a decrease in the productivity level of sheep so that it detriments to the farmers. The purpose of this study was to compare the type, the prevalence and the intensity of endoparasites in each age group. Samples were taken as many as 7 fresh feces from each group of lamb, virgin, and adult from a total of 21 sheeps. Fecal samples were examined and prepared using the E.P.M Mc Master Protocol while endoparasites were identified under microscopical examination. The results showed 5 species of gastrointestinal endoparasites, i.e *Eimeria* sp., *Dicrocoelium* sp., *Haemonchus* sp., *Strongyloides* sp., and *Capillaria* sp. *Eimeria* sp. showed a prevalence of 85.7% in commonly category and intensity of 2216.6 or super infection category. *Haemonchus* sp. had a prevalence of 85% (common), an intensity of 671.4 (very severe), and the *Strongyloides* sp. had a prevalence of 85% (common) and an intensity of 200 (very severe). The lambs were found to be infected by *Eimeria* sp. with a prevalence of 100% (common) and an intensity of 2871.4 (super infection). *Haemonchus* sp. had a prevalence of 100% (common) and an intensity of 1200 (super infection), *Strongyloides* sp. had a prevalence of 71% (usual) and an intensity of 860 (very severe) and the *Dicrocoelium* sp. had a prevalence of 14% (often) and an intensity of 400 (very severe). The adult sheeps were found to had *Eimeria* sp. with a prevalence of 100% (common) and an intensity of 371.4 (very severe). The *Haemonchus* sp. had a prevalence of 14% (often) and an intensity of 200 (very severe). *Strongyloides* sp. had a prevalence of 57% (frequent) and an intensity of 200 (very severe) and the *Capillaria* sp. had a prevalence of 14% (often) and an intensity of 100 (severe). The prevalence and intensity of endoparasites in sheeps at Arjuna Farm was classified as very high.

1. Introduction

Ruminant production is severely hampered by gastrointestinal parasites because of the accompanying morbidity, mortality, treatment, and control expenses [1]. Parasitic infections can cause acute and chronic disease. Such infections are common even when the sheeps appear healthy. In contrast to adult sheeps who have clinical signs, the lambs that are infected with endoparasites in their bodies will nonetheless appear healthy. Clinical symptoms such as diarrhea, dehydration, loss of appetite, failure to gain weight and accompanied by anemia are known to be associated with endoparasitic infestation on the ruminants [2]. Low production is witnessed as a result of inadequate feed consumption and stunted development [3]. Countless animal products are lost due to protozoan parasites as well, which is a burden in all parts of the world [4].

Newborn mammals have immature immune systems that cannot protect themselves from infectious diseases. The variation in the risk of infection depends on how effectively the host immune system defends itself against particular parasites. Determining the relationship between age and infection risk is critical in identifying the part of the host population that contributes to parasite dynamics [5]. Lambs commonly have lower rates of gastrointestinal (GI) parasitic infections than adult sheeps [6]. However, lambs are more susceptible to helminths infection than adults, so the immune response of lambs is more important than adults. The GI parasitic infections in young animals (<12 months) were higher followed by older (more than 5 years) and adults (1-5 years) [7].

Endoparasite infection in sheep can be found in several areas with various species of parasites in every age group, including young, adolescent, and adult ruminants. Parasitic diseases that are often found in ruminant animals such as coccidiosis and nematodiasis. Gastrointestinal nematode species are often found in some ruminant animals, including *Haemonchus* sp., *Strongyloides* sp., *Bunostomum* sp., *Trichuris* sp., *Gaigeria* sp., *Chabertia* sp., *Oesophagostomum* sp., Coccidians, *Nematodirus*, *Trichostrongylus*, and *Fasciola* [8]. Common endoparasites in buffalo including *Entamoeba* spp., *Moniezia* spp., *Haemonchus* spp. and Coccidians [6]. While in cattle, 3 genera of endoparasites were reported namely *Haemonchus* (Nematoda), *Fasciola* (Trematoda) and *Paramphistomum* (Trematoda) [9]. Parasite-infected animals may have negative economic effects [10]. Inadequate body development, decreased milk output, adult ruminant with sub-optimal body weights, organ damage, and lower fertility and metabolism are all contributing factors to the economical decline in ruminant industry. The goal of the current study is to identify and evaluate endoparasites in sheep of various ages in order to provide a broad description of the infection that may be taken into account for long-term management.

2. Materials and Method

2.1. *Fecal sampling*
Sheep feces were taken directly in the morning at 8 AM from a total of 21 individuals, composed from each age group i.e lambs (< 7 months), adolescent sheeps (8–12 months) and adult sheeps (>12 months) at Arjuna Farm, Deli Tua District, Deli Serdang Regency, North Sumatra, Indonesia. Approximately 10 g of fresh fecal samples were collected, put in a plastic bag, tied with rubber and labeled with a code. The samples were stored in cold condition prior GI endoparasite identification at the Laboratory of Parasitology, Regional I Veterinary Investigation and Testing Center Medan.

2.2. *Sample preparation*
The preparation of fecal samples followed the protocol of E.P.G (Egg Per Gram) Mc. Master or a flotation technique to yield a significant amount of helminth eggs using a saturated solution [11]. Approximately 2 g of fecal sample were suspended into 58 mL of saturated salt solution, mixed well and filtered (100-mesh). The filtrates were pipetted and checked under compound microscope with magnification of 10× for endoparasite identification and count.

2.3. *Data analysis*
The data were subjected to descriptive analysis with the parameters of parasitic GI helminths, species prevalence and intensity (infestation) [12].

3. Results and Discussion
Based on the identification results, endoparasites in the fecal samples of sheeps at various ages consisted of 5 species as presented in Table 1.

Table 1. Endoparasites of sheeps at various ages in Arjuan Farm

Phylum	Species	Age group		
		Lambs (<7 months)	Adolescents (± 8-12 months)	Adults (>12 months)
Protozoa	*Eimeria* sp.	√	√	√
Platyhelminthes	*Dicrocoelium* sp.	-	√	-
Nemathelminthes	*Haemonchus* sp.	√	√	√
	Strongyloides sp.	√	√	√
	Capillaria sp.	-	-	√

The assemblage of endoparasites between age class of sheeps was not entirely different. In lambs, we only documented three endoparasites species i.e *Eimeria* sp, *Haemonchus* sp, and *Strongyloides* sp. *Capillaria* sp was only found in adult sheeps. The difference in the number of endoparasite species found in each age group of sheep can be seen that the lambs is less susceptible to parasites than the adolescent and adult sheeps. In contrast to adolescent and adult sheeps which consume food like forage, lambs are still quite dependent on their mothers during the lactation period, which may account for this low susceptibility. Infections with intestinal parasites are frequently brought on by the grazing method [13]. The treatment of the livestock management also greatly affects the number of parasites found in this study. Unsanitary surroundings, negligent handling, direct contact with animals, and an intense climate of diseased sheep with a wide range of gastrointestinal parasites [8]. Meanwhile, the prevalence of endoparasites is presented in Table 2.

Table 2. Prevalence of endoparasite in sheeps at various ages in Arjuna Farm

Endoparasites	Age group					
	Lambs (<7 months)		Adolescents (± 8-12 months)		Adults (>12 months)	
	Prevalence	Category	Prevalence	Category	Prevalence	Category
Eimeria sp.	85.7%	Almost always	100%	Always	100%	Always
Dicrocoelium sp.	-	-	14%	Often	-	-
Haemonchus sp.	100%	Always	100%	Always	14%	Often
Strongyloides sp.	85%	Almost always	71%	Usually	57%	Frequently
Capillaria sp.	-	-	-	-	14%	Often

Based on Table 2, there is a different prevalence of endoparasites among age group in sheeps. *Eimeria* sp. was documented as the most prevalent endoparasite in sheeps followed by *Haemonchus* sp. in lambs and adolescent sheeps and *Strongyloides* sp. in lambs. This finding is most likely to occur because adolescent and adult ruminants have a higher intensity of eating compared to young ruminants [14]. Adult sheeps consumed more food and their feeding period were more frequent than lambs. Oocysts found in feces had an easier time adequately sporulating due to high feeding intensity and neglected cage hygiene. Oocysts will then sporulate once they are outside of the host body. Under harsh climatic circumstances, the infectious spore of the oocyst *Eimeria* sp. may persist for a very long period. In contrary, Leon et al [15] stated that the presence of *Eimeria* sp. showed no correlation between its prevalence and age group in ruminants. The second and third most prevalent endoparasites namely *Haemonchus* sp. and *Strongyloides* sp. were most likely to occur due to the different immunity traits among age group [5]. Young livestock are more susceptible to parasitic infestations compared to adult livestock, because young livestock have immunity that is susceptible to parasitic worms while adult livestock at least have been infested and are able to survive the infestation. The infestation or intensity of endoparasites is presented in Table 3.

Table 3. Intensity of endoparasites in sheeps at various ages in Arjuna Farm

Endoparasites	Age group					
	Lambs (< 7 months)		Adolescents (± 8-12 months)		Adults (>12 months)	
	Intensity	Category	Intensity	Category	Intensity	Category
Eimeria sp.	2216.6	Super infection	2871.4	Super infection	371.4	Very severe
Dicrocoelium sp.	-	-	400	Very severe	-	-
Haemonchus sp.	671.4	Very severe	1200	Super infection	200	Very severe
Strongyloides sp.	200	Very severe	860	Very severe	200	Very severe
Capillaria sp.	-	-	-	-	100	Severe

Based on Table 3, the intensity of endoparasites particulary *Eimeria* sp. was regarded as *super infection* in lambs and adolescent sheeps. *Haemonchus* sp. also showed a *super infection* in adolescent sheeps. In contrary, the category of endoparasite intensity in adult sheeps showed no *super infection* with only *very severe* and *severe* categories. Shaw et al [16] stated that intensity is the average number of types of parasites found against the host infected with the parasite. This demonstrates that the endoparasite intensity in each age group vary owing to age considerations, maintenance practices, cage hygiene, feeding, and routine sheep health monitoring. The effect of forage might end up being the primary source of parasites. The ability of the parasite to adapt around the cage environment is very suitable to infest the animals in the cage. Physical factors include temperature, climatic conditions and humidity greatly affect the development and survival of parasite eggs [17]. The geographic distribution of gastrointestinal endoparasites varies according to factors such as climate, temperature, humidity, and livestock density. Gastrointestinal endoparasites often dwell in several areas. The likelihood of endoparasite infestations is thought to be high in North Sumatra because of the temperature and humidity, which are thought to be factors in the development of endoparasite larvae.

4. Conclusion

Five species of endoparasites were identified from the fecal samples of sheeps at various ages in Arjuna Farm, North Sumatra i.e. *Capillaria* sp., *Dicrocoelium* sp., *Eimeria* sp., *Haemonchus* sp., and *Strongyloides* sp. Based on the age class, the lambs were less susceptible to the endoparasites as indicated from the number of endoparasite species ($N = 3$) but heavily infested with *Eimeria* sp. The presence of Eimeria sp. in all age groups of sheep needs to be addressed in order to reduce the prevalence and intensity of attacks of this species on sheep in the study area.

References

[1] Nwosu C O, Madu P P and Richards W S 2007 *Vet Parasitol* **144** 118
[2] Gregory N and Grandin T 2007 *Animal Welfare and Meat Production* (Cambridge: CABI Publishing)
[3] Pedreira J, Paz-Silva A, Sanchez-Andrade R, Suarez J L, Arias M, Lomba C, Diaz P, Lopez C, Diez-Banos P and Morrondo P 2006 *Prev Vet Med* **75** 56
[4] Hendrix C M and Robinson E D 2017 *Diagnostic Parasitology for Veterinary Technicians* (United States: Elsevier Inc)
[5] Combrink L, Glidden C K, Beechler B R, Charleston B, Koehler A V, Sisson D, Gasser R B, Jabbar A and Jolles A E 2020 *Biol Lett* **16** 20190811
[6] Khan T, Khan W, Iqbal R, Maqbool A, Fadladdin Y A J and Sabtain T 2022 *Braz J Biol* **83** e242677
[7] Maharana B R, Kumar B, Sudhakar N R, Behera S R and Patbandha T K 2016 *J Parasit Dis* **40** 1174
[8] Gadahi J A, Arshed M J, Ali Q, Javaid S B and Shah S I 2009 *Vet World* **2** 51
[9] Azlan R and Tanjung M 2021 *IOP Conf Ser Earth Environ Sci* **912** 012097

[10] Charlier J, Waele V D, Ducheyne E, van der Voort M, Velde F V and Claerebout E 2016 *Ir Vet J* **69** 14
[11] Colville J 1991 *Diagnostic Parasitology for Veterinary Technicians* (United States: American Veterinary Publishing)
[12] Margolis L, Esch G W, Holmes J C, Kuris A M and Schad G A 1982 *J Parasitol* **68** 131
[13] Min B R, Hart S P, Miller D, Tomita G M, Loetz E and Sahlu T 2005 *Vet Parasitol* **130** 105
[14] Nurcahyo R W, Ekawasti F, Haryuningtyas D, Wardhana A H, Firdausy L W, Priyowidodo D and Prastowo J 2021 *IOP Conf Ser Earth Environ Sci* **686** 012063
[15] Leon J C P, Delgado N U and Florez A A 2019 *Vet World* **12** 48
[16] Shaw A K, Sherman J, Barker F K and Zuk M 2018 *Proc R Soc B Biol Sci* **285** 20182147
[17] Soulsby E J L 1982 *Helminths, Arthropods, and Protozoa of Domesticated Animals* (London: Baillière Tindall)

Soil Permeability in Various Vegetation Stands at USU Campus Padang Bulan Medan

N Anggraini [1*] and OKH Syahputra [1]

[1] Department of Forest Management, Faculty of Forestry, Universitas Sumatera Utara, Medan 20155, Indonesia

*Email: novitaanggraini@usu.ac.id

Abstract. The University of Sumatera Utara (USU) campus is one of the green campuses with a reasonably high diversity of tree species. The ability of stands on the USU Campus to pass water (soil permeability) must be considered for flood prevention. Soil permeability studies in field surveys were carried out by taking soil samples—determination of soil samples for soil permeability properties based on the constant head method. Soil chemical analysis was carried out for soil organic fraction content and physical properties in soil fraction and texture. The analysis showed that saga stands (SS) showed the highest permeability among mahogany stands (MS), cape stands (CS), glodokan tiang stands (GTS), and oil palm stands (OPS).

1. Introduction

Water is a natural resource that is very important for human life. The water balance that occurs on earth is regulated through the hydrological cycle [1]. The hydrological cycle is a continuous process in which water on the ground will return to the atmosphere through evaporation or to the sea through surface and subsurface flows [2]. In detail, the hydrological cycle is also described in [3], where rainfall that falls to the earth's surface will wet the plant surface through interception and reach the soil. Then penetrate the soil surface through infiltration, becoming groundwater, flowing into canals, lakes, rivers, seas, and lower places. Furthermore, it undergoes evaporation (evapotranspiration), and the water vapor from the evaporation forms clouds that will condense into rainfall. The process of water movement into the soil is an important aspect that needs to be known to know the ability of the land to pass water.

The soil quality to pass water or air, which is measured based on the amount of flow through the unit of soil that has been saturated in advance per unit of time, called permeability. The physical properties of the soil strongly influence permeability. Changes in water temperature slightly affect permeability. In water-saturated soils, the permeability varies between wide limits, from less than 25 cm per year in dense clay soils to several thousand meters per year in gravel formations. Each type of soil has different permeability capabilities. Several soils for agriculture include andepts, inceptisols, and ultisols that need to know their permeability capabilities for various purposes such as designing drainage channels and washing soil salinity [4].

Soil permeability is a property that states the rate of movement of a liquid in the soil through a medium with macro and micro pores vertically and horizontally. Soil permeability indicates the ability to pass water. All types of soil are permeable, where water is free to flow through the empty spaces (pores) that exist between soil grains. The value of soil permeability increases, and also the infiltration rate [5].

The University of North Sumatra (USU) campus is one of the green campuses with diverse tree species, totaling 121 species from 37 families [6]. The ability of stands on the USU Campus to pass water needs to be studied as consideration for flood prevention. Based on this, measuring soil permeability in various tree stands on the USU Campus is necessary.

2. Methods

2.1 Study Area

USU Campus (latitude 3°56'18.2"N and longitude 98°65'6.27"E) that located in Padang Bulan, Medan Baru District, Medan City was chosen as the study area. Medan city had an altitude of 2.5-37.5 meters above sea level. The minimum temperature ranges from 23.2° C-24.3°C, and the maximum temperature ranges from 30.8°C to 33.2°C. The average humidity in Medan City is around 84-85%. Wind speed is 0.48 m/s while monthly evaporation is around 104.3 mm, with inceptisol soil type [7].

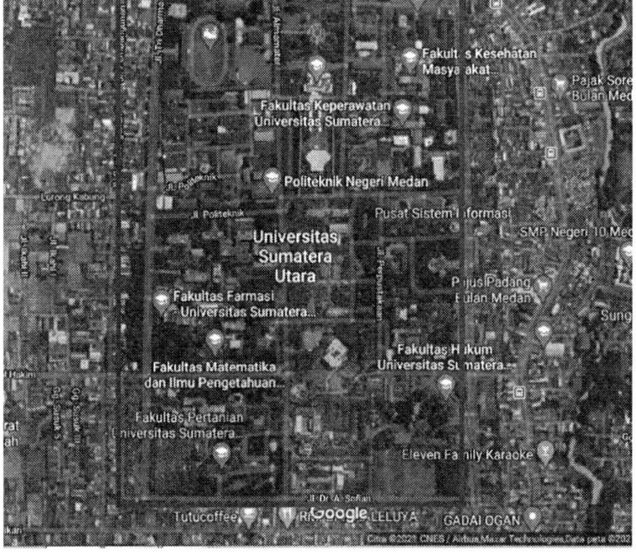

Figure 1. USU Campus Padang Bulan Medan (source : googleearth.com)

2.2 Soil Permeability and Soil Physical Properties Measurement

Determination of the sampling was done by purposive sampling with dividing the points in several types of stand (Table 1). Each stand was measured in three times. The observation point was determined using GPS. Soil sampling was carried out at each stand. [8] explains that intact soil sampling is carried out using rings or tubes to determine soil permeability properties based on the *constant head method* and soil permeability measurements are carried out through the following stages: (a) soil samples in the ring are immersed in water in an immersion bath to a depth slightly below the top of the ring for more than 12 hours or until the soil appears wet; (b) after the saturation process is complete, the top of the ring containing the soil sample is connected to the empty ring, using adhesive tape. During the grafting process, the soil sample remains in the immersion water; (c) furthermore, the soil sample is transferred to the measuring instrument, then water is flowed into the device by keeping the water level above the soil sample constant; (d) measuring the volume of water that comes out through the soil mass with the first measurement after 6 hours and subsequent measurements at one hour intervals.

The USU campus has an area of approximately 120 hectares and includes a campus that provides a fairly large green area dominated by various types of trees and land uses (Figure 2). The mahogany tree (*Swietenia macrophylla*) is the most dominant tree species on the USU Campus [6]. The sampling location under mahogany is in the Tridharma Forest. Tridharma forest is an educational forest with a monoculture type which is dominated by mahogany trees. This forest is also used as a *tracking path* for walking, *jogging*, as well as a place to relax. The cape (*Mimosups elengi*)is located on the street next to Medan State Polytechnic (POLMED), the cape dominates the greenery between the sidewalk and the road drainage. The glodokan tiang is at Jl. Biotechnology in front of the Faculty of Forestry. The oil palm is located on Jl. Alumni and the saga is on Jl. University (Table 1).

Table 1. Sampling Location

No	Location Code	Location Coordinates	Type of Stand
1	Mahogany Stand (MS)	3.559358 N, 98.653381 E	*Swietenia macrophylla*
2	Cape Stand (CS)	3.562558 N, 98.653566 E	*Mimosups elengi*
3	Glodokan Tiang Stand (GTS)	3.558999 N, 98.653370 E	*Polyalthia longifolia*
4	Oil Palm Stand (OPS)	3.562758 N, 98.657540 E	*Elaeis guineensis*
5	Saga Stand (SS)	3.556710 N, 98.660102 E	*Adhenanthera pavonina*

Figure 2. Vegetation Stand Types in USU Campus Padang Bulan

The observed physical properties of the soil, namely: soil structure, soil texture, soil porosity, *bulk density*, and soil moisture content, will be tested at the Central Laboratory, Faculty of Agriculture, University of Sumatera Utara. Soil structure was analyzed *by feeling* method . Soil texture was analyzed by hygrometer method [4]. Soil porosity, bulk density, soil moisture content using the gravimetric method [9].

3. Result and Discussion

The soil type in the study area is inceptisol [7]. Inceptisols are young parent soils caused by their relatively rapid formation due to weathering of the material. Inceptisols have a low clay content of less than 8% at a depth of 20-50 cm. Inceptisol soil is classified as moderately weathered and washed soil and has a soil solum of 1-2 meters thick, black or gray to dark brown, the texture of sand, dust, and clay, crumb soil structure, loose consistency, pH 5.0 to 7.0, moderately high organic matter (10% to 31%), moderate to high nutrient content, moderate to high soil productivity [10]. Table 2 shows each sample's soil fractions of sand, silt, and clay.

Table 2. Soil Texture Content at the Research Site

No	Location Code	Sand (%)	Dust (%)	Clay (%)	Soil Texture
1	MS	79	13	8	Loamy sand
2	CS	69	23	8	Sandy loam
3	GTS	72	20	8	Sandy loam
4	OPS	60	32	8	Sandy loam
5	SS	60	32	8	Sandy loam

The soil texture at the study site is dominated by sandy loam with a sand content of less than 75 %, namely TT, TGT, TSW, and TS (Table 2). Characteristics of soil with a sandy loam texture is a slightly rough taste, forming a ball that is rather hard but crumbles easily and sticks. Meanwhile, TM has a loamy sand texture with less than 75% sand content. Clay-sand soils are characterized by an apparent coarse taste, forming balls that break easily and are very slightly attached [8].

Table 3. Average Soil Organic Matter and Permeability at the Study Area

No	Location Code	Organic Matters (%)	Permeability (cm/hour)
1	MS	2.59	4.26
2	CS	2.40	4.19
3	GTS	2.83	4.63
4	OPS	1.90	4.12
5	SS	3.36	4.65

The results of soil permeability measurements (Table 3) showed that all types of stands were categorized at a moderate level based on [11] classification. It is due to the soil texture, which is dominated by sandy loam (Table 2). The pore size and the relationship between the pores will determine whether the soil has low or high permeability [12]. Soil with a sandy loam texture has pores that are pretty good at absorbing water.

The highest organic matter content was found in saga stands (TS). The more organic matter in the soil, the more water content in the field capacity, as a result of the increase in medium-sized pores (meso) and the decrease in macro pores, so that water holding capacity increases and has an impact on increasing the availability of water for plant growth [13]. In addition to the organic content factor, the impact of tillage affects the permeability capacity of the soil. Compaction by rain, animals, or heavy equipment can drastically reduce the ability of the soil to absorb water by closing the soil pores [14].

4. Conclusion

Soil permeability of each stand is different according to the factors that influence it. The content of texture fractions (sand, silt, and clay), organic matter, and land use tillage affect the size of the soil permeability. Soil permeability under saga stands (SS) give a higher value than mahogany stands (MS), cape stands (CS), glodokan tiang stands (GTS), and oil palm stands (OPS).

Acknowledgement

This work was funded by TALENTA grant 2021 from University of Sumatera Utara (No.139/UN5.2.3.1/PPM/SPP-TALENTA USU/2021).

References

[1] Soedjoko S, Suyono, dan Suryatmojo H. 2016. *Hidrologi Hutan: Dasar-Dasar, Analisis, dan Aplikasi.* Yogyakarta: UGM Press.
[2] Sharp J dan Sawden P1984. Elements of Hydrology. In J. Sharp, & P. Sawden, *Basic Hydology* (pp. 13-15). Oxfort, UK: Butterworth-Heinemann.
[3] Anggraini N, Slamet B, Ahmad AG. 2021. Buku Ajar Hidrologi Hutan. USU Press
[4] Siregar NA, Sumono, Munir AP. 2013. Kajian Permeabilitas Beberapa Jenis Tanah di Lahan Percobaan Kwala Bekala USU melalui Uji Laboratorium dan Lapangan. Jurnal Rekayasa Pangan dan Pertanian 1(4) : 138-143.
[5] Maro'ah S. 2011. Kajian Laju Infiltrasi dan Permeabilitas Tanah pada Beberapa Model Tanaman (Studi Kasus Sub DAS Keduang Wonogiri). Skripsi. Universitas Sebelas Maret. Surakarta. Jawa Tengah.
[6] Susilowati A, Rangkuti AB, Rachmat HH, Iswanto AH, Harahap MM, Elfiati D, Slamet B, Ginting IM. 2021. Maintaining Tree Biodiversity in Urban Communities on the University Campus. Biodiversitas 22 (5): 2839-2847.
[7] BPS Medan. 2020. Kota Medan Dalam Angka 2020. Badan Pusat Statistik Kota Medan.
[8] Mulyono A, Lestiana H dan Fadilah A. 2019. Permeabilitas Tanah Berbagai Tipe Penggunaan Lahan di Tanah Aluvial Pesisir DAS Cimanuk, Indramayu. Jurnal Ilmu Lingkungan 17(1) : 1-6.
[9] Irawan T, Yuwono SB. 2016. Infiltrasi pada berbagai tegakan hutan di arboretum Universitas Lampung. Jurnal Sylva Lestari. 4 (3). pp. 21-34
[10] Ketaren SE, Mabun P, Marpaung P. 2014. Klasifikasi Inceptisol Pada Ketinggian Tempat yang Berbeda di Kecamatan Lintong Nihuta Kabupaten Hasundutan. Jurnal Online Agroekoteknologi. 2(4): 1451-1458
[11] Uhland, R.E. and A.M. O'neal. 1951. Soil Permeability Determinations For Use In Soil and Water Conservations. SCS-TP101, 36 pp., Iiius, New York
[12] Dariah, A., Yusrial, dan Mazwar. 2006. Penetapan Kondukstivitas Hidrolik Tanah dalam Keadaan Jenuh: Metode Laboratorium: Sifat Fisik Tanah dan Metode Analisisnya. Balai Besar Litbang Sumberdaya Lahan Pertanian. Badan Penelitian dan Pengembangan Pertanian, Departemen Pertanian
[13] Scholes, M.C., Swift, O.W. Heal, P.A. Sanchez, JSI. Ingram and R. Dudal. 1994. *Soil Fertility research in response to demand for sustainability.* In The biological managemant of tropical soil fertility (Eds Woomer, Pl. and Swift, MJ.) John Wiley & Sons. New York.
[14] Lee, R. 1990. *Hidrologi Hutan.* Gadjah Mada University Press. Yogyakarta.

Diversity of Hymenoptera in the Forest Area with Special Purpose (KHDTK) Aek Nauli, North Sumatra

D A S Siahaan*, K Berliani, R Y Pangaribuan, D Agustin, Yulinar, F Mutiara and T S U B Tarigan

Department of Biology, Faculty of Mathematics and Natural Sciences, Universitas Sumatera Utara, Medan, North Sumatra 20155, Indonesia

*E-mail: donialdo@usu.ac.id

Abstract. As one of the four largest insect orders, Hymenoptera are important pollinators, parasitoids, and predators in the agricultural sector. The objective of this research was to study the diversity and abundance of hymenoptera in the Forest Area with Special Purpose/ KHDTK Aek Nauli, Simalungun Regency, North Sumatra, Indonesia. The insect sampling was conducted during the day and night using yellow and light trap technique. The results obtained a total of 13 species of Hymenoptera consisting of 45 ind. The relative abundance was recorded from 2.22 to 22.22 with the highest abundance from *Vespula* sp (Vespidae) and *Sceliphron curvatum* (Spechidae). The highest Importance Value Index (IVI) of Hymenoptera species was documented from *Brachymyrmex* sp. (Formicidae) at 115.56 and *Polybia fastidiosuscula* (Vespidae) at 106.67. The Shannon's diversity (*H'*) of Hymenoptera ranged between 0.89 and 1.83 among sites or categorized as low-to-medium level of biodiversity with an evenness index (*E*) from 0.64 to 1.00. This finding may be noted as our first attempt in documenting the beta diversity of insects in the forest region.

1. Introduction

The wide-spread and closely-placed rows of trees that make up a forest constitute an ecosystem. The forest also contains an ecosystem in the form of stands, each of which has its own unique characteristics in terms of structure, age class, species composition, and interconnected processes. Rivers, fish, animals, and grasslands are also part of forests, including plantation forests, protected forests, urban forests, industrial forests, and non-industrial forests [1].

Hymenoptera is the third largest insect order after Lepidoptera and Coleoptera. This order consists of phytophagous insects, social insects, bees, solitary wasps and parasitoids [2]. According to their role in nature, parasitoid species make up around 80% of the Hymenoptera genus. Predators, pollinators, and phytophages are additional species. One of the key themes in ecological research to examine the composition and composition of species and their abundance in an ecosystem is species diversity, particularly with regard to Hymenoptera [3]. The variety of Hymenoptera species that inhabit different habitats might vary, notably in terms of species richness and abundance. The order Hymenoptera is divided into two sub-orders, namely the sub-order Symphyta and the sub-order Apocrita [4]. Many members of the suborder Symphyta are plant pests. The suborder Apocrita has the most predatory species and parasitoids that are widely used in the Integrated Pest Managements (IPM) [5].

The KHDTK Aek Nauli is a forest region located in North Sumatra which is still limited for its information regarding the insect diversity. The goal of this research was to investigate the diversity of Hymenoptera, some of which may be significant for the conservation of forests while others may become the first documentation of prospective insects in the future.

2. Method

2.1. *Study area*

The research was conducted within the Forest Area with Special Purpose (KHDTK) Aek Nauli, Sibaganding Village, Girsang Sipangan Bolon District, Simalungun Regency, North Sumatra, Indonesia. The forest area is structured by primary and secondary forest with an area of 1,900 ha, altitudes from 1,100–1,750 m a.s.l. The forest is mainly composed by the pine vegetation and other tropical tree and shrub species with some sites implemented for an *ex situ* conservation of Sumatran Elephant (*Elephas maximus sumatranus*). The forest region is officially managed by the Forestry Research, Development and Innovation Agency (BP2LHK) Aek Nauli under the Indonesian Ministry of Environment and Forestry.

2.2. *Insect collection*

A total of 3 sampling sites were purposively selected in the forest area near the pine stands (Site-1), deer cage (Site-2), and elephant's conservation area (Site-3). The sampling was conducted on May 2022 within three consecutive days from 08:00 AM to 17:00 PM. Light trap was installed at each site for insects that prefer light while yellow trap was installed for insects interested in yellow surface with a pan filled with detergent water as a preserving agent [6]. Trapped insects were collected in a bottle that was filled with 96% alcohol. The specimens were identified at the Laboratory of Animal Systematics, Universitas Sumatera Utara based on some identification manuals and an online tool [4,7,8].

2.3. *Data analysis*

Shannon's diversity index (*H'*) and evenness index (*E*) of Hymenoptera species were calculated using PAST4.03 [9]. Importance Value Index (IVI) of each species was calculated as the sum of relative abundance (*Ra*) and relative frequency of occurrence (*Fo*).

3. Results and Discussion

We identified a total of 13 morphospecies (Hymenoptera) in the forest area of KHDTK Aek Nauli from three sampling sites namely Site-1 (Pine stands), Site-2 (Deer cage), and Site-3 (Elephant's conservation area). The species list and total number of Hymenoptera individuals are presented in Table 1. We documented ants (Formicidae), bees (Apidae), and wasps (Eurytomidae, Pompilidae, Sphecidae, Tiphiidae, Vespidae) as the Hymenoptera community in the study area. The highest number of Hymenoptera species was recorded at pine stands or Site-1 (27 ind), followed by Site-2 or deer cage (14 ind), and Site-3 or elephant site (4 ind). This may due to the existence of trees as the important habitats for Hymenoptera community in the forest region. In a study led by Budiaman et al [10], the Hymenoptera diversity found in pine plantation forest, West Java is considerably high even after the thinning of timber that tend to reduce the species composition but not with the species abundance and the number of individuals.

In general, *Sceliphron curvatum* and *Vespula* sp, the two wasp species were recorded with the highest abundance insect population in the forest region. Wasps including the parasitoid ones as the part of canopy arthropods play a significant role in the balance of insect population [11]. In terrestrial ecosystems, parasitoids are essential because they greatly influence the population of herbivores [12]. According to Dyer et al [13], low levels of herbivory are linked to higher levels of plant development, which in turn increases the diversity of wildlife. Parasitoid insects are incredibly diverse, comprising nearly around 10% of all known insect species at the moment, in part because of their high level of host specialization [14]. In addition, the highest Importance Value Index (IVI) was obtained from the ant species namely *Brachymyrmex* sp. (115.56), followed by a wasp species, *Polybia fastidiosuscula* (106.67). Based on the diversity profile among sampling sites, the Shannon's diversity index (H') ranged between 0.89 to 1.83 or classified as low-to-medium level of biodiversity. Meawhile, the evenness index (E) of Hymenoptera community ranged between 0.64 to 1.00 indicating no species dominance in the community.

Table 1. Composition, relative abundance, relative frequency of occurrence, and IVI of Hymenoptera in KHDTK Aek Nauli

No.	Family	Species	*N* (Ind.) Site-1	Site-2	Site-3	Total	*Ra* (%)	*Fo* (%)	IVI
1.	Formicidae	*Brachymyrmex* sp.	4	2	1	7	15.56	100	115.56
2.		*Carebara diversa*	2	-	-	2	4.44	33.33	37.77
3.		*Camponotus vagus*	5	-	-	5	11.11	33.33	44.44
4.		*Myopopone* sp.	1	-	-	1	2.22	33.33	35.55
5.	Apidae	*Melipona* sp.	2	-	-	2	4.44	33.33	37.77
6.		*Trigona* sp.	-	-	1	1	2.22	33.33	35.55
7.	Eurytomidae	*Eurytoma* sp.	-	1	-	1	2.22	33.33	35.55
8.	Pompilidae	*Auplopus* sp.	1	-	-	1	2.22	33.33	35.55
9.	Sphecidae	*Sceliphron curvatum*	-	10	-	10	22.22	33.33	55.55
10.	Tiphiidae	*Tiphia* sp.	-	-	1	1	2.22	33.33	35.55
11.	Vespidae	*Polybia fastidiosuscula*	1	1	1	3	6.67	100	106.67
12.		*Vespa velutina*	1	-	-	1	2.22	33.33	35.55
13.		*Vespula* sp.	10	-	-	10	22.22	33.33	55.55
	Total		27	14	4	45	100	-	-

Note: *Ra* = Relative abundance, *Fo* = Frequency of occurrence, IVI = Importance Value Index

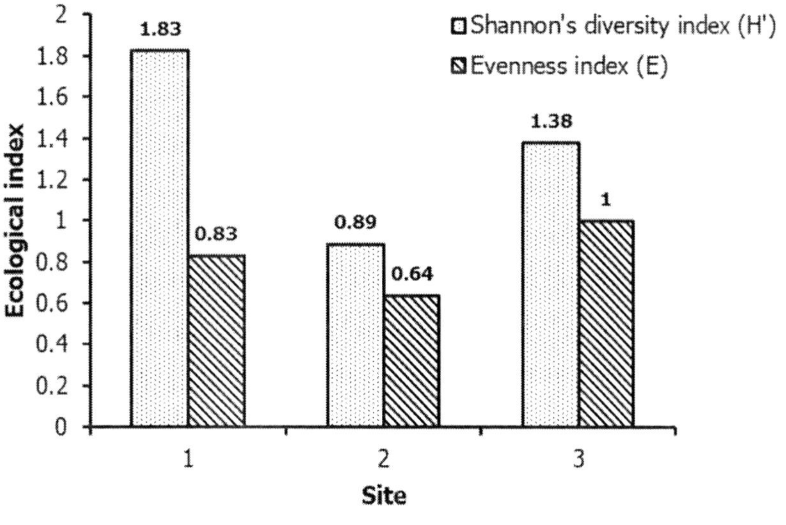

Figure 1. Diversity profile of Hymenoptera community across sampling sites in KHDTK Aek Nauli

4. Conclusion

The Hymenoptera community in the forest area of KHDTK Aek Nauli was documented for the first time revealing a total of 13 species comprising 7 families namely Formicidae (ants), Apidae (bees), Eurytomidae (wasps), Pompilidae (wasps), Sphecidae (wasps), Tiphiidae (wasps), and Vespidae (wasps).

References

[1] Siregar P G, Supriatna J, Koestoer R H and Harmantyo D 2018 *Biotropia* **25** 103
[2] Ovawanda E A, Witjaksono and Trisyono Y A 2016 *Jurnal Perlindungan Tanaman Indonesia* **20** 15
[3] Arnan X, Bosch J, Comas L, Gracia M and Retana J 2010 *Insect Conserv Soc* **4** 200
[4] Goulet H and Huber J T 1993 *Hymenoptera of the World: An Identification Guide to Families* (Ottawa: Research Branch Agriculture Canada Publication)
[5] Viggiani G 2000 *Crop Protection* **19** 665
[6] Ikhsan Z, Hidrayani, Yaherwandi and Hamid H 2020 *Biodiversitas* **21** 1020
[7] Kalshoven L 1981 *The Pest of Crops in Indonesia* (Jakarta: Ichtiar Baru – Van Hoeve)
[8] Triplehorn C A and Johnson N F 2004 *Borror and DeLong's Introduction to the Study of Insect* (Stanford: Cengage Learning)
[9] Hammer Ø, Harper D A T and Ryan P D 2001 *Paleontol Electron* **4** 9
[10] Budiaman A, Haneda N F and Suhada A 2020 *IOP Conf Ser Earth Environ Sci* **468** 012017
[11] Godfray H C J 1994 *Parasitoids: Behavioral and Evolutionary Ecology* (Princeton: Princeton University Press)
[12] Azhar A, Hartke T R, Bottges L, Lang T, Larasati A, Novianti N, Tawakkal I, Hidayat P, Buchori D, Scheu S and Drescher J 2022 *Agr Forest Entomol* (DOI: 10.1111/afe.12512)
[13] Dyer M I, Turner C L and Seastedt T R 1993 *Ecologic Appl* **3** 10
[14] Santos A M C and Quicke D L J 2011 *Entomolog Sci* **14** 371

Forest structure change after forest fire in peatland of Central Kalimantan

M A Qirom[1*], D Rachmanadi[1], F Lestari[2] and S Andriani[3]

[1]Research Center for Ecological and Etnobiology, National Research and Innovation Agency, Jl. Raya Jakarta – Bogor Km. 46, Cibinong-Bogor 16911, West Java, Indonesia
[2]Research Center for Plant Conservation, Botanic Gardens, and Forestry, National Research and Innovation Agency, GedungKusnoto, Ir. H. Juanda No. 18 Bogor 16122, West Java, Indonesia
[3]Institution for Environment and Forestry Standard Instrument Implementation of Banjarbaru, Ministry of Environment and Forestry Jln. Ahmad Yani Km 28.7 Banjarbaru-South Kalimantan- Indonesia

[*]E-mail: muhammad.abdul.qirom@brin.go.id

Abstract. The hydrological system of tropical peatland has been changed due to drainage. The condition caused prone to peatland burning. Forest fire affected the vegetation composition in peatland. Repeated fire causes more severe impact on vegetation and the change the peatland characteristics. The aim of the study is to obtain species composition on peatland after fire in Central Kalimantan. The research was conducted on several plots of vegetation observation on peatland that burned in 2015 (three years after fire and forth fire frequency). The plot sizes were depending on vegetation stage (seedling, sapling, poles, and trees) of 2 Ha or 50 observation plots. The measurements of parameter were species, density, frequency, and important value index. The species composition was described by ecological indices such as Shannon wiener index, Simpson index, and evenness index. The results showed that species diversity and ecological index were very low. That condition indicated that the fire has changed the species composition and the natural regeneration has been obstructed. *Combretocarpus Rotundatus* and *Cratoxylon glaucum* were presented on overall tree stage as pioneer species, but the climax species didn't found on three years after forest fire. Based on the results, the repeated fires caused an extreme decrease of natural regeneration so the ability of recovery is not running well. The parameters can be used in determining the priority of revegetation program on large scale.

1. Introduction

Vegetation cover of a forest is determined by its vegetation structure composer. The destruction of vegetation structure will cause the decrease of vegetation cover quality. This will lead to the increase of undergrowth that can inhibit the process of forest regeneration and raise soil surface temperature [1]. The peat swamp forest ecosystem, like other forest ecosystems, also experiences dynamics of changes in vegetation cover caused by various factors, including exploitation, land conversion and forest and land fires. Peat swamp forest is one of Indonesia's wetland ecosystems and plays a global role in climate change. Wetland ecosystems are known to have large carbon stores, which are 20-30% of the world's organic carbon. As a fragile ecosystem, this large store of carbon has the implication of being released back into the atmosphere in the form of carbon and methane when the climate is warmer and the wetland ecosystem is drier [2].

The area of peat swamp forest owned by Indonesia is approximately 14 million Ha [3] and is the largest tropical peat swamp forest ecosystem in the world. This ecosystem stores 55 Gt of carbon with an estimated peat thickness of 4.5 m [4]so its management and conservation is very important in climate change adaptation and mitigation. Peat swamp forest also produces various species of wood so that this ecosystem does not escape exploitation, both legally and illegally. Exploitation causes changes in vegetation structure but can recover in the next logging rotation even though the dominance of the species changes [5]. Difficult accessibility is causing humans to take various ways to facilitate the process of exploitation and utilization of other forest products, one of which is by making canals with various dimensions. The construction of the canal will certainly have an impact on the hydrological status of peatlands as a wetland ecosystem. Then massive land conversion such as the construction of one million hectares of agricultural land in Central Kalimantan caused a drastic change in its hydrological status and caused the peat ecosystem to dry up so that it was very vulnerable to fires [6].

The fires that then occur almost every year in Central Kalimantan drastically also cause changes in forest structure and vegetation cover. Fires cause the loss of all species of vegetation in every layer of the forest structure. It is very different from structural change due to exploitation which generally only removes commercial vegetation species. This study attempts to quantify changes in vegetation structure in peat swamp forest ecosystems that have experienced repeated forest fires. The resulting data will provide an overview of the ability to recover of peat swamp forest ecosystems and how human intervention is needed for the recovery process.

2. Materials and Method

2.1. *Location and time*

The study was conducted in Forest Research of Tumbang Nusa (FRTN), Central Kalimantan. FRTN position is 6^0Southand 130^0East. The land cover are various such as secondary forest, natural forest after fire, shrub, and revegetation area but the dominant land cover is burned area. The research was carried out on the burned area of 2015. The characteristic area has been burned on 1997, 2002, 2009, and 2015. The observation of species composition was accomplished on August and December 2018 or three years after fires. The observation time was represented the dry and rainy seasons.

2.2. *Methods*

2.2.1. Category, characteristic, plot size and shape, and number of plot . The land and strata of category was determined to accelerate and clarify work in the field. The strata of plants category were divided into 4 stage: seedlings, saplings, poles, and trees [7,8]. It's stratum affected to individual dimension and size of observation plots. The land characteristics to determine the study area was fire frequency [9]. The criteria were chosen because the more frequent the fires, the poorer the peatland quality. The locations of plot were on burned area of 2015 or three years after fires. The plot sizes were depend on plants stage such as: seedling with 4 m^2(2 m x 2 m), area of 25 m^2(5 m x 5 m) for saplings, the plot size of poles is 10 m x 10 m, and trees with 400 m^2 or 20 m x 20 m [7]

The shape of observation plots were rectangular with diverse size of vegetation stage [8]. The parameters of vegetation stage divided to4 criteria such as seedling with tree height < 1.5 m, sapling with tree height > 1.5 m and DBH < 10 cm, poles with diameter 10-20 cm, and trees with diameter ≥ 20 cm. In the fields, the overall of observation plot was 100 m x 200 m (2 Ha) so the total plots measurement was 50 plots which divided into 400 m^2 for each plot. The plots were assigned accordance to the location characteristic based on preliminary surveys or recent imagery. These plots divided into 10 blocks as replication. It's represented of the whole of burnt area (Figure 1).

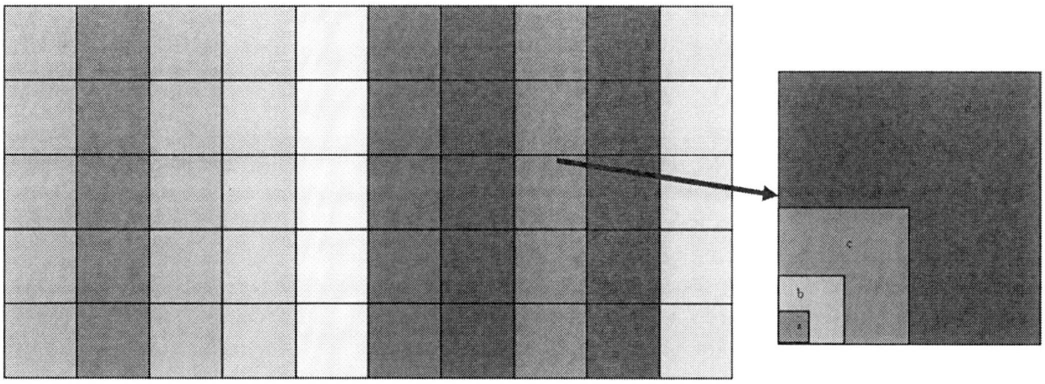

Remarks: difference color showed replicant; a: seedling, b: sapling; c: pole; d: trees

Figure 1. Observation plots layout

2.2.2. Measurement and data processing. The parameter of measurement were species, number of species, plant height, and number of plant. The identification of species was based on local name from local identifier. The data base of list species was used to identify scientific name [10]. The ecological parameters were used for data processing of measurement. The parameters were included density, frequency, dominance and important value index. These parameters are formulated as follows:

$$K = \frac{K_{indv}\ of\ each\ plot}{total\ of\ plot\ area};\ KR = \frac{K_{ind}}{\sum K_{ind}} x100\%\ ; F_{ind} = \frac{number\ of\ individual\ found}{total\ plot\ area}; \quad (1)$$

$$FR = \frac{F_{ind}}{sum\ of\ species\ frequency} x\ 100\%\ ;\ DR = \frac{D_{species}}{sum\ of\ dominance\ of\ total\ species} x100\%; \quad (2)$$

$$INP_1 = KR + FR\ ;\ INP_2 = KR + FR + DR$$

K is density (individual/ha), KR is relative density (%), F_{ind} is frequency of one individual, FR is relative frequency (%), D is dominance (m^2/ha), DR is relative dominance (%), INP_1 is important value index for vegetation with diameter less than 10 cm (%), INP_2 is Important value index for vegetation with diameter more than 10 cm (%).

2.2.3. Data analysis. The performance of species composition after fire was assessed by several ecological indices. The indices can describe the stand quality as seen from diversity, dominance, and evenness of stand composition. These indices were formulated as follows [8]:

$$H' = -\sum \frac{ni}{N} \log\left(\frac{ni}{N}\right) \qquad \text{Shanon-Wiener Index [8]} \quad (3)$$

Category of Shannon-Wiener index $H > 3$ high diversity, $1 \leq H \leq 3$ medium diversity, $H < 1$ low diversity

$$DI = \sum \left(n.i/N\right)^2 \qquad \text{Simpson Index [8]} \quad (4)$$

Category $0 < DI \leq 0.5$ low dominance; $0.5 < ID \leq 0.75$ medium dominance; $0.75 < ID \leq 1$ high dominance

$$E = \frac{H'}{\mathrm{Log}\, S} \qquad \text{evenness index [8]} \quad (5)$$

Category of evenness index $0 < E \leq 0.5$ distressed community, $0.5 < E \leq 0.75$ unstable community, $0.75 < E \leq 1$ stable community

3. Result and Discussion

3.1. Diversity and density of species

The results of the vegetation analysis showed that the species that composed the stand were very limited with the main species *Co. rotundatus* and *Cr. Glaucum* (Figure 2). These species are always present at the seedling and sapling levels. Those species are pioneer species that will be present in areas with open conditions either due to fire or other disturbances such as illegal logging and land conversion. The diversity of species 3 years after burning is very low because the stand only consists of 3 species. This condition is different when compared to the results of measurement of vegetation diversity after burning in peatland areas in South Sumatera. In this area, the composer species of the stand are more diverse and the stand structure is relatively complete. This is indicated by more tree species in the stand and regeneration at pole level found at the site. However, the stand is incomplete because the tree level is not found in the observation plot [11]. This condition is influenced by the lower frequency of fires at that location than that occur at the study site.

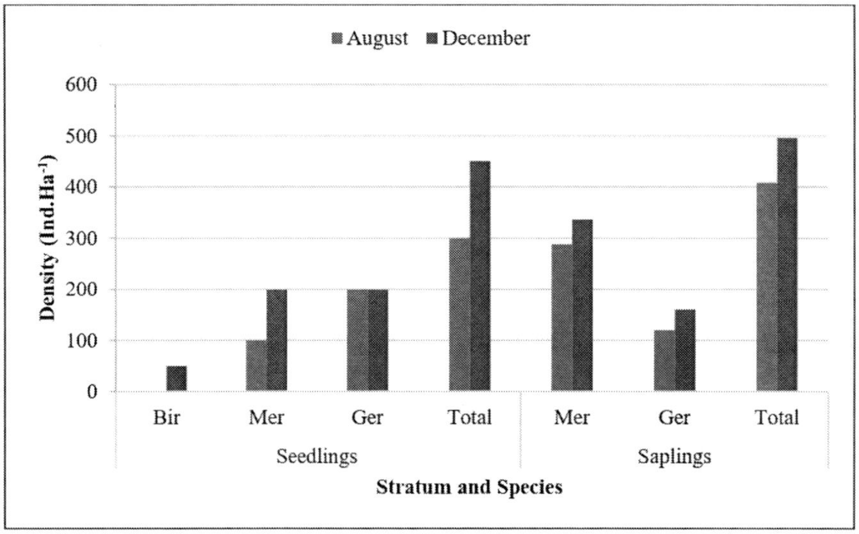

Remarks: Bir: Birayang, Mer: *Co. rotundatus*, Ger: *Cr. glaucum*

Figure 2. Diversity and density of species after three years of fires

The structure of the stand is incomplete. This can be seen from the loss of plant species at the pole and tree level. This condition indicates that the research location is experiencing disruption of the natural regeneration process. The characteristics of natural recovery vary widely. Only a small part of secondary forest is able to have a complete species composition (34%) [12]. This natural recovery requires human intervention so the stands will have a complete species composition. The speed of recovery varies greatly and not all phases will recover completely. The recovery measures were species abundance and vegetation composition, the surrounding land cover plays a very important role in ensuring the sustainability of land recovery related to seed sources and their distribution. The speed of recovery is related to management intensity, the success of natural regeneration is determined by the strength of an ecosystem to changes that occur. Stands remain after damage as a determining factor and a driving force for success as a source of seeds for natural regeneration [13]. The results of the study at Forest Research of Tumbang Nusa (FRTN) showed that the distance from the seed source/dwelling stand and forest determines the level of species diversity in disturbed areas in peat swamp forest [14].

Table 1. Distribution of the species of each stratum in the whole observation plot

Measurement time	Stratum	Species	Observation plot										
			1	2	3	4	5	6	7	8	9	10	Total
August	Seedling	*Co. rotundatus*				1	1						2
		Cr. glaucum		2		1	2	3			4	3	15
	Saplings	*Co. rotundatus*	5	7	3	5	4	1	2		4	5	36
		Cr. glaucum		2		1	2	3			4	3	15
December	Seedling	Birayang				1							1
		Co. rotundatus				3				1			4
		Cr. Glaucum	1					3					4
	Saplings	*Co. rotundatus*	5	7	3	8	7	1	2	1	3	5	42
		Cr. Glaucum	1	2		1	3	3		2	4	4	20

The condition of the research area is in a very severe disturbed condition. It can be shown that in all the observation plots no regeneration was found at all regeneration levels (Table 1). In addition to conditions of low diversity, individual density per certain area is very rare. The density was less than 0.06 individuals.m^{-2} in the second measurement (December 2018), and in the August 2018 measurement, the species density was only 0.04 individuals.m^{-2} (Figure 2 and Table 1). This density is still lower than the density of stands in the 1997 burned areas in Central Kalimantan [9,15]. The low density and diversity of this species indicates that this area cannot achieve natural succession. This is caused by several factors, namely the very low density remaining stands, both seeds and seed banks in the soil, damage to environmental conditions due to changes in the physical and chemical properties of the soil, as well as changes in the hydrological conditions of the peat [16].

The addition new species after 4 months of observation is very little, only one new plant species. This condition shows the severity of changes in soil conditions due to forest and land fires. Fires have caused changes in the microtoprography of peatlands which make it very difficult for new species to adapt to environmental changes [17]. These changes mainly occur in very extreme fluctuations in ground water level. In the context of revegetation of degraded peatlands due to forest fires, areas with changed conditions with natural regeneration and very limited species addition, human intervention is absolutely necessary. These interventions are in the form of selecting species that are very tolerant to extreme conditions [18], environmental manipulation, and regulation of peat hydrology. Indigenous peat swamp species especially Pioneer species are highly recommended for degraded peatlands[14,15,18,19]. This is due to the native species of peat swamps which have the ability to adapt very quickly to environmental changes, especially inundation that last for a long time [20].

3.2. Ecological index

In general, the ecological indices obtained are very low (Figure 3) for the characteristics of diversity, dominance, and evenness of species. These conditions indicate that the burnt area is in a very disturbed condition. Based on the ecological indices, it shows that the area has very low species diversity ($H'<1$), dominance at the intermediate level ($DI<0.75$), and depressed community with $E<0.5$. The low ecological index occurred at all levels of regeneration (seedlings, saplings, and total regeneration).

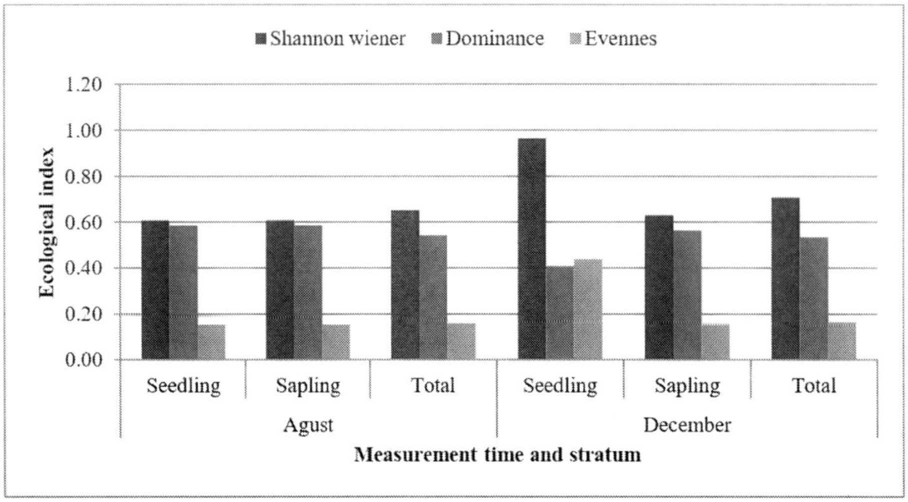

Figure 3. Ecological index of seedling and sapling in two measurement time

Ecological index shows an increase in the index value with burning time or at the second measurement. This condition indicates that the natural regeneration process is taking place but the rate of natural regeneration is very slow. This condition can also be seen from the very high community similarity index value between August and December measurements. The community similarity index reached 79.7% at the seedling level and the similarity index reached 97.7% at the sapling level. The magnitude of the index value indicates that the time of 4 months period did not change the composition of the plant species of the stand.

In areas of extremely severe disturbance, natural recovery takes more than 100 years. The speed of recovery varies greatly depending on the severity of stand damage, intensity and severity of disturbance, availability of seed sources, seed vectors [21], and environmental conditions. Human intervention is needed in encouraging the interaction between these factors to run optimally. This condition occurs in the research area. Based on the parameters of species diversity and ecological index, this area requires human intervention to accelerate the revegetation process. These interventions are in the form of adding prospective new species, selecting suitable species, and improving environmental conditions.

4. Conclusions

Forest fires have changed the structure and composition of the standing species. These changes are very extreme with a decrease in species composition reaching more than 95% of secondary forest species and loss of plant strata at the level of poles and trees. This condition causes burnt areas to lose the ability to recover naturally, so human intervention is absolutely necessary to accelerate the revegetation process in burnt areas.

Acknowledgement

This research was funded in collaboration with the Peat and Mangrove Restoration Agency. The authors would like to thank the management of the Institution for Environment and Forestry Standard Instrument Implementation of Banjarbaru for the permission and facilitation in conducting research at the KHDTK Tumbang Nusa. The authors also thank the engineering technicians for their assistance in field measurements.

References

[1] Dupuy J M and Chazdon R L 2008 *For. Ecol. Manage.* **255** 3716
[2] Mitsch W J, Nahlik A, Wolski P, Bernal B, Zhang L and Ramberg L 2010 *Wetl. Ecol. Manag.* **18** 573
[3] Anda M, Ritung S, Suryani E, Hikmat M, Yatno E, Mulyani A and Subandiono R E 2021 *Geoderma* **402** 115235
[4] Jaenicke J, Englhart S and Siegert F 2011 *J. Environ. Manage* **92** 630
[5] Daryono H 2000 *Kondisi hutan setelah penebangan dan pemilihan jenis pohon yang sesuai untuk rehabilitasi dan pengembangan hutan tanaman di lahan rawa gambut* (Banjarmasin: Balai Teknologi Reboisasi Banjarbaru)
[6] Boehm H D V and Frank J 2009 *Peatland topography and hydrology of ex-MRP and Sebangau National Park measured with Airborne Laser Scanning- (LIDAR)-Technology* (Hokkaido University and State Ministry of Research and Technology)
[7] Badan Standardisasi Nasional (BSN) 2011 *Pengukuran dan penghitungan cadangan karbon – Pengukuran lapangan untuk penaksiran cadangan karbon hutan (ground based forest carbon accounting)* (Jakarta)
[8] Krebs C J 2014 Species diversity measures *Ecological Methodhology* (Boston, United States: Addison-Wesley Educational Publishers, Inc.)
[9] Blackham G V, Webb E L and Corlett R T 2014 *For. Ecol. Manage* **324** 8
[10] Thomas A 2013 *Panduan Lapangan Identifikasi Jenis Pohon Hutan* (Palangka Raya)
[11] Kunarso A, Angga T, Syabana A and Rachmanadi D 2022 *J. Galam* **2** 125
[12] Rozendaal D M A, Bongers F, Aide T M, Alvarez-Dávila E, Ascarrunz N, Balvanera P, Becknell J M, Bentos T V and et al 2019 *Sci. Adv.* **5** 3114
[13] Jakovac C C, Peña-Claros M, Kuyper T W and Bongers F 2015 *J. Ecol.* **103** 67
[14] Rachmanadi D, Faridah E, Sumardi and Meer P J Van Der 2017 *J. Ilmu Kehutan.* **11** 224
[15] Blackham G V, Andri Thomas, Webb E L and Corlett R T 2013 *Biol. Conserv.* **167** 215
[16] Vijayakumar D B P, Raulier F, Bernier P, Paré D, Gauthier S, Bergeron Y and Pothier D 2016 C *For. Ecol. Manage* **360** 170
[17] Lampela M, Jauhiainen J, Kämäri I, Koskinen M, Tanhuanpää T, Valkeapää A and Vasander H 2016 *Catena* **139** 127
[18] Lampela M, Jauhiainen J, Sarkkola S and Vasander H 2017 *For. Ecol. Manag* **394** 52
[19] Lampela M, Jauhiainen J, Sarkkola S and Vasander H 2018 *For. Ecol. Manage* **429** 217
[20] Rachmanadi D, Faridah E, Meer P Van Der, Qirom M A, Agro J and Bulaksumur N 2021 *J. Galam* **1** 123
[21] Tamin R P, Ulfa M and Saleh Z 2021 *Wahana For J Kehutan* **16** 181

Optimization of green open space in Ambon City

F M Selanno[1], M R Sitanala[2], P Santoso[3] and H Arinah[4*]

[1]Faculty of Forestry, Universitas Gadjah Mada, Jl. Agro, Bulaksumur No.1, Kocoran, Caturtunggal, Kec. Depok 55281, Sleman, Special Region of Yogyakarta
[2]Department of Forestry, Faculty of Agriculture, Pattimura University, Ambon, 97233
[3]Department of Bioresources Technology and Veterinary, Vocational College Universitas Gadjah Mada Jl. Yacaranda Sekip Unit 2, Depok 55281, Sleman, Daerah Istimewa Yogyakarta
[4]Faculty of Forestry, Universitas Sumatera Utara, Medan 20155, North Sumatra, Indonesia

[*]Email: hafizaharinah@usu.ac.id

Abstract. The increase in population in Ambon City has resulted in increased space requirements and impacts the conversion of green areas into built-up areas. This indirectly impacts the decline in environmental quality, marked by an increase in the concentration of greenhouse gas (GHG) emissions in the air. The Green Open Space (GOS) arrangement program involves efforts to reduce GHG emissions in Ambon City. However, in reality, the allocation of space to develop green open space in Ambon City still does not meet the 30% standard or is still lacking 737.8 ha. This study aims to obtain information about the arrangement of green open spaces in Ambon City. The interview method was conducted in a structured manner with a questionnaire. SWOT analysis was used to analyze internal factors and external factors. The results showed that the critical strategies in optimizing Ambon City's green open space were to improve the quality of green open space by planting plants that have high GHG emissions absorption potentials, such as *Samanea saman*, *Pterocarpus indicus*, and *Acacia mangium*. Conduct harmonization and socialization with other institutions such as the Ambon City Satpol PP and green open space users to increase the supervision of green open space. Using the basic green coefficient as the primary condition for obtaining a Building Permit (IMB). The present study recommended that future research on green open spaces in Ambon City should focus on other influencing elements including geography, climate, socio-culture, and so on.

1. Introduction

The strategic position of Ambon City as the Capital of Maluku Province has become a magnet for residents to come and live (urbanization). This urbanization process has caused various problems, such as the increasing concentration of the population [1]. In its development, Ambon City was affected by the urbanization process, as seen from [2], which stated that the population of Ambon City in 2017 was 444,956 people, or an increase compared to the 2016 population data.

The population that continues to increase in Ambon City causes most green areas to turn into built-up areas. This is evidenced by [3] research, which explains that the built-up area in Ambon City Center has experienced an area increase of 1322 ha in 2015. It is predicted that this land-use change process will continue to increase along with high activity, population, and adequate facilities and infrastructure. If not controlled, it will disrupt the ecological balance and decrease environmental quality [4].

The decline in environmental quality, marked by increasing concentrations of greenhouse gas (GHG) emissions in the air, is slowly causing climate change. As a country with many small islands and a tropical climate, Indonesia is highly affected by climate change conditions [5]. One of the small islands in Indonesia whose existence is very threatened by various impacts of climate change namely Ambon Island. The real threats felt by the fishing community on Ambon Island include changes in the fishing ground, chaos in the fishing season, rising seawater temperatures, rising sea levels, increasing storm intensity, and high sea waves [6]. The Ambon City Local Government (Pemda) took appropriate and strategic steps to reduce the main variant of climate change, namely GHG emissions, through the Green Open Space (GOS) arrangement program.

Various types of plants in Ambon City Green Open Space can reduce GHG emissions in the air. This is explained in [7] research, which states that the Samanea saman plant species in Ambon City Green Open Space can absorb carbon dioxide gas emissions of 191,378 tons/ha. The above green space capability should be followed by adequate space allocation. However, in reality, the allocation of space for the arrangement of green open space in Ambon City still does not meet the standard of 30% or lacks 737.8 ha [8]. Therefore, it is necessary to optimize Ambon City's green open space to improve the quality of the urban environment

2. Materials and Methods

The research was conducted in Ambon City for three months, from December 2018 to February 2019. The data collected consisted of primary data and secondary data. Preliminary data was obtained through field observations, measuring the area of green open space (calculated using Google Earth), and interviews. In contrast, secondary data resulted from a literature study, including data on the condition of Ambon City Green Open Space, regional statistical information, and other supporting data from books, journals, reports, and the internet.

A structured interview technique with a list of questions (questionnaire) is used to obtain information from critical respondents (key informants). Determination of key respondents (key informants), such as the Ambon City Environmental Service (DLH), the Ambon City Planning and Development Agency, and one of the lecturer staff (Academics) in the forestry department of Pattimura University Ambon, using a purposive sampling technique with the consideration that the respondents are the parties related to the arrangement of green open space.

The data analysis used in this research is SWOT analysis (strengths/strengths, weakness/weaknesses, opportunities/opportunities, and threats/challenges/threats) to determine internal and external factors [9]. Internal factor analysis was conducted using an internal strategic factor analysis matrix (Internal Strategic Factors Analysis Summary / IFAS). In contrast, external factor analysis used an external strategic factor matrix (External Strategic Factors Analysis Summary / EFAS). Based on the IFAS and EFAS matrices, a SWOT matrix is prepared. Each indicator in the SWOT matrix will determine the primary strategy for optimizing Ambon City's green open space.

3. Result and Discussion

3.1 The existing condition of Ambon City green open space

The arrangement of Ambon City's green open space is directed at urban areas. It aims to create a harmonious and balanced urban ecosystem and can meet the standards of 30% of the city area (10% private green open space and 20% public green open space). The results of field observations and measurements show that currently, Ambon City has an area of 554.11 ha of green open space and is classified into 13 types of green open space (Table 1 and Figure 1).

Table 1. The area of each type of green open space in Ambon City

No	GOS Type	Nusaniwe sub-district (ha)	Sirimau sub-district (ha)	Baguala sub-district (ha)	Teluk Ambon sub-district (ha)	Amount (ha)	green open space needs		
							Private GOS (10%)	Public GOS (20%)	Total (30%)
Public		38,06	6,90	0,63	2,45	46,04		909,50	
1	Urban Park		3,74						
2	Green Line Road	0,28	1,67	0,40	2,31				
3	road median		0,16	0,23	0,09				
4	Grave		1,00						
5	Road Island		0,33		0,05				
6	Siwalima Tourism	35,78							
Private		2,44	17,69	327,78	160,85	508,77	454,75		
1	Mangrove			59,71	2,91				
2	Field		0,22	1,71	1,12				
3	Military area			40,53					
4	protected forest			203,58					
5	Empty land	0,63	6,67	21,52	25,51				
6	Pertamina's park	1,06							
7	Ukim park	0,76							
8	Park		0,07	0,75					
9	Graveyard		3,74						
10	Brimob area		1,47						
11	Urban Forest		2,00						
12	Forest		3,53						
13	airport land				131,31				
					total				1364,25

Figure 1. Existing conditions of green open space in Ambon City

The existing condition of Ambon City's green open space, which is only 554.11 ha, does not meet the city's green open space needs of 1364.25 ha (30%), so it is necessary to optimize the area to minimize urban environmental problems. According to [10], optimizing green open space is necessary to improve green open space quality.

3.2 Ambon City green open space optimization strategy

Efforts to optimize Ambon City's green open space require a strategic plan. Strategic planning is supported by a good analysis stage of strengths, weaknesses, opportunities, and threats [9]. SWOT analyzed these strategic factors to produce a strategy and policy alternatives. The strategic factor analysis (IFAS and EFAS) is described as follows:

3.2.1 Internal factor. The analysis of internal factors, including factors of strength and weakness, can be seen in Table 2.

3.2.2 External factor. Analysis of external factors, including opportunity and threat factors, can be seen in Table 3.

Table 2. Internal factors in the optimization of the green open space in Ambon City

	Internal Factors	Value	Rating	Score
	strength			
1	GOS benefits and functions	0,165	4	0,658
2	Regulations related to the development of green open space	0,152	4	0,608
3	Use of green base coefficient	0,152	3	0,456
	amount	0,468		1,722
	Internal Factors			
	Weakness			
1	Limitations of the Regional Revenue and Expenditure Budget (APBD)	0,190	2	0,380
2	Poor arrangement	0,177	2	0,354
3	Overlapping of GOS management between agencies	0,165	3	0,494
	amount	0,532		1,228
	Total	1,00		2,95

Table 3. External factors in the optimization of the green open space in Ambon City

	External Factor	Value	Rating	Score
	Opportunity			
1.	Green community (environmental care community)	0,165	2	0,329
2.	Availability of undeveloped land outside the city center	0,153	3	0,459
3.	Private sector involvement in funding for green open space development	0,165	2	0,329
	Jumlah	0,482		1,117
	External Factor			
	Threat			
1.	The intensity of supervision on the existence of green open space is still low	0,176	2	0,353
2.	Acceleration of infrastructure development and public service facilities	0,176	2	0,353

3.	The absence of technical regulations for the development of green open space	0,165	3	0,494
	Amount	0,518		1,200
	Total	1,00		2,317

Table 2 and Table 3 show the magnitude of the value of internal and external factors. The value of the difference in internal factors (strengths and weaknesses) is 0.49, while the difference in external factors (opportunities and threats) is -0.08. A positive value for internal factors indicates the dominance of the strength factor over the weakness factor. In contrast, a negative value for the external factor means that the opportunity factor cannot dominate the threat factor. By looking at the value of internal factors (IFAS) and external factors (EFAS), then the position of the matrix is in quadrant II (Figure 2).

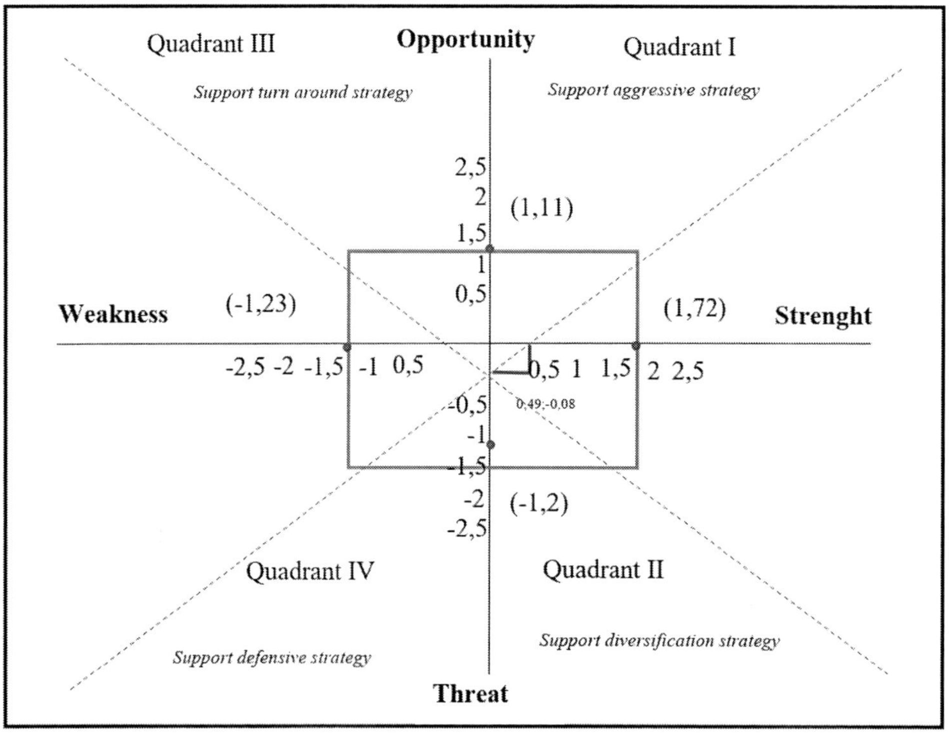

Figure 2. Position of IFAS-EFAS quadrants

According to [9], the position of quadrant II shows a condition where there is a very big threat. Still, on the other hand there is strength, so the strategy applied is a diversification strategy (ST). The application of the diversification strategy in optimizing the green open space of Ambon City resulted in three main strategies (Table 4).

3.3 Improve the quality of green open space

The existing condition of Ambon City Green Open Space which does not meet the standard of 30% of the city area (Table 1), has caused various environmental problems, such as increased GHG emissions, floods/landslides, global warming, and loss of biodiversity. One of the efforts that can be made to overcome these problems is by planting trees. Trembesi tree species (Samanea saman), linggua tree (Pterocarpus indicus), and acacia tree (Acacia mangium) have been proven to have high carbon absorption capacity [7], so that the strategy of optimizing green open space through planting these types is the right step.

3.4 Conduct harmonization and socialization to improve supervision of green open space

Ambon City's development and development to fulfill national goals impact green open space availability. The current condition of Ambon City's green open space continues to change its shape into settlements and other supporting infrastructure. Research by [3] states that the conversion of green areas in Ambon City has increased due to the high demand for residential land and the increase in the city's economic activity. To minimize this, the government and the community synergy is needed [11]. The synergy between the government and the community can be done by increasing harmonization and socialization between related institutions such as the Satpol PP and area users.

3.5 Use of green base coefficient

Using the basic green coefficient is a form of government support for managing green open space. According to [10], government support through related regulations will result in a green open space management system free from destruction and conversion of area functions. In its implementation, the Ambon City Regional Government (Pemda) requires using the basic green coefficient to obtain a building permit (IMB). This can be used as the primary strategy for optimizing Ambon City's green open space.

Table 4. Matrix SWOT analysis optimization of green open space Ambon City

Internal / **External**	**Strength (S)** 1) Benefits and functions of green open space 2) Regulations related to the development of green open space 3) Use of green base coefficient	**Weakness (W)** 1) 1) Limited Regional Revenue and Expenditure Budget (APBD) 2) Poor arrangement 3) Overlapping of green open space management
Opportunity (O) 1) Green Community (environmental care community) 2) Development of Green Open Space on undeveloped land outside the city center 3) Private sector involvement in funding for green open space development	**Strategy (SO)** 1) Optimizing regulations to support sustainable green open space management 2) Planning the development of green open space on undeveloped land outside the city center 3) involving green communities, communities, and other parties in the development of Green Open Space	**Strategy (WO)** 1) Involving regional banks, private banks, and other parties in improving the management of Ambon City's green open space through nursery and pot-making activities 2) Improving the management of green open space in Ambon City so that the Ambon City green open space development program can run well

Threat (T)	Strategy (ST)	Strategy (WT)
1) The intensity of supervision on the existence of Green Open Space is still low 2) Acceleration of infrastructure development and public service facilities 3) There is no technical regulation on the development of green open space	1) It is improving Green Open Space's quality by planting plants with the potential for high GHG emission absorption, such as Samanea saman, Pterocarpus indicus, and Acacia mangium. 2) Harmonization and socialization with other institutions such as the Ambon City Satpol PP and green open space users to increase supervision of green open spaces 3) Using the basic green coefficient as the primary condition for obtaining a Building Permit (IMB)	1) Increase collaboration between government, private, and community in planning green open spaces such as city parks so that their existence is more valuable and avoids destruction and conversion of functions

4. Conclusion

Based on the SWOT analysis results, three main strategies were formulated to optimize the green open space of Ambon City. The main strategies are: (1). Improving green open space quality by planting plants with high GHG emissions absorption potentials, such as *Samanea saman*, Pterocarpus indicus, and Acacia mangium; (2). Conduct harmonization and socialization with other institutions such as the Ambon City Satpol PP and green open space users to increase the supervision of green open space; (3). The basic green coefficient is the primary condition for obtaining a Building Permit (IMB).

References

[1] Harahap F R 2013 *Jurnal Society* I **1**
[2] BPS 2018 Kota Ambon dalam Angka Kota Ambon Badan Pusat Statistik Kota Ambon
[3] Metekohy E F, Mononimbar W, Tarore C R 2016 *Spasial Perencanaan Wilayah Kota* **3** 1
[4] Dahlan E N 2008 *Media Konservasi* **13** 2
[5] Peace 2007 *Indonesia dan Perubahan Iklim Status Terkini dan Kebijakannya* (Jakarta : PT Pelangi Energi Abadi Citra Enviro)
[6] Subair, Kolopaking L M, Adiwibowo S, Pranowo M B 2014 *Jurnal Komunitas* **6** 57
[7] Selanno F M, Purwanto R H, Santoso P 2021 *Journal of Tropical Biology* **9** 3
[8] Tim Swakelola P2KH Kota Ambon 2013 *Laporan Penyususnan Masterplan Ruang Terbuka Hijau Perkotaan Kota Ambon* (Ambon: Tim Swakelola P2KH Kota Ambon)
[9] Rangkuti F 2015 *Analisis SWOT Teknik Membedah Kasus Bisnis* (Jakarta : Gramedia Pustaka Utama)
[10] Astriani N 2014 *Fiat Justisia Jurnal Ilmu Hukum* **8** 2
[11] Hamrun and Prianto A L 2017 *Kebijakan Pengelolaan Ruang Terbuka Hijau di Kota Makassar*

Assessing the local farmer's perspective on Tapanuli Orangutan in Agroforest Landscape of Batang Toru Ecosystem, Indonesia.

Hamid Arrum Harahap[1], Yonariza[2], Endrizal Ridwan[3]
[1,2]Faculty of Agriculture, Universitas Andalas, Limau Manis, Padang, Indonesia
[3] Faculty of Economy, Universitas Andalas,Limau Manis, Padang, Indonesia
*Corresponding Author: yonariza@gmail.com

Abstract. Crop raiding by great apes presents a challenge to conservation due to enhancing human and wildlife conflict, and it is one of the main threats to the existence of newly identified species of Tapanuli Orangutan (Pongo Tapanuliensis). Our study aims to understand communities' socioeconomics, farmers' knowledge, perception, attitude, action against Tapanuli Orangutan. Nearly 33% of our respondents showed negative action against Tapanuli Orangutan. Farmer's claimed that they drive the Orangutan out of their agroforests using tools such as air rifles, sharp tools, as well as stones and sticks. Farmer's knowledge of Tapanuli Orangutan as a forest engineer is very low. Seventy three per cent of the respondent did not know any ecosystem services provided by Orangutans. We found out that cultural services of orangutans are the most understandable services by local farmers (14%), followed by regulation services (10%) and provisioning services as they consume the bushmeat of Orangutan (3%). Simultaneously, we found that knowledge, perceptions and attitude affect farmer actions by 62.2%. Understanding the farmer's action against the Tapanuli Orangutan will provide vital information for the stakeholders. We suggested further research to determine farmers' willingness to accept compensation for their crop-raiding by an Orangutan and the resolution conflict scheme of humanorangutan conflict. Using the economic valuation of orangutans, some strategic approaches may be taken to solve the conflict.

1. Introduction

Since the Tapanuli Orangutan *(Pongo tapanuliensis)* was discovered as new species separated from the Sumatran Orangutan in 2017 (1), the world is racing with time to save this most critically endangered great apes from diverse threats to their population and their habitat. Tapanuli Orangutan was scientifically reported in the last two decades (2), and it was still considered part of the Sumatran Orangutan population in the southern part of Lake Toba. Like other great ape species, the Tapanuli Orangutan is also on the verge of extinction. The cause of the decreasing population of orangutans is habitat loss, habitat degradation, habitat fragmentation, hunting and captivity (3). Twenty years later, (4) specifically mention the causes of threat to the Tapanuli Orangutan sustainability are habitat loss, hunting and conflict killing, extractive industries such as gold mining and power generation from geothermal and hydro energy. The growing threat of the Tapanuli Orangutan population in recent decades has resulted in a smaller living space for the remaining Tapanuli Orangutan. Even though the Tapanuli orangutan is the late great ape discovered in the 21st century, threats to its small population could make them the first great ape to become extinct (5).

The remaining Tapanuli orangutans live in three subpopulations in the fragmented forests of the Batang Toru ecosystem. Based on the decree and forest area designation No.579/Menhut-II / 2014 issued by the Ministry of Environment and Forestry, the Batang Toru area is mostly protected forest, followed by nature reserves and production forest. The rest consists of areas for other purposes (APL), covering 36.8%, and water bodies, covering 0.2%. The Batang Toru landscape spans 249,169 hectares, but the remaining orangutan habitat counts only around 132,040 ha thereof, according to the land cover analysis. It is no wonder that the orangutans have a high encounter in the human settlement such as in agroforest.

1.1 Orangutan and Ecosystem Services

People (stakeholders) receive and manage ecosystem services in a range of forms, rates, magnitudes, and sizes (6). Given the interdependence of natural resources, integrated and coordinated efforts across stakeholders will be necessary to plan and manage forests through a payment mechanism for biodiversity conservation (7), including Orangutan protection. Orangutan *(Pongo spp.)* plays an important role as a forest ecosystem engineer, and by diverse ecosystem services they provide. In 1994, scientist introduced the concept of an ecosystem engineer where organisms participate in changing or supporting an ecosystem's physical (8). The primate is considered an ecosystem engineer (9). Orangutan plays an important role in supporting and maintaining the ecosystem's sustainability by turning the plant they eat into new seed in the forest structure (10–13) Not only dispersing the seed, but the Orangutan also maintains the forest's health by germinating the seed (11,14). The success of germination increases and shortens the time it takes for up to 50% to reach germination when the seeds were swallowed earlier by orangutans (10).

Besides helping the sustainability of ecology, Orangutans play a vital role through cultural ecosystem services such as recreational value and traditional knowledge; in recreational value, for example, Zander et al. (15) estimated that the voluntary contribution of visitors to Orangutan conservation is around USD 6.6 million per year in Sarawak. In addition, these activities also generate between USD 13 and USD 23 million into the local economy annually through their spending on local businesses. Another economic valuation for Orangutan ecotourism through a willingness to pay was conducted in Bukit Merah Orangutan Island. Mohd-Asri et al. (16) found that the visitor's WTP was around USD 13, determined through the bidding process. Using cost benefit analysis, Fauzi et al. (17) conducted economic valuation of orangutan tourism in Bukit Lawang, Sumatra, Indonesia. The economic valuation obtained is around USD 513,308, where the profit earned from orangutan tourism is around USD 119,788 per year. Orangutans play an important role in the ecosystem, not only supporting the environment and sustainable ecosystems but also supporting local communities and the national economy.

Apart from the economic value of Orangutans have from recreation, the Dayak people in Borneo Island have a wide range of traditional knowledge about orangutans. They believe that orangutans are reincarnations of respected community members like their ancestors (18). Although orangutans are revered in the Iban Dayak community's traditions, the capitalization and increase in illicit logging, as well as habitat damage due to conversion to monoculture plantations such as oil palm and rubber, have decimated the population in recent decades (19). But what would happen if local people do not have or lack value traditional knowledge or taboos to an orangutan, and at the same time, capitalism is also active in the site? What would be suitable conservation strategies? The paper will explore the case of the Tapanuli orangutan.

1.2 Human-Orangutan Interaction in Batang Toru Ecosystem, Tapanuli, North Sumatra.

Despite the services provided by an orangutan, human-orangutan conflict is still a challenge in ape conservation, including the Batang Toru ecosystem, the settlement among the Tapanuli people. Meijaard et al. (4) recently claimed that in 2020, two adult females were killed in the first six months. This continues, with two baby Tapanuli orangutans caught from the wild, reported so far in 2020, with one of them confiscated from their owners and the other being released illegally to evade the law repercussions. In 2019, a male orangutan was seriously injured by humans (20). Another male orangutan was caught and removed twice due to complaints about crops with the local community (21). In 2013, an adult male was killed after being injured for several days around the resident's plantations (22).

Human-orangutan conflicts can be purely human-driven when the animals are hunted for economic value. Or, broader wildlife-driven conflicts can arise, for example, when orangutans eat agricultural commodities such as fruits or crops grown by locals. When aiming to mitigate human-orangutan conflicts, understanding the different dimensions of such conflicts is crucial to reducing conflict between humans and orangutans (23–25) Even a few deaths in a

IOP Conf. Series: Earth and Environmental Science 1115 (2022) 012021 doi:10.1088/1755-1315/1115/1/012021

year can impact the small population's long-term survival, and this is due to the slow birth rate despite its lengthy lifetime. Hence, a low extrinsic death rate (i.e., >1% every year) poses a major threat to the Tapanuli orangutan's small population (26). Together with this, conflict with local communities represents the major threat to the long-term growth, stability, and persistence of the Tapanuli orangutan population. Therefore, it is necessary to analyze local and socio-culturally specific information in Batang Toru communities (4) and the attitudes and social behaviour towards orangutans. This state urgently needs attention to ensure the survival of the only existing Tapanuli orangutans. It is very important to understand the drivers of farmer action against the Orangutan. Our study focuses on the socio-economic of society and explores farmers' knowledge of Orangutan services, their perceptions and attitude against Orangutans. Furthermore, we use these findings to formulate recommendations and the conservation strategy of Tapanuli Orangutan.

2. Material and Methods.

2.1 Study Area

Our research focuses on local farmers in the Batang Toru Ecosystem (BTE). The BTE is a fragmented forest complex in Sumatra, Indonesia (See Figure 1). Administratively, the BTE spread across three regencies in North Sumatra Province, these are South Tapanuli, Central Tapanuli and North Tapanuli Regencies. This ecosystem consists of lowland forest to upland forest, where most of the forest is at an altitude of 800 m asl to 1781 m asl. These fragmented forests are divided into three main areas, west block, east block and nature reserve. Batang Toru ecosystem consists of several statuses such as protection and conservation forest, nature Reserve, production forest and other uses area. Around 27,906 households live around the Batang Toru Ecosystem (27). They depend on the sustainability of the forest and ecosystem services for their livelihoods. The main local livelihoods depend on water irrigating rice fields and also agroforestry for various crops such as Durian (*Durio zibethinus*), Petai (*Parkia speciosa*), and Salak *(Salacca zalacca)*.

Figure 1 Batang Toru Ecosystem Map.

2.2 Research Design

We use an explorative research design, and we employ a survey technique. Our questionnaire is a close-ended one to explore the knowledge, perceptions and attitudes of local farmers towards the Tapanuli Orangutan. Farmers' knowledge about the ecosystem services provided by Orangutans with the option yes and no type of questions; 'I know' and 'I do not know'. Furthermore, our interviews are conducted with open questions regarding the benefits that people feel obtained from the presence of orangutans. Local farmers' perceptions of orangutans also use the options: 'beneficial' and 'detrimental'. The attitude of local farmers regarding orangutans using the options: 'pleased' and 'displeased'. The action of local farmers when meeting orangutans directly in their fields were obtained using the 'do nothing' and 'expelled with tools' option. We also conduct open-ended questions to determine how the community expels orangutans and the community's views on the existence of orangutans in the Batang Toru Ecosystem. Furthermore, we collected information about community demographics, socio-economy, ownership of durian trees (the preferred crop of orangutans), perceptions of orangutans as pests and also their belief in the spirit animal.

2.3 Sampling technique

We interviewed 150 respondents, mostly heads of households, during December 2020. The respondents spread across six villages on the edge of the Batang Toru forest. The villages were selected based on two categories, namely agroforestry and interaction with the Tapanuli orangutan. These six villages were chosen to represent the demographic and administrative diversity of the Batang Toru Ecosystem. The ethnic of respondents consisted of the Batak Toba, Batak Angkola and Nias. The religions practised by the respondents are Islam, Protestant Christianity and Catholic Christianity. The selected villages represent three blocks and three districts in the Batang Toru Ecosystem (See Table.1).

Before the survey, we secured fieldwork permits from the local government and visiting respondents who were selected through the permission of the village head. The sample was determined using a population list in the village database by determining the 30 selected people. We explain the purpose of the interview to each respondent. We started the interview after getting permission from the respondents. Either at home or in their fields when they are resting from work. After that, the interview was conducted by reading the questions that had been compiled. Interviews usually last about 30-45 minutes.

Table 1 Distribution of samples by district, Villages and forest block.

No	District	Village	Forest Block	Number of respondents
1	North Tapanuli	Lobu Pining	East Block	30
2	North Tapanuli	Sidua Bahal	West Block	15
3	North Tapanuli	Sitolu Bahal	West Block	15
4	Central Tapanuli	Huta gurgur	West Block	30
5	South Tapanuli	Simaninggir	West Block	30
6	South Tapanuli	Aek Nabara	West Block, Sibual-buali Nature reserve	30

2.4 Data Analysis
We used statistical analysis with IBM SPSS Statistics 26. In evaluating the influence of knowledge, perception and Attitude against Tapanuli Orangutan, descriptive analysis, correlation, partial t-test and simultaneous test were used. To ensure correlation and interaction between variables, ANOVA was used. We tested our independent variable, such as farmer knowledge, perception and attitude, and their influence on the dependent variable is farmer's behaviour against Orangutan. We tested socio-economic variables such as owning a durian tree, education and belief in spiritual animals on their correlation with farmer's knowledge, perception, attitude and action toward Orangutan. We believe that the socioeconomics variables will affect farmers behaviour against orangutans, and we believe that farmer's knowledge, perception, and attitude significantly influence the farmer's behaviour against orangutans, as previously studied by scientists (24,28–30).

3. Result and Discussion

3.1 Farmer Demographics
Our study found that most respondents were male (n = 99; 97%). Protestant Christianity (n = 44, 45%) was the religion most practiced by respondents and followed by Muslim (n = 44, 45%), while the least number of respondents was Catholic (n = 10, 10%). For ethnicity, most respondents in thisstudy were Batak ethnic (n = 85, 83.3%), and the rest were ethnic Nias (n = 17, 16.7%). The age range of our respondents varied where the most were 41-50 years old (n = 34, 34%) and 50-60 years(n = 23, 24%). Only a few of our respondents claimed to be highly educated at university (n = 3, 3%). The majority of respondents admitted that their level of education is a primary school and junior high school (n = 59, 58%). 28.7% (n=43) of respondents admitted to having attended forestrytraining, and the remaining 71.3% (n=107) said they had never attended forestry training at all. A total of 50.6% of farmers in this study admitted owning durian trees (n = 76). Regarding hunting tools, 75% (n = 76) of respondents stated they did not own hunting tools, and 25% (n = 26) stated they had hunting tools. More than 50% of our respondents (n=80) believe that a spirit embodied inanimal bodies such as Orangutan Tapanuli (*Pongo tapanuliensis*), Tiger (*Panthera tigris sumatrae*) and Siamang (*Symphalangus*

syndactylus).

Furthermore, we asked what farmers in durian farming most hated crop raider (animal). Surprisingly we found out that farmers mentioned the Tapanuli Orangutan (n=36). 53% (n=19) stated that they hate orangutans because when orangutans raided the durian, they were forbidden tokill or hurt them. The government protects it. A resentment arises because they thought the government prioritized the Tapanuli Orangutan over the farmer's economy. Our analysis also foundthe crop raiders of durian that farmers hated the next most were the long-tailed macaque (*Macaca fascicularis*) where (n = 23), the Squirrel (*Sciuridae*) where (n = 18), and the Sun bear (*Helarctos malayanus*) where (n = 11).

Table 2 Factors associated with farmers' behaviour toward orangutan.

Correlations					
		Knowledge	Perception	Attitude	Action
Owning durian tree	Pearson Correlation	-.376**	-.329**	-.156	-.334**
	Sig. (2-tailed)	.000	.000	.056	.029
	N	150	150	150	150
Education	Pearson Correlation	.318**	.241**	.139	.288**
	Sig. (2-tailed)	.000	.003	.090	.056
	N	150	150	150	150
Believe in spiritual animal	Pearson Correlation	-.196*	-.178*	-.364**	-.384**
	Sig. (2-tailed)	.016	.029	.000	.000
	N	150	150	150	150

**. Correlation is significant at the 0.01 level (2-tailed).
*. Correlation is significant at the 0.05 level (2-tailed).

3.2 Knowledge toward Tapanuli Orangutan
Orangutans provide a variety of services to ecosystems and communities in Tapanuli. Of our total respondents, 27% (n=40) said that they received benefits from the services provided by Orangutans. The role of the Tapanuli Orangutan, according to a local farmer, can be seen in figure 2 below.

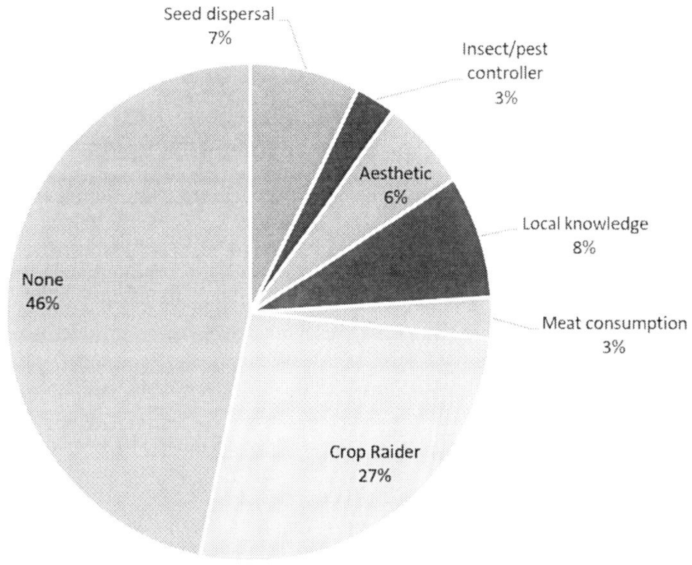

Figure 2 Orangutan services according to a local farmer.

Figure 3 explains that most of the respondents (73%) said Orangutan are crop raider and do not provided services to them, Hence, 27% of them stated that orangutans are detrimental to their pest. Meanwhile, ecosystem services mostly found in local farmers are cultural ecosystem services where orangutans are considered to provide aesthetics and contribution to the local knowledge of thecommunity. The other ecosystem service farmers know about orangutans is regulatory services, namelyseed dispersal and insect/pest controller at 10%. These are the services provided by the Orangutan as a forest engineer. Surprisingly, 3% of respondents considered Orangutans to be instrumental in provisioning services, and they said that they still see Orangutans as bushmeat for consumption if they are found in their agroforest land.

We tested what factors influenced farmers' knowledge of orangutan role as forest engineers (Table 2).These factors are durian tree ownership, education, and belief in spirit animals. Our statistics found thatowning durian trees and education where ($p=0.000$) are correlated with the farmers' knowledge on Orangutan role as forest engineers on the 1% significance level ($p < 0.01$). Meanwhile, belief in animalspirits does not correlate and does not influence farmer's knowledge of Orangutan role as forest engineers ($p=0.016$)

3.3. Perception on Tapanuli Orangutan

Our statistical test results show that only 10% (n=15) of farmers perceived that Orangutans benefit themthrough their ecosystem services. The remainder, 90% (n=135), perceived that Orangutans are detrimental to them—this percentage indicating that lack of knowledge affects the wrong perception ofmost of the farmers. We tested factors that influenced farmers' perception of orangutans (see Table.2).All three factors (owning durian tree, education, and belief in spiritual animals) only owning durian treecorrelate with farmer's perception ($p=0.000$). We found out that farmer's who own durian tree tends toperceive negative or detrimental perception toward Orangutan.

3.4 Attitude against Tapanuli Orangutan

Most farmers are negatively affected by Tapanuli orangutan, where the farmers see the Tapanuli orangutan as an unpleasant creature. Our results show that 73% of farmers admit that they are unpleased to see the Tapanuli Orangutan. There are various reasons for their displeasure, such as crop-raiding by the Tapanuli orangutan, detrimental to farmers. We conducted statistical tests on what factors influenced farmers' attitudes towards Tapanuli orangutans. Of all the tested factors, only belief in spiritual animals correlated with the farmer's attitude towards the Tapanuli orangutan (p = 0.000). We found that farmers who believe in spiritual animals tend to show a positive attitude towards the Tapanuliorangutan.

3.5 Action against Tapanuli Orangutan

We studied farmers' actions when meeting orangutans on their agroforest lands. The question we ask is, "how do you do when you see the Tapanuli Orangutan in your agroforest?". This is a closed-ended question with an option to answer: 'Do nothing' and 'Expelled with tools'. The results of our study foundthat as many as 32% of farmers (n = 48) answered that they expelled Orangutans with tools. We continued with open-ended questions about what tools are used to evict Orangutans. The most commonanswers were air rifles, machetes, wood and stones. We conducted statistical tests to find out what factors correlated with the actions of farmers on the Orangutan Tapanuli. From the three factors we tested, farmers who believed in spiritual animals tended to do nothing or leave Tapanuli orangutans alone when they were in their agroforest land (p = 0.000). Meanwhile, farmers who owned durian (p =0.026) and education (p = 0.059) had a negative correlation with the actions of farmers on the Tapanuliorangutan where (p <0.01).

We conducted other statistical tests to determine what variables influence farmers' actions on orangutans. Hence, we conducted a partial t-test with three independent variables (X), namely farmers' knowledge, farmers' perceptions and farmers' attitudes towards the dependent variable (Y), the actionsof farmers towards the Orangutan Tapanuli. The results of our statistical analysis for the partial t-test can be seen in Table 3 below:

Table 3 Partial t-test.

Coefficients[a]

Model	Unstandardized Coefficients		Standardized Coefficients	t	Sig.	Collinearity Statistics	
	B	Std. Error	Beta			Tolerance	VIF
(Constant)	.029	.098		.294	.770		
Knowledge	.478	.107	.341	4.469	.000	.444	.250
Perception	.063	.122	-.041	-.521	.603	.423	.362
Attitude	.645	.060	.617	10.794	.000	.793	.261

a. Dependent Variable: Action against Orangutan

The results of our partial t-test show that of the three variables tested, two of them have a positive and significant effect on farmers' actions against Orangutans. These two variables are knowledge and attitude, where the significance value is <0.05. In the knowledge variable, it was obtained a significancevalue of 0.000 <0.05. This value indicates that the hypothesis is accepted, which means that knowledgehas a positive and significant effect on farmers' actions against Orangutans. Our research shows that farmers who are aware of the ecosystem services that Orangutans play as forest engineers tend to havepositive actions against the Tapanuli Orangutan than farmers who do not know the role of forest engineers.

Whereas in variable perception, a significance value of 0.603 was obtained. Because the significance value is 0.603> 0.05 probability, the hypothesis is rejected, meaning perception does not have a positiveand significant effect on farmers' actions against orangutans. This study indicates that the perceptions given by farmers to orangutans, whether the farmers perceive orangutans to be beneficial or detrimental,did not have a positive and significant effect on their actions.

The attitude variable has a significant value smaller than the probability where the significance value is 0.000 <0.05. Therefore, the alternative hypothesis is accepted, where attitude significantly affects farmers' actions against Orangutans. Our statistical results show that farmers with a pleasing attitude towards Orangutans tend to do nothing when meeting Orangutans on their durian fields. On the other hand, farmers who are unpleased when they see orangutans tend to evict orangutans with tools on theirdurian fields. Therefore, an attitude has a positive and significant effect on farmers' actions against orangutans.

Table 4 F test.

ANOVA[a]						
Model		Sum of Squares	df	Mean Square	F	Sig.
1	Regression	20.303	3	6.768	80.094	.000[b]
	Residual	12.337	146	.084		
	Total	32.640	149			

a. Dependent Variable: Action against Orangutan

b. Predictors: (Constant), Knowledge, perception, attitude

To find out the relationship of all independent variables to the dependent variable, we performed a simultaneous F test (see table 4). We tested whether the variables of knowledge, perception, and attitudesignificantly affected the dependent variable of farmer action against orangutans simultaneously. The results of our study indicate that the independent variables (knowledge, perception and attitude) have a significant influence simultaneously on the dependent variable (action against Orangutans). Our F tableresult is 2.67. Ftable = k (3); n-k (147). We found out that F counts 80.094> F table 2.67. The hypothesis is accepted where the independent variables significantly affect the dependent variable. Therefore, knowledge, perception and attitude simultaneously influence the actions of farmers against orangutans.

Table 5 Rsquare test.

Model Summary[b]					
Model	R	R Square	Adjusted R Square	Std. Error of the Estimate	Durbin-Watson
1	789[a]	.622	.614	.291	1.635

a. Predictors: (Constant), Knowledge, Perception, Attitude

b. Dependent Variable: Action against Orangutan

The next statistical test we ran was the Rsquare test to quantify the effect of the independent variables (knowledge, perception and attitudes) on the dependent variable (action against Orangutans). Our statistical results found that the Rsquare value was 62.2% which means the independent variable influenced the dependent variable by 62.2%. This percentage implied that 37.8% of the variables otherthan knowledge, perception and attitude were not included in this study which affected the dependent variable (action against Orangutan).

Deforestation and changes in land use have led orangutans to enter agroforests and agricultural lands and raid the crops. Meijaard et al. (24) conducted extensive research on human-orangutan conflicts driven by deforestation. Crop raiding by orangutans has been observed throughout Kalimantan,although it appears to be more serious in some places in central East Kalimantan, eastern and western Central Kalimantan, and southern West Kalimantan. Orangutans that wandered into respondents' gardens elicited a wide range of responses.

Nearly half of credible respondents who had seen an orangutan near the community and encountered conflict reacted by attempting to chase or scare it away, while 5% killed or attempted to kill it. An annual report from YOSL-OIC (31), an organization with human-orangutan conflict response in Leuser and Batang Toru ecosystem, reported that they have already rescued 24 orangutans from 2016-2017. In the West-South Coast of Aceh, the human-orangutan conflict has increased the number of orangutans rescued. Sumatran orangutans in the Leuser ecosystem have experienced a significant increase in confiscation and evacuation cases during 2012-2016, where there is extensive land clearingtaking place for oil palm plantation in Orangutan habitat (31).

There is no prior study about crop raiding of Tapanuli Orangutan in agroforest. Over the past few years,conflicts against the Tapanuli Orangutan have persisted even though they are the most endangered apespecies globally. Our research found that farmers are fully aware of the Tapanuli Orangutan as a protected species by the government. However, the public's knowledge of orangutan ecosystem servicesas a whole is still low, especially Orangutan services as forest engineers. A small number of respondentsadmitted that they consider orangutans as animals with powers and are considered haunted. If they disturb or harm the orangutans, they are afraid of bad luck. We found that respondents who adhere to this taboo tend to have a more friendly attitude towards orangutans when these creatures enter their agroforest lands. However, we only found this in respondents from South Tapanuli, and we did not findthe same taboos applied in the Central and North Tapanuli.

We also noticed that not everyone held on to this taboo, especially when the orangutan encountered their agroforest land. Whether the farmers will drop the orangutan away with tools or harm them to outof the agroforest land, the taboo is no longer enforced. The taboos with an orangutan in South Tapanulihas similarity in the Sentarum Lake, Kalimantan. The Dayak people in this area regard the orangutan as an ancestor, and it is forbidden to harm or kill one. Certain Dayak communities also think that orangutans taught their ancestors' important lessons, such as giving birth and using medicinal herbs (18). The Iban Dayak people, for example, have a unique story regarding orangutans;their great-grandfathers informed them that orangutans aided them in driving away adversaries duringthe ethnic battles of the 1900s, this leaving a positive memory on the orangutan. Nevertheless, we understand that the Batak people in Tapanuli only saw the orangutan as haunted animals rather than ancestors like the Dayak people. Because of that, the approach to conserving Tapanuli orangutan mustbe different from that in Kalimantan.

Understanding the human-orangutan conflict in Tapanuli cannot be mere a crop-raiding issue, and changing farmers' behaviour is never be enough, and it wouldn't be addressed the roots of the problem. It must be viewed with a bigger lens, as claimed by Wich et al. (32) where thefrequency of killing and conflict increases might be linked with deforestation and land conversion. Tangible cooperation and conservation effort from various parties is a must. The role of government and industry, and local farmers is very vital if we want to see the Tapanuli Orangutan population survivefor the long term. Companies must have systematic conflict mitigation and responsible practices such as building biological corridors to connect and reconnect the Batang Toru forest blocks and agroforests (5,33).

The two biggest groups of farmers who see Orangutans are farmers who do not know the services of Orangutans at all and farmers who consider Orangutans as crop raiders. Our results support the study by Hocking & Humle (34) stated that when orangutans eat fruit trees or plants in plantations/agriculture around the forest, they frequently get shot or killed, and surviving babies end upbeing trafficked or in a cage. Especially in Batang Toru Ecosystem, the remaining Tapanuli orangutans are threatened by conflict, killing and hunting, and lowland habitat loss (35). When farmers see orangutans in the forest, they are fine with it and do not disturb them. Nevertheless, when farmers see orangutans in their gardens, they will drive them away immediately, making them uncomfortable using tools/air guns. Although Orangutans are may not the primates that cause the mostharm to agriculture, they are one of the most annoying primates for durian farmers. Every orangutan raids the durian, the farmer, will decide whether to leave them or expel them to injure them. And if theychoose the latter option, they will feel bad because the government protects orangutans.

Another resentment also arises because they think the government cares much more about and protectsorangutans than the community's livelihoods. For example, in Aek Nabara, one of the village leaders said that when orangutans eat durian, the farmer suffers losses, and there is no solution from the government. Conversely, when orangutans eat durian and farmers injure them, the government responds. If the conflict continues and there is no solution, farmers and orangutans will be disadvantaged together. Farmers will lose their profits economically, while Orangutans will be pushedtowards extinction. Abram et al. (36) has addressed an extensive human-orangutan killing and conflicts in Borneo, where many populations are at extinction risk locally. Borneo Orangutan populations are expected to go extinct in several regions of Kalimantan over the next decade. The fate of the Tapanuli Orangutan population living with less than 800 individuals is no safer than Borneo's counterpart.

The urgency to save the last population of the Tapanuli Orangutan is even more urgent, considering theremaining population is getting smaller. However, orangutan conflicts with local farmers still occur inthe Batang Toru ecosystem. If there is no change in farmer actions, the conflict will disrupt the population growth of the Tapanuli Orangutan (37). Our study provides an overviewof how to understand conflict with orangutans through the knowledge, perceptions and attitudes of localfarmers. Local farmers still consider Orangutans crop raiders as pests for durian plantations rather thanforest engineers.

Several approaches have been discussed and carried out to prevent and mitigate the human-orangutan conflict. Campbell-Smith et al. (25) advise that easy-to-install netting for lucrative tree crops is developed and that farmers' perceptions of orangutans as a harmful species must be changed to preventconflict. The United States Agency for International Development (USAID) issued an article that summarized a field report on the human-orangutan conflict in Sumatra. It recommended that Human- orangutan conflict mitigation strategies include involving local figures and community champions in campaigns to motivate people to help, incorporating human-wildlife conflict mitigation into village planning priorities, conducting population monitoring, and integrating conflict mitigation with improved forest management (31).

Though our findings enrich knowledge in understanding local farmers' perspectives regarding the presence of a new species of Orangutan in agroforests of Tapanuli, similar attention has been addressedpreviously to other orangutan species in understanding human-wildlife conflict in anthropocentric landscapes. For example, the local economy from agriculture, including agroforests, is one of the indicators in evaluating the sustainability of the Bornean Orangutan habitat and its challenges Siregar et al. (19). Human-orangutan conflict driven by crop-raiding requires a peaceful mechanism to resolve it (38,39). The peaceful mechanism means not only saving orangutans from conflict but also saving the durian harvest from people whom orangutanshave raided.

Therefore, we noticed the importance of mitigation schemes/approaches that benefit local people. We recommend that stakeholders understand appropriate steps to mitigate orangutan-human conflicts. It isimperative to link the ecosystem services that orangutans provide to the environment in general and human life in particular. It is necessary to change the farmer's perspective towards orangutans, no longeras crop raiders or pests. As a result, the risk of human–orangutan violence is reduced, and farmers' financial security is improved. Incentivising the local farmer's on their durian loss might be a good way to reduce conflict with Orangutans in the agroforest landscape, as Siregar et al. (19) suggest to initiate payment for environmental services or other incentive mechanisms to conserve and protect the remaining forests and the Orangutan's habitat. Special attention needs to be paid to the human-orangutan conflict resolution scheme through incentivesfor durian loss for farmers, which must be included in the government's Orangutan conservation actionand plan strategy. It is necessary to understand local community perspectives and see the present and future opportunities in placing local communities at the core of conservation and acknowledging their role (38).

IOP Conf. Series: Earth and Environmental Science 1115 (2022) 012021 doi:10.1088/1755-1315/1115/1/012021

4. Conclusion

In ensuring the long-term survival of the Tapanuli Orangutan, it is necessary to minimize or eliminate conflicts between Orangutans and humans, especially local farmers. Conflicts between the Tapanuli Orangutan and local farmers still exist in the Batang Toru Ecosystem. Farmers consider orangutans to be a disadvantage as a crop raider rather than as forest engineers. Farmers still lacked knowledge of theservices and roles of the Tapanuli orangutan. Besides that, farmers were also far more likely to perceivethe Tapanuli orangutan as detrimental, and they had an unpleasant attitude towards the Tapanuli. Farmers who own durian trees tend to know the presence of orangutans as crop raiders, where many of them perceive that orangutans are detrimental to their durian trees. The education factor also correlateswith the durian tree ownership factor related to farmers' knowledge and perceptions of orangutans. However, farmers who still hold spiritual values and believe in taboos on orangutans tend to have positive attitudes and friendly actions towards orangutans in agroforest lands.

Adequate payment for ecosystem services mechanism might likely fit with orangutan conservation andeconomic empowerment by giving an incentive for local farmer's loss on durian eaten by the Orangutan. Hence, there is a cost when excluding the local community from conservation and conflict resolution. Nevertheless, agroforestry has been practised in Batang Toru for hundreds of years, and it will likely last and continue since agroforests as an important source of income.Peaceful co-existence in the agroforest landscape between local farmer's and Orangutan are urgently needed. More social science research is needed to advise policymakers better and concentrate conservation efforts. Incorporating economic mechanisms would lead to improved long-term conservation goals forcritically endangered species such as Tapanuli Orangutan as the local farmer's activities in theagroforest landscape continue.

Reference

1. Nater A, Mattle-Greminger MP, Nurcahyo A, Nowak MG, de Manuel M, Desai T, et al. Morphometric, Behavioral, and Genomic Evidence for a New Orangutan Species. Current Biology. 2017;27(22):3487-3498.e10.
2. Meijaard E. A Survey of some Forested Areas in South and Central Tapanuli, North Sumatra; new Chances for Orang Utan Conservation. 1997.
3. Rijksen HD, Meijaard E. Our Vanishing Relative. The status of wild orang-utans at the close of the twentieth century. A Tropenbos Publication; 1999.
4. Meijaard E, Ni'matullah S, Dennis R, Sherman J, Onrizal, Wich SA. The historical range and drivers of decline of the Tapanuli orangutan. PLoS One [Internet]. 2021;16(1 January):1–23. Available from: http://dx.doi.org/10.1371/journal.pone.0238087
5. Wich SA, Fredriksson G, Usher G, Kühl HS, Nowak MG. The Tapanuli orangutan: Status, threats, and steps for improved conservation. Conserv Sci Pract. 2019;1(6):e33.
6. Maynard S. A methodology for ecosystem services assessments across multiple scales. 2016;(October):1–371. Available from: http://hdl.handle.net/1885/111380
7. Yonariza, Andini BA, Mahdi, Maynard S. Addressing knowledge gaps between stakeholders in payments for watershed services: Case of Koto Panjang hydropower plant catchment area, Sumatra, Indonesia. Ecosyst Serv [Internet]. 2019;39(April):100995. Available from: https://doi.org/10.1016/j.ecoser.2019.100995
8. Jones CG, Lawton JH, Shachak M. Organisms as ecosystem engineers. Oikos. 1994;69(3):373–86.
9. Chapman CA, Bonnell TR, Gogarten JF, Lambert JE, Omeja PA, Twinomugisha D, et al. Are Primates Ecosystem Engineers? Int J Primatol. 2013;34(1):1–14.
10. Tarszisz E, Tomlinson S, Harrison ME, Morrogh-Bernard HC, Munn AJ. Corrigendum: Gardeners of the forest: Effects of seed handling and ingestion by orangutans on germination success of peat forest plants [Biological Journal of the Linnean Society, 123, 1, (2018), 125-134]doi 10.1093/biolinnean/blx133. Biological Journal of the Linnean Society. 2018;124(2):278–278.
11. Nielsen NH, Jacobsen MW, Graham LLLB, Morrogh-Bernard HC, D'Arcy LJ, Harrison ME. Successful germination of seeds following passage through orang-utan guts. J Trop Ecol. 2011;27(4):433–5.
12. Tarszisz E. The ecophysiology of seed dispersal by Orangutans in Bornean peat swamp forest. 2016;275.
13. Tarszisz E, Tomlinson S, Harrison ME, Morrogh-Bernard HC, Munn AJ. An ecophysiologically informed model of seed dispersal by orangutans: Linking animal movement with gut passage across time and space. Conserv Physiol. 2018;6(1):1–15.
14. Ancrenaz M. Seed spitting and seed swallowing by wild orangutans (Pongo pygmaeus. 2014;(June).
15. Zander KK, Pang ST, Jinam C, Tuen AA, Garnett ST. Wild and valuable? Tourist values for orang-utan conservation in Sarawak. Conservation and Society. 2014;12(1):27–42.
16. Mohd-Asri NF, Kamaluddin SN, Dharmalingam S, Idris WMR, Md-Zain BM. Valuing ecotourism in Bukit Merah Orang Utan island, Malaysia based on visitors' experience. Biodiversitas. 2021;22(3):1543–9.
17. Fauzi A, Kusmana C, Santoso N. Economic valuation of the orangutan conservation tourism in Bukit Lawang Langkat District , North. 2019;16(4):8–19.
18. Yuliani EL, Adnan H, Achdiawan R, Bakara D, Heri V, Sammy J, et al. The roles of traditional knowledge systems in orang-utan Pongo spp. and forest conservation: A case study of Danau Sentarum, West Kalimantan, Indonesia. Oryx. 2018;52(1):156–65.
19. Siregar PG, Supriatna J, Koestoer RH, Harmantyo D. Borneo Orangutan Pongo pygmaeus pygmaeus habitat sustainability and its challenges. IOP Conf Ser Earth Environ Sci. 2020;481(1).
20. Mongabay. Terjebak di Kebun Warga, Orangutan Tapanuli Ditemukan Luka-luka. 2019 Sep 28;Mongabay.co.id. Available from: https://www.mongabay.co.id/2019/09/28/terjebak-di-kebun-warga-orangutan-tapanuli-ditemukan-luka-luka/

21. MoEF. Penanganan Konflik Orangutan di Adiankoting Sumatera Utara [Internet]. 2020. p. Kementerian Lingkungan Hidup dan Kehutanan. Available from: http://ksdae.menlhk.go.id/info/7489/penanganan-konflik-orangutan-di-adiankoting-sumatera-utara.html
22. Mongabay. Dijerat Pemburu di Tapsel, Orangutan Penuh Luka Berbelatung. Mongabay.co.id [Internet]. 2013 Nov 22; Available from: https://www.mongabay.co.id/2013/11/22/dijerat-pemburu-di-tapsel-orangutan-penuh-luka-berbelatung/
23. Davis JT, Mengersen K, Abram NK, Ancrenaz M, Wells JA, Meijaard E. It's Not Just Conflict That Motivates Killing of Orangutans. PLoS One. 2013;8(10).
24. Meijaard E, Buchori D, Hadiprakarsa Y, Utami-Atmoko SS, Nurcahyo A, Tjiu A, et al. Quantifying killing of orangutans and human orangutan conflict in Kalimantan, Indonesia. PLoS One. 2011;6(11):1–10.
25. Campbell-Smith G, Sembiring R, Linkie M. Evaluating the effectiveness of human-orangutan conflict mitigation strategies in Sumatra. Journal of Applied Ecology. 2012;49(2):367–75.
26. Marshall AJ, Lacy R, Ancrenaz M, Byers O, Husson SJ, Leighton M, et al. Orangutan population biology, life history, and conservation: Perspectives from population viability analysis models. Orangutans: Geographic Variation in Behavioral Ecology and Conservation. 2009;311–26.
27. Tarigan J, Roshetko J, Martini E, Ekadinata A. Non-timber forest products as a source of livelihood diversification for local communities in the Batang Toru Orangutan conservation program. 2010;15.
28. Campbell-Smith G, Simanjorang HVP, Leader-Williams N, Linkie M. Local attitudes and perceptions toward crop-raiding by orangutans (Pongo abelii) and other nonhuman primates in Northern Sumatra, Indonesia. Am J Primatol. 2010;72(10):866–76.
29. Hockings KJ, McLennan MR. Problematic Primate Behaviour in Agricultural Landscapes: Chimpanzees as 'Pests' and 'Predators.' 2016;137–56.
30. Wallace GE, Hill CM. Crop Damage by Primates: Quantifying the Key Parameters of Crop-Raiding Events. PLoS One. 2012;7(10).
31. USAID. Story from the Field – On the Front Lines of Human-Orangutan Conflict Mitigation [Internet]. Lestari-Indonesia. 2017 [cited 2022 Sep 22]. Available from: https://www.lestari-indonesia.org/en/story-field-front-lines-human-orangutan-conflict-mitigation/
32. Wich SA, Gaveau D, Abram N, Ancrenaz M, Baccini A, Brend S, et al. Understanding the Impacts of Land-Use Policies on a Threatened Species: Is There a Future for the Bornean Orang-utan? PLoS One. 2012;7(11):1–10.
33. Kuswanda W, Harahap RH, Alikodra HS, Sibarani R. Characteristics of the Tapanuli Orangutan Habitat in the Conflict Area of Batang Toru Landscape, North Sumatra, Indonesia. Forest and Society. 2021;5(April):90–108.
34. Hockings K, Humle T. Best Practice Guidelines for the Prevention and Mitigation of Conflict Between Humans and Great Apes. IUCN; 2009.
35. Nasution A, Perwitasari-Farajallah D, Utami-Atmoko SS. Declining orangutans population in the unprotected forest of batang toru. Trop Life Sci Res. 2018;29(2):77–87.
36. Abram NK, Meijaard E, Wells JA, Ancrenaz M, Pellier AS, Runting RK, et al. Mapping perceptions of species' threats and population trends to inform conservation efforts: The Bornean orangutan case study. Divers Distrib. 2015;21(5):487–99.
37. Kuswanda W, Harahap RH, Alikodra HS, Sibarani R. Characteristics and community perception on mitigation of Human-Tapanuli Orangutans conflict around Dolok Sipirok Natural Reserve, South Tapanuli. IOP Conf Ser Earth Environ Sci. 2021;713(1).
38. Ancrenaz M, Dabek L, O'Neil S. The costs of exclusion: Recognizing a role for local communities in biodiversity conservation. PLoS Biol. 2007;5(11):2443–8.
39. Campbell-Smith G, Campbell-Smith M, Singleton I, Linkie M. Apes in space: Saving an imperilled orangutan population in Sumatra. PLoS One. 2011;6(2).

Species checklist of Orthoptera in the Forest Area with Special Purpose (KHDTK) Aek Nauli, North Sumatra

D A S Siahaan*, Nursal, I R Alfakih, Y Dzawita, F Zahra, M J Harianja and A Hartanto

Department of Biology, Faculty of Mathematics and Natural Sciences, Universitas Sumatera Utara, Medan, North Sumatra 20155, Indonesia

*E-mail: donialdo@usu.ac.id

Abstract. We present the species checklist of Orthoptera in the Forest Area with Special Purpose/ KHDTK Aek Nauli, Simalungun Regency, North Sumatra, Indonesia. The checklist is provided as a basic information on the grasshopper diversity and species assemblage in the forest region. KHDTK Aek Nauli is classified into a tropical montane forest with an elevation of 1,100–1,750 m a.s.l dominated by the pine species and intended as an *ex situ* conservation for Sumatran Elephant (*Elephas maximus sumatranus*). Sampling of grasshoppers was conducted for four consecutive days using two methods namely pit fall trap and sweeping net with a total of 12 sampling points. Approximately of 21 orthopteran species from 97 individuals belonging to suborder Ensifera (12 species) and suborder Caelifera (9 species) were documented. The highest Importance Value Index (IVI) orthopteran species was obtained from *Paratettix curtipennis* (102.92), followed by *Teleogryllus emma* (79.04), *Oxya intricata* (69.67), and etc. All grasshoppers were commonly identified for Asiatic region to worldwide distribution.

1. Introduction

Orthoptera are typical grassland insects that regularly contribute to the regional abundance, biomass, and biodiversity [1]. Grasshoppers may be found in a variety of habitats, including plantations, open vegetation, woodlands, and agricultural fields. Most Orthoptera are regarded as significant agricultural pests because of their role as food for other organisms farther up the food chain [2]. Species diversity and ecological study on Orthoptera, particularly their ecological roles, are still largely understudied in Southeast Asia [3]. The two main suborders of Orthoptera are Caelifera and Ensifera. Based on the length and thickness of their femora, the suborders are distinguished. Caelifera includes those whose rear femora are longer and thicker than their front femora, while Ensifera includes those whose femora are almost the same in length and thickness [4].

The Orthoptera, which include crickets, grasshoppers, and katydids, are the largest group of terrestrial insects with more than 27,000 species known to exist [5]. They are usually studied as essential parts in grasslands and agroecosystems, however little is known about their role in the forest ecosystems [6]. In addition, Orthoptera are also significant elements of the understorey fauna in a forest ecosystem involved in the food chain [7]. This study aims to analyze the community assemblage and alpha diversity of Orthoptera in the forest region or KHDTK Aek Nauli, North Sumatra. The finding will be a valuable source of knowledge for initiating a comprehensive inventory in the future and possibly as a part of the species conservation initiatives.

2. Method

2.1. Study area
The research was conducted within the Forest Area with Special Purpose (KHDTK) Aek Nauli, Sibaganding Village, Girsang Sipangan Bolon District, Simalungun Regency, North Sumatra, Indonesia. The forest area is structured by primary and secondary forest with an area of 1,900 ha, altitudes from 1,100–1,750 m.a.s.l. The forest is mainly composed by the pine vegetation and other tropical tree and shrub species with some sites implemented for an *ex situ* conservation of Sumatran Elephant (*Elephas maximus sumatranus*). The forest region is officially managed by the Forestry Research, Development and Innovation Agency (BP2LHK) Aek Nauli under the Indonesian Ministry of Environment and Forestry.

2.2. Sampling sites, techniques and grasshoppers identification
A total of 12 sampling points (2°43'3"N, 98°56'25"E) were randomly selected in the forest area following some accessible paths. The sampling was conducted on May 2022 within four consecutive days in the morning from 08:00-10:00 AM and from 14:00-16:00 PM. A number of pit fall traps was adjusted in each site while sweeping net was employed at adjacent shrubs. The obtained Orthoptera specimens from direct sweeping and overnight traps were put in bottles filled with 75% EtOH (v/v). The specimens were identified at the Laboratory of Animal Systematics, Universitas Sumatera Utara based on some identification manuals and an online tool [4,5,8,9].

2.3. Data analysis
The abundance, alpha diversity, and species richness of Orthoptera were analyzed descriptively. Shannon's diversity and evenness of species were calculated using PAST4.03 [10]. Importance Value Index (IVI) of each species was calculated as the sum of relative abundance (*Ra*) and frequency of occurrence (*Fo*).

3. Results and Discussion
The species assemblage of Orthoptera in the forest or KHDTK Aek Nauli was revealed from two suborders namely Caelifera and Ensifera with a higher species proportion from Ensifera than Caelifera (Figure 1A). In terms of species composition, Gryllidae was observed as the most speciose family (10 species), followed by Acrididae (8 species), Tettigoniidae (3 species), and Tetrigidae (1 species) as can be seen in Figure 1B. In total, 21 species were documented with 97 individuals from a 4 days-survey in the forest area (Table 1).

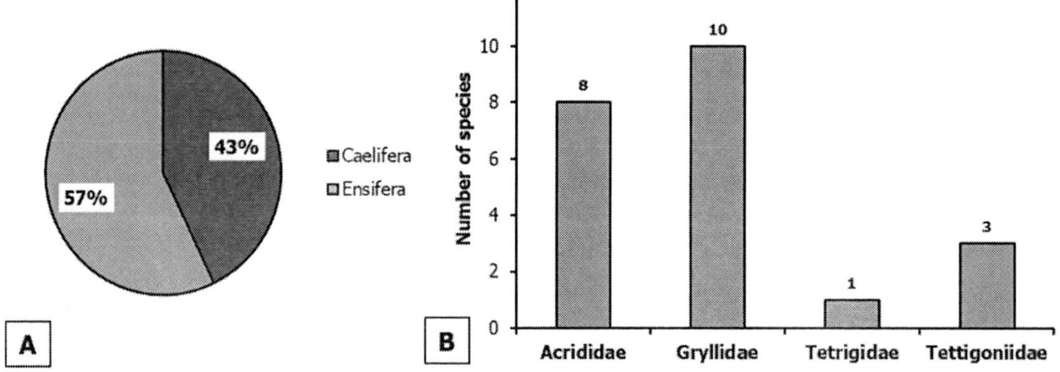

Figure 1. Composition of Orthoptera based on the percentage of suborders (A) and number of species assemblage from each family (B)

Table 1. Composition and species list of grasshoppers in KHDTK Aek Nauli

No.	Suborder	Family	Genus	Species	*N* (ind.)
1.	Caelifera	Acrididae	*Oxya*	*O. chinensis*	8
2.				*O. intricata*	2
3.			*Phlaeoba*	*Phl. angustidorsis*	1
4.				*Phl. antennata*	3
5.				*Phl. formosana*	3
6.				*Phl. infumata*	1
7.			*Valanga*	*V. nigricornis*	9
8.			*Velarifictorus*	*Velarifictorus* sp	2
9.		Tetrigidae	*Paratettix*	*P. curtipennis*	2
10.	Ensifera	Gryllidae	*Acheta*	*A. domesticus*	2
11.			*Loxoblemmus*	*L. equestris*	11
12.				*L. parabolicus*	19
13.			*Nisitrus*	*N. hughtani*	7
14.				*N. vittatus*	2
15.			*Teleogryllus*	*T. albipalpus*	1
16.				*T. emma*	1
17.				*T. mitratus*	1
18.				*Teleogryllus* sp	12
19.		Tettigoniidae	*Conocephalus*	*C. cognatus*	1
20.				*C. maculatus*	1
21.			*Mecopoda*	*M. elongata*	8

Table 2. Relative abundance, frequency of occurrence and Importance Value Index (IVI) of each species in KHDTK Aek Nauli

No.	Species	*Ra* (%)	*Fo* (%)	IVI
1.	*Acheta domesticus* (Linnaeus, 1758)	8.25	50.00	58.25
2.	*Conocephalus cognatus* (Redtenbacher, 1891)	2.06	8.33	10.40
3.	*Conocephalus maculatus* (Le Guilou, 1841)	1.03	8.33	9.36
4.	*Loxoblemmus equestris* (Saussure, 1877)	3.09	16.67	19.76
5.	*Loxoblemmus parabolicus* (Saussure, 1877)	3.09	16.67	19.76
6.	*Mecopoda elongata* (Linnaeus, 1758)	1.03	8.33	9.36
7.	*Nisitrus hughtani* (Robillard and Tan, 2021)	9.28	33.33	42.61
8.	*Nisitrus vittatus* (Haan, 1844)	2.06	8.33	10.40
9.	*Oxya chinensis* (Thunberg, 1815)	2.06	8.33	10.40
10.	*Oxya intricata* (Stål, 1861)	11.34	58.33	69.67
11.	*Paratettix curtipennis* (Hancock, 1912)	19.59	83.33	102.92
12.	*Phlaeoba angustidorsis* (Bolivar, 1902)	7.22	25.00	32.22
13.	*Phlaeoba antennata* (Brunner von Wattenwyl, 1893)	2.06	8.33	10.40
14.	*Phlaeoba formosana* (Shiraki, 1930)	1.03	8.33	9.36
15.	*Phlaeoba infumata* (Brunner von Wattenwyl, 1893)	1.03	8.33	9.36
16.	*Teleogryllus albipalpus* (Chopard, 1961)	1.03	8.33	9.36
17.	*Teleogryllus emma* (Chopard, 1961)	12.37	66.67	79.04
18.	*Teleogryllus mitratus* (Burmeister, 1838)	2.06	8.33	10.40
19.	*Teleogryllus* sp	1.03	8.33	9.36
20.	*Valanga nigricornis* (Burmeister, 1838)	1.03	8.33	9.36
21.	*Velarifictorus* sp	8.25	41.67	49.91

Note: *Ra* = Relative abundance, *Fo* = Frequency of occurrence, IVI = Importance Value Index

Eleven genera of Orthoptera were documented namely *Acheta, Conocephalus, Loxoblemmus, Mecopoda, Nisitrus, Paratettix, Phlaeoba, Teleogryllus, Oxya, Valanga* and *Velarifictorus* (Table 1). Based on the relative abundance, *Paratettix curtipennis, Teleogryllus emma,* and *Oxya intricata* was observed as the most abundant species, followed by their frequency of occurrence and IVI. The species found in this study are identified and confirmed as common in the Asiatic region. The Shannon's diversity index or the alpha diversity of Orthoptera in this study was categorized into moderate level of diversity (H' = 2.592) with an evenness (E) value of 0.636.

The majority of Orthoptera are numerous and well adaptable to environmental changes in their habitat although ultimately studied in agroecosystems. In addition, the number of orthopteran species are commonly lower in the forest ecosystems. A total of 12 orthopteran species were documented from three forests in Southern Cameroon using a combination of sweep net, quadrat method and pit fall traps [7]. The occurrence of *P. curtipennis* was reported from a study in the agricultural areas of Batu City, East Java, Indonesia [11]. The species, *T. emma* was claimed as the first record for Riau Province, Indonesia [12]. *Oxya intricata* was also documented from a study in two sites i.e corn field and forest ecosystem of Banyumas Regency, Central Java, Indonesia [13]. The difference in the number of species richness, abundance and diversity across region may due to the limited number of sampling efforts, time of sampling, type of microhabitats, and sampling techniques [14]. To discover a greater variety of orthopteran species in the forest ecosystems, several authors proposed doing an extensive investigation in diverse microhabitats and the forest canopy [15].

4. Conclusion

The community assemblage of Orthoptera in the forest region or KHDTK Aek Nauli, North Sumatra was composed of 21 species from 11 genera. The alpha diversity was categorized as moderate level of species diversity with

References

[1] Welti EAR, Prather RM, Sanders NJ, de Beurs KM and Kaspari M 2020 *J Anim Ecol* **89** 1286
[2] Latchininsky A, Sword G, Sergeev M, Cigliano MM and Lecoq M 2011 *Psyche* **2011** 578327
[3] Tan MK, Choi J and Shankar N 2017 *Zootaxa* **4238** 127
[4] Johnson, Norman and Charles A 2004 *Borror and DeLong's Introduction to the Study of Insects* (Florence: Brooks/Cole)
[5] Cigliano MM, Braun H, Eades CC and Otte D 2017 (http://orthoptera.speciesfile.org/)
[6] Joshi PC, Lockwood JA, Vashishth N and Singh A 1999 *J Orthoptera Res* **8** 17
[7] Ngoute CO, Kekeunou S, Lecoq M, Fiemapong ARN, Nyobe PCAU and Bilong CFB 2020 *J Orthoptera Res* **29** 25
[8] Tan MK 2012 *Orthoptera in Bukit Timah and Central Catchment Nature Reserves (Part 1): Suborder Caelifera* (Singapore: Raffles Museum of Biodiversity Research)
[9] Tan MK 2012 *Orthoptera in Bukit Timah and Central Catchment Nature Reserves (Part 2): Suborder Ensifera* (Singapore: Raffles Museum of Biodiversity Research)
[10] Hammer Ø, Harper DAT and Ryan PD 2001 *Paleontol Electron* **4** 9
[11] Abrori M, Leksono AS and Gama ZP 2021 *Biotropika* **9** 19
[12] Syahlan S, Yolanda R and Lubis RR 2017 *Scripta Biologica* **4** 99
[13] Prakoso B 2017 *Biosfera* **34** 80
[14] Leksono AS, Yanuwiadi B, Khotimah A and Zairina A 2022 *Biodiversitas* **23** 75
[15] Floren A, Riede K and Ingrisch S 2001 *Ecotropica* **7** 33

Study of peat-mangrove ecosystem in East Kalimantan Province

S Surahman[1], A Ahmad[2], T Butarbutar [2] and M A Palgunadi[3*]

[1]Study Program of Food Crops Production Technology, Faculty of Vocational Studies, Hasanuddin University, Makassar, Indonesia
[2]Deutsche Gesellschaft fur Internationale Zusammenarbeit (GIZ) Gmbh, Indonesia
[3]Department of Forest Conservation, Forestry Faculty, Hasanuddin University, Makassar 90245, Indonesia

[*]Email: muhammad.alriefqi@gmail.com

Abstract. Indonesia's peatland cover is about 11 million ha of land in tidal swamp areas and 3.9 million hectares in swamp and coastal areas. Based on these data, there are possibility interconnection between peat and mangrove ecosystems in the one landscape. Therefore, the purpose of this study is to analyze the characteristics of interconnection between the peat and mangrove ecosystems in Indonesia, specifically in East Kalimantan. Vegetation characteristics such as structure and species composition will be measured, as well as soil physical characteristics including analysis of water content, bulk density, particle density and porosity. characteristics of soil chemical properties include analysis of pH, EC, ash content, bases (K, Na, Ca, Mg), micro elements (Fe, Mn, Cu, Zn), C-total, N-total and S-total. The results obtained are East Kalimantan is located in the coastline area which is still influenced by high salinity tides. The level of species diversity and community stability in the transition area is in the low to moderate limit with a Shannon index value of 2.24. Species with high salinity tolerance dominate the transition area (true mangrove vegetation species).

1. Introduction

The peat ecosystem is one of the ecosystems that has a high complexity in terms of species composition, vegetation structure and fauna diversity. The health and function of peat ecosystems depend on the interaction between forest, peat, and hydrology [1]. Most tropical peat forests are found in lowland sub-coastal areas and are formed from broadleaf rainforest vegetation and higher plant associations. In several case studies, peat found in the lowlands is often found to be associated with mangrove ecosystems [2].

Mangrove ecosystems exist on different substrates, namely clay, sand or peat. Although peat can form in various geological settings around the world, mangrove peat can only form in salty and brackish mangroves along coastlines and tropical and subtropical islands [3]. Mangrove ecosystems that make up peat are widely reported [4]. Mangrove peat in other parts of the world is common and has been studied extensively in Belize, Puerto Rico, Honduras & Panama and Florida [5,6]

In the tropics, peatlands usually form a large dome located behind the mangrove forest along the coastal plain and along major rivers. During the period of development, the peat level in this system is getting higher compared to the surrounding land. The increase in peat during this period came from rainwater that replaced tidal and river floods as the main water source, and the domed peat surface became increasingly nutrient-poor and acidic. Peat-forming vegetation is a type of tropical rain forest, peat swamp forest that is detached from stressful environmental conditions such as anoxia, acidity, and low nutrient availability [2]. The findings of these peat deposits under mangrove ecosystems explain the unusual pattern of mangrove development in the Southeast Asia region, as these coastal forests usually grow in areas with dominant mineral soils and the deposition of peat soils under mangroves is rarely reported elsewhere in the region. this region.

The substrate beneath the mangrove forest in the Pelican Cays complex is dominated by peat which consists mostly of mangrove roots. Leaves and wood account for less than 20% of the peat mass. At Cat Cay, the peat depth ranges from 0.2 m along the shoreline to 1.65 m in the center of the island, indicating that the island has expanded both horizontally and vertically through biogenic processes underground. Mangrove roots thus play an important role in soil formation, vertical accretion, and stability of these mangrove islands. The species composition of fossil roots changed markedly with depth: *Rhizophora mangle* (red mangrove) was an early colonizer at the coral base, followed by *Avicennia germinans* (black mangrove), which increased in abundance and extended radially from the center of the island. The center of the Avicennia stands eventually dies, leaving behind a shallow pool of no vegetation. Thus, peat maintains a record of mangrove development, succession, and decline in response to changes in sea level and concomitant hydroedafic conditions that control mangrove distribution, formation, growth, and mortality on oceanic islands in Belize [5].

Mangrove peat is generally limited to areas with salty and brackish water located along tropical and subtropical coastlines and islands [3]. Peat with a rich organic matter content can be found in areas with such conditions, for example in South Florida [7] and elsewhere in the Caribbean and western Pacific. It is estimated that more than 23 million hectares (62%) of tropical peatlands globally are located in Southeast Asia. This peat occupies most of the lowlands around the coastal and sub-coastal areas [2].

Based on the problems above, there is a possibility of ecological interconnection between peat and mangrove ecosystems, so a more in-depth study is needed. Through GIZ PROPEAT, it is hoped that it will be able to carry out technical activities in collaboration between the Government of Indonesia and the Government of Germany which focuses on peat and mangrove management activities in Indonesia. This activity was carried out in the province of East Kalimantan. The study of the characteristics of the peat and mangrove ecosystem will provide data and information to support the process of preparing the management plan and protection of the peat and mangrove ecosystem (RPPEG) in East Kalimantan. This study aims to analyze the profile of peat and mangrove ecosystems as a landscape. Furthermore, the results of this study are expected to describe the interconnection between peat and mangrove ecosystems.

2. Material and Method

2.1. Study site

The research was conducted at the PROPEAT (Peatland Rehabilitation and Management Project) project locations in East Kalimantan Province covering areas; Batu Putih, Hole Klatak and Sepinang (Figure 1).

2.2. Vegetation measurements

The variables observed in this study were species composition, base area diameter/area, vegetation structure (profile diagram and canopy projection), and percentage of cover. To obtain data on these variables, plot with size of 20 cm x 20 cm was established and each individual species in the plot will be recorded. All species in the plot will be taken samples of leaves, flowers and fruit for later analysis at the Hasanuddin University Forest Conservation Laboratory and the Indonesian Institute of Sciences to determine the family, genus and species level.

Tree species with a circumference of more than 15 cm will be measured in diameter, while tree species with a circumference of less than 15 cm and herbaceous species will be counted. The profile diagram and the projection of the canopy will be drawn in a 10 m wide strip which is placed lengthwise in the center of the plot in the direction of intersecting the contour. Furthermore, the percentage of species cover will be measured using the Brawn-Blaquet method.

2.3. *Soil measurements*

Preliminary Investigation and sampling of Soil samples were taken from cross-sections of soil profiles, i.e. complete soil samples are taken with a ring sampler at a specified depth, such as soil profiles in cross-sections every 10 cm from the soil surface to the depth of the water table. Soil chemistry and biological analysis were performed in combination with soil sampling. Use a peat drill to subsample soil samples from several locations at depths of 0-20 cm, 20-40 cm, and 40-60 cm. The samples are drilled with peat, then placed in a plastic bag and in a sample box to keep the samples safe during transport from the collection site to the laboratory. Physical properties of peat soils were analyzed using intact soil samples. Peat soil sampling methods use peat borers, ring samplers and box samplers [8].

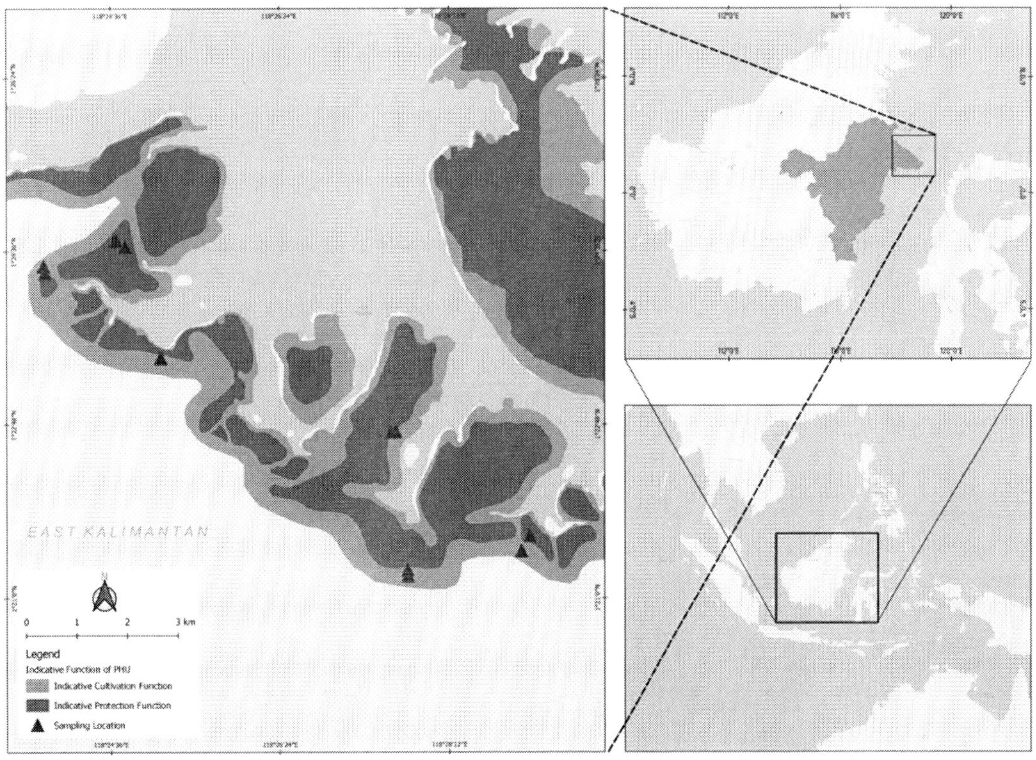

Figure 1. Map of determining observation plots in East Kalimantan

2.4. *Data analysis*

The measurement data obtained in the field are used to calculate the base area (LBDS), density, frequency. The formulas used are:

$$LBDS = 1/4\pi D^2 \quad (1)$$

$$\text{Density (K)} = \text{(Number of individuals in the plot)/(Area of the plot)} \quad (2)$$

$$\text{Frequency (F)} = \text{(Number of subplots found of a type)/(Sum of all subplots) x 100\%} \quad (3)$$

The LBDS value, density, percentage of cover and the results of the analysis of the dominance index are the basis for determining the dominance of species in the ecosystem. Furthermore, the results of the depiction of the stand structure (profile diagram and projection of the crown) will be digitized again using the CorelDraw program. The Dominance Index is calculated using the Simpson's dominance index formula:

$$D = (ni/N)2 \tag{4}$$

Descriptions: D = Simpson's Dominance Index,
ni = Number of Individuals of each species,
N = Number of Individuals of all species

The diversity index (H') describes the state of the population of organisms mathematically to make it easier to analyze information on the number of individuals of each type in a community. For this reason, calculations are carried out using the Shannon-Wiener equation.

$$H^I = -\sum \left[\frac{ni}{N}\right] \mathbf{log} \left[\frac{ni}{N}\right] \tag{5}$$

Descriptions: H = Shannon diversity index;
Ni = Number of individuals per species;
N = Number of individuals of all species.

3. Result and Discussion

3.1. Vegetation characteristics

Vegetation observations were carried out in 3 locations, namely Batu Putih village, Lubang Klatak and Sepinang islands, Berau Regency. In general, the three observation locations are in the coastal area, which is still affected by the tides. 9 Observation plots with an area of 20m x 20m were then purposively placed in the mangrove ecosystem area and the mangrove-peat ecosystem transition. A total of 15 species were found in the observation plot of the mangrove-peat ecosystem transition area. The dominant species were from the family Rhizophoraceae, Moraceae, Pteridaceae, Meliaceae and Combretaceae (Table 1).

The species with the highest density was Rhizophora mucronata at 122/0.02Ha, followed by Acrostichum aureum (109/0.02Ha), Xylocarpus sp. (100/0.02Ha), Lumnitzera littorea (Jack) Voight (63/0.02Ha), Ficus sp.4 (41/0.02Ha), Ficus sp.2 (30/0.02Ha), and Ceriops tagal (Press) C.B.Rob. (24/0.02Ha). Furthermore, the species with the largest number of Base Areas (LBDs) were Rhizophora mucronata at 299333.25 cm2/0.02Ha and Lumnitzera littorea (Jack) Voight at 162703.31 cm2/0.02Ha.

The species with the smallest occurrence frequency was found at 20% (only found in 1 observation plot) and the largest occurrence was found at 60% at the research site. The species with the highest occurrence rate were Acrostichum aureum and Xylocarpus sp. with a frequency of 60%. The species with the smallest occurrence rate were Cinnamomum cuspidatum Miq., Ceriops tagal (Pers.) C.B.Rob., Liana sp.13, Sapindaceae, Sp.75, and Rutaceae with each frequency value of 20%.

The diversity of species in the transitional area is at a low to moderate borderline with an index value of 2.24 [10]. Species of mangrove vegetation seem to dominate this ecosystem with a dominance index value of 0.1. Species Rhizophora mucronata, Xylocarpus sp., Lumnitzera littorea (Jack) Voight and ferns (Acrostichum aureum) dominates the composition of the mangrove-peat transitional area in the 3 research sites in East Kalimantan. The dominance of mangrove species in the transitional area is most likely influenced by tides and high salinity levels so that the species that can survive are species with high salt tolerance [9].

The vegetation structure is projected in the form of profile diagrams and canopy projections (Figures 2, 3, and 4). Based on the diagram, in general, transitional ecosystems have 1-2 strata of canopy with a medium to high coverage area of 55-85%. The average tree height in the first strata is

10-17 m, which is dominated by species from the Moraceae, Meliaceae and Rhizophoraceae families. The second canopy layer with an average height of 5-10 m is dominated by species from the family Combretaceae, Lauraceae, Sapindaceae and Rutaceae.

The highest canopy cover value was found in Batu Putih village, in a transitional area dominated by the *Lumnitzera littorea* (Jack) Voight mangrove stands, reaching 70-85% (Figure 7). Furthermore, a more open canopy cover was found in the village of Sepinang with a cover area of about 55% and dominated by species from the Moraceae family or banyan tribe.

Table 1. Composition of vegetation types in the mangrove-peat transitional ecosystem in East Kalimantan, covering area; Batu Putih Village, Lubang Klatak and Sepinang (K = Density; LBDs = Base Area; F = Frequency of Attendance)

Family	Species	K (0,02Ha)	LBDs (cm²)	F (%)
Combretaceae	Lumnitzera littorea (Jack) Voight	63	162703,31	40
Rhizophoraceae	Ceriops tagal (Pers.) C.B.Rob.	24	29513,81	20
Pteridaceae	Acrostichum aureum	109	0	60
Moraceae	Ficus sp.4	41	21468,26	40
Moraceae	Ficus sp.2	30	6581,10	40
	Liana sp.10	11	0	40
Lauraceae	Cinnamomum cuspidatum Miq.	9	51,50	20
	Liana sp.1	13	0	40
	Liana sp.2	12	0	40
Rhizophoraceae	Rhizophora mucronata	122	299333,25	40
Meliaceae	Xylocarpus sp.	100	135365,61	60
	Liana sp.3	7	0	20
Sapindaceae	Sapindaceae	11	193,49	20
	Sp.1	6	339,62	20
Rutaceae	Rutaceae	18	972,64	20

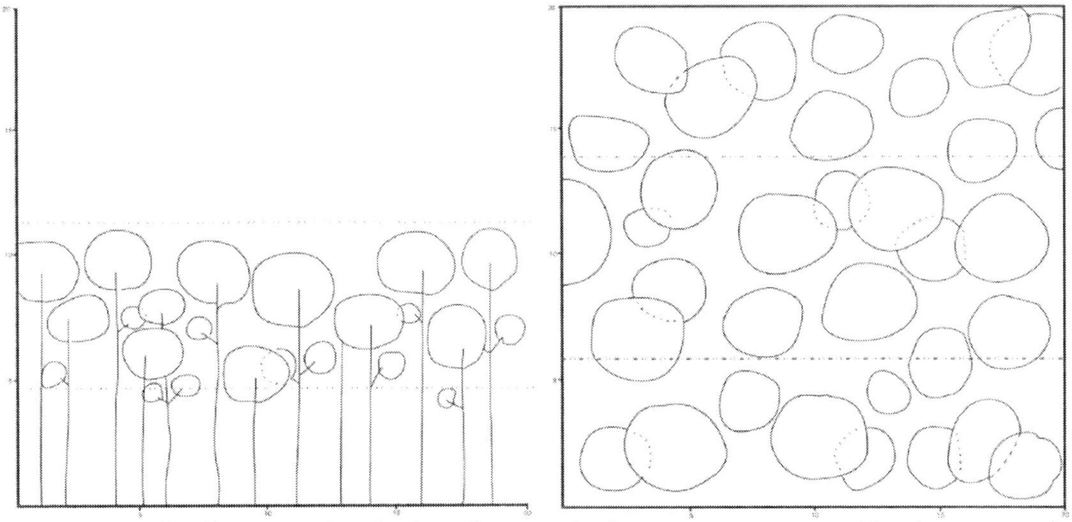

Figure 2. Profile diagram and projection of canopy in the mangrove-peat transitional ecosystem in Batu Putih Village, East Kalimantan (Scale 1:200)

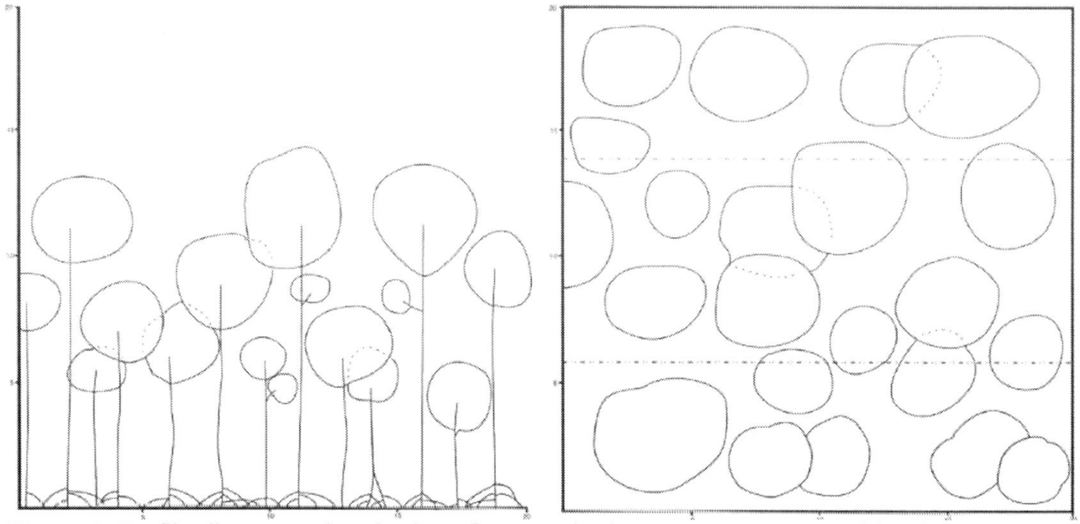

Figure 3. Profile diagram and projection of canopy in the mangrove-peat transitional ecosystem of Lubang Klatak, East Kalimantan (Scale 1:200)

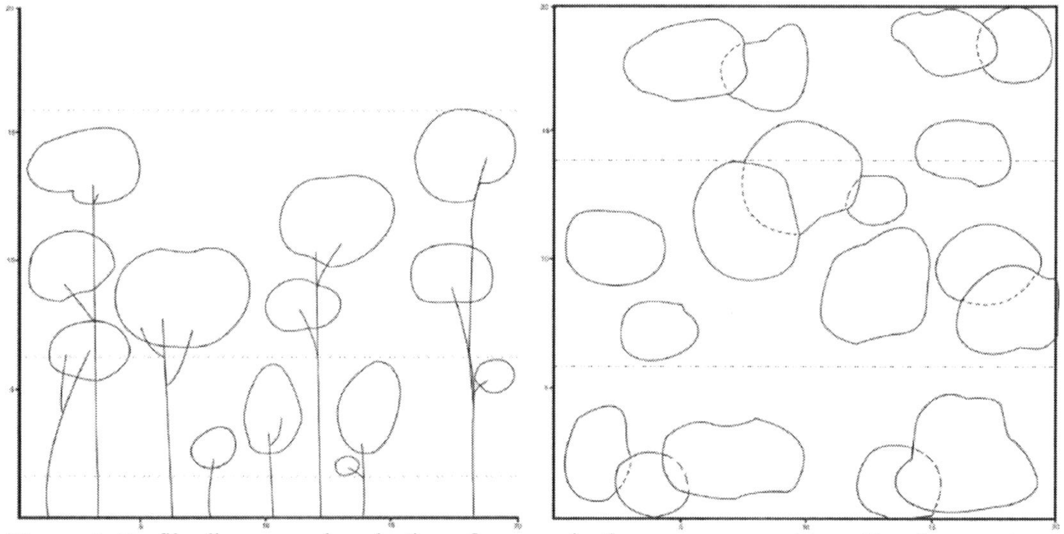

Figure 4. Profile diagram and projection of canopy in the mangrove-peat transitional ecosystem of Sepinang, East Kalimantan (Scale 1:200)

3.2. Peat characteristics

Peat depth is an important physical property to be analyzed. Based on the depth criteria, peat can be classified as shallow (< 50-100 cm), medium (100-200 cm), deep (200-400 cm), very deep (400-800 cm), and very deep (> 800 cm). [12]. The results of measuring the depth of peat in the field using a peat drill (eijkel camp).

The average value of fiber content can indicate the level of maturity of the peat. Based on the maturity level class, peat is divided into three types, namely fibric, hemic and sapric. Peat maturity level is based on the content of peat fiber remaining on hand: fibric peat >3/4, hemic between 3/4-1/4, and sapric <1/4 [8]. Peat maturity analysis at the study site is classified as fiber maturity

The results of bulk density analysis at the study site showed an average BD value of 0.53 g/cm3. This is in accordance with the opinion of Tie and Lim [11] that very low BD of peat soil, which is <0.1 g/cm3 is found in fibric (raw) peat located at the bottom, while coastal peat and peat located in river flow paths. has a relatively higher BD, which is >0.2 g/cm3 due to the influence of mineral materials, but it is still far from the BD of mineral soils which ranges from 0.7-1.4 g/cm3.

A high porosity value indicates that the number of pores in the soil is very large, which makes the soil more porous. Porosity is also closely related to the level of soil density, the denser the soil, the smaller the porosity of the soil and vice versa. The results of the porosity analysis in the study area of East Kalimantan Province showed an average porosity value of 54.65%

Peat soil can function as a very effective water storage because it has a highwater holding capacity. In a saturated state, the water content of peat soil can reach 450-3000% of its dry weight, because the water consumed for plants is available throughout the year and is present in excess [12]. The results of the analysis of soil water content in the study location of East Kalimantan Province showed an average soil water content value of 64.8%.

The carbon content in peat soil is very large in peat soil because it is generally formed from the process of accumulating organic matter. The organic fraction of peat soil in Indonesia is more than 95%, less than 5% the rest is the inorganic fraction. The organic fraction consists of about 10% to 20% humic compounds, mostly consisting of non-humic compounds which include lignin, cellulose, hemicellulose, wax, tannin, resin, suberin, and a small amount of protein. Meanwhile, humic compounds consist of humic acid, himatomelanate and humin [13,11]

The chemical characteristics of peat soils in Indonesia are very diverse and are determined by several factors such as: mineral content, thickness, types of peat constituent plants, mineral types in the substratum (at the bottom of the peat), and the level of peat decomposition. Polak [15] made of wood species dominates the peat in Sumatra and Kalimantan. Therefore, the organic matter composition is mostly lignin which generally exceeds 60% of the dry matter, while the content of other components such as cellulose, hemicellulose, and protein generally does not exceed 11%.

Peat soil acidity level is relatively high, which is in the range of pH 3 - 4. Research on oligotropic peat in Berengbengkel, Central Kalimantan has a pH range of 3.25 - 3.75, this peat has a quartz sand substrate [16,17]. The level of acidity of peat soil is closely related to the content of organic acids, namely humic acid and fulvic acid, peat soils in Indonesia are mostly acidic to very acidic with a pH <4.0. [18,19]

The conductivity is usually much lower than that of inorganic soils, from 40-100 mS/cm [20] for inland peat in Central Kalimantan to 140-320 mS/cm for coastal peat in West Kalimantan [21], which the latter may reflect the entry of marine aerosols onto the peat surface. The average value of DHL content in the East Kalimantan Study site was 24.76 mS/cm. The high DHL value in this area is caused by the high dissolved salts in the soil (Ca, Mg, K, Na). The position of the area which is a tidal area also plays a role in increasing the value of DHL.

Ash content of shallow peat with a thickness of >1 m or about 5%, middle to deep peat with a thickness of 1-3 m or between 11-12%, and deep peat with a thickness of >3 m or about 15% [22]. The higher the minerals contained in the peat soil, the higher the ash content. The ash content of oligotrophic peat is approximately 2%, mesotrophic peat between 2-7%, & eutrophic peat (high mineral content) >14% [22]. The average value of Ash Content in the East Kalimantan Study site is 4.55%.

The average value of C-Organic content in the East Kalimantan Study site is 37.80%. The higher the organic matter added, the more organic acids content will increase the organic C content. According to Hillel [23], the distribution of dead roots and roots continues, especially root hairs can also stimulate the activity of microorganisms that can contribute organic matter to the soil.

The average value of N-Total content in the East Kalimantan Study site is 1.15%. The process of decomposition of soil organic matter will produce compounds containing nitrogen. The C and N ratio is basically a comparison value between the C and N content in the soil. The average value of the ratio of C and N at the study site in East Kalimantan is 32.96. Organic matter has a C/N ratio > 30, there will be a N immobilization process, where the N produced from the mineralization process is used by micro-organisms for their daily needs [24].

The average values of the Ca, Mg, K, and Na bases in the East Kalimantan Study site were 9.85%, 3.20%, 1.26% and 2.50%, respectively. This value is in the medium to high range. The thicker the peat, the lower the ash content, the lower the Ca and Mg content and the more acidic the soil reaction [25]. The low base content is accompanied by a high cation exchange capacity (CEC) value, causing the availability of bases to be low. The low base content of inland peat is closely related to its formation process which is more influenced by rainwater [26]. Base saturation (KB) of inland peat soils is generally very low. Bereng bengkel inland peat soil in Central Kalimantan has a KB value <100% (Bogor Agricultural Institute Team, 1974), as well as the KB value for lowland peat in Riau [22].

East Kalimantan is located on the shoreline area which is still affected by the tides of sea water with high salinity. The level of species diversity and community stability in the transition area is at a low to moderate borderline with a Shannon index value of 2.24. Species with high salinity tolerance dominate the transition area (species of true mangrove vegetation) with 15 species identified. Furthermore, the vegetation structure tends to be more uniform with an average of 1-2 strata of canopy layers and a high cover of 70-85%.

Landscape of the study area in East Kalimantan is in the coastal area of the sea (coastal). The formation of peat in this area is strongly influenced by the tides of sea water, causing the mangroves to move forward closer to the coast [27]. This condition causes the mangrove peat ecosystem in East Kalimantan to be dominated by accumulation of peat with alluvial deposits (peat) and a relatively uniform/homogeneous mangrove vegetation community. The results of the assessment of physical properties that describe the characteristics of the peat depth recorded in this area are relatively uniform, which is included in the category of shallow (thin) peat. The peat formed in this area is classified as peat with a fibric maturity level. Other physical properties such as bulk density, porosity and water content show that this area is very porous and muddy. Conditions like this allow the vegetation on it, especially mangroves, to develop properly.

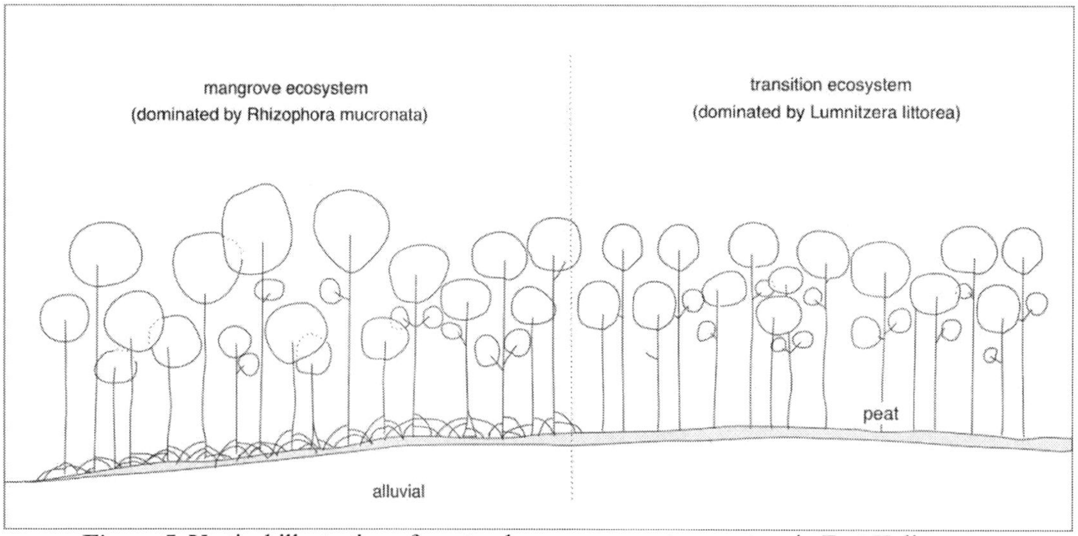

Figure 5. Vertical illustration of a natural mangrove peat ecosystem in East Kalimantan

The results of the assessment of chemical properties in the mangrove and peat ecosystem areas showed a very acidic soil reaction (pH). The low pH in this area is related to the condition of the area being inundated and mixed with the alluvial layer. The DHL value associated with salinity levels in this region is also relatively higher. In addition, the content of bases in this region also has the potential to increase the salinity value. The high salinity is thought to be caused by the high concentration of Fe in the soil solution where Fe cations are the result of the oxidation of pyrite mineral materials.

4. Conclusions

Peat and mangrove are an ecological unit whose management cannot be separated from each other and must be managed in an integrated and sustainable manner. Peat as a 'sponge' that absorbs, stores and supplies fresh water reserves maintains the stability of the mangrove ecosystem, and vice versa, mangroves as a barrier from tides/estuaries can stabilize dissolved salts so that the potential for peat to become saline can be minimized. Based on the built logical framing, this study recommends several concrete considerations as follows: (1) Incorporate mangrove peat issues into the preparation of the peat ecosystem protection and management plan (RPPEG) with the direction of protection function or limited use; (2) The government and stakeholders need to look for innovations in community empowerment activities so that the utilization of peat or mangrove ecosystems runs according to environmental/ecosystem conditions; (3) There needs to be a good and sustainable socialization about the benefits of peat as an ecological unit; (4) For mangrove planting, species such as *Bruguiera gymnorrhiza* (L.) Lamk., *Rhizophora mucronata*, lumnitzera sp. as well as several other true mangrove species.

Acknowledgements

We would like to extend our sincere gratitude to GIZ and Faculty of Forestry, Hasanuddin University for providing experimental equipment and convenience during the whole research process.

References

[1] Wijedasa L S et al 2017 *Global Change Biology* **23** 977
[2] Rieley J and Page S 2016 *Tropical peatland ecosystems* 3
[3] Cameron C C, Esterle J S and Palmer C A 1989 *Int. J. Coal Geol.* **12** 105
[4] Joosten H 2012 *Natur und Landschaft* **87** 50
[5] McKee K L and Faulkner P L 2000 *Atoll Research Bulletin* **468** 45
[6] Macintyre I G, Toscano M A, Lighty R G and Bond G B 2004 *Atoll Research Bulletin* 510 1
[7] Odum W E 1988 *Annual Review ofEcology and Systematics* 19 147
[8] Sumawinata B, Djajakirana G and Handayani L *2015 Penilaian Sifat Fisik, Kimia dan Biologi Gambut*
[9] Ball M C 1988 *Functional Plant Biology* **15** 447
[10] Shannon C E and Weaver W 1949 *The mathematical theory of communication* (The University of Illinois Press: Urbana, IL, USA)
[11] Tie Y L and J S Lim 1991 *Proc. International Sysposium on Tropical Peatland*
[12] Darwiati W and Nurhaedah M 2010 *Jurnal Mitra Hutan Tanaman* **5** 27
[13] Stevenson F J 1994 *Humus Chemistry.Genesis, Composition, and Reactions* 443 (New York: John Wiley andSons.Inc.)
[14] Tan K H 1994 *Environmental Soil Science* 304 (New York: Marcel Dekker Inc.)
[15] Polak B 1975 *Mod. Quatern. Res. SE Asia* **2** 71
[16] Halim A 1987 *Pengaruh Pencampuran Tanah Mineral dan Basa dengan Tanah Gambut Pedalaman Kalimantan Tengah dalam Budidaya Tanaman Kedelai* (Bogor, Indonesia: Fakultas Pascasarjana)
[17] Salampak 1999 *Peningkatan Produktivitas Tanah Gambut yang Disawahkan dengan Pemberian Bahan Amelioran Tanah Mineral Berkadar Besi Tinggi* (Bogor, Indonesia:

Program Pascasarjana, Institut Pertanian Bogor)
[18] Andriesse J P 1974 *Dept. of Agric. Res.Of the Royal Trop. Inst.Comm.* **63** 869
[19] Miller R W and Donahue R L 1990 *An Introduction to Soil and Plant Growth* Prentice Hall International Edition
[20] Kurnain A, Notohadikusumo T, Radjagukguk B, and Hastuti S 2001 *Proceeding on Peatlands for People: Natural Resources Functions and Sustainable Management* 168-78
[21] Suryanto 1994 *Improvement of the P Nutrient Status of Tropical Ombrogenous Peat Soils from Pontianak, West Kalimantan, Indonesia* 216 (Belgia: Universiteit Gent)
[22] Suhardjo H and Widjaja-Adhi I P G 1976 *Soil Res. Inst. Bogor* **3** 74
[23] Hillel D 1996 *Introduction to Soil Physics* (Armhest, Massachusetts: Department of Plant and Soil Sciences, University of Masschusetts)
[24] Tisdale S L, Nelson W L and Beaton J D 1985 *Soil Fertility and Fertilizers* (New York: 4th Edition,Macmillan Publishing Company)
[25] Driessen P M and Rochimah L 1976 *Proceedings ATA 106 Midterm Seminar, Soil Research Institute* 56
[26] Leiwakabessy F M 1978 *Seminar Pemantapan Usaha-usaha Pembangunan di Daerah Transmigrasi oleh JTKI-PPSM* (Bogor: Institut Pertanian Bogor. 1976. Kriteria Penilaian Kandungan Unsur dan KemasamanTanah Daerah Pasang Surut)
[27] Middleton B A and McKee K L 2001 *Journal of Ecology* 818

IOP Conf. Series: Earth and Environmental Science 1115 (2022) 012024 doi:10.1088/1755-1315/1115/1/012024

Identification of ornamental plant species in the Martelu Purba Nature Reserve, North Sumatra

R Rambey[1,2*] and J Permana[1,2]

[1]Department of Forestry, Faculty of Forestry, University of Sumatera Utara. Jl. Tridharma Ujung No. 1 Kampus USU, Medan 20155, North Sumatra, Indonesia.
[2]JATI - Sumatran Forestry Analysis Study Center. Jl. Tri Dharma Ujung No. 1 Kampus USU, Medan 20155, North Sumatra, Indonesia

*Email: ridahati.rambey@usu.ac.id

Abstract. Indonesia has a fairly high biodiversity, both flora and fauna. One of the biodiversity in Indonesia is various species of ornamental plants found in the wild. One of the locations that has a diversity of ornamental plant species is the Martelu Purba Nature Reserve. The purpose of this study was to identify the species of ornamental plants found in the Martelu Purba Nature Reserve. The method used in this research is the roaming method, namely by exploring the Martelu Purba nature reserve and then noting the species of plants that have the potential to be ornamental. The results obtained indicate that there are 19 species of ornamental plants at that location. The species of ornamental plants include the family Araceae, Orchidaceae and species of ferns. There are plants that live Epiphyte and several other species grow terrestrially.

1. Introduction

Indonesia has a diverse range of flora and fauna and many species are still largely undiscovered [1]. Habitat fragmentation and loss are the two main causes of species decline worldwide [2]. A nature reserve forest is a nature reserve area which due to its natural condition has the uniqueness of plants and/or animals and their ecosystem, which need to be protected and development takes place naturally [3]. The Martelu Purba nature reserve is administratively located in Purba Tongah Village and Tiga Runggu Village, Purba District, Simalungun Regency, North Sumatra Province. Based on the geographical location, the Martelu Purba Nature Reserve is located at coordinates 20 53' – 20 54' North Latitude and 980 42' – 980 43' East Longitude. The Martelu Purba Nature Reserve area is located at an altitude of up to 1,320 masl.

One of the potentials in the Martelu Purba Nature Reserve is ornamental plants. Ornamental plants are plants that have a certain beauty and attractiveness value. In addition, it also has economic value for indoor and outdoor decoration purposes [4]. The term 'ornamental plants' refers to all plants of ornamental value, which generally have beautiful flowers or a special plant architecture [5]. One of the ornamental plants is a fern which is widely used as an ornamental plant even though it does not have flowers, but has diverse leaves and is no less beautiful than flowering plants. The beauty of the leaves and sorus has attracted the attention of ornamental plant lovers so that ferns are sought after [6].

As stated by Majanah and Saputri [7] ornamental plants are plants that are cultivated or planted or grow wild because they have a beauty value both in flowers, leaves, and all parts of the plant.

Ornamental plants are usually found in the yard/home garden and grow wild in the forest. Ornamental plants in addition to providing beauty can also be used as medicine.

Various species of plants exist in the Martelu Purba Nature Reserve, but there has been no previous research in this location. Ornamental plants such as ferns, orchids and taro are ornamental plant species that can be developed in the context of conservation and utilization of the area. The species of ornamental plants in the location of the Martelu Purba Nature Reserve have not been well documented, therefore there needs to be a scientific study. This study aims to identify plant species that have ornamental potential in the Martelu Purba Nature Reserve and determine their conservation status.

2. Material and Method

This research was conducted from April 2020 to October 2020. This research was conducted in the Martelu Purba Nature Reserve, Purba District, Simalungun Regency, North Sumatra. The tools used in this research are data collection and data analysis tools. Data collection tools include GPS (Global Position System), measuring tape, knife, string of raffia, camera, phiband, haga hypsometer and stationery. The data analysis tool used is a computer, Microsoft Excel. While the materials used in the study were tallysheets, plastic bags and location maps.

Primary data was obtained from direct observation in the field using the roaming method. The species of plants that have the potential to be ornamental are then recorded, observed and commented on. Species identification is carried out with local species identifiers and identification books.

3. Results and Disscussion

3.1. Identification of ornamental plants in the Martelu Purba Nature Reserve

Based on Table 1, it can be seen that in the Martelu Nature Reserve there are 19 species of Ornamental Plant Species, which consist of; *Asplenium nidus, Nephrolepis biserrata, Coelogyne fuliginosa, Anoectochillus reinwardtii Blume, Rhapidopora* sp, *Caladium* sp, *Dendrobium connatum, Dennstaedtia scandens, Dianella ensifolia, Cyathea contaminans, Curculigo latifolia, Oberonia complanata, Alocasia arifolia, Gleichenia longissima, Angiopteris avecta, Eria stelata, Huperzia phlegmaria* and *Tacca chantrieeri*. The species of ornamental plants found in the Martolu Purba Nature Reserve can be seen in Table 1.

Table 1. Species of ornamental plants in the Martelu Purba Nature Reserve

No	Local name	Scientific name	Family	Life pattern
1	Paku sarang burung	*Asplenium nidus*	Aspleniaceae	Epiphyte
2	-	*Nephrolepis biserrata*	Nephrolepidaceae	Epiphyte
3	Anggrek Pohon	*Agrostophyllum* sp	Orchidaceae	Epiphyte
4	Anggrek Permata	*Anoectochillus reinwardtii Blume*	Orchidaceae	Terresterial
5	Rapidopora	*Rapidhopora* sp	Araceae	Epiphyte
6	Keladi	*Caladium* sp	Araceae	Terresterial
7	Anggrek	*Dendrobium connatum*	Orchidaceae	Epiphyte
8	Pakis	*Dennstaedtia scandens*	Dennstaedtiaceae	Terresterial
9	Anggrek tanah	*Cymbidium* sp	0rchidaceae	Terresterial
10	Syngonium	*Syngonium podhophllum*	Araceae	Terresterial
11	Paku Tiang	*Cyathea contaminans*	Cyatheaceae	Terresterial
12	Sungkit	*Curculigo latifolia*	Hypoxidaceae	Terresterial
13	Anggrek oberonia	*Oberonia complanate*	Orchidaceae	Epiphyte
14	Alocasia	*Alocasia arifolia*	Araceae	Terresterial
15	Paku andam	*Gleichenia longissimi*	Gleicheniaceae	Terresterial
16	Pakis gajah	*Angiopteris avecta*	Marattiaceae	Terresterial

17	Paku huperzia	*Huperzia phlegmaria*	Lycopodiaceae	Epiphyte
18	Bunga kelelawar	*Tacca chantrieeri*	Diosscoreales	Terresterial
19	Anggrek	*Eria stelata*	Orchidaceae	Terresterial

The most abundant ornamental plant species in the Martelu Purba Nature Reserve are from the Orchidaceae family as many as 6 species (31.57%) and Araceae as many as 4 species (21.05%). Other families such as Aspleniaceae, Nephrolepidaceae, Denstadetiaceae, Chateaceae, Hypoxidaceae, Gleicheniaceae, Marattiaceae, Lycopodiaceae and Diosscoreales each number one species. Based on the nature of its growth as many as 12 species (63%) of ornamental plants live terrestrially and as many as 7 species of living efifit (37%).

Table 2. Percentage of species abundance

No	Family	Total	Percentage (%)
1	Aspleniaceae	1	5
2	Nephrolepidaceae	1	5
3	Orchidaceae	6	31.57
4	Araceae	4	21.05
5	Denstadetiaceae	1	5
6	Chateaceae	1	5
7	Hypoxidaceae	1	5
8	Gleicheniaceae	1	5
9	Marattiaceae	1	5
10	Lycopodiaceae	1	5
11	Diosscoreales	1	5

Syngonium podophyllum

Alocasia arifolia

Agrostophyllum sp

Anoectochillus reinwardtii

Rhapidopora sp

Oberonia complanata

Dendrobiun connatum

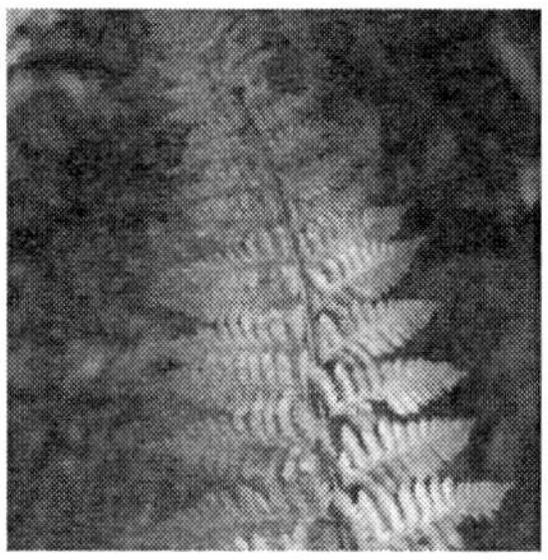
Dennstaedtia scandens

Asplenium nidus *Huperzia phlegmaria* *Tacca chantrieeri* *Cyathea contaminans*

Figure 1. Documentations of ornamental plants in Martelu Purba Nature Reserve

Based on the research above, many ornamental plants of the same type are found, such as the species of ferns, orchids, and the species of taro which are also included in ornamental plants. The most common type of fern is bird's nest fern. Ferns are more commonly found in mountainous areas than in the lowlands [8]. In his research stated that ferns can live in humid places, in general the number of species of ferns in mountainous areas is more than in the lowlands, this is due to the high humidity, the amount of water flow, the presence of fog, even the amount of rainfall affects the kind. In addition to the difference in height, there are also differences in the variety of trees at the three locations, so that the variety of trees will also affect abiotic factors which in turn affect the diversity of ferns. At the research site, there were 7 species of ferns, both terrestrial and epiphytic. It is different from [9] that there are 12 species of epiphytic fern in Mount Gede Pangrango National Park.

Likewise orchid species, rare orchid species are also found in the Martelu Purba Reserve, *Anoectochillus reinwardtii Blume,* not so many species of orchids are found as ornamental ferns, considering the species of plants Orchids are very difficult to find and very rare and there are also many ornamental plant lovers who are willing to look for rare ornamental plants in the wild growing forest. However, in Indonesia there is no quota issued for orchids, meaning that orchid species cannot be applied internationally. For conservation purposes, species of commercial crops should be cultivated, to avoid over exploitation of nature. According [10] orchid is one of the plants that has a figure or appearance that is quite beautiful. Orchid plants present various strands, shapes, and patterns of flowers. Orchids can grow in dark forests, on open slopes, on steep rocks at the foot of mountains.

The ornamental plants of the taro type are also included in the species of ornamental plants, but not all species are included in the species of ornamental plants. The ornamental species of taro plants that are found are Caladium, Anthurium, Alocasia arifolia, this type is also commonly found in the ancient Martelu nature reserve considering that this nature reserve is included in a humid and dark nature reserve. Based on [11] state that a plant is classified as a flower ornamental plant if the plant has attractive flowers caused by attractive colors, beautiful and charming shapes, fragrant smells, or by their special size. Leaf ornamental plants have their own charm in the leaves caused by the shape, color and composition of leaves with beautiful stems. Stem ornamental plants rely on the beauty of their stems on display. The beauty of the stem that is usually displayed is its shape or color.

4. Conclusion

There are 19 species of ornamental plants found in Martelu Purba Nature Reserve. The most abundant ornamental plant species in the Martelu Purba Nature Reserve are from the Orchidaceae family as many as 6 species (31.57%) and Araceae as many as 4 species (21.05%). Other families such as Aspleniaceae, Nephrolepidaceae, Denstadetiaceae, Chateaceae, Hypoxidaceae, Gleicheniaceae, Marattiaceae, Lycopodiaceae and Diosscoreales each number one species. Based on the nature of its growth as many as 12 species (63%) of ornamental plants live terrestrially and as many as 7 species of living epiphyte (37%).

References

[1] Kuswanto L, Chusna N A, Purnomo E, Krisantini and Ahmad M U 2022 *Identification and documentation of wild plant species with ornamental potentials at Mount Prau* (Indonesia : Ornamental Horticulture) **28** 110
[2] Shijin W 2021 *Meteorological Applications* **28** 1987
[3] UU No 41. 1999 Konservasi Sumberdaya Alam Hayati
[4] Lakamisi H 2010 *Jurnal Ilmiah Agribisnis dan Perikanan* **3** 55
[5] Zheng T, Li P, Li L and Zhang Q 2021 *J Horticulture Research* **8** 65
[6] Sukarsa S, Hidayah H A and Chasanah T 2011 *Majalah Ilmiah Biologi BIOSFERA: A Scientific Journal* **28** 23
[7] Majanah dan Saputri 2019 Jurnal Jeumpa **6** 1
[8] Surfiana, Samsul K and Muslich H 2018 *Prosiding Seminar Nasional Biotik* **1** 452
[9] Hasanah U, Hadisunarso and Praptosuwiryo T N 2021 *Biodiversitas* **22** 4968
[10] Puspitaningtyas D M 2017 *Biodiversitas* **18** 341
[11] Mutmainnah 2015 *Prosiding Semirata* 632

IOP Conf. Series: Earth and Environmental Science 1115 (2022) 012025 doi:10.1088/1755-1315/1115/1/012025

Forest fire analysis in Habinsaran Sector PT Toba Pulp Lestari industrial forest plantation area 2014 to 2021

S H Sidabukke[1*], Fahmi Rasyid[2], M Sianipar[3], R H M Panjaitan[3] and F R Aulin[4]

[1]Faculty of Agriculture, Department of Forestry, Universitas Simalungun, Pematang Siantar 21142, North Sumatra, Indonesia
[2]S2 Students of Natural Resource and Environmental Management Study Program, Post Graduate School of Universitas Sumatera Utara.
[3]PT Toba Pulp Lestari, Medan 20231, North Sumatra, Indonesia
[4]Faculty of Forestry, Universitas Sumatera Utara, Medan 20155, North Sumatra, Indonesia

[*]Email: simon201265@gmail.com

Abstract. Forest fires are one of the impacts of the increasing level of pressure on forest resources. Impacts related to forest or land fires are environmental damage and pollution, such as damage to flora and fauna, soil and water. Forest fires impact on the environment and biological ecosystems. The aim of becoming a reference for forest managers in their efforts to prevent forest fires at PT Toba Pulp Lestari. The analysis tested in this study is an analysis of a descriptive quantitative. The interview data contained in the questionnaire will be analyzed to determine the level of forest fires. The results obtained from this study are the age of plants that burn the most are 2-3 years old plants as much as 96 times. And the age of the plant that rarely burns is the plant that is 1-2 years old as much as 16 times. The most common factor causing forest fires from 2014 to 2021 was due to being burned by unknown people 187 times, followed by community farming activities for 31 times. The number of incidents and the area of forest fires occurred 221 times and an area of 163.23 ha. The highest period of fires occurred in February and June while the lowest occurred in December and January.

1. Introduction

Forest is a community consisting of individuals who are neatly arranged and have a reciprocal relationship between plants and plants, plants and animals and the relationship between all living things and the environment which is an ecosystem unit that cannot be separated from one another [1]. Until now, the threat of forest sustainability is still dominated by the danger of fire which causes changes in forest ecosystems [2]. Thus any disturbance to the forest, such as forest fires or damage by natural events, has damaged the forest ecosystem, although naturally damaged forests will return to forests with their original conditions marked by the formation of secondary forests to primary forests [3].

Fire is a term used to describe the physical phenomena produced by the combination of the flow of oxygen and fuel, characterized by the presence of heat, light and fire. Meanwhile, forest fires are the burning of materials in the forest such as trees, shrubs, weeds and other materials in the forest that spread widely and uncontrollably [4]. If this is allowed to continue, it will endanger the diversity of existing flora species. The destruction of one type of flora will result in the loss of the characteristics of all types of flora contained in the forest area [5].

According to [6] forest fires are divided into 3 types based on differences in the way the fire spreads and the position of the fire from the ground, namely (1) Ground fire, the fire burns organic material that is under the litter on the forest floor. In this type of fire spreads slowly and is often difficult to detect, the damage caused is the greatest when compared to the damage caused by other fires. From ground fires usually progress to surface fires, depending on the moisture in the organic layer on the forest floor; (2) Surface fire, the fire usually burns litter on the ground surface, understorey vegetation, such as grasses, herbs, shrubs and tree saplings. This type of fire is most common in timber stands; (3) Crown Fire, the fire spreads from the canopy of trees/shrubs to the crowns of other trees/shrubs. This study aims to provide information about the problems and impacts of forest fires on the environment and biological ecosystems with the aim of becoming a reference for forest managers in their efforts to prevent forest fires at PT Toba Pulp Lestari, especially the Habinsaran sector.

2. Methods

2.1. Time and location of the study
This research was carried out in each estate of the Industrial Plantation Forest PT. Toba Pulp Lestari, Tbk, Habinsaran Sector, Habinsaran sub-district, Toba Regency, North Sumatra Province. This research was conducted in November 2021.

2.2. Research tools and materials
The tools and materials used in this research include a map of the location of the industrial plantation forest of PT Toba Pulp Lestari in the Habinsaran sector and stationery. The materials used in this research include industrial forest areas that have experienced forest fires, the object of research and questionnaires.

2.3. Research procedure

2.3.1. Data collection method. Data collection techniques are carried out by collecting primary and secondary data, namely (1) Observation, data collection was carried out by direct observation. Observations were made based on the results of the determination of forest fire prone areas. The location of the field check is also based on information from the technical agency that handles forest fires. Locations with high hotspot density and land cover indicating fire areas are the locations for field checks; (2) Interview, this research was conducted by survey method. This method aims to collect data through direct interviews. The interview was intended to find out the history, causes, impacts and local wisdom about forest fires. Interviews and discussions were conducted using a questionnaire with the respondents. The required data information includes: Are there frequent fires in the location, the background of the occurrence of forest fires, the frequency of forest fires, the causes of forest fires and community activities.

2.3.2. Analysis of interview results. After collecting interview data, a descriptive quantitative approach will be analyzed. The interview data contained in the questionnaire will be analyzed to determine the level of forest fires at PT Toba Pulp Lestari in the Habinsaran sector from 2014 to 2021.

3. Result and Discussion

3.1. Forest fires
Forest fire is an event where fire devours vegetated fuel that occurs in forest areas that spreads freely and uncontrollably [7]. Forest fires are one of the impacts of the increasing level of pressure on forest resources. Impacts related to forest or land fires are environmental damage and pollution, such as damage to flora and fauna, soil and water [8].

IOP Conf. Series: Earth and Environmental Science 1115 (2022) 012025 doi:10.1088/1755-1315/1115/1/012025

Forest fires occur due to the availability of three main things, namely sources of fire, fuel, and wind. Forest conditions affect the density of vegetation and the types of land use that exist in a place. Forest conditions, vegetation density and land use are related to the availability of fuel for fires in one place [9].

3.2. *Forest fire incidence by age*

The resistance of each tree species to forest fires is different due to differences in bark thickness, age and physiological characteristics of the tree. Tree resilience is meaningless if fires occur for a long time, even if the intensity is small [10]. The recapitulation of forest fires based on age in the PT. Toba Pulp Lestari Habinsaran Sector from 2014 to 2021 as follows.

Table 1. Forest fire occurrences based on plant age in 2014-2021

Year	0-1 years	1-2 years	2-3 years	> 3 years
2014	52	4	5	27
2015	7	2	1	10
2016	24	30	10	11
2017	10	4	0	1
2018	3	0	0	0
2019	0	0	0	0
2020	0	0	0	0
2021	0	0	0	0
Total	96	40	16	49

The results in the table show that the age of plants that burned the most were 0-1 years old plants with 96 times, followed by plants > 3 years old with 49 times. And the age of the plant that rarely burns is the plant that is 2-3 years old as much as 16 times. From the table it can also be seen that the number of forest fires has decreased from year to year. Temperature of 54°C was sufficient to kill the cambium [11]. Fire also causes damage to the surrounding trunk, the tree will languish and die at a temperature of 65°C to 69°C [12].

The thickness of the bark will affect the length of time the tree will die at a certain temperature. The speed at which trees die depends on the temperature of the fire, the higher the temperature the faster the trees die. This has been investigated by Jawad who concluded that for a temperature of 100°F trees will die within 1 hour, while for a temperature of 130°F trees will die in just a few minutes [13]. While the exact temperature range to cause the death of a tree is not known [14].

The resistance of trees to fire varies according to the thickness of the bark, the ease of burning of the bark, the age of the tree, the moisture content of the tree and the special properties possessed by each type of tree [15]. The consequences of forest fires on trees and saplings are in the form of injuries, dead trees and saplings, reducing increment and growth, damaging plant rejuvenation and disrupting other functions of trees. The level of forest management also determines the level of damage if the forest burns [16].

3.3. *Causes of forest fires*

Forest fires are caused by a combination of the three elements of fire, namely fuel, oxygen and heat. The unification of the three elements is due to the presence of a heat source that functions as a trigger for forest fires [17]. The causes of forest fires in the PT. Toba Pulp Lestari from 2014 to 2021 as follows.

Table 2. Recapitulation of data on the causes of forest fires in HTI areas in 2014-2021

No	Causative Factor	Total (Freq)
1	Community farming activities	31
2	Burned by unknown people	187
3	Cooking activities in the area	0
4	Allegedly burned by the community (Claim area)	0
5	Hunting activity	2
6	Unknown	1
	Total	221

Table it shows that the most common factor causing forest fires in 2014 to 2021 was because they were burned by unknown people 187 times, followed by community farming activities for 31 times. This is in accordance with the statement [18] which states that the causes of forest fires are different for each forest area. The causes of forest fires are divided into three main factors, namely intentional factors, accidental factors and natural factors [19]. Forest fires can occur intentionally or unintentionally, while the main factor for forest fires is intentional, 90% of forest fires occur because of intentional behavior from the community. Some community activities that trigger forest fires are farming activities, plantations, etc [20].

The trigger for the fire is the result of a chemical combustion process that is intentionally made by humans for a specific purpose. Various activities that are often carried out by communities around the forest using fire, cannot be separated from their relationship with livelihoods or additional activities [21]. Fires often appear from forested areas to roadsides for the purpose of burning grass that is already thick in fields and roadsides. This custom has been around for a long time. What is even more ironic is that burning has been routinely carried out by owners of idle land, both on dry and wet land, with the sole aim of preventing land from becoming forest or showing ownership [22].

The main cause of forest fires is the human factor that originates from activities or problems with the traditional farming system of the population. This is due to the high demand for land for socio-economic needs and forest area boundary demarcation activities have not been carried out [23]. Human activity factors and biophysical factors interact with each other. Biophysical factors only provide conditions for fuel availability and flammability in an area. As for whether or not an area is burned, it is largely determined by human activities carried out on land with certain characteristics [24].

One of the unwise human actions related to fire hazards is shifting cultivation. Fires often appear from forested areas to roadsides for the purpose of burning grass that is already thick in fields and roadsides. This custom has been around for a long time. What is even more ironic is that burning has been routinely carried out by owners of idle land, both on dry and wet land, with the sole purpose of preventing land from becoming forest or showing ownership. They don't think that the smoke that is generated has endangered the health of others and interferes with vehicle transportation [25].

3.4. Number of fire incidents and land area

The area of forest fires in each region is very diverse, there are several factors that affect the area of forest fires such as length of exposure to sunlight, rainfall, number of rainy days, humidity, air pressure, etc. From these factors, each estate has a different area of fire. The impacts caused by forest fires are quite large, namely ecological damage, decreased biodiversity, decreased economic value of forests and soil productivity, changes in the micro or global climate, and the smoke can interfere with public health and also disrupt transportation, both land, river, sea and air [26]. The number of incidents and the extent of fires in the PT. Toba Pulp Lestari Habinsaran Sector from 2014 to 2021 as follows.

Table 3. Number of incidents and area of forest fires in 2014-2021

Year	Description	
	Number of events	Area
2014	93	75.94
2015	20	2.89
2016	88	68.63
2017	15	6.94
2018	5	8.83
2019	0	0
2020	0	0
2021	0	0
Total	221	163.23

Table 3 shows that the highest number of occurrences and the area of forest fires occurred at Habinsaran estate as much as 221 times and covering an area of 163.23 ha. The nature of forest fires is characteristic of the way the material starts to burn, the development of the flame, the spread of the flame and the appearance of the flame [27]. This is due to the factors that affect forest fires. According to [28] one way to find out the nature of forest fires is by means of a case study, which is to collect information from a fire regarding the spread of fire, the size of the fire, the high and low of the flame, the source of the fire, wind motion, weather, topography, etc.

Each forest fire has a different speed of fire spread and the size of the fire, this is due to differences in the factors that can influence it [12]. These factors are important to be considered and known for consideration in efforts to prevent and control forest fires, including fuel, fuel moisture content, topography, wind, microclimate in the forest and sources of forest fires [19].

3.5. Forest fires by month of occurrence

There is an annual cycle in rainfall and vegetation growth, so each area has a clear fire season when forest fires often occur. The length of this fire season varies from region to region, there may also be two fire seasons a year and sometimes throughout the year [10]. The recapitulation of forest fires based on the month of occurrence in the PT. Toba Pulp Lestari Habinsaran estate from 2014 to 2021 as follows.

Table 4. The forest fire incidents by month of occurrence (2014-2021)

Month											
Jan	Feb	Mar	Apr	May	Jun	Jul	Aug	Sep	Oct	Nov	Dec
0	41	39	1	1	26	40	27	20	24	2	0

Table 4 shows that the most frequent forest fires occurred in February as many as 41 times and in July as many as 40 times. And forest fires do not occur in December and January and rarely occur in April and May. This is in accordance with the statement [8] that the most vulnerable periods in North Sumatra are in June, July and August. By knowing the weather conditions that are suitable and which are not suitable for forest fires, forest fire guards will be able to predict when the fire hazard increases or decreases [27].

All fire typologies, the time of fire generally occurs in June to August, while in September the hotspots have started to decrease, because the rain has started to fall [24]. Several reasons why fires occurred during that month were: a) the community had finished cutting the bush which usually took about 40 days, which is around April and May, and since June the cut bush had dried up, b) during that period there is a very hot and scorching dry season, which encourages people to burn shrubs or grass / litter that has been cut down previously, c) if the burning has started, then a series of unexpected and often uncontrolled fires occurs, especially on unused land that is ownership is not recorded or the owner resides outside the location/in the city (land of guntai) [12].

The average number of fires occurred in July. Communities in general carry out land clearing activities during the dry season, resulting in a high number of hotspots. The number of fires identified every mid-year ie in June and July the number of hotspots increases and reaches its peak. This corresponds to the peak of the dry season [29]. Climate is one of the natural factors that can cause forest and land fires, because climatic conditions can affect the level of dryness of surface fuels, the amount of oxygen available, and the speed of fire spread [4].

4. Conclusion

The plants that burned the most were 0-1 years old, 96 times. And the age of the plant that rarely burns is the plant that is 2-3 years old as much as 16 times. The most common factor causing forest fires from 2014 to 2021 was due to being burned by unknown people 187 times, followed by community farming activities for 31 times. The highest number of occurrences and the area of forest fires occurred as many as 221 times and covering an area of 163.23 ha. The highest fire period occurs in February and June. And the lowest or not occurs in December and January.

References

[1] Kusumaningtyas R and Chofyan I 2013 *Jurnal Perencanaan Wilayah dan Kota* **13** 1
[2] Rasyid F 2014 *Jurnal Lingkar Widyaiswara* **1** 47
[3] Cahyono S A, Warsito S P, Andayani W and Darwanto D H 2015 *Jurnal Sylva Lestari* **3** 103
[4] Wibowo K A 2019 *Jurnal Studi Sosial dan Politik* **3** 69
[5] Suryani A S 2012 *Aspirasi: Jurnal Masalah-masalah Sosial* **3** 59
[6] Yusuf A, Hapsoh H, Siregar S H and Nurrochmat D R 2019 *Dinamika Lingkungan Indonesia* **6** 67
[7] Adiputra A 2018 *Geografi, Edukasi dan Lingkungan (JGEL)* **2** 1
[8] Maranatha E I and Kusmayadi I M 2020 *Jurnal Kajian Jurnalisme* **3** 153
[9] Ratnaningsih A T and Prastyaningsih S R 2017 *Wahana Forestra: Jurnal Kehutanan* **12** 37
[10] Itsnaini N, Sasmito B, Sukmono A and Prasasti I 2017 *Jurnal Geodesi Undip* **6** 62
[11] Akbar A, Sumardi S, Hadi R, Purwanto P and Sabarudin M S 2011 *Jurnal penelitian hutan tanaman* **8** 287
[12] Pasai M 2020 *Jurnal Pahlawan* **3** 36
[13] Jawad A, Nurdjali B and Widiastuti T 2015 *Jurnal Hutan Lestari* **3** 88
[14] Pinem T 2016 *GEMA TEOLOGIKA: Jurnal Teologi Kontekstual dan Filsafat Keilahian* **1** 139
[15] Suhendri S and Purnomo E P 2017 *Journal of Governance and Public Policy* **4** 174
[16] Anggraini N and Trisakti B 2011 *Jurnal Penginderaan Jauh dan Pengolahan Data Citra Digital* **8** 11
[17] Mapilata E, Gandasasmita K and Djajakirana G 2013 *Majalah Ilmiah Globe* **15** 178
[18] Widodo R B 2014 *Jurnal Pembangunan Wilayah dan Kota* **10** 127
[19] Saputro J G J, Handayani I and Najicha F U 2021 *Jurnal Manajemen Bencana (JMB)* **7** 27
[20] Sugianto C A 2015 *Techno Com* **14** 336
[21] Putra I K, Saharjo B H and Wasis B 2019 *Jurnal Ilmu Pertanian Indonesia* **24** 151
[22] Murtinah V, Edwin M and Bane O 2017 *Jurnal Pertanian Terpadu* **5** 128
[23] Zulkifli I and Kamarubayana L 2017 *Agrifor: Jurnal Ilmu Pertanian dan Kehutanan* **16** 141
[24] Agustiar A B, Mustajib M, Amin F and Hidayatullah A F 2020 *Profetika: Jurnal Studi Islam* **20** 124
[25] Irwandi J and Ismail B 2016 *Agrifor: Jurnal Ilmu Pertanian dan Kehutanan* **15** 201
[26] Harafani H and Wahono R S 2015 *Journal of Intelligent Systems* **1** 82
[27] Cabucci M O and Maulina P 2021 *Jurnal Komunikasi dan Kajian Media* **5** 205
[28] Pratiwi T A, Irsyad M, Kurniawan R, Agustian S and Negara B S 2021 *CESS (Journal of Computer Engineering, System and Science)* **6** 139

[29] Aflahah E, Hidayati R, Hidayat R and Alfahmi F 2019 *Jurnal Pengelolaan Sumberdaya Alam dan Lingkungan (Journal of Natural Resources and Environmental Management)* **9** 405

Identification of the generative phases of *Amorphophallus gigas* in the Sabungan Village and Langga Payung Village, Sungai Kanan District, North Sumatra Province, Indonesia

R Rambey[1,2*], Rahmawaty[2,3], A Rauf[3,4] and E S M Nababan[3,5]

[1]Doctoral Program in Natural Resources and Environment Management, Graduate School, Universitas Sumatera Utara. Jl. Sivitas Akademika No. 9, Medan 20222, North Sumatra, Indonesia
[2]Faculty of Forestry, Universitas Sumatera Utara, Jl. Tri Dharma Ujung No. 1 Kampus USU, Medan 20155, North Sumatra, Indonesia
[3]Natural Resources and Environment Management, Graduate School, Universitas Sumatera Utara. Jl. Sivitas Akademika No. 9, Medan 20222, North Sumatra, Indonesia
[4]Faculty of Agricultue, Universitas Sumatera Utara, Medan 20155, North Sumatra, Indonesia
[5]Faculty of Mathematics and Natural Sciences, Universitas Sumatera Utara, Jl. Bioteknologi No. 1 Kampus USU, Medan 20155, North Sumatra, Indonesia

[*]Email: ridahati.rambey@usu.ac.id

Abstract. *Amorphophallus gigas* is one of Sumatra's endemic *Amorphophallus* species that is protected in Indonesia. *Amorphophallus gigas* is one type of *Amorphophallus* with the highest flowers in the world. The conversion of land to other uses is one of the causes of the reduced population of *Amorphophallus gigas*. One of the natural habitats of *Amorphophallus gigas* was found in the Sabungan Village and Langga Payung Village, North Sumatra. The purpose of this study was to determine the distribution of the generative phase of *Amorphophallus gigas* in the Sabungan Village and Langga Payung Village, to determine the types of plants found around the place of growth and to determine the morphology of the generative phase of *Amorphophallus*. The method used in this research is the roaming method and then records the coordinates of the generative phase of *Amorphophallus*, records the types of pollinating insects around it and observes the morphology of *Amorphophallus gigas*. The results showed that there were 11 individuals in the generative phase of *Amorphophallus gigas* in the Sabungan Village and Langga Payung Village, where 10 individuals grew on rubber land cover and 1 individual grew on oil palm land cover. A total of 2 individuals were found in bud condition, 1 individual was in bloom, 2 individuals were wilted and 6 individuals were found at the seed formation stage. There are 12 types of insects found around the generative phase of the *Amorphophallus*.

1. Introduction

The genus *Amorphophallus* has more than 200 species in the world from West Africa to tropical Asia and northern Australia [1]. Indonesia has about 25 species or about 12.5% of the total species in the world. The distribution of carrion flowers are: 8 species in Sumatra, 5 species in Java, 3 species in Kalimantan and 1 other species in Sulawesi [1]. Sumatra Island is an area with the highest level of endemism compared to other islands in Indonesia [2]. This plant grows from the lowlands to 1000 m above sea level, with temperatures between 25-35°C, while the rainfall is between 300-500 mm per month during the growth period [3].

Amorphophallus gigas is one of the *Amorphophallus* species protected by the government according to the Regulation of the Minister of Environment and Forestry No. 20 of 2018. *Amorphophallus gigas* is one of the *Amorphophallus* endemic in North Sumatra [4] and is the highest *Amorphophallus* flower in the world [5]. One of the locations where *Amorphophallus gigas* is located is the Sabungan Village and Laggapayung village, Sungai Kanan District, North Sumatra.

Sabungan Village and Langga Payung Village are located in Sungai Kanan District, North Sumatra Province. This village has flat and hilly land conditions with a height of 30 to 50 meters above sea level. It is in the Sungai right sub-watershed and is included in the Asahan Barumun watershed [6]. There are several types of land cover in this village, namely rubber land cover and oil palm land cover. Changes in rubber land to oil palm continue to occur because according to the community, palm oil is a promising commodity. However, the practice of clearing oil palm land using herbicides causes some to continue to experience extinction such as *Amorphophallus*. In rubber plantations, *Amorphophallus gigas* continues to grow under stands because rubber plantations are usually left under bushes to resemble secondary forests and are rarely cleaned with herbicides.

There are 3 growth phases of *Amorphophallus*, namely the generative phase, the dormant phase and the vegetative phase. The vegetative phase is the phase where the leaves emerge. The generative phase of *Amorphophallus* is the flowering phase until the formation of seeds. This vegetative phase begins at the beginning of the rainy season with the production of a single large leaf and lasts for 6-12 months, followed by a dormant phase for 1 to 4 years before entering the flowering cycle. Flowering cycles are generally irregular [7].

Research on the bioecology of species and their distribution patterns is an alternative that can be done to support conservation efforts of these species [8]. Land conversion and taking carrion flower bulbs are a threat to the existence of carrion flower populations. Public knowledge of the scarcity status of carrion flowers is also still lacking). very low resulting in carrion flowers not getting much attention. In addition to human factors, natural factors also determine the regeneration of carrion flowers. In damaged environmental conditions, *Amorphophallus* is difficult to self-pollinate because there is no natural pollination by the help of animals around it. In a good environment, *Amorphophallus* can self-pollinate naturally. So that pollination techniques are also a challenge in ex situ conservation of this *Amorphophallus* species. According to [9] Self-pollination is almost impossible, but sometimes it works naturally with the help of pollinators. Pollen from other plants is usually used for artificial pollination.

The objectives of this study were (1) to determine the distribution of the generative phase of *Amorphophallus gigas* in the Sabungan Village and Langga Payung Village, North Sumatra, (2) to determine the types of pollinators found around *Amorphophallus gigas* and to determine the morphology of the generative phase of *Amorphophallus gigas*.

2. Methodology

This research was carried out in 2022 in Sabungan Village and Langga Payung Village, Sungai Kanan District, South Labuhanbatu Regency, North Sumatra Province. The method used in observing the generative phase of *Amorphophallus* is the cruising method. Explorations were carried out on the banks of the channel and riverbanks on several land covers, namely open land, oil palm land cover and rubber land cover and interviews with local communities where *Amorphophallus* was commonly found. To observe the types of animals that are around *Amorphophallus* by direct observation of the types and numbers later. The placement of the plots was done intentionally, namely only on *Amorphophallus* in the generative phase in various land covers. The morphological measurements of amorphpohallus in the generative phase were flower height, spata diameter, male flower length, female flower length and flower diameter, spadic diameter.

3. Result and Disscussions

Based on the results of the research conducted, there were 5 individuals of *Amorphophallus gigas* in the generative phase, namely 10 individuals on rubber land cover and 1 individual on oil palm land cover. *Amorphophallus gigas* grows wild in several land covers in Simandiangin Hamlet in several land covers such as rubber land cover and oil palm land cover. This is in line with [11] that *Amorphophallus* is abundant under tree stands such as agroforestry, secondary forest and open land. Details of *Amorphophallus gigas* found in two villages can be seen in Table 1. The distribution of generative phase of *Amorphophallus gigas* in the Sabungan Village and Langga Payung Village can be seen in Figure 1.

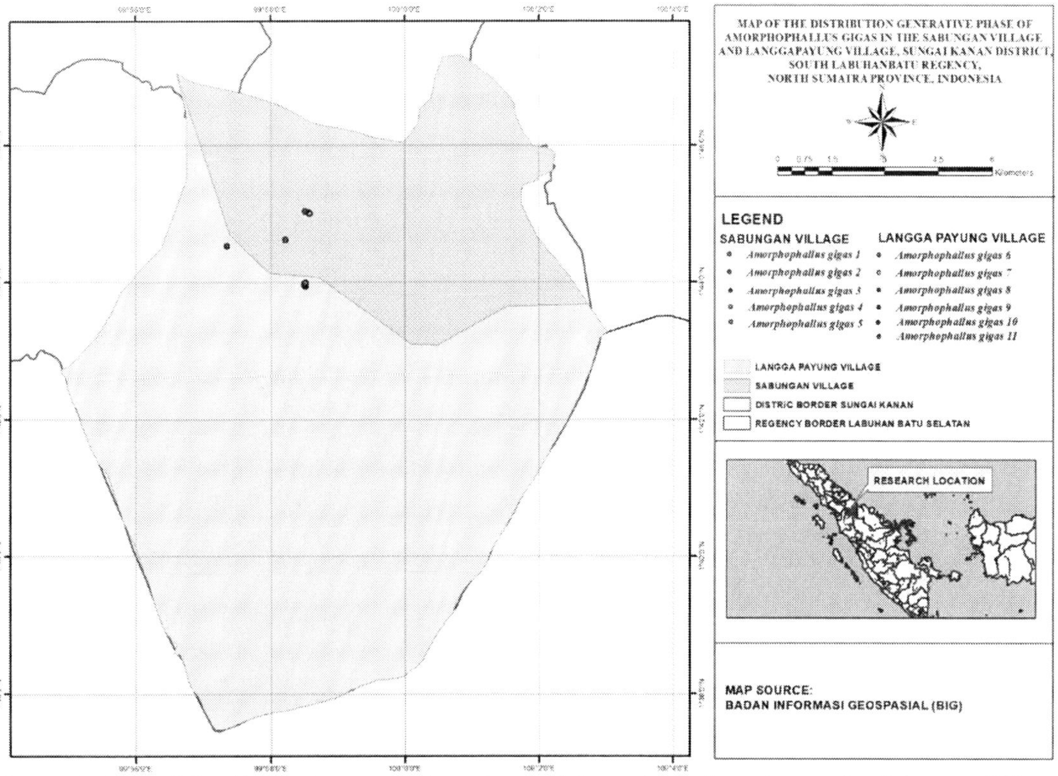

Figure 1. Map of the distribution generative phase of *Amorphophallus gigas* in the Sabungan Village and Langga Payung Village

Table 1. Description of the generative phase of *Amorphophallus gigas* found in the Sabungan Village and Langga Payung Village, North Sumatra

Phase	Location	Description
Amorphophallus gigas 1	Rubber land cover (Sabungan Village)	This flower was found in bloom with a height of 191 cm with a diameter of 32.48 cm. The height of the peduncle is 111, the height of the female flower is 13 cm, the height of the male flower is 4.5 cm, the height of the spadix is 62.5 cm. Peduncle diameter 5.25 cm, spadix diameter 10.50 cm. The spata is purple, the spadix is yellow, the peduncle is green with round white spots like tinea versicolor. The color of young female flowers is dark purple, while the color of male flowers is yellow. The male flower is on top of the female flower wrapped in a spatha.

Amorphophallus gigas 2	Rubber land cover (Sabungan Village)	*Amorphophallus gigas* was found in a green seed condition. The total height from peduncle to seed is 176 cm, peduncle height is 147 cm, peduncle diameter is 6.36 cm, seed height is 29 cm with a diameter of 7 cm.
Amorphophallus gigas 3	Oil palm land cover (Sabungan Village)	This *Amorphophallus* is found in almost ripe seed condition which is characterized by a yellow seed color. The total height from peduncle to seed was 102 cm, peduncle height was 78 cm, peduncle diameter was 3.18, seed height was 24 cm and seed diameter was 6.36 cm. Seeds are arranged in large numbers attached to the top of the peduncle.
Amorphophallus gigas 4	Rubber land cover (Sabungan Village)	The condition of the seeds is almost yellowish. The total height from the base of the peduncle to the seed was 134 cm, the height of the peduncle was 104 cm, the diameter of the peduncle was 4.76 cm, the height of the seed was 30 cm and the diameter of the seed was 7 cm.
Amorphophallus gigas 5	Rubber land cover (Sabungan Village)	This flower is found in bud condition, this flower is in the process of blooming. The total height of the buds is 93 cm.
Amorphophallus gigas 6	Rubber land cover (Langga Payung Village)	The total height is 195 cm, the peduncle height is 166 with a diameter of 5.09 cm, the seed height is 29 cm and the diameter of the cob is 6.68 cm. The color of the seeds starts to yellow. Diameter of seeds 0.5 cm
Amorphophallus gigas 7	Rubber land cover (Langga Payung Village)	Total height 162 cm, peduncle height 120 cm with a diameter of 8.59 cm, seed height 42 cm with a diameter of 9.55 cm, yellowish green seed color.
Amorphophallus gigas 8	Rubber land cover (Langga Payung Village)	Total height 121 cm, peduncle height 111 with a diameter of 10.50 cm, seed height 20 cm with a seed diameter of 9.55 cm. New seeds are formed after the spatha and male flowers wither. This means that pollination in nature is perfectly formed. The seeds are still small and purple
Amorphophallus gigas 9	Rubber land cover (Langga Payung Village)	Total height 127 cm, peduncle height 109 cm with a diameter of 7.00. The height of the female flower is 13 cm with a seed diameter of 5.09 and the height of the male flower is 5 cm with a diameter of 3.82 cm. Flowers (spatha) wilt but complete pollination to form purple seeds. Male flowers begin to wither.
Amorphophallus gigas 10	Rubber land cover (Langga Payung Village)	Total height 87 cm, peduncle height 72 cm with a diameter of 3.50. The height of the female flower is 11 cm and the height of the male flower is 4 cm with a seed diameter of 3.82 cm, the diameter of the male flower is 3.18 cm. Flowers (spatha) wilt but complete pollination to form purple seeds. Male flowers begin to wither.
Amorphophallus gigas 11	Rubber land cover	The total height is 114 cm with an oeduncle diameter of 5.41 cm, a peduncle height of 79 cm, an appendix (bud) height of 35 cm and a bud diameter of 6.36 cm. dark green buds.

(Langga Payung Village)

Figure 2. The generative phase of *Amorphophallus gigas* in various land covers

3.1. Pollinators found around Amorphophallus gigas

There are 12 types of pollinators around the generative phase of *Amorphophallus gigas*, including: *Lucilia* sp., *Sithopilus oryzae, Plagiodera* sp., *Vespa affinis, Sylvicola fenestralis, Sylvicola fenestralis, Araneus* sp., *Monomorium minimum, Psyllobora vigintiduopunctata, Psyllobora vigintiduopunctata, Carabus* sp., *Aedes albipictus* and *Blatella* sp.

Table 2. Pollinators found around *Amorphophallus gigas*

Scientifict name	Family	Found on
Lucilia sp.	Muscidae	Ag1, Ag5
Sithopilus oryzae	Tenebrionidae	Ag1, Ag6, Ag 9, Ag 10
Plagiodera sp.	Chrysomelidae	Ag5
Vespa affinis	Vespidae	Ag5
Sylvicola fenestralis	Anisopodidae	Ag1
Araneus sp.	Araneidae	Ag7. Ag8
Monomorium minimum	Formicidae	Ag1, Ag2, Ag3, Ag4, Ag6, Ag8
Psyllobora vigintiduopunctata	Coccinellidae	Ag5
Carabus sp.	Carabidae	Ag5
Aedes albipictus	Culicoedea	Ag1, Ag2 Ag6, Ag8, Ag9, Ag10. Ag 11
Panesthia angustipennis	Blaberidae	Ag7

Based on [12] the presence of carrion flowers that have a distinctive smell can invite several types of attacks, namely beetles (*Onthophagus* sp.) and carrion flies (*Lucilia* sp.) These insects help the pollination process which will then develop into fruit. The fruit of the carcass flower becomes food for the finches (*Pignonotus aurigaster*), so to maintain the existence of the corpse flower, the finches are needed as a seed dispersal agent from the carcass flower. Based on [13] reported that insect visitors to *A. paeoniifolius* inflorescences in the Bogor Botanical Gardens were flies (*Calliphora vomitoria*), meat flies (*Sarcophaga* sp.), house flies (*Musca domestica*), stingless bees (*Trigona* sp.), hoverflies (Syrphidae) and sap-eating beetles (*Epuraea haptoncurina*).

According to [14] the success of conservation depends on whether a plant species can regenerate naturally and achieve its life cycle in the long term. Therefore, it is very important to assess a species' reproductive strategy before deciding whether in situ and/or ex situ conservation should be undertaken. The role of pollinators in the reproduction of species is indispensable in plant reproduction for perfect pollination to occur.

In addition to cultivation techniques, pollination techniques are also a challenge in ex situ conservation of this *Amorphophallus* species. Self-pollination is almost impossible, but sometimes it works naturally with the help of pollinating animals. Pollen from other plants is usually used for artificial pollination [10].

4. Conclussions

Amorphophallus gigas is a wild plant that grows as an understorey in the Sabungan Village and Langga Payung Village. There are 11 individuals in the generative phase in the Sabungan Village and Langga Payung Village consisting of 1 blooming flower, 2 withering flowers and 8 with seeds. The generative phase of *Amorphophallus* was found in rubber land cover and oil palm land cover. There are 12 species of insects found around the generative phase of *Amorphophallus gigas*.

References

[1] Yuzammi 2015 *Rencana Aksi Konservasi Bunga Bangkai* (Kementerian Lingkungan Hidup dan Kehutanan)

[2] Santosa E, Lian C L, Sugiyama N, Misra R S, Boonkorkaew P and Thanomchit K 2017 *PLoS ONE* **12** 6

[3] Sumarwoto 2005 *Biodiversitas* 6 185

[4] Rambey R, Rangkuti A B, Onrizal, Susilowaty A, Wijayanto N and Siregar I Z 2020 *IOP Conf. Series: Earth and Environmental Science* **782** 2021

[5] Steinecke H 2016 *Amorphophallus gigas blüht erstmals im Palmengarten. Der Palmengarten.* (German : University Library J. C. Senckenberg) **80** 47

[6] Wikipedia.org 2022

[7] Poerba Y S and Yuzammi 2008 *J Biodiversitas* **9** 103

[8] Hetterscheid W L A and Claudel C 2014 *Botanical Studies* **55** 2
[9] Nursanti, Wulan C and Felicia M R 2019 *Jurnal Silva Tropika* **3** 2
[10] Puspitaningtyas D M and Ariati S R 2016 *Pros Sem Nas Masy Biodiv Indon* **2** 219
[11] Mursyidin D H and Hernanda M A 2021 *Biodiversitas* **22** 2821
[12] Jintan, Yuzammi, Suwastika N and Pitopang R 2015 *Jurnal of Natural Science* **4** 17
[13] Hanfayani T, Yuzammi, Hadiah JT. 2020 *Biodiversitas* **21** 5835
[14] Tang J R R, Liu E , Zhang Y, Schinner L J, Sun W and Chen G 2020 *BMC Genetics* **21** 102

IOP Conf. Series: Earth and Environmental Science 1115 (2022) 012027 doi:10.1088/1755-1315/1115/1/012027

Effect of gamma iradiated seeds of Jabon Merah *(Neolamarckia macrophylla (Wall.) Bosser)* to genetic diversity

S H Larekeng[1,2*], N Qalbi[3], A Rachmat[4], I Iswanto[1], M Restu[1]

[1]Biotechnology and Tree Breeding Lab., Faculty of Forestry, Universitas Hasanuddin, Makassar, Indonesia.

[2]Research and Development Center of Natural Heritage and Biodiversity, Institute of Research and Community Service, Universitas Hasanuddin, Makassar, Indonesia.

[3]Forestry Vocational School of Makassar, Indonesia.

[4]Research Center for Genetic Engineering, National Research and Innovation Agency, Indonesia

Corresponding author's e-mail: sittihalimah@unhas.ac.id

Abstract. Jabon merah (*Neolamarckia macrophylla* (Wall.) as one of the potential industrial forest plantations to meet the need for wood that has a moderate level of genetic diversity. This study aims to analyze the morphological and genetic characters and analyze the stability of genes in explants from red jabon seeds induced by gamma rays at doses of 0 gray, 5 gray, 10 gray, 15 gray, 30 gray, 45 gray and 60 gray, 75 gray, 90 gray, 120 gray and 240 gray. The induced seeds were cultured *in vitro*. The subculture was started with an initialization on MS0 medium and repeatedly cultured. Finally, the LD50 value was determined using curve fit analysis, explant morphological analysis with SPSS, genetic analysis of explants using RAPD markers and SSR/Microsatellite. The DNA analysis with 9 SSR/Microsatellite and RAPD marker primers showed that most of the induction doses produced more fragments than the control.

Mutagenic,*Neolamarckia macrophylla*, RAPD, SSR.

1. Introduction

Jabon merah (*Neolamarckia macrophylla* (Wall.) Bosser), as a potential species of wood-producing plant, has a limited natural distribution. The distribution of jabon merah covers the Central - Eastern part of Indonesia, including in Sulawesi, Maluku, North Maluku and Papua [1,2]. With a high development potency and a limited distribution area, there is a possible risk of decreasing genetic variation which have a major impact on the sustainability of the species. [3] suggested that the loss of genetic variation and the occurrence of inbreeding contributed greatly to the risk of extinction. Species with low genetic diversity have an impact on the reduced ability of the species to survive environmental changes during evolution and lead to increased susceptibility to disease, pests, parasites and weeds [4].

Mutation breeding as an alternative to improve plant genetic quality can be a source of extensive genetic diversity. According to Bramasto [5], mutation breeding has great potential, especially in generating new diversity that increases heterozygicity to obtain plants with characters that are more adaptive to environmental changes with a high level of productivity. Gamma ray irradiation as a form of induced mutation has been tested on many types of agricultural and forestry crops.

The genetic variation of jabon merah have been carried out. Arif et al. [6] found that the genetic diversity of the jabon merah in the Sulawesi provenance is at a moderate level, so a combination of in situ and ex situ genetic conservation is needed. Rinaldi [7] showed the coefficient of genetic variation in jabon merah from Sulawesi and Maluku in the research forest of Parung Panjang Bogor have tall characters was 8.8% (medium) and 3.53% (low) for diameter characters, indicating that genetically there is moderate diversity for high characters, while diameter characters have low diversity or tend to be uniform. From this study, information was obtained that the genetic variation of the Sulawesi jabon merah provenance was still classified in the low to medium category.
The molecular marker approach has a very high ability to capture the diversity of characters between individuals [8]. The markers used are RAPD markers which are commonly used in research related to genetic variation [9] and microsatellite markers show higher polymorphic and allele variability and have been tested on jabon merah [6].

2. Materials and Methods

2.1. Plant Material and Research Location

The fruits of Jabon Merah were collected from seed orchard at the class of identified seeds stand in Luwu Regency, South Sulawesi, Indonesia. Induction of Gamma irradiation was carried out at P3TIR BATAN, Jakarta, Indonesia. Seed germination testing was performed at BPTH Region II, Makassar, South Sulawesi.

2.2. Gamma Irradiation and in Vitro Propagation

Induction of Gamma irradiation was performed using several doses: 0, 5, 10, 15, 30, 60, 120 and 240 Gray by irradiation Gamma Chamber 4000 A.

2.3. Polymerase Chain Reaction

A total of 9 universal SSR and RAPD markers used in this study. DNA amplification using Polymerase Chain Reaction (PCR) was performed in a total volume of 15 µL containing 20 ng of template DNA, 10 µM of each primer (forward and reverse), 6.25 µL of KAPA PCR ready mix (KAPA Biosystem). The PCR amplification was programmed to initial denaturation at 94 °C for 3 min, cycle 35 times under the following conditions: denaturation for 30 sec at 94 °C, annealing for 50 sec at 55 °C, and elongation for 1 min at 72 °C. Final extension was held for 5 min at 72 °C. Amplicons were migrated on 8% non-denatured polyacrylamide gels and used a 100 bp-DNA ladder for molecular weight size. The electrophoresed gel was stained with ethidium bromide and visualized under ultraviolet light in UV Transilluminator and documented for allele detection:

Table 1. List of primers for SSR and RAPD markers

Molecular marker	Primer	Sequens of basa	Reference
SSR/Microsatelite	NCA C11	F: GTTGGCTTGGTTTCTCCATC R: ACATTACCCACAACCCCATC	[10]
	NCA C15	F: TGCGTGTATTGTGCGTGTAG R: TAACACTTAGGGCCGCTTG	

	NCA G01	F: CATTCCGCGTACGTGTTACC R: TGAGAGATCCCGACGACAC	
RAPD	OPY 09	AGC AGC GCA C	[11]
	OPP 08	ACA TCG CCC A	
	OPK 20	GTG TCG CGA G	
	OPA15	TTC CGA ACC C	
	OPA 02	TGC CGA GTC G	
	OPD 03	GTC GCC GTC A	

2.4.Data Analysis

Analysis of molecular data was done by looking at the results of the advent of DNA amplification using a primers. The obtained data in scoring was present (1) or not present (0) ribbon of each individual at a certain size in each primer. From these data, it can be seen the number and percentage of polymorphic loci.

3. Result and Discussion

The results of the study on the M1 population showed that the dose of gamma-ray irradiaton affected the number of explants. Gamma rays irradiation which applied to plants will affect the morphological characteristics due to damage to DNA. This damage induces mutations at both the chromosome level and free radical genes. Physical mutagenesis produces any DNA changes of point mutation, insertion, deletion that can affect phenotypic variation [12].
Molecular techniques such as molecular marker offers an opportunity to help detect the changes by examining both genetic variants in the population and detect the desired mutants in a more robust way [13]. This molecular marker can served as complements for morphological markers [14]. Molecular markers can reveal differences among irradiated and non-irradiated individual plants at the DNA level and, thus, provide a more direct, reliable and efficient tool for germplasm conservation and management [15].

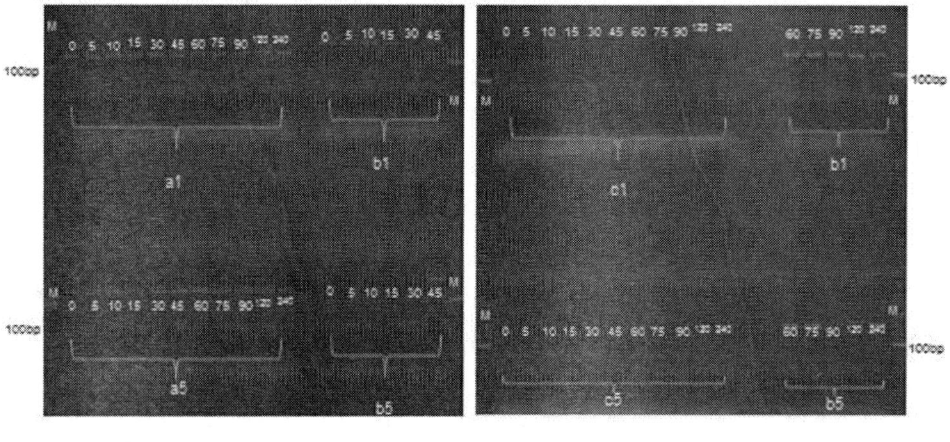

Figure 1. Banding pattern of Jabon merah mutant lines based on SSR: *M = Marker ; 0,5,10,15,30,45,60,75,90,120,240 = gamma irradiaton dose (gray); a1= primer NCA C11; a5= primer NCA C11; b1= primer NCA C15; b5= primer NCA C15; c1= primer NCA CG01; c5= primer NCA CG01*

Amplification using 3 SSR primers as recapitulated in Figure 1 and Table 2., showed that the tested SSR primers produced monomorphic fragments with a size of 200–300 bp. Wuryaningsih [10] conducted an analysis of genetic variation in jabon merah with NCA C11, NCA C15 and NCA CG01 primers produced polymorphic fragments. This could be due to differences in the individual trees used as sample sources. Induced mutation technique could improve one or more characters without changing the basic character of native varieties. Induced mutation through seed becomes the most effective breeding approaches in developing varieties according to the desired characters [4,12].

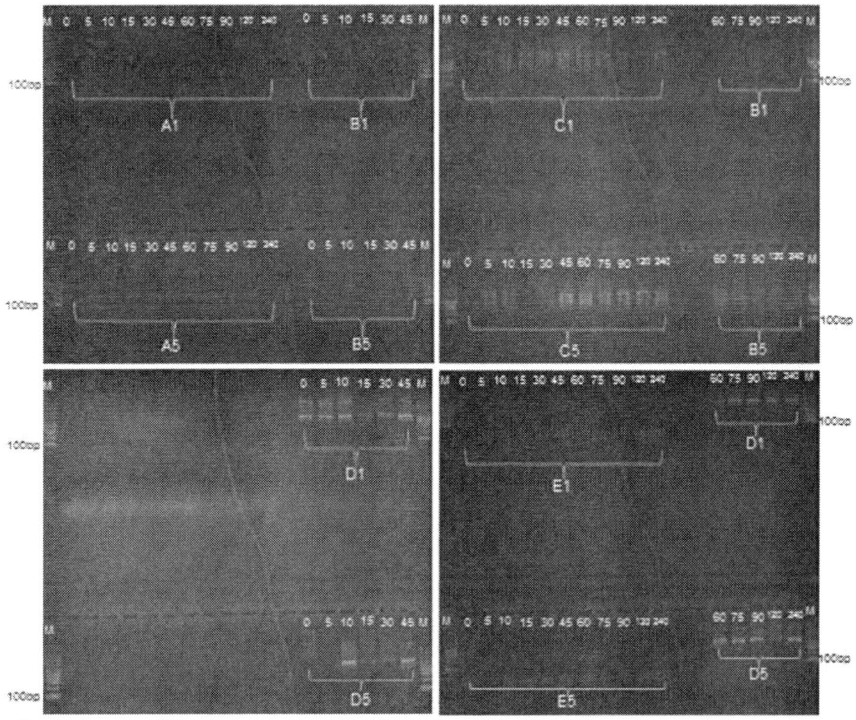

Figure 2. Banding pattern of Jabon merah mutant lines based on RAPD markers.

M=Marker a1= primer NCA C11; a5= primer NCA C11; b1= primer NCA C15; b5= primer NCA C15 c1= primer NCA CG01; c5= primer NCA CG01.

Individual genetic diversity of a species is geographically different even though in one population.The genetic variation caused by each individual plant has a unique DNA shape [16]. The genetic diversity of a species in a population can be influenced by the environment, natural conditions, population size, methods of reproduction and natural selection [17]. In addition, genetic diversity can also be influenced by mutations, gene flow and genetic drift [18].
Amplification DNA using RAPD primers on mutagen samples, most of them produced polymorphic bands with fragment sizes ranging from 200-900 bp showed in Figure 2, and amplification DNA fragments produced using 9 primers showed in Table 2.

Table 2. Recapitulation of the number of fragments using SSR and RAPD primer

No	Primers	Dose /number of fragment										
		0	5	10	15	30	45	60	75	90	120	240
1	NCA C11	1	1	1	1	1	1	1	1	1	1	1
2	NCA C15	1	1	1	0	0	1	1	1	1	1	1
3	NCA CG01	0	0	0	0	0	0	0	0	0	0	1
4	OPK 20	3	1	1	2	1	3	3	3	2	2	1
5	OPA 15	3	1	2	0	3	1	3	3	4	4	4
6	OPA 02	1	1	1	0	3	2	2	2	1	1	2
7	OPY 09	3	0	5	4	4	3	2	2	2	3	2
8	OPP 08	0	2	3	1	1	3	3	3	3	5	2
9	OPD 03	0	0	0	3	3	2	2	2	1	2	2

Genetic stability is one of the measurement indicators in plant subcultures, in-vitro culture to determine the true-to-type quality of the tested explants. In this study, the genetic stability of the explants was assessed by comparing the DNA fragments in subculture 1 with DNA fragments in subculture 5. The results of amplification with 9 primers resulted in DNA fragments as shown in Figure 2. In this study, genetic stability analysis was carried out on the fragments obtained. produced by RAPD primers, namely OPK 20, OPA 15, OPA 02, OPY 09 and OPD 03, like research by Zulfiana and Makmur [19,20]. The fragments produced by SSR/Microsatellite primers were monomorphic so genetic stability analysis was not performed. Primer OPP 08 also only produced one fragment so that stability analysis was not carried out on the primer. Recapitulation of the number of fragments produced in the primary OPK 20, OPA 15, OPA 02, OPY 09 and OPD 03 at each radiation dose as shown in Table 3.

The percentage of stable fragments obtained as shown in table 3 indicated a fairly large value (greater than 50%) in 3 primers, namely OPK 20, OPA 15 and OPD 03. The results obtained that genetically the subcultured explants relatively stable, some species show subculture frequencies ranging from 3 to 9 times still showing good growth. Konar et al [21] showed the highest shoot multiplication in *Hibiscus sabdariffa* was obtained in subculture 8. Harianja et al [22] also stated that subculture in Medicago sativa can still be carried out until subculture 3 because it remains genetically stable.

Table 3. Recapitulation of stable fragments using RAPD primers.

No	Primers	Type of fragment	Number of fragment at dose (gray)											Total	Percent (%)
			0	5	10	15	30	45	60	75	90	120	240		
1	OPK 20	stable fragment	2	1	1	1	1	2	3	1	2	0	1	15	60,0
		Total of fragment	3	2	1	3	1	3	3	3	2	2	2	25	
2	OPA 15	stable fragment	2	0	2	0	2	1	1	1	4	4	3	20	60,6
		Total of fragment	3	1	3	3	3	1	3	4	4	4	4	33	
3	OPA 02	Fragmen stabil	0	1	1	0	1	1	1	1	1	0	1	8	44,4
		stable fragment	1	1	1	1	4	2	2	2	1	1	2	18	
4	OPY 09	Fragmen stabil	1	0	2	2	2	2	1	1	0	1	1	13	33,3
		Total of fragment	4	2	5	4	4	3	3	3	4	4	3	39	
5	OPD 03	stable fragment	0	1	2	1	1	2	2	2	2	2	2	17	54,8
		Total of fragment	2	2	3	2	2	3	3	3	4	5	2	31	

4. Conclusion

The results of DNA analysis using 9 primers for SSR/Microsatellite and RAPD markers showed that most of the gamma-irradiation doses given produced more fragments than control, which is indicate that the genetic diversity of the induced explants at that dose was greater than the control.

5. Acknowledgments

The authors gratefully acknowledge financial support from the Ministry of Research, Technology and Higher Education of the Republic of Indonesia and National Research and Innovation Agency, Indonesia.

6. References

[1] Setyaji T, Arif N, S. S, K S D, S.K. D And Sumaryana 2014 *Budidaya Intensif Jabon Merah (Anthocephalus Macrophyllus) Si Kecil Dari Jawa Timur* (Jakarta, Indonesia: IPB Press)

[2] Shi S, Larekeng S H, Lv P, Nie Y, Restu M And Yang H 2020 The Complete Chloroplast Genome Of *Neolamarckia Macrophylla* (Rubiaceae) *Mitochondrial DNA Part B* **5** 1611–2

[3] Chung M Y, Son S, Herrando-Moraira S, Tang C Q, Maki M, Kim Y D, López-Pujol J, Hamrick J L And Chung M G 2020 Incorporating Differences Between Genetic Diversity Of Trees And Herbaceous Plants In Conservation Strategies *Conserv. Biol.* **34** 1142–51

[4] Cires E, Sanna M, Vázquez V M And Prieto J A F 2018 Genetics And Conservation Of Rare And Endemic Plants: The Case Of Genista Sanabrensis (Fabaceae) In The Iberian Peninsula *Mediterr. Bot.* **39**

[5] Bramasto Y, Putri P, Zanzibar P K, Zanzibar P K, Danu M And Danu M 2016 Pemanfaatan Teknik Irradiasi Sinar Gamma Untuk Meningkatkan Viabilitas Benih Sengon *J. Hutan Trop.* **4** 14–20

[6] Arif A, Larekeng S H, Restu M, Cahyaningsih Y F And Mukti J 2019 A Genetic Diversity On Jabon Merah (*Anthocephalus Macrophyllus* Roxb.) From Three Different Provenances In South Sulawesi *IOP Conf. Ser. Earth Environ. Sci.* **270**

[7] Rinaldi S P 2018 *Keragaman Pohon Induk Dan Parameter Genetik Pertumbuhan Awal Uji Keturunan Jabon Merah (N.Macrophylla) Asal Sulawesi Dan Maluku* (Bogor)

[8] EL-Bakatoushi R And Ahmed D G A 2018 Evaluation Of Genetic Diversity In Wild Populations Of Peganum Harmala L., A Medicinal Plant *J. Genet. Eng. Biotechnol.* **16** 143–51

[9] Akzad M Uhamma. B, Nuraeni S And Larekeng S H 2021 Detecting DNA Polymprhism On Mulberry (*Morus* Sp.) Using RAPD And ISSR Markers *Plant Cell Biotechnol. Mol. Biol. ISSN 0972-2025* **22** 106–11

[10] Wuryaningsih A 2017 *Variation Microsatellite DNA Of Jabon Merah (Anthocephalus Macrophylus) On Conservation Site Genetic Resources In South Sulawesi. (Tesis)* (Gadjah Mada University)

[11] Larekeng S ., Dermawan R, Iswoyo H And Mustari K 2019 RAPD Primer Screening For Amplification On Katokkon Pepper From Toraja, South Sulawesi, Indonesia *IOP Conference Series: Earth And Environmental Science* Vol 270 P 012023

[12] Singh D P, Singh A K And Singh A 2021 Mutation Breeding *Plant Breed. Cultiv. Dev.* 279–90

[13] Abdulhafiz F, Kayat F And Zakaria S 2018 Effect Of Gamma Irradiation On The Morphological And Physiological Variation From In Vitro Individual Shoot Of Banana Cv. Tanduk (*Musa* Spp.) *J. Plant Biotechnol.* **45** 140–5

[14] Wahyudi D, Hapsari L And Sundari 2020 RAPD Analysis For Genetic Variability Detection Of Mutant Soybean (*Glycine Max* (L.) Merr) *J. Trop. Biodivers. Biotechnol.* **5** 68–77

[15] Werden L K, Sugii N C, Weisenberger L, Keir M J, Koob G And Zahawi R A 2020 Ex Situ Conservation Of Threatened Plant Species In Island Biodiversity Hotspots: A Case Study From Hawai'I *Biol. Conserv.* **243** 108435

[16] Yelni G, Syarif Z, Kasim M And Dewi H P 2019 Meningkatkan Keragaman Genetik Bawang Putih (Allium Sativum L.) Melalui Mutasi Irradiasi Gamma *J. Sains Agro* **4** 1–13

[17] Engels J M M, Ebert A W And Carvalho M 2021 Plants A Critical Review Of The Current Global Ex Situ Conservation System For Plant Agrobiodiversity. I. History Of The Development Of The Global System In The Context Of The Political/Legal Framework And Its Major Conservation Components

[18] Cai M, Wen Y, Uchiyama K, Onuma Y And Tsumura Y 2020 Population Genetic Diversity And Structure Of Ancient Tree Populations Of *Cryptomeria Japonica* Var. Sinensis Based On RAD-Seq Data *Forests* **11** 1–16

[19] Zulfiana A S, Larekeng S H And Restu M 2020 Amplification Of Rapd Primers Of Sugar Palm (Arenga Pinnata Merr) From South Sulawesi *Eur. J. Mol. Clin. Med.* **7** 4514–21

[20] Makmur M F, Larekeng S H And Restu M 2020 Genetic Diversity Of Eight Types Of Bamboo Based On Random Amplified Polymorphic DNA (RAPD) Markers *Plant Arch.* **20** 2333–7

[21] Konar S, Adhikari S, Karmakar J, Ray A And Bandyopadhyay T K 2019 Evaluation Of Subculture Ages On Organogenic Response From Root Callus And SPAR Based Genetic Fidelity Assessment In The Regenerants Of *Hibiscus Sabdariffa* L. *Ind. Crops Prod.* **135** 321–9

[22] Harianja D N, Karti P D M H And Prihantoro I 2021 Morfologi Mutan Alfalfa (*Medicago Sativa* L.) Hasil Iradiasi Sinar Gamma Pada Cekaman Kering *J. Ilmu Nutr. Dan Teknol. Pakan* **19** 59–65

The carbon storage capacity of plantation forests and agroforestry pattern in South Sulawesi in the context of mitigating climate change

S A Paembonan[1*], A S Soma[2] and D A Rampisela[3]

[1]Laboratory of Silviculture, Faculty of Forestry, Hasanuddin University Makassar, Makassar 90245, Indonesia
[2] Laboratory of Watershed Management, Faculty of Forestry, Hasanuddin University Makassar, Makassar 90245, Indonesia
[3] Laboratory of Soil Science, Faculty of Agriculture, Hasanuddin University Makassar, Makassar 90245, Indonesia

*Email: paembonansa@gmail.com

Abstract. Comparison between the ability of carbon storage for various types of land use, both inside and outside forest areas, needs to be done in order to predict its ability to deal with climate change mitigation and adaptation programs in Indonesia. Local baseline data need to be known because each vegetation growth is strongly influenced by local environmental conditions. This study was conducted to compare the ability of carbon storage in pine plantations and agroforestry patterns at high altitudes. The research method was carried out purposively by determining sample plots that were considered representative of the type of land use. The results showed that there was a significant difference between the ability to store aboveground carbon between pine plantations of 86.80 tons/ha and agroforestry patterns of 56.14 tons/ha. Similarly, the comparison of the aboveground and underground carbon storage capacity, there was a significant difference, namely for pine stands 78.99% to 21.01%, while agroforestry pattern was 53.58% to 46.42%. Overall, both types of land use can be relied on to address climate change mitigation programs in Indonesia.

1. Introduction

Plantation forests consist of various plant species according to the program and purpose of planting as well as the location of planting [1]. One of the most abundant species of plantation forest in South Sulawesi is the pine plantation and the largest is found in Tana Toraja and North Toraja Regencies. Based on data from the Forestry Service in 2011, pure pine forest covers an area of 7,909.25 ha and mixed gardens in the form of agroforestry covers an area of 38,356.00 ha. Pine plantations in Tana Toraja district are planted artificially inside and outside forest areas and dominate forest land cover in Tana Toraja.

Plantation forests provide a variety of products, both wood and non-timber products. This product provides significant economic returns for the community as a source of community income [2]. Agroforestry pattern also provides a variety of products because it is a mixture of woody tree species, fruit trees, commodity plants and seasonal crops so that it allows the community to obtain multiple benefits, both in the form of building materials, fruit products and commodity crops for economic purposes and annual crops for family consumption [3].

Ecologically, community plantation forests and agroforestry make a significant contribution in terms of their ability to store and absorb carbon, both on a plot scale and in a landscape scale [4-6]. Therefore, in the context of mitigating climate change, it is necessary to compare the capacity of carbon storage and sequestrations between plantation forests and agroforestry in the highlands of Tana Toraja.

2. Methods

This research was conducted from April to July 2022 in Tana Toraja regency to compare the potential for carbon storage in pine plantation forests and agroforestry under the same environmental conditions. The research location in Tana Toraja Regency is represented by Mengkendek District, which is geographically located at 3° 10' 55" latitude and 119° 52' 57" east longitude (Figure 1). The topography of the Mengkendek district is a hilly area and the altitude above sea level is 800 – 1200 mdpl with slopes varying and ranging from 15 to 55 %.

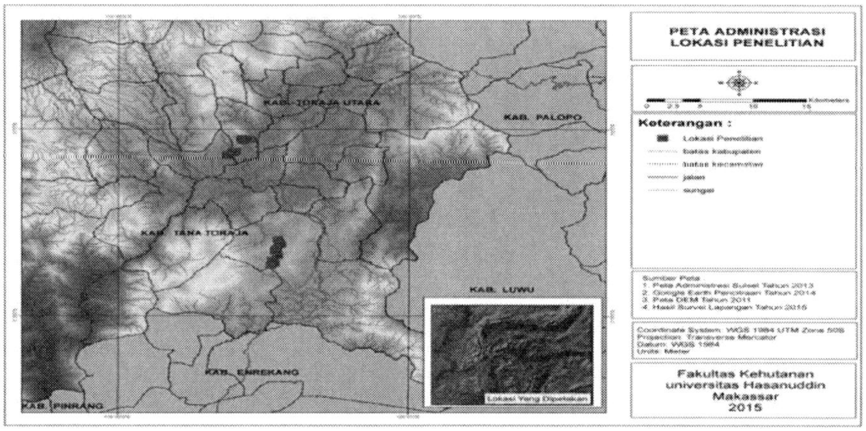

Figure 1. Research location in Tana Toraja Regency

This research uses the purposive sampling method on agroforestry patterns which were widely developed in Tana Toraja Regency and mature pine trees. The data collected were tree dimension measurements, undergrowth data, necromass and litter and soil organic carbon. Tree data was collected for biomass estimation by non-destructive sampling and for undergrowth and necromass by destructive sampling.

The research sample plots with a size of 25 m x 25 m (Figure 2) totaling 20 plots were used to measure the dimensions of tree growth, consisting of 10 plots each for agroforestry patterns and 10 plots for pine plantations. Within each main plot, 5 sub-plots which the size of 1 m x 1 m were systematically made for data collection of undergrowth and litter [4,6]. Soil samples to measure soil organic carbon were taken at a depth of 20 cm and analyzed in a silviculture laboratory [7,8].

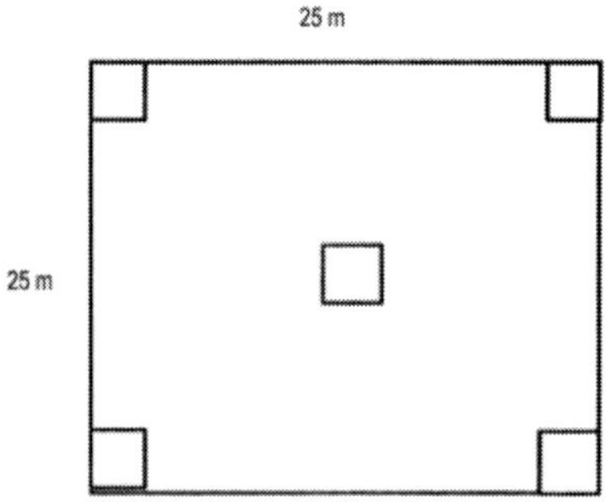

Figure 2. Sketch of sample plots and subplots

2.1. Data analysis

Primary data collection was carried out through direct observation and measurement of research objects in the field and in the laboratory. To estimate tree carbon biomass, the allometric equation model developed by [9] is used as follows:

$$W = 0{,}11\,\rho\,D^{2{,}62} \qquad (1)$$

where :
W : Tree Biomass (Kg)
D : Diameter of the tree at breast height (cm)
ρ : Density of Wood (*Mg/m3, kg/dm3, g/cm3*)

To estimate the biomass of coffee, cocoa, bamboo, banana and palm tree species, we use the formula by [4]. Likewise, to estimate the biomass of undergrowth and litter, the Hairiah and Rahayu [4] formula is used. Based on the Indonesian National Standard (SNI) [10] the calculation of tree carbon uses the following formula:

$$C = W \, x \, 0.47 \qquad (2)$$

where :
C : Carbon (kg)
W : Total Biomass (kg)
0.47 is the percentage value of carbon organic content,

2.2. Calculation of soil organic carbon

To estimate the C content of the soil, calculations were carried out using the following formula [6] :

$$Sc = Sd \, x \, \rho \, x \, \%\, C\, Organic \qquad (3)$$

where:

Sc	: soil carbon content, expressed in grams (g/cm2);
Sd	: depth of soil sample, expressed in centimeters (cm)
ρ	: bulk density, expressed in grams per cubic centimeter (g/cm3);
% C organic	: percentage of carbon content.

3. Results and Discussion

3.1. Pinus merkusii plantation forest

In general, the productive cycle of pine trees is around 20-25 years which is an economical cycle, while the technical cycle is adjusted to the needs of the industry, where the utilization of pine for the purpose of making pulp and paper industry is about 5-15 years. Therefore, in this study, it was only limited to pine stands with the age of 20 years so that they could be compared with agroforestry pattern which is a mixed crop.

Forests are able to absorb CO_2 from the atmosphere through photosynthesis and store carbon in the biomass [1,11,12]. Each forest ecosystem and the species in it have different abilities to absorb CO_2 and produce biomass [6,13,14]. Based on the results of estimation on samples obtained in the field, the values of biomass and carbon are obtained in Table 1.

Table 1. Value of biomass and carbon storage in 20 years old stand of *Pinus merkusii*

Elements	Biomass (tons/ha)	Carbon (tons/ha)	Relative Value (%)
Trees	181.68 ± 5.79	85.39 ± 2.72	
Undergrowths	1.35 ± 0.15	0.63 ± 0.07	
Litter	1.66 ± 0.11	0.78 ± 0.05	
Aboveground Total	**184.69 ± 5.75**	**86.80 ± 2.70**	**78.99**
Soil Organic Carbon		23.09 ± 5.04	21.01
Total		**109.89 ± 5.47**	**100.00**

The results of the calculation of the carbon values above and below the soil surface for pine stands obtained a ratio of 78.99 % to 21.01 %. The value of CO_2 absorption is directly proportional to the value of carbon storage. The greater the CO_2 absorbed through the process of photosynthesis, the more carbon is stored in the tree. Plants absorb CO_2 from the air and convert it into organic compounds in the form of biomass [11,15]. The greater the value of CO_2 absorption, the greater the value of a tree's carbon storage. The amount of carbon stored in trees is linearly proportional to the increase in tree biomass and reaches a stable condition when the age of the tree has reached the cutting period [1,16,18].

3.2. *Agroforestry pattern*

Total species of plants in the agroforestry area were 25 species. All of these plants consists of 6 species of trees, thats are: *Elmerrilla ovalis, Casuarina junghuhniana, Albizzia falcatana, Tectona grandis, Palaquium rostrata, Pinus merkusii* and *Toona sureni*, 9 species of fruit trees are: *Lansium domesticum, Durio zibethinus, Ceiba pentandra, Artocarpus integra, Persea Americana, Musa paradisiasa, Cocos nucifera L., Nephelium lappaceum L. and Garcinia mangostana,* 3 species of commodity crops, namely: *Theobroma cacao*, *Coffea* sp and *Syzygium aromaticum and* 3 species of palm and bamboo, namely: *Pigafetta filaris, Arenga pinnata and Bambusa* sp., medicinal and spice plants, consisting of lombok, turmeric and galangal.

3.3. *Potential of stored carbon in trees*

Compared to undergrowth, litter and soil, trees are a component that has a very large potential for storing carbon in agroforestry systems (Table 2). This is due to the more varied types of stands that make up the agroforestry pattern, the tree diameter varies greatly according to the age of the plant and the varying spacing due to the large number of seasonal plants that require sunlight [6,17].

Table 2. Average value of biomass and carbon in agroforestry

Elements	Biomass (tons/ha)	Carbon (Tons/Ha)	Relative Values (%)
Trees	114.43 ± 6.11	53.78 ± 2.87	
Litter	3.98 ± 0.76	1.87 ± 0.36	
Undergrowth	1.04 ± 0.14	0.49 ± 0.06	
Aboveground Total	**119.45 ± 6.24**	**56.14 ± 2.34**	**53.58**
Soil Organic Carbon		48.63 ± 2.47	46.42
Total		**104.77 ± 2.82**	**100.0**

The results of the calculation of the carbon values above and below the soil surface for agroforestry pattern obtained 53.58 % to 46.42 %, respectively. The value of litter carbon storage was different in each plot due to the stand density of each plot. The denser a stand and canopy, the higher the level of competition for sunlight so that a lot of litter is produced because the trees will release their branches and leaves due to lack of light for the photosynthesis process [1]. Factors that affect the carbon potential in litter are tree age, canopy density and the amount of fallen litter as well as its quality. Litter quality is influenced by environmental factors (climate, soil fertility), plant species and season [7,19].

Undergrowth is one of the components in the forest ecosystem that grows to occupy the lowest strata of the tree and obtain sunlight for its metabolism through the gaps between trees. Undergrowth is a plant that has a diameter of < 3 cm, including seedlings, sprouts, ferns, grasses, climbing plants and mosses [3,6,17]. Hairiah and Rahayu [4] stated that the amount of carbon stored in undergrowth varies between fields, depending on the diversity and density of existing trees. The denser the tree crowns that make up a land, the undergrowth plant biomass will decrease due to the lack of sunlight reaching the forest floor for plant growth.

From the calculation of carbon potential, it can be seen that the carbon potential of undergrowth is relatively smaller than the potential of carbon storage in trees and litter. This is because the denser the tree canopy, the lower the carbon stock of the undergrowth due to the reduced sunlight reaching the forest floor [3,4,6]. The quality of the growing place affects the development of undergrowth as well as the growth of trees in stands, where stands that have good growing quality will provide better nutrients needed by plants so that the growth of understorey will be better too [7].

3.4. Potential organic carbon content in soil

In estimating the biomass and carbon contained in the soil, the method used is the composite method. The value and amount of soil carbon content varies with each land use. From the results of the study in Table 2. it can be seen that the potential for carbon storage of community forests using agroforestry patterns, both from above ground and underground carbon.

Organic C is part of the soil which is a complex and dynamic system, sourced from plant and/or animal residues found in the soil which continuously changes shape, because it is influenced by biological, physical and chemical factors and chemistry [7]. Land use systems, past and present land management intensity and organic carbon supply in soils are also determined from forest productivity and unidirectional decomposition of mineral soils. If the soil is denser (the greater the BD), the soil will contain a greater mass at a certain depth, so that the carbon content is higher [7,8].

4. Conclusions

There is a significant difference between above and below ground carbon storage between pine plantations and agroforestry systems in Tana Toraja district. The comparison of carbon storage in aboveground to below ground pine plantations is 78.99 % to 21.01%, while for agroforestry systems it is 53.58 % to 46.42%. Differences in carbon storage in pine forests and agroforestry systems are determined by plant species, stand density, age of stands and variations in the types of stands. The contribution of pine plantations and agroforestry to their ability to store carbon is quite good in the context of adaptation and mitigation of climate change.

Acknowledgements

The authors would like to thank the Directorate of Research and Community Service, Ministry of Education and Culture of the Republic of Indonesia and the Chancellor of Hasanuddin University for the approval and research funding provided under decree number 071/E5/PG.02.00/2022. This article is one part of the research results from a series of research that will be carried out for 2 years.

References

[1] Paembonan S A 2014 *Hutan Tanaman dan Serapan Karbon* (Makassar: Masagena Press) 168
[2] Yuningsih L, Lensari D and Milantara N 2018 *Jurnal Sylva Tropika* **2** 3
[3] Paembonan S A, Putranto B, Millang S and Nurkin B 2019 *IOP Conference Series: Earth and Environmental Science* **270** 012035

[4] Hairiah K and Rahayu S 2007 *Pengukuran karbon tersimpan di berbagai macam penggunaan lahan Bogor* (Malang: University of Brawijaya) 77
[5] Siringoringo H H 2014 *Jurnal Penelitian Hutan dan Konservasi Alam* **11** 13
[6] Hairiah K, Ekadinata A, Sari R R and Rahayu S 2011 *Pengukuran Cadangan Karbon: dari tingkat lahan ke bentang lahan* (Malang: University of Brawijaya)
[7] Vinh T V, Marchand C, Linh T V K, Jacotot A, Nguyen, T N and Allenbach M 2022 *Geophysical Monograph* **267**
[8] Rusdiana O and Lubis R S 2012 *Jurnal Silvikultur Tropika* **3** 1
[9] Kettering Q M, Coe R, Van N, Ambagau M Y and Palm C A 2001 *Journal of Forest Ecology and Management* **146** 199
[10] Badan Standar Nasional Indonesia (SNI) 2011 *Badan Standarisasi Indonesia SNI*
[11] Toochi E C 2018 *Forestry Research and Engineering: International Journal* **2** 148
[12] Stas S M, Rutishauser E, Chave J, Anten N P and Laumonier Y 2017 *Forest Ecology and Management* **389** 27
[13] Cameron C, Kennedy B, Tuiwawa S, Goldwater N, Soapi K and Lovelock C E 2021 *Environmental Research* **192** 110213
[14] Setiawan G, Syaufina L and Puspaningsih N 2015 *Jurnal Pengelolaan Sumberdaya Alam dan Lingkungan* **5** 141
[15] Kellogg W W and Schware R 2018 *Climate change and society: consequences of increasing atmospheric carbon dioxide* (New York : Routledge)
[16] Firyadi, Widiatmaka, Iswati A, Ardiansyah M and Mulyanto B 2018 *Jurnal Pengelolaan Sumberdaya Alam dan Lingkungan* **8** 178
[17] Analuddin K, La Ode K, La Ode M Y H andi S, Idin S, La S, Rahim S, La Ode A F and Kazuo N 2020 *Biodiversitas Journal of Biological Diversity* **21** 4
[18] Bakar S N 2022 *Inovasi Tata Kelola Lingkungan Hidup dan Kehutanan* (Malang: Universitas Brawijaya)
[19] Zhu L, Hu Y and Zhao P 2020 *Functional Plant Biology* **47** 558

IOP Conf. Series: Earth and Environmental Science 1115 (2022) 012029 doi:10.1088/1755-1315/1115/1/012029

Analysis of the species and derived bamboo (Bambuceae) product in Berastagi, Kabanjahe and Tiga Panah of Karo District, North Sumatra

Samsuri[1,2*], Margaretta Tampubolon[1] and Anita Zaitunah[1]

[1]Forest Management Department, Forestry Faculty, Universitas Sumatera Utara, Medan, Indonesia 20155

[2]Sumatran Landscape Study Center (Puskalanskap), Universitas Sumatera Utara, Medan, Indonesia 20155

[*]Email: samsuri@usu.ac.id

Abstract. Bamboo forests have ecological and economic functions. Bamboo is a commodity with bright prospects if it is widely developed in forestry. Bamboo standing stock and its derived product are important to be measured. The research aimed to identify bamboo species and analyze the benefit type of bamboo in North Sumatra Province. The sample analysis was carried out by analyzing the existing bamboo growing stock. The stratification interview technique with bamboo garden owners regarding the use of bamboo is based on local wisdom. A square plot of 50 m x 50 m was measured to get a standing stock of bamboo. The research found *Gigantochloa sp.* dominates the bamboo forest. The bamboo products are on a large scale, but they have not been sent out of town because they are very cheap and expensive in transportation. Many bamboo farmers have switched professions because of the low price of bamboo.

1. Introduction

Forests help balance the environment to prevent erosion, slow global warming, produce oxygen and provide water, among other things. Forest areas can make a microclimate mainly in forests and surrounding areas. Forests also provide a habitat for biological and wildlife diversity. The bamboo plant is one of them [1]. According to forestry ministerial regulation number: P.35 / Menhut-II / 2007, nonwood forest products (NWFP); except for forest wood, it consists of forest products, both vegetable and animal, as well as derivative and cultivated products. NWFP is a component of the forest ecosystem that plays a variety of roles in both the natural environment and human life. Sandalwood, gaharu, sago, rattan, sugar palm, breadfruit, eucalyptus, various medicinal plants, essential oils, bamboo, natural silk, jernang, incense and honey are examples of NWFP that the public used and commercialized [2].

Bamboo plants cover about 75% of the bamboo forest, which is a homogeneous forest. Bamboo plants have promising economic potential in addition to being easy to cultivate. However, bamboo cultivation remains unappealing because bamboo is regarded as a less commercial plant. Bamboo plants are considered populist plants in rural areas because they provide a primary or secondary source of income [3]. Bamboo is Gramineae (grass) family, which are made up of a number of stems (reeds) that grow gradually and clump together, beginning with bamboo shoots, young stems and adults at the age of 4-5 years. Bamboo stalks are cylindrical and have hollow sections and hard walls. Bamboo roots are made up of rhizomes (rhizomes). The bamboo growth is ascending direction and sometimes climbing. The bamboo stalks' end is usually hung and the leaves appear to wave when they grow tall [4].

IOP Conf. Series: Earth and Environmental Science 1115 (2022) 012029 doi:10.1088/1755-1315/1115/1/012029

Bamboo is a Gramineae family (grasses) and is also mentioned as giant grass. The bamboo reaches 3-4 years of age as mature bamboo [5]. Recently data show about 2,104,000 ha of bamboo forest in Indonesia. Bamboo has a lot of promise as a natural resource because of its multiple applications as a plant and ease of harvesting. For present demands, bamboo can take the place of industrial wood [6]. Bamboo is a controlled plant that can be found around homes, farms, fields and the boundaries of rice fields. The most effective technique to generate high-quality bamboo with a high selling price is through bamboo plantations. Bamboo has various functions that are Juvenile bamboo can be a delicious cooked vegetable. Bamboo clumps can protect the environment from hazardous wind [7]. Many crafts employ bamboo as raw materials, including plaits, baskets, mats, musical instruments and even architectural materials [8]. Bamboo is an essential resource for the lives of people in rural Indonesia since it benefits for welfare community. In modern industry, bamboo has been extensively utilized, from low-tech to high-tech. Due to its widespread rural growth and versatility as a plant for rural communities, bamboo is more affordable than other building materials. [9]. The underground rhizomes' branching roots, which constitute the morphology of bamboo, may be recognized. The trunk is a segmented and book-filled reed. Branches typically appear at nodes. Leaf veins run parallel along the bamboo strands [10].

Many nations are embracing bamboo as a new, renewable energy source since it produces more biomass per unit than most other plant species. Additionally, because bamboo is inexpensive, it is frequently utilized as a raw material for buildings, furnishings for the home, tools for transportation, bamboo ladders, crafts and many other things [1]. Simple home furnishings, kitchen furniture, building supplies and other items, as well as bamboo coating, bamboo lamination and contemporary paper sectors, are only a few of the uses and advantages of bamboo. This research is expected to provide references and benefits to the Karo district community in determining bamboo species so that subsequent efforts can be made to maximize bamboo production and discover the benefits of bamboo more effectively and efficiently. (1) Identify bamboo species in Berastagi, Kabanjahe and Tiga Panah sub-districts, Karo District, Sumatera Utara Province and (2) analyze the benefits of bamboo species in Berastagi, Kabanjahe and Tiga Panah sub-districts, Karo District, Sumatera Utara Province.

2. Materials and Methods

This research was conducted in September 2020. The research was carried out precisely in Karo district, Sumatera Utara Province, which can be seen on the map below (Figure 1) with an administrative area of 2,127.25 km^2 consisting of 17 districts, ten villages and divided into 259 villages. Data processing and analysis were carried out at the Forest Management Laboratory, Forestry Study Program, Faculty of Forestry, University of North Sumatra. The descriptive analysis was also applied in this study

The Geographic Positioning System and Maps Camera Mobile Application were used for field surveys. Other tools are personal computers as hardware; meanwhile, the software tool is Microsoft Excel, Arcmap 10.3, Sentinel Application Platform and ERDAS Imagine 9.2. We also use supporting tools that are a camera, a questionnaire (tally sheet), stationery and a printer. The materials used in the study consisted of (1) Spatial data, which are the Administrative map of Karo district, Sumatera Utara Province and (2) non-spatial data collected from a field survey.

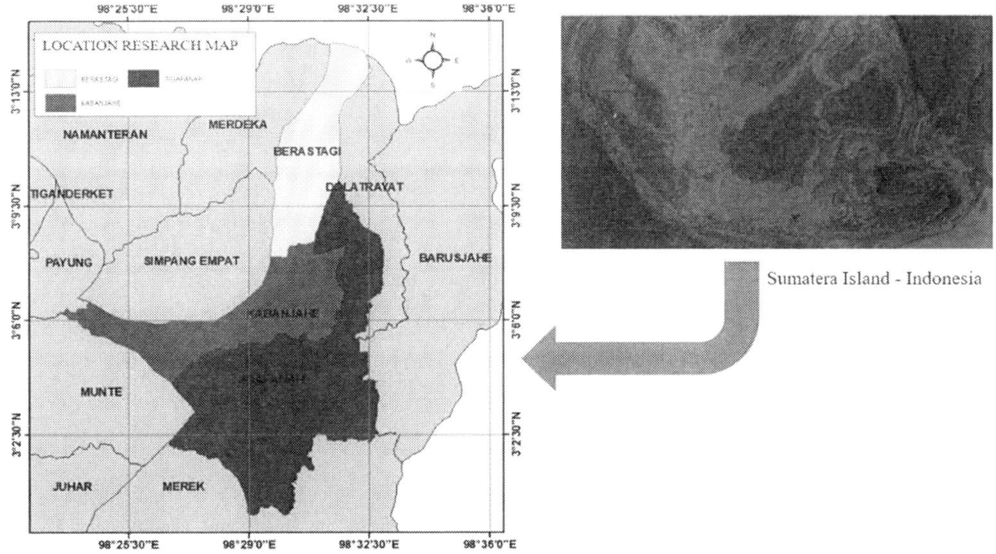

Figure 1. Site research, Karo district - Sumatera Utara

2.1. *Data collecting*

Sentinel-2 satellite images were downloaded from the site *www.eartexplore.usgv.gov* which is needed according to the analysis's purpose. Delineation of the area or potential area for bamboo was carried out to obtain data around the research location, vegetation cover near settlements, rice fields and rivers.

In this study, the two-stage sampling method was carried out (1) selecting samples in satellite imagery (map) and (2) selecting samples based on the selected samples in the first stage. In the second stage, a sample of sub-districts with bamboo potential was selected within three sub-districts: Berastagi sub-District, Kabanjahe sub-District and the Tiga Panah sub-District. After knowing the potential in 3 sub-districts, each sub-district took a sample of 3 villages. In the Berastagi sub-district, bamboo's potential is in the villages of Doulo, Raya and Berastagi House. In the Kabanjahe sub-district, the potential for bamboo is in Lepar village, Stadium street and the village of Sumber Mufakat; and in Tiga Panah sub-District that has the potential for bamboo is in the village of Jalan Kabanjahe, Mulawari Village.

To identify the bamboo species in that location using a google map and based on the knowledge of the garden owner and the surrounding community. Field survey activities are designed to collect data and validate land use classifications. Field data obtained through direct field observations include documentation of field conditions, marking the position of points in the field and input into tally sheets collected from permanent sample plots and field surveys.

2.2. *Farmer interviewing*

Researchers interviewed bamboo garden owners in Berastagi, Kabanjahe and Tiga Panah sub-districts, Karo District, Sumatera Utara Province. Regarding the use of bamboo species based on local wisdom. Apart from garden owners, he also conducted interviews with bamboo users, bamboo craftsmen and bamboo product sellers. The species of bamboo plants found at the research location were recorded and interviews with the community about using bamboo plants were conducted. The data collected was then entered into a Microsoft Excel table. Following the research flow, the observations on the bamboo plants' species and how to use them were recapitulated in the questionnaire.

2.3. Bamboo standing stock

Sample plots are designed to collect data and validate land use classifications. Field data is obtained through direct field observations that are documentation of field conditions, marking the position of points in the field. Field surveys are also designed to collect data and validate land use classifications. They were plotted on a grid of 50 m x 50 m. As shown in Figure 2, create 27 50 m x 50 m plots in a bamboo plant area as illustrated in Figure 2.

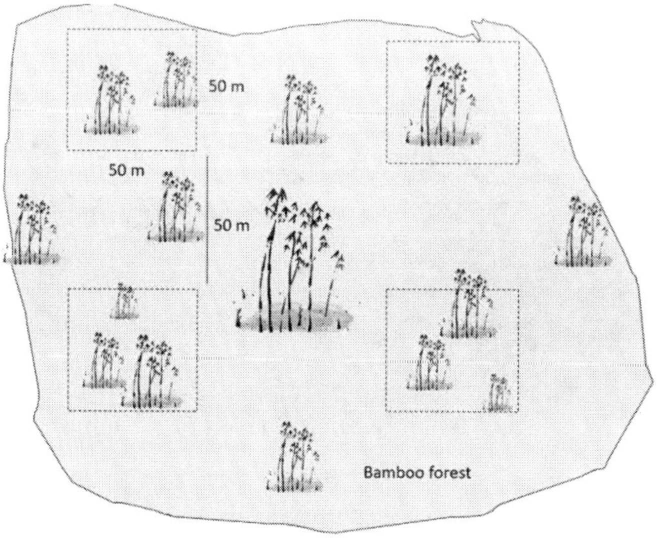

Figure 2. Layout sampling plot

The sample plot procedures are as follows: (a). Count the number of clumps of bamboo and (b). Count how many bamboo culms are in small and large clumps. The number of bamboo stands was calculated using data from the field survey. The following equations were used to analyze the data.

$$Y = \frac{\sum_{1}^{n} yi}{total\ plot\ area} \quad \text{.......................................} (1)$$

Note :
Y = Total bamboo clump per ha
yi = the number of bamboo clump per plot (i=1, 2, 3….etc)
The number of bamboo stands was calculated using data from the field survey

3. Result and Discussion

Bamboo plants grow in clumps most of the time. It can grow as a shrub or solitary stem. The bamboo plants are sympodial bamboo plants, which have stems that gather in one clump due to the rhizome branching in the soil and collecting more. Bamboo's morphology can be seen in its characteristics: underground rhizomes' roots form a branching system. The trunk is made up of books and segments. The reed midrib is a modified leaf that is attached to each segment of the reed midrib, reed midrib and ligula. Branches are frequently located at nodes. Parallel leaf veins can also be seen in bamboo leaf strands. The leaf blade is joined to the midrib by a leaf stalk; meanwhile, the midrib is attached to the ligula by a midrib [10].

Bamboo plants can grow in a day about 15-18 cm and reach the highest in 4-6 months. Bamboo culms are made up of 50% parenchyma fibers, 40% fibers and 10% connecting cells. Bamboo biomass productivity is higher than most other plant species, so many countries are turning a renewable energy source from bamboo. Bamboo is inextricably linked to Indonesian culture because it is easy to work with, shape and light. Bamboo is broadly utilized as a raw material for building houses, household equipment, transportation instruments, bamboo ladders, crafts and many other things because of its low cost. It is recognizable that bamboo has a significant role in Indonesian society [1].

Bamboo is a large tree-like family plant that grows by using the root of the root rhizome, which is segmented with one shoot in each segment, alternating with each subsequent vertebra; growth using rhizomes allows bamboo to grow with sympodial, monopodial and amphipod clumps. Bamboo has a wide range of applications and benefits, from simple home furnishings, kitchen furniture, building materials and other equipment to bamboo coating, bamboo lamination and modern paper industries.

3.1. Lands cover

Image classification resulted in 11 land cover types, comprising primary dryland forest, secondary dryland forest, plantation forest, dryland agriculture, mixed dryland farming, rice fields, shrubs, settlements, vacant land and water bodies. Land cover grouping. The largest change in the land cover area in Karo Regency occurred in dryland agriculture, which became a wetland, followed by forest land, which became dryland agriculture [7].

Bamboo plants dominate more than 75% of the forest cover in the bamboo forest, which is a homogeneous forest. If developed on a large scale, bamboo is a commodity with promising prospects in the forestry sector. Bamboo plants have a high economic potential as well as being simple to cultivate. On the other hand, bamboo cultivation remains unappealing because bamboo is regarded as a less commercial plant. Bamboo is a popular plant in rural areas, both economically and socially, because it is a primary or secondary source of income [3].

Bamboo plants dominate more than 75% of the forest cover in the bamboo forest, which is a homogeneous forest. Bamboo is a commodity with promising prospects in the forestry sector if it is developed on a large scale. Bamboo plants have a high economic potential and are easy to cultivate. However, bamboo cultivation remains unappealing because bamboo is regarded as a less commercial plant. Bamboo plants are a populist plant in rural areas, both economically and socially.

3.2. Species identification

Based on the measurement plot data, three bamboo species were found as in Table 1 Apus bamboo is the dominant species in the study area. Apus bamboo is a species that is widely planted by the community because it has the potential for high economic value. Tali bamboo is one of the species of bamboo that is often used in Indonesia, namely the species of Tali bamboo, Apus bamboo andong bamboo, Betung bamboo, Duri bamboo and Hitam bamboo. Because of their high potential, these bamboo species are frequently used. However, when compared to wood, bamboo has technical flaws (physical, mechanical and chemical properties) that make it unsuitable for industrial use. This necessitates knowledge of bamboo plants in order to obtain high-quality bamboo plants [11].

Table 1. Bamboo species found in research

No	Species	Number of bamboo/clump
1	Bambu Apus (*Gigantochloa apus*)	54
2	Bambu Tali (*Gigantochloa sp*)	26
3	Bambu Regen (*Gigantochloa purrens* Widjaja)	20

3.3. Bamboo potency

Bamboo plants belong to the Poaceae (grass) family of plants. Bamboo itself is a plentiful resource with a wide variety of species. Arundinaria, Bambusa, Dendrocalamus, Dinochloa, Gigantochloa, Melocanna, Nastus, Phyllostachys, Schizostachyum and Thyrsostachyum are the ten most well-known bamboo genera in Indonesia. As one of the world's tropical countries, Indonesia has potential bamboo resources.

Bamboo can be found in both rural and forested areas of Indonesia. Except for coastal land, bamboo can be planted in any soil type [10]. The identification of bamboo species in the Karo district sample plot with the help of a google map and based on the knowledge of the bamboo forest owner and the surrounding community [12]. The identification showed that the most bamboo species planted by the Karo community was Apus bamboo (*Gigantochloa apus*). Most people plant bamboo Apus with the consideration that the main income from bamboo is making baskets. Usually, those who buy bamboo from the community are neighbors bamboo craftsmen and no one has said that the bamboo plant is sent outside the city. This bamboo apus species is very much found in the residents' yard at the back of the house

3.4. Usage of bamboo

Interview the community and bamboo farmers to find out their perceptions and knowledge of bamboo stands. The data obtained from the interviews included the benefits and uses of bamboo. The respondent's character is gender, age, education level, ras, income, bamboo owner and bamboo forest area. Respondents are 53 persons, comprised of 19 females and 34 males. The number of male respondents who were interviewed was more than that of women, with ages greater than 20 years. Based on the level of education, the respondents were mostly elementary school graduates. Socially, the Karo tribe dominates the number of respondents, followed by the Toba Batak tribe. The income range of more than IDR 1,100,000 dominates the respondents' income level. Based on bamboo ownership status, 27 people still own their bamboo land and the remaining 26 people are bamboo buyers. The planted area under management for growing bamboo is mostly less than 1 hectare. The bamboo harvest is used and used as an alternative job.

In terms of socio-cultural practices, Indonesian people have long used bamboo in artistic and traditional activities. Bamboo plants had their own history during Indonesia's independence struggle in 1945. Spiked bamboo was a traditional weapon used in the anti-colonialist fight. As a result, it is not surprising that many of Indonesia's struggle monuments are made of bamboo. The yellow bamboo plant, for example, is planted in Pekalongan City's Monument Park 45 as a symbol of struggle. Meanwhile, the pointed bamboo monument serves as a focal point in the Sharp Bamboo Monument Park in Surabaya and the Digulis Park in Pontianak. It is clear that bamboo is important in Indonesian society [13].

3.5. Social economy aspects

Bamboo plants benefit the community's economic life, so they play an important role in people's lives in rural Indonesia. Bamboo has been widely used up to this point, from the most basic technology to high technology on an industrial scale, to make it easy to distribute. Bamboo is the least expensive building material because it grows in many rural areas and is a versatile plant for rural communities. [9].

Data on the use of bamboo in the research location were obtained from interviews with selected respondents. In this study, Respondents are mostly farmers, with only a small proportion of traders and private employees. Economically, the price of raw bamboo in the market is also low; namely, bamboo is only sold for 1,000 - 1,500 / meter, 7,000 / stem and 18,000 / bunch and even if it is taken directly to farmers, the bamboo price is lower. The amount of bamboo usage per day ranges from 1.5 meters x 150 meters. Bamboo is used by each family and reaches 1050 meters per week and 31,500 meters of use for one month.

Bamboo species identification indicates the species of bamboo commonly used: *lemang* bamboo, blanked bamboo, regen bamboo and rope bamboo. The preference for bamboo plants is bamboo rope. Bamboo is harvested to supply local needs and suppliers of bamboo in traditional markets and factories. The bamboo harvest consists of young bamboo (bamboo shoots) and old bamboo. Young bamboo is a food source for the community, while old bamboo is used as raw material for the handicraft industry and building materials. Bamboo as a food ingredient is the raw material for making household food and for the food industry. Table 2 shows the types of food made of young bamboo and old bamboo as building materials and crafts.

Table 2. Most products use bamboo as raw material

No	Product with bamboo raw material	Productivity (unit/day)
1	Karo instrument music (Keteng-keteng)	0.1
2	Bamboo charcoal	10
3	Bamboo wall	0,3
4	Broom bamboo handle	10
5	Chicken cage	0.5
6	Handycraft	5-7
7	Basket	15-18
8	Lemang food container	Base on ordered
9	Bamboo chair	0.2
10	Small farmhouse	1
11	Small tourism tower	10
12	Bamboo salad	Based on the growth stage
13	Broom of bamboo fiber	10
14	Flute music instrument	3
15	Winnowing	10
16	Bamboo ladder	3
17	Bamboo pillar	20
18	Flower stick	20
20	Banner stick	20
21	Hat traditional	5-7
22	Table covered	5

The number of products from bamboo is very diverse. It is just that the demand is minimal and the price is low, making bamboo farmers sell their bamboo land and some are diverting bamboo land into orange land and others. Furthermore, the selling price of bamboo has also decreased significantly. This resulted in many bamboo farmers changing their professions to become orange farmers. In the field, it is also said that only this species of *bamboo Tali* can be used as material for making baskets and hats from bamboo. Because basket income is also seasonal, many people sell bamboo land at very low prices, only during the growing season for oranges and tomatoes. When it is not in season, people's income will be cut off if they only rely on bamboo.

Because there are not many bamboo enthusiasts, it is difficult for the community to process it. In the field, many people complain because, in addition to the decreasing selling price of bamboo, those interested in bamboo products can already compete with products from abroad. Almost all household appliances that were previously made of bamboo are replaced with plastic. For example, Tampi, Dipper, Plate, Glass and even Flute now there are made of plastic. Many people also complain that they do not know how to process bamboo because there are no handicraft teachers around where they live.

4. Conclusion

The dominant species of bamboo that grows in Berastagi, Kabanjahe and Tiga Panah, Karo Regency, Sumatera Utara Province, is Apus bamboo. The bamboos that many craftsmen need are Lemang Bamboo, Regen Bamboo and Blangke Bamboo. Some products in the research area are produced on a large scale, but they have not been sent out of town because they are very cheap and expensive in transportation. Many bamboo farmers have switched professions because of the low price of bamboo. The bamboo forest is running out over time.

Acknowledgments

The author would like to thank the Research Institute at the Universitas Sumatera Utara and the Directorate of Research and Community Service, which funded this research through the 2020 Higher

Education Leading Basic Research Scheme (PDUPT) contract number 99/UN5.2.3.1/PPM/KP-DPRP/2021. The author also thanks Universitas Sumatera Utara for the support as well.

References

[1] Mayasari A and Suryawan A 2012 *J. Info BPK Manado* **2** 139
[2] Tang M, Malik A and Hapid A 2019 *J. Warta Rimba* **7** 19
[3] Alamsyah R, Affandi O and Batubara R 2012 *https://media.neliti.com/media/publications/157760-ID-analisis-potensi-ketersediaan-dan-pemasa.pdf*
[4] Widnyana K 2002 *Bambu Dengan Berbagai Manfaatnya* (Denpasar: Fakultas Pertanian Universitas Mahasaraswati) 1
[5] Hingmadi D 2012 *Keanekaragaman Ciri Morfologi Jenis-Jenis Bambu (Bambusa Sp.) di Kelurahan Teunbaun Kecamatan Amarasi Barat* (Kupang: Fakultas MIPA-Biologi Universitas PGRI-NTT) 1
[6] Arsad E 2015 *J.Riset Industri Hasil Hutan* **7** 45
[7] Prastiyo S 2010 *Identifikasi Potensi Dan Pemasaran Produk Dari Hutan Rakyat Bambu* (Medan: Universitas Sumatera Utara)
[8] Ekayanti N W 2016 *J. Bakti Saraswati* **5** 132
[9] Putro D S, Jumari and Murningsih 2014 *J. Biologi* **3** 2 71-79
[10] Nasution E Z 2018 *Keanekaragaman Jenis Dan Pemanfaatan Bambu (Bambusa Sp) oleh Masyarakat Sekitar Hutan Kawasan Taman Nasional Batang Gadis* (Medan: Fakultas Kehutanan Universitas Sumatera Utara)
[11] Arhamsyah 2009 *J. Riset Industri Hasil Hutan* **1** 30
[12] Samsuri, Zaitunah A, Ahmad A G and Ramadhan M F 2021 *IOP. Conference Series: Earth and Environmental Science* **912**
[13] Damayanto I P 2018 *J. Arsitektur Lansekap* **4** 211

IOP Conf. Series: Earth and Environmental Science 1115 (2022) 012030 doi:10.1088/1755-1315/1115/1/012030

Phylogenetic reconstruction of tree species on the University of North Sumatra Campus, based on the *rbcL* gene

A B Rangkuti[1,2*], A Mawarni[1] and R R rangkuti[1]

[1]Forestry Studies Program, Faculty of Forestry, Universitas Sumatera Utara, Jl. Tri Dharma Ujung No.1 USU Medan Campus 20155
[2]Center of Excellence for Mangrove, Universitas Sumatera Utara, Medan 20155, Indonesia

*Email: baiqunirangkuti@usu.ac.id

Abstract. A phylogenetic tree is an arrangement that can describe kinship data in the form of a branched tree. In the field of forestry, identification using morphological data is difficult because it takes a long time and the fruit and flower seasons are uncertain. Another alternative that can be used to create a phylogenetic tree is to use secondary data. This study aims to explore the availability of GeneBank sequence data for tree species at the Universitas Sumatera Utara and create a phylogenetic tree to see their relationships based on rbcL markers. The results showed that the highest sequence of trees on the USU campus using the rbcL gene was 86.8%. In the reconstruction of the phylogeny tree, the boostrap value is above 70%, and tree species are grouped according to the same family.

1. Introduction

The kinship studied through a phylogenetic approach is based on the level of similarity that exists. The kinship of a plant can be interpreted as the relationship between one plant and another, where the total similarity between species determines the pattern of the plant group [1]. In the phylogenetic approach, a group of organisms each has members with many similar characters, traits, or traits that are considered to have a very close kinship with their, and it is thought that they have been passed down by one ancestor who is believed to be the elder of the species [2].

In the forestry sector, identification using morphological data is difficult because it takes longer and the fruit and flower seasons are uncertain [3]. On the other hand, identification using molecular data such as DNA or protein also takes a long time because it uses laboratories and other sophisticated equipment. Another alternative used to create a phylogenetic tree is to use secondary data. The data used is the nucleotide sequence of the entire tree with the marker rbcL obtained from the NCBI website. The method commonly used in the analysis of kinship is phylogenetic which can be used to interpret the relationship between species with phylogenetic reconstruction [4]. A phylogenetic tree is an arrangement that can describe kinship data in the form of a branching tree [5]. The Neighbor-Joining, Parsimony, Maximum Likehood, and UPGMA models were used in phylogenetic making to determine the similarity between sequences [6].

This study aims to explore the availability of GeneBank sequence data for tree species at the Universitas Sumatera Utara and create a phylogenetic tree to see their relationships. The phylogenetic analysis was able to describe the cladistics and the evolution of kinship between species so as to facilitate the conservation of tree species at the USU Padang Bulan Campus.

2. Research method

2.1. Research time and place
This research was conducted in June-August 2021. Data analysis and processing were carried out at the Forestry Biotechnology Laboratory,Universitas Sumatera Utara.

2.2. Research methodology
This sresearch is divided into two stages: sequence data collection in NCBI and phylogenetic reconstruction. Data collection includes sequencing from GeneBank. Phylogenetic reconstruction includes sequence data editing, sequence alignment and phylogenetic tree creation.

2.3. Data collection
The data used in making the phylogeny tree is secondary data in the form of DNA sequences from the NCBI website. Tree sequences were collected using the rbcL primer. Sequence data with the most rbcL primers were found, so they were used to construct a phylogenetic tree.

2.4. Phylogenetic tree reconstruction

2.4.1. Sequence data editing. MEGA X software is used to edit the sequence data. MEGA X program is used to determine the level of similarity between sequences. Thus, the MEGA X program can read DNA sequences, both nucleotide base sequences, and proteins and the alignment of one sample with another sample. That way, sequences that have kinship relationships can be identified and identified by occupying the closest branch [7].

2.4.2. Alignment. Sequence alignment is done to avoid misreading between one sequence and another so that all sequences are cut to align. Sequence alignment can be done using "ClustalW" on MEGA X software [8]. The data that has been aligned can then be used to construct a phylogenetic tree. The purpose of alignment is to sequence and determine the level of homology of each base pair in the DNA [9].

2.4.3. Making a Phylogenetic Tree. The phylogeny tree in this study used the Neighbor-Joining method.

3. Results and Discussion

3.1. Types of trees on the USU campus with sequences available at GeneBank
Results identification [10] found 121 tree species belonging to 36 families and 88 genera. However, only 115 species on the USU Campus have data sequences in GeneBank.

3.2. DNA sequence
DNA sequence is the most basic information on a gene that contains and carries compounds in the form of instructions that contain the information needed to form a living thing [11]. This DNA sequence is usually used to determine the identity of living things by comparing the DNA sequence with other known DNA sequences [12]. The sequence can be known when it has been amplified using a primary pair. Amplification plays a very important role in determining the phylogeny of a species. However, the final result depends on the primary selection to be used [13].

The results of genetic data tracing from NCBI, the most widely used primer for tree species phylogeny at USU Padang Bulan Campus was rbcL, which was 105 species with the largest percentage of 87%, while the least used primer was psbM-trnD with a percentage of 2.5%. Not all primers were available in GeneBank, such as primers psbK-I and coxl which were completely absent from Gene Bank (Table 1).

Table 1. Number of primers found on the NCBI portal

No	Primary name	Number of types	percentage
1	*mat*K	102	84.3%
2	*rbc*L	105	86.8%
3	ITS	41	33.9%
4	ITS2	77	63.6%
5	*psb*A-trnH	86	71.1%
6	*atp*F-atpH	13	10.7%
7	*ycf*5	6	4.9%
8	*psb*KI	-	-
9	*psb*M-trnD	3	2.5%
10	*rps*16	35	28.9%
11	coxl	-	-
12	*pulse*1	8	6.6%
13	*trn*LF	22	18.2%
14	*rpo*B	32	26.4%
15	*rpo*C1	35	28.9%

Based on the consideration of sequence availability, the primer used for phylogenetic tree reconstruction in this study was rbcL primer because it had the highest number of tree sequences for the USU Campus (105). However, several types of trees on the USU Campus are also not available for their rbcL sequences, namely burahol, ki damar, fir balloon, keben, duku, and kere parasol.

3.3. Reconstruction of the phylogenetic tree

In analyzing a phylogenetic tree, it is called cladistics (clades or groups of descendants derived from a common ancestor), usually represented by a branching system [14]. The farther the branching distance and the more branches, the further the level of resemblance to the ancestor will be. According to Dharmayanti [15] explains that the bootstrap value can also affect the results of a reliable phylogeny tree. The bootstrap value is the value used to test how well the model is used. High or low bootstrap values can affect the results of the phylogeny tree. If the bootstrap value is high, it can be said that the level of similarity of a species is similar to its ancestor. On the other hand, if the bootstrap value is low,

According to Dharmayanti [15], the Neighbor-Joining Tree method selects sequences that when combined will provide the best estimate of the branch length that most closely reflects the real distance between the sequences. The phylogenetic tree was tested statistically using the bootstrap method with 1000 replications. Hall [16] stated that bootstrap values of 100 to 1000 replicates were used to estimate the confidence level of a phylogenetic tree. In addition, according to Hasibuan et al [17] that the greater the bootstrap value used, the higher the confidence of the reconstructed tree topology based on the distribution of characters in the data is strongly influenced by random effects.

Based on the classification of clades on the phylogeny, found similarities between species, namely family similarity. So it can be concluded that the clades are grouped by family of each species. However, it is undeniable that all species are not in the same family, because there are several species that are separated from species that have the same family.

In Figure 1, the resulting phylogenetic tree mostly has bootstrap values above 70. According to Subari et al [18], a good phylogenetic tree is a phylogenetic tree that has a bootstrap value above 70. The presence of a higher dominant bootstrap value makes it more certain that the phylogeny tree can be trusted, and vice versa. If the dominant bootstrap value is low, then the phylogenetictree cannot be trusted.

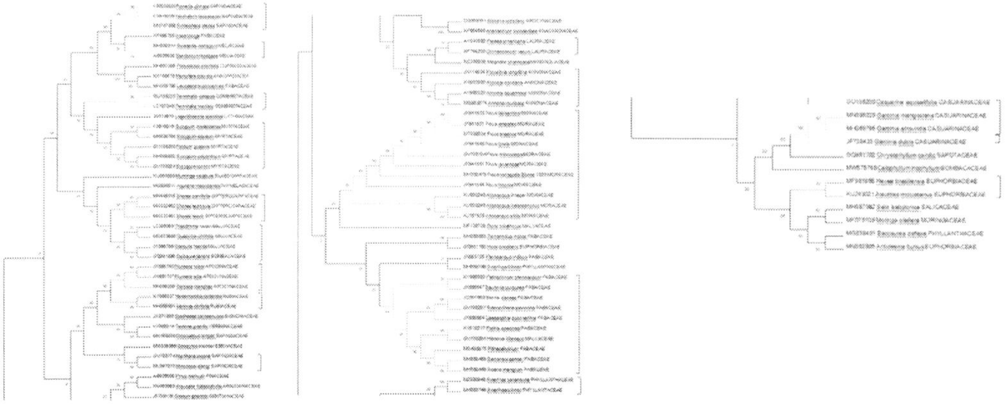

Figure 1. Phylogenetic tree species on the USU campus

When viewed from the phylogenetic results, it can be concluded that the grouping occurs because each species was in the same family. This can be seen in the species belonging to the Anacardiaceae, Meliaceae, Apocynaceae, Annonaceae, Ebenaceae, Sapindaceae, Clusiaceae, Sapotaceae, Calophyllaceae, Euphorbiaceae, Fabaceae, and Moraceae families.

4. Conclusion

The results of the exploration of sequence data on GeneBank for tree species on the USU Campus, as many as 86.8% were found using the rbcL marker. Reconstruction of the phylogeny tree using the neigbor joining method produces a boostrap value above 70, this indicates a fairly good classification result. Grouping occurs based on the same family.

References

[1] Arrijani 2003 *Biodiversitas* **4** 83
[2] Cahyandari R and Nursolihah R 2015 *Statistika* **15** 73
[3] Lestari D A, Azrianingsih R and Hendrian 2018 *Journal of Tropical Biodiversity and Biotechnology* **3** 1
[4] Hidayat T and Pancoro A 2008 *Jurnal AgroBiogen* **4** 35
[5] Lubis K 2014 *Jurnal Pengabdian Kepada Masyarakat* **20** 66
[6] Bhambri P and Gupta O P 2012 *Development of Phylogenetic Tree Based on Kimuras's Method*
[7] Fitmawati, Juliantari O P and Sofiyanti N 2017 *Applied Science and Technology* **1** 127
[8] Irawan P D, Tallei T E and Kolondam B J 2016 *Jurnal Ilmiah Sains* **16** 43
[9] Tindi M, Mamangkey N G F and Wullur V 2017 *Jurnal Pesisir dan Laut Tropis* **1** 32
[10] Susilowati A, Rangkuti A B, Harahap M M, Iswanto A H, Rachmat H H and Ginting I M 2020 *Identifikasi Jenis-Jenis Pohon di Kampus Universitas Sumatera Utara* (Medan: USU Press)
[11] Sinaga E 2010 *Biologi Molekuler Regulasi Ekspresi Gen* (Jakarta: Universitas Nasional)
[12] Lokapirnasari W P, Sahidu A M, Nurhajati A M, Supranianondo K and Yulianto A B 2017 *Jurnal Veteriner* **18** 76
[13] Prakoso S P, Wirajana I N and Suarsa I W 2016 *Indonesian Journal of Legal and Forensic Sciences* **2** 45
[14] Brinkman F S L and Leipe D D 2001 *Phylogenetic Analysis. In: Bioinformatics: A Partical Guide to the Analysis of Genes and Proteins. Baxevanis A D dan B F F Ouellette (Eds).* (United States: John Willey and Sons)
[15] Dharmayanti N L P I 2011 *WARTAZOA* **21** 1
[16] Hall B G 2001 *Phylogenetic Trees Made Easy: A How-To Manual for Molecular Biologists*

(United State: Sinauer Associates)
[17] Hasibuan F E, Mantiri F R and Rumende R R H 2017 *Jurnal Ilmiah Sains* **17** 59
[18] Subari A, Razak V and Sumarmin R 2021 *Jurnal Biologi Tropis* **21** 89

Identification of Araceae in Pondok Buluh Training Forest, Simalungun Regency, North Sumatra Province

H Manurung[1,2*], M Hasibuan[1] and R Rambey[1,2]

[1]Forestry Studies Program, Faculty of Forestry, Universitas Sumatera Utara, Jl. Tri Dharma Ujung No.1 USU Medan Campus 20155
[2]JATI-Sumatran Forestry Analysis Study Center, Deli Serdang 20353, Indonesia

*Email: harisyah_manurung@usu.ac.id

Abstract. The Pondok Buluh Training Forest is located in Simalungun Regency, North Sumatra, which has a fairly high diversity of flora, one of which is a plant from the Araceae family. Araceae is a family of taro that has potential for ornamental plants, medicinal plants, food and animal feed, therefore it is important to conduct research on Araceae species in this location to determine the diversity and potential of Araceae. From the results of the research, it was found that there were 13 species of Araceae which were included in 6 genera, namely Aglaonema, Alocasia, Diffenbachia, Homalomena, Schismatoglottis and Syngonium. The species are *Aglaonema crispum, Alocasia arifolia, Alocasia* sp., *Alocasia longiloba, Alocasia sanderiana, Alocasia scabriuscula, Dieffenbachia seguine, Diffenbachia amoena, Homalomena cordata schott, Homalomena pendula, Homalomena rubescens, Schismatoglottis calyptrata*, and *Syngonium podophyllum*. The potential of Araceae plants is as a medicinal plant, ornamental plant, vegetable pesticide and animal feed.

1. Introduction

Indonesia is a country that is rich in biological and natural resources diversity of flora and fauna species found throughout the archipelago. This natural wealth must be protected and preserved so that it can be used by the community for future interests [1]. One of the diversity of flora types is plants from the Araceae family that grow a lot in Indonesia. The Araceae family consists of 110 genera, which includes 3,200 species [2]. Research on the characteristics of Araceae in Indonesia is still limited, while Indonesia has a high level of Araceae diversity. The lack of information about the benefits of Araceae causes people to not be interested and do not cultivate plants belonging to the Araceae tribe. A study on the characteristics of Araceae needs to be carried out to determine the character possessed by Araceae as the latest information.

Araceae, better known as taro, is a plant that is very familiar but few people know it in depth. The main characteristic of the Araceae family of plants is the inflorescence arranged in the form of a cob (spadix) surrounded by a spathe. Araceae consists of 105-110 genera, 2500-3700 species, generally found in the tropics, namely Southeast Asia (including Indonesia, Malaysia, Brunei Darussalam, Philippines, Singapore), America and Papua New Guinea [3].

Araceae plants are widely known by the public as ornamental plants found in home gardens, but these plants can also be found living wild in the forest. Plants from the Araceae family are one of the many plant groups found in Indonesia [4]. Based Mayo et al [3] stated that this family consists of 105 genera and 3300 species.

Araceae are found in tropical areas, such as Malaysia, Indonesia, the Philippines, Singapore and Brunei Darussalam. Indonesia has 31 genera of Araceae spread over several regions, including Sumatra, Java, Kalimantan, Sulawesi and Irian Jaya. The Araceae tribe is one type of herbaceous plant that lives on land (terrestrial), in water (aquatic) and propagates on trees (epiphytes) [5]. Araceae plants are widely known by the public as ornamental plants found in home gardens, but these plants can also be found living wild in the forest. Tropical climates such as Indonesia is a very suitable habitat for this plant. About 90% of genera and 95% of Araceae species live in the tropics [6].

According to [7] Araceae are hairless herbaceous plants or shrubs, mostly bulbous or rooted in rhizomes, varied leaf shapes, have sheaths, flowers with lobes and androgynous. Araceae are generally found in 3 habitats, namely living on land, in water, and living in epiphytes. Most of these tribes live on land, both on forest floors, river banks and rocks.

The characteristics of Araceae plants are inflorescences arranged in the form of a cob (spadix) surrounded by a spathe of unisexual or bisexual inflorescence types, and can grow throughout the year [7]. Araceae plants have benefits that are not widely known by the public both from an economic and scientific perspective. The use of Araceae plants by the community includes ornamental plants, food sources and medicines. Some of them are used as alternative food ingredients, for example from *Colocasia esculenta* (L.) Schott (taro), *Amorphophallus paeoniifolius* (Dennst.) Nicolson (suweg) , and *Xanthosoma sagittifolium* (L.) Schott (taro) as ornamental plants because of their beautiful leaves and flowers, for example the types of Anthurium and Alocasia; and efficacious as an antibacterial, antioxidant, and anticancer drug, namely *Typhonium flagelliforme* Blume (rat taro) [8].

The Pondok Buluh Education and Training Forest area with a forest area of 1,272.70 Ha located in Simalungun Regency is the place chosen by the researchers to conduct their research because this forest area has never been researched on Araceae plants and has a wealth of biological natural resources, this forest also has a dense and extensive forest cover where as is known this araceae plant can also be found living wild in the forest, so it is necessary to know the diversity of araceae plant species in it. The objectives of this research are as follows: (1) To determine the types of Araceae plants found in the Pondok Bulu Training Forest, Simalungun Regency, North Sumatra Province, and (2) to determine the potential of Aracaea plants found in the Pondok Buluh Training Forest.

2. Methodology

Observations of Araceae plant species were carried out by roaming, namely by looking for locations overgrown by Araceae and then observing and identifying the species. Plant identification was carried out by matching the observed Araceae plants with the Araceae plant identification book. Identification of Araceae plants was carried out by measuring the leaf blade and petiole length using a ruler and tape measure. Observation of leaf shape was carried out by matching and comparing the types of Araceae leaf blades based on the Araceae identification book. Then the types of Araceae produced are described descriptively. To find out the use of aracae, in-depth interviews with field guides were conducted. After knowing the benefits, then classifying the benefits of each of the Araceae tribes.

3. Results and Disscussion

Based on the results of research that has been carried out, it was found that 13 species of the entire Araceae family were *Aglaonema crispum, Alocasia arifolia, Alocasia* sp., *Alocasia longiloba, Alocasia sanderiana, Alocasia scabriuscula, Dieffenbachia seguine, Diffenbachia amoena, Homalomena cordata* schott, *Homalomena pendula, Homalomena rubescens, Homalomena cordata, and Syngonium podophyllum.*

3.1. Alocasia longiloba

Alocasia longiloba is a herbaceous plant with a height of up to 150 cm and has a rhizome. Beautiful patterned leaves, big arrow shape. Petiole purple brown, pink to green. Cob cylindrical, up to 13 cm long. This species is native to Southeast Asia and southern China. Now the keris taro has been widely distributed in various tropical countries. This type can be cultivated by means of tuber cuttings and seedling separation. This species has been cultivated by the community as an ornamental plant, both indoors and outdoors. Based on the results of research by [9] this species in Bali has not been widely used, but has the potential of ornamental plants. In Malaysia this type is used in treating fever and scabs. In addition, this type can also be used to treat fever and edema by boiling the roots and drinking the boiled water [10].

3.2. *Homalomena cordata*

Homalomena cordata is a herbaceous plant with a height of up to 80 cm. heart-shaped leaves (cordatus). The leaves of H. cordata are about 80 cm long, yellowish green with fine green stripes. This species originates from Java and is now widespread throughout Indonesia and Malaysia. Habitat grows in the forest up to an altitude of 200 meters above sea level. This species can be used as a medicinal and ornamental plant. Rhizome efficacious eliminate colds and strengthen bones. Leaves can be used as food wrappers. Based on the results of research by [11] this type can be used as an anti-swelling drug. Based on [12] *Homalomena cordata* schott is an Araceae plant that is used as a medicinal and ornamental plant. Rhizome efficacious eliminate colds and strengthen tendons and bones. Leaves can be used as food wrappers.

3.3. *Homalomena pendula*

Homalomena pendula is a herbaceous plant with a height of up to 1 m. Stem erect, petiole 85 cm long, green to dark green. The leaves are hearty, green to yellowish green on the upper surface, pale green to dark green on the lower surface. This species is widely distributed in Malaysia, Thailand, Myanmar to Bangladesh. *H. pendula* can be cultivated by propagation of seeds and rhizomes. This type can be used as an ornamental plant in the yard or city park.

3.4. *Dieffeenbachia seguine*

Dieffenbachia seguine is a herbaceous plant with a height of up to 2 m. the stem is green, the bark is yellowish white to green and gummy. The broad leaves are round to oval, the color varies from poppy green, green with light green spots, to white or yellow. This species comes from the Caribbean Islands to the tropics of South America. Now, happy leaves have spread widely in tropical areas of Africa and Asia, including Indonesia. This species can be cultivated by stem cuttings and saplings. This type can be used as an ornamental plant in the yard, room and decoration. In Africa, the impact of this plant part is used as arrow poison.

3.5. *Alocasia scabriuscula*

Alocasia scabriuscula has a characteristic on the petiole that is attached directly to the base of the leaf and the petiole is green. Alocasia has the upper and lower surfaces of smooth, waxy leaves (pruinosus) with dark green on the upper surface of the leaves and light green on the lower surface. Alocasia has an elongated tuber shape. *Alocasia scabriuscula* has potential as a medicinal plant. Based on the results of research by [13] this plant has medicinal properties to reduce bee stings, which is taken from the sap. Ancient people believed this species as a natural medicine that can cure bee stings.

3.6. *Syngonium podophyllum*

Syngonium podophyllum is a herbaceous plant that has arrow-shaped leaves (sagitttatus) with flat leaf edges (integer), tapered leaf tips (acuminatus), grooved leaf base (emarginatus) and pinnate leaf spines (penninervis). The upper and lower surfaces of the leaves are smooth (laevis). Based on the results of research by [14], several Araceae species that are used by the surrounding community in Cisoka Village are mostly as ornamental plants. Usually planted in front of the yard. Traditionally used as a stage decoration in wedding parties or commemoration of the independence day of the Republic of Indonesia, one of which is *S. podophyllum.*

3.7. *Schismatoglottis calyptrata*

Schismatoglotis calyptrata is a herbaceous plant with a height of about 60 cm and forms dense colonies. The petiole is smooth, about 50 cm long. The leaves are oblong lanceolate, scalloped at the base, usually light green in color. This species is spread from southwest China, Indochina, to Maluku and Vanuatu. This type can be cultivated by propagation of seeds, stem cuttings and separation of tillers. This species can be used as a medicinal and ornamental plant [12]. stated that Schismatoglottis calyptara can be used as a medicinal plant and ornamental plant in the home garden. This plant is efficacious as a remedy for back pain. In West Java, the leaves are often used as fish feed. Meanwhile, based on the research results of [6] In Bali this type is used as a ceremonial material. However, in Malaysia, the roots and leaves are used as food or can be consumed

3.8. *Homalena rubescens*

Homalena rubescens is a herbaceous plant with a height of up to 80 cm. The leaves of *H. rubescens* are heart-shaped (cordatus) with a tapered leaf tip (acuminatus) with a curved leaf base (emarginatus). The leaves of H. rubescens range from 10-23 cm long and 6-18 cm wide. The petiole and leaf veins on the underside of the leaf surface are reddish green while the color of the leaf veins on the upper surface of the leaf is green. Based on the research results of [15] the leaf stalk of H. rubescens can be used as pig feed by boiling it first. Then the leaves can be used as a means of wrapping spices.

3.9. *Dieffenbachia amoena*

Dieffenbachia amoena is a popular ornamental plant commonly grown in gardens. Its beauty comes from the shape of the canopy and also the color of the leaves which vary in green with light green or yellow spots. This species has the potential as an ornamental plant. Based on the results of [16], this plant is resistant indoors, although for a certain period it needs to be treated also in open spaces. The color of the leaves tends to darken when placed indoors or in the shade, but becomes bright bright in the sun. Propagation is generally done by cuttings. The sap of the leaves and stems of Dieffenbachia contains calcium oxalate which forms needles in the cells and can cause itching and spasms of the lips and tongue.

3.10. *Alocasia sanderiana*

Alocasia sanderiana is a plant from the Araceae family. *Alocasia Sanderiana* is also known as the keris plant because of the resemblance of the tip of its leaves to the wavy blade of a kris (also known as a kris plant). It is a tropical perennial with shiny erect leaves, V-shaped and very curved. Plant height can reach 6 feet [17]. Alocasia sanderiana has potential as an ornamental plant. Based on the results of research by [18] most of the Alocasia species in the Philippines are ornamental plants or have potential ornamental value and are currently threatened by habitat destruction. One of these species is Alocasia sanderiana whose distribution is limited and only known from a few natural places today. This species is threatened on various levels and requires protection. However, it is not included in the CITES list.

3.11. *Alocasia* sp.

Alocasia sp. is a herbaceous plant with beautiful leaves and large arrows. Petiole purple brown, pink to green. The cob is cylindrical in shape, up to 13 cm long. sheath shaped like a canoe, green to white, 7-17 cm long. The back of the leaves is red-purple. This plant has the potential as an indoor and indoor ornamental plant. This type has a beautiful shape so it has many enthusiasts to cultivate it. The right and left sides of *A. greenvelvet* form a wave groove with a slightly pointed end. This plant has been traded in the market at a price of 40,000.

3.12. *Alocasia arifolia*

Alocasia arifolia is a herbaceous plant that has heart-shaped leaves and smooth leaf surfaces. *A arifolia* has arrow-shaped leaves, dark green with a thin striping on the leaves, large leaves growing to a length of 20-90 cm on long stalks. Based on the results of research by [19], the results of phytochemical tests carried out on lust showed that: that the chemical content contained in this plant is flavonoids, terpenes, and alkaloids. With the high content of secondary metabolites, it can be concluded that Birah can be used as an ingredient natural pesticides. Birah can be used as a vegetable pesticide, which is to function as an insect repellent.

3.13. Aglaonema crispum
Aglaonema crispum has fibrous roots; stem: light green, Stem contains a lot of water. The leaves are oval in shape, the tips of the leaves are pointed, the color of the leaves is green with a pattern, there are also red leaves. The leaf spines are pinnate. *A. crispum* has potential as anornamental plant. Based on the results of [20], leaves derived from crossing Aglaonema species so that they have diverse and interesting patterns are called hybrids. Leaf performance resulting from this cross produces a stunning leaf pattern. Hybrid type Aglaonema plants have a variety of leaf patterns so that flower hobbyists "never get tired of seeing it" and also the selling value of this hybrid type Aglaonema plant is higher than the Aglaonema species. From a business point of view, hybrid Aglonema leaves which have a unique style are in great demand even though the price is high.

3.14. Potential of araceae plant
The potential utilization of several species of araceae in the Pondok Buluh Training Forest in detail can be seen in Table 1.

Table 1. Potential of Araceae plants

Scientific name	Potential utilization
Homalomena cordata	Medicinal and ornamental plants
Homalomena pendula	Medicinal and ornamental plants
Alocasia longiloba	Ornamental plants
Dieffeenbachia seguine	Ornamental plants
Alocasia scabriuscula	Medicinal plant
Syngonium podophyllum	Ornamental plants
Schismatoglottis calyptrata	Medicinal and ornamental plants
Homalomena rubescens	Feed
Dieffenbachia amoena	Ornamental plants
Alocasia sanderiana	Ornamental plants
Alocasia sp.	Ornamental plants
Alocasia arifolia	Pesticides
Aglaonema crispum	Ornamental plants

From the results of the study, it is known that 3 species of Araceae have potential as medicinal and ornamental plants, 7 species of Araceae have potential as ornamental plants, 1 species have potential as medicinal plants, 1 species have potential as vegetable pesticides and 1 species has potential as animal feed.

4. Conclussions
From the results of the study, it can be concluded that there are 13species of Araceae plants in the Pondok Buluh Training Forest consisting of namely Aglaonema, Alocasia, Diffenbachia, Homalomena, Schismatoglotris and Syngonium. Based on its potential 3 species of Araceae have potential as medicinal and ornamental plants, 7 species of Araceae have potential as ornamental plants, 1 species have potential as medicinal plants, 1 species have potential as vegetable pesticides and 1 species has potential as animal feed.

References
[1] Sriastuti W, Herawatiningsih R and Tavita E 2018 *Jurnal Hutan Lestari* **6** 147
[2] Wilyasari L, Yulianty, Zulkifli and Nurcahyani E 2020 *Jurnal Ilmiah Biologi Eksperimen dan Keanekaragaman Hayati* **7** 35

[3] Mayo S J, Bogner J and Boyce P C 1997 *The Marga of Araceae* (Belgium: Royal Botanical Gardens, Kew)
[4] Widiyanti D, Mukarlina and Turnip M 2017 *Jurnal Protobiont* **6** 207
[5] Kurniawan A, Asih N P S, Yusammi P C and Boyce 2013 *Garden's bulletin Singapore* **65** 157
[6] Asih N P S and Kurniawan A 2019 *Widya Biologi* **10** 135
[7] Van Steenis C G G J 2008 *Flora* (Jakarta: Pradnya Paramita)
[8] Hutasuhut M 2020 *Jurnal Biolokus* **3** 288
[9] Sulaiman B and Mansor M 2002 *Proceedings of The 4th IMT-GT UNINET Conference* 216
[10] Lemmens R H M J and Bunyapraphatsara N 2003 *Plant Resources of South East Asia : Medicinal and Poisonous Plants* (Bogor: Prosea Foundation)
[11] Kurniawan A, Asih N P S, Yusammi P C and Boyce 2013 *Garden's bulletin Singapore* **65** 157
[12] Munawaroh E, Yuzammi, Mar'atus and Suhendar S 2017 *Tumbuhan Berpotensi sebagai Tanaman Hias* (Bogor: LIPI Press)
[13] Elsi Y, Satriadi T and Istikowati W 2020 *Jurnal Sylva Scienteae* **3** 193
[14] Mutaqin A Z, Fatharani M, Iskandar J, Partasasmita R 2018 *Biodiversitas* **19** 640
[15] Asih N P S, Warseno T and Kurniawan A 2014 *Prosiding Seminar Nasional Biodiversitas* **1** 521
[16] Ulfa S W 2019 *Best Journal (Biology Education Sains and Technology)* **2** 15
[17] Selvakumar P, Kaniakumari D and Loganathan V 2016 *Journal of Pharmacognosy and Phytochemistry* **5** 114
[18] Medecilo M and Madulid A 2010 *Philippine Journal of Science* **142** 145
[19] Sihombing F 2016 *Eksplorasi Tumbuhan Beracun di kawasan Cagar Alam Dolok Saut, Desa Pansur Natolu, Kecamatan Pangaribuan, Kabupaten Tapanuli Utara* (Medan: Fakultas Pertanian, Program Studi Kehutanan, Universitas Sumatera Utara)
[20] Apriansi M and Suryani R 2019 *Jurnal Agroqua* **17** 141

eDNA envisaged conservation of IUCN threatened taxa of the tropical mangrove ecosystems

S Kannan[1*], S Balamurugan[2], P Ragavan[3], B Deivasigamani[1], A K S Wee[4], S G Salmo III[5], M Basyuni[6,8] and T Kajita[7]

[1]CAS in Marine Biology, Annamalai University, Porto Novo, India.
[2]National Institute of Oceanography, Goa, India.
[3]Ministry of Environment, Forest and Climate Change, New Delhi, India.
[4]Faculty of Science & Engineering, University of Nottingham (Malaysia Campus), Jalan Broga, Malaysia.
[5]Institute of Biology, University of the Philippines, Diliman, Philippines
[6]Department of Forestry, Universitas Sumatera Utara, Medan 20155, Indonesia
[7]Iriomote Station, Tropical Biosphere Research Center, University of the Ryukyus, Japan
[8]Center of Excellence for Mangrove, Universitas Sumatera Utara, Medan 20155, Indonesia

[*]Email: oceanoactin@gmail.com

Abstract. eDNA sampling is a more affordable, quick, and consistent method to understand the aquatic ecosystem, especially the ocean, with freely associated DNA in water suitable for estimating diversity, composition, and behaviour. Furthermore, it helps to propose management plans to use sustainable ocean resources. The world's mangroves are currently under threat, and they also provide habitat for International Union Conservation of Nature (IUCN)-listed threatened species (e.g., Dugong, Bengal Tiger, etc.). These species are inextricably linked to the mangrove ecosystem, forming a sustenance chain that can be disrupted if lost. On the other hand, the more reliable methods to monitor threatened species are not clearly established and mostly depend on the non-standardized method, which requires practical and taxonomic expertise and limits the data-driven conservation acts. The implementation of eDNA in ocean data collection and monitoring of endangered species is still in a progressive stage and the studies on the eDNA of threatened species from mangrove ecosystems are few. Environmental DNA surveys have also mapped the distribution of endangered European eel (*Anguilla anguilla*) during spring and documented threatened sharks and rays across eastern Indonesia. As a hallmark, UNESCO recently launched a project to understand the richness of biodiversity. The main objective is to monitor fishes, including the IUCN red-listed species, through eDNA collected across the selected marine World Heritage sites. The recent observations ensure that eDNA might be used to monitor the IUCN-designated threatened taxa found in the tropical mangrove ecosystem. The outcomes of this eDNA technique are crucial for providing information regarding conservation priorities.

1. Introduction

Tropical countries have vast and varied coastal wetlands, viz., saltmarshes, tidal flats, mangroves, seagrasses, lagoons, and coral reefs. Besides having a robust framework of laws and policies, governing the wetland conservation it reamins difficult [1]. However, conservation methods can be improved given the rising expenses, threats and the mediocre results of restoration or rehabilitation. Protecting habitat conversion, optimizing adaptive potential (i.e., populations' or species' capacity to quickly adapt to changing environments with little disruption), and keeping the ecosystem healthy by conducting non-invasive ecosystem monitoring (e.g., eDNA metabarcoding) are all important goals to safeguard the existing coastal wetlands and sustain the provisional ecosystem services offered by them rather than the short-term increase in area by feeble restoration/rehabilitation efforts. Mangrove forests are a unique ecosystem and a critical interface between terrestrial, estuarine, and near-shore marine ecosystems [2].

Despite its importance in ecological and economic values, global mangroves are gradually disappearing. It is important to have well-functioning and undamaged marine ecosystems that are essential for various organisms. However, anthropogenic activities, like extensive use of coastal land, alien species, harbor/boat activities, ballast water, and coastal pollution, threaten the condition, marine ecosystem function, and the existing biodiversity at a local to global scale. The management and guarding of the marine ecosystem are thus the highest priority. An adequate monitoring of marine ecosystems is there by essential, and has a long tradition, both with respect to variables looked at and methods used [1].

2. Methods

In recent decades, the use of eDNA approaches has overtaken aquatic biodiversity monitoring as it overcomes the limitations of conventional biomonitoring methods, which support non-invasive sampling, covering a wide range of taxonomic groups, with high sensitivity and possible automation. In fish ecology research, eDNA has been used in species diversity and conservation to monitor threatened and invasive species and analyse their abundance and distribution [3]. Thomsen P F et al [4] investigated marine fish biodiversity using the eDNA obtained directly from seawater samples. Tina E. Berry [5] and his team gave a new insight into identifying a wide range of different phylogenetic organisms at the species level using eDNA from plankton samples collected over five years on the same site. Thus, eDNA sampling is a more affordable, quick, and consistent method to understand the aquatic ecosystem, especially the ocean, with freely associated DNA in water suitable to estimate the diversity, composition, and behaviour and to propose management plans to use sustainable ocean resources.

3. Result

3.1. IUCN threatened taxa of the tropical mangrove ecosystems

The world's mangroves are under threat, and the following mangroves viz., *Pelliciera rhizophorae, Bruguiera hainesii, Sonneratia griffithii, Avicennia bicolor, Camptostemon philippines, Heritiera fomes* & *Heritiera globose* are classified under threatened category [6].

The mangroves also provide habitat for IUCN-listed threatened species such as the Dugong (*Dugong dugon*), Pygmy Three-toed Sloth (*Bradypus pygmaeus*), Bengal Tiger (*Panthera tigris tigris*), False Water Rat (*Xeromys myoides*), Proboscis Monkey (*Nasalis larvatus*), Smooth Coated Otter (*Lutrogale perspicillata*), Western Red Colobus (*Procolobus badius*), Fishing Cat (*Prionailurus viverrinus*), Sambar Deer (*Rusa unicolor*), Philippine Cockatoo (*Cacatua haematuropygia*), Mangrove Finch (*Camarhynchus heliobates*), Madagascar Teal (*Anas bernieri*), Mangrove Hummingbird (*Amazilia boucardi)*, Illidge's Ant Blue Butterfly (*Acrodipsas illidgei*), Mangrove Skimmer (*Orthetrum poecilops*), Hawksbill Turtle (*Eretmochelys imbricate*), American Crocodile (*Crocodylus acutus*), Four-Toed Terrapin (*Batagur baska*), Sharptooth Lemon Shark (*Negaprion acutidens*), Roughnose Stingray (*Pastinachus solocirostris*), Tiger Tail Seahorse (*Hippocampus comes*) and Fin-Joined Goby (*Gobulus birdsongi*) [6]. The following species found in the mangroves are listed as critically endangered and facing an extremely high risk of extinction. These endangered species are inextricably linked to the mangrove ecosystem, forming a sustenance chain that can be disrupted if lost. On the other hand, the more reliable methods to monitor threatened species are not established and mostly depend on the non-standardized methods, which require practical and taxonomic expertise [5].

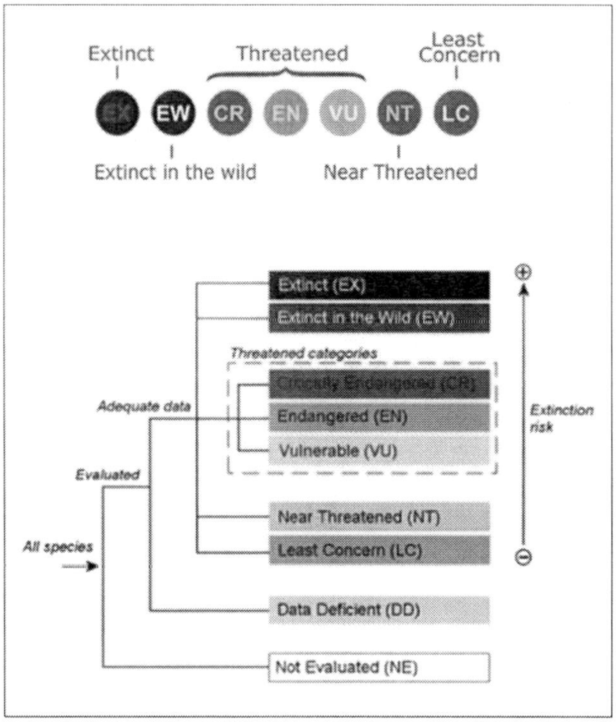

Figure 1. IUCN's structure of the categories [6]

Table 1. Critically endangered species found in the tropical mangroves

Critically endangered species	Group
Bruguiera hainesii	Mangroves
Sonneratia griffithii,	Mangroves
Bradypus pygmaeus	Mammals
Cacatua haematuropygia	Birds
Camarhynchus heliobates	Birds
Eretmochelys imbricate	Reptiles
Batagur baska	Reptiles
Gobulus birdsong	Fishes

3.2. Studies used eDNA to identify endangered or threatened species in mangrove and coastal waters
Studies over the past two decades have reported a rapid decline of many marine species and their habitats at an alarming rate [8,9]. Previous results showed that marine species thriving in Southeast Asia face high pressure. However, based on the IUCN red list, the relative proportion of threatened species is comparable to other geographical areas [10]. Identifying these threatened species in any ecosystem is challenging, especially in the mangrove ecosystem, as sampling and taxonomic biases are inherent in assessing the diversity of these species.

Many studies have implemented eDNA to detect the occurrence and abundance of IUCN-listed endangered marine species. The eDNA study conducted in the Merbok Estuary, Peninsular Malaysia, identified seven IUCN-endangered fish species (*Carcharhinus amblyrhynchos-EN, Carcharhinus melanopterus-NT, Brevitrygon walga-NT, Gymnura poecilura-NT, Anguilla bicolor-NT, Harpadon nehereus-NT,* and *Scomberomorus commerson-NT*) and two invasive species (*Oreochromis niloticus* and *Poecilia vivipara*). Among them, *Carcharhinus amblyrhynchos-EN*, a black-tail reef shark, numbers declined to an undetectable level because of over-catch [11]. Likewise, environmental DNA has proven applications in monitoring threatened taxa, such as the discovery of previously unrecorded elasmobranch species in Singapore's waters [12], mapping the distribution of endangered European eel [13] and documenting rare and threatened sharks and rays across eastern Indonesia [14]. Similarly, the critically endangered large-tooth sawfish (*Pristis pristis*) inhabit in the waters of northern Australia was identified using eDNA method (COI gene) [15]. Also, the critically endangered small tooth sawfish (*Pristis pectinata*), was identified in Chandeleur Islands, Louisiana using eDNA. The small tooth sawfish that thrive throughout the Atlantic water is now restricted to the southwest of Florida and the Bahamas [16]. Also, the eDNA method was used to assess the endangered *Neophoca cinerea* (Australian sea lion) [17]. However, the use of eDNA-based methods in tropical mangrove ecosystems is still in its early stages.

4. Conclusion

Mangrove ecosystem biota can now be evaluated using environmental DNA (eDNA). By characterizing the Spatio-temporal distribution of eDNA, metabarcoding provides a non-destructive way to monitor coastal biodiversity, particularly that of the mangroves. Given the availability of such biodiversity data, coastal ecosystem resilience can be increased through efficient data-driven coastal management. As a hallmark, UNESCO launched a project to understand the richness of biodiversity based on the eDNA method to monitor fish and IUCN red-listed species from the selected marine World Heritage sites. The recent observations ensure that eDNA might be used to monitor the IUCN-designated threatened taxa found in the tropical mangrove ecosystem, and the outcomes of this eDNA technique are crucial for providing information regarding conservation priorities. Using environmental DNA as a key technique for biomonitoring in the mangrove environment may provide solutions for the sustainable development of coastal biological resources.

Acknowledgements
We extend our gratitude to Professor TK for his assistance in the production of this publication. We also acknowledge the support of JSPS Core to Core, Tropical Biosphere Research Center, University of the Ryukyus, Japan.

References

[1] Ragavan P, Kathiresan K, Mohan P M, Ravichandran K, Jayaraj R S C and Rana T S 2020 *Wetlands Ecol Manage* **29** 5
[2] Polidoro B A, Carpenter K E, Collins L, Duke N C, Ellison A M, Ellison J C, Farnsworth E J, Fernando E S, Kathiresan K and Koedam N E 2010 *PLoS ONE* **5** 4
[3] Xing Y, Gao W, Shen Z, Zhang Y, Bai J, Cai X, Ouyang J and Zhao Y 2022 *Front Environ Sci* **73**
[4] Thomsen P F, Kielgast J, Iversen L, Wiuf C, Rasmussen M, Gilbert M Gilber T P, Orlando L and Willerslev E 2011 *Mole Ecol* **21** 11
[5] Berry T E, Saunders B J, Coghlan M L, Stat M, Jarman S, Richardson A J, Davies C H, Berry O, Harvey E S and Bunce M 2019 *PLoS genetics* **15** 2
[6] IUCN Red list categories and criteria, version 3.1, second edition *IUCN Library System* [https://portals.iucn.org/library/node/10315]
[7] Ii S, Kitade H, Ishida S, Imai Y, Watanabe Y and Wada S *PLoS computational biology* **16** 6 e1007943
[8] Sarah C 2013 Endangered Species Associated with Mangroves *Mangrove Action Project* [https://mangroveactionproject.org/wcontent/uploads/2018/10/Endangered-Species]
[9] Waycott M, Duarte C M, Carruthers T J B, Orth R J, Dennison W C and Olyarnik S 2009 *Proc Natl Acad Sci* **106** 30
[10] Selig E R, Longo C, Halpern B S, Best B D, Hardy D, Elfes C T, Scarborough C, Kleisner K M and Katona S K 2013 *PLoS ONE* **8** 4 e60284
[11] Zainal A D H, Lavoué S, Abdul R M, Jamaludin N A and Mohammed A N A 2021 *Scientific Reports* **11** 1
[12] Ip Y C A, Tay Y C, Chang J J M, Ang H P, Tun K P P, Chou L M, Huang D and Meier R 2020

Environmental DNA **3**
[13] Weldon L, O'Leary C, Steer M, Newton L, Macdonald H, and Sargeant S L 2020 *Environmental DNA* **2** 4
[14] Moore A M, Jompa J, Tassakka A C M A R, Yasir I, Ndobe S, Umar W, Gold Z and Barber P H 2021 *Aquaculture, Aquarium, Conservation and Legislation* **14** 2
[15] Simpfendorfer C A, Kyne P M, Noble T H, Goldsbury J, Basiita R K Lindsay R, Shields A, Perry C and Jerry D R 2016 *Endang Spec Res* **30**
[16] Humphreys E M 2021 Environmental DNA (eDNA) Surveys for the Smalltooth Sawfish, Pristis pectinata, in the Chandeleur Islands (The University of Southern Mississippi Louisiana: USA)
[17] Berry T E, Osterrieder S K, Murray D C, Coghlan M L, Richardson A J, Grealy A K, Stat M, Bejder L and Bunce M 2017 *Ecol Evoln* **7** 14

Phylogenetic analysis of Shorea Genus in University of North Sumatra Campus, Indonesia, based on *mat*K genes

A B Rangkuti[1,2*], L Irmayanti[3] and S S H Harahap[1,2]

[1]Forestry Studies Program, Faculty of Forestry, Universitas Sumatera Utara, Jl. Tri Dharma Ujung No.1 USU, Medan 20155, North Sumatra, Indonesia
[2]Center of Excellence for Mangrove, Universitas Sumatera Utara, Medan 20155, Indonesia
[3]Major of Forestry, Faculty of Agriculture, Khairun University, North Maluku, Indonesia

*Email: baiqunirangkuti@usu.ac.id

Abstract. Shorea is one of the genera in the Dipterocarpacea family with a very large number of species. Shorea tree species have high economic and ecological value. Currently more than 100 species of Shorea are in the endangered category. Taxonomic aspects as well as information on biological characteristics are one of the important bases of species conservation or restoration activities. The University of North Sumatra campus is one of the green open spaces that can be used as an ex-situ conservation location. This study aims to determine the types of trees of the Shorea genus on the North Sumatra University campus and their kinship based on phylogenetic analysis. It is necessary to identify the Shorea species on the USU campus using the current taxonomic method using phylogeny analysis. Sequences of the identified Shorea species were collected from the NCBI (National Center for Biotechnology Information) web, then edited the data and constructed a phylogenetic tree using Meng X.1. In this study found 4 types of Shorea, namely *Shorea parvifolia* (Meranti Sarang Punai), *Shorea laevis* (Bangkirai), *Shorea macrophylla* (Tengkawang Tungkul) and *Shorea leprosula*. *Shorea leprosula* species are separated from other Shorea groups. This indicates that *Shorea leprosula* is closely related to other Shorea. Meanwhile, *Shorea macrophylla* and *Shorea parvifolia* are closely related because they are in one cluster, with a boostrap value of 85%.

1. Introduction

Dipterocarpaceae is a tree family that dominates lowland forests, especially in Indonesia. The family also has an important role in fulfilling wood supply, both domestically and abroad [1]. Shorea is one of the largest genera in the Dipterocarpaceae family, of which 196 species have been found in Indonesia and are generally found on the islands of Kalimantan and Sumatra and only 138 species have been documented [2]. Based on data from the International Union Conservation Nature (IUCN) Red list [3] more than 100 species of Shorea fall into the critical category (Critically Endangered).

Taxonomic aspects as well as information on biological characteristics are one of the important bases of species conservation or restoration activities [4]. Misidentification of species occurs quite often which results in unfavorable results when it is associated with tree breeding activities and procurement of quality seeds in the context of restoration programs [5]. Identification activities with morphological data

IOP Conf. Series: Earth and Environmental Science 1115 (2022) 012033 doi:10.1088/1755-1315/1115/1/012033

are difficult because it will take a long time and the fruit and flower seasons are not the same [6,7]. A phylogenetic tree is an arrangement that can describe kinship data in the form of a branching tree [8].

The construction of phylogenetic trees is the most important and interesting thing in the study of evolution. There are several methods for constructing phylogenetic trees from molecular data (nucleotides or amino acids) [9]. Phylogenetic analysis of a nucleotide or amino acid sequence family is an analysis to determine how the family was inherited during the evolutionary process. Evolutionary relationships between sequences are described by locating sequences as outer branches of a tree. The branch relationships in the interior of the tree reflect the degree to which the different sequences are related. Two very similar sequences will be located as neighboring outside of the branches and connected in a common branch [10]. Barcoding studies of plant loci generally use matK (maturase K) because of the high success rate of amplification for many species and easy sequencing [11].

The University of North Sumatra campus is one of the green open spaces in the city of Medan and has great potential as a conservation location for tree species that have a threatened status such as the genus Shorea. Based on research by Arida et al 2021 [12], there are as many as 121 tree species on the USU campus. It is very interesting to conduct further research related to species and their phylogenetic studies, where there are still very limited studies conducted on the USU 1 campus. The purpose of this study was to determine the types of plants of the genus Shorea and their kinship found on the USU campus.

2. Methods

Tree data collection was carried out by taking an inventory of the types of Shorea that exist on the USU campus. Once found, identification of species is carried out, finding out local names, synonyms, benefits and conservation status. For phylogeny analysis data, the Shorea type sequence data found was taken from the NCBI web (Table 1). Data editing and phylogeny creation were carried out using MEGA X.1 software, phylogeny trees using the Neighbor joining (NJ) method.

Table 1. Sequence data information based on NCBI web

No	Types	Locus/ Accession	Sequence length	Title
1	*Shorea leprosula*	MH332586	771 bp	Generating DNA sequence data with limited resources for molecular biology: lessons from a barcoding project in Indonesia [13]
2	*Shorea parvifolia*	MF418826	824 bp	Plant DNA Barcodes and assesment of phylogenetic community structure of a tropical mixed dipterocarp forest in Brunei Darussalam (Borneo) [14]
3	*Shorea Laevis*	MH332600	771 bp	Generating DNA sequence data with limited resources for molecular biology: lessons from a barcoding project in Indonesia [13]
4	*Shorea macrophylla*	AB246441	972 bp	Comprehensive molecular phylogeny of the sub family Dipterocarpoideae (Dipterocarpaceae) based on Chloroplast DNA sequences. [15]

5	*Tectona grandis*	KJ012800	719 bp	A molecular phylogeny for the trees of puerto rico. [16]

3. Results and Discussion

This research found 4 species of the genus Shorea on the campus of the University of North Sumatra, namely *Shorea parvifolia* (Meranti Sarang Punai), *Shorea laevis* (Bangkirai), *Shorea macrophylla* (Tengkawang Tungkul) and *Shorea leprosula*. The uses of the Shorea tree species found include: construction materials, panels, furniture, veneers, plywood, bridge furniture, ship raw materials, etc. While the conservation status starts from Endangered (EN), Vulnerable (VU), Least Concern (LC) and Near Threaned (NT), the data can be seen in Table 2.

Table 2. Types of Shorea found on the North Sumatra University Campus

No	Types	Local Name	Synonym	Utilization	Conservation Status
1	*Shorea parvifolia* Dyer	Meranti Sarang Punai (Meranti Merah)	*Shorea scutulata* King *Shorea gentilis* Parijs	Building Materials, Panels, Furniture, Crates, Boxes and Veneers	Endangered (EN) IUCN Redlist since 1998 [3]
2	*Shorea laevis*	Bangkirai (Meranti Kuning)	*Hopea laevifolia*, *Shorea laevifolia*, *Shorea rogersiana*	Building Materials, Panels, Furniture, Crates, Boxes and Veneers	Vulnerable (Vu) IUCN Redlist since 2017 [17]
3	*Shorea macrophylla*	Tengkawang Tungkul (Meranti Merah)	-	Furniture, Light construction, Plywood, Fruit	Least Concern (LC) IUCN redlist since 2019 [18]
4	*Shorea leprosula*	Meranti Bunga, Meranti Batu (Meranti Merah)	-	Construction Materials, Plywood, furniture, bridges, boats and sills.	Near Threaneted (NT) Since 2019 [18]

Compilation of phylogeny trees is considered objective if using outgroup species [19], in this study using teak species (*Tectona grandis*). Based on the resulting phylogeny tree (Figure 1), it shows that the *Shorea leprosula* species is separate from the other Shorea groups. This indicates that *Shorea leprosula* is distantly related to other Shorea. Meanwhile, *Shorea macrophylla* and *Shorea parvifolia* have a close relationship because they are in the same cluster. The results of this study are in line with Kamiya et al [20] that the clusters/groups in the phylogenetic tree of the type *Shorea leprosula* are separated from those of *Shorea macrophylla*.

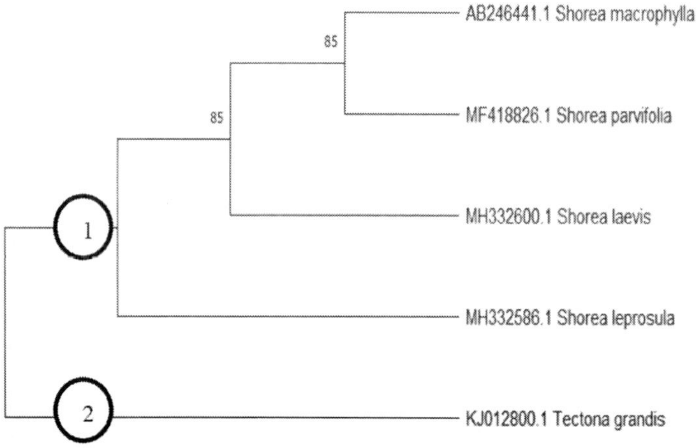

Figure 1. The phylogeny tree of the Genus Shorea found on the USU Campus

Species that are in one group or one cluster in the phylogeny tree can be estimated at their evolutionary level [21,22]. The evolutionary relationships among the sequences of each species are described by placing the sequences as the outer branches of a tree. The branch relationships in the interior of the tree reflect the degree to which the different sequences are related. Based on this, it can be said that the genetic range of *Shorea leprosula* is different from other types of Shorea used in this study, especially *Shorea macrophylla* and *Shorea parvifolia*. In terms of the growth rate of the early phase (seedlings), *Shorea leprosula* has a slower growth compared to *Shorea macropylla* [23].

4. Conclusion

There are 4 species of Shorea found on the USU campus, namely *Shorea parvifolia* (Meranti Sarang Punai), *Shorea laevis* (Bangkirai), *Shorea macrophylla* (Tengkawang Tungkul) and *Shorea leprosula*. The phylogeny analysis showed that the *mat*K primer was able to classify Shorea species within 1 clade, which means that they have a close genetic distance.

Reference

[1] Tsumura Y, Kado T, Yoshida K, Abe H, Ohtani M, Taguchi Y, Fukue Y, Tani N, Ueno S, Yoshimura K, Kamiya K, Harada K, Takeuchi Y, Diway B, Finkeldey R, Na'iem M, Indrioko S, Ng K K S, Muhammad N and Lee S L 2011 *Journal of Plant Research* **124** 35
[2] Ashton P S 1982 *Flora malesian* **9** 237
[3] [IUCN] International Union for The Conservation of Nature 1998 *IUCN Red List of Threatened Species*
[4] Schmeller S, Bauch B, Gruber B, Juskaitis R, Budrys E, Babij V, Lanno K, Sammul M, Varga Z and Henle K 2008 *Biodiversity Conservation* **17** 3623
[5] Soberon J and Medellin R A 2007 *Conservation Biology* **21** 1366
[6] Lestari D A, Azrianingsih R and Hendrian H 2018 *Journal of Tropical Biodiversity and Biotechnology* **3** 1
[7] Rangkuti A B, Susilowati A, Arinah H, Irmayanti L, Siregar I Z and Siregar U J *2021 IOP Conf. Series: Earth and Environmental Science* **912** 012084
[8] Lubis K 2014 *Pengabdian Kepada Masyarakat 20* 66
[9] SAITOU N and IMANISHI T 1989 *Mol Biol Evol* **6** 514
[10] MOUNT D W 2001 *Phylogenetic prediction. In: Bioinformatic, Sequence and Genome Analysis* (New York Press) 237

[11] Rangkuti A B, Susilowati A, Siregar UJ, Irmayanti L and Siregar I Z 2021 *Journal of Sylava Indonesiana* **04** 45
[12] Susilowati A, Rangkuti A B, Rachmat H H and Ginting I D A M 2021 *Biodivers J Biol Divers* **22** 2839
[13] Dean G H, Asmarayani R, Ardiyani M, Santika Y, Triono T, Mathews S and Webb C O 2018 *Applications in Plant Sciences* **6** e1167
[14] Heckenhauer J, Salim K A, Chase M W, Dexter K G, Pennington R T, Tan S, et al 2017 *PLoS ONE* **12** e0185861
[15] Gamage D T, de Silva M P, Inomata N, Yamazaki T and Szmidt A E 2006 *Genes Genet Syst* **81** 1
[16] Muscarella R, Uriarte M, Erickson D L, Swenson N G, Zimmerman J K, et al 2014 *PLoS ONE* **9** e112843
[17] [IUCN] International Union for The Conservation of Nature 2017 *IUCN Red List of Threatened Species*
[18] [IUCN] International Union for The Conservation of Nature 2019 *IUCN Red List of Threatened Species*
[19] Muzzazinah 2017 *Prosiding Seminar Nasional Pendidikan Biologi dan Biologi* (Jurusan Pendidikan Biologi, Fakultas MIPA, Universitas Negeri Yogyakarta)
[20] Kamiya K, Harada K, Tachida H and Ashton P S 2005 *American Journal of Botany* **92** 775
[21] Harnelly E, Thomy Z and Fathiya N 2018 *Biodiversitas* **19** 1074
[22] Dharmayanti N L P 2011 *Wartazoa* **21** 1
[23] Fawzi N I and Juliansyah 2019 *Wahana Forestra : Jurnal Kehutanan* **14** 1

Bioinformatics analyses of embryoid genes in oil palm tissue culture (*Elaeis guineensis* Jacq.)

R P Astari[1,3], M Basyuni[2,5*], L A P Siregar[4] and R I M Damanik[4]

[1]Doctoral Program of Agricultural Science, Faculty of Agriculture, Universitas Sumatera Utara, Medan, North Sumatra, Indonesia
[2]Department of Forestry, Faculty of Forestry, Universitas Sumatera Utara, Medan, North Sumatra, Indonesia
[3]PT. Socfin Indonesia, Jl. K. L. Yos Sudarso No. 106, Medan 20244, North Sumatra, Indonesia
[4]Faculty of Agriculture, Universitas Sumatera Utara, Medan, North Sumatra, Indonesia
[5]Centre of Excellence for Mangrove, Universitas Sumatera Utara, Medan 20155, Indonesia

*Email: m.basyuni@usu.ac.id

Abstract. Oil palm (*Elaeis guineensis* Jacq.) is a beneficial tropical crop that produces oil vegetables, those needs continue to increase. Clone planting on a large scale is one of the possibilities to increase oil palm production. Identification of genotype oil palm embryoid can optimize the efficiency of the culture process. Twenty-six projected embryoid genes with code amino acid from 850 to 936 bp from oil palm tissue culture were collected from the NCBI database. This study clarified the variation and role of physicochemical characteristic, possible peptide transfer and subcellular localization of the oil palm (*Elaeis guineensis* Jacq.) embryoid genes. The similarity of genes in embryoids on phylogenetic trees showed the potential to detect the role of specific embryoid genes to help the breeders optimize cloning processes.

1. Introduction

Elaeis guineensis Jacq (Oil palm plant) is tropical crop that produces important oil vegetable [1,2]. This plant originated in West Africa and spread to several countries [3]. Oil palm is a beneficial crop in the world. It is the highest commodity export agriculture as a source of foreign exchange in Indonesia [4]. It is a perennial including monocot plant, monoecious, belonging to the Arecaceae family [5]. Oil palm breeding with conventional genetic is a long-term process and there is variability in the needs selection of the plant. Propagation with tissue culture enables a true-to-type plant with the characters likely the parents [6,7].

Tissue culture propagation used by the elite oil palm is one of the potential technologies to help assemble uniform superior oil palm seeds with high production. Planting oil palm tissue culture clones significantly increased production (30%) compared to commercial Dura x Pisifera *(DxP)* seeds in a large-scale trial area [8-10]. The successfully of tissue culture processes depends some Factors. There are growth condition of material [11], type of media, room condition [12] and the genotypes of donor plants. Among these factors, the genotype is factor affecting the efficiency of the culture process [13].

The response of molecular cellular oil palm embryoid gene is not well understood. Genes of embryoid are genes associated with callogenesis and embryogenesis in culture. Furthermore, in the NCBI database, there is a lot of data on oil palm embryoid tissue culture [14]. This study used mRNA sequence oil palm tissue culture to identify the genes embryoid. This present report thus proposes to identify twenty-six gene embryoids of oil palm tissue culture with the bioinformatics method.

2. Materials and Methods

2.1. Materials

Twenty six (26) calculated of embryoid genes in oil palm tissue culture (*Elaeis guineensis*) from NCBI database. The NCBI reference of mRNA sequence was used with the number accessions:

1. *E. guineensis* embryoid clone with number accession: EY410244.1; pOP-EO07352.
2. *E. guineensis* embryoid clone with number accession: EY411811.1; pOP-EO06767.
3. *E. guineensis* embryoid clone with number accession: EY410308.1; pOP-EO07377.
4. *E. guineensis* embryoid clone with number accession: EY412800.1; pOP-EO06600.
5. *E. guineensis* embryoid clone with number accession: EY412967.1 pOP-EO06712.
6. *E. guineensis* embryoid clone with number accession: EY409141.1 pOP-EO06539.
7. *E. guineensis* embryoid clone with number accession: EY410753.1 pOP-EO06866.
8. *E. guineensis* embryoid clone with number accession: EY409959.1 pOP-EO06745.
9. *E. guineensis* embryoid clone with number accession: EY409112.1 pOP-EO06590.
10. *E. guineensis* embryoid clone with number accession: EY409046.1; pOP-EO06599.
11. *E. guineensis* embryoid clone with number accession: EY412020.1; pOP-EO06859.
12. *E. guineensis* embryoid clone with number accession: EY408968.1; pOP-EO06793.
13. *E. guineensis* embryoid clone with number accession: EY412574.1; pOP-EO06797.
14. *E. guineensis* embryoid clone with number accession: EY408659.1; pOP-EO06785.
15. *E. guineensis* embryoid clone with number accession: EY411917.1; pOP-EO07861.
16. *E. guineensis* embryoid clone with number accession: EY410807.1; pOP-EO06585.
17. *E. guineensis* embryoid clone with number accession: EY410867.1; pOP-EO07461.
18. *E. guineensis* embryoid clone with number accession: EY411546.1; pOP-EO07373.
19. *E. guineensis* embryoid clone with number accession: EY410329.1; pOP-EO06601.
20. *E. guineensis* embryoid clone with number accession: EY413071.1 pOP-EO06776.
21. *E. guineensis* embryoid clone with number accession: EY409815.1 pOP-EO06489.
22. *E. guineensis* embryoid clone with number accession: EY408618.1; pOP-EO06603.
23. *E. guineensis* embryoid clone with number accession: EY412776.1; pOP-EO06880.
24. *E. guineensis* embryoid clone with number accession: EY409295.1; pOP-EO06504.
25. *E. guineensis* embryoid clone with number accession: EY410845.1; pOP-EO06716.
26. *E. guineensis* embryoid clone with number accession: EY413689.1; pOP-EO06782.

2.2. Physicochemical characteristics of the embryoid genes in E. guineensis clones

The structure and physicochemical characteristics of embryoid gene in oil palm tissue culture (*E. guineensis*) were analyzed with Protparam online with by web.expasy.org/protparam. The analyzed factor designated molecular weight, number of amino acids, theoretical isoelectric values, coefficients of extinction, half-life period, coefficients of instability, index of aliphatic and hydropathicity values [15].

2.3. Promising peptide transfer and subcellular prediction of embryoid genes

Possible peptide transfer analyzed used online by Target P-2.0 program with the link (web:services.healthtech.dtu.dk/service/TargetP-2.0). It is to the estimated location of one of the signal peptides of secretory pathway, mitochondrial peptide target, chloroplast peptide transit peptide, thylakoid transfer peptide and other signals. The estimated for sub-cellular location was analyzed by online website (www.psort.org) to prediction the subcellular locations of embryoid genes in oil palm tissue culture reported [16].

2.4. Phylogenetic analysis of 26 embryoid genes

Phylogenetic analyzed twenty six sequence of genes embryoid in oil palm tissue culture from NCBI database. Phylogenetic of amino acid genes embryoid in oil palm tissue culture was performed using MEGA version 11.0.11 design by a neighbor-joining method with bootstrap investigation 1000 times to measure the similarity of embryoid genes in the phylogenetic map [17].

3. Results and Discussions

3.1. Physical and chemical attributes of embryoid genes of E. guineensis clones

Several physicochemical attribute embryoid genes of *E. guineensis clone* showed in Table 1. The *E. guineensis* tissue culture used of twenty six clones. The length of the genes embryoid is varies from 850 to 936 bp. The heterogeneity related to several amino acids, in molecular weight, number of amino acids, theoretical isoelectric values, coefficients of extinction, half-life period, coefficients of instability, index of aliphatic and hydropathicity values of genes embryoid in oil palm tissue culture. Analyzed showed that there are variations in physicochemical characters [16,18].

It is the variation to known that the approximate half-life value was varied among the embryoid gene oil palm clone (min 1.2h; max 30h) in Table 1. *Ganoderma boninense* genes from *Elaeis guineensis* Jacq (min 1.2h; max 7.2 h) [19], actin genes in mangrove plant also have varied half-life (min 1.3h; max 30h) [15], Oxidosqualenecyclase (OSC) genes in *R.tribe* with half-life 4.4 h [20], polyprenol gene reductase in *H. brasiliensis* (1.2-30 h) [21].

Table 1. Physicochemical characteristic of the embryoid genes in oil palm tissue culture

Variant	G1	G2	G3	G4	G5	G6	G7
Length of mRNA/bp	855.00	899.00	881.00	876.00	858.00	901.00	906.00
Total of amino acids	855.00	899.00	881.00	876.00	858.00	901.00	906.00
Molecular mass	70829.58	74073.33	72543.52	71709.95	70136.34	75825.18	77794.48
Theoretical isoelectric values	5.12	5.1	5.02	5.09	5.09	5.05	5.11
Number of atoms	9140	9495	9360	9116	8867	9567	10071
Coefficient of extinction	10125	11375	10375	12000	12375	15125	11250
Half-life point	1.2	30.0	7.2	1.2	30.0	30.0	1.2
Coefficient of instability	40.27	34.2	41.92	42.52	30.59	50.47	52.28
Index of aliphatic	29.24	27.36	30.53	25.46	23.54	26.42	25.61
Hyropathicity values	0.708	0.71	0.738	0.724	0.718	0.885	0.63

Noted: G1-G7: accession number of gene embryoid oil palm tissue culture (*Elaeis guineensis* Jacq.).

Table 1. Continued

Variant	G8	G9	G10	G11	G12	G13	G14
Length of mRNA/bp	913.00	881.00	872.00	850.00	916.00	886.00	883.00
Total of amino acids	913.00	881.00	872.00	850.00	916.00	886.00	883.00
Molecular mass	76460.37	72856.39	72979	72203.45	76413.17	73312.05	75019.3
Theoretical isoelectric values	5.00	5.09	5.04	5.07	5.04	5.09	5.03
Number of atoms	9795	9293	9099	9146	9585	9381	9422
Coefficient of extinction	12250	12125	15750	13500	16000	12000	16375
Half-life point	7.2	1.2	1.2	1.2	1.2	1.2	1.2
Coefficient of instability	40.46	46.99	60.69	47.55	47.27	46.65	62.65
Index of aliphatic	26.4	26.56	20.99	22.82	26.75	29.23	28.77
Hyropathicity values	0.718	0.743	0.829	0.755	0.936	0.793	1.02

Noted: G8-G14: accession number of gene embryoid oil palm tissue culture (*Elaeis guineensis* Jacq.).

Table 1. Continued

Variant	G15	G16	G17	G18	G19	G20
Length of mRNA/bp	879.00	860.00	873.00	852.00	935.00	853.00
Total of amino acids	879.00	860.00	873.00	852.00	935.00	853.00
Molecular mass	74665.65	74105.51	74308.11	70541.19	75665.98	72209.23
Theoretical isoelectric values	5.00	5.06	5.1	5.09	5.16	5.11
Number of atoms	9148	9408	9544	9004	9898	9308
Coefficient of extinction	19375	14500	11875	12000	8375	10750
Half-life point	30.0	30.0	1.2	30.0	1.2	30.0
Coefficient of instability	45.85	45.49	43.64	42.13	29.21	36.8
Index of aliphatic	19.57	29.07	25.89	29.34	35.29	25.44
Hyropathicity values	0.996	0.934	0.706	0.823	0.712	0.644

Noted: G15-G20: accession number of gene embryoid oil palm tissue culture (*Elaeis guineensis* Jacq.).

Table 1. Continued

Variant	G21	G22	G23	G24	G25	G26
Length of mRNA/bp	871.00	886.00	936.00	859.00	853.00	859.00
Total of amino acids	871.00	886.00	936.00	859.00	853.00	859.00
Molecular mass	75060.76	73033.34	79411.77	71446.05	71301.04	71421.16
Theoretical isoelectric values	5.1	5.11	5.12	5.06	5.08	5.09
Number of atoms	9651	9383	10318	8967	9044	9114
Coefficient of extinction	11875	11125	10500	14500	12750	12125
Half-life point	1.2	30.0	30.0	1.2	30.0	1.2
Coefficient of instability	45.13	35.89	39.77	50.7	43.88	47.96
Index of aliphatic	22.96	29.68	26.92	24.56	23.92	28.29
Hyropathicity values	0.63	0.758	0.609	0.858	0.741	0.801

Noted: G21-G26: accession number of gene embryoid oil palm tissue culture (*Elaeis guineensis* Jacq.).

IOP Conf. Series: Earth and Environmental Science 1115 (2022) 012034 doi:10.1088/1755-1315/1115/1/012034

3.2. Prediction of peptide localization and subcellular embryoid genes in oil palm

The prediction of peptide potential transit of the embryoid genes in oil palm tissue culture analyzed showed in Table 2. There are four the options peptipe transfer: signal peptipe of secretory pathway, mitochondrial target peptipe, chloroplast transite peptide, thylakoid luminal transfer peptipe dan other signal. The result showed, there 20 from 26 genes with potential transit peptide maximum embryoid oil palm tissue culture from 0.7138 to 0.9999 on other signal.

There is also 4 genes embryoid with predictied transit peptide maximum on signal peptipe of secretory pathway (0.6750-0.9671). Reported result in *H. brasiliensis* drought genes with the potential transit peptide in the secretory pathway (0.026-0.38) [22].The potential transit peptide maximum on chloroplast peptide (C26: 0.8665). The mitochondrial target peptipe maximum in C22: 0.6324. Prediction transfer peptide in thirteen clones of *Ganoderma. boninense* from 0.090 to 0.274 [19], rubber protein dehydration induced with mitochondria transfer peptide from 0.020 to 0.342 [21].

Table 2. The prediction of transfer peptide of the embryoid genes in oil palm tissue culture

Variant (Sample Type) Likelihood	Reliability				
	Other	Signal peptide of secretory pathway	Mitochondrial target peptide	Chloroplast transit peptide	Thylakoid luminal transfer peptide
G1	0.7653	0.0009	0.0001	0.0086	0.225
G2	0.9724	0.0002	0.0031	0.0165	0.0078
G3	0.9999	-	-	-	-
G4	0.7138	0.1433	0.1423	0.0002	0.0004
G5	0.7356	0.1064	-	0.1563	0.0015
G6	0.9953	-	0.0041	0.0005	0.0001
G7	0.8068	0.0008	0.1715	0.0096	0.0112
G8	0.9997	0.0002	-	-	0.0001
G9	0.0298	0.9671	0.0028	0.0002	-
G10	0.9982	0.0017	-	-	-
G11	0.9303	0.0005	0.0023	0.0664	0.0006
G12	0.9234	0.0001	0.0606	0.016	-
G13	0.9919	0.0001	0.0031	0.005	-
G14	0.9955	-	-	0.0043	0.0001
G15	0.9974	0.001	0.0015	-	-
G16	0.9120	0.0405	0.0354	0.0052	0.0068
G17	0.2644	0.7349	0.0005	-	0.0001
G18	0.9365	0.0002	0.0404	0.0175	0.0055
G19	0.9901	0.0084	0.0014	-	-
G20	0.3232	0.675	0.001	0.0008	0.0001
G21	0.9856	0.0115	0.0008	0.0014	0.0006
G22	0.2912	0.0002	0.6324	0.0758	0.0004
G23	0.9958	0.0021	0.002	-	-
G24	0.998	0.0013	0.0007	-	-
G25	0.2857	0.6919	0.0217	0.0001	0.0006
G26	0.1212	-	0.0002	0.8665	0.012

The location of subcellular embryoid genes of oil palm tissue culture was showed in Tabel 3. The result showed embryoid gene of oil palm culture was detected in fourteen locations subcellular: nucleus, cytoplasmic, c.stroma, c. thylakoid membrane, c.t thylakoid space, membrane endoplasmic reticulum, e. reticulum (lumen), golgy-body, membrane of plasma, micro-body (peroxisome), membrane inner of mitochondrial, matrix space of mitochondrial, intermembrane space of mitochondrial and outside. There are fourteen genes in embryoid oil palm tissue culture detected inside cytoplasm, membrane of endoplasmic reticulum and micro body (peroxisome). It was the same the location subcellular of the genes generally inside cytoplasm, membrane of endoplasmic reticulum and micro body (peroxisome) [19]. There are two genes detected highest subcellular location in nucleus 0.96 (C13 and C26), showed data in Table 3. It is the same with rubber gen protein dehydration induced subcellular maximum in nucleus 0.96 [21].

Table 3. Subcellular localization of the embryoid genes in oil palm tissue culture

Variant	Nuc	Cy	Cs	Ctm	Cts	Er(m)	Er(l)	Gb	Pm	M(p)	Mim	Mms	Mis	Ost
G1	nt	nt	nt	0.297	nt	nt	Nt	0.300	0.790	0.519	nt	nt	nt	nt
G2	0.700	nt	nt	0.100	nt	nt	Nt	nt	nt	0.572	nt	0.100	nt	nt
G3	nt	0.650	nt	nt	nt	nt	Nc	nt	nt	nt	nt	0.100	nt	nt
G4	nt	nt	nt	nt	nt	nt	Nt	nt	nt	0.563	0.486	0.807	nt	nt
G5	nt	nt	nt	nt	nt	0.685	0.100	0.460	0.640	nt	nt	nt	nt	nt
G6	0.700	nt	nt	0.100	nt	nt	Nt	nt	nt	0.542	nt	0.100	nt	nt
G7	nt	nt	nt	0.313	nt	0.300	Nt	0.400	0.600	nt	nt	nt	nt	nt
G8	nt	0.450	nt	0.100	nt	nt	Nt	nt	nd	0.174	nt	0.100	nt	nt
G9	nt	nd	nt	nt	nt	0.100	Nt	nt	0.685	0.120	nt	nt	nt	0.820
G10	nt	0.650	nt	nt	nt	nt	Nt	nt	nt	0.062	nt	0.100	nt	nt
G11	nt	0.650	nt	0.100	nt	nt	Nt	nt	nt	0.006	nt	0.100	nt	nt
G12	0.700	nt	nt	nt	nt	nt	Nt	nt	nt	nt	0.103	0.426	0.103	nt
G13	0.960	nt	nt	0.100	nt	nt	Nt	nt	nt	0.140	nt	0.100	nt	nt
G14	nt	nt	0.851	0.405	0.405	nt	Nt	nt	nt	nt	nt	0.100	nt	nt
G15	0.600	nt	nt	0.100	nt	nt	Nt	nt	nt	nt	nt	0.100	nt	nt
G16	nt	nt	nt	0.468	nt	0.300	Nd	0.400	0.600	nt	nt	nt	nt	nt
G17	nt	0.450	nt	nt	nt	nt	Nt	nt	nt	0.333	nt	0.437	0.121	nt
G18	nt	0.450	nt	0.100	nt	nt	Nt	nt	nt	0.498	nt	0.360	nt	nt
G19	nt	nt	nt	nt	nt	0.685	0.100	0.460	0.640	nt	nt	nt	nt	nt
G20	nt	nt	nt	nt	nt	0.100	0.100	nt	0.460	0.166	nt	nt	nt	nt
G21	nt	nt	nt	0.311	nt	0.300	Nt	0.400	0.600	nt	nt	nt	nt	nt
G22	nt	0.450	0.200	0.280	nt	nt	Nt	nt	nt	0.300	nt	nt	nt	nt
G23	nt	nt	nt	nt	nt	0.685	0.100	0.460	0.640	nt	nt	nt	nt	nt
G24	nt	nt	nt	nt	nt	0.550	0.100	nt	nt	0.315	nt	nt	nt	0.100
G25	nt	nt	nt	nt	nt	0.100	0.100	nt	0.460	nt	nt	nt	nt	0.100
G26	0.960	nt	nt	0.280	nt	nt	Nt	nt	nt	nt	0.300	0.360	nt	nt

Note: Nuc: Nucleus; Cy: cytoplasma; Cs: stroma of chloroplast; Ctm: membrane of chloroplast thylakoid; Cts: thylakoid space of chloroplast; Er (m): endoplasmic reticulum (membrane); Er(l): endoplasmic reticulum (lumen); Gb: golgi body; Pm: plasma membrane; M(p): microbody (peroxisime); Mim: mitochondrial inner membrane; Mms: mitochondrial matrix space; Mis: mitochondrial intermembrane space; Ost:outsite; nc: not clear; nt: not detected.

3.3. Phylogenetic construction

Phylogenetic analyzed to verify the homology amongst the embryoid gene oil palm tissue culture [18]. The similarity among embryoid gene of (*Elaeis guineensis* Jacq.) tissue culture varies from 41% to 100%, a clustering was formed (Figure 1). The phylogenetic tree characterizes two branches for 26 embryoid genes. The first branch consist 14 embryoid genes (C6, C10, C15, C25, C5, C12, C24, C2, C11, C1, C19, C4, C9 and C17) and another branch consist 12 embryoid genes C7, C20, C22, C3, C8, C13, C26, C18, C14, C16, C21 and C23) (Firuge 1).

The similarity of genes in embryoid on phylogenetic trees showed the potential to detect the role of specific embryoid genes which is important genes in tissue culture process (callus and embryo induction). Therefore an efficient tissue culture process to produce the superior seed to optimize increasing oil palm productivity.

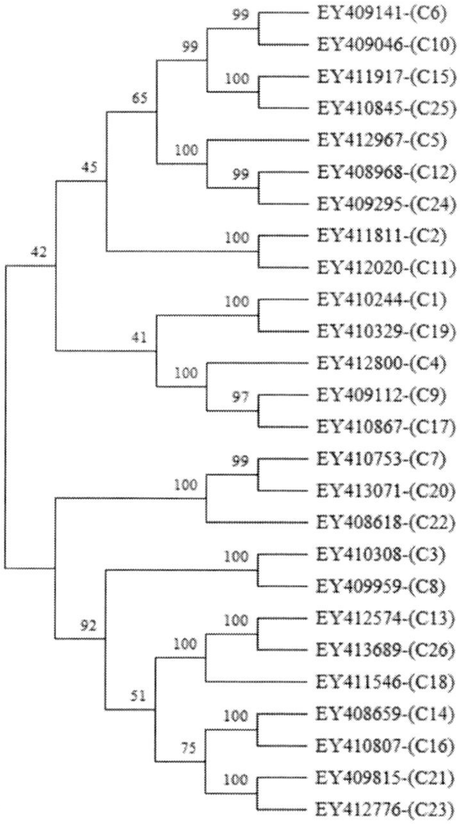

Figure 1. Clustering of twenty six genes embryoid *Elaeis guineensis* Jacq clone with the similarity varies from 41-100 %. Dendogram designed by Neighbor-Joining method used CLUSTAL alignment [22] with bootstrap 1000 repetition [17].

4. Conclusions

There are 26 nucleotides of gene embryoid of tissue culture oil palm in the database Gene-Bank NCBI at a gene length of 850-936 bp. This study successfully clarified the variable of physicochemical characters, possible peptide transfer and subcellular localization and phylogenetic homology gene embryonic of palm oil tissue culture (*Elaeis guineensis* Jacq.). The embryoid gene was the potential to detect the role of specific embryoid genes for cloning processes efficiently.

Acknowledgments
This study was funded by the *Penelitian Disertasi Doktor* scheme No. 51/UN5.2.3.1/PPM/KP-DRTPM/TI/2022, from the Directorate General of Research, Technology and Community Service, the Ministry of Education, Culture, Research and Technology of the Republic of Indonesia.

References

[1] Bourgis F, Kilaru A, Cao X, Ngando-Ebongue G F, Drira N, Ohlrogge J B and Arondel V 2011 *Proc. Natl. Acad. Sci. U. S. A.* **108** 12527
[2] Zou J, Zhang Q, Zhu Z, Gao L, Zheng Y and D Li 2019 *Scientia Horticulturae* **245** 125
[3] Kushairi A, Mohd Din A and Rajanaidu N 2011 *MPOB* **1** 47
[4] Ho Y W, Tan C C, Soh A C, Wong G dan Wong S P 2009 *Proc National Conference On Oil Palm* 86
[5] Chia G S, Lopes R, Cunha R N V, Rocha R N C and Lopes M T G 2009 *Acta Amaz* **39** 249
[6] Baudouin L and Durand-Gasselin T 1991 *Oléagineux* **49** 313
[7] Hashim A T, Ishak Z S K, Rosli M, Abdullah O, Ooi S, Husri M N and Bakar D A 2018 *Somatic Embryogenesis* (Springer International Publishing: USA)
[8] Cochard B, Durand-Gasselin T, Amblard P, Konan E K and Gogor S 1999 *Proceedings of 1999 PORIM International Palm Oil Congress*
[9] Ong-Abdullah M, Ordway J M, Jiang N, Ooi S E, Kok S Y, Sarpan N et al 2015 *Nature* **525** 533
[10] Kushairi A, Tarmizi A H, Zamzuri I, Ong-Abdullah M, S Kamal, Ooi S E and Rajanaidu N 2010 *International Seminar on Advances in Oil Palm Tissue Culture*
[11] Caswell K, Leung N and Chibbar R N 2000 *Plant Cell Tiss. Org. Cult.* **60** 69
[12] Saharan V, Yadav R C, Yadav R N and Chapagain B P 2004 *Afr. J. Biotechnol.* **3** 256
[13] Gandonou C H, Errabii T, Abrini J, Idaomar M, Chibi F and Senhaji S 2005 *African Journal of Biotechnology* **4** 1250
[14] Ting N C, Jansen J, Nagappan J, Ishak Z, Chin C W, Tan S G, Cheah S C and Sigh R 2013 *PLoS One* **8** e53076
[15] Basyuni M, Wasilah M and Sumardi 2017 *J. Phys. : Conf. Ser.* **801** 012013
[16] Basyuni M, Wati R, Sagami H, Oku H and Baba S 2018 *J. Phys.: Conf. Ser.* **978** 012044
[17] Felsenstein J 1985 *Evolution* **39** 783
[18] Basyuni M, Hayati R, Sihaloho M A, Slamet B, Bimantara Y, Habsyah T S, Hanafiah S S and Julianti E 2019 *IOP Conf. Series: Earth and Environmental Science* **305** 012036
[19] Basyuni M, Purba A, Putri L A P, Hayati R, Chalil D and Syahputra I 2018 *Journal of Physics: Conf. Series* **1235** 012071
[20] Basyuni M and Wati R 2017 *J. Phys. : Conf. Ser.* **801** 012011
[21] Pasaribu S A, Basyuni M and Hayati R 2019 *IOP Conf. Series: Earth and Environmental Science* **305** 012042
[22] Pratomo B, Lisnawita, Nisa T C and Basyuni M 2021 *IOP Conf. Series: Earth and Environmental Science* **713** 012053

Etlingera elatior extract inhibits early developmental stage of *Fasciola gigantica* egg *in vitro*

A R Wulandari, A Nurlaelasari, D Prasetyo, W Pawestri, D M Nuraini, M Cahyadi and P H Hamid*

Department of Animal Science, Faculty of Agriculture, Universitas Sebelas Maret, Surakarta, Central Java 57126, Indonesia

*Email: pennyhumaidahhamid@staff.uns.ac.id

Abstract. The use of plants as natural anthelmintic is quite promising for parasite control. In this report, we evaluate the activity of *Etlingera elatior* to *Fasciola gigantica* eggs *in vitro*. The eggs were collected from naturally-infected bovine in the Ampel Abbatoir, Boyolali, Central Java. The eggs were exposed to *E. elatior* ethanolic extract in different concentrations i.e., 5%, 2.5% and 1.25%. Fluconix 340 in the equal concentrations was positive control of anthelmintic and distilled water served as negative treatment. The eggs were placed in the dark until 9 days and observed under the light microscope on days 5 and 9. The results showed that *E. elatior* in the concentration of 5%, 2.5% and 1.25% significantly inhibited the initial development of *F. gigantica* compared to the negative control (water) on days 5 and 9 treatments ($P<0.05$). The *E. elatior* effectively inhibits morulation and larva formation. Fluconix 340 5%, 2.5% and 1.25% were possessing higher ovicidal activities compared to *E. elatior* since day 5 of treatment. *E. elatior* showed effective inhibition to the initial development of *F. gigantica* egg *in vitro*.

1. Introduction

Fasciola sp. is an endemic trematode in Indonesia causing liver damage to ruminants. The worm adult stage is in the liver parenchyma and gallbladder of infected ruminants. *Fasciola* sp. causes general gastrointestinal disturbances, the inefficiency of metabolic nutrients and thus economic losses. The disease has a major impact on the economy on livestock production, either directly or indirectly. Control of Fasciolosis is mandatory by considering the disease impact and the zoonotic importance of the disease. The global prevalence of *Fasciola* sp. infection is estimated at 17 million people from 61 countries and 180 million people are at risk of infection because they are in endemic areas. In Indonesia, a 23.57% prevalence rate was reported in the Pontianak City Slaughterhouse, West Kalimantan [1], while a prevalence of 17-48% was recorded in the Special Region of Yogyakarta Province between 2016-2018 [2-4]. The latest published data in 2020 shows that the rate of the disease in Aceh Province, Central Java Province, West Sumatra Province, Lampung Province and Bali Province was 56.3%, 33.33%, 23.81, 30% and 15%, respectively [5-9].

Control of fasciolosis now mostly relies on anthelmintics. In general, ruminants infected by *Fasciola* sp. are treated using anthelmintic drugs which are widely distributed in the market. The common anthelmintic drugs that are widely used belong to several classes e.g., Benzimidazole, Intermectin, Nitroxynil and Triclabendazole, the last as the most recommended anthelmintic for Fasciolosis. Drugs of this type are widely used because of their effectiveness against many classes of gastrointestinal parasites and their affordability in developing countries [10]. However, the use of certain types of anthelmintics without rotation may cause resistance [11-13].

The use of natural materials can be an alternative solution as an anthelmintic which is environmentally friendly and based on biological resources abundantly available in Indonesia. *Etlingera elatior* is known as "Kecombrang" often used as herbs and traditional salad food in Indonesia. *E. elatior* has healing properties, antimicrobial effects and insecticidal properties [14-16]. The use of *E. elatior* ethanolic extract as an anthelmintic against *F. gigantica* has never been studied. In this study, we evaluate the potential use of *E. elatior* in the early developmental stage of *F. gigantica* eggs.

2. Materials and Methods

2.1. Eggs collection

Fasciola gigantica eggs were collected from the gallbladder of the naturally infected bovine. The bovine is slaughtered at the Ampel Slaughterhouse, Boyolali, Central Java. All the gallbladders were transported on ice to the Division of Biology, Integrated Laboratory, Universitas Sebelas Maret, Indonesia. The gallbladder was dissected and the bile was filtered using a 100-mash filter to separate the worm eggs from debris. The eggs were sedimented by using tap water in a 1000 ml beaker glass for 30 minutes. After 30 minutes, the liquid was drained slowly, leaving worm deposits on the bottom of the beaker glass. The sedimentations were further repeated until the eggs were clean from cell debris. *Fasciola* sp. eggs were finally washed by using a centrifuge at 5,500 rpm for 10 minutes [17].

2.2. Parasite identification

The *F. gigantica* eggs were identified morphologically under a light microscope. Eggs were taken as much as 50 ul then placed on an object glass and observed. Examination of eggs by using a magnification of 400 times. The eggs are then categorized into normal, developing and damaged eggs [18]. In this study, developed eggs were described as morulated eggs on day 5 and initial development of larva on day 9.

Biomolecular identification was also carried out to identify whether the eggs used were *F. gigantica*, not *Fasciola hepatica* or hybrid eggs. The DNA from pooled eggs was isolated by using a DNeasy isolation kit (Qiagen, Germany). Briefly, DNA isolation was carried out by pooled eggs that contain a minimum of 100 eggs into a 1.5 ml microtube. 180 µl of Buffer ATL and 20 l of Proteinase K were added. The solution was homogenized using a vortex after being incubated for 24 hours at 56°C. The eggs were vortexed occasionally during the incubation process. Then the 200 µl of buffer AL was added and then vortexed. The eggs were incubated for 10 minutes at 56°C and thereafter added by 200 µl ethanol and transferred to a DNeasy spin column. The samples were centrifuged at 10,000 rpm for 1 minute. The lysate was added with 500 µl of Buffer AW1, centrifuged then the filtrate was discarded. 500 µl of AW2 buffer was added, then centrifuged 2 times and then the filtrate was discarded. The DNeasy spin column was transferred to a 1.5 ml new microtube and 100 µl of Buffer AE was added to elute the DNA. The DNA was stored in a freezer at -20°C until use.

F. gigantica identification was performed by duplex PCR T100 Thermal cycler machine (BioRad, Singapore). PCR mixes were as follows: 12.5 µl BioLine HS (BioLine, USA), 8.5 µl of nuclease-free water, 1 µl of DNA samples and 1 µl of each primer. The PCR reaction was pre-denaturation at 95°C for 3 minutes, denaturation at 95°C for 30 seconds, annealing at 52°C for 30 seconds, extension at 72°C form 2 minutes and the last step is the final extension at 72°C for 2 minutes [19]. The PCR results were loaded onto 1.5% agarose gel (Nacalai Tesque, Japan) and visualized in UV light. The identification of DNA multiplication for Fasciola species was descriptive.

2.3. Preparation of *Etlingera elatior* extract

E. elatior was extracted from fresh flowers collected in Cilacap, Central Java, Indonesia. *E. elatior* flowers were identified in the Department of Plant Systematic, Universitas Gadjah Mada no.

072/S.T.b./IV/2022. 10 kg of fresh flowers were cleaned, cut into small pieces and dried in the sun. The dried *E. elatior flowers* were then mashed and macerated using 96% ethanol as a solvent mixed with 2% citric acid for 5 days in the dark. The process was continued with re-maceration again 3 times. The re-maceration results were mixed with the first filtrate and then evaporated using a rotary evaporator until a thick *E. elatior* flower extract was obtained [20]. All the procedures were performed in LPPT, Universitas Gadjah Mada, Indonesia.

2.4. Eggs treatment
The isolated eggs were soaked in 1 ml of *E. elatior* extract solution with concentrations of 5%, 2.5% and 1.25%. The drug control of egg immersion was Fluconix 340 (Interchemie, Holland) with the same concentrations. The negative control of the study was aquadest [17]. All concentrations were performed in triplicates. To observe changes in eggs, observations were made under a light microscope on days 5 and 9.

2.5. Statistical analysis
The result was analyzed by GraphPad Prism 4.0 (GraphPad Software Inc., USA) for graphical presentation and statistics.

3. Results and Discussions
Various herbs were reported to possess anthelmintic activity against *Fasciola* sp. eggs [17,21,22]. Natural plants harboring anthelmintic activity can be used as fluke control. The large abundance of herbs in the tropical region will guarantee the availability and ease for extraction and further analyzed for the bioactive substances.

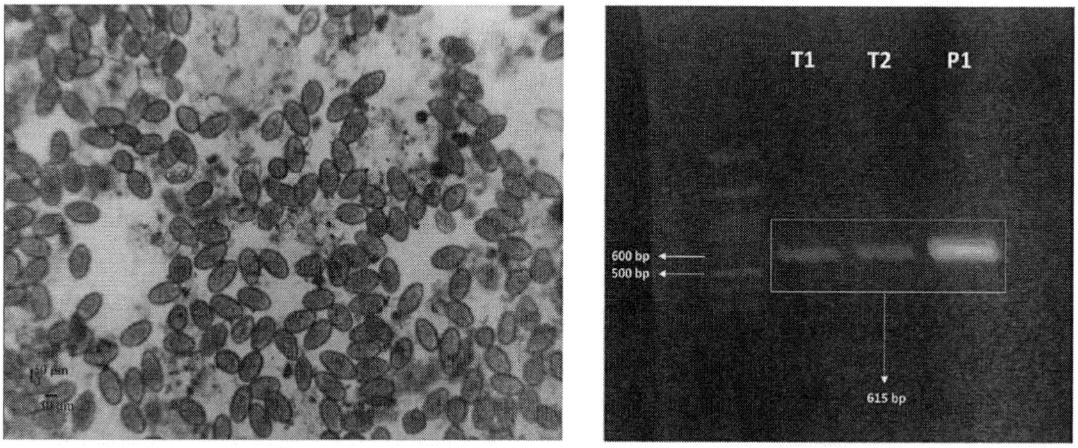

A B

Figure 1. Isolation and identification of *F. gigantica* eggs. A. *F. gigantica* eggs collected and observed under light microscopy. B. Duplex-PCR results from the pooled sample of eggs. T1, T2 are pooled eggs samples in comparison to adult tissue of *F. gigantica*, P1.

Eggs from the gallbladder were recovered successfully (Fig. 1. A) and showed as highly viable eggs. Each egg is surrounded by an intact cell wall and yellowish inside morphologically. Although well-known as *F. gigantica* endemic countries, some reported presence of hybrid species in Indonesia [23]. The duplex PCR with specific primers aimed to evaluate eggs homogeneity of *Fasciola* species to be used in all experiments. Egg-pooled samples were recognized as *F. gigantica* eggs with 615 bp from PCRs results (Fig. 1. B). The species were *F. gigantica* only since no double bands were observed to represent hybrid or *F. hepatica* eggs in the population [19].

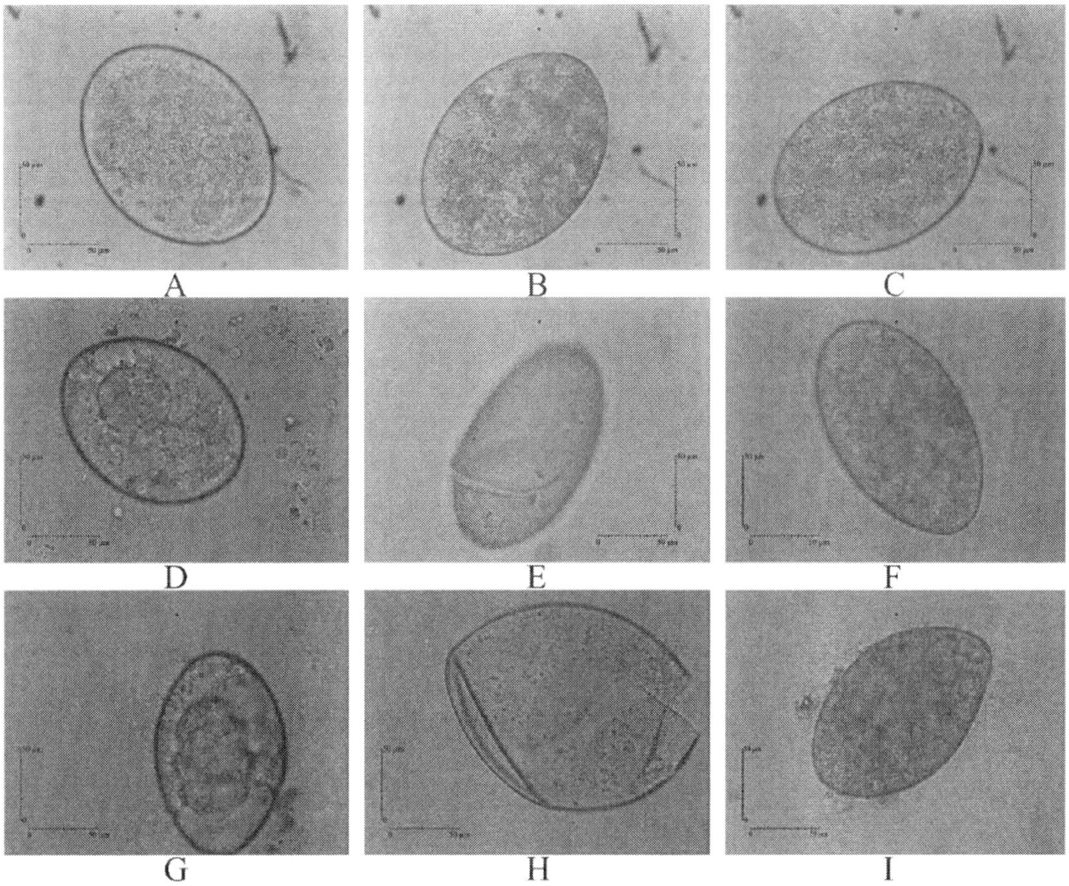

Figure 2. Inhibition of *F. gigantica* eggs development by *E. elatior*. Day 0: water (A), Fluconix 340 5% (B), *E. elatior* 5% (C); day 5: water (D), Fluconix 340 5% (E), *E. elatior* 5% (F); day 9; water (G), Fluconix 340 5% (H), *E. elatior* 5% (I).

F. gigantica egg showed a morulated embryo on day 5 incubated in water as control (Fig. 2). The larva is developed on day 9 after incubation in normal water (Fig. 2). Total larva developed on day 9 reached 76% of egg population (Fig. 3). Ethanolic extract of *E. elatior* showed inhibition of *F. gigantica* development in the early stage. *F. gigantica* eggs treated by *E. elatior* at 5% concentration showed significantly lesser developed eggs on days 5 (P=0.0065) and 9 (P=0.00047) compared to negative controls (Fig. 3). *E. elatior* at 5% showed only 30% development to morula on day 5 and 26% of the egg population is developed to the initial larval stage on day 9. Eggs development was still inhibited in *E. elatior* at 2.5% and 1.25% by 60% and 67% only eggs developed on day 9, respectively. The inhibition of *E. elatior* extract led to several eggs were remain in blastomeres stage without morula formation on day 5 and larval development on day 9. Besides, ovicidal effect was also observed by degraded eggs wall. The Fluconix 340 5% act stronger in the ovicidal activity in significant numbers compared to control (P=0.0001). The Fluconix 340 5% has a stronger ovicidal effect compared to *E. elatior* 5% (P=0.0078). No development was observed on eggs soaked in Fluconix 340 5% on day 5 and only 3% larvated on day 9 (Fig. 3). Eggs degradation was observed on day 5 (Fig. 2). Eggs were developed in higher numbers at Fluconix 340 at 2.5% and 1.25% on day 5 i.e., 13% and 20% respectively. Fluconix 340 2.5% and 1.25% on day 5 possessing higher ovicidal activity compared to *E. elatior* with the same concentration (P=0.0078, P=0.025).

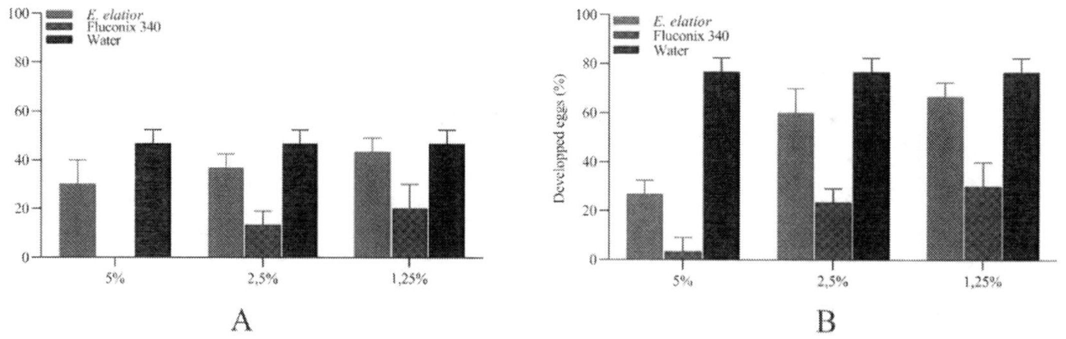

Figure 3. Inhibition of *E. elatior* to *F. gigantica* eggs in the early-developmental stages, day 5 (A) and 9 (B).

The *E. elatior* is also known as "Karondo" were used as a traditional remedy for helminthiasis for human in Sulawesi [24]. The herb is also utilized as a fresh salad in a mix with chili sauce in almost all areas in Indonesia since easily cultivated or naturally grown. No toxicity was even reported so far in mammals while consuming *E. elatior*. Natural herbs are expected to have less toxicity and unwanted side-effect compared to chemical compounds. *E. elatior* in this study showed significant inhibition to egg until 1.25% as the lowest concentration used. It was reported that *E. elatior* essential oil contains long-chain alcohols and mono- and sesquiterpenes by GC-MS [16]. Another study showed that embryonated *F. gigantica* eggs are more susceptible to ethanolic extract of *Moringa oleifera* leaf than the non-embryonated eggs [17]. Other plants possessing ovicidal activities *in vitro* to *Fasciola* eggs are *E. uniflora* L. *H. procumbens*, *P. guajava* L., *S. adstringens* and *Z. officinale [25, 26]*. Almost all of the plants tested contain phenolic compounds, tannins and terpenes [26]. The significant results of this study may indicate further experiment necessary which is focus on lower concentrations and bioactive analysis of the ethanolic extract. Additionally, the experiments will also be performed on egg hatch assay and its flukicide efficacies.

4. Conclusions

E. elatior in the concentration of 5%, 2.5% and 1.25% significantly inhibits the initial development of *F. gigantica* compared to the negative control (water) on days 5 and 9 treatments ($P < 0.05$). The *E. elatior* effectively inhibits morulation and larval formation during initial development. *E. elatior* showed effective inhibition to the initial development of eggs and may act as eco-friendly anthelmintic to *F. gigantica*.

References

[1] Tantri N, Setyawati T R and Khotimah S 2013 *J Protobiont* **2** 102
[2] Hamid P H, Kristianingrum Y P, Prastowo J and da Silva L M R 2016 *American Journal of Animal Veterinary Science* **11** 119
[3] Nugraheni Y R, Priyowidodo D, Prastowo J, Rohayati E S, Sahara A and Awaludin A 2018 *Ilmu Peternakan Terapan* **1** 46
[4] Rinca K F, Prastowo J, Widodo D P and Nugraheni Y R 2019 *Vet. World* **12** 593
[5] Hambal M, Sayuti A and Dermawan A 2013 *Jurnal Medika Veterinaria* **7** 49
[6] Indrasanti D, Samsi M, Sufriyanto S, Nafi M I and Anandita I 2020 *Prosiding Semnas LPPM Unsoed* **10** 151
[7] Ridwan M, Suwanti L T, Suprayogi T W, Mufasirin, Kusnoto and Hastutiek P 2021 *Media Kedokteran Hewan* **32** 105
[8] Valinata S, Susilo J, Pramono B, Karmil F, Hamzah A, Gani F A, Jalaluddin M and Harris A 2020 *Jurnal Medika Veterinaria* **14** 74

[9] Winaya I B O, Adi A A A M, Berata I K, Kardena I M, Adnyana I B W and Suardana I B K 2020 *Jurnal Veteriner* **21** 654
[10] McKellar Q A and Jackson F J T I P 2004 *Trends Parasitol* **20** 456
[11] Fairweather I, Brennan G P, Hanna R E B, Robinson M W and Skuce P J 2020 *Int. J. Parasitol. Drugs Drug Resist* **12** 39
[12] Olaechea F, Lovera V, Larroza M, Raffo F and Cabrera R 2011 *Vet. Parasitol* **178** 364
[13] Cabada M M, Lopez M, Cruz M, Delgado J R, Hill V and White Jr A C 2016 *PLoS Negl. Trop. Dis* **10** e0004361
[14] Ghasemzadeh A, Jaafar H Z E, Rahmat A and Ashkani S 2015 *BMC Complement. Altern. Med* **15** 335
[15] Juwita T, Pakpahan W H P, Puspitasari I M, Saptarini N M and Levita J 2020 *Pakistan Journal of Biological Sciences* **23** 1193
[16] Bezerra-Silva P C, Dutra K A, Santos G K N, Silva R C S, Iulek J, Milet-Pinheiro P and Navarro D M A F 2016 *PLoS One* **11** e0150008
[17] Hegazi A G, Megeed K N A, Hassan S E, Abdelaziz M M Toaleb N I, Shanawany E E E and Aboelsoued D 2018 *Vet. World* **11** 215
[18] Fairweather I, McShane D D, Shaw L, Ellison S E, O'Hagan N T, York E A, Trudgett A and Brennan G P 2012 *Vet. Parasitol* **183** 249
[19] Le T H, Nguyen K T, Nguyen N T B, Doan H T H, Le X T K, Hoang C T M and De N V 2012 *J. Clin. Microbiol* **50** 2720
[20] Rasyadi Y, Agustin D, Aulia G, Merwanta S and Hanifa D 2021 *Parapemikir: Jurnal Ilmiah Farmasi* **10** 54
[21] Alvarez-Mercado J M, Ibarra-Velarde F, Alonso-Diaz M A, Vera-Montenegro Y, Avila-Acevedo J G and Garcia-Bores A M 2015 **11** 45
[22] Pereira C A J, Oliveira L L S, Coaglio A L, Santos F S O, Cezar R S M, Mendes T, Oliveira F L P, Conzensa G and Lima W S 2016 *Vet. Parasitol* **228** 160
[23] Hayashi K, Ichikawa-Seki M, Allamanda P, Wibowo P E, Mohanta U K, Sodirum, Guswanto A and Nishikawa Y 2016 *Parasitol. Int* **65** 424
[24] Samarang S, Isnawati R and Murni M 2015 *Jurnal Penyakit Bersumber Binatang* **2**
[25] Moazeni M and Khademolhoseini A A 2016 *J. Parasit. Dis* **40** 662
[26] Marques L T, Guedes R A, Rodrigues W D, Archanjo A B, Severi J A and Martins I V F 2020 *Parasitology* **50**

IOP Conf. Series: Earth and Environmental Science 1115 (2022) 012036 doi:10.1088/1755-1315/1115/1/012036

Community's local wisdom and its relationship with environmental conservation efforts in Bangka Belitung, Indonesia

Henri*, **F Fatansyah, Alita, Y Lestari, A Sonia, J E Putri and Rahmasari**

Department of Biology, Universitas Bangka Belitung, Bangka, Indonesia

*Email: biology.henry@gmail.com

Abstract. This study aims to identify and examine the role of the community's local wisdom inenvironmental conservation or conservation efforts and to understand the history, values and behavior of local wisdom in Bangka Belitung. The study uses semi-structured interviews and descriptive analysis. The results show that local wisdom consists of three moral values: humans are not the center of nature; protecting the environment is a local job; and a holistic wayof thinking. The Malay community's efforts in environmental conservation in Bangka Belitung are to use local wisdom by saving the environment from unused tin mines with the term origin or impact. The impact of local wisdom on environmental conservation is that environmental sustainability will be maintained by no longer converting forest functions into mining land. The local wisdom of the Malay community on Belitung Island include the tradition of conserving plants or what is commonly referred to as *kelekak* and the *pemalian* forest. In the beginning, *kelekak* was an expanse of wilderness. Then, the community cut down the trees that grew in the wilderness to be used as *huma* and then abandoned or commonly referred to as *bebak*. *Huma* is one of the categories of the forbidden forest because of the influence of supernatural things thatcan prevent the unrestricted use of the forest. This local wisdom is a way for the community to conserve the environment.

1. Introduction

Indonesia is a mega biodiversity country with very high biodiversity and genetic resources. Indonesia as a biodiversity hotspot country, currently pays very little attention to the sustainable use of this diversity [1]. This makes the threat of biodiversity extinction one of the highest in the world. Bangka Belitung is one of Indonesia's provinces with an area of ± 81,725.14 km2 with various environmental problems. One of them is land conversion, especially forests reaching up to ± 5,400 ha, into tin mining areas. This causes environmental degradation, land damage and the narrowing of productive community land [2].

Local wisdom is important in people's lives in adapting to nature and becoming a cultural heritage for the utilization and management of natural resources with knowledge or ideas, customary norms and cultural values contained in the concept of community thinking [3]. Suggested that local knowledge and community management practices strongly support environmental conservation efforts [4]. Environmental management through traditional ecological knowledge plays an important role since it refers more to individual practices, knowledge, values and beliefs in developing an environment historically, conceptions and perceptions of local communities [5].

The Indonesian state recognizes local wisdom in managing the environment, as stipulated in Law no. 32/2009 on Environmental Protection and Management (UUPPLH). Furthermore, Article 1 Paragraph 30 explains that local wisdom is the noble value applied in the community's life too, among others, protect and manage the environment sustainably. Local wisdom is also one of the principles in environmental protection and management, as mentioned in Article 2 Letter (l) of the Law on Environmental Protection and Management (UUPPLH) and clarified in its explanation. The principle of local wisdom is that in the protection and management of the environment must pay attention to the noble values applied in the community's life [6].

Local wisdom could not be separated from elements of local rules, local traditions, customs, or cultural values, which are evidence of the harmonious relationship between society and the environment [7]. Local wisdom is local ideas that are wise, full of wisdom, valuable, embedded and followed by community members. In addition, local wisdom becomes an important basis for decision-making in reducing the potential for environmental damage [8].

A community can hold their respective local wisdom as a reference to living and survive longer in modern life. Modern life is currently carrying out exploitation activities with high intensity which causes widespread environmental damage [9]. The community can use local wisdom as local knowledge to survive in an environment that is integrated into belief systems, norms and culture and is expressed in traditions and myths that have been held for a long time. Local wisdom is also a manifestation of the behavior of a particular community or society so that it can coexist with nature/environment without destroying it [10]. Therefore, it is necessary to identify and examine the role of the community's local wisdom in environmental conservation or conservation efforts and to understand the history, values and behavior of local wisdom in Bangka Belitung.

2. Method

This research was conducted directly by communicating with several community leaders who understand the local wisdom of the people in Bangka Belitung. This study uses a snowball sampling technique and semi-structured interviews for data collection. In addition, this research is also based on a literature study through books, articles and official documents. This study is descriptive exploratory research to describe the state of a phenomenon, variable, or symptoms instead of testing certain hypotheses [11].

3. Results and Discussions

3.1. Local wisdom of the Malay Community in Bangka Island

The Malay community in Bangka has local wisdom to save the environment of Bangka Regency from the presence of tin mines, namely, the existing tin has been discarded/eliminated, in which the local village community calls *diasal* or *ampak*. All areas (villages and hamlets) in Bangka Island that have been *diasal* are Malay community settlements.

The history of this *diasal* tradition began during the Dutch colonial period and the entry of ethnic Chinese from China to Bangka, where there was the exploitation of tin sand in Bangka. Meanwhile, those that have never been exploited are used for the area of *diasal*. *Diasal* is a condition where the tin is so light, has no value and cannot be sold, as is the case with ordinary sand. The community believes that *ampak* or *diasal* tin is a result of the violation of actions that are prohibited in mining [6].

During British rule, where Bangka Island was divided into three divisions based on the division of exploration or tin mining areas. The northern part covers the areas of Jebus, Belinyu, Sungailiat, Merawang, (the area of Bangka Regency). The western part includes Mentok, Belo, Kotawaringin, while eastern parts covering almost half of Bangka Island, namely Pangkalpinang, Sungai Selan, Bangka Kota, Paku, Permis and Toboali. This map explains that the entire island of Bangka, including Bangka Regency contains a natural wealth of tin. However, with previous people's consideration of environmental sustainability for the future, there are some *diampak* areas while others are not *diampak* with considerations such as agricultural land and farming [6].

There are three moral values contained in this local wisdom. First, humans are not the center of nature, because ecologically, living things, including humans and their behavior and other abiotic communities, are interrelated and have reciprocal relationships and influence each other. Second, the understanding that protecting the environment is a total job since it involves the existence of mankind and other living things on this earth. Third, value is a holistic way of thinking, shown by the understanding that the goal of punishment is not limited to humans who live and currently exist. In addition, it includes generations that will live in the future for humans, non-human beings and ecosystems and habitats that support human life [12].

Their local wisdom maintains environmental sustainability since the activity of converting forest functions into mining land in some areas is reduced. This is because the tin has been *ampak*, so it can discourage people from mining tin in that area. When viewed in areas where tin is not *ampak*, the environment in Bangka Belitung tends to experience a lot of damage due to people's mining activities being abandoned [12].

For the time being, *diampak* tradition is rarely done anymore for some reason. Many people are tempted by the profits from tin mining, although it damages the environment. It can help the economy of the village community, especially if the company processes it. Thus, it is suspected that *ampak* is a form of community resistance when they have not felt the benefits of tin mining in the new order era and to provide land for living and farming in the colonial period.

3.2. Community's local wisdom in Belitung Island

3.2.1. Kelekak, Plant Conversion Tradition. *Kelekak* is an area of fruit trees usually reserved for family members from generation to generation. It is defined that the previous generation guarantees food availability for their descendants when they die. This local wisdom is an agrarian tradition handed down by the ancestors of the Belitung people, including the people on Bangka Island. The term *kelekak* itself is a term from the next generation, not from the person who planted it. Meanwhile, in Bangka, *kelekak* is often connoted with the meaning *kelak kek ikak* which means "later for you."

Kelekak was originally a stretch of wilderness. Then the community cut down the trees that grew in the wilderness to serve as *huma* as a place to plant rice. After the huma is harvested, it is left by the owner for several years. The owner of the *huma* grows various types of long-term crops as a sign that the *huma* left behind is already owned by a certain person so that other people are no longer working on it. Thus, the ownership status of the abandoned *huma* becomes the right of the first person to work on it. The *huma* left for some time was intended to restore soil fertility naturally, because agricultural activities carried out then did not use fertilizers. During the period of abandonment (*bera*), they look for new arable land or return to previously cultivated land if the period of returning soil fertility is considered sufficient.

The *huma* left behind is known as *bebak*. This area is commonly overgrown with wild plants such as reeds and *keramonting* (*Rhodomyrtus tomentosa*), can be reused as a village and can also be allocated as settlements, plantations and tombs. Meanwhile, the rest of the area is planted with various edible fruit-producing perennials such as mango (*Mangifera indica*), durian (*Durio zibethinus*) and rambutan (*Nephelium lappaceum*). Fruit trees planted in this *bebak* land are then called *kelekak*.

The future contains ecological, social and economic values. Ecologically, planting fruit trees can be interpreted as an effort to build green open lands. The land that has been planted with various fruit trees can function to absorb water so that the area does not dry out quickly during the dry season and does not get flooded during the rainy season. The social value is to maintain the relationship between family members who own the *kelekak* tree. By remembering the *kelekak* tree, the family tree can be re-woven because the descendants of the person who planted the *kelekak* tree can be traced. With *kelekak*, there are times when those who have been separated in other areas can gather when there are those who intend to enjoy the fruit of the *kelekak* tree planted by their ancestors. Meanwhile, economic benefits can be obtained from the sale of fruit from the *kelekak* tree. The trees planted are edible fruit trees so they have economic value [13].

Another function of *kelekak* is to test the patience in waiting for the time needed for fruit trees planted by parents or ancestors (elders) to bear fruit. Although the *kelekak* plants are planted intentionally, the planting pattern is not carried out regularly, both from the distance and the type of fruit trees planted. Therefore, the fruit plants in it grow irregularly and have various types. These fruit plants are cared for in such a way by the person who planted them or their families. The long waiting time is one of the factors causing the fruit trees that are used at a later date is not intended for the grower. Even though they have given energy and attention to care for them, the results of this plant are not to be enjoyed alone but to be passed on to their descendants in the future. This behavior shows that the elders are very concerned about the needs of the next generation of their descendants [13]. *Kelekak* a traditional plantation on Bangka Island is a traditional representing the intimacy between nature and people on Bangka Island. A traditional in the late 19th Century was strictly controlled by the Dutch goverment to encourage forest preservation [14].

The *kelekak* plant has a meaning that the Belitung people have understood the concept of conservation inherited by their ancestors. This refers to the definition of conservation which is defined as the management of living natural resources. The use of which is done wisely to ensure the continuity of their supply, while maintaining and enhancing the quality of diversity and its value. With the *kelekak* tradition, the Belitung people already have wisdom related to ecology by conserving the land, even though it is reduced to an agricultural scale. The goal is a forest that has been cleared to be used as huma. However, when left for several years, it does not become critical land since the Belitung people plant the former huma land with various fruit plants. *Kelekak* has an ecological function in maintaining mangrove forests so that there is no degradation caused by deliberate destruction. Mangrove forests that are maintained and protected from damage will be passed on to the next generation and ensure the next generation's survival from environmental disasters.

3.2.2. Forbidden/ Pemalian Forest. A forbidden (*pemalian*) forest is defined as the forest that cannot be utilized for any reason. The *pemalian* forest is a meeting forest between village shamans in the supernatural world. The *pemalian* forest is difficult to identify along with the growing settlements, except for each village shaman. Traditionally, it is forbidden to use the *pamalian* forest area because this area is the abode of spirits and ancestral spirits. With the prohibition forest, the forest area intended to be used as a *huma* or *ume* (Belitung language) must first obtain the approval of the village shaman.

The opening of *huma* in ancient times was usually done in groups. Before carrying out their activities, the group of rice growers in *huma* first asked the village shaman for permission to ensure that the area could be used traditionally. The village shaman and farmer groups will review the forest in question. Then, the village shaman asked the direction of the *huma* location he wanted to manage, whether the requested area was to the west, east, north, or south. After the direction of the requested area is clear, the village shaman then 'communicates' with the designated forest dwellers to ascertain whether or not the forest area is allowed to be used as *huma*. This means that the village shaman does not necessarily agree with or reject the proposal of the rice farmer group to clear the forest. The village shaman answers after receiving instructions from the 'ruler' of the forest area through a certain ceremony or ritual.

The village shaman communicates with the spirits of forest dwellers by making *kesalanta* a sign that the shaman has spoken to forest dwellers. The *kesalanta*, which has been enchanted or given a spell by the shaman, is placed in a coconut shell on a three-legged wooden so that the shell is balanced and the contents do not spill. When the *kesalanta* in the coconut shell is placed on the three-legged wood, the shaman says: "*kalo mimang nak deizinkan, berik tande desinek tige ari agik kamek kan dateng*" (if you want it to be allowed, put a sign here, we will come in three days).

The directions will be given by *seniang* at the place of coconut shells placed in the forest area that is 'requested' to be processed. After waiting for three days, if the 'weird' signs are found on the coconut shells, it shows that the area is included in the forbidden forest. The peculiar signs include the discovery of three green leaves in a coconut shell or the carcass of an animal. On the other hand, if the sign is not present, the forest may be cleared for use as *huma*.

When the forest begins to be cleared, the place where the coconut (*Cocos nucifera*) shell is served as a container to put the salad becomes the starting point for the opening of the *huma*, with directions that have been previously informed to the village shaman. Groups of farmers who clear land also have their local wisdom in dividing land to be used as *huma*. The basis for the division of land is adjusted to the age level, starting from the youngest person to the older person. The area of land is divided according to the needs and ability of a person or his family to manage it.

Based on this age group, the person who has the *huma* furthest from the starting point is the oldest person. It departs from the idea that young people do not step over their father's 'head', because in ancient times, there was only one road to clear the forest to reach the land's end. On the other hand, if the place where young people live is further away, when they want to cut down the forest, they will pass through the land of people who are older than themselves. This can be interpreted as stepping over the 'head' of an older person and is a taboo subject.

The location of the land that will be used as a *huma* is not protested by the farmer group even though the land is due to this rule. Suppose they ask for forest areas to be cleared and used as human rights, not one group. In that case, one group of farmers and another group of farmers should not see each other when carrying out their agricultural activities. Therefore, between one *huma* and another, there must be vacant land or forest that should not be used, which is used as a barrier (*riding* in Belitung's term) between one land and another. The size (width) of this *riding* also has no exact provisions; only an estimate adapted to its function.

In addition, forest *riding* is also used as a crossing for wild animals in the forest area so that they do not pass through the huma and damage the plants in it. With this function, the width of the riding forest can be assumed to be as wide as the humpback. If the hummingbird is not disturbed, that is the width level. However, if the humpback is still afraid to cross the riding forest, then its territory must be widened again. The number of *riding* forests that he has documented in Belitung is six, namely Inding, Ibul, Lais, Mengkuang, Gelanggang, Kebang and Mawai, all of which are located on Seliu Island. Because belief in spirits is still so high, this forest *riding* is also beneficial in the occult. It means that the riding forest is also prepared as a traffic lane for supernatural spirits from one place to another. Considering the benefits of forest riding as well as a crossing path for spirits, forest *riding* is also categorized as a forbidden forest [15].

4. Conclusions

People in Bangka Belitung use local wisdom to preserve the environment. The Malay community in Bangka Island uses the *ampak* tradition where they use the *diampak* land as agricultural land and cultivate crops so that the land is not used for mining. Other local wisdom in Bangka Island are *kelekak* (tradition of conserving plants) and forbidden or *pemalian* forest. *Kelekak* means the tradition of conserving plants, which is the use of land from the former manufacture of *huma* to plant fruit crops. In terms of ecology, it has the function of maintaining mangrove forests and avoiding degradation. Meanwhile, the forbidden/*pemalian* forest is a forest that is believed by the community as a place for spirits or ancestral spirits to reside. Thus, the development of *huma* there must have a permit from the village shaman. These three local pearls of wisdom of the community can preserve the environment by balancing the use of natural resources and efforts to preserve or conserve the environment sustainably.

Acknowledgments

We would like to thank the Department of Biology Universitas Bangka Belitung and the community and traditional leaders who have provided an understanding of the local wisdom needed in this research.

References

[1] Sutarno and Setyawan A D 2015 *Pros Sem Nas Masy Biodiv Indon* **1** 1
[2] Ginting P, Budi R and Khalid K 2014 *WALHI in collaboration with Milieudefensi* (Amsterdam)
[3] Nurdin B V and Ng K S F 2013 *Procedia-Social and Behavioral Sciences* **91** 113
[4] Richeri M, Cardoso M B and Ladio A H 2013 *Ecologia Austral* **23** 184
[5] Henri H, Hakim L and Batoro J 2018 *Jurnal Ilmu Lingkungan* **16** 49
[6] Rahayu P 2016 *Jurnal Hukum IUS QUIA IUSTUM* **23** 320
[7] Mulyadi A, Dede M and Widiawaty M A 2022 *Jurnal Geografi Gea* **22** 55
[8] Dede M, Nurbayani S, Ridwan I, Widiawaty M A and Anshari B I 2021 *IOP Conf. Ser.: Earth Environ. Sci* **683** 012134
[9] Amirudin A, Sumarmi and Sufia R 2016 *Jurnal Pendidikan* **1** 726
[10] Sumarmi and Amirudin 2014 *Pengelolaan Lingkungan Berbasis Kearifan Lokal* (Malang: Aditya Median Publishing)
[11] Zank S and Hanazaki N 2012 *Evidence-Based Complementary and Alternative Medicine* **2012** 563570
[12] Hasmonel, Rahayu D P and Faisal 2020 *Jurnal Dinamika Hukum* **20** 33
[13] Elvian A, Sulaiman R, Purnama A, Indradjaja A, Ahkap A S, Ziwar A, Usman A, Rahayu D P, Imam F, Rahayu F A, Sudana A 2017 *Kelekak Sejarah Bangka* (Bangka: Dinas Kebudayaan dan Pariwisata kabupaten Bangka)
[14] Nuraeny E, Susanto D and Kurniawan K R 2018 *E3S Web of Conferences* **67** 04018
[15] Fithrorozi 2011 *Ngenjungak Republik Kelekak* (Yogyakarta: Kaukaba Dipantara)

IOP Conf. Series: Earth and Environmental Science 1115 (2022) 012037 doi:10.1088/1755-1315/1115/1/012037

Post-modification of ultra-low molar ratio urea-formaldehyde adhesive with different types and levels of organic acid hardener

M A R Lubis[1,2*], A H Iswanto[3,4,10*], L S Hua[5], L Kristak[6], P Antov[7], R Putri[8], I Juliana[8] and E Mardawati[2,9]

[1]Research Center for Biomass and Bioproducts, National Research and Innovation Agency, Cibinong, West Java 16911, Indonesia
[2]Research Collaboration Center for Biomass and Biorefinery Between BRIN and Universitas Padjadjaran, National Research and Innovation Agency, Cibinong, West Java 16911, Indonesia
[3]Department of Forest Product, Faculty of Forestry, Universitas Sumatera Utara, Medan, North Sumatera 20155, Indonesia
[4]JATI-Sumatran Forestry Analysis Study Center, Universitas Sumatera Utara, Medan North Sumatera 20155, Indonesia
[5]Laboratory of Biopolymer and Derivatives, Institute of Tropical Forestry and Forest Product, Universiti Putra Malaysia 43400 UPM Serdang, Selangor, Malaysia
[6]Faculty of Wood Sciences and Technology, Technical University in Zvolen 96001 Zvolen, Slovakia
[7]Faculty of Forest Industry, University of Forestry 1797 Sofia, Bulgaria
[8]PT. Greenie Alam Indonesia, Tanggerang, Banten 15310, Indonesia
[9]Department of Agro-industrial Technology, Faculty of Agro-industrial Technology, Universitas Padjadjaran, Sumedang, West Java 45363, Indonesia
[10]Center of Excellence for Mangrove, Universitas Sumatera Utara, Medan 20155, Indonesia

*Email: marl@biomaterial.lipi.go.id; apri@usu.ac.id

Abstract. Ultra-low (UL) molar ratio urea–formaldehyde (UF) adhesive is a class of UL formaldehyde-emitting adhesive systems for wood-based panels. However, the UL molar ratio exhibits poor reactivity, low cohesion strength, and inferior adhesion. This study aimed to enhance the performance of UL molar ratio UF adhesive via post-modification with the addition of different types and levels of organic acid hardener, such as acetic acid (AA), maleic acid (MA) and citric acid (CA) at 3 and 5% level based on the theoretical solids content of UF adhesive. The solids content, gelation time, viscosity, pH, and free-formaldehyde content decreased with the addition of a higher level of organic acids. This showed that the incorporation of organic acid enhanced the reactivity of the UL molar ratio UF adhesive via cross-linking of carboxylic acid groups and methylol groups or the free-formaldehyde as confirmed by the results of gelation time, pH, and free-formaldehyde content. Post-modification with 5% CA resulted in the lowest solids content, gelation time, viscosity, pH, and free-formaldehyde content. The adhesion strength values of plywood bonded with modified UF adhesive were greater than that of control UF adhesive. In addition, the FE of plywood bonded with modified UF adhesive was lower than that of control UF adhesive. This study showed that 5% CA could be added as a hardener of UF adhesive.

1. Introduction

The biggest problem in the manufacturing of wood composites is the free-formaldehyde content of wood adhesives and formaldehyde emission (FE) from wood-based boards [1]. Urea-formaldehyde (UF) is the primary glue used in the production of wood composites, with an anticipated annual global consumption of around 11 million tons, followed by melamine and phenolic adhesives [2-4]. The domination of UF adhesives is because of their excellent adhesion properties, high reactivity, short press time, low curing temperatures, relatively low cost, and ease of handling [2-4]. The primary disadvantage of UF adhesive is its low water resistance, which restricts the interior use of wood-based panels coated with UF. Similarly, FE is recognized as one of the key causes of sick building syndrome [1,5]. The FE is for the most part connected with unfavorable human medical conditions, such as nausea, skin sensitization, genotoxicity, respiratory irritation, leukemia, and nasopharyngeal cancer [6–8]. Formaldehyde is likewise a known dermal allergen in people and a poisonous air toxin to the California Air Resource Board (CARB) in 1992 [5].

In 2007, the CARB also backed a harmful airborne control action to reduce the FE from fiberboard, particleboard, and plywood [5]. External and internal factors influence the FE of wood-based paneling. The principal environmental variables are relative humidity and temperature [9,10]. Other external factors include heating, vibration, sanding, sawing, and grinding. The FE of wood-based boards decreases with age [9,10]. The internal factors include wood species, adhesive type, adhesive content and hardener used [9-14]. Indoor air pollution is reliant upon relative stickiness, room temperature, intrinsic wood-based board properties, air speed at the emanating surface, air conversion scale, the proportion of producing surface-to-room aspects, and the removal of formaldehyde [9,10].

Reducing the FE has been a prominent area of research as of late. Experiments in this field include the use of formaldehyde scavengers, the reduction of the formaldehyde/urea (F/U) molar ratio, and the substitution of non-toxic reactive compounds for formaldehyde [2-8]. Massive F/U molar ratio reduction in UF adhesive has resulted in reduced FE at the expense of sluggish reactivity, lower adhesion strength, and crystal formation in the adhesive [15]. The high crystallinity in the linear polymer structure prevented the creation of a polymer network in low molar ratio UF adhesive [16]. This oddity likely resulted from hydrogen bonding between the straight designs of polymer particles, similar to polypeptide crystallization [17,18]. Consequently, it is essential to improve the performance of low molar ratio UF adhesive by reducing gelation time and increasing bonding strength.

UF adhesive is for the most part prepared to utilize an initial F/U mole ratio of 2.0 to 4.0 [19-21]. Because urea has four NH reactive sites, greater initial F/U mole ratios are possible. Tetra-hydroxymethyl urea is still not reported elsewhere. Urea and formaldehyde combine to generate urea and methylene glycol anion. Mono-hydroxymethyl urea is made by adding urea and methylol at 7.8-8.5 pH. Di- or tri-hydroxymethyl urea is then made [21]. The arrangement of mono-, di- and tri-hydroxymethyl ureas of 1.0 to 1.6 F/U mole ratio UF resins are counted for around 20-25%, 60-65%, and 10-15%, respectively [22-24].

The build-up of methylene-bridged urea derivatives occurs during the condensation reaction. This reaction normally occurs at acidic pH of 4.6-6.0 [25,26]. Likewise, the reaction also occurs at alkaline pH of 12-13 [27,28]. The ether linkages, methoxymethylene, and water are formed in further condensation reaction of urea and hydroxymethyl forms [25-28]. A few examinations have revealed the formation of UF condensation products, which is proportional to the final F/U molar ratio.

Commercially, UF adhesives are produced using a F/U mole ratio of 1.2-1.6, which emits a greater amount of formaldehyde [22-24]. Low molar ratio UF adhesive is produced using a F/U mole ratio of 0.9-1.0 by several researchers [25-28]. This paper introduced a F/U mole ratio of 0.8 as an ultra-low molar ratio because it consists of low free-formaldehyde content and eventually lower formaldehyde emission. However, lowering the F/U mole ratio results in a lower cohesion and adhesion strength of the UF adhesive.

Therefore, this work aims to prepare an ultra-low (UL) molar ratio UF adhesive at 0.8 F/U mole ratio via an alkaline-acid two-stage reaction, which is considered to release ultra-low FE from the wood-based panels. The post modification of UL molar ratio UF adhesive is done by mixing with different types of organic acids such as acetic acid (AA), maleic acid (MA), and citric acid (CA) at two levels of 3 and 5% based on the solids content of ultra-low molar ratio UF adhesive. The basic and chemical properties of the adhesive are measured according to the published methods and the adhesion performance as well as the FE of UF adhesive is determined in plywood based on the Japanese Agricultural Standard (JAS) No. 233.

2. Experimental

2.1. Materials

For the manufacture of UL molar ratio UF adhesive, formalin (37% HCHO), urea granules (99%), sodium hydroxide (NaOH, 20% w/v) and formic acid (HCOOH, 20% w/v) were used. As a control hardener, ammonium chloride (NH4Cl, 20% w/v) was used. The organic acid hardener consisted of acetic acid (AA, 20% w/v), maleic acid (MA, 20% w/v) and citric acid (CA, 20% w/v).

2.2. Synthesis and post modification of ultra-low molar ratio UF adhesive

An addition-condensation reaction was used to make UL UF adhesive (Figure 1). A beaker glass containing 200 g formalin was swirled at 300 rpm. Using NaOH, the pH was adjusted to 8.0–8.2 and the temperature to 40 °C. Later, 74.7 g of first urea was added. For 1 h, the solution's pH and temperature were held at 8.0 and 90 °C to maintain the additional reaction. After that, the pH was adjusted to 4.9–5.2 using HCOOH and the condensation temperature was set to 80 °C until 250–275 mPa.s was reached. 112.2 g of second urea was added to make a 0.8 F/U mole ratio. After dissolving the urea, the temperature was set to 272°C and the pH was adjusted with NaOH. 24 hours passed before examining the glue. Based on UF's theoretical solids content, 3% NH4Cl was added to the control glue.

The post-modification of UL molar ratio UF adhesive was done before the application of UF as a plywood adhesive. The post-modification of UF adhesive was performed by mixing AA, MA, and CA at two different contents of 3 and 5% based on the theoretical solids content of UF adhesive. The mixing process was performed at room temperature under 200 rpm of agitation.

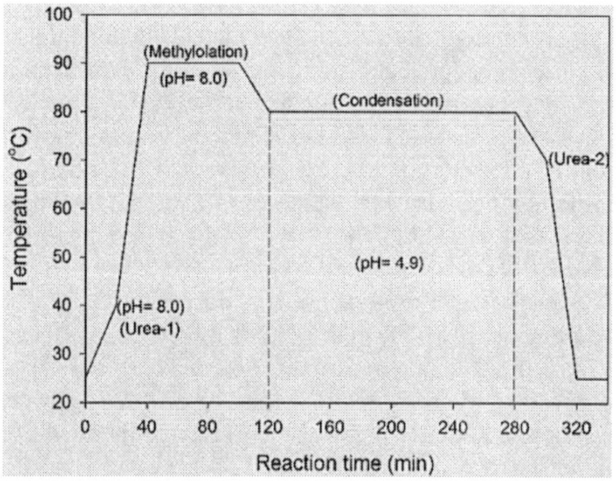

Figure 1. Reaction condition of ultra-low molar ratio UF adhesive

2.3. *Characterization of UF adhesives*

Using procedures that are already well established, this work examined the solids content, pH, gelation duration, and viscosity of control and modified UF adhesives at a mole ratio of 0.8 F/U [29,30]. The solids content of adhesives was determined by drying UF adhesives in an oven heated to 105 °C for three hours and dividing the weight upon drying by the initial weight. At 100 °C, the gelation time of adhesives was determined using a gel time meter (Techne GT-6, Coleparmer, United States). Using a digital pH meter (Laqua, Horiba, Japan), the pH of adhesives was measured. The viscosity of the control and modified UF adhesives was determined using a rotational rheometer (RheolabQC, AntonPaar, Austria) with spindle No. 27 at 100 rpm and 25 °C. Each experiment was carried out thrice.

The free-formaldehyde (HCHOfree) content of adhesives was determined based on a standard procedure [31]. The dilution of 10 grams of UF glue was done using distilled water. The sample was mixed with 7 drops of thymolpthalein indicator and afterward, 50 mL of sodium sulfite was introduced to convert the color from white to blue. The sample was titrated immediately with sulfuric acid till the blue color turned white. The experiment was performed three times.

Using ATR-FTIR spectroscopy (Spectrum Two, Perkin Elmer, United States) with a 4 cm–1 resolution, the functional groups of liquid and cured adhesive were scanned between 4000 and 400 cm–1. A minimum-maximum method was applied to normalize every spectrum [29,30].

2.4. *Preparation of UF-bonded plywood*

Plywood (300 mm × 300 mm × 6 mm) was fabricated from rubber (*Hevea brasiliensis*) wood veneers and the UL molar ratio UF adhesive. The glue spread was 180 g/m^2. 20 minutes of pre-pressing at 0.8 MPa was followed by four minutes of hot pressing at 140 °C and 0.8 MPa was applied to produce plywood. Before analysis, plywood was conditioned at room temperature for one week [32].

2.5. *Evaluation of UF-bonded plywood*

Dry tensile shear strength (TSS) and FE were assessed based on the procedure of JAS standard No. 233 [32]. Six specimens were utilized for determining the TSS using a universal testing machine (AGX 10 kN, Shimadzu, Japan) with a cross-head speed of 2 mm/min. UV–visible spectrophotometer (Optizen 3220 UV, Mecasys, Korea) was used to determine the FE values of plywood according to the desiccator method at a wavelength of 412 nm. Plywood properties were statistically analyzed by comparing the mean values using analysis of variance and post-hoc test at $\alpha = 0.05$ using SPPS 17 software (SPSS Inc., Chicago, United States).

3. Results and Discussion

The solids content of wood adhesive indicates the total solids in the adhesives that provide more adhesive coverage on the bonded products. Lower solids content will cause shrinkage, which may lead to cracking or other quality issues. The solids content of control UF adhesives was 62.3% (Figure 2). The solids content decreased with the addition of organic acids into a range of 59.5 – 60.4%. The addition of higher levels of organic acids further reduced the solids content of UF adhesive. This is due to the 5% level of organic acids containing more water compared to the 3% level. This was in agreement with the published works that revealed a decrease in the solids content of UF adhesive mixed with a higher level of NH_4Cl hardener [33,34]. Statistical analysis showed no significant differences in solids content at different types and levels of the organic acids.

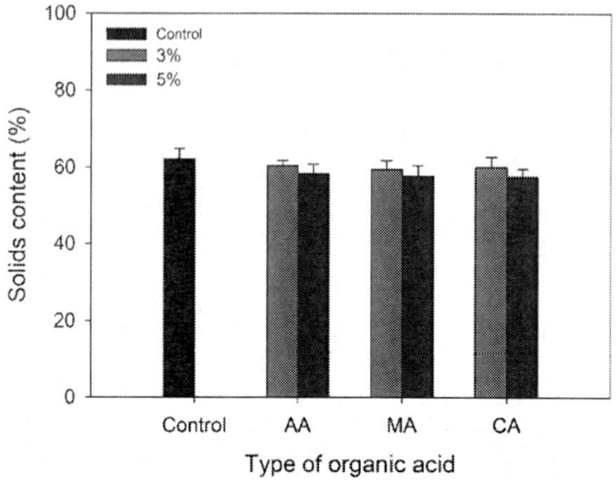

Figure 2. The solids content of ultra-low molar ratio UF adhesive mixed with different types and levels of organic acids

The gelation time of wood adhesive indicates the time needed for the formation of a gel from adhesive polymers [19-21]. Branched can frame joins between the chains, which leads to logically bigger polymers. That reaction is characterized as a gel point, where the framework loses smoothness and viscosity turns out to be exceptionally huge. The gelation time of control UF adhesives was 220.5 s (Figure 3a). The gelation time decreased by 6.8 – 30.0% with the addition of organic acids into a range of 154 – 205 s. The addition of higher levels of organic acids further reduced the gelation time of UF adhesive. This is due to the 5% level of organic acids containing more carboxylic groups (-COOH) compared to the 3% level. This was following the result of pH, showing that the addition of 5% of organic acid decreased the pH from 8.1 to 5.3 (Figure 3b). The addition of 5% CA resulted in the lowest pH because CA has three –COOH groups, while MA has two –COOH groups and AA has one –COOH group. The –COOH can react with the free-formaldehyde ($HCHO_{free}$) and the methylol group ($-CH_2OH$) to form a cross-linked polymer of UL molar ratio of UF adhesive [25-28]. Statistical analysis showed significant differences in gelation time and pH of UF adhesive at different types and levels of the organic acids. The addition of 5% CA into UL molar ratio UF adhesive provided significant results compared to others.

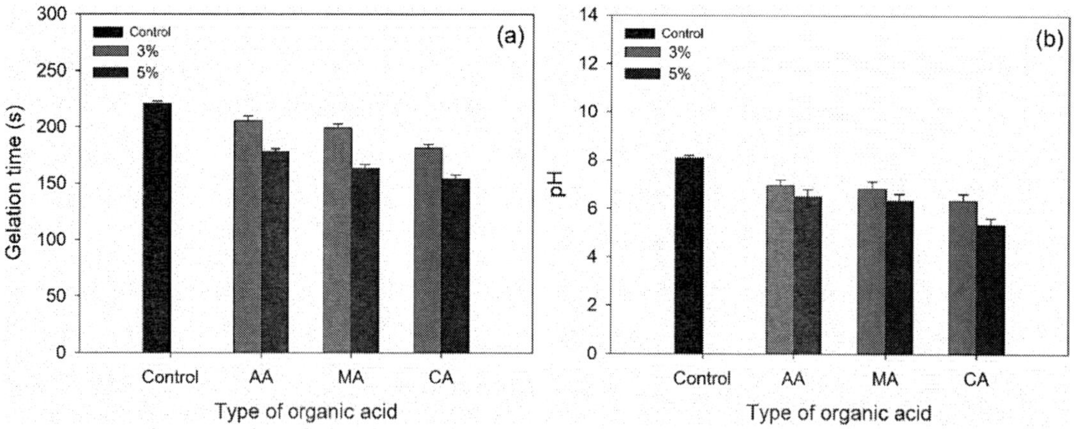

Figure 3. Basic properties of ultra-low molar ratio UF adhesive mixed with different types and levels of organic acids. (a) gelation time, (b) pH

Viscosity is the proportion of resistance to gradual deformation by pressure. For liquid adhesives, viscosity corresponds to the thickness of the liquid [19-21]. Water has a viscosity of 1 mPa.s at 20°C. Fluids that have a viscosity not as much as water are mobile fluids and those with a viscosity higher than water are viscous. Viscosity is the main property in the assembling of wood-based composites. It is essential to keep the viscosity below a certain threshold to ensure the distribution and penetration of wood adhesives. The viscosity of control UF adhesives was 236 mPa.s (Figure 4). The viscosity decreased with the addition of organic acids into a range of 178 – 199 mPa.s. The addition of higher levels of organic acids further reduced the viscosity of UF adhesive. This is due to the 5% level of organic acids containing more water compared to the 3% level. This was in agreement with the published works that showed a decrease in the viscosity of adhesive mixed with a higher level of NH_4Cl hardener [33,34]. Statistical analysis showed significant differences in the viscosity of adhesive at different types and levels of organic acids. The addition of 5% CA into UL molar ratio UF adhesive provided significant results compared to others.

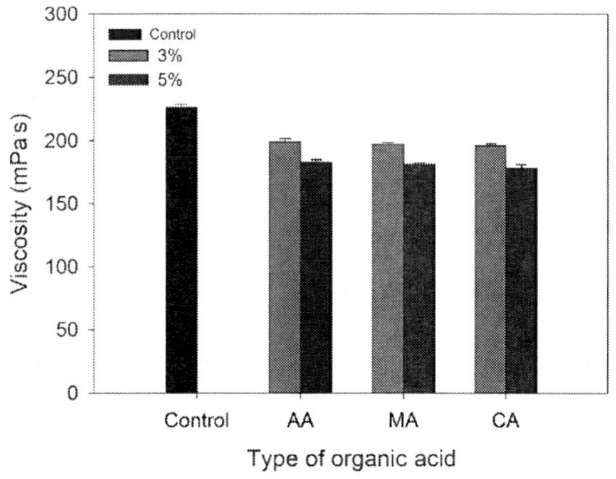

Figure 4. The viscosity of ultra-low molar ratio UF adhesive mixed with different types and levels of organic acids

The $HCHO_{free}$ content is defined as the unreacted formaldehyde that exists in a finishing solution. Formaldehyde resins are generally used in polymer and textile industries as adhesives, finishes, molded objects, etc. These resins are obtained by condensation reactions with formaldehyde. The performance, strength, and water resistance ability of such resins are affected by the $HCHO_{free}$ content [6–8]. In addition, formaldehyde emission is strictly regulated by various standards in formaldehyde resin-based adhesives [1]. Hence, the determination of the $HCHO_{free}$ in resins is important. The $HCHO_{free}$ of control UF adhesives was 0.25% (Figure 5). The $HCHO_{free}$ decreased with the addition of organic acids into a range of 0.18 – 0.23%. The addition of higher levels of organic acids further reduced the $HCHO_{free}$ of UF adhesive. This is due to the 5% level of organic acids containing more carboxylic groups (–COOH) compared to the 3% level. The addition of 5% CA resulted in the lowest $HCHO_{free}$ value because CA has three –COOH groups, while MA has two –COOH groups and AA has one –COOH group. The –COOH can react with the free-formaldehyde ($HCHO_{free}$) and the methylol group ($–CH_2OH$) to form a cross-linked polymer of UL molar ratio of UF adhesive [25–28]. Statistical analysis showed significant differences in the $HCHO_{free}$ of UF adhesive at different types and levels of the organic acids. The addition of 5% CA into UL molar ratio UF adhesive provided significant results compared to others.

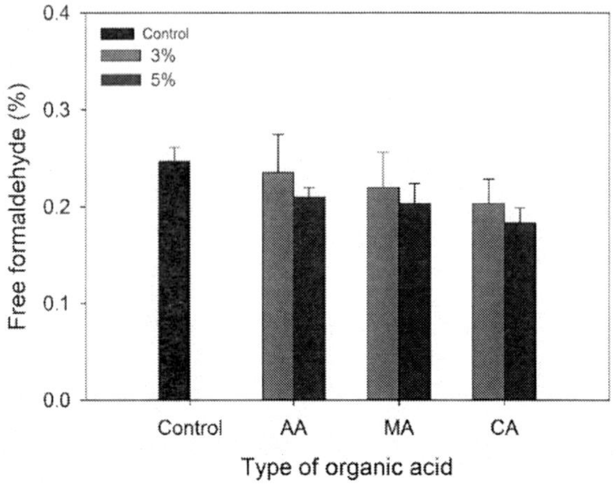

Figure 5. Free-formaldehyde content of ultra-low molar ratio UF adhesive mixed with different types and levels of organic acids

Functional groups of ultra-low molar ratio UF adhesive mixed with different types and levels of organic acids were analyzed using FTIR spectroscopy (Figure 6). All absorption bands were interpreted based on the published works [19-21]. Regardless of the types and levels of organic acids, the functional groups of liquid UF adhesive were built of the –N–H_2, C–H, C=O, –N–H of primary amide at a wavenumber of 3310, 2960, 1635, and 1600 cm^{-1}, respectively. No differences in FTIR spectra were found at the liquid UF adhesive mixed with 3 and 5% levels of organic acids.

After curing, the functional groups altered slightly owing to the formation of the three-dimensional network from the reaction of the UF adhesive and the hardener. The intensity of the functional groups was markedly reduced after curing. The –N–H_2 at 3310 cm^{-1} was still detected with smaller intensity compared to the liquid adhesives. The C–H, C=O, and N–H of primary amide were slightly moved to the wavenumber of 2900, 1670, and 1550 cm^{-1}, respectively, after the curing process. Theoretically, the –COOH of organic acids can react with the $HCHO_{free}$ and the –CH_2OH group to form a cross-linked polymer of UL molar ratio of UF adhesive [25-28]. This probably altered the functional groups of UF adhesive after curing.

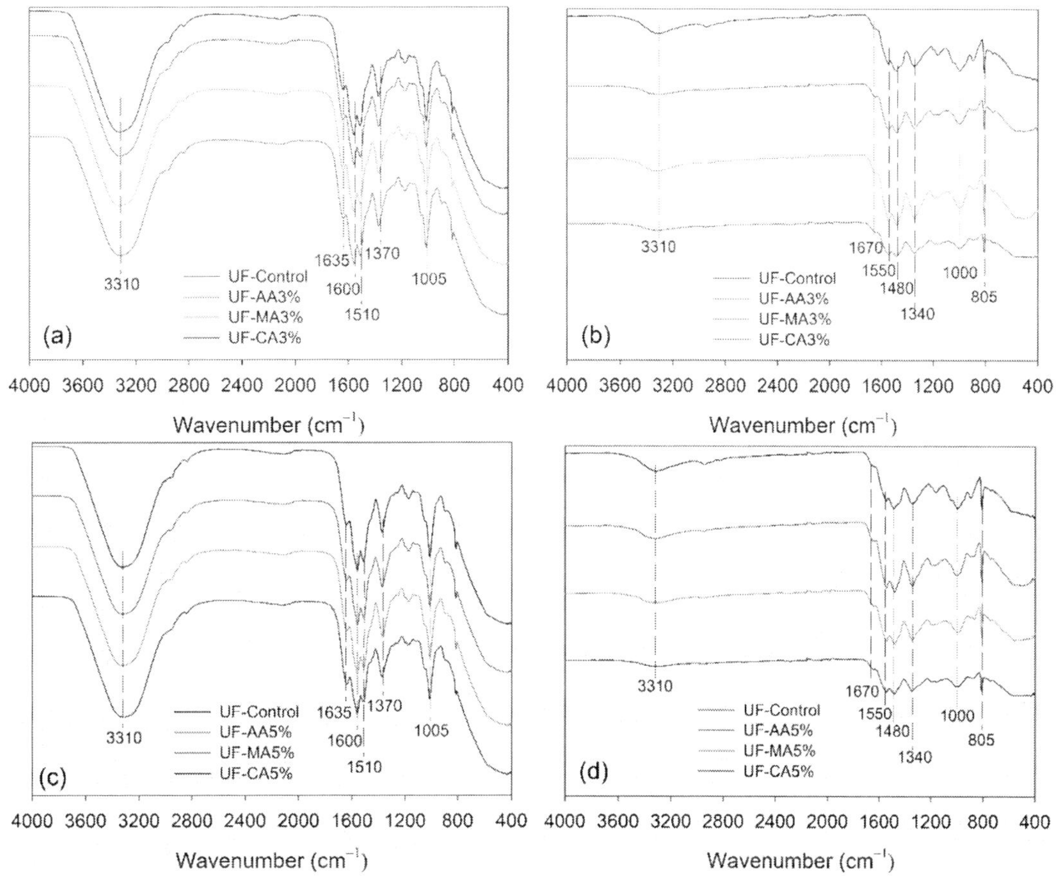

Figure 6. FTIR spectra of ultra-low molar ratio UF adhesive mixed with different types of organic acids at a 5% level. (a) liquid adhesive mixed with 3% organic acids, (b) cured adhesive mixed with 3% organic acids, (c) liquid adhesive mixed with 5% organic acids, (d) cured adhesive mixed with 5% organic acids

Plywood was prepared using UL molar ratio UF adhesive mixed with different types and levels of organic acids. Dry TSS and FE of plywood bonded with UF mixed with different types and levels of organic acids were statistically investigated (Table 1). Regardless of the type and level of the organic acid, the TSS values of modified UF adhesive were greater than that of control UF adhesive. Statistical analysis reveals that the addition of a 5% level of CA could enhance the TSS of plywood by 16.81% from 0.94 MPa to 1.13 MPa. There were no significant differences in TSS value between control UF adhesives and modified UF with a 3% level of AA. A similar trend was obtained for modified UF with a 5% level of AA and a 3% level of MA, as well as for modified UF mixed with a 5% level of MA and a 3% level of CA. Based on JAS No. 233, the TSS values of plywood bonded with UL molar ratio UF adhesives have met the minimum requirement of 0.7 MPa for interior application.

The FE values of UF-bonded plywood mixed with different types and levels of organic acids were similar to the trend of TSS values. Regardless of the type and level of the organic acid, the FE values of modified UF adhesive were lower than that of control UF adhesive. Statistical analysis reveals that the addition of a 5% level of CA could reduce the FE of plywood by 59.65% from 0.57 mg/L to 0.23 mg/L. There were no significant differences in TSS value between control UF adhesives and modified UF with a 3% level of AA. A similar trend was obtained for modified UF with a 5% level of AA and a 3% level of MA, as well as for modified UF mixed with a 5% level of MA and a 3% level of CA. According to

JAS No. 233, the FE values of plywood bonded with UL molar ratio UF adhesives meet the minimum requirement of F*** (maximum 0.7 mg/L) and F **** (maximum 0.4 mg/L).

Table 1. The TSS and FE of plywood bonded with the UL molar ratio UF adhesive at different types and levels of organic acids

Type of adhesive	TSS (MPa)	F-test	P-value	FE (mg/L)	F-test	P-value
UF-control	0.94 ± 0.06 (d)**	14.02	0.0028	0.57 ± 0.05 (d)	6.28	0.0463*
UF-AA3%	0.99 ± 0.05 (d)	13.60	0.0031	0.54 ± 0.03 (d)	6.49	0.0434
UF-AA5%	1.03 ± 0.04 (c)	13.23	0.0034	0.53 ± 0.04 (c)	6.63	0.0416
UF-MA3%	1.03 ± 0.06 (c)	13.24	0.0033	0.42 ± 0.06 (c)	7.60	0.0409
UF-MA5%	1.06 ± 0.07 (b)	12.95	0.0037	0.31 ± 0.05 (b)	8.61	0.0326
UF-CA3%	1.08 ± 0.06 (b)	12.77	0.0038	0.32 ± 0.04 (b)	8.54	0.0332
UF-CA5%	1.13 ± 0.07 (a)	12.33	0.0043	0.23 ± 0.06 (a)	9.43	0.0272

*The value is statistically different if less than 0.05 of the significance level
**The same letter is not significantly different at a significance level of 0.05

4. Conclusions

Reducing the FE of UF-bonded products has recently been a hot research subject. As the primary wood adhesive, UF is typically synthesized with a starting F/U mole ratio of 2.0 and a final F/U mole ratio of 1.0. An alkaline-acid two-stage reaction was used to successfully manufacture an ultra-low (UL) molar ratio UF glue with a mole ratio of 0.8 F/U. The UL molar ratio UF adhesive was post-modified by mixing it with various organic acids such as AA, MA, and CA at 3 and 5% based on the solids content of the UL molar ratio UF adhesive. The addition of more organic acids reduced the solids content, viscosity, gelation time, pH, and free-formaldehyde content. This demonstrated that the addition of organic acid increased the reactivity of the UL molar ratio UF adhesive via cross-linking of carboxylic acid groups and methylol groups or the free-formaldehyde, as evidenced by pH, gelation time and free-formaldehyde concentration. The lowest solids content, viscosity, gelation duration, pH, and free-formaldehyde content were obtained after post-modification with a 5% CA. Plywood glued with modified UF adhesive had higher TSS values than control UF adhesive. Furthermore, plywood glued with modified UF adhesive had lower FE values than control UF adhesive. This study found that a 5% CA might be used as a UF adhesive hardener.

Acknowledgments

This study was supported by PT. Greenie Alam Indonesia through a research collaboration with Research Center for Biomass and Bioproducts, National Research and Innovation Agency (No. B-714/V/KS.00.00/10/2021 and No. 001/Greenie/PKS/X/2021) with a title "Penelitian dan Pengembangan Perekat Ramah Lingkungan untuk Produk Panel Komposit Non-Kayu Berbasis Limbah Pelepah Pinang".

References

[1] Kristak L, Antov P, Bekhta P, Lubis M A R, Iswanto A H, Reh R, Sedliacik J, Savov V, Taghiyari H R, Papadopoulos A N, Pizzi A and Hejna A 2022 *a review Wood Mater. Sci. Eng.* **0** 1
[2] Kumar RN P 2019 *Urea formaldehyde resins* ed A Pizzi (Hoboken, NJ, USA: Wiley-Scrivener)
[3] Dunky M 1998 Urea–formaldehyde (UF) adhesive resins for wood *Int. J. Adhes. Adhes.* **18** 95
Pizzi A and Valenzuela J 1994 Theory and practice of the preparation of low formaldehyde emission uf adhesives *Holzforschung* **48** 254
[4] Salthammer T 2019 Formaldehyde sources, formaldehyde concentrations and air exchange rates in European housings *Build Environ* **150** 219
[5] World Health Organization 2008 *Sick Building Syndrome Symptoms* (Textos Complet)

[6] Jiang C, Li D, Zhang P, Li J, Wang J and Yu J 2017 *Build Environ* **117** 118
[7] Zhou X, Dong X, Wang X and Wang F 2022 *Build Environ* **207** 108419
[8] Roffael E, Behn C and Dix B 2012 *Eur J Wood Wood Prod* **70** 911
[9] Roffael E, Schneider T and Dix B 2015 *Eur J Wood Wood Prod* **73** 597
[10] Park B D, Kang E C, Lee S M and Park J Y 2016 F *J Korean Wood Sci Technol* **44** 600
[11] Park S, Jeong B and Park B-D 2021 *Forests* **12** 1037
[12] Mantanis G I, Athanassiadou E T, Barbu M C and Wijnendaele K 2018 *Wood Mater Sci Eng* **13** 104
[13] Lubis M A R, Jeong B, Park B-D, Lee S-M and Kang E-C 2019 *J Korean Wood Sci Technol* **47** 579
[14] Park B and Kim J 2008 *J Appl Phys* **108** 2045
[15] Wang H, Cao M, Li T, Yang L, Duan Z, Zhou X and Du G 2018 *Polymers (Basel)* **10**
[16] Singh A P, Causin V, Nuryawan A and Park B 2014 *Eur Polym J*
[17] Dunker A K, John W E, Rammon R, Farmer B and Johns S J 1986 *J Adhes* **19** 153
[18] Lubis M A R and Park B-D 2021 *J Adhes* **97** 1271
[19] Wibowo E S, Lubis M A R and Park B D 2021 *J Adhes Sci Technol* **0** 1
[20] Lubis M A R and Park B-D 2020 *J Korean Wood Sci Technol* **48** 1
[21] Kim M G 2001 *J Appl Polym Sci* **80** 2800
[22] Kim M G 2000 *J Appl Polym Sci* **75** 1243
[23] Kim M G 1999 *J Polym Sci Part A Polym Chem* **37** 995
[24] Steinhof O, Kibrik É J, Scherr G and Hasse H 2014 *Magn Reson Chem* **52** 138
[25] Steinhof O, Scherr G and Hasse H 2016 *Magn Reson Chem* **54** 457
[26] de Jong J I and de Jonge J 1952 *Recl des Trav Chim des Pays-Bas* **71** 643
[27] Chiavarini M, Bigatto R and Conti N 1978 *Die Angew Makromol Chemie* **70** 49
[28] Yadav S M, Lubis M A R and Park B-D 2021 *Wood Mater Sci Eng* **0** 1
[29] Yadav S M, Lubis M A R, Wibowo E S and Park B 2020 *Polym Bull*
[30] TAPPI 2001 Analysis of formaldehyde in aqueous solutions and of free formaldehyde in resins *TAPPI T 600 C*
[31] Japanese Agricultural Standard (JAS) for Plywood 2003 *Japanese Agric Stand*
[32] Nuryawan A, Singh A P, Zanetti M, Park B and Causin V 2017 *Int J Adhes Adhes* **72** 1
[33] Jeong B and Park B-D 2019 *Wood Sci Technol* **53** 665

Emission factor for emission reduction estimation in reduced impact logging

I W S Dharmawan[1] and M Ridwan[2]*

[1]Research Center of Ecology and Ethnobiology, National Research and Innovation Agency (BRIN), Bogor, Indonesia.
[2]Centre for Climate Risk and Opportunity Management in Southeast Asia Pacific (CCROM-SEAP), Bogor Agricultural University (IPB), Bogor, Indonesia.

E-mail: *salifa03@yahoo.co.id or mhdridwan2008@gmail.com

Abstract. Indonesia has a forest area of 125.92 million hectares and Production Forest has the largest area, which is approximately 55% of the total forest area of Indonesia. To ensure sustainability of forest management, the use of a logging system that provides minimal damage is highly desirable. Reduced Impact Logging (RIL) as one of the logging systems can minimize the impact of damage due to logging. It is important to provide the emission factor for emission reduction estimation in RIL to support achievement of Indonesian NDC target for net sinker in 2030. This research aims to estimate the emission factor of stands stock in forest concession areas that have applied conventional logging systems and RIL. The circular plot was used to conduct a vegetation survey in concessionaries areas in Central Kalimantan Province and the data collected were then analyzed further for carbon stock and emission factor for emission reduction. The results of the research in study sites showed that emission factor for emission reduction estimation in RIL compare to conventional logging (CL) amounted 30.03 tonC/ha and equivalent 110.2 tonCO2eq/ha or 55.71%. Thus, this value can be used as emission factor to estimate the emission reduction locally in RIL compare to the CL in Central Kalimantan.

1. Introduction

Indonesia has an area size reaches 125.92 million hectares, which include: conservation forest (HK) with an area of 27.43 million hectares, protected forest (HL) with an area of 29.66 million hectares, limited production forests (HPT) with an area of 26.79 million hectares, a fixed production forest (HP) with an area of 29.22 million hectares and converted production forests (HPK) with an area of 12.82 million hectares [1]. Production Forest has the largest area, which is approximately 55% of the total forest area of Indonesia.

In managing natural forest production, the use of a harvesting system that has an impact of minimal environmental damage is highly expected [2]. Reduced Impact Logging (RIL) as one of the logging systems can minimize the impact of damage due to logging [3]. Reduced impact logging is a timber harvesting technique that has a low impact/ minimum through a comprehensive approach between planning, implementation, monitoring and evaluation [4]. This technique is improvement of road, logging, and skidding practices that are currently running. Reduced impact logging implementation has the goal of minimizing negative influences on the environment, improving harvesting efficiency through emphasis on volume of logging waste, harvesting costs and increasing timber production, creating a growing space that is conducive to standing up so that the growth of trees and non-timber forest products will be better, increasing income, health and safety work and make pre-conditions for sustainable forest management. It is expected to reduce forest damage through the application of good RIL at the location of timber harvesting in the forest concessionaries area. The implementation of good RIL can increase timber production and support sustainable forest sustainability including biodiversity [5]. Sustainable forest management, namely efforts to reduce emissions through sustainable forest management with sustainable innovations to provide low environmental impact, including the application of the Reduced Impact Logging (RIL) technique, as well as the application of other sustainable forest management techniques, namely reducing the damaging impacts of timber harvesting activities and applying certification PHPL (Sustainable Production Forest Management) [6].

To ensure sustainability of forest management, the use of a logging system that provides minimal damage is highly desirable. Related to the issue of climate change and global warming, then one way to maintain the ecological function of the forest is to treat and maintain forest vegetation from possible damage (deforestation and degradation).

World attention to the importance of the existence of forests in climate change mitigation can be seen from the release of the Clean Development Mechanism (CDM) and REDD (Reducing Emissions from Deforestation and Forest Degradation) in carbon trading. The issue of REDD has been discussed with intensively at COP-13 (Conference on Parties-13) in Bali. Reduced impact logging is a good technique to support reducing of carbon emissions as well as opportunities to benefit from reductions in carbon emissions, e.g., voluntary carbon markets, REDD+, Nationally Determined Contributions to the UN Paris Climate Agreement [7]. This technique can have improved natural forest management and has potency to contribute in climate change mitigation, but this activity still has high uncertainty [8]. To support reducing uncertainty in the RIL technique for climate change mitigation, it is important to provide the emission factor for emission reduction estimation in RIL to support achievement of Indonesian NDC target for net sinker in 2030. This research aims to estimate the emission factor of stands stock in forest concession areas that have applied conventional logging systems and RIL.

2. Material and Methods

2.1. Research location

Research activities have been carried out in the area of Sinamang Forest Area 1 covering an area of 44,970 ha, and the Sinamang Forest Area 2 covering an area of 38,450 ha, which are located in Sanaman Mantikei District, Katingan Regency, Central Kalimantan Province. The map of research area can be seen in Figure 1.

Figure 1. The map of research area.

2.2. Plot establishment

The materials and tools used in this research include a tape measure, phi-band, machete, chainsaw, wood saw, scale, sack, rope, plastic, marker, pencil, camera, calculator, and computer. In the survey of biomass measurement, data were collected by random sampling using a circular plot of SNI 7724:2011 [9]. The plots were made of 155 plots with a distribution of: a) Baseline condition, 32 plots for vegetation survey in forest area before logging; b) Sinamang Forest (Area 1; 44970 ha), 62 plots divided into 31 plots for CL and 31 plots for RIL; c) Sinamang Forest (Area 2; 38450 ha), 61 plots divided into 29 plots for CL and 32 plots for RIL.

The circular plot shape and the size of the radius can be seen in Figure 2. Circle A with radius 1.13 m for seedlings (diameter < 2 cm), understorey and litter; Circle B with radius 2.82 m for saplings (diameter 2 – 10 cm); Circle C with radius 5.64 m for poles (diameter 10 – 20 cm), and Circle D with radius 11.29 m for trees (diameter > 20 cm). The plot did not put in timber collecting location, logging road, main road, sub-main road to avoid the bias data.

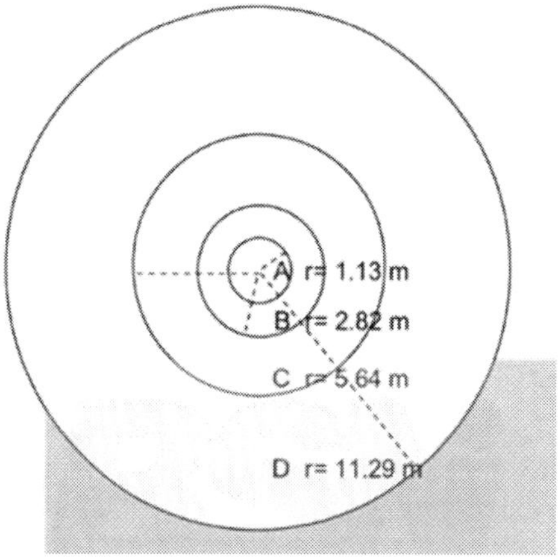

Figure 2. Circular plot shape (BSN, 2011).

2.3. Data analysis

Data analysis was carried out to obtain the value of carbon stocks in the baseline area, conventional logging/CL area and RIL logging area. The flow of calculating the value of carbon stocks to get the percentage reduction in RIL emissions to CL can be seen in Figure 3.

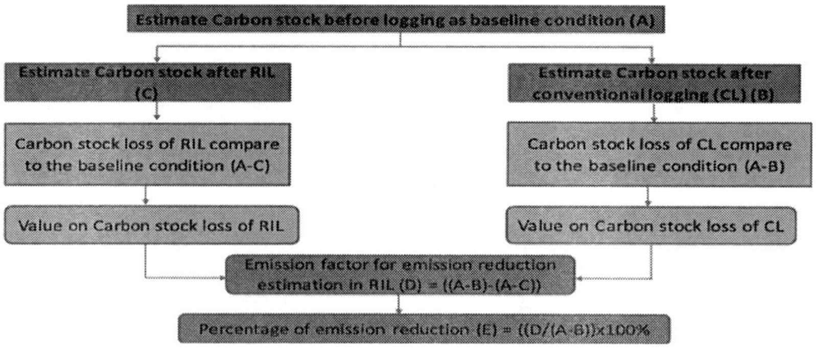

Figure 3. Flowchart of estimation on carbon stock and percentage of emission reduction.

The biomass of aboveground biomass (sapling, pole and tree) was estimated using the allometric equations for secondary forest [10] as follow:

$$W = 0{,}11 * \varrho * D^{2{,}62} \quad (1)$$

where W is the aboveground biomass (kg), ϱ is the wood density (g/cm³), and D is the diameter at breast height (cm).

The total aboveground biomass for each plot was calculated by adding up all of the estimated aboveground biomass values in the plot expressed in ton. The seedling and understorey biomasses were also calculated and added to the aboveground biomass (sapling, pole and tree) to get the total biomass in each plot.

The carbon contained in the total biomass was approximately 47% or ton biomass = 0.47 tonC [11]. The total carbon was estimated as the amount of carbon aboveground biomass, seedling biomass and understorey biomass. Their values were equivalent to the carbon dioxide values in the atmosphere by multiplying the carbon stocks by a factor (44/12).

To calculate the emission factor value in RIL implementation, the following formula can be used [12]:

$$EF = [(A - B) - (A - C)] \quad (2)$$

where EF is the Emission Factor (tonC/ha), A is the carbon stok in baseline condition/before logging (tonC/ha), B is the carbon stok in conventional logging/CL (tonC/ha), and C is the carbon stok in reduced impact logging/RIL (tonC/ha).

For estimation of emission reduction percentage in RIL implementation, the following formula can be used [12]:

$$\% \; ER = [(EF / (A - B)) \times 100\%] \quad (3)$$

where $\% \; ER$ is the percentage of emission reduction (%), EF is the Emission Factor (tonC/ha), A is the carbon stok in baseline condition/before logging (tonC/ha), and B is the carbon stok in conventional logging/CL (tonC/ha).

3. Results and Discussion

3.1. Plot distribution

Placement of the distribution of research plots with consideration not to overlap with timber collecting location, logging road, main road, sub-main road; distance from timber collecting location, logging road, main road, sub-main road at least 25 m; distance between plot at least 25 m; and distance from river or sub-river at least 25 m. Distribution of study plot can be seen in Figure 4 and Figure 5.

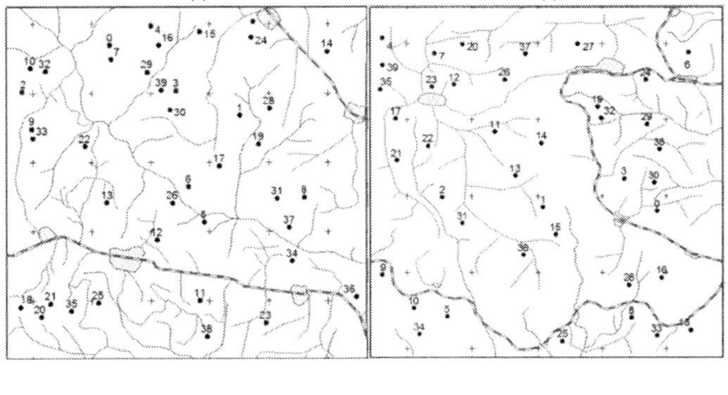

Figure 4. Plot distribution in Sinamang Forest Area 1. (A) in CL, (B) in RIL.

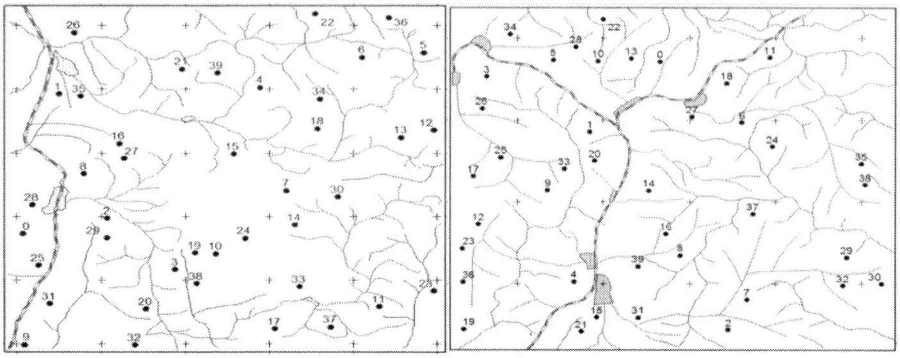

Figure 5. Plot distribution in Sinamang Forest Area 2. (A) in CL, (B) in RIL.

3.2. Carbon stock

The dynamics of the distribution of carbon stocks at the research plot locations varied greatly, both in Sinamang Forest Area 1 and Sinamang Forest Area 2 (Figure 6 and Figure 7). Overall, the potential carbon stock in the Sinamang Forest Area 1 is higher than that of the Sinamang Forest Area 2. The location of the Sinamang Forest Area has the highest carbon stock of 450 tonsC/ha or equivalent 1,651.5 tonCO_2eq/ha. Meanwhile, the Sinamang Forest Area 2 location has the highest carbon stock of 320 tonC/ha or equivalent 1,174.4 tonCO_2eq/ha. The high carbon stock in Sinamang Forest Area 1 is strongly dominated by forest stands from the Dipterocarpaceae, Euphorbiaceae, Myrtaceae, and Fabaceae families (Figure 8). The high carbon stocks in Sinamang Forest Area 2 are strongly dominated by forest stands of the Dipterocarpaceae, Myrtaceae, and Annonaceae families (Figure 8).

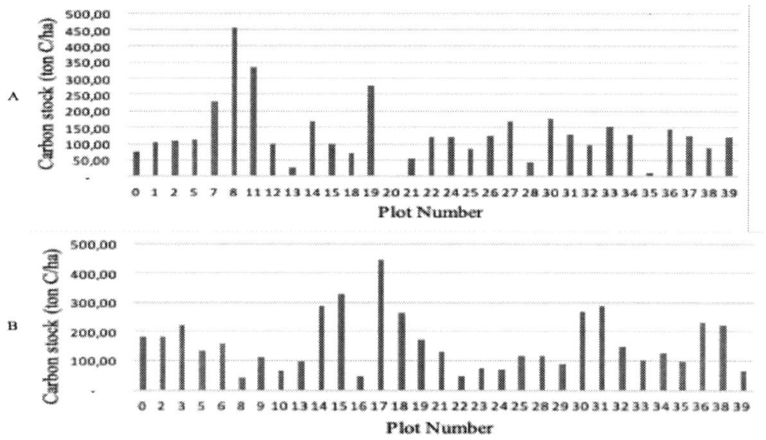

Figure 6. Distribution of carbon stock in Sinamang Forest Area 1; (A) in CL, (B) in RIL.

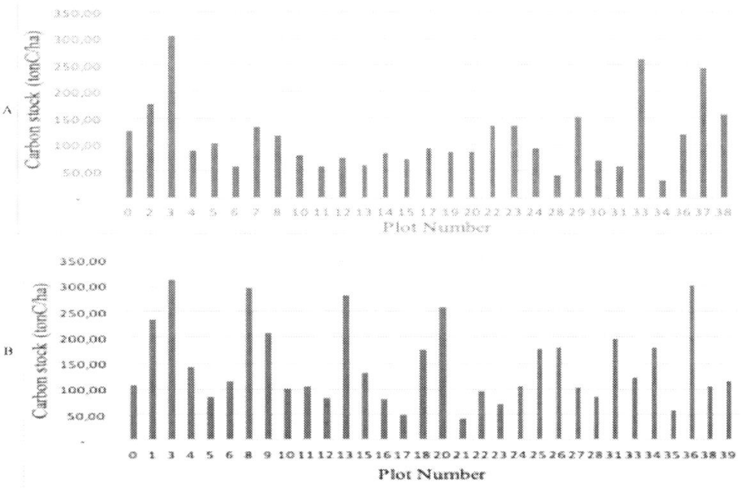

Figure 7. Distribution of carbon stock in Sinamang Forest Area 2; (A) in CL, (B) in RIL.

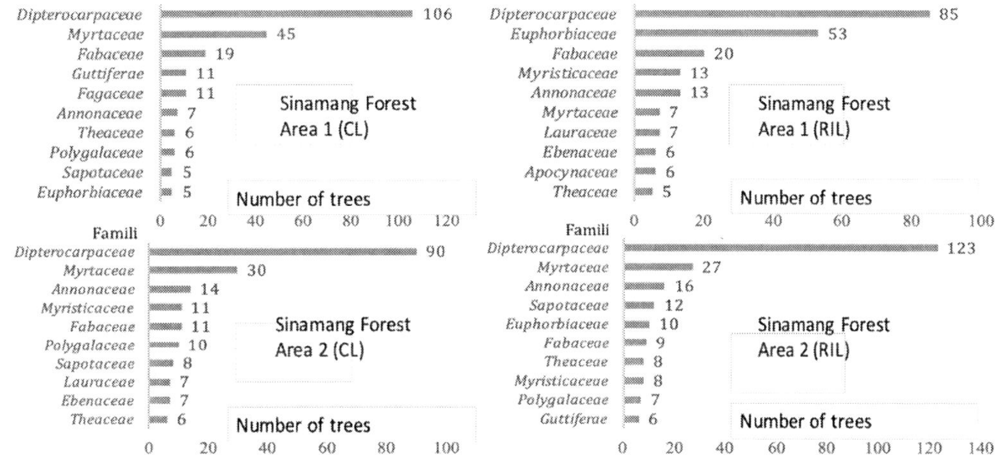

Figure 8. Trees family dominant.

Each forested area usually has a vegetation composition such as the level of seedlings & undergrowth, saplings, poles and trees. There were no significant changes in plant families based on the number of individual trees at the location before and after felling, in both the conventional and RIL systems, there was no significant change and was still very much dominated by the Dipterocarpaceae, Myrtaceae, Fabaceae and Euphorbiaceae families [13]. At the research site before felling, the species *Shorea* sp., *Syzigium* sp. and *Dipterocarpus borneensis* were found in 5 species with the highest Importance Value Index and were always found at the growth rate of saplings, poles and trees. At the conventional logged-over and RIL locations, the distribution of the three species was also found with various variations in Importance Value Index values but still in 5 species with the highest Importance Value Indexes both at the growth rate of saplings, poles and trees [13]. The Dipterocarpaceae family is

IOP Conf. Series: Earth and Environmental Science 1115 (2022) 012038 doi:10.1088/1755-1315/1115/1/012038

the most dominant family and is most commonly found in natural and secondary forest areas on the islands of Kalimantan and Sumatra. Myrtaceae is a family that is often found in logged-over forests [14, 15, 16]. Overall, the potential carbon stock in the baseline condition is 168.14 tonC/ha or equivalent 617.1 tonCO_2eq/ha with the largest proportion being tree carbon stock as much as 82.78% and the smallest proportion being seedling carbon stock and undergrowth as much as 0.50% (Table 1.)

Table 1. Potency of aboveground carbon stock in baseline condition.

No	Vegetation Type	Carbon Stock (tonC/ha)	Proportion (%)
1	Seedling and understorey	0.85	0.50
2	Sapling	10.49	6.24
3	Pole	17.62	10.48
4	Tree	139.19	82.78
	Total	168.14	100.00

3.3. *Emission factor*

In Sinamang Forest Area 1, the carbon stock in conventional logging and reduced impact logging are 121.51 tonC/ha or equivalent 445.9 tonCO_2eq/ha and 149.34 tonC/ha or equivalent 548.1 tonCO_2eq/ha, respectively. This shows that RIL activities can reduce emissions by 27.83 tonC/ha and equivalent 102.1 tonCO_2eq/ha or 59.68% compared to the baseline condition (conventional logging system). In Sinamang Forest Area 2, the carbon stock in conventional logging and reduced impact logging are 105.86 tonC/ha or equivalent 388.5 tonCO_2eq/ha and 138.08 tonC/ha or equivalent 506.8 tonCO_2eq/ha, respectively. This shows that RIL activities can reduce emissions by 32.22 tonC/ha and equivalent 118.2 tonCO_2eq/ha or by 51.73% (Table 2). In overall, the results of the research in study sites showed that average of emission factor for emission reduction estimation in RIL compare to conventional logging (CL) amounted 30.03 tonC/ha and equivalent 110.2 tonCO_2eq/ha or 55.71%.

Table 2. Emission factor for emission reduction estimation in RIL.

No	Location	Carbon Stock in Baseline Condition (Before Logging) (tonC/ha)	Carbon Stock (tonC/ha)		Emission Factor for Emission Reduction Estimation in RIL (tonC/ha)	Percentage of Emission Reduction (%)
			After Conventional Logging	After Reduced Impact Logging		
		A	B	C	D = ((A-B)-(A-C))	E = ((D/(A-B))x100%
1	Sinamang Forest (Area 1)	168.14	121.51	149.34	27.83	59.68
2	Sinamang Forest (Area 2)	168.14	105.86	138.08	32.22	51.73
	Average	168.14	113.69	143.71	30.03	55.71

The potential of RIL in reducing emissions in Indonesia is quite large when viewed from the size of the timber extraction area. The results of this study indicate that good RIL technique can reduce the loss of carbon stocks/reduce emissions quite high by 59.68% in Sinamang Forest Area 1 and 51.73% in Sinamang Forest Area 2 (Table 2). This results similar to the Elias statement that RIL can reduce carbon loss by up to 51 percent as compare to the conventional logging methods [17]. In the East

Kalimantan Province, the results also showed that the application of environmentally friendly low-carbon emission logging (RIL-Carbon) techniques can reduce forest damage by almost 50%, and reduce carbon emissions by almost 30% compared to conventional logging (as Business As Usual/BAU) currently running [18].

In the other site in Amazonian Forest, the RIL activities can reduce the loss of carbon stock by 23% [19]. According to [20] and [21] also reported that by applying reduced-impact logging techniques as compared to conventional logging operations could reduce carbon losses by 30%. According to [21] reported that reduced impact logging in the tropical forests would keep at least 0.16 gigatons of carbon in each year.

Although RIL activity to support emission reduction still has several challenges, but countries with big area for timber extraction could implement this technique to achieve large portions of their NDCs with input of low costs [3]. Reduced impact logging in addition to playing an important role in reducing emissions, overall RIL also has shown to reduce damage to soils and residual stands up to 50% when compared with conventional logging [20, 22-24].

According to [25] also showed the results that RIL activities could reach to 41% less damage to residual stands when compared to the conventional logging. Tropical forests carbon stock can be kept by reducing deforestation as well as reducing carbon loss by following Reduced impact logging (RIL). Along with that fact, scaled-up application of RIL practices could ensure sustainable tropical forest management and enhanced carbon stock in the tropical forests which in turn will help mitigate climate change [26].

3.4. Implication of Reduced Impact Logging (RIL)

The improvement of the current conventional forest harvesting system (as BAU) with the application of environmentally friendly harvesting systems such as RIL is expected to make a significant contribution to the role of natural production forest management in reducing carbon emissions in the forestry sector, given the higher percentage of natural production forests than protected forest and conservation forest. The government's interest as a regulator is that the application of RIL technique can demonstrate the role of production forests in climate change and will also facilitate the monitoring function because RIL can only be implemented if the inventory and planning data are carried out correctly. In the context Indonesian policy related sustainable forest management and its contribution to the emission reduction, the reducing forest degradation due to excessive logging and encroachment in the production forests has been put as other priority.

Indonesia has issued mandatory certification systems (Ministerial Regulation No. 30/ 2016 by the Minister of Environment and Forestry) that push the adoption of sustainable management practices in the production forests especially in reduced impact logging/RIL. With this policy, many forest concessions have improved their management practices. Under LCCP (Low Carbon Compatible with Paris Agreement), by 2050 all forest concession companies have implement sustainable harvesting practices. At present, the forest concessions that have received the good forest certification reach about 76% [27].

4. Conclusion

The results of the research in study sites showed that average of emission factor for emission reduction estimation in RIL compare to conventional logging (CL) amounted 30.03 tonC/ha and equivalent 110.2 tonCO_2eq/ha or 55.71%. Thus, this value can be used as emission factor to estimate the emission reduction locally in RIL compare to the CL in Central Kalimantan.

Production forest activities based on the reduced impact logging technique in theory and practice are proven to be more environmentally friendly than conventional logging techniques. The remaining forest stands after reduced impact logging is better than conventional logging, using a parameter of carbon stock after logging according to the stock difference method. Forest concessionaries in Indonesia need to produce timber using the reduced impact logging technique to reduce emissions on a forest company scale and help reduce national emissions of land based sector according to the NDC.

Acknowledgements
The authors would like to thank to the APHI (Asosiasi Pengusaha Hutan Indonesia) during research study.

References

[1] Direktorat Jenderal Pengelolaan Hutan Produksi Lestari 2020 *Rencana strategis tahun 2020 – 2024 (Strategic plan year 2020-2024)* (Jakarta, Indonesia: Ministry of Environment and Forestry)

[2] Muhdi, Elias, Murdiyarso D, and Matangaran J R 2012 Kerusakan tegakan tinggal akibat pemanenan kayu reduced impact logging dan konvensional di hutan alam tropika (Studi kasus di areal IUPHHK PT. Inhutani II, Kalimantan Timur) (Remaining stands damage due to reduced impact logging and conventional timber harvesting in tropical natural forest) (Case study in the IUPHHK PT. Inhutani II Area, East Kalimantan) *Jurnal Manusia dan Lingkungan* **19** p 303-311

[3] Ellis P W, Gopalakrishna T, Goodman R C, Putz F E, Roopsind A, Umunay P M, Zalman J, Ellis E A, Mo K, Gregoire T G, and Griscom B W 2019 Reduced-impact logging for climate change mitigation (RIL-C) can halve selective logging emissions from tropical forests *Forest Ecology Management* **438** p 255-266

[4] Suhartana S, Yuniawati, and Rahmat 2014 *Pemanenan kayu ramah lingkungan di hutan tanaman lahan kering di Sumatera, Kalimantan dan Jawa Barat (Friendly timber harvesting in dryland plantation forest in Sumatera, Kalimantan and West Java) In Pemanfaatan Sumberdaya Terbarukan untuk Kesejahteraan Manusia dan Kelestarian Lingkungan (Renewable resources utilization for human wellbeing and environmental sustainability)* Suwinarti W, Kusuma I W, Erwin, Ismail (Eds.) Proceeding Seminar Nasional MAPEKI 16 (National Seminar Proceeding MAPEKI 16) p 402-406 (Jakarta, Indonesia: Masyarakat Peneliti Kayu Indonesia)

[5] Bicknell J E, Struebig M J, Edwards D P, and Davies ZG 2014 Improved timber harvest techniques maintain biodiversity in tropical forests *Current Biology* **24** p 1119-1120

[6] Ministry of Environment and Forestry 2019 *Roadmap Nationally Determined Contribution mitigation* (Jakarta, Indonesia: Ministry of Environment and Forestry)

[7] United Nations Framework Convention on Climate Change 2015 *COP 21 Climate Agreement* (Paris, French: United Nations Framework Convention on Climate Change)

[8] Griscom B W, Adams J, Ellis P W, Houghton R A, Lomax G, Miteva D A, Schlesinger W H, Shoch D, Siikamäki J V, Smith P, Woodbury P, Zganjar C, Blackman A, Campari J, Conant R T, Delgado C, Elias P, Gopalakrishna T, Hamsik M R, Herrero M, Kiesecker J, Landis E, Laestadius L, Leavitt S M, Minnemeyer S, Polasky S, Potapov P, Putz F E, Sanderman J, Silvius M, Wollenberg E, and Fargione J 2017 Natural climate solutions *Proceedings of the National Academy of Sciences* **114** p 11645–11650

[9] Badan Standardisasi Nasional [BSN] 2011 *Standar Nasional Indonesia (SNI) 7724:2011 tentang Pengukuran dan penghitungan cadangan karbon –Pengukuran lapangan untuk penaksiran cadangan karbon hutan (ground based forest carbon accounting) (Indonesian National Standard (SNI) 7724:2011 on Measurement and estimation of carbon stock – ground based forest carbon accounting)* (Jakarta, Indonesia: National Standardization Agency)

[10] Ketterings Q M, Coe R, Van Noordwijk M, Ambagau Y, and Palm C 2001 Reducing uncertainty in the use of allometric biomass equations for predicting above-ground tree biomass in mixed secondary forest *Forest Ecology Management* **146** p 199-209

[11] Manuri S, Brack C, Rusolono T, Noor'an F, Verchot L, Maulana S I, Adinugroho W C, Kurniawan H, Sukisno D W, Kusuma G A, Budiman A, Anggono R S, Siregar C A, Onrizal, Yuniati D, and Soraya E 2017 Effect of species grouping and site variables on aboveground biomass models for lowland tropical forests of the Indo-Malay region *Annals Forest Sciences* **74** p 1-14

[12] Ridwan M, Suparna N, Dharmawan I W S, Nugraha A, and Widyantoro B 2020 *Petunjuk praktis penghitungan penurunan emisi dari kegiatan RIL (Practical guidance of emission reduction estimation from RIL/Reduced Impact Logging) activity)* (Jakarta, Indonesia: PT Wana Aksara)

[13] Dharmawan I W S, Ridwan M, and Suparna N 2021 Jenis tanah, komposisi dan keanekaragaman jenis tegakan pada pengusahaan hutan alam secara konvensional dan RIL (Soil type, composition and species diversity in the natural forest logging of conventional and RIL) *Jurnal Ilmu Lingkungan* **19** p 555-564
[14] Kartawinata K, Purwaningsih, Partomihardjo T, Yusuf R, Abdulhadi R, and Riswan S 2008 Floristic and structure of a lowland *dipterocarp* forest at Wanariset, East Kalimantan. *Reinwardtia* **12** p 301-323
[15] Samsoedin I 2009 Dinamika keanekaragaman jenis pohon pada hutan produksi bekas tebangan di Kalimantan Timur (Dynamic of tree species diversity at logged over production forest in East Kalimantan) *Jurnal Penelitian Hutan dan Konservasi Alam* **6** p 69-78
[16] Samsoedin I, and Heriyanto N M 2010 Struktur dan komposisi hutan pamah bekas tebangan ilegal di kelompok hutan Sei Lepan, Sei Serdang, Taman Nasional Gunung Leuser, Sumatera Utara (Structure and species composition of lowland disturbed forest at Lepan River Forest Complex, Sei Serdang, Gunung Leuser National Park, North Sumatera) *Jurnal Penelitian Hutan dan Konservasi Alam* **7** p 299-314
[17] IPB University 2021 *Prof Elias explains RIL-C approach in reducing carbon emissions due to tropical forest logging* https://ipb.ac.id/news/index/2021/07/prof-elias-explains-ril-c-approach-in-reducing-carbon-emissions-due-to-tropical-forest-logging/02fad66becf2a11436987d4192fbf020
[18] Yasman I, Nurrochmat D, Septiani Y, and Lasmini 2013 *Policy paper: Peran pengelolaan hutan produksi alam dalam perubahan iklim (REDD+, Pengelolaan Hutan Lestari dan RIL-C) (Role of natural production forest management in climate change (REDD+, Sustainable Forest Management and RIL-C))* (Jakarta, Indonesia: The Nature Conservancy, Program Terestrial Indonesia)
[19] Mazzei L, Sist P, Ruschel A, Putz F E, Marco P, Pena W, and Ferreira J E R 2010 Above-ground biomass dynamics after reduced impact logging in the Eastern Amazon *Forest Ecology Management* **259** p 367-373
[20] Pinard M, and Putz F 1996 Retaining forest biomass by reducing logging damage *Biotropica* **28** p 278–295
[21] Putz F E, Zuidema P A, Pinard M A, Boot R G, Sayer J A, Sheil D, Sist P, Elias, and Vanclay J K 2008 Improved tropical forest management for carbon retention *PLoS Biology* **6** p 1368-1369
[22] Elias P S 1997 *Reduced impact timber harvesting in the tropical natural forest in Indonesia; Forest harvesting case study 11* (Rome, Italy: Food and Agriculture Organization of the United Nations)
[23] Winkler N 1997 *Environmentally sound forest harvesting: Testing the applicability of the FAO models code in the Amazon in Brazil; Forest harvesting case study 8* (Rome, Italy: Food and Agriculture Organization of the United Nations)
[24] Pinard M, Barker M G, and Tay J 2000 Soil disturbance and post-logging forest recovery on bulldozer paths in Sabah, Malaysia *Forest Ecology Management* **130** p 213–225
[25] Klassen A 2002 *Impediments to the adoption of reduced impact logging in the Indonesian corporate sector In Applying Reduced Impact Logging to Advance Sustainable Forest Management* Enters T, Durst P B, Applegate C, Kho P C S, Man G (Eds.) (Bangkok, Thailand: Food and Agriculture Organization of the United Nations, Regional Office for Asia and the Pacific)
[26] Rahman M, Islam M, Islam R, and Sobuj N A 2017 Towards sustainability of tropical forests: Implications for enhanced carbon stock and climate change mitigation *Forest Environmental Sciences* **33** p 281-294
[27] Ministry of Environment and Forestry 2021 *Indonesia long-term strategy for low carbon and climate resilience 2050* (Jakarta, Indonesia: Ministry of Environment and Forestry)

IOP Conf. Series: Earth and Environmental Science 1115 (2022) 012039 doi:10.1088/1755-1315/1115/1/012039

Bolt-bearing strength parallel and perpendicular to the grain of four Indonesian hardwoods

E Herawati*, R Hartono and T N Harahap

Faculty of Forestry, Universitas Sumatera Utara. Kampus USU 2 Bekala, Deli Serdang 20353, North Sumatra, Indonesia

*E-mail: evalina@usu.ac.id

Abstract. Bearing strength is an important factor in determining the load-carrying capacity of a bolt-timber connection besides the bolt-bending yield strength. This study aims to determine the bolt-bearing strength parallel and perpendicular to the grain using three bolt diameters (6, 9 and 12 mm) in four hardwoods species with different specific gravities, namely terap (*Artocarpus odoratissimus*), durian (*Durio zibethinus*), mindi (*Melia azedarach* L) and rubberwood (*Hevea brasiliensis* Muell. Arg). The bolt-bearing strength was determined as described in ASTM D5764 using half-hole testing. The moisture contents ranged from 13.49–14.77% and specific gravities were 0.33–0.60. The higher the specific gravity, the higher the bearing strength. The bearing strength values parallel to the grain were higher than those perpendicular to the grain. The bearing strengths parallel to the grain were 3–18% and 2–24% lower than those based on the Indonesian National Standard (SNI) 7973 and National Design Specification (NDS) for Wood Construction 2015, respectively. In contrast, the experimental bearing strength of mindi was 8% higher than the SNI. Except for mindi, which was 22% exceed SNI and 9% above NDS, bearing strengths perpendicular to the grain were 3-32% and 14-41% less than values based on SNI and NDS, respectively.

1. Introduction

In addition to the bolt's bending yield strength, bearing strength is an important factor in determining the load-carrying capacity of a bolt-timber connection. Both the National Design Specification (NDS) for Wood Construction 2015 [1] and the Indonesian National Standard (SNI) 7973: 2013 Design Specifications for Wood Construction [2] state that the wood density affects bearing strength in parallel and perpendicular to the grain, but in perpendicular to the grain also affected by the diameter of the bolt. Meanwhile, according to another test standard, especially Eurocode 5 [3], in a parallel test of bearing strength, the diameter of the bolt is also taken into calculation, but from many studies, it is known that the diameter of the bolt has no significant effect, so it is proposed to replace it with the dowel material properties [4].

Previous studies have shown that the formula contained in the referenced standard is more suitable for softwood than for hardwood [4,5]. Some studies on the dowel bearing strength using hardwood showed a lower value than the value found in the referenced standard [5–9]. Besides the limited hardwood database in formulating the standard, this is also related to the wood moisture content (MC) in the test, where usually test for tropical hardwood is carried out in air-dried MC conditions which are usually above 12% to 20%. An increase in MC will cause a decrease in the wood-bearing strength of about 2–3% per 1% MC [10,11]. If tested at 12% MC, the experimental findings exceeded the standard values, such as the bearing strength of merbau wood, which was 11% more than NDS 2015 and 23% above the SNI 7973 [9].

A previous study evaluated the bolt-bearing strength of six wood species from natural forests, with a specific gravity range of 0.4-0.8 [9]. Meanwhile, much of the timber traded comes from plantations or community forests. The woods were used by society for a variety of objects, including construction materials and furnishings [12]. So it is necessary to conduct a study related to the characteristics of wood from community forests. Among the many species planted by the community, four of them terap (*Artocarpus odoratissimus*), mindi (*Melia azedarach* L), durian (*Durio zibethinus*) and rubberwood (*Hevea brasiliensis*) had been studied for their physical and mechanical properties [13]. Since the woods are also used for structural purposes, their application as a connection is also very feasible. So it is basic to know the bearing strength of the bolt if it will be used in building components that require bolted connections.

This study was a series of studies on the fundamental characteristics of wood, in particular its physical and mechanical characteristics up until making wood connections using bolts. At various bolt diameter sizes, the evaluated bearing strength was not limited perpendicular to the grain but also parallel to the grain. This study's objective was to examine the bolt-bearing strength of four Indonesian hardwoods that were sourced from community forests in Binjai Regency, North Sumatra. It is expected that the information on the bolt-bearing strength would be taken into consideration when designing wood connections.

2. Materials and Method

This study investigated four hardwood species, including rubberwood (*Hevea brasiliensis*), durian (*Durio zibethinus*), mindi (*Melia azedarach* L) and terap (*Artocarpus odoratissimus*). The specific gravity of wood ranged from 0.35 to 0.60. The logs are obtained from community sawmills in the Binjai Regency, North Sumatra, Indonesia. The bolts used consist of three diameter sizes, namely 6 mm, 9 mm and 12 mm.

Bolt-bearing strength testing was carried out in two directions of loading, namely parallel and perpendicular to the grain. The test was carried out with reference to ASTM D5764-97a (Reapproved 2013) Standard Test Method for Evaluating Dowel-Bearing Strength of Wood and Wood-Based Products [14]. The wood blocks were dried to an air-dry moisture content before being cut into samples for bearing strength testing. The size of the test samples used was 5 x 5 x 5 cm, each with 10 replications so the number of test samples made was 240 units. The bearing strength test was carried out on a Universal Testing Machine (UTM) of the Tensilon merk with a capacity of 50 kN. The test was conducted using the half-hole method, where the load was evenly distributed along the bolt. The load was applied until half of the diameter of the bolt was inserted in the test sample or reached the maximum load with a loading speed of 1 mm/min (Figure 1). By fitting a straight line to the initial linear section of the load-displacement curve, the bearing yield load was obtained. Next, deform this line by 5% of the fastener's diameter to offset it and then choose the load at which the offset line crosses the load-displacement curve.

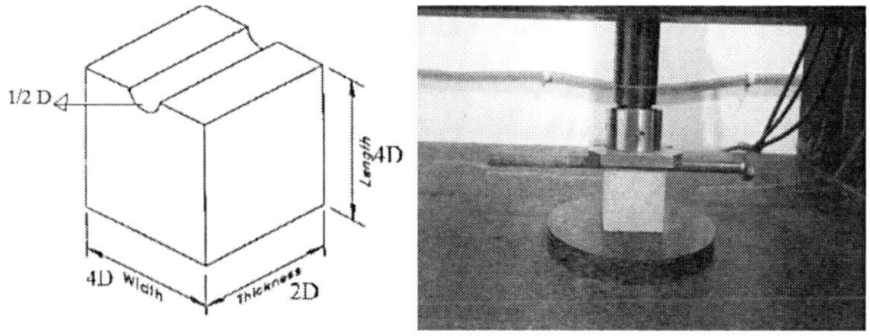

Figure 1. Specimen configuration and the testing apparatus

The following formula was used to determine the bearing strength (Fe):

$$F_e = \frac{P_{0.05D}}{Dt} \quad (1)$$

Where $P_{0.05D}$ is the load at the 5% offset diameter, D is the bolt's diameter and t is the wood sample thickness. Additionally, each wood's density and moisture content were assessed using the ASTM D2395 [16] and D4442 [15] standards, respectively. Furthermore, the data were analyzed using descriptive statistics and compared with reference standards.

3. Results and Discussion

3.1. Bolt-bearing strength parallel to the grain

Figure 2 shows the load vs deformation graphs for three bolt sizes during the parallel to the grain bearing strength test. The graph shows that a larger bolt diameter and higher specific gravity (SG) of the wood resulted in a higher load being reached during the test. The graphic pattern for testing all wood species was almost the same, after reaching a certain load, the graph tended to be flat because there was an increase in deformation at a relatively constant load. The condition of the sample in this case experienced compaction at the bottom of the bolt. This pattern was also shown in *Eucalyptus grandis* [6] and light and heavy red meranti wood where the specific gravity was 0.43 and 0.62 respectively [9]. Determination of the load used to calculate the bearing strength was based on the yield load determined by the 5% offset diameter method from the load-deformation curve of the test results. The intersection of the lines to obtain the yield load was generally located at the point before the maximum load.

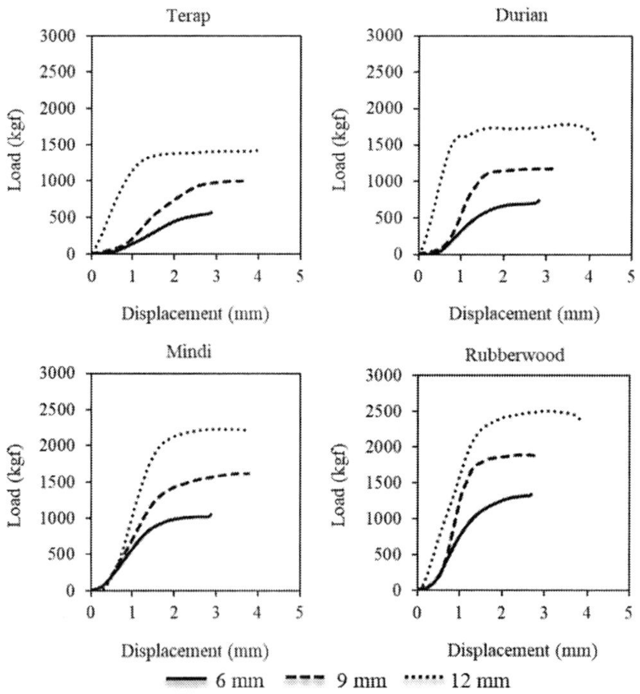

Figure 2. Load vs displacement curves for parallel to the grain bolt-bearing strength tests

The density and moisture content of each sample was obtained after testing was done to assess the wood bearing strength. The moisture content of terap, durian, mindi and rubberwood ranged from

13.49 to 14.77%, where this MC was in the range of air-dried moisture content. Meanwhile, the wood specific gravities used were 0.35, 0.40, 0.49 and 0.60 respectively. While the other woods are in strength class III, terap wood is in strength class IV [13,17].

Furthermore, Table 1 presents the parallel bearing strength value based on experimental data, SNI 7973 and NDS 2015. With an increase in the wood specific gravity, the value of the parallel to grain bearing strength increases., it can be seen that the bearing strength of terap wood with the lowest specific gravity has the lowest value as well. Meanwhile, rubber wood has the highest bearing strength value according to its specific gravity. In general, the bolt bearing strength increased from 6 mm to 9 mm in diameter but decreased at a diameter of 12 mm. This decrease could be due to the larger surface area of the wood hole, thus lowering the strength limit of the wood. This is also in line with the trend of the results of studies conducted by [8,9]. The coefficient of variation in the parallel to the grain bearing strength test was generally the lowest in rubberwood (4–10%) and the highest in terap wood (18–25%), the values were still within the range of normal values for testing the wood mechanical properties.

Table 1. The average bearing strengths parallel to the grain (Fes ∥) values based on the current experimental findings, SNI and NDS

Wood species	MC (%)	SG	F_{es} ∥ (MPa)					
			6 mm	9 mm	12 mm	Average	SNI 7973	NDS 2015
Terap	13.49	0.35	21(25)	22(18)	18(20)	21	25	27
Durian	14.49	0.40	24(16)	28(24)	23(21)	25	28	31
Mindi	14.77	0.49	35(10)	42(12)	34(13)	37	34	38
Rubberwood	14.11	0.60	42(7)	41(10)	39(4)	41	42	46

Coefficients of variation (%) are given within parentheses

For dowels measuring in ≤ ¼ in (6 mm), the bearing strength parallel to the grain in SNI 7973 and also NDS 2015 only has one value, which means that even though the bolt sizes are different, the bearing strength is calculated the same. The study of wood bearing strength with reference to Eurocode 5 also shows that actually, the effect of bolt diameter on the parallel to grain bearing strength test is so small, that an equation solely based on wood density would be sufficient [5].

Based on the bearing strength value on SNI 7973 and NDS 2015, it can be seen that the greater the specific gravity, the greater the bearing strength value as well as the results obtained from this study. The bearing strength values determined from this study generally show a lower tendency when compared to the strength values based on SNI 7973 and also NDS 2015. This difference ranged from 3 to 18% and 2-24%, except for the bearing strength of mindi wood which was between the SNI and NDS values.

3.2. Bolt-bearing strength perpendicular to the grain

Figure 3 shows the load vs deformation graphs for three bolt sizes during the bearing strength test perpendicular to the grain. The perpendicular-to-grain bearing strength test, such as the parallel-to-grain bearing strength test, demonstrated that the load attained during the test was higher with the larger bolt diameter and the wood-specific gravity (SG). Meanwhile, the graphic pattern on the test of all wood species showed an increasing tendency of the load during the test. This was different from the parallel-to-grain test where the graph pattern tended to flatten after reaching a certain load. This curve is in line with the curve obtained from the study of the wood-bearing strength perpendicular to the grain of *Eucalyptus grandis* [6]. The condition of the sample, in this case, experienced compaction at the bottom of the bolt so it gave a tendency to increase strength due to the compaction. However, the intersection of the lines to obtain the yield load, in general, can still be determined even though the load and deformation continue to increase.

IOP Conf. Series: Earth and Environmental Science 1115 (2022) 012039 doi:10.1088/1755-1315/1115/1/012039

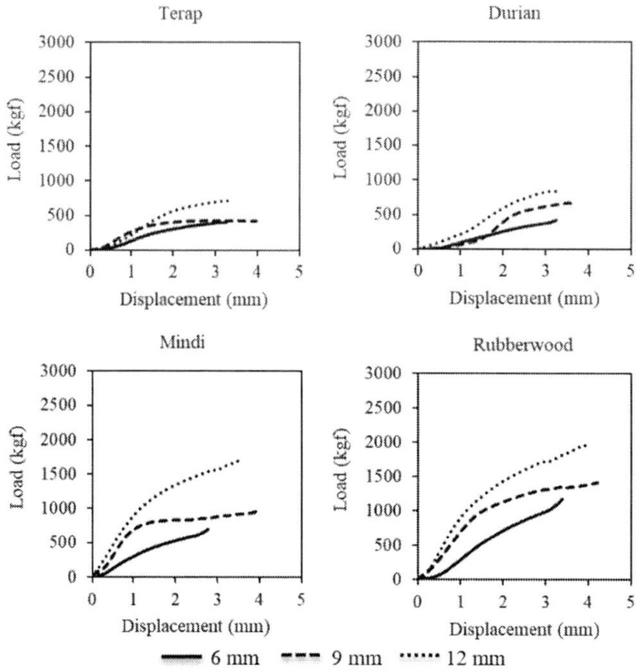

Figure 3. Load vs displacement curves for perpendicular to the grain bolt-bearing strength tests

According to testing results, SNI 7973 and NDS 2015, the value of the perpendicular bearing strength is shown in Table 2. Similar to the parallel-to-grain test, the perpendicular-to-grain test generally demonstrated that the bearing strength of wood increases with its specific gravity. The larger the bolt diameter, the smaller the value of the bearing strength perpendicular to the grain, although the load achieved during the test was higher with the larger the diameter of the bolt. And this was consistent with the values in the referenced standards SNI 7973 and as well NDS 2015, according to which the bolt diameter also has an impact on the bearing strength perpendicular to the grain in addition to specific gravity. In comparison to the bearing strength parallel to the grain, the bearing strength perpendicular to the grain had a lower value. This was matched with the wood's measurements for compression strength parallel and perpendicular to the grain, where the parallel value was higher than the perpendicular value [18].

Table 2. The average bearing strengths perpendicular to the grain (Fes ⊥) values based on the current experimental findings, SNI and NDS

Wood species	F_{es} ⊥ (MPa)								
	6 mm			9 mm			12 mm		
	Experiment	SNI 7973	NDS 2015	Experiment	SNI 7973	NDS 2015	Experiment	SNI 7973	NDS 2015
Terap	13(35)	16	18	10(28)	13	15	9(13)	11	13
Durian	13(22)	19	22	13(15)	16	18	12(17)	14	16
Mindi	25(23)	26	30	20(27)	21	24	23(11)	18	21
Rubberwood	31(18)	35	40	28(19)	29	33	24(27)	25	28

Coefficients of variation (%) are given within parentheses

The values of the bearing strength perpendicular to the grain in this study ranged from 3–31 MPa for a bolt diameter of 6 mm, 10–28 MPa for a bolt diameter of 9 mm and 9–24 MPa for a bolt diameter of 12 mm. The coefficient of variation ranged from 11–35% which indicated that the values obtained were quite diverse but still in the range of values that were common in the testing of wood mechanical properties. Compared to the values parallel to the grain test, the values perpendicular to the grain test were more diverse from the coefficient of variation. The bearing strength values obtained were generally lower than the values listed in the SNI 7973 and 2015 NDS referenced standards except for mindi wood. Except for mindi, which was 22% exceed SNI 7973 and 9% above NDS 2015, bearing strengths perpendicular to the grain were 3-32% and 14-41% less than values based on SNI 7973 and NDS 2015, respectively.

4. Conclusion

Terap, durian, mindi and rubberwood all had moisture contents between 13.49 and 14.77% and specific gravities between 0.35 and 0.60. The higher the specific gravity of the wood, the higher the bearing strength. Similarly, the larger the bolt diameter, the greater the value of the load achieved during the test, but the smaller the value of the bearing strength, especially the bearing strength perpendicular to the grain. The bearing strength values parallel to the grain were higher than those perpendicular to the grain. In comparison to those based on SNI 7973 and NDS 2015, the bearing strengths parallel to the grain in this investigation were 3-18% and 2-24% lower, respectively. The experimental bearing strength of Mindi, however, was 8% greater than that of SNI 7973. Except for mindi, which was 22% above the SNI 7973 and 9% above the NDS 2015 value, bearing strengths perpendicular to the grain were indeed 3-32% and 14-41% below those calculated on SNI 7973 and NDS 2015, respectively.

Acknowledgement

The TALENTA USU Research Grant on Basic Research Scheme of contract no 4167/UN5.1.R/PPM/2019 provided funding for the study. The authors appreciate Arfan, Dedy and Hardin's assistance with the laboratory work.

References

[1] American Wood Council 2014 *National Design Specification (NDS) for Wood Construction 2015 edition* (Leesburg: American Wood Council)
[2] National Standardization Agency of Indonesia 2013 *Indonesian National Standard (SNI) 7973:2013 Design Specifications for Wood Construction* (Jakarta: National Standardization Agency of Indonesia)
[3] British Standards Institution 2004 *BS EN 1995-1-1:2004 Eurocode 5: Design of Timber Structures Part 1-1: General-Common rules and rules for buildings* (London: British Standards Institution)
[4] Yurrita M and Cabrero J M 2018 *Construction and Building Materials* **173** 238
[5] Sandhaas C, Ravenshorst G J P, Blass H J and van de Kuilen J W G 2013 *Eur J Wood Prod* **71** 599
[6] Sosa Zitto M A, Köhler J and Piter J C 2011 *Eur J Wood Prod* **70** 433
[7] Glišović I, Stevanović B and Kočetov-Mišulić T 2012 *Wood Research* **57** 639
[8] Sadiyo S, Susanto D and Pratiwi N E S 2017 *Jurnal Teknil Sipil* **24** 153
[9] Herawati E, Sadiyo S, Nugroho N and Karlinasari L 2017 *International Wood Products Journal* **8** 233
[10] Rammer D R and Winistorfer S G 2001 *Wood and Fiber Science* **33** 126
[11] Cousin A and Salenikovich A 2012 *Proc of World Conf on Timber Engineering* ed Quenneville P (Auckland: New Zealand Timber Design Society) 473
[12] Center for R and D of Wood Products 2008 *A Handbook of Selected Indonesian Wood Species* (Bogor: Center for R & D of Wood Products, ISWA and ITTO)

[13] Herawati E, Hartono R and Sinaga H M M 2021 *IOP Conf Ser Earth Environ Sci* **912** 012024
[14] American Society for Testing and Material 2013 ASTM D5764−97a (Reapproved 2013) *Standard Test Method for Evaluating Dowel-Bearing Strength of Wood and Wood Based Products* (West Conshohocken: ASTM International)
[15] American Society for Testing and Material 2016 *ASTM D4442–16: Standard Test Methods for Direct Moisture Content Measurement ff Wood and Wood-Base Materials* (West Conshohocken: ASTM International)
[16] American Society for Testing and Material 2014 *ASTM D–2395–14: Standard Test Methods for Specific Gravity of Wood and Wood-Based Materials* (West Conshohocken: ASTM International)
[17] Oey D S 1990 *Specific gravity of Indonesian woods and its significance for practical use* (Bogor: Forest Products Research Institute. Announcement No.1)
[18] Leijten A J M 2016 *Construction and Building Materials* **105** 528

IOP Conf. Series: Earth and Environmental Science 1115 (2022) 012040 doi:10.1088/1755-1315/1115/1/012040

The effect of extraction methods towards antioxidant activity of ethanol extract of *Picria fel-terrae* Lour. Herbs

A Dalimunthe[1*], D Pertiwi[2], M Muhmmad[2], V E Kaban[3], N Nasri[3] and D Satria[2]

[1]Department of Pharmacology, Faculty of Pharmacy, Universitas Sumatera Utara. Jl. Tri Dharma No.5, Medan, North Sumatra, Indonesia.
[2]Department of Pharmaceutical Biology, Faculty of Pharmacy, Universitas Sumatera Utara. Jl. Tri Dharma No.5, Medan, North Sumatra, Indonesia.
[3]Departement of Pharmacy, Sekolah Tinggi Ilmu Kesehatan (STIKes) Senior Medan. Jl. Jamin Ginting KM. 8,5 No. 13, Medan 20141, North Sumatra, Indonesia

[*]Email : aminah@usu.ac.id

Abstract. Antioxidants are molecules that can inhibit or prevent the oxidation reactions of other molecules that generate free radicals. Antioxidants can be synthesized in the body (endogenous) or obtained from food intake (exogenous). Modern pharmacological investigations show that the extract of *Picria fel-terrae* Lour. Herbs. Effective as an antioxidant because of the high content of flavonoids and phenols. This study aims to analyze the antioxidant activity, total phenol content and total flavonoid content in ethanol extract (EE). The powder was extracted by Soxhletation, percolation and reflux method with absolute ethanol. Antioxidant activity was analyzed by 1,1-diphenyl-2-picrylhydrazyl (DPPH), FRAP, ABTS and CUPRAC. Total phenol content was measured using the Folin-Ciocalteau method and total flavonoid content was measured using $AlCl_3$ reagent using colorimetric method. The best antioxidant activity and the highest total phenolic content were indicated by the soxhlet extraction method with IC_{50} 23.52 ± 0.17 g/mL and 160.77 ± 0.95 mg GAE/g, and the highest total flavonoid content was indicated by the reflux method (1.08 ± 0. 02 mg QE/g). In conclusion, the ethanolic extract of *Picria fel-terrae* Lour. Herbs Lour with reflux extraction method can be further developed as an antioxidant.

1. Introduction

Oxidative stress causes degenerative diseases that lead to death, such as cancer, atherosclerosis, diabetes, and heart disease [1,2]. In living things, oxidation is a crucial process [3]. Free radicals that are produced by metabolism or environmental factors continuously interact with biological systems. Oxidation reactions release free radicals, which can cause cell damage. Free radicals are charged molecules that assault our cells, disrupt cell membranes, and cause damage to the body's nucleic acids, proteins, and enzymes as a natural result of human metabolism. Oxidative stress refers to any type of free radical damage that might harm a cell's structure and function [4]. Free radicals are produced by cellular metabolism, radiation, ozone, cigarette smoke, hyperoxia, and heavy metal exposure. Free radicals in excess can destroy macromolecules including DNA, lipids, and proteins [5]. Antioxidants can be utilized to treat diseases induced by oxidative stress [6]. Antioxidants are chemicals that can block or prevent other molecules from oxidizing and producing free radicals [6,7]. Antioxidants can be made in the body (endogenous) or consumed through diet (exogenous). Antioxidants can be divided into two categories: enzymatic and non-enzymatic antioxidants. Superoxide dismutase (SOD), glutathione peroxidase (GPX), and catalase are three enzyme antioxidants that play a key function (CAT). Many natural exogenous anti-oxidants are now being produced to suit the body's antioxidant needs. This is due to the fact that natural antioxidants have fewer adverse effects than synthetic antioxidants. Non-enzymatic antioxidants, such as glutathione, vitamin C, vitamin E, carotenoids, uric acid, thiols, and polyphenols, are examples of natural antioxidants found in plants [8,9].

Indonesia is a country rich in spice plant resources. Poguntano (*Picria fel-terrae* Lour. Herbs) from the family *Scrophulariaceae*, which is one form of spice that has the potential to be developed. The essential oil can be found in the fruit, stems, roots, and leaves of this plant. This gives the whole plant a pleasant smell Poguntano (*Picria fel-terrae* Lour. Herbs) with a high content of flavonoid components). Diuretic, antipyretic, hepatoprotective, cardioprotective, antidiabetic, antioxidant, anti-inflammatory, anthelmintic, and analgesic activity are the traditional uses of *Picria fel-terrae* Lour. Herbs [11,12].

Extraction technique is a method for obtaining high yield extracts by reducing the difficulty of the extraction process [13]. Using an extraction method with a ratio of time and solvent requirements greatly determines the efficiency of extraction, in addition to the use of temperature, which has a significant impact on the quality (biological activity in vitro or in vivo) and quantity (amount or yield) of the extract obtained [14,15]. Cold extraction (maceration and percolation) and hot extraction (reflux and soxhletation) are the most popular extraction processes [16]. Based on this background, we want to see how different extraction procedures (maceration, percolation, and reflux) affect antioxidant activity, total phenolic content, and total flavonoid content when using ethanol as a solvent.

2. Materials and Methods

2.1. *Preparation of extract*
The air-dried and powdered of *Picria fel-terrae* Lour. Herbs (100 g) was extracted by maceration, percolation and reflux extraction method with ethanol absolute (Merck). The filtrate was collected and then evaporated under reduced pressure to give a viscous form and then dried to dry on waterbath [7].

2.2. *DPPH radical scavenging assay*
The free radical scavenging activity was measured by the 1,1-diphenyl-2-picrylhydrazyl (DPPH•) method. 0.2mM solution of DPPH• in methanol was prepared and 100 µL of this solution was added to various concentrations of *Picria fel-terrae* Lour. Herbs extracts. After 60 mins, absorbance was measured at 516 nm. All assays were carried out in triplicate [17,18].

2.3. *CUPRAC radical scavenging assay*
The sample concentration was added to a 0.01M $CuCl_2.2H_2O$ solution, 0.0075 ethanolic neukuproin, and pH 7. 1M ammonium acetate buffer, incubated for 1 hour, and evaluated at 450 nm using the Cupric Ion Reducing Antioxidant Capacity (CUPRAC) method [19].

2.4. *FRAP radical scavenging assay*
The Ferric Reducing Antioxidant Power (FRAP) method was used to test free radical scavenging activity using 0.2 M phosphate buffer and K3Fe (CN)6 1 percent incubation for 20 minutes, adding TCA, and 0.1 percent FeCl3, and measured absorbance at 720 nm [20].

2.5. *ABTS radical scavenging assay*
The 2,2-azinobis-3-ethylbenzothiazoline-6-sulfonic acid (ABTS) free radical scavenging method was carried out by mixing the sample with an ABTS and potassium persulfate solution, which was incubated for 12 hours and detected at a wavelength of 734 nm [21].

2.6. *Determination of total phenolic content*

The total phenol concentration (TPC) of the sample was determined using a folin reagent. Briefly, 100 µL of EE (500 µg/mL) were mixed with 7.9 mL of distilled water and 0.5 mL of Folin-ciocalteu's reagent (1:10 v/v) and mixed with vortex for 1 min. After mixing, 1.5 mL of 20% aqueous sodium bicarbonate was added, and the mixture was allowed to stand for 90 mins within intermittent shaking. The absorbance was measured at 775 nm using a spectrophotometer. Total phenolic concentration is expressed as gallic acid equivalent in mg per gram of extract. The methanol solution was used as a blank. All assays were carried out in triplicate [17,22].

2.7. *Determination of total flavonoid content*

The amounts of total flavonoids in the extracts were measured spectrophotometrically as previously reported. Briefly, 2 mL of extract in methanol was mixed with 0.10 mL of 10% aluminium chloride ($AlCl_3.6H_2O$), 0.10 mL of sodium acetate ($NaC_2H_3O_2.3H_2O$) (1M), and 2.80 mL of distilled water. After incubation of 40 mins, absorbance was measured at 432 nm using a spectrophotometer. To calculate the concentration of flavonoids, we prepared a calibration curve using quercetin as standard. The flavonoid concentration is expressed as quercetin equivalents in mg per gram of extract. All assays were carried out in triplicate [22,23].

3. Results and Discussion

3.1. *Antioxidant activity, total phenolic content and total flavonoid content of Picria fel-terrae Lour. Herbs*

Lour extract of *Picria fel-terrae* Lour. Herbs. When examined utilizing the DPPH, CUPRAC, ABTS, and FRAP procedures, spice was found to have excellent antioxidant activity. The activity of the DPPH approach with the soxhletation extraction method employing ethanol as a solvent was high. The sample concentration required to capture 50% of free radicals was calculated using the IC_{50} value. As antioxidants, phenolic compounds including flavonoids work by reducing free radicals [24]. The amount and kind of hydroxyl groups that a chemical contains determines how effective it is as an antioxidant [25]. The antioxidant testing method using the DPPH method is the method that gives the best results because the DPPH method has more stable radical properties than the elements of other antioxidant methods, making it easier to convert free radicals into non-radical compounds. The extraction method using the hot soxhlet method is an iterative extraction method that can be more certain that the extraction process is perfect and the metabolite compounds are more completely attracted by the occurrence of a colour change in the extraction process from coloured to colourless, which ensures that the extraction process has been completed [16,26]

Table 1 shows the results of antioxidant tests using *Picria fel-terrae* Lour. Herbs ethanol extract with IC_{50} values. Herbs employs the DPPH, CUPRAC, ABTS, and FRAP methodologies, as well as a variety of extraction techniques. The IC_{50} value metric is used to measure antioxidant activity, with the lower the IC_{50} value, the stronger the antioxidant activity. [7,26,27].

Table 1. Result of antioxidant activity (IC_{50}) with DPPH, FRAP, ABTS, total phenolic and total flavonoid content

NO	Extraction Method	DPPH (µg/mL)	FRAP (µg/mL)	ABTS (µg/mL)	CUPRAC (µg/mL)	Total phenolic (mg GAE/g)	Total flavonoid (mg QE/g)
1.	Reflux	31.78 ± 0.06	149.70 ± 0.37	123.99 ± 0.16	32.74 ± 0.03	140.78 ± 0.96	1.80 ± 0.02
2.	Soxhletation	23.52 ± 0.17	137.69 ± 0.22	85.80 ± 0.14	25.31 ± 0.10	160.77 ± 0.95	0.08 ± 0.01
3.	Percolation	33.91 ± 0.01	154.75 ± 0.44	216.57 ± 0.44	28.30 ± 0.08	125.79 ± 0.92	0.58 ± 0.01

Parameter values of IC50 can be classified Very strong (IC50 50 μg/mL), strong (IC50 50 - 100 μg/mL), moderate (IC50 101 - 150 μg/mL), and weak antioxidant activity (IC50> 150 μg/mL) was measured [29]. Table 1 shows the reflux extraction method giving total phenol values of (140.78 ± 0.96 mg GAE/g), soxhletation (160.77 ± 0.95 mg GAE/g), and percolation (125.79 ± 0. 92 mg GAE/g), total flavonoids extracted by reflux (1.80 ± 0.02 mg QE/g), soxhletation (0.08 ± 0.01 mg QE/g), and percolation (0.58 ± 0.01 mg Q/g).

When employing a polar solvent as a solvent, phenol is a chemical that is soluble in polar solvents, giving it a high solubility value. Because not all phenol concentrations in the extract are part of flavonoids, the overall phenol value is higher than the total flavonoid value, as shown in Table 1. Chemical polyphenols include phenolic acid, melanin, quamarine, flavonoids, lignin, and tannin, which are all metabolites found in plants [30].

The ability of the sample to diminish free radical activity was used to test antioxidant activity. The IC_{50} value is utilized as the parameter for the free radical scavenging test. This is the value achieved as a result of the ability to absorb 50% of free radicals [22]. The DPPH method is used to assess antioxidants because it functions as a synthetic radical that takes hydrogen atoms, transforming DPPH into non-radical molecules [17]. The CUPRAC antioxidant test method can oxidize antioxidant molecules in a sample to generate a bisneocuproin-copper (I) complex, which changes color from blue to yellow visibly [31]. The ABTS test method is a method for determining a sample's ability to act as an indicator [7,31].

The FRAP test method is based on antioxidant chemicals' capacity to reduce iron (III)-tripyrydyl-triazine compounds to iron (II)-tripyrydyl-triazine compounds to iron (II)-tripyrydyl-triazine compounds to iron (II)-tripyrydyl-triazine compounds to iron (II)-tripyrydyl-triazine compounds [21]. Because not all antioxidants can decrease Fe(II) molecules at the right time for incubation, the FRAP test is a poor test method in this case.

4. Conclusion

The soxhlet extraction procedure employing the DPPH method yielded the highest antioxidant activity and phenolic content. The reflux extraction method revealed the highest overall flavonoid concentration.

Acknowledgments

This research was funding by Ministry of Education, Culture, Research and Technology through Penelitian Dasar Unggulan Perguruan Tinggi research grant 2022

References :

[1] Sen S, Chakraborty R, Sridhar C, Reddy Y S R and De B 2010 *International journal of pharmaceutical sciences review and research* 3 **91**
[2] Valko M, Leibfritz D, Moncol J, Cronin M T D, Mazur M and Telser J 2007 *The international journal of biochemistry & cell biology* **391** 44
[3] Harahap U, Dalimunthe A, Hertiani T, Muhammad M, Nasri and Satria D 2021 *AIP Conference Proceedings* **2342**
[4] Aher V D, Wahi A, Pawdey A M and Sonawane A 2011 *Int J Curr Pharm Res* **3** 8
[5] Birben E, Sahiner U M, Sackesen C, Erzurum S and Kalayci O2012 *World Allergy Organ J* **5** 9
[6] Al-Dalaen S M and Al-Qtaitat AI 2014 *Am J Biosci Bioeng*, **2** 5 60
[7] Dalimunthe A, Hasibuan P A Z, Silalahi J, Sinaga S F and Satria D 2018 *Oriental Journal of Chemistry* **34** 1149
[8] Partap S 2012 *Journal of Pharmacognosy and Phytochemistry* **1** 4
[9] Gomes E C, Silva A N and de Oliveira M R 2012 *Oxidative Medicine and Cellular Longevity* 756132
[10] Pacôme O A, Bernard DJYH N, Sékou D, Joseph D A, David N J, Mongomaké K and Hilaire K T 2014 *Research Journal of Pharmaceutical, Biological and Chemical Sciences* 5 1453

[11] Auliafendri N, Rosidah, Yuandani, Suryani S and Satria D 2019 *Open Access Macedonian Journal of Medical Sciences* 7 24
[12] Harahap U, Yuandani, Marianne, Agustya H M, Azizah D U and Alfiah S W 2018 *Scientia Pharmaceutica* **86** 34
[13] Quispe-Condori S, Foglio M A, Rosa P T V and Meireles M A A 2008 *The Journal of Supercritical Fluids* **46** 27
[14] Gan C Y and Latiff A A 2011 *Food chemistry* **124** 1277
[15] Hmidani A, Bouhlali E D T, Khouya T, Ramchoun M, Filali-zegzouti Y, Benlyas M and Alem C 2019 *Scientific African* **5** 00143
[16] Desmiaty Y, Elya B, Saputri F C, Dewi I I and Hanafi M 2019 *Jurnal Ilmu Kefarmasian Indonesia* **17** 227
[17] Rosidah, Yam M F, Sadikun A and Asmawi M Z 2008 *Pharmaceutical Biology* **46** 616
[18] Satria D, Silalahi J, Haro G, Ilyas S, Hasibuan P A Z 2017 *Asian Pacific journal of cancer prevention: APJCP* **18** 399
[19] Hidayat M A, Chassana R I, Ningsih I Y, Yuwono M and Kuswandi B 2019 *European Food Research and Technologt* **245** 2063
[20] Diniyah N, Alam M B and Lee S H 2020 *Arabian Journal of Chemistry* **13** 55208
[21] Sridhar K and Charles A L 2019 *Food Chemistry* **275** 41
[22] Nazliniwaty H, Avriyanti O, Pertiwi D, Satria D and Muhammad M 2021 *American Institute of Physics Conference Series* 2342 080010
[23] Senguttuvan J, Paulsamy S and Krishnamoorthy K 2012 *J Appl Pharm Sci* **2** 149
[24] Lubis M F, Syahputra H, Illian D N and Kaban V E 2022 *J Res Pharm.* **26** 565
[25] Gurning K et al 2021 *J. Pendidik. Kim* **13** 137
[26] Nazliniwaty H et al 2021 *American Institute of Physics Conference Series* 2342 080010
[27] Molyneux P 2014 *Songklanakarin J. sci. technol* **26** 211
[28] Sitorus P, Hasibuan P A Z and Satria D 2017 *Asian J. Pharm. Clin. Res* **10** 243
[29] Blois M S 1958 *Nature* **181** 1199
[30] Moein S and Moein M R 2010 *Journal of Medicinal Plants Research* **4** 517
[31] Bener M, Sen F B and Apak R 2018 *Talanta* **187** 148
[32] Untea A, Lupu A, Saracila M, and Panaite T 2018 *Bulletin UASVM Animal Science and Biotechnologies* **75** 111

Comparison of microwave assisted extraction and ultrasound assisted extraction on recovery of residual palm oil from spent bleaching earth

Taslim*, **R Manurung, Iriany, O Bani, A Pranata and A Nugraha**

Department of Chemical Engineering, Faculty of Engineering, Universitas Sumatera Utara, Medan 20155, Indonesia

*Email: taslim@usu.ac.id

Abstract. Recovery of oil contained in spent bleaching earth (SBE) will have a good impact on the environment and economy. The process can be carried out using Microwave Assisted Extraction (MAE). This study aimed to determine best operation parameters for residual palm oil recovery using MAE and compare it to Ultrasound Assisted Extraction (UAE). The oil extraction was carried out using hexane as solvent at variable microwave power (100, 180, and 300 W), SBE to solvent ratio (1:4, 1:5, 1:6), and extraction time (50,70 and 80 min). The highest oil yield was achieved at microwave power of 300 W, SBE to solvent ratio of 1:5 and extraction time of 80 min at 20.97%. The main components of the oil produced were C16 (56.65 %wt) and C18.1 (28.35 %wt). The free fatty acid content, moisture, iodine value, deterioration of bleachability index (DOBI) and peroxide value of the resulting oil were 10.91%, 1.52%, 56.71 g I2/100 g, 1.372 and 7.419 meq/kg respectively.

1. Introduction

The bleaching of palm oil aims to improve the appearance, color, odor, and stability of palm oil as well as to remove soap, heavy metals, and phosphorus [1,2]. Generally, bentonite-based bleaching earth is used as an adsorbent in the bleaching process. Approximately 5 – 20 kg of fresh bleaching earth/ton of vegetable oil is required for the process [2]. However, this process will produce a large amount of solid waste which is commonly known as spent bleaching earth (SBE) [1]. In Indonesia, more than 100,000 tons of SBE are produced during the production of 18 million tons of palm oil annually [3]. The condition becomes even more problematic due to rapid industrial growth and increase in waste production [4]. SBE is categorized as dangerous waste [5] and is usually disposed by landfilling. However, this kind of treatment can cause water and soil pollution as well as contributing to economic losses and fire hazards due to oil residue [5,6]. Generally, the oil residue contained in SBE is around 20-40% but in some cases, it can reach 72% [7]. Thus, it is necessary to carry out oil recovery process.

The recovered oil in SBE can be used for the manufacture of biodiesel [2] or lubricant [8]. It can also be added to plant or animal waste to produce compost [6]. Generally, the oil recovery is carried out by solid-liquid extraction. However, this method requires long extraction time, lots of SBE, and lots of solvents while the resulting yield is small [9,10]. In recent years, a new extraction method has been developed, namely microwave assisted extraction (MAE) [11].

MAE is a method of oil extraction by utilizing microwaves which can be converted into heat. It is better than conventional extraction which relies on conduction and convection and requires longer extraction time. Microwave extraction has been proven to be effective in accelerating oil extraction from the matrix to solvent using less solvent [9,11]. It is therefore necessary to explore and develop the MAE of palm oil residue from SBE. This paper aimed to determine best operation parameters for palm oil residue extraction by MAE and compare it to UAE.

2. Materials and methods

2.1. *Materials*
SBE was obtained from one of the palm oil factories in Medan, Indonesia. Hexane (p.a., Merck, Germany) was purchased from a chemical shop in Medan.

2.2. *Microwave assisted extraction*
SBE was weighed and mixed with hexane at SBE to solvent ratio (m/v) of 1:4, 1:5, and 1:6. The mixture was then placed in a 20-L microwave. The microwave was operated at power of 100, 180, and 300 W for 50, 70, and 80 min. After completion, the mixture was let to cool to room temperature. Afterwards, SBE was separated from the mixture by filtration using filter paper. The filtrate, which contained both oil and solvent, was heated at 80°C for 2 h to remove the solvent. The oil remained was weighed and put into a sample bottle and analyzed for its free fatty acids, moisture, iodine value, deterioration of bleachability index (DOBI), and peroxide value using Near-Infrared Spectroscopy (NIRS). Oil yield is calculated using the equation (1):

$$Yield\ of\ oil = \frac{Weight\ of\ Oil}{Weight\ of\ SBE} \times 100\% \qquad (1)$$

The parameter value that generated highest oil yield using MAE was selected as reference for UAE and the results were compared.

2.3. *Oil composition analysis by GC-MS*
As much as 2 drops of sample (oil) was put into a vial by pipetting, then 2 mL 0.5 N methanolic NaOH solution was added. The mixture was heated for 1 h at 80 – 85°C then added with 1 mL boron trifluoride. The mixture was reheated for 10 min at 80 – 85°C then added with 2 mL hexane and 1 mL 37 %wt. HCl and stirred. The mixture was left to form 2 layers. The top layer was taken and put into a vial to be injected into the GC-MS (Agilent) and the results were recorded.

3. Results and Discussion

3.1. *Effect of SBE to solvent ratio on oil yield*
In figure 1, increasing SBE to solvent ratio from 1:4 to 1:5 increased oil yield. This is because the concentration gradient between the solvent and solid becomes larger which acts as a driving force in the mass transfer process [12]. As a result, the adsorbate will have a greater tendency to leave the matrix (feed) into the solvent through dissolution and diffusion [11]. However, further increase of SBE to solvent ratio to 1:6 decreased oil yield. This is due to higher energy requirement and longer reflux time [13]. In addition, excess solvent will inhibit the heating in SBE matrix due to absorption of microwave energy by solvent. The same result was reported by Dejkajorn et al [9] on the extraction of soybean oil from SBE in which highest yield was obtained at SBE to solvent ratio of 1:15 and the yield decreased when the ratio was increased to 1:30. Excessive use of solvents will result in wasted solvents and higher operating costs.

Microwave power is another important parameter during extraction process. Microwave power will heat the SBE and produce a driving force in SBE matrix which triggers the oil diffusion out of the SBE matrix into the solvent [9]. In figure 1, extraction yield increased with the increase of microwave power. Increasing the power will decrease the viscosity and surface tension of the solvent so that mass transfer increases. It will also increase the capacity of the solvent to dissolve the adsorbate from the active site in the matrix [11]. On the other hand, increasing power will cause more solvent to evaporate [14].

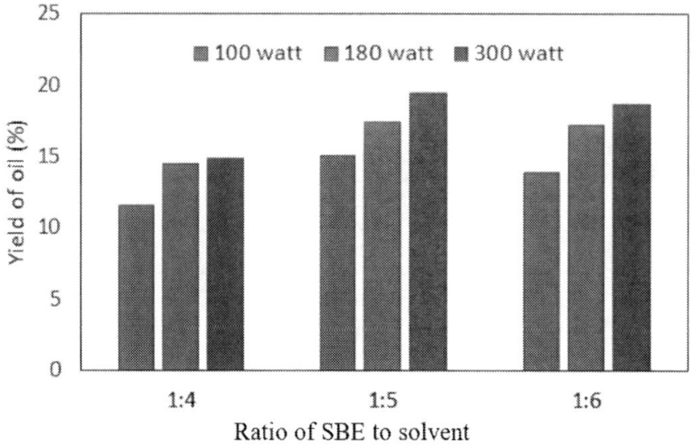

Figure 1. Effect of operating conditions on oil yield

3.2. Effect of extraction time on oil yield

Extraction time is another important parameter that affects yield, quality, energy, and costs [11]. In figure 2, increasing extraction time increased oil yield. However, the increase was insignificant. This was because the oil in the SBE was limited [15]. Further addition of extraction time will thus produce a fixed extraction yield.

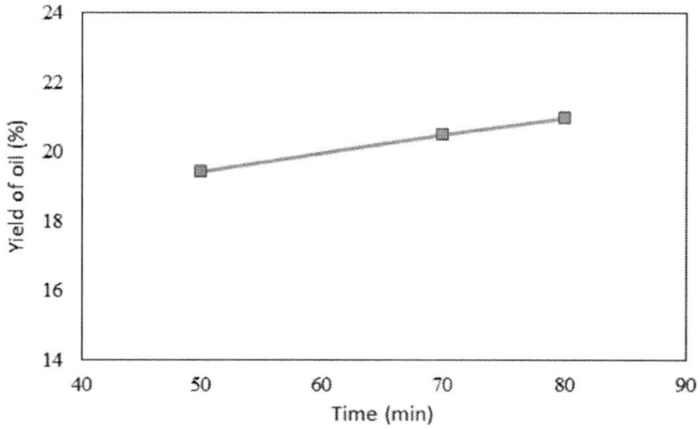

Figure 2. Effect of extraction time on oil yield

3.3. Comparison of MAE and UAE on palm oil residue recovery

A comparison of residual palm oil extraction methods was carried out using the MAE and UAE methods. The extraction process with UAE (Elmasonic S 300 H) with a frequency of 37 kHz was carried out under the best conditions of MAE, namely SBE to solvent ratio of 1:5 and extraction time of 80 min. The comparison of residual palm oil extracted from SBE using MAE and UAE is in table 1. The yield of oil produced with MAE (20.97%) was higher than that of UAE (18.35%). This is because microwave energy is much better at breaking up solid particles in a liquid than ultrasonic wave energy. Dahmoune et al. [16] explained that there was a change in the surface structure of starfish powder after the extraction process using MAE and UAE. MAE gives this highly crushed surface indicating that microwaves play an important role in breaking down cell walls in contrast to UAE which plays an insignificant role in breaking down starfish cell walls.

Table 1. Comparison of MAE and UAE methods on residual palm oil produced

Characteristic	MAE	UAE
Yield of oil (%)	20.97	18.35
FFA (%)	10.91	13.50
Moisture (%)	1.52	2.28
IV (g I_2/100 g)	56.71	51.27
DOBI	1.37	1.13
PV (meq/kg)	7.42	8.79

The effect of extraction method was evaluated on the quality of oil produced. The quality includes free fatty acid (FFA), moisture, iodine value (IV), deterioration of bleachability index (DOBI), and peroxide value (PV), and was analyzed using Near Infrared Spectroscopy (NIRS) as in table 1. Free fatty acids of oil produced by MAE (10.91%) were lower than that by UAE (13.50%). The increase in free fatty acids was caused by oil hydrolysis due to factors such as lipase enzyme, heat, and moisture. In addition, moisture of oil produced by MAE was also lower, in which humidity correlates with free fatty acids. The heating caused by MAE can inhibit lipase and moisture from reacting with oil [12]. IV is used to measure the level of oil saturation. IV of oil produced by MAE (56.71 g I_2/100 g) was greater than that by UAE (51.27 g I_2/100 g). High IV indicates good oil quality. PV indicates the degree of oxidation that occurs in the oil. PV of oil produced by UAE (8,797 meq/kg) was higher than that of MAE (7,419 meq/kg). This is because ultrasonic waves can produce many reactive radicals. These reactive radicals will then react to form H_2O_2 and other oxidizing agents [17,18]. Meanwhile, DOBI of oil produced by MAE and UAE was low and not significantly different i.e., 1.37 and 1.13. The standard DOBI value for Crude Palm Oil (CPO) ranges between 2.5 and 4.0 [19]. Color is also another parameter related to oil quality. In figure 3, color of oil produced by MAE is yellow while that by UAE is orange. The yellow and orange colors of the oil indicated that the oil is less sensitive to the Maillard reaction in which this reaction can cause oil to darken [12].

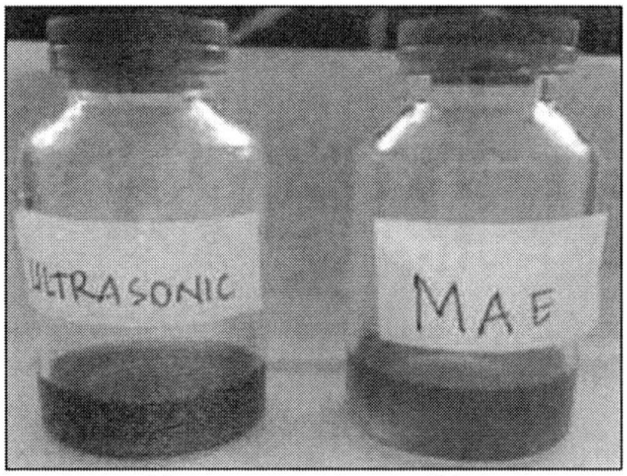

(a) (b)

Figure 3. Oil produced by the extraction method (a) UAE (b) MAE

3.4. GC-MS analysis

The compositions of oil produced by MAE and UAE were analyzed using GC-MS as in Table 2. The main component of oil extracted from SBE was dominated by C16 and C18.1, similar to crude palm oil (CPO). This indicates that extraction by MAE and UAE does not change the composition of the resulting oil.

Tabel 2. Oil composition

Composition	CPO (%)	MAE (%)	UAE (%)
C8	-	0.82	-
C10	0.01	0.09	-
C12	0.19	0.44	0.18
C14	1.02	1.59	1.13
C16	41.48	46.65	44.07
C18.0	4.41	6.00	4.49
C18.1	40.70	36.39	39.13
C18.2	11.24	4.15	10.17
C18.3	0.66	-	0.38
C20	0.20	1.20	0.35

Note: C18.0, C18.1, C18.2, and C18.3 stand for stearic acid, oleic acid, linoleic acid, and linolenic acid respectively

4. Conclusion

Oil recovery from SBE using MAE gave better results than UAE. Oil recovery yield of 20.97% were achieved at microwave power of 300 W, SBE to solvent ratio of 1:5 and extraction time of 80 min. In addition to the higher oil yield, MAE also produced better oil quality compared to UAE. However, both extraction methods did not cause significant changes to the composition of resulting oil.

References

[1] Majid R A and Mat C R C 2017 *J. Oil Palm Res.* **29** 579
[2] Sedghamiz M A, Raeissi S, Attar F, Salimi M, and Mehrabi K 2019 *Fuel* **237** 515
[3] Hindryawati N, Daniel D, Erwin E and Maniam G P 2017 *Jurnal Bahan Alam Terbarukan* **6** 68
[4] Lee C G, Seng C E and Liew K Y 2000 *J Am Oil Chem Soc* **77** 1219
[5] Yuan Z, Shen Y, Yuan H, Sui A and Zhu N 2020 *J Clean Prod* **258** 120622
[6] Wafti N S A, Yoo C K, Lin S W, Yaw T C.S and Abdullah L C 2011 *J Oil Palm Res* **23** 999
[7] Loh S K, Cheong K Y and Salimon J 2017 *Appl Clay Sci* **140** 59
[8] Amiri A and Ghanbarzadehlak M 2013 *ISWA World Congress Proceed* **1** 1
[9] Dejkajorn C, Suttiarporn P, Sukkathanyawat H, Wichianwat K, Sriphan S and Srisurat T 2021 *E3S Web Conf* **302** 01009
[10] Sedghamiz M A, Attar F and Raeissi S 2019 *Process Saf. Environ Prot* **124** 121
[11] Sonar M P and Rathod V K 2020 *Chem. Data Collec* **30** 100545
[12] Zhong J, Wang Y, Yang R, Liu X, Yang Q and Qin X 2018 *Ind. Crops Prod* **120** 1
[13] Ibrahim N A and Zaini M A A 2018 *Chin J Chem Eng* **26** 2516
[14] Chen Q, Dong W, Wei C, Hu R and Long Y 2020 *Ind. Crops Prod* **151** 112405
[15] Megawati, Fardhyanti D S, Sediawan W B and Hisyam A 2019 *Indus Crops Prod* **131** 315
[16] Dahmoune B, Houma-Bachari F,Chibane M, Akrour-Aissou C, Guégan J P, Vives T, Jéhan P, Dahmoune F, Mouni L, Ferrières V and Hauchard D 2021 *Chem Eng Process: Process Intensif* **163** 108359
[17] Kumar K, Srivastav S and Sharanagat V S 2021 *Ultrason. Sonochem* **70** 105325
[18] Vinatoru M, Mason T J and Calinescu I 2017. *TrAC - Trends Anal Chem* **97** 159
[19] Ifa L, Wiyani L, Nurdjannah N, Ghalib A M T, Ramadhaniar S and Kusuma H S 2021 *Heliyon* **7** e07230

IOP Conf. Series: Earth and Environmental Science 1115 (2022) 012042 doi:10.1088/1755-1315/1115/1/012042

Experimental Study of Solar Cold Energy Storage with Additional PCM and Without PCM

Calvin [1,2], Samar [1], H Ambarita[1*]

[1]Sustainable Energy Research Centre, Faculty of Engineering, Universitas Sumatera Utara, Medan, Sumatera Utara, Indonesia.
[2]Mechanical Engineering Department, Faculty of Engineering, Universitas Sumatera Utara, Medan, Sumatera Utara, Indonesia.

*himsar@usu.ac.id

Abstract. This Increased energy efficiency and comfort conditions in the building environment are increasingly needed. Therefore, phase change materials (PCM) are devoted to reducing the energy consumption of buildings. This research is an experiment with a solar-powered CES working system with dimensions of 740 mm x 410 mm x 310 mm which is used to supply cold air to a room with dimensions of 4.8 m × 3.7 m × 3.3 m. The main component of the CES cooling system consists of a combination of a thermoelectric system and an actively applied PCM. In active applications, forced convection is the dominant heat transfer mechanism between PCM and air in CES where some additional electrical devices are required. In this test, the results of CES without PCM were obtained, on the first and second days there were differences in the outlet temperature for the room and the inlet temperature for the room which were almost the same with an average of 1.43°C and 1.5°C. While CES with PCM on the first and second days with an average of 7.36°C and 3.85°C. For COP values obtained CES without PCM on the first and second days has an average value of 0.25. While the value of COP CES with PCM on the first and second days was 0.39 and 0.33.

1. Introduction

Increased energy efficiency and comfort conditions in the building environment are increasingly needed. Therefore, phase change materials (PCM) are devoted to reducing the energy consumption of buildings [1]. One way to reduce energy requirements is the use of thermal energy storage (TES) [2, 3]. Among various TES applications, the accumulation of hot and cold temperatures from -50°C to 120°C has a higher market potential when using phase change materials (PCM) [4, 5]. Phase change material (PCM) which is LTHES (latent heat thermal energy storage) when used as cold energy storage (CES), offers a promising alternative to conventional cooling systems. PCM has a large thermal energy storage capacity in a temperature range close to the point of transformation and exhibits almost isothermal behavior during the charging and discharging process [6].

PCM technology can be implemented into buildings, either passively or actively [7]. In active applications, forced convection is the dominant heat transfer mechanism between PCM and indoor air where some additional electrical or mechanical devices are required [8]. The use of PCM as cold energy storage (CES) can be integrated with a thermoelectric system. Thermoelectric cooling (TEC) has several advantages compared to other cooling systems such as compact in size and light in weight, high reliability, no mechanical moving parts so it is not noisy, no working fluid so it is environmentally friendly, and thermoelectric cooling can be powered DC (direct current) which means it is easy to integrate with PV (photovoltaic) panels [9].

The cooling system under study is solar cold energy storage. In this cold energy storage system, PCM and TEC are used which are powered by solar panels. This research focuses on the temperature that occurs in cold energy storage and the performance value of the cold energy storage working system.

2. Method and Experimental Setup

The test equipment and problem formulation are described in the following sections.

2.1. System description

In this study, the object of research observed was a test room located in the city of Medan, North Sumatra with dimensions of 4.8 m x 3.7 m x 3.3 m which was conditioned by air conditioning based on solar cold energy storage. A schematic representation of the analyzed test chamber with the walls of the test chamber containing a window with geometric dimensions and experimental settings is shown in Figure 1.

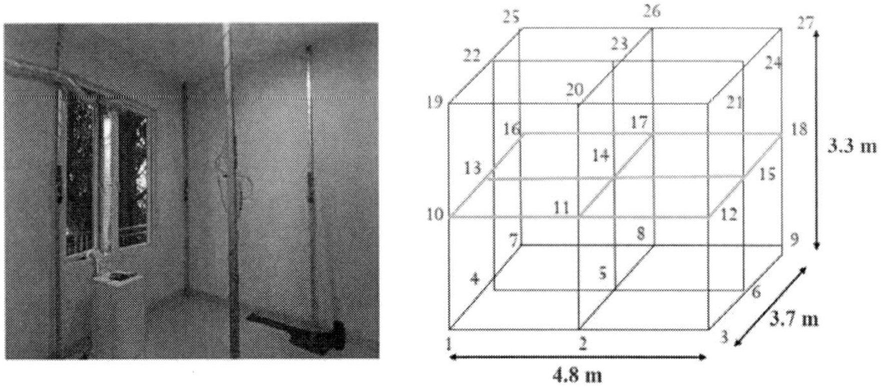

Figure 1. Test room (a) Setup Experimental, (b) schematic representation of the test room

As seen in Figure 2, the size of the styrofoam solar cold energy storage cooler is 74 cm x 41 cm x 31 cm with 4 sets of Peltier as a cooling source plus an ice gel PCM and 6 DC fans (8 x 8) cm for air circulation in in the test room, where 4 fans are installed in the test room on the styrofoam box as output to the test room, and 2 more fans are installed in the air conditioner as input from the test room with electrical power supplied from solar panels. The solar cold energy storage cooler has three openings, two openings for the air inlet and one for the air outlet.

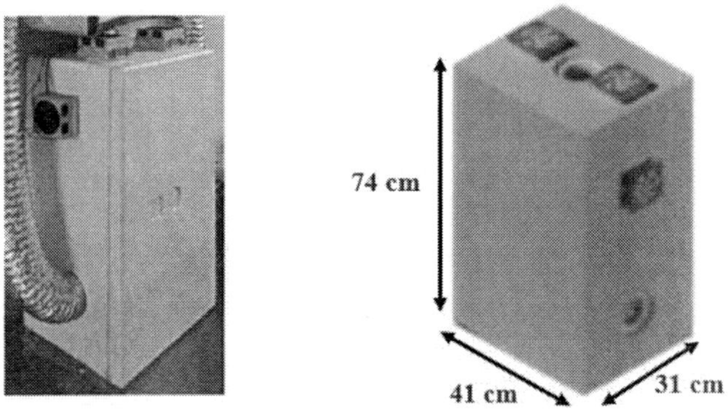

Figure 2. Solar Cold Energy Storage (a) Setup Experimental, (b) schematic representation of Solar Cold Energy Storage

2.2. Procedure

Two different conditions were carried out during the experiment to analyze and compare the measurements.

a. Mode 1: Cold energy storage without PCM.
b. Mode 2: Cold energy storage with PCM.

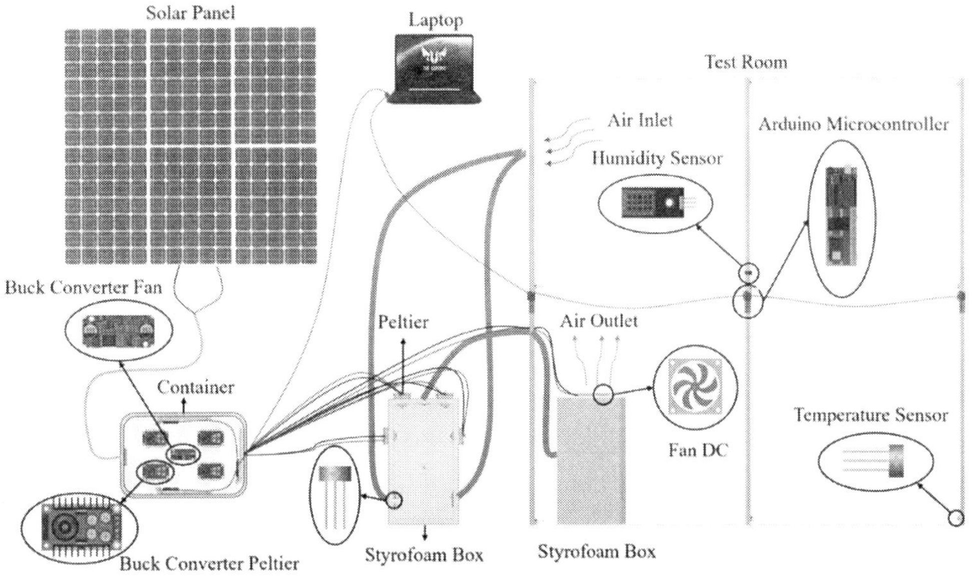

Figure 3. Test Scheme

Tests carried out on October 2-03 2021 for mode 1 and January 11-12 2022 for mode 2 with the same styrofoam box. In Figure 3 it can be seen that solar panels are used as an energy source to run the cooling device. It is known that solar radiation energy absorbed by solar panels is quite volatile, so a buck converter device is used to reduce the excess voltage according to the working voltage of the cooler, which is 12 V. The cooling device used is a cooling system based on solar cold energy storage. Thermoelectric and PCM are installed in the cooling device.

The workings of the cooler starts with the air from the room being sucked from the top of the room into the cooler box. The heat from the room entering the cooler is absorbed and then flowed back into the room using a DC fan so that it produces air output with a lower temperature into the room.

Next, the Arduino Uno sensor starts to run to be able to acquire real-time data stored on the SD-card. After that, the experimental settings are run whether there is a problem with the sensor using a laptop. After that, the Arduino microcontroller reads the temperature for the set time, the microSD is removed and reads in Microsoft Excel on the computer. For solar cold energy storage coolers, there are 14 temperature sensor points shown in Figure 4 and 1 temperature sensor around the CES.

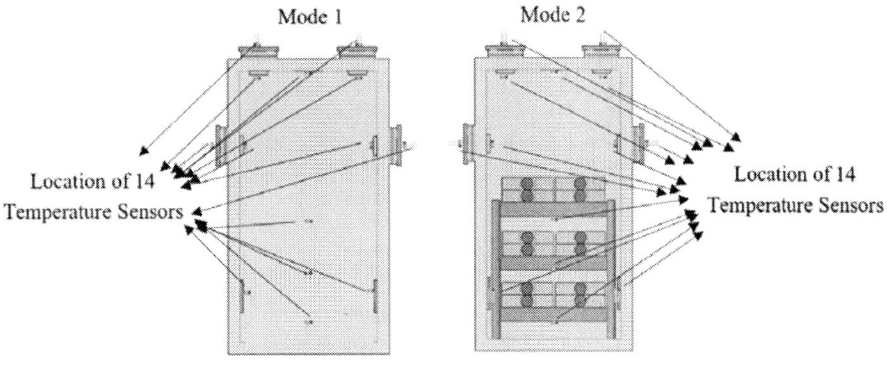

Figure 4. Location of temperature sensor on the CES cooler

2.3. Equation used

Comparing the work of a cooling system, means talking about the coefficient of performance (COP). From the air side, the cooling capacity (Q_c) and COP of the refrigeration unit can be evaluated by the following equation:

$$Q_c = c_{p,udara}\dot{m}\left(T_{in,udara} - T_{out,udara}\right) \quad (1)$$

Where $\dot{m}$ is the mass flow rate of air from the refrigeration unit, $T_{in,udara}$ and $T_{out,udara}$ are the inlet and outlet temperatures of the air flow through the refrigeration unit. $c_{p,udara}$ is the heat capacity of the air passing through the refrigeration unit which is taken at the average reference temperature $T_{in,udara}$ and $T_{out,udara}$.

$$COP_{unit} = \frac{Q_c}{P_{unit}} \quad (2)$$

Where Q_c is the cooling capacity generated from the cooling unit and P_{unit} is the total electrical power consumed by the thermoelectric and DC fan.

For the DC fan work function equation used:

$$v = -0{,}0184V^2 + 0{,}4636V - 0{,}7807 \quad (3)$$

Where v is the speed of the DC fan and V is the voltage supplied to the DC fan.

2.4. Tools and materials used

The measuring instrument for the test is shown in Figure 5 and the specifications are given in Table 1.

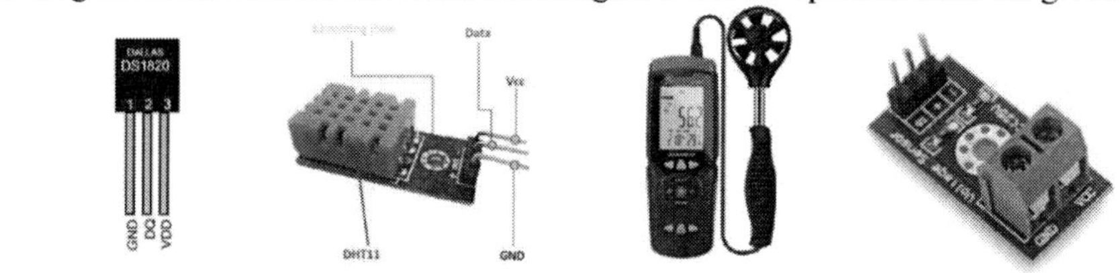

Temperature Sensor **Humidity Sensor** **Anemometer** **voltage sensor**

Figure 5. Measuring Tools

Table 1. Measuring Equipment Specifications

Equipment	Model	Range	Accuracy/resolution
Temperature Sensor	DS18B20	-55 +125 °C	± 0,5 °C
Humidity Sensor	DHT11	20 – 90% RH	±5% RH
Anemometer	GT8907	0 – 45m/s	± 3%.
Voltage Sensor	-	0 – 25 V	0,00489 V

The Peltier module used is the "TEC-12706" peltier module with specifications that can be seen in Table 2. For the Peltier work graph, it can be seen in Figure 6.

Table 2. Peltier Module Specification "TEC1-12706"

Hot Temperature Side (°C)	25°C	50°C
Q_{max} (Watts)	50	57
Delta T_{max} (°C)	66	75
I_{max} (Amps)	6.4	6.4
V_{max}(Volts)	14.4	16.4
Module Resistance	1.98	2.3

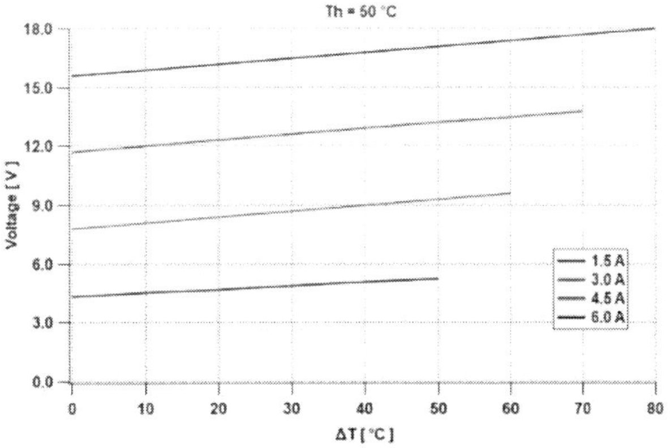

Figure 6. Peltier Work Function Graph

There is a DC Fan (8 X 8) cm with a power of 2.6 Watts. For the DC fan function graph in Figure 7, it appears that the Exp data points obtained through measurements using an anemometer in the voltage range of 2-12 V with 0.5 V increments. While the equation curve is the result of polynomial regression of the Exp data points so that equation (3) which is in section 2.3.

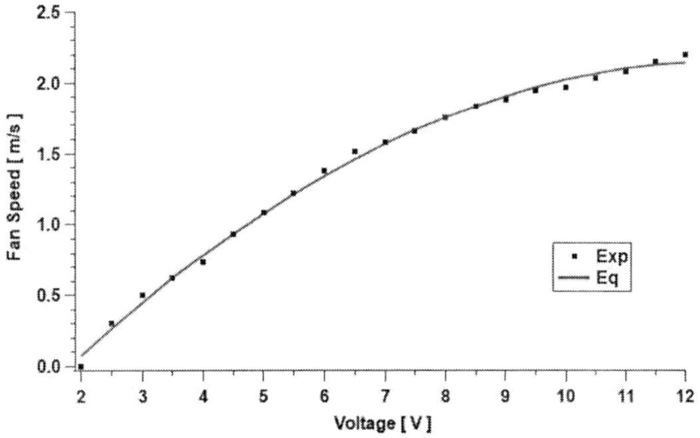

Figure 7. DC Fan Work Function Graph

The material used for the test is a material that changes the ice gel phase in 24 containers of 400 ml with thermophysical properties which can be seen in table 3 [10].

Table 3. Thermo-Physical Properties of Ice Gel

Melting temperature	-5 °C
Freezing temperature	-5 °C
latent heat	507952 J/ kg
Specific heat of liquid	4024 J/kg °C
Solid specific heat	1787 J/kg °C
Density	1000 kg/m^3
Liquid thermal conductivity	0,55677 W/m K @20 °C
Solid thermal conductivity	1,96 W/m K @-5 °C

3. Results

Experimental testing for the CES cooling system was conducted in Medan, North Sumatra. The test was conducted to determine the characteristics of the CES cooling system.

In this study, the temperature data in the CES box was analyzed using experimental data on working hours for 2 days in the mode 1 test (without PCM) from 02-03 October 2021. The graph can be seen as follows:

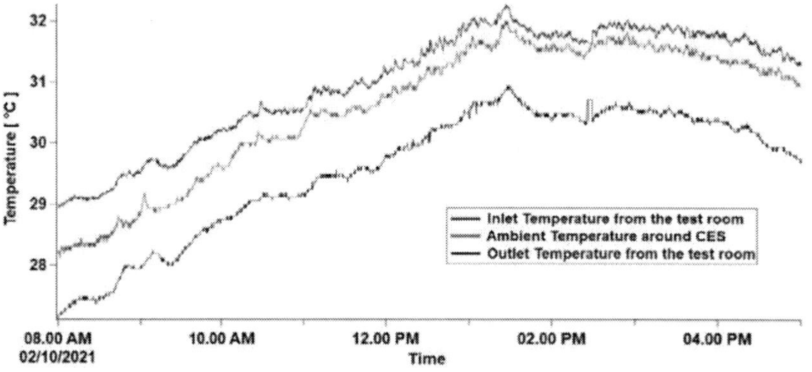

Figure 8. Temperature Graph without using PCM (02 October 2021)

It can be seen in Figure 8, from 08:00 to 17:00, the average inlet and outlet temperatures are 31°C and 29.57°C with the ambient temperature around CES having an average of 30.6°C. Based on the test results, the average difference between the outlet temperature for the room and the inlet temperature for the room is 1.43°C. This means that the performance of the peltier functions well to lower the air temperature of the room.

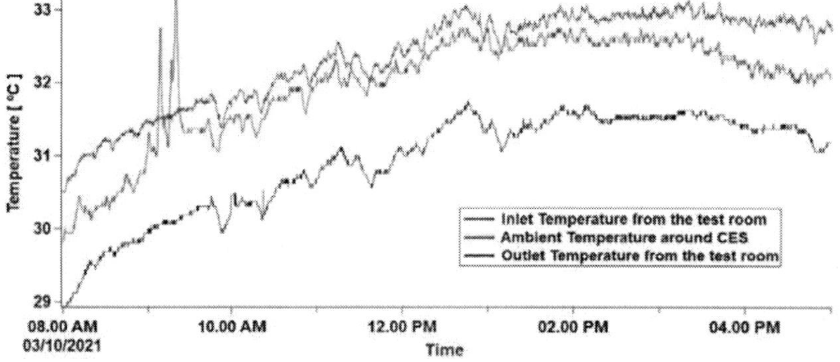

Figure 9. Temperature Graph without using PCM (02 October 2021)

It can be seen in Figure 9, from 08:00 to 17:00, the average inlet and outlet temperatures are 32.36°C and 30.85°C with the ambient temperature around CES being 32°C on average. Based on the test results, the average difference between the outlet temperature for the room and the inlet temperature for the room is 1.5°C. This means that the performance of the peltier functions well to lower the air temperature of the room.

In mode 2 testing, temperature data were analyzed using experimental data during 2-day working hours on 11-12 January 2022 using PCM. The graph can be seen as follows:

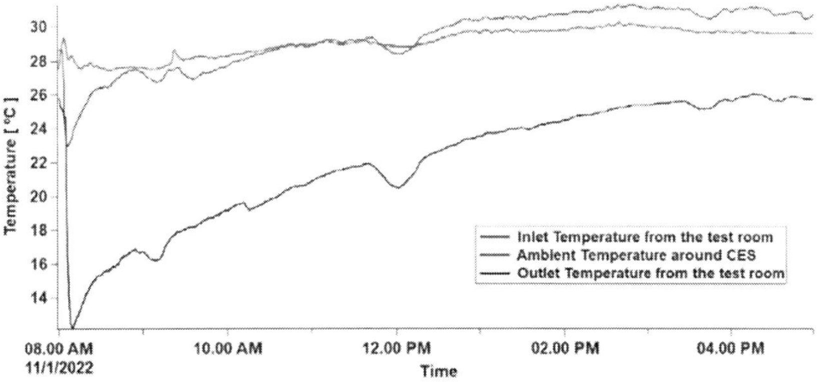

Figure 10. Temperature Graph using PCM (11 January 2022)

It can be seen in Figure 10, namely the inlet and outlet temperatures using PCM. Based on the test results, the difference in the outlet temperature for the room and the inlet temperature for the room can reach 11.63°C at 08:11. Where the outlet temperature can be maintained below 20°C is for 2.5 hours from 08:05 to 10:35. At 17:00 the outlet temperature difference for the room and the inlet temperature for the room becomes 5°C with an average of 7.36°C. With an average inlet and outlet temperature of 29.34°C and 22°C with an average ambient temperature of CES 29.12°C. This means that the performance of the PCM with the peltier functions optimally to reduce the air temperature from the room.

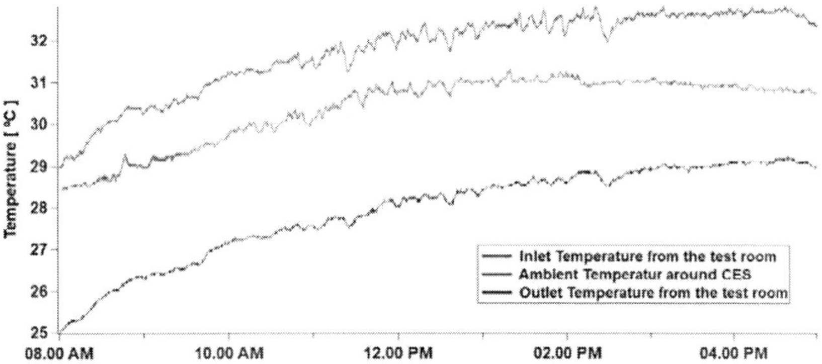

Figure 11. Temperature Graph using PCM (12 January 2022)

Seen in Figure 11, namely the inlet and outlet temperatures using PCM. Based on the test results, using PCM which was still placed in the box on the previous day's conditions, namely 11 January 2022, the difference in the outlet temperature for the room and the inlet temperature for the room has decreased by an average 3.85°C. With an average inlet and outlet temperature are 31.77°C and 27.9°C with an average ambient temperature of CES is 30.32°C. This means that the performance of the PCM is less able to absorb air heat in the ongoing test. But the performance of the PCM with the peltier still works well to lower the air temperature from the room.

The following chart compares the air temperature without PCM and using PCM is as follows:

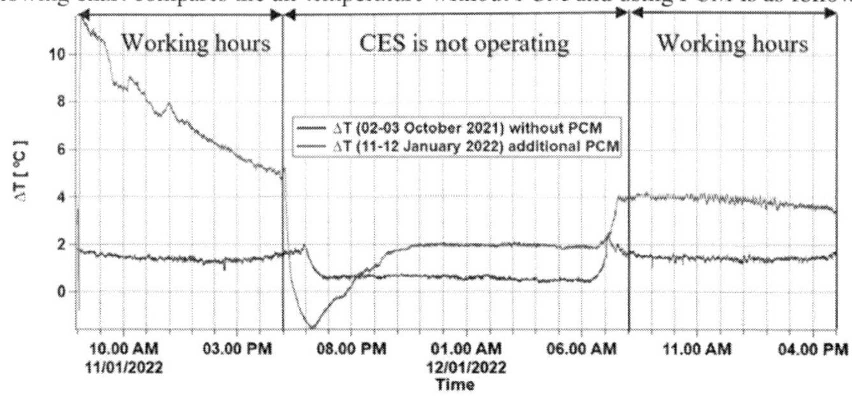

Figure 12. Comparison Graph of the Result Temperature difference on each day

Seen in Figure 12, which is a comparison graph of the difference in temperature produced on each day. This result is obtained from the temperature difference (ΔT) = inlet temperature (Tin) – outlet temperature (Tout). It can be seen from 08:00 WIB to 17:00 WIB on October 2, 2021, the temperature difference is 1.43°C, when CES is not operating the temperature difference decreases again because the peltier and fan do not work because solar radiation at that time is no longer there and the next day at 08:00 WIB to 17:00 WIB on October 3, 2021, the temperature difference is 1.5°C, and the performance of the peltier functions optimally to supply cold temperatures for the room.

On January 11, 2022 at 08:00 WIB to 17:00 WIB, the temperature difference is higher, this proves that by adding PCM, the outlet temperature given for the room is better along with the decrease in cooling power from 11.63°C to 5°C , when the CES is not operating the temperature difference decreases again, it can be seen at 18:00 WIB the temperature difference becomes -1.53°C at its lowest point because the cold air produced by the PCM spreads through the temperature sensor around the 2 inlet openings in the cooler box and for the next day at 08:00 WIB to 17:00 WIB on January 12, 2022, the temperature difference will be in the range of 4°C - 3.5°C, this is because the PCM is not able to absorb air heat in the ongoing test. Based on this, it can be concluded that by adding PCM, it can absorb heat better and produce better cold temperatures for the room than just depending on the peltier without PCM support.

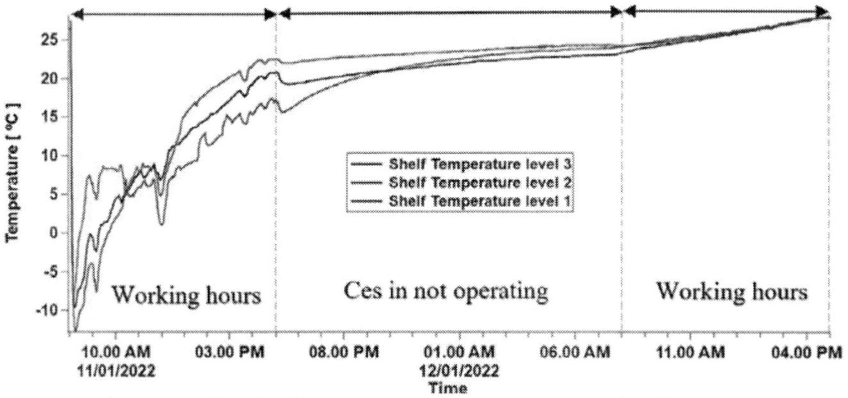

Figure 13. Graph of Temperature in the area near PCM for 2 days

Seen in Figure 13, the temperature in the area near the PCM. Based on the test results for 2 days, starting from 08:00 WIB in the morning on January 11, 2022 until 17:00 WIB in the afternoon on January 12, 2022, it can be seen that the shelf temperature difference at the three levels, at the first level is seen from the bottom of the shelf, the temperature rises faster because it is closer to forced convection from the fan, while the second level is further from the fan and the third level takes longer to rise because it is further exposed to forced convection from the fan than the first and second levels. It can be seen that the temperature of this level has stabilized again at 17:00 pm on January 12, 2022, because the PCM is no longer able to absorb air heat in the test with an average temperature of 28°C.

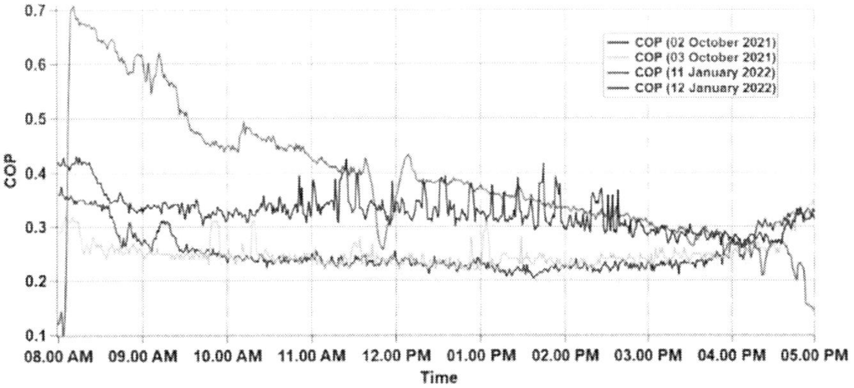

Figure 14. Comparison graph of COP values for each day on the CES system

Seen in Figure 14, which is a comparison graph of the COP (Coefficient Of Performance) value on each day. On October 2, 2021, there is a COP value with an average value of 0.25 and on October 3, 2021 there is a COP value with an average value of 0.25. On January 11, 2022, there is a good COP value with the highest value at 08:11 WIB, which is 0.7 and the COP value continues to decrease along with the ability to absorb heat from PCM which is already less able to absorb heat, then at 17:00 WIB the COP value has reached 0.14 with an average value of 0.39. On January 12, 2022, there was a COP value with an average value of 0.33 because the PCM entered on the previous day was not able to absorb heat properly anymore. It can be concluded that by adding PCM to the cold energy storage system, it can produce a better COP value than not using PCM.

4. Conclusions

In this study, tests have been carried out to evaluate the performance of the CES cooling system comparatively through testing mode 1 (without PCM) and testing mode 2 (with PCM). Several conclusions can be drawn from the test results:

- In the CES test without PCM, on the first and second days there were differences in the outlet temperature for the room and the inlet temperature for the room which were almost the same with an average of 1.43°C and 1.5°C. Meanwhile, CES and PCM on the first day have a difference of 11.63°C at 08:11 WIB where the outlet temperature can be maintained below 20°C for 2.5 hours starting from 08:05 WIB to 10:35 WIB and at 17:00 WIB the difference has become 5°C with an average of 7.36°C. On the second day, the temperature difference has decreased by an average of 3.85°C.
- Based on the results of the mode 2 test, it was found that the PCM can be used for 33 hours before finally reaching an average temperature of 28°C.
- The value of COP CES without PCM on the first and second days with an average value of 0.25. While the value of COP CES with PCM on the first day, with a range value of 0.14-0.7 with an average value of 0.39. For the COP value on the second day, with an average value of 0.33.

IOP Conf. Series: Earth and Environmental Science 1115 (2022) 012042 doi:10.1088/1755-1315/1115/1/012042

Acknowledgement
The researcher would like to thank Mr. Samar for the support of the tools and equipment used to this research in mechanical engineering, university of Sumatera Utara.

References
[1]. Pigliautile, I., Chàfer, M., Pisello, A. L., Pérez, G., & Cabeza, L. F. (2020). Inter-building assessment of urban heat island mitigation strategies: Field tests and numerical modelling in a simplified-geometry experimental set-up. *Renewable Energy*, 147, 1663–1675.
[2]. Souayfane, F., Fardoun, F., & Biwole, P. H. (2016). Phase change materials (PCM) for cooling applications in buildings: A review. In *Energy and Buildings* (Vol. 129, pp. 396–431).
[3]. Sarbu, I., & Sebarchievici, C. (2016a). Solar Heating and Cooling Systems: Fundamentals, Experiments and Applications. *Solar Heating and Cooling Systems: Fundamentals, Experiments and Applications*, 1–424.
[4]. Kenisarin, M. and Mahkamov, K., 2016. Salt hydrates as latent heat storage materials: Thermophysical properties and costs. *Solar Energy Materials and Solar Cells*, 145, pp.255-286.
[5]. Rasta, I.M. and Suamir, I.N., 2018. The role of vegetable oil in water based phase change materials for medium temperature refrigeration. *Journal of Energy Storage*, 15, pp.368-378
[6]. Gholamibozanjani, G., & Farid, M. (2020). Application of an active PCM storage system into a building for heating/cooling load reduction. *Energy*, 210.
[7]. Kozak, Y., Abramzon, B., & Ziskind, G. (2013). Experimental and numerical investigation of a hybrid PCM-air heat sink. *Applied Thermal Engineering*, 59(1– 2), 142–152.
[8]. Chiu, J. N. W., Gravoille, P., & Martin, V. (2013). Active free cooling optimization with thermal energy storage in Stockholm. *Applied Energy*, 109, 523–529.
[9]. Zhao, D., & Tan, G. (2014). Experimental evaluation of a prototype thermoelectric system integrated with PCM (phase change material) for space cooling. *Energy*, 68, 658–666.
[10]. Leonardo, H., Samar, & Ambarita, H. (2021). Simulation of Cold - Energy Storage Distribution Temperature Using Phase - Change Material. 01, 28–31.

IOP Conf. Series: Earth and Environmental Science 1115 (2022) 012043 doi:10.1088/1755-1315/1115/1/012043

Tree damage index in Taman Merdeka, Binjai City, North Sumatra Province

A G Ahmad*, M M Harahap and N I Putratama

Faculty of forestry, Universitas Sumatera Utara, Medan, Indonesia

*Email: alfan1@usu.ac.id

Abstract. Tree damage is one of the important indicators for tree health monitoring. Data on tree damage conditions will be very useful for city park managers to determine tree care measures. The purpose of this study was to determine the index of tree damage in Taman Merdeka, Binjai City, North Sumatra Province. The assessment of the three parameters (location of damage, type of damage, and level of damage severity) on the tree damage indicators refers to the Forest Health Monitoring method developed by the USDA Forest Service (1997). A total of 86 (eighty six) individual trees from 15 (fifteen) tree species growing in Taman Merdeka – Binjai City were successfully measured for their damage conditions. The results of this study indicate that overall the trees that grow in Taman Merdeka Binjai City are in good condition. This can be seen from the absence of trees that have a heavy damage index. The results of this study showed that as many as 47,67% of tree species growing in Taman Merdeka – Binjai City had a moderate damage index, dan as many as 52,33% had a low damage index. Five species of trees are known to have moderate damage, namely mahoni (*Swietenia macrophylla*), saga (*Adenanthera pavonina*), dadap merah (*Erythrina christagali*), ketapang kencana (*Terminalia mantally*), and mangga (*Mangifera indica*).

1. Introduction

Binjai is one of the cities in the province of North Sumatra, Indonesia. The city is located about 22 kilometers west of Medan City (the provincial capital of North Sumatra (Medan). The area of Binjai is 90,23 km^2 bordering with Deli Serdang District and Langkat District [1]. As a satellite city that continues to grow, Binjai has green open space in the form of an urban park, namely Taman Merdeka Binjai. The existence of this urban park in addition to affecting the beauty of the city, also has an important role in the sustainability of the city [2]. Trees play an important role in maintaining the sustainability of urban park. Trees are an essential component of urban nature, creating a vibrant aesthetic, contributing critical ecosystem services, and cultivating a sense of community safety Healthy Trees-Healthy People [3].

Development that is more directed towards physical development has ruled out the existence of green open spaces. This affects the physical condition of the tree by showing various signs of damage. Tree health needs to be maintained because if there is damage it will cause a decrease in the ability of trees to absorb carbon and other types of pollution so that it will have an impact on environmental quality [4].

Taman Merdeka - Binjai as one of the green open spaces that maintains the environmental health of Binjai City needs to be preserved. This can be done by maintaining the species of trees that compose it. One form of supervision that can be applied is to conduct a tree damage survey. This data is needed as a decision-making material for managers to take steps to maintain and prevent further damage. Thus the risk of accidents due to tree damage can be minimized. The purpose of this study was to obtain data about the damage that occurred to the trees in Taman Merdeka, Binjai.

2. Materials and methods

This research was conducted at the Taman Merdeka Binjai. Identification of tree diversity is done by collecting data on various species of trees found in city parks. Making a tree distribution map is done by taking the coordinates of each tree using GPS.

Damage recorded on each tree is a maximum of three damage. When multiple damage occurs in the same place then only the worst damage is recorded. Parameters used as an indicator of tree damage are location, type of damage and level of damage severity .

Damage assessment uses various criteria based on the FHM (Forest Health Monitoring) method [5]. Damage assessment data becomes the basic data in calculating the tree damage index value. The calculation uses equation 1.

$$\mathrm{LDI} = \sum_{i=1}^{n}(\mathrm{xi.yi.zi}) \quad (1)$$

Note: LDI: Level Damage Index, xi: damage type value, damage position value, zi: damage severity value. If the LDI is lower it is mean more healthy tree. This LDI is used to determine the class of damage tree. The final result is a tree damage index value which is divided into four classes with the following categories;

Healthy : 0 – 5
Low Damage : 6 – 10
Moderate Damage : 11 – 15
Heavy Damage : 16 – 20

3. Result and Discussion

This research determine the tree damage based on the data of damage type, damage position and damage severity. This damage will affect the physiological function of the tree, reduce the rate of tree growth and can cause tree death.

3.1. Composition and diameter structure of tree species

The results of data collection in the field can be seen as many as 15 species of trees growing in the Taman Merdeka, Binjai city with a total of 86 individual trees. Mahoni (*Swietenia macrophylla*) and saga (*Adenanthera pavonina*) are tree species with highest number of individual compositions, which are 32,56% and 31,40% respectively, as presented in Table 1.

Table 1. Composition and diameter structure of tree species

No.	Species	Number	Percentage	Diameter (cm)		
				Min	Max	Average
1	Mahoni (*Swietenia macrophylla*)	28	32,56%	24,00	106,00	60,76
2	Saga (*Adenanthera pavonina*)	27	31,40%	23,00	81,00	42,69
3	Akasia (*Acacia mangium*)	3	3,49%	28,00	115,00	45,33
4	Dadap merah (*Erythrina christagali*)	5	5,81%	24,00	28,00	25,57
5	Ketapang kencana (*Terminalia mantali*)	6	6,98%	25,00	33,00	28,25
6	Talok (*Muntingia calabura*)	1	1,16%	22,00	22,00	22,00
7	Krei Payung (*Filicium deficiens*)	1	1,16%	40,00	40,00	40,00
8	Mangga (*Mangifera indica*)	6	6,98%	21,00	41,00	27,92
9	Ketapang (*Terminalia catappa*)	1	1,16%	21,00	21,00	21,00
10	Matoa (*Pometia pinnata*)	1	1,16%	21,00	21,00	21,00
11	Tanjung (*Mimisop elengi*)	2	2,33%	24,00	26,00	25,00
12	Beringin Putih (*Verigata benzamine*)	1	1,16%	25,00	25,00	25,00
13	Angsana (*Pterocarpus indicus*)	1	1,16%	85,00	85,00	85,00
14	Pulai (*Alstonia scholaris*)	2	2,33%	28,00	29,00	28,50
15	Karet Hutan (*Hevea brasliensis*)	1	1,16%	48,00	48,00	48,00
	Total	**86**	**100,00%**			36,40

Based on tree diameter data, it can be seen that the tree species that grow in Taman Merdeka Binjai city have all been included in the tree category. It is known that there are 5 species of trees that have sindividual trees with diameters above 40 cm, namely mahoni (*Swietenia macrophylla*), saga (*Adenanthera pavonina*), akasia (*Acacia mangium*), krei payung (*Filicium decipiens*) dan angsana (*Pterocarpus indicus*).

3.2. Damage type

The results of observations of the visible signs of damage, it was found that there were 9 type of tree damage in Taman Merdeka Binjai City. There are three type of tree damage that are commonly found, namely: open wounds, loss of apical dominance or dead terminal, and broken or dead, as presented in Table 2. The number of types of tree damage found at Taman Merdeka Binjai is the same as the amount of damage to the Saga tree on the USU campus [6]. The tree damage found in Taman Merdeka Binjai was less than the damage found on the Mahoni Tree on The USU campus [7].

Table 2. Type of tree damage at Taman Merdeka Binjai City

No.	Damage type	Percentage (%)
1	Cancer	5,02
2	Conks, Fruiting bodies, and other indicator of advanced decay	10,00
3	Open wounds	30,00
4	Resinosis / gummosis	2,30
5	Brooms on roots or bole	8,10
6	Loss of apical dominance, dead terminal	20,00
7	Broken or dead	20,00
8	Damage foliage or shoots	1,10
9	Discoloration of foliage	3,48
	Total	**100,00**

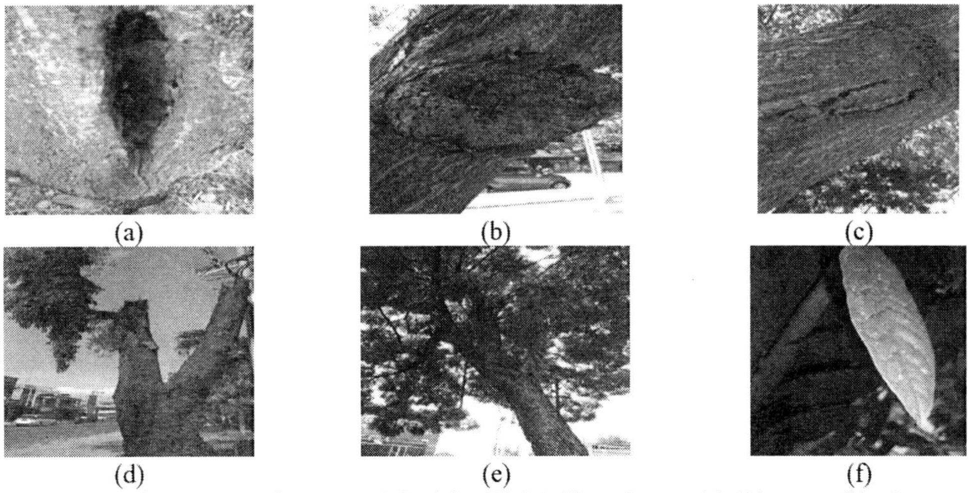

Figure 1. Tree damage type in Taman Merdeka Binjai City: Cancer (a), Liver rot (b), Open wounds (c), Brum on root or stem (d), Loss of dominant tip (e), Leaf color change (f)

3.3. Damage location

Based on Forest Health Monitoring (FHM) the location of damage to trees is divided into nine of damage locations. The results of field observations showed that the most tree damage in Taman Merdeka Binjai City was found at location 4 (lower and upper boles), as presented in Figure 2.

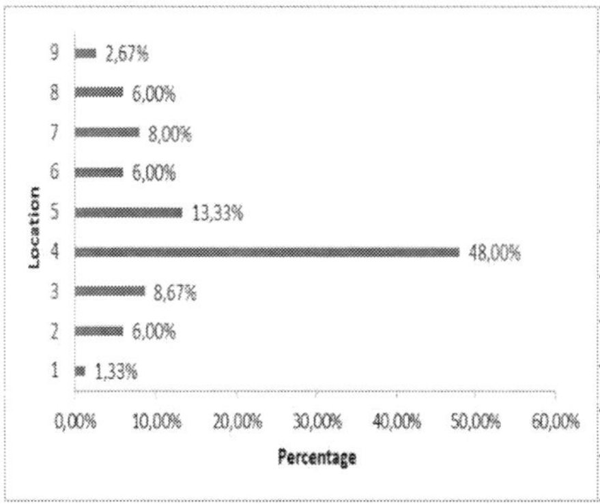

Code	Location
0	No damage
1	Exposed roots and "stump" (12 inch (30 cm) in height from ground level)
2	Roots and lower bole
3	Lower bole (lower half on the trunk between the "stump" and base of the live crown)
4	Lower and upper boles
5	Upper bole (upper half on the trunk between "stump" and base of the live crown)
6	Crown stem (main stem within the live crown area, above the base of the live crown)
7	Branches (woody stems other than main stem)
8	Buds and shoots (the most recent year's growth)
9	Foliage

Figure 2. Location of tree damage in Taman Merdeka Binjai City

3.4. *Tree damage index*

The level of tree damage in Taman Merdeka Binjai City is presented in more detail in Table 3. Generally, the trees in Taman Merdeka Binjai City are in good condition. No trees were seriously damaged. As many as 52,33% of trees found low damage, and 47,67% found moderate damage.

Tabel 3. Tree damage index in Taman Merdeka Binjai City

No.	Species	Tree Damage Level Index							Number of Individu
		low damage	% Total species	% Each Species	Moderate damage	% Total	% Each species	Heavy damage	
1	Mahoni (*Swietenia macrophylla*)	13	28,89%	46,43%	15	36,59%	53,57%	0	28
2	Saga (*Adenanthera pavonina*)	7	15,56%	25,93%	20	48,78%	74,07%	0	27
3	Akasia (*Acacia mangium*)	3	6,67%	100,00%	0	0,00%	0,00%	0	3
4	Dadap merah (*Erythrina christagali*)	1	2,22%	20,00%	4	9,76%	80,00%	0	5
5	Ketapang kencana (*Terminalia mantali*)	5	11,11%	83,33%	1	2,44%	16,67%	0	6
6	Talok (*Muntingia calabura*)	1	2,22%	100,00%	0	0,00%	0,00%	0	1
7	Krei Payung (*Filicium deficiens*)	1	2,22%	100,00%	0	0,00%	0,00%	0	1
8	Mangga (*Mangifera indica*)	5	11,11%	83,33%	1	2,44%	16,67%	0	6
9	Ketapang (*Terminalia catappa*)	1	2,22%	100,00%	0	0,00%	0,00%	0	1
10	Matoa (*Pometia pinnata*)	1	2,22%	100,00%	0	0,00%	0,00%	0	1
11	Tanjung (*Mimisop elengi*)	2	4,44%	100,00%	0	0,00%	0,00%	0	2
12	Beringin Putih (*Verigata benzamine*)	1	2,22%	100,00%	0	0,00%	0,00%	0	1
13	Angsana (*Pterocarpus indicus*)	1	2,22%	100,00%	0	0,00%	0,00%	0	1
14	Pulai (*Alstonia scholaris*)	2	4,44%	100,00%	0	0,00%	0,00%	0	2
15	Karet Hutan (*Hevea brasliensis*)	1	2,22%	100,00%	0	0,00%	0,00%	0	1
	Total species	45			41			0	86
	Percentage	**52,33%**			**47,67%**			**0%**	

4. Conclusion

In general, the condition of trees in the Taman Merdeka Binjai City is still in good condition. However, to optimize the function of tree, it is necessary to make better maintenance efforts. One of the is through pruning. In addition, public education is also needed so that current tree damage can be reduced.

References

[1] *Statistic of Binjai Municipality 2022* (Binjai Municipality in Figures 2022)
[2] Cheisura A 2004 *Landscape Urban Planning* **68** 129
[3] Norman-Burgdolf H and Rieske L K 2021 *Urban Forestry and Urban Greening* **60**
[4] Nuhamara S T and Kasno 2001 *Forest Health Monitoring to Monitor the Sustainability of Indonesia Tropical Rain Forest Volume I* (Bogor: SEAMEO BIOTROP)
[5] Mangold R 1997 *Forest Health Monitoring Field Methods Guide* (Washington: USDA Forest Service)
[6] Susilowati A, Ahmad A G, Rangkuti A B and Harahap M M 2020 *IOP Conf. Series: Earth and Environmental Science* **782**
[7] Susilowati A, Ahmad A G, Silalahi V and Rachmat H H 2018 *IOP Conf. Series: Earth and Environmental Science* **203** 012028

IOP Conf. Series: Earth and Environmental Science 1115 (2022) 012044 doi:10.1088/1755-1315/1115/1/012044

Identifying management strategies for the development of wildlife-based ecotourism: A study case of Sulawesi bear cuscus (*Ailurops ursinus*) inhabited the Educational Forest of Hasanuddin University, Maros

A U B Putri [1], R I Maulany [2] and A Achmad [2]

[1] Graduate Student of Forestry Department, Forestry Faculty, Hasanuddin University. Makassar
[2] Laboratory of Forest Conservation and Ecotourism, Forestry Faculty, Hasanuddin University. Makassar

*Email: risma.maulany@unhas.ac.id

Abstract. The educational forest of Hasanuddin University is intended to be developed as ecotourism spot with various physical and biological tourism attractions including bear cuscus (*Ailurops ursinus*). However, the concrete management strategies required have not been fully explored and examined. This research aimed to identify and formulate management strategies of wildlife-based tourism for bear cuscus in the area by examining bio-ecological of the species and analyzing community's perceptions and attitudes towards bear cuscus and its conservation in the area. Data on the biology and ecology of *A. ursinus* and its habitat were gained from literatures while the perceptions were obtained directly from questionnaire distribution to all relevant stakeholders as respondents. SWOT analysis was used and further analyzed by using AHP (Analysis Hierarchy Process) to determine quadrants and priority levels. There were 11 strengths and 4 weaknesses found as internal factors and 5 opportunities and 3 threats for the external. The IFAS matrix score were 1.82 with the weighted total value of 3.29 and the weakness' score of 1.47. For the EFAS matrix, the total opportunities score was 2.95 and the total threats score of 1.00. The high priority strategy to be implemented in the context of ecotourism management was located in Quadrant I which indicates that the opportunity for the bear cuscus' ecotourism is favorable.

1. Introduction

Cuscus is known as marsupial mammals from the family of Phalangeridae [1] and has limited distribution in the world. The species of this family can only be found in the eastern parts of Indonesia (Papua, Maluku, and Sulawesi), Australia, and Papua New Guinea with the total of 6 genera (Ailurops, Phalanger, Spilocuscus, Stigocuscus, Wyulda, and Trichosurus) [2,3]. In Indonesia, only four genera existed: Ailurops, Phalanger, Spilocuscus, and Stigocuscus [4].

Sulawesi bear cuscus (*Ailurops ursinus*) of the Genus Ailurops is one of the endemic marsupials in the Wallacea region. The species is strictly distributed in Sulawesi and small islands adjacent to the mainland such as Peleng, Muna, Buton, and Togean [1,4,5]. Based on the assessment conducted by the International Union for Conservation and Nature (IUCN), the species has been categorized as *Vulnerable* due to decreasing number of mature individuals in the wild [1].

Currently, rapid population declines from many places were reported due to high threats from illegal hunting and trading as well as habitat loss and destruction, reasons behind the activities were related to the use of species for food and protein intake resources, rituals and special occasions by local people, also to be sold as pets in the local markets [6-8]. However, this species is not regulated under CITES (Convention on International Trade of Endangered Flora and Fauna Species). Meanwhile, at the regional level, previously all species under the Family of Phalangeridae were protected under the Government Regulation No. 7/1999 regarding the preservation of flora and fauna, but then in the revised version enacted by the Ministry Regulation No. 92/2018 on list of Indonesian's protected species, *A. ursinus* is no longer mentioned [9]. This means both at the national and international levels, the protection for the species is not well guaranteed.

Wildlife-based ecotourism is one way of non-consumptive utilization of the animal species for sustainable use of natural resources. Wildlife tourism offers various experiences not only to observe but also to have direct involvement with the animals in real-life situation which can be difficult as they considered threatened, endangered, or even rare and with wildlife-based tourism this can be made possible [10-13]. Although wildlife ecotourism has not been well developed in Indonesia, but it has been proven in some developed countries to have positive long-term impact for the conservation of wildlife and their habitats [12,14].

In wildlife-based ecotourism management, the role of the community in supporting conservation efforts is very important. The community, especially those who live in the vicinity of the wildlife habitat, needs to participate directly to maximize the potential benefits of this type of tourism [15]. The form of community participation can be seen from the perceptions and attitudes shown towards the presence of these wild animals [16]. Perceptions and attitudes regarding ecotourism management are required in order to build a positive attitude to the sustainability of tourism development as well as to change perceptions of the community into more positive direction towards wildlife to enable more effective conservation particularly for problematic keystone species and species in conflict with human [17].

The educational forest of Hasanuddin University (Unhas) is identified as one of the habitats of *A. ursinus* in the region [18]. The forest is intended to be developed as ecotourism spot with various physical and biological tourism attractions including bear cuscus (*A. ursinus*). Some studies on the species have been conducted focusing on daily behavior, nesting habitat, type of diets, diet preferences, and potential of diet's distribution across the forest [19-24]. However, the concrete management strategies required have not been fully explored and examined. In particular, no information was gathered on the community's views on both bear cuscus and its conservation, as well as on the idea of developing wildlife-based ecotourism. Therefore, this research aimed to identify and formulate management strategies of wildlife-based tourism in particular bear cuscus in the educational forest of Unhas by examining the bio-ecological of the species and analyzing the community's perceptions and attitudes towards bear cuscus and its conservation in the area.

2. Materials and methods

2.1. Study Area

This study was conducted in four months between September to December 2020. The study site was located in the Educational Forest of Hasanuddin University, Maros Regency (South Sulawesi) around 65 km from the capital city of South Sulawesi Province (Makassar) (Figure 1). Geographically, the area lies between 119°44''34' - 119°46''17'E and 04°58''07' - 05°00''30'S. The average rainfall occurred in the area was 162.5 mm with daily temperature ranging from 18.4°C to 38.3°C and relative humidity of 76.25%.

IOP Conf. Series: Earth and Environmental Science 1115 (2022) 012044 doi:10.1088/1755-1315/1115/1/012044

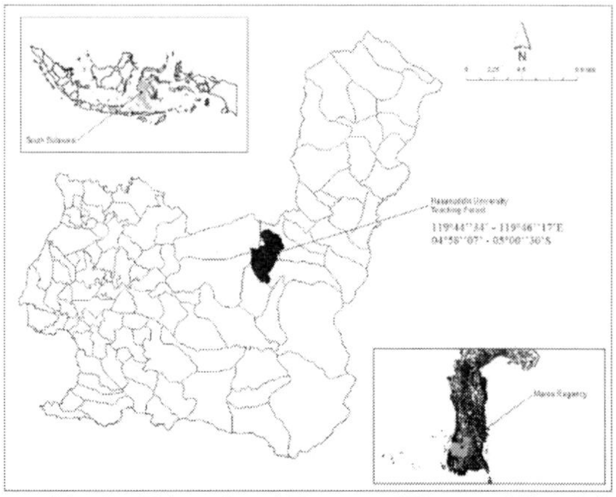

Figure 1. Study site located in the Educational Forest of Hasanuddin University, Maros Regency

2.2. Data Collection and Analysis

This study consisted of three parts: 1) examining biological and ecological conditions of Sulawesi bear cuscus (*A. ursinus*) and its habitats; 2) analyzing public perceptions towards *A. ursinus*; and 3) identifying and formulating management strategies for the development of wildlife-based ecotourism with Sulawesi bear cuscus as the main attraction. Data on the biology and ecology conditions of *A. ursinus* as well as the habitats were gained through exploration of various literatures while the perceptions were obtained directly from the field through questionnaire distribution to all relevant stakeholders as respondents (local community nearby the site, the management of educational forest, government institutions related with the issues, and potential tourists).

The local community sampled nearby the area was 30 respondents selected based on the encounter opportunities and interactions with *A. ursinus*. Information from the management of the educational forest was gained from the main actor (the manager). The government institution interviewed was represented by 7 respondents. Potential tourists were also surveyed to overview the development of wildlife-based ecotourism in the area (34 respondents).

Once the data on biological-ecological conditions and the habitat of *A. ursinus* and public perceptions were compiled and analyzed, the next step was to identify and formulate management strategies by using SWOT analysis. In this particular section, the first step was to identify internal and external factors influencing the management which was continued by scoring and rating the factors. Then, the results were analyzed to later composing matrices of IFAS (Internal Factors Analysis Strategy) and EFAS (External Factors Analysis Strategy). Then, each factor identified was further analyzed by using AHP (Analysis Hierarchy Process) method to determine the priority level. There were 3 respondents who were academicians involved and are experts in the relevant field. This was used to determine quadrants and build the SWOT matrix.

3. Result

3.1. Eco-biology Condition and Habitats of A. ursinus

The educational forest of Hasanuddin University (Unhas) is a designated forest area established for specific purposes granted by the Ministry of Forestry to the Forestry Faculty of Unhas in 1980 based on the decree N0. 86/Menhut II/2005. With the total of 1,400 ha, the area has been used as the main learning source and conducting practicals for students from the faculty. Apart from that, limited recreational activities have been also applied in the area.

The site has become the habitats for some endemic Sulawesi fauna including bear cuscus (*Ailurops ursinus*). Based on the literatures, the individuals of the species were commonly found in Block 37-38 of the educational forest or nearby Rompegading Village. Number of individuals recorded were less than 10 individuals per km2 with more than 50% of daily activities of each individual contributed for resting. The species was also known to be restricted in the secondary and pine forests with close distance movement (78m2-357m2). As an arboreal animal, *A. ursinus* utilized 10 species of trees for nesting and 9 species of foraging sources. Potential trees with >15 m height for both nesting and foraging can be considered sufficient for the above population with around 49 individuals/0.6 ha of diet trees and 89 individuals/0.6 ha (Table 1).

Table 1. Bio-ecological conditions and habitat of *A. ursinus* in the Educational Forest of Unhas

Criteria Assessed	Indicators	Sources
Bio-ecology Conditions of *Ailurops ursinus*		
Number of population	7 - 8 individuals/ km^2	Achmad et al. (2016); Jathi, (2020)
Dominant daily behavior	Resting behaviour 97% (52-85%)	Mangalla (2014); Jathi (2020)
Dominant activity budget	Resting behaviour 22% (85-92.22%)	
Species movement	78 m^2 – 357 m^2	Achmad et al. (2016); Fauzan (2020)
Homer range	Secondary Forest and Pine Forest of Teaching Forest Area	Fauzan (2020)
Habitat conditions of *Ailurops ursinus*		
Status of the area	Forest Area for Specific Purpose	The decree of Ministry of Forestry No.86 / Menhut II / 2005 concerning changes to the Decree of the Director General of Forestry No. 63 / Kpts / BS / 1/1980 dated March 31, 1980
Habitat conditions	Natural habitat	Alamsyah (2015); Fauzan (2020)
Land cover	Mixed forest (primary and secondary forest) Secondary Forest and Pine Forest	
Forest ecosystem type	400 - 700 m above sea level included in lowland forest (<600 asl) and sub montane forest (600 - 900 asl)	Ground check
Vegetation density at tree level	89 ind/0.6 ha	Hidayat (2015); Aziz (2020)
Diet tree density (tree diameter)	49 ind/ 0.6 ha	
Species of Diet tree	9 species (*Mangifera indica, Palaquim obovatum, Dracontomelon dao, Diospyrus celebica, Spondias pinnata, Ficus sp., Cinnamomum sp., Arthrophyllum diversifolium* and Liana)	
Species of nesting tree	10 species (*Aleurites moluccana, Arthrophyllum diversifolium, Bauchanania arborescens,*	Alamsyah, (2015); Fauzan (2020)

Criteria Assessed	Indicators	Sources
	Cinnamomum inner, Dracontomelon dao, Diospyrus celebica, Ficus sp, Myristica fragrans, Palaquium obovatum, and *Pinus merkusii*)	
Position of *A.ursinus* on the tree stratum	Stratum c	
Nesting tree characteristics	Tree with the height between 15 m to 25 m, diameter > 0,5 m, tree canopy width of 86,54 m²-455,93 m²	
Distance from water sources	No influences of individual cuscus distances to water sources	Dimomonmau (2000) in Fauzan (2020)

3.2. Public Perception and Attitudes towards A. ursinus

3.2.1. Characteristics of Respondents. To obtain information on public perceptions and attitudes towards A. ursinus and its conservation, 30 respondents living nearby the educational forest were questioned along with 7 respondents from the government agencies. There were also 34 prospective tourists interviewed to gain information on the potentials of wildlife-based ecotourism in the area. The manager of the educational forest of Unhas was also included as key informant. In general, the community nearby the area mostly only completed their elementary schools (n=22 respondents). In contrast, 76% of the respondents from prospective tourists were graduated from colleges (n=26 respondents). All respondents of the government agencies and the educational forest held undergraduate or master's degrees (Figure 2a). Around 70% of the local people interviewed worked as farmers. While the prospective tourists had various occupations but were dominated by college students (38.2%) (Figure 2b). All respondents from government agencies and the educational forest were civil servants.

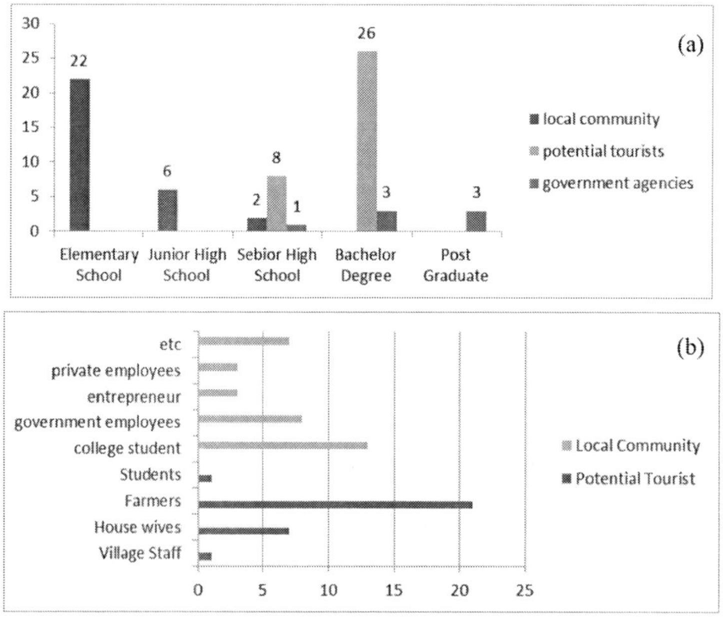

Figure 2. Profile of respondents based on (a) educational level; (b) types of occupation

IOP Conf. Series: Earth and Environmental Science 1115 (2022) 012044 doi:10.1088/1755-1315/1115/1/012044

3.2.2. Local Community Nearby the Study Area. From the interview sessions with the local community, it was revealed that more than 50% of the respondents had basic knowledge on bear cuscus as an endemic species, changes in its population over time, and the impact on the existence of bear cuscus. However, there was 83.3% of the respondents had no knowledge on the important of bear cuscus to the environment. More than half of the respondents also had limited understanding on conservation including the conservation of bear cuscus (Figure 3a).

In term of attitudes of local community nearby the habitat of bear cuscus, there were 96.7% of the locals had stated that they never involved in hunting activities on bear cuscus. Even though more than 90% of respondents had an understanding that bear cuscus can be consumed and hunted. All respondents have been recorded to agree with conservation activities but only 83.3% agreed to be involved in the activities. In general, most of the respondents (96.7%) approved with the idea of having bear cuscus as part of ecotourism activities in the area (Figure 3b).

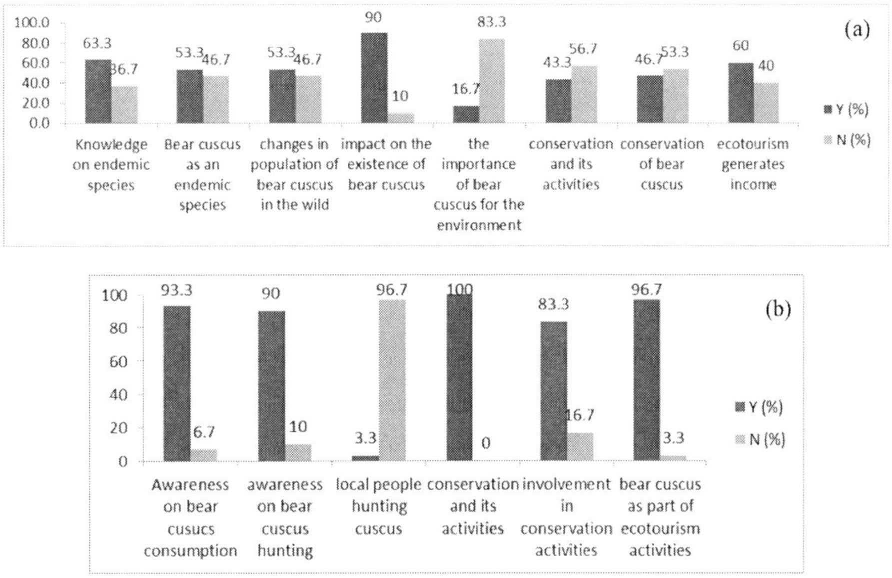

Figure 3. Local community's perceptions (a) and attitudes (b) towards knowledge on bear cuscus existence and conservation in the Educational Forest of Hasanuddin University, Maros

3.2.3. Potential Tourists. In relation with the development of wildlife-based ecotourism with the bear cuscus as the centre of attractions, some questions were also asked to potential tourists on basic knowledge related with the species and its conservation. Almost the majority of respondents were well equipped with on basic knowledge of bear cuscus (>90%) although not all of the respondents had sufficient information regarding the changes of its current population (only 70.6%). Surprisingly, around 85.3% had known the educational forest as one of the habitats for bear cuscus in the region [Figure 4a]. For the attitudes of prospective tourists, it can be seen that all respondents agree with the conservation of bear cuscus and ecotourism development plan (100%). However, only around 3-8% of the respondents had disagreed on the idea of making the bear cuscus and animal as tourism objects [Figure 4b].

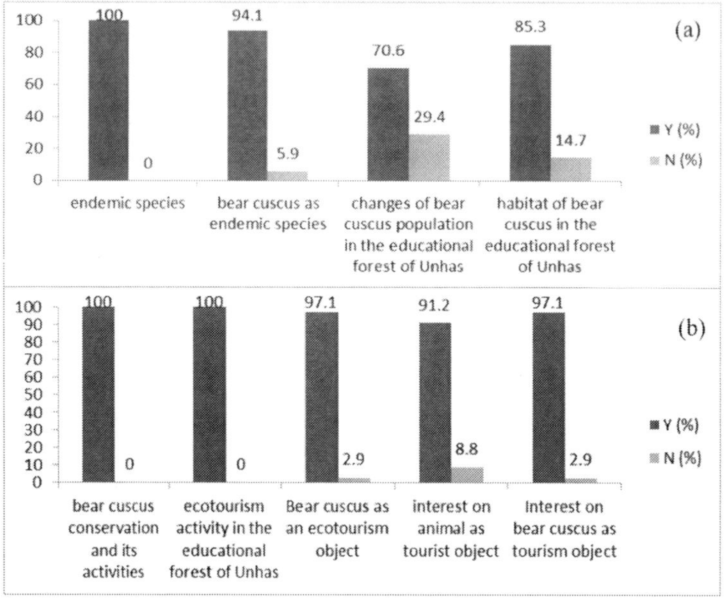

Figure 4. Perceptions (a) and Attitudes (b) of potential tourists on bear cuscus, its conservation and ecotourism development

3.2.4. Government Institutions. To portray on the stakeholders, some staff of relevant government institution were interviewed. In term of knowledge about endemic animals, 100% of respondents had fully understanding on the matter but only 75% had been aware that bear cuscus as endemic animals. Only 62.5% of the respondents had knowledge on changes occurred in wild bear cuscus population and around 87.5% have not heard the educational forest of Unhas as one of the habitats for bear cuscus in the province [Figure 5a].

Different with both the locals and potential tourists, all staff interviewed had agreed to be involved and contributed to the conservation of bear cuscus (100%). While 87.5% have agreed the idea of ecotourism development plan in the educational forest of Unhas with only 50% supported the activities and made the bear cuscus as an object of ecotourism [Figure 5b]

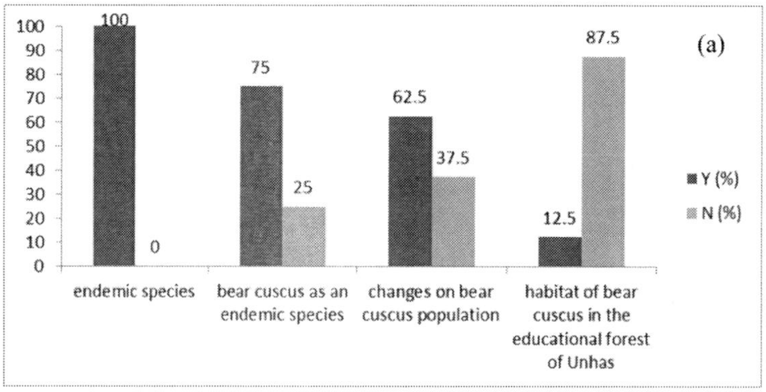

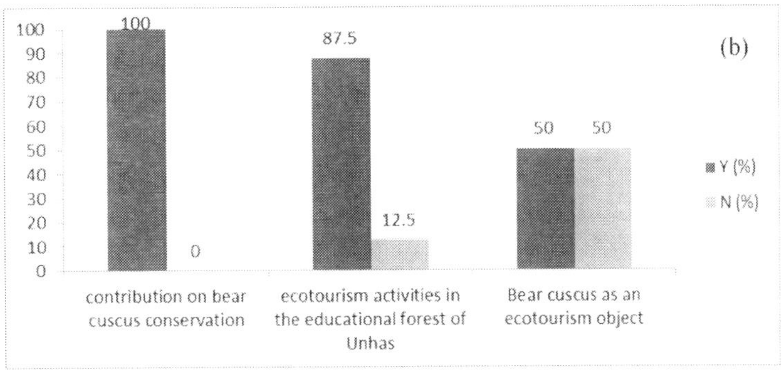

Figure 5. Perceptions (a) and (b) Attitudes of government institutions towards bear cuscus, its conservation and ecotourism development

3.2.5. Management of the Educational Forest. In order to further implement the idea of developing ecotourism in the educational forest of Unhas with bear cuscus as the main attraction, supporting data related with the management of the area were needed [Table 2]. The data were gained from a deep interview with the manager of the educational forest of Unhas. In general, the management of the Educational Forest of Unhas had agreed on the idea of ecotourism development in the area including making bear cuscus as the main attraction. This also strengthen with some facts that some contribution have been made by the management which also involved various intensive research, internal management policy, the availability of human resources, including facilities and infrastructure. However, there were no networking established and no collaboration have been made in accordance with the plan.

Table 2. Supporting data by the Educational Forest of Unhas, Maros

Indicators	Y	N
Contribution on conservation activities of Sulawesi bear cuscus	√	
Attitude on ecotourism activities in the Educational Forest of Unhas	√	
Attitude on the idea of bear cuscus as the main object of ecotourism	√	
The availability of internal policies on ecotourism management	√	
Networking and collaboration with other stakeholders specific on bear cuscus		√
The availability of human resources	√	
The availability of basic facilities and infrastructures for ecotourism	√	
Specific records on number of tourists visit bear cuscus in the Educational Forest of Unhas	√	
The availability of previous programs for the local community to support the development of ecotourism	√	

3.3. Ecotourism Management Strategies

In this research, ecotourism management strategy of cuscus was determined by using a SWOT analysis. The formulation process of management strategies was consisted of three stages: 1) identification stage of internal factors (IFAS) and external factors (EFAS); 2) analysis stage; and 3) decision-making stage. In the first stage, internal factors were used to measure the extent of the strengths and weaknesses of the eco-biological conditions and habitat of bear cuscus as well as the perceptions and attitudes of forest managers. While then, the external factors were used to summarize and evaluate on opportunities and threats related with the perceptions and attitudes of all involved actors (local community nearby, potential tourists and government agencies) in the development of ecotourism. There were 11 strengths

IOP Conf. Series: Earth and Environmental Science 1115 (2022) 012044 doi:10.1088/1755-1315/1115/1/012044

and 4 weaknesses found as internal factors (Table 3). For the external factors, 5 opportunities and 3 threats were identified (Table 4).

Table 3. Internal Factors Identified and Internal Factors Analysis Strategy (IFAS) Matrix of Wildlife-based Ecotourism Development in the Educational Forest of Unhas

	Internal Factors	Weight (B)	Rating (N)	Score (BXN)
	Strength factors			
S1	Stable bear cuscus population	0.10	3.7	0.35
S2	Resting behavior as the dominant daily behavior performed by bear cuscus	0.09	3.3	0.29
S3	Data on the species of diet plants consistently consumed by bear cuscus	0.07	3.3	0.25
S4	Data on nest tree species consistently used by bear cuscus including its nest tree criteria	0.11	3	0.33
S5	Information on the habitat inhabited by bear cuscus	0.07	4	0.30
S6	Sufficient foraging plants based on the vegetation survey	0.07	3.3	0.25
S7	No land cover change occurred in the habitat of bear cuscus	0.07	4	0.30
S8	The availability of internal policy regarding ecotourism management in the Educational Forest of Unhas	0.10	3	0.31
S9	The availability of human resources for ecotourism management in the Educational Forest of Unhas	0.10	3	0.31
S10	Sufficient facilities and infrastructures for supporting ecotourism activities in the Educational Forest of Unhas	0.10	3	0.31
S11	Data on visitors to the Educational Forest of Unhas	0.10	3	0.31
	Total	1.00		3.29
	Weakness factors			
W1	Limited number of bear cuscus individuals found in the area	0.22	1	0.22
W2	Irregular homerange of bear cuscus	0.31	1	0.31
W3	No networking and collaboration established with the management of the Educational Forest of Unhas on bear cuscus ecotourism	0.23	2	0.47
W4	No specific infrastructures built for bear cuscus wildlife-based ecotourism (e.g. observation deck/tower)	0.23	2	0.47
	Total	1.00		1.47
	Score Strengths - Weaknesses> IFAS =			1.82

Table 4. External Factors Analysis Strategy (EFAS) Matrix of Wildlife-based Ecotourism Development in the Educational Forest of Unhas.

	External Factors	Weight (B)	Rating (N)	Score (BXN)
	Opportunity factors			
O1	High percentage of local community agreed to make bear cuscus as an object of ecotourism	0:21	4	0.82
O2	Positive response of the community to get involved in ecotourism management	0.21	4	0.82
O3	No hunting on bear cuscus occurred by the local community	0.21	4	0.82

	External Factors	Weigh t (B)	Rating (N)	Score (BXN)
	Opportunity factors			
O4	High interests of potential tourists for ecotourism and in particular bear cuscus as the main attraction	0.22	4	0.87
O5	Support from government agencies towards the development of ecoutourism	0.16	3.7	0.61
	Total	1.00		3.95
	Threat factors			
T1	The local community been aware the possibility of bear cuscus to be hunted and consumed	0.33	1	0.33
T2	Low public perception on the importance of bear cuscus for the environment	0.33	1	0.33
T3	Low community perception on the concept of ecotourism	0.33	1	0.33
	Total	1.00		1.00
	Score of Opportunity - Threats> EFAS =			2.95

After identifying the internal and external factors, the weight of the internal and external factors were calculated in order to determine the quadrant which later also decide the priority focus of bear cuscus ecotourism management strategies. The calculation of the factor weight was done by tabulating the IFAS - EFAS (Internal - External Strategic Factor Analysis Summary) score [Tables 3 and 4].

The results of the identification have shown that the management of bear cuscus ecotourism in the Educational Forest of Unhas had 11 advantages as summarized in Table 2. The highest ranking gained from the greatest strength aspects which were the stable population of bear cuscus and the presence of consistent bear cuscus' nest tree species information. While the weaknesses were found to be limited individuals of bear cuscus and the irregularity of its home range.

Meanwhile, the management strategies based on external conditions were identified by five opportunities and three threats, each with total threats scores lower than the opportunity score [Table 3]. The results of the IFAS matrix analysis were 1.82, including the weighted total value with an average score of 3.29 and weaknesses of 1.47. For the EFAS matrix analysis, the total opportunities score was 2.95 from the total score value of 3.95 and the total threats score of 1.00. This shows that these factors need to be considered, especially in the quality of conditions for ecotourism purposes.

Determination of the priority of strategies for the management of bear cuscus ecotourism in the Educational Forest of Unhas can be seen by using a graph gained from SWOT analysis, which combines the two weighted values of the values obtained from the internal factor analysis (IFAS) for the horizontal axis and the analysis of external factors for the vertical axis (EFAS) [Figure 6].

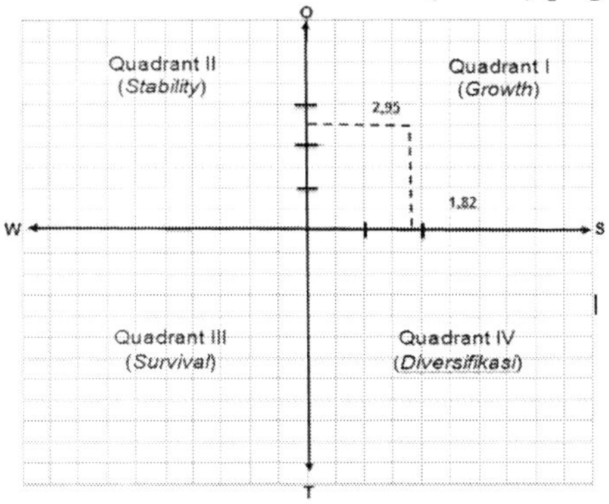

Figure 6. Quadrant SWOT Analysis for prioritizing management strategies for future ecotourism development in the Educational Forest of Unhas

Based on the quadrant position formulation in Figure 6, the high priority strategy to be implemented in the context of ecotourism management was located in Quadrant I or laid between internal strengths and external opportunities (growth strategy). The quadrant describes a strong internal condition with a supportive environment so that the appropriate direction, goals and strategies was progressive or in the S - O position. The growth strategy in the quadrant can be interpreted as having the opportunity to grow and develop by exploiting the existing potentials and opportunities. This gives an indication that the opportunity for the bear cuscus' ecotourism management strategy is in a favorable condition, where in addition to have very large strengths (3.29) from weaknesses (1.47), it also has a great opportunity (3.95) compared to the existing threats (1.00).

4. Discussion

The Educational Forest of Unhas as one of main habitats of Sulawesi bear cuscus (*Ailurops ursinus*) has an infinite potentiality to be wildlife-based ecotourism site. The area lays on 1,400 ha dominated by secondary tropical lowland forest where it has become an ideal place for the population of Sulawesi bear cuscus. Sequential studies on eco-biological of the species in the area have been conducted since 2014 and from the studies, there were no changes found on the population of the bear cuscus. This will benefit the idea of wildlife watching activities for wildlife-based ecotourism development. Combined with the resting behaviour as dominant characteristics of the animals and slow motion behavior, have made this species to be easily spotted and followed.

However, it was revealed that the population of the species in the area was too small and sparsely distributed (irregular home range). Living solitarily in higher canopy cover and spends most of its time above the ground [25] can make difficulties for wildlife watching activities if the population number is continuously very low. Animal populations are important information in ecotourism attractions because animal population information can determine the ease of encounter and distribution of animals [26]. This will become a challenge for both tourists and management in the implementation of wildlife-based ecotourism in the area.

Apart from population and home range issues, it has been highlighted in the findings that internally, there were some concerns related with the development of wildlife-based ecotourism. No networking and collaboration established in the management of the species were also need to be addressed by the internal party. To support the wildlife-based ecotourism development plan, specific infrastructure for bear cuscus tourism had to be initiated to support the ecotourism in the area. Other concerns are related to possibilities of bear cuscus to be hunted and or consumed by the locals due to low public perceptions on the importance of this animal as well as the concept of ecotourism.

Therefore, after the SWOT analysis, several alternative programs and activities are recommended to optimize the management strategies of bear cuscus in the area. The SO strategy is used as a priority strategy as its position in Quadrant I. The alternative program and activities proposed are:

4.1. S - O strategy

4.1.1. Managing habitat of bear cuscus. Sufficient information on diet and nesting plants as well as habitat type in the area have already been identified and therefore, this can be used to support the management in conducting habitat enrichment and protection. Furthermore, as found in the results of SWOT analysis, management is required to guarantee the species welfare. The welfare of the species could guarantee the lifelong of both conservation and the business of ecotourism [27]. In order to optimize the development plan, the welfare of bear cuscus inhabited the area becomes important. This can be conducted by improving eco-biological conditions and its habitat. Efforts to improve the habitat quality of bear cuscus are needed by increasing the productivity of natural food species in order to elevate the bear cuscus population. This habitat quality improvement program is expected to increase population density in the future which will later also increase the chances of encountering with that animal.

4.1.2. Take advantage on community support in maintaining eco-biological conditions and habitat of bear cuscus through direct and active involvement in ecotourism activities. High support for both conservation of bear cuscus (A. ursinus) and ecotourism by the local community nearby the site and potential tourists is one of the strength shown from the analysis. The support in particular those who live nearby the site can become a vital capital to maintain the eco-biological conditions and habitat of bear cuscus as well as in safeguarding the conservation of the species in the future. The support provided by the community through ecotourism practices can become an effective tool for wildlife conservation [28].

The community must actively take part in ecotourism development and management by becoming actors in the development of wildlife ecotourism. Community-based management can be an effort to improve environmental conditions in the area. However, in order to achieve the goal, direct involvement in any activities can be implemented including a series of capacity building. Ecotourism-based interventions provide benefits to the local and thus, has triggered more positive perceptions on even towards conflicting wildlife species [17]. This can also elevate the low perceptions occurred in the community as well as educating the community to later on increase the awareness and change their attitudes.

4.1.3. Cooperating with research institutions, government agencies and NGOs. One of the highlight was to extend more collaborations with research institutions, government agencies and NGOs in the field of research and funding. Collaboration can be carried out to explore the potential in the ecotourism area, development of facilities and infrastructure as well as conservation campaigns regarding the preservation of flora and fauna in the area. A study in Sempu Island (East Java) on the opportunities of ecotourism business also revealed that in order to minimize negative impacts generated from the ecotourism, collaboration must be established within the community and to other involving stakeholders [29].

4.2. W - O strategy
There is a need for habitat protection and fostering to increase the population of bear cuscus. Take advantage of community support in preventing factors from decreasing bear cuscus population through intensive monitoring to avoid hunting activities towards bear cuscus. Habitat management is required to protect the habitat of bear cuscus from any destructions and also to prevent hunting efforts. As it is informed that the species has become a target for wildlife trading to Java in large numbers [9]. This was also worsened by weak law enforcement in tackling illegal hunting and trade of bear cuscus [30]. It has become a crucial issue in conserving the species where the species has not been protected by any laws or regulations in particular at the national level. In fact, according to some informants, permissions have been released to one commercial breeder so it can be used a breeding stock supply [31]. There is a need for finding funding sources through expanding network and collaboration in order to build adequate facilities and infrastructure for bear cuscus.

4.3. S – T Strategy
Developing conservation education activities for the community to change the perception not only to the nearby community but also in the wider context so that the protection and conservation of bear cuscus can be achieved. In the future, it will unavoidable to have free impacts from wildlife-based ecotourism in particular to the conservation of wildlife. Therefore, communicating the issues to the local community as well as to the potential tourists as a form of education are urgently required to anticipate the tourism activities that might harm the animal [31-33]. Increasing the level of awareness among people in regards to the importance of wildlife and its role towards the environment. Develop training activities and dissemination of the concept of ecotourism to all relevant stakeholders.

4.4. Strategy W - T
Establishing several discussion forums for the community and government to gain continuous support in the management of bear cuscus ecotourism. It is proven that building communication among

stakeholders and elevates the level of relationship, for example: by considering the tourists involved in the ecotourism activities as partners including discussing any constraints existed and providing consistent message in conducting interactions with wildlife, have been known as successful wildlife-based tourism [34].

5. Conclusion

The high priority strategy to be implemented in the context of ecotourism management was located in Quadrant I which indicates that the opportunity for the bear cuscus' ecotourism is favorable. To be the target area of wildlife-based ecotourism, the educational forest of Hasanuddin University (Unhas), have to recognize carefully internal and external factors in developing an ecotourism plan. Though the journey to accomplish the idea could still be a long road but identifying those factors may provide guidance to prepare better for the development of ecotourism in the area. And to the fact, that the strategies proposed were constructed from not only strengths, weaknesses, but also from viewing the availability of opportunities and threats by examining bioecological of the species and analyzing the community's perceptions and attitudes towards bear cuscus and its conservation in the area could provide a comprehensive and complete picture of what should become a priority in developing the plan.

References

[1] Salas L, Dickman C, Helgen K and Flannery T 2019 *Ailurops ursinus (Temminck, 1824) The IUCN Red List of Threatened Species*
[2] Sinery A S, Boer C and Farida W R 2012 *Biodiversitas* **13** 86
[3] Helgen K M and S M Jackson 2015 Family Phalangeridae (Cuscuses, Brush-tailed Possums and Scaly-tailed Possum) In: Wilson, DE, RA Mittermeir (Eds). *Handbook of the mammals of the world Vol. 5: Monotremes and Marsupials* (Barcelona: Lynx Editions)
[4] Widayanti R, Wijayanto H, Wendo W D and Kunda R M 2015 *Jurnal Veteriner* **16** 227
[5] Whitten T and Henderson G S 2012 *Ecology of Sulawesi* (USA: Tuttle Publishing)
[6] Kusumaningrum E N, Supriatna J, Abinawanto and Bowolaksono A 2018 *Biodiversitas* **19** 2140
[7] Brodie J F, Giordano A J, Zipkin E F, Bernard H, Azlan J M and Ambu L 2015 *Society for Conservatyion Biology* **9** 110
[8] Saroyo S 2011 *Jurnal Bioslogos* **1** 26
[9] Shepherd C R, Sy E Y, Janssen J and Morgan J 2018 *Journal of Indonesian Natural History* **6** 30
[10] Liu X 2013 *Non-consumptive wildlife tourism and community-based conservation: a case study in Yunnan, China* (Colombus: The Ohio State University)
[11] Steven R, Pickering C and Castley J G 2011 **92** 2287
[12] Tisdell C A and Wilson C 2012 *Nature-based tourism and conservation: New economic insights and case studies* (United Kingdom: Edward Elgar Publishing).
[13] Cooper C, L Larson, A Dayer, R Stedman and D Decker 2015 *The Wildlife Society* **79** 446
[14] Motlagh E Y, Hajjarian M, Zadeh O H and Alijanpour A 2020 *Land Use Policy* **94** 1
[15] Kiper T 2013 *Role of ecotourism in sustainable development. In: Ozyavuz, M (eds) Advances in Landscape Architecture* (Turkey: Intechopen)
[16] Stem C J, Lassoie J P, Lee D R, Deshler D D and Schelhas J W 2010 *Society & Natural Resources* **16** 387
[17] Vanelli K, Hampton M P, Namgail T and Black S A 2019 *Human Dimensions of Wildlife* **24** 180
[18] Achmad A, Ngakan P O, Maulany R I and Asrianny A 2016 *Proc of the National Seminar on Biology*. (Makassar: Faculty of Mathematics and Natural Sciences, Hasanuddin University)
[19] Mangalla R R 2014 *Daily behaviour of Bear Cuscus (Ailurops ursinus) in the educational forest of Hasanuddin University* Undergraduate Thesis (Makassar: Hasanuddin University)
[20] Alamsyah R 2015 *Resting habitat preferences of Bear Cuscus (Ailurops ursinus) in the Educational Forest of Hasanuddin University, Maros, South Sulawesi* Undergraduate Thesis (Makassar: Hasanuddin University)
[21] Hidayat B R 2015 *Types and potentials of Bear Cuscus feed plants (Ailurops ursinus) in*

Hasanuddin University Educational Forest Undergraduate Thesis (Makassar: Hasanuddin University)

[22] Aziz A 2020 *Preferences and potential of Bear Cuscus (Ailurops ursinus) diet plant species in the Field Laboratory of Forest Conservation, the Educational Forest of Hasanuddin University, Maros, South Sulawesi* Undergraduate Thesis (Makassar: Hasanuddin University)

[23] Fauzan 2020 *Resting tree characteristics of Bear Cuscus (Ailurops ursinus) at various vegetation cover in the field laboratory of Forest Conservation, the Educational Forest of Hasanuddin University, Maros, South Sulawesi* Undergraduate Thesis (Makassar: Hasanuddin University)

[24] Jathi I 2020 *Daily Behaviour of Bear Cuscus (Ailurops ursinus) in the field laboratory of Forest Conservation, the Educational Forest of Hasanuddin University, Maros, South Sulawesi* Undergraduate Thesis (Makassar: Hasanuddin University)

[25] Talumepa P A A, Wungow R S H, Poli Z and Rimbing S C 2016 *Journal Zootek* **36** 174

[26] Kuswanda W, Sitomorang R O P, Berliani K, Barus S P and Silahi J 2018 *Elephant conservation and ecotourism: A Model from KHDTK Aek Nauli* (Bogor: PT Penerbit IPB Press)

[27] Williams C L, Mazzola S M and Pastorino G Q 2017 *Annual Research & Review in Biology* **19** 1

[28] Stronza A L, Hunt C A and Fitzgerald L A 2019 *Annual Review of Environment and Resources* **44** 229

[29] Purnomo H, Sulistyantara B and Gunawan A 2013 *Jurnal Penelitian Sosial dan Ekonomi Kehutanan* **10** 247

[30] Flannery T F and Helgen V 2016 *Ailurops melanotis. The IUCN Red List of Threatened Species 2016*: e.T136218A21949526

[31] Newsome D, Moore S A and Dowling R K 2012 *Natural area tourism: Ecology, impacts and management* (Toronto: Channel View Publications)

[32] D'cruze N, Machado F C, Matthews N, Balaskas M, Carder G, Richardson V and Vieto R 2017 *Nature Conservation* **22** 1

[33] Bach L and Burton M 2017 *Journal of Sustainable Tourism* **25** 181

[34] Ballantyne R, Packer J and Hughes K 2009 *Tourism Management* **30** 658

IOP Conf. Series: Earth and Environmental Science 1115 (2022) 012045 doi:10.1088/1755-1315/1115/1/012045

The use of chicken manure, lowland peat and highland peat briquettes on the growth of *Macadamia integrifolia* in an arid area of Pematang Tambun Village, Simalungun Regency

B Utomo[1*], F A Lubis[1], A Dalimunthe[1], P Tamba[1], D C G Silalahi[1], S P Lesmana[1] and L Fauziah[2]

[1]Faculty of Forestry, Universitas Sumatera Utara, Kampus USU Padang Bulan, Medan 10155, North Sumatra, Indonesia
[2]Faculty of Agriculture, Universitas Sumatera Utara, Kampus USU Padang Bulan, Medan, 10155 North Sumatra, Indonesia

[*]Email: budiutomo@usu.ac.id

Abstract. The catchment area of Lake Toba continues to experience a decline in quality due to frequent fires in this area. Various recovery measures in the form of reforestation are continuously being carried out but often fail due to the difficult terrain due to the steep slope and the rapid loss of water due to gravity. The water immediately dissipates to a lower elevation leading to the lake. As a result, this condition leaves an expanse of barren area that is only overgrown with weeds. This study aimed to see whether lowland and highland peat soil can be used as a water retaining material that has the same quality as manure and to obtain the right dose of this water retaining material to support the growth of *Macadamia integrifolia* plants. The study was designed using 3 types of water retaining material briquettes, namely: chicken manure, lowland peat and highland peat, each of which was divided into several treatments, namely: 0, 0.5, 0.75, 1.0, 1.25 and 1.5 kg/plant. The difficulty of the steep terrain makes it difficult for planting actions. Therefore, the technology of making briquettes is considered to be a solution. In addition, the conversion into briquettes gives a positive value to peat where acidic peat water can be removed so that the toxic effects of peat can be reduced. The results showed that the three types of briquettes significantly increased plant growth in terms of plant height, leaf area and crown area of *M. integrifolia*. Although the best plant growth performance was shown by the briquette dose of 1.5 kg/plant, the dose of 0.75 kg/plant seemed to have given no significant growth when compared to a dose of 1.0-1.5 kg/plant.

1. Introduction

Lake Toba, which is located in North Sumatra Province at an altitude of about 905 meters above sea level, is the largest lake in Indonesia and the largest volcano-tectonic lake in the world. The catchment area of Lake Toba has an area of 379,940.35 ha with a land area of 275,412.10 ha (72.49%) and a water area of 104,528.25 ha (27.51%) [1]. The Lake Toba catchment area has always been dominated by steep slopes and steep cliffs, so the government designated it as a protected forest area. Meanwhile, on the flat and sloping side, which is located adjacent to the lake, the community uses it as an area for farming rice, onion, etc. The problem that arises is that the slopes of the mountains and steep cliffs continue to be bare and only overgrown by weeds. This triggers a landslide hazard and creates an arid area due to the lack of trees to bind humus and rainwater. This also causes the water level of Lake

Toba to continue to shrink from time to time. Various reforestation efforts have been carried out by various parties but do not seem to have produced encouraging results. Every year forest and land fires often occur due to various factors, such as: land burning which causes the fire to spread to forest areas, human behavior that throws cigarette butts in any place, burning garbage, etc.

Efforts to replant forest land have been repeatedly carried out by the government and various other related parties. However, the results are still far from expectations because the plant mortality rate in the field is very high, reaching 90%. This happens because planting activities are rarely followed by maintenance actions, especially watering plants. This is because the terrain is very steep, so it is very dangerous for workers to carry out maintenance actions.

The steep and steep slope of the land makes it very easy for rainwater to be lost to the lake area by the high gravity of the earth. Shallow soil solum increases the acceleration of groundwater loss so that cultivated tree plants immediately lose their ability to get water from the soil. This is what causes cultivated plants planted in this forest area to die.

To increase the survival percentage of seedlings planted on sloping land like this, technology is needed. If on flat and sloping land, usual maintenance actions such as watering and fertilization can be carried out, then for this case of catchment area will be different. We tried to use 3 types of water-retaining materials, namely chicken manure, lowland peat and highland peat. The use of peat material is based on the consideration that its availability is quite abundant in the North Sumatra region. Lowland peat spreads from the east coast of North Sumatra which borders Aceh Province to the border area of Riau Province. Lowland peatlands in Sumatra have a total area of 6,436,649 Ha of which 261,234 Ha (4.06 %) of the total area is in North Sumatra [2]. Meanwhile, highland peat is located in Humbang Hasundutan Regency which is located at an altitude of 1400 m above sea level covering an area of 2,998 Ha [3].

Problems arise when we faced to transporting this material in sloping and steep land. This tough terrain does not allow the use of transport vehicles to the planting site. Therefore, manure and peat material must be transformed into a medium that is easy to carry. The alternative is to turn it into briquettes.

One of the potential species for rehabilitation in the Lake Toba catchment is macadamia (*Macadamia integrifolia*). Based on the results of the trial planting in Sipisopiso Village (1,100 masl) which is located 10 km from the research location, this species is able to adapt to *Imperata* land with low nutrient availability, high temperature, low humidity, and high light intensity. This plant does not require intensive care, does not drop leaves, does not die easily when a forest floor fire occurs (surface fire) and is easy to sprout again after burning [4]. Tolerance to fire is an important character because fire is a frequent occurrence in this area, is non-invasive, liked by the community and has high economic value. This also underlies the use of this plant as a sample plant. This study aimed to compare the possibility of lowland and highland peat can be used as a water retaining material that has the same quality as manure. In addition, this study aims to obtain the right dose of this water-retaining material to support the growth of *M. integrifolia* plants.

2. Materials and Method

This research was conducted in the Lake Toba Catchment Area, precisely in the community forest area of Pematang Tambun Raya Village, Pematang Sidamanik District, Simalungun Regency, North Sumatra. This research was conducted in March-September 2021.

The materials used in this study were *M. integrifolia* seedlings aged 6-8 months, chicken manure, peat originating from the lowlands and peat from the highlands. The tools used in this research are manual hydraulic press, stationery, ruler, califer, scales, hoe, camera, SPSS software and Microsoft Excel. The study was designed using a non-factorial randomized block design method. Each plant received a different dose of briquettes, both for manure, lowland peat and upland peat, namely: control (without application of manure briquettes, treatment of manure briquettes, lowland peat briquettes and highland peat briquettes respectively doses of 0.25 kg, 0.5 kg, 0.75 kg, 1.0 kg, 1.25 kg, and 1.50 kg per plant. The research was conducted with the following stages:

IOP Conf. Series: Earth and Environmental Science 1115 (2022) 012045 doi:10.1088/1755-1315/1115/1/012045

The seeds of the *M. integrifolia* plant used in this study were seeds originating from the Simalungun Regency area, precisely in the PT Inhutani IV Nursery. The *M. integrifolia* seeds used were the result of generative propagation through seeds. The seeds used are seeds aged 6-8 months and have good health and physical condition.

Water with tapioca flour is boiled in a saucepan until it produces a clear color. Furthermore, this mixture is cooled and mixed with manure until evenly distributed, then shaped and allowed to stand until it curdles. After that this dough is printed using a manual press.

The size of the planting hole used in this study was 30 cm x 30 cm with a depth of 30 cm. The spacing used in this study was 5 m x 5 m. The spacing is made so wide that after the research is complete the plants can be allowed to grow up for the purpose of reforestation and future production.

Before making observations on each research parameter, the initial height and diameter data were collected first. Observations were made after the adaptation of plants in the field that had been treated, ie 1 week after planting. Seedlings need time to adapt after transplanting. Seedlings from the nursery will be very risky if they are planted directly in the field because the conditions in the nursery and the planting location are very different [5].

The observed variables were: (1) the decrease in the weight of the water retaining material observed in the laboratory, namely by weighing 1000 g for each water retaining material after being saturated with water and drained for 24 hours until no water dripped, then pressed and then drained. weighed again, as well as the measurement of the volume [6]. (2) increase in plant height, (3) increase in stem diameter, (4) leaf area, and (5) canopy area. Specifically, the observation of plant height and stem diameter was observed every 2 weeks while other parameters were carried out at the end of the study.

3. Results and Discussion

3.1. Reduction of weight and volume of water retaining material before and after becoming briquettes

From the results of research in the laboratory, it was found that the three water-retaining materials studied (chicken manure, lowland peat and highland peat) could be reduced in weight and volume by pressing them into briquettes. The results of the calculation of peat weight and volume reduction are presented in the following figure (Figure 1).

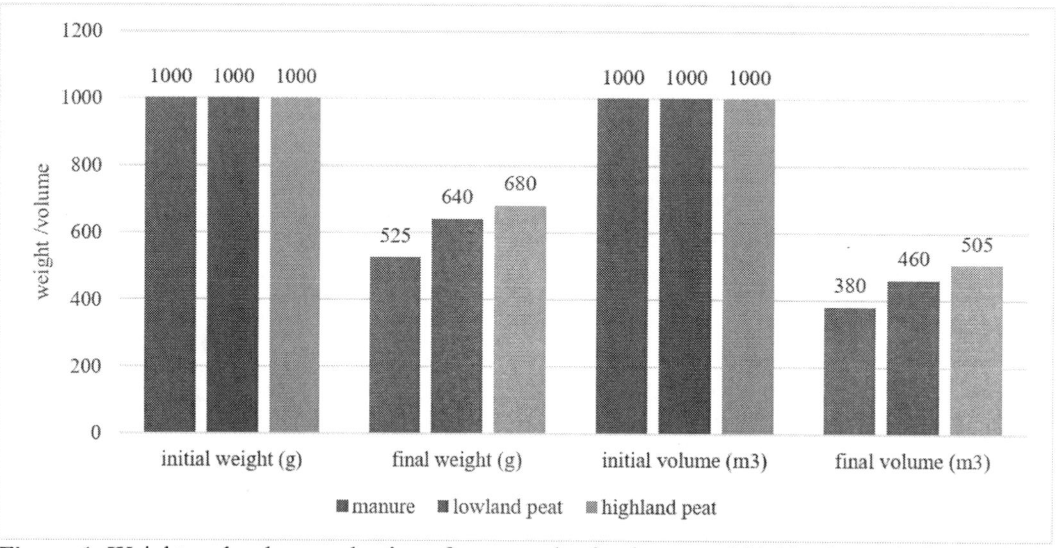

Figure 1. Weight and volume reduction of manure, lowland peat and highland peat before and after pressing into briquettes

The use of a press machine can significantly reduce the weight and volume of briquettes due to the loss of water and compaction of pore spaces [7]. The reduction in weight and volume of manure was lower than that of lowland peat and upland peat. Likewise, the decrease in weight and volume of lowland peat is lower than that of highland peat. This could be due to the coarser particle size of upland peat compared to lowland peat and manure. The low air temperature on highland peat taken from an altitude of 1,400 masl causes the weathering of peat in the highlands to take place slowly. This causes organic peat material in the highlands to be more resistant to weathering processes [8].

In this study, none of the planted *M. integrifolia* plants died even though these plants grew in dry conditions without watering. This proves that this plant is able to adapt and survive in extreme conditions in the catchment area of Lake Toba. The type of *M. integrifolia* is able to adapt in *Imperata* fields with low nutrient availability, high temperature, low humidity, and high light intensity [9]. Therefore, this *M. integrifolia* plant is good for use as a reforestation plant in arid and fire-prone areas because of the superiority of *M. integrifolia* plants that are resistant to fire [10].

3.2. Plant height and stem diameter

The results of measuring the increase in the average height of *M. integrifolia* plants and the addition of stem diameter until the age of 4 months after planting can be seen in Table 1.

Table 1. Effect of briquettes on the increase in plant height and stem diameter of M. *integrifolia* at 16 weeks of age

Treatment	Plant Height (cm)			Stem Diameter (cm)		
(kg)	Manure	Lowland peat	Highland peat	Manure	Lowland peat	Highland peat
Kontrol	10.51 a	10.51 a	10.51 a	0.73	0.73	0.73
0.25	18.93 ab	18.23 ab	15.16 ab	0.89	0.91	0.81
0.5	25.61 c	23.61 bc	20.23 b	1.02	0.96	0.86
0.75	28.17 cd	23.87 bc	20.81 b	1.11	1.02	0.92
1	29.28 d	26.26 cd	22.19 bc	1.09	1.19	1.07
1.25	31.22 e	27.32 d	25.12 c	1.05	1.09	1.11
1.5	30.25 de	26.25 cd	24.20 bc	1.04	1.06	1.01

Table 1 shows that the application of briquettes significantly increased plant height growth in all types of water retainers, but had no significant effect on increasing stem diameter. The increase in plant height varies based on the material used. In the use of chicken manure, plant height growth showed the best performance when compared to material derived from lowland peat and highland peat.

3.3. Leaf area and plant canopy area

On the parameters of leaf area and plant canopy area up to 4 months of age, the observations showed a significant effect for the three water-retaining materials studied. These differences can be seen in Table 2.

Table 2. Effect of manure briquettes, lowland peat and upland peat on leaf area and canopy area of *M. integrifolia* at 16 weeks of age

Treatment	Leaf Area (cm)			Crown Area (cm)		
(kg)	Manure	Lowland peat	Highland peat	Manure	Lowland peat	Highland peat
Control	16.05 a	16.05 a	16.05 a	107.02 a	107.02 a	107.02 a
0.25	25.15 b	23.41 bc	18.13 ab	195.22 b	188.34 b	170.25 b
0.5	33.95 bc	25.88 bc	19.82 ab	251.34 c	196.32 bcd	180.18 bc
0.75	38.97 cd	28.27 bcd	24.33 bc	272.55 cd	195.18 bcd	185.08 bc
1	45.60 d	28.30 bcd	24.67 bc	282.01 cde	206.46 cd	195.22 bc
1.25	45.86 de	30.05 cd	28.76 cd	306.07 de	209.01 d	196.91 bc
1.5	46.44 e	32.21 d	29.88 d	318.32 e	206.34 cd	201.17 c

Based on Table 2, it can be seen that the three water-retaining materials studied gave a positive response in supporting plant growth for both leaf area and plant crown area observed up to 16 weeks of age. However, when compared between the three, the use of manure gave the best plant growth response, which was indicated by the highest leaf area and plant crown area.

In plants treated with lowland peat, the leaf area of the plant was not as good as that of plants treated with chicken manure, but there was a trend of increasing growth with the use of lowland peat as a water retaining material. Similarly, the response of plants to the treatment of highland peat was also shown. Considering that *M. integrifolia* is a tree-shaped plant with an annual age, the response of plants to the given water retaining material tends to be slower than that of annual plants. The increase in plant growth was mostly supported by improvements in the physical properties of the soil. However, peat is reported to be able to supply 64.1% amino acid nitrogen for plants [11]. These physical properties provide a better growth space for plants, retain and provide water in the rhizosphere longer than without water-retaining materials [12].

In contrast to peat which only acts as a soil water retaining material, manure has a low C/N value so that the nutrients contained in it, especially N, P and K are more available and help increase plant growth [13]. This resulted in a better response of *M. integrifolia* plants to manure treatment than lowland peat and highland peat. The N element stimulates faster leaf development, while the P element is used for the development of meristem tissue. The development of meristem tissue causes the cells to elongate and enlarge, so that plant parts such as leaves and shoots will be longer and wider and will affect plant leaf area [14,15]. Nutrients P and K are needed in the growth of stems and branches and function in the formation of carbohydrates so as to produce a large number of leaves [16]. Leaves are important organs in photosynthesis that produce substances needed by plants. The more leaves, the higher the photosynthesis that occurs, [17]. Plants that have a larger leaf area will be more efficient in absorbing sunlight which is useful in the photosynthesis process and affects plant fresh weight [18].

4. Conclusion

The use of chicken manure, lowland peat and highland peat packaged in the form of briquettes significantly increased plant growth in plant height, leaf area and canopy area of *M. integrifolia* up to 16 weeks of age. The best plant growth response was shown by the treatment of chicken manure followed by lowland peat and highland peat.

Acknowledgment

On this occasion the author would like to thank the University of North Sumatra for the Talent Grant in year 2021 so that the author could carry out the superior research of this college.

References

[1] Sihotang H, Purwanto M Y J, Widiatmaka and Sambas B 2012 *Journal of Natural Resources*

and Environmental Management **2** 65
[2] Ritung S, Wahyunto, Nugroho K, Sukarman, Hikmatullah, Suparto and Tafakresnanto C 2011 *Indonesian Peatland Map Scale 1:250,000 (Indonesian Peatland Map at The Scale 1:250,000)* (Bogor: Indonesian Center for Agricultural Land Resources Research and Development)
[3] BBSDLP 2022 Center for Agricultural Land Resources 2020 Identification and Characterization of Upland Peatlands in Humbang Hasundutan Regency, North Sumatra. https://bbsdlp.litbang.pertanian.go.id/
[4] Aswandi and Kholibrina CR 2017 *Restoration of the Lake Toba Ecosystem* (Medan: Pioneer Media Development)
[5] Directorate General of Plantations 2006 Guidelines for the Cultivation of Macadamia (*Macadamia Integrifolia*) (Jakarta)
[6] Aransiola E F, Oyewusi T F, Osunbitan J A and Ogunjimi L A O 2019 *Energy Reports journal* **5** 909
[7] Mallika T, Thanchanok P, Kasidet P and Prasong W 2015 *Energy Procedia* **79** 890
[8] Jeb S F and Nazim G 2021 *IOP Conference Series Earth and Environmental Science* **384**
[9] Emmanuel J Z, Kadmiel M, Shonil A B, Kauêde S, Andrew E, William R, Yoseph N A 2021 *PLOS ONE* **16**
[10] Aswandi, Pratiara, and Kholibrina C R 2017 *Policy Brief* **11**
[11] Guennadi S, Lech S and Irina S 2019 *Soil Science Society of American Journal* **59** 188
[12] Oleg Misnikov 2020 *E3S Web of Conferences Vth International Innovative Mining Symposium* **174** 01015
[13] Oagile D and Namasiku M 2010 *Journal of Soil Science and Environmental Management* **1**
[14] Hartatik W, Husnain And Kadiyani R W 2015 *Journal of Land Resources* **9**
[15] Damanik M M B, Hasibuan B E, Fauzi, Sarifuddin and Hanum H 2011 *Soil Fertility and Fertilization* (Medan: USU Press)
[16] Sucipto 2010 *Embryo Journal* **7** 67
[17] Wahida R S, Nadira H L and Hernusye 2011 *Application of Chicken Manure on Three Varieties of Sorghum (Sorghum bicolar (L). moench)* (Padang: Andalas University)
[18] Utomo S A and Purnamasari R T 2017 *Journal of Agrotechnology Merdeka Pasuruan* **1** 22

The use of oil palm fruits waste brickets and manure in relationship as groundwater resistance materials to increase plant growth in critical land

B Utomo[1], A Prasetyo[2], M N Alfin[1], A Dalimunthe[1] and P Tamba[1]

[1]Faculty of Forestry, Universitas Sumatera Utara, Kampus USU Padang Bulan, Medan 20155, North Sumatra, Indonesia
[2]Students of Faculty of Forestry, Universitas Sumatera Utara, Kampus USU Padang Bulan, Medan 20155, North Sumatra, Indonesia

[*]Email: budiutomo@usu.ac.id

Abstract. The catchment area of Lake Toba geologically has soil properties that are sensitive to erosion because the land is hilly or mountainous with very steep slopes, very shallow soil solum (< 30 cm) to shallow (30-60 cm). Even though the level of rainfall in this area is quite high, the slope of the land causes the soil to easily lose ground water by the earth's gravity. Similarly, the level of erodibility is also high. The purpose of this research is to take advantage of the high potential of oil palm bunches waste which allows it to be used as a water retaining material compared to manure. To facilitate transportation, these two types of organic water retaining materials are used as briquettes. The study was structured using a non-factorial Randomized Block Design (RAK). Where each type of organic matter used 6 treatments with 6 replications, namely without using briquettes (P0) Doses of 0.25 kg, 0.50 kg, 0.75 kg, 1.00 kg and 1.25 kg, with 6 replications. The results shows that the waste briquettes of oil palm fruit bunches have the ability to hold water and increase the growth of *Macadamia integrifolia* plants which is the same as the ability to use manure.

1. Introduction

Lake Toba is located in North Sumatra Province and has an altitude of about 900 meters above sea level. This lake is the largest lake in Indonesia and the largest volcano-tectonic lake in the world. Lake Toba has an area of about 110,000 ha with a catchment area of about 280,000 ha. The Lake Toba catchment area is mostly forest area and only a small part has a sloping topography that can be planted with agricultural crops [1]. As a forest area that deserves to be protected as a lake catchment area, this area should be overgrown with dense trees. But in reality, almost 80% of the forest area in this area is bare and only overgrown with weeds and weeds.

The catchment area of Lake Toba has soil types that are sensitive to erosion, the land is hilly or mountainous with very steep slopes, the soil solum is very shallow (< 30 cm) to shallow (30 cm - 60 cm). In addition, the catchment area of Lake Toba also has high rainfall (although only less than 4 months a year). Thus naturally the catchment area of Lake Toba has a high potential for erosion hazard [2]. The steep terrain has resulted in reforestation programs often failing to grow because it is not possible to carry out maintenance on the planted tree seedlings. Even planting this area requires a lot of effort because of the steep cliffs and high slope. As a result, until now the forest area in the catchment area of Lake Toba is still barren and arid. Meanwhile, the clearing of areas on more sloping plains continues for residential needs, the tourism industry and agricultural businesses, resulting in the air temperature in the vicinity being no longer as cool as in the past.

Figure 1. The description of the catchment area of Lake Toba which is arid and barren, not many trees are able to grow and develop under conditions of lack of water due to the slope

The water requirement factor is vital in the soil, especially in conditions of steep slopes where water loss can occur quickly so that plant roots will immediately lose water if they are not able to immediately absorb it. Therefore, efforts to maintain water in the soil longer are important so that plants can meet their minimum water requirements from the rhizosphere.

Various reforestation efforts have been carried out since 2003, starting with the government's initiative to plant a million hectares through the National Movement for Forest and Land Rehabilitation. This has been followed by a series of other reforestation programs to date, both by the government and other stakeholders. However, the success of these programs is still far from expectations (<10%). An integrative approach is needed based on the findings of the research that has been done [3]. Modifications in an effort to increase the ability to store water in the soil, especially in the rhizosphere, have been carried out recently, with the result that the role of organic matter is very important as a water-retaining material, especially in areas prone to rapid groundwater loss [4].

Indonesia is now the largest palm oil producer in the world. With this capacity, the number of empty palm oil bunches that become unused waste is also abundant. This waste has the potential to be investigated for its ability to help increase the growth of reforestation plants in the catchment area of Lake Toba in relation to its ability to improve the physical properties of the soil in the rhizosphere [5]. As a comparison, organic materials derived from chicken manure were also used. The plant used as an indicator plant is macadamia (*Macadamia integrifolia*) which has been proven to be able to grow in the catchment area of Lake Toba. This study aims to determine whether the waste of palm oil bunches modified as briquettes is able to work as a water reservoir that can increase plant growth in the catchment area of Lake Toba which easily loses ground water due to steep topography.

2. Method

2.1. Tools and materials

This research was conducted in the catchment area of Lake Toba, precisely in Si Pintuangin Village, Dolok Pardamean, Simalungun Regency, North Sumatra. This research was conducted from February to June 2021.

The tools used in this study were tag labels, tally sheets, manual hydraulic presses, stationery, hoes, SPSS software, rulers, and cameras. The materials used in this study were palm oil bunches (Elaeis guineensis Jacq.), tapioca flour, water, and 8 month old M. integrifolia seedlings.

2.2. Research procedure

In the implementation of the research carried out with the following stages:

2.2.1. Briquette and manure preparation. Oil palm bunches waste briquettes (Elaeis guineensis Jacq.) are made by reducing the water content using a manual press machine, then the oil palm tanks are chopped according to the specified size, after that the chopped results are mixed and stirred with adhesive (starch starch) with a ratio of waste: adhesive (10 kg: 1 kg) in the container until homogeneous, then the briquettes are printed using an oval mold, then the briquettes are dried with the help of sunlight for 3 days. The same thing was done in the treatment of manure.

2.2.2. Seed preparation. The M. integrifolia seedlings taken from PT Inhutani IV Nursery in Simalungun Regency. The M. integrifolia seedlings used were the result of generative propagation through seedlings. The seedlings used are seedlings that have an age of around 8 months and have good health and physical condition as indicated by seedlings that are not attacked by pests and diseases.

The size of the planting hole used in this study was 30 cm x 30 cm with a depth of ± 25 cm. The spacing used in this study was 4 x 5 m, the mini spacing was made so that after the completion of the research the plants could continue to be maintained to grow and produce in that place. The placement of waste briquettes of oil palm bunches and manure was carried out simultaneously with the planting of *M. integrifolia* plants. After the planting hole was formed, the seeds of *M. integrifolia* were put into the soil and between the soil containing the roots of the seeds were put in the waste briquettes of oil palm bunches according to the treatment. The same was done for the treatment of manure.

The study was designed using a non-factorial Randomized Block Design method. While the waste briquettes of empty oil palm fruit bunches used were 6 treatments with 6 replications. The levels of treatment are as follows:

P0 = control; P1 = 0.25 kg of oil palm bunch waste briquettes; P2 = 0.50 kg of oil palm bunch waste briquettes; P3 = Dose of 0.75 kg oil palm bunch waste briquettes; P4 = dose 1.0 kg of oil palm bunch waste briquettes; P5 = Dose of 1.25 kg of oil palm bunch waste briquettes. Each treatment was repeated 6 times. Thus in the treatment of waste oil palm briquettes, 36 *M. integrifolia* plants were used. In the manure treatment, the same treatment was also used, namely as many as 36 *M. integrifolia* plants. Thus overall this study used 72 sample *M. integrifolia* plants.

2.2.3. Observation variable. The observed variables were: (1) the decrease in the weight of the water retaining material observed in the laboratory, namely by weighing 1000 g for each water retaining material after being saturated with water and drained for 24 hours until there was no water dripping, then pressed and then drained and then weighed again, as well as the measurement of the volume [9]. (2) increase in plant height, (3) increase in stem diameter, (4) leaf area, and (5) canopy area. Specifically, the observation of plant height and stem diameter was observed every 2 weeks while other parameters were carried out at the end of the study.

3. Results and discussion

3.1. Weight loss of both types of water retaining materials

Based on research conducted in the laboratory, the waste of oil palm bunches is compacted by pressing using a press. A total of 1000 g of palm oil bunch waste is pressed to remove the water content. With the release of water content, the weight of this waste decreases. The same is done for the volume of waste from oil palm bunches. A total of 1000 g of waste after compaction is calculated volume. For the same thing is done on the treatment of manure. The results of these two organic materials are then compared and presented in Figure 2.

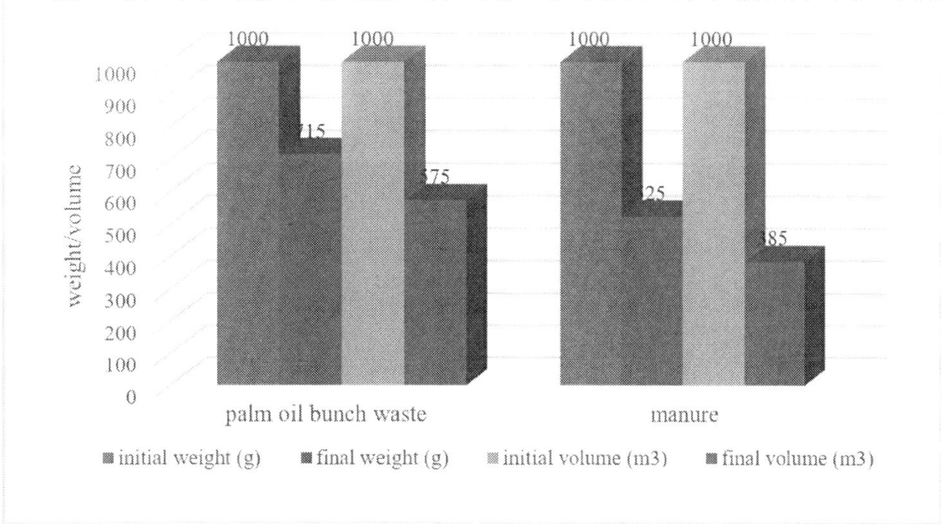

Figure 2. Weight and volume measurement of oil palm tanks to facilitate the transportation process in the use of tanks

Based on Figure 2, it can be seen that the reduction in the weight of the waste of oil palm bunches was lower than the decrease in the weight of the manure. Similarly, the volume caused by pressing treatment. The reduction in the weight of the waste of oil palm bunches was only 28.5%, while in manure the weight loss was 47.5%. Likewise, the decrease in the volume of waste from oil palm bunches decreased by 42.5%, while in manure the decrease reached 62.5%. This decrease was due to the high fiber content in palm oil bunch waste which is more resistant.

3.2. Plant growth

3.2.1. Plant height. Whether or not the effect of adding organic matter into the soil in supporting plant growth must be tested on plants. The response shown by plants is considered to represent how well an organic material supports plant growth through the effect of improving physical properties given. Modification of the shape into briquettes is intended to facilitate the transportation of this organic material to the field. The results of measuring the height increase of M. integrifolia plants up to the age of 4 months after planting in the field can be seen in Figure 3.

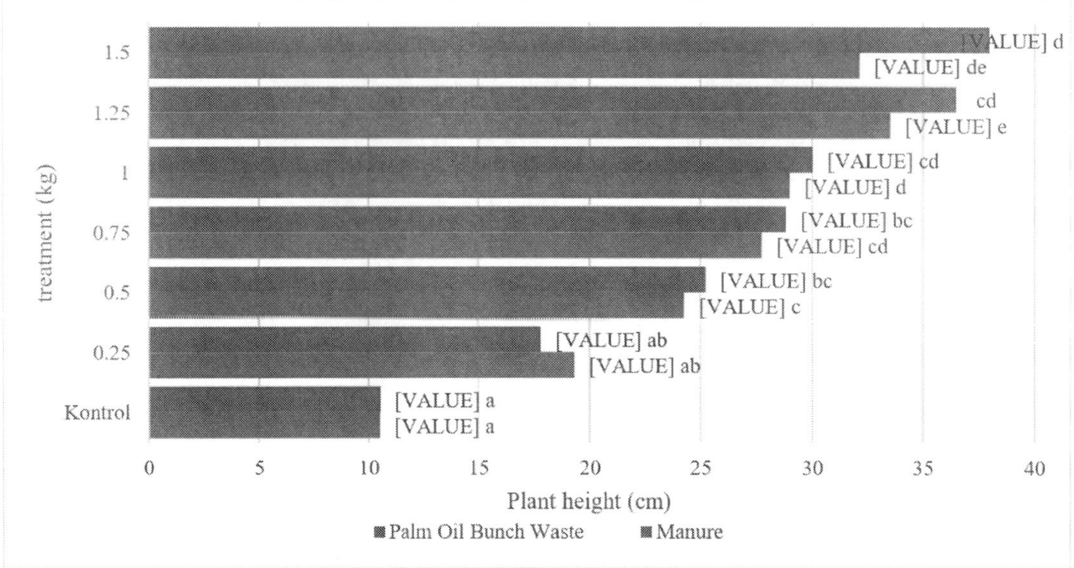

Figure 3. Plant height growth of *M. integrifolia* at 4 months of age after treatment

From Figure 3, it can be seen that the use of both organic materials, both waste briquettes of oil palm bunches and manure briquettes, has a significant effect in increasing plant height growth until the plant is 4 months old. Even at the same age, the response of giving oil palm bunches waste briquettes was better than the response of plants given manure briquettes.

The use of oil palm bunches waste briquettes as much as 1.5 kg/plant resulted in the best plant height (37.98 cm. However, the use of 1 kg/plant has given the same results. This can be interpreted that the use of waste briquettes of oil palm bunches of 1 kg/plant has sufficient to support plant growth. While the application of 1.25 kg of manure briquettes/plant resulted in the best plant height growth. The use of oil palm bunches waste has a positive effect in supporting the increase in plant height which can compete with plant responses. indicated by manure briquettes.

3.2.2. Stem diameter. Stem diameter in plants is considered the best predictor of survival percentage and seedling growth in the field. Larger diameter also indicates a large root system and stem volume [6]. The results of measuring the average diameter of macadamia plants can be seen in Figure 4.

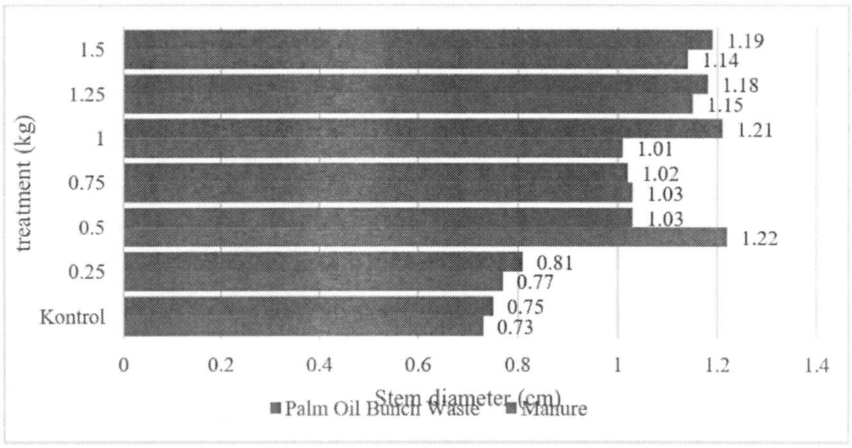

Figure 4. The growth of diameter *M. integrifolia* plants at the age of 4 months after treatment

Figure 4 shows that until the age of 4 months in the field the increase in stem diameter between all treatments was not significantly different. This could be due to the longer research period because *M. integrifolia* is a long-lived perennial tree. This plant takes a long time to show a response to the enlargement of its stem diameter due to the treatment given. Therefore, the study time of 4 months was considered insufficient to provide a response to differences in stem diameter.

3.2.3. Leaf area. In the parameter of leaf area to plant age of 4 months, the observation turned out to show a significant effect for the two water retaining materials given, both oil palm bunch waste briquettes and manure briquettes. These differences can be seen in Figure 5.

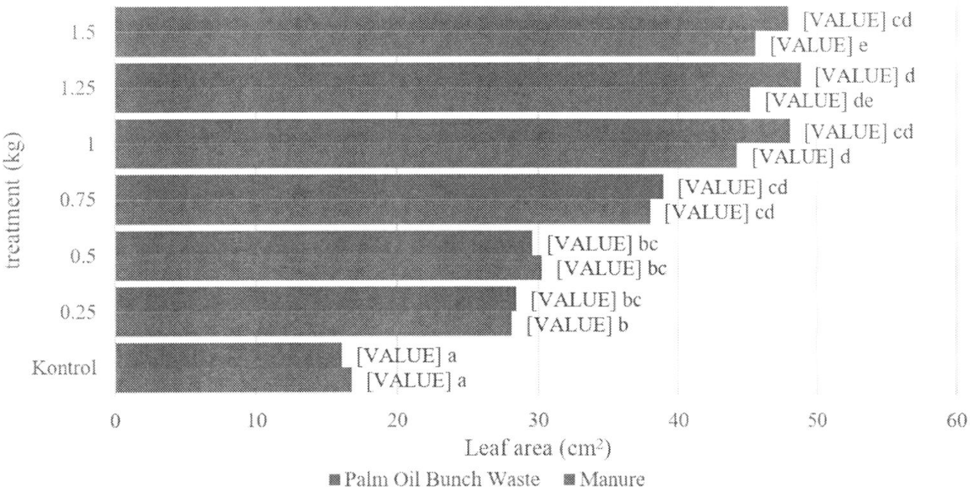

Figure 5. Leaf area of *M. integrifolia* at 4 months after treatment

The provision of water-retaining material in the form of waste oil palm bunches briquettes and manure briquettes had a significant effect on increasing leaf area of *M. integrifolia* at 4 months of age in the field. Oil palm bunches waste briquettes gave a good response in supporting the growth of plant leaf area which was indicated by leaf area that matched the response of plants given manure briquettes. Plants that were given these two water-retaining materials showed a more healthy visual appearance and more fertile than control.

3.2.4. Crown area. On the parameters of the crown area up to the age of 4 months, the observations showed a significant effect for the two types of water-retaining materials studied. These differences can be seen in Figure 6.

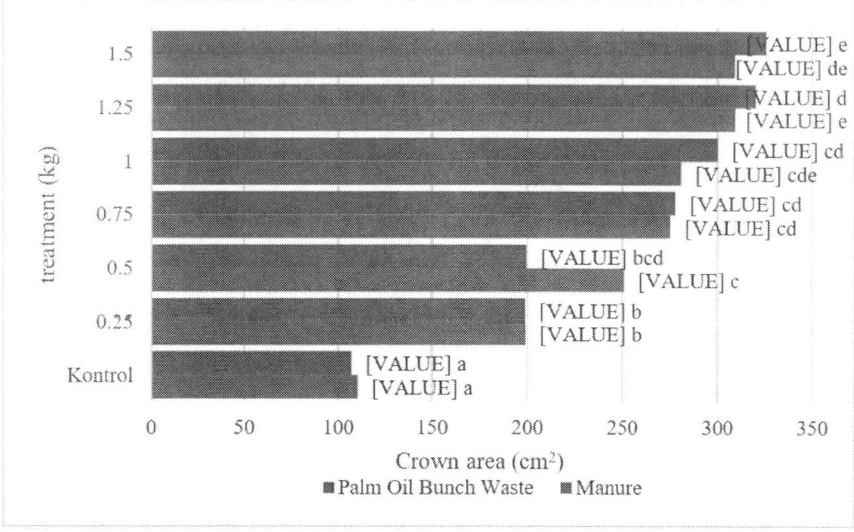

Figure 6. *M. integrifolia* plant crown area growth for 4 months after treatment

In line with the increase in leaf area, the plant crown area was also significantly different between all treatments on the two water-retaining materials given. The more briquettes given, both waste oil palm bunches and manure briquettes, the better the response indicated by the increase in the area of the crown area. The application of waste briquettes of oil palm bunches increased the area of the plant canopy which matched the response of plants to manure briquettes. This gives a breath of fresh air that the waste of oil palm bunches can be used as a water retaining material or organic fertilizer which has the same effect as manure.

Plant height is a plant size that is often observed as an indicator of growth as well as a parameter to measure the influence of the environment or the treatment applied because plant height is the easiest growth measure to see [7]. Based on the graph of the average increase in plant height of *M. integrifolia*, the height of macadamia plants continued to increase in both types of water-retaining materials. This shows the positive effect of the briquettes in retaining water in the soil which in turn gives a positive response to increased plant growth.

The addition of organic fertilizers to the soil in addition to playing a role in increasing its nutrient content also plays a role in improving the state of the structure, aeration, soil water holding capacity, influencing or regulating soil temperature conditions and providing substances resulting from overhaul that can help plant growth. [8]. Organic matter in oil palm bunches waste generally contains complex nutrients (macro and micro) although in small amounts, besides physically this organic material can improve the structure and stability of soil aggregates, increase water absorption and storage capacity, so that soil microbial activity takes place smoothly. aims to support the decomposition of organic matter into available nutrients for plants [9].

Fertilizers with controlled release forms can optimize the absorption of nutrients (one of them nitrogen) by plants, because these fertilizers can control the release of nutrients according to the time and amount needed by plants [10].

The number of leaves is related to the height of the plant where the taller the plant, the more leaves are formed. The increasing number of leaves results in more light being absorbed by plants for the photosynthesis process, this is very influential in plant growth and development [11].

Planting media that is able to provide good pore space for root growth will accelerate the number of leaves. This is because the number of leaves allows it to absorb a lot of sunlight so that the photosynthesis process also takes place more quickly [12].

The physical measurements of macadamia plants treated with oil palm bunches waste briquettes and manure briquettes showed differences, namely there was an increase in leaf area and plant crown area compared to controls. This indicates the alleged ability to retain water in the rhizosphere and provide it to meet the growth needs of macadamia plants.

The area of the canopy affects the process of photosynthesis and the absorption of solar energy. The effect of lack of water during the vegetative stage is the development of smaller leaves, which can reduce light absorption [13].

Water has a great influence on plant growth. Water conditions that are not easily available can interfere with plant physiological processes which will put pressure on plants, over time the plants begin to wither, even death [14]. Plants that experience water shortages generally have a smaller size than plants that grow normally [15].

Lack of water also reduces chlorophyll synthesis and reduces the activity of some enzymes (eg nitate reductase). The decrease in chlorophyll content when plants lack water is related to the activity of photosynthetic apparatus and decreases the rate of photosynthesis of plants [16]. Photosynthesis is a very important metabolic process in plants, the things that must be met in photosynthesis are light, CO2, O2, chlorophyll and water. Water is very influential on the turgidity of the stomata guard cells, if there is a lack of water, the turgidity of the cells will decrease and will cause the stomata to close [17].

4. Conclusions

The use of waste oil palm bunches briquettes can increase the growth of *M. integrifolia* plants in the field. Plant response was shown by significant differences in the increase in plant height, leaf area and plant canopy area. This response was also able to compensate for the response of plants treated with manure briquettes. These two water-retaining materials show a balanced plant performance in increasing plant growth.

Refferences

[1] [LTEMP] Lake Toba Ecosystem Management Plant 2006 *Guidelines for the Management of the Lake Toba Area Ecosystem* (Medan: The Coordinating Board for the Management of the Lake Toba Catchment Ecosystem)
[2] Sihotang H, Purwanto M Y J, Widiatmaka and Sambas B 2012 *Journal of Natural Resources and Environmental Management* **2** 65
[3] Aswandi A, Kholibrina C R and Lumbantobing S 2018 *Sumatrana Forestry Research Journal* **2** 39
[4] Utomo B, Silalahi D C G and Dalimunthe A 2021 *3rd International Conference on Natural Resources and Technology IOP Conf Ser Earth Environment science* **912** 012066
[5] Budianto, Kusdarini E, Amrullah N H and Aidawiyah A 2020 Kelapa sawit sebagai basis indonesia penghasil utama biofuel dan karbon aktif
[6] Nur A, Kurniawan and E Jalaluddin 2020 *LPPM UMJ National Research Seminar*
[7] Sitompul S M and Guritno B 1995 *Plant Growth Analysis* (Yogyakarta: Gadjah Mada University Press)
[8] Punomo J, Mulyadi, Amin I and Suhardjo H 1992 *Journal of Soil and Agroclimate* **8** 61
[9] Haryanti A, Norsamsi, Sholiha P S F and Putri N P 2014 *Conversion Journal* **3**
[10] Banyo Y and Song 2013 *Leaf Chlorophyll Concentration as an Indicator of Water Deficiency in Plants* (Manado)
[11] Mamonto R, Rombang J and Lasu M 2019 *Journal of Forestry Unsrat* **1**
[12] Djaenudin D, Saefoedin B and Suhardjo H 2001 *Journal of Agricultural Research and Development Agency* **20** 32
[13] Ai N S and Lenak 2014 *Leaf curling in monocot plants when water shortage* (Manado: Biology FMIPA Sam Ratulangi University)
[14] Gardner F P, Pearce R B and Mithcell R L 1991 *Physiology of Cultivated Plants. Herawati Susilo translation* (Jakarta: UI Press)
[15] Fitter A H and Hay R K M 1981 *Environmental Physiology of Plants* (Yogyakarta: Gadjah

Mada University Press)
[16] Neneng I 2020 *BirEx Journal* **2** 14
[17] Ai N S 2012 *The Evolution of Photosynthesis in Plants* (Manado: Sam Ratulangi University. Manado)
[18] Utomo S A and Purnamasari R T 2017 *Journal of Agrotechnology Merdeka Pasuruan* **1** 22

Potential and value of utilization of community forest logging waste in Cenrana Baru Village, Cenrana District, Maros Regency

A Mujetahid*, M Dassir, A V Faradiba and Heriah

Faculty of Forestry, Hasanuddin University, Makassar, Indonesia

*Email: mujetahid@unhas.ac.id

Abstract. This study aims to determine the potential and value of using logging waste in community forests in Cenrana Baru Village, Cenrana District, Maros Regency. This research was carried out from August 2019 – January 2022 in a community forest. Based on the results of the study, it was found that the largest waste potential of the average volume of waste in acacia trees is 0.28 m^3 (30.39%), stump waste 0.06 m^3 (11.21%), waste above branch free 0.07 m^3 (8.89%), branch waste 0.04 m^3 (4.79%), while in the candlenut tree the largest waste is the main trunk 0.80 m^3 (35.87%), stump waste 0.14 m^3 (6.75%), branch waste 0.13 m^3 (6.41%), waste above branch free 0.09 m^3 (4.83%). Community forest logging waste is used by the community for making garden fences and making charcoal from acacia wood, this waste has a price value of IDR 15,000 per stem. while in the candlenut tree, the largest waste is the main trunk 0.80 m^3 (35.87%), stump waste 0.14 m^3 (6.75%), branch waste 0.13 m^3 (6.41%), the above waste is free to branch 0.09 m^3 (4.83%).

1. Introduction

Until now, the utilization of wood in Indonesia has not been effective and efficient because the amount of wood used is generally still low compared to the volume of wood that is cut. Tree parts such as stumps, branches, twigs and defective trunks are generally left in the forest and become waste [1]. Harvesting waste is a part of the trunk or part of a tree that is allowed to be cut but is not utilized based on the current utilization pattern and is left in the forest. This harvesting waste can come from stumps, branch-free stems, upper stems and branches. The results of research conducted by Mujetahid [2] show that utilization of waste will increase the supply for the needs of the wood processing industry which is still lacking. The need for logs for the industry in 2015 reached ± 63.4 million m^3 but log production was only ± 35.1 million m^3 resulting in a shortage of supply of 28.3 m^3[3]. Nurrochmat [4], stated that the installed capacity of the wood processing industry is estimated to lack the need for logs to reach ± 40 million m3/year.

Research on logging waste is mostly done on the wood waste that occurs in industrial plantation forests, but research on logging waste using wood processing in community forests has not been widely carried out so not much information is available. Research is needed on the potential and utilization of community forest logging waste. One of the potential community forests in South Sulawesi Province, especially Maros Regency, has a large enough potential that is around 15,000 ha, based on data from the Forestry and Plantation Service of Maros Regency in 2011. The potential of wood will have economic value if it has been removed from the forest to be industrial or industrial.

IOP Conf. Series: Earth and Environmental Science 1115 (2022) 012047 doi:10.1088/1755-1315/1115/1/012047

The purpose of this research is to determine the potential of community forest logging waste and the value of utilizing the waste generated in Cenrana Baru Village, Maros Regency.

2. Research Methods

This research was conducted from August 2019 - January 2022 in the Community Forest area in Cenrana Baru Village, Cenrana District, Maros Regency, South Sulawesi Province. The tools and materials used in this study were a rolling meter and tape meter used to measure the length and circumference of the logs and the diameters that could be utilized or in the form of waste from logging, calculators, writing instruments, tally sheets and haga meters. The materials used are acacia and candlenut trees which are ready to be cut down. Data collection is done through observation and recording in the field. The data collected was then processed using the formula for measuring the diameter and calculating the volume of waste with a total sample of 20 trees.

2.1. Data analysis

The data collected is then analyzed quantitatively using the formula:

2.1.1. Diameter measurement. Measure the diameter of the log using the equation below:

$$D = \frac{Dp+Du=\frac{1}{2}(d1+d2)+\frac{1}{2}(d3+d4)}{2} \quad (1)$$

Where:

D = Diameter of the log (cm)

Dp = The average diameter of the base bounty in multiples of one full cm, obtained from the shortest diameter (d1) and the longest diameter (d2) through the center of the bounty (cm).

Du = The average diameter of the tip bounty in multiples of one full cm, obtained from the shortest (d3) and longest (d4) diameters through the center of the bounty as well (cm).

2.1.2. Waste volume calculation. The calculation of the volume of waste consists of the waste stump, main stem, branches and above without branches using the Brereton metric formula, according to the following instructions for measuring and determining the contents of logs in Indonesia [5]:

$$V = 0.7854 \text{ x } D^2 \text{ x } L \quad (2)$$

Where:

V = volume or volume of logs (m^3)

L = log length (m)

D = diameter of log (cm)

0.7854 is the number of = x 3.1416

To calculate the total volume of wood, the following formula is used:

$$VTK = VKpb + VLpb \quad (3)$$

Where :

VTK = Total volume per tree (m^3)

VKpb = Volume of felled wood (m^3)

VLpb = Volume of felling waste (m^3)

To calculate the total volume of waste, the following formula is used:

$$VTL=VLtg + VLbu + VLcb + VLdbc \quad (4)$$

Where:
VTL = total volume waste per tree (m^3)
VLtg = volumestump waste (m^3)
VLbu = volume main stem waste (m^3)
VLcb = volume branch waste (m^3)
VLdbc = volume waste above branch free (m^3)

Calculation of the percentage of waste wood

$$PL = \frac{VTL}{VTK+VTL} x\ 100\% \tag{5}$$

Where:
PL = Percentage of waste wood in percent (%)
VTL = total volume of waste per tree (m^3)
VTK = total volume of usable wood per tree (m^3)

3. Results and Discussion

3.1. Potential logging waste

Forest harvesting always leaves the wood in the forest which is called harvesting waste. Logging has the potential to produce waste in the form of stumps and broken stems. Cutting the canopy (topping) and cutting branches and twigs (debranching) have the potential to produce waste in the form of branches and twigs. Meanwhile, bucking has the potential to produce waste in the form of short pieces, as well as defective wood waste such as rotting wood waste and rotten liver/*gerowong*. The definition of waste in this study, namely the part of the wood or tree that is cut down and still possible to use because it has a diameter that still meets the requirements for sawn wood for ingredients for local villagers' houses or raw material for furniture by industries in the city. If it is not utilized as much as possible by the residents, a lot of wood waste that is left or not utilized is left to rot in the forest which results in even greater carbon emissions resulting from the decaying process of wood waste at the logging location [6]. The general definition of waste as described previously is that felling waste is a part of wood from a tree due to defects, damage, or breakage, too small a diameter and a length that does not meet the requirements for a particular purpose. Based on this understanding, the following research results were obtained. The stages of forest harvesting that have the potential to produce the largest waste wood are tree felling activities [7].

3.2. Logging process

The felling team consisting of one chainsaw operator and one helper plays an important role in feller activities. Based on the experience carried out by loggers with the ability to cut trees and divide trunks an average of five trees per day. This community forest logger uses a logging team consisting of one chainsaw operator assisted by one helper. The helper is usually appointed by just anyone because it does not require special skills. In this study, the person who acts as a helper is the owner of the wood itself. This helper is in charge of helping determine the direction of the tree's fall and cleaning up disturbing things around the tree. Determination of the direction of fall is carried out by the operator by taking into account the direction of skidding, the place for the stem and avoiding breakage to a minimum.

The next stage is the division of stems which is carried out by stem-by-stem management, namely the size of the cutting starting from the base to the tip while taking into account the quality of the wood on the branches that can be used for carpentry. Before dividing the trunk, measurements were made on the stem according to the specified size using wood that had been previously oiled. Logs are cut according to the marking path that has been made into a specific bearing or sortiment. The wage system used at the time of felling is per tree and per sortimen size. For the calculation per tree, one tree is calculated IDR. 100,000 - IDR. 200,000 depending on the size of the tree, while for the count per size sortiment, for a pole with a size of 10 cm as much as IDR. 10,000/pole. For blocks with a size of 4 x 6 cm as much as IDR. 4,000/beam and size 3 x 12 cm as much as IDR. 7,000/beam and for the size of the board with a size of 2.5 x 20 cm as much as IDR. 5,000/board. Experienced chainsaw operators can produce high felling productivity [8]. Matangaran and Anggoro [9], stated that logging has the potential to produce waste in the form of the stump and broken stems, while the division of stems has the potential to produce waste in the form of short pieces, as well as defective wood waste such as rotted and rotten wood waste or gerowong.

3.3. Arrear waste

The measurement of the height of the stumps above the ground level up to the highest limit of the former ballast notch created by the volume of stump waste that occurred from the logging of acacia and candlenut wood for each tree can be seen in Table 1.

Table1. Length, diameter and volume of acacia and candlenut stumps in community forest logging activities in Cenrana Baru Village, Cenrana District, Maros Regency

Tree Name	Arrears Height	Diameter (cm)	Tree Volume (m^3)	Solid Waste Volume (m^3)	Percentage (%)
Acacia	39	56.45	1.10	0.10	7.68
	54	44.72	0.97	0.08	2.98
	21	41.79	0.66	0.03	9.84
	59	37,40	0.47	0.06	6.61
	24	40.67	0.47	0.03	8.36
	30	40.76	0.67	0.04	7.12
	45	36.79	0.60	0.05	7.91
	29	45.49	1.03	0.05	4.29
	31	42.66	1.00	0.04	15.96
	65	55.82	1.56	0.16	41.30
Amount			8.52	0.64	
Average	39.7	44.26	0.85	0.06	11.21
Hazelnut	58	68.58	2.68	0.21	8.00
	50	72.38	3.32	0.21	6.18
	31	60.83	2.15	0.09	4.18
	35	52.71	1.23	0.08	6.18
	40	63.32	2.22	0.13	5.68
	46	59.15	1.99	0.13	6.34
	38	53.74	1.40	0.09	6.15
	80	62.69	1.71	0.25	14.47
	28	53.92	1.55	0.06	4.13
	51	69.25	3.13	0.19	6.13
Amount			21.38	1.43	
Average	45.7	61.65	2.14	0.14	6.75

Table 1 shows the results of felling acacia and candlenut with an average diameter of 44.26 cm and 61.65 cm and an average stump height of 39.7 cm and 45.7 cm, respectively. The average volume of acacia waste is 0.06 m^3 and candlenut 0.14 m^3. This indicates that the chainsaw operator performs logging with an average stump height of 35 cm. The determination of the height of the stump only considers the technical ease of the feller. Cowl waste that is left behind or not used is left as waste because the community and local industry do not yet have the knowledge to process cowpea waste into useful raw materials.

3.4. Main trunk waste

The main stem waste was measured in length and the diameter of the tip and base. This waste occurs as a result of making the wrong notch and trimming the base, tip and middle of the stem that is defective/damaged. Dimensions measured were base diameter, tip diameter and stem length.

Table 2. Length, diameter and volume of acacia and candlenut main trunk waste in community forest logging activities in Cenrana Baru Village, Cenrana District, Maros Regency

Tree Name	Length (m)	Diameter(cm)	Tree Volume (m^3)	Main Stem Waste Volume (m^3)	Percentage (%)
Acacia	6.1	74.97	1.10	0.55	49.60
	7.25	74.20	0.97	0.23	23.53
	5.25	53.22	0.66	0.14	21.92
	4.3	56.02	0.47	0.17	37,10
	3.25	52.21	0.47	0.12	24.66
	4.1	51.97	0.67	0.14	20,19
	4	30.22	0.60	0.18	30.01
	4.6	67.09	1.03	0.21	20.57
	4	35,30	1.00	0.28	28.04
	11	166.93	1.56	0.75	48.32
Amount				8.52	2.77
Average	5.39	66.21	0.85	0.28	30.39
Hazelnut	7.2	98.95	2,676	1.02	38.11
	10.23	154.04	3,324	1.42	41.63
	6.07	90,70	2,153	0.49	22.25
	8.08	75.00	1,235	0.55	44.58
	12.32	127.55	2,216	0.95	41.82
	5.6	89.35	1,992	0.42	20.57
	6.05	80.81	1,401	0.53	36.73
	9.19	116.96	1,705	0.72	41.02
	6.18	84.04	1.546	0.41	26.07
	10.1	155.75	3,130	1.48	45.94
Amount			21.379	8.00	
Average	8.10	107.32	2,138	0.80	35.87

Table 2 shows the average main stem length is 5.39 m in acacia and candlenut trees 8.10 m, the average diameter is 66.21 cm and 107.32 cm, the average volume is 0.29 m^3 and 0.80 m^3. Even distribution of the base of the tree is minimal because the wage system is calculated per sortimen where the loggers tend to cut the logs to even out the diameter, so the logger operator maximizes the length of the main trunk so that the wages obtained are even greater. This is in accordance with the statement of Indrawati [10] which states that the amount of main stem waste from logging activities is due to the large diameter of the main stem waste and the length of the waste. Another factor causing the high percentage of waste in addition to the diameter and length of the wood is how large the volume of unused wood is compared to the total volume of trees.

3.5. *Waste above branch free*

The waste above is free of branches, namely the main stem from the top of the stump or buttress to the first branch. This waste is in the form of short pieces or logs. The dimensions measured include the diameter of the base, the diameter of the tip and the diameter of the base.

Table 3. Length, diameter and volume of waste on free acacia and candlenut branches in Community Forest Logging Activities in Cenrana Baru Village, Cenrana District, Maros Regency

Tree Name	Length (m)	Diameter(cm)	Tree Volume (m3)	Main Stem Waste Volume (m3)	Percentage (%)
Acacia	6.53	13.54	1.10	0.09	8.51
	4.3	11.40	0.97	0.04	4.54
	3.68	7.29	0.66	0.02	2.33
	3.3	9.08	0.47	0.02	4.52
	5.2	9.87	0.47	0.04	8.50
	8	7.87	0.67	0.04	5.78
	7.4	14.65	0.60	0.12	20.92
	7	17.64	1.03	0.17	16.57
	8.5	13.79	1.00	0.13	12.74
	3.9	15,13	1.56	0.07	4.49
Amount				8.52	0.75
Average	5.78	12.03	0.85	0.07	8.89
Hazelnut	4.3	14.33	2.68	0.07	2.59
	2.4	15,19	3.32	0.04	1.31
	3.1	15.61	2.15	0.06	2.75
	3.1	19.52	1.23	0.09	7.51
	2.83	18,50	2.22	0.08	3.43
	7.53	20.06	1.99	0.24	11.95
	3.72	16.82	1.40	0.08	5.89
	4.1	14.49	1.71	0.07	3.96
	3.26	17.52	1.55	0.08	5.08
	3.78	20.06	3.13	0.12	3.82
Amount			21.38	0.93	
Average	3.81	17.21	2.14	0.09	4.83

Table 3 shows that the average length of waste above free branches is 5.78 m and 3.81 m, with a diameter of 12.03 cm and 17.21 cm, a volume of 0.07 m^3 and 0.09 m^3. Branch waste occurs because in branch cleaning activities not all branches are taken. After all, they are considered to have small diameters. The amount of waste above branch-free shows that logging activities are caused by the large diameter of the waste above branch-free and the length of the waste, as well as the large volume of total wood that can be utilized.

3.6. *Branch waste*

Branch waste is part of the stem from the first branch to the crown which is an extension of the main stem. Dimensions measured include base diameter, tip diameter and stem length.

Table 4. Length, diameter and volume of acacia and candlenut branch waste in community forest logging activities in Cenrana Baru Village, Cenrana District, Maros Regency

Tree Name	Length (m)	Diameter (cm)	Tree Volume (m^3)	Main Stem Waste Volume (m^3)	Percentage (%)
Acacia	12.9	28.06	1.10	0.08	7.33
	14.7	31.91	0.97	0.09	9.15
	12	24.49	0.66	0.03	4.73
	5.4	11.50	0.47	0.01	1.68
	10.9	15.92	0.47	0.03	6.54
	5.6	14.27	0.67	0.02	2.29
	9.4	24.62	0.60	0.03	5.22
	8.3	15,19	1.03	0.02	1.74
	10.1	16.72	1.00	0.06	5.14
	14.2	34.17	1.56	0.06	4.05
Amount				8.52	0.43
Average	10.35	21.68	0.85	0.04	4.79
Hazelnut	11.4	28.66	2.68	0.08	3.06
	23.8	46.66	3.32	0.30	8.99
	12.41	27.55	2.15	0.09	4.36
	11.3	32.42	1.23	0.11	8.55
	9.6	35.06	2.22	0.10	4.73
	21.8	46.21	1.99	0.25	12.32
	15.9	31.91	1.40	0.10	6.93
	11.5	31.82	1.71	0.10	6.04
	7.8	30.89	1.55	0.07	4.39
	16.4	43.44	3.13	0.15	4.68
Amount			21.38	1.34	
Average	14.19	35.46	2.14	0.13	6.41

Table 4 shows that the average length of branch waste is 10.35 m and 14.19 m, with a diameter of 21.68 cm and 35.46 cm, a volume of 0.04 m^3 and 0.13 m^3. This waste is so large that the logger operator also cuts the tip of the tree into several long pieces. The occurrence of branch waste is caused by felled trees generally have many shoots/tips, but have a small diameter so that a lot of them become waste. The results of this study are in line with the research of Erajaa et al [11], which said that the majority of waste wood due to logging is wood that comes from the tree canopy and has a small diameter.

3.7. Waste volume recapitulation

The results of measuring the diameter and length of the potential for acacia and candlenut logging waste can be obtained by recapitulating the composition of the average volume and percentage of four types of waste as shown in Table 5.

Table 5. Recapitulation of average volume and percentage of waste in logging activities of community forest acacia tree species in Cenrana Baru Village, Cenrana District, Maros Regency

Tree Name	Waste Composition	Average Volume (m^3/tree)	Percentage (%)
Acacia	Arrear Waste	0.06	11.21
	Main Trunk Waste	0.28	30.39
	Waste Above Branch Free	0.07	8.89
	Branch Waste	0.4	4.79
Total		0.45	55.28
Hazelnut	Arrear Waste	0.14	6.75
	Main Trunk Waste	0.80	35.87
	Waste Above Branch Free	0.09	4.83
	Branch Waste	0.13	6.41
Total		1.16	53.86

Table 5 shows that the largest waste wood comes from the main stem waste with an average waste percentage of 30.39% and 35.87%, the average volume is 0.28 m^3 and 0.80 m^3 of the total waste. This happens because the length does not meet the requirements requested so it is left in the forest as waste. The occurrence of this waste is also caused by topping activities, loggers tend to cut the ends of the wood at the boundary where the relative diameter is still close to the base diameter, as a result, there is a log residue between the topping limit and the first branch in the form of waste wood.

Second order of wood stump waste, namely 11.21% and 6.75%, volume 0.06 m^3 and 0.14 m^3 of the total waste, while in candlenut wood the waste sourced from above is free of branches with an average percentage of 10.20% and volume 0.129 m^3. The occurrence of this waste is caused by topping activities, loggers tend to cut the ends of the wood at the limit where the relative diameter size is still close to the base diameter, as a result, there are residual logs between the topping limit and the first branch in the form of waste wood and the large percentage of the waste is because the wood located on the branch has a small diameter and long. Because operators who are not able to process small size wood will be left in the felling plot in the form of waste because they only use wood with a certain diameter and are used by residents only.

The third order of waste sourced from the above section is free of branches with an average percentage of 8.89% and a volume of 0.07 m^3 of the total waste in acacia wood while candlenut wood comes from branches with an average percentage of 6.41% and a volume of 0.13 m^3. The large percentage of the waste is because the wood located on the branches has a small diameter and length. Because operators who are not able to process small size wood will be left in the felling plot in the form of waste because they only use wood with a certain diameter size and are used only by residents. Because operators who have not been able to process small sizes of wood will be left in the felling plot in the form of waste because they only use wood with a certain diameter size and are only used by residents and loggers tend to cut the ends of the wood at the limit where the relative diameter is still close to the diameter of the base.

The fourth order of acacia wood waste is sourced from branch waste with an average percentage of 4.79% with a volume of 0.04 m^3, The large percentage of the waste is because the wood located on the branches has a small diameter and length. Because the operator is not able to process small size wood, it will be left in the felling plot in the form of waste because it only uses wood with a certain diameter size, while candlenut wood is sourced from above branch-free waste with a percentage of 4.83 with a volume of 0.09%. This happens because the logger operator does the logging without considering the size of the waste, but based on the position of the felling which makes logging easier, the cutting time is faster because the logging fee is based on the volume of the felled wood. When applying the lowest possible logging, the time used is longer.

3.8. Waste utilization

Harvesting of forest products can be defined as a series of forestry activities that convert trees or other biomass into forms that can benefit the economic and cultural life of the community. Timber harvesting has several purposes that are not only economically beneficial but not ecologically damaging, so a system for its utilization is needed. Based on the results of research in the field, the potential for acacia wood waste that is utilized by local communities can be seen in table 6.

Table 6. Utilization of acacia wood logging waste in Cenrana Baru Village, Cenrana District, Maros Regency

No.	Logging Section that Becomes Waste	Waste Average Volume (m^3/tree)	Waste Utilization	Parties who use waste
1.	Arrears	0.06	Not used	Not used
2.	Main Trunk	0.28	Garden Fence	Residents
3.	Above Branch Free	0.07	Charcoal and Garden Fence	Residents
4.	Branch	0.04	Charcoal and Garden Fence	Residents

Table 6 shows that residents in Cenrana Baru Village, Cenrana Subdistrict, use logging waste for daily needs and market it to residents and the city. Residents use the waste from the main stem and branch-free as material for fence posts, waste from above is branch-free and branch waste is the raw material for making charcoal and firewood. Branch waste and branch-free waste above can be utilized so these two wastes are referred to as logging waste. Wood waste that has a minimum length of 2 m with a diameter of < 30 cm, the community uses for making garden fences and wood charcoal for IDR. 15,000 per stem. The community also uses charcoal for their daily needs and is traded at a price per sack weighing 50 kg for IDR. 100,000,-. This is in line with research [8], which suggests that people use harvesting waste as a trading commodity and as fuel for their own needs. The candlenut wood waste is not utilized by the community.

4. Conclusion

Based on the results of data analysis obtained from observations and measurements of community forest logging activities in Cenrana Baru Village, Maros Regency, several conclusions can be obtained as follows are the largest potential waste volume of waste in acacia trees is main trunk waste of 0.28 m^3 (30.39%), stump waste 0.06 m^3 (11.21%), waste above branch free 0.07 m^3 (8 .89%), branch waste 0.04 m^3 (4.79%), while in candlenut tree the largest waste was the main trunk 0.80 m^3 (35.87%), stump waste 0.14 m^3 (6.75%), branch waste 0.13 m^3 (6.41%), waste above branch free 0.09 m^3 (4.83%). Waste that can be used as raw material for making garden fences is the main stem, while branch waste and above-free branches are used as charcoal from acacia wood, this waste has a price value of IDR. 15,000 per stem.

References

[1] Ujang S, Matangaran JR and Harmawan F 2013 *Jurnal Manusia dan Lingkungan* **21**
[2] Mujetahid 2010 *Jurnal Perennial* **6**
[3] Ministry of Environment and Forestry, 2016. Statistik Produksi Kehutanan 2015. Kementrian

Lingkungan Hidup dan Kehutanan. Jakarta
[4] Nurrochmat R R 2016 *Prediksi keseimbangan supply-demand hasil hutan kayu Indonesia*
[5] Directorate General of Forestry Production Development 2004 *Metode Pengukuran Kayu Bulat Rimba Indonesia* (Jakarta: Departemen Kehutanan)
[6] Yuniawati and Suhartana 2014 *Jurnal Ilmu Lingkungan* **12** 21
[7] Budiaman A and Pradata AA 2013 *Journal of Tropical Forest Management* **19** 194
[8] Dulsalam, Suhartana S, Sukadaryati and Yuniawati 2014 *Pemanenan kayu di hutan tanaman* (Bogor: Forda Press)
[9] Matangaran J R and Anggoro R 2012 *Jurnal Perennial* **8** 88.
[10] Indrawati B 2009 *Excess demand kayu dan dampaknya pada illegal logging di Indonesia* (Jakarta : Ubhara Jaya)
[11] Erajaa S, Halme P, Kotioho JS, Markkanen A, Toivaren T 2010 *Silva Fennica* **44** 203

IOP Conf. Series: Earth and Environmental Science 1115 (2022) 012048 doi:10.1088/1755-1315/1115/1/012048

The availability of Perum Perhutani potential forest area to support the PT. PLN's biomass cofiring sustainability: A Mini Review

A Ibrahim[1, 2], **D Hermawan**[2*] **and J Sutiawan**[3]

[1] Directorate of Commercial, Perum Perhutani, Jakarta, Indonesia.
[2] Department of Forest Products, Faculty of Forestry and environment, IPB University, Bogor, Indonesia
[3] Faculty of Forestry, Universitas Sumatera Utara, Medan, Indonesia

*Email: dedehe@apps.ipb.ac.id

Abstract. High-rank coal reserves continue to decrease, while what is widely available is in the form of low-rank coal with low calorific value and high sulfur. Biomass can be a solution for providing fuel to replace low-rank coal (4,000 kcal/kg). To ensure the sustainability of the supply of biomass cofiring, it is necessary to have sufficient land available for energy plantations/forestry plantations (HTE). This research is aimed at assessing the availability of land in forest areas in Perum Perhutani for energy crop activities by using the right types of plants to supply biomass cofiring. Perhutani has entered into a cooperation agreement with PT. PLN to meet the need for cofiring-biomass PLTU. Currently on the island of Java, there are 11 PLTUs with a total electricity capacity of 14,645 MW. For 5% biomass cofiring, 189,849 ha of HTE land is required. Perum Perhutani has established a business portfolio of Biomass Plants covering an area of 67,356 ha. This area is equivalent to the effective area of 48,496 ha of HTE which supplies 3,756 MW of electricity (26% of the total capacity of the Java PLTU). The types of plant biomass used were Gamal (*Gliricidia sepium*) and Kaliandra (*Calliandra callothyrsus*).

1. Introduction

The Indonesian government continues to anticipate energy scarcity in Indonesia by encouraging the use of new and renewable energy (EBT), the Ministry of Energy and Mineral Resources informs that high-rank coal reserves continue to decrease, and prices tend to rise. Meanwhile, the national coal reserves still quite widely available today are low-rank coal, with low heating value and high sulfur. Therefore, biomass is a solution for providing fuel with low ash and sulfur content with a heat content that can match low-rank coal (above 4,000 kcal/kg). Nationally, the current energy mix is still dominated by fossil energy. Indonesia has a very large NRE potential but has not been managed properly [1].

Currently, Indonesia already has around 65.8 thousand MW of power plants, but the use of EBT has only reached 9,000 MW (only around 13-14%). By 2025, the Indonesian government targets the achievement of utilization to reach 23% as mandated in the RUEN. In addition to the EBT mix target, the issue of high Greenhouse Gas emissions from coal-fired power plants encourages efforts to reduce GHG emissions from steam power plants so that the emission reduction targets declared in the Paris Agreement can be achieved. Several action plans related to reducing GHG emissions from coal-fired

IOP Conf. Series: Earth and Environmental Science 1115 (2022) 012048 doi:10.1088/1755-1315/1115/1/012048

power plants are: (1) a moratorium on the construction of coal-fired power plants, (2) a phase-out of coal-fired power plants in 2050, and (3) an increase in the efficiency of coal-fired power plants by implementing low-carbon technology through biomass co-firing [1].

2. Wood chip as a renewable energy source

Wood Chip which can also be in the form of sawdust is a renewable energy source from sawdust and other wood materials. Currently, various countries in the world are competing and switching to using biomass as a substitute for coal to reduce greenhouse gas emissions and increase the use of renewable energy. Wood chips and wood pellets are included in the category of carbon-neutral fuel because wood as a source of biomass comes from photosynthesis using CO2 from the atmosphere. The carbon will be released back into the atmosphere when used as an energy source. This shows that on the balance sheet, there is no addition of CO2 in the atmosphere or carbon neutral and environmentally friendly [2].

3. Co-firing of PLTU PLN coal

The potential of the domestic good biomass market is very prospective regarding the number of needs. To meet the need for 1% co-firing at PLTU in Indonesia 17.470 tons of biomass per day or 5 million tons of wood pellets per year are needed. PLN needs 109 PLTU units with a total MW of 18,275. It requires 17,470.80 tons/day of biomass or 5,101,473.60 tons/year of biomass (wood pellets). Assuming the calorific value of wood biomass is 4000 kcal/kg, total biomass production is 5 tons/ha/year, and wood biomass harvesting can be carried out every three months for 20 years. So the demand for wood biomass supply and wood biomass land area to replace 1% and 5% of coal energy supply in all existing PLTUs are 0.58 million tons and 2.89 million tons, respectively, and a land area of 38,508 ha and 192,539 ha [3].

Perhutani is targeting the opportunity to provide feedstocks to fulfill the co-firing of PLTU in Java. Co-firing for PLTU is an alternative effort to reduce the use of coal by replacing some of the coal with biomass fuel while still paying attention to the quality of the fuel as needed. In addition to supporting the contribution of renewable energy to the national energy mix, this program also has a positive impact on the development of a productive people's economy (circular economy) through the creation of a populist electricity ecosystem that involves active community participation in its implementation (www.esdm.go.id).

Although this is a very big opportunity, Perhutani must consider the suitability of the biomass price by PLN with the cost of procuring biomass by Perhutani, where Perhutani, as one of the state-owned enterprises, must also benefit from the sale of the biomass. On January 28, 2022, Perum Perhutani signed the Cooperation with PT. PLN for the fulfillment of coal biomass co-firing at two PLTUs owned by a subsidiary of PT. PLN, namely PLTU Sluke in Rembang (PT. PJB) and PLTU Pelabuhan Ratu in Sukabumi (PT. Indo Power). The biomass supply for the two PLTUs is fulfilled with 12,000 tons of sawdust biomass per year. The target of 12,000 tons per year is a very minimal quantity because of the bid price from PT. PLN, whose economic value is still much lower when compared to selling wood pellet products to the export market.

4. Solutions to land productivity problems

Perum Perhutani, based on PP72/2016, was assigned to carry out management activities for the use of state forests covering an area of approximately 2.4 million hectares on the island of Java. However, with PP No. 23 of 2020 concerning Forestry Governance, the government will issue a new policy on KHDPK (Forest Areas with Special Management). As a consequence, the government will withdraw forest land (+ 1.2 million hectares) that Perum Perhutani now manages to become KHDPK. One of the things that the government considers in issuing this policy on KHDPK is the government's belief that Perum Perhutani's productivity in land use has not been maximized, so it is deemed necessary to transfer its management to other parties.

The development of the biomass business in Perhutani is also driven by initiatives to improve the low productivity of forest land managed by Perhutani due to the dominance of forest land use using long-cycle business commodities. The community around the forest does not like long-cycled commodities because people do not want to wait a long time to be able to enjoy the harvest. This situation can be a triggering factor for increasing social conflicts that disrupt forest sustainability. Economic analysis shows that the biomass industry is a quick-yield commodity and can also be categorized as a Green Economy activity. With the inclusion of Perhutani in the biomass business development industry, it is hoped that this activity can be an effort to improve the productivity of land use to obtain new sources of income for Perhutani.

5. Biomass Production Plan for wood chips/sawdust for co-firing

In 2022, in the context of developing the biomass industry, Perhutani will build 1 (one) factory for the production of wood pellets and wood chips and 2 (two) factories for PLN's sawdust co-firing products (Table 1). The Biomass Industry in Rembang and Sukabumi will develop Saw Dust products for Cofiring PLTU Rembang and PLTU Sukabumi, each with a production capacity of 12,000 tons per year. Raw Materials for each industry, including Gamal Wood, and Kaliandra Wood, amount to 20,000 tons/year [4]. To meet the needs of 20,000 tons of raw materials, an area harvested annually is 330 hectares or 660 hectares of effective area. At least to meet one industrial sawdust for PLN co-firing, an HTE area of 1,000 hectares is required.

Table 1. Perhutani biomass industry development plan

Total generating capacity (MW)	5% cofiring land requirement (ha)	HTE (ha) Perum Perhutani	KPH Location
600	7778		
1600	20741		
1800	23334		
625	8102		
945	12250		
315	4083		
1050	13612	3050	Sukabumi
990	12834	8817	Purwakarta, Sumedang, Indramayu
660	8556		
630	8167	3998	Saradan
1320	17112	11430	Semarang, Telawa
1320	17112	4858	Purwodadi, Pati
630	8167	5747	Mantingan, Blora, Cepu
700	9074	10757	Jatirogo, Padangan, Bojonegoro
660	8556	37	Madura
800	10371		
14645	189849	48694	

Perhutani has designated 73,000 ha of forest land for Energy Plantation Forest as a source of energy wood for biomass industry activities (Table 2). The energy plants planted are species that can be coppice or ditrubus in the long term (replanting at the age of 15 years). Namely Gamal (Glicidia sepium/ gliriside) and Red Calliandra (Calliandra callothyrsus) have high-calorie content equivalent to light coal.

IOP Conf. Series: Earth and Environmental Science 1115 (2022) 012048 doi:10.1088/1755-1315/1115/1/012048

Table 2. Total area of HTE Perhutani

Cluster Jawa Tengah		Cluster Jawa Timur		Cluster Jawa Barat	
KPH	Area (Ha)	KPH	Area (Ha)	KPH	Area (Ha)
Blora	1925	Bojonegoro	2332	Indramayu	845
Cepu	1742	Jatirogo	3436	Purwakarta	4704
Mantingan	2651	Padangan	10421	Sukabumi	4346
Pati	731	Saradan	6084	Sumedang	8507
Purwodadi	1260				
Semarang	7178				
Telawa	1193				
Total	16680	Total	22273	Total	18402

Perhutani also sees the opportunity to use felling waste from tool wood and sap-producing wood such as pine. Logging waste from Pine replanting activities currently has a potential of around 5,000 ha/year. Forms of wood waste that can be used as raw material for wood pellets, for example, are sortimen used square, cut shoots, branches, plant twigs. In addition, wood from thinning pine trees still has a great potential of around 10,000 ha/year. Wood pellets made from pine wood are known to have high calorific value and meet market standards. Perhutani also sees the potential for waste outside of Perhutani's production which is quite potential, namely Sengon plywood industry waste, teak wood processing industry, and other types of wood.

6. Perhutani biomass development policy

Biomass development in Perhutani is intended as a new portfolio for quick business yields and to support the National Energy Mix target of 23% EBT in 2025. Perhutani has designated 73,000 ha of Energy Plantation Forest as a source of energy wood for biomass industry activities. If Perhutani can supply the need for feedstocks (biomass raw materials) nationally, the progress of meeting the national energy mix target will be able to increase significantly.

The development of the Biomass industry is integrated with the management of Energy Plantation Forests (HTE), intended so that the biomass industry activities always get a sustainable supply of feedstock [5-7]. The implementation of HTE uses an agroforestry cropping pattern system with a composition of 70% energy crops and 30% agricultural plant partnership area, following the wishes of forest farmers. This method is believed to have advantages because it involves the community actively in securing energy crops in the 70% cluster. Upstream, the community can enjoy waste feedstock in the form of leaves for animal feed, and also feedstock stands can be used as a place for bees to fly. To build a Social Safeguard System for the biomass business, the agroforestry farmer institutions are intensively supported by Perhutani. In addition to planting on Perhutani land (nuclear land), farmers are encouraged to plant feedstock on their land (plasma land). Perhutani will purchase the results as an additional supply of biomass feedstock. Biomass industry activities are also used as an economic safeguard system for the preservation of energy plantations (Figure 1).

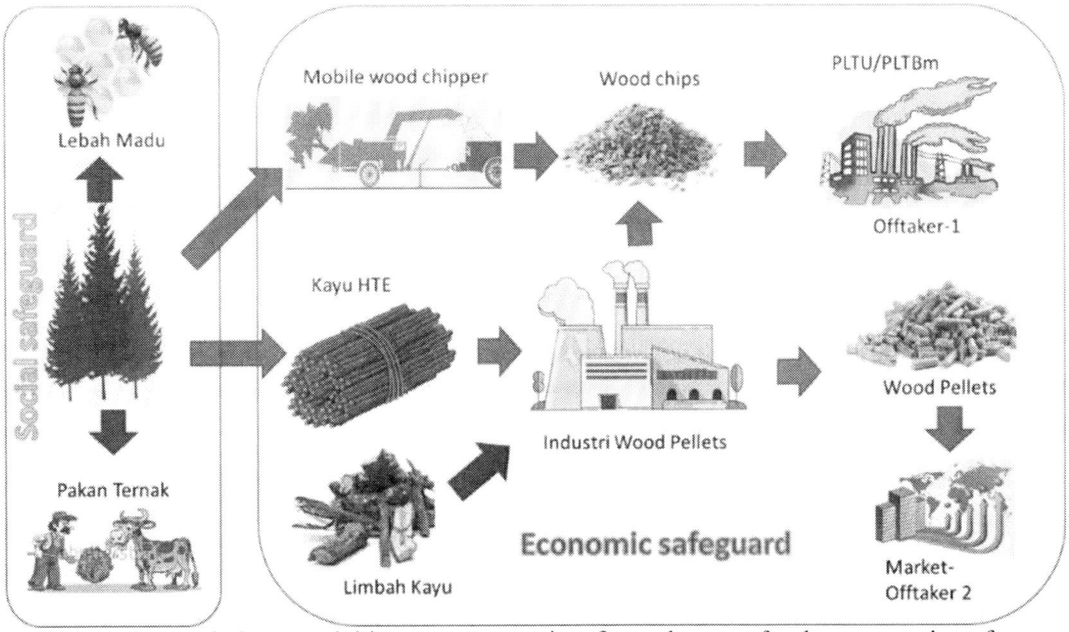

Figure 1. Biomass industry activities as an economic safeguard system for the preservation of energy plantations

1. The National Energy Policy targets 23% of new and renewable energy by 2025. This policy has encouraged all related elements of society to explore all potential bioenergy resources to achieve the energy mix target.
2. The development of the biomass business in Perhutani is driven by initiatives to take advantage of opportunities to use less productive land or land specifically designated to plant wood as an energy source. The energy wood produced must be managed sustainably through energy gardens to meet the continuous supply of renewable energy products based on biomass.
3. Perhutani's biomass industry development portfolio has implemented a cropping pattern using an agroforestry system, with a composition of 70% energy crops and 30% agricultural plant partnership area following the wishes of forest farmers. This method is believed to have advantages because it involves the community actively in securing energy crops in the 70% cluster.
4. Market opportunities for international and domestic biomass products are still very open, so in the production strategy and marketing strategy to pay attention to the nature and characteristics of the market so that Perhutani Biomass production can compete in local and global markets.
5. As a wood-pellet exporter to South Korea, Indonesia still ranks number 3 in market share after Malaysia and Vietnam. Indonesia has the opportunity to achieve a larger market share because wood pellets from Indonesia must be more competitive, especially with Malaysia. Wood pellet production from Indonesia must be able to supply in larger quantities from a more efficient production process to offer prices that sell cheaper.
6. Perhutani is targeting the opportunity to provide feedstocks to fulfill the co-firing of PLTU in Java as an alternative effort to reduce coal use by replacing some coal with biomass fuel while still paying attention to the fuel quality as needed. In addition to supporting the contribution of renewable energy to the national energy mix, this program also has a positive impact on the development of a productive people's economy (circular economy) by creating a populist electricity ecosystem that involves active community participation in its implementation.

7. Although this is a very big opportunity, Perhutani must consider the suitability of the biomass price by PLN with the cost of procuring biomass by Perhutani, where Perhutani as one of the SOEs must also benefit from the sale of the biomass

7. Conclusion

Currently on the island of Java, there are 11 PLTUs with a total electricity capacity of 14,645 MW. For 5% biomass cofiring, 189,849 ha of HTE land is required. Perum Perhutani has established a business portfolio of Biomass Plants covering an area of 67,356 ha. This area is equivalent to the effective area of 48,496 ha of HTE which supplies 3,756 MW of electricity (26% of the total capacity of the Java PLTU). The types of plant biomass used were Gamal (Gliricidia sepium) and Kaliandra (Calliandra callothyrsus). HTE uses an agroforestry cropping pattern with a composition of 70% energy crops and 30% agricultural crops. This method is quite effective to involve the community. In addition, farmers are also encouraged to plant feedstock on their own land (plasma land) whose results will be purchased by Perhutani. The economic safeguard system activities for the conservation of energy plantations are quite positive for the development of a productive people's economy (circular economy) through the creation of a populist electricity ecosystem that involves active community participation in its implementation.

References

[1] Peraturan Pemerintah No. 79 Tahun 2014 tentang Kebijakan Energi Nasional (Jakarta: Lembaran Negara RI)
[2] Gunawan I, Damayanti A, Muharyani N, Amin Y and Pari G 2020 *IOP Conf Ser Earth Environ Sci* **415** 012025
[3] Ideas Semesta Energi 2021 *Studi Kelayakan Pengembangan Industri Biomassa Terpadu di Perum Perhutani Divisi Regional Jawa Tengah* (Jakarta: PT. Ideas Semesta Energi)
[4] Hendrati R L 2020 *Pemuliaan Tanaman Hutan Tropis Penghasil Biomassa Kayu Untuk Kemandirian Energi Nasional* (Jakarta: IPB Press)
[5] Sidabutar V T P 2018 *Jurnal Ilmu Kehutanan* **12** 99
[6] Yuldashev F, Illukpitiya P, Tegegne F and Ekanem E 2020 *International Wood Products Journal* **11** 173
[7] Yulianti M, Adinugroho W C and Krisnawati H 2020 *IOP Conf Ser: Mater Sci Eng* **935** 012049

Dynamics of environmental change and affecting factors in the Malili lake complex, South Sulawesi

Rahmawati[1,2*], J Jompa[3], Budimawan[3] and A A Arief [3]

[1]Doctoral Program of Agricultural Science,Graduate School Hasanuddin University, Makassar, Indonesia
[2]Muhammadiyah University of Makassar, Makassar, South Sulawesi 90221, Indonesia
[3]Faculty of Marine Science and Fishery, Hasanuddin University, Makassar, Indonesia

[*]Email: rahmawatiralim@gmail.com

Abstract. The exploitation of the Malili Lake Complex in South Sulawesi as a natural resource that is not accompanied by wisdom has caused various major problems, such as damage to natural resources, loss of resources and the emergence of various wastes, which result in a decrease in the quality of the environment.The objectives of this research are: to examine in depth the dynamics of the quality of the aquatic environment and the influencing factors. Testing of environmental parameters measured in the field using a tool, namely the Water Quality Checker. Parameter measurements carried out in the field during June-August 2021 by taking six observation stations, namely: Salonsa Beach, Dream Beach, Malili River, Karebbe River, Sorowako Pier and Timampu Pier. The condition of the waters in has been polluted by the presence of domestic and industrial waste and the activities of the people who live in the vicinity.

1. Introduction

Lake ecosystems are the foundation of human life in fulfilling their present and future needs, because lake ecosystems provide productive natural resources both as a source of raw water for drinking and daily needs, sources of protein, minerals and energy, transportation media, and as a source of energy. tourist area. The importance of maintaining lake ecosystems with a lake revitalization program is regulated in the Regulation of the Minister of Environment and Forestry 2020, namely: The strategy of the Ministry of Environment and Forestry in supporting National Priorities for Food, Water, Energy and Environment Security is by establishing and protecting protected areas, strengthening Forest Management Unit (KPH), forest and watershed rehabilitation, and lake revitalization. In addition, the Ministry of Environment and Forestry carries out environmental quality improvement through forest fire prevention, environmental and forestry law enforcement, waste and waste management, and monitoring of water, seawater and air quality.

The typology of lakes in Indonesia varies widely and most of the lakes in Indonesia are natural lakes. The number of lakes in Indonesia reaches 840 large and small lakes. On the island of Sumatra there are 170 lakes with a maximum area of 3,700 km^2, on the island of Kalimantan 139 lakes with a maximum area of 1,142 km^2, on the islands of Java and Bali as many as 31 lakes with a total area of 62 km^2, on the island of Sulawesi there are 30 lakes with an area of 1,599 km^2, and on the island of Sulawesi there are 30 lakes with an area of 1,599 km^2. Papua Island has 127 lakes with an area of more than 600 km^2. Based on this information, it is known that Sumatra has the most many lakes and

IOP Conf. Series: Earth and Environmental Science 1115 (2022) 012049 doi:10.1088/1755-1315/1115/1/012049

Sulawesi has the largest average lake area. Sumatra has the largest lake, Lake Toba with a body area of nearly 1,200 km^2, the maximum depth measured in 2002 was 505 m (529 m in 1933) with the lake water level at 902.5 m above sea level [8]. In Sulawesi also found a number of tectonic lakes with oligotrophic water bodies, cliffs around the lake and steep lakebed walls such as Lake Matano, Towuti, Mahalona, Poso, and Lindu.

The utilization of the Malili Lake Complex for development activities affects the quality of the surrounding waters. Development which is a process of change to improve the standard of human life, is also inseparable from the activities of utilizing natural resources such as lake ecosystems. This activity often causes changes that have an impact on the availability of natural resources and have an impact on the quality of the environment. The higher the rate of development, the higher the level of utilization of natural resources and the greater the changes that occur in the environment. Exploitation of lakes as natural resources that are not accompanied by wisdom causes various big problems, such as damage to natural resources, loss of resources and the emergence of various wastes, which result in a decrease in environmental quality [9]. The aim of this research are: examine in depth the dynamics of the quality of the aquatic environment and influence factors in the Malili Lake Complex, South Sulawesi.

2. Methodology

This research was conducted during June-August 2021 at the Malili Lake Complex (Lake Matano, Lake Towuti and Lake Mahalona) South Sulawesi. The location selection was done deliberately with the consideration that this area has abundant potential for natural resources, exotic natural scenery, and the diverse socio-economic life of the local community.

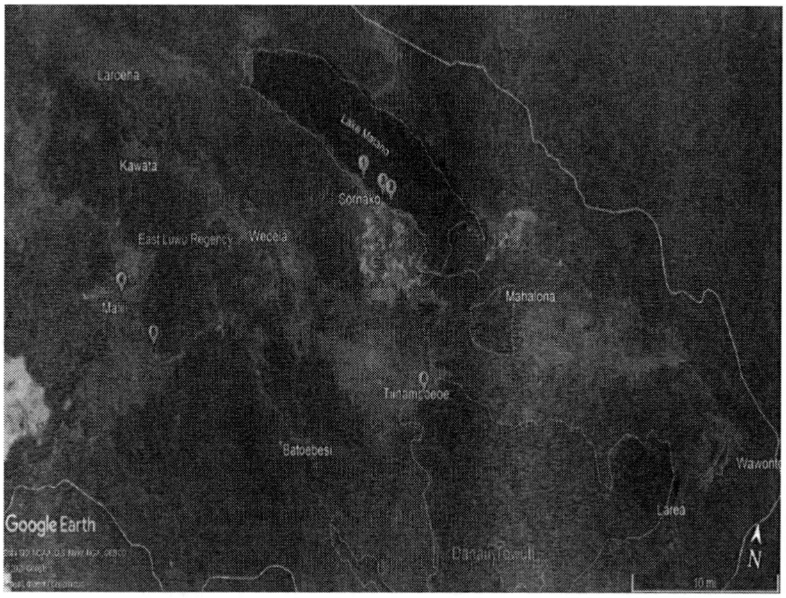

Figure 1. Malili Lake complex, South Sulawesi

Based on Figure 1, it can be determined that the environmental quality parameter observation stations are follows at Table 1:

Table 1. Water sampling Station in Malili Lake complex, South Sulawesi

Number	Place	Coordinate Position	Description
1	Salonsa Beach	5-2.5157036 E121.34891	Tourist and industrial area, densely populated settlements,near from Matano Lake
2	Impian Beach	5-2.523985 E 121.369876	Tourist and industrial area, densely populated settlements,near from Matano Lake
3	Malili River	5-2.637646 E 121.111499	Tourist area, Hydroelectric power plant,originates in three lakes: Matano, Towuti and Mahalona Lake
4	Karebbe River	5-2.671308 E 121.160296	Tourist area, Hydroelectric power plant,originates in three lakes: Matano, Towuti and Mahalona Lake
5	Sorowako Pier	5-2.520754 E 121.361144	Shipping and tourist area, near from Matano Lake
6	Timampu Pier	5-2.659556 E 121.426263	Shipping and tourist area, between Mahalona and Towuti Lake

Testing of environmental parameters measured in the field using a tool, namely the Water Quality Checker. Parameter measurements carried out in the field during June 2021 by taking six observation stations, namely: Salonsa Beach, Impian Beach, Malili River, Karebbe River, Sorowako Pier and Timampu Pier. The use of this measuring instrument is very important to see the water quality at the six observation stations. The six observation stations are still in one stream from Lake Matano, Mahalona and Lake Towuti . Before use, first prepare stationery such as pencils, small blackboards, paper, raffia rope. The parameters to be tested in full are presented in Table 2 below:

Table 2. Result test of water quality with using water quality checker in Malili Lake Complex, South Sulawesi

No	Parameter	Description	Unit
1	Turbidity	The level of turbidity or sedimentation in the water	NTU
2	Amonia/NH_3-N	A gaseous chemical compound with a very pungent odor	mg/l

Further testing was carried out at the Water Quality Laboratory, Faculty of Engineering, Hasanuddin University, Makassar. Water sampling was carried out in the morning, when the weather was still cool and the water was still calm. Water samples were put in plastic bottles, 1.5 meters each and then labeled 1 Salonsa Beach, 2 for Dream Beach, 3 for Malili River, 4 for Karebbe River, 5 for Sorowako Pier, 6 for Timampu Pier. Next, put in styrofoam coated with ice cubes on the bottom and top, ready to be taken to the test site. The parameters (10 parameters), tested are presented in full in Table 3 below:

Table 3. Description of parameter with laboratory tests in the Malili Lake Complex, South Sulawesi

No	Parameter	Description	Unit
1	Temperature	Hot or cold conditions	mg/liter
2	Total Dissolved Solid (TDS)	Amount of dissolved solids in water	mg/liter
3	Acidity (pH)	The degree of acidity of the lake	-
4	Dissolved Oxygen (DO)	Amount of Oxygen in the water	mg/liter
5	Biochemical Oxygen Demand (BOD)	Amount of biochemically degraded in water	mg/liter
6	Chemical Oxygen Demand (COD)	Measurement of Oxygen equivalent of organic and inorganic materials	mg/liter
7	Nitrat (NO_3)	The amount of nitrogen in the water	mg/liter
8	Phospate (PO_4)	Phosphorus forms that can be utilized by plants	mg/liter

3. Results

The study was conducted by testing a sample of 10 parameters, including: temperature, Total Dissolve Solid (TDS), degree of acidity (Ph), Dissolved Oxygen (DO), Biochemical Oxygen Demand (BOD), Chemichal Oxygen Demand (COD), Nitrate (NO_3), Phosphate (PO_4) Codencity (Condensity), Turbidity (Turbidity), and Ammonia (NH_3-N) at six observation stations namely: Salonsa Beach, Dream Beach, Malili River, Karebbe River, Sorowako Pier and Timampu Pier. The results of measurements and tests for each parameter can be seen in the following Table (Table 3-14) and Graph (Figure 2-12):

Table 4. Results test of temperature in the Malili Lake Complex, South Sulawesi

No	Observation Station	Results Test (°C)	
		Test 1	Test 2
1	Salonsa Beach	29,00	28,50
2	Impian Beach	29,20	28,30
3	Malili River	29,50	28,20
4	Karebbe River	29,10	28,20
5	Sorowako Pier	29,70	28,30
6	Timampu Pier	29,70	28,40

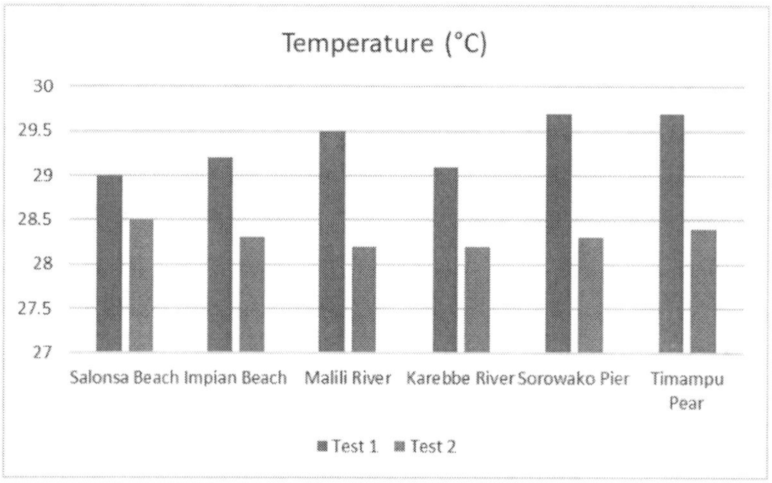

Figure 2. Result test of temperature

Table 5. Results test of TDS in the Malili Lake Complex, South Sulawesi

No	Observation Station	Results Test (g/l)	
		Test 1	Test 2
1	Salonsa Beach	14,19	17,98
2	Impian Beach	17,63	18,01
3	Malili River	17,22	18,06
4	Karebbe River	17,01	16,97
5	Sorowako Pier	14,23	17,69
6	Timampu Pier	17,89	14,89

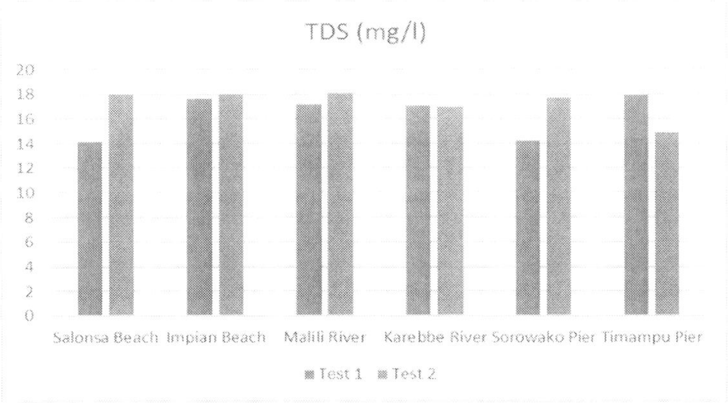

Figure 3. Results test of TDS

Table 6. Results test of pH in the Malili Lake Complex, South Sulawesi

No	Observation Station	Results Test	
		Test 1	Test 2
1	Salonsa Beach	7,10	8,00
2	Impian Beach	7,10	7,90
3	Malili River	7,10	7,70
4	Karebbe River	7,00	7,70
5	Sorowako Pier	7,00	7,60
6	Timampu Pier	7,10	7,30

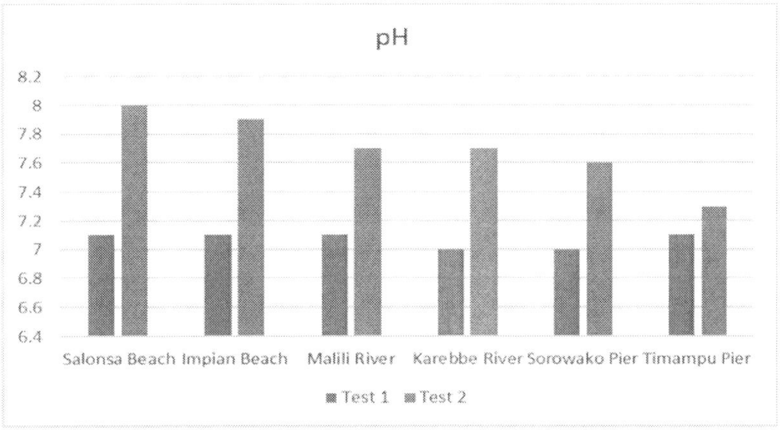

Figure 4. Results test of pH

Table 7. Results test of DO in the Malili Lake Complex, South Sulawesi

No	Observation Station	Results Test(mg/l)	
		Test 1	Test 2
1	Salonsa Beach	10,40	10,00
2	Impian Beach	11,50	11,20
3	Malili River	9,00	10,20
4	Karebbe River	10,80	10,50
5	Sorowako Pier	12,60	11,50
6	Timampu Pier	11,40	11,10

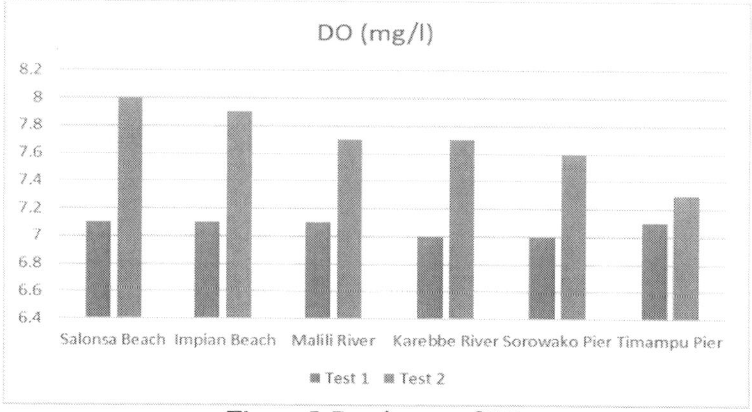

Figure 5. Results test of DO

Table 8. Results Test of BOD in the Malili Lake Complex, South Sulawesi

No	Observation Station	Results Test (mg/l)	
		Test 1	Test 2
1	Salonsa Beach	9,00	9,20
2	Impian Beach	8,00	5,80
3	Malili River	4,30	8,40
4	Karebbe River	3,90	6,30
5	Sorowako Pier	9,12	10,40
6	Timampu Pier	8,50	7,80

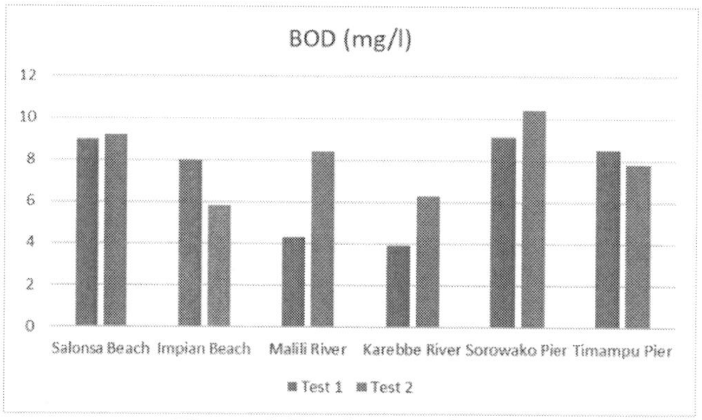

Figure 6. Results test of BOD

Table 9. Results test of COD in the Malili Lake Complex, South Sulawesi

No	Observation Station	Results Test (mg/l)	
		Test 1	Test 2
1	Salonsa Beach	16,00	20,00
2	Impian Beach	16,00	12,00
3	Malili River	8,00	16,00
4	Karebbe River	8,00	12,00
5	Sorowako Pier	20,00	24,00
6	Timampu Pier	16,00	16,00

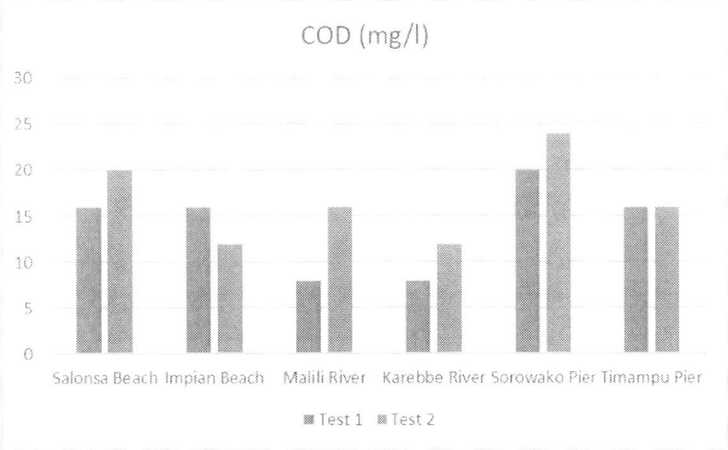

Figure 7. Results test of COD

Table 10. Results test of NO_3 parameter testing in the Malili Lake Complex, South Sulawesi

No	Observation Station	Test Result (mg/l)	
		Test 1	Test 2
1	Salonsa Beach	21,01	22,05
2	Impian Beach	20,13	25,30
3	Malili River	21,62	16,51
4	Karebbe River	19,53	16,54
5	Sorowako Pier	20,21	24,28
6	Timampu Pier	19,57	10,94

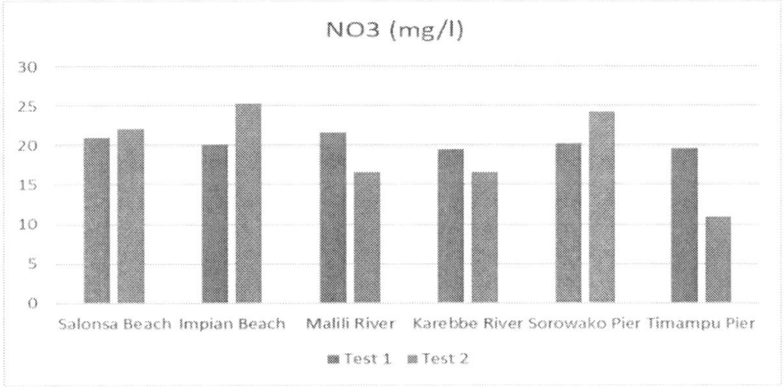

Figure 8. Results test of NO_3

Table 11. Results test of PO_4 in the Malili Lake Complex, South Sulawesi

No	Observation Station	Results Test (mg/l)	
		Test 1	Test 2
1	Salonsa Beach	1,07	1,75
2	Impian Beach	0,97	1,73
3	Malili River	1,51	1,03
4	Karebbe River	1,78	1,21
5	Sorowako Pier	1,66	1,15
6	Timampu Pier	1,03	1,50

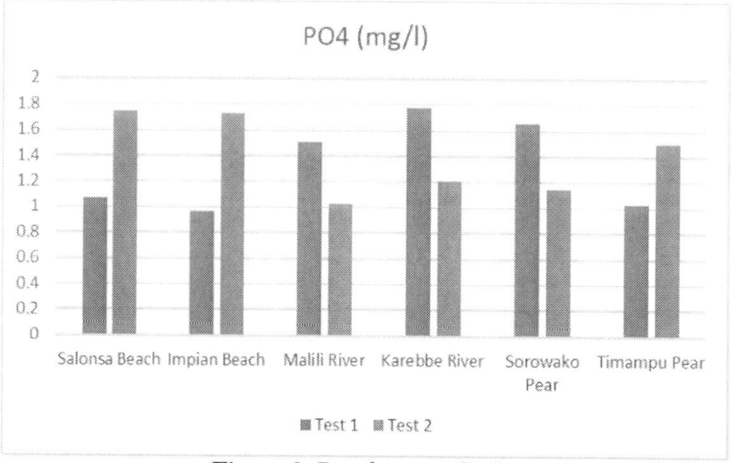

Figure 9. Results test of PO_4

Table 12. Results test of turbidity in the Malili Lake Complex, South Sulawesi

No	Observation Station	Results Test (mg/l)	
		Test 1	Test 2
1	Salonsa Beach	1,57	1,88
2	Impian Beach	0,90	0,70
3	Malili River	1,80	1,90
4	Karebbe River	0,60	0,70
5	Sorowako Pier	2,40	1,70
6	Timampu Pier	1,98	2,40

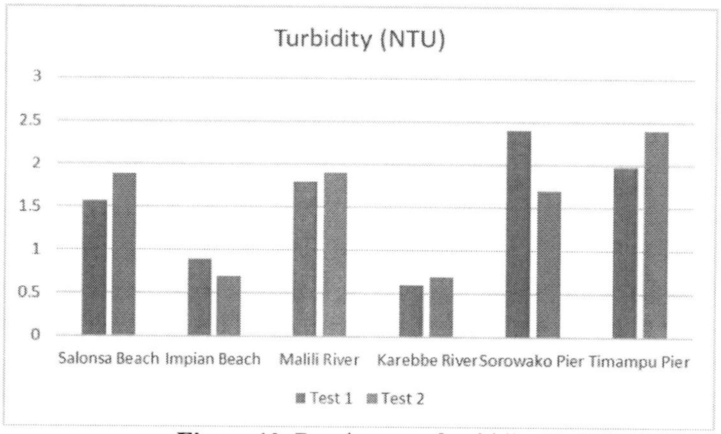

Figure 10. Results test of turbidity

Table 13. Results test of NH_3-N in the Malili Lake Complex, South Sulawesi

No	Observation Station	Results Test (mg/l)	
		Test 1	Test 2
1	Salonsa Beach	0,46	0,46
2	Impian Beach	0,38	0,37
3	Malili River	0,30	0,30
4	Karebbe River	0,30	0,29

5	Sorowako Pier	0,38	0,38
6	Timampu Pier	0,33	0,33

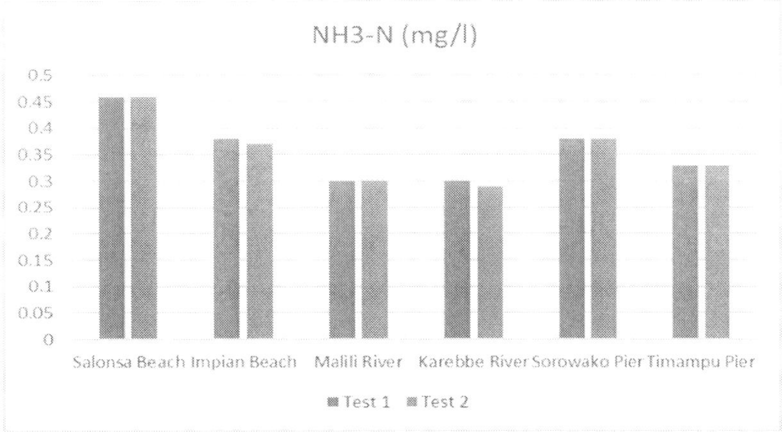

Figure 11. Results test of NH_3-N

Tabel 14. Range of values and means based on parameters and stations

No	Parameter	Range of Values (Unit) Minimum/ Station	Maximum/ Station	Averages	Quality Requirements Based on PP RI No. 82 year 2001	Description
1	Temperature	28,20 ° C / Malili River and Karebbe River	29,00 ° C / Salonsa Beach	28,84 ° C	Deviation 3	Not accordance
2	TDS	14,19 mg/l / Salonsa Beach	18,06 mg/l / Malili Beach	16,81 mg/l	10,00 mg/l	Not accordance
3	Ph	7,00 / Karebbe River and Sorowako Pier	8,00/ Salonsa Beach	7,38	6,00-9,00	Accordance
4	DO	10,00 mg/l / Salonsa Beach	12,60 mg/l / Sorowako Pier	10,85 mg/l	3,00-4,00 mg/l	Not accordance
5	BOD	3,90 mg/l Karebbe River	10,40 mg/l / Sorowako Pier	7,56 mg/l	3,00-6,00 mg/l	Not Accordance

No	Parameter	Range of Values (Unit)		Averages	Quality Requirements Based on PP RI No. 82 year 2001	Description
		Minimum/ Station	Maximum/ Station			
6	COD	8,00 mg/l / Malili and Karebbe River	24,00 mg/l / Sorowako Pier	16,00 mg/l	25,00-50,00 mg/l	Not accordance
7	NO_3	10,94 mg/l / Timampu Pier	25,30 mg/l / Impian Beach	19,80 mg/l	10,00-20,00 mg/l	Not Accordance
8	PO_4	0,97 mg/l / Impian Beach	1,78 mg/l / Karebbe River	1,37 mg/l	0,20-1,00 mg/l	Not accordance
9	Turbidity	0,60 NTU / Sungai Karebbe	2,40 NTU / Dermaga Timampu	1,54 NTU	10,30 NTU	Not accordance
10	NH_3-N	0,29 mg/l / Sungai Karebbe	0,46 mg/l / Pantai Salonsa	0,36 mg/l	0,50 mg/l	Accordance

Sumber: Results Test with Using Water Quality Checker and Laboratory, 2021.

4. Discussion

4.1. Temperature, pH and Dissolved Oxygen (DO)

Water temperature is influenced by the intensity of sunlight that enters the waters. Temperature in addition to affecting the specific gravity, viscosity and density of water, also affects the solubility of gases and elements in water. Light that enters the water will experience absorption and change into heat energy. Temperature plays a role in controlling water conditions, affecting the life and growth of aquatic biota [22]. High temperatures cause an increase in the rate of evaporation so that the concentration of pollutants will increase [11].

Based on research that has been carried out at the Malili Lake Complex, South Sulawesi at six observation stations, the lowest temperature value was 28.20 °C in the Malili River and Karebbe River; the highest temperature is 29 °C at Salonsa Beach. The required temperature value is in not accordance with the standard value based on Government Regulation (PP) of the Republic of Indonesia Number 82 of 2001, namely deviation 3. This is reinforced by the statement of Sitorus, 2009 which states that the temperature range for aquatic organisms is around 20-30 °C. The different water temperatures at each station are thought to be due to differences in water depth. Shallow waters will receive a higher intensity of sunlight than shallow waters will be higher than in deeper waters [12].

Water pH is one of the factors that can affect water productivity [12]. The pH value indicates that H+ ions and OH- ions are present in the water [17]. Based on research that has been conducted at the Malili Lake Complex, South Sulawesi at six observation stations, the lowest pH value was 7.00 in the Karebbe River and Sorowako Pier; The highest pH is 8.00 at Salonsa Beach. The required pH value is in accordance with the standard value based on Government Regulation (PP) of the Republic of Indonesia Number 82 of 2001 which is 6.00-9.00. The results of this study are supported by the statement [2], which states that the lower the pH of the waters, the higher the mobility of heavy metals, the higher the pH of the waters causing the balance of Ammonium and Ammonia in the water to be disturbed.

The pH value is influenced by several parameters including biological activity, temperature, oxygen content and ions. From biological activity, CO2 gas is produced which is the result of respiration. The more CO2 produced from respiration, the reaction moves and releases H+ ions which causes the pH of the water to drop. On the other hand, photosynthetic activity that requires a lot of CO2 causes the pH of the water to rise. The pH value in natural waters ranges from 4-9.

Based on research that has been carried out at the Malili Lake Complex, South Sulawesi at six observation stations, the lowest DO value is 10.00 mg/l at Salonsa Beach; The highest DO is 12.60 mg/l at Sorowako Pier. The required DO value is not in accordance with the standard value based on Government Regulation (PP) of the Republic of Indonesia Number 82 of 2001 which is 3.00-4.00 ml. In general, the condition of the waters at the study site has begun to be polluted by easily biodegradable organic matter.

4.2. Total dissolved solid (TDS)

TDS is a solid consisting of organic and inorganic compounds that are soluble in water, minerals and their salts [7]. TDS is strongly influenced by flowing water, so a decrease in water discharge can increase the TDS value [15].

Based on research that has been conducted in the Malili Lake Complex, South Sulawesi at six observation stations, the lowest TDS value was 14.19 mg/l at Salonsa Beach; The highest TDS is 18.06 mg/l in the Malili River. The required TDS value is not accordance with the standard value based on Government Regulation (PP) of the Republic of Indonesia Number 82 of 2001 which is 10.00 mg/l. This shows that the waters contain a lot of metallic solids and have started to be polluted by mining activities in the vicinity. The TDS value is strongly influenced by weathering of rocks, run off from the soil and anthropogenic influences, in the form of domestic and industrial waste [5]. High TDS affects the taste and smell of water and in general, levels above 300 mg/L will be seen by consumers [6].

4.3. Dissolved oxygen (DO) and amonia (NH_3-N)

Based on research that has been carried out at the Malili Lake Complex, South Sulawesi at six observation stations, the lowest DO value is 10.00 mg/l at Salonsa Beach; The highest DO is 12.60 mg/l at Sorowako Pier. The required DO value is not in accordance with the standard value based on Government Regulation (PP) of the Republic of Indonesia Number 82 of 2001 which is 3.00-4.00 ml. This means that oxygen absorption is quite high at the research site. Dissolved oxygen is produced through the photosynthesis process of aquatic plants and from the air enters through a diffusion process which slowly penetrates the water surface [21].

The lowest value of Ammonia (NH_3-N) content is 0.29 mg/l in the Karebbe River; the highest is 0.46 mg/l at Salonsa Beach in accordance with the standard value based on Government Regulation (PP) of the Republic of Indonesia Number 82 of 2001 which is 0,50 ml.In water with a high dissolved oxygen content, the ammonia content is relatively small so that ammonia increases with increasing depth [17].

4.4. Dissolved Oxygen (DO), Biochemical Oxygen Demand (BOD) and Chemical Oxygen Demand (COD)

The DO value fluctuates depending on organic pollutants (BOD) and the natural purification of rivers and lakes. The concentration of DO in water depends on the level of sinking and solubility [1]. The concentration of pollutant released can reduce the DO content due to consumption by microbes to degrade organic matter / deoxygenation. DO in surface runoff, was found to be higher than groundwater [6]

Based on research that has been carried out at the Malili Lake Complex, South Sulawesi at six observation stations, the lowest BOD value was 3.90 mg/l in the Karebbe River; The highest BOD is 10.40 mg/l at Sorowako Pier. The required BOD value is in accordance with the standard value based on Government Regulation (PP) of the Republic of Indonesia Number 82 of 2001 which is 3.00-6.00 ml. Meanwhile, the lowest COD value was 8.00 mg/l in the Malili River and the Karebbe River; the highest value of 24.00 mg/l at Sorowako Pier is not in accordance with the standard value based on Government Regulation (PP) of the Republic of Indonesia Number 82 of 2001 namely: 25.00-50.00

mg/l. Utilization of rivers and lakes as toilets and household waste disposal causes an increase in BOD and COD concentrations, which are indicator parameters of organic waste pollution originating from community activities along the rivers and lakes [23].

BOD is the Oxygen index number required by biodegradable pollutants. BOD indicates the amount of dissolved oxygen required by microorganisms in the decomposition of organic matter during aerobic conditions [19]. The COD content is influenced by the degradation of organic and inorganic materials originating from community activities. Excess COD content in waters will be the same as BOD content and affect DO which causes a decrease in water quality [18]. The higher the COD concentration in the lake due to urban wastewater and agricultural runoff, it will actively affect aquatic life [20]. High levels of BOD can cause water quality problems such as severe damage, oxygen depletion and death of fish in the water [3].

The need for BOD and COD is an indicator that indirectly supports the health of the environment and plant biological life. Biochemical oxygen demand (BOD) and chemical oxygen demand (COD) in water are indirect indicators of the health and capacity of water to support biological life [13].

4.5. Nitrate (NO_3)

Based on research that has been conducted at the Malili Lake Complex, South Sulawesi at six observation stations, the lowest NO_3 value was 10.94 mg/l at Timampu Pier; The highest NO_3 is 25.30 mg/l at Pantai Impian. The required NO_3 value is in not accordance with the standard value based on Government Regulation (PP) of the Republic of Indonesia Number 82 of 2001, namely 10.00-20.00 mg/l. This shows that the condition of the waters has been polluted by human activities. This statement is supported by **[5]**, which states that the levels of Nitrate-Nitrogen in natural waters almost never exceed 0.1 mg/l, if more than 5 mg/l describes the occurrence of pollution originating from human activities and animal feces. Nitrate levels for drinking water purposes should not exceed 10 mg/l/.

4.6. Phosphate (PO_4)

Phosphate/Phosphorus is widely used as fertilizer, soap or detergent, ceramic industrial material, lubricating oil, beverage and food products, catalysts and so on [5]. lowest PO_4 0.97 mg/l at Impian Beach ; The highest PO_4 is 1.78 mg/l in the Karebbe River. The required PO_4 value is in not accordance with the standard value based on Government Regulation (PP) of the Republic of Indonesia Number 82 of 2001, namely 0.20-1.00 mg/l. This shows that the use of phosphates, especially agricultural fertilizers at the research site is still controlled because it has not polluted the surrounding environment.

The high value of phosphate in water is not toxic to aquatic animals and does not interfere with human health [7]. Then Effendi [5], stated that phosphorus is not toxic to humans, animals and fish. The presence of phosphate in the waters is very important, especially in the formation of protein and metabolism for organisms. However, the high phosphate causes eutrophication in the form of an explosion in the number of algae (blooming) which is bad for the cultivation of biota.

4.7. Turbidity

Turbidity describes the optical properties of water, namely the amount of light that is absorbed and emitted by the materials contained in the water [2,5]. Based on research that has been carried out at the Malili Lake Complex, South Sulawesi at six observation stations, the lowest turbidity value is 0.60 NTU in the Karebbe River; highest turbidity 18.80 NTU at Salonsa Beach. The required turbidity value is in not accordance with the standard value based on Government Regulation (PP) of the Republic of Indonesia Number 82 of 2001, namely 10,30 NTU. High turbidity can result in disruption of the osmoregulation system, such as respiration and visibility of aquatic organisms, and can inhibit light penetration into the water [5].

5. Conclusion

Based on the research that has been done, it is concluded that in general the condition of the waters in the Malili Lake Complex in South Sulawesi has been polluted by the presence of domestic and industrial waste and the activities of the people who live in the vicinity.. Several parameters used as research samples were above the predetermined threshold, including: temperature, TDS, DO, BOD, COD, NO_3, PO_4, turbidity; while other parameters are within the normal threshold, such as: pH, NH_3-N.

Acknowledgments

Thank you very much to LPDP Ministry of Finance Republic Indonesia as the main sponsor of this research.

References

[1] Sameer A and Keshari A K 2021 *Water Science and Technology* **84**
[2] Barus T A 2001 *Pengantar Limnologi Studi Tentang Ekosistem Sungai dan Danau* (Medan: Program Studi Biologi USU FMIPA)
[3] Penn C J, Mullins G L, Zelazny L W, Warren J G and McGrath J M 2004 *Journal of Environmental Quality* **33** 1431
[4] Connor R 2015 *The United Nations World Water Development Report 2015: Water for a Sustainable World* (UNESCO Publishing)
[5] Effendi H 2003 *Study of Water Quality for Management of Aquatic Resources and Environment.* Kanisius (Yogyakarta)
[6] Emmanuel O and Thomas 2021 *Water Science* **35** 11
[7] Fardiaz S 1992 *Polusi Air dan Udara* (Yogyakarta: Kanisius)
[8] Haryani G S and Hehanussa 1999 *Proceedings of the Annual Scientific Meeting MLI*
[9] Haryani G S 2013 *Lake as a Foothold for Sustainable Freshwater Fish Resource Management* (Indonesian Institute of Sciences: Research Professor Inauguration Oration)
[10] Ngoye E and Machiwa J F 2004 *Physics and Chemistry of the Earth* **29** 1161
[11] Reddythota D and Mosisa T T 2022 *Hindawi Journal of Environmental and Public Health*
[12] Rugebregt, Marsy J, Caleb M and Syafrizal 2020 *Journal of Environmental Science* **18** 589
[13] Mohammed S, Nyade P K, Abdul-Hamid M and Hodgson I O A 2021 *Water Science* **35** 39
[14] Salam A 2020 *Internet of things for Water Sustainability In Internet of things for Sustainable Community Development* (Cham: Springer)
[15] Salmani M H and Jajaei E S 2016 *Journal of Hydrology* **535** 148
[16] Shen Y, Cao H, Tang M and Deng H 2017 *Northeast China Water* **9**
[17] Sulistyorini, Iin S, Muli E and Adriana S A 2016 *Jurnal Hutan Tropis* **4** 64
[18] Suparjo M 2009 *Journal of Fisheries Science* **4**(2) 38
[19] Tanjung R H R, Hamuna B and Alianto 2019 *Journal of Ecological Engineering* **20** 87
[20] Wang W, Liu X, Wang Y, Guo X and Lu S 2016 *Environmental Science and Pollution Research* **23** 4886
[21] Wardhana W A 2001 *Impact of Environmental Pollution* Third Edition (Yogyakarta: Penerbit Andi)
[22] Warman 2015 *Lais River Estuary Water Quality Test for Fisheries in North Bengkulu* **13** 24
[23] Yetti E, Soedharma D and Haryadi S 2011 *Evaluation of the Water Quality of the Rivers in the Upper Brantas Watershed of Malang*

IOP Conf. Series: Earth and Environmental Science 1115 (2022) 012050 doi:10.1088/1755-1315/1115/1/012050

Analysis of methyl compounds variation in the crude *Pongamia pinnata* oil from Banten, West Java and North Java populations

Jayusman[1], L Hakim[2], A Dalimunthe[3] and V Yuskianti[4]

[1]Research Center for Plant Conservation, Botanic Garden and Forestry, National Research and Innovation Agency, Bogor 16122, Indonesia
[2]Wood Technology Department, Faculty of Forestry, Universitas Sumatera Utara, Medan, North Sumatra 20155, Indonesia
[3]Silviculture Department, Research Center for Ecology and Ethnobiology, National Research and Innovation Agency, Bogor 16911, Indonesia
[4]Research Center for Ecology and Ethnobiology, National Research and Innovation Agency, Bogor 16911, Indonesia

[*]Email: luthfi@usu.ac.id.

Abstract. Identification of methyl compounds from crude oil is an important baseline provenance to support the development of renewable bioenergy resources in Indonesia. *Pongamia pinnata*, known as *Malapari*, has great potential to be an alternative to renewable bioenergy resources and an environmentally friendly substitute for fossil fuels. Our analysis of four provenances of crude *Pongamia* oil has identified eight (8) methyl compounds with a relative value of >1%, i.e. palmitoleate, lenoleate, eicostrienoic, erucate, docosadienoic, nervonate, eicosapentaenoate and arachidate. The value of the individual methyl compounds varied in each provenance. The unsaturated fatty acid content of *Pongamia* crude was higher and needs to be balanced to maintain the physical properties of *Pongamia* biodiesel. In addition, the quality of the tested *Pongamia* crude oil from all provenances that show high acidity and water content in the crude oil but exceed the standard for biodiesel (SNI 04-7182-2006) shows a need for improvement advanced process. The variations of the methyl compounds found in each provenance can be used as a selection basis for the development of products in the pharmaceutical and agricultural sectors.

1. Introduction

Indonesia has set the goal to increase the use of biofuels to 5% by 2025. It is well known that biofuels are very environmentally friendly due to the exhaust gases released into the atmosphere when they are burned are absorbed by plants for photosynthesis. To achieve the national target launched by the government, a lot of information such as the diversification of raw material sources, the properties of crude oils and their processing, and their production capacity need to be studied [1].

Pongamia pinnata known as Malapari is one of the alternative renewable resources for biodiesel in Indonesia. It has wide natural distribution, including Java, Sumatra, Kalimantan, Bali, East Nusa Tenggara, Maluku to Papua, indicating its suitability to plant in many areas in Indonesia. Many studies have been conducted, such as its natural distribution on the Java Island [2] and their genetic relationship [3,4]. Other information related to crude oil production, such as the extraction method of

P. pinnata seeds to produce crude oil [2,5-8] physico-chemical characterization [9]; Methylated crude oil [5,10]; their biodiesel tests [11,12] and a comparison of the thermal braking efficiency (BTE) between the biodiesels with the blends B50, B65 and B80 [13] were reported. The cost analysis for biodiesel production was also carried out [14]. *P. pinnata* oil is not only useful as a biodiesel, Usharani et al. [15] reported that *P. pinnata* oil is capable to stop bleeding when it is applied to a wound, anthelmintic, and giving good results in leprosy, piles, liver pain, chronic fever, ulcers and rheumatism arthritis scabies.

Methyl compounds play an important role in the production of biodiesel. Analysis of methyl compounds in crude *P. pinnata* oil, referred to as Pongamia oil, is therefore important to support its use as biodiesel. Several studies analyzing constituents of Pongamia crude oil have been reported [10,16-17]. To complete the baseline information of *P. pinnata* as biodiesel, this study was conducted. The aim of this study specifically was to identify the composition of methyl compounds in Pongami crude oil as a basis for further processing of the crude oil to meet the biodiesel standard in Indonesia (SNI 04-7182.2006) [18]. Analysis of Pongamia crude oil from multiple provenances can also be used to select for biodiesel methyl compounds in the species. From the result, we then recommend several strategies for strengthening and formulating the final processing of raw Pongamia oil.

2. Materials and Methods

2.1. Sample collection

Samples for analysis of Crude Pongamia Oil were collected from four provenances in Java, viz. H. Alas Purwo (AP) in East Java Province; Batu Karas (BK) in West Java Province; Pulau Haundelem Ujung Kulon (UK-01) and Pantai Cilintang Ujung Kulon (UK-02) in Banten Province. Two provenances from Ujung Kulon are more widespread than the AP and BK provenances. Detailed information on the four provenances, including their geographical location, rainfall and temperatures, is shown in Figure 1.

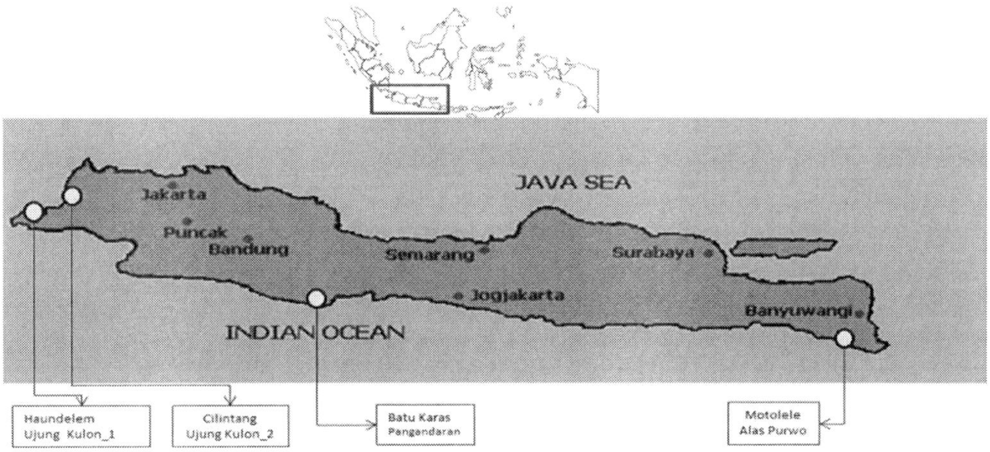

Description	Provenances			
	Alas Purwo (AP) North Java Province	BK (Batu Karas) West Java Province	Ujung Kulon (UK)_1 Banten Province	Ujung Kulon (UK)_2 Banten Province
Latitude	8° 26' 45" S	07° 41' 15" S	6° 52' 17" S	6° 43' 40" S
Longitude	114° 20' 16" E	108° 39' 32" E	105° 02' 32" E	104° 8' 33" E
Altitude (m asl)	2 - 8	3 - 6	3 - 5	2 – 7
Rain Fall (mm/year)	2079	2987	3250	3250
Temperature (°c)	24 - 30	23 - 30	24 - 30	24 - 30

Figure 1. Detailed information and map of four provenances of *Pongamia pinnata* used in this study

2.2. *Methods*
The sample fruits collected from the field were dried in the sun until their pericarp can be opened to remove the seed. The seed obtained from the dried fruit is then dried further to reduce its water content (8-12%). Pongamia crude oil was obtained by extracting the seed using a Screw Press Expeller (SPE) machine with a capacity of 30 kg/hour. To increase crude oil recovery, the SPE machine was pressed up to 5 times with the addition of rice hulls. The extracted crude oil was separated according to origin. Analysis of methyl compound content was performed at Gadjah Mada University's Integrated Testing and Development Laboratory (LPPT), Yogyakarta. Analysis of the acid number of Pongamia crude oil using volumetric technique, while the water content using distillation technique

2.3. *Data analysis*
Three steps of analysis were carried out. The first step was to identify the methyl content which have a relative value > 1%; Secondly, we continued the analysis of the methyl content, acid number and water content of each provenance and in the last step was to compare the result with the Indonesian government biodiesel standard (SNI 04-7182. -2006) [18]

3. Result and Discussion

3.1. *Identification of methyl compound*
Analysis of four provenances of crude Pongamia oil identified eight (8) methyl compounds that have a value >1%. Below you will find explanations of the identified methyl compounds and their variations in the four provenances examined.

3.1.1. Methyl of palmitoleate and lenoleate compound. Our study found that the methyl palmitoleate content of the four provenances varied between 12.02% and 14.5%, with the highest value being in the UK provenances (UK 01 and UK 02) (Figure 2A), while the value for methyl lenoleate varied from 38.18-61.20 with the highest content in Alas Purwo (AP) and the lowest in UK 02 (Figure 2B). The largest methyl component in crude pongamia oil is methylene lenoleate. Lenoleic acid belongs to the group of unsaturated fatty acids, whereas palmitoleate to the group of saturated fatty acids. Biodiesel feedstocks are derived from vegetable oils including P. pinnata and tend to be low in saturated fat. This result indicates the need to reduce the unsaturated fatty acids and maintain a balance between saturated and unsaturated fatty acids.

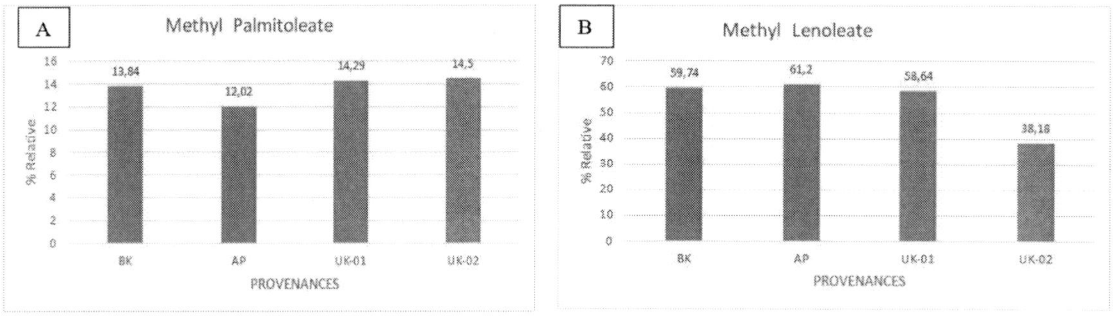

Figure 2. Histogram of palmitoleate (A) and lenoleate (B) methyl compound from crude Pongamia pinnata oil in four provenances

3.1.2. Methyl of eicostrienoic acid and erucate compound. The methyl eicostrienoic content of the four provenances varied from 1.47% – 2.41% with the AP provenance showing the highest (Figure 3A). The methyl erucate value ranging from 0.89 – 1.40, showing the highest in UK-02 and the lowest in AP provenance (Figure 3B).

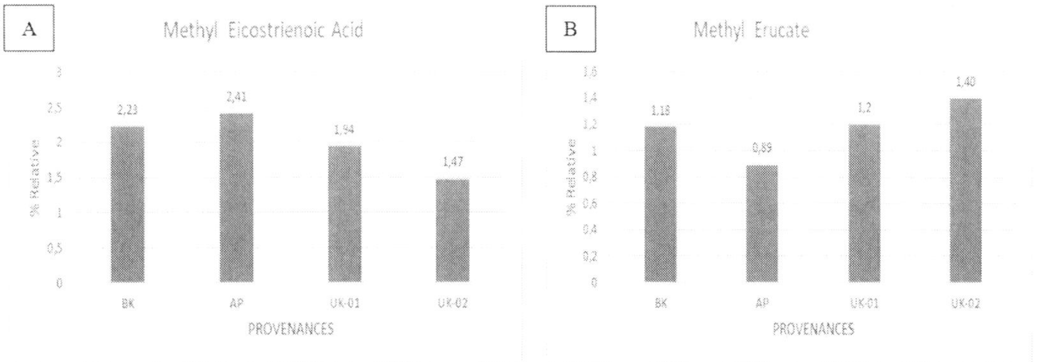

Figure 3. Histogram of methyl eicostrienoic acid (A) and methyl erucate (B) compound from crude Pongamia pinnata oil in four provenances

3.1.3. Methyl of docosadienoic and nervonate compound. Docosadienoic acid methyl content in four provenances varied from 9% – 12.26 % with the highest in AP (Figure 4A), whereas methyl nervonate ranged from 1.1 to 4.95 with the highest in UK-02 and the lowest in UK-01 (Figure 4B).

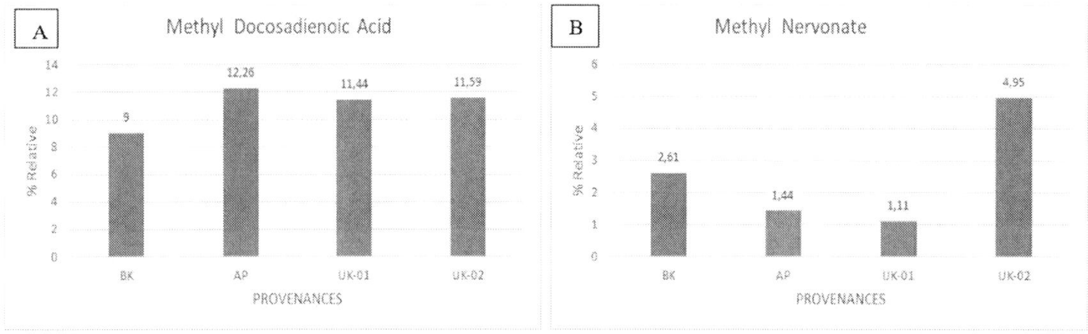

Figure 4. Histogram of docosadienoicacid methyl (A) and nervonate methyl (B) compound from crude *Pongamia pinnata* oil in four provenaces

3.1.4. Methyl of eicosapentaenoate and arachidate compound. The value of eicosapentaenoate methyl in four provenances varied from 0.13 % – 1.99 % with the AP showing the highest, and UK-01 as the lowest (Figure 5A), whereas the value of methyl arachidate ranged from 3.94 to 8.18 with UK-01 as the highest and UK-02 as the lowest (Figure 5B).

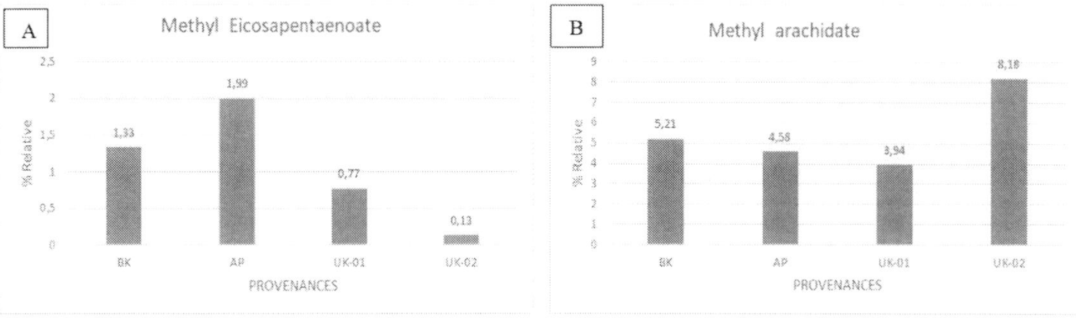

Figure 5. Histogram of eicosapentaenoate methyl (A) and arachidate methyl (B) compound from crude *Pongamia pinnata* oil in four provenances

The variation of methyl compound in crude oil was also observed in the study by Prakash et al. [19] who reported the characterization of the oil and the variation of the methyl compound in *Melletia pinnata* (synonym of *Pongamia pinnata*). Pongamia crude oils have many properties for different applications. For example, the properties of raw pongamia oil have been used in agriculture, especially for bioinsecticides, pharmaceutical products, and raw materials for the synthesis of fatty acid methyl esters [8,15-17].

3.2. Acid value and water content

Analysis using the volumetric technique in the four origin of crude Pongamia oil showed that the acid value varied between 7.41 and 13.84 (Figure 6A). This result is higher than the maximum requirement in the biodiesel standard (SNI 04-7192-2006), which was set at 0.6% [18]. This result was also different from Arpawi study (7), who found an acid number of 0.05%. These differences may be caused by the different extraction method used as our study using the Screw Press Expeller (SPE) method in the extraction process. Oil with a high acid number is considered as poor quality. This is because free fatty acids greater than 1% form a soap emulsification makes them difficult to separate in biodiesel [15]. In addition, the acid number of *P. pinnata* crude oil in our study was also higher than the acid number of used cooking oil (4.03), and new cooking oil (0.10) [7].

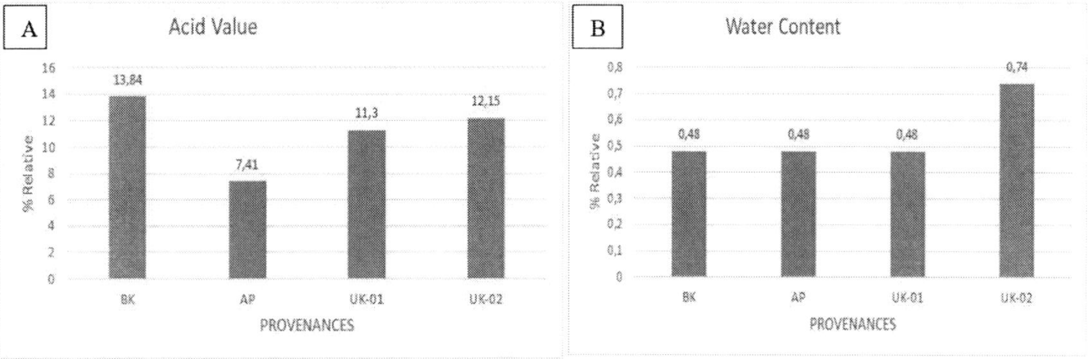

Figure 6. Histogram of acid value (A) and water content (B) of crude *Pongamia pinnata* oil in four provenances

The water content of *P. pinnata* crude oil from the four provenances varies ranging from 0.48 to 0.74 (Figure 6B). This result was lower than that previously examined using five populations of *P. pinnata* from Java island (ranging from 1.24 to 2.7) [5] and higher than Arpiwi [7] (0.01%). These results are also higher than the required standard for biodiesel in the SNI 04 7182 2006 (0.05%) [18]. The water content of biodiesel is expected to be minimal or even zero because it will affects the optimization of fuel combustion in the engine as well as a good medium for the growth of microorganisms. The high water content value obtained in our study may be affected by the storage of the pressed oil and its extraction method.

Our extraction method using screw press expeller (SPE) produced high water content. The high water content in the processed *P. pinnata* seed determines the magnitude of the physicochemical properties of the vegetable oil in the seed, particularly the viscosity and yield produced. The acid number and water content in Pongami crude oil from four provenances in this study (Figure 6) did not meet the Indonesian standard for biodiesel (0.6 mg KOH/g and 0.05%) [18]). High water content in biodiesel can cause corrosion when it reacts with sulfur, reducing heat of combustion and causing foam [8]), while high acid number, indicating the amount of free fatty acids in biodiesel, is corrosive and can cause soot or scale in diesel engine injectors.

3.3. Molecular formula and properties of methyl compound

The analysis of methyl compound in the crude pongamia oil has successfully identified eight (8) methyl compounds from four provenances *P. pinnata* in Indonesia. Detailed information of the identified methyl compound including its molecular formula and properties is shown in Table 1.

Table 1. Synonyms, molecular formula and properties of eight identified methyl compound from crude pongamia oil obtained in this study

Names & synonyms	Molecular formula	Properties
Methyl palmitoleate/ methyl cis-9-hexadecenoate	$C_{17}H_{32}O_2$	Used to prevent obesity
Methyl lenoleate/ Methyl (9Z,12Z)-octadeca-9,12-dienoate	$C_{19}H_{34}O_2$	Used to reduce hydration, skin barrier
Methyl eicostrienoic / methyl 11-cis,14-cis,17-cis-eicosatrienoate	$C_{21}H_{36}O_2$	Human Endogenous Metabolite
Methyl erucate/ Erucic acid methyl ester	$C_{23}H_{44}O$	Used for the production of mustard alcohol, erucic acid amide and other derivatives
Methyl docosadienoic / Methyl docosa-13,16-dienoate	$C_{23}H_{42}O_2$	Strongly inhibits the secretion of ghrelin by isolated mouse gastric cells
Methyl eicosapentaenoate /Eicosapentaenoic acid methyl	$C_{21}H_{32}O_2$	An anticancer compound by inducing apoptosis
Methyl arachidate/ Arachidic acid methyl ester	$C_{21}H_{42}O_2$	Used as Nutritional Supplement in Pharmaceutical product
Methyl nervonate/ Methyl (15Z)-tetracos-15-enoate	$C_{25}H_{48}O_2$	Ester version of the free acid which may be more suitable for the formulation of fatty acid-containing diets and dietary supplements.

Source: [20, 21]

The variation in methyl compound found in this study shows the importance to conduct tree improvement and conservation program to the species. A research on its natural distribution, for example in Bali, found lower number of trees in the southern and western part of the island, indicating a need for further action for conserving the species [22]. Analysis on five populations of *P. pinnata* in Java Island using RAPD markers shows different level of genetic diversity where the population in Baluran was considered as a potential diversity center of malapari in Java, whereas the population in Kebumen was found to have the lowest genetic diversity [23]. The existence of substantial variation and diversity can be utilized for genetic resource conservation and tree improvement programs [24]. In addition, the quality and genetic diversity of seed sources is also very important on industrial plantation forest programs for bioenergy and land restoration in Indonesia [25].

4. Conclusions

Basic information on the properties of Pongamia crude oil should follow the Indonesian Biodiesel Standard (SNI 04 7182 2006). However, the results of our study revealed high levels of unsaturated fatty acids as indicated by linoleic acid content, high acid number and high water content; these were higher than the maximum requirements for the biodiesel standard (SNI 04 7182 2006). Other studies to maintain some important biodiesel physical properties as specified in the standard are required. Our study also successfully identified eight (8) methyl compounds from crude Pongamia oils collected from four provenances of *Pongamia pinnata* in Indonesia. The variation in the content of methyl compound found from each provenance has great potential as a basis for selection for different purposes, such as a bioinsecticide in agriculture and pharmacy.

Acknowledgments

We would like to express our sincere thanks to the Ministry of Environment and Forestry Republic of Indonesia for providing fund to support our research "Breeding Research of *Pongamia pinnata* for Renewable Energy" (Contract Number: MAK 5395.001.001.099.F.524111- DIPA 2020).

References

[1] Aminah A, Supriyanto S, Siregar I Z and Suryani A 2017 *Jurnal Penelitian Hasil Hutan* **35** 255
[2] Agarwal A K and Rajamanoharan K 2009 *Applied Energy* **86** 106
[3] Arpiwi N L 2015 (Udayana University: Annual Report)
[4] BSN 2006 SNI 04-7182-2006 Biofuel Badan Standardisasi Nasional
[5] Bobade S N and Khyade V B 2012 *Research Journal of Chemical Sciences* **2** 16
[6] Doddabasava R P 2014 *Int. J. Sci. Res.* **3** 128
[7] Febritasari F, Arpiwi N L and Wahyuni I G A S 2016 *Jurnal Metamorfosa* **III** 74
[8] Thirugnanasambandham K 2019 *Chemistry and Functionality* (Springer, Cham: US)
[9] Jayusman 2016 *Prosiding Seminar Nasional Sumberdaya Genetik* 24 November 2016
[10] Jayusman 2019 *Wana Benih* **20** 1
[11] Kesari V, Das A and Rangan L 2010 *Biomass and Bioenergy* **34** 108
[12] KNRT 2006 Kementrian Negara Riset dan Teknologi KNRT Jakarta
[13] Mukta N, Murthy I Y L N and Sripal P 2009 *Industrial Crops & Products* **29** 536
[14] Soudagar M E M, Nik-Ghazali N N, Akram N, Al-Rashid M A, Badruddin I A, Khan H and Ukkund S J 2020 *AIP conference proceedings* **2247** 030003
[15] Afzal A, Kareemullah M and Razak R K A 2018 *Journal of Engineering Research* **6** 1
[16] Naik P L and Katpatal D C 2013. *International Journal of Science and Research* **2** 2319
[17] Khayoon M S, Olutoye M A and Hameed B H 2012 *Bioresource Technology* **111** 175
[18] Usharani K V, Naik D and Manjunatha R L 2019 *J Pharmacogn Phytochem* **8** 2181
[19] Prakash S R, Aravindhkumar R, Madhav R and Vinoth B 2016 *Int Res J Eng Technol* **3** 512
[20] https://www.chemsrc.com/en/cas/506-37-6_1107311.html
[21] https://pubchem.ncbi.nlm.nih.gov/
[22] Arpiwi N L, Wahyuni I G A S, Muksin I K and Sutomo 2018 *Biodiversitas* **19** 1607

[23] Aminah A, Supriyanto, Suryani A and Siregar I Z 2017 *Biodiversitas* **18** 677
[24] Divakara B N and Das R 2011 *Journal of Forestry Research* **22** 193
[25] Hasnah T, Leksono B, Sumedi N, Widyarini E, Adinugraha H A, Baral H and Artati Y 2020 *ICUE 2020 on Energy, Environment, and Climate Change, Asian Institute of Technology*, Thailand

IOP Conf. Series: Earth and Environmental Science 1115 (2022) 012051 doi:10.1088/1755-1315/1115/1/012051

Season, basal media and plant growth regulators effect in wood plant in vitro propagation: a comprehensive review

Jayusman[1], L Hakim[2] and A Dalimunthe[3]

[1]Research Center for Plant Conservation, Botanic Garden and Forestry, National Research and Innovation Agency, Bogor 16122, Indonesia
[2]Wood Technology Department, Faculty of Forestry, Universitas Sumatera Utara, Medan, North Sumatra 20155, Indonesia
[3]Silviculture Department, Research Center for Ecology and Ethnobiology, National Research and Innovation Agency, Bogor 16911, Indonesia

[*]Email: luthfi@usu.ac.id

Abstract. Wood plant in-vitro propagation is currently widely used and is not limited to the preparation of superior seeds but has been widely utilized for the production of phytopharmaceuticals and the production of a valuable and specific tree for the benefit of man. Studies focused on the influence of explant collection seasons, selection of basic media, and plant growth hormone. The results of the study showed the magnitude of the variables that must be considered in the propagation of tissue cultures even in the same species and types of cultures. The period of January and March is the best time for explant retrieval because it provides the best explant response in bud breakage, lowest browning, and best bud induction. The basic media of MS (Murashige & Skoog), as well as its various modifications and the basic medium of WPM (Wood Plant Medium), tend to be widely chosen as the basic media at various levels of subcultures in many woody plant species. The application of growth regulators shows a tendency to use BAP (benzyl amino purine) and NAA (naphthalene acetic acid) whose application is separately or in combination with a certain concentration according to the species and subculture phase. Other plant growth regulators (PGRs) that are also quite widely used are the IBA (indole butyric acid), IAA (indole acetic acid), and G3 (gibberellic acid) groups which are used in the multiplication and rooting stages. The information from the study can be used as basic information at the stage of optimizing tissue culture activities in different species, explanatory, and regeneration types. The success of the optimization stage will be used as an improvement to the previous woody plant tissue culture protocol so that optimization of success is achieved. In vitro propagation is always associated with the number of resources used, so efforts to improve in vitro protocols are a very important activity to evaluate that will lead to increased effectiveness to reduce resource inputs, and be able to realize low-cost wood plant tissue culture propagation.

1. Introduction

Plant tissue culture is an in vitro aseptic culture of cells at the level of plant tissues and organs under controlled nutritional and environmental conditions to provide an environmental culture conducive to its growth and multiplication. These conditions include the proper supply of nutrients, pH medium, adequate temperature, and proper environment. For the success of plant tissue culture, it is necessary to provide aseptic conditions, selection of plant tissues free of microbes, proper sterilization, proper nutrition in tissue culture media, and taking appropriate measures in browning and somaclonal variation.

Acclimatization of field conditions becomes a crucial stage at the end of tissue culture [11]. Tissue culture has shared advantages when compared to conventional propagation, namely (a) rapid propagation of genetically similar plants that have desired properties, (b) being able to propagate one plant cell to multiply into thousands of plants in a very short time without being limited by the season, (c) plant propagation can be done without seeds or pollinators needed to produce seeds, (d) regeneration of genetically modified plant cells from plant cells helps nurseries to select superior clones of ornamental plants in large quantities and (e) tissue cultured plants can reduce the possibility of transmission of diseases, pests, and pathogens.

Application of plant tissue cultures include anther or pollen cultures for haploid production, fusion protoplast-somatic cultures, cell cultures, tissue culture screening and selection, in vitro gene banks, plant stock banks, DNA transformation systems, clone propagation, direct organogenesis, indirect organogenesis, armpit bud propagation, somatic embryogenesis, embryogenesis rescue, pathogen-free plants, somatic variations, mutation induction, in vitro grafting, in vitro grafting, biosynthesis in bioreactors (production of secondary metabolites). Another tissue culture utilization opportunity is to be able to cross distantly related species to regenerate hybrid novels through protoplast fusion, production of double monoploid crops with kolkisin treatment from haploid cultures to reach homozygous strains faster in breeding programs, seed production of certain plants that require a high level of genetic conservation for seed production, as a better scope and future for the production of important phytopharmaceuticals, as well as helping the successful production of valuable, specific plant chemicals that have not yet been discovered. Successful propagation of tissue cultures of woody plants has been widely reported, among others, in the species *Tectona grandis* [34], *Santalum album* [29], *Acacia sinuate* [18] and *Aquilaria malaccensis* Lam [30]. Separate screenings related to the influence of seasons [3,5,6,8,21] basic media [16,35] and growth-stimulating substances [13,14,34] have been widely reported as well as the results of comprehensive screening related to the three factors mentioned above [1,2,4,9,12,13,18,20,22,23,25-30].

Efforts to increase the propagation of finger cultures have been widely reported, including rapid tissue culture for the preparation clones [19], [33], chances of in vitro propagation in woody plants [7, 17], in vitro for clonal propagation [20] In vitro regeneration testing for shoot multiplication [37], Phloroglucinol and silver nitrate enhances axillary shoot proliferation in nodal explants [23], Effect of glutamine for high frequency in vitro regeneration [30], mapping barriers in tissue culture, [17,31], and low-cost tissue culture [32], Browning Treatment in Tissue Culture [27], Influence of explant collection period, anti-browning strategy and growth regulators composition [28]. Optimization in tissue culture techniques to obtain the most suitable protocol [29], is very dynamic and continuous must be carried out because it is strongly influenced by plant characteristics, the technology used, and activity targets.

Determination of current status of plant tissue culture has been reviewed [36], identification of factors influencing in vitro establishment [25] and identification of tissue culture problems [17]. The purpose of the review in this research is to prepare user information for the improvement and improvement of tissue culture propagation protocols in woody plants and is not limited to producing genetically superior seedlings but also for selection activities related to the properties of plants to be glorified. Information on the results of the study can be used as a basis for pioneering and realizing low-cost tissue culture through efficiency and optimization of all stages of tissue culture in the target species.

2. Methodology

2.1. Manuscript review

The results of review research journal manuscripts related to the effects of the explant retrieval season, the use of basic media, and plant growth regulator in woody plant cultures were subsequently tabulated for easy analysis and the results are listed in Appendix 1.

2.2. Bibliometrical analysis

Two steps of analysis were carried out. The first step was to identification of trends in seasonal influences, use of basic media and PGRs (plant growth regulators); Second, perform data comparisons and determine the basis for improving tissue culture techniques on aspects of season, basal media and PGRs.

3. Results and Discussion

3.1. Seasonal effect

The regeneration system of tissue culture plants depends on many factors but the explant retrieval season has been identified as one of the main factors that should get serious attention in tissue culture. The influence of seasons, explant type, explant sterilization, genotype, leptin and melatomin [2-6,8,12,15], as well as the combined influence of seasons, explants and PGRs has been widely reported results [1] and Role explant collecting season have been identified [21]. Studies related to the influence of explant harvesting seasons on the success of tissue culture on woody plants are listed in Figure 1.

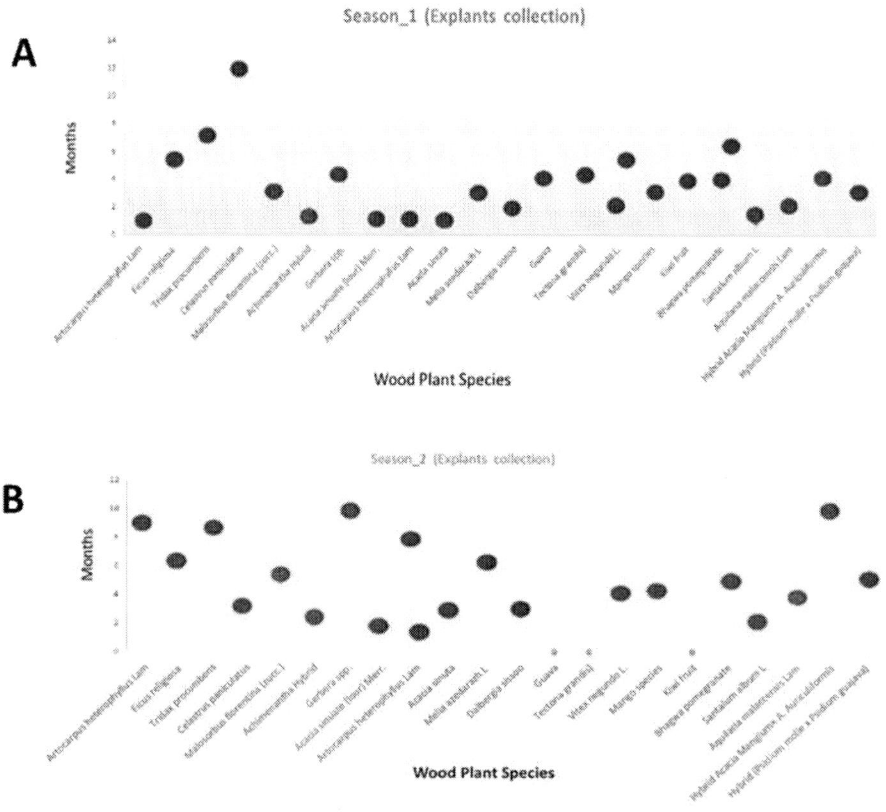

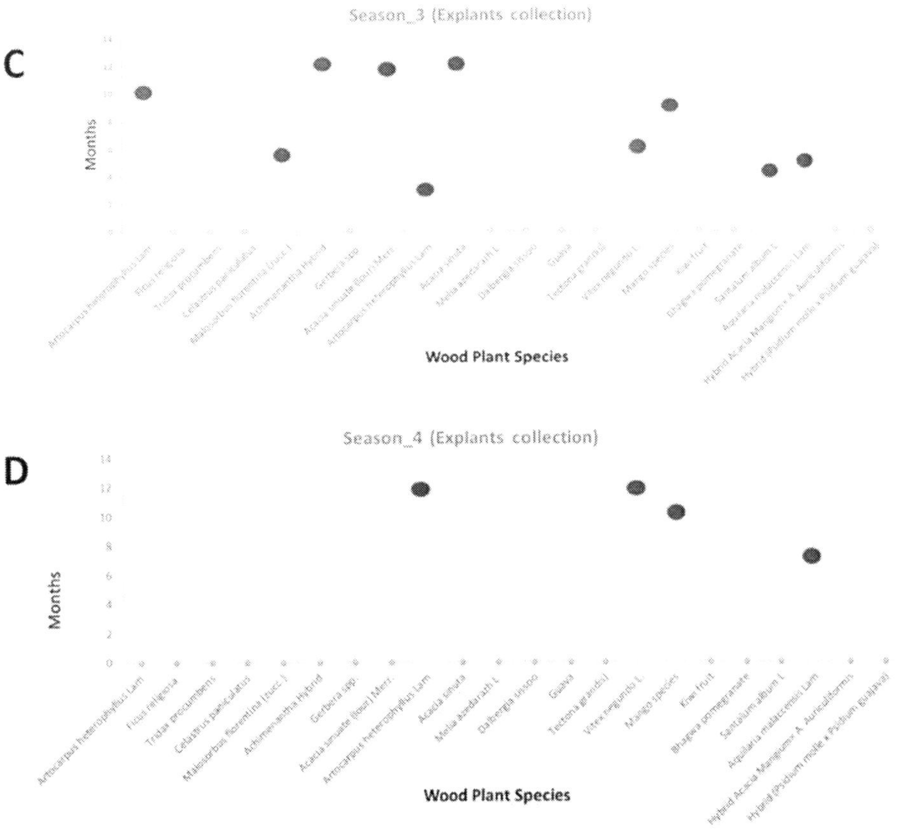

Figure 1. Scatter diagram of variations of the seasons of explant collection in January-March (A), March – July (B), March – December (C) and July – December (D)

Explant retrieval is the first step in the tissue culture stage and the retrieval time greatly affects the quality of the explant obtained. Based on a certain model system it is seen that totipotent explants can be bred from any part of the plant. The selection of explant retrieval can take into account the source of the explant from a healthy mother plant, from the field or seedbed in the house, and the right position and age of the explant because it allows the growth of the explant to be more. Explant survival increases taking into account the retrieval time as stated in Figure 1-A, which is the period January – March. Other option options could consider the periods in Figures 1-B, 1-C, and 1-D. The status of the tissue culture protocol applies specifically and must still be accompanied by optimization to ensure the basic information can be verified to ensure its validity [10,38].

Improper use of explants will increase the risk of microbial contamination also increases. The meristematic tips of plants such as the tip of the shoot, the tip of the auxiliary shoot, and the tip of the root are commonly used because it has a high potential for a high level of cell division and produces the necessary growth regulatory hormones including auxins and cytokinin. Some explants collected from the tips of the roots are difficult to isolate and contaminated with the soil microflora that forms a close bond with the root system or is even born inside the roots. Soil particles that are also attached are very difficult to remove without injuring the roots that cause microbial attacks.

3.2. Basal media

Tissue culture basal media is known to be quite widely known including (a) Murashige and Skoog (MS), (b) white basal media, (c) basal medium Vacant Went (VW), (d) Basal medium Nitsch and Nitsch, (e) Basic medium Schenk and Hildebrandt, (f) Basal medium Woody Plant Medium (WMP) and (g) basal medium N6. The importance of optimization in the selection of basic media can be seen in Appendix 1, which shows the basic media MS in the tissue culture of woody plants, albeit in the general designation of MS media is for herbaceous plants while the basic media of WPM and Medium base of Schenk & Hildebrandt are intended for woody plants. The compilation of the use of basal media in the tissue culture of woody plants is shown in Figure 2.

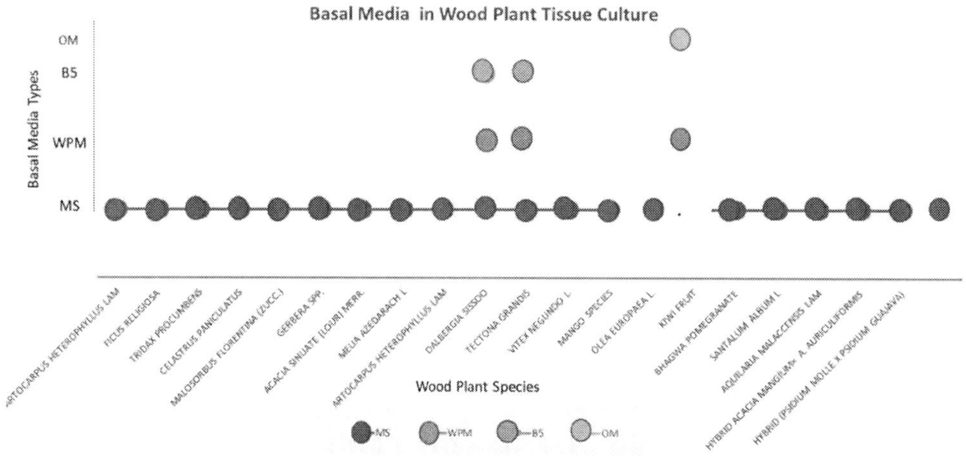

Figure 2. Scatter diagram of the number of uses of PGRs in tissue culture of woody plants

The basic protocols based on MS media [16] in the field of research and industry was carried out with many modifications. However, there are still some species and cultivars that have not responded to the protocol. The publication of protocols by many scientists regarding species that are not reproduced by other laboratories or do not survive under continuous production. This condition encourages scientists to review the basic principles including in modern plant tissue culture carried out under aseptic conditions with filtered air (free from microorganisms).

Plant tissue culture media contains all the nutrients (macronutrients, micronutrients, vitamins), supporting organic components, Plant growth regulators, carbon sources, and some gelling agents in a solid medium needed for normal growth and development of plants. MS media are most widely used for vegetative propagation of many plant species tissue culture. The pH factor of the base medium is also important because it affects both plant growth and the activity of growth regulators. The composition of the basic medium and its combination with plant hormones and nitrogen sources has a profound effect on the beginning of the explanatory response [26].

3.3. Plant growth regulators

Observations of plant growth regulators (PGRs) used in woody plant tissue culture activities vary greatly because they are influenced by the type, composition, concentration, regeneration, and subculture stage. PGRs usage mapping is shown in Appendix 2. Based on the mapping, a scattering diagram based on species and the use of PGRs used in explant regeneration, induction phase, multiplication phase, and root phase were carried out, the results of which are shown in Figure 3.

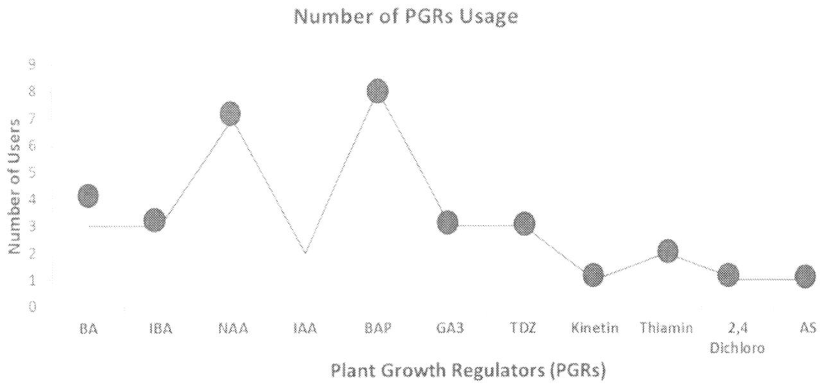

Figure 3. Scatter diagram of the number of uses of PGRs in tissue culture of woody plants

Figure 2 shows that BAP and NAA are the most preferred choices in woody plant tissue culture. The selection of BAP and NAA added in the base medium is strongly related to the breakdown of the bud in the induction phase. It has been reported to be able to support very high explant survival rates and high regeneration frequencies and requires the least number of days. Information related to callus induction, proliferation, and regeneration using a combination of BAP and NAA in various plants. Different reports suggest that the addition of BAP plus 2,4-phenoxy acetic acid (2.4 D) can cause necrosis (coagulative characterized by gelatin substance formations in dead tissue) which is very high in explants. Propagation in the culture media had different forms of callus cultures like cell suspension culture, protoplast culture, embryo culture, anther culture and pollen culture. Each type is preferred for various aspects in the culture of plants. Identification of different culture media [26], plant tissue culture media and practices [35], Effect various plant growth regulator [34] and combination effect PGRs, Season and type cutting explant [9] the results have been made available.

PGRs play an important role in determining the developmental pathways of plant cells and tissues in the culture medium. Auxins, cytokinins and gibberellin are the most commonly used growth regulators. The type and concentration of hormones used depend primarily on the plant species, the tissue or organ being cultured, and the purpose of the experiment. Auxins and cytokinins are the most widely used growth regulators in plant tissue culture and their number determines the type of culture created or regenerated. High concentrations of auxin generally favor root formation, while high concentrations of cytokinins promote the regeneration of shoots. The balance of auxins and cytokinins leads to the development of a mass of undifferentiated cells known as a callus. Review PGRs effect has been reported among others effect of various plant growth regulators [34], effect of season and growth regulators on in vitro [13] and Effect of auxins (IBA and NAA) and season on rooting [14], Effect of plant growth regulators, type of cutting and season [9].

3.4. Improvement of the tissue culture protocol

The tissue culture protocol is not generally applicable but is very specific and is influenced by many factors and the approach that is often carried out is to optimize all stages of tissue culture. Screening of factors that influence the success of tissue culture is not limited to the season and time of taking explants, growing media, and plant growth regulators but can be supplemented by other major determining factors such as environmental factors as lighting. Sing [31] reported problems and their countermeasures in plant tissue culture (a) explant selection, (b) Sterilization of culture media and containers, (c) sterilization of explants, (d) Influence of ultraviolet-c radiation in explant sterilization, (e) role of endophytic organisms in contamination, (f) Preparation of medium, nutrition, and hormonal imbalances, (g) contamination, (h) Construction of tissue culture laboratories:, (i) role of antibiotics:, (j) browning of explants:, (k) somaclonal variations, (l) somaclonal variations, (m) contamination in subcultures and (n) hardening and acclimatization.

Optimization of these components must be carried out simultaneously and continuously to produce the most suitable protocol. Improvements in invitro techniques include invitro techniques in the field of breeding [11], rapid in vitro multiplication [19] and rapid in vitro cloning [33]. Availability of information supporting the improvement includes the results of a review on problems and its remedy [17], improved micropropagation protocol [22] A modified and improved protocol development for in vitro clonal propagation [29] and bioreactor for the production of secondary metabolites [24].

3.5. The need for low-cost tissue culture

The commercial application of tissue culture technology is limited due to high production costs, therefore the most challenging aspect today is to reduce production costs, thereby improving production efficiency [32]. Micropropagation protocols have been developed for many woody plant species but due to the high cost of production then production is still on a small scale through micropropagation and efforts to overcome these limitations, so low-cost strategies began to be developed around the world. Low-cost tissue culture is very useful not only for farmers but also for regular large-scale commercial propagation. Components of plant tissue culture that require attention include basic media and their complementary compositions (plant growth hormone, vitamin and mineral nutrients), explants, equipment, and infrastructure. These components are closely related to cost reduction efforts.

Improvement and improvement of tissue culture protocols will be an important part of increasing the number of resources invested. Naik et al [32] mentioned several future tissue culture targets that include (a) the need for low-cost tissue culture technology, (b) adoption of low-cost options, (c) low-cost sterilization operations, (d) low-cost gelling agent alternatives for plant-based tissue cultures, (e) the use of liquid media and physical matrices, (f) low-cost nutritional alternatives and (g) low-cost carbon and energy alternatives. The success of propagation through tissue culture does not stop at the stage of availability of superior seedlings but must also pay attention to the demands of low-cost operations.

4. Conclusions

Propagation of woody plant tissue cultures has several tendencies related to the selection of explant harvesting seasons in January – March, but some other seasons and times can be used as alternative options. Basic media shows MS Media is the main choice followed by WPM media. The selection of PGRs shows the choice is directed at the BAP and IAA. The combination of media and the combination of PGRs is most widely carried out and requires foresight in determining the best combination. Seriousness is needed in realizing low-cost tissue culture through the continuous collaboration of researchers and related partners. Considering that tissue culture activities require a fairly long time, optimization is a keyword in streamlining all components that determine the success of the wood plant tissue culture target.

Acknowledgments

We would like to express our sincere thanks to the Ministry of Environment and Forestry Republic of Indonesia for providing funds to support our research "Breeding Strategy of *Toona sinensis* Roem. & *Toona sureni* Merr." for raw material of furniture wood industry. Contract Number: MAK 5391.DDA.002.051.T- DIPA 2021.

References

[1] Siwach P, Gill A R and Kumari K 2011 *African Journal of Biotechnology* **10** 4879
[2] Yari M B, Gholami M, Ashari E M and Kheradmand H 2014 *International Journal of Agriculture and Crop Sciences* **7** 624
[3] Tosca A, Arcara L and Frangi F 1999 *Plant cell, tissue and organ culture* **59** 77
[4] Mishra M, Sharma M R and Rajan S 2019 *Green Farming* **10**
[5] Singh S K, Khawale R N and Singh S P 2002 *Indian Journal of Horticulture* **59** 233

[6] Yadav K and SinghN 2011 *Analele Universitati din Oradea Fascicula Biologie* **18**
[7] Lu Y, Huang L, Wang H and Shahid M Q 2021 (https://doi.org/10.21203/rs.3.rs-690162/v1)
[8] Gorka K B, Szczesna E, Molik M and Zieba D A 2010 *Small ruminant research* **91** 231
[9] Bharathy P V, Sonawane P C and Sasnu V P 2004 *Indian Journal of Horticulture* **61** 338
[10] Roh M S and Wocial M 1988 *International Symposium on the Development of New Floricultural Crops* **252** 129
[11] Smith M K and Drew R A 1990 *Functional Plant Biology* **17** 267
[12] Yadav K, Lal D and Singh N 2011 *Journal of Agricultural Technology* **7** 1355
[13] Mannan M A, Nasrin H and Islam M M 2006 *Khulna University Studies* **7** 83
[14] Swamy S L, Puri S and Singh A K 2002 *New Forests* **23** 143
[15] Papafotiou M and Martini 2007 *International Symposium on New Floricultural Crops* **813** 503
[16] Murashige T and Skoog F 1962 *Physiol Plant* **15** 473
[17] Singh C R 2018 *Asian J. Biol. Sci.* **11** 165
[18] Vengadesan G, Ganapathi A, Anand R P and Selvaraj N 2003 *In Vitro Cellular & Developmental Biology-Plant* **39** 409
[19] Husain M K and Anis M 2009 *Acta physiologiae plantarum* **31** 765
[20] Arya I D, Nautiyal S and Arya S 2013 *International Journal of Biotechnology Research* **1** 58
[21] Devi S and Yadav N R 2022 *The Pharma Innovation Journal* **11** 1244
[22] Tiwari S K, Tiwari K P and Siril E A 2002 *Plant Cell, Tissue and Organ Culture* **71** 1
[23] Stephen M, Nagarajan S and Ganesh D 2010
[24] Yancheva S, Georgieva L, Badjakov I, Dincheva I, Georgieva M, Georgiev V and Kondakova V 2019 *Journal of Central European Agriculture* **20** 321
[25] Chandra R, Padaria J C and Srivastava A 2004 *Indian Journal of Plant Physiology* **9** 136
[26] Moreira R M and Schuch M W 2018 *Agronomy Science and Biotechnology* **4** 1
[27] Wang R, Song Q, Kong L and Zhang Y 2020 *IOP Conference Series: Materials Science and Engineering* **740** 012194
[28] Kumar K, Arora P K, Brar J S, Bhatia D and Kumar A 2019 *Indian Journal of Horticulture* **76** 273
[29] Bhargava P, Ravindra N and Singh G 2018 *Tropical Plant Research* **5** 193
[30] Borpuzari P P and Kachari J 2018 *Journal of Medicinal Plants Studies* **6** 9
[31] Singh CR 2018 *Asian J. Biol. Sci.* **11** 165
[32] Naik R, Bhushan A, Gupta R K, Walia A and Gaur A 2020 *International Journal of Biochemistry Research and Review* **29** 66
[33] Arora K, Sharma M, Srivastava J, Ranade S A and Sharma A K 2010 *Agroforestry Systems* **78** 53
[34] Mishra J P, Bhadrawale D, Yadav U, Mohammad N and Shirin F 2018 *Trop. Plant Res.* **5** 152
[35] Phillips G C and Garda M 2019 *In Vitro Cellular & Developmental Biology-Plant* **55** 242
[36] Tefera A A 2019 *Journal of Natural Sciences Research* **9** 20
[37] Zarei H K, Salehi H and Jowkar A 2017 *International Journal of Horticultural Science and Technology* **4** 51
[38] Malik C P and Wadhwani C 2009 *African Journal of Biotechnology* **8** 3239

IOP Conf. Series: Earth and Environmental Science 1115 (2022) 012051 doi:10.1088/1755-1315/1115/1/012051

Appendix 1. Summary of season, basal media and plant growth regulators in wood plant tissue culture propagation

Species	Treatment			Explant & regeneration type	References
	Season	Basal Media	Plant Growth Regulators (PGR's)		
Artocarpus heterophyllus Lam	Sept tp May (Januari Best)	MS	-NAA-0.5 mgl-1 with different levels (1.0, 1.5 & 2.0 mgl-1) -BA (benzil adenin) at 1.0 and 1.5 mgl-1to study the proliferation and multiplication of shoots	Shoot culture	[13]
Ficus religiosa L	May, June	MS	2.26 μM 2,4-dichlorophenoxyacetic acid and 3% sucrose	Shoot culture	[1]
Tridax procumbens	July and September	MS	BAP (1 mg l-1)	Shoot culture	[38]
Celastrus paniculatus	December, March	MS	benzil amino purine (BAP) (2.0 mg / l) + NAA (0.5 mg / l).	Callus induction	[6]
Malosorbus florentina (zucc.)	March, May, June	MS	0,7 - 1 mg / l BA with 0.1 mg /l IBA. Shoot culture in MS + 1 mg /l BA + 0.1 mg /l IBA or NAA & MS 1 mg /l BA + 0.1 mg /l IBA + 1 mg /l kinetin + 0.1 mg /l IAA + 0.05or 0.2 mg /l TDZ	Shoot culture	[15]
Picea abies (L.)	Spring - Fall	MS	2-iso-pentenyl adenine (2iP) and Zeatin (Z)	Shoot culture	[37]
Acacia sinuate Merr.	Jan to Dec & Dec – March	MS	PM 6-benzyl adenine (BA), 2.5 PM thidiazuron (TDZ), and 135.7 PM adenine sulfate (AS) at the end of the first transfer following initial culture (60 day after inoculation). Gibberellic acid (GA3) at 1.8 PM promoted shoot elongation.	Shoot culture	[18]
Melia azedarazh L	March – Juni Best	MS	PM 6-benzyl adenine (BA) shoot lengh, 2.5 PM thidiazuron (TDZ), and 135.7 PM adenine sulfate (AS) at the end of the first transfer following initial culture (60 day after inoculation). Gibberellic acid (GA3) at 1.8 pM promoted shoot elongation.	Shoot cultured	[19]
Dalbergia sissoo	February - March	MS & WPM, B5	BAP (2.5-10.0 μM) and auxin NAA at varying concentrations and cultured on full strength MS media supplemented with BAP (0.0-12.5 μM) separately or in combination with NAA	Shoot culture	[20]
Tectona grandis	April – May the best season	MS	Explant colleting 22.2μM benzyl amino purine and then serially transferred to fresh medium after12, 24, 48 and 72 h gave maximum culture establishment (76.8%)	Shoot Culture	[22]
Vitex negundo L.	March-May & Dec-Peb the best season for explant collecting	MS	BA at 1mg/l was effective (PG) at 100mg/l and silver nitrate (20 mg/l) in addition to BA (1 mg/l) Among the different concentrations of IBA (0.1, 0.5 and 1.0 mg/l) tested, 0.5 mg/l IBA was more effective for	inducing roots	[23]
Mango species	March April – Sept-Oct the best	MS	3 mg/l indole acetic acid + 1 mg/l kinetin + 200 mg/l adenine sulphate + 1 mg/l thiamine + 1 mg/l riboflavin + 100 mg/l casein hydrolysate+ 400 mg/l glutamine + 100 mg/l PVP + 100 15mg/l ascorbic acid + 60 mg/l sucrose	Shoot culture	[25]
Olea europaea L.	Winter gave phenolic oxidation & spring recomendation	MO & WPM	Salt and Thiamin (Vitamin) *No information of PGRs*	Explant treatement	[26]

Kiwi fruit	April - lower browning	MS	survival rate of callus induction in the treatment of 75% alcohol 20s + 0.1% mercury for 5min was significantly higher than other treatments, *No information of PGRs*	Callus induction	[27]
Bhagwa pomegranate	April-July	MS	A combination of growth regulators (BAP, NAA) and growth substances (AgNO3, Adenine sulphate) were also tested for shoot bud initiation The significantly highest response (85.17%) was achieved on MS medium supplemented with BAP (1.0 mg l-1) + NAA (0.25 mg l-1)	Shoot culture	[28]
Santalum album L	Januari, Peb-April the best	MS	Highest shoot multiplication with longest shootlets (2.90 cm) was achievedon MS medium containing 0.5 mg L-1BAP and 5.0 ml L-1 IBA after 30 days of culture.	Shoot culture	[29]
Aquilaria malaccensis Lam	Peb – April, May_July	MS	Micro shoots treated with 50.0 ml L-1IBA for 48 h produced roots on growth regulator free. GA3 in combination with adenine GA31.0+ adenine 0.2 without BAP recorded as highest 1.52 ± 0.23 shoot elongation In combine effect MS+BAP 0.5+NAA 0.5+glutamine 20 mg/l was recorded maximum number of shoots 2.41 ± 0.74	Non-embrionic and shoot culture	[30]
Hybrid *Acacia mangium* × *A. auriculiformis*	April to October is the most optimum period tocollect explants for bud induction	MS	he concentration of 6-benzylaminopurine (6-BA) was 0–1 mg/L, the germination rate Therefore, the best bud induction medium combination for A. mangium × A. auriculiformisis 1/8 MS + 6-BA 0.5 mg/L, with a maximum germination rate of 92.55%	Induction culture	[7]
Hybrid *Psidium molle* × *Psidium guajava*	March – Mei (low contaminant)	MS, Juglans basal medium, Woody plant medium, Date palm basal medium and Banana multiplication medium	combinations of auxin and cytokinin tried, 2-ip 0.5mg/l + NAA 0.5mg/l containing media gave 63.43 per cent survival of explants, high regeneration frequency (32.12%) and required only 12.7 days for callusing	Leaf induction culture	[4]

IOP Conf. Series: Earth and Environmental Science 1115 (2022) 012051 doi:10.1088/1755-1315/1115/1/012051

Appendix 2. Summary of plant growth regulators in wood plant tissue culture propagation

Species	Plant Growth Regulators (PGRs)												
	BA	IBA	NAA	IAA	BAP	GA3	TDZ	Kinetin	Thiamin	2.4-D	AS	2-ip	Zeatin
Artocarpus heterophyllus Lam			√										
Ficus religiosa										√			
Tridax procumbens					√								
Celastrus paniculatus			√		√								
Malosorbus florentina (zucc.)	√	√	√	√			√						
Picea abies (L)												√	√
Acacia sinuate (lour) Merr.	√					√	√						
Melia azedarach L	√					√	√				√		
Dalbergia sissoo			√		√								
Tectona grandis					√								
Vitex negundo L.		√											
Mango species				√				√	√				
Olea europaea L.									√				
Kiwi Fruit													
Bhagwa pomegranate			√		√								
Santalum album L		√			√								
Aquilaria malaccensis Lam					√	√							
Hybrid *Acacia mangium* × *A. auriculiformis*			√		√								
Hybrid *Psidium molle* × *P. guajava*			√										

Potential analysis of Nipa Palm (*Nypa Fruticans*) in Singkil

T M Ibrahim, E Julianti, T Supriana and Delvian

Doctorate Program of Natural Resources and Environmental Management, Universitas Sumatera Utara, Medan, North Sumatra, Indonesia 20155

*Email: t.miftahibrahim@gmail.com

Abstract. Nipa palm (*Nypa fruticans*) is the plant that grouped in mangrove forest ecosystems which has a great economic potential when developed with a good strategy. The potential of nipa palm in Singkil has not been mapped properly, so that the objective of this research was to analyze the potential of nipa palm in Singkil. Data in this research were obtained directly from observations in the form of ground checks at several research locations and aerial photo data taken directly in the field using UAV vehicles. The results of data obtained through observation and documentation are then processed before being used to validate the results of image data processing. The use of UAV data provides a solution for mapping and monitoring nipa palm with high accuracy as well as a solution to the limited availability of nipa palm data in Singkil. Nipa palm was found spread along the estuary of the Lae Singkil River with total area 768.08 Ha. In the aerial photo taken in Kampong Pulosarok covering an area of 37.01 Ha, it was found that the land covering of nipa palm was 20.12 Ha or 54%. Whereas, in the aerial photo taken in Kampong Kilangan covering an area of 34.41 ha, it was found that the land covering of nipa palm was 22.16 ha or 64%. From the data, it can conclude that the nipa palm potential in Singkil is large.

1. Introduction

Mangrove forest is a collection of trees that grow in the area around the coastline which is affected by tides and in places where organic matter accumulates and mudslides. Mangrove forest is an ecosystem that is unique because of the inundation cycle activity by the tides. In general, the existence of mangrove forests worldwide is decreasing, including in Indonesia. The cause of the reduction in mangrove land is usually due to conversion to agricultural land and salt production, coastal industrialization, conversion to aquaculture and others [1]. Data from the Aceh Singkil Forestry and Plantation Office in 2012 states that Aceh Singkil has a mangrove ecosystem area of 6,514 hectares [2].

Nipa palm (*Nypa fruticans*) is the plant that spread throughout the archipelago and grouped in mangrove forest ecosystems because they grow along rivers that are affected by tides. This plant has a great economic potential when developed with a good strategy because almost all parts of this plant can be used for food, industrial raw materials and renewable energy. In addition, Nipa is a plant that is friendly to the environment and is very suitable for conservation plants because it has the ability to maintain soil stability, protect against erosion, and reduce the strength of high -sea storms and high sea waves in coastal areas [3].

Nipa palm have high potential to be developed as a source of food and new and renewable energy, namely by tapping the sap from fruit bunches. The resulting sap is then converted into sugar in solid or liquid form or fermented into bioethanol and purified so that it can be used as fuel [4–8]. The problem is, the high potential of nipa palm has not been utilized optimally. In fact, if this potential is optimized, it will become a large source of income for foreign exchange which certainly has a positive effect on social, economic and environmental aspects. However, exploitation of nipa palm that does not care to ecological factors can also endanger sustainability.

Aceh Singkil is one of the regencies in Aceh Province which has large area of nipa palm. The regency with an area of approximately 1,857.88 km^2 is dominated by land areas and partly in the form of islands. Singkil is a sub-district in Aceh Singkil located at the estuary of the Lae Singkil River. Nipa palm are spread along the estuary of the Lae Singkil River, which is downstream from the Lae Cinendang River and Lae Soraya River. So far the potential of nipa palm in Singkil has not been mapped properly, so that the objective of this research was to analyze the potential of nipa palm in Singkil.

2. Research Methods

This research was conducted from May to June 2022 located in Aceh Singkil Regency, Aceh Province. The tools used in this study are Global Positioning System (GPS), smartphones equipped with the DJI Go 4 and Pix4D Capture applications, DJI Phantom 4 UAVs, tally sheets, and stationery. The data processing is carried out with software ArcGIS version 10.3, Quantum GIS version 2.14.6 and Agisoft Photoscan. The primary data used are obtained directly from observations in the form of ground checks at several research locations based on GPS tracking and aerial photo data taken directly in the field using Unmanned Aerial Vehicle (UAV). The results of field data obtained through observation and documentation are then processed before being used to validate the results of image data processing. The process of acquiring UAV images or aerial photograph is carried out through several stages, namely the preparation stage, the acquisition/shooting stage, and the data processing stage. The height chosen is 120 meters high and the resulting image resolution is 2.01cm/pixel. The vehicle used is a DJI Phantom 4 UAVs drone with a camera resolution of 20 MP with the ability to fly for 130 minutes per battery. The aerial photo data obtained is then processed in the Agisoft Photoscan Pro software to processing align photo, build mesh and build orthomosaic [9–12].

3. Results and Discussion

3.1. Characteristics of the studied area

Aceh Singkil Regency is located at 20°02'-20°27'30" North Latitude and 97°04'-97°45'00" East Longitude, with altitude ranging from 6 to 74 meters above sea level. The total area of Aceh Singkil Regency is 1.857,88 km2 or about 18% of the area of Aceh Province. Aceh Singkil Regency has administrative boundaries which include the North bordering the City of Subulussalam, the South bordering the Indonesian Ocean, the East bordering the Province of North Sumatra and the West bordering Trumon, South Aceh Regency [13].

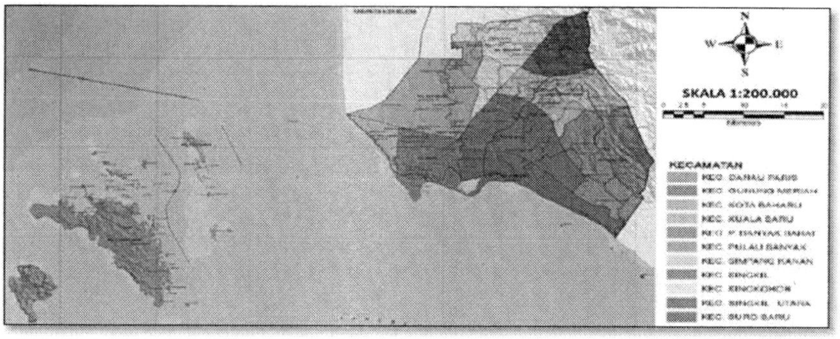

Figure 1. Subdistricts distribution in Aceh Singkil

IOP Conf. Series: Earth and Environmental Science 1115 (2022) 012052 doi:10.1088/1755-1315/1115/1/012052

According to data from Meteorological, Climatological, and Geophysical Agency Aceh in 2020, the average air temperature in Aceh Singkil Regency was a minimum of 27.57^{0}C and a maximum of 29.73^{0}C. The average humidity was minimum 70.89%, and maximum 82.62%. The average air pressure minimum and maximum was 1003.39 mb and 1006.21 mb, respectively. Meanwhile, the rainfall was 16.8 mm in January (minimum) and 388.9 mm in may (maximum) with minimum rainy days of 4 times a month and a maximum of 27 times [13]. These environmental conditions are very suitable as a condition for the growth of nipa palm plants.

3.2. Nipa palm distribution

Aceh Singkil Regency is divided into 11 subdistricts, 5 of which are coastal areas with mangroves. The 5 subdistricts are Singkil, North Singkil, Kuala Baru, Banyak Island and West Banyak Island. However, nipa palm was not found in all mangrove areas but only spread along the estuary of the Lae Singkil River, which is downstream from the Lae Cinendang River and Lae Soraya River. Figure 2 shows the data on the distribution of nipa palm in Singkil measured using GPS tracking. The total area of nipa palm in Singkil was 768.08 Ha.

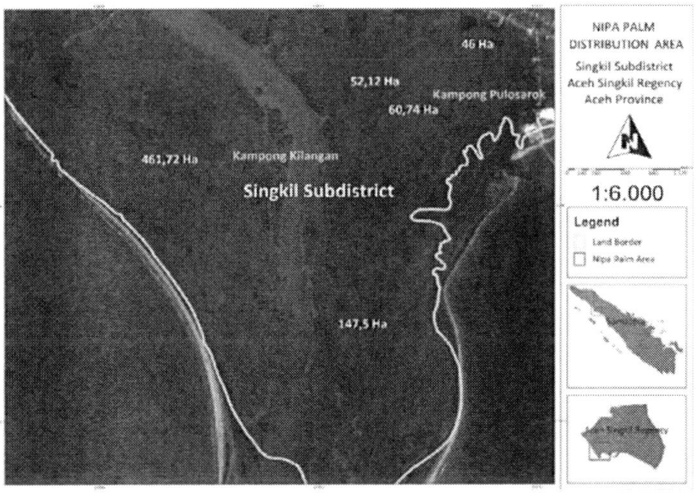

Figure 2. Nipa palm distribution in Singkil

3.3. Land cover analysis of nipa palm using Unmanned Aerial Vehicle (UAV)

Remote sensing is the technology that is able to prevent and overcome the damage to the distribution of nipa palm by mapping nipa palm. Remote sensing is a science and art that studies an object and to obtain information about a phenomenon through the process of analyzing data using a tool without direct contact with the object being analyzed. This technology has developed rapidly and currently uses satellites to record data and process it to get information [14].

Unmanned Aerial Vehicle (UAV) can be used to analyze land covering, especially in coastal areas. This land cover analysis can be used to identify the potential of a plant in an area [10]. The use of this UAVs has the advantages that it makes easier for us to analyze the potential of the land. In addition, the costs incurred are also cheaper than using satellite imagery with a comparable level of image resolution. However, the use of UAVs also has disadvantages such as when bad weather or other natural disturbances can produce blurry photos that can cause land identification errors. In addition, the use of UAVs on surface waters also has a greater risk of falling and will be damaged by water. The aerial photo data obtained from the UAV which have processed in the Agisoft Photoscan Pro software can be seen at Figure 3 and Figure 4.

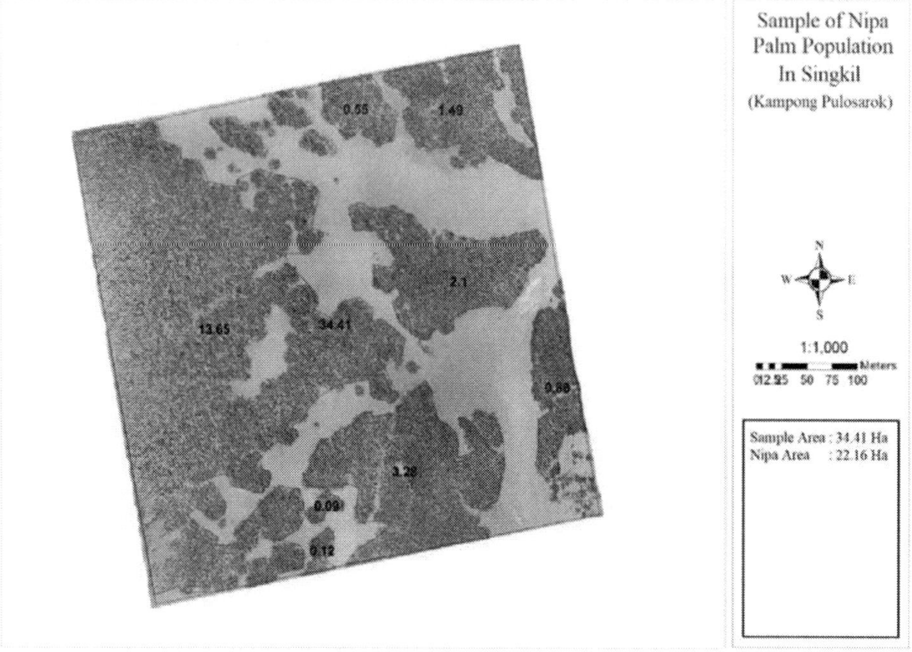

Figure 3. Sample of nipa palm population in Kampong Pulosarok, Singkil taken by UAV

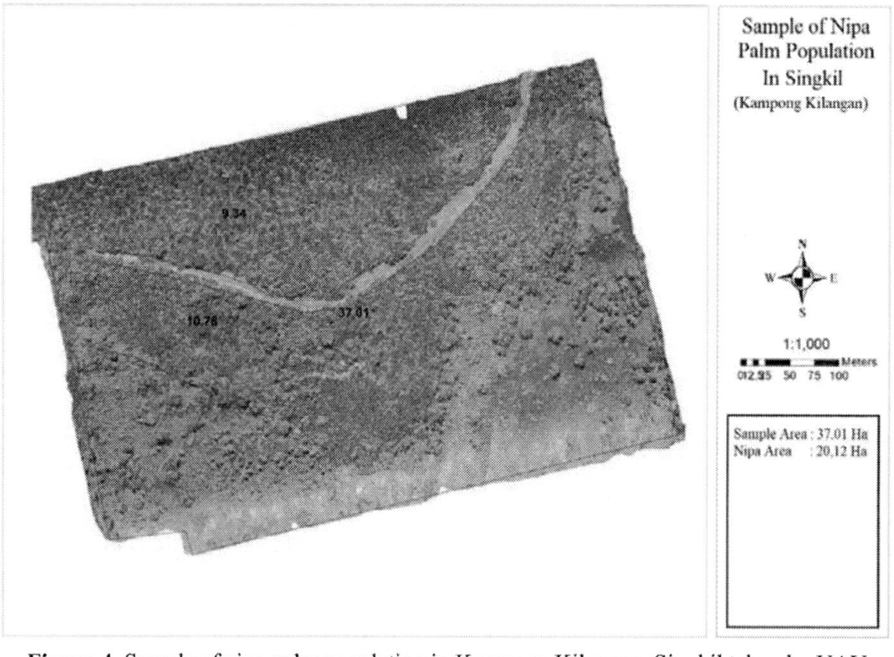

Figure 4. Sample of nipa palm population in Kampong Kilangan, Singkil taken by UAV

In the aerial photo taken in Kampong Pulosarok covering an area of 37.01 Ha, it was found that the land covering of nipa palm was 20.12 Ha or 54% (Figure 3). Whereas, in the aerial photo taken in Kampong Kilangan covering an area of 34.41 ha, it was found that the land covering of nipa palm was 22.16 ha or 64% (Figure 4). The area of the nipa palm was obtained by identifying aerial photographs obtained and then validated by field observations. In most locations, nipa palms grow in clusters without other plants, but there are also a small number of nipa palms that grow side by side with other mangrove plants.

From the observations and data obtained, we can conclude that the nipa palm potential in Singkil is large. The potential benefits will be even greater when nipa palm can be utilized into products that have greater added value. Whether the product is produced from leaves, fruit or sap tapped from fruit bunches. The utilization carried out must pay attention to sustainability certainly, so that it still maintain the preservation of nature and the environment.

4. Conclusion

Nipa palm was found spread along the estuary of the Lae Singkil River with total area 768.08 Ha In the aerial photo taken in Kampong Pulosarok covering an area of 37.01 Ha, it was found that the land covering of nipa palm was 20.12 Ha or 54%. Whereas, in the aerial photo taken in Kampong Kilangan covering an area of 34.41 ha, it was found that the land covering of nipa palm was 22.16 ha or 64%. From the data, it can conclude that the nipa palm potential in Singkil is large.

Acknowledgments

Thanks to the Indonesian Endowment Fund for Education (LPDP) funded this research.

References

[1] Susanto A H, Soedarti T and Purnobasuki H 2018 *Vegetos* **31** 09
[2] Dina E H, Firmadiana and Basri H 2021 *J Ilm Mhs Pertan* **6** 86
[3] Hossain M F and Islam M A 2015 *Am J of Agric and For* **3** 156
[4] Hidayat I W 2015 *Pros Sem Nas Masy Biodiv Indon* **1** 109
[5] Hidayat I W 2018 *IOP Conf Ser: Earth Environ Sci* **166** 012045
[6] Tamunaidu P, Kakihira T, Miyasaka H and Saka S 2011 *Green Energy and Technology* **66** 159
[7] Tamunaidu P and Saka S 2012 *Green Energy and Technology* **108** 121
[8] Tamunaidu P, Matsui N, Okimori Y, and Saka S 2013 *Biomass and Bioenergy* **52** 96
[9] Basyumi M, Yuntha B and Rizka A 2021 *Mengenal Drone dalam Sistem Informasi Geografis dan Aplikasinya Dalam Penelitian Kehutanan* (Medan: USU Press)
[10] Adade R, Aibinu A M, Ekumah B and Asaana J 2021 *Environ Monit Assess* **193** 154
[11] Sarron J, Malézieux E, Sané C A B and Faye E 2018 *Remote Sens* **10** 1
[12] Giordan D, Adams MS, Aicardi I, Alicandro M, Allasia P, Baldo M and Troilo F 2020 *Bull Eng Geol Environ* **79** 3437
[13] (BPS) BPS-Statistics of Aceh Singkil Regency 2022 "*Aceh Singkil Regency in Figures 2022*" (Singkil: BPS-Statistics of Aceh Singkil Regency)
[14] Ghosh S M and Behera M D 2021 *Comput Geosci* **150** 104737

IOP Conf. Series: Earth and Environmental Science 1115 (2022) 012053 doi:10.1088/1755-1315/1115/1/012053

Seed characteristics of *Curculigo latifolia* and its prospect to *in vitro* propagation

W Muslihatin[1,2,3], Y S W Manuhara[1*] and S Werbrouck[2]

[1] Department of Biology, Faculty of Science and Technology, Universitas Airlangga, Surabaya, Indonesia
[2] Laboratory for Applied In Vitro Plant Biotechnology, Faculty of Bioscience Engineering, Ghent University, Ghent, Belgium
[3] Department of Biology, Faculty of Science and Data analytics, Institut Teknologi Sepuluh Nopember, Surabaya, Indonesia

*Email: yoshepine-s-w-m@fst.unair.ac.id

Abstract. The propagation of *Curculigo latifolia* only occurs naturally in the wild and it is difficult to be domesticated via seed. It is suspected that the seeds are recalcitrant and require a special medium and care. In this study, we evaluated the structure of seed of *C. latifolia* using a stereo microscope, SEM and tetrazolium for seed viability test. The characteristics of the seeds observed in this study were the shape and color of the seeds, viability of the seeds with tetrazolim test. Seeds from the fresh fruit were cultured on MS0 medium to determine the ability of the seeds to germinate. This study showed that seeds aged 8-9 weeks after flowering were viable seeds with a red color indicated on all seed surfaces through the tetrazolium test.The structure of *C. latifolia* embryo is classified as an axial embryo with hypogeal germination type. Seeds of *C. latifolia* have potential to be used as explants for *in vitro* culture

1. Introduction

Curculigo latifolia is locally known as Lemba or marasi is an herb from the family of Hypoxidaceae that so far only naturally grows in the wild. Besides in Kalimantan, the plant can also be found in Sumatra, Bangka, Papua, and Java. Lemba is a Kalimantan's landrace plant having high potential for cultivation because of its high economic value and commercial prospect owing to its diverse purposes. Fibers of leaves of *C. latifolia* are utilized for making fishing net and rope. In Eastern Kalimantan especially among *Dayak* tribe, the fibers of leaves of *C. latifolia* are used for traditional cloth (*Ulap Doyo*) [1]. The plant is famous for its unique sweet proteins, namely, curculin and neoculin that have been proven to be 500 to 9000 times sweeter than sucrose by weight [2], and can be used as a low-calories sweetener for diabetic and obese people. *C. latifolia* has potential as an alternative sweetener to replace sugar cane.

C. latifolia is a monocotyledonous plant, growing erect, clumped with a height of less than 2 m. In the wild, multiplication in this plant use rhizome and could not domestification via seed. It is suspected that the seed are recalsitrant. There are many seeds in a fruit. The seed has hard and thick coat. The seeds coat has many important functions, one of them is to save the seeds from bad environmental condition. The germination of some species depends on the coat. Nyman punctured the seed coat of *P. Sylvestris* [3]. Seeds of different species have evolved to vary enormously in their structural and anatomical complexity and size (the weight of a seed varies from 0.003 mg for orchids to over 20 kg for the double coconut palm (*Lodoicea maldvica*) [4]. So far, there is no report that describes the detail structure of seeds of *C. latifolia*. The aim in this study is to describe the characteristics of *C. latifolia* seeds and its prospect to germinate in in vitro condition

2. Material and Method

2.1. Samples of seeds

The seeds were obtained from the fresh fruit of *C. latifolia* which grew in Pangandaran, West Java. The fruit was selected by the characteristics of fresh white color, the color of the seeds visible from the outside with the age of the fruit 8-9 weeks after flowering. Fruits size were measured using a digital caliper. Seeds were removed from the fruit by hand. Seeds were cleaned from fruit flesh and rinsed with water.

2.2. Observation of seed structure

Size of seed were measured using a digital caliper. Whole seeds were observed with a stereo microscope with a 40x magnification. Observations included seed coat color and seed shape. The seeds were observed for the structure of the cotyledon using SEM with 100x magnification to determine the seed diameter, seed coat thickness and cotyledon diameter.

2.3. Observation of seed viability with tetrazolium

35 seeds of *C. latifolia* without seed coat were thoroughly soaked in 0.1% 2,3,5- trifenil tetrazolium klorida (w/v) and incubated at 35°C for 4 hours. Furthermore, the seeds were observed under a stereo microscope with 40x magnification. Observations included seed color which indicated the viability of seeds and determined endosperm and embryo staining.

2.4. In vitro propagtion

The seeds were planted on MS basal medium without plant growth regulator (MS0), 6 g/L agar and 3% sucrose. The medium pH was adjusted at 5.8. The cultures was incubated on white fluorescence lamp on 16 h/8 h light/dark photoperiod. The cultures were incubated for 4-5 weeks under temperature of 25 ± 2 °C.

3. Result and Discussion

In flowering plants, a fruit is a ripened ovary and seeds is a result of ovule development after fertilization. The inflorescences of *C. latifolia* are bunches, up to 5 cm long, the flowers emerge from the rhizome and look as if they are growing from the ground. The color of the flowers is bright yellow, and small. A flower in an inflorescence is composed of one pistil with one stigma, six stamina, and 6 petals (Figure 1). Fruit of *C. latifolia* was oval with the length of 2-3 cm approximately. The surface color of the fruit that was harvested 2-4 weeks after flowering appeared whitish; the fruit appearance turned greenish after 4–7 weeks of flowering and white or pale yellow after 8-10 weeks. Fruit contains many of seeds, approximately 35-45 seeds.

Figure 1. Morphology of *C. latifolia*. A). Whole plant; B). Inflorescence; C) and D). Fruits

A seed may be defined as the unit of reproduction which has the embryo and commonly food-laden endosperm enveloped by a seed coat derived from the integument of the ovule. Based on observations, it is clearly found that the shape of seed is ellipsoid, 2 - 4 mm long and with diameter of 1,5 -2,2 mm globose, with black seed coat, glossy, rough with wavy lines, beak knob-like,curved inside and hard (Figure 2). According to previous study [5], the seed coats of the fruits were pink or soft brown from 2 to 4 weeks and black from 6 to 8 weeks of flowering. The seeds and the seed coats were soft even 8 weeks after flowering, suggesting that the seeds were immature. After 10 weeks, the fruits matured, and seed coats hardened and black adequately to allow crushing by hand. In this study, it is showed that seeds having age 8-9 weeks was hard seed coat.

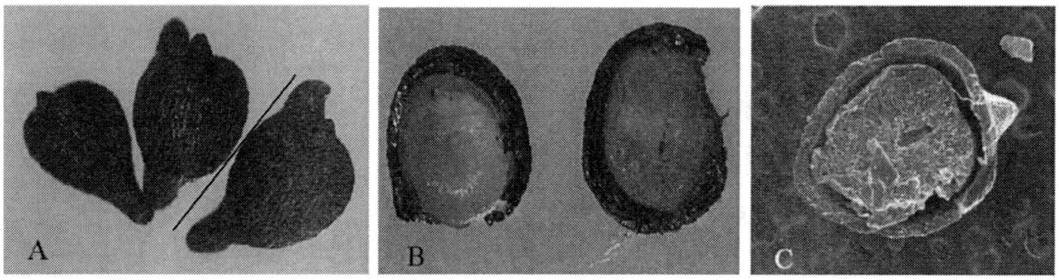

Figure 2. Morphology of *C. latifolia* seed. A). Whole seed; B). Endosperm; C). Longitudinal section of seed showed coat seed thickness. (A and B showed atenuation of 40x with stereo microscopy and C 100x with SEM) (Bar: 2 mm)

From Figure 2, it is found that the endosperm of *C. latifolia* has a white and shiny color. Observation of endosperm diameter can be known by using SEM. From the measurements, it can be seen that the diameter of the cotyledon is 1.55 – 1.55 mm and the thickness of the seed coat is 0.11 – 0.15 mm. *C latifolia* is monocotiledone plant. They contain one cotyledon.

The viability of the seed is a measure of how many seeds are alive and could develop into plants which will reproduce themselves, given at the appropriate conditions. Based on tetrazolium test, it is shown that the seeds had viable properties which were indicated by a red color on the entire surface of the endosperm and pink or lighter color on the embryo (Figure 3). The percentage of germination of *C. latifolia* was low (41.440%). This phenomena was effected by dormancy. One of the method to break the dormancy was puctured the seed coat.

Tetrazolium tests have been carried out on several of seeds, for example, barley (*Hordeum vulgare*), *Cattleya labiata* and *C. tigrina*, gherkin (*Cucumis anguria*), papaya (*Carica papaya*), tambor or guapuruvu (*Schizolobium parahyba*) [6-10], and on several cereal crops [11].

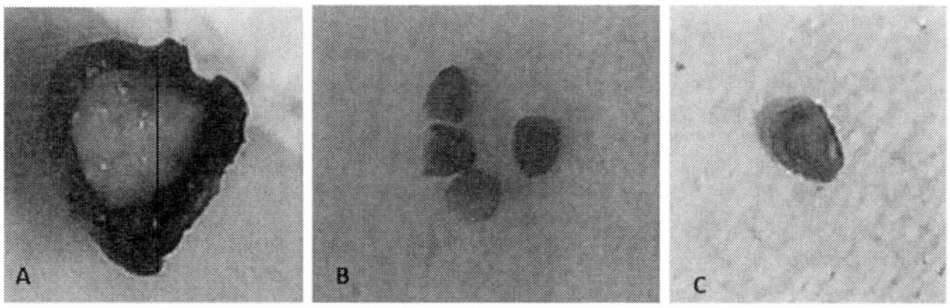

Figure 3. A. Transversal section of *C. latifolia* seed; B. Seed with tetrazolium test; c). Endosperm with peripheral embryo (Bar: 1,5 mm)

The tetrazolium (TZ) viability test uses a colorless solution of 2,3,5-triphenyl tetrazolium chloride, which is used as an indicator of the reduction process that occurs in living cells. This indicator will be imbibited by the seed, then interact with the reduction process in living cells and receive hydrogen from the dehydrogenase process. Through the hydrogenase of 2,3,5-triphenyl tetrazolium chloride, triphenyl formazan is formed which is red, stable and is an insoluble substance produced in living cells. This is what makes the difference between the red color of the live seed and the colorless part of the dead seed. The evaluation of the TZ test was carried out based on the staining pattern on the seed tissue, which would be categorized into viable and non-viable seeds. Viable seeds are seeds that show staining throughout the seed tissue required for normal germination development, while non-viable seeds show deficiencies and/or imperfect staining patterns, so that it can indicate the development of seeds into abnormal sprouts.The mature angiosperm embryo consists of an embryonic axis with a single cotyledon (monocots) [12]. Based on observations from the endosperm, it can be seen that the embryo of *C. latifolia* is classified as an axial embryo. This result is supported by Black et al. [12], it states that axial embryos range from small to occupying the entire seed interior, and may be central and straight, curved, coiled, bent or folded. The characteristics of seeds are summarized in Table 1.

Seeds with the above characteristics can be used as explants for in vitro propagation. Previous study [13] reported that the seeds of *C. latifolia* did not response on medium MS. In this study, the seeds were succesfully cultured on MS medium with free plant growth regulator and showed that the seeds could grow normally. Seed germination pattern showed the same hypogeal pattern with other monocot seeds. Hypogeal Germination is a type of seed germination that the cotyledons remain in the soil due to the epicotyl's length, which allows for faster growth and development. Seeds take 30-40 days to germinate. The average height of sprouts is 0.3-0.4 cm with 1 number of leaves. Furthermore, the stem will elongate and branch to form many petioles and leaves (Figure 4).

IOP Conf. Series: Earth and Environmental Science 1115 (2022) 012053 doi:10.1088/1755-1315/1115/1/012053

Table 1. Characteristic of seed *C. latifolia*

Fruit and seed	Characteristcs
Fruit shape	Oval
Fruit color	White glossy
Number of seeds on fruit	35-45 seeds
Seed shape	Ellipsoid, globose, black, glossy, rough with wavy lines, beak knob-like, curved inside.
Seed size	2-4 mm long
Seed coat color	Black and hard (mature)
Seed coat thick	110,7 - 151,5 um
Cotyledon diameter	1,45 mm
Endosperm color	White
Type of embrio	Axial embryo
Type of germination	Hypogeal
Germination time	30-40 days after planted
Percentage of seed germination	41,40%
Height of sprouts	0.3-0.4 cm

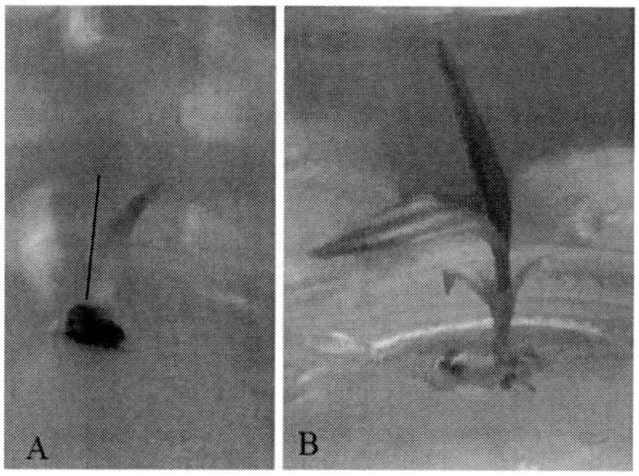

Figure 4. Propagation of seed of *C. latifolia* on in vitro condition. A). Hypogeal type on in vitro germination ; B). Plantlet of *C. latifolia* (Bar: 0,4 cm)

Micropropagation of *C. latifolia* has been reported by Babaei et al [14], who used shoot tips of Malaysian Lemba as explants on MS basal medium that was supplemented with different concentrations and combinations of thidiazuron (TDZ: 0, 0.5, 1, 1.5, and 2 mg/L) and indole-3-butyric acid (IBA: 0, 0.25, and 0.5 mg/L) for shoot regeneration and scalp induction. Furthemore, Farzinebrahimi et al. [13], successfully induced callus from several explants of *C. latifolia* from Malaysia. They used MS basal medium containing 3% sucrose that was solidified with 2.5 g/L Gelrite and supplemented with various concentrations of IBA either alone or in combination with BAP (0.5–4 mg /L). In addition, the previous study [15] succeeded in inducing callus and its regeneration from leaf tissues for plant multiplication. This study succesfull to grow seeds in MS0 medium showed that viable seeds in large numbers had the potential to be used as explants for in vitro propagation. The results will contribute to *C. latifolia* propagation for green stock.

4. Conclusion
C. latifolia seeds at the age of 8-9 weeks had good characteristics and were viable based on the tetrazolium test. *C. latifolia* seeds with the characteristics discussed above have the potential to be used as explants in in-vtro propagation.

Acknowledgements
This research was supported by LPDP-Ministry of Finance of The Republic Indonesia and PUSLAPDIK-Ministry of Education, Culture, Research and Technology of The Republic of Indonesia.

References
[1] Raden I, Nugroho C Syahrani 2017 *Biodiversitas* **18** 1367
[2] amashita H, Theerasilps S, Aiuchis T, Nakayas K, Nakamurag Y and Kuriharal Y 1990 *The Journal of Biological Chemistry* **265**
[3] Nyman B 1961 *Nature* **19** 11219
[4] Sliwinska E and Bewley D 2013 Overview of seed development, anatomy and morphology In *Seeds: the ecology of regeneration in plant communities* (pp 1–17 CABI)
[5] Okubo S, Asakura T, Okubo K, Abe K, Misaka T, Akita T and Abe K *Journal of Plant Physiology* **165** 1964
[6] Ribeiro De Souza Grzybowski C, de Castro Ohlson O, Carvalho Da Silva R and Panobianco M 2012 *Revista Brasileira de Sementes* **34** 47
[7] Hosomi S T, de Souza T B, Custódio C and Neto N B M 2017 *Australian Journal of Crop Science* **11** 1320
[8] de Paiva E , Torres S B, de Almeida J N, Sá F V da S and Oliveira RRT 2017 *Revista Ciencia Agronomica* **48** 118
[9] Carvalho S M C, Torres B, Sousa E C, Sousa D M M, Pereira K T O, Paiva EP de, Matias J R and Santos B R V dos 2018 *Journal of Agricultural Science* ***10***335
[10] Espitia-Camacho M, Araméndiz-Tatis H and Cardona–Ayala C 2020 *Actualidad & Divulgación Científica* **23**
[11] França-Neto J de B and Krzyzanowski FC 2019 *Journal of Seed Science* **41** 359
[12] Black M, Bewley J D and Halmer P 2006 *The Encyclopedia of Seeds Science Technology and Uses CAB International* (Wallingford: UK)
[13] Farzinebrahimi R, Mat Taha R, Rashid K A, Ali Ahmed B, Danaee M and Rozali SE 2016 *Evidence-Based Complementary and Alternative Medicine* 6429652
[14] Babaei N, Psyquay Abdullah N A, Saleh G and Abdullah L T 2014 *The Scientific World Journal* **9** 275028
[15] Umar A H, Ratnadewi D, Rafi M, Sulistyaningsih Y C and Hamim H 2021 IOP Conf Series: Earth and Environmental Science **948** 012051

Study of Vitality Change in Teak Community Forest Management (Case Study on Teak Community Forest in Natar Sub-district, South Lampung Regency, Lampung Province)

Arum Nurcahyani[1], Rahmat Safe'i[2*)], Afif Bintoro[3] dan Hari Kaskoyo[4]

[1, 3, 4] Department of Forestry, Faculty of Agriculture, University of Lampung
[2, 3] Master of Forestry Study Program, Department of Forestry, Faculty of Agriculture, University of Lampung

*)Corresponding author: rahmat.safei@fp.unila.ac.id

Abstract. Community forests are forests that are owned, built and managed by the people. Community forests can provide direct and indirect benefits. So that the health of community forests has various cropping patterns that can be assessed from the parameters of forest health, namely vitality. One of the plants that is in great demand in the timber industry is teak (*Tectona grandis*) because of its very high quality. This study aims to determine the changes in the value of vitality indicators and health value status of teak community forests in Rulung Helok Village, Natar District to better understand the importance of forest health in realizing sustainable forest management. The method used in this research is Forest Health Monitoring (FHM). This research was conducted twice in June 2020 and January 2022. The results shows that the change in the value of vitality indicators in the value of measurement 1 and measurement 2 experienced changes in vitality indicators which tended to increase. The assessment of each teak forest health indicator with an average measurement of 1 is 2.44 in the good category (cluster plots 2, 6, 7, and 8) the medium category is absent and the bad category (cluster plots 1, 3, 4, and 5) and the average of the 2nd measurement is 1.75 in the good category (cluster plots 2, 4, and 6), the medium category (cluster plot 8), and the bad category (cluster plots 1, 3, 5, and 7). The status of changes in the 1st and 2nd measurements of the health of teak community forests obtained an average category of medium. This is due to the occurrence of tree damage in each cluster-plot.

1. Introduction

Community woods are those that the populace owns, cultivates, and maintains. Community forests can provide direct and indirect benefits. Direct benefits are in the form of income, community welfare and sources of industrial raw materials. As for the indirect benefits in the form of preserving ecological functions (maintaining ecological stability, supporting ecosystems, and protecting the environment) that the existence of community forests is important to maximize its benefits.

Community forests have agroforestry cropping patterns with various cropping patterns, namely cropping patterns, single cropping patterns, mixed cropping patterns and agroforestry cropping patterns. Monoculture is a homogeneous or similar cropping pattern so that there is only one type of tree, and polyculture is a combination of woody and non-woody forestry plant species so as to produce a diversity of tree species. The cropping pattern of agricultural theory works on forestry and food crops, so that the health of community forests has various cropping patterns this can be studied using a parameter of forest health, namely vitality indicator. One of the plants that is in high demand in the wood industry is teak (*Tectona grandis*) because of its very high quality[13]. One of the community forests in Lampung Province which is the main teak plantation is located covers a 60 hectare in Natar District, South Lampung Regency, Lampung Province.

South Lampung's Natar district is home to the Teak Community Forest is one of the community forests that focuses on developing woody plants for industrial raw materials. Because the wood products are marketed to Madiun, East Java for industrial purposes [16]. The teak community forest applies a monoculture cropping pattern. The manager's goal in choosing this pattern is to maximize forest products in the form of good quality and quantity of wood. By evaluating forest health indicators, one may ascertain the state of the community-managed teak community forests in South Lampung district's Natar sub-district. Utilizing tree damage condition characteristics, one of the signs is the vitality indicator. Calculating damage location, damage kind, and harm severity yields vitality numbers [18]. A tree is considered healthy when it is able to perform its physiological functions and is resistant to pests and other external factors.

Declining tree health can be caused by various factors, including human activities, both biotic and abiotic variables [4]. Trees are damaged by disease, pests, weeds, fire, weather, animals or human activities. Tree damage can impede several forest functions, including conservation forests. The gradual destruction of trees will have an impact on how trees grow and develop in the forest, which will have an impact on the health of seat belts [7]. The value of changes in the health of teak community forests on vitality indicators in the community forests of Natar district is one of the objectives of this study to determine changes in the value of vitality indicators and the health value status of teak community forests in Rulung Helok Village, Natar District to better understand the importance of forest health in achieving sustainable forest management.

2. Method

This research was conducted in January 2022, located in the teak people forest (*Tectona grandis*) Natar district, South Lampung Regency, Lampung Province. The map of the research location for productivity indicators in forest health can be seen in the following Figure 1.

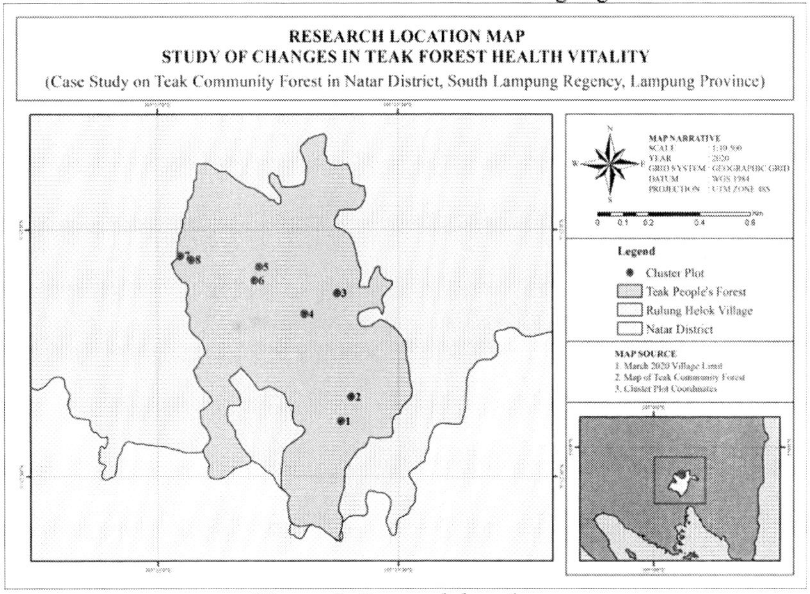

Figure 1. Research location map

2.1. Tools and Materials

The tools to be used in data collection include GPS (Global Positioning System), plastic labels, 50 meter roll meter, 150 cm tape meter, pralon pipe (Ø = 1 inch), digital camera, compass, calculator and laptop [17]. The materials used are tally sheets and forest health books.

2.2. Data Retrieval Method

The object of this research is the type of teak plant contained in the Forest Health Monitoring (FHM) cluster plot in the teak community forest with 8 cluster plots or 32 plots with an area of 32 ha of observational cluster plots. Determination of the number of cluster plots is done by purposive sampling method which is a sampling technique using certain considerations [16].

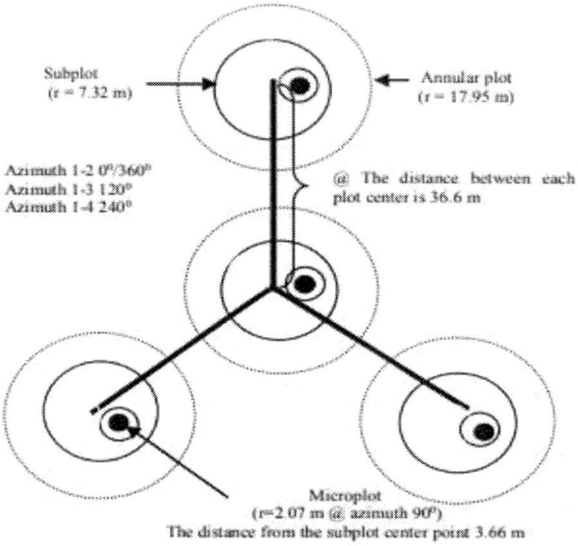

Figure 2. Diagram of plot cluster design based on Forest Health Monitoring (Source: Mangold 1999 [10]).

The Forest Health Monitoring (FHM) method or double plots in the shape of a circle were used to measure and collect data on vitality markers. A system of monitoring forest health conditions called Forest Health Monitoring (FHM) was developed by the USDA for monitoring forest health [20]. In addition, the health of teak community forests will have a good impact on the fulfillment of forest functions, preferably teak community forests as production forests [19]. Therefore, forest health can be used as a form of control to carry out the main functions of the forests and also to create the sustainability of teak people's forests [2]. In order to assess alterations in the value of the condition of the local forests based on vitality indicators in teak community forests, it is crucial to carry out this research.

Ecological measures of forest health, such as vitality, are used to assess the health of forests. Tree damage condition (CLI), crown condition, and other variables are used as criteria for forest health indicators (VCR). Based on the criteria for damage with forest health monitoring approaches, the tree damage assessment comprises the location of the damage, the type of damage, and the level of tree damage in each tree (Forest Health Monitoring) [12]. Damage that satisfies the criteria in accordance with pre-established requirements is the sort of damage assessed [18].

The damage index (IK) at the index cluster-level damage are used to examine tree damage (CLI), the plot-level damage index (PLI) and the tree level (TLI). The following equation was used to calculate the tree damage index for each cluster.

$$IK = [\text{X location x } Y \text{ damage x } Z \text{ severity}] \quad (1)$$

The amount of the weighted value, x, y, and z, varies depending on how much each component affects the growth and resistance of the tree. The final value of the health state of the teak woods was used to assess the status of the community teak forests in Natar District. You can use the following formula to determine the final value of forest health:

$$CLI = \frac{\sum PLI}{\sum Plot} \quad (2)$$

$$PLI = \frac{\sum TLI \text{ in the plot}}{\sum \text{Trees in the plot}} \quad (3)$$

$$TLI = [IK1] + [IK2] + [LK3] \quad (4)$$

$$NKHRj = \sum (NT \text{ x } NS) \quad (5)$$

Based on the value of the crown appearance rating, the state of the header was examined (Visual Crown Ratio/VCR). The value of the crown appearance rating (Visual Crown Ratio VCR) was obtained from the assessment of the five condition parameters of the title. The VCR value of a tree is worth between 1 to 4 which is based on the criteria assessment that has been developed.

3. Result and Discussion

If destroyed forest stand trees are not immediately dealt with, they will deteriorate and perhaps die . As a result, the condition of the trees that make up the stand can provide insight into the health of forest ecosystems. A healthy tree is one that can perform its physiological tasks, has a high level of ecological resistance against pest disturbances and other environmental conditions, and is not susceptible to disease [2]. Based on the type of damage and the location of the damage, the state of the trees was assessed. According to the research that has been done, the damage was discovered in the following areas: roots, stems, branches, shoots, shoots, and leaves. Results of tree damage assessment as health vitality indicator parameter forest through tree damage index in each cluster plots are in Table 1 and Figure 3.

Table 1. Results of Assessment of Changes in Tree Damage

Cluster Plot	CLI Measurement to-1	CLI Measurement to-2	CLI Change
CL1	1.857	2.353	0.496
CL2	1.886	2.237	0.351
CL3	1.816	2.195	0.379
CL4	1.570	2.054	0.484
CL5	1.526	1.996	0.470
CL6	1.284	1.291	0.007
CL7	1.330	2.172	0.842
CL8	1.260	1.649	0.389

Source: processed from the results of field data

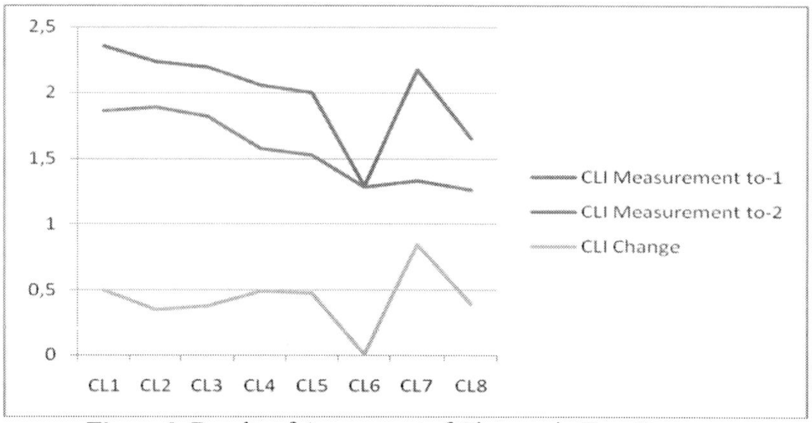

Figure 3. Results of Assessment of Changes in Tree Damage
Source: processed from the results of field data

The information in Table 1 and Figure 3 reveals that the cluster plots that have bad tree damage conditions are cluster plots 1 and 2 with CLI values of 2.353 and 2.237. Meanwhile, the other clusters, namely 3, 4, 5, 6, 7 and 8 still show tree damage which is not too worrying because it is still in good condition, with a CLI value below 3.00. The higher level of tree damage in clusters 1 and 2 compared to other plot clusters was due to attacks or disturbances of abiotic and biotic factors that attacked every part of the plant so that plants suffered greater damage than clusters 3, 4, 5, 6, 7 and 8. Disease, pests, weeds, fire, the weather, animals, etc. can all cause damage. This tree damage may prevent the forest from performing its current functions [17]. Tree damage will have a negative effect on wood quality and community woodlands health [18].

Vitality indicator parameters have an influence on tree growth. This is due to the presence or absence of defects or damage to the tree that is the cause tree growth becomes stunted or abnormal. If the damage is not repaired, it will eventually result in the death of the tree since the quality of the tree has reduced and its physiological activities have been disrupted [3]. Defines tree damage as the condition in which specific trees sustain harm due to infections, insects, air pollution, natural circumstances, and often performed human activities [14]. Based on the location of the tree damage and the type and level of the severity of the damage to the tree, the state of the tree can be assessed.

The main photosynthetic process occurs in the tree's crown, which has a significant impact on a tree's ability to grow and is closely tied to the tree's health. The header condition might therefore be one of the elements used to define the state of the forest. Regarding the results the assessment of vitality indicators through crown conditions is shown in Table 2 and Figure 4.

Table 2. Results of Assessment of Changes in Crown Conditions

Cluster Plot	VCR Measurement to-1	VCR Measurement to-2	VCR Change
CL1	3.713	2.487	1.226
CL2	3.872	4.630	-0.758
CL3	3.650	1.035	2.615
CL4	3.600	3.767	-0.167
CL5	3.550	0.935	2.615
CL6	3.600	1.287	2.313
CL7	3.700	0.932	2.768
CL8	3.650	1.155	2.495

Source: processed from the results of field data

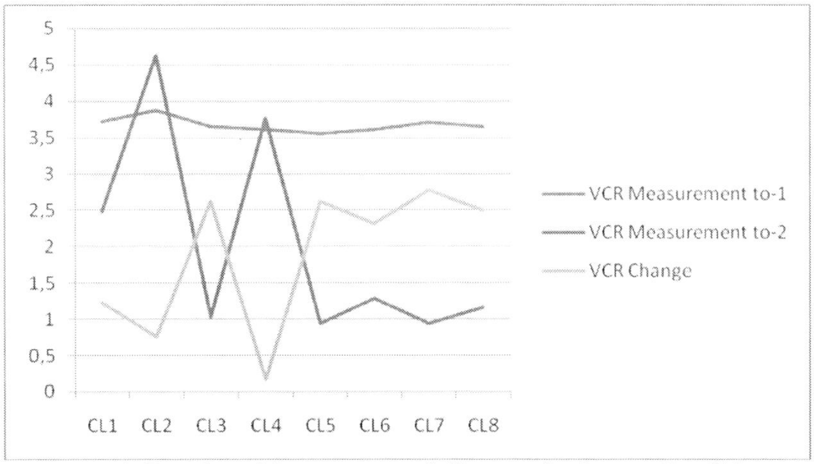

Figure 4. Results of Assessment of Changes in Crown Conditions
Source: processed from the results of field data

Based on the data in Table 2 and Figure 4 it is known that the cluster plots 2, 4, 1, 6, 8 and 3 have good crown conditions, while cluster plots 5 and 7 have bad header conditions. A high VCR number for a tree in terms of assessing the health of a forest implies that photosynthesis is occurring well and that the results are better in carbohydrate form, whereas low VCR values indicate a damaged crown, and light may not show the best photosynthetic outcomes [3]. This demonstrates how the crown's size provides information about the overall health of the tree. The five elements of the crown measurement have a significant impact on both the high and low values of the VCR [11]. Therefore, the condition of the dense crown and the width of the crown, as well as the large crow life ratio indicate that the tree's speed is fast, it will lead to better tree health condition. Meanwhile, the low density value of a tree can illustrate that the tree has a thin crorw which can be caused from disease, insects or environmental factors that affect it. This means that such conditions indicate that the condition of the forest is less able to withstand disturbances and shows that the forest ecosystem is less stable.

The health of the teak community forest can be divided into three categories: good, medium, and bad. Based on the threshold value of teak community forest health in Table 3 and Figure 5.

Table 3. The result of changes in the final health value of teak community forests

Cluster Plot	Final Value of Teak Community Forest Health		Forest Health Condition Category	
	Measurement to-1	Measurement to-2	Measurement to-1	Measurement to-2
CL1	1.850	1.350	Bad	Bad
CL2	2.930	2.700	Good	Good
CL3	1.540	0.730	Bad	Bad
CL4	1.920	2.850	Bad	Good
CL5	1.650	1.190	Bad	Bad
CL6	2.840	2.570	Good	Good
CL7	3.420	0.730	Good	Bad
CL8	3.380	1.880	Good	Medium
Average	2,44	1,75	Medium	

Source: processed from the results of field data

IOP Conf. Series: Earth and Environmental Science 1115 (2022) 012054 doi:10.1088/1755-1315/1115/1/012054

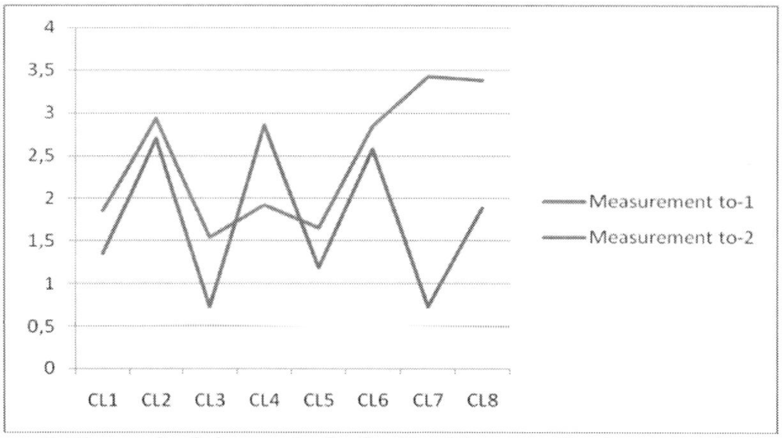

Figure 5. The result of changes in the final health value of teak community forests
Source: processed from the results of field data

The eventual health value of teak community forests tends to decline, as seen in Table 3 and Figure 5. With the exception of cluster plot 4, all cluster plots showed this decline. Teak community forests' health value significantly decreased in cluster plot 7, where there was a 2.69 difference in value. The value of the teak community forest's health likewise fell as a result of the second measurement's CLI value increasing while the VCR value decreasing. The VCR value has risen for cluster plot 4. As a result, only cluster plot4 saw a rise in the finger community forest's final health value. The more significant the weight and score values for each aspect of the higher the final value of the ecological indicator of forest health, the better [18].

Three codes that, respectively, describe where the harm is, what kind of damage it is, and how severe it is were used to classify tree damage. Damaged trees were removed in up to three damage locations, beginning at the spot with the lowest code, to meet the severity criterion. A lower damage type code number will get you a greater priority [23]. The final evaluation in a tree damage assessment is the cluster-level damage index. The tree damage status displayed in the CLI is one of the factors that can be used to gauge the stand's health [11]. Changes in environmental circumstances, both biotic and abiotic, may lead to increased tree damage. According to [5], The environment's (biotic and abiotic) negative interactions with plants can interfere with and lead to significant crop damage.

This change in environmental factors causes each tree to be more sensitive to pest disturbances. Biological factors, including: attack by pests, diseases, or other organisms that can cause damage. Abiotic factors such as: timber theft, natural disasters and land clearing [8]. Tree damage can therefore impact how trees function physiologically, slow down their growth, and even kill them [15]. Tree damage can also be caused by human activities, such as open wounds caused by managers using sharp objects during management activities. A wound that gradually opens will allow pathogens to enter the trunk and cause damage. Tree damage can be minimized through management practices such as fertilization, thinning, weeding, and weeding [1].

Types of tree trunk damage are further damage to trees caused by plant pests that make holes in plant parts. A wound or group of wounds that have peeling skin or visible wood inside and no further signs of deterioration are referred to as open wounds. Open wounds can allow pathogens to enter the tree, thereby reducing the tree's health. Termite houses are nesting places for termites, which use wood for food because wood contains cellulose. According to [21], cellulose compounds are organic compounds that are abundant in nature but cannot be digested by humans or other higher organisms, and termites can easily digest cellulose due to the presence of enzymes that secrete enzyme in their intestines trichomonas parasite therefore, termites eat wood [22].The size of the crown can be used to gauge the tree's general health. A broad and dense crown denotes rapid growth. Small and sparse canopies are a sign of bad growing site conditions, additional effects, such as storms or leaf disease, or competition with other trees or low humidity [9]. Impede healthy tree growth, whereas the trunk is crucial for producing processed wood [20].

Difference in VCR values can be caused by a number of factors, one of which is facility maintenance performed by managers. Plant care with fertilization can improve the quality of the crown through the growth of new leaves that make up the crown. This fertilization activity can also increase the value of soil CECs, which support tree growth and development [19]. In addition, larger distances make plants more adaptable to the environment [18], so plants grow better, resulting in a thicker crown. The ratio of crown length to tree height is known as the crown lifespan ratio. The quantity of plant parts that make up that crown, or the amount of sunlight that prevents sunlight from reaching the forest, is included in density. The quantity of light obstructed through trees is therefore represented by crown density [17]. A tree's high density can be explained by the fact that it has a dense crown of leaf, which meets the tree's requirement for photosynthetic energy [6].

Low soil pH levels and high tree damage levels can also contribute to bad forest health. The trees that make up the forest must therefore be in good health for it to be in good condition. Infertile site quality carrying capacity indicates that the land cannot support optimal stand growth if the tree stands are unhealthy [17].

4. Conclusion s

The results of this study indicate that the value of the results of the 1st and 2nd measurement changes in vitality indicators which tend to increase. The health assessment of teak community forests with an average of the 1st measurement is 2.44 with a good category (cluster plots 2, 6, 7, and 8) medium category is none and bad category (cluster plots 1, 3, 4, and 5) and average of the 2nd measurement was 1.75 in good category (cluster plots 2, 4, and 6), medium category (cluster plot 8), and bad category (cluster plot 1, 3, 5, and 7). The status of change in the 1st and 2nd measurement of the health of teak community forests obtained an average of medium category. This change can be caused by the occurrence of tree damage in each cluster-plot. The value of changes in vitality indicators in teak community forests can be used as a basis for management and decision making that pays attention to ecological indicators that play an important role in improving forest health or causing a decrease in forest health conditions managed by farmers.

References

[1] Anwar, P. S., Safe'i, R and Darmawan, A. 2022. Landscape Characteristics on Forest Health Measurement Plots in Several Forest Functions. *IOP Conference Series: Earth and Environmental Science* **950** 1–10

[2] Ardiansyah, F and Safe'i, R. 2021 Analysis of Changes in Health of Coastal Mangrove Forest on the East Coast of Lampung. *IOP Conference Series: Earth and Environmental Science* **755** 1–7

[3] Arwanda, E, R and Safe'i, R. 2021. Assessment of Forest Health Status of Panca Indah Lestari Community Plantation Forest (Case Study in Bukit Layang Village, Bakam District, Bangka Regency, Bangka Belitung Province). *IOP Conference Series: Earth and Environmental Science* **918** 1–11

[4] Ajijah, L, N., Safe;i, R., Yuwono, S, B and Kaskoyo, H. 2022. Forest Health Analysis Based on Flora Biodiversity Indicators in Gapoktan Harapan Sentosa KPHL BatuTegi, Lampung. *IOP Conference Series: Earth and Environmental Science* **995**

[5] Akamani, K. and Hall, T, E. 2015. Determinants of The Process and Outcomes of Household Participation in Collaborative Forest Management in Ghana: a Quantitative Test of a Community Resilience Model. *J Environ Manag* **147** 1–11.

[6] Barkey, R, A and Nursaputra, M. 2019. The Detection of Forest Health Level as an Effort to Protecting Main Ecosystem in Watershed Management in Maros Watershed, South Sulawesi. *IOP Conference Series: Earth and Environmental Science.* **270** 1–6

[7] Doria, C., Safe'i, R., Iswandaru, D and Kaskoyo, H. 2021. Fauna Biodiversity as One of Repong Damar Forest Health Indicators. *IOP Conference Series: Earth and Environmental Science* **886**

[8] Gautam, S., Timilsina, S and Shrestha, M. 2021. The Effects of Forest Management Activities on Genetic Diversity of Forest Trees. *International Journal of Social and Environmental Issues (IJSEI)* **2** 110-118

[9] Hutchison, C, Gravel, D, Guichard, F and Potvin, C. 2018. Effect of diversity on growth, mortality, and loss of resilience to extreme climate events in a tropical planted forest experiment. *Scientific Reports.* **8** 81–10

[10] Mangold, R. 1999. *Forest Health Monitoring: Field Method Guide*. USDA Forest Service. New York.

[11] Maulana, I, R., Safe'i, R and Rochmah, S, F. 2021. Analysis of The Vitality of The Agathis Dammaratree at Situ Gunung Resort, Gunung Gede Pangrango National Park. *IOP Conference Series: Earth and Environmental Science* **912**

[12] Maulana, I, R., Safe'i, R., Febryano, I, G and Rahmat, A. 2022. The Relationship Between The Health of Mangrove Forests and The Level of Community Welfare. *IOP Conference Series: Earth and Environmental Science* ***1027***

[13] Martinah, V., Marjena., Ruchaemi, A and Ruhiyat, D. 2015. Pertumbuhan Hutan Tanaman Jati (*Tectona grandislinn.f.*) di Kalimantan Timur. *Jurnal Agrifor.* 14(2); 287-292

[14] Pratama, M, R., Safe'i, R., Kaskoyo, H and Febryano, I, G. 2022. Forestry Value for Health Status: An Ecological Review. *IOP Conference Series: Earth and Environmental Science* **995**

[15] Puspita, E, N., Safe'i, R and Kaskoyo, H. 2021. Forest Health Study in Efforts to Preserve Community Forest Agroforestry Patterns in Kubu Batu Village, Gedong Tataan District, Pesawaran Regency, Lampung Province. *IOP Conference Series: Earth and Environmental Science* **886**

[16] Rochmah, S, F., Safe'i, R., Bintoro, A and Rahmat, A. 2021. The Effect of Forest Health on Social Conditions of The Community. *IOP Conference Series: Earth and Environmental Science* **739**

[17] Safe'i, R., Ardiansyah, F., Banuwa, I, S., Maulana, I, R and Muslih, A, M. 2021. Analysis of Internal Factors Affecting The Health Condition of Mangrove Forests in The Coastal Area of East Lampung Regency. *IOP Conference Series: Earth and Environmental Science* ***912***

[18] Safe'i, R., Arwanda, E, R., Doria, C and Taskirawati, I. 2021. Health Assessment of Vegetation Composition in The Reclamation Area of PT Natarang Mining, Tanggamus Regency, Lampung Province. *IOP Conference Series: Earth and Environmental Science* **886**.

[19] Safei, R., Darmawan, A., Irawati, A, R., Pangestu, A, Y., Arwanda, E, R and Syahiib, A, N. 2022. Cluster Analysis on Forest Health Conditions in Lampung Province. *International Journal of Design & Nature and Ecodynamics*. **17** 257–262

[20] Safe'i, R., Maulana, I., Ardiansyah, F., Banuwa, I, S., Harianto, S, P., Yuwono, S, B and Apriliyani, Y. 2022. Analysis of Damage to Trees in The Coastal Mangrove Forest of East Lampung Regency. *Int J Sustain Dev Plan* **17** 307–12

[21] Safe'i, R., Rezinda, C, F, G., Banuwa, I, S., Harianto, S, P., Yuwono, S, B., Rohman, N, A and Indriani, Y. 2022. Factors Affecting Community-Managed Forest Health. *Environment and Ecology Research*. **10** 467–474

[22] Thirumalai, C., Chandhini, S, A and Vaishnavi, M. 2017. Analysing The Concrete Compressive Strength Using Pearson and Spearman. *International Conference of Electronics, Communication and Aerospace Technology* **2** 215–8.

[23] Widodo, L, R., Safe'i, R., Winarno, G, D., and Yuwono, S, B. 2022. Plant Success Rate in Program Forest and Land Rehabilitation in Unity Batutegi Forest Management. *IOP Conference Series: Earth and Environmental Science* ***1030***.

IOP Conf. Series: Earth and Environmental Science 1115 (2022) 012055 doi:10.1088/1755-1315/1115/1/012055

Comparison of Crown Health Assessment Using Forest Health Monitoring and Remote Sensing Techniques (Case Study: KTH Lestari Jaya 8, KPHL Kota Agung Utara, Lampung)

Anggi Feriansyah[1], Rahmat Safe'i[2*], Arief Darmawan[3], Hari Kaskoyo[4]

[1,3,4] Department of Forestry, Faculty of Agriculture, University of Lampung
[2,3] Master of Forestry Study Program, Department of Forestry, Faculty of Agriculture, University of Lampung

Corresponding Author : rahmat.safei@fp.unila.ac.id

Abstract. Tree crown is one of the main components that must be monitored to assess the health condition of the forest. The crown condition is described by five ecological parameters namely : live crown ratio, foliage transparency, crown density, crown diameter, and the dieback . The method used to assess the canopy condition was the Forest Health Monitoring (FHM) method and remote sensing with the Normalize Difference Vegetation Index (NDVI) transformation.. This study was conducted with the aim of being able to determine the condition of tree crown in protected forests using the FHM and remote sensing methods so that a comparison of the assessment results of the two methods can be carried out. This research was carried out at HKm Beringin Jaya, a managed area of KTH Lestari Jaya 8 which is located in Sumberejo District, Tanggamus Regency. The results of this study are that the Visual Crown Ratio value obtained ranges from 2.50 - 3.25. NDVI value describing the condition of the tree crown ranges from 0.75-0.83. The conclusion obtained from this study is that with the FHM technique, the health condition of the tree crown obtained is moderate to good, while with remote sensing techniques, the condition of the tree crown is in a very good condition.

1. Introduction

The tree crown is one of the main components that must be monitored to assess the health condition of the forest [1]. The crown of a tree is where the process of photosynthesis occurs and is the part of the tree that can describe the physical condition of the tree so that it has a close relationship with the health of the tree [2-3]. There are several methods or ways to assess the condition of the crown, one of the methods that can be used to assess the crown condition is the Forest Health Monitoring (FHM) method and remote sensing with the Normalize Difference Vegetation Index (NDVI) transformation.

FHM methods begins with determining the sampling index of the forest to look for its health status, then the assessment of the crown condition is carried out based on ecological parameters [4-23]. The crown condition is described by five ecological parameters namely : live crown ratio, foliage transparency, crown density, crown diameter, and the dieback [5]. The condition of the wide and dense tree crown describes the good physiological condition of the tree and the function of the forest is going well as well, on the contrary, the condition of the small and not dense tree crown describes the physiological condition of the tree that is not good so that it can cause damage to the tree which will make the function of the forest disturbed [6-7].

IOP Conf. Series: Earth and Environmental Science 1115 (2022) 012055 doi:10.1088/1755-1315/1115/1/012055

Another method of assessing the crown condition is by remote sensing. Remote sensing provides visual data such as imagery containing information about the appearance of the earth's surface [8]. The remote sensing method that can be used in the assessment of the condition of the crown is with the NDVI transformation. NDVI transformation is one of the image processing methods by calculating the bands on the image used to obtain the greenish level, then an analysis is carried out to determine the distribution of vegetation density [9]. This study was conducted with the aim of being able to determine the condition of tree crown in protected forests using the FHM and remote sensing methods so that a comparison of the assessment results of the two methods can be carried out.

2. Research Materials and Methods

2.1 Research Location

This research was carried out at KTH Lestari Jaya 8 KPHL Kota Agung Utara which is located in Sumberejo District, Tanggamus Regency. The Location of the Study is shown by Figure 1.

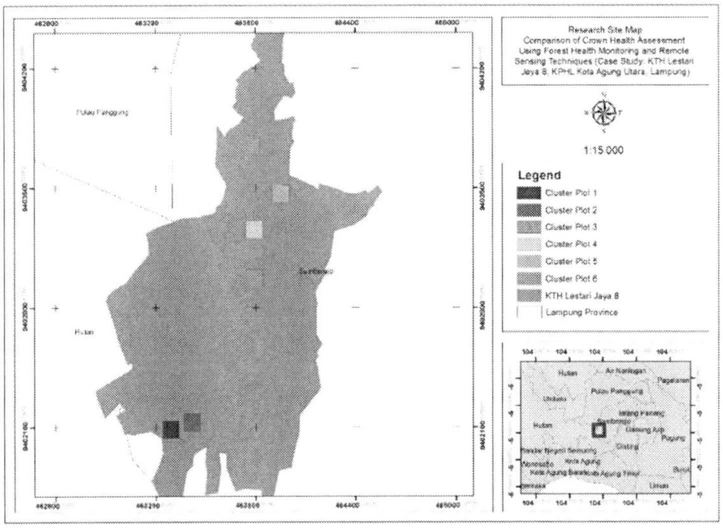

Figure 1. Research site map

2.2. Materials and Tools

The equipment used in this study is a tally sheet, rollmeter, compass, magic card, hagameter, camera, binocular and laptop that has installed microsoft excel software, ESA SNAP, ENVI 5.3 and ArcGIS 10.3. Meanwhile, the research is a tree that is included in the FHM plot cluster and sentinel-2 imagery data downloaded via the USGS web.

2.3. Data Collection

Data collection in this study includes:

1. Field Data, including data on the condition of tree crown contained in the cluster plot. The total cluster plot created is 6 (24 measuring plots).
2. Spatial Data, namely sentinel-2 imagery data (downloaded in < http://earthexplorer.usgs.gov >) Path/Row 124/64 time acquisition April 2022, administrative map of Lampung Province, administrative map of Tanggamus Regency (downloaded in < http://tanahair.indonesia.go.id >) and research area map (source: KTH Lestari Jaya 8).

2.4. Data Analysis

2.4.1. Field Data
The crown condition data obtained include parameters : live crown ratio (LCR), crown density (Cden), foliage transparency (FT), crown diameter (CDW,CD90°), and dieback (CDB). Assessment of the crown condition using criteria as presented in table 1.

Table 1. Crown condition criteria.

Parameters	Criteria		
	Good (value = 3)	Keep (value = 2)	Bad (value = 1)
Live crown ratio	≥ 40%	20 - 35%	≤ 15%
Crown density	≥ 55%	25 - 50%	5 - 20%
Foliage transparency	0 - 45%	50 - 70%	≥ 75%
Crown diameter	≥ 10.1	2.5 - 10 m	≤ 2.4 m
Dieback	0 – 5%	10 - 25%	≥ 30%

Source : [27]

The VCR value of a tree has a value between 1-4 depending on the magnitude of the observation value of each parameter of the crown condition. The following can be seen the assessment on individual VCR trees in Table 2.

Tabel 2. Tree Individual VCR Values

VCR Value	Criteria
4 (High)	The entire crown condition parameter is valued at 3, or only 1 parameter has a value of 2, no parameter is worth 1.
3 (Moderate)	More combinations between values 3 and 2 in the crown parameter, or all are worth 2, but no parameter is worth 1.
2 (Low)	At least 1 parameter is worth 1, but not all parameters
1 (Very Low)	All crown condition parameters are 1 value

Source : [27]

The plot-level VCR value is derived from the average VCR value of each individual tree in one plot, formulated as follows.

$$\text{VCR Plot} = \frac{\text{VCR value tree 1+VCR value tree 2+VCR value...n}}{\text{The number of trees in one plot}} \quad (1)$$

The VCR values of each plot cluster are formulated as follows.

$$\text{VCR cluster plot} = \frac{\text{VCR plot 1+VCR plot 2+VCR plot 3+VCR plot 4}}{4} \quad (2)$$

2.4.2. Spatial Data
The processes in this remote sensing method include: sentinel-2 image download, image atmospheric correction, image geometric correction, supervised classification, and NDVI transformation. To perform the NDVI transformation, the formula is used :

$$NDVI = \frac{\text{NIR}-\text{RED}}{\text{NIR}+\text{RED}} \quad (3)$$

Information
NIR : Near Infrared Band (Band 8)
RED : Band Red (Band 4)

Based on the NDVI transformation, the lowest value will be obtained, namely -1 to the highest, which is 1. Furthermore, a classification is carried out based on the range of NDVI values as presented in Table 3 below:

Table 3. NDVI value categories

Categories	NDVI value
Very Bad	-0.10 – 0.12
Bad	0.12 – 0.22
Moderate	0.22 – 0.42
Good	0.42 – 0.72
Very Goodt	0.72 – 0.92

Source : [20]

3. Results and Discussion

3.1. Forest Health Monitoring

The state of the crown condition can be described based on the obtained VCR value [10]. The VCR value ranges between the smallest which is 1 and the largest is 4. The greater the VCR value obtained describes the better crown condition. The spread of tree VCR values is presented in the following figure 2.

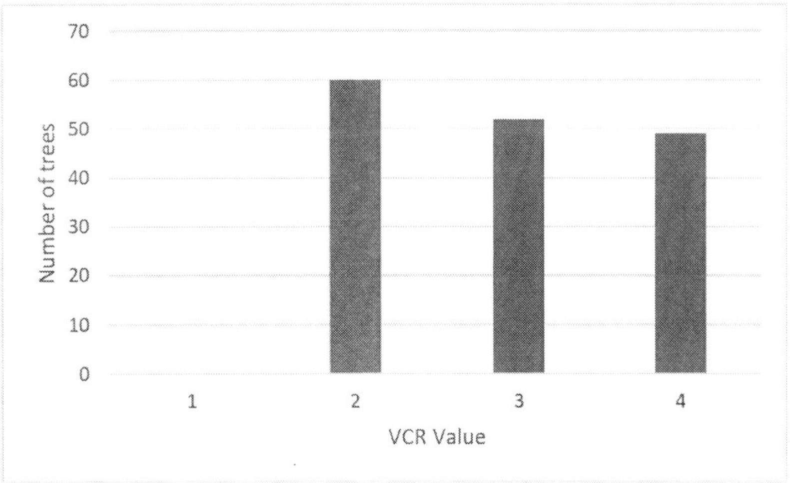

Figure 2. Graph of the values of the tree VCR.

In this study, the number of trees in the study plot amounted to 161 trees. Based on figure 2 above, it is known that there are 60 trees obtaining a value of 2 (low), 52 trees obtaining a value of 3 (moderate) and 49 trees obtaining a value of 4 (good). The difference in VCR values obtained is influenced by various factors, including plant treatments carried out by the community [6]. The VCR values obtained in this study are shown by Table 4.

Table 4. VCR value

Cluster Plot	VCR
1	2.50
2	3.25
3	3.08
4	3.24
5	2.83
6	2.63

Based on the data above, the VCR value obtained with the smallest value is found in cluster plot 3 with a VCR value of 2.50. Meanwhile, the largest VCR value can be found in plot cluster 6 with a value of 3.25. A wide, dense crown can describe a rapid growth rate, while a small, non-bushy crown describes slow growing conditions [11]. The condition of the tree crown can be affected by tree damage that occurs due to disruption of physiological processes caused by diseases, insects and other abiotic causes. In addition, climate change can also cause damage disturbances to trees [12-13-14].

In addition, the crown condition can also affect the population of living things, where the crown provides habitat for fauna to live and also provides food availability [15-25]. The crown condition has an influence on the nutrients contained in the soil, where theengorging that has a healthy crown condition will produce a greater amount of litter. This litter is then decomposition and will become humus which is useful for increasing the infiltration capacity of the soil so that it can provide organic matter that is very important for plant growth [16-17]. The state of crown conditions is influenced by environmental factors around the forest such as nutrient availability, sunlight, water, and growing space for tree crown [18-28]. A high VCR value can illustrate that the tree or stand has a crown with lush foliage, so that the need for photosynthesis to support tree growth can be fulfilled [6].

Forest utilization activities carried out by surrounding communities must not exceed the carrying capacity of the forest because it can result in forest areas experiencing degradation or deforestation [19]. Regular forest maintenance is indispensable to prevent the spread of pests and diseases that can lead to a decrease in the condition of the tree crown [24-26]. With the data of the crown condition, it can be a reference for community to be able to carry out sustainable forest management.

3.2. Remote Sensing

Assessment of the crown condition by remote sensing using the NDVI transformation method. The image data used is sentinel-2 imagery. The Sentinel-2 imagery presents data that can be used by various sectors from agriculture to forests. This imagery can be used for land monitoring, detection of land cover changes, as well as with land use [20]. NDVI has arange of values between -1 and 1 which represents the density of vegetation cover. If the value is close to 1, it means dense vegetation whereas if it is less than zero it represents water and clouds [21]. The values of the NDVI transformation are presented in Table 5.

Table 5. NDVI value

Cluster Plot	NDVI
1	0.82
2	0.83
3	0.81
4	0.80
5	0.80
6	0.75

Based on the data above, the NDVI value obtained in the range of 0.75 - 0.83. From these values, it is known that the condition of the tree crown assessed by remote sensing technique has good conditions. The NDVI value that is getting closer to 1 illustrates the better the health condition of the tree crown. The NDVI method can show vegetation-related parameters, including green foliage biomass, and green foliage areas whose values can be estimated to perform vegetation analisis [9]. A healthy forest will acfive perform photosynthesis. This process will absorb most of the red waves of sunlight and reflect the near infrared waves higher. Meanwhile, less healthy or stressed forests will reflect more red waves and less on near infrared waves [22].

4. Conclusions

The conclusion obtained from this study is that the VCR value obtained using the FHM technique ranges from 2.50 - 3.25. From this value, it is known the health condition of the tree crown in moderate to good condition. Meanwhile, in the assessment using the remote sensing method, the NDVI value describing the condition of the tree crown ranges from 0.75-0.83. From these values it is known that the state of health of the tree crown is in good condition.

References

[1] Stakėnas V, Varnagirytė-K I and Kabašinskas A 2021. A methodological approach for the assessment of basic crown parameters in scots pine stands. *J. Baltic Forestry* **27** 55

[2] Cavalli J P and Finger C A G 2017 Modelling of upper crown exposed to light of Cedrela fissilis (Vell.) open-grown trees by a non-destructive method *J. Forestry* **90** 312-317

[3] Agustini L and Irianto R S B 2016. Hubungan Antara Kondisi Tajuk Eucalyptus pellita F. Muell dan Infeksi Penyakit Busuk Akar *J. Penelitian Hutan Tanaman* **13** 1-11

[4] Safe'i R, Darmawan A., Irawati A R, Pangestu A. Y, Arwanda E. R and Syahiib A N 2022. Cluster Analysis on Forest Health Conditions in Lampung Province *International Journal of Design & Nature and Ecodynamics* **17** 257–262

[5] Puspita E N, Safe'i R, and Kaskoyo H. 2021. Forest health study in efforts to preserve community forest agroforestry patterns in Kubu Batu Village, Gedong Tataan District, Pesawaran Regency, Lampung Province. *IOP Conference Series: Earth and Environmental Science* **886** 1-7

[6] Maulana I R, Safe'i R, Febryano I G, Kaskoyo H, and Rahmat A 2022 The Relationship Between The Health of Mangrove Forests and The Level of Community Welfare *IOP Conference Series: Earth and Environmental Science* **1027** 1-15

[7] Arwanda E R and Safe'i R 2021 Assessment of forest health status of Panca Indah Lestari Community Plantation Forest (case study in Bukit Layang Village, Bakam District, Bangka Regency, Bangka Belitung Province). *IOP Conference Series: Earth and Environmental Science* **918** 1-11

[8] Furoida K, Saifuddin A and Fakhri M 2021 Analisis Kerusakan Hutan Mangrove Berdasarkan Klasifikasi NDVI Pada Citra Sentinel (Studi Kasus Kecamatan Sayung, Kabupaten Demak) *Seminar Nasional Geomatika* **6** 189-195

[9] Syifa P E., Widiasari A, Karim R. A, Somantri L and Ridwana R 2021. Pemanfaatan Citra Sentinel-2 Untuk Analisis Kerapatan Vegetasi Di Wilayah Gunung Manglayang. *J. Pendidikan Geografi Undiksha* **9** 133–143

[10] Maulana I R, Safe'i R and Rochmah S F 2021 Analysis of the vitality of the Agathis dammara tree at Situ Gunung Resort, Gunung Gede Pangrango National Park *IOP Conference Series: Earth and Environmental Science*, **912**

[11] Pratama R M, Safe'i R, Kaskoyo H and Febryano I G 2022 Forestry Value for Health Status: An Ecological Review *IOP Conference Series: Earth and Environmental Science* **995**

[12] Safe'i R, Maulana I R, Ardiansyah F, Banuwa I S, Harianto S P, Yuwono S B, and Apriliyani,

Y 2022 Analysis of Damage to Trees in the Coastal Mangrove Forest of East Lampung Regency. *International Journal of Sustainable Development and Planning* **17** 307–312

[13] Anwar P S, Safe'i R and Darmawan A 2022 Landscape Characteristics on Forest Health Measurement Plots in Several Forest Functions *IOP Conference Series: Earth and Environmental Science* **950**

[14] Safe'i R, Arwanda E R, Doria C, and Taskirawati I 2021 Health assessment of vegetation composition in the reclamation area of PT Natarang Mining, Tanggamus Regency, Lampung Province. *IOP Conference Series: Earth and Environmental Science,* **886**

[15] Doria C, Safe'i R, Iswandaru D, and Kaskoyo H 2021 Fauna biodiversity as one of Repong Damar forest health indicators *IOP Conference Series: Earth and Environmental Science* **886**

[16] Irawan T and Yuwono S B 2016 Infiltrasi pada berbagai tegakan hutan di arboretum universitas lampung *J. Sylva Lestari* **4** 21-34

[17] Riyanto, Indriyanto and Bintoro A 2013 Produksi seresah pada tegakan hutan di blok penelitian dan pendidikan taman hutan raya wan abdul rachman provinsi lampung *Jurnal Sylva Lestari* **1** 1-8

[18] Safe'i R, Ardiansyah F, Banuwa I S, Yuwono S B, Maulana I R and Muslih, A. M. 2021. Analysis of internal factors affecting the health condition of mangrove forests in the coastal area of East Lampung Regency. *IOP Conference Series: Earth and Environmental Science,* **912**

[19] Sukristiyono S, Purwanto R H, Suryatmojo H and Sumardi S 2021. Analisis Kuantitas dan Kualitas Air dalam Pengembangan Pemanfaatan Sumber Daya Air Sungai di Kawasan Hutan Lindung Sungai Wain *J. Wilayah Dan Lingkungan* **9** 239–255

[20] Kawamuna A, Suprayogi A and Wijaya A 2017 Analisis Kesehatan Hutan Mangrove Berdasarkan Metode Klasifikasi Ndvi Pada Citra Sentinel-2 (Studi Kasus : Teluk Pangpang Kabupaten Banyuwangi) *Jurnal Geodesi Undip* **6** 277-284

[21] Simarmata N, Wikantika K, Tarigan T A, Aldyansyah M, Tohir R K, Fauziah A and Purnama Y 2021 Analisis Transformasi Indeks NDVI, NDWI dan SAVI Untuk Identifikasi Kerapatan Vegetasi Mangrove Menggunakan Citra Sentinel Di Pesisir Timur Provinsi Lampung *J. Geografi* **19** 69–79

[22] Noviantoro P N, Sasmito B, and Prasetyo Y 2017 Analisis Perubahan Kerapatan Hutan Menggunakan Metode NDVI dan EVI pada Citra Satelit Landsat 8 Tahun 2013 Dan 2016. Area Studi : Kabupaten Semarang. *J Geodesi Undip* **6** 21–27

[23] Ajijah L N, Safe'i R, Yuwono S B, and Kaskoyo H 2022 Forest Health Analysis Based on Flora Biodiversity Indicators in Gapoktan Harapan Sentosa KPHL BatuTegi, Lampung *IOP Conference Series: Earth and Environmental Science* **995**

[24] Widodo L M, Safe'i R, Winarno G D and Yuwono S B 2022 Plant Success Rate in Program Forest and Land Rehabilitation in Unity Batutegi Forest Management *IOP Conference Series: Earth and Environmental Science* **1030**

[25] Sanjaya F A, Safe'i R, Winarno G D, and Setiawan A 2022 Health Assessment of Conservation Forest-Based on Biodiversity Indicator Tree Types (Case Study at Elephant Respont Unit Margahayu Way Kambas National Park). *IOP Conference Series: Earth and Environmental Science,* **950**

[26] Safe'i R and Upe A 2022 Mapping of Tree Health Categories in Community Forests in Lampung Province *IOP Conference Series: Earth and Environmental Science* **995**

[27] Safe'i R and Tsani M K 2016 *Kesehatan Hutan* (Yogyakarta : Plantaxia)

[28] Safe R, Farshilia C, Rezinda G, Banuwa I S and Harianto S P 2022 Factors Affecting Community-Managed Forest Health. *Environment and Ecology Research* **10** 467–74

Characteristics of instant porridge from cassava flour fortified anchovy flour, and yellow pumpkin flour from Indonesian Local Resources

M Nurminah*, Y Pratiwi and Z Lubis

Department of Food Technology, Faculty of Agriculture, Universitas Sumatera Utara, Medan 20155, North Sumatra, Indonesia

*Email: miminurminah@usu.ac.id

Abstract. This study aimed to produce modern functional food products, obtain the bestquality instant porridge formula, and determine the effect of the formulation of cassava flour and anchovy flour and the addition of yellow pumpkin flour for health. This study used a completely randomized design (CRD) with 2 factors, namely the ratio of cassava flour to anchovy flour (Factor I) which was 97.5%: 2.5%; 95%:5%; 92.5%:7.5%; 90%:10% and the addition of yellow pumpkin flour (Factor II) ie 0%; 5%; 10%; 15%. Several parameterscarried out in this study consisted of proximate tests (moisture content, ash content, fiber content, fat content, protein content, carbohydrates) and organoleptic tests (color, aroma, taste, texture, and general acceptance). Results obtained on all test parameters, it can be determined that instant porridge with a ratio of cassava flour and anchovy flour (90%: 10%) and the addition of yellow pumpkin flour (15%) was the best treatment of instant porridge product.

1. Introduction

Pumpkin or *Cucurbita moschata* has the popular Indonesian name of *waluh* plant which is easily available in various places, especially in Indonesia. This plant is one of the various sources of provitamin A that has the potential to be exploited for its content because it ranges from 181 SI of carotene but the development of pumpkin is also not widely found. Carotene has similarities with other carotenoids in the form of pure pigments that are soluble in fat and have an effect on the heating and oxidative stages [1].Pumpkin is a plant that has great potential to be used as a source of nutrition. In addition to being a source of provitamin A, the fruit also has various nutritional contents, namely protein, polysaccharides,carotenoids, amino acids, and minerals [2].

The increasing public need for easy-to-prepare food requires research to find ways to process ingredients that apart from being easy to prepare, must also pay attention to aspects of price and nutritional value. Efforts that are currently being carried out are making ready-to-eat products. Ready-to-eat products area type of product that is easy to prepare as a morning meal and can be consumed at any time with or without milk. Ready-to-eat products for breakfast are generally made from cereals. This product consistsof two types, namely those that require preheating for serving and those without the need for preheatingor the product is ready to be consumed directly. Products that can be prepared easily are products that are widely used and favored by consumers. The utilization of sweet potato into flour can reduce import of wheat flour to Indonesia [3]. Therefore the manufacture of instant porridge can be diversified using composite flour. Composite flour used for making instant porridge is a local food ingredient that is commonly found in Indonesia and has a high nutritional value such as orange sweet potato, carrot, and red bean.

Making porridge generally requires ingredients from wheat flour. In minimizing the continuous use of flour, other materials are needed that are used as substitutes for the basic ingredients. One of them is yellow pumpkin flour. In maximizing the nutritional value of the porridge, it is possible to use pumpkin composition ingredients, namely with other ingredients such as anchovy flour. The nutritional content In anchovies per 100 g of ingredients that can be consumed includes "77 kcal energy, 16 g protein, 1 g fat, 0 g carbohydrate, and contains high enough calcium (500 mg/ 100 g)", so that it can provide an increase calcium and protein in the type of food. And also because anchovy has an affordable price than other animal protein sources and has a large supply.

It is hoped that the substitution of anchovy flour and yellow pumpkin can be used as an ingredient for instant porridge, and can increase the level of sensory character that influences public acceptance. Therefore, the implementation of the research is useful in looking at the effects of anchovy and pumpkin flour used on the physical and sensory characteristics of instant porridge, in seeing the formulations of anchovy and pumpkin meal used appropriately to match the results of the desired chemical and physical characters and interested.

2. Methodology

2.1. Materials and tools

In this study, the ingredients used were cassava, anchovies, pumpkin, and salt. The reagents used are ethanol, aquadest, n-hexane pa, H_2SO_4 0.325 N, concentrated H_2SO_4, H_2SO_4 0.02 N, NaOH 0.1 N, NaOH 1.25 N, NaOH 40%, CuSO4, K2SO4, chloroform, oxalic acid, sodium bicarbonate, Mensel indicator and 96% alcohol. The equipment used in this study were stove, boiling pan, knife, spoon, stainless steel pan, basin, plastic drain, blender, 60 mesh filter, digital scale, vacuum sealer, mixer, plastic food box, mask, cloth gloves, gloves. plastic hand, baking sheet, medium plastic container, polyethylene plastic, aluminum foil, cling wrap, Whatman 41 filter paper, filter paper, tissue paper, erlenmeyer, measuring cup, measuring flask, funnel, vacuum pump, beaker glass, and porcelain cup,

2.2. Research implementation

2.2.1. Making cassava flour. Cassava was washed, then cut and sliced. The sliced cassava was arranged in a baking dish and dried in a cabinet drier at 60 °C for 6 hours. Next, it cooled to room temperature and was blended and filtered using a 60 mesh sieve, then it was packaged with polyethylene plastic and sealed.

2.2.2. Making anchovy flour. The anchovy head was removedo, then washed and drained. The anchovy was soaked for 1 hour and washed again. The anchovy was steamed for 20 minutes, then mashed until smooth. Next, dried in the oven, at 60 °C for up to 5-6 hours. After that, the anchovy was grinded. The anchovy was then sifted with 60 mesh to obtain anchovy flour.

2.2.3. Making pumpkin flour. The pumpkin was peeled and cleaned from the skin and seeds. Reduce the size of the pumpkin and slice it up to 0.2 cm thick. The production of pumpkin flour by dried at a temperature of 50 °C for up to 24 hours aims to produce flour with a better water content. Grind and sieve using a 60 mesh sieve

2.2.4. Instant porridge making. The processing of instant porridge was carried out by preparing ingredients with various formulations, then mixing the ingredients by adding 1% salt, then adding water with a formulation ratio of 3: 1 to ingredients (150 g), and cooking until thickened. The slurry was dried on a baking sheet with a thickness of ± 3 mm and oven at 60°C for up to 12 hours. Next, blended until smooth and sieved the size of 60 mesh. Dried porridge was wrapped in polyethylene plastic.

2.2.5. Data analysis. This study used a factorial Completely Randomized Design (CRD) consisting of two factors, namely factor I: Comparison of cassava flour and anchovy flour (T): T1 = 97.5% cassava flour: 2.5% anchovy flour T2 = 95% cassava flour : 5% anchovy flour,T3 = 92.5% cassava flour : 7.5% anchovy flour, T4 = 90% cassava flour : 10% anchovy flour, factor II addition of yellow pumpkin flour (Y): Y0 = 0% pumpkin flour, Y1 = 5% pumpkin flour, Y2 = 10% pumpkin flour, Y3 = 15% pumpkin flour. The number of treatment combinations (Tc) treatment combinations was 4 x 4 = 16, and it carried three repetitions for research accuracy, so the total number of samples was 48 samples. If the results obtained are significantly different and very significant, then it is continued using the mean difference test, using theLeast Significant Range (LSR) test. The analysis carried out in this study was water content, ash content,protein content, fat content, crude fiber content, carbohydrate content, water absorption and organoleptictests of color, aroma, taste, texture and general acceptance.

3. Result and Discussion

Characteristics of instant porridge with factor the comparison of cassava flour with anchovy flour and the addition of yellow pumpkin flour at several stages of treatment can be seen in Table 1 and Table 2.

Table 1. The effect of the comparison of cassava flour with anchovy flour on the quality characteristics of instant porridge

Parameters	Cassava flour formulation with anchovy flour (%)			
	T_1 (97.5%:2.5%)	T_2 (95%:5%)	T_3 (92.5%:7.5)	T_4 (90%:10%)
Water content (%bb)	3.1784	3.4149	3.5147	3.7861
Protein content (%bk)	9.8434	10.0763	10.3469	10.9925
Fat content (%bk)	6.6966	6.7847	6.8352	7.2354
Crude fiber content (%bk)	3.1227	2.8936	2.7042	2.5417
Carbohydrate content (%bk)	16.4522	17.2700	17.6288	18.8793
Color (hedonic)	4.8917	4.9786	5.3583	5.6917
Aroma (hedonic)	4.9036	4.9893	5.3429	5.2679
Taste (hedonic)	5.8679	5.6571	5.7571	5.4929
General acceptance (hedonic)	5.8905	5.8381	5.8369	5.6988

Note: This figure is obtained from the average of 3 replications. The difference in letters following the numbersin the same line, namely the 5% level with lowercase letters representing a significant effect and the 1%level with capital letters representing very significant through the LSR test.

Table 2. The effect of adding yellow pumpkin flour to the quality characteristics of instant porridge

Parameter	Addition of pumpkin flour (%)			
	Y_0 (0%)	Y_1 (5%)	Y_2 (10%)	Y_3 (15%)
Water content (%bb)	3.1905	3.3409	3.5714	3.7913
Protein content (%bk)	9.8815	10.1557	10.4562	10.7657
Fat content (%bk)	6.5340	6.7670	7.0044	7.2465
Crude fiber content (%bk)	2.4875	2.6808	2.9425	3.1515
Carbohydrate content (%bk)	16.3212	17.0240	17.9786	18.9065
Color (hedonic)	5.0786	5.2381	5.2774	5.3262
Aroma (hedonic)	4.9583	5.0869	5.1095	5.3488
Taste (hedonic)	5.7202	5.7012	5.6476	5.7060
General acceptance (hedonic)	5.8524	5.8452	5.7679	5.7988

Note: This figure is obtained from the average of 3 replications. The difference in letters following the numbers in the same line, namely the 5% level with lowercase letters representing a significant effect and the 1% level with capital letters representing very significant through theLSR test.

3.1. Water content

The comparison of cassava flour with anchovy flour (Table 1) gave a very significant difference ($P<0.01$). The addition of yellow pumpkin flour (Table 2) gave a very significant difference ($P<0.01$). The interaction of the comparison of cassava flour with anchovy flour and the addition of yellow pumpkin flour gave a very significant difference ($P<0.01$). In comparison, the use of more anchovy flour will increase the moisture content of instant porridge because anchovy flour has a high protein content. These results are in line with Lehninger [3] which explains that the increase in water content dueto the hydroxyl combination of proteins can cause proteins to bind water molecules.

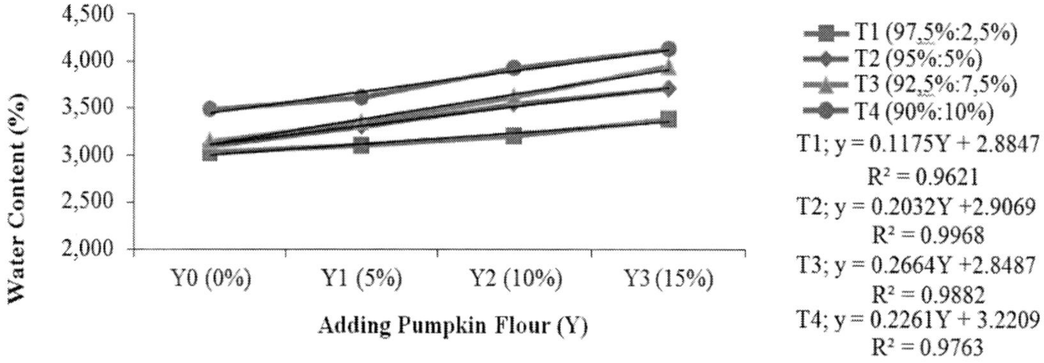

Figure 1. The interaction of the comparison of cassava flour with anchovy flour and the addition of yellow pumpkin flour on the moisture content of instant porridge

The addition ofpumpkin flour can also increase the moisture content of instant porridge, this is due to the high fiber content of pumpkin flour as its raw material. The binding of water tothe fiber has difficulty releasing the product even through heating [4]. The interaction relationship between the comparison of cassava flour with anchovy flour and the addition of yellow pumpkin flour to the moisture content of instant porridge can be seen in Figure 1.

3.2. Protein content

The comparison of cassava flour with anchovy flour (Table 1) gave a very significant difference ($P<0.01$). The addition of yellow pumpkin flour (Table 2) gave a very significant difference ($P<0.01$). The interaction of the comparison of cassava flour with anchovy flour and the addition of yellow pumpkin flour gave a very significant difference ($P<0.01$). The use of a lot of anchovy flour will produce a high proteincontent of instant porridge, explainingthat anchovy is a small and economical fish variant, similar to other marine fish variants that this fish has a high protein content, where the protein is composed of various essential amino acids [5], which include methionine and lysine. In addition, the addition of yellow pumpkin flour can increase the protein contentof instant porridge due to its high protein content of 11.2248% [6]. The interaction of the comparison of cassava flour with anchovy flour and the addition of yellow pumpkin flour on the protein content of instant porridge can be seen in Figure 2.

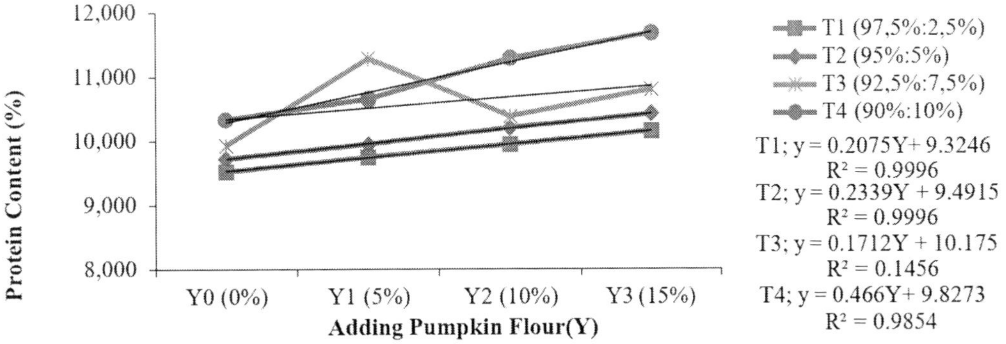

Figure 2. The interaction of the comparison of cassava flour with anchovy flour and the addition of yellow pumpkin flour on the protein content of instant porridge

3.3. Fat content

The comparison of cassava flour with anchovy flour (Table 1) gave a very significant difference ($P<0.01$). The addition of pumpkin flour (Table 2) gave a very significant difference ($P<0.01$). The interaction of the comparison of cassava flour with anchovy flour and the addition of pumpkin flour gave a very significant difference ($P<0.01$). The use of a lot of anchovy flour will produce a high fat content of instant porridge. This is because the use of anchovy flour has a higher fat content (9.5170%)than cassava flour (1,2357%) and pumpkin flour (2,0437%). So that the more use of anchovy flour, the fat content of instant porridge also increases.

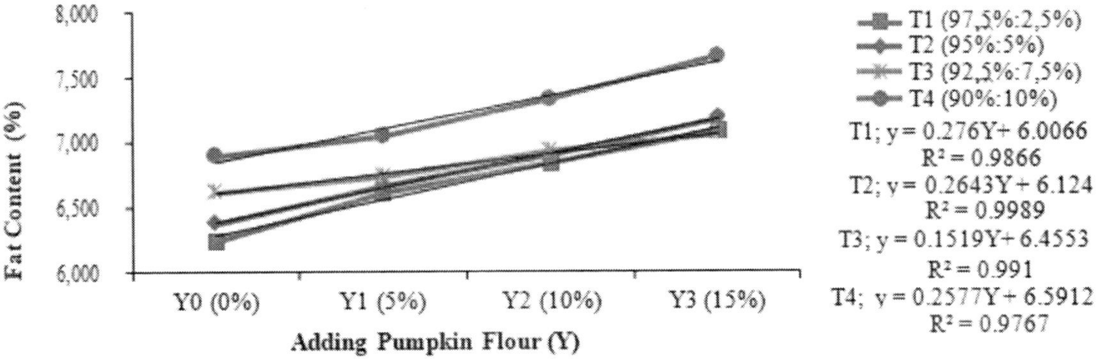

Figure 3. The interaction of the comparison of cassava flour with anchovy flour and the addition of yellow pumpkin flour on the fat content of instant porridge

The function of fat is to protect body organs, maintain body temperature, provide a delicious and filling taste, maintain protein stability, as an alternative to transporting dissolved fat vitamins to produce essential fatty acids, form various cells, and especially asa source of energy [7]. The interaction of the comparison of cassava flour with anchovy flour and the addition of yellow pumpkin flour on the protein content of instant porridge can be seen in Figure 3.

3.4. *Crude fiber content*

The comparison of cassava flour with anchovy flour (Table 1) gave a very significant difference ($P<0.01$). The addition of yellow pumpkin flour (Table 2) gave a very significant difference ($P<0.01$). The interaction of the comparison of cassava flour with anchovy flour and the addition of yellow pumpkin flour gave a very significant difference ($P<0.01$). The use of a small amount of anchovy flour will result in a low crude fiber content of instant porridge, which explains that the low fiber content of anchovy flour causes the fiber content to decrease in line with the increased anchovy flour formulation for instant porridge [5]. The addition of more yellow pumpkin flour will produce higher instant porridge, there was an effectof adding fruit such as pumpkin as a functional ingredient that has great potential as an increase in fibercontent and active compounds in the form of -carotene and carotenoids to the resulting product [8]. The interaction of the comparison of cassava flour with anchovy flour and the addition of pumpkin flour to the crude fiber content of instant porridge can be seen in Figure 4.

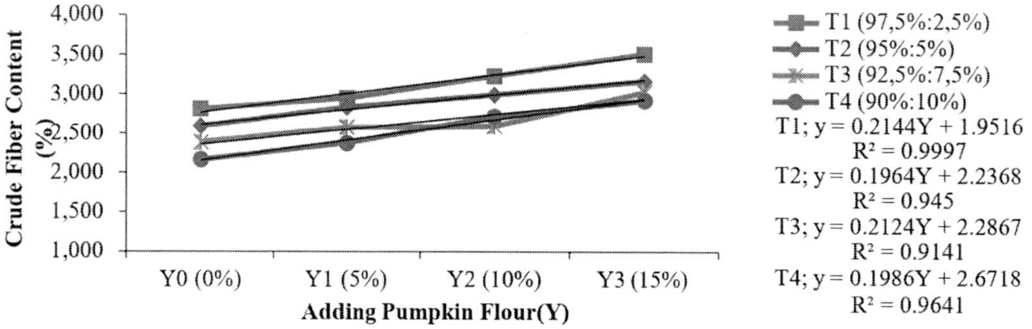

Figure 4. The interaction of the comparison of cassava flour with anchovy flour and the addition of yellow pumpkin flour on the crude fiber content of instant porridge

3.5. *Carbohydrate content*

The comparison of cassava flour with anchovy flour (Table 1) gave a very significant difference ($P<0.01$). The addition of pumpkin flour (Table 2) gave a very significant difference ($P<0.01$). The interaction of the comparison of cassava flour with anchovy flour and the addition of pumpkin flour hada significantly different effect ($P<0.05$). The use of a lot of anchovy flour will produce a high carbohydrate content of instant porridge. This is because the carbohydrate content of anchovy flour is the highest compared to other flours, which is 62.6783%. So that its use is increasingly affecting the carbohydrate content of the instant porridge obtained. Carbohydrates also function as energy producers in the body. Calculation of carbohydrate content is done by difference, then other content affects the carbohydrate content. The low level of these other nutrients results in a high content of carbohydrate ingredients [9]. The interaction of the comparison of cassava flour with anchovy flour and the addition of pumpkin flour on the carbohydrate content of instant porridge can be seen in Figure 5.

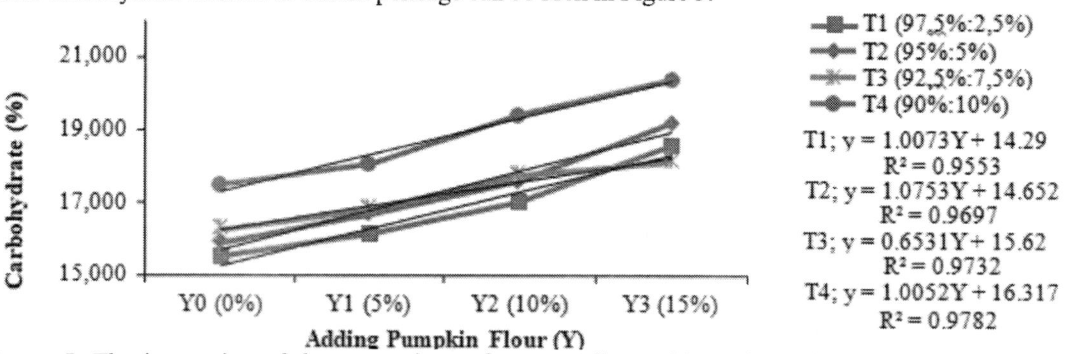

Figure 5. The interaction of the comparison of cassava flour with anchovy flour and the addition of yellow pumpkin flour on the carbohydrate content of instant porridge

3.6. Hedonic of color

The comparison of cassava flour with anchovy flour (Table 1) gave a very significant difference ($P<0.01$). The addition of pumpkin flour (Table 2) gave no significant difference ($P>0.05$). The interaction of the comparison of cassava flour with anchovy flour and the addition of pumpkin flour gave a very significant difference ($P<0.01$). Judging from the hedonic value of the color obtained, it shows that the hedonic value of the instant porridge color obtained has increased in line with the increase in the amount of anchovy flour used in the formulation. This is because anchovy flour has a brownish yellow color, so if you add more flour, the more contrast or darker the color of the instant porridge will be, which will affect the attractivenessand interest of the panelists [10]. The more pumpkin and anchovy flour used resulted in an increase in the panelists' appetite because the instant porridge was brownish yellow, it explained that pumpkin flour has the main color, namely yellow-orange and is able to produce an effecton the color of the final product [11]. The interaction of the comparison of cassava flour with anchovy flour and the addition of pumpkin flour to the hedonic color of instant porridge can be seen in Figure 6.

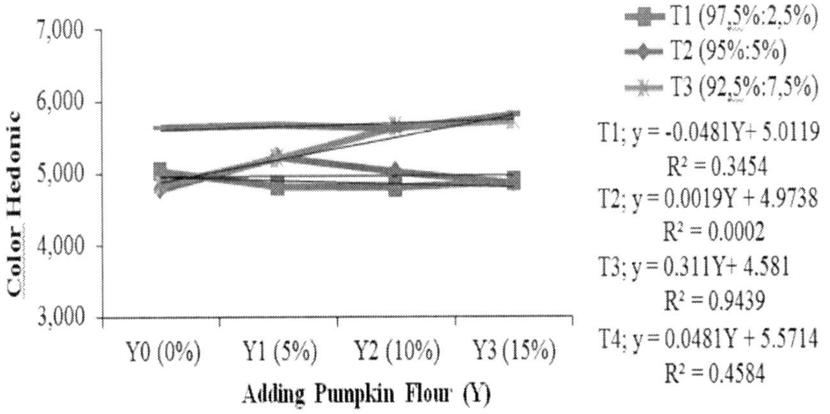

Figure 6. The interaction of the comparison of cassava flour with anchovy flour and the addition of yellowpumpkin flour to the hedonic color of instant porridge

3.7. Hedonic of aroma

The comparison of cassava flour with anchovy flour (Table 1) gave a very significant difference ($P<0.01$). The addition of pumpkin flour (Table 2) gave a very significant difference ($P<0.01$). The interaction of the comparison of cassava flour with anchovy flour and the addition of pumpkin flour gave a very significant difference ($P<0.01$). Judging from the hedonic value of the aroma obtained, it shows that the hedonic value of the aroma of instant porridge obtained increased in line with the increasein the amount of anchovy flour used in the formulation up to the T3 treatment, which was 7.5%. However, the hedonic value of instant porridge aroma decreased in T4 treatment with the use of anchovyas much as 10%. This means that the panelists prefer the aroma of instant porridge with the maximum use of anchovy flour as much as 7.5%. The amount of addition of anchovy flour will increase the aromaof anchovy in instant porridge, which is a little fishy. Fish has a distinctive smell that is fishy due to thepresence of nitrogen compounds in the form of "guanidine, Trimethyl Amine Oxide (TMAO), and imidazole derivatives".

The addition of more pumpkin flour resulted in an increase in the hedonic value of the instant porridge aroma. The reason is because pumpkin flour contains chemical compounds, namely flavonoids, the element "aliphatic alcohol and carbonyl compounds, hexenal, 2- hexenal. 3-hexen and 2,3 butanodiene" identified pumpkin aroma [12]. So that the amount of flour addedin making instant porridge increases the preferences and tastes of the panelists. The interaction of the comparison of cassava flour with anchovy flour and the addition of pumpkin flour to the hedonic aroma of instant porridge can be seen in Figure 7.

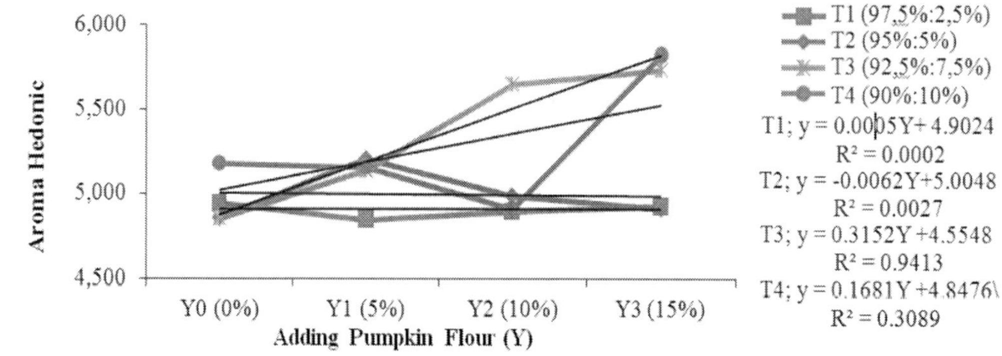

Figure 7. The interaction of the comparison of cassava flour with anchovy flour and the addition of yellow pumpkin flour to the hedonic aroma of instant porridge

3.8. Hedonic of taste

The comparison of cassava flour with anchovy flour (Table 1) gave a very significant difference ($P<0.01$). The addition of pumpkin flour (Table 2) gave no significant difference ($P>0.05$). The interaction of the comparison of cassava flour with anchovy flour and the addition of pumpkin flour had a significantly different effect ($P<0.05$). Judging from the hedonic value of the taste obtained, it shows that the hedonic value of the instant porridge taste obtained is less favored by the panelists, it explained that the panelists were not accustomed to assessing the dominant taste of biscuit products for the taste of anchovies because various other products found in the community had not usedthis taste even though the protein content of the fish was high [13]. The result of anchovy substitution leads to a bitter taste and a strong anchovy characteristic that is not liked by the assessors. What affectsthe bitter taste is the hydrolysis of various amino acids that exist from the effect of Maillard on processedfish meal or biscuits.

The amount of fortification of anchovy flour can create a product with a savory taste and a stronger anchovy taste [10]. In addition, the use of pumpkin flour alsoproduces a savory final product taste but tends to taste a bit bitter. This is because the taste produced bypumpkin flour tends to be strong and is obtained through the flavonoid substances present in the Pumpkin [7]. The interaction of the comparison of cassava flour with anchovy flour and the addition of pumpkin flour to the hedonic flavor of instant porridge can be seen inFigure 8.

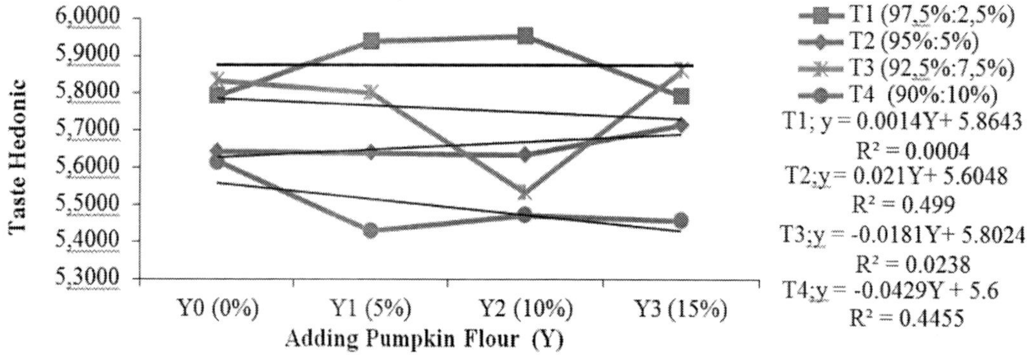

Figure 8. The interaction of the comparison of cassava flour with anchovy flour and the addition of yellow pumpkin flour to the hedonic taste of instant porridge

IOP Conf. Series: Earth and Environmental Science 1115 (2022) 012056 doi:10.1088/1755-1315/1115/1/012056

3.9. Hedonic of general acceptance

Comparison of cassava flour with anchovy flour (Table 1) gave a significantly different effect (P<0.05). The addition of pumpkin flour (Table 2) gave no significant difference (P>0.05). The interaction of the comparison of cassava flour with anchovy flour and the addition of pumpkin flour gave no significant difference (P>0.05).Judging from the hedonic value of the general acceptance of instant porridge the highest or most preferred by the panelists, namely the T1 treatment of 5.8905 where the percentage of anchovy flour used in the formulation was the least (2.5%). Sensory testing uses various test parameters as a whole from the value given by consumers to all uses of sensory attribution in seeing the level of acceptance of the rater on the product as a whole. Like the product according to the manufacturing process, sensory, social, taste, preparation stage, and various other aspects. There are differences in assigning value to food quality according to tastes and preferences [7].

4. Conclusion

The addition of more pumpkin flour will increase the value of water content, protein content, fat content, crude fiber content, carbohydrates content, color hedonic, hedonic aroma of instant porridge. Meanwhile, the hedonic value of taste and acceptance generally decreased. Instant porridge with a ratio of cassava flour and anchovy flour (90%: 10%) and the addition of yellow pumpkin flour (15%) was the best treatment of instant porridge product.

References

[1] Prabasini H, Ishartani D and Rahadian D 2013 *J. Food Sci* Tech. **2** 93
[2] Suwanto, Suranto and Purwanto E 2015 *J. UNS Graduate.* **3** 61
[3] Lehninger A L 1995 *Basics of Biochemistry* (Jakarta: Erlangga Publisher)
[4] Sarofa U, Mulyani U and Wibowo Y A 2013 *J. Veterans NationalDevelopment University* **5** 58
[5] Takzim F, Tamrin and Rezeki S 2018 *J. Food Tech Sci.* **3** 1
[6] Purnamasari, Ismawati A, Pradana B T and Jonathan J G 2006 *J. Food and Nutrition* **4** 27
[7] Cahyaningtyas F I, Basito and Anam C 2014 *J. Sci Tech* **3** 13
[8] Murdiati A and Amaliah 2013 *Healthy Food Preparation Guide For All* (Jakarta: Kencana Prenadamedia Group)
[9] Sugito, Manalu W, Astuti D A, Handharyani E and Chairul 2006 *J. Animal Husbandry and Veterinary Sci* **12** 68
[10] Ambarita B U 2020 Pengaruh fortifikasi tepung ikan teri asam (*Stolephorus sp.*) dalam pembuatan stik terhadap penerimaan konsumen [Effect of fortification of sour anchovy flour (*Stolephorus sp.*) in the manufacture towards consumer acceptance] *Skripsi* (Pekanbaru: Fakultas Perikanan dan Kelautan-Universitas Riau)
[11] See E F, Wan N W A and Noor A A A 2007 *ASEAN Food Journal* **14** 123
[12] Berger R G 2007 *Flavor and Fragrance: Chemistry, Bioprocessing and Sustainability Chemistry and Materials Science* (USA: Springer Publisher)
[13] Ramadhan R, Nuryanto and Wijayanti H S 2019 *J Nutrition College* **8** 264

Agent-based modelling application for social forestry policy evaluation

L Abdulah[1], Subarudi[2], N S Lestari[1*], M A Qirom[1], M Yulianti[1], D Wicaksono[1], H Krisnawati[1], Y Lisnawati[1] and S Astana[3]

[1] Research Center for Ecology and Ethnobiology, National Research and Innovation Agency (BRIN), Jl. Raya Jakarta-Bogor Km 46 Cibinong, West Java, Indonesia 16911
[2] Research Center for Population, National Research and Innovation Agency (BRIN), Jl. Gatot Subroto No. 10, Jakarta Selatan, DKI Jakarta, Indonesia 12710
[3]Research Center for Behavioural and Circular Economics, National Research and Innovation Agency (BRIN), Jl. Gatot Subroto No. 10, Jakarta Selatan, DKI Jakarta, Indonesia 12710

*Email: nurul.silva.lestari@brin.go.id

Abstract. Social forestry (SF) as a community involvement program in forest areas needs to be frequently evaluated, particularly on policy aspects considering the dynamics of program implementation. Policy issues related to SF program have not been well identified. As a result, the implementation of SF policies may not achieve the target and recurrently lead to new problems. This research aims to develop the best scenario to support the achievement of SF target but is limited to SF areas adjacent to oil palm plantations. The actor-based modelling was developed using Netlogo 6.2 analytical tool. The results showed that the implementation of SF program should include the economic welfare improvement of local communities as a main consideration. The best scenario to achieve SF target is providing full access for local communities to manage forest areas and developing a mechanism to market the commodities by considering domestic wood consumption. The model simulation indicates that this management approach would be able to maintain forest cover as well as increase the community's income. Therefore, information on wood consumption level needs to be promoted as an investment opportunity in the forestry sector, which has not been well considered previously in SF program. This strategy can be used to increase the participation of local communities in SF program by providing sufficient economic benefit as well as improving the inclusion of local communities in sustainable forest management.

1. Introduction

Social forestry (SF) is a government program that aims to solve tenurial problems and improves the community's welfare as well as ensure forest sustainability. Moreover, this program is also expected to increase community participation in forest management [1]. The livelihood sources in SF areas are forest timber and non-timber utilization in production and protected forests [2]. However, the implementation of SF program remains a big challenge. The allocated SF area has not been optimally utilized by local communities living in surrounding forest areas. Several issues related to SF program include policy implementation, complicated licensing process, access to funding, forest products marketing, profit sharing, insufficient institutional capacity and variability of forest resource characteristics [3,4]. As a result, SF program attractiveness tends to be low [5,6].

Local communities participating in SF program are still prioritizing economic benefits over other benefits [6]. Therefore, they use this priority as a main consideration for selecting activities to be applied in SF areas. This study only includes SF areas near oil palm plantation areas. This is related to the allocation of SF areas, which in many cases, were appointed in forest areas that have been utilized by local communities to avoid tenurial conflicts. Many of them plant oil palm in such areas because of its high potential in increasing communities' revenue [7,8].

SF issues can also be seen using an institutional approach [9], including three important elements: agent, regulation and environment. Agent is local communities/farmers participating in SF program to get benefits from land utilization which may affect the environment. Regulation refers to SF regulations that are applied based on utilization rights. Meanwhile, environment refers to forest areas allocated for SF programs. These elements are used to develop agent-based modelling, which enables the integration between information and knowledge [10]. Agent based modelling is useful to study behavior [11], even though the information is incomplete. Such information can be supported by knowledge related to the issue. Therefore, system analysis requires strong basic knowledge of particular topics [10]. In this study, the basic knowledge includes SF objectives and strategies. Considering the importance of SF program, this study aims at evaluating SF policies and developing the best SF scenario using agent-based modelling to support the achievement of SF program target, particularly in those located near oil palm plantation areas.

2. Method

2.1. Data collection

The data used to develop models was collected from literature review of relevant scientific publications about social forestry. This includes research articles from journal and conference proceeding articles [2,7,8,12]. Additional supporting data were also derived from Global Forestry Statistics [13] and BPS-Statistics Indonesia [14].

2.2. Data analysis

Agent behavior-based modelling emphasizes how an agent thinks, communicates and interacts with other agents in a particular system [15]. The steps of model development using the Netlogo 6.2 analytical tool are as follows:

a) Identify problem, objective and model boundaries
   In this model, the problem is the performance of SF policies in supporting sustainable forest management which consider ecological, economic and social aspects. The aim of the model is to evaluate SF program. Meanwhile model boundaries only include forest conversion due to land and commodity competition [16].
b) Conceptual model formulation
   Local communities who live surrounding forest area have the willingness to participate in government program as long as they get the benefits [17]. For instance, oil palm is more likely to be planted by local communities because of its economic benefits [18]. In addition, forest areas tend to be converted into non forest areas when the required plant species under SF program have no economic benefit [19,16].

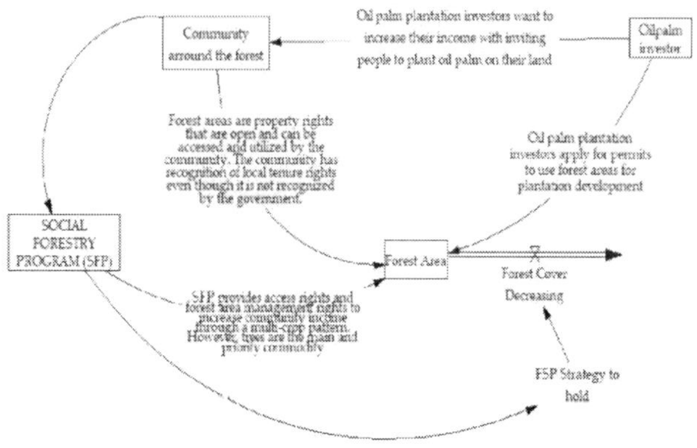

Figure 1. Conceptual model

In this conceptual model, the participation level of local communities in SF program is influenced by commodity price. In SF areas surrounded by oil palm plantations, local communities tend to compare the economic benefits of commodities that would be developed in SF areas. This may lead to forest conversion when the profits from planting non forest tree species such as oil palm is higher than forest tree species. Meanwhile, correlation between forest area status and investor profit is related to the land cost. Investors are more likely to get more lower land cost by involving smallholders as their partner to expand oil palm plantation areas [12].

c) Mathematical model formulation
Mathematical model in this study applied logical model. For example, if local communities have SF permits and require to plant forest tree species and other horticultural plants using their own resources, they are more likely to receive investor offer to plant oil palm which has a more certain market. Another example is when the communities have information about wood products supply and demand, and there is a certainty that wood price would be higher than production cost, they tend to plant trees in SF areas.

d) Model simulation
We developed three model scenarios in this study. First is SF program under the business as usual scenario. The second is SF program which includes the synergy and collaboration of relevant stakeholders. Lastly, SF program which consider wood consumption to improve tree planting in SF areas.

e) Model evaluation
The models were evaluated by comparing three scenarios and its actual implementation in the field.

f) Model application
The models were used to evaluate the achievement SF program target.

3. Results and discussion

3.1. Current status of SF program

The progress of SF implementation can be assessed using the extent of area allocated for SF as an indicator. According to Minister Regulation No. 83 of 2016, SF consists of 5 schemes: village forest (*hutan desa*/HD), community forest (*hutan kemasyarakatan*/HKm), community plantation forest (*hutan tanaman rakyat*/HTR), forestry partnership (*kemitraan kehutanan*/KK) and customary forest (*hutan adat*/HA). In addition, social forestry program within state-owned forestry company (Perum Perhutani) areas in Java is under the permit for social forestry utilization (*Izin Pemanfaatan Hutan Perhutanan* Sosial/IPHPS) scheme. Village forest is forest area within village territory which is managed by village to be utilized for community welfare. Community forest on the other hand, is only managed by a group of people. Meanwhile, community plantation forest is plantation forest that is located in production forest managed by community groups by applying silvicultural practices to improve the quality of production forest as well as ensuring forest sustainability. Forestry partnership requires cooperation between local communities and forest management units or forest concessionaires. Meanwhile customary forest is SF program implemented in customary law areas. Based on these definitions, the ownership of SF area is not handed over to local communities. They only have access to manage the forest but cannot have the right to transfer the land.

The progress of SF permits has been impressive since the stipulation of Minister regulation No 83/2016. Forest areas which have been granted SF program is around 4.1 million ha in 2020. The largest area is village forest with the total area of 1.5 Mha [20]. The detail of SF areas is presented in the following table.

Table 1. National accumulation of SF areas in the period of 2007-2020

No	Schemes	Area (in thousand ha)						
		2007 - 2014	2015	2016	2017	2018	2019	2020
1	HKm	154	175	177	276	433	433	757
2	HD	78	142	238	589	940	940	1,552
3	HTR	199	201	216	244	270	270	352
4	KK	19	21	45	72	75	75	425
5	IPHPS	-	-	-	6	19	19	26
6	HA	-	-	13	22	22	22	950

Source: (MoEF, 2020)

3.2. Impacts of SF on forest cover and community welfare under business as usual scenario

Business as usual scenario in this study refers to the conditions in which SF participants are required to plant particular tree species and assistance provision is insufficient. Under this scenario, the model simulation showed that the implementation of SF program is unlikely to increase forest cover. This is because the options of commodity to be planted in SF areas is limited. In a case where SF areas are located near to oil palm plantations, SF participants are more likely to plant oil palm in order to improve their incomes. As a result, there is no difference on forest cover change before and after the implementation of SF program, in which forest cover tend to be decreased (Figure 2).

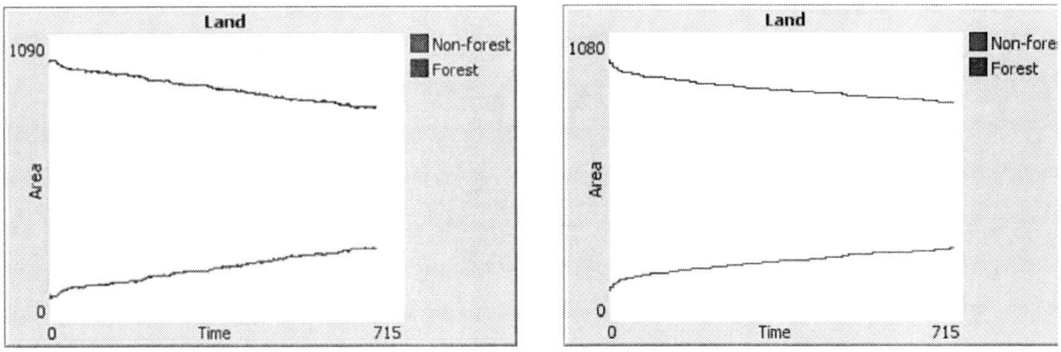

(a) (b)

X = Simulation time in months; Y = area

Figure 2. Forest cover (a) before social forestry program, (b) after social forestry program

Based on this model, local communities need assistance in managing forest areas under SF program. Land and commodity competitions often occur in forest management. Commodity competition may affect land use pattern, such as oil palm development [7]. The total oil palm plantation area in Indonesia in 2021 is about 15.02 million ha, in which 8.42 million ha (55.8%) managed by private companies, 6.08 million ha (40.3%) managed by smallholders, and 0.60 million ha is managed by state companies [21]. This indicates that oil palm plantation has become an important income source for communities.

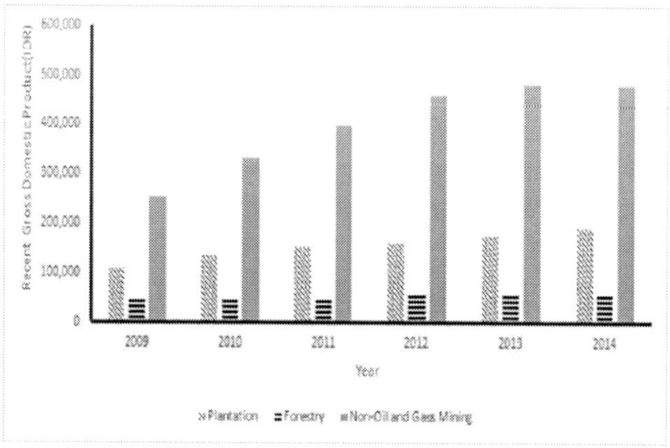

Figure 3. Gross Domestic Product (GDP) by Sector

Considering the extent of plantation areas, this sector has significant contribution to the country's revenue. Meanwhile the GDP of forestry sector has been the lowest since 2009 compared to plantation and non oil and gass mining (Figure 3). This indicates that policies and regulations that have been implemented in forestry sector including SF program could not significantly contribute to GDP improvement. Whereas in every 1,000 ha of oil palm plantation, economic benefit would increase by 4.02% [22].

Model simulation showed that SF implementation under business as usual scenario may still increase the revenue of oil palm investor, similar to before SF implementation. Meanwhile, the revenue of SF participants may decrease due to unprofitable commodity development (Figure 4). This can be a threat to forest sustainability, in which about 2.8 million ha of oil palm plantation is located in forest areas [7], although the economic benefit for smallholders is still questionable [8].

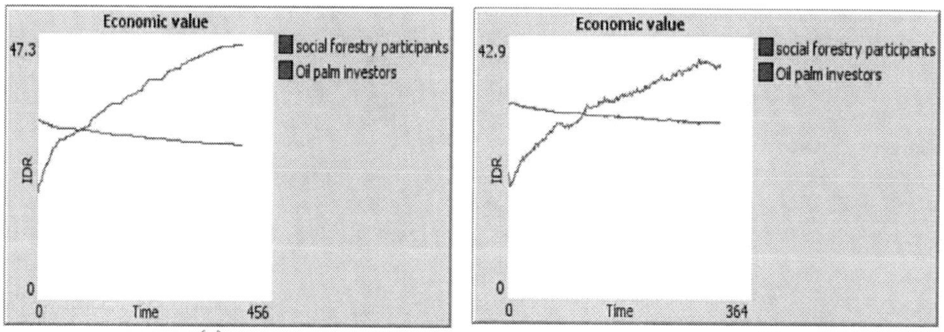

(a) (b)

X = simulation time in months; Y = land values in million IDR

Figure 4. Economic welfare (a) before social forestry program, (b) after social forestry program

The level of farmers' revenue is highly influenced by market price. The economic value of lands planted with oil palm is higher than other land covers. The farmers who claimed the land would get 0.5 million IDR/ha, marketing intermediaries would have 0.7 million IDR/ha, while land clearing workers would be paid 11.5 million IDR [12]. The income of oil palm smallholders in dryland is around 29.6 million IDR/ha/year, with production cost of 4.2 million IDR/ha/year. Hence, the net benefit of oil palm plantation is about 25.4 million IDR/ha/year, or approximately 2 million IDR/ha/year. Whereas the income of oil palm smallholders in wetland is 28.7 million IDR/ha/year, with production cost of 9 million IDR/ha/year. The net benefit in wetland is smaller than in dryland which is about 19.6 IDR/ha/year [23].

Meanwhile, SF program can only contribute to about 20 million IDR/ha/year to the income of SF participants [24]. This income was generated annually because the amount was not evenly distributed each month. A main challenge for improving SF participant incomes is lack of capital to support small business. HTR program development is often followed by an instant institutional establishment. This may hamper the fulfilment of administration requirements. As a result, land utilization could not be optimized. Moreover, lack of human resource capacity and program assistants of HTR program also may hinder the implementation of SF program. In addition, legal aspect also still becomes an issue that needs to be addressed [25].

3.3. Improving the effectiveness of SF implementation

3.3.1. The collaboration of SF management. SF program needs to be reviewed under a broader perspective. This program should be implemented as an integrated measures of several directorates under the Ministry of Environment and Forestry. HTR program is under the management of Directorate General of Sustainable Forest Management (PHL), while HKm is under the management of Directorate General of Social Forestry and Environmental Partnership (PSKL) [26], which indicates that SF management needs comprehensive and inter-sectoral approaches. Therefore, collaboration among institutions is an essential part of SF implementation.

To address this issue, SF program needs to be mainstreamed across directorates under the Ministry of Environment and Forestry. The implementation of this program should refer to the main objectives of social forestry as stated in the Minister Regulation No 83 of 2016: the improvement of community welfare and forest sustainability. These objectives should be translated into integrated action plans which involve various stakeholders.

Model simulation showed there was no significant difference in in the implementation of SF under business as usual scenario and under collaboration among institutions scenario terms of the improvement of SF participants welfare in SF program. Meanwhile the income of oil palm investors tend to be fluctuated (Figure 5). This is because the community income from wood plantation depends of the rotation cycle. The shorter the rotation, the quicker farmers will obtain the economic benefits. In addition, other issues that need to be considered in wood industry sector is wood quality, yield and durability. In this scenario, marketing strategy to develop SF commodities also still lack. Even though the SF participants are allowed to develop profitable commodities which involve various institutions, technical assisstance and marketing strategies are still required to achieve SF target in improving community livelihood. This also may be caused by additional new SF participants under collaboration scenario which would affect per capita income, hence there is no difference in SF participants incomes between business as usual and collaboration scenarios.

IOP Conf. Series: Earth and Environmental Science 1115 (2022) 012057 doi:10.1088/1755-1315/1115/1/012057

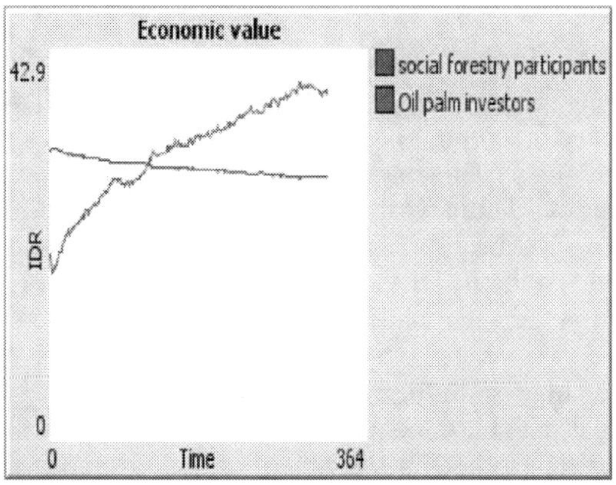

(a)

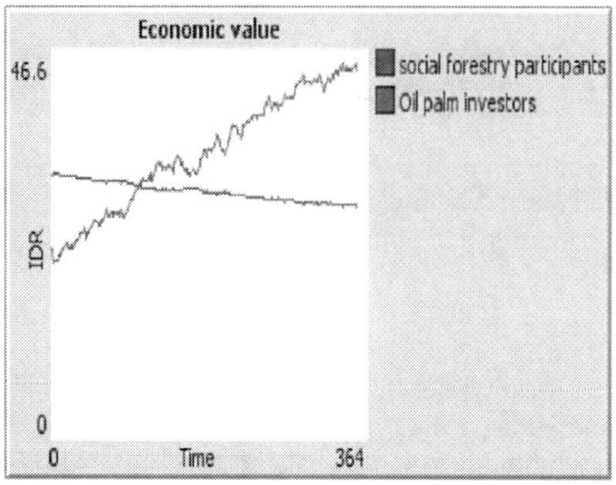

(b)

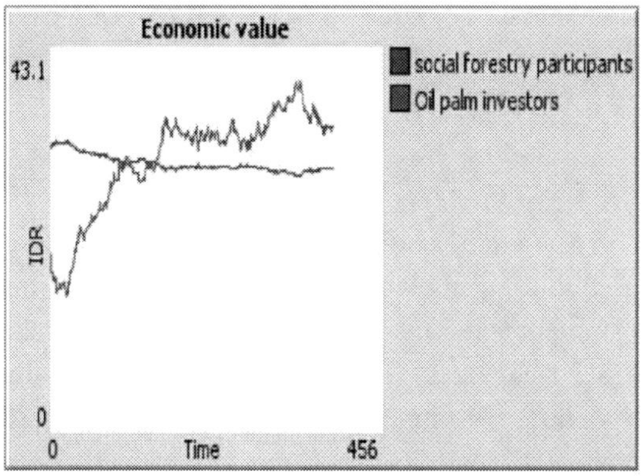

(c)

X = S*imulation time in months*, Y = L*and values in million IDR*

Figure 5. Economic welfare (a) under business- as usual SF Program), (b) under collaboration scenario), (c) under wood product consumption scenario

3.3.2. The inclusion of wood consumption factor as a strategy to improve SF program. As a result of sectoral forest management, forest productivity assessment is merely based on tree growth and not considering wood demand. Food and Agriculture Organization (FAO) noted that wood consumption per capita is 0.15 m^3/year. Meanwhile log production in 2012-2016 reached about 28 million m^3 and other wood products such as plywood, veneer and sawn wood was about 23 million m^3. This number was measured based on industrial capacity at above 6000 m^3 (Figure 6) [27].

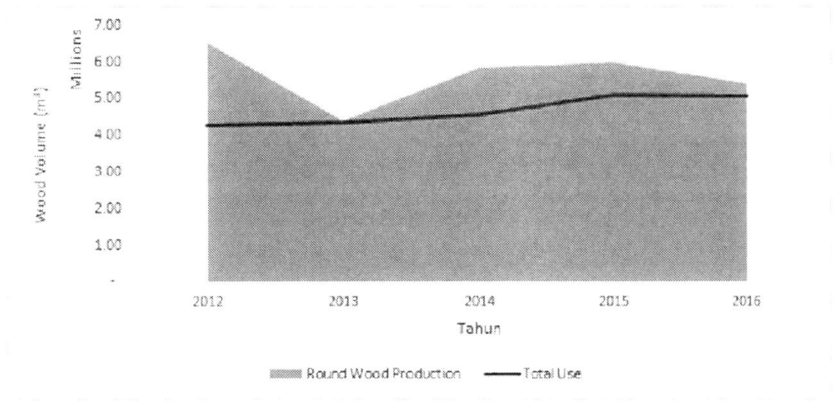

Figure 6. Log production and consumption

Considering these potentials, model simulation in this study includes wood consumption as a variable to develop a scenario for SF implementation. Under this scenario, a mechanism to sell logs produced by SF participants to wood-processing industries is available, following the production capacity of industries. In model simulation, wood consumption refers to wood products such as sawn wood, plywood and other laminated veneer lumber products. The production of such wood products has been increasing while log production has been declining in the past few years. This indicates that the increase in wood products demand has not been followed by the increase in wood supply. This can be an opportunity for SF program to support the improvement of wood production at the national level, while at the same time improve community livelihood and sustainable forest management.

The amount of domestic wood consumption is derived from wood production and imports substracted by wood products export [28]. It means that the volume of wood products export is more likely to affect domestic wood consumption. The model simulation showed that the increase in wood demand has a positive trend on SF participants' income in a long term. In addition, the gap on the obtained economic benefit between SF participants and oil palm investor is also smaller compared to the two other scenarios (Figure 5). This can be achieved when planting wood commodities is more attractive to local communities than planting oil palm in terms of economic benefit. However, wood price needs to be taken into consideration because it can also affect forest conversion rate. The decrease in wood price by 10% would reduce forested area by 0.09% [29].

Under SF program with wood product consumption scenario, there is an opportunity that farmer incomes would increase, assuming that most consumption is from domestic market. Based on the extent of forest area, wood consumption in Indonesia (436.48 m^3/ha) is generally far below those in the USA (979.52 m^3/ha) and China (2,131.05 m^3/ha). Indonesia's wood consumption to forest area ratio is only about 230.17 m^3/ha. This number is considered low compared to other countries, even the lowest in the Southeast Asia region [13].

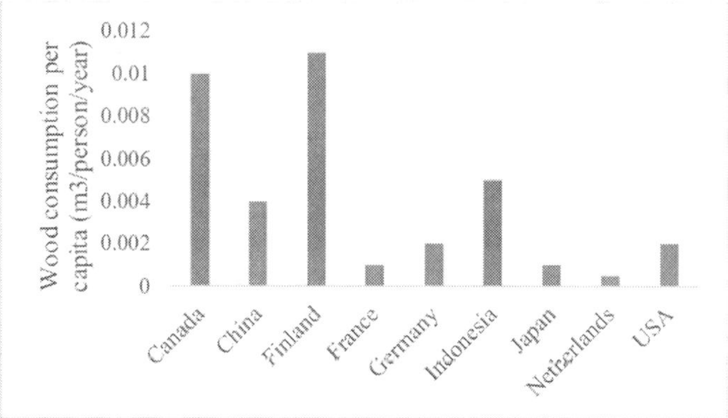

Figure 7. Per capita wood consumption (FAO, 2107)

Figure 7 shows that Indonesia's per capita wood consumption was lower than Canada and Finland but higher than China, Germany, the USA, France and the Netherlands. Based on national GDP, Indonesia's wood consumption is about 0.00465 m^3/person/year [13]. With a total population of 265 million and population growth of 1.33% [30], it was estimated that wood demand reached 1.2 million m^3 in 2019 and will be likely to increase up to 1.4 million m^3, assuming that GDP growth would be similar to 2019 growth. This simulation also shows that SF program implementation is not only about forest management but also depends on other sectors. In order to support the achievement of SF program target, commodity management and business development are required to support the improvement community welfare.

4. Conclusion

The achievement of SF targets to improve community welfare and sustainable forest management needs to be supported by appropriate regulations, particularly in SF commodities marketing. Collaboration among institutions which highly consider the interests of SF participants is required. In addition, the provision of technical assistance and forest products marketing strategy is also essential to ensure that SF program can significantly improve the community welfare. To improve its effectiveness, forest products consumption data, which represents wood demand in Indonesia, can be included as a basis for developing a strategy for SF implementation. This can attract more SF participants to support the improvement of wood production which may also result in the increase in livelihoods, as well as to avoid forest conversion within SF areas.

References

[1] MoEF. 2021. The State of Indonesia's Forests. The Ministry of Environment and Forestry

[2] Susilo, Y.S., 2019. Dampak Perhutanan Sosial Terhadap Pendapatan Masyarakat. *ISEI Economic Review*, *3*(1), pp.16-27.

[3] Rakatama, A. and Pandit, R., 2020. Reviewing social forestry schemes in Indonesia: Opportunities and challenges. *Forest policy and economics*, *111*, p.102052.\

[4] Hakim, I., Irawanti, S., Murniati, S., Widiarti, A., Effendi, R., Muslich, M. and Rulliaty, S., 2010. Social Forestry Menuju Restorasi Pembangunan Kehutanan Berkelanjutan. *Bogor, Indonesia: Pusat Penelitian dan Pengembangan Perubahan Iklim dan Kebijakan*.

[5] Laksemi, N.P.S.T. and Sulistyawati, E., 2019. Sustainable social forestry in Bali (a case study at Hutan Desa Wanagiri). *Jurnal Sylva Lestari*, *7*(2), pp.150-163.

[6] Dewi, I.N., 2018. Kemiskinan masyarakat sekitar hutan dan program perhutanan sosial. *Buletin Eboni*, *15*(2), pp.65-77.

[7] Ma'ruf, A., Suradiredja, D., Marhaento, H., Santoso, H., Saif, I., Zamroni, S., & Saputra, W.

2019. Hutan Kita Bersawit "Gagasan Penyelesaian Untuk Perkebunan Kelapa Sawit dalam Kawasan Hutan. *Jakarta. KEHATI.*

[8] Obidzinski, K., Dermawan, A. and Hadianto, A., 2014. Oil palm plantation investments in Indonesia's forest frontiers: limited economic multipliers and uncertain benefits for local communities. *Environment, Development and Sustainability*, *16*(6), pp.1177-1196.

[9] Purnomo, H., Suyamto, D. and Irawati, R.H., 2013. Harnessing the climate commons: an agent-based modelling approach to making reducing emission from deforestation and degradation (REDD)+ work. *Mitigation and Adaptation Strategies for Global Change*, *18*(4), pp.471-489.

[10] Svitek, M., 2015. Towards complex system theory. *Neural Network World*, *25*(1), p.5.

[11] Abdulah, L. 2019. Model interaksi pelaku hutan rakyat dalam perdagangan kayu: pendekatan simulasi model berbasis agen. *Jurnal Penelitian Hutan Tanaman*, *16*(9), 1689–1699. https://doi.org/10.1017/CBO9781107415324.004

[12] Purnomo, H., Dewayani, A.A., Achdiawan, R., Ali, M., Komar, S., Okarda, B. and Juniwaty, K.S., 2016. Tata kelola rantai nilai sawit dan kebakaran hutan dan lahan.

[13] FAO. 2017. *Forest Products*. Food And Agriculture Organization of The United Nations.

[14] Badan Pusat Statistik [BPS]. 2016a. *Statistik Produksi Kehutanan*. Badan Pusat Statistik.

[15] Purnomo, H., 2019. *Pemodelan dan simulasi untuk pengelolaan adaptif sumber daya alam dan lingkungan*. PT Penerbit IPB Press.

[16] Bert, F., North, M., Rovere, S., Tatara, E., Macal, C. and Podestá, G., 2015. Simulating agricultural land rental markets by combining agent-based models with traditional economics concepts: The case of the Argentine Pampas. *Environmental Modelling & Software*, *71*, pp.97-110.

[17] Ilham, Q. P., Purnomo, H., & Nugroho, T. 2019. Model of multi-stakeholder forest management: A system study of Protected Forest Management Unit in Solok, Indonesia. *IOP Conference Series: Earth and Environmental Science*, *285*(1). https://doi.org/10.1088/1755-1315/285/1/012009

[18] Darwis, I. 2015. Skripsi : Dampak Keberadaan Perusahaan Kelapa Sawit Terhadap Kesejahteraan Sosial Masyarakat Di Desa Bulu Mario Kabupaten Mamuju Utara (Vol. 53, Issue 9). Universitas Hasanuddin. https://doi.org/10.1017/CBO9781107415324.004

[19] Zunariyah, S., 2012. Dilema ekspansi perkebunan kelapa sawit di Indonesia: Sebuah tinjauan sosiologi kritis. *Universitas Sebelas Maret Institutional Repository*.

[20] MoEF. 2020. *Capaian Izin Perhutanan Sosial Tahun 2007 - 2020*. Direktorat Jenderal Perhutanan Sosial dan Kemitraan Lingkungan.

[21] MoA. 2022. Luas Areal Kelapa Sawit Menurut Provinsi di Indonesia, 2017-2021. retrieved on 20 July 2022.

[22] Purnomo, H., Okarda, B., Dewayani, A.A., Ali, M., Achdiawan, R., Kartodihardjo, H., Pacheco, P. and Juniwaty, K.S., 2018. Reducing forest and land fires through good palm oil value chain governance. *Forest policy and economics*, *91*, pp.94-106.

[23] Lifianthi, Oktarina, S. and Aryani, D., 2014. Perbandingan kontribusi pendapatan dan pengeluaran konsumsi petani plasma kelapa sawit di dua tipologi lahan di Sumatera Selatan. *Prosiding Seminar Nasional Lahan Suboptimal 2014 Tema "Pengembangan Teknologi Pertanian Yang Inklusif Untuk Memajukan Petani Lahan Suboptimal, September*, 139–146. https://doi.org/979-587-529-9

[24] Subarudi, S., 2014. Analisis Kelayakan Sosial, Finansial Dan Pasar Produk Hutan Tanaman Rakyat: Studi Kasus Di Kabupaten Dompu, Nusa Tenggara Barat. *Jurnal Penelitian Sosial Dan Ekonomi Kehutanan*, *11*(4), p.29100.

[25] Santika, T., Meijaard, E., Budiharta, S., Law, E.A., Kusworo, A., Hutabarat, J.A., Indrawan, T.P., Struebig, M., Raharjo, S., Huda, I. and Ekaputri, A.D., 2017. Community forest management in Indonesia: Avoided deforestation in the context of anthropogenic and climate complexities. *Global Environmental Change*, *46*, pp.60-71.

[26] Herawati, T., 2010. Komparasi Kebijakan Hutan Tanaman Rakyat dan Hutan Kemasyarakatan

dalam Perspektif Teori Hak Kepemilikan. In *Prosiding Seminar Nasional MAPEKI (Masyarakat Peneliti Kayu Indonesia) XIII. https://doi. org/10.13140/2.1* (Vol. 4768).

[27] MoEF. 2017. *Statistik lingkungan hidup dan kehutanan 2016*. The ministry of environment and forestry

[28] Lim, B., Brown, S., Schlamadinger, B., Sokona, Y., Abel, K., Boer, R., Breed, W., Ford-Robertson, J., Karjalainen, T., Nabuurs, G.J. and Masera, O., 1999. Evaluating approaches for estimating net emissions of carbon dioxide from forest harvesting and wood products.

[29] Kim, T.J., Wear, D.N., Coulston, J. and Li, R., 2018. Forest land use responses to wood product markets. *Forest policy and economics*, *93*, pp.45-52.

[30] [BPS] Badan Pusat Statistik. 2020. *Statistik Indonesia Dalam Infografis 2019*. Badan Pusat Statistik.

Land capability analysis using LCLP software in cangkringan micro watershed model

S Fadhilla[1a*], A Kusumandari[2], Senawi[3], Y S Lubis[4], A W Siregar[5], L Hakim[6a]

[1] Department of Forest Resource Conservation, Faculty of Forestry, Universitas Sumatera Utara, Kampus 2 USU Bekala, Pancur Batu, Kabupaten Deli Serdang, Indonesia 20353
[2] Department of Forest Resource Conservation, Faculty of Forestry, Universitas Gadjah Mada, Jl. Agro no.1, Yogyakarta, Indonesia 55281
[3] Department of Forest Management, Faculty of Forestry, Universitas Gadjah Mada, Jl. Agro no.1, Yogyakarta, Indonesia 55281
[4] Research Center for Biomass and Bioproduct, National Research and Inovation Agency (BRIN), Jl. Raya Jakarta-Bogor Km. 46 Cibinong, Kab. Bogor, Jawa Barat 16911
[5] Study Program of Forestry PSDKU USK Gayo Lues, Faculty of Agriculture, Universitas Syiah Kuala, Jl. Blangkejeren-Blang Pidie Km. 20, Gayo Lues, Aceh, Indonesia 24655
[6] Departemen of Forest Product Technology, Faculty of Forestry, Universitas Sumatera Utara, Kampus 2 USU Bekala, Pancur Batu, Kabupaten Deli Serdang, Indonesia 20353
[a] JATI-Sumateran Study Analysis Centre, Faculty of Forestry, Universitas Sumatera Utara, Kampus 2 USU Bekala, Pancur Batu, Kabupaten Deli Serdang, Indonesia 20353
[*]Corresponding author: surifadhilla@usu.ac.id

Abstract. Each land in a watershed area has different capabilities depending on the characteristics of the land in the area. Land capability analysis is needed in land management and land use, the aim is that land use can be carried out optimally and reduce the impact of land damage. Determination of land capability class using the matching method, by matching and comparing land characteristics or land limiting factors with land capability class criteria according to the USDA classification system. This study was started by conducting field observations and taking soil samples, then analyzed using Land Capability and Landuse Planing (LCLP) software and GIS to get a map of the land capability class. The results showed that the land capability in site are consists of class II-IV and VI-VII. The limiting factors for land management are erosion (e), permeability (P), texture (t) and slope (L)

1. Introduction

Land use must consider the ability of the land, because otherwise it will result in a decrease in soil productivity, increasing critical land and land degradation [1-3]. Each land has land capability wich is the quality of the soil for extensive cultivation activities and the cultivation of livestock feed pastures with the condition that there is no damage to the land in long term [4-5]. Land capability evaluation is process to evaluate land potency according to its capability for sustainable land uses [6]. Land evaluation and land use planning needed as base for making policy regarding land use, so that resources planning and development land resources can be guaranteed sustainability for now and future [7]. As one of the requirements in land use planning is to carry out a land suitability assessment, by analayzing the current lan use wether or not it is in accordance with the land capability class and then determining the appropriate land use for the area[8].

Land capability classification is an assessment of land components systematically and grouping them into several categories based on the characteristics and obstacles to their sustainable use [9]. The parameters of land capability such as the types of soil, its depth and texture, underlying geology, topography, hydrology and land use. These parameters become the limit in managing land to become productive lan for cultivation crops [10]. Land is classified into three main categories, namely: class, sub-class and capability unit or management unit. The land capability classification based on the intensity of the land constraining factors. Land is grouped into 8 classes, used Roman letters I to VIII. Class one to four have characteristics suitable for agricultural activities, while class five to eight are not suitable for agricultural business and directed to serve as procteted areas [11-12]. In carrying out soil conservation, land capability analysis and land suitability maps are needed to determine the recommended conservation action needed [13]. Land suitability clasification in land management practices is useful for ensuring that the use of exiting agricultural land is in accordance with the potential of the land so that sustainable agricultural land use will be obtained [14]. In watershed planning, it is very important to identify the land capability class as a determination of appropriate land management measures [11]. This study aims to analyze land capability using Land Classification and Land Use Planning (LCLP) software and Geography Information System (GIS) in Cangkringan Micro Watershed Model.

2. Material and Method

2.1. Research Location

The location of study are lies between a latitude of 110° 26.53' and 110° 28.06' South and longitude of 7° 34.17' and 7° 41,23' East. The location is part of sub-watershed whose management is in the micro watershed model unit, in cangkringan sub-district, Sleman district. The condition of the slope is relatively flat between 2%-8%, but in the upstream the slope is very steep and flat in the downstream. The shape of Cangkringan micro watershed model is elongated which has a slow peak flood time, meaning that if it rains upstream the water/flood will not reach downstream quickly. The condition of drainage density in Cangkringan watershed is experiencing drought, this happens because the river upstream is part of cold lava flow of Mount Merapi.

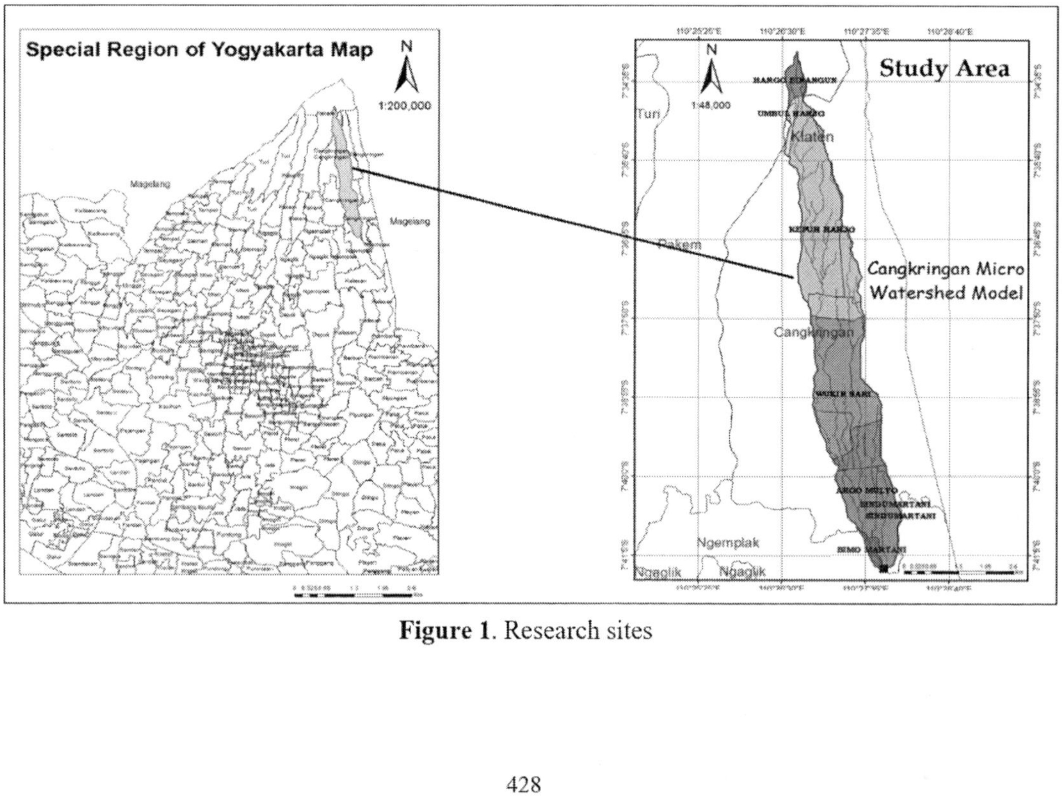

Figure 1. Research sites

2.3. *Data Collection*
Field observations in this study were carried out to obtain the conditions and characteristics of the land, then soil samples were taken fo analysis in the evaluation of land research objects usually use land unit, in this study there were 32 land units obtained from the overlay of soil type maps, land slope maps, rainfall distribution maps, and land use/land cover types maps. Field observations carried out include soil texture, drainage, soil depth, percentage of gravel and rock, measurements of topographic properties (slope), to observe hazard (flood, land slide and erosion) and also to collect soil sample. Soil analaysis was carried out to obtain soil characteristic data in the form are soil structure and texture, permeability and soil organic matter.

2.4. *Data Analysis*
Determination of land capability class using the matching method, by matching and comparing land characteristics or land limiting factors with land capability class criteria according to the USDA classification system (Table 1). This land capability classification system divides land into a number of categories sorted according to the number and intensity of inhibiting factors that affect plants, these categories are sorted from highest to lowest [15]. Inhibiting factors that affect the ability of the land are soil texture (top and bottom layers), slope, erodibility, erosion rate, soil depth, permeability, drainage, percentage of rock/gravel and flood threat [16]. Data analysis was performed using Land Classification and Land Use Planning (LCLP) software, and then mapped using ArcGIS.

Table 1. USDA land capability classification system [15-16]

No	Limiting factors	Land Capability Class							
		I	II	III	IV	V	VI	VII	VIII
1	Surface slope	A	A	B	C	A	D	E	E
2	Soil erodibility (KE)	KE1, KE2	KE3	KE4, KE5	KE6	(*)	(*)	(*)	(*)
3	Erosion	e0	e1	e2	e3	(**)	e4	e5	(*)
4	Soil depth	k0	k1	k2	k2	(*)	k3	(*)	(*)
5	Soil texture	t1,t2, t3	t1,t2, t3	t1,t2, t3,t4	t1,t2,t3,t4	(*)	t1,t2, t3,t4	t1,t2, t3,t5	t5
6	Permeability	P2,P3	P2,P3	P2,P3,P4	P2,P3,P4	P1	(*)	(*)	(*)
7	Drainage	d1	d2	d3	d4	d5	(**)	(**)	d0
8	Rock/gravel	b0	b0	b1	b2	b3	(*)	(*)	b4
9	Flood threat	O0	O1	O2	O3	O4	(**)	(**)	(*)

Description : (*) can have any properties
(**) doesn't apply as a parameter

The result of analysis using LCLP software will classify land based on land capability class, land capability sub-class based on limiting factors, division, sub-division and conservation action. The results of this study will provide an overview of land capability for the type of land use in general. To determine land suitability for specific uses, for example for plantations, agriculture or housing, further analysis is needed to use the land suitability analysis.

3. Result and Discussion

3.1. *Land characteristics in Cangkringan Micro Watershed Model*
Overall the land in Cangkringan Micro Watershed Model is easy to cultivate or plant, because it has almost little gravel or rocks and there is no threat of flooding. The topography is predominantly flat to slightly steep, the soil type is classified as a slightly erosion sensitive class (association of andosol,

regosol, litosol and latosol soil types) with moderate to moderate erosion sensitivity values which have the potential for the magnitude of erosion that occurs. The erosion class is dominated by the light class covering area of 670.33ha (49.51%), the medium class covering area of 545.37ha (40.28%), the heavy erosion class covering area of 92.83ha (6.85%) and the very heavy erosion class covering area of 44.37ha (3.36%). The drainage class is classified as good because the air circulation in the soil is quite good and there are no gray or brown spots on the soil layer.

3.2. Land capability classification

The land characteristic data that has been collected is processed into input data of land capability parameters in the LCLP software, which is then analyzed to obtain the results of the distribution of land capability classes on each land unit in Cangkringan micro watershed model. The results of the land capability classification show that the Cangkringan micro watershed model has various land capability classes consisting of 6 land capability classes, namely class II covering area of 757.47 ha (55.94%), III covering area of 405.03 ha (29.91). %), IV covering area of 143.42 ha (10.59%), VI covering area of 15.66 ha (1.16%), VII covering area of 30 ha (2.22%), and VIII covering area of 2.43 ha (0.18%). Land in class II and III can be processed for intensive agricultural and forestry cultivation activities, while land in class IV, VI, VII and VIII cannot be used for intensive agriculture, only designated for certain uses such as limited production forests, limited plantations, limited grass plants, conservation forest and protected forest. The result of analysis using LCLP software show in table 2.

Table 2. Land capability class in Cangkringan Micro Watershed Model

No	Land capability class	Land capability sub-class	Limiting factor	Division	Area (ha)
1	II	II-e	Erosion	Land can be cultivated	141.77
		II-t	Soil texture		233.11
		II-P	permeability		382.58
2	III	III-e	Erosion	Land can be cultivated	405.03
3	IV	IV-L	Surface slope	Land can be cultivated	107.28
		IV-e	Erosion		36.14
4	VI	VI-L	Surface slope	Land can't be cultivated	1.58
		VI-e	Erosion		14.08
5	VII	VII-L	Surface slope	Land can't be cultivated	30.00
6	VIII	VIII-e	Erosion	Land can't be cultivated	2.43

Based on the results of the land capability class analysis, the Cangkringan land is dominated by class II – IV with a total area of 96.44%, which means that the use of land to be cultivated as agricultural land is quite extensive. The higher the land capability class, the worse the land quality, which means the risk of land use that can be applied is increasingly limited. Classes I – IV are suitable for agricultural businesses, while classes V – VIII are not suitable for agricultural businesses or require very high costs for their management [16]. In class IV, if the land is used for agriculture, the efforts that need to be made to maximize land productivity require appropriate and appropriate conservation actions. Land management efforts for agriculture are very supportive on Cangkringan, but it is necessary to consider the characteristics of the land and the factors that become obstacles in the management. The land capability class map can be seen in figure 2.

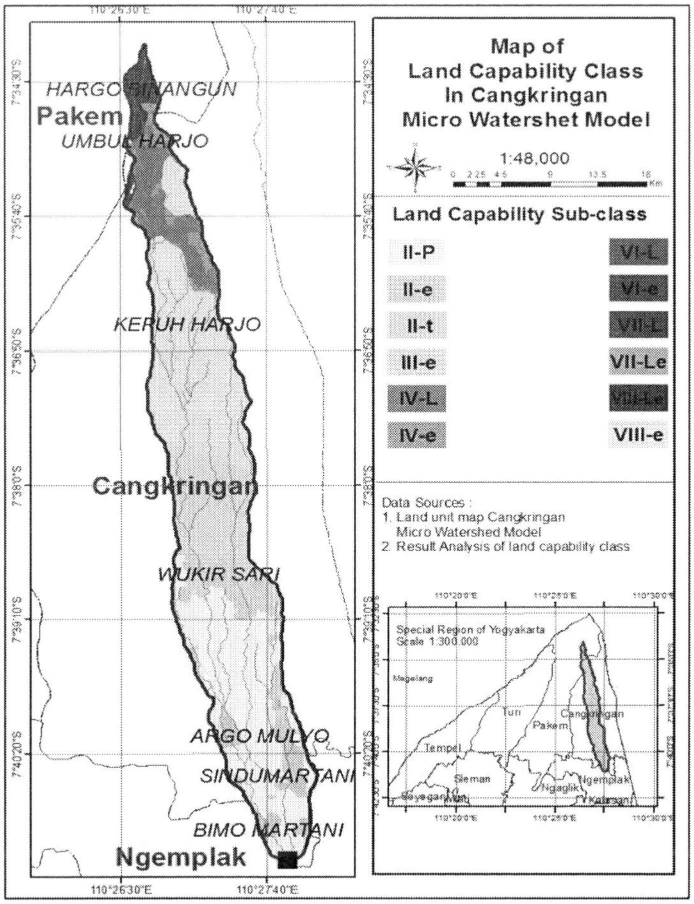

Figure 2. Land capability class and the limiting factor in Cangkringan micro watershed model. II – VIII are land capability class P, e, t and L/Le are limiting factors P : permeability, e : erosion, t : texture, L/Le : slope.

3.3. Limiting factor
Land capability class II has the main limiting factor in the form of soil with a coarse texture in the form of sand, the land distribution is located on a flat slope, making it suitable for all types of land use businesses except extensive agriculture with light to moderate soil conservation measures. The main limiting factor in capability class III is moderate erosion rate with fast permeability conditions, but coarse soil texture conditions with dominance of sandy soil. Land with ability class III is spread on a flat slope. The condition of the land in this class still allows for extensive agricultural activities by applying moderate soil conservation measures. Ability class IV has the main limiting factors in the form of moderate erosion and slopes. The level of erosion occurs due to poor soil cover conditions and is supported by slightly steep slope conditions. The condition of the land in this class still allows for marginal agricultural activities, but must apply heavy soil conservation efforts.

Class VI has the main limiting factor in the form of heavy erosion and some have steep land conditions. This condition make the land unsuitable for agricultural activities and is directed to limited pastures for grazing and forestry crops or forms of high density gardens accompanied by heavy soil conservation measures. Class VII has the main limiting factor of steep slope conditions, which makes the land unusable for agricultural activities and is directed to be used as limited production forest with dense soil cover and conservation forest. Class VIII has the main limiting factor of very heavy erosion, due to land use conditions with little land cover which is located on a very steep slope. The condition of the land in this class causes the land to be directed to be used as a protected forest with heavy soil conservation efforts.

4. Conclusion

Land capability in Cangkringan Micro Watershed Model are consists of class II-IV and VI-VII, class II covering area of 757.47 ha (55.94%), III covering area of 405.03 ha (29.91). %), IV covering area of 143.42 ha (10.59%), VI covering area of 15.66 ha (1.16%), VII covering area of 30 ha (2.22%), and VIII covering area of 2.43 ha (0.18%). The limiting factors are erosion (e), permeability (P), texture (t) and slope (L). Land in class II and III can be processed for intensive agricultural and forestry cultivation activities, while land in class IV, VI, VII and VIII cannot be used for intensive agriculture, only designated for certain uses such as limited production forests, limited plantations, limited grass plants, conservation forest and protected forest.

Acknowledgments

The authors are thankful to watershed management laboratory, GIS laboratory of Forest Faculty Universitas Gadjah Mada and our colleague Deni Susanto, Febri Arif, Pinky Novitasari, Faezal Prandeka and Marya Tiara.

References

[1] Cholidah N N Z and Masruroh H 2021 Analysis of land capability as a land use direction using Geographic information systems (GIS) in Nganjuk Regency *Jurnal Integrasi dan Harmoni Inovatif Ilmu-Ilmu Sosial* **1** 1167–81

[2] Kusumandari A and Nugroho P 2015 Land capability analysis based on hydrology and soil characteristics for watershed rehabilitation *Procedia Environmental Science* **28** 142–147

[3] Asnur P, Thoha A S and Elfiati D 2019 Mapping of degraded soil properties in Gunung Leuser National Park *Journal of Sylva Indonesia* **2** 70–79

[4] FAO 1983 *Guidelines: Land evaluation for rain fed agriculture* Soils Buletin 52 Food and Agriculture Organization of the United Nation p273

[5] Gad A 2015 Land capability classification of some western desert Oases, Egypt, using remote sensing and GIS *The Egyptian Journal of Remote Sensing and Space Sciences* **18** S9–S18

[6] Maryanti S 2013 Land capability evaluation of reclamation area in Indonesia coal mining using LCLP software *Procedia Earth and Planetary Science* **6** 465–473

[7] Osok R M, Talakua S M and Supardi D 2018 Determination of land capability class and directions for watershed land rehabilitation Wai Batu Merah, Ambon City, Maluku Regency *AGROLOGIA* **7** 32–41

[8] Akinci H, Ozalp A Y and Turgut B 2013 Agricultural land use suitability analysis using GIS and AHP technique *Computers and Electronics in Agriculture* **97** 71–82

[9] Duwila R, Tarore R C and Takumansang E D 2019 Analysis of land capability in Sulabesi Island Sula Island District *Journal Spasial* **6** 703–713

[10] Panhalkar S S 2011 Land capability classification for integrated watershed development by applying remote sensing and GIS techniques *ARPN Journal of Agricultural and Biological Science* **6** 46–55

[11] Silpa T J M and Nowshaja 2016 Land capability classification of Ollukara Block Panchayat using GIS *Procedia Technology* **24** 303–308

[12] Arsyad 2012 *Soil and Water Conservation* (Bogor: IPB Press) p 105
[13] Feudis M D, Falsone G, Gherardi M, Sperenza M, Vianello G and Antisari L V 2021 GIS-based soil maps as tools to evaluate land capability and suitability in a coastal reclaimed area (Ravenna, northern Italy) *International Soil and Conservation Research* **9** 167–179
[14] Kilic O M, Ersayin K, Gunal H, Khalofah A and Alsubeie 2022 Combination of fuzzy-AHP and GIS techniques in land suitability assessment for wheat (*Triticum aestivum*) cultivation *Saudi Journal of Biological Science* **29** 2634–44
[15] Senawi 1999 *Evaluation and land use Textbooks* (Yogyakarta: Faculty of Forestry Universitas Gadjah Mada) p 20
[16] Hardjowigeno S and Wisiatmaka 2011 *Land Suitability Evaluation and Land Use Planning* (Yogyakarta: UGM Press) p 122

IOP Conf. Series: Earth and Environmental Science 1115 (2022) 012059 doi:10.1088/1755-1315/1115/1/012059

The influence of biochar with pH, CEC, C organic on paddy field

A Sipayung* **and R Damanik**

Department of Agrotechnology, Universitas Sumatera Utara, Prof. A. Sofyan No. 3 Kampus USU, Medan 20155, Indonesia

*Email: sipayungantoniomarro@usu.ac.id

Abstract. An organic material called biochar used for soil conditioners and having a role to restore and improve the quality of fertility of degraded soil. It was got from the incomplete of combustion process. The purpose of this study was to evaluate the influence of biochar application in terms of pH, CEC, C Organik a modification on paddy crops. The Factorial Split Plot Experimental Design was used for this reseach with the factors were the treatment of planting as a main plot with two stages (Jajar Legowo 2:1 dan Jajar Legowo 4:1) and the sub-plot was the biochar application with three levels, namely: B0 (no application), B1 (25 ton/ha), and B2 (50 ton/ha). The research carried out at Tanjung Garbus Kampung's Village, Pagar Merbau Sub- district, Deli Serdang Regency with an elevation of ± 50 Meters above sea level. The paddy field had irrigation system. The biochar application has no statistically affected paddy production as well as any significant influence on pH, CEC, and C Organic.

1. Introduction

Because of the population in Indonesia going increase at rate of 2% per year that affected the need of rice. Rice is main food sources in Indonesia. The pattern of consumption of non-rice to rice had changed. On the other hand, the decline of fertile paddy land due to land conversion for purposes other than agriculture, also the phenomenon of irrigated paddy productivity tends to decrease [1].

Makarim et al [2] stated there were some causing of making the unproductivity of rice in paddy fields, namely, a) low fertilizer efficiency; b) ineffective pest control has not been effective; c) the use of seed fertilizers and pesticides that are lacking in quality and the varieties selected are less adaptive; d) soil physical properties are not optimal. He added that the way to increasing the national grain production was by optimizing the rice productivity. Normally the paddy production is 6 to 7 ton/ha. However, the average production is 4.7 ton/ha.

Efforts are made to obtain maximum production from superior varieties of rice that must be planted on fertile land, nutrients must be available, sufficient irrigation, the soil should contain organic material, integrated pest control, and crop processing must be done well (7-12 ton/ha). Paddy field land requires fertile mud soil with a thickness of 18-22 cm. Soil acidity between pH 4.0-7.0. In paddy field land, inundation will change the pH of planting to neutral (7.0). In principle, calcareous soil with a pH of 8.1 to 8.2 does not damage paddy crops [3].

The physical, chemical, biological, and morphological characteristics of the soil become tangible caused by paddy soil preparation. The condition of the natural soil changes to the state of the artificial land and deviates from the state desired by other crops. Paddy fields are the cultivation of plants that use the most water. It takes a lot of water to overflow the soil, to inundate plantations, and to be able to flow from one plot to another. Because of the paddy soil is reductive (anaerobic), the paddy soil is one of the main producers of methane gas. The other main producer is swamped [4]. Land with fine-medium texture (sandy clay, clay, dusty clay, clay loam, sandy clay loam, dusty clay loam, very fine sandy loam, loam, the dusty loam, and dust) suitable to be used as paddy fields [5].

The previous of soil analysis at paddy field experiment showed that 5.3 of pH and 1.73% for c organic content. According to [6] stated that 3% is a normal number for Organic. Therefore, the soil is less fertile, and the productivity need to improve. The previous production were 5 tons and 4.8 ton respectively.

Gani [7] mentioned about the biomass carbonization process that obtaining the solid material called biochar. Biochar also called charcoal deriving from plants is a porous charcoal substance. A good habitat for soil microbes for instance bacteria provided by soil which contain biochar. Biochar can overcome several problems with the soil in the cultivation process and provide additional options for managing the soil. These soil problems are easy to lose nutrients and moisture, for instance.

According to Glaser [8] the biomass heating process in a limited oxygen resulting a black charcoal or called biochar. An organic material using as a soil ameliorant has stable properties called biochar. In order to restore the soil management and improving the quality of soil fertility, biochar became more popular rather than other organic source material.

According to Lehmann et al [9] the approach to reducing the greenhouse gas emissions and increasing the crop productivity for agricultural activities by the biochar application. Through artificial pyrolysis producing biochar either for renewable energy sources or carbon sequestration. He added soil fertility and crop fertility can increase using the biochar.

The potential use of biochar in Indonesia is quite large because of its abundant raw materials, for example, wood, coconut shell, rice husks, and mangrove plants. The making of charcoal is quite well known in Indonesia but has not been utilized properly as a soil enhancer. In general, charcoal making is carried out for export purposes. The use of soil-based materials to make up agricultural remains that are difficult to decompose is one alternative that can be taken to accelerate the improvement of the quality of the physical properties of the land in the use of land as a food source so that production can be increased [10].

Straw, wood, manure, and other organic waste are the biomass sources for producing the biochar. A lot of raw material for making biochar potentially attractive for soil amendments. Charcoal could be a promising solution to improve soil condition which stable relatively of organic form. The efficient way to achieve agronomics benefit is by the biochar amending to the soil including to reduce the plant diseases [12].

The use of rice husk biochar in Inceptisol paddy soil causes changes in the organic C content of the soil. The higher the dose of biochar, the organic C content of the soil will also increase. Likewise with the fineness of biochar where the smoother the soil organic C content will increase as well. The data showed that there was an increase from 1.09% to 1.41% for soil organic C-content [14].

The results of research conducted by [15] showed that biochar residues significantly affected grain per panicle and yield per ha. The average total grain of per panicle for biochar application of 10 tons/ha is 112.67 grains/panicle with a production of 6.07 tons/ha. Whereas for treatment without the application of biochar residues, a total grain of permits was obtained, 101.01 grains/panicle with a production of 5.45 tons/ha. Many people use charcoal husk for treatment factors as well as for experiment with varying results. It can be functioned as natural fertilizer base on the material of manufacture. [19] mentioned that only a few of Indonesia peasant communities using husk fornatural fertilizer while it is important for soil.

Kuzyakov [17] stated that the simultaneous application of organic and inorganic substances into the soil, causing positive and negative impacts on the soil C-organic content. The negative impact that occurs is the decrease in C-organic content, presumably caused by the saturation process of carbon sequestration in the soil. The input of organic material that cannot be stabilized by the soil mineral association will affect the balance of the carbon portion and cause the process of saturation of carbon sequestration. In this case, it can be said that the application of chemical fertilizers in a low dose of

(40% of the recommended dose) causes the dissociation of C-organic from straw-Biochar compost because of insufficient energy sources for the metabolic activity of soil microorganisms carry out decomposition and mineralization.

2. Material and Method

The research was carried out Tanjung Garbus Kampung village, Pagar Merbau sub-district, Deli Serdang Regency (3°31'34.1"N and 98°51'16.1" E) with an area of 650 m^2 and altitude of ± 50 meters. Soil analysis was carried out at PT. Socfin Indonesia Dolok Masihol Estate. Soil analysis took place before the experiment has been taken and showed that the texture is sandy loam. The study was conducted from December 2017 to March 2018.

The materials used in this study were biochar from paddy husks, inorganic fertilizers (Urea, ZA, TSP, MOP, NPK). The tools used in this study are laboratory instruments such as analytical scales, Kjeldahl flasks, 10 ml measuring pipettes, 1000 ml measuring cups, 100 ml measuring cups, 50 ml measuring cups, distillation, destruction, 100 ml Erlenmeyer, analytical balance, burette, shaker, filter paper, funnel, suction rubber, 50 ml measuring flask, 250 ml Erlenmeyer, 50 ml Erlenmeyer, shake-bottle, and soil drying oven.

3. Result and Discussion

The soil analysis taken before and after the paddy planting. The soil anlaysis conducted in the Socfindo Laboratory. The soil analysis taken after the paddy harvesting divided into three treatments, namely (B0/control, B1 25.5 ton/ha, and B2 50.5 ton/ha).

Table 1. Preliminary soil properties

No	Parameter	Preliminary Soil Analysis
	Texture	
1	Sand %	60.98
2	Dust %	24.83
3	Clay %	14.02
4	pH H_2O	5,3
5	C Organik %	0,092361111
6	pH MOP	04.02
7	CEC	0,896527778
8	K-exch (me/100mg)	00.05
9	Ca-exch (me/100mg)	05.53
10	Mg-exch (me/100mg)	03.33
11	Na-exch (me/100mg)	00.34

Table 2. Soil properties after treatment

No	Parameter	Biochar Application		
		B0	B1	B2
		0 ton/ha	25.5 ton/ha	50.5 ton/ha
	Texture			
1	Sand %	44.13	51.42	41.05
2	Dust %	38.4	13.88	41.61
3	Clay %	17.47	34.7	17.35
4	pH H_2O	4.5	4.3	4.5
5	C Organic %	0.89	0.55	0.53
6	pH KCl	3.8	3.91	3.72
7	CEC	8.93	8.56	13.25
8	K-exch (me $100mg^{-1}$)	0.27	0.25	0.17
9	Ca-exch (me $100mg^{-1}$)	1.09	1.1	1.67
10	Mg-exch (me $100mg^{-1}$)	1.14	0.92	0.89
11	Na-exch (me $100mg^{-1}$)	0.13	0.13	0.27

The soil analysis carried out at the laboratory and the sample take after the harvest. It showed that biochar's treatment with a control level (B0) had a pH of 4.5 and followed by the application of biochar with a level of 25.5 tons ha^{-1} (B1) pH 4.3 and biochar application with a level of 50.5 tons/ha (B2) pH 4.5. As well as the soil C organic content for biochar applications for the level control (B0), 25.5 tons/ha (B1) and 50.5 tons/ha (B2) of 0.89%, 0.55%, and 0.52% respectively. Based on the soil analysis, the cation exchange capacity (CEC) of biochar's application showed that control levels (B0), 25.5 tons/ha (B1) and 50.5 tons/ha (B2) is 8.93 me/mg, 8.56 me/mg, and 13.25 me/mg respectively.

Table 3. Biochar properties from paddy husk

Biochar (Paddy Husk)	C-Organik (%)	N-Total (%)	P_2O_5(%)
	49,07%	0,84%	0,32%
	K_2O (%)	Lignin	C/N
	3,25%	32,49%	58,42%

Giving paddy husk for biochar's did not give an increase in the pH of the soil after planting paddy crops. The soil analysis has taken place previously. Initially the soil pH was 5.3 after planting and harvesting. The measuring of soil pH from the block of each level of biochar's application B0 (control), B1 (25 ton/ha), B2 (50 ton/ha) were 4.5, 4,3 and 4,5 respectively. In basic alkaline reactions after submergence, a decrease insoil pH occurs due to the presence of OH-ions produced by the CO_2 reduction reaction with H_2O. The process of inundation will affect the increase in the degree of soil acidity (pH) in paddy field land starting in the second week until the fourth week and gradually decreasing in the following weeks towards equilibrium. So, the provision of biochar's is not very influential on the pH of the paddy soil.

Several factors causing the changes of pH in paddy soils into iron, sulfate to sulfide, carbon-dioxide to methane, and ammonium buildup. It is because not significantly affect to pH of paddy soil [18].

The paddy husk effect of biochar's had slight difference of the cation exchange capacity (CEC) on soil after harvesting. Provision of biochar's residues from paddy husk can change the soil chemical properties by increasing soil CEC levels. According to [20] biochar's as a soil enhancer has recalcitrant properties, is more resistant to oxidation, and is more stable in the soil so that it has a long-term effect on improving soil fertility quality (CEC). Biochar's stay long in the soil, and it can be used as soil enhancer in addition to improve the soil physio-chemical properties as well as carbon storage. Therefore, the biochar's application from paddy husk in the first season's paddy soils has not been shown to improve soil fertility quality

The effect of paddy husk for biochar did not provide an increase in C-organic content on the soil after paddy planting. [17] states that the simultaneous application of organic and inorganic substances into the soil, causing positive and negative impacts on soil C-organic content. The negative impact that occurs is the lowing of C-organic content, allegedly caused by the process of saturation of carbon sequestration in the soil. The input of organic material that cannot be stabilized by the soil mineral association will affect the equilibrium of the carbon portion and cause the process of saturation of carbon sequestration.

The biochar's application increases the P levels by the P-Bray method. P (phosphate) available in the soil before planting is 5,632 ppm, however, the P available after planting was 11,341 ppm (B0/control), 10,284 ppm (B1/25.5 tons/ha) and 12,413 ppm (B2/50.5 tons/ha). The highest biochar dose is B2 50.5 tons/ha increasing P-availability. By following [7] the addition of biochar to the soil increases the availability of main cations and phosphorus, total N, and the cation exchange capacity of the soil (CEC) which ultimately increases yield. The amount of biochar added influencing the crop productivity. According to [13] several mechanisms causing the availability of phosphor, namely, (1) the occurrence of soil mechanical disturbance during tillage, (2) the diffusion from soil because of adsorption and orthophosphate desorption from the sediment depends on the relative amount of P in soil solution and (3) plankton and fauna activities that occur due to the possibility of P transfer from soil to standing water due to active changes by paddy soil microorganisms.

According to [16], stated that biochar's can be cheap soil conditioners including a way to protect the environment from agricultural waste. It can be functioning as soil conditioners, and it gets from the incomplete combustion process. The use of biochar also implements to the area that lacks water or less rainfall and it could catch the limited water, therefore, available for crops. Another function of biochar's in agricultural land is to mitigate the greenhouse effect, such as N_2O and CH_4.

4. Conclusion

Giving biochar for paddy did not give any impact for soil. It is becaused biochar as organic matter that has long process for soil improvement. The application of biochar for paddy had slight difference of the cation exchange capacity (CEC) on soil after harvesting. Provision of biochar's residues from paddy husk can change the soil chemical properties by increasing soil CEC levels. The Biochar did not increase the C-organic content on soil. It took after the paddy harvesting.

Acknowledgement

We would like to thanks Prof. Ir. T. Sabrina, M.Agr.Sc., Ph.D from Department of Agrotechnology, Facultyof Agriculture, Universitas Sumatera Utara for her guiding regarding this study.

References:

[1] Balai Litbang Pertanian 2008 *Pengelolaan tanaman terpadu dan sumberdaya terpadu pada sawahirigasi* (Jakarta: Departemen Pertanian)
[2] Makarim A K, Udin S N and Unang G K 2000 *Teknologi produksi padi sawah. Bogor: Pusat Penelitian dan PengembanganTanaman Pangan*
[3] Simanjuntak C P S, Ginting J and Meirani 2015 *Jurnal Online Agroekoteknologi* **3** 1416
[4] Notohadiprawiryo T 1992 *Ceramah Ilmiah Sehari bertema Pencetakan lahan sawah sebagai salah satu alternatif kebijaksanaan dalam pengembangan tata guna lahan* (Yogykarta: Fakultas Pertanian, Universitas Pembangunan Nasional)
[5] Djaenuddin D, Marwan H, Subagyo H, Mulyani A and Suharta N 2003 *Kriteria Kesesuaian lahanuntuk komoditas pertanian* (Badan Litbang Pertanian)
[6] Suriadikarta D A and Simanungkalit R D M 2006 *Pupuk organik dan pupuk hayati* (Bogor: Balai BesarPenelitian dan Pengembangan Sumberdaya Lahan Pertanian) 1
[7] Gani A 2010 *Multiguna arang-hayati biochar* (Sinar Tani Edisi 13-19 Oktober 2010)
[8] Glaser B 2001 *Die Naturwissenschaften* **88** 37
[9] Lehmann J, Gaunt J and Rondon M 2006 *Mitig Adapt Strat Glob Chang* **11** 395
[10] Lehmann J 2007 *Frontiers in Ecology and the Environment* **5** 381
[11] Ding Y, Liu, Y X, Wu W X, Shi D Z, Yang M and Zhong Z K 2010 *Water Air Soil Pollut* **213** 47
[12] Elad Y, David D R, Harel Y M, Borenshtein M, Kalifa H B, Silber A and Graber E R 2010 *Phytopathology* **100** 913
[13] Rogers G S, Payne L, Milham P and Conroy J 1993 *Plant and Soil* **156** 231
[14] Salawati, Basir M, Kadekoh I and Taha AR 2016 *Jurnal Agroland* **23** 101
[15] Mawardiana, Sufardi and Husen E 2013 *Jurnal Manajemen Sumberdaya Lahan* **2** 255
[16] Neneng N L 2014 *Jurnal Sumberdaya Lahan* **8** 57
[17] Kuzyakov Y 2002 *J. Plant Nutr. Soil Sci* **165** 382
[18] Setyorini D and Abdulrachman S 2009 *Pengelolaan Hara Mineral Tanaman Padi* (Balai Besar Penelitian Tanaman Padi) 109
[19] Siringoringo H H and Siregar C A 2011 *Pengaruh aplikasi arang terhadap pertumbuhan awal michelia montana blume dan perubahan sifat kesuburan tanah pada tipe tanah latosol* (Bogor: Pusat LitbangKonservasi dan Rehabilitasi)
[20] Steiner C, Teixeira W G, Lehmann J, Nehls T, de Macêdo J L V, Blum W E and Zech W 2007 *Plant and soil* **291** 275

Analysis of income from pine resin tapping on Forest Farmers Groups in partnership with UPT KPH Region II Pematang Siantar

P R Purba[1*], Rahmawaty[2] and A Susilowati[2,3]

[1]Magister Student of Forestry Study Program, Faculty of Forestry, Universitas Sumatera Utara, Medan 20155, Indonesia.
[2]Faculty of Forestry, Universitas Sumatera Utara, Medan 20155, Indonesia.
[3]Center of Excellence for Mangrove, Universitas Sumatera Utara, Medan 20155, Indonesia

[*]Email: riandopolman@gmail.com

Abstract. The forestry partnership pattern is a social forestry scheme which is an empowerment program that prioritizes the principles of agreement, equality, mutual benefit, local specificity, trust, transparency and participation in its implementation. One of the partnership patterns adopted by UPT. KPH Region II Pematang Siantar is pine resintapping. The purpose of this study was to analyze the factors that influence the income of pine resin tappers in farmer groups partnering with UPT. KPH Region II Pematang Siantar. The research was conducted in a qualitative descriptive manner using primary and secondary data. A total of 70 members of the forest farmer group (KTH) from the Mantap Cooperative of Sibaganding Village, were taken as respondents in this study. The method used is Analytical Hierarchy Process (AHP) using expert choice software. The data collection method is Focus Group Discussion (FGD). From the results of weighting with AHP, it was obtained that the determinants of the income of the pine resin tappers of the Mantap Cooperative were the price (0.41), followed by the number of trees being tapped (0.26), the weather (0.13), the distance from the tapping location (0.10) and gender (0.08). The value of the level of inconsistency in this study is 0.07.

1. Introduction

Forestry Partnership (KK) is a collaboration between local communities and forest managers, holders of business permits for forest use/forest services, permits for borrowing and using forest areas, or holders of business permits for primary forest products. This partnership or cooperation can be carried out in all types of forest, be it production forest, protected forest, or conservation forest. One of the managers holding this forestry partnership activity is the Technical Implementation Unit (UPT) of Regional Forest Management Unit (KPH) II Pematang Siantar, Forestry Service of North Sumatra Province.

The technical implementing unit of the Pematang Siantar Region II Forest Management Unit, the Forestry Service of North Sumatra Province has several farmer groups that have received permits for pine resintapping. There are 7 (seven) forest farmer groups (KTH) that have made a memorandum of understanding of cooperation (NKK) as partners engaged in the pine resin tapping business, namely the Mantap Cooperative, Gapoktan karya Bersama, KTH Luhur, KTH Wahana, KTH Mandiri, KTH

Mekar and KTH Gorbus Nauli. All of these partners are in Simalungun Regency. It is hoped that the existing forestry partnership will open up opportunities and business networks for tapping pine resin in North Sumatra in order to increase tappers' income.

Initial studies conducted on KPH partners in Pematang Siantar region II showed that pine resin production had a fluctuating trend every month, and not all KTHs were able to produce. The Mantap cooperative is a partner capable of producing 19.20 tons of resin per month. However, the production also shows a fluctuating trend. Several factors are thought to influence the production of resin in these partners. Several factors that are thought to influence people's income from tapping pine resin are the tapping area [1], tapping technique [2], tree age [3], and the use of stimulants [4]. Information related to the factors that affect the income of farmers tapping pine resin in UPT. KPH Region II Pematang Siantar has not been obtained so far. It is necessary to do this in relation to the income of tapper farmers after partnering with UPT. KPH Region II Pematang Siantar. To analyze the factors obtained, the AHP (Analytical Hierarchy Process) method is often used for this purpose.

According to Marsono [5] Analytical Hierarchy Process (AHP) is a method for solving a complex unstructured situation into several components in a hierarchical arrangement, by assigning a subjective value to the relative importance of each variable, and determining which variable has the highest priority in order to affect the outcome of the situation. According to Rahmawaty et al [6], the stages in AHP are, namely the preparation of the problem hierarchy, paired comparison assessment, priority and consistency. According to Mustika [7] there are four working principle of the AHP method is identification of causal factors, hierarchical arrangement, prioritization, consistency and priority weighting. The causal factors can be expanded into criteria in decision making. The arrangement of the hierarchy is to determine which are faced which are the goals of the system as a whole. Meanwhile, according to Tominanto [8] the arrangement of the hierarchy is by determining the goals which are the overall system goals at the top level. The purpose of this study was to analyze the priority level of the factors that affect the income of the pine resin tappers of Mantap Cooperative, Sibaganding Village, Girsang Sipangan Bolon District, Simalungun Regency.

2. Methods

2.1. Research location
This research was carried out at the Mantap Cooperative. The Mantap cooperative is located in Sibaganding Village, Girsang Sipangan Bolon District, Simalungun Regency.

2.2. Data collection
The data used in this study are primary and secondary data. Primary data includes general data on farmer households including name, age, address, main job, side job, gender, last education, marital status and number of family members in one house. While secondary data is in the form of supporting research data which includes the general condition of the research location, both the physical environment, socio-economic community, statistical data on population identity, data on the Mantap Cooperative both active and inactive members in the institution as well as other data related to research conducted sourced from libraries or related institutions.

2.3. Method of taking research objects
To determine the best determinant of income for pine resin tappers, the Analytical Hierarchy Process (AHP) method using expert choice software was used in this study. The data collection method is carried out trough a systematic process of collecting data and information about a particular and specific problem through group discussions is Focus Group Discussions (FGD). The sources used are based on experience, expertise and positions. In this study, the resource person used was Management of the KTH Cooperative Mantap, the Forestry Extension / Companion Extension and the Head of UPT. KPH Region II Pematang Siantar.

IOP Conf. Series: Earth and Environmental Science 1115 (2022) 012060 doi:10.1088/1755-1315/1115/1/012060

2.4. *Data analysis technique*

To analyze the determination of the best factor priority using the Analytical Hierarchy Process (AHP) method, the procedure is as follows: hierarchical arrangement in the AHP hierarchical structure chart, making a pairwise comparison matrix between criteria, determining the priority weights of criteria, measuring logical consistency, Making pairwise comparison matrices and priority weights between alternatives in relation to criteria and measuring their logical consistency and global prioritization.

The criteria used in this study are (a). Weather includes sub-criteria: good, moderate and bad, (b). Gender includes male and female sub-criteria, (c). Prices include high, medium and low sub-criteria, (d). Number of trees that were tapped and (e). The distance from the house to the wiretapping location with sub-criteria is far and near. The alternatives used in this research are tapper conditions, resin production and price stability.

3. Results and Discussion

Based on the results of the data analysis obtained, it is known that the weight of the criteria for weather, sex, price, number of trees that were tapped and the distance from the tapping location resulted in different values and rankings (Table 1). Based on Table 1, it is known that the price criteria are in the first place with a value of 0.41 followed by the number of trees being tapped with a value of 0.26, weather with a value of 0.13, the distance to the tapping location is 0.10 and gender has a weight of 0.08 The value of the level of inconsistency is 0.07. The alternative weighting is presented in Table 2.

Table 1. Criteria weights of AHP output Mantap Cooperative

No	Criteria	Value of Criteria	Ranking
1	Weather	0.13	3
2	Gender	0.08	5
3	Price	0.41	1
4	Number of trees tapped	0.26	2
5	Distance to the wiretapping location	0.10	4
	Inconsistency = 0.07		

Table 2. Alternative weights of AHP output Mantap Cooperative

No	Alternative	Value of Alternative	Ranking
1	Tapping condition	0.34	1
2	Resin production	0.32	2
3	Price stability	0.32	2
	Inconsistency = 0.06		

Based on Table 2, the condition of tappers is at the top rank with a value of 0.34, aspects of resin production and price stability following with the same value of 0.32. The level of inconsistency of this alternative criteria is 0.06. The value of inconsistency in the weighting of criteria and alternatives is < 10%. This indicates that the respondents' answers about comparisons between elements are considered consistent. If the inconsistency value is >10%, the research needs to be evaluated and re-interviewed with informants regarding the comparison of elements [5].

The price criterion is a priority factor affecting the income of forest farmer group of mantap cooperative. At the farmer level, the price of pine resin is around IDR 8500/kg. The selling price of pine resin in the market is in the price range of IDR 16,000 – IDR 20,000.00. The number of trees tapped also affects farmers' income. Therefore, there is a need for socialization about good operational standards of pine resin tapping. Standards related to pine resin tapping have been stated in SOP No.2./jasling/uhhbk/hpl.2/1/2020 issued by the director of environmental services and non-timber forest products.

Weather factors greatly affect the production of pine resin produced. The highest production of pine sap is obtained in the dry season. The results of the research by [9] showed that there were differences in the productivity of pine resin during the summer and the rainy season. During the summer, the productivity of pine resin at BKPH Ponorogo Barat is around 1.10 kg/tree to 1.85 kg/tree. This value range is higher than the productivity during the rainy season, which is only around 0.81 kg/tree to 1.53 kg/tree. In addition, the annual production of pine resin is also influenced by the number of tapping trees in a year, the more the number of tapping trees, the higher the resin production. In general, the annual productivity at BKPH Ponorogo Barat is around 1.91 kg/tree/year to 3.39 kg/tree/year.

Distance, land topography and road conditions to the tapping location also affect farmers' income. In conducting wiretapping, the taps of the Mantap cooperative still carry out conventional activities. As a result, there are still many locations that have not been tapped due to road conditions that have not been cleaned and tidied up, the absence of temporary resin collection points and there is also no work hut to rest.

Gender is known to affect the productivity of tapping. In general, the tappers involved in tapping pine resin in the Mantap cooperative are male. Several previous studies stated that men tend to get more resin than women. In this study, the tapper condition became the alternative with the highest weight. Tapping conditions which include physical health, experience or skills, and education affect the amount of pine resin obtained. The tapping staff in the cooperative KTH has not focused on tapping activities because tapping pine resin is still a side job. As a result, the potential of the existing sap has not been properly exposed.

In addition to gender, the age of farmers is also known to affect the potential for wiretapping. Physical strength at productive age is considered to be better and tougher and has strong energy compared to old age. In addition, farmers who have productive age are easier to accept innovations compared to old farmers. Based on the results of the study, it is known that the Mantap cooperative group is dominated by farmers with an age group between 30 - 39 (42.85%). This age is classified as productive age. The education level of pine resin tappers from the members of the Mantap Cooperative is the majority equivalent to high school graduates (75.71%), followed by junior high school education (20%) and elementary school education totaling 4.28%.

The production of resin is an alternative that has the second highest weight. The production of pine resin with high quantity and quality is very influential on the income of tappers. The pine resin tapping activity at the Mantab cooperative is carried out throughout the year. The average production every month is 362 kg and in a year it is obtained about 4,345 kg. The selling price of pine resin at the tapping location is IDR 8,500 per kg. The amount of money received by members of the Mantap Cooperative from the results of tapping pine sap for a year is IDR 30,600,000 to IDR 51,000,000 with an average of IDR 36,933,714.

Price stability affects the income of farmers tapping pine resin. Over the past five years, the market price of pine resin has been relatively stable at a price range of IDR 16,000 to IDR 21,000 per kilogram. Meanwhile, at the farmer level, the price of pine resin in the past year has been around IDR 8,500. According to Rahmawaty et al [10,11] states that natural resource management often requires alternative management decisions. The decision makinginstrument must be rational, careful and justifiable. AHP is one technique that can be usedin making these decisions.

4. Conclusions

The priority factors that influence the income of farmers tapping the pine resin of the Mantab cooperative are price with a weight value of 0.41, followed by the number of trees tapped with a value of 0.26, weather 0.13, distance from tapping locations 0.26 and gender with a value of 0.08. The alternative weight of the tapper condition is in first place with a value of 0.34, resin production is in second place with a value of 0.32 and price stability is in third place with a value of 0.32 with an inconsistency value of 0.06.

Reference

[1] Ningrum K D A 2006 *Analisis pendapatan penyadap getah Pinus merkusii Jungh et de Vriese dan hubungannya dengan tingkat partisipasi masyarakat dalam kegiatan penyadapan getah di bkph Karangkobar Kph Banyumas Timur* (Bogor : Institut Pertanian Bogor)
[2] Sukadaryati 2014 *Jurnal Penelitian Hasil Hutan* **32** 62
[3] Idris M M and Soenarno 1983 Aspek teknis eksploitasi hutan pinus di Pulau Jawa *Proceeding Simposium Pengusahaan Hutan Pinus* Pusat Penelitian dan Pengembangan Hasil Hutan Bogor
[4] Soenarno, Lempang M and Aksar M 2000 *EBONI* **15** 1
[5] Marsono 2014 *Penerbit In Media* Bogor **912** 012048
[6] Rahmawaty, Abdul R and Idasari S 2013 *Proceedings of the International Symposium on Tropical Forest Ecosystem Science and Management 2013*
[7] Mustika 2017 *Computer Science Journal Collection (CLICK)* **4** 1
[8] Tominanto 2012 *Infokes Journal* **2** 1
[9] Wijayanto A, Nurmadina, Darwitono and Wardhana T W 2019 *Jurnal Silva Tropika* **3** 99
[10] Rahmawaty 2011 *Indian Journal of Ecology* **38** 159
[11] Rahmawaty, Teodoro R V and Myrna G C 2011 (Germany : *Academic Publishing GmbH and Co. KG, Saarbrucken*)

Efficacy of teak and pine litter extract with several solvent against wood termites (*Cryptotermes cynocephalus* light)

S Nuraeni[1], M Rajab[1*], N Azizah[1], A Nurfadilla[1] and K S Madjid[2]

[1]Faculty of Forestry, Hasanuddin University, Makassar, South Sulawesi, Indonesia
[2]Faculty of Mathematics and Sciences, Hasanuddin University, Makassar, South Sulawesi, Indonesia

*E-mail: marwanrajab032@gmail.com

Abstract. Termites are social insects that play an important role in the decomposition and nutrient turnover. In addition, termites are also categorized as pests that eat wood or materials composed of cellulose. Some form of control one of them has been done by making biopesticides. Various types of raw materials have been used from agricultural, plantation and forestry crop groups. Termite mortality is one indicator to determine extract bioactivity by calculating the percentage of the number of termites that died after being given treatment at a certain time. Compound content bioactive substances such as alkaloids, phenolics and other chemicals that can be used to kill and controlling insects in the environment. This study examines the effectiveness teak litter extract and pine litter extract with various solvents against wood termites through experimental research activities in the laboratory. The content of teak litter extract levels and pine litter obtained by 2.5%. Based on the termite mortality value test, the highest toxicity in the ethyl acetate extract, which is at week 3, it has reached 100% at a concentration of 1% which is. In addition, based on testing the retention value of the test paper, The highest percentage of weight loss was in the control, followed by the n-hexane fraction 49.55%. Whereas in testing the level of termite consumption, the highest percentage was in the aquades fraction.

1. Introduction

Termites are social insects that play an important role in the decomposition and rotation of elements nutrient. In addition, termites are also categorized as pests that eat wood (xylophagus) or materials composed of cellulose [1]. Based on the analysis of economic losses [2] attacks termites to 36 public junior high schools in Pekanbaru city of equivalent to USD 36,483,57. Two years Furthermore, termite attacks on housing in the Mijen area of Semarang City have experienced many problems damage to the door frame by 33% with a light damage intensity of 77.2% [3].

To overcome this problem, several forms of control have been carried out, one of which is by making biopesticides. Biopesticide is an insecticide made from from plant parts that have strong toxic compounds for insects [4] effort Control in this way is ideal. This is because the materials for making Insecticides are abundant in nature. In addition, biopesticides are natural materials that are easily biodegradable in nature so that it does not pollute the environment and is relatively safe for humans [5].

Various types of raw materials have been used for the manufacture of termite control biopesticides from agricultural, plantation and forestry crop groups. Such as tobacco, salak, sangitan leaves, tuba leaves and lemongrass [6-9]. Termite mortality is one indicator to determine extract bioactivity by calculating the percentage of the number of termites that died after being given treatment at a certain time. The effectiveness of agricultural ingredients, namely kitchen lemongrass which has concentration of 25% can produce a termite mortality of 97.33%. Different from extract ingredients salak Madura with mortality of 81.3% and Tobacco 77.3%. While on the tuba leaves resulted in the highest termite mortality effect in the treatment of extracts with a concentration of 6% with a mortality percentage of 93.3%. Of the several biopesticide materials that have been tested, they have mortality effectiveness varies. However, if you look at the availability of materials that are still small, for use on a large scale.

Some of the results of these experiments indicate that there are extracts of plant parts that are toxic to pests. Extracts from plants, such as wood, bark, leaves, flowers, fruit or seeds, believed to have the potential to prevent the growth of fungi or resist the presence of destructive insects such as termites [10]. contain bioactive compounds such as Alkaloids, Phenolics and other chemical substances that can be used to kill and control insects in the environment.

Based on data [11], the most widely distributed plants in Indonesia are teak and pine. Both have an area of 1,100,534 ha and 476,126 ha respectively and produce litter with an amount of about 2.81 tons/ha and 6.857 tons/ha. In addition, when viewed from the component chemically, pine litter produces secondary metabolites in the form of terpenoid components (monoterpenes - pinene and -pinene) which are toxic to other plants as well as insects [12] and allelopathic substances as bioherbicides in weeds [13]. Therefore, research on the potential of teak and pine litter was carried out as a biopesticide material to determine the effectiveness of termite control for the existence of wood.

2. Method

2.1. Extract making

Making litter extract begins with cleaning teak litter, wide litter cut into pieces to facilitate the grinding process, then the ground teak powder is ground add 100 g. The teak powder is then filtered through a 40-60 mesh sieve. then macerated with a ratio of 1:10 between teak litter powder and acetone extract was obtained. And the results of the filtrate were thickened using a rotary vacuum evaporator. That matter This applies to both teak litter and pine litter. Then To find out levels of extractive substances obtained from the extraction and fractionation were calculated using formula [14] as follows:

$$\text{Extractive Substance Level} = \frac{\text{Wa}}{\text{Wb}} \text{ x } 100\% \quad (1)$$

Information:
Wa = weight of extractive solids (g)
Wb = dry weight of powder (g)

2.2. Fractionation

The step of the fractionation process to separate the extract is carried out by extracting fractionated using ethyl acetate solvent and water, then shaken and allowed to separation between ethyl acetate solvent and water then the ethyl acetate fraction obtained is then put into a glass and tightly closed and carried out until the solution is clear. Yield residue fractionation with n-hexane then fractionated again with n-hexane solvent and shaken and until the separation occurs, after the separation of the ethyl acetate fraction is separated and stored in a tightly closed glass. Furthermore, this fractionation is also carried out until the solution is clear Then each fraction was evaporated to obtain an extract from each solvent that used.

2.3. Concentration making

The process of making the concentration of pine litter extract and teak litter based on a comparison between the weight of the extract and the solvent with a concentration of 0.5%, 1% (v/v). The solvent used is aquades, N-Hexane, acetone and ethyl acetate so that 9 treatments are obtained, then the solution allowed to become clear which is assumed that all compounds have been extracted Then furthermore, a control is also prepared, namely a control test that is treated with a solution commercial termiticide with active ingredient Deltamethrin 25 g.

2.4. *Testing and preparation of cellulose test paper*

The test method refers to the cellulose pads method that has been carried out [15]. Stages for Testing cellulose test paper, namely preparing test paper, the paper used is Whatman no. 41 with a paper diameter of 125 mm then the paper is dipped until all parts submerged in the extract and then aerated to remove the solvent that is then put in a desiccator for ± 15 minutes after drying, the test paper is weighed to knowing the initial weight then the number of test samples used in the study was 9 fruit is a combination of 2 concentrations (0.5 and 1%), and termiticide as a control. As for The calculation of the percentage of weight loss of the test paper was carried out after 28 days of feeding, by using the following equation:

$$\text{Weight Loss} = \frac{B0-B1}{B0} \times 100\% \quad (2)$$

Information :
B0 = dry weight of test paper desiccator before feeding (g)
B1 = dry weight of test paper desiccator after feeding (g)

2.5. *Test Paper Feeding against Termites*

The steps for feeding the test paper against termites, namely the preparation of termite colonies *C. cynocephalus* which will be used as many as 450 healthy and active worker termites later put into each plastic container (50 termites per treatment. Next, do observation for 28 days or 4 weeks to determine the mortality of each treatment. Then the extractive toxicity test on termite survival was carried out for feeding. Furthermore, the calculation of the percentage of termite mortality using the formula [14] as follows :

$$\text{Termite Mortality} = \frac{M1}{M2} \times 100\% \quad (3)$$

Information :
M1 = Number of termites that died in the i-th test sample (tails)
M2 = Number of termites at the beginning of the feed (50 heads)

Meanwhile, the rate of termite consumption was calculated based on research [16]. Consumption of termite food the average is also calculated by the formula:

$$\text{Termite Feeding Consumption} = \frac{B0-B1}{N} \quad (4)$$

Information :
B0 = Weight of test sample before feeding (g)
B1 = Weight of the test sample after being fed (g)
N = Number of initial worker termites.

2.6. *Data analysis*

Data analysis was carried out to calculate the difference in the effect of solvent treatment and concentration by calculating the number of termite mortality every day for 28 days after application and the results obtained were compared with treatments using existing termiticides on the market (the active ingredient is Deltamethrin).

3. Results and Discussion

3.1. *Extraction of teak litter and pine litter*

The extract content obtained from 200 g of teak powder and pine litter powder produces a paste as much as 10 g. So that the levels of extractive substances obtained from extraction with ethanol solvent yielded 2.5% ethanol extract.

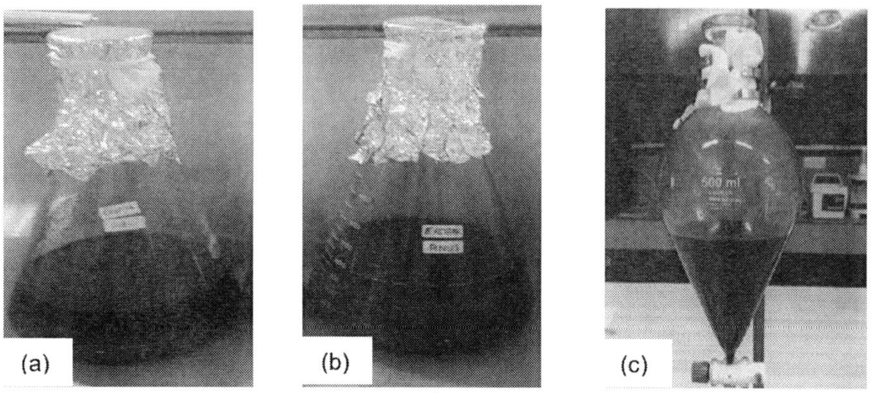

Figure 1. Extraction results of teak litter (a) pine litter (b) and (c) pasta

The levels of extractive substances in teak litter and pine litter are in the medium category, where the levels of wood extractives are classified as high class if the extractive content is greater than 4%, medium class if the extractive conten between 2-4% and low grade if the extractive content is less than 2% [17]. This extractive substance plays a role in the natural durability of mahogany wood against biological attack in the form of compounds polyphenols, terpenoids and tannins [18]. Meanwhile, the size of the extract yield value shows the effectiveness of the extraction process [19].

3.2. Fractionation
The acetone extract produced in this study is brown, the soluble extractive substance in n-hexane solvent it is clear yellow, distilled water is clear and brown black in acetyl acetate. This is thought to depend on the type of each solvent on the type of dissolved compounds. Differences in the color of the fractionated results obtained with different solvents [20].

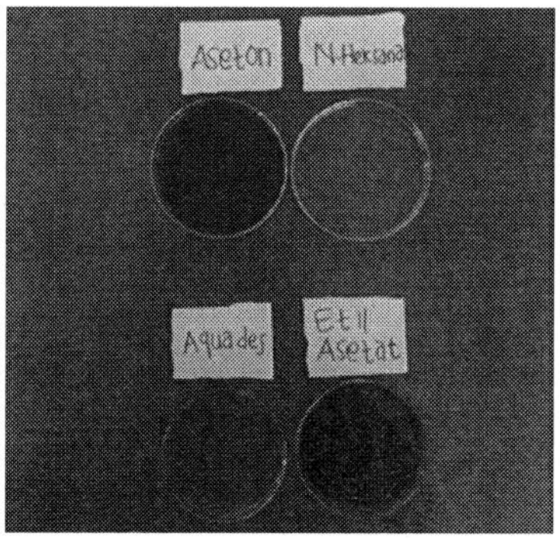

Figure 2. Differences in color from fractionation with different solvents

3.3. Test paper retention
The indicator used to see the level of substance retention by weighing the test paper before immersion and after immersion. The greater the value of the extractive substances absorbed in the the test paper shows that the test paper has a high content of extract substances. Percentage value retention of test paper based on research results.

Table 1. Retention of cellulose test paper containing teak and pine litter extract after bait for the wood termite *C. cynocephalus*.

Type of Extractive Substance	Concentration (%)	Active Ingredient Absorbed (%)
Acetone	0,5	45.01
	1	44.82
N-hexane	0,5	46.80
	1	49.55
Aquadest	0,5	47.67
	1	49.50
Ethyl Acetate	0,5	44.36
	1	45.02
Termiticide	-	38.78

The lowest retention in the test sample was 38.78% termiticide, while the highest was N- hexane 49.55% (Table 1). The retention of the test paper ranged from 49.55% - 38.78% This proves that the higher the concentration of extractive substances, the more extracts absorbed in the extract cellulose test paper and this means more toxins are added [21]. So experience decrease in the percentage of absorption of the test paper. These conditions allow termites to refuse to eat it so that it will inhibit the activity of eating termites.

3.4. Termite consumption rate

Another indicator that can be used to see the anti-termite properties of teak litter extractive substances and pine litter is the rate of termite consumption of the paper test sample. consumption rate, indicated by the amount of termite consumption after being fed. This can be seen in Table 2. teak and pine litter extract after feeding against the wood termite *C. cynocephalus*.

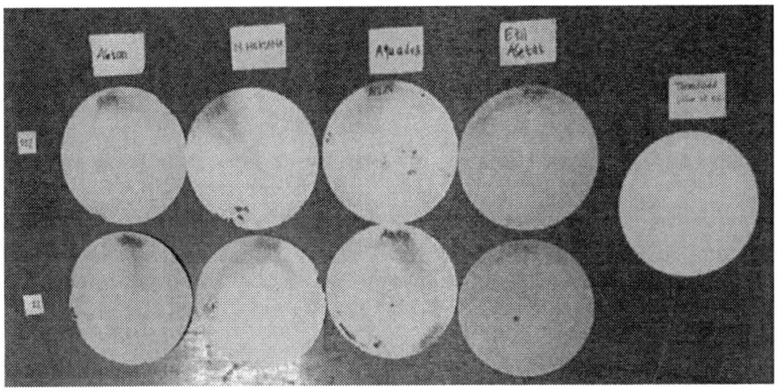

Figure 3. Termite consumption for 28 days after application

The lowest weight loss in the test sample was 1.9 mg, namely in the test sample that was given 1% Aquades extractive substance. The rate of weight loss can be affected by concentration as well as absorption and retention of extracts [19].

Table 2. Consumption rate of *C. cynocephalus* wood termites. on cellulose test paper containing teak and pine litter extract

Type of Extractive	Concentration (%)	Termite Eating Consumption (mg)
Acetone	0.5	1.7
	1	1.7
N-hexane	0.5	1.8
	1	1.8
Aquades	0.5	1.8
	1	1.9
Ethyl Acetate	0.5	1.7
	1	1.7
Termiticide	-	1.1

3.5. Drywood termite mortality (C. cynocephalus).
The indicators used in determining the effectiveness of the anti-termite extractive substances are termite mortality was calculated up to 28 Days after application. The percentage value of termite mortality is presented in Figure 4.

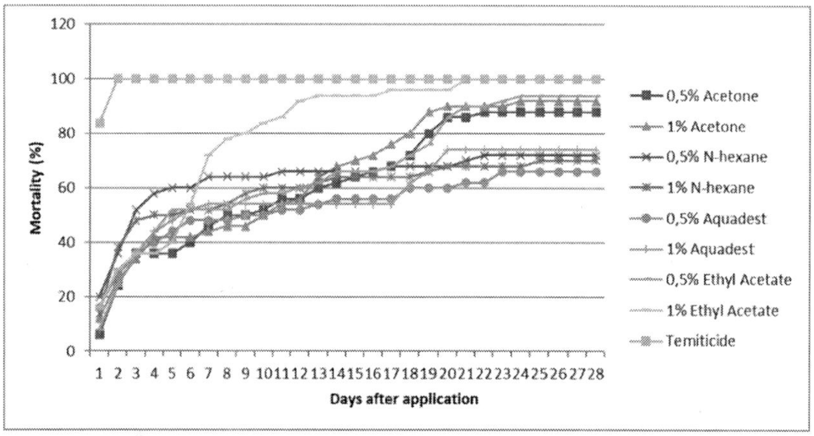

Figure 4. Termite mortality after application of teak and pine litter extracts in several solvents

In Figure 4 it can be seen that the highest mortality was ethyl acetate at a concentration of 1%. Termite mortality has occurred 94% on day 13 and at week 3 (day 21 after application) has reached 100%. Meanwhile, the positive control, namely termiticide as a comparison of extractive substances, has also been tested experienced 100% mortality in the first week. The mortality rate and weight loss of the test sample can be interpreted if the higher the level of termite mortality, the lower the weight loss that occurs. This can happen because The four extractive substances contain various compounds that are antifeedant, such as: nicotine, acetoginin, flavonoids, saponins [23] and others that can cause termite death and prevent damage to the test sample. The difference in weight loss of the test sample is thought to be due to each type of extractive substances have different compounds so that they can reduce the high mortality rate different too.

4. Conclusion

The extract content of teak litter and pine litter obtained was 2.5%. The highest toxicity occurred on day 13 using 1% ethyl acetate solvent. i.e. termite mortality was 94% and reached 100% on the 21st day after application. The value of the retention of test paper, the highest percentage of weight loss was followed by the n-hexane fraction 49.55%.

References

[1] Titarsole J, Maail R S and Franz J J 2019 *Jhppk* **3** 2621
[2] Endasura M, Hakim L and Afifuddin Y 2014 *Economic loss analysis and mapping The distribution of termite attacks on public junior high schools in Pekanbaru City* (Medan: Faculty Agriculture Universitas Sumatera Utara)
[3] Savitri A, Martini and Yuliawati S 2016 *Journal Public health* **4** 100
[4] Utari N, Diba F and Sisillia L 2018 *Journal Tengkawang* **8** 75
[5] Kardinan A 2011 *Agricultural Innovation Development* **4** 262
[6] Sholehah D N 2011 *Agrovigor* **4**
[7] Zulyusri, Desyanti and Usmal M 2013 *Semirata Proceedings of the Faculty of Mathematics and Natural Sciences* 521
[8] Hidayatullah S, Rizaldy A A, Gracia H and Syahidah 2017 *Journal of Tropical Wood Technology* **15** 167
[9] Sufyan, Jayuska and Destiarti L 2018 *Chemistry Journal Equator* **7** 47
[10] Arif A M, Natsir U and Fatmawaty S 2012 *Journal Parrenial* **3** 15
[11] Perum PERHUTANI 2012 *Statistics of Perum PERHUTANI 2006-2010* (Jakarta: Directors of Perum PERHUTANI)
[12] Cahyanti L D, Sumarni T and Widaryanto E 2013 *Journal of Environmental Science* **7** 48
[13] Bilkis F G, Triastinurmiatinigsih and Ismanto 2018 *Journal of Forestry Science* **15** 26
[14] Syafii W, Sofyan K and Febrianto F 1994 *Research report; possible use Indonesian tropical wood extractive substances as preservatives* (Bogor: Bogor (ID) Inter Center University, Bogor Agricultural Institute)
[15] Yanti H, Syafii W and Darma I G K T 2012 *Journal Tengkawang* **2** 82
[16] Rudi and Nandika 1999 *Feeding Consumption and Survival of Subterranean Termites (Coptotermes curvignathus) Holmgren (Isoptera: Rhinotermitidae) in laboratory testing IPB Bogor*
[17] Lestari S B and Pari G 1990 *Research Journal Forest Products* **7** 96
[18] Oemry S and Lubis L 2015 *Agroecotechnology* **3**
[19] Fatmawati F 2022 *Efficacy of Leaf Extract and Shell of Pangi Seed (Pangium edule Reinw.) against wood rot fungus Schizophyllum commune* (Doctoral dissertation, University of Hasanuddin)
[20] Marcelinda A, Ridhay A and Prismawiryanti P 2016 *Natural Sciences: Journal of Science and Technology* **5**
[21] Rahmawati F 2015 *Optimization of the use of thin layer chromatography (TLC) in separation Pulai leaf alkaloid compounds* (*Alstonia scholaris LR Br*) (Doctoral dissertation, Universitas Islamic State of Maulana Malik Ibrahim)
[22] Mariana E and Ariyanti E 2013 *War Jungle* **1**
[23] Luth F 2020 *Scientific Journal of Agriculture* **8** 8

Antioxidants and chemical compounds of hot water extract agarwood leaves *Wikstroemia tenuiramis* Miq

R Batubara[1,2*], S M Lubis[1], K Aulia[1], R H Harahap[1], M R Mahendra[1], M S Nasution[1] and O Affandi[1]

[1]Faculty of Forestry, Universitas Sumatera Utara, Medan 20155, North Sumatra, Indonesia
[2]Center of Excellence for Mangrove, Universitas Sumatera Utara, Medan 20155, Indonesia

[*]Email: ridwanti@usu.ac.id

Abstract. One type of agarwood from the *Wikstroemia* genus is *Wikstroemia tenuiramis* Miq. The leaves have been used as raw materials for agarwood leaf tea. The aim of the research was to determine the antioxidant activity of *W. tenuiramis* Miq. agarwood leaf tea in ready-to-drink formulations and chemical compounds contained in hot water extract of agarwood leaf type *W. tenuiramis* Miq. The method of making tea by boiling for 7 minutes. Analysis of antioxidant activity test using the DPPH method. Gas Chromatography - Mass Spectrometry (GC-MS) instruments was used to analyze chemical compounds in hot water extracts. The antioxidant test results that hot water extracts of agarwood leaves had strong antioxidant activity. The results of the GC-MS analysis contained 33 compounds, with the main compounds being Hexadecanoic acid and methyl ester.

1. Introduction

The findings of study on the chemical constituents of agarwood leaves, such as flavonoid chemicals, tannins, steroids and triterpenoids, justify the usage of agarwood leaves as brewed tea by the community [1,2]. Phytochemical screening of ethanol, methanol and agarwood leaf water extracts in several species showed the presence of flavonoids, tannins and saponins [3,4].

Agarwood leaf infusion and kombucha contain phenolic chemicals, flavonoids and tannins [5]. Sugar, color, tannin, gum and starch are examples compound inof extractive compounds that dissolve in water [6]. The existence of tannins in the extraction results contributed to the antioxidant activity of agarwood leaves because tannins are a class of phenolic compounds.

The results of the phytochemical test of the aqueous fraction of the thick extract of agarwood leaves contain flavonoid, phenolic and triterpenoid compounds and their antioxidant activity is 76,95 μg/ml [7]. The aqueous agarwood leaf extract has antioxidant activity [8]. Agarwood leaf *A. malaccensis* hot water extract contains flavonoids, triterpenoids, tannins and glycosides with potent antioxidant activity [9].

Based on various literature and the development of its application in society, agarwood leaves are a contemporary drink that has been sold in the form of tea bags or ready-to-drink tea. One of the types developed is ready-to-drink tea from the leaves of agarwood *W. tenuiramis* Miq. Ready-to-drink tea is extracted with hot water by boiling it until it boils. There has been no research report on the antioxidant activity and chemical compounds contained in this agarwood leaf hot water extract. Based on this, it is necessary to conduct research on antioxidant activity and the content of chemical compounds in hot water extract of agarwood leaves *W. tenuiramis* Miq.

2. Research Method

2.1. Research time and place
This research was conducted from February to June 2022. The material of agarwood leaves is *W. tenuiramis* Miq. taken from Siantona Village, Mandailing Natal Regency. The antioxidant test was carried out at the Phytochemical Laboratory, Faculty of Pharmacy, Universitas Sumatera Utara, while the extraction was carried out at USU's Links and the GC-MS test was carried out by the Medan Customs Laboratory.

2.2. Simplicia making
Agarwood leaves are washed with water until clean, then drained and then dried in a drying cabinet at a temperature of around 40°C. After drying, the simplicia is powdered in a blender, then stored in a clean and tightly closed container.

2.3. Agarwood leaf extraction with hot aquades solvent
Extracts were made by boiling agarwood leaf powder in hot water, with the ratio of hot water and powder being 4.5 g of agarwood leaf powder and 250 ml of hot water. The stew is filtered and added with sugar as much as 15 g then packed in ready-to-drink packages in 250 ml bottles.

2.4. Antioxidant activity analysis with DPPH
The test procedure refers to [10,11] methods. The comparison is vitamin C. Antioxidant test using the 1,1-diphenyl-2-picrylhydrazyl (DPPH) method for further calculation of IC50 (*Inhibition Concentration 50*) as an indicator of antioxidant analysis.

2.5. Identification of active compound extract with GC-MS
Hot water extract of agarwood leaves was used to identify chemicals using Gas Chromatography-Mass Spectrometry (GC-MS). 0.5 ml of the extract was put into the GC-MS apparatus. The device is a GC-MS 7890B.

3. Result and Discussion

3.1. Antioxidant of hot water extract of agarwood leaf type W. tenuiramis Miq.
The results of antioxidant testing of hot water extract of agarwood leaf *W. tenuiramis* Miq. with the comparison of ascorbic acid (vitamin C) can be seen in Table 1. Hot water extract of agarwood leaves *W. tenuiramis* Miq. which are formulated to be ready to drink with the addition of sugar and water are in a strong category.

The results of previous studies that the antioxidant of agarwood leaf type *W. tenuiramis* and cultivated agarwood leaf type *A. mallacensis* had IC50 values of 30,482 and 41,130 μg/ml, in this study, ethanol extract was tested [12]. The antioxidant activity of agarwood leaf, fresh leaves and rotting leaves of type *W. tenuiramis* Miq was 27,78 and 28,80; 25,35 and 28,59 g/m, respectively, in ethanol solvent and hot water [1], in this study, the viscous extract, both ethanol extract and hot water extract, were tested.

Table 1. Antioxidant activity of hot water extract of agarwood leaves *W. tenuiramis* Miq. and ascorbic acid

Material	Antioxidant Activity (µg/ml)
Agarwood Leaf Hot Water Extract	65.090 ± 0.010
Ascorbic acid	2.600 ± 0.020

3.2. Identification of compounds by Gas Chromatography Mass Spectrometry (GC-MS)
Based on the spectrogram data acquired and the fragmentation patterns of each chemical as shown in Figure 1, the compounds present in the hot water extract of agarwood leaves were identified. Up to 33 different compounds were found in the hot water extract of *W. tenuiramis* Miqagarwood .'s leaves. Table 2 lists thc names of the compounds. The major ingredient is methyl ester of hexadecanoic acid. In addition, 9-Octadecenoic acid (Z)-, methyl ester, is the second major substance. A class of acid compounds includes 9-Octadecenoic acid (Z)-, methyl esters and hexadecanoic acid, methyl esters.

The results of the Py GC-MS test revealed that a family of acidic compounds, specifically Octadecanoic acid (CAS) Stearic acid, was the predominant compound in both the old and young leaves of *W. tenuiramis* Miq [12]. The major ingredient with the highest concentration in agarwood leaves from Laru Village, type *A. malaccensis* Lamk, is octadecanoic acid (CAS) stearic acid [13]. The primary substances present in the ethanol extract of agarwood leaves of the *A. malaccensis* Lamk species are neophytadiene and hexadecanoic acid, methyl ester compounds [14].

The results of the GC-MS test also support the results of the antioxidant test, where the dominant compound is a class of acid compounds, namely Hexadecanoic acid compounds, methyl esters. Similar results were also reported by [13] which notes that Octadecanoic acid Stearic acid, the main ingredient with the highest concentration in agarwood leaves from Laru Village, had a significant part in the leaves' potent antioxidant activity.

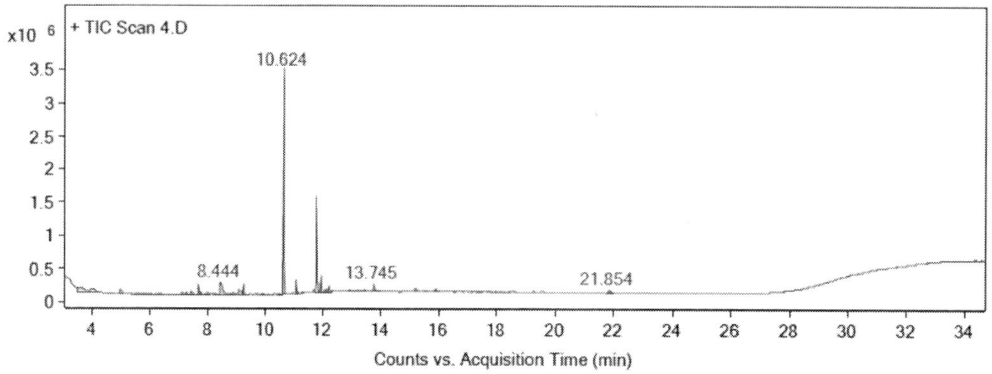

Figure 1. Results from the GC-MS chromatogram *W. Tenuiramis* Miq hot water extract of agarwood leaf

Tabel 2. Compounds in agarwood leaves extracted in hot water

No	Retention Time (Rt)	Compound Name	Formula
1.	3.604	2,5,5-Trimethyl-[2-13C]-1-pyrroline-Noxide	$C_7H_{13}NO$
2.	4.011	Nonane	C_9H_20
3.	4.602	Cyclododecene, 1- [(phenylsulfinyl)methyl]-	$C_{19}H_{28}OS$
4.	4.990	1,2-Benzenediol	$C_6H_6O_2$
5.	5.636	1,3-Benzenediol	$C_6H_6O_2$
6.	5.876	Methyl 3-hydroxy-2-(ptoluenesulfonyloxy)methyl-10-tetrahydropyranyloxydecanoate	$C_{24}H_{38}O_8S$
7.	6.338	Decanoic acid	$C_{10}H_{20}O_2$
8.	7.114	Ethanone, 1-(3-hydroxyphenyl)-	$C_8H_8O_2$
9.	7.409	.beta.-D-Glucopyranose, 1,6-anhydro-	$C_6H_{10}O_5$
10.	7.668	Dodecanoic acid, methyl ester	$C_{13}H_{26}O_2$
11.	7.723	Heptanoic acid, 3-oxo-, methyl ester	$C_8H_{14}O_3$
12.	7.982	Dodecanoic acid	$C_{12}H_{24}O_2$
13.	8.444	1,3,4,5- Tetrahydroxycyclohexenecarbo xylic Acid	$C_7H_{12}O_6$
14.	8.850	Octhyl Decanoate	$C_{18}H_{36}O_2$
15.	9.072	Myo-Inositol, 4-C-methyl-	$C_7H_{14}O_6$
16.	9.220	Myristic acid, methyl ester	$C_{15}H_{30}O_2$
17.	9.848	Cyclohexanone, 2,2-dimethyl-	$C_8H_{14}O$
18.	10.476	9-Octadecenoic acid (Z)-, methyl ester	$C_{19}H_{36}O_2$
19.	10.624	Hexadecanoic acid, methyl ester	$C_{17}H_{34}O_2$
20.	11.067	Hexadecanoic acid, ethyl ester	$C_{18}H_{36}O_2$
21.	11.270	1-Methythio-7-methyl-4-oxo8,10,12,13-tetraoxapentacyclo[5.5.1.0(2,6).0(3,11).0(5,9)]tridecane	$C_{11}H_{12}O_5S$
22.	11.769	9-Octadecenoic acid (Z)-, methyl ester	$C_{19}H_{36}O_2$
23.	11.917	Methyl stearate	$C_{19}H_{38}O_2$
24.	12.120	9,12-Octadecadienoic acid (Z,Z)-, methyl ester	$C_{19}H_{34}O_2$
25.	12.212	Ethyl (9E)-9-Octadecanoate	$C_{20}H_{38}O_2$
26.	12.933	9-Octadecenoic acid, methyl ester, (E)-	$C_{19}H_{36}O_2$
27.	13.191	2-Methylene-1,5-pentanediol	$C_6H_{12}O_2$
28.	13.450	Hexadecanoic acid, 10-hydroxy-, methyl ester	$C_{17}H_{34}O_3$
29.	13.745	heptadecanoic acid, 2-methoxyethyl ester	$C_{20}H_{40}O_3$
30.	15.186	1-hexacosanol	$C_{26}H_{54}O$
31.	15.500	4-Hydroxy-2,2,6,6-1etramethyl-1-(1- phenylsulfanyl)hex-5-enoyloxy)piperidine	$C_{21}H_{33}NO_2$ S
32.	15.870	8-Octadecenoic acid, methyl ester, (E)-	$C_{19}H_{36}O_2$
33.	21.854	6,10,14-Hexadecatrienoic acid, 3,7,11,15-tetramethyl-, methyl ester, [R-(E,E)]-	$C_{21}H_{36}O_2$

Furthermore the hot water extract of agarwood leaves included 33 different chemicals, more than 15 compounds are of the acid group. Also found compound 1,2-Benzenediol, with the chemical formula $C_6H_4(OH)_2$ commonly called pyrocatechol; o-benzenediol; catechol; 2-hydroxyphenyl is commonly used as a flavoring agent. The flavor is a combination of characteristic ingredients in the form of taste and aroma sensations [15], from the hedonic test of *W. tenuiramis* Miq agarwood leaf tea highly favored by consumers of the taste and aroma. Another compound found was 1,3-Benzenediol, or resorcinol, which is a polyphenolic compound. The compound 2,5,5-Trimethyl-[2-13C]-1-pyrroline-Noxide is a pyrroline compound that contributes to flavor.

4. Conclusion

Strong antioxidant activity can be seen in agarwood leaf hot water extract used in ready-to-drink tea formulations. Antioxidant power is affected by chemical constituents in agarwood leaf hot water extract found in ready-to-drink tea formulations.

Acknowledgment

This research was funded by Pre-Start Up Innovation 2021.

References

[1] Surjanto, Batubara R, Hanum T I and Julianti E 2019a *IOP Conf Ser Earth Environ Sci* **305** 012061
[2] Surjanto, Batubara R, Hanum T I and Pulungan W 2019 *IOP Conf Ser Environ Sci* **260** 012101
[3] Kamonwannasit S, Nantapong N, Kumkrai P, Luecha P, Kupittayanant S and Chudapongse N 2013 *Microbiol Antimicrob* **12** 1
[4] Khalil A S, Rahim A A, Taha K K and Abdallah K B 2013 *J Appl Ind Sci* 2013 **1** 78
[5] Nurmiati and Wijayanti E D 2018 *Journal Cis-Trans (JC-T)* **2** 6
[6] Sjostrom E 1998 *Kimia Kayu Edisi 2* (Yogyakarta: Gadjah Mada University Press) Yogyakarta
[7] Hasanah M, Apriyanti D and Patmayuni D 2020 *Jurnal Penelitian Sains* **22** 25
[8] Wil N N A N, Omar N A M, Brahim N A and Tajuddin S N 2014 *Journal of Chemical and Pharmaceutical Research* **6** 688
[9] Batubara R, Wirjosentono B, Siregar B A, Harahap U and Tamrin 2021 *IOP Conference Series: Earth and Environmental Science* **830**
[10] Prakash A 2001 *Analytical Progress* **19** 1
[11] Molyneux P 2004 *Songklanakarin J Sci Technol* **26** 212
[12] Batubara R, Hanum T I, Afandi O and Wahyuni H S 2020 *Biodiversitas* **21** 4616
[13] Batubara R, Surjanto, Hanum T I, Handika A and Afandi O 2020 *Biodiversitas* **21** 1588
[14] Batubara R, Wirjosentono B, Siregar A H, Harahap U and Tamrin 2021 *Biodiversitas* **22** 2884
[15] BPOM RI 2016 Peraturan Kepala BPOM RI No 22 Tahun 2016 *tentang Persyaratan Penggunaan Bahan Tambahan Pangan Perisa*

Mini review: Extracellular enzymes and proteins produced by nematophagous fungi

L D S Hastuti[*], K Berliani, M B Mulya, A Hartanto and S Pahlevi

Department of Biology, Faculty of Mathematics and Natural Sciences, Universitas Sumatera Utara, Medan, North Sumatra 20155, Indonesia

[*]Email: liana.hastuti@usu.ac.id

Abstract. While the advanced use of nematophagous and nematode-trapping fungi has been promoted and proven to be effective in the field to control root-know nematodes, the prospect of enzymes produced by these group have remained unclear and not extensively summarized. The aim of this review was to describe the current progress and possibly suggestion for further investigation regarding the extracellular enzymes and proteins produced by the fungi. We identified some studies published between 1985 and 2022 from the Scopus database by using a combination of keywords or Boolean operators: "(Nematophagous OR Nematode-trapping) AND (Fungi OR Fungus) AND Enzyme" in the "Article title, Abstract, Keywords" section. The procedure yielded a total of 77 research articles regarding the context with its peak of publication in 2007 with 6 articles. Members of Orbiliaceae or nematode-trapping fungi were predominantly mentioned in the articles among other fungal taxa such as *Arthrobotrys oligospora, Duddingtonia flagrans,* and *Monacrosporium thaumasium*. Proteases, serine proteases, and chitinases were the most investigated enzymes produced by nematophagous fungi, with a topic emphasizing on their utilization and role in biological control against nematodes.

1. Introduction

Nematophagous fungi is a group of microfungi which predate on vermiform living nematodes by using specialized structures or hyphal modifications and through secretion of bioactive metabolites (bio-nematicides), extracellular enzymes and proteins [1–3]. Earlier studies on nematophagous fungi were still focused on the predation mechanism, the emergence of trapping structures, physiological adaptation, and the identification or taxonomy of nematophagous fungi, particularly Orbiliaceae [4]. With the advancement of molecular biology, there have been more publications on the genetic engineering of genes that contribute to the production of metabolites and enzymes that can be transferred to other hosts in order to improve their nematicidal activity compared to isolated samples or consortia application against animal- or plant-parasitic nematodes [5].

A review by Anders et al [6] summarized the adhesion mechanisms of nematophagous fungi on nematodes in 1992. Following that publication, another review focusing on three nematophagous fungi namely *Arthrobotrys oligospora, Paecilomyces lilacinus*, and *Pochonia chlamydosporia,* promoted particular fungal taxa to be applied as potential biological control agents [7]. When nematophagous fungi were introduced in the agricultural fields, the prospect of fungal chitinases were highlighted as key components in promoting plant defense mechanism against root-knot nematodes [8]. Meanwhile, there have been a number of reviews that have concentrated on the molecular processes that result from the interaction between nematophagous fungi, pathogens, and plants [9–11].

Fungi are notably, the most superior agents in producing organic acids and extracellular enzymes, however reports on this topic, especially from nematophagous fungi have not been widely summarized. In this review, we aim to provide a contribution to the subject by conducting a bibliographic analysis of the global scientific literature, taking into consideration the advancement of research verifying the extracellular enzyme and protein produced by nematophagous fungi.

2. Methodology

The first step to conduct this study was to download some articles from a database. We utilized a document search engine in Scopus (https://www.scopus.com/search/form.uri?display=basic#basic) using the following search terms and respective Boolean operators: "(Nematophagous OR Nematode-trapping) AND (Fungi OR Fungus) AND Enzyme" in the "Article title, Abstract, Keywords" section (accessed on August 1st, 2022). This step allowed a selection on the number of published articles and types of studied enzymes and proteins. There was no limitation on the date of publication. For every article recovered from the database, the following information was generated: (1) Trend of published articles per year, (2) Percentage of Orbiliaceae and non-Orbiliaceae species mentioned in the number of articles, (3) Rank of fungal taxa that appeared most in the articles, and (4) Type of extracellular enzyme and protein produced by the fungi.

3. Results and Discussion

By using the standard search procedure, a total of 161 articles were found in Scopus database. The collection was divided into research articles ($N = 147$), review articles ($N = 8$), book chapters ($N = 2$), short survey ($N = 2$), a conference paper, and a letter. An early research article was published in 1985 with the context of observing the initial stage of predation and development of trapping device by *Arthrobotrys oligospora* against nematodes [12]. The analysis was then conducted only on 77 research articles which reported the diversity of extracellular proteins, enzymes, and identity of nematophagous fungal species while excluding some errors such as duplicate articles and articles out of context. The peak of published articles was observed in 2007 ($N = 6$). The average number of published articles is 2 articles ($\bar{x} = 2.02$) for each year. In this case, there is a fluctuative trend of publication from 1985 to 2022 with fewer publications published in the last ten years as can be seen in Figure 1.

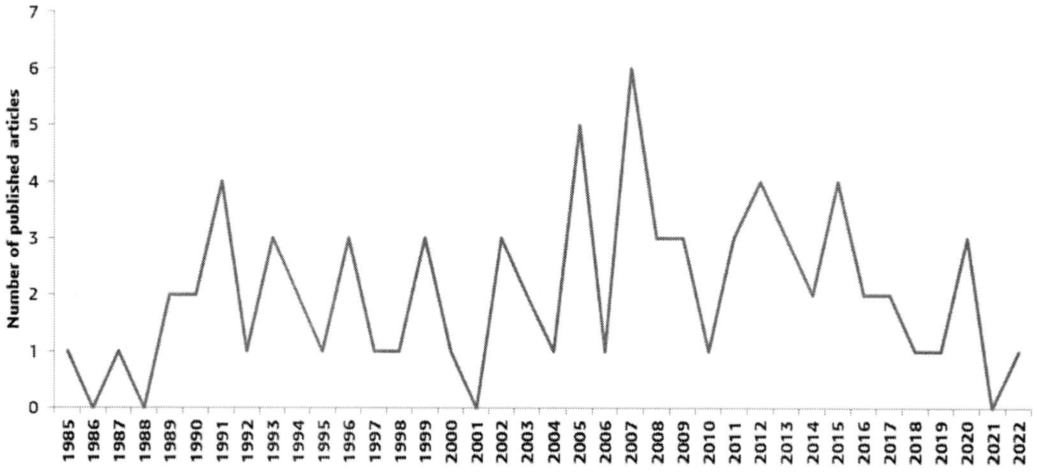

Figure 1. Temporal distribution of published articles ($N = 147$) in the context of extracellular enzymes and proteins produced by nematophagous fungi

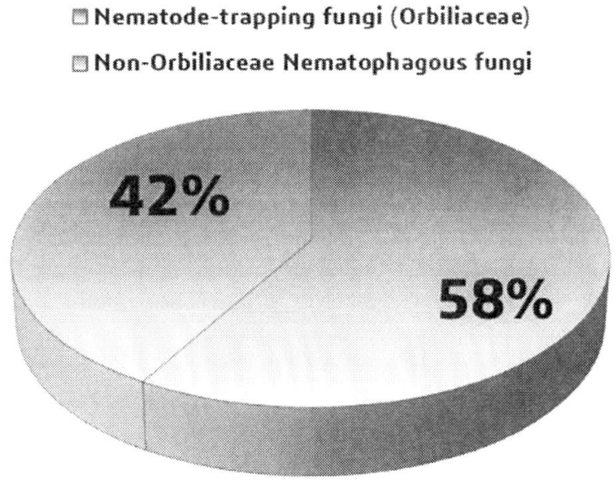

Figure 2. Percentage of Orbiliaceae and non-Orbiliaceae fungi in the collection of articles (N = 77) mentioning their extracellular enzymes and proteins

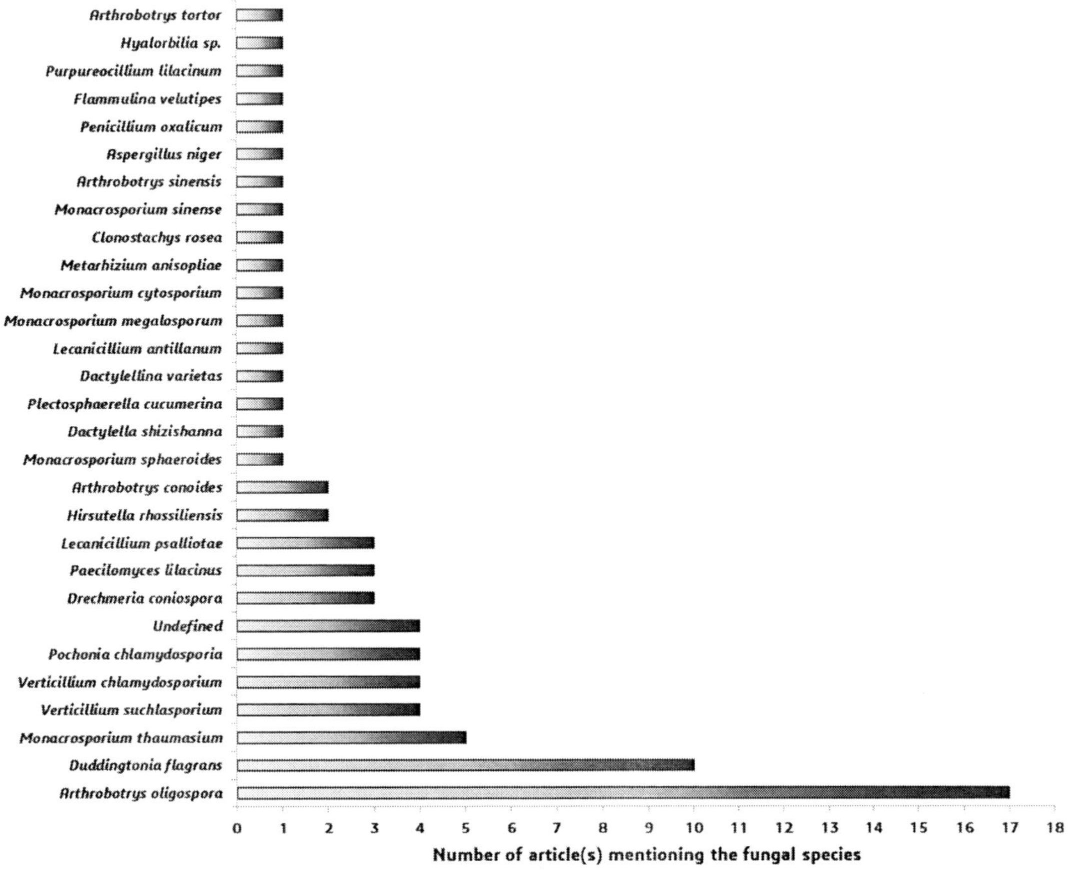

Figure 3. Distribution of article(s) mentioning fungal taxa (N = 77) in the context of extracellular enzymes and proteins

The percentage of articles that discussed the study of enzymes and proteins from non-Orbiliaceae nematophagous fungi is almost equal to the number of articles mentioning Orbiliaceae fungi or nematode-trapping fungi (Figure 2). A total of 29 fungal taxa was documented as nematophagous fungi (Figure 3). The members of Orbiliaceae or nematode-trapping fungi were predominant in this review such as *Arthrobotrys conoides, A. oligospora, A. sinensis, A. tortor, Dactylella shizishanna, Dactylellina varietas, Duddingtonia flagrans, Hyalorbilia* sp, *Monacrosporium cytosporum, M. megalosporum, M. sinense, M. sphaeroides,* and *M. thaumasium*. Four articles were categorized as "Undefined" because they contained information on various strains but did not provide the species identity. Based on the result, *Arthrobotrys oligospora* was mentioned the most in publication ($N = 17$), followed by *D. flagrans* ($N = 10$), *A. thaumasium* ($N = 5$), non-Orbiliaceae fungi such as *Verticillium suchlasporium* ($N = 4$), *V. chlamydosporium* ($N = 4$), *Pochonia chlamydosporia* ($N = 4$), etc.

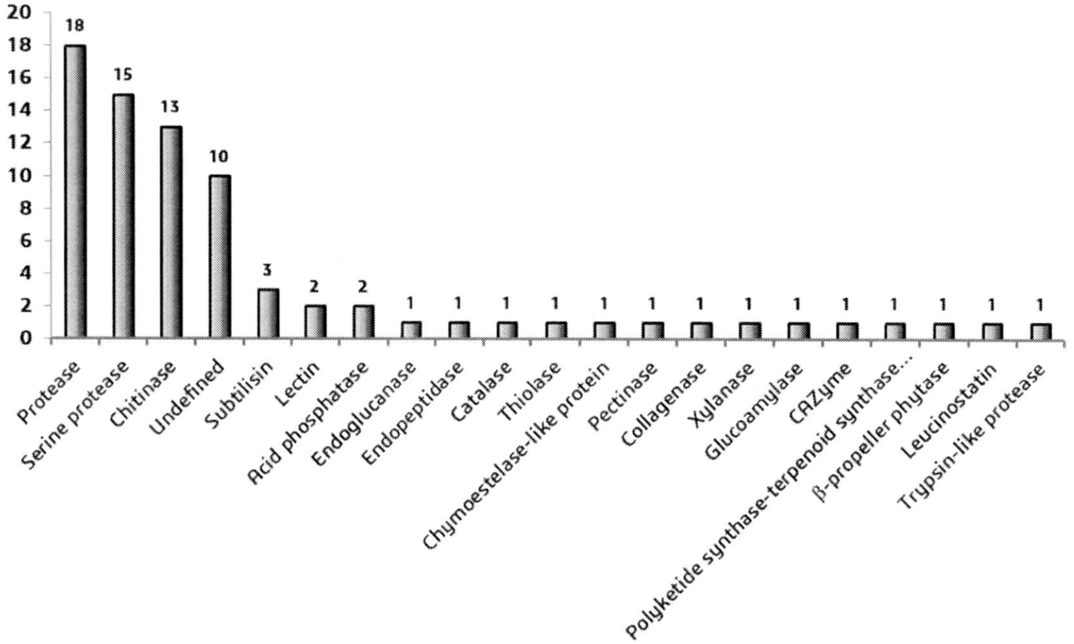

Figure 4. Type of extracellular enzyme and protein studied from nematophagous fungi that was mentioned in the collected articles (N=77)

According to the categories of enzymes and proteins reported, proteases are the most often reported enzymes, followed by serine proteases, chitinases, etc. The term "Undefined" articles were contextualized without specifying the type of enzymes/ proteins or simply assaying crude enzymes without any precise determination. Proteases and serine proteases are hydrolytic enzymes that disintegrate both the nematode defensive component and the intracellular compartment of the eggs and larvae [13]. Chitinases are hydrolytic enzymes that are involved in the crucial digestion of nematode cell walls, particularly in egg parasitism by nematophagous fungi [14]. *Verticillium suchlasporium* (non-Orbiliaceae) released proteases and endopeptidases that was identified during its penetration into nematode eggs and was capable of degrading some cyst nematode proteins [15]. *Arthrobotrys oligospora* (Orbiliaceae) also produced a number of serine proteases as cuticle-degrading enzymes, which were observed during the contact between fungal trapping devices and nematodes [16]. In general, the production of protease and its relatives was significant in the recognition of nematodes, whether parasitic or not, and in demonstrating the heterotrophic lifestyle of nematophagous fungi.

Another protein discovered in this review was lectin, which is defined as a multifunctional protein engaged in cellular-to-cellular communication. This protein was connected as an extracellular polymer layer on the fungal cell wall during the capture of nematodes by *A. oligospora* [17]. The presence of lectin was later recognized to be closely related to Deuteromycetous and Basidiomycetous fungi, which may explain why some Basidiomycete members, such as *Flammulina velutipes*, also demonstrated nematicidal properties [18,19]. In addition, subtilisins are serine proteins first discovered in *Bacillus subtilis*, although their synthesis has been shown in other fungal taxa such as *V. chlamydosporium* [20], *Metarhizium anisopliae* [20], *Arthrobotrys oligospora* [21], members of Pezizomycotina [22], and members of Cordycipitaceae, Clavicipitaceae and Ophiocordycipitaceae [23].

Some other unique enzymes, but with a limited number of publications, have also been found in nematophagous fungi, and will be discussed further. The initial article in 1985 mentioned acid phosphatase, which was probably involved in the interaction between *A. oligospora* and nematodes [12]. Cruz et al [24] corroborated the findings by suggesting that acid phosphatase may be synthesized by nematophagous fungus, in this case, *Duddingtonia flagrans*, when it interacts with nematodes. Nematophagous fungi also produced collagenase, a specific protease capable of digesting fibrous or structural protein in most humans and animals. In a laboratory test, *Arthrobotrys tortor* isolated from a cold environment, was found to release collagenase [25]. The possibility of finding a suitable agent for producing collagenase is centered on its rapid turnover and prospective applications in medicines (wound and scar treatment) and the leather industry (fur and hide tanning) [26]. Glucoamylase, a carbohydrase from *A. conoides*, was identified as having some promising qualities for application in the livestock sector and food science due to its high raw starch digestibility [27]. A novel phytase, specifically β-propeller phytases (BPP), was described for the first time from a fungal source, a nematophagous fungus, *A. oligospora*, which is also potential in the livestock sector as feed processing agent [28]. In addition, *Pochonia chlamydosporia* was shown to synthesize not just proteases but also carbohydrate-active enzymes (CAZyme), which could be significant for nematophagous activity and a target for strain enhancement [29].

4. Conclusions

The analysis on research articles concerning the extracellular enzymes and proteins produced by nematophagous fungi revealed a declining trend in the number of studies for the past ten years. Members of Orbiliaceae, particularly *Arthrobotrys oligospora* and *Duddingtonia flagrans* appeared mostly during the search. Proteases, serine proteases, and chitinases were the most common enzymes studied for their biocontrol capabilities, although other unique enzymes were also noted in this review.

Acknowledgments

This project is funded by the Indonesian Ministry of Higher Education, Research and Innovation, *Direktorat Riset dan Pengabdian Kepada Masyarakat* (DRPM) year 2018 under scheme of *Penelitian Dasar Unggulan Perguruan Tinggi* (PDUPT) No. 45/UN5.2.3.1/PPM/KP DRPM/2018.

References

[1] Berhanu M, Waktole H, Mamo G and Terefe G 2022 *BMC Microbiol* **22** 159
[2] Degenkolb T and Vilcinskas A 2016 *Appl Microbiol Biotechnol* **100** 3799
[3] Minglian Z, Minghe M and Keqin Z 2003 *Mycologia* **96** 16
[4] Yang Y, Yang E, An Z and Liu X 2007 *Proc Natl Acad Sci U S A* **104** 8379
[5] Xie M, Zhang Y J, Zhang X L, Peng D L, Yu W B and Li Q 2016 *J Invertebr Pathol* **138** 86
[6] Tunlid A, Jansson H B and Nordbring-Hertz B 1992 *Mycol Res* **96** 401
[7] Morton C O, Hirsch P R and Kerry B R 2004 *Nematology* **6** 161
[8] Gortari M C and Hours R A 2008 *Mycol Prog* **7** 221
[9] Li J, Zou C, Xu J, Ji X, Niu X, Yang J, Huang X and Zhang K Q 2015 *Annu Rev Phytopathol* **53** 67

[10] Ying S H and Feng M G 2019 *Virulence* **10** 429
[11] Tong S M and Feng M G 2019 *Appl Microbiol Biotechnol* **103** 577
[12] Veenhuis M, Nordbring-Hertz B and Harder W 1985 *Antonie Van Leeuwenhoek* **51** 385
[13] Muszewska A, Stepniewska-Dziubinska M M, Steczkiewicz K, Pawlowska J, Dziedzic A and Ginalski K 2017 *Sci Rep* **7** 9147
[14] Soares F E D F, Sufiate B L and de Queiroz J H 2018 *Agric Nat Resour* **52** 1
[15] Lopez-Llorca L V 1990 *Can J Microbiol* **36** 530
[16] Tunlid A and Jansson S 1991 *Appl Environ Microbiol* **57** 2868
[17] Rosen S, Ek B, Rask L and Tunlid A 1992 *J General Microbiol* **138** 2663
[18] Rosen S, Kata M, Persson Y, Lipniunas P H, Wirkstrom M, van den Hondel C A M J J, van den Brink J M, Rask L, Heden L O and Tunlid A 1996 *Euro J Biochem* **238** 822
[19] Ferreira J M, Carreira D N, Braga F R and Soares F E F 2019 *3 Biotech* **9** 410
[20] Segers R, Butt T M, Keen J N, Kerry B R and Peberdy J F 1995 *FEMS Microbiol Lett* **126** 227
[21] Ahman J, Johansson T, Olsson M, Punt P J, van den Hondel C A M J J and Tunlid A 2002 *Appl Environ Microbiol* **68** 3408
[22] Li J, Yu L, Yang J, Dong L, Tian B, Yu Z, Liang L, Zhang Y, Wang X and Zhang K 2010 *BMC Evol Biol* **10** 68
[23] Varshney D, Jaiswar A, Adholeya A and Prasad P 2016 *BMC Evol Biol* **16** 720
[24] Cruz D G, Silva C P, Carneiro C N B, Retamal C A, Thiebaut J T L DaMatta R A and Santos C P 2009 *J Invertebr Pathol* **102** 238
[25] Tosi S, Annovazzi L, Tosi L, Iadarola P and Caretta G 2002 *Mycopathologia* **153** 157
[26] Wanderley M C D A, Neto J M W D, Filho J L D L, Lima C D A, Teixeira J A C and Porto A L F 2017 *Braz J Microbiol* **48** 13
[27] Shetty P 2016 *Appl Biochem Microbiol* **52** 176
[28] Hou X, Shen Z, Li N, Kong X, Sheng K, Wang J and Wang Y 2020 *Microb Cell Factories* **19** 84
[29] Aranda-Martinez A, Lenfant N, Escudero N, Zavala-Gonzalez E A, Henrissat B and Lopez-Llorca L V 2016 *Environ Microbiol* **18** 4200

Visitor valuation for natural tourism development in Simarjarunjung, North Sumatra, Indonesia

A Purwoko*, **P Patana and S A Ginting**

Faculty of Forestry, Universitas Sumatera Utara, Medan, North Sumatra, Indonesia 20155

*Email: agus9@usu.ac.id

Abstract. Ecotourism is a type of tourism that became popular around the 1990s. Indonesia has a variety of unique nature that is good for the development of ecotourism. Most foreign tourists come to Indonesia because of its natural beauty and uniqueness. The increase in the local and national economy is obtained from visits and expenses incurred by tourists, so it must continue to be improved and maintain satisfaction. This study aims to analyze the characteristics, level of visitor satisfaction, and the factors that influence the motivation to visit tourist attractions in the Simarjarunjung area. This study uses descriptive and explanatory methods with a quantitative approach. Score and categorization of satisfaction level in each aspect studied, namely natural scenery, access/affordability, security, facilities, cleanliness, comfort, toilet facilities, road infrastructure, and promotion. Scores for all aspects are also obtained as an overall picture of the level of visitor satisfaction for each category of push and pull factors. From the regression analysis obtained the factors that have a significant effect on the motivation of visitors along with the direction and coefficient value. Aspects of satisfaction and motivation as well as the factors that influence them become important policy briefs for stakeholders to develop nature tourism.

1. Introduction

Forests have an important meaning in human life because they have ecological, social, and economic functions [1-3]. In its development, forests were under such strong pressure that the economic function of the forest became more dominant while the ecological function and social function of the forest received less attention [4,5]. Forest management requires the organization of forest and area management at the site level to manage effectively and efficiently [6,7].

Ecotourism is a type of tourism that became popular around the 1990s [8,9]. Indonesia is a country with a variety of natural uniqueness, Indonesia has a fairly good potential for the development of types of ecotourism. Most foreign tourists come to Indonesia because of the beauty of nature. Natural resources provide increase the competitiveness of Indonesian tourism. In the World Economic Forum, Indonesia ranks 14th in potential natural resources in 2017 [10]. Observing the positive impact of ecotourism in increasing tourism competitiveness, community-based regional ecotourism development is needed. It is necessary to improve environmental conditions, carry out conservation/ reforestation/planting efforts within the location of the ecotourism area, and build ecotourism infrastructure. Increasing the capacity building of ecotourism managers involving the community is also important in the development of ecotourism [11].

Indonesia is one of the countries that have a lot of tourism and ecotourism potential, one of which is in North Sumatra [12]. One of the attractions in North Sumatra is the beautiful charm of Lake Toba

IOP Conf. Series: Earth and Environmental Science 1115 (2022) 012064 doi:10.1088/1755-1315/1115/1/012064

[13]. One of the most famous and in-demand areas is Simarjarunjung Nature Tourism [14]. In general, natural tourist destinations, as well as all-natural tourist attractions in Simarjarunjung offer beauty and scenery in the form of beautiful scenery with a variety of tourist attractions provided by each tourist actor [15].

Tourist satisfaction is indicated by the desire of the tourist to come back to the destination he visited [16]. Tourist products in a tourism attraction play an important role to satisfy its visitors. If the performance of the tourist product (or result) is higher than the traveler's expectations, 2 then a positive assessment can be realized. However, if the performance (or result) of such a tourist product is below the expectations of tourists, then a negative assessment occurs. When there is a positive assessment, the traveler will evaluate the experience positively, and when there is a negative assessment, the traveler will evaluate the experience negatively. So the satisfaction of tourists is seen from the level of consumer feelings whether they feel happy or not in choosing and visiting [17]. The purpose of this study is to analyze the motivation of the visiting community and analyze the level of tourist satisfaction with tourist attractions in the Simarjarunjung Area.

2. Materials and Methods

This research was conducted in the Simarjarunjung natural tourism area, Simalungun Regency, North Sumatra, Indonesia.

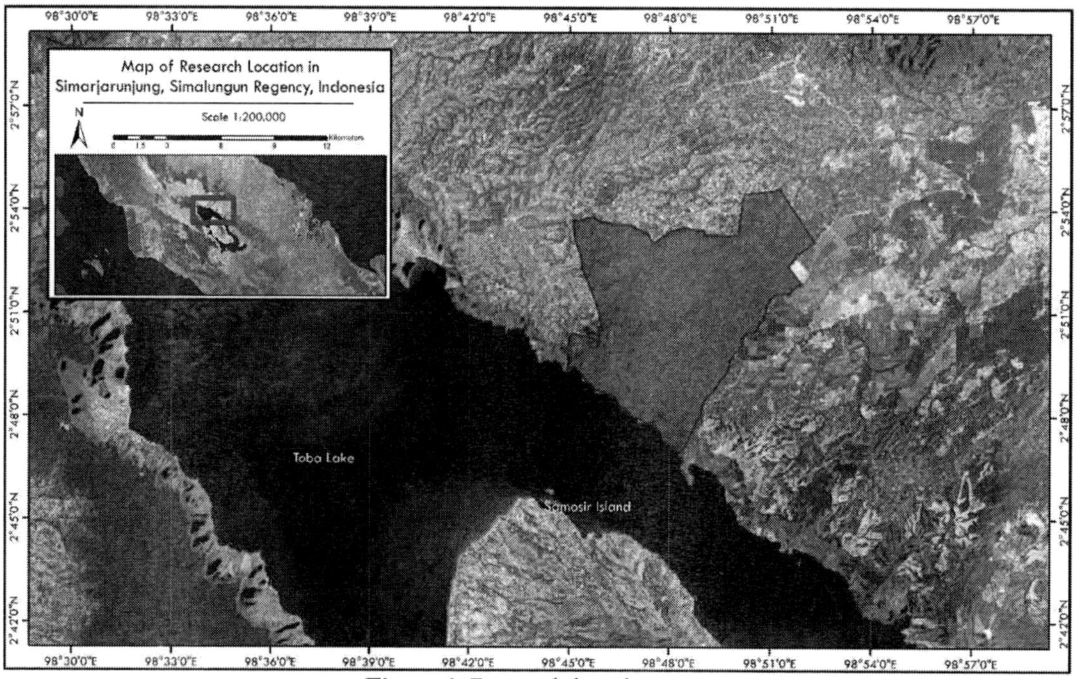

Figure 1. Research location map

2.1. *Population and data retrieval techniques*

The population determined in this study is people who visit natural attractions in Simarjarunjung district, Simalungun Regency. This study used the accidental sampling technique where the technique conducted interviews with visitors who were encountered by chance.

2.2. *Data analysis*

Technical data analysis discusses sampling methods, data retrieval techniques, variable measurement methods and, method of analyzing determinants of motivation. The complete technical analysis data is described as below.

2.2.1. Sampling methods. The visitor sampling method used in this study is to use the accidental sampling technique, which is a technique that is a determination based on coincidence or anyone who is visiting and can be met by researchers. To determine the sample size of visitors, sample measurements were carried out using the Slovin formula [18-20]. The error tolerance limit is 10% because if the number is greater than 100 people, it is taken between 10-15% or 20-25% [21]. The population of visitors to natural attractions in the Simarjarunjung area, North Sumatra in this study was 31,200 in 2019 (Simalungun Regency Tourism Office, 2020). Using the 10% error tolerance limit percentage, the number of research samples needed is 99.68 rounded to 100 visitors.

2.2.2. Data collection methods. The data collection method in this study used interviews with questionnaire tools to obtain as much data and information as possible about the issues raised about visitor motivation and satisfaction.

2.2.3. Motivation measurement methods. The method of measuring visitor motivation data uses the Guttman scale. The Guttman scale is unequivocal and consistent with providing answers to yes- and no questions [22]. The affirmative assessment is given a score of 1 and the dissenting answer is given a score of 0. After scoring, it is calculated by percentage (%). Answer the question to find out the motivation of the respondent by using the following formula:

$$P = \frac{a}{b} \text{ x } 100\% \tag{1}$$

Information:
P = Percentage (%)
a = Number of yes answers
b = Number of respondents
The calculation results are interpreted in the form of percentages into descriptive criteria of percentages and then interpret them using qualitative sentences. The division of descriptive criteria of percentages is divided into two categories according to the Guttman scale i.e. if admission is 51%-100% then the category is 'yes' while if it is 0%-50% the category is 'no'.

2.2.4. Satisfaction data measurement method. The method of measuring visitor satisfaction data in this study is a Likert scale used to measure a person's attitudes, opinions, and satisfaction with the measured social phenomena.

2.2.5. Method of analyzing determinants of motivation. The method of analyzing the factors affecting visitor motivation uses multiple linear regression analysis with the following basic model:

$$Y = a + b_iX_i + e \tag{2}$$

Information:
Y = Motivation
a = Constants
b_i = Regression coefficient of the i-th variable
X_i = Estimator variables (sight, access, security, facilities, infrastructure, cleanliness, convenience means of toilet facilities and promotion)

3. Results and Discussion

3.1. Visitor characteristics of natural attractions in the simarjarunjung area

Based on gender, the most visitors were women, which was 56%, while men were 44%. Age categorization is only divided into 2, namely children and adolescents (≤ 20) and adults age (>20). The results showed that the majority of visitors came from adults (57%), while the rests were from children and adolescents (43%). From the educational aspect, the majority of visitors have a high school education (SLTA) or equivalent (68%), while 26 5 of them are educated at the college level.

Table 1. Visitor description based on city/region origin

	City of Origin	Percent	Percentage	Cumulative Percent
1	Medan	26.0	26.0	26.0
2	Siantar	29.0	29.0	55.0
3	Kaban Jahe	5.0	5.0	60.0
4	Binjai	3.0	3.0	63.0
5	Tebing Tinggi	11.0	11.0	74.0
6	Langkat	2.0	2.0	76.0
7	Lubuk Pakam	8.0	8.0	84.0
8	Sibolga	8.0	8.0	92.0
9	Berastagi	3.0	3.0	95.0
10	Merek	5.0	5.0	100.0
	Total	100.0	100.0	

Based on Table 1, it can be seen that the highest percentage of visitors comes from the nearest city, namely Pematang Siantar, followed by the city of Medan, Tebing Tinggi, and Lubuk Pakam. This illustrates that natural tourism visitors in this area come from urban areas, but are still limited to local cities around the location of natural attractions. This data deserves attention for managers to segment services and promotion of natural tourism that need to be done to increase tourist visits. It is also necessary for policymakers to increase their support so that this tourist attraction can attract larger visits and from the origin of the wider area, including international tourists.

3.2. *Tourist satisfaction visiting*

Tourist satisfaction visiting discussed includes aspects natural landscape, access or availability, security, facilities available, road infrastructure, clearlyness, comfort in the natural tourist area, bathing, washing, and latrine toilet facilities and promotion on social media. More details are presented below.

3.3. *Natural landscape*

Simarajarunjung natural tourism is known to many people because it has beautiful scenery, a stretch of trees located on the side of the road, and the beauty of the panorama of Lake Toba. If the natural scenery offered by Simarjarunjung is in a good category such as no scattered garbage and the environment is made beautiful, visitors will be comfortable.

Table 2. The natural scenery of Simarjarunjung tourism

Criterion	Frequency	Percentage (%)	Score	Number of scores
Very Dissatisfied	0	0	1	0
Dissatisfied	0	0	2	0
Intermediate	9	9	3	27
Satisfied	23	23	4	92
Very Satisfied	68	68	5	340
Total	100	100		459

In Table 2, it can be seen that the interpretation criteria on the score in the natural scenery section of the Simarjarunjung natural tourist attractions are 459, namely in the category of 'very satisfied. From the table, someone travels to Simarjarunjung because Simarjarunjung has a natural beauty that is still natural, where many trees make the atmosphere there beautiful and cool, this certainly makes visitors feel comfortable and satisfied. Indicators related to enjoying the beauty of nature are motivation and satisfaction with tourist visitors [24].

3.3.1. Access or affordability. Access/affordability in tourist attractions is something that can affect visitor satisfaction. Simarjarunjung Nature Tourism itself makes it easy for visitors to access one attraction to another by providing clear directions. The visitors have no difficulty accessing tourist destinations. This can make visitors feel comfortable when traveling.

Table 3. Access or affordability of Simarjarunjung natural tourism area

Criterion	Frequency	Percentage (%)	Score	Number of Scores
Very Dissatisfied	2	2	1	2
Dissatisfied	11	11	2	22
Intermediate	31	31	3	93
Satisfied	51	51	4	204
Very Satisfied	5	5	5	25
Total	100	100		346

In Table 3, it can be seen that the criteria for interpretation of the score in the access/affordability section in the Simarjarunjung natural tourist attractions are 346, namely in the category of 'satisfied'. Access conditions 30 roads to tourist attractions, and the access road must be related to public infrastructure [25]. The condition of public roads and access roads determine the accessibility of a tourist attraction. This accessibility is a very important requirement for tourist attractions that make it easy for tourists to visit the tour.

3.3.2. Security. Everyone when making a tourist visit certainly wants a pleasant experience, visitors do not want any security disturbances around tourist attractions. The better the security of a tourist attraction, the more people who visit will be.

Table 4. Security in the natural tourist area of Simarjarunjung

Criterion	Frequency	Percentage (%)	Score	Number of Scores
Very Dissatisfied	0	0	1	0
Dissatisfied	12	12	2	24
Intermediate	23	23	3	69
Satisfied	34	34	4	136
Very Satisfied	31	31	5	155
Total	100	100		384

In Table 4, it can be seen that the interpretation criteria for the score in the security section are 384 and are in the category of 'satisfied'. There have never been riots or other dangerous things so visitors feel satisfied with security when they are in simarjarunjung nature tourism. The safety of travel is very important because if the attraction is not safe it can be a danger to tourists [26].

3.3.3. Facilities available. Facilities are physical equipment available at tourist attractions to make it easier for visitors when carrying out activities in tourist attractions. One of the factors that make visitors feel satisfied while traveling in Simarjarunjung is the facilities provided by Simarjarunjung, such as the availability of toilets, parking lots, prayer rooms, playgrounds, and others. The better the availability of facilities in tourist attractions, the more satisfied visitors will be in tourist attractions.

Table 5. Facilities available in the Simarjarunjung natural tourism area

Criterion	Frequency	Percentage (%)	Score	Number of Scores
Very Dissatisfied	1	1	1	1
Dissatisfied	13	13	2	26
Intermediate	29	29	3	87
Satisfied	35	35	4	140
Very Satisfied	22	22	5	110
Total	100	100		364

In Table 5 it can be seen that the criteria for interpretation of the score in the section of facilities available in the Simarjarunjung natural tourist attractions are 364, namely in the category of 'satisfied'. Tourist facilities have a positive and significant effect on tourist satisfaction [27]. In addition, facilities are one of the important things to developing tourism. The availability of these tourist facilities is very important to be available in tourist destinations.

3.3.4. Road infrastructure. Road infrastructure in Simarjarunjung natural tourism also affects the satisfaction of visitors who do tours. Road infrastructure includes how the road conditions are in the Simarjarunjung natural tourist area. If the road is in good condition, visitors can visit tourist attractions in the Simarjarunjung Area easily, and vice versa.

Table 6. Road infrastructure in Simarjarunjung natural tourist area

Criterion	Frequency	Percentage (%)	Score	Number of Scores
Very Dissatisfied	11	11	1	11
Dissatisfied	24	24	2	48
Intermediate	31	31	3	93
Satisfied	27	27	4	108
Very Satisfied	7	7	5	35
Total	100	100		295

In Table 6, it can be seen that the criteria for interpretation of scores in the road infrastructure section of the Simarjarunjung natural tourist attractions are 295, namely in the 'medium' category. Road infrastructure has a significant influence on tourists' interest in visiting, namely the presence of good road conditions can make it easier for tourists to visit tourist attractions [28].

3.3.5. Cleanliness. The cleanliness of Simarjarunjung's natural tourist attractions also affects visitors. Cleanliness is one of the factors that are very much a concern for general visitors to tourist attractions, especially natural tourism. If the environment is clean, it can have a comfortable impact on visitors. Cleanliness is also a factor that can determine the interest of visitors to come back to visit Simarjarunjung natural tourism.

Table 7. Cleanliness in the natural tourist area of Simarjarunjung

Criterion	Frequency	Percentage (%)	Score	Number of Scores
Very Dissatisfied	3	3	1	3
Dissatisfied	16	16	2	32
Intermediate	30	30	3	90
Satisfied	35	35	4	140
Very Satisfied	16	16	5	80
Total	100	100		345

In Table 7, it can be seen that the criteria for interpretation of the score in the hygiene section at the Simarjarunjung natural tourist attractions are 345, namely in the 'satisfied' category. The large number of visitors who come, in addition to having a positive impact on income, also hurts environmental conditions, one of which is cleanliness. The cleanliness of the tourist facilities greatly affects the desire of tourists to visit again [29].

3.3.6. Comfort in the natural tourist area. The comfort of traveling is one of the factors that determine visitor satisfaction. It is undeniable that every individual wants comfort wherever they are. Visitors who feel comfortable with facilities, safety, or cleanliness while in tourist attractions can affect their desire to return to visiting simarjarunjung nature tourism.

Table 8. Comfort in the natural tourist area of Simarjarunjung

Criterion	Frequency	Percentage (%)	Score	Number of Scores
Very Dissatisfied	2	2	1	2
Dissatisfied	24	24	2	48
Intermediate	30	30	3	90
Satisfied	34	34	4	136
Very Satisfied	10	10	5	50
Total	100	100		326

In Table 8, it can be seen that the criteria for interpretation of scores in the hygiene section in Simarjarunjung natural attractions are 326, namely in the 'medium/medium' category. Comfort and safety are vital factors in tourism destinations [30]. Comfort is also something that will be very important in building the progress of the tourism destination. The level of comfort of tourists can be seen in the condition of the facilities of the tourist attraction destination. If the facilities and access to tourist attractions are complete and controlled, the level of comfort for returning visits will be as high as possible.

3.3.7. Bathing, washing, and latrine (BWL) facilities. The facilities of BWL facilities in tourist attractions are things that must be a priority in tourist destinations. Not only its availability but also paying attention to cleanliness and feasibility. The available BWL facilities are decent, clean, and comfortable will greatly affect the satisfaction of visitors who carry out tourist activities.

Table 9. BWL facilities in the Simarjarunjung natural tourism area

Criterion	Frequency	Percentage (%)	Score	Number of Scores
Very Dissatisfied	7	7	1	7
Dissatisfied	17	17	2	34
Intermediate	36	36	3	108
Satisfied	19	19	4	76
Very Satisfied	21	21	5	105
Total	100	100		330

In Table 9, it can be seen that the interpretation criteria for hygiene scores amount to 326 and are in the 'medium/medium' category. Tourist facilities are actually not directly related to tourism but are often part of the needs of tourists, namely by meeting the needs of tourists during their stay for a while in the tourist area visited.

3.3.8. Promotion on social media. Promotion is one of the treatments of products, services, and goods to attract potential consumers. Promotion of a product or service can be done in various ways, one of which is by using social media platforms. In today's era, many people have a high level of dependence on social media. Promotion is an activity to communicate that provides an explanation that can be convincing and a meaningful target market that if you do a promotion, the decision to do tourism will increase [31]. Therefore, social media can be used if Simarjarunjung nature tourism wants to promote it so that it is easier and faster to be accessed by the public.

Table 10. Promotion in simarjarunjung natural tourism area

Criterion	Frequency	Percentage (%)	Score	Number of Scores
Very Dissatisfied	12	12	1	12
Dissatisfied	18	18	2	36
Intermediate	41	41	3	123
Satisfied	26	26	4	104
Very Satisfied	3	3	5	15
Total	100	100		290

In Table 10, it can be seen that the interpretation criteria for hygiene scores amount to 290 and are in the 'medium/medium' category. This condition is quite unfortunate, considering that cleanliness is a universally necessary factor in every tourist activity [32]. Similar studies also report the same thing where overall cleanliness is an important factor in the management of tourist attractions [33].

3.4. Motivational factors of tourist visitors

Motivation is the underlying factor of a person in determining the attitudes and deeds he will do. In this case, motivation is an important factor for a person in making decisions regarding the tourist attractions to be visited. The higher the motivational drive of visitors created, the more the decision to visit from visitors will increase [23].

3.4.1. Driving factors. A person's motivation to visit a tourist attraction is determined by several driving factors such as visitors who have previously been to a tourist attraction, come to a tourist attraction for assignments (school/college), come to a tourist attraction for recreation, and gather with friends or family.

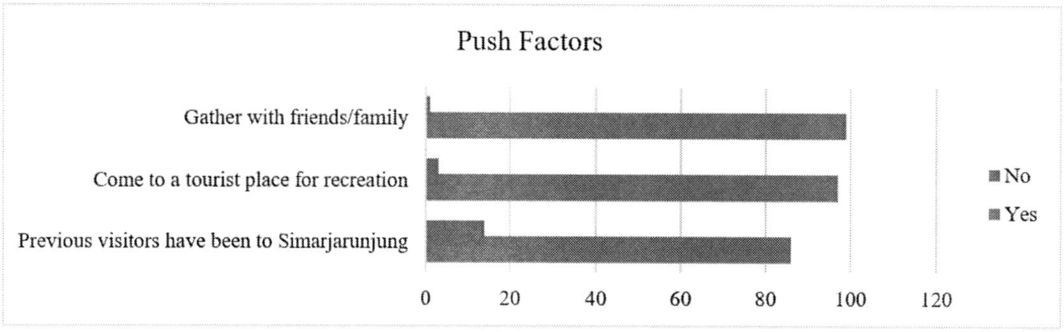

Figure 2. Factors driving natural tourism visits

3.4.2. Attracting factor. A person's motivation is also determined by several factors such as the attractiveness of the Simarjarunjung natural attractions, the affordable costs of tourist attractions, access to the simarjarunjung natural attractions, and the existence of good photo spots.

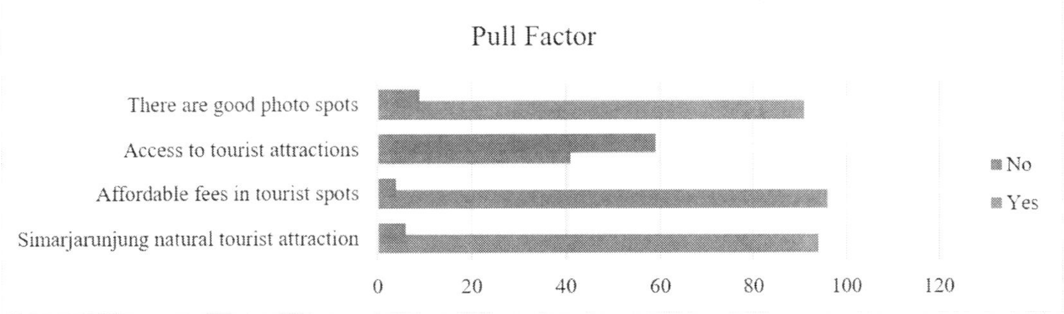

Figure 3. Factors attracting natural tourist visits

3.5. Factors affecting motivation

This study uses multiple linear regression analysis to predict how far the value of the dependent variable changes, namely visitor motivation, if the value of the estimating variables increases or decreases.

Table 11. The results of the analysis of variant (Anova)

	Model	Sum of Squares	df	Mean Square	F	Sig.
1	Regression	3,894	9	,433	9,241	,000[b]
	Residual	1,966	42	,047		
	Total	5,861	51			

a. Dependent variable: motivasi
b. Predictors: (constant), promotion, sight, comfort, facilities, access, cleanliness, infrastructure, security, toilet facilities

Based on the test results with Test F in Table 11, it is known that the calculated F value is 9.243, greater than the F-table value (1.98). The significance value is close to zero. This shows that the independent variable simultaneously affects the dependent variable, namely comfort. The results of the F test have not been able to determine exactly the independent variables that have a direct and significant effect on the dependent variable, namely motivation. Therefore, this analysis is continued with the T-Test to be able to determine independent variables that have a real influence on dependent variables partially.

Table 12. Partial test results (T-test)

		Unstandardized Coefficients		Standardized Coefficients		
	Model	B	Std. Error	Beta	t	Sig.
1	(Constant)	2,941	,359		8,189	,000
	View*	,155	,055	,262	2,843	,007
	Access*	-,102	,038	-,255	-2,678	,011
	Security*	,224	,037	,674	6,015	,000
	Facilities*	-,081	,033	-,242	-2,442	,019
	Infrastructure*	,070	,030	,227	2,290	,027
	Hygiene	,042	,030	~~,133~~	~~1,400~~	~~,169~~
	Comfort	-,017	,037	~~-,047~~	~~-,456~~	~~,651~~
	BWLfacilities	,082	,036	,292	2,274	,028
	Promotion	,006	,031	~~,019~~	~~,201~~	~~,842~~

Dependent variable: visitor motivation
*) significant

IOP Conf. Series: Earth and Environmental Science 1115 (2022) 012064 doi:10.1088/1755-1315/1115/1/012064

The value of the coefficient of determination of 66.4% which means that 66.4% of the motivation for tourist visits to the Simarjarunjung Natural Tourism Area can be simultaneously explained by variations of nine independent variables, namely scenery, comfort, facilities, access, cleanliness, infrastructure, security, BWL facilities, and promotion. While the remaining 33.6% is explained by other variables outside the model that were not included in the study. Aspects attached to objects have great potential to be used as new and additional independent variables in research. Based on Table 12 above, shows the variables of sight, facilities, access, infrastructure, security, and BWL facilities. has a significance value below 0.05. This shows that these variables have a significant influence on visitor motivation

The presence of a positive and significant influence on the variables of scenery, infrastructure, security, and BWL facilities on the motivation to visit the Simarjarunjung Natural Tourism Area is by the supporting theoretical assumptions. In this case [31] explained that the factors that influence the motivation to visit senior tourists to tourist attractions consist of driving and pulling factors. The driving factors are ego enhancement, self-esteem, seeking knowledge, relaxation, and socialization. The attracting factors are in the form of cleanliness and safety, facilities, events, costs, and natural scenery. Various similar studies also report on the same subject, among which reported that seeking novelty and the search for knowledge is one of the attractions that move people to travel [35].

Empirically, the fact that the location of the Simarjarungjung natural tourism district does have a very beautiful view in the form of a view of Lake Toba from the top of the hill and a view of the ridges of the mountains surrounding Lake Toba. Charming. The microclimate of the neighborhood which still has a forest ecosystem and good vegetation cover offers an attraction to relax. The supporting infrastructure and security and BWL facilities in the object are very adequate. Therefore, some of these driving factors can generate various motivations for visitors and can be a factor causing the emergence of visitors' desire to visit again. This explanation strengthens the importance of natural attraction factors such as scenery and the support of artificial factors such as infrastructure, security, and BWLoilet facilities for the development of natural tourism objects.

Theoretically, ease of access and availability of facilities for tourists are very important to support various tourist activities. The study report states that facilities greatly influence how tourists choose destinations for vacation [36]. Other studies have also concluded that tourist attractions and service facilities together have a significant effect on tourist satisfaction [37]. Likewise, with access, research [38] states the same thing, where the increase in distance that generally leads to an increase in the cost of transportation access is a significant factor that affects the total cost of tourism. Many studies on tourism define the transport aspect as accessibility that affects tourist activities. Transportation is seen as a link between the visitor's home area and the location of the tourist attraction.

However, in this study, there are different facts, where the variable coefficient of the facility is marked negatively. The empirical explanation that may be the cause of this phenomenon is the cost charged to each additional item of facilities such as photo spots, artificial rides, lodging, BWL facilities, parking, documentation, and others. The cost of manpower on each item of this object facility may burden local visitors, the majority of whom are from the lower middle economic group, thus causing a negative relationship with motivation. Visitors are more focused on enjoying its natural attractions in the form of views of Lake Toba, microclimate, natural scenery of the mountains, and the like. No adequate explanation has been found for the negative phenomenon of regression coefficients in access variables. It is possible to have different perceptions of visitors in interpreting this variable, so further research is needed by improving the measurement indicators on this variable to get more thorough results.

Stakeholders related to the management and development of natural tourism objects in the Simajarungjung Area must prioritize the factors that support visitors' activities and strive to improve them. On the other hand, things that have the potential to cause trade-off effects also need to be minimized so that they do not have an impact on reducing tourist visits to this region. With the absence of natural tourism around the residence, the community and the local government will greatly benefit [39]. This is because the growing natural tourism produces good social and economic impacts for managers, governments, visitors, and the local community. The existence of this tourism provides business opportunities for the surrounding community [35].

4. Conclusion

Visitor satisfaction for each aspect of natural tourism is in the range of 'medium/medium' to 'very satisfied. The aspect that gets the assessment category is 'very satisfied' is only the natural scenery (score of 459). Aspects that received a 'satisfied' assessment were safety (384), facilities (364), and cleanliness (score 345). Other aspects get the 'medium/medium' assessment category is access/affordability (score 326), comfort (score 326), road infrastructure (score 295), BWL facilities (score 330), and promotion (score 290). The average score for all aspects is 348 which can be classified into the category of 'satisfied' assessment. The motivation of visitors to travel to Simarjarunjung natural tourism is classified as a driving factor and a pulling factor. The driving factors that motivate visitors include the desire to re-visit (86%), academic/educational activities (97%), and recreation with friends or family (99%). Attracting factors that motivate visitors include the attraction of natural attractions (94%), affordable costs 96(%), easy access (59%), and the presence of good photography spots (91%). Simultaneously all predictor variables have a significant effect on the motivation of visitors to nature tourism. Factors that partially have a significant and positive effect on visitor motivation include scenery, security, infrastructure, and BWL facilities. Access and facilities have a significant effect but have a negative coefficient on visitor motivation.

Acknowledgments

The author would like to thank the Indonesian Ministry of Education, Culture, Research, and Technology for funding this research with Agreement/Contract Number 059/E5/PG.02.00/PT/2022. Thanks are also conveyed to the managers and visitors of tourist attractions in the Simarjarunjung Area who have supported this research.

References

[1] Lewin H A, Robinson G E, Kress W J, Baker W J, Coddington J, Crandall K A and Zhang G 2018 *Proceedings of the National Academy of Sciences* **115** 4325
[2] Mori A S, Lertzman K P and Gustafsson L 2017 *Journal of Applied Ecology* **54** 12
[3] Halsey M and White R 2017 *Crime, ecophilosophy and environmental harm* (Routledge: Green Criminology)
[4] Puettmann K J, Wilson S M, Baker S C, Donoso P J, Drössler L, Amente G and Bauhus J 2015 *Forest Ecosystems* **2** 1
[5] Arora-Jonsson S 2014 *In Women's Studies International Forum* **47** 295
[6] Bettinger P, Boston K, Siry J and Grebner D L 2016 *Forest management and planning* (Academic press)
[7] Keenan R J 2015 *Annals of forest science* **72** 145
[8] Liu S and Li W Y 2020 *Sage Open* **10**
[9] Gezon L L 2014 *Journal of Sustainable Tourism* **22** 821
[10] Hakim L 2017 *AIP Conference Proceedings* **1908** 030008 AIP Publishing LLC
[11] Bansal S P and Kumar J 2013 *Ecotourism for community development: A stakeholder's perspective in Great Himalayan National Park. In Creating a sustainable ecology using technology-driven solutions* (IGI Global)
[12] Butarbutar R and Soemarno S 2013 *Journal of Indonesian Tourism and Development Studies* **1** 97
[13] Chesner C A 2012 *The Toba caldera complex Quat Int* **258** 5
[14] Purwoko A, Slamet B and Pohan N Z P 2022 *IOP Conference Series: Earth and Environmental Science* **977**
[15] Purwoko A and Sahputra A 2022 *IOP Conference Series: Earth and Environmental Science* **977**
[16] Antón C, Camarero C and Laguna-Garcia M 2017 *Tourism* **20** 238

[17] Del C G, Andreu L and Gallarza G M 2014 *International Journal of Culture, Tourism and Hospitality Research* **8** 420
[18] Arikunto S 2012 Prosedur Penelitian Suatu Pendekatan Praktik (Jakarta: Rineka Cipta)
[19] Susanti A, Soemitro R A A, Suprayitno H and Ratnasari V 2019 *Journal of Infrastructure and Facility Asset Management* **1**
[20] Hasibuan C F 2020 *IOP Conference Series: Materials Science and Engineering* **909**
[21] Panjaitan U I 2015 *Analisis Potensi dan Strategi Pengembangan Obyek Wisata Alam Air Terjun Teroh – Teroh Desa Rumah Galuh Kecamatan Sei Bingai, Kabupaten Langkat, Sumatera Utara* [Undergraduate Thesis] Universitas Sumatera Utara
[22] Sugiyono 2011 Metode Penelitian Bisnis, Edisi kedua (Bandung: Alfabeta)
[23] Kempiak J, Hollywood L, Bolan P and McMahon-Beattie U 2017 *International Journal of Heritage Studies* **23** 375
[24] Kim H, Lee S, Uysal M, Kim J and Ahn K 2015 *Journal of Travel and Tourism Marketing* **32** S76
[25] Abdullah S, Razak A A and Jaafar M 2014 *SHS Web of Conferences* **12** (01096)
[26] Boakye K A 2012 *Tourism Management* **33** 327
[27] Coban S 2012 *The effects of the image of destination on tourist satisfaction and loyalty: the case of Cappadocia*
[28] Le-Klähn D T, Roosen J, Gerike R and Hall C M 2015 *Tourism Geographies* **17** 738
[29] Marcelina S D, Febryano I G, Setiawan A and Yuwono S B 2018 *Jurnal Belantara* **1** 45
[30] Salman A and Hasim M S 2012 *Asian Social Science* **8** 48
[31] Familmaleki M, Aghighi A and Hamidi K 2015 *International Journal of Economics and management sciences* **4** 1
[32] Zˇabkar Y, Brencic M M and Dmitrovic T 2010 *Tourism Management* **31** 537
[33] Buhalis D 2000 Tourism Management **21** 97
[34] Jang S C and Wu C F 2005 *Tourism Management* **27** 306
[35] Sangpikul A 2018 *International Journal of Culture, Tourism and Hospitality Research* **12** 106
[36] Švec R, Martinát S, Pícha K, lusáček P, Knotek J, Schaefer JC, Březinová and Navrátil J 2021 *Environment, Development and Sustainability*
[37] Snyman S 2014 *Koedoe: African Protected Area Conservation and Science* **56** 1
[38] Prideaux B 2000 *Tourism Management* **21** 53
[39] Purwoko A, Latifah S and Siahaan C 2021 *IOP Conf Series: Earth and Environmental Science* **912** 012050

IOP Conf. Series: Earth and Environmental Science 1115 (2022) 012065 doi:10.1088/1755-1315/1115/1/012065

Institutional models of natural resource management and utilization: lesson learning from Simarjarunjung tourism area, Lake Toba, Indonesia

A Purwoko*, **M Ulfa and R Kusnadi**

Forestry Study Program, Faculty of Forestry, Universitas Sumatera Utara, Medan, North Sumatra, Indonesia.

*Email: agus9@usu.ac.id

Abstract. Nature tourism is an effort to utilize natural resources that have been designated as tourist attraction objects to be targeted for tourism potential. One of the national strategic natural tourism areas around Lake Toba is Simarjarunjung. This area has an ecosystem of forest areas with very beautiful views that are managed by the community. This study aims to examine how the institutional management of natural tourism objects and the use of natural resources in the Simarjarunjung area of North Sumatra. This study uses a descriptive method with a qualitative approach. The results of this study present various characteristics of the institutional aspects of nature tourism management. Various advantages and disadvantages exist in the form of management institutions, such as the Association of Forest Farmers Groups (Gapoktan), Tourism Aware Groups (Pokdarwis), and individuals/volunteers. The potential of natural resources that are utilized and managed in the form of views of Lake Toba, the natural charm of the mountains, the charm of pine stands, agroforestry ecosystems, and a cool microclimate. The results of the comparison of various institutional models and the use of natural resources can be used as a reference in the development of natural tourism by managers and related stakeholders.

1. Introduction

Nature tourism is defined as tourism featuring nature as its main attraction [1]. Nature tourism or nature-based tourism is also often interpreted synonymously with terms such as ecotourism, sustainable tourism, green tourism, alternative tourism, and responsible tourism. Ecotourism is a form of tourism that is inspired primarily by the natural history of an area, including its original culture [2,3]. Ecotourism visits relatively undeveloped areas in a spirit of appreciation, participation, and sensitivity. However [4-6] defines ecotourism differently, i.e. a specific type of tourism, distinguished from nature tourism and outdoor recreation with the aim of its conservation and development. The general key for ecotourism is natural areas, ecotourism in terms of benefits for conservation, local culture, benefits for local communities, and the educational value of ecotourism for tourists and residents [7].

Lake Toba as a super-priority destination for Indonesian tourism has tremendous potential. In [8,9], The Toba Caldera is one of the most remarkable volcanic features formed over a long period. In this area, there are many tourist objects with various attractions and attractions. One of the natural attractions being developed is the Simarjarunjung Hill area (02°50'09.8" N; 98°45'32.6" E) at an altitude of 1479 masl [10]. This area is a magnet for tourists because this area is one of the best places to see the view of Lake Toba. Before the beginning of 2017, the area of the Bukit Indah Simarjarunjung tourist attraction was 2 hectares [11]. This nature tourism has beauty and landscapes in the form of beautiful scenery with various kinds of tourist attractions provided by each tourist actor [12,13]. This tourist attraction has now developed into a natural tourist area consisting of several nature-based tourism objects.

Tourism institutions are institutions that are responsible for the management of a natural tourism object. The formation of institutions in the community will make it easier to take advantage of the potential of natural resources with the availability of natural resources that will overcome sustainable economic development [14,15]. The institutional function provides behavioral guidelines to improve the economy by managing the potential of natural resources [16,17]. The importance of an institution in terms of overcoming obstacles emphasizes how social resources can be grown and developed by requiring institutions that have responsibility for the realization of strategies and policies [18,19].

Two important things must be done on the natural resources owned, namely utilization and supervision [20-22]. The utilization of natural resources is intended solely to improve the welfare of the community. Maximum supervision in the management of natural resources can prevent negative impacts [23,24]. The availability of natural resources that are managed optimally will be able to overcome the problems of economic development and sustainable development [25]. The importance of a good institution, among others, is to overcome various obstacles and emphasize that social resources can be grown and developed. Effective community institutions are expected to provide convenience in managing and increasing the potential of natural resources [14,15]. Credible community institutions are expected to raise awareness and increase awareness of the potential for nature tourism [26,27]. The purpose of this study is to find out how the institutional aspects in the management of natural tourism objects and how to utilize the potential of natural resources in the Simarjarunjung area.

2. Materials and Methods

2.1. Research locations

The study was conducted in the Simarjarunjung nature tourism area in Nagori (Village) Parik Sabungan, Dolok Pardamean District, Simalungun Regency, North Sumatra Province (Figure 1).

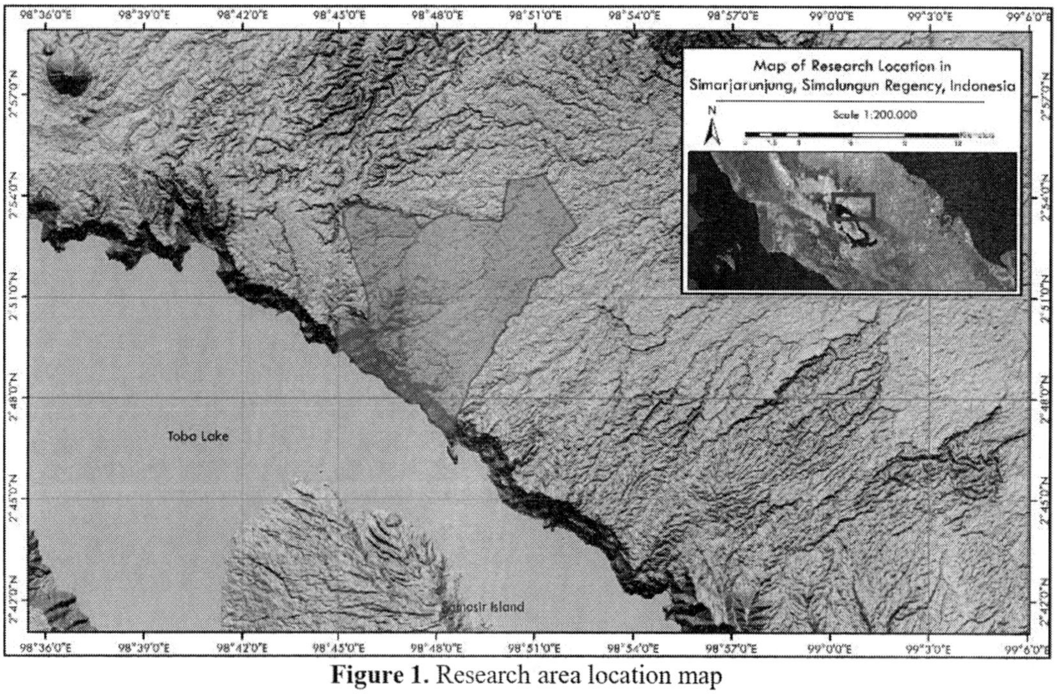

Figure 1. Research area location map

2.2. Data sources

The data was collected in the form of primary data and secondary data. Primary data were obtained from field observations in the form of observations and interviews with stakeholders (community, managers, community leaders, and visitors). Secondary data was collected to complement primary data through literature studies on the results of previous studies related to research, local government reports (sub-district and or village offices), textbooks, and so on.

This research was descriptive and qualitative with a research focus on institutional management in the Simarjarunjung nature tourism area. The number of resource persons for this study was 23 people consisting of the village head, village secretary, management of natural tourism objects, local community leaders, and resource persons representing tourists visiting each Simarjarunjung nature tourism object.

2.3. Data collection method

Data collection methods used in this study were as follows:

2.3.1. Survey method namely data collection which was done through observation and interview methods. The direct observation method was carried out in the Simarjarunjung nature tourism area, Simalungun Regency on holidays and weekdays. The interview guide used at the time of the interview is a questionnaire so that the questions will be more structured and focused.

2.3.2. The library method is a way to obtain data by reading and browsing the literature, reports, scientific works, and the results of previous research related to research.

2.3.3. Documentation was used to complete and at the same time add to the accuracy, and correctness of data or information collected from existing documentation materials in the field and can be used as material in checking the validity of the data.

2.4. Data analysis

The institutional theory used was the institutional theory according to Meyer and Rowan which describes the relationship between the organization and its environment, how and why the organization runs a structure and process, and what are the consequences of the institutional process that is carried out.

2.4.1. Institutional structure analysis. Institutional structure analysis examines the order that forms a group in society, and the arrangement, and organization of interrelated elements in an object or organized system.

2.4.2. Analysis of institutional forms. An institutional form is a set of rules, Syst, ems, and incentive structures that shape organizational behavior. Institutional forms often refer to formal and non-formal rules in society or social systems.

2.4.3. Norm analysis. Norms are behavior within a community group and certain regional boundaries. Norms will develop along with the social agreements of the community, often also referred to as social rules. The theory of norms used is the theory according to Hans Kelsen which is the legal norms and is layered in a hierarchy (organization). A higher norm applies, originates, and is based on a higher norm. Legal norms can be formed in writing or unwritten by the competent institutions that form them, while moral, customary, religious, and other norms occur unwritten, grow and develop from the habits that exist in society.

3. Results and Discussion

3.1. Nature tourism management institution

The Nature Tourism Management Institutions in the Simarjarunjung nature tourism area, Parik Sabungan, Dolok Pardamean District, Simalungun Regency, North Sumatra Province at the Village Level are as follows.

3.1.1. Tourism aware group (Pokdarwis). Increasing the role of the community in tourism development requires various empowerment efforts. This empowerment effort is intended so that the community can play a more active, optimal role and at the same time receive positive benefits from development activities carried out in improving their welfare. The forum for community development whose profession is related to tourism management in Pariksabungan Village is the Tourism Aware Group (Pokdarwis). This group becomes a top-down coaching medium for the local government and other relevant stakeholders, as well as a bottom-up channel for channeling aspirations from the community [28].

Pokdarwis in the Simarjarunjung area has a simple organizational structure led by a chairperson who is elected by the members. The chairpersons are assisted by a secretary, treasurer, and several section heads consisting of security and order, cleanliness and beauty, tourist attractions and memories, public relations, human resource development, and business development. The Pokdarwis management also has coaches and advisors from elements of the village, sub-district, and local community leaders. People who join this group are those who have reasons or incentives to carry out the tourism profession in groups [18].

3.1.2. Farmers group association (Gapoktan). Members of Pokdarwis Setia Tani own and/or work in the tourism sector, but they are members of the Forest Farmers Group (KTH). There are 2 KTH, namely KTH 'Green' which manages Dolok Sipigul nature tourism, and KTH 'Setya Tani' which manages Pinus Hills nature tourism. As an institution rooted in the community, apart from being the main target for various forestry extension activities, currently KTH is also the main actor in forestry development activities in the field.

In the Simarjarunjung nature tourism area, the existing KTH formed a Farmers Group Association (Gapoktan) called Karya Bersama. Gapoktan is a combination of a collection of farmer groups of the same type of business. In addition, Gapoktan is a communication medium for exchanging information between farmer groups that are members. The purpose of establishing Gapoktan is to provide assistance, training, and facilitate natural tourism management activities so that people understand the meaning of environmental conservation and the importance of community-based nature tourism [29].

3.2. Institutional management of object management

In entrepreneurship, people tend to join groups. Groups will be able to manage natural resources better. In groups, group members have reasons or incentives to carry out tasks in groups [30]. Therefore, Pokdarwis Setia Tani cadres formed the Forest Farmers Group (KTH) to manage the natural tourism potential in this area. The 'Green' Forest Farmers Group was formed to manage the natural tourism object of Dolok Sipigul, and the KTH 'Setya Tani' was formed to manage the natural tourism object of Pinus Hills. As a representation of the community in managing forest ecosystems, KTH is the main target of various forestry extension activities. Currently, KTH is also a partner of the government and the main actor in forestry sector development at the site level.

IOP Conf. Series: Earth and Environmental Science 1115 (2022) 012065 doi:10.1088/1755-1315/1115/1/012065

Table 1. Institutional management of natural tourism objects in the Simarjarunjung area

No	Nature Tourism	Tourism Management Institutions
1.	Dolok Sipigul	The natural tourism object of Dolok Sipigul is managed jointly by the Forest Farmers Group (KTH) which has a management structure of the KTH 'Green), has agreed institutional forms and norms, and is applied in the management of tourism objects.
2.	Pine Hills	The natural tourism object of Pinus Hills is managed jointly by the Setya Tani Group which has a management structure for the Forest Farmers Group "Hijau" and has agreed to and implemented institutional forms and norms so that tourism development continues.
3.	Simarjarunjung Beautiful Hill	The natural tourism object of Bukit Indah Simarjarunjung is managed privately because the land is claimed to be privately owned. The management structure is only a family that works together to manage tourism objects. This tour has an organization called the Cooperation 'Jasa Destinasi Wisata Danau Toba' (Lake Toba Tourism Destination Service).
4.	Dolok Situangin	The natural tourism object of Dolok Sipintuangin is managed privately on the land which is claimed as private property. This tour does not have a management structure, it is only managed by the owner's family.
5.	Bukit Indah Simarjarunjung Annex	The natural tourism object of Bukit Indah Simarjarunjung Annex is managed independently on land that is claimed as private property. This tour has no organizational structure, only managed by the owner's family.

Not all tourism objects in the Simarjarunjung area are managed by community institutions (Table 1.). Only the natural attractions of Dolok Sipigul and Pinus Hills are managed by organized community groups. The rest (the natural attractions of Bukit Indah Simarjarunjung, Dolok Sipituangin, and Bukit Indah Simarjarungjung Annex) are managed individually by the party who claims to control the land, this is also as reported [31].

3.2.1. Dolok Sipigul nature tourism. Dolok Sipigul Nature Tourism is managed by KTH 'Hijau', which was established on November 28, 2017 by the letter of determination of Pangulu Nagori Parik Sabungan, Dolok Pardamean District, Simalungun Regency. The highest power and decisions are held by the 'Rapat Anggota Tahunan' (Annual Members Meeting). The concession permit for the manager of Dolok Sipigul's natural tourism services has been obtained from the Ministry of Environment and Forestry under the guidance of the Forest Management Unit (Kesatuan Pemangkuan Hutan or KPH) II Pematang Siantar. The permit refers to Indonesian Government Regulation No. 3 of 2008, stipulating the Regulation of the Minister of Environment and Forestry No. P.31/MenLHK/Setjen/Kum.1/3/2016 concerning Guidelines for Business Activities for Utilizing Nature Tourism Environmental Services in Production Forests. If it has developed properly, the existence of a concession permit for the management of Dolok Sipigul natural tourism objects can later provide income for the state as a form of contribution to the use of environmental services.

The structure of the KTH 'Green' includes a protector or builder consisting of the Head of Forest Management Unit KPH II Pematang Siantar, Pangulu (head) of Nagori Parik Sabungan, and the chairman of the Gapoktan 'Karya Bersama'. Advisors are elements of management that come from local community leaders. The core management structure consists of a chairman, secretary, treasurer, rubber transportation coordinator, cooperation coordinator, and members. The management of KTH 'Hijau' has been going well because it always holds member meetings to share the schedule of cooperation activities. At first, the people who joined the KTH 'Hijau' did not have the same thought and there were still misunderstandings. The coaches and advisors in this case play a role so that KTH 'Hijau' members have the same thoughts to jointly build and develop the Dolok Sipigul nature tourism object to what it is today. From the development aspect, the Dolok Sipigul nature tourism object is hampered by the financing factor.

Communities that are members of the KTH 'Hijau' must always participate in forest management activities for nature tourism by the KTH 'Hijau' work program, including:

1. Manage and utilize non-timber forest products in the form of pine tree sap in forestry partnership cooperation areas by applicable regulations.
2. Managing and utilizing forest areas with an agroforestry pattern with a combination of forestry plants and crops of lemongrass, coffee, corn, ginger, and other agroforestry crops. Honeybee cultivating is also carried out in forestry partnership cooperation areas by applicable regulations.
3. Implement forest protection in areas within and around the location of forestry partnership cooperation which includes prevention of forest fires, forest area encroachment, illegal logging of trees, and hunting of wild animals.
4. Maintain the conduciveness of the implementation of activities in the area of forestry partnership cooperation.
5. Carry out other obligations by the agreement and applicable laws and regulations
6. Resolve problems according to applicable rules or in a family manner if there is a land dispute with the community or other parties.

Regarding forest management, the objectives of the KTH 'Hijau' by the basic laws and basic rules or organization are as follows:

1. Gather all forest communities that manage land in forest areas to get definite legal protection from the government, in this case, the Ministry of Environment and Forestry through Forest Management Unit (KPH) II Pematang Siantar, Forestry Service Affairs of North Sumatra Province.
2. Obtaining economic and social benefits from pine tree stands and agroforestry activities while still paying attention to sustainability aspects to improve the welfare of communities around the forest.
3. Fostering the spirit and awareness of the nation and state to create good citizens.
4. Implement forest protection in areas within and around the location of forestry partnership cooperation in the form of forest fire prevention, forest area encroachment, illegal logging of trees, and hunting of wild animals.
5. Carry out plant enrichment in unproductive forest areas in collaboration with the government and other community organizations to promote Nagori Parik Sabungan, Dolok Pardamean District, and Simalungun Regency.

3.2.2. Pine Hills nature tourism. The natural tourism object of Pinus Hills in Nagori Parik Sabungan, Dolok Pardamean District, Simalungun Regency is a pine forest that is managed through a social forestry program so that it can be accessed and utilized by the community. Areal around the Pine Hills has the potential for natural beauty that is very well managed for tourism. Nearly one hundred hectares of forest land are used by farmer groups to carry out various activities including tourism activities [8,19].

Pinus Hills natural tourism object is institutionally managed by KTH Setya Tani which was formed in 2018. This organization consists of people who live around Pinus Hills Nature Tourism and have a passion for developing tourism. The establishment of the Setya Tani KTH is specifically carried out by the community who will become members and play a role in managing Pinus Hills nature tourism. The community already knows and is aware of what to do and also the problems faced to develop tourism. They also realize that the community is one of the important elements in tourism and the community must be involved as one of the stakeholders. The current community acts as a host to provide good service to tourists.

The structure of KTH Setya Tani includes a coach, chairman, secretary, treasurer, field coordinator, and members. The structure can evolve according to the needs of the organization. The purpose of KTH Setya Tani is to bring together communities around the forest to manage and develop the existing potential to obtain economic benefits while maintaining ecological functions. The organizational structure of KTH Setya Tani is as follows:

3.2.3. Bukit Indah Simarjarunjung nature tourism. The natural tourism object of Bukit Indah Simarjarunjung is institutionally managed by the Cooperation 'Jasa Destinasi Wisata Danau Toba'. The establishment of this cooperation institution aims as a forum to involve the community so that they are aware of the huge potential for nature tourism in this region which can have a positive impact on the economy. This is because the tourism sector has an important role in encouraging and accelerating the economic growth of the community, especially in reducing the number of unemployed [32,33]. If it has developed, natural tourism like this also has the potential to be a source of foreign exchange for the country [34].

The organizational structure of the Cooperation 'Jasa Destinasi Wisata Danau Toba' includes a chairman, secretary, treasurer, and members. The structure refers to regulations related to the establishment of cooperatives and the need for the management of tourism objects in the field. Cooperation is built by carrying out the values of togetherness and cooperation [35]. Until now, this cooperative does not yet have a permit from the government in its activities to manage nature tourism. Further strengthening and administrative processes are still needed to immediately obtain a local government cooperative permit.

3.2.4. Penatapan (View Point) Dolok Sipintuangin. Penatapan Dolok Sipintuangin is a tourist attraction in the form of hills that are a viewpoint to see Lake Toba from a height, where residents call it a 'penatapan'. From this height, visitors can enjoy the view of Lake Toba, the mountain's microclimate, and also intragenic spots for the public and visitors who want to take pictures. A tourist area of 0.8 hectares, Penatapan Dolok Sipintuangin offers a contemporary-style photo spot with a modern design that looks very stunning. Natural Tourism Penatapan Dolok Sipintuangin has not been managed in groups and tourism management is those who own land that is used for tourism. So this tour does not have a management structure. Institutionally, this tourist attraction is claimed to be privately owned, so the natural tourism of Penatapan Dolok Sipintuangin does not have a group-based management institution. The managers of these tourist objects are those who own land that is used as a tourist attraction, so they do not have a management structure. Natural tourism manager Penatapan Dolok Sipintuangin finds it difficult to manage tourism because, during the Covid-19 pandemic, managers have difficulty in developing tourism due to the drastic decline in visitors.

3.2.5. Bukit Indah Simarjarunjung Annex. Bukit Indah Simarjarunjung Annex was formed in 2018, formed from the separation process with BIS. This tourism object is not managed as a group, but individually by a party who claims to own land assets in the tourism object. BIS Annex and BIS natural tourism objects were previously merged into one. In 2018 there was a separation due to problems related to land. This obstacle takes place after a year of this natural tourist attraction running. For a long time, the problem was not resolved until finally each of them carried out land acquisition for access to tourist sites. From then on they became two separate objects from each other. The manager hopes that there will be cooperation with the Government for tourism development in the form of promoting tourism to the community.

3.3. Utilization of Natural Resources in the Simarjarunjung Nature tourism area

Utilization of natural resources uses an approach of 4 categories of ownership and credit of natural resources, namely public property (open access), state property, private or individual property, and communal property [36,37]. Each natural tourism object in the Simarjarunjung area has different natural resources (Table 2) [38].

Table 2. Forms of the utilization of natural resources in the Simarjarunjung Nature Tourism Area

No	Tourism site	Weld Status	Forms of Utilization of Natural Resources
1.	Dolok Sipigul Nature Tourism	KTH cooperation	• The natural charm of the mountains • Aesthetics of pine stands • Pine stands microclimate • Agroforestry Ecosystem
2.	Pine Hills Nature Tourism	KTH cooperation	• The natural charm of the mountains • Pine stands microclimate • Aesthetics of pine stands
3.	Bukit Indah Simarjarunjung Nature Tourism	Independent	• Lake Toba View • The natural charm of the mountains • Pine stands microclimate
4.	Dolok Sipintuangin Nature Tourism	Independent	• Lake Toba View • The natural charm of the mountains • Microclimate of pine stands.
5.	Bukit Indah Simarjarunjung Nature Tourism Annex	Independent	• Lake Toba View • The charm of the mountains • Pine stands microclimate

The form of utilization of natural resources in the Simarjarunjung Nature Tourism Area is explained clearly in the section below.

3.3.1. Dolok Sipigul. The natural resources of Dolok Sipigul tourism are state-owned and open access but have been given exclusive rights in the form of forest service management for nature tourism activities to KTH Hijau. The natural resources in Dolok Sipigul Nature Tourism are utilized and managed together. Efforts made by Dolok Sipigul tourism actors from the city government, village government, community, and all related parties, in this case, are in the form of utilizing forest services for business development in the tourism sector. This sector is supported by the potential of Parik Sabungan Village which is rich in agricultural resources so it has superior potential in the agro-tourism sector [39]. The most important assets to attract tourist visits are authenticity, uniqueness, comfort, natural beauty, and local wisdom.

Natural resources are elements of the natural environment, both physical and biological that are needed by humans to meet their needs and improve their welfare. For tourism, these natural resources have a very important role in the management of Dolok Sipigul tourism objects. This is because the management of natural tourism cannot be separated from its natural potential, especially in Dolok Sipigul where the agricultural sector is a potential sector to become a tourist attraction. The next thing to do is control efforts so that the potential of these natural resources can be used sustainably. Based on the statement [40], tourism currently exploits the natural and cultural resources of the community so there is a need for control to maintain natural and cultural resources for the benefit of tourism activities [41].

This nature tourism also has a camping ground which is an arena for camping for visitors. The camping ground is generally in demand by tourists who come to this area for various types of special activities including research, organizational/scout activities, and enjoying the beautiful scenery at night at tourist attraction locations. The rural atmosphere and beautiful natural environment are the main attraction for visitors to enjoy nature by camping. The management of the camping ground still needs to be developed as a natural tourist attraction by the manager with government support, so that the tourist attraction can be better known and visited by many tourists.

3.3.2. Pine Hills. The natural resources in the Pine Hills forest tour are the beauty of pine tree stands and beautiful hills that attract tourists to travel in the Pine Hills forest tour. Natural resources will not run out because they can be reproduced and the quantity and quality of the resources are to their management. The process of utilizing natural resources is still being developed by the Pokmas Setya Tani which is still in the development process.

The utilization of natural resources in the natural attractions of Pinus Hills is managed in cooperation. One of the efforts that can be done is by utilizing exotic natural resources as tourism objects and utilizing agricultural resources through the use of non-timber forest products to improve welfare. The development of tourism in this location is an effort that is expected to be a shortcut to accelerating economic growth. The community manages existing natural resources such as forest plants that can be used for food, medicine, crafts, and other purposes. The technology used daily by the community consists of simple tools made of iron and wood such as hoes, axes,

3.3.3. Bukit Indah Simarjarunjung. The potential of natural resources that can generally be utilized in the highlands is tourism potential. The highlands in this area have great potential that can be developed because the natural scenery at the location is very beautiful. The potential natural resources in this object include the main view of Lake Toba, the natural charm of the mountains, and the microclimate of pine stands. The utilization of natural resources in the Bukit Indah Simarjarunjung natural tourism area is managed privately because the land is claimed as private property. Therefore, its contribution to the welfare of the surrounding community is still minimal. Local governments also have limited access and roles in their development. In fact, according to article 17 paragraph (1) of the Indonesian Repiblik Law on the Local Government states that local governments can utilize, maintain, control the impact and cultivate it. This condition is different from other objects that get a lot of support from relevant government agencies because they are managed by community institutions that formally have concessions from the government for tourism management.

3.3.4. Dolok Situangin. The potential of natural resources that exist in the natural tourism object of Penatapan Dolok Si Pintuangin is the charm of the beautiful view of Lake Toba, the natural charm of the mountains, and the microclimate of pine stands. Not all of this potential is utilized properly, it is still only limited to the beauty of the charm of Lake Toba. The potential for microclimate has not been utilized properly, even now they complain about changes in the microclimate around tourism objects. The perceived change in microclimate is characterized by an increase in the average air temperature, an increase in the amount of rainfall, an increase in the number of rainy days, and a decrease in wind speed. The use of natural resources for tourism in Dolok Si Pintuangin has implemented sustainable principles. Efforts to increase the manager's income through tourist visits are carried out by continuing to strive for environmental balance to be maintained. This is by the conception that the most important elements that become the attraction of an ecotourism destination are its natural conditions, unique, rare, and endemic flora and fauna conditions, conditions of natural phenomena, and customary and cultural conditions [42].

The manager of the natural tourism object of Dolok Si Pintuangin hopes for cooperation and support from various parties in the management and development of tourism and the use of natural resources to preserve existing natural resources, including local governments. However, from the institutional aspect of management where this tourism object is claimed and managed as private property, it causes a lack of cooperation and support from the expected parties.

3.3.5. Bukit Indah Simarjarunjung Annex. The Bukit Indah Simarjarunjung Annex natural tourism object utilizes the potential of natural resources in the form of the beautiful view of Lake Toba from a height, the beautiful natural charm of the mountains, and the microclimate of pine stands. From this tourist attraction, you can immediately see the view of Lake Toba from a very good angle. In addition, this tour has a microclimate that includes cool air, warm morning sun, and sufficient rainfall. The existence of this kind of natural gift is not always available in other locations so its use must be maintained so that it continues to be sustainable. BIS Annex natural tourism objects are also claimed and managed as private property so that the use of natural resources is also managed privately. This causes a lack of collaboration and support from various relevant stakeholders in its development.

(a) (b) (c) (d)

Figure 2. Some natural attractions in the Simajarunjung area (a) Dolok Sipigul (b) Pinus Hills (c) Simarjarunjung Hills (d) Dolok Sipintuangin viewpoint

Tourist attractions can be in the form of tourist objects and tourist attractions (Figure 2). A tourist attraction is a tourist attraction that is static and tangible. In addition, service facilities within a destination include accommodation, restaurants, and facilities [43]. The development of tourism objects and attractions which are the main drivers of the tourism sector requires the cooperation of all stakeholders consisting of the community, government, private sector, non-government organizations, and other parties. By its duties and authorities, the government is a facilitator who has a role and function in making and determining all policies related to the development of tourism objects and attractions [44,45]. Meanwhile, the community is the party that should be the main actor as well as those who most feel the benefits of managing the potential of renewable natural resources.

The different characteristics and motivations of natural tourism visitors that have been known through this study should be an important reference source for managers to provide the right services according to the segmentation and motivation of visitors. The visitor satisfaction that has been measured in this study should be the most effective mirror for managers to determine what steps should be taken to develop. Likewise, the results of the study are related to factors that affect visitor motivation significantly and positively and should be a reference for the policies of all relevant stakeholders in developing this tourist area. Collaborative development of natural attractions is highly recommended so that the results are better and more efficient. The application of the principles of sustainable management is also required to continue to be respected by all parties involved in mini-style tourism activities, both managers and visitors.

4. Conclusions

Natural tourism management institutions in the Simarjarunjung area have various institutional management models. Some of them are in the form of Joint Farmers Groups and Tourism Aware Groups (Dolok Sipigul and Pinus Hills). Some of them are managed privately as land owners. The self-managed natural attractions are Bukit Indah Simarjarunjung Nature Tourism, Dolok Si Pintuangin, and Bukit Indah Simarjarunjung Annex. Natural resources in the Simarjarunjung Tourism Area have different natural resources, including natural charm, mountains, pine stands, agroforestry plants, and microclimate. Natural resources in the Simarjarunjung Tourist Area have different natural resources including views of Lake Toba, the natural charm of mountains, aesthetics of pine stands, agroforestry ecosystems, and the microclimate of pine forests. The community also uses nontimber forest products and agro-commodities as supporting activities for natural tourism. The use of these natural resources is managed to improve the community's economy.

Acknowledgment
The authors would like to thank the Ministry of Education, Culture, Research and Technology for funding this research. Special thanks to the Simarjarunjung Nature Tourism manager and visitors who have provided data and field guidance. We also express our deepest gratitude to other parties who cannot be mentioned for their cooperation during this research.

References
[1] Western Australian Tourism Commission and Western Australia 1997 *Department of Conservation and Land Management 1997 Nature-based tourism strategy for Western Australia* (Perth, W.A: Western Australian Tourism Commission: Department of Conservation and Land Management)
[2] Weaver D, Faulkner B and Lawton L 1998 *Nature-based tourism in Australia and Beyond: a preliminary investigation* (CRC Tourism Work in Progress Report Series, Report 1, Griffith University)
[3] Ziffer K 1989 *Ecotourism: The Uneasy alliance In Conservational International* (Washington, DC: Ernst & Young, Working Papers on Ecotourism)
[4] Krüger O 2005 *Biodivers Conserv* **14** 579
[5] The International Ecotourism Society (TIES) 2018 *What is Ecotourism?* (Washington DC: TIES)
[6] Stronza A L, Hunt C A and Fitzgerald L A 2019 *Annual Review of Environment and Resources* **44** 229
[7] Fennell D A 2001 *Issues Tour* **4** 403
[8] Chesner C A 2012 *Quaternary International* **258** 5
[9] Hidayat T W and Irwan N 2019 *Jurnal Ilmu Administrasi Publik* **7** 88
[10] Mambrasar Y M, Damayanto I P G P and Atikah T D 2018 *IOP Conf Series: Earth and Environmental Science* **308** 012083
[11] Damanik C G B 2018 *JOM FISIP* **5** 2
[12] Badan Pusat Statistik Kabupaten Simalungun 2019 *Kecamatan Dolok Pardamean* (ISBN: 978-602-5512-65-0)
[13] Joseph E M and Amanda L S 2011 *Journal of Environmental Management* **9** 1950
[14] Nuraeni Y 2018 *Jurnal Akuntansi Manajerial (Managerial Accounting Journal)* **3** 42
[15] Simarmata M M, Sudarmanto E, Kato I, Nainggolan L E, Purba E, Sutrisno E and Karim A 2021 *Ekonomi Sumber Daya Alam* (Yayasan Kita Menulis)
[16] Nikoyan A, Kasim S, Uslinawati Z and Yani R 2020 *Fakultas Kehutanan dan Ilmu Lingkungan Universitas Halu Oleo* **16** 34
[17] Barak D R 2006 *How Community Institutions Create Economic Advantage: Jewish Diamond Merchants in New York*
[18] Soleh A 2017 *Jurnal Sungkai* **5** 32
[19] Suaib H H 2017 *Suku Moi: nilai-nilai kearifan lokal dan modal sosial dalam pemberdayaan masyarakat* (An1mage)
[20] Lestari E and Satria A 2015 *Buletin Ilmiah Marina Sosial Ekonomi Kelautan dan Perikanan* **1** 67
[21] Listiyani N, Hayat M A and Mandala S 2018 *Jurnal Media Hukum* **25** 217
[22] Insarullah I 2020 *Tadulako Law Review* **5** 49
[23] Ploeg F V D 2011 *Journal of Economic Literature* **49** 366
[24] Olalekan R M, Omidiji A O, Williams E A, Christianah M B and Modupe O 2019 *MOJ Ecology and Environmental Sciences* **4** 114
[25] Nerini F F, Tomei J, To L S, Bisaga I, Parikh P, Black M and Mulugetta Y 2018 *Nature Energy* **3** 10
[26] Sri F and Ayu S K 2020 *International Conference on Climate Smart Sustainable Agriculture* 1
[27] Wolfe D A, Jaffe P G and Jetté J L 2003 *American Psychological Association* D12
[28] Andiani N D and Widiastini N M A 2017 *Jurnal Kewirausahaan dan Bisnis (JKB)* **20**
[29] Trisnanto T B, Fitriani F and Fatih C 2017 *Society, Culture and Politics* **30** 59

[30] Restu J, Rinaldy D, Suparmoko M, Setyo S, Moersidik and Harry W 2017 *Environmental value losses as impacts of natural resources utilization of in coal open mining* (MATEC Web of Conferences)
[31] Purwoko A, Slamet B and Pohan N Z P 2022 *IOP Conf Series: Earth and Environmental Science* **977** 012111 IOP Publishing
[32] Suhel S and Bashir A 2018 *Economic Journal of Emerging Markets* **10** 32
[33] Puah C H, Jong M C, Ayob N and Ismail S 2018 *International Journal of Business and Management* **13**
[34] Alegre J, Pou L and Sard M 2018 *Current Issues in Tourism* **22** 1138
[35] Faedlulloh D 2015 *IJPA-The Indonesian Journal of Public Administration* **2**
[36] Hidayat 2011 *Universitas Negeri Medan: Sejarah Citra Lekha XV* **1** 19
[37] Katherine B H 2007 *Urban Geography* **28** 113
[38] Purwoko A and Sahputra A 2022 *The 5th International Conference on Agriculture, Environment, and Food Security IOP Conf Series: Earth and Environmental Science* **977**
[39] Ramadi F 2016 *Strategi Pengelolaan Objek Wisata Istana Kota Rebah Sungai Carang oleh Dinas Pariwisata dan Kebudayaan Kota Tanjungpinang* (Tanjung Pinang: Universitas Maritim Raja Ali Haji)
[40] Elwijan F S and Damayanti M 2017 *Wilayah dan Lingkungan* **5** 71
[41] Andrea G and Janet H K 2012 *Community-based tourism and local culture* **10** 173
[42] Sunarko, Eno M, Subyanto P C and Sudarsono A 2003 *Pemanfaatan Sumber Daya Alam di Indonesia*
[43] John F K 1998 *Conservation and Utilization of Natural Resources in the East Usambara Forest Reserves: Conventional Views and Local Perspectives Wageningen* (Agricultural University the Netherlands)
[44] Devi A H and Soemanto R B 2017 *Sosiologi Dilema* **32**
[45] Chenglin M, Debin F, Liyan S and Qiaoling L 2010 *Conservation and Recycling* 153

Anatomical properties of branches and twigs mangrove woods

L E Panggabean[1], A Nuryawan[1,2*], I Risnasari[1,2,4] and Susilawati[3]

[1]Faculty of Forestry, 2nd Campus in Kuala Bekala, Universitas Sumatera Utara, Deli Serdang 20355, Indonesia
[2]Integrated Research Laboratory, Universitas Sumatera Utara, Medan 20155, Indonesia
[3]Department of Physics, Faculty of Mathematics and Natural Science, Universitas Sumatera Utara, Medan 20155, Indonesia
[4]Center of Excellence for Mangrove, Universitas Sumatera Utara, Medan 20155, Indonesia

*Email: arif5@usu.ac.id

Abstract. In order to optimize the utilization of mangrove woods, particularly branches and twigs parts, this research was conducted to investigate anatomical properties of three species of mangrove woods, namely api-api (*Avicennia marina*), bakau minyak (*Rhizopora apiculata*), and berembang (*Sonneratia caseolaris*). The properties were comprised of both microtome sections and macerations dimensions. Procedure for microtome followed Sass method, resulted in three sections for each part, namely cross-section (L/longitudinal), radial/R section, and tangential/T section. Technical procedure for maceration based on Forest Product Laboratory (FPL) method, resulted in measurement of fiber dimension and derivative of fiber dimension. In term of microtome, the anatomical features between branches and twigs for each species in L, R, T sections are similar, but their quantitative anatomical observations between twigs and branches have differences in the vessel, rays and fiber parameters statistically. In term of fiber dimension and its derivatives (Runkel Ratio, Felting Power, Mulhstep Ratio, Coefficient Rigidity, Flexibility Ratio) both branches and twigs showed same quality class, namely 3rd class with scored 150-175 points.

1. Introduction

Indonesia is an archipelagic country that has 16,766 islands [1] with a coastline of approximately 95,000 km [2]. This condition showed that Indonesia has had large territorial waters. Both ocean and coastal zone are part of the territorial waters, these two zones have richness and uniqueness that can be utilized in the economic sector, including in the forestry sector. The mangrove forest is the primary ecosystem that supports life in the ocean and coastal areas. Mangrove forests provide many benefits such as environmental services (coastal protection from erosion, reducing abrasion, stabilizing coastal zones, reducing tidal waves), animal habitats, and sources of livelihood for coastal communities [3].

Despite the fact that mangrove ecosystem has many advantages, its existence is currently being degraded due to logging beyond the limits of sustainability and changing functions caused by development activities. Mangrove forest is a forest with a "fragile" system type which means it is susceptible to environmental changes. Under uncontrolled exploitation, it will be extremely easy to have severe degradation [4]. The damaged mangrove ecosystem will take a long time and much efforts to recover. Therefore, it is necessary to manage natural resources that are optimal environmentally so that the existing natural resources constantly sustainable [5].

In order to optimize the use of mangrove wood as an effort to support the sustainable use of mangrove resources, this study used branches and twigs of three types of mangrove wood as research material. So far, the community has used the two only as a result of handicrafts, even though they are considered wood waste as a result of pruning the canopy already intertwined with one another [6]. Further, to find out the extent to which branches and twigs can be used as a product that has added value from the economic and ecological aspects, it is necessary to identify the basic characteristics of the mangrove wood. The use of wood species that have known basic properties can be used as a source of information, so that their use is in accordance with the characteristics of the wood, considering that the basic nature of one wood can be different from another.

This study aimed to compare anatomical structures and fiber quality of three species of mangrove woods, particularly branches and twigs parts, originated from species of api-api (*Avicennia marina*), bakau minyak (*Rhizopora apiculata*), and berombang (*Sonneratia caseolaris*).

2. Materials and Methods

2.1. Place and time of research

This research was conducted at Laboratory of Forest Products Technology, Universitas Sumatera Utara, 2nd Campus Deli Serdang, Indonesia, from March to June 2022.

2.2. Sample selections

The woods raw materials were taken in Lubuk Kertang, by exploring three types of mangrove trees that predominant in the mangrove forest area. The mangrove wood used is ± 10 years old and obtained from stands that are close to the mainland [7]. Branches and twigs of wood were taken by pruning the lower branches of the tree canopy.

After obtaining stands that qualified, samples of branches and twigs of each species are collected. Number of branch and twig were 5 (five) for each, with a length of 20-30 cm for branch, and a length of 10-20 cm for twig, respectively. During collecting, the samples were then kept in wet fiber sacks [8]. In the laboratory, plastic bags containing sample are labeled according to the species and stored in a refrigerator for a few days until further observation [9].

2.3. Observation of the anatomical features

Observation of microscopic characteristics was carried out on the microtome incision according to the working procedure of Sass (1961) [9]. Branches and twigs without bark were cut 5 cm for each and then boiled for 72 hours. Further, the wood was soaked with a 1:1 glycerin alcohol to soften the sample. Samples were hand-sectioned approximately 10-12 μm thick using a razor blade for the longitudinal, radial, and tangential section. The best specimens were stained with safranin, dehydrated with graded alcohol, and placed on microscopic slides for measurement.

Anatomical features were observed based on the standards from the International Association of Wood Anatomists (IAWA) were used for hardwood identification [10] and classification of wood anatomy from the book of Atlas Kayu Indonesia. The parameters observed included the arrangement and distribution of the vessels/ pore of cells, parenchyma, and the arrangement of the radial.

2.4. Fiber analysis

Procedure for microscopic anatomical structure of the wood included the maceration based on Forest Product Laboratory method. The maceration samples were manufactured by slicing wood into some matchstick sizes, putting them in a test tube, adding glacial acetic acid and H_2O_2 into the test tube until the wood slices sinked. The test tube containing matchstick wood then was boiled on a waterbath at a temperature of ± 95°C.

After the wood changed into white and soften, the tube is shaken for 10 minutes until the fibers are separated individually. The fibers were filtered with filter paper and washed with distilled water (aquadest) until the fibers were acid free. Safranin is dripped sufficiently on the washed fiber, left for 3 hours, dehydrated with graded alcohol. Preparat is made by taking the fiber using a pipette and placing onto a glass object and covered with a cover glass. Observations were made using a light microscope equipped with a measurement application.

The value of the derivative of the fiber dimensions was determined based on the measurement results of the fiber dimensions according to Silitonga *et al.* (1972) [11]. The fiber dimensions measured were fiber length, fiber diameter, cell wall thickness, and lumen diameter. Measurements were made on number of 100 fibers. Fiber wall thickness were calculated based on Equation 1.

Fiber wall thickness (W) $= \frac{D-1}{2}$ ....(1)

Fiber derivative was calculated based on fiber morphology. The derived values were calculated based on Equation 2 to 6 as follow:

Runkel Ratio (RR) $= \frac{2W}{l}$ ....(2)

Felting Power (FP) $= \frac{L}{D}$ ....(3)

Mulhsteph Ratio (MR) $= \frac{(D^2-l^2)}{D^2} \times 100\%$ ....(4)

Coefficient Rigidity (CR) $= \frac{W}{D}$ ....(5)

Flexibility Ratio (FR) $= \frac{l}{D}$ ....(6)

Where:
D is width (μm); w is fiber wall thickness (μm); L is Length (μm); and l is lumen width (μm)

The derived values of the fiber dimensions were classified according to the classification table of Pasaribu and Silitonga (1977) as presented in Table 1 regarding the quality and grade of wood fiber as raw material for pulp and paper [12].

Table 1. Indonesian wood fiber assessment criteria

No	Remarks	Quality Class					
		I		II		III	
		Requirements	Score	Requirements	Score	Requirements	Score
1	Length (μm)	> 2000	100	1000-2000	50	<1000	25
2	*Runkle ratio* (RR)	< 0,25	100	0.25-0.50	50	0.51-1.0	25
3	*Felting power* (FP)	> 90	100	50-90	50	< 50	25
4	*Muhlsteph ratio* (MR)	< 30	100	30-60	50	61-80	25
5	*Flexibility ratio* (FR)	>0,80	100	0.50-0.80	50	< 0.50	25
6	*Coefficient of Rigidity* (CR)	< 0,10	100	0.10-0.15	50	> 0.15	25
	Interval value	450-600		225-449		<225	

2.5. *Statistical analysis*

Differences in anatomical structure between branches and twigs of mangrove wood were tested using statistical tests T-Test or Z Test to find out the difference between the two average sample parameters.

If the data is unnormally distributed, the statistical analysis used is the Mann-Whitney test. All tests were performed with a 5% confidence interval (p<0.05).

3. Result and Discussion

3.1. Api-api (Avicennia marina (Forssk.) Vierh)

3.1.1. Macroscopic observation of wood. Physical characteristics between the branches and twigs of Api-api (A. marina) do not have big differences. Observations showed that branches and twigs of wood are still not formed between sapwood and heartwood. The branches and twigs of Api-api have a smooth texture and have a distinctive decorative pattern. The cross section of the wood looks shiny when exposed to light with a rough touch impression. Twigs and branches of *A. marina* is classified as having a moderate hardness and a special smell when it is fresh.

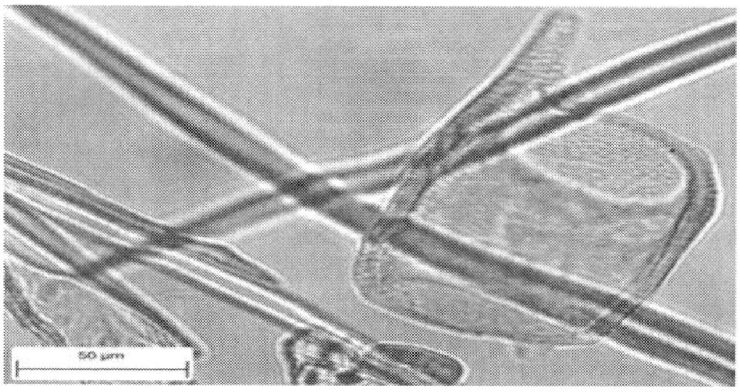

Figure 1. Vessel and fiber from maceration brances *A. marina*

3.1.2. Anatomical structure. The results of anatomical observations of twigs and branches of *A. marina* were adjusted according to the IAWA classification [10]. The anatomical results of the branches and twigs of *A. marina* showed that both of woods had clear growth rings which were indicated by the presence of parenchyma marginal bands, porosity vessel is wood ring porous, vessel arrangement is partially solitary and partially multiple radial vessels of 2-6 cells. The number of solitary vessels in *A. marina* twigs is 48%, which means that the vessels are partly solitary and partly multiple, while in the branches is 41%, which means that the arrangement of vessels is mostly multiple. *A. marina* has outline of rounded vessel. Perforation plates is simple, while shape of intervessel pits circular to oval (Figure 1). The results of microscopic observations of branch and twigs *A. marina* wood namely transverse / cross section, radial section and tangential section are presented in Figure 2 and Figure 3.

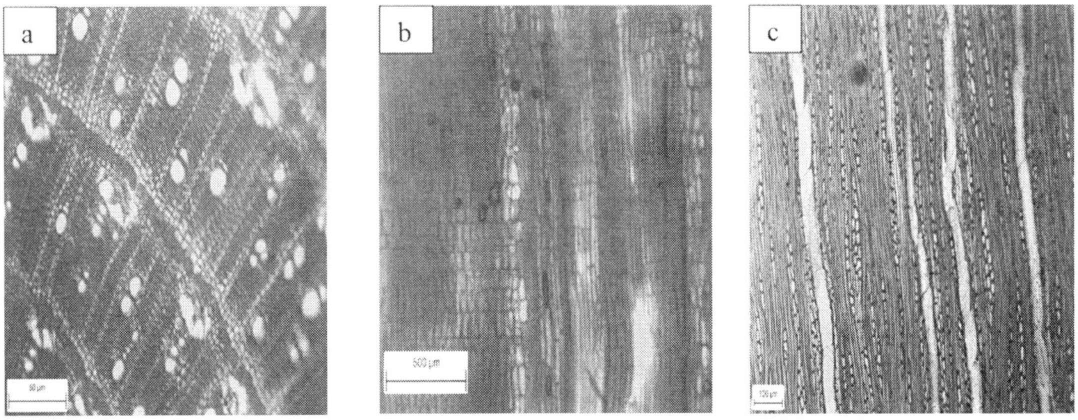

Figure 2. Anatomical features of twig of *A. marina* wood speciecs: a) longitudinal section b) radial section and c) tangensial section

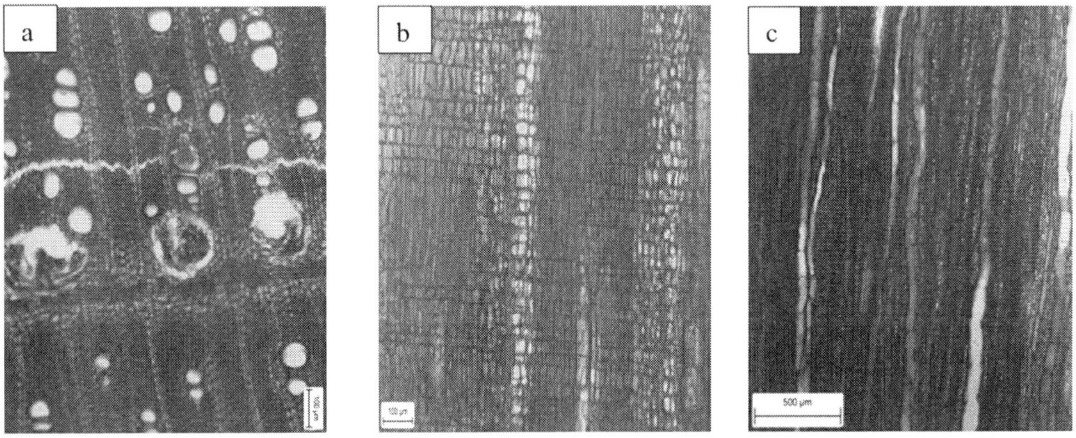

Figure 3. Anatomical features of branch of *A. marina* wood species a) Longitudinal section b) radial section and c) Tangential section

Characteristics of the axial parenchyma in the *A. marina* observed in the transverse section (Figure 2 and Figure 3) has a paratracheal parenchymal arrangement axial with parenchyma scanty paratracheal. In the tangential sections of the *A. marina* twigs (Figure 1c and Figure 2c) Showed rays type multiseriate and ray width 1 to 4 cells. The rays can be observed in the radial section as in Figure 1c and Figure 2c. The whole arrangement is square ray cell.

The ray's frequency of twigs and branches of *A. marina* is quite large, namely 14 per mm^2 and 11 mm^2, respectively. The ray's width of the twigs is classified as narrow, namely 21.12 ± 7.63 µm while in the branch, the width of the rays is quite narrow, namely 32.25 ± 12.3 µm. The height of the rays on the observation of twigs and branches of *A. marina* was classified as very short, namely 763.33 ±172.64 and 862.12 ±159.54 µm. In the transversal section, the strands extending tangentially near the growth ring of *A. marina* wood are concentric phloem strands interspersed with xylem zones [10]. Research by Schmitz [13] suggested that phloem bands in *A. marina* wood are united in parenchyma tissue and bordered by sclereids on the outside which play a role in metabolism repair and water retention. Phloem in *A. marina* wood plays an important role in drought stress, can tolerate salinity stress and affects wood density, if the proportion of phloem area is low, the wood density will increase [14].

IOP Conf. Series: Earth and Environmental Science 1115 (2022) 012066 doi:10.1088/1755-1315/1115/1/012066

3.2. Bakau minyak (Rhizophora apiculata)

3.2.1. Macroscopic observation of wood. The macroscopic characteristics of the R. apiculata wood particularly the twigs and branches showed that there was no sapwood and heartwood. Bakau minyak twigs and branches have had a smooth texture and have a plain pattern. The cross section of the wood looks shiny when exposed to light with a rough touch impression. Bakau minyak wood is classified as having a moderate to soft hardness and a special smell when it is fresh.

3.2.2. Anatomical structure. The results showed that the twigs and branches of R. apiculata had growth ring boundaries indistinct with diffuse-porous porosity and vessels in radial pattern. Group of vessels was partl solitary in 2-6 cells or clusters. In the twigs parts the number of solitary vessels was 42% while in branches was 45%. The twig and branch vessels of R. apiculata have had the outline of vessel as rounded with perforation plates with scalariform of less than ten bars (Figure 6). The results of microscopic observations of branch and twigs R. apiculata wood namely transverse / cross section, radial section and tangential section are presented in Figure 4 and Figure 5.

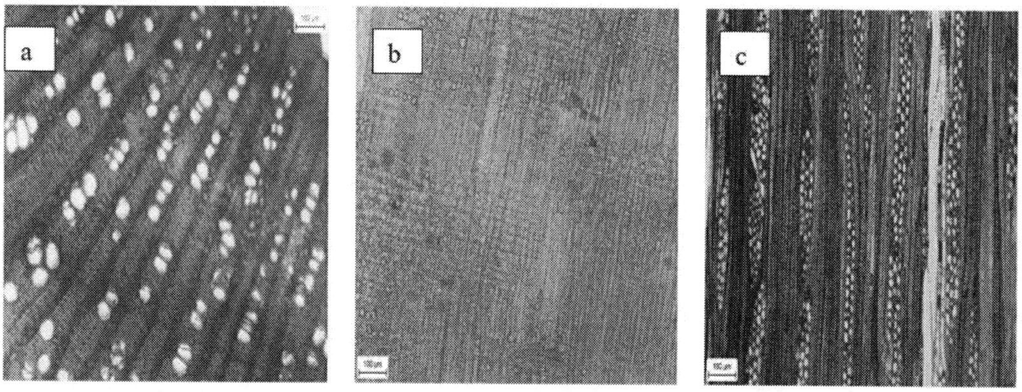

Figure 4. Anatomical features of twig of *R. apiculata* wood speciecs: a) longitudinal section b) radial section and c) tangensial section

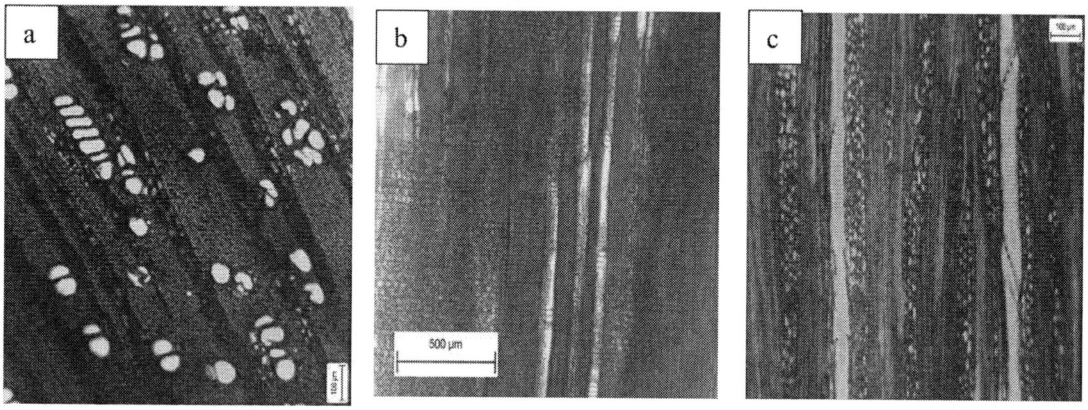

Figure 5. Anatomical features of branch of *R. apiculata* wood speciecs: a) longitudinal section b) radial section and c) tangensial section.

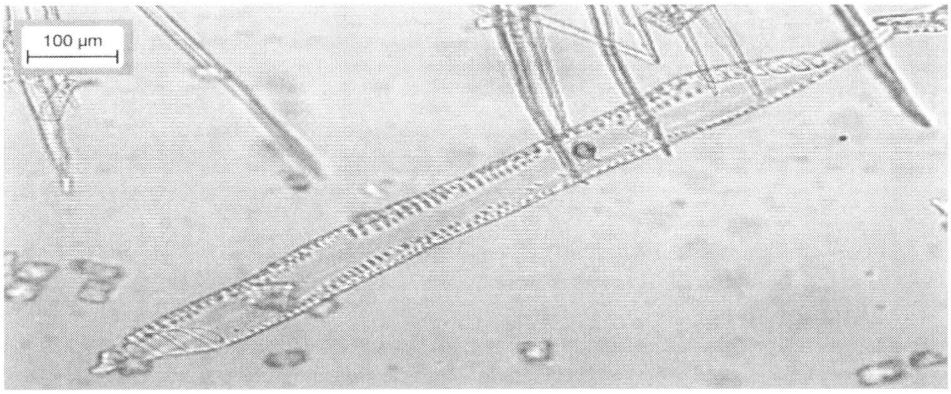

Figure 6. Branch Vessel *R. apiculata*

The results of quantitative data on vessel measurements of *R. apiculata* wood showed differences between the twigs and branches, namely vessels diameter and frequency. The vessels diameter of the branches is larger than the vessels diameter of the twigs. The diameter of the twig vessel of *R. apiculata* is 43,58 ± 8,02 µm which is classified as very small. Meanwhile, the branch has a vessel diameter of 59,72 ± 12,80 µm which is quite small. The frequency of vessels in the branches is less than the frequency of vessels in the twigs. In twigs, the frequency of vessels is 78 cells per mm^2 which is very high, while in branches, 39 cells per mm^2 are quite a lot. Vessel cell height in the twigs and branches were 400,11 ± 102,77 µm and 433,98 ± 176,99 µm, both of which were classified as moderate. Vessel contents in the twigs and branches were not found to have deposits or tilosis.

Characteristics of the axial parenchyma in *R. apiculata* wood observed in the cross section (Figure 4a and Figure 5a) has a paratracheal parenchymal arrangement with axial parenchyma scanty paratracheal. The results of the tangential incision on the twigs of *R. apiculata* (Figure 4c and Figure 5c) showed rays type multiseriate with ray width one to three cells. The arrangement of the rays in the radial section (Figure 4b and Figure5b) shows the body of the rays is square cells with mostly two to four rows of upright procumbent marginal cells.

The number of twigs of *R. apiculata* per mm^2 is 10 cells so it is quite a lot, the width of the rays is 28,77 ± 14,65 µm is classified as narrow. The height of the rays is 1078,76 ± 284,22 µm which is short. Meanwhile, in the branch, the rays of 7 cells per mm^2 is quite rare, the width of the rays is 42,34 ± 13,92 µm which is quite narrow, and the height of the rays is 1199,35 ± 344,78 which is short.

3.3. Berembang (Sonneratia caseolaris)

3.3.1. Macroscopic observation. The macroscopic characteristics of the *S. caseolaris* tree in the twigs and branches showed that the two have not yet formed between sapwood and heartwood. The wood is light brown-white with integrated straight wood grain directions. Twigs and branches of slender wood have had a smooth texture and a plain pattern. The cross section of the wood looks shiny when exposed to light with a rough touch impression. *S. caseolaris* wood is classified as having a soft hardness when pressed in the transverse, radial and tangential sections. If smelled it has no characteristic odor when it is fresh.

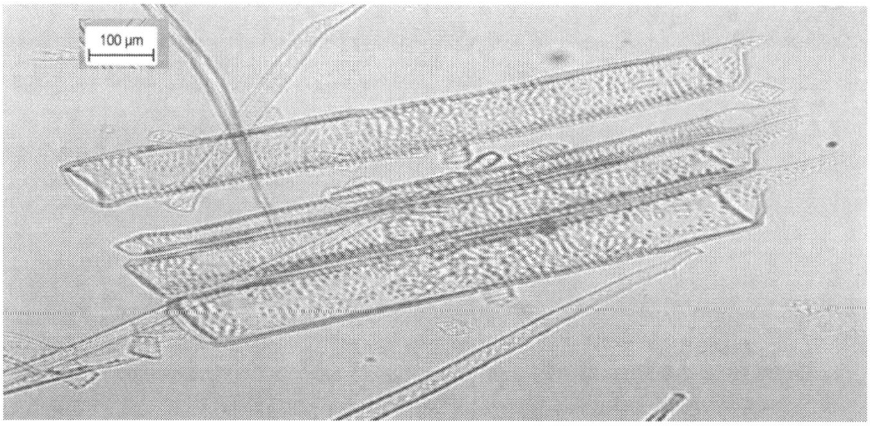

Figure 7. Branch vessel *S. caseolaris*

3.3.2. Anatomical structure. The results showed that the twigs and branches of *S. caseolaris* had growth ring boundaries indistinct, diffuse-porous porosity, vessels in radial pattern. Vessel grouping partly is solitary, composed of partly in 2-4 cells. The number of solitary vessels in twigs was 48% while in branches was 46%. The vessel of *S. caseolaris* have a simple perforation and Intervessel pits scalariform to opposite (Figure 7).

In the longitudinal section (Figure 8a and 9a) it can be observed that *S. caseolaris* wood is axial parenchyma absent. The results of tangential section on the twigs and branches of *S. caseolaris* showed rays type exclusively uniseriate. The arrangement of the rays in the radial section (Figure 8b and 9b) shows the body of the rays in the form of square cells. Frequency rays of *S. caseolaris* per mm^2 is 23 cells on the twigs and 17 cells on the branches, respectively, so it's quite a lot. The width of the branch rays is 16.85 ± 3.62 µm and the branch is 18.46 ± 4.62 µm which is classified as narrow. The rays of the branches are 321.48 ± 101.35 µm and those of the branches 489.94 ± 154.16 are both classified as unusually short.

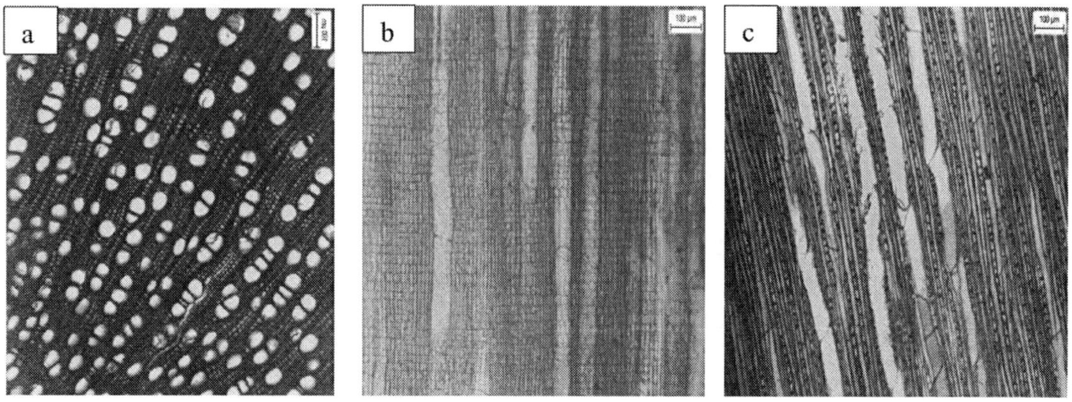

Figure 8. Anatomical features of twig of *S. caseolaris* wood speciecs: a) longitudinal section b) radial section and c) tangensial section

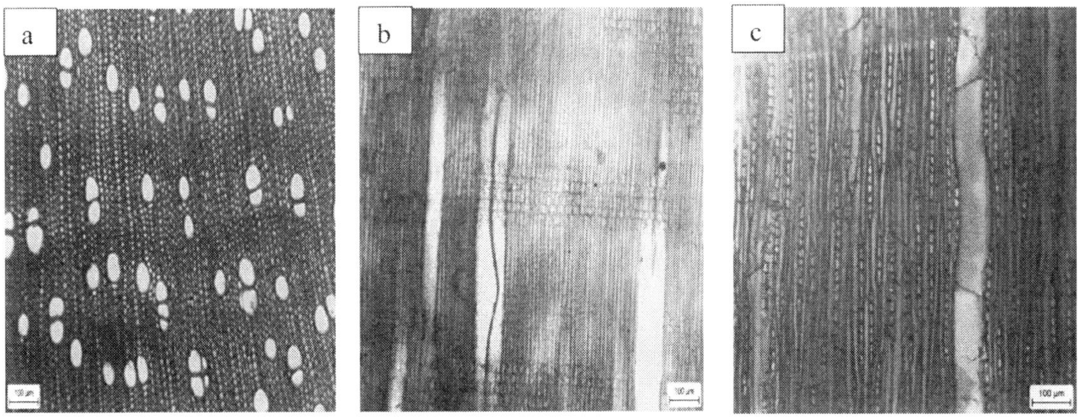

Figure 9. Anatomical features of branch of *S. caseolaris* wood speciecs: a) longitudinal section b) radial section and c) tangensial section

The results of measurements of diameter and pore height of *S. caseolaris* wood showed that the branch section had a larger vessel diameter and a longer pore height than the twig section. The diameter of the vessel on the twig is 42.49 ± 7.95 μm which is classified as very small, while the vessel has a diameter of 52.07 ± 6.55 μm which is relatively small. In twigs, the length of the vascular cells is 300.74 ± 74.45 μm, which is relatively short. While the length of vessel cells in the branch section is 384.83 ± 176.99 μm is classified as medium. The frequency of vessels in the twigs and branches of *S. caseolaris* is 188 and 66 cells per mm^2, respectively, that are very high. The contents of the rays on the twigs and branches were not found to have deposits or tilosis.

Quantitative microscopic characteristics between twigs and branches of mangrove wood in this study, contained categorical similarities on several parameters. However, statistically on the T-test there was a significant difference ($p<0.05$) between twigs and branches on the average parameter value even though they were grouped in the same class (Table 2).

Tabel 2. Statistical analysis of the difference in the mean of quantitative anatomical characteristics that have the same category

Species	Frequency Vessel	Vessel Length	Frequency Rays	Rays Width	Rays Length
Api-api putih (*A. marina*)	0,569	0,587	0,06 *	-	0,041 *
Bakau minyak (*R. apiculata*)	-	0,412	-	-	0,184
Berembang (*S. caseolaris*)	195 x 10^{-6} *	-	6 x 10^{-6} *	0,177	35 x 10^{-6} *

Remarks: * Significant at p < 0,05
- Twigs and branches have different categories

The frequency of branch vessels and twigs of *S. caseolaris* were both categorized as many, statistically the frequency of vessels in the twigs is more than the branches. The frequency of the rays of *A. marina* branches and twigs is quite large. *A. marina* twigs has a higher number of rays than those of branches of *S. caseolaris* wood in both parts. It was classified as having a very large number of rays because the number of rays on the twigs is more than those of the branches. In *A. marina* and *S. caseolaris* wood, the rays of the branches are longer than that of the twigs.

In this study, the types of mangrove wood that had good anatomical characteristics were the branches and twigs of *A. marina* and *R. apiculata* because they had fewer vessels and rays (parenchyma cells) than those of the branches and twigs of *S. caseolaris*. The less number of vessels and wood parenchyma will make the wood have greater density and strength in the producing of paper and high quality wood [15].

3.4. Fiber quality

The wood cells were separated individually through the maceration process. Measurement has been conducted to obtain the average fiber dimension. Important quantitative of fiber dimension of twigs and branches of *A. marina, R. apiculata* and *S. caseolaris* were tabulated in Table 3. Fiber lengths were categorized according to the classification recommended by the IAWA [16].

Table 3. Average value of fiber dimension

Wood Species	Position	Fiber Length (μm)	Fiber Width (μm)	Lumen Width (μm)	Fiber Wall Thickness (μm)
Api-api putih	Twigs	649.47±105.37	18.16±3.07	4.16±1.37	7.00 ±1.22
(*A. marina*)	Branch	665.63±230.69	15.80±2.41	3.67±0.84	6.06±1.09
Bakau minyak	Twigs	885.17±140.54	20.06±3.82	4.95±1.71	7.56±1.57
(*R. apiculata*)	Branch	1100.05±212.38	24.61±3.80	6.83±2.28	8.89±1.62
Berembang	Twigs	672.83±114.11	19.84±2.87	5.84±1.92	7.00±1.21
(*S. caseolaris*)	Branch	689.90±127.40	18.07±3.00	5.10 ±3.88	6.46±2.14

The fiber length of twigs and branches in *A. marina* and *S. caseolaris* species has an average fiber length of 649-689 μm including the very short category (fiber length is 500-700 μm). Twigs and branches of *R. apiculata* have longer fibers than those of *A. marina* and *S. caseolaris*. Twigs of *R. apiculata* have an average fiber length of less than 900 μm including the short category and branches of *R. apiculata* have an average fiber length in the medium category (900-1600 μm).

Statistically, *A. marina* and *S. caseolaris* did not have a significant difference in fiber length between branches and twigs (p>0.05) (Table 4). While the average fiber length of *R. apiculata* had a significant difference ($3.31 \times 10^{-14} < 0.05$) between the twigs and branches. The length of branch fiber *R. apiculata* is greater than of twigs. The length of the fiber greatly affects the quality of the wood product produced, in this study the fibers of mangrove branches and twigs meet the fiber length requirements in the manufacture of pulp and paper. Suansa and Al-Mefarrej (2020) stated that the longer the fiber (> 400 μm) the better the paper [8].

Table 4. Analysis of differences in average fiber length, fiber dimensions, lumen dimensions and fiber wall thickness in twigs and branches of three mangrove woods

Species	Fiber Length	Fiber Width	Lumen Width	Fiber Wall Thickness
Api-api putih (*A. marina*)	0,05	$2{,}94 \times 10^{-8}$ *	0,024*	$4{,}4 \times 10^{-8}$*
Bakau minyak (*R. apiculata*)	$3{,}31 \times 10^{-14}$ *	$6{,}23 \times 10^{-15}$*	$2{,}83 \times 10^{-12}$*	$1{,}42 \times 10^{-8}$*
Berembang (*S. caseolaris*)	0,31	$8{,}52 \times 10^{-5}$*	2×10^{-6}*	0,036*

Remarks: * Significant at p < 0,05

The average diameter of the twigs and branches of these three types of mangrove wood ranges from 15-24 μm, according to the Wagenfeur classification [17], the diameters of the twigs and branches of A. *marina*, R. *apiculata* and *S. caseolaris* are in the medium category (fiber diameter is between 10–24 μm) [18]. The lumen diameter of the twigs and branches of *A. marina* is in the very small category, while the lumen diameter of *S. caseolaris* in both parts is in the small category. The lumen diameter of the twigs of *R. apiculata* is in the very small category, on the branches it is in the small category [19]. According to Wheeler *et al.* (1989) a fiber is categorized as thin-medium if the lumen diameter is three or equal to two fiber wall thicknesses and is categorized as very thick if the lumen diameter is almost covered by the fiber diameter [10]. So the fiber walls of both parts of *R. apiculata* and *S. caseolaris* belong to the medium category. While the twigs and branches of A.*marina* are included in the very thick category. *A. marina* fibers with thick fiber walls can increase resistance to wind and tidal damage during storms [14].

Fiber diameter, lumen diameter and fiber wall thickness in these three types of wood are in the same category but statistically the average value is significantly different between branches and twigs for each type of wood where $p < 0.05$ (Table 4). Overall, the lumen diameter and a large fiber wall thickness was narrow, according to Mutmmainnah [20] if the lumen diameter is narrower and the fiber wall thickness is getting bigger in paper making, it will produce paper with low breakout and tensile strength [20].

3.4.1. Fiber derivative. Fiber cells are mechanical reinforcement in wood so that the stems, branches and twigs are not easily broken when exposed to the wind due to the secondary wall thickening process consisting of wood/lignin material [21]. In this research, the fiber dimensions obtained were calculated and assessed for the derivatives of the fiber dimensions for each type (Table 5 and Table 6). Fiber dimension is one of the important properties of wood to find the derivative value of wood dimensions. This value is used as a basis for estimating a type of wood as raw material for pulp and paper [22].

Table 5. Fiber derivative

Wood Species	Position	Runkel Ratio	Felting Power	Muhlsteph Ratio	Rigidity Coeffisien	Flexibility Ratio
Api-api putih (*A. marina*)	Twigs	3,37	35,76	94,75	0,39	0,23
	Branch	3,30	42,13	94,60	0,38	0,24
Bakau minyak (*R. apiculata*)	Twigs	3,05	44,13	93,91	0,38	0,25
	Branch	2,60	44,70	92,30	0,36	0,28
Berembang (*S. caseolaris*)	Twigs	2,40	33,92	91,34	0,35	0,29
	Branch	2,51	38,17	91,89	0,36	0,28

Table 6. Indonesian wood fiber assessment criteria to evaluate as raw material for pulp

Wood Species	Position	Scoring Result						
		FL	RR	FP	MR	RC	FR	Class (Total Score)
Api-api putih (*A. marina*)	Twigs	III (25)	III (25)	III (25)	III (25)	III (25)	III (25)	III (150)
	Branch	III (25)	III (25)	III (25)	III (25)	III (25)	III (25)	III (150)
Bakau minyak (*R. apiculata*)	Twigs	III (25)	III (25)	III (25)	III (25)	III (25)	III (25)	III (150)
	Branch	II (50)	III (25)	III (25)	III (25)	III (25)	III (25)	III (175)
Berembang (*S. caseolaris*)	Twigs	III (25)	III (25)	III (25)	III (25)	III (25)	III (25)	III (150)
	Branch	III (25)	III (25)	III (25)	III (25)	III (25)	III (25)	III (150)

The value of Runkle Ratio (RR) of branches and twigs of *A. marina, R. apiculata* and *S. caseolaris* (Table 5) is 2,40-3,37 classified in quality class III. In all samples the RR value showed poor fiber quality, twigs and branches of *S. caseolaris* had a lower runkle ratio than A. *marina* and R. *apiculata*. According to Frianto and Rojidin (2014) the smaller the Runkle Ratio value, the thinner the fiber cell wall so that it will produce better fiber quality [24].

The assessment of wood pulp raw materials based on fiber length (FL) and dimensional derivatives of branch and twig fibers of *A. marina, R. apiculata* and *S. caseolaris* (Table 6) confirmed that all samples obtained a total score of 150-175 so they were classified as quality class III. Fibers with a quality value of III such as branch and twig fibers of *A. marina*, *R. apiculata* and *S. caseolaris* have thick fiber walls and narrow lumens, where in the formation of pulp the fibers are difficult to flatten with bonds between fibers and weaving are not good and produce paper sheets with tear strength, and low tensile strength [23].

The value of Muhlstep Ratio (MR) of branches and twigs of A. *marina*, R. *apiculata* and *S. caseolaris* (Table 5) is 91,23-94,75 that belongs to the quality class III, based on the classification of muhlsteph numbers Kasmudjo (1994) the sample belongs to the class quality 4 is medium fiber quality and poor fineness. The *S. caseolaris* branch has the highest Flexibility Ratio (FR) of 0.29 and the *A. marina* branch has the lowest 0.23. According to Sunarti et al [25] wood as a raw material with a lower Muhlstep number value will produce better pulp quality, the density of the pulp sheet produced will be better with better strength properties. A high coefficient of flexibility will have a thin wall thickness and easily deform so that it will produce pulp sheets with better strength.

The value of Felting Power (FP) in the sample is 33,92 – 44,70 which is included in the quality class III. FP shows the number of bonds between fibers [24]. The sample has a low inter-fiber bond so that it affects the folding, tensile and breaking strength of the paper [26]. The value of FP required for good quality paper with fiber as raw material from hard wood is 40-60 [27]. Fibers of *A. marina*, twigs and branches of *R. apiculata* had FP values > 40. Coefficient of rigidity (CR) values in the sample were 0.35-0.39 and classified in quality class III. CR is the ratio between fiber wall thickness and fiber diameter, CR has a positive correlation to pulp yield and density [28].

In this research, branches had better quality than twigs, but both showed weaknesses in structural use as well as paper-making materials because they had high vessels or parenchyma and poor fiber quality. Utilization of twigs and branches of *A. marina, R. apiculata* and *S. caseolaris* may be directed in the manufacture of energy wood. In the research, Suansa and Al-Mefarej [8] suggested that branched wood has a high percentage of sapwood so that its chemical components contain large amounts of holocellulose which have the potential to produce products such as biochar, biofuel or bioenergy. A study has examined the physical properties of the branches of mangrove wood as the main material for making briquettes and pellets, because mangrove wood was previously known to have a high calorific value. The results show that branch wood has varying density values but tends to be high and will be advantageous if it is used in the manufacture of energy-dense wood because it can increase its high calorific value [29].

4. Conclusions

The observation of anatomical features of *A. marina*, *R. apiculata* and *S. caseolaris* have different variations both macroscopically and microscopically. Each type shows almost the same anatomical features between branches and twigs. Statistically, the results of quantitative microscopic anatomy observations between branches and twigs have differences in the vessel, rays and fiber parameters. The microscopic characteristics of branch wood have a better quantitative value than twigs. Branches or twigs both show weakness against structural use as well as paper-making materials because they have poor vessel, parenchymal and fiber characteristics, perhaps their use can be directed to energy wood production. In this research, the dimensions of the fiber and its derivatives (Runkel Ratio, Felting Power, Mulhstep Ratio, Coefficient Rigidity, Flexibility Ratio) both branches and twigs in all samples have the same dimensional derivative value, namely a score of 150-175 with quality III.

Acknowledgment
This research was funded by Directorate Research and Community Service, Ministry of Education, Culture, Research, and Technology Republic of Indonesia through Research Scheme of PDUPT (Fundamental Research of Competitive Higher Education) 2021-2023, year of 2022 to AN. Authors also thank to the late Mrs. Irawati Azhar as one of the members of this research.

References
[1] Statistics Indonesia 2021 *Statistics of Marine and Coastal Resources Year of 2021 BPS* (Central Bureau of Statistics)
[2] Wardhani M K 2011 *Marine journal* **4** 60
[3] Eddy S, Iskandar II, Ridho MR and Mulyana A 2019 *Indobiosains* **1**
[4] Saidah S and Sofia L A 2016 *Tropical Forest Journal* **4** 265
[5] Takarendehang R, Calvyn F S, Erly K, Deslie K, Indri S, Unstain N J W and Rembet 2018 *Journal of Tropical Coasts And Seas* **2** 45
[6] Batubara B, Mohammad B, Heri I, Bejo S, Arida S, Deni E, Mariah U and Ryzki P 2021 *Journal of Abdi Moestopo* **4** 12
[7] Imiliyana A, Muryonom M and Purnobasukim H 2012 *Journal Department of Biology, Faculty of Mathematics and Natural Sciences Sepuluh Nopember Institute of Technology*
[8] Suansa N I and Al-Mefarrej 2020 *Resouce Bioresource* **15** 479
[9] Sass J E 1961 *Botanical Microtechnique Iowa, the United State of America* (The Iowa State University Press)
[10] Wheeler, Elisabeth A, Pieter B and Peter E Gasson, eds 1989 *IAWA* **219**
[11] Silitonga T, Siagian R, and Nurachman A *Special publication* **12** 1
[12] Pasaribu and Silitonga 1977 Experiment Processing of Broadleaf Wood and Mixed Wood as Raw Material for Pulp and Paper Report No 100 Result Research Institute Forest Bogor
[13] Schmitz N, Robert E M, Verheyden A, Kairo J G, Beeckman and Koedam N 2008 *Avicennia Marina? Annals of Botany* **101** 49
[14] Santini N S, Schmitz N and Lovelock C E 2012 *Trees Springer* 0729
[15] Zhao X, Pingping G, Zhaolin Z, and Haixin P 2019 *Bioresources* **14** 1980
[16] Aprianis Y and Rahmayanri S 2009 *Journal of Forest Products Research* **27** 11
[17] Wagenfuehr R 1984 *Anatomic Des Holzes Veb Fachhicl Verlag Lepzig*
[18] Supartini and Erwin 2020 *Journal of Dipterocarp Ecosystem Research* **6** 25
[19] Pasaribu G, Sahwalita S and Sipayung B 2008 *Journal of Forest Products Research* **26** 16
[20] Muthmainnah M 2019 *Forestsains* **17** 6
[21] Sunaryo 2002 *Biology News* **6**
[22] Bogidarmanti R and Mindawati N 2011 *Materials in Prosiding MAPEKI*
[23] Kasmudjo 1994 *Methods of Determining the Proportion of Cell Types and Dimensions of Wood Cells Fostering Foundation for the Faculty of Forestry* (Yogyakarta: Gajah Mada University)
[24] Frianto D and Rojidin A 2014 *Seminar Nasional Mapeki XVII Medan* **11**
[25] Sunarti S, Praptoyo H and Nirsatmanto A 2016 *Journal of Forest Plant Breeding* **10** 135
[26] Sutiya D, Istikowati W T, Rahmadi and A Sunardi 2012 *Journal of Bioscientiae* **9** 8
[27] Jayusman J and Hakim L 2021 *Bioresources* **16** 4769
[28] Marbun S D, Wahyudi I, Suryana J and Nawawi DS 2019 *Journal of the Korean Wood Science and Technology* **47** 617
[29] Nuryawan A, Syahputra R S, Azhar I and Risnasari I 2021 *Iop Conference Series: Earth and Environmental Science* **891** 012005

Recycle energy: waste oil as alternative eco-friendly fuel

L Hilda[1*], R Lubis[1], Misdawati[2], Syafiruddin[3], G R Arafah[4] and A Y Halomoan[4]

[1]IAIN Padangsidimpuan, North Sumatra, Indonesia
[2]UNIVA Medan, North Sumatra, Indonesia
[3]UGN Padangsidimpuan, North Sumatra, Indonesia
[4]Universitas Brawijaya, Malang. East Java, Indonesia

[*]Email: lelya.hilda@gmail.com

Abstract. Used oil is a dangerous and toxic material whose use is strictly prohibited because it damages the environment and human life. Therefore, special handling is needed to dispose of the waste or reuse it with other functions (renewable energy). Researchers are interested in recycling used oil as a fuel that can help overcome the depletion of fossil fuels. This research is an experimental study conducted in a laboratory. Used oil which for a lot of time as waste that is not utilized will affect the environment and also health because it is a pollutant of Hazardous and Toxic Materials (B3). But before being reused, the harmful metals Pb, Fe, Al and Mn are removed by a filtering process, acidification to remove impurities and bind metals that contaminate the oil. After that, it is reused as a very efficient fuel in terms of price compared to using LPG or kerosene.

1. Introduction

High energy demands in Indonesia are still dominated by non-renewable energy with high pollution, derive from fossil, crude oil, and coal. The non-renewable energy gives implied meaning that this kind of energy cannot be reproduced this energy increasingly depleting so it is necessary to look for alternative energy to meet the needs. The energy alternate is clean and environmentally friendly [1,2].

In this century, energy has become one of the primary needs of mankind, this is highly used for cooking, fuel, industry, etc. but very unfortunate that crude oil holds the market which is non-renewable energy whose existence is dwindling. This means that the availability of alternative fuels is very necessary [3,4].

Fuel oil or BBM (*Bahan Bakar Minyak*) dominates national energy needs, with a total consumption of 52.50% - followed by coal with 21.52% of total consumption, gas: 19.04%, water: 3.73%, geothermal: 3.01%, and renewable energy is the least consumption with amount 0.2% of the total [5]. The increase in the use of national fuel oil is due to an increase in the transportation sector, as well as diesel power generation centers.

The energy needs are exponential increasing and prices are getting higher become a homework for the nation, if this is not getting solved then another problem will keep coming, like the problems for the lower middle class people to meet their daily fuel needs. High demands from nation community, also means an increase in energy demand for the industry. But industrial processing will produce waste as a by-product, high processing means high of produce waste, and one concern is used oil that used as a lubricant [6].

The increasing use of motorized vehicles causes the need for engine oil to be higher too, because motorized vehicles must change the oil gradually in order to maintain engine durability. By-product of waste oil is source of pollution in the environment because it contains heavy metals and hydrocarbon compounds in used oil as hazardous and toxic waste (B3) [7].

The lack of processing technology and research in Indonesia caused Indonesia, cannot to cultivate the waste oil, merely utility the waste oil as a lubricant, and even considered it worthless (low price selling, IDR.1,000/litre). Utilization that has not been maximized but its availability is getting higher, can be said that oil is a source of environmental pollutants, which must be considered in handling [8].

Waste oil is a serious problem for environmental and human health matter because due to the presence of very dangerous contaminants, become mandatory to manage based on an efficient collection system and proper processing. Waste oil is classified as dangerous waste, it is not only because of the abundant quantity, but waste oil content heavy metals, polychlorinated biphenyls (PCBs) and polycyclic aromatics hydrocarbons (PAHs)) which are harmful to green ecosystem and human healthy. On the other hand, waste oil can be considered a valuable resource, in the sense that it is possible to recover profitable energy or materials for further use. Thus, the gross calorific value of used oil is estimated to be around 40 MJ/kg and in 1 liter of motor oil can be produced using only 1.6 liters of used oil in recycling [9].

In the European Union about 5.7 megatons of lubricant (oil) are consumed every year. From this oil, and almost 2.7 megatons are potentially collectible. Therefore, the right way of waste oil processing, based on a proper collection system and processing process, it is a good opportunity to get the highest margin from cheap of raw material. At this moment, in the European Union that about 75% of waste oil is recycled & reused, and the amount of waste will decrease (reducing) [10].

Waste oil have been used for firing, but the utility is outside of general consumption, and only used as fuel for limestone and aluminum smelting. Insufficient waste treatment will cause a buildup of waste and that is a disaster, from the research results of used oil as the best stove fuel which has the lowest viscosity (10.58 cts) as a result of heating to a temperature of 100°C. The use of used oil as fuel for stoves is more efficient than using kerosene [3].

This study aims to eliminate heavy metals as a source of Hazardous and Toxic Materials (B3), the use of which is strictly prohibited if there is no process to produce new eco-friendly materials. Removal of heavy metals in used oil produces used oil as an eco-friendly fuel.

2. Materials and Method

The procedure for reducing or removing heavy metal in waste oil, by; filtering, is done by separating dissolved solids. Heavy metal content of waste oil is conducted by filtering. Zeolite sand is the filter medium used in this study [11]; precipitation is the process of converting a solution into a fluid from the contaminant and the contamination will turn into a solid. Clay or clay is a depositional medium used in this study. The clay content is silica (SiO_2) and fine aluminum oxide (Al_2O_3), which has a high absorption rate that can reduce the levels of ions and heavy metals and the ability to bind impurities [11]; neutralization reaction is a reaction between acids and bases that produce products in the form of water (H_2O), by releasing positive ions (H^+) and negative ions (OH^-). Acid and base reagents used in this study are HCl and NaOH. The addition of HCl in this study is useful for reducing and binding compounds olefins, aromatics, and non-hydrocarbon compounds present in used oil, while sodium hydroxide is useful for neutralizing acidity in used oil [12].

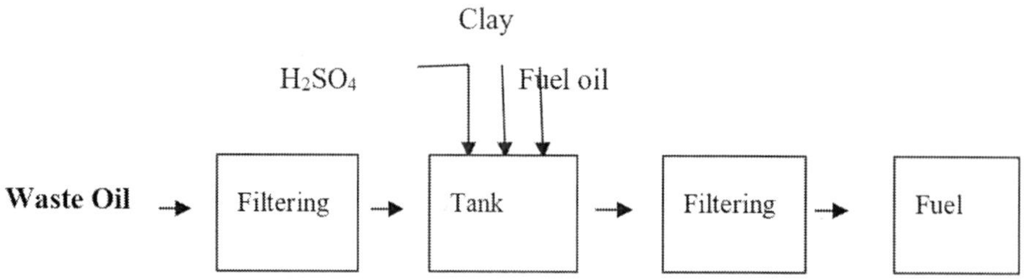

Figure 1. Scheme of waste oil processing into fuel

3. Result and Discussion

3.1 Result

3.1.1. Heavy metal determination of Pb, Al, Fe, and mn in waste oil. All of the result determine from experimental laboratory, and this was done in PPKS (*Pusat Penelitian Kelapa Sawit*).

Table 1. Heavy metal determination of pb, al, fe, and mn in used oil

Metal	A = Waste Oil = 100%	B = 90%	C = 80%	D = 70%	E = 60%
Pb	2.41	0.3	0.23	0.21	0.1
Al	31.61	12.83	8.65	6.59	4.79
Fe	19.91	19.47	18.21	16.56	8.12
Mn	4.12	3.43	2.84	2.75	0.64

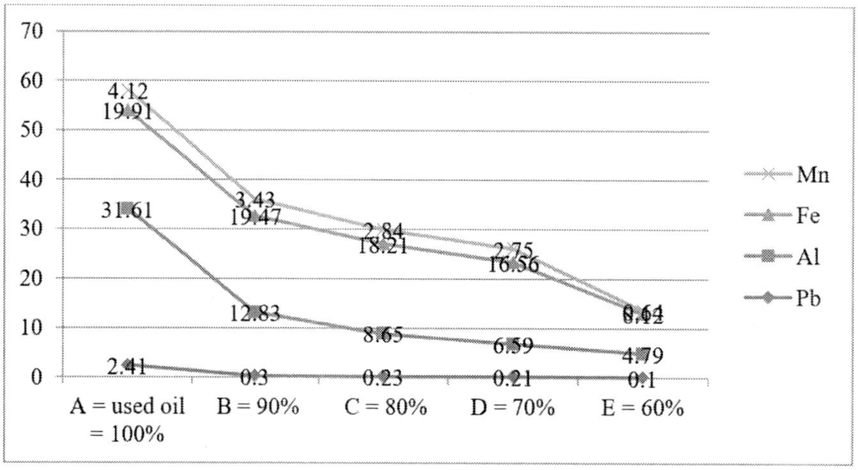

Figure 2. Heavy metal content rate of Pb, Al, Fe, and Mn in waste oil with several treatments

Pb Metal

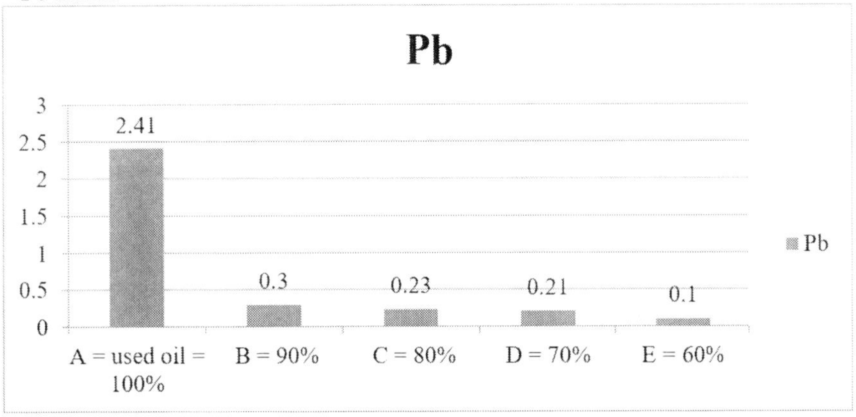

Figure 3. Pb concentration in waste oil with several treatments

The result shows that new oil or unused oil gives the value of Pb content as 0 ppm. Whereas waste oil that has not been touched by any treatment has a Pb rate of 3.12 ppm and the result will decrease with the addition of acetic acid as a metal binder. And the reaction was stopped by adding 25 ml of acetic acid to 50 ml of used oil.

The existence of Pb in the surrounding is very cautious because Pb is a kind of dangerous and toxic atom (B3). The maximum lead content in the air is 2.0 mg/l. based on PP RI No. 41 of 1999 [13]. While that which has been obtained from waste oil exceeds the threshold. the lead threshold value for clean water and drinking water is based on the Minister of Health of the Republic of Indonesia No. 416 of 1990 which is equal to 0.05 mg/l.

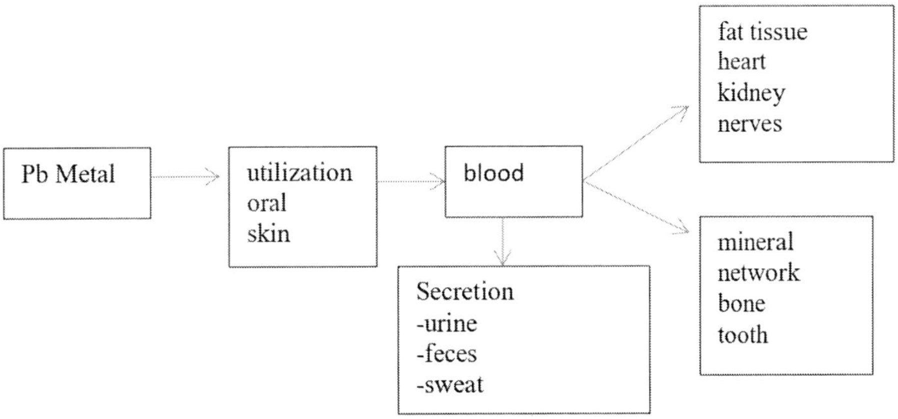

Figure 4. Accumulation of Pb metal in human organ

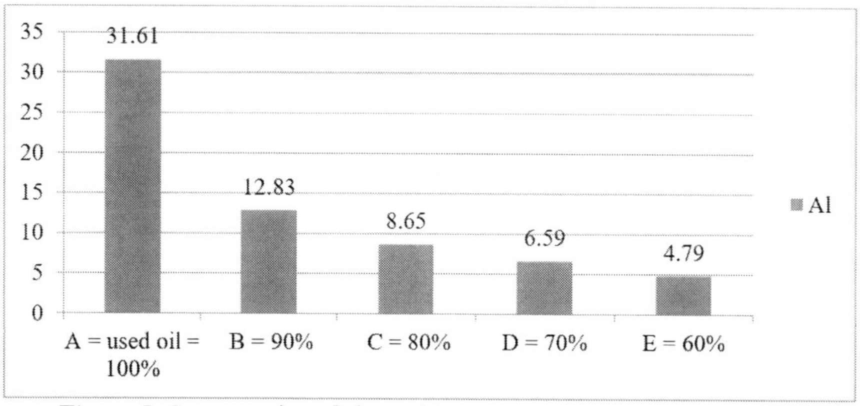

Figure 5. Concentration of al metal in waste oil with several treatments

Exposure of Al metal to human' skin will clog pores. The high Al concentration will caused central nerve damage, memory loss, lethargy, trembling.

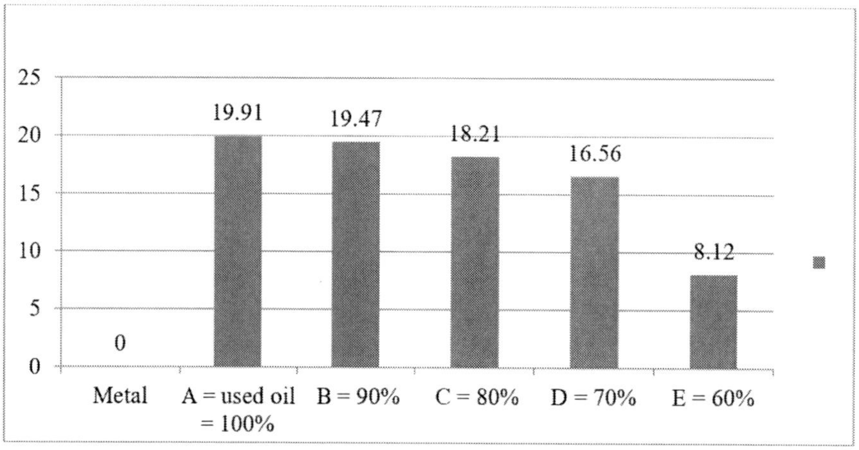

Figure 6. Fe metal content in waste oil with several treatments

Fe is an essential metal or metal that is needed by humans, but if it exceeds the specified threshold, it will have a harmful effect on humans.

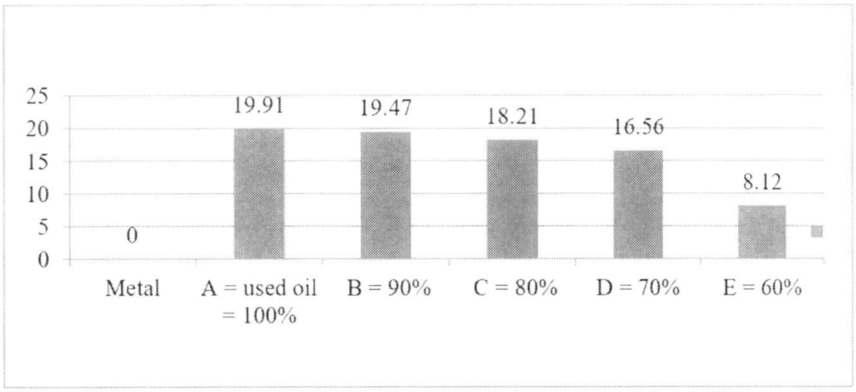

Figure 7. Mn metal content in waste oil with several treatments

Mn metal same as Fe metal that essential but if it's exceed the allowed threshold will cause chronic poisoning in humans so that it has an impact on causing weakness in the legs, dull face and continuous effect (slow speech and hyperreflexia) for human that poisoning on Mn [13-15].

Table 2. Viscosity determination

Viscosity	I	II	Average
Viscosity (new)	2.03	2.04	2.03
No Treatment	0.43	0.44	0.44
Viscosity A	0.40	0.40	0.40
Viscositas B	0.35	0.35	0.35
Viscositas C	0.35	0.35	0.35
Viscositas D	0.35	0.37	0.37

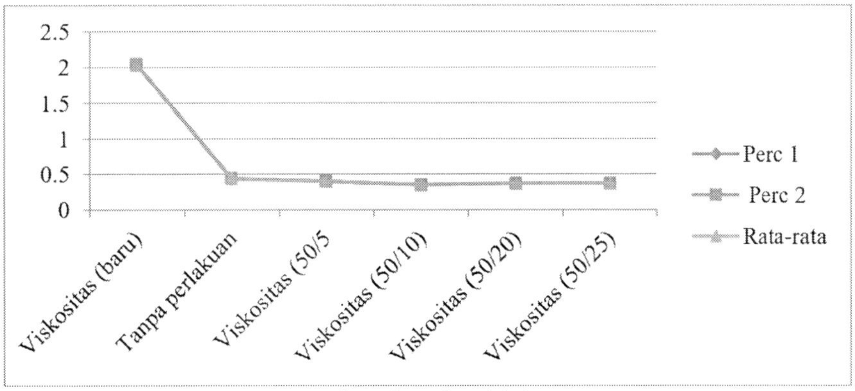

Figure 8. Viscosity of multiple treatments

Viscosity will increase correlate with concentration and will decrease with high temperature. This is due to the movement of the liquid particles which gets faster when the temperature is increased and the viscosity decreases. This graph is performed at T = 100°C with different concentrations.

3.1.2. Oil as fuel. After all treatment done by reducing and removing dangerous metal content, and measuring viscosity, then continued testing the waste oil as fuel. Next trial is using the waste oil as fuel for drying lemongrass. Looking to industry, known that businessmen use LPG gas to dry lemongrass – in all the needs are five 3 kg gas cylinders with a price of IDR 25,000/cylinder, so the total modal need is 125,000 for the fuel. While using waste oil as fuel requires 5 litres, 1 liter is IDR 2,000, so from an economic point of view it is very profitable to use waste oil that is already environmentally friendly.

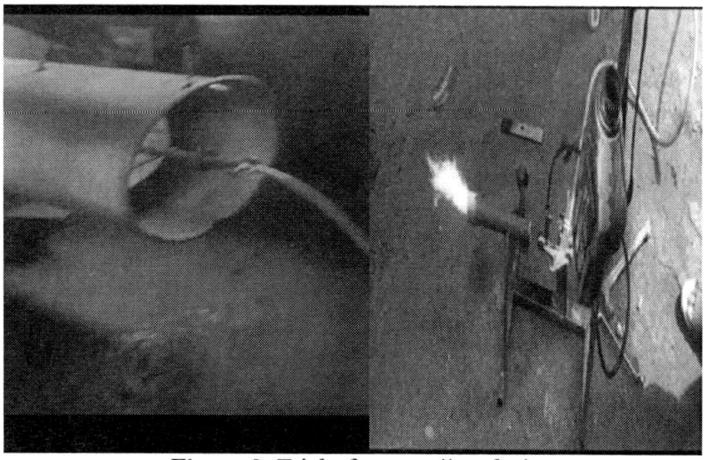

Figure 9. Trial of waste oil as fuel

The result of Song et al [4] said that Agricultural and forestry wastes (AFWs) are produced globally in enormous quantities and are considered an important resource to be used to reduce dependence on fossil fuels. Biofuels and/or chemicals can be produced. Utilization of biomass waste as raw material is the main alternative to fossil fuels or chemicals, because they are abundant and relatively inexpensive. AFW is considered as a promising alternative for supplied biofuels and/or chemicals by thermochemical liquefaction.

4. Conclusion

The results showed that the levels of Pb, Fe, Al and Mn metals decreased significantly after filtering and adding acid with the best concentration of 60% and obtained results that tended to decrease with low content and were allowed to use. Likewise, the viscosity decreases, this indicates that the concentration with the addition of acid will be lower. After testing it as a fuel, it provides a very high efficiency with a ratio of 1: 5: 7 (2.5 liters of used oil required 1 LPG cylinder (3 kg) and liter of kerosene) at a price of 5000: 25000: 35000).

References

[1] Afin A P and Kiono B F T 2021 *Jurnal Energi Baru Dan Terbarukan* **2** 144Wardhani M K 2011 *Marine journal* **4** 60
[2] Jaelani A, Firdaus S and Jumena J 2017 *International Journal of Energy Economics and Policy* **7** 4 193
[3] Kusnadi A, Djafar R and Mustofa M 2020 *Jurnal Teknologi Pertanian Gorontalo (JTPG)* **5** 49
[4] Song C, Zhang C, Zhang S, Lin H, Kim Y, Ramakrishnan M, Du Y, Zhang Y, Zheng H and Barceló, Damià 2020 *Science of the Total Environment*, **749** 141972
[5] Herdito T and Lutfi M 2021 Pemanfaatan Limbah Oli Bekas Menjadi Bahan Bakar High Speed Diesel (HSD) **7** 2
[6] Agarwal A K 2007 *Progress in Energy and Combustion Science* **33** 233
[7] Dahlan M, Setiawan A and Rosyada A 2014 *Jurnal Teknik Kimia Universitas Sriwijaya* **20** 38
[8] Asidu L O A D, Hasbi M and Aksar P 2017 *Jurnal Mahasiswa Teknik Mesin*, **2** 1

[9] Boston J A, Moreno J, Espada J J, Serano D P and Dufour J 2017 *LCA* **125** 315
[10] Azharuddin D 2020 *Jurnal Austenit* **12** 48
[11] Dahlan M, Setiawan A and Rosyada A 2014 *Jurnal Teknik Kimia Universitas Sriwijaya* **20** 38
[12] Muzhaffar I, Nurwulan, F and Pfis M 2021 *eProceedings of Engineering* **8** 1884
[13] Raharjo W P 2004 *Mekanika* **3** 23
[14] de Groot K 2018 *Bioceramics of calcium phosphate*
[15] Nainggolan A F, Taringan M A P and Khair H 2017 *Jurnal Teknik Lingkungan Unand* **14** 1

IOP Conf. Series: Earth and Environmental Science 1115 (2022) 012068 doi:10.1088/1755-1315/1115/1/012068

Literature Review: Identification Of Sustainable Ecotourism In Tangkahan

Onan Marakali Siregar[1], Afrila Mulyati Siregar[1], Andriansyah[1]

[1] Business Administration Studies Program, Faculty of Social and Political Sciences, Universitas Sumatera Utara

*Email: onan@usu.ac.id

Abstract. Ecotourism development is an effort to develop sustainable and environmentally friendly tourism so that it can be enjoyed for a more extended period. Ecotourism does not only offer travel but creates an element of learning in it. As a result, managing sustainable tourism destinations related to governance in carrying out tourism activities is increasingly important. In this paper, the approach used in the literature review is collected from several publishers such as emerald, Wiley, research gate, science direct, and Thomson and then studied systematically, which is related to this paper, totalling 50 articles. The purpose of this paper is to present a conceptual framework for ecotourism that combines the concept of 4a's tourism as a framework for identifying and developing tourist destinations. This paper explores the essence of ecotourism related to sustainable tourism through the 4A component of tourism. The findings show that ecotourism contributes differently to other tourism depending on the sub-topics raised. It can be underlined that the literature review is long-term research integrated between ecotourism management and stakeholders such as the government, residents and consumers needed to carry out ecotourism literacy for the environment.

1. Introduction

Indonesia is one of the countries in Southeast Asia that has diverse tourism. One of them is ecotourism, a sub-component of sustainable tourism development developed from the nature tourism market segment, which includes a mixture of culture and rural tourism [1]. The success of ecotourism management must have a relationship with the government and residents in maintaining and finding the best way to manage ecotourism based on sustainable nature. The synergy between managers and local governments is the main thing that becomes a vital spotlight in ecotourism.

At first, the concept of ecotourism development was first coined by the World Conservation Union in 1983, and then the International Ecotourism Society re-explained the concept of ecotourism. ICS explained that ecotourism is a tourism concept responsible for protecting nature and the lives of surrounding communities [2]. Various countries widely use ecotourism to promote sustainable tourism [3]. Ecotourism areas have become an important destination not only for environmental protection but in providing employment, recreation and learning media [4]. However, an appropriate and conceptual interpretation of ecotourism can explain that all problems in tourism can be solved by ecotourism, such as economic development, poverty reduction, poverty preservation and cultural preservation, to educate tourists about all problems related to nature [5][6].

This paper aims to produce a systematic literature review on ecotourism in Tangkahan to find the best strategy for developing sustainable ecotourism. Writing a literature review aims to collect systematically, document, examine and critically analyze research related to ecotourism which will produce sustainable ecotourism.

The writing in this study was chosen by reviewing previous literature described to find the right idea. The writing of this research will produce a systematic literature design for further research. Conclude various types of research that discuss sustainable ecotourism, constraints and socio-economic impacts on the community. Linking ecotourism with sustainable tourism requires the identification of tourist objects, one of which is the identification of 4A: attraction, accessibility, amenities and ancillary. Several components of the tourism market are the main characters in developing tourism. Understanding and implementing tourism by the manager regarding the 4A (four) components so that they can attract, fulfil desires, and make it easier for visitors or tourists. Then the identification is related to how sustainability can be carried out in Tangkahan, which is the object of this research.

The author assumes that in the development of ecotourism, identification must be carried out clearly and integrated with public policies from local or central governments to produce sustainable ecotourism that is of interest to tourists but also has minimal social and natural impacts like the Tangkahan tourist attraction in North Sumatra. Tangkahan, which is one of the ecotourism areas in North Sumatra, is located in the buffer zone of the Gunung Leuser National Park (TNGL). Administratively, Tangkahan is located at an altitude of about 130-200 meters above sea level. The Tangkahan ecotourism area is located in the middle of the Buluh and Batang Serangan rivers. The air temperature in this ecotourism area is around 21.1°-27.5°C. The topography of the Tangkahan ecotourism area is a hill with a stunning landscape. Based on data from the Tangkahan Tourism Institute, Tangkahan ecotourism is located on the border of an oil palm plantation area belonging to PTPN II Kuala Sawit. In the east, it is bordered by the hamlet of Kuala Buluh. This ecotourism is flanked by two villages, Namo Sialang Village and Sei Serdang Village, located in Batang Serangan District, Langkat Regency, North Sumatra. Tangkahan is a natural tourism object, and Sumatran elephant captivity offers the concept of ecological education tourism in North Sumatra. The location, about 3 hours from the centre of Medan City, offers attractive and attractive ecotourism to tourists.

1.1 Identify 4A Ecotourism

Managing tourist destinations must-have elements in its implementation. According to [7], a good tourist attraction has 4A elements in the tourism industry in attracting tourists, namely:

a. Attraction

   Tourist destinations must have an attraction; that is why someone travels. This tourist attraction is related to the attractions offered by the tourist attraction. Attractions are places of interest that tourists visit, usually because of their inherent cultural values such as history, natural beauty, culture, lodging or artificial entertainment opportunities that aim to attract tourists [8]. Tourist attraction refers to all types of natural and artificial resources that have unique and aesthetic values that encourage and motivate tourists to visit them [9]. The form of natural attraction comes from nature, for example, natural scenery, diversity of flora and fauna, jungle, the beauty of the sea and so on. While those from non-natural (artificial), for example, ancient relics, cultural arts, museums, agro-tourism, water tourism, adventure tourism, recreational parks, and entertainment venues.

b. Accessibility

   An essential component that makes travel activities possible accesses. Accessibility is all forms of facilities and conveniences presented to tourists and potential tourists that can be used as a trajectory to get to tourist destinations [10]. Improved accessibility ensures that the flow of visitors to tourist destinations is well connected [11]. A destination must be accessible if it is to facilitate tourist visits. However, if access becomes too easy, this can result in too much demand and lead to congestion, making the destination less attractive to tourists. While more intrepid travellers may be willing to put themselves at great expense to see some of the world's more exotic places, most tourists will not be attracted to a destination unless it is relatively easy to reach.

c. Amenities
Amenities (Facilities) are essential services that meet the needs of tourists. Amenities are elements in a destination or related to a destination that allows tourists to stay in the destination to enjoy or participate in the attractions it has to offer [12]. Amenity is closely related to the presence of accommodation facilities to spend the night at tourist destinations and places to eat/restaurants/cafes and everything that allows tourists to meet their consumption needs. Tourists also need other necessities during their trips, such as public toilets, rest areas, parking lots, health clinics, and religious facilities should also be available at a destination. Of course, these facilities also need to assess the situation and conditions of the destination for the tourists' needs. Not all amenities have to be close together and in the central area of the destination. Natural destinations and historical relics should be far from commercial amenities, such as hotels, restaurants and rest areas [13].

d. Ancillary
Ancillary tourism can be explained as the availability of the organization or person who manages the destination. An ancillary is a form of tourism organization that aims to manage and market tourist destinations. In tourism, a destination will perform its duties as an organization with responsibilities like a company. Tourism institutions aim to manage all facilities and infrastructure to develop tourist objects in the tourism area [14]. Managing destinations so that they can provide benefits to related parties such as the government, surrounding communities, tourists, the environment and other stakeholders.

1.2 Nature-Based Tourism Governance

Good governance is now regarded as the driving force for success in all life [15]. Governance can be described as the processes, rules and behaviours that provide how power is located and exercised, especially regarding the principles of openness, participation, accountability, effectiveness and coherence. Establishing governance must focus on stakeholders, both internal and external. The government is one of the actors in the implementation of governance. Governance is a way for the community to regulate itself to make and implement the decisions reached by considering the attitude of mutual understanding, agreement and action. Governance, including its social, political and economic dimensions, operates at every level of human enterprise, in households, villages, municipalities, regions or the world [16].

Today's tourism management tends to lead to sustainable management. The management in question is management that pays attention to and maintains the ecosystem in tourist destinations. Other parties involved in governance vary in the context of tourism depending on the context of the tourism concept being carried out. Tourist destinations are complex linkages between ecosystems and social systems in tourist areas [17].

Wood [18] explains that the management of nature-based and ecotourism tourism must include: 1) Providing an adequate budget to preserve popular tourist areas and allocating tourism costs for conservation. 2) Tourism businesses must pay impact fees that fund infrastructure for solid waste treatment, sewage treatment capacity, electricity and water, as their guests will need these services. Tourism properties should receive clear incentives to conserve electricity and water. 3) An impact monitoring program must follow the environmental impact assessment program. These programs should be linked to business licensing and certification systems. 4) Well-managed camping trails and areas should provide marked rules for low-impact use. 5) Published and recognized rules or regulations for public lands should detail the types of vehicles allowed in certain zones, speed limits, fines for dumping or pollution, clear rules for limiting off-road driving, and supervision of the number of visitors allowed in the designated areas—categorized, and instructions on how and when to observe wildlife. The guidelines must state what is inappropriate. 6) Training programs are needed to provide opportunities for residents to run their businesses. 7) Affordable housing programs must ensure that residents are not excluded from their communities due to inflated real estate prices.

1.3 Sustainable Ecotourism

The shift in the trend of mass tourism to tourism awareness programs, where ecotourism is a development sub-sector whose primary goal is to protect the surrounding environment. Ecotourism is the fastest growing tourism sector in the global tourism industry, which mainly focuses on preserving nature and promoting environmental and cultural understanding, appreciation and conservation [19]. Ecotourism has been recognized as one of the vitals to promote sustainable development.

Environment, economy, and society, as well as the effectiveness of implementing sustainable tourism plans and management [20]. That sustainability depends on how well planning is formulated relative to the particular characteristics of an area. The sustainable development approach can be applied to any scale of tourism development, from large resorts to special interest tourism of limited size. In an academic discourse in the 1980s, ecotourism was advocated as a viable form of tourism development, especially for developing countries [21][22][23][24][25]

IUCN (now the World Conservation Union), ecotourism is environmentally sound travel and visits to relatively undisturbed natural areas in order to enjoy and appreciate nature (and any accompanying cultural features-both past and present) that promote conservation, have low negative visitor impact and provide the socio-economic benefits of active engagement with residents [18]. Developing destination and regional principles assess the potential of unique tourism resources and competitive advantages to develop environmentally friendly places. In general, as Wang discussed, ecotourism is considered a viable form of tourism development because [26]: 1)Promote the application of sustainable tourism principles in nature. 2) Socio-cultural areas and arrangements; considering sustainability and adhering to the triple-bottom-line concept. 3) Emphasize a trip's educational component, thereby improving the visitor experience and helping change their behaviour and lifestyle to become more responsible travellers.

2. Methodology

This paper is a conceptual paper with a literature review type. The literature review is a research method for assessing, evaluating and interpreting certain studies, specific topics, or phenomena of concern [27]. Overall, the systematization of this field uses data from the main study of interest rather than direct experimentation. This paper reviews articles from online scientific journal portals, namely: ResearchGate, Elsevier, Science Direct, and Emerald, according to the keyword "sustainable ecotourism", "ecotourism", "sustainable tourism", and "tourism 4a". A search for publications between 2012 and 2022 was completed to identify relevant studies exploring ecotourism theory thoroughly.

A search on each scientific journal portal resulted in 4 ResearchGate articles, 14 articles on Elsevier, 1 Science Direct, and 3 Emerald articles, a total of 21 articles. Then classify these articles into two categories: articles that only mention "ecotourism" and articles related to ecotourism with sustainability. The author then summarizes, analyzes and synthesizes critically, and in-depth eight articles from articles deemed appropriate, which are then quoted or reviewed as in Table 1.

Table 1. Empirical Research Used in Systematic Reviews

Authors	Year	Elements
Jaafar and Maideen	2012	• Appreciate nature. • Conservation.
Mearns	2013	• Public. • Conservation. • Tourist. • Governance.
Ştefan P., Gheorghe	2014	• Natural areas.

		• Ecotourism interpretation. • conserve nature. • tourist satisfaction. • Marketing.
The International Ecotourism Society,	2015	• Conscientious. • Sensitivity towards, and appreciation of, local cultures and biodiversity. • Support for local conservation efforts. • Sustainable benefits to local communities. • Local participation in decision-making. • Educational components for both the traveler and local communities.
Winkler and Zimmermann	2015	• Economic Benefits. • Environmental Protection. • Education. • Cultural Reception. • Community empowerment. • Economic Benefits.
Xu and Chan	2016	• Self-enhancement. • Relaxation, and knowledge. • Escaping from routine. • Life, the scenery of the destination. • Information and convenience. • Various activities for fun
Kamri and Radam	2018	• Challenge excursion. • Social trip. • Nature tour. • Getaway outing.
Chow, Cheng, and Cheung	2019	• Relaxation. • Escape from daily life. • And mental health.
Forje, Tchamba, and Eno-Nku	2020	• Reference to where ecotourism occurs (natural areas). • Conservation. • Culture. • Benefits to locals. • Education.
Choi, Oh, and Chon	2021	• Diversity. • Connectivity. • Learning. • Participation. • Polycentric governance.

Carvache-Franco et al	2022	• Self-development and Interpersonal relationships. • Building personal relationships. • Escape. • Nature. • Terrestrial nature. • Rewards.

3. Result

This literature review aims to describe the results of ecotourism studies that have been carried out and sustainability so that they can be used as further research literature. Systematic reviews in tourism research are still minimal [36]. The results of a literature review depend on what has been done, the findings, and turning reports that are sourced from the actual interpretation of research [37]. Ecotourism research indicates significant issues for further research. Ecotourism research reviews differ from other tourism depending on the subtopics raised. The literature review is long-term research that integrates ecotourism managers and stakeholders such as the government, residents and consumers needed to carry out ecotourism literacy for the environment. Tangkahan tourism object is an ecotourism ecology that should have implemented 4A in its marketing strategy. The role of each stakeholder in the tourism industry, especially the nature-based tourism industry, will make a tourist attraction called sustainable tourism. The implementation of sustainable tourism must first identify the selected tourism objects for development priorities.

4. Recommendation

The conclusion above can be identified that ecotourism research in the future requires the involvement of every stakeholder in its development. Ecotourism will be very attractive and maintained if all stakeholders own tourism literacy. The need for conducting ecotourism research by identifying the characteristics of tourism in advance and understanding the socio-economic impact will make this tourism sustainable tourism. Ecotourism conditions related to 4A can be evaluated from various aspects of the tourism industry. Further studies are expected to expand the research sample to see that the Tangkahan tourism object is one of the conservation objects owned by Indonesia and is of interest to domestic and foreign tourists.

Further research is recommended for the international ecotourism context by identifying sources of satisfaction for tourists from different countries and cultures, drawing on emerging differences. Future research can also focus on classifying service attributes according to other criteria such as gender, age, travel destination, length of visit and so on to form an appropriate tourism marketing strategy for Tangkahan.

References

[1] P. K. Perera, "Marketing forest-based ecotourism in Sri Lanka: predicting the ecotourism behavior and defining the market segment through a behavioral approach," 2011.

[2] X. . Li, "Study of interests demand of the main stakeholders: taking daming mount national nature reserve in Guangxi as a case," *Ecol. Econ*, vol. 10, 141–14, 2011.

[3] R. Buckley, C. Cater, Z. Linsheng, and T. Chen, "SHENGTAI LUYOU: Cross-Cultural Comparison in Ecotourism," *Ann. Tour. Res.*, vol. 35, no. 4, pp. 945–968, 2008, doi: https://doi.org/10.1016/j.annals.2008.07.002.

[4] T. C. H. Tao and G. Wall, "Tourism as a sustainable livelihood strategy," *Tour. Manag.*, vol. 30, no. 1, pp. 90–98, 2009.

[5] H. M. Donohoe and R. D. Needham, "Ecotourism: The evolving contemporary definition," *J. Ecotourism*, vol. 5, no. 3, pp. 192–210, 2006.

[6] F. ISMAIL, A. IMRAN, N. KHAN, and M. I. QURESHI, "Past, present and future of

ecotourism, a systematic literature review from last decade," *Estud. Econ. Apl.*, vol. 39, no. 4, 2021, doi: 10.25115/eea.v39i4.4592.
[7] O. A. Yoeti, *Perencanaan dan Pengembangan Pariwisata*. Jakarta: Pradnya Paramita, 1997.
[8] D. Jerab, M. Alper, and A. Baslar, "The impact of core competencies on competitive advantages in Istanbul tourists companies," *Soc. Sci. Res. Netw.*, 2011.
[9] Erislan, "Tourist attraction and the uniqueness of resources on tourist destination in west Java, Indonesia," *Rev. Integr. Bus. Econ. Res.*, vol. 5, no. 1, pp. 251–266, 2016.
[10] O. A. Yoeti, *Perencanaan dan Pengembangan Pariwisata*. Jakarta: Pradnya Paramita, 2008.
[11] J. Gržinić and D. Saftić, "Approach to the development of destination management in Croatian tourism," *Manag. J. Contemp. Manag. Issues*, vol. 17, no. 1, pp. 59–74, 2012.
[12] I. G. Pitana and I. K. S. Diarta, *Pengantar Ilmu Pariwisata Edisi I*. Yogyakarta: CV Andi Offset, 2009.
[13] Isdarmanto, *Dasar-Dasar Kepariwisataan dan Pengelolaan Destinasi Pariwisata*. Yogjakarta: Gerbang Media Aksara dan STiPrAm, 2017.
[14] B. Dewandaru, A. N. Rahmadi, and N. Susilaningsih, "Pengaruh Attraction, Accesibility, Amenity Dan Ancillary Terhadap Kepuasan Wisatawan Pada Kawasan Wisata Besuki Kediri," in *Conference on Economic and Business Innovation (CEBI)*, 2021, pp. 498–508.
[15] G. Walter, N. P. Awazi, and J. Kimengsi, "Ecotourism governance in sub-Saharan Africa: a systematic review," *Environ. Res. Commun.*, Mar. 2022, doi: 10.1088/2515-7620/ac5f1c.
[16] M. Sudders and J. Nahem, *Governance Indicators: A Users' Guide*. Oslo: United Nations Development Programme, 2007.
[17] M. Paskova, G. Wall, D. Zejda, and J. Zelenka, "Tourism carrying capacity reconceptualization: Modelling and management of destinations," *J. Destin. Mark. Manag.*, vol. 21, p. 100638, 2021.
[18] M. Wood, *Ecotourism: Principles, Practices and Policies For Sustainability*. UNEp, 2002.
[19] J. R. Timmons, *Trouble in paradise: Globalization and environmental crises in Latin America*. Routledge, 2003.
[20] G. Wall, "Is Ecotourism Sustainable?," *Environ. Manage.*, vol. 483–491, 1997.
[21] C. Ashley and J. Mitchell, *Tourism and poverty reduction: Pathways to prosperity*. Routledge, 2009.
[22] J. Butcher, "Can ecotourism contribute to tackling poverty? The importance of 'symbiosis,'" in *Tourism and the Millennium Development Goals*, Routledge, 2013, pp. 219–232.
[23] E. Cater, "Ecotourism in the Third World: problems and prospects for sustainability.," *Ecotourism a Sustain. option?*, pp. 69–86, 1994.
[24] S. Snyman, "The impact of ecotourism employment on rural household incomes and social welfare in six southern African countries," *Tour. Hosp. Res.*, vol. 14, no. 1–2, pp. 37–52, 2014.
[25] A. Wondirad, "Ecotourism development challenges and opportunities in Wondo Genet and its environs, southern Ethiopia," *J. Place Manag. Dev.*, vol. 13, no. 4, pp. 465–491, 2020.
[26] A. Wondirad, D. Tolkach, and B. King, "Stakeholder collaboration as a major factor for sustainable ecotourism development in developing countries," *Tour. Manag.*, vol. 78, p. 104024, 2020.
[27] B. Kitchenham, "Procedures for performing systematic reviews," *Keele, UK, Keele Univ.*, vol. 33, no. 2004, pp. 1–26, 2004.
[28] M. Jaafar and S. A. Maideen, "Ecotourism-related products and activities, and the economic sustainability of small and medium island chalets," *Tour. Manag.*, vol. 33, no. 3, pp. 683–691, 2012.
[29] C. C. Ştefan P., Gheorghe S., "NEW APPROACHES CONCERNING THE DEVELOPMENT OF THE ECOTOURISM IN ROMANIA," *Manag. Econ. Eng. Agric. Rural Dev.*, vol. 14, no. 2 ISSN 2284-7995, 297–302, 2014.
[30] T. Winkler and F. Zimmermann, "Ecotourism as community development tool–development of an evaluation framework," *Curr. Issues Tour. Res.*, vol. 4, no. 2, pp. 45–56, 2015.

[31] J. B. Xu and S. Chan, "A new nature-based tourism motivation model: Testing the moderating effects of the push motivation," *Tour. Manag. Perspect.*, vol. 18, pp. 107–110, 2016.
[32] T. Kamri and A. Radam, "Motivation of Visiting Bako National Park," *Asian J. Qual. Life*, vol. 3, no. 9, p. 123, 2018.
[33] A. S. Y. Chow, I. N. Y. Cheng, and L. T. O. Cheung, "Self-determined travel motivations and ecologically responsible attitudes of nature-based visitors to the Ramsar wetland in South China," *Ann. Leis. Res.*, vol. 22, no. 1, pp. 42–61, 2019.
[34] G. W. Forje, M. N. Tchamba, and M. Eno-Nku, "Determinants of ecotourism development in and around protected areas: The case of Campo Ma'an National Park in Cameroon," *Sci. African*, vol. 11, p. e00663, 2021.
[35] Y. E. Choi, C.-O. Oh, and J. Chon, "Applying the resilience principles for sustainable ecotourism development: A case study of the Nakdong Estuary, South Korea," *Tour. Manag.*, vol. 83, p. 104237, 2021.
[36] M. Weed, "Sports Tourism Research 2000–2004: A Systematic Review of Knowledge and a Meta-Evaluation of Methods," *J. Sport Tour.*, vol. 11, no. 1, pp. 5–30, Feb. 2006, doi: 10.1080/14775080600985150.
[37] C. M. Hall *et al.*, "Accommodation Consumers and Providers' Attitudes, Behaviours and Practices for Sustainability: A Systematic Review," *Sustainability* , vol. 8, no. 7. 2016. doi: 10.3390/su8070625.

Utilization of effluent biodigester as a bioactivactor for the decomposition of rice husk (*Oryza sativa*) into compost

S Afrianti[1,5], Irvan[2*], T B Nur[3] and Delvian[4]

[1]Doctoral Program in Natural Resources and Environment Management, Universitas Sumatera Utara, Medan 20222, Indonesia.
[2]Department of Chemical Engineering, Faculty of Engineering, Universitas Sumatera Utara, Medan 20155, Indonesia
[3]Department of Mechanical Engineering, Faculty of Engineering, Universitas Sumatera Utara, Medan 20155, Indonesia
[4]Department of Natural Resources and Environment Management, Universitas Sumatera Utara, Medan 20222, Indonesia.
[5]Department of Agrotechnology, Faculty of Agro Technology, Universitas Prima Indonesia, Medan 20117, Indonesia

[*]Email: irvan@usu.ac.id

Abstract. Accelerate the biodegradation of rice husks into compost requires a bioactivator. The bioactivator used is the effluent from a biogas reactor which integrates with the fermenter. The composting process for rice husk (*Oryza sativa)* adds a bioactivatorwith two treatments. Treatment A uses an effluent bioactivator Biogas reactor; treatment B uses an effluent bioactivator from the University of North Sumatra pilot plant. This study aimed to compare the effect of variations in composting treatment and B on the results of composting rice husks. From the results of the study, the two treatments at temperatures were not too much different; the average of treatment A was 33 ^{0}C, and B was 33.20 ^{0}C; the average value of pH in Treatment A was 7.1 and B was 7.3, while the average value of Content moisture in treatment A was 55.59% and in B was 55.16%. Meanwhile, the value of nutrients and macros is not too much different between treatments A and B. The organic C in A is 14.12%, and B is 14.36%. The N-Total value in treatment A is 1.08%, and B is 1.12%, the value of $P2O5_A in_{0.77}$ and B is 0.69%, the value$_A$is 0.57% and B is 0.60%, the value of CaO in treatment A is 0.17%, and B is 0.22%, the value of MgO in treatment A is 0.29%, and B is 0.34, the value of Cu in treatment A is 9 ppm, and B is 9.1 ppm, the value of Zn in treatment A is 50 ppm, and B is 52 ppm.

1. Introduction

Rice is one of the leading agricultural products in agrarian countries, including Indonesia. The culture of Indonesian society cannot separate rice as a staple food, so Indonesia has become one of the countries with an enormous consumption of rice. High rice consumption also increases rice production and grows every year, as recorded in the annual report [1]. In 2005 rice production in Indonesia was 54 million tons. Production from 2005 to 2006 increased to 54.45 million tons; this increase occurred continuously every year starting from 2007 to 2011 was 57.15; 60.33; 64.40 66.41 million tons of rice.

Producing rice from the rice will have solid waste in rice husks. The composition of the rice mill generally yields 5 – 8% bran, and 20 – 22% husk, 72% rice. Rice husk is solid waste, or some people call it a by-product of rice milling; if rice production recorded at the Research and Development Center in 2011 was 66.41 tons of rice, rice husk produced was 47.81 tons.

So far, farmers are very limited in using rice husk waste, considered a product that does not have high economic value. In contrast, its use has been limited to fuel for land clearing, fuel for brick making, ornamental plant media and minimal local needs. One of the reasons rice husks are not used is because they are hard, while on the other hand, rice husks have a high carbon nutrient value, so, with this high carbon element, it can be used as the primary raw material for composting.

Like other agricultural organic waste, Rice husk waste contains carbon (C) elements and requires microorganisms for the decomposition process into compost. The content of this carbon nutrient is a substance needed by microorganisms in their growth. The value of this carbon content will decrease if it approaches mature compost because the carbon content has been decomposed so that it becomes a CO2 compound which is released into the air. Rice husk has lignin and cellulose bonds, making it difficult to decompose. One of the polysaccharide compounds is lignin and cellulose; in general, these compounds are found in all plants, which are constituents of plant cell walls.

Decomposition of organic materials or overhauling complex compounds into simpler compounds with the help of microorganisms is called compost. This composting process can occur naturally or with human assistance. Composting naturally can occur by *open dumping* or accumulating the waste in nature, while human-assisted composting utilizes technology, one of which is using a bioactivator. It Will occur more quickly; the bioactivator serves as a medium for the process of accelerating the composting process.

In previous studies, biogas reactors have been made using organic restaurant waste as raw materials. This process produces effluent that can be used as a bioactivator for compost decomposition. Various efforts have been made to accelerate the composting time, one of which is by utilizing a bioactivator, in this study uses *effluent* from the biogas production process to become a bioactivator in the decomposition of rice husks into compost.

2. Exsperimental Section

2.1. Materials

The research materials used are (1). Effluent biogas reactor as a bioactivator, (2). Rice husks, as well as materials for compost analysis, include (3). Solution of $K_2Cr_2O_7$ 0.1 N, (2). H_2SO_4(p), (3). H_3PO_4 85%, (4).NaF 4%, (5). Solution; $C_6H_4OHCOOH$.

2.2. Tools

The tools used are (1). Digital thermometer (2).digital pH meter (3). Shakers (4).Plastics tube (shake bottle) (5). Analytical balance (6).Beaker glass (7).Composter (8).Scales.Ovens (10).Desiccator (11).Cup (12). Filter Paper (13). Conductivity Meter.

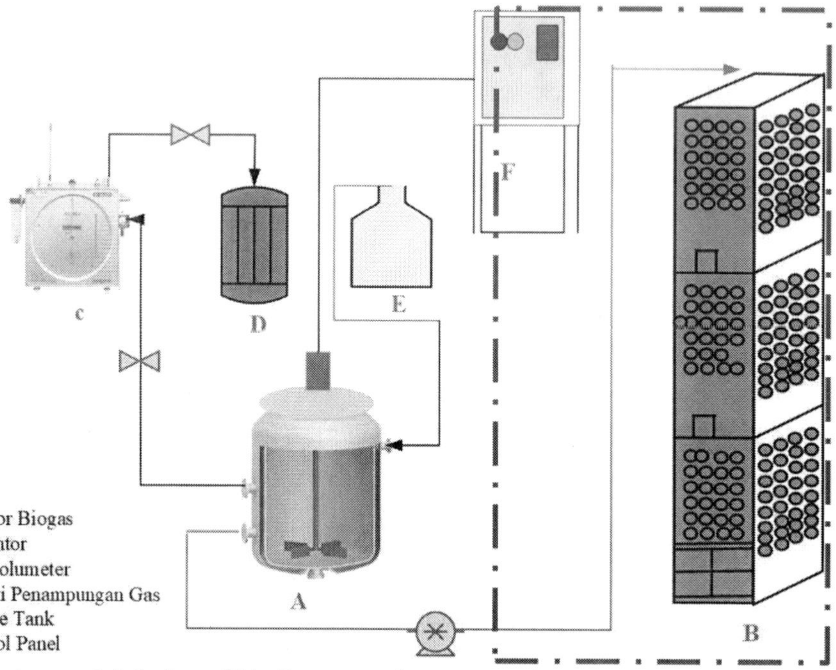

Figure 1. Combination model design of biodigester and composter in managing organic waste into biogas and compost

Figure 2. Combination of biodigester and composter in managing organic waste into biogas and compost

2.3. Methods

This study carried out two treatments. The first treatment uses the *activator effluent* from the reactor biogas as a bioactivator. The second treatment uses a starter from the pilot plant of the Pusdiklat LPPM University of North Sumatra as a bioactivator. A starter and effluent biogas reactor are pumped every day to the fermenter in the form of a liquid phase. Everyday observations will be made on the pH, moisture content and moisture content. At the end of the composting or after the compost has matured, a test of micro and macronutrients in the compost will be carried out using the Kjeldahl and AAS methods. This study also analyzed the characteristics of the effluent used. The parameters analyzed were COD using the open reflux method, TS, VS, TSS and VSS pH using the standard method.

2.4. Process description

This research is an anaerobic digestion process by designing an integrated combination of biodigester and composter, the type of reactor designed is a stirred tank flow reactor to produce biogas. In the initial stage, restaurant waste mixed with methanogenesis starter with a composition of 1:10 solids and water was put into the feed tank and then stirred until homogeneous. The mixture is fed into the reactor with a working volume of 10 litres at 55 °C.

The process begins with treatment without giving an organic load (batch) until the pH and alkalinity are stable (2000 – 4000 ppm). After a stable condition, a continuous stage carries out. Gradual organic loading was carried out at a temperature of 55°C, pH 7, and the alkalinity was stable (2000 – 4000 ppm). It is kept stable so that the reactor's microbial life conditions can develop properly. To prevent a decrease in pH and alkalinity can be added $NaHCO_3$ into the reactor from the beginning of the process to the end of the operation.

Analyzes pH, alkalinity, COD, TS, VS, TSS, and VSS. Analysis of pH, M-Alkalinity, TS, VS, TSS and VSS was carried out every day, while the study of COD and gas was carried out every three days until the end of the process. Analysis of pH, M-Alkalinity, and COD used samples that came out of the overflow in the reactor. In contrast, the TS, VS, TSS, and VSS analysis used pieces at sampling points along the reactor.

The biogas formed is measured through a gas meter connected to the reactor, and the analysis carried out is the gas composition of CH_4, CO_2 and H_2S using a gas detector.The effluent reactor is flowed to the composter to function as an activator for composting. The raw material for composting is rice husk. The composting water content is maintained until the optimum composting water content is 55%-65%. Temperature measurements are carried out two times a day; pH measurements are carried out once a day, water content measurements are taken carried out once a day, and measurement of water holding capacity was carried out once in 10 days. The N, P, K and C/N ratio levels were analyzed after the compost was ripe. This paper only displays the results of the study of the fermenter. In contrast, the results of the analysis and observations on the biogas reactor are displayed in other articles or journals that are separate from this journal.

3. Results and Discussion

The value of the analysis of the characteristics of the biogas effluent starter in treatments A and B is pH at A 8, B 7.5, the value of *Chemical Oxygen Demand* (COD) A is 31,000 mg/ LB is 34,000 mg/L, the value of *Total Solid* (TS) in treatment A is 22,000 mg/L, and B is 27,500 mg/L.*Volatile Solid* (VS) in treatment A was 12,000 mg/L and B was 10,900 mg/L, the value of *Total Suspended Solids* (TSS) in treatment A was 10,000 mg/L and B was 15,300 mg/L, and the importance of *Volatile Suspended Solid* (VSS)in treatment A was 17,000 mg/L and B 11,300 mg/L.

Table 1. Results of analysis of biogas stater effluent characteristics

No	Parameter	Test Methode	units	Value	
				A	B
1	Ph	APHA 4500-H	-	8	7,5
2	*Chemical OxygenDemand* (COD)	Spektrofotometri	mg/L	31.000	34.000
3	*Total Solid* (TS)	APHA 2540B	mg/L	22.000	27.500
4	*Volatile Solid* (VS)	APHA 2540E	mg/L	12.000	10.900
5	*Total Suspended Solid* (TSS)	APHA 2540B	mg/L	10.000	15.300
6	*Volatile Suspended Solid* (VSS)	APHA 2540E	mg/L	17.000	11.300

*A. starter from *the pilot plant* LPPM Education and Training Center, University of North Sumatra
B. Biogas Reactor Effluent

3.1. Composting temperature observation profile

Temperature is an indicator that shows the activity of decomposing microorganisms during the composting process. The movement of decomposing microorganisms in the composting process affects the temperature changes in the compost heap. Microorganisms generate heat during the process of overhauling organic matter. When the decomposition process of organic matter at the beginning of the composting phase is faster, the heat generated will increase faster. At the end of the composting process, the organic matter decomposed by decomposing microorganisms has been exhausted, decreasing the compost pile's temperature. There will be a substantial temperature increase for the first 3-5 days in the aerobic composting process, and the compost pile's temperature can reach 30-60°C. This temperature range is the best for the growth of microorganisms, where decomposing microorganisms can multiply up to three times so that the activity of overhauling organic matter becomes faster.

In general, compost undergoes three stages of the composting process: the acclimation stage, the thermophilic stage, and the compost maturation stage. The composting process begins with the acclimation stage, which is adjusting the temperature of the compost material; at this stage, there is an increase in the temperature of the mixture of organic materials used as compost. The next stage is the thermophilic stage. The microorganisms involved in this stage can live at a temperature of 40-60°Cand aim to degrade organic matter quickly by consuming carbohydrates and proteins. At this stage, the decomposition process begins to slow down and is marked by the achievement of the peak temperature so that the organic matter is more easily decomposed. The last stage is the compost maturation stage. At this stage, the thermophilic microorganisms will die due to increased temperature above 60°C and are replaced by mesophilic microorganisms. Suppose the maximum temperature has been reached and all the activities of the microorganisms involved during the decomposition process have stopped. In that case, the temperature will drop again until it gets to the initial temperature range. At this stage, the compost has been formed and is ready to use.

The continuity of the decomposition process of rice husk waste into compost in the treatment using biogas reactor effluent and starter from the USU LPPM pilot plant can be seen from temperature changes. Temperature observations are carried out twice a day, morning and evening; the table above shows the average between morning and evening temperature measurements, which have been done every day during composting.

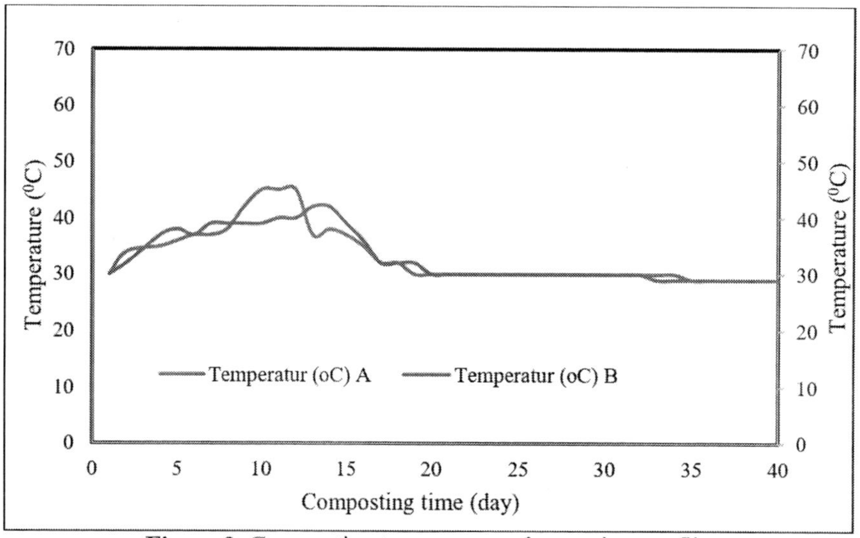

Figure 3. Composting temperature observation profile

Figure 3 above is the result of observing the Composting Temperature Observation Profile in two treatments. The treatment uses a biogas effluent reactor (A), while the second treatment uses a pilot plant starter is symbolized by (B). After 24 hours of treatment A and B in the process of decomposition of rice husks into compost, the temperature increased, the value of the increase in treatments A and B. increased from A are 30°C, the second day became 34 °C, then continued to increase to 34 °C, 35 °C, 36 °C, 37 °C, until it reaches the highest temperature 45 °C, namely at the time of composting on day 8, the same thing also happened in treatment B where the initial temperature was 30 °C. It rose to 32 °C, 37 °C, and 38 °C until the highest temperature was 40 °C on day 10. The increase in temperature in the decomposition process of rice husk compost into compost indicates that the decomposer microbes are active in the fermenter. The temperature profile described in this study is also the same as reported in the research results [2]. If a material contains high organic matter and is added to a material containing cellulose, the material will generate heat.

The description of the temperature of the decomposition of rice husks into compost decreased on the 8th day; we can see in treatment A that on the 8th day, the temperature was 45 °C to 39 °C, 37 °C, 35 °C, 32 °C, then constant on the 8th day. The 18th to 40th day, namely with a temperature value of 30 °C.this illustrates that composting process has been completed, according to research [3] which says that after the process of increasing the temperature in the composting process, the temperature will slowly decrease, so that the occurrence of degradation is slow due to the depletion of the availability of nutrients for microorganisms in composting raw materials.

In composting, temperature fluctuations illustrate that the decomposition of microbes in the composter is active; according to [4] heat is one of the products produced during decomposition in the composting process. The occurrence and increase in temperature in composting are significant because it helps the hygienic process, namely to kill pathogenic bacteria; the composting process will take place in a combination process between mesophilic and thermophilic temperatures [5].

3.2. *Moisture content observation profile*

Content is one of the parameters that affect the composting process [6]. According to [7] in the composting process, microbes need water to support their growth, affecting the biodegradation of organic materials such as rice husks. In this study, the source of water content comes from the effluent of the biogas reactor and is flowed continuously during the process so that the water content can be maintained between 55% - 65%.

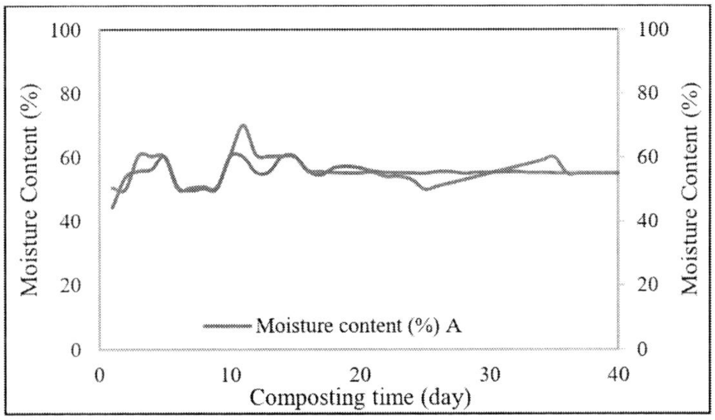

Figure 4. Moisture content profile observation results

In the flow of the biogas reactor effluent to the fermenter, one of its functions is to keep the water content of the fermenter which is around 55%. - 65%, the volume of water content every day is equal to the feed that is fed into the biogas reactor; for one day, the feeding is carried out twice a day, in the morning and evening, from the picture above the average daily water content is between 55% - 65%, This is following research [8], that the average water content in the composting process is between 55% - 65%.

The importance of water content in composting must be highly maintained because for the sake of sustainable microbial growth to degrade organic material into compost; it also illustrates that the oxygen level needed by microbes in the degradation process is trapped in the cavity of the organic material substrate particles so that the water content must be maintained. If the water content is less than 12%, it will interfere with the composting process, can cause microbial activity to stop, and the degradation process does not go well; this is following the report [9], saying that water content is significant in the composting process.

3.3. pH profile observations

This study checked the compost once a day to determine the pH profile of the compost every day, In Figure 5. In the pH observation profile above, the pH value varies every day, and the pH value varies with time composting; the research above has two treatments, namely treatment A uses an effluent bioactivator from a biogas reactor while treatment B uses a bioactivator from the pilot plant, everyday therapies A and B change, as seen in treatment A the pH value from the first day is 8.3; 8.3 then dropped to 7.8; and so on then increased again with a maximum weight of 8.5 increase, as well as treatment B from the first day with a pH value of 8.5; 8.4;8; 7 then rose again to 8.5, this is following the research report that [10], but the pH value will rise again to pH 9, this indicates the occurrence of organic acid concentration in the composting process [11]. Asam The acid in composting serves as one of the parameters for the increase in the microbial population in the composter, so the rise in pH will increase continuously because acid is a source of nutrition for microbes [12].

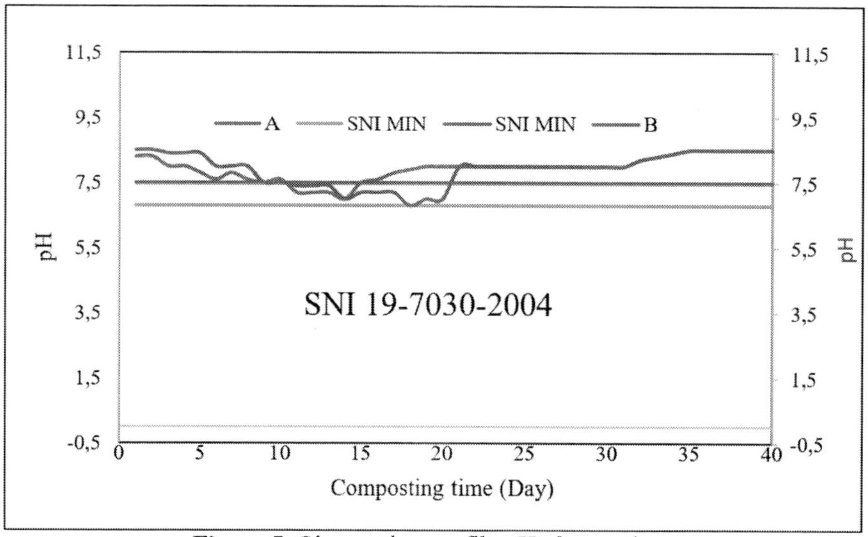

Figure 5. Observation profile pH observations

Changes in pH during the decomposition process of rice husks into compost are caused by microbial activity in the composting process [13], a decrease in pH in the composting process indicates that ammonium evaporation occurs, and the process of releasing hydrogen ions as a result of the nitrification process [14].

3.4. Results of analysis of macro and micro nutritional composts

To determine the quality of compost, an analysis of the nutrient content of compost was carried out with two treatments, namely treatment A and treatment B, then the results of the two treatments were compared with SNI 29-7030-2004; below are the results of the analysis Nutrients Compost of rice husks using effluent bioactivator from a biogas reactor.

Table 2. Results of nutrition analysis of macro and micro nutritional composts

NO	PARAMETER	Unit	Test Methode	Value		SNI 19-7030-2004
				A	B	
1	Temp.	(°C)	SNI 06-6989.23.2005	30	30	Water soil temp.
2	*Moisture Content*	*(%)*	SNI 03-1971-1990	55	54,89	≤ 50
3	pH		SNI 03-6787-2002	7,35	7,55	6,80- 7,49
4	C-organik	%	Gravimetri	14,12	14,36	9,8 – 32
5	N-Total	%	Kjeldahl	1,08	1,12	≥ 0,4
6	P_2O_5	%	Spectrofotometri	0,77	0,69	≥ 0,2
7	K_2O	%	AAS	0,57	0,60	≤ 25,5
8	CaO	%	AAS	0,17	0,22	≤ 25,5
9	MgO	%	AAS	0,29	0,34	≤0,6
10	Cu	ppm	AAS	9	9,1	≤ 100
11	Zn	ppm	AAS	50	52	≤ 500
12	Pb	ppm	AAS	Td*)	Td*)	≤ 50

Information: Td*) = Not detected

Based on Table 2, can be seen the results of elemental analysis of compost nutrients from rice husks; the compost above is the result of two treatments, where treatment A is composted using a bioactivator from a biogas effluent reactor while treatment B is abioactivator used from the USU pilot plant, from the table above we can see that the results of treatments A and B are not very good. Much different, when compared with SNI 19-7030-2004, which is above the standard, treat A with a temperature value of 30 ^{0}C, the value of *Moisture Content* is 55 while the middle of SNI 19-7030-2004 is ≤ 50, pH value of 7.35 is still within the threshold SNI is 6.80 – 7, 49, while the value of organic C is 14.12%, the value of SNI is 9.8 – 32, the N-Total value is 1.08% SNI 0.4; the importance of P_2O_5 is 0.77 the value of SNI 0.2; the value of K_2O is 0.57 %, the value of SNI is 25.5%, the value of Cao is 0.17 %, the value of SNI is 25.5%, as well as the value of macro and micronutrients are on the threshold of SNI 19-7030 -2004.

In general, macronutrients are the content of NP K, where the role of nitrogen is for the formation of plant cells, plant tissues and organs [15]; nitrogen has the primary function as the synthesis of chlorophyll, protein, and amino acids [16]. Therefore the element N is needed in large quantities in plants, especially in the vegetative phase; in this study, it can be seen in the table above that the N value is 1.08% while the SNI value is 0.4, meaning that the N value in this study meets the SNI standard. No. 19-7030-2004.

The element P_2O_5 plays a role in the growth of roots, seeds, fruit and flowers [17]. The part of the P_2O_5 on plant roots is to improve the root structure so that the absorption of plant roots to nutrients will be better [18], ; the P_2O_5 in this study in treatment A was 0.77 while the treatment B is 0.69 these two values are in the standard SNI 19-7030-2004 with a value of ≥ 0,2.

K2O functions to regulate physiological processes in plants such as accumulation, photosynthesis, translocation of carbohydrate transport, opening and closing of stomata, or controlling the distribution of water in tissues and cells [19]. Lack of K2O plants, Can cause leaves to burn and fall [20]. The table above shows that the value A K2O treatment is 0.57%, while treatment B is 0.60, and the value for treatments A and B is above the SNI quality standard, which is ≤ 25,5.

Magnesium is an activator that transports energy for several enzymes in plants [21]. The element magnesium is very dominant in its presence in leaves, especially for chlorophyll availability [22]. Sufficient magnesium in plants is needed to facilitate photosynthesis [23]. Magnesium is an essential component in forming chlorophyll and enzymes in various protein synthesis processes [24]. The value of MgO in treatments A and B was below the SNI quality standard threshold, namely ≤ 0,6.

The element CaO plays a role in plant cell growth. CaO is a component that regulates the development and maintenance of plant cell walls. The part of CaO is essential for plant root growth. Even if there is a lack of CaO in plants, it will disrupt the formation and development of roots and result in inhibited nutrient absorption. CaO functions in the process of cell division and elongation and regulates the distribution of photosynthetic products. The value of CaO in compost at treatments A and

B was 0.17 % and 0.22 %; this value was below the quality standard of SNI SNI 19-7030-2004, namely ≤ 25,5.

Acknowledgments
The authors would like to thank all those who have supported the research and writing of this paper; I hope this article will be useful for many people.

Reference

[1] Puslitbang 2015 *Laporan Tahunan Penelitian dan Pengembangan Perkebunan*
[2] Wei Y, Wu D, Wei D, Zhao Y, Wu J, Xie X, Zhang R and Wei Z 2019 *Bioresource Technology* **271** 66
[3] Trisakti B, Mhardela P, Husaini T, Irvan and Daimon H 2018 *IOP Conference Series: Materials Science and Engineering* **309**
[4] Sun C, Wei Y, Kou J, Han Z, Shi Q, Liu L and Sun Z 2021 *Journal of Cleaner Production* **299** 126928
[5] Alves D, Villar I and Mato S 2019 *Journal of Environmental Management* **239** 30
[6] Li M X, He X S, Tang J, Li X, Zhao R, Tao Y Q, Wang C and Qiu Z P 2021 *Chemosphere* **264**
[7] Bao H, Wang J, Zhang H, Li J, Li H and Wu F 2020 *Journal of Hazardous Materials* **385** 121595
[8] Trisakti B, Mhardela P, Husaini T, Irvan and Daimon H 2018 *IOP Conference Series: Materials Science and Engineering* **309**
[9] Lalander C, Ermolaev E, Wiklicky V and Vinnerås B 2020 *Science of the Total Environment* **729**
[10] Cao Y, Wang X, Liu L, Velthof G L, Misselbrook T, Bai Z and Ma L 2020 *Journal of Environmental Management* **264**
[11] Voběrková S, Maxianová A, Schlosserová N, Adamcová D, Vršanská M, Richtera L, Gagić M, Zloch J and Vaverková M D 2020 *Science of the Total Environment* **723**
[12] Wu D, Wei Z, Mohamed T A, Zheng G, Qu F, Wang F, Zhao Y and Song C 2022 *Chemosphere* **286**
[13] Wang W, Zhang L and Sun X 2021 *Bioresource Technology* **320**
[14] Wong J W C, Mak K F, Chan N W, Lam A, Fang M, Zhou L X, Wu Q T and Liao X D 2001 *Bioresource Technology* **76** 99
[15] Bernard S M and Habash D Z 2009 *New Phytologist* **182** 608
[16] Hörtensteiner S and Feller U 2002 *Journal of Experimental Botany* **53** 927
[17] Ginandjar S, Hakim L A, Subandi M, Rahmadi A and Hakim M T A 2021 *IOP Conference Series: Earth and Environmental Science* **739**
[18] Singana H, Moreno P S and Hussauin D M 2006 Bacterial Biofertilizers For Sustainable Crop Production: A Review Related papers Bact erial Biofert ilizers for sust ainable crop product ion: A review Dat aset (www.arpnjournals.com)
[19] Alaoui-Sossé B, Genet P, Vinit-Dunand F, Toussaint M L, Epron D and Badot P M 2004 *Plant Science* **166** 1213
[20] McCauley A, Jones C and Jacobsen J 2011 *Nutrient Management Module* **9** 1
[21] Soetan K O, Olaiya C O and Oyewole O E 2010 *African Journal of Food Science* **4** 200
[22] Afrousheh M, Ardalan M, Hokmabadi H and Afrousheh M 2010 *Scientia Horticulturae* **124** 141
[23] Cakmak I and Yazici A M 2010 *Better Crops* **94** 23
[24] Hörtensteiner S and Kräutler B 2011 *Biochimica et Biophysica Acta - Bioenergetics* **1807** 977

IOP Conf. Series: Earth and Environmental Science 1115 (2022) 012070 doi:10.1088/1755-1315/1115/1/012070

Lignin utilization from black liquor as raw material lignin resorsinol formaldehyde adhesives (LRF)

H Manurung* **and I Azhar**

Faculty of Forestry, Universitas Sumatera Utara, Medan, North Sumatra, Indonesia

*Email: harisyah_manurung@usu.ac.id

Abstract. Black liquor is the result of pulp processing which contains several chemical compounds such as lignin, carbohydrate degradation products, resins and fatty acids. The main objective of the study was to determine the amount of lignin content contained in black liquor from the kraft process and its potential as a raw material for adhesives. Several research results state that the lignin contained in black liquor can be used as an adhesive. Lignin was isolated from black liquor by means of repeated deposition (reprepitation). The lignin obtained was then tested by FTIR, then the lignin was used as an adhesive with the addition of resorcinol and formaldehyde with an applicative composition of lignin resorcinol formaldehyde adhesive on a resin with a mole ratio of lignin (L) : resorcinol (R) : formaldehyde (F) = 1: 0.5 : 2 ; 1 : 0.3 : 2 ; with an additive content of 1.5% of the solid resin. Then tested for specific gravity, pH, content of non-volatile solids, viscosity, and gelatination. The results of the study were in the form of black liquor, dark brown to blackish in color with the form of a slightly viscous concentrated liquid. In testing with a pH meter, it showed that the initial pH of the black liquor was 12, this indicates that the waste pulp in the form of leachate is alkaline. Indicators on pH and specific gravity alone do not meet PF adhesive standards. The shape, appearance, viscosity (thickness), residual evaporation and gelatinization period are in accordance with PF adhesive standards. The results obtained from the ratio of different moles of resorcinol do not differ much from the test results, but it can be seen from the data that the specific gravity of the LRF adhesive (1:0,5:2) is closer to the standard for PF adhesive, although the distance between the two values is not much different. Overall this adhesive has the potential and needs further research for its application.

1. Introduction

The unitary state of the Republic of Indonesia, which is geographically located on the equator, has trees that grow three times faster than in countries in cold regions, so that there is extensive forest available as a source of raw materials. the middle of Asia is developing into a new economic giant that will become the world's largest pulp and paper market in the future. Indonesia is ranked 11th in the world for the paper industry and ninth in the world for the pulp industry [1].

Pulp mills have residual waste products in the form of solid and liquid waste. Generally, liquid waste is in the form of black liquor, which contains several chemical compounds that are harmful to the environment. Black liquor is a liquid left over from pulp cooking. Black liquor from pulp processing also contains several chemical compounds such as lignin, carbohydrate degradation products, resins and fatty acids. Among these compounds, lignin is the largest component of black liquor which is about 46% of the total solids.

In general terms, lignin can be described as a three-dimensionally cross-linked polymer formed biologically from phenol-propane-type monomer units [2]. Several research results state that the lignin contained in black liquor can be used as an adhesive [3]. Recently, the academic and industrial interest in lignin as a renewable resource for many valuable applications has been on the rise [4]. In the manufacture of lignin adhesive from black liquor, it is possible to use resorcinol formaldehyde. Resorcinol formaldehyde has advantages, including less odor when compared to phenol formaldehyde, faster hardening at low temperatures, more active than phenol formaldehyde, resistance to weather effects, high humidity, boiled water, high temperature and biodeterioration as well as exterior use, for shipping, structural and for marine construction [5].

2. Materials and Methods

2.1. Materials

The materials used is black liquor from PT. Toba Pulp Lestari, Sulfuric acid 2N, NaOH, 0.1N, filter paper, aquadest, resorcinol, and formaldehyde. The tools used in this study were: erlenmeyer, universal pH indicator, filter paper, aluminum foil, stirring rod, gloves, mask, electric scale, oven, desiccator, centrifuge, petri dish, dropper pipette, measuring cup, 100 mesh filter, viscometer, test tube, pH-meter, aqua glass, nails, rubber band, digital scale, 30 cm x 30 cm print, Teflon cloth, hot press, 0.9 cm thick iron plate, saw, caliper and pycnometer tube.

2.2. Lignin isolation

Isolation of lignin from the remaining pulp cooking solution was carried out by inserting 200 ml of the remaining cooking solution into an erlenmeyer and dripping with 2 N sulfuric acid, slowly (1 ml per minute) until the residual solution showed pH 2. This cooking solution was then centrifuged at a speed of 2500 rpm for 25 minutes. The lignin precipitate formed was dissolved with 0.1 N NaOH solution and filtered. Furthermore, the lignin solution was deposited again with the addition of sulfuric acid, carried out again as above (reprepitation). The precipitated lignin obtained was then washed with hot water and finally with cold water until the washing water was no longer acidic. This can be determined by calculating the pH of the washing water. This lignin is then dried in an oven at a temperature of about 50 °C, then crushed and sieved using a 100 mesh sieve. The obtained lignin is ready to be used as an adhesive raw material for adhesive lignin. Yield is expressed in weight percent (grams) per volume of the remaining cooking solution (ml) [5].

2.3. Lignin analysis

2.3.1. Yield. The lignin yield from black liquor kraft pulp with the following formula:

$$\text{Yield (\%)} = \frac{output}{input} \times 100\% \tag{1}$$

2.3.2. Functional group (FT-IR). Identification of Functional Groups with FT-IR Spectrophotometer Lignin isolate (as much as 1 mg of lignin flour) was added with KBr (as much as 150 mg), then made into pellets which were then analyzed by FT-IR spectrophotometer (IR Spectrophotometer Shimadzu FTIR-8201 PC) with a number wave 4000 cm^{-1} to 400 cm^{-1}.

2.3.3. Manufacturing of adhesive. The application composition of the lignin resorcinol formaldehyde adhesive is the resin with the lignin mole ratio (L): resorcinol (R): formaldehyde (F) = 1 : 0.5 : 2 ; 1 : 0.3 : 2 ; with an additive content of 1.5% of the solid resin. The use of resorcinol variations in this preparation, basically wanted to see which effectiveness was the best. This resorcinol is also used to assist formaldehyde in reacting with lignin [5].

2.3.4. Specific gravity. The procedure for testing specific gravity by weighing an empty pycnometer (W1). Then put distilled water into the pycnometer until it is full, then close it, so that there are no air bubbles in it. The outside of the pycnometer containing distilled water was cleaned and dried, then weighed (W2). Removed water from the pycnometer, cleaned and dried. Then put the adhesive test sample into the pycnometer until it is full and close it, so that there are no air bubbles. Cleaned and dried the outside of the pycnometer containing the test sample, then weighed (W3). Calculated, specific gravity with the equation:

$$\text{Specific gravity} = \frac{W3-W1}{W2-W1} \tag{2}$$

2.3.5. pH. The pH measurement was carried out with a pHmeter, through the steps of standardizing the pH meter using a buffer solution of pH 7 and pH 10. Then pour the sample into a 200 ml cup to taste and measure the pH of the sample [6].

2.3.6. Non-volatile solid content. The test for the content of non-volatile solids was carried out by weighing the sample as much as 1.5 grams and placing it in a cup (W1). Then dry for 3 hours at a temperature of (105±2) °C, cool in a desiccator and weigh (W2). The remaining evaporation is calculated by the equation:

$$\text{Residual evaporation (\%)} = (W2/W1) \times 100\% \tag{3}$$

2.3.7. Viscosity. The viscosity of the adhesive was measured using a viscometer, with the steps of measuring the adhesive mass to be used as much as ± 6 g. Then the volume of the same adhesive was measured as much as ± 5 ml. Then the adhesive is slowly inserted into the viscometer. After the adhesive enters the viscometer, then the adhesive is sucked through the measuring flask to the upper limit of the capillary tube. Then the adhesive from the top of the capillary tube is allowed to drop to the bottom of the capillary tube. Then the time is calculated. The calculated time is the time required for the adhesive to descend from the top to the bottom of the capillary tube [6].

2.3.8. Gelatination. The test procedure for the gelatinization mass is weighed ± 10 g of the test sample and put it in a test tube, then close it. Then heated on a water bath at a temperature of 100 °C, the surface of the sample is placed 2 cm below the surface of the water. Then the time required for the test sample in the gelatin tube was observed by tilting the test tube and it was seen that the test sample was no longer flowing [6]. The test results obtained from the research were then compared with the quality of the liquid Phenol Formaldehyde adhesive. The quality of liquid phenol formaldehyde for plywood adhesives can be seen in Table 1.

Table 1. The quality of liquid phenol formaldehyde fplywood adhesives

No	Parameter	Unit	Requirements
1	Form	-	Liquid
2	Appearance	-	Blackish red and free from dirt
3	pH (25 °)	-	10.0-13.0
4	Viscosity (25 °)	cps	130-300
5	Specific gravity (25 °)	-	1.165-1.200
6	Non-volatile solid content	%	40-45
7	Gelatination (100 °C)	minute	>30

IOP Conf. Series: Earth and Environmental Science 1115 (2022) 012070 doi:10.1088/1755-1315/1115/1/012070

3. Results and Discussions

3.1. Source of lignin

The raw material in the form of black liquor used from PT. Toba Pulp Leatari, Tbk with the raw material used in the pulping process is *Eucalyptus* sp. Eucalyptus plants belong to the family Myrtaceae, the genus Eucalyptus with the species *Eucalyptus* sp. Commonly known species include *Eucalyptus alba, Eucalyptus deglupta, Eucalyptus grandis, Eucalyptus plathyphylla, Eucalyptus saligna, Eucalyptus umbellate, Eucalyptus camadulensis, Eucalyptus pellita, Eucalyptus tereticornis.*

The taxonomy of eucalyptus plants is as follows:

Kingdom	: Plantae
Division	: Spermatophyta
Class	: Myrtales
Family	: Mytaceae
Genus	: Eucalyptus
Species	: *Eucalyptus* sp. [7]

3.2. Yield

Based on the results of the isolation of lignin from black liquor, the percent yield of lignin obtained was 12.907%. This output value is obtained from the isolation of lignin from black liquor weighing 77.44 g compared to the volume or input of raw materials of 600 ml. The yield obtained from black liquor is obtained from *Eucalyptus* sp.

Figure 1. Lignin from black liquor

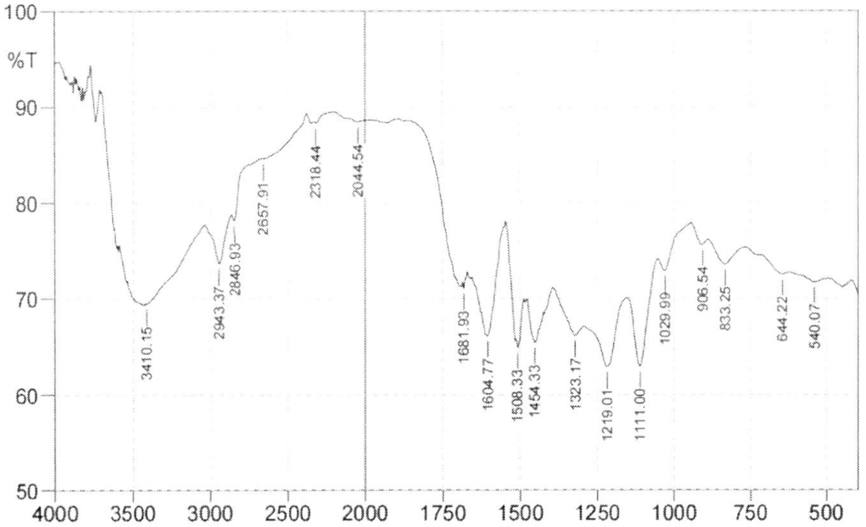

Figure 2. FT-IR results of lignin functional groups

3.3. Functional groups

Identification of Lignin Functional Groups using an FT-IR Spectrophotometer (IR Spectrophotometer Shimadzu FTIR-8201 PC) with a wave number of 4000 cm^{-1} to 400 cm^{-1}. The results can be seen from Figure 2 and Table 2. In Table 2, it can be seen that there is a stretch of the O-H and N-H functional groups at a frequency of 3410.15 cm^{-1} which is in the standard range of 3300 cm^{-1} -3600 cm^{-1}. The -CH3, -CH2-, C-H, C-H aldehyde strains were found at frequencies of 2943.37 cm^{-1} and 2846.93 cm^{-1} which were in the standard range of 3000 cm^{-1} -2700 cm^{-1}. In the -C≡C- strain, C≡N was found at a frequency of 2318.44 according to the standard range of 2400 cm^{-1}-2100 cm^{-1}[8].

Table 2. Functional groups in lignin

Frequency (cm^{-1}) on research	Absorption band standard range (cm^{-1})	Function group description
3410.15	3300-3600	apart O-H, N-H
2943.37 and 2846.93	3000-2700	apart -CH3, -CH2-, C-H, C-H aldehyde
2318.44	2400-2100	apart -C≡C-, C≡N
1681.93	1900-1650	regang C=O (acids, aldehydes, ketones, amides, esters, anhydrides)
1604.77 and 1508.33	1675-1500	apart C=C (aromatic and aliphatic), C=N
1454.33 and 1323.17	1475-1300	C-H bending
906.54 and 833.25	1000-650	C=C-H, Ar-H bending

3.4. Specific gravity

For the adhesive, the density was measured using a pycnometer. The average density value obtained for LRF adhesive with a ratio (1:0.3:2) is 1.02 g/ml, while for LRF adhesive with a ratio (1:0.5:2) is 1.09 g/ml.

3.5. pH
The value of the degree of acidity LRF (1:0.5:2) is classified as alkaline with a value of 8.74 and LRF (1:0.3:2) is also classified as alkaline with a value of 9.30. In the manufacture of lignin resorcinol formaldehyde, NaOH is added which functions as a catalyst to accelerate the reaction between lignin, resorcinol and formaldehyde. According to Pizzi [9], an increase in pH above 3 causes the reaction process to be faster. This happens because the NaOH catalyst causes the activation of lignin with resorcinol and formaldehyde. The results obtained by LRF (1:0.5:2) of 8.74 and LRF (1:0.3:2) of 9.30 are not in accordance with the standard phenol formaldehyde which requires a pH of 10.0-13.0. With respect to the polar character of lignin, two factors of great importance are the value of pH and lignin-polysaccharide complexes. lignin's solubility in water decreases with decreasing pH [10].

3.6. Non-volatile solid content
The average value of the non-volatile solids content with the difference in the use of resorcinol did not differ much. The results obtained by LRF (1:0.5:2) are 44.66% and LRF (1:0.3:2) are 43.1%. This result is in accordance with the Phenol formaldehyde standard which requires 40-45%.

3.7. Viscosity
The viscosity or viscosity of the LRF adhesive (1:0.5:2) is 187 cps and the LRF (1:0.3:2) is 262 cps. This viscosity result is in the range of the standard phenol formaldehyde which requires between 130-300 cps. The viscosity of the LRF resin was made relatively unaffected by the increase in moles of resorcinol, and was attempted to be thinner than the PF adhesive, in order to have a longer pot-life.
This parameter is not the measured quantity, but the final target of the reaction. This condition was created because according to Maloney [11], a resin with a high solids content with an appropriate viscosity will make it able to penetrate the wood pores well and form an optimum bond, resulting in satisfactory adhesion. Facts in the field often show that adhesives with high viscosity, shorter pot-life and will harden faster than runny ones, so the adhesive quality is relatively low.

3.8. Gelatination
The gelatinization period is the time required for the adhesive to thicken or become a gel so that it cannot be added with other materials and is ready to be glued [5]. The average value of gelatinization time for LRF adhesive (1:0.5:2) was 44.66 minutes and for LRF adhesive (1:0.3:2) was 45 minutes. This result is in accordance with the standard of phenol formaldehyde which requires >30 minutes. Gelatin time represents the pot-life of the resin. The LRF resin produced had a longer gelatinization time than the PF adhesive used as a comparison. This suggests that resorcinol is bound to methylated lignin under alkaline conditions to form branched chains [12]. This results in limited crystallization of LRF, so that the resin is more almorphic than PF. All characteristics of lignin product is presented in Table 3.

Table 3. Test results data with PF Comparison

No	Test	Unit	LRF 1:0.5:2	LRF 1:0.3:2	Phenol Formaldehyde
1	Form	-	Liquid	Liquid	Liquid
2	Appearance	-	blackish red	blackish red	blackish red and free from dirt
3	pH	-	8.74	9.30	10.0-13.0
4	Viscocity	Poise	1.87	2.62	1.30-3.00 poise or 130-300 cps
5	Specific gravity	-	1.09	1.02	1.165-1.200
6	Residual evaporation	%	44.66	43.1	40-45
7	Gelatination	minute	44.66	45	>30

Based on the results obtained that the content for pH and specific gravity alone does not meet the standards of PF adhesive. The shape, appearance, viscosity (thickness), residual evaporation and gelatinization period are in accordance with PF adhesive standards. The results obtained from the ratio of different moles of resorcinol do not differ much from the test results, but it can be seen from the data that the specific gravity of the LRF adhesive (1:0.5:2) is closer to the standard for PF adhesive, although the distance between the two values is not much different.

4. Conclusions

The results of the study were in the form of black leachate, dark brown to blackish in color with the form of a slightly viscous concentrated liquid. In testing with a pH meter, it showed that the initial pH of the black leachate was 12, this indicates that the waste pulp in the form of leachate is alkaline. The results obtained are that the content for pH and specific gravity does not meet the standards of PF adhesive. The shape, appearance, viscosity (thickness), residual evaporation and gelatinization period are in accordance with PF adhesive standards. The results obtained from the ratio of different moles of resorcinol do not differ much from the test results, but it can be seen from the data that the specific gravity of the LRF adhesive (1:0.5:2) is closer to the standard for PF adhesive, although the distance between the two values is not much different.

Acknowledgments

The authors are thankful to PT. Toba Pulp Leatari, TbkResearch institute TALENTA University of North Sumatra for fiscal year 2021 No: 6789/UN5.1.R/PPM/2021 and Sri Megawati Lubis who have helped.

References

[1] ICN [Indonesian Commercial Newsletter] 2011 *Profil Industri Pulp Dan Kertas* (http://datacon.co.id/Pulp-2011Industri.html)
[2] Sjöström E 1993 *Wood Chemistry: Fundamentals and Applications* (Academic Press: San Diego)
[3] Maulana H 2018 *Pemanfaatan Lignin dari Lindi Hitam sebagai Bahan Pembuatan Perekat* (Universitas Muhammadiyah Surakarta: Surakarta)
[4] Lauwaert J, Stals I, Lancefield C S, Deschaumes W, Depuydt D, Vanlerberghe B, Devlamynck T, Bruijnincx P C A and Ververckmoes A 2019 *Separation and Purification Technology* **221** 226
[5] Ruhendi S, Koroh D N, Syamani F A, Yanti H, Nurhaida, Saad S and Sucipto T 2007 *Analisis Perekatan Kayu* (IPB University: Bogor)
[6] SNI [Standar Nasional Indonesia] 1992 *Standar Nasional Indonesia untuk Kayu Lapis* (SNI 01-2704-1992).
[7] Departemen Kehutanan 1994 *Eucalyptus Pedoman Teknis Penanaman Jenis-jenis Kayu Komersial* (Badan Litbang Departemen Kehutanan: Indonesia)
[8] Dachriyanus 2004 *Analisis Struktur Senyawa Organik secara Spektroscopy* (Universitas Andalas: Padang)
[9] Pizzi A 1994 *Advanced Wood Adhesives Technology* (Marcel Dekker: New York)
[10] Hubbe M A, Raimo A, Michael P, Miyuru K and Jonas K 2019 *BioResources* **14** 2300
[11] Maloney T M 1993 *Modern Particleboard and Dry-Process Fiberboard Manufacturing* (Miller Freemann Inc.: USA)
[12] Santoso A 2003 *Sintesis dan Pencirian Resin Lignin Resorsinol Formaldehida untuk Perekat Kayu Lamina* (IPB University: Bogor)

Partial purification and characterization of keratinase produced by keratinolytic fungi, *Earliella scabrosa* A2 and *Aspergillus flavus* A11

C Gozali, D Suryanto* and A Hartanto

Department of Biology, Faculty of Mathematics and Sciences, Universitas Sumatera Utara, Medan, North Sumatra 20155, Indonesia

*E-mail: dwisuryanto@usu.ac.id

Abstract. Chicken feather waste originating from the poultry industry and its processing is a potential source of protein and amino acids that can still be used as functional compounds. This study aimed to study the keratinolytic activity of local fungal isolates in vitro as a keratin- degrading agent. Keratinase secreted by keratinolytic fungal isolates were produced and characterized in 1% (w/v) Feather Meal Broth (FMB) composed of pure chicken feathers. Fungal isolates used in this study were A2 and A11 which were later identified based on ITS- rDNA genetic similarity to *Earliella scabrosa* and *Aspergillus flavus,* respectively. Crude enzymes were precipitated using an increasing ammonium sulphate salt gradient. The enzyme activity of *Earliella scabrosa* A2 and *Aspergillus flavus* A11 were 19 and 7.5 U/mL respectively. The crude enzymes showed an increase of activity during 80% of precipitation for isolate A2 with 10 U/mL and for isolate A11 with 18.6 U/mL. The specific activity after dialysis were 61.99 U/mg for isolate A2 and 75.11 U/mg for isolate A11. The optimum condition of keratinase activity for isolate A2 were at pH 7 and 30 °C while isolate A11 showed its optimum activity at pH 8 and 45°C.

1. Introduction

The increase in chicken farming has resulted in an increase in chicken feather waste produced from thechicken slaughterhouse industry and from other chicken slaughterhouses. In the chicken slaughterhouse industry, chicken feather waste is something that needs special handling because it has a huge impact on environmental pollution. Chicken feathers are waste that still has the potential to be utilized because it still contains a crude protein nutrient content of 80-91% of dry matter, most of which is keratin [1].

Keratin is a biopolymer composed of fibrous protein which is difficult to dissolve and has a strong structure so it is very difficult to be degraded or hydrolyzed due to the presence of disulfide bonds and cross bonds. Keratin has a large number of cysteine residues, which have a thiol group (-SH), resultingin strong, covalent disulfide bonds that cleave the polypeptide bonds of the chains together and also cross-link the matrix molecules [2-4]. Keratin cannot be digested using ordinary protease enzymes soit requires keratinase enzymes in the degradation process.

Keratinase enzymes can be obtained by isolating microorganisms that live in nature, one of whichis obtained from waste soil around chicken slaughter. Keratinase includes protease enzymes which are extracellular enzymes, which are classified as serine proteases. Microorganisms that can be used to produce keratinase enzymes include the genera Bacillus, Streptomyces and Actinomycetes [5]. The largest group of organisms that can utilize keratin as the sole source of carbon and nitrogen are keratinolytic fungi [4].

Keratinolytic fungal isolates have been isolated and are known to be able to degrade chicken feather waste due to the activity of the keratinase enzyme. However, the isolates of the keratinolytic fungi have not yet known the enzyme activity and the exact incubation time to produce enzymes with high activity. The objective of this research was to examine the in vitro keratinolytic activity of local fungal isolates as a keratin-degrading agent. Therefore, it is necessary to determine the timing of enzyme production and partial purification of enzymes as well as characterization of enzymes including characterization of pH and temperature of the isolates of the keratinolytic fungi.

2. Materials and Method

2.1. Subculture of Keratinolytic Fungal Isolates
Keratinolytic fungal isolates were obtained from stock cultures of the Microbiology Laboratory FMIPA USU. Keratinolytic fungal isolates A2 and A11 were rejuvenated by cutting the hyphae using a cork borer and then inoculated into Potato Dextrose Agar media. Fungal isolates were incubated for6 days at room temperature.

2.2. Measurement of Keratinase Enzyme Activity
Keratinolytic fungal isolates were rejuvenated on Potato Dextrose Agar (PDA) media and incubated for 6 days. Furthermore, the mycelium of the keratinolytic fungi was cut to a size of 1 x 1 cm. The keratinolytic fungi pieces were put into Feather Meal Broth (FMB) media. FMB media containing pieces of keratinolytic fungi were incubated for 15 days using an orbital shaker at 120 rpm at room temperature. Liquid cultures were measured for crude extract enzyme activity every 3 days for 15 days. The incubation results were filtered using Whattman filter paper no. 1 then the filtrate was centrifuged at 10,000 rpm for 15 minutes. The supernatant from the centrifugation was taken and stored at 4°C. The supernatant was referred to as crude extract enzyme, which was then measured using the modified method [6].

2.3. Keratinase enzyme production
The production of the keratinase enzyme was carried out in the same steps as measuring the enzyme activity of the crude extract. Keratinolytic fungal isolates A2 were put into production medium (FMB) and then incubated under agitation for 12 days, while A11 fungal isolates were incubated under agitated conditions for 9 days at room temperature. The incubation results are then separated from the remaining chicken feathers by filtering using Whattman filter paper no. 1 and then centrifuged to separate the pellet and supernatant. The supernatant obtained was taken and stored at 4°C for further partial purification of the keratinase enzyme, while the protein content was measured using the Lowry method.

IOP Conf. Series: Earth and Environmental Science 1115 (2022) 012071 doi:10.1088/1755-1315/1115/1/012071

2.4. Keratinase activity measurement

Measurement of keratinase enzyme activity was determined using the modified method [6]. 0.5 ml of keratinase enzyme was dissolved with 0.5 ml of 0.1 M potassium phosphate buffer solution (pH 7), then 0.5 ml of 0.5% keratin solution was added. The mixture was incubated for 15 minutes at 31°C andthen 10% TCA was added 13 mL after incubation. The mixture was filtered using a 0.45 µl syringe filter. The filtrate was measured at 280 nm. Control was carried out by adding 1 mL of TCA solution before adding 0.5% keratin solution. One unit of keratinase activity (U/ml) was defined as an increase in absorbance of 0.01. The initial stage of preparation is washing the feathers with water and detergent. The feathers were then soaked in acetone to remove the lipid content in the feathers for 24 hours. The chicken feathers that have been soaked in acetone are then dried at 40°C using an oven for 72 hours. Dried chicken feathers were cut into small pieces of 2 cm and then ground until smoothusing a mortar and then sieved [7].

2.5. Determination of protein levels

Protein content of the enzyme keratinase was measured based on the Lowry method [8]. 1 mL of enzyme was mixed with 5 mL of Lowry's solution A. The mixture was homogenized and incubated for 10 minutes. Then 0.5 mL of Lowry B solution was added. The mixture was homogenized and incubated for 30 minutes. The mixture was measured at a wavelength of 750 nm. Bovine Serum Albumin (BSA) standard curve was created to calculate the protein concentration of the sample. BSA standard curves were made with a series of standard concentrations of 0, 5, 10, 15, 20, 40, 60, 80, 100, 150, 200, 250 and 300 mg/mL. A standard curve is created by relating the standard concentration to the corrected absorbance.

2.6. Keratinase precipitation

The crude extract enzyme obtained from the previous enzyme production stage was precipitated with six levels of saturation of the ammonium sulfate salt, namely 30%, 40%, 50%, 60%, 70% and 80%. The ammonium sulfate salt used in the precipitation process according to the EnCor Biotechnology Inc protocol. Ammonium sulfate salt with a saturation level of 30% was added to 50 mL of crude extract enzyme slowly and stirred using a magnetic stirrer under cold conditions for 2 hours. Furthermore, the crude extract enzyme which was added with 30% ammonium sulfate was centrifuged at 10,000 rpm for 15 minutes at 4°C. The pellets and the supernatant obtained were separated and then re-precipitated with a saturation level of 30% by adding ammonium sulfate salt to the supernatant from the previous centrifugation and centrifuged again under the same conditions until pellets and supernatants were obtained at a saturation level of ammonium sulfate up to 80%.

Supernatant were tested for enzyme activity, while the pellets obtained were first dissolved in 10 ml of potassium phosphate buffer solution pH 7. Measurement of the keratinase enzyme activity of pellets and supernatants was measured by the method used in the modified [6] study. Protein content was measured using the Lowry method.

2.7. Keratinase dialysis

The dialysis process was carried out after enzyme precipitation using ammonium sulfate. Dialysis was started by cutting a cellophane bag (cut off 10 kD) 15 cm long and then immersed in 100 ml of phosphate buffer solution for 15 minutes. Cellophane plastic was tied at one end and the other end was opened to fill 9 ml of the enzyme precipitated with ammonium sulfate, then tied tightly.

Cellophane plastic that already contains the enzyme precipitation is put into a beaker glass containing 600 ml of cold phosphate buffer solution and stirred using a magnetic stirrer for 1 hour, then the phosphatebuffer solution is replaced and stirred using a magnetic stirrer for 1 hour, then the buffer solution is replaced again. and stirred using a magnetic stirrer for 4 hours. Keratinase activity was measured usingthe modified [6] method, while protein content was measured using the Lowry method.

2.8. Keratinase characterization based on pH and temperature

Keratinase enzymes are characterized based on different pH and temperature. Characterization based on pH was carried out by adding 0.5 ml of 0.5% keratin solution into a tube containing 0.5 ml of buffer solution and 0.5 ml of dialyzed keratinase enzyme. The mixture was homogenized and incubated for 15 minutes at 31°C with different pH of 4, 5, 6, 7, 8 and 9. To adjust the pH of the solution, sodium acetate buffer solution was used for pH 4 and 5, pH 6 and 7 potassium phosphate and Tris-HCl for pH 8 and 9. Enzyme activity was measured using the modified method [6].

Enzyme characterization was based on temperature with different temperatures, namely 25, 30, 35, 40, 45 and 50°C. The 0.5% keratin solution was put into a tube containing a buffer solution at the optimum pH of the enzyme from each isolate, then the dialyzed enzyme was added. The mixture was homogenized and incubated at a predetermined temperature for 15 minutes. Enzyme activity was measured using the method of [6]. The initial stage of preparation is washing the feathers with water and detergent. The feathers were then soaked in acetone to remove the lipid content in the feathers for 24 hours. The chicken feathers that have been soaked in acetone are then dried at 40°C using an oven for 72 hours. Dried chicken feathers were cut into small pieces of 2 cm and then ground until smooth using a mortar and then sieved [7].

3. Results and Discussion

3.1. Keratinase enzyme activity

Measurement of the keratinase enzyme activity of *Earliella scabrosa* A2 and *Aspergillus flavus* A11 isolates was carried out every three days for 15 days. *Earliella scabrosa* A2 isolate showed the highestactivity on the 12th day of incubation with enzyme activity of 19 U/mL, while *Aspergillus flavus* A11 isolate showed the highest activity on the 9th day of incubation with an activity of 7.5 U/mL shown in Figure 1.

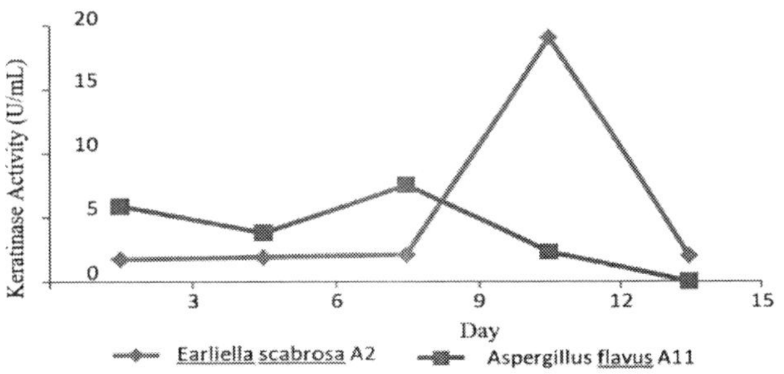

Figure 1. Profile of keratinase enzyme activity.

The incubation period is one of the important parameters for the production of keratinase enzymes by microorganisms under fermentation conditions [9,10] successfully tested 5 species of fungi, namely *Aspergillus flavus*, *A. fumigatus*, *A. niger*, *A. nidulans* and *A. terreus* which were able to grow and degrade chicken feathers by releasing compounds containing cysteine and keratinase. The proteolytic activity and biomass production of *Aspergillus flavus* K-03 increased gradually in basal media (chicken feathers) from the 8th to 15th day of incubation. Research conducted by Preczeski et al [11] showed that the highest keratinase enzyme activity was found in feather meal media at 113.5 U/mL on the 9th incubation day. The enzymes produced by keratinolytic fungi depend on the composition of the substrate, because the structure of keratin is not only composed of keratin itself.

3.2. Dialysis of keratinase enzyme

The results of precipitation of the keratinase enzyme produced by the two different isolates showed that the enzyme activity in the highest pellet contained a different fraction in each isolate (Figure 2), namely the 70% fraction for Aspergillus flavus A11 isolates and 80% fraction for *Earliella scabrosa* A2 isolates with activities of 18.6 U/mL and 10 U/mL, respectively. It shows enzyme activity (U/mL) that at a concentration of 70% ammonium sulfate the enzyme produced by *Aspergillus flavus* A11 precipitated well while the enzyme produced by isolate *Earliella scabrosa* A2 precipitated well at a concentration of 80% ammonium sulfate shown in Figure 2.

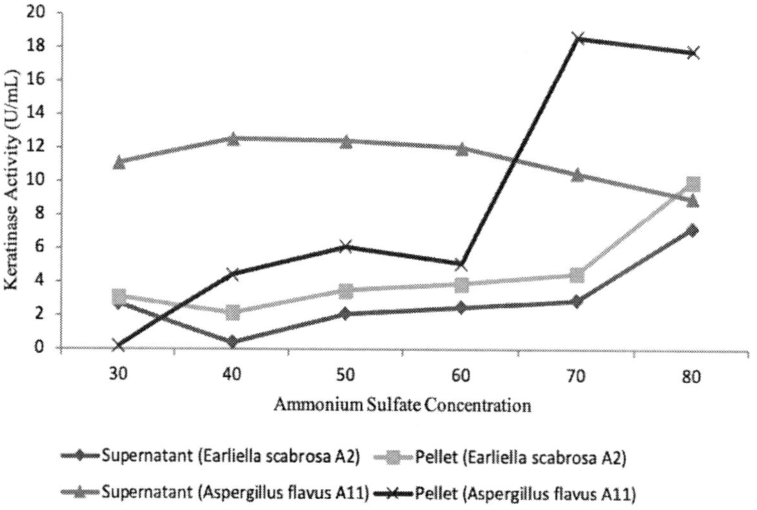

Figure 2. Keratinase activity of supernatants and pellets of keratinolytic fungal isolates precipitation

Precipitation of crude extract enzymes using ammonium sulfate salt causes water molecules contained in enzymes to bind to salt ions. The water that is bound to the enzyme will be released and the enzyme or protein will settle. Enzyme activity will increase because contaminants that block the active site of the enzyme can be released and the active site of the enzyme can bind to the substrate [12]. The salts and organic solvents used in the precipitation process increase the ionic strength and reduce the dielectric constant thereby increasing the protein–protein interactions that stand out from the protein–water interactions, resulting in precipitated protein aggregates [11].

Precipitation with 30-40% ammonium sulfate fraction had the same activity pattern. This indicates that this fraction has not been able to precipitate the keratinase enzyme. Precipitation with ammonium sulfate concentration fraction of 70% can precipitate protein from isolate *Aspergillus flavus* A11 and at fraction 80% can precipitate protein from isolate *Earliella scabrosa* A2.

Dialysis using a permeable membrane showed a decrease in the specific activity of the enzyme from both types of keratinolytic fungal isolates. *Earliella scabrosa* A2 isolate decreased from specific activity of 171.48 U/mg to 90.49 U/mg after precipitation with ammonium sulfate salt with a fraction of 80% and continued to decrease to 61.99 U/mg. *Aspergillus flavus* A11 isolate showed an increase from specific activity of 26.91 U/mg to 105.98 U/mg after precipitation with ammonium sulfate salt with a fraction of 70% and continued to decrease to 75.11 U/mg (Table 1).

Table 1. Purification of keratinolytic fungal keratinase enzymes

Isolate	Volume (mL)	Enzyme	Total Activity (U)	Total Protein (mg)	Specific Activity (U/mg)	Yield	Result (%)
A2	300	Crude Extract	5.700	33.24	171.48	1	100
	60	Ammonium Sulphate (80%)	600	6.63	90.49	0.527	10.5
	35	Dialysis	168	2.71	61.99	0.361	2.94
A11	300	Crude Extract	2.250	83.61	26.91	1	100
	60	Ammonium Sulphate (70%)	1.116	10.53	105.98	3.93	49.6
	35	Dialysis	164.5	2.19	75.11	2.79	7.3

Enzyme purification generally coincides with the purification steps, the specific activity of the enzyme can be increased with crude extract enzymes and a decrease in total protein. The results of this study showed the same pattern where there was a decrease in the specific activity of the enzymes isolates *Earliella scabrosa* A2 and *Aspergillus flavus* A11 at each enzyme purification step, while *Aspergillus flavus* A11 isolates increased at the ammonium sulfate precipitation stage and decreased at the dialysis stage. Another factor that is thought to affect the decrease in the specific activity of the enzyme after enzyme purification is the damage or loss of some of the protein structure that occurs in the precipitation step using ammonium sulfate.

3.3. Effect of pH on keratinase activity

Enzyme characterization with different pH showed that *Earliella scabrosa* A2 had the highest activity at pH 7 with an activity of 4 U/mL and *Aspergillus flavus* A11 had the highest activity at pH 8 with an activity of 5.4 U/mL (Figure 3). Most of the keratinase enzyme groups are classified as alkaline or neutral proteases, but there are several enzymes keratinase which has high enzyme activity at extreme pH [13].

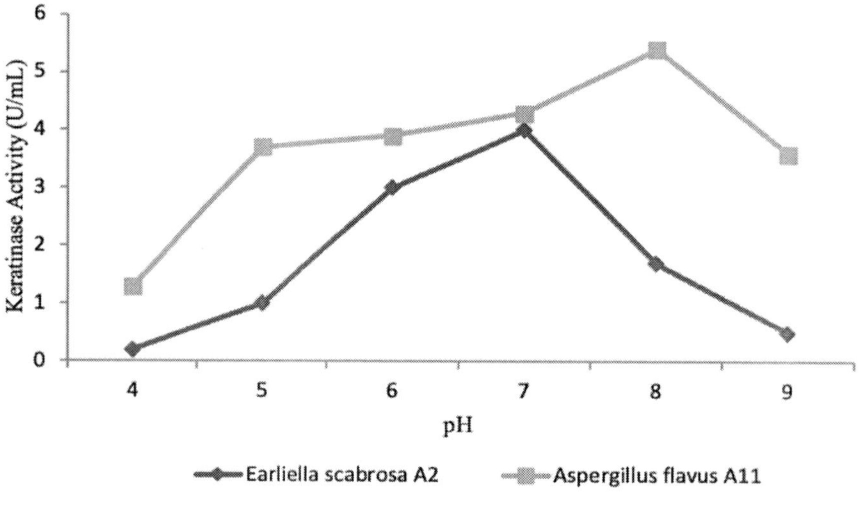

Figure 3. Effect of pH on keratinase activity

Research conducted by Kim [10] on *Aspergillus flavus* isolates showed that these isolates had the highest activity at pH 9, depending on the buffer solution used. The enzyme activity of the isolate was quite stable at pH 8-11 and could maintain its activity more than 80% after 24 hours at room temperature. Most of the keratinase enzyme groups are classified as alkaline or neutral proteases. However, there are several keratinase enzymes produced by keratinolytic fungi that have high enzyme activity at a maximum pH of 8 [13,14].

3.4. Effect of temperature on keratinase activity

The temperature characterization of the dialyzed enzymes was carried out with a temperature range between 25-50°C. The results of temperature characterization on isolates of *Earliella scabrosa* A2 at a temperature of 30°C with an activity of 14.5 U/mL while isolates of *Aspergillus flavus* A11 at a temperature of 45°C with an activity of 9.3 U/mL (Figure 4).

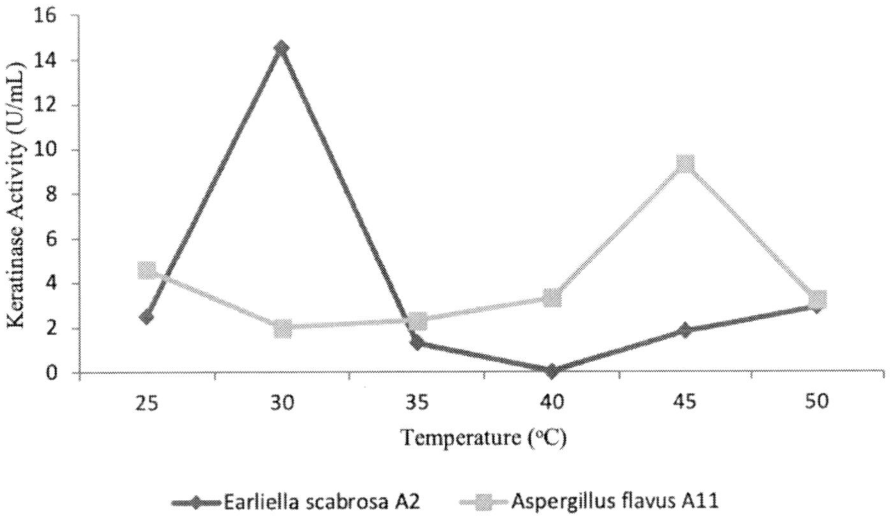

Figure 4. Effect of temperature on keratinase activity

High temperatures can affect enzyme activity because temperature can affect enzymes and substrates to form products. This is because the enzyme and substrate have enough energy to collide with each other to form products. In addition, high temperatures can also affect the performance of enzymes because enzymes can be denatured and even damaged at high temperatures. The enzymes produced by keratinolytic fungi have their respective optimum temperatures. Keratinase had the highest activity at 45°C and at 65°C the enzyme activity decreased drastically [15]. Research conducted by [16] found the highest *A. terreus* keratinase activity found in temperature 35-40°C.

4. Conclusions

Earliella scabrosa A2 and *Aspergillus flavus* A11 fungi isolates had crude extract enzyme activity of 19 U/mL and 7.5 U/mL, respectively. The specific activity of *Earliella scabrosa* A2 after precipitation with ammonium sulfate at 80% concentration was 10 U/mL, while the *Aspergillus flavus* A11 isolate precipitated at 70% concentration had a specific activity of 18.6 U/mL. The enzyme dialysis results from *Earliella scabrosa* A2 isolate had a specific activity of 61.99 U/mg and *Aspergillus flavus* A11 isolate had a specific activity of 75.11 U/mg. The keratinase activity of *Earliella scabrosa* A2 isolates had optimum conditions at pH 7 and a temperature of 30°C, while *Aspergillus flavus* A11 isolates at pH 8 and 45°C.

Acknowledgements
The authors gratefully acknowledged that this research was supported by TALENTA 2020, *Lembaga Penelitian Universitas Sumatera Utara* with a Contract Number: 4142/UN5.1.R/PPM/2020, 27th April 2020.

References

[1] Ketaren S 2008 *Pengantar Teknologi Minyak dan Lemak Pangan* (Jakarta: UI Press)
[2] Peng Z, Mao X, Zhang J, Du G and Chen J 2019 *Microbial Cell Factories* **18** 1
[3] McKittrick J, Chen P Y, Bodde SG, Yang W, Novitskaya E E and Meyers M A 2012 *JOM* **64** 449
[4] Moorthy K, Prasanna I, Vimalan S, Lavanya V, Selvi A T, Mekala T and Thajuddin N 2011 *Biosciences Biotechnology Research Asia* **8** 633
[5] Mini K D, George S M and Mathew J 2015 *International Journal of Current Microbiology andApplied Series* **4** 535
[6] Cai C and Zheng X 2009 *J Zhejiang Univ Sci B* **9** 60
[7] Marcondes N R, Taira C L, Vandresen D C, Svidzinski T I E, Kadiwaki M K and Peralta R M 2008 *Microb Ecol* **56** 13
[8] Plummer D T 1978 *An Introduction to Practical Biochemistry 2nd Edition* (New Delhi: Tata McGraw-Hill Pubishing Company Ltd)
[9] El-Ghonemy D H and Ali T H 2017 *Journal of Applied Pharmaceutical Science* **7** 172
[10] Kim J D 2003 *Mycobiology* **31** 209
[11] Preczeski K P, Dalastra C, Czapela F F, Kubeneck S, Capini T, Camargo A F, Zanivan J, Bonatto C, Stefanski FS, Venturin B, Fongaro G and Treichel H 2020 *Front Bioeng Biotechnol* **8** 1
[12] Roe S 2001 *Protein Purification Techniques* Second Edition (New York: Oxford UniversityPress)
[13] Brandelli A Dariot D J and Riffel A 2010 *Appl Microbiol Biotechnol* **85** 1735
[14] Sangali S and Brandelli A 2000 *Journal of Applied Microbiology* **89** 735
[15] Kim J D 2007 *Mycobiology* **34** 219
[16] Koutb M, Morsy F M, Bagy M M K and Hassan E A 2012 *Journal of Advanced Laboratory Research in Biology* **3** 210

Potential of cellulolytic fungi isolated from the intestine of Asiatic rhinoceros beetle (*Oryctes rhinoceros*)

A F Nasution[1], E Munir[1,2*], D Suryanto[1] and Y Yurnaliza[1]

[1]Department of Biology, Faculty of Mathematics and Natural Sciences, Universitas Sumatera Utara, Medan, North Sumatra 20155, Indonesia
[2]Center of Excellence for Mangrove, Universitas Sumatera Utara, Medan 20155, Indonesia

*E-mail: erman@usu.ac.id

Abstract. Cellulose is a potential source of fermentable sugars and a promising cost-effective source for biofuel production. This study aimed to obtain potential cellulolytic fungi from the intestine of Asiatic rhinoceros beetle (*Oryctes rhinoceros* L.). The cellulolytic fungi were isolated using Potato Dextrose Agar (PDA) media using spread plate method. Qualitative and quantitative screening tests were carried out using 1% of CMC (w/v) as a sole carbon source. Twelve isolates were recovered with varying cellulolytic capabilities. The highest cellulolytic activity was obtained from KOr-15J with an index of 1.93 while KOr-23J produced the lowest cellulolytic index (1.01) based on the qualitative screening. The highest cellulase activity assayed using DNS (3,5-Dinitrosalicylic acid) method revealed KOr-21J with an activity of 0.018 U/mL while three isolates (KOr-16J, KOr-19J and KOr-22J) produced the lowest activity (0.003 U/mL). This study provides multiple verifications of the potential ability of cellulolytic fungi and their cellulase activity in the intestine of Asiatic rhinoceros beetle.

1. Introduction

Cellulase is an enzyme system consisting of three main types of enzymes, namely the endo-β-1,4-glucanase complex (carboxymethylcellulase/CMCase, Cx cellulase, endocellulase), the exo-β-1,4-glucanase complex (avicellase, cellobiohydrolase, C1 cellulase) and 1,4-glucosidase or cellobiase. These enzymes can degrade cellulose by breaking the bonds of 1,4 glycosides which produce oligosaccharides derived from cellulose and glucose. These enzymes are classified into 3 groups depending on their specificity in hydrolyzing cellulose, namely endo-1,4-β-glucanase (CMCase), exo-1,4-β-glucanase and D-glucosidase. Cellulases hydrolyze cellulose by breaking the 1,4-glycosidic bonds in cellulose, cellodextrin, cellobiose and their derivatives into simple sugars or glucose from the ends of the cellulose chain [1,2].

The utilization of cellulases extracted from microorganisms is versatile. In addition to food and industry, the use of cellulases from bacteria can provide a solution to the waste management by reducing the amount of cellulosic wastes such as leaf litter in the final disposal area, agricultural waste and seaweed on the beach which can be modified into processed organic fertilizers. This is because organic waste is a part of an environmental issue that has the potential to become a pollutant factor once abandoned and not managed properly. All of these waste materials can also be modified to produce various beneficial materials for biogas production using cellulolytic enzymes.

Some insects can produce cellulolytic enzymes. Most of the enzymes produced in insects are derived from various microorganisms in the intestinal environment. Asiatic rhinoceros beetle (*Oryctes rhinoceros* L.) is a weevil-type insect that inhabits the oil palm plantation and mainly consumed oil palm trees or celluloses as their food. This indicates that there is a role for microorganisms that can hydrolyze cellulose in the intestine of *O. rhinoceros*. A previous study on finding cellulose-degrading microorganisms from other coleopterans have been reported revealing a number of fungal species i.e. *Fusarium culmorum* and *Penicillium crustosum* [3]. Eleven cellulolytic fungi have been isolated from the digestive tract of *O. rhinoceros* larvae which showed their potential ability based on the qualitative results [4]. Based on a study by Goyari et al [5], three isolates of cellulolytic fungi were isolated from forest soil showed the highest cellulolytic activity and FPase activity, based on the molecular identification of 3 cellulolytic isolates classified in *Talaromyces verruculosus*, *Trichoderma gamsii* and *Trichoderma atroviride*. Therefore, this research is aimed to isolate potential cellulose-degrading fungi from the intestine of *O. rhinoceros* and to specify its cellulose-degrading ability on the CMC medium.

2. Materials and Method

2.1. Isolation of intestinal fungi from Oryctes rhinoceros

Two adult beetles (3-5 cm long) were collected as samples for the isolation of cellulolytic fungi. The intestine part was removed and cut into small pieces. The intestinal fragments were put into sterile physiological saline solution and diluted until 10^8. Approximately 0.1 mL of diluted solution was inoculated into Potato Dextrose Agar (PDA, Merck) medium using the spread plate technique. The isolation media were incubated at 37°C for 3 days. The fungal isolates were characterized based on their colony appearances including form, elevation, margin and color [6,7].

2.2. Qualitative screening of cellulase-producing isolates

The isolates were screened for their cellulolytic activities on 1% CMC agar medium containing (g/L): 1 g of CMC (Sigma-aldrich); 0.02 g of $MgSO_4.7H_2O$ (Merck); 0.075 g of KNO_3 (Merck); 0.05 g of K_2HPO_4 (Merck); 0.002 g of $FeSO_4.7H_2O$ (Merck); 0.004 g of $CaCl_2.2H_2O$ (Merck); 0.2 g of yeast extract (Merck); 1.5 g of agar and 0.1 g of glucose (Merck). The culture media were incubated at 37°C for 3-5 days. Further testing was carried out by pouring a 0.1% (W/v) of congo red solution into the culture and left for 30 min. The immersed plates were rinsed with NaCl (Merck). A formation of clear zone around fungal colonies indicate the cellulolytic activity by the isolates [8,9].

2.3. Crude enzyme production

Three fungal plugs from each isolate were inoculated into a 50-mL conical flask containing 25 ml of CMC liquid medium. The fermentation medium was incubated at 37°C and agitated at 120 rpm for 3-5 days. The crude enzymes were harvested by sampling 5 mL of fermentation medium and centrifuged at 10,000 rpm for 10 min. The supernatants or culture filtrates were used for further enzyme or cellulase assay [10].

2.4. Cellulase assay

The cellulase (endo-β-1,4-glucanase) activity was estimated from a reaction between DNS (3,5-dinitrosalicylic acid) the glucose content in a reaction mixture. One mL of crude cellulases was put into a 1-mL reaction tube containing 1 mL of 1% CMC. The blank was prepared from a reaction mixture between 1 mL of 1% CMC and 1 mL of sterile distilled water added with 1 mL of 20 mM phosphate buffer (pH = 7). The mixture was incubated at 50°C for 10 min later added with 2 mL of DNS and heated at 100°C for 5 min. The solution was cooled for 20 min and the absorbance was measured using a spectrophotometer at 540 nm. One activity unit (U/mL) is expressed as the total amount of reduced sugars (1000 μmoL) in the reaction mixture per 30 min [11,12].

3. Results and Discussion

Twelve fungal isolates were isolated from the intestine of *O. rhinoceros* showing some distinct colony characteristics (Table 1). All isolates produced circular forms with a dominant flat elevations. The margins of the colony were entire (N = 8), filamentous (N = 2), curled (1 isolate) and umbonate (1 isolate). The typical colors of fungal isolates are ranging from white to dark green (top view) and from white to green (bottom view) as represented in Figure 1.

Table 1. Colony characteristics of the fungal isolates from intestine of *O. rhinoceros*

Isolate code	Form	Elevation	Margin	Color colony	
				Top view	Bottom view
KOr-12J	Circular	Flat	Filamentous	White	White
KOr-13J	Circular	Flat	Filamentous	White	White
KOr-14J	Circular	Flat	Entire	White	White
KOr-15J	Circular	Flat	Entire	Brown	Brown
KOr-16J	Circular	Flat	Entire	Dark green	Green
KOr-17J	Circular	Flat	Entire	White	White
KOr-18J	Circular	Flat	Entire	White	White
KOr-19J	Circular	Raised	Curled	Grey	White
KOr-20J	Circular	Raised	Entire	White	White
KOr-21J	Circular	Flat	Entire	White	Yellow
KOr-22J	Circular	Flat	Entire	Brown	Brown
KOr-23J	Circular	Flat	Umbonate	Green	Yellow

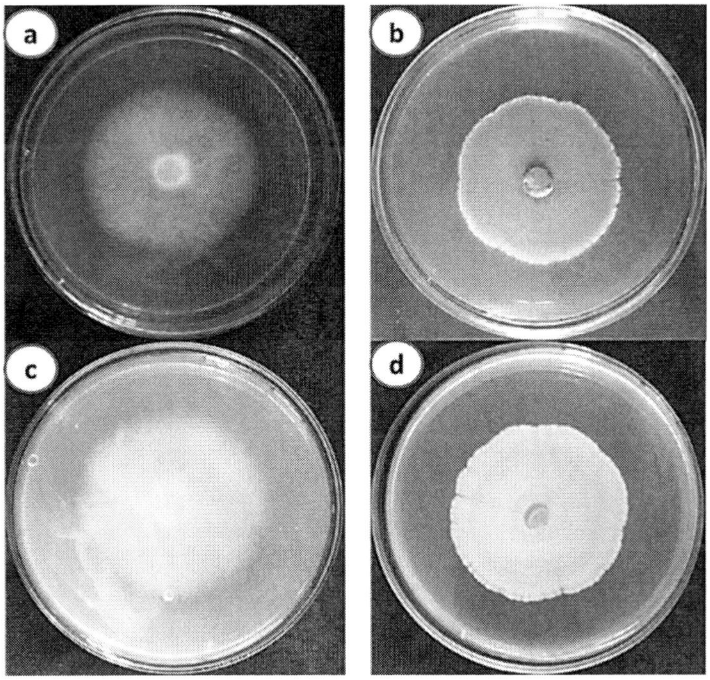

Figure 1. Colony of cellulolytic fungi from the intestine of *O. rhinoceros* beetle on CMC medium. (a) KOr-12J top view, (b) KOr-16J top view, (c) KOr-12J bottom view and (d) KOr-16J bottom view

The qualitative screening using 1% CMC in solid medium showed that all isolates could utilize cellulose as indicated by the formation of a clear area around the colony after immersion with 0.1% (w/v) of congo red solution (Figure 2). The cellulolytic activity expressed as cellulolytic index was varying from 1.01 (KOr-23J) to 1.93 (KOr-15J) or presented in Figure 3.

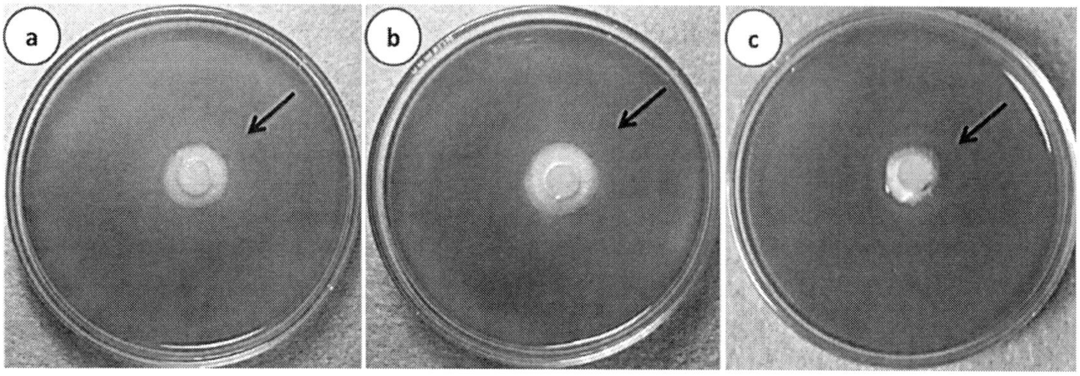

Figure 2. Qualitative screening on CMC medium with congo red (0.1%), (a) isolate KOr-12J, (b) isolate KOr-13J and (c) isolate KOr-16J

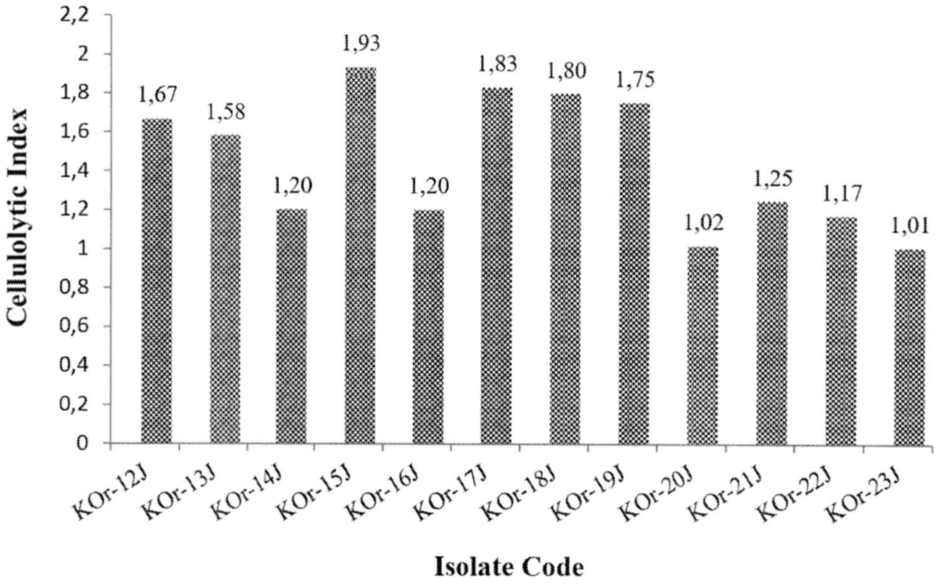

Figure 3. Cellulolytic activity from the intestine of *O. rhinoceros* beetle

The presence of clear zone (Figure 2) shows the cellulolytic activity of the extracellular cellulase secreted by twelve isolates of cellulolytic fungi after immersion with congo red dye solution. The dye will bind specifically to polysaccharides or in this case, the endo-β-1,4-glucanase bonds. Belmont-Montefusco et al succeeded in isolating cellulolytic fungi from the digestive tract of Amazonian aquatic insects, with a total of 175 isolates. Majority of their isolates (70%) produced a higher cellulolytic index (>2.0) from 7 genera including *Cladosporium, Gliocephalotrichum, Penicillium, Pestalotiopsis, Talaromyces, Trichoderma* and *Umbelopsis* [13]. Pathania et al have isolated cellulolytic fungi from different soils in Punjab, India. Five fungal isolates were successfully isolated with codes B, W1, W2, W3LG and W3DG showing the maximum hydrolysis zone and identified as *Aspergillus* sp [14].

The cellulase production was on its peak on the 5th day of incubation as indicated from the consistency change of CMC liquid medium. The medium changed from cloudy and viscous solution from day 0 to a clear and transparent solution on day 5. Twelve isolates of cellulolytic fungi were evaluated for their endo-β-1,4-glucanase activity by measuring the reducing sugar content using the DNS method. Based on the results (Figure 4), the isolate, KOr-21J produced the highest enzyme activity (0.018 U/mL), followed by KOr-15J (0.017 U/mL). The lowest cellulase activity (0.003 U/mL) was obtained from three isolates (KOr-16J, KOr-19J and KOr-22J). This is in line with a study by Nagah et al [15], who reported 8 fungal strains isolated from several Egyptian environments. The best medium in producing extracellular cellulase was CMC with different duration of incubation (3, 6 and 9 days). The most prominent strains were identified as the members of *Aspergillus* sp and *Penicillium* sp [15]. In addition, a study by Azzaz et al [16] showed the ability CMC-ase produced by *Aspergillus niger* with an enzyme activity of 0.076 U/mL.

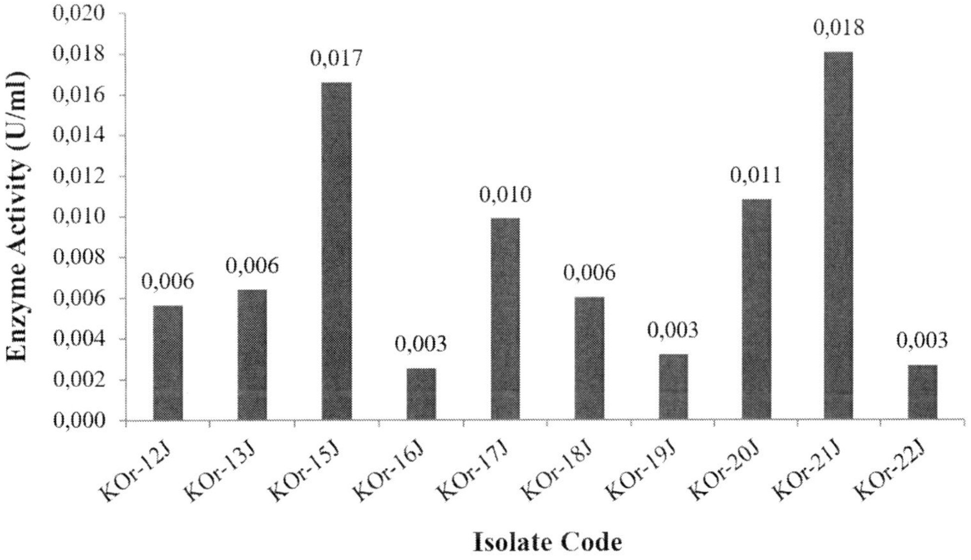

Figure 4. The cellulase or CMCase (endo-β-1,4-glucanase) activity of cellulolytic fungi from the intestine of *O. rhinoceros* bettle

4. Conclusion

The study obtained twelve isolates of cellulolytic fungi from the intestine of *Oryctes rhinoceros*. All isolates could hydrolyze cellulose, forming a clear zone on medium CMC treated with congo red (qualitative test). The isolate, KOr-15J produced the highest cellulolytic index (1.93) while KOr-22J fungal isolate produced the lowest cellulolytic index (1.01). The highest cellulase activity was obtained from KOr-21J (0.018 U/mL). The results of this study support the prospect of celluloytic fungi from insects in the conversion of cellulose-based materials into its derivative products in the future.

Acknowledgment

This study was funded by Talenta Research Grant of Universitas Sumatera Utara under the scheme of National Collaboration Research for World Class University Grantee.

References

[1] Ganguly A, Das S and Dey A 2012 *Journal of Pharmacy and Biological Science* **2** 1
[2] Munafiah I, Chasanah E, dan Fawzya Y N 2011 *Prosiding ISISM International Seminar of Indonesian Society for Microbiology* (Bogor-Perhimpunan Mikrobiologi Cabang Bogor)
[3] Hidayat A R and Isnawita 2021 *LenteraBio* **10** 176
[4] Nasution F A, Munir E and Suryanto D 2021 *IOP Conference Series: Earth and Environmental Science* **9** 1
[5] Goyari S, Devi S S, Kalita C S and Talukdar C N 2014 *SpringerPlus* **3** 700
[6] Saini A, Aggrawal K N and Yadav A 2016 *Bioengineering and Bioscience* **4** 88
[7] Shil K R, Mojumder S, Sadidam F F, Uddin M and Sikdar D 2014 *Brazilian Archives of Biology and Technology an International Journal* **57** 927
[8] Ram L, Kaur K and Sharma S 2014 *International Journal of Pharmaceutical Science Investment* **3** 12
[9] Ruiz-C C, Avendan R, Leyva-E E, Barboza-C G, Chaverri P and Chavarria M 2018 *Scientific Reports* **8** 1
[10] Murtiyaningsih H and Hazmi M 2017 *Agritrop* **15** 293
[11] Lokapirnasari P W, Nazar S D, Nurhajati T, Supranianondo K and Yulianto B A 2015 *Veterinary World* **8** 367
[12] Jannah A, Aulanni'am, Ardyati T and Suharjono 2018 *J. Chem* **18** 514
[13] Belmont-Montefusco L E, Nacif-Marcal L, Assuncao de N E, Hamada N and Nunes-Silva G C 2020 *Acta amazonica* **50** 346
[14] Pathania N, Pachouri C U and Kumar A 2015 *International Journal of Advanced Research* **3** 842
[15] Nagah M, Abdel-aziz S M, Al-Sherbiny M G, Moghannem M A S and Shawky T B 2016 *Middle East Journal of Applied Sciences* **6** 579
[16] Azzaz H H, Murad A H, Kholif M A, Hanfy H M and Gawad A H M 2012 *Research Journal of Microbiology* **7** 23

Histological analysis of somatic embryogenesis from shoot explant of oil palm (*Elaeis guineensis* Jacq.)

N H Panggabean[1,3*], I Nurwahyuni[1*], Elimasni[1] and M Basyuni[2,4]

[1]Department of Biology, Faculty of Mathematics and Science, Universitas Sumatera Utara, Jl. Bioteknologi No.1 Medan, North Sumatra 20155, Indonesia
[2]Department of Forestry, Faculty of Forestry, Universitas Sumatera Utara, Jl. Tri Dharma Ujung No. 1 Medan, North Sumatra 20155, Indonesia
[3]STKIP Asy – Syafi'iyah Internasional Medan, Jl. Karya Tani No.1 Medan, North Sumatra 20146, Indonesia
[4]Centre of Excellence for Mangrove, Universitas Sumatera Utara, Medan 20155, Indonesia

*Email: isnaininurwahyuni@yahoo.co.id

Abstract. The goal of this research was to investigate the ontogenic path followed by the somatic embryo from oil palm (*Elaeis guineensis* jacq.) with in vitro culture and to characterize somatic embryo from shoot of explants. Explant were injected into a medium for the induction of calluses, histological analysis were perfomed on embryogenic callus and somatic embryo. Embryogenic callus with nodular structure and still proliferate because the cells are meristematic and very actively to dividing. The samples were dehydrated in an alcohol series, FAA-fixed and paraffin-infused for this. For staining the sample, it was done by inserting the sample in 1% safranin dye for 12 hours, then rinsed with distilled water until the color completely disappeared, then continued with dehydration in ethanol and added 2% fast green solution for 30 seconds. Oil palm somatic embryos consist of 3 phases, namely globular, skutelar and coleoptilar, where somatic embryos in the globular phase will be larger than nodular callus. globular embryo size > 200 μm with the characteristics of this phase, namely the formation of epidermal tissue, this tissue is the result of differentiation from the meristematic surface layer called protoderm. This layer is a periclinal cleavage at the transition from the proembryonic to globular phase. This method of multiplication allowed the original cell mass to proliferate, resulting in new calli lineages, particularly the embryogenic yellowish nodular lineage, which gave rise to somatic embryos.

1. Introduction

In comparison to organogenesis, somatic embryogenesis is the recommended in vitro regeneration method for palms, as this morphogenetic process may increase the number of regenerated plantlets [1]. *Elaeis guineensis* Jacq. Sometimes known as the African oil palm, is a tropical plantation crop produced for the extraction of palm oil and palm kernel oil. It belongs to the Arecaceae family and is a monocotyledonous plant, which comprises around 2.000 palm species [2–4]. The average yearly yield per hectare is as high as 4, 27 tons. Asia, Malaysia, Indonesia, West and Central Africa, South America and North and Central America are the world's top producers of oil palm, with Indonesia being the top producer and exporter of palm oil (more than 50 percent production and export) [5,6].

Oil palm trees grow in tropical climates, where they may grow and produce the most fruit. Oil palm trees thrive in temperatures between 30 and 32 degrees Celsius. Oil palm species are badly harmed by temperatures below 20 degrees Celsius, while native African oil palm species are also harmed by temperatures exceeding 40 degrees Celsius. The cross-pollinated oil palm tree has a ten-year reproduction cycle. Seeds are mainly used to cultivate oil palms. Seed propagation, on the other hand, cannot be employed economically in a true-to-type manner. Furthermore, oil palm seed germination is poor and under natural conditions, it can require more time to germinate [7]. For clonal proliferation of superior varieties, the development of an efficient plant regeneration method for oil palms is critical. Because uniform trees composed of clones of highly productive genotypes can be created, oil palm clones have the potential to increase output and productivity [8]. During the development of methods for the regeneration of new plants from tissue culture, morphological and histological aspects of course need to be analyzed in order to obtain information about the explant response in vitro, such as callus development and proliferation of somatic embryos, as well as morphogenetic characterization. Based on [9] developed a step-by-step procedure for date plant regeneration via somatic embryogenesis, including histological analysis of each phase. They detailed the calli's origins and development, as well as histochemical changes that occurred throughout in vitro culture and might be used as markers for somatic embryogenesis competency.

Somatic embryogenesis takes a long time, from callus induction through embryo formation to plantlet formation and the most essential stages of somatic embryogenesis to be analyzed and monitored in oil palm seed production are callus induction and embryo formation, which have a low success rate [10]. Studies have shown that the indirect somatic embryogenesis method of oil palm in vitro propagation includes a number of stages, including callus induction, embryogenic callus or suspension cultures, somatic embryo development, maturation and germination [11,12]. This study used histological analysis to evaluate the ontogenic route taken by an oil palm somatic embryo cell differentiation.

2. Methods

Somatic embryogenesis has evolved into a useful marker for the spread of elite genotypes of oil palm. Somatic embryo formation can take place either directly or indirectly. In the first instance, determined cells that are already programmed for embryo differentiation acquire embryogenic competence and produce somatic embryos without going through the callus formation stage [6].

2.1. Callus morphological

Observation of the structure and color of primary callus was observed at the growth period of 8 weeks after inoculation. Based on the results of morphological observations of the structure and color of the primary callus, it was yellowish white to brownish white, clustered around the explants. White callus is embryonic tissue that does not contain chloroplasts but has high starch grains. While for embryogenic callus the color of callus was yellow, dense, nodular consisting of milky white embryos with a smooth callus surface and observation were made at week 16 after inoculation.

2.2. Histological Embryogenic callus

Histological analysis was carried out using embryogenic callus. Embryogenic callus with nodular structure and still proliferate because the cells are meristematic and very actively to dividing. Sample of embryogenic callus were fixed using formalin 40%, alcohol 70%, glacial acetic acid (1: 8:1) and soaked for 24 hours. Dehydration stage, followed by immersing the sample gradually in graded alcohol series to absolute alcohol, then continued in xylol solution (3 hours for each process). in the infiltration stage, the samples will be put in paraffin with a ratio of xylol and liquid paraffin 3:1, 2:2, 1:3 (3 times for 3 hours) and 0:4 for 24 hours, respectively. Then proceed in the pure paraffin immersion for one day. The next step of embedding and slicing. For staining the sample, it was done by inserting the sample in 1% safranin dye for 12 hours, then rinsed with distilled water until the color completely disappeared, then continued with dehydration in ethanol and added 2% fast green solution for 30 seconds.

3. Results and Discussion

The development of oil palm shoot explants in vitro culture begins with callus formation which differentiates into embryogenic callus with nodular structure. Nodular callus will differentiate to produce somatic embryo. Oil palm somatic embryos consist of 3 phases, namely globular, skutelar and coleoptilar, where somatic embryos in the globular phase will be larger than nodular callus. globular embryo size > 200 μm with the characteristics of this phase, namely the formation of epidermal tissue, this tissue is the result of differentiation from the meristematic surface layer called protoderm. This layer is a periclinal cleavage at the transition from the proembryonic to globular phase.

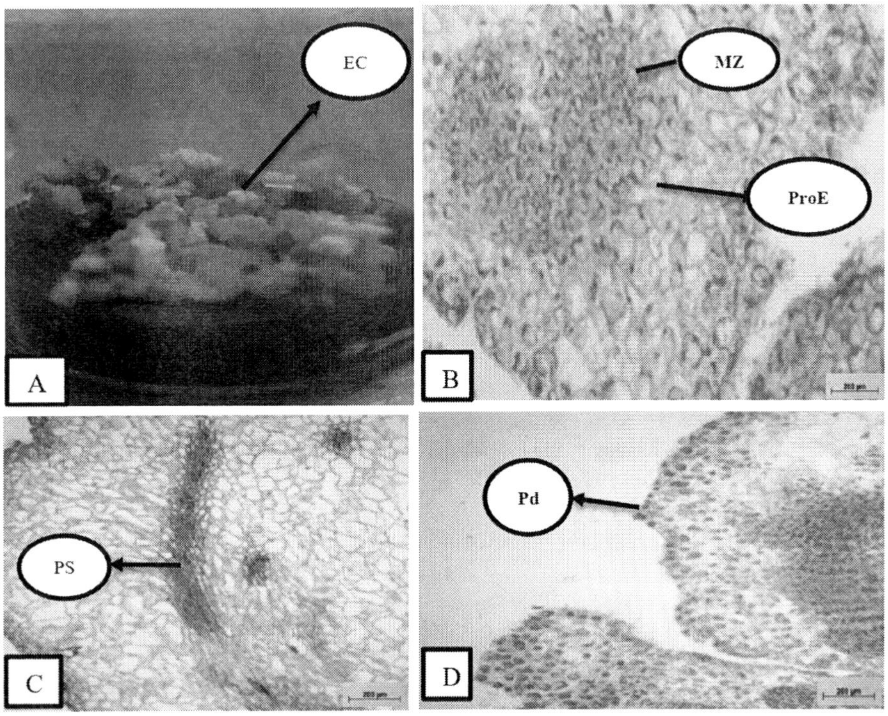

Figure 1. (A) Morphological Primary Callus Of Oil Palm (B,C,D) Histological Analysis Somatic Embryogenesis of Oil Palm Globular Phase, (EC) Embryogenic Callus; (ProE) embryogenic aggregates and proembryo cells; (MZ) Merismatic Zone; (Ps) Procambial Strands; (Pd) Protodermis (Magnification 40 x 10)

A meristematic zone (MZ) formed of small meristematic cells was discovered during the formation of nodular calli. Following that, these meristematic cell masses evolved into somatic embryos. Small, distinct groupings (two or four celled) or proembryo aggregates with thicker cell walls were created by some of the rapidly proliferating embryogenic calls (ProE). Most of the walls of the somatic embryo exhibit continuous development with complete polarization, as demonstrated by the presence of procambium (PS) strands and somatic embryo development begins with the continuous proliferation of proembryos and matures into embryos that are distinguished by the presence of a well-defined protodermis (Pd). In palm family, many reports have documented that the first cell division is observed in cells adjacent to the vasculer tissue [13–15]. Based on [16] There is evidence suggesting the origin of callus from procambial cells due to the early cell proliferation near the vascular bundle and the absence of parenchymal cell differentiation signals in the early ontogenic callus phase. Procambium cells, which typically develop into vascular tissue when driven by the growth regulator auxin present in the growing medium, are reprogrammed to increase cell proliferation and callus formation [17].

The use of several approaches that allow the events that occur when plant species are cultivated in vitro to be described has shown to be more valuable in the study of somatic embryogenesis through histological observations. For example, knowing and observing the cell and tissue growth of plant material during the various stages of somatic embryogenesis, as well as identifying the more sensitive and/or embryogenic tissue and/or cell indicators early in the process, are all achievable with this method. Furthermore, histological investigations allow researchers to clarify features of the somatic-embryogenic cell transition that are currently unknown, particularly in palm palms.

4. Conclusion

In comparison to organogenesis, the most favoured in vitro regeneration method is the somatic embryogenesis approach for palms, as this morphogenetic process may improve the number of regenerated plantlets. Seeds are mainly used to cultivate oil palms. Seed propagation, on the other hand, cannot be employed economically in a true-to-type manner. Furthermore, Oil palm clones have the potential for better yield/productivity because it is feasible to generate homogeneous trees made up of identical clones of highly productive genotypes. Because it is possible to grow uniform trees made of identical clones of highly productive genotypes under natural settings, oil palm clones have the potential to produce crops with higher yields and productivity. Morphological and histological aspects should be investigated during the construction of a protocol for plantlet regeneration to elucidate the in vitro response; using histological analysis, it is feasible to know.

Acknowledgment

The work was funded by the Indonesian Ministry of Education, Culture, Research and Technology in BPPDN (*Beasiswa Pendidikan Pascasarjana Dalam Negeri*) program year 2019.

Reference

[1] Steinmacher D A, Krohn N G, Dantas A C M, StefenonmV M, Clement C R and Guerra M P 2007 *Annals Of Botany* **699** 709
[2] Weckx S, Dirk I and Ludo M 2019 *Frontiers in plant science* **1** 17
[3] Kong E Y Y, Biddle J, Foale M and Adkins S W 2020 *Industrial Crops and Products* **147** 112
[4] Basyuni M, Prayogi, Putri L A P, Syahputra I, Siregar E S, Risnasari I, Wati I and Arifiyanto D 2018 in *ICBS Medan*
[5] Panggabean N H, Nurwahyuni I, Elimasni 2021 *IOP Conf Series Materials Science and Engineering* **1156** 1
[6] Panggabean N H, Basyuni M, Nurwahyuni I and Elimasni 2021 *3rd International Conference on Natural Resources and Technology* **912**
[7] Thuzar M, Vanavichit A,Tragoonrung S and Jantasuriyarat C 2012 *African Journal of Biotechnology* **82** 14761
[8] Sane F, Bertossi A, Gassama-Dia Y K, Trouslot M F, Duval Y and Borgel A 2006 *Annals Of Botany* **98** 301
[9] Meira F S, Luis Z G, Cardoso I M and Pereira JE 2019 *Flora* **250** 59
[10] Romyanon K, Mosaleeyanon K and Kirdmanee C 2015 *Scientia Horticulturae* **195** 1 7
[11] Marbun C L M, Mathius N T, Reflini and Utomo C 2015 *Procedia Chemistry* **14** 122
[12] Gomes H T, Bartos P M C, Pereira J E S 2017 *Plant Cell Tiss, Organ Cult* **131** 269
[13] Fernando S C, Verdeil J L, Hocher V, Weerakoon L K and Hirimburegama K 2003 *Plant Cell Tissue Org. Cult* **72** 281
[14] Sane D, Bertossi F, Gassama Y K, Sagna M, Trouslot M F, Duval Y and Borgel A 2006 *Ann. Bot* **98** 301

[15] Thuzar M,Vanavichit A, Tragoonrung S and Jantasuriyarat C 2011 *Acta Physiol Plant* **33** 123
[16] Gomes HT, Bartos P M C and Pereira S 2017 *Plant Cell Tiss. Organ Cult* **131** 269
[17] Meira F S, Luis Z G, Cardoso I M and Pereira J E S 2019 *Flora* **250** 59

IOP Conf. Series: Earth and Environmental Science 1115 (2022) 012074 doi:10.1088/1755-1315/1115/1/012074

Utilization of UAV (Unmanned Aerial Vehicle) technology for mangrove species identification in Belawan, Medan City, North Sumatera, Indonesia

A S Thoha[1]*, O A Lubis[1], D L N Hulu[1], T Y Sari[1], M Ulfa[1] and Z Mardiyadi[2]

[1]Faculty of Forestry, Universitas Sumatera Utara, Medan 20155, North Sumatra, Indonesia

[2]Faculty of Forestry, University of Papua, Manokwari, Indonesia

*Email: a.siddik@usu.ac.id

Abstract. The mangrove forests in Indonesia are starting to decrease over time because there is still a lack of knowledge among people around the coast about the importance of mangroves. Kampung Nelayan, Medan Belawaan, is a coastal community area whose life is very dependent on coastal, mangrove, and water resources. Unmanned Aerial Vehicle (UAV) technology has the potential to provide a fast, cost-effective, and efficient mangrove mapping technique. It is very useful because mangrove areas are located in remote areas, where field measurements are difficult, time-consuming, and expensive. The objective of this study is to analyze mangrove species using UAV imagery with Object-Based Image Analysis (OBIA) classification. The object-based classification result for the overall accuracy is 82.94% where there are 7 classes of mangrove species based on the classification process, including: *Avicennia alba, Avicennia officinialis, Avicennia Marina, Rhizopora apiculata, Nypah fruticans, Scyphipora hydrophylacea, Bruguiera gymnorriza*. There are also two classes for non-mangrove, consisting of the water body and non-mangrove. The largest area of mangrove species in the research site is Avicennia Marina with a percentage of 33.86% covering an area of 7.80 Ha. The second-largest mangrove species with a percentage of 21.88% is Avicennia officinalis with an area of 5.04 Ha.

1. Introduction

Indonesia is an archipelagic country with mangrove forests overgrown along its coastline. Mangrove forests in Indonesia are immensely wide. However, the mangrove forests are starting to decrease over time due to a lack of public awareness and the conversion of land into ponds [1]. Mangrove forests can be considered the only type of forest that can grow in areas experiencing high tides. This is because the mangrove plant community can tolerate salt contained on the seaside [2].

Kampung Nelayan, Medan Belawan, is a coastal community area whose life is very dependent on mangrove ecosystem [3]. Mangrove ecosystems provide various ecosystem services including food provision, providing good habitat for fauna species, regulating for climate, and protecting from diasters [4]. The extent of degradation of mangrove forests is increasing as a result of massive exploitation by people who are very dependent on the mangrove ecosystem for their lives.. Unsustainable coastal resource management can threaten ecosystems and fishermen's catches, thereby threatening their livelihoods [5]. Consequently, mangrove restoration is needed by considering the ecological and economic aspects in its implementation [6].

Vegetation analysis is an important step taken to determine the number of mangrove vegetation found in the area since vegetation is the main element in soil and water conservation efforts. Therefore, it is necessary to carry out a vegetation analysis regarding the presence of plants to study the vegetation structure and composition of plant species. Vegetation analysis aims to determine the composition of the type (composition) of plants and the form (structure) of vegetation in the analyzed area [7].

Remote sensing provides the effective and efficient method for mapping of mangrove ecosystem. Remote sensing is very useful in measurements in mangrove ecosystems because it can overcome somes constrains such as difficult access, expensive costs and long time. Remote sensing ultimately supports the sustainable management of mangrove ecosystems [8]. The data taken from the image aims to detect changes in vegetation that are related to spatial resolution, temporal resolution and spectral resolution [9].

Unmanned Aerial Vehicle (UAV) is also widely applied in government programs such as forest stand mapping, tree volume estimation, land cover assessment, pest monitoring, and harvest planning. In addition, non-governmental organizations are also interested in using drones for work related to conservation, such as monitoring wildlife, monitoring land use changes and monitoring illegal activities [11]. The commercial availability of UAVs is increasing and the prices are more economical, making UAVs increasingly widely used as a tool for environmental and ecological analysis. The UAV function can be applied for monitoring agricultural land, forest fires impact measurement, distribution of moss in the poles and mapping mangroves ecosystem [12].

Cameras mounted on UAVs are very useful in small scale forest inventories with low cost and high flexibility. Image characteristics of UAVs that have very high spatial resolution have the potential to identify mangrove species. In recent years, object-based identification methods combined with high-resolution imagery are applied for mapping of mangrove species [14].

Generally, vegetation mapping with remote sensing imagery uses automatic classification by distinguishing various types of objects.. Objects in the form of mangrove vegetation will be distinguished based on the suitability of the image, color or wavelength reflected by the object [13]. An object-based classification is an approach that is widely used and the classification process does not only consider the spectral but also the spatial aspects of the object [15]. Therefore, the use of UAV and object-based image analysis will be indispensable in identifying species in the mangrove ecosystem so that survey work becomes faster, cheaper and easier. The objective of this study was to analyze the type of mangrove using UAV imagery with the classification of Object-Based Image Analysis.

2. Materials and Methods

2.1. Research site

This study was conducted in the mangrove forest area of Medan Belawan, North Sumatera. The data collection was conducted from March to April 2022. Aerial photo analysis was carried out in the Laboratory of Forest Resources Conservation, Faculty of Forestry, Universitas Sumatera Utara. The location of the research site is presented in Figure 1.

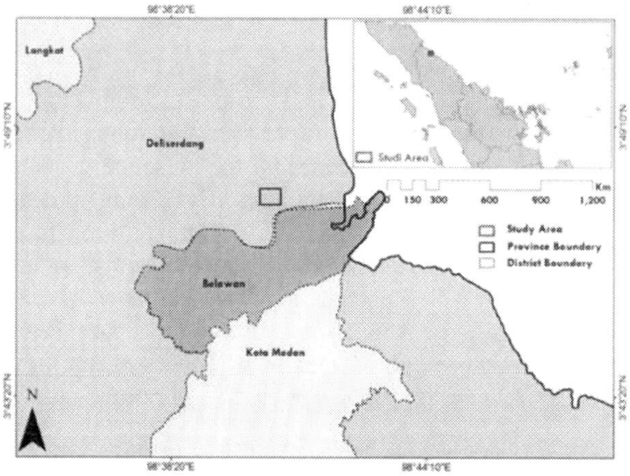

Figure 1. Research site map

2.2. Tools and materials

The hardware used in this study is personal computer (PC) and a drone type of DJI Mavic 2 Pro. The software used was ArcGIS 10.8, Agisoft Metashape, eCognition, Pix4D Capture, and Microsoft Excel, The material used in this study was high-resolution aerial photography of the Belawan mangrove forest area. The field survey tools were in the form of Avenza Maps software, maps of the Belawan mangrove forest area, stationery, digital cameras and ropes to make measuring plots for observing tree species.

2.3. Data collection

The data collection was in the form of aerial photo acquisition with flight missions and plotting for analysis of vegetation types. The acquisition of aerial photographs required several stage namely preparation, data acquisition and the data processing. In the preparation stage, identification of the area to be mapped is carried out in the form of a grid, flight path, flight height, overlap, flight speed and determination of pixel resolution or ground sample distance (GSD). Flying height in data acquisition is 80 m and 80 % of photo overlap. The UAV or Drone Type used was the DJI Mavic 2 Pro. The determination of a flight mission for the identification area of mangrove species used Pix4D Capture, an Android and IOS-based application used to estimate an area, number of paths, GSD and duration of the flight.

Figure 2. Display on the flight path application, camera location and image overlap

2.4. Data analysis

The processing of aerial photos into orthomosaic used Agisoft Metashape software with steps including importing photos and flight path reconstruction, Align photos, merging Dense Point Clouds, construction, texture models, DEM development, Orthophoto development and exporting orthomosaic in GeoTIFF format. The results of processing aerial photos in orthomosaic can be seen in Figure 3.

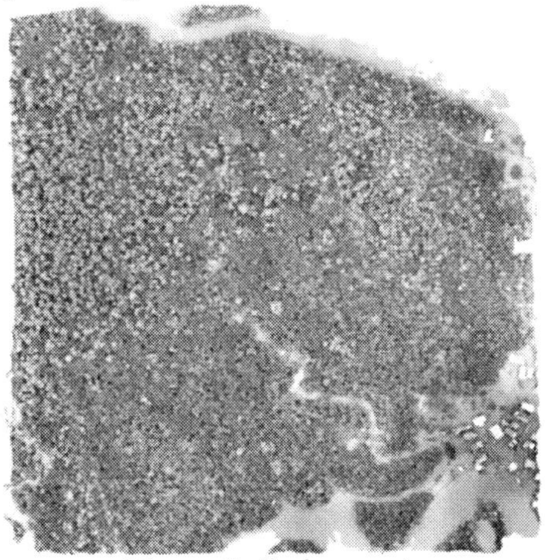

Figure 3. *Orthomosaic*

2.4.1. Object-Based Image Analysis (OBIA). The image classification that is functioned to map the mangrove composition used Object-Based Image Analysis / OBIA. The OBIA approach is considered superior to pixel-based classification because it not only considers spectral but also spatial aspects [16]. The OBIA method is a classification method developed by segmenting and analyzing objects or image classification processes based on their spatial, spectral and temporal characteristics to produce image objects or segments which are then used for classification [17].

2.4.2. Segmentation. Image segmentation is the first step of Object-Based Image Analysis and is used to produce images of objects that are more homogeneous among themselves than those with nearby regions [18]. The most widely used segmentation is multiresolution, a bottom-up region merging technique, which is used to generate image objects. In this study, the segmentation algorithm used was the multiresolution segmentation (MRS) algorithm. This algorithm is done with a single-pixel and combining neighboring segments until a heterogeneity threshold is reached. In the segmentation process using the MRS algorithm, there are three important parameters, including shape, compactness and scale. All classification object-based image processing procedures were performed using the eCognition Developer 64 software (Trimble Germany GmbH, Munich, Germany).

2.4.3. Field survey. The sampling method used in the identification of mangrove species was the stratified systematic sampling method, where the species recorded were taken in sample plots made with an area of 20 x 20 m in the form of a square. The research area was 23 ha with a sampling intensity of 1%. The distance between sampling plots was 100 m with a total of 10 sampling plots made. In the sample plots, all the names of tree species were recorded, specifically vegetation with a trunk diameter of more than 10 cm.

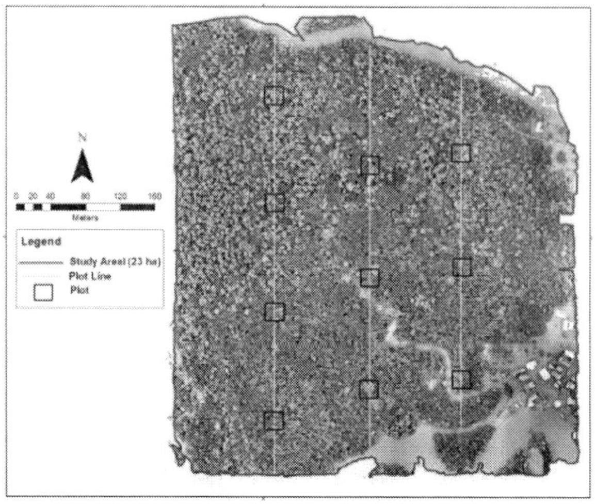

Figure 4. Map of sampling plots

2.4.4. Accuracy test. Overal accuracy is used to determine the accuracy of the calculation from the map with groundcheck data. Determination of accuracy using confusion matrix. Overall accuracy using the following formula:

$$\text{Overall Accuracy} = \frac{\sum_{i=1}^{r} xii}{n} 100\%$$

Xii : Diagonal values of the i-th row and i-th column of the contingency matrix
X+I: Number of pixels in the i-th column
Xi+: Number of pixels in the i-th row
N : Number of pixels in the sample [20]

Overall accuracy is defined as the total class classified divided by the total class. The determination of the training area is carried out by various combinations to get the best results in the accuracy test. The combinations in the training area determination can be done in small, medium or large sizes. Sampling points for accuracy test are carried out through field verification.

The calculation of accuracy in this study is based on the guidelines for processing satellite image data. In the land cover classification guidelines from satellite imagery, the accuracy value used is not less than 75% [19].

3. Results and Discussion

3.1. Aerial photo interpretation

The results of the aerial photo analysis with a total of 888 photos at an altitude of 60 produced orthomosaic with a Ground Sampling Distance (GSD) of 1.3 cm. The objects from orthomosaic visuals had clearer details and shapes. The crown of vegetation and tree shadows were clearly visible from UAV data, which could assist the visual classification process easily. Likewise, the appearance of the bottom of shallow waters could be seen clearly.

Drone mapping results show that the color and hue of the mangrove vegetation area will look greener and darker because the mangroves are usually located in wet sediments and characteristic of the water absorbs a lot of electromagnetic waves rather than reflecting them. The texture of mangrove vegetation will look rougher than other lower-level vegetation. The results of the mapping show that mangrove vegetation is scattered in all parts of the study area.

3.2. Segmentation

Multiresolution segmentation was introduced by [21] where this algorithm is a region merging technique that groups areas that have similarities and adjacent pixels into objects. In this study, the segmentation algorithm used was the multiresolution segmentation (MRS) algorithm with three parameters namely shape, compactness and scale. This algorithm was chosen because it is considered to best describe the real conditions. The formation of objects in the segmentation process was carried out with 1 level. This study was carried out on a 125 scale with a shape of 0.1 and a compactness of 0.5. The number of objects obtained from the segmentation process was 189,622 objects. The segmentation results are presented in Figure 5.

Figure 5. Segmentation results with a scale of 125

3.3. Mangrove type classification

The results of the classification using multiresolution segmentation are presented in Table 1. The determination of the training area refers to the survey results, leaf shape and different color hues from orthomosaic. At the classification stage, it is done by looking at the pixel values that correspond to the training area. Pixel values that are in the range of pixel values in the training area will be classified into a predetermined class. The distribution and area of OBIA classification results can be seen in Figure 6.

Tabel 1. The results of the classification of mangrove and non-mangrove species with multiresolution segmentation

Image Appearance	Image Classification	Field Condition	Mangrove and Non-Mangrove Classisification
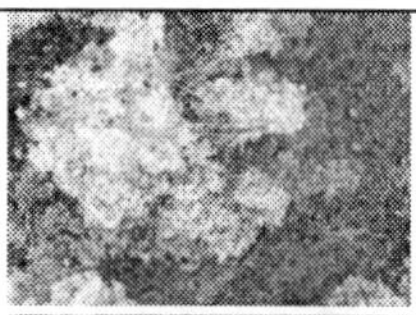			*Avicennia alba*
			*Avicennia officinalis*
			*Avicennia Marina*
			*Nypah fruticans*
			*Rhizopora apiculata*
			*Bruguiera gymnorriza*
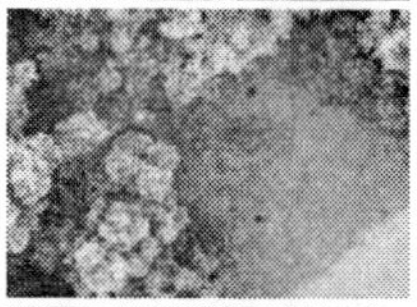			*Scyphipora hydrophylacea*
			Non-Mangrove (Settlement)
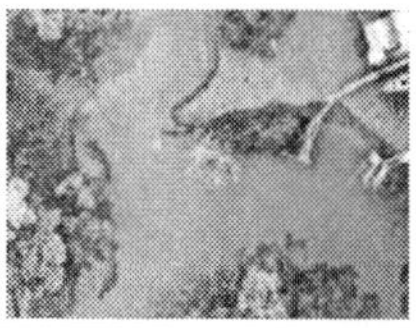			Water Body

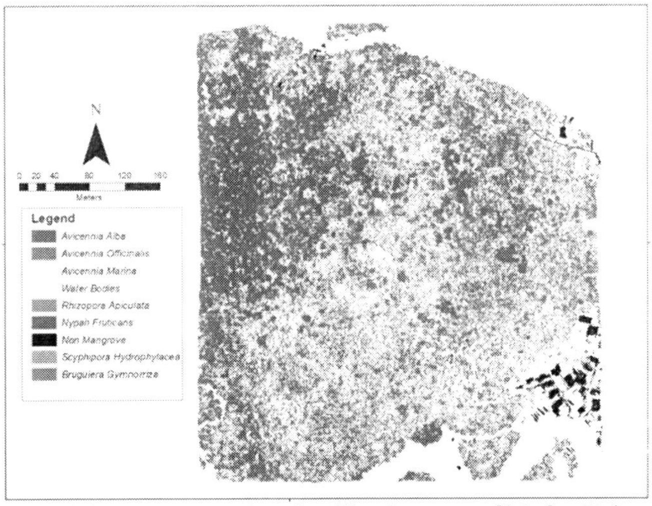

Figure 6. Mangrove species classification map of Medan Belawan

Table 2. Results of land cover identification using the object-based image analysis method

Class	Area (ha)	Percentage (%)
Avicennia alba	3.40	14.77
Avicennia officinalis	5.04	21.89
Avicennia Marina	7.80	33.87
Water body	4.91	21.31
Rhizopora apiculata	0.29	1.27
Nypah fruticans	1.07	4.65
Non-Mangrove	0.29	1.28
Scyphipora hydrophylacea	0.02	0.09
Bruguiera gymnorriza	0.19	0.87
Total Area	23.00	100

The results of class identification using OBIA with the multiresolution segmentation (MRS) algorithm (Table 2) showed the largest area of mangrove species in the research site, specifically *Avicennia Marina* with a percentage of 33.86% covering an area of 7.80 Ha. The second-largest mangrove species with a percentage of 21.88% is *Avicennia officinalis* with an area of 5.04 Ha.

3.4. Accuracy test (confusion matrix)

After the classification process is complete, it is necessary to conduct the accuracy test. Based on Table 3, it can be seen that in class 1 there are 17 points that are classified correctly, and 1 point is in another class. Furthermore, in class 2 there are 16 entry points in the correct class and 6 points outside the supposed class. Finally, in class 3 there are 26 points that are classified correctly and 1 point is not in the correct class and so on.

The segmentation result is validated by the species identification from field survey with the sample unit in a plot size of 20 x 20 m. We identify the dominant canopy mangrove species seen from above (UAV) in the sample plot. In one plot we found more than 1 mangrove species.

Table 3. Confusion matrix of mangrove species classification

Class	1	2	3	4	5	6	7	8	9	*Grand Total*	*Producer Accuracy*	*User Accuracy*
1	17		1							18	94.44	77.27
2	2	16	4							22	72.72	80.00
3		1	26							27	96.29	70.27
4				13						13	100.00	86.66
5	2		3	1	12	1				19	63.15	100.00
6						10				10	100.00	90.90
7							4			4	100.00	100.00
8		1						2		3	66.66	100.00
9	1	2	2	1					7	13	53.84	100.00
Grand Total	22	20	36	15	12	11	4	2	7	129		
Overall Accuracy	82.94											
Kappa Accuracy	84.51											

Class Description: (1) *Avicennia alba,* (2) *Avicennia officinalis,* (3) *Avicennia Marina,* (4) *Water Body,* (5) *Rhizopora apiculate,* (6) *Nypah fruticans,* (7) *Non-Mangrove,* (8) *Scyphipora hydrophylacea,* (9) *Bruguiera gymnorriza*

Errors of classification can occur because the classes have similar color and hue of other classes. For example, The classification error is caused because certain classes have a color and hue that is close to the same as the color and hue of other classes. In Table 4, errors in classification occur in class 1 where there are 17 correct points and 1 point is incorrectly classified and enters class 3. Class 1 and Class 3 are located on adjacent pixels so that they have almost the same color and hue. Visually there are similarities in shape and color between *Avicennia alba, Avicennia officinalis* and *Avicennia Marina*. Even in the *Rhizopora apiculata* class, apart from being included in other types, it is also included in the water body class.

The study area is a secondary forest area that has been planted in the past. The results of field observations still found formations according to the natural zoning of mangroves. It is evident that the results of segmentation method found that most of the *Avicenia spp* were closest to the sea. Generally, the identified species distribute randomly within the study area.

The canopy of *Avicennia,* which height dominated in almost every research site, becomes an obstacle in the process of making the training area. The results showed that the *Avicennia* species dominate. This is because this type of wood has a low selling price, making this plant not hunted by local residents, in contrast to *Rhizopora* and *Bruguiera* whose wood has a fairly high selling price.

The study finding related to the previous studies which found that the aerial photo interpretation using object-based image analysis can produce very accurate classification. Identification of species in mangrove vegetation adjusts to the image based on the color or wavelength reflected by the object. [13,23]. In this case show that data from UAV systems is reliable to identify mangrove species. A study by [14] and [24] found that drone mapping of mangrove species in recent years using UAV obtained the reliable accuracy.

Mapping using UAV technology is very reliable in mangrove ecosystem, especially in analyzing ecological aspects of mangrove ecosystem. This study found that it can be seen in detail the dominant mangrove species and other types that can be used as sources of seeds, food and habitat for animals and aquatic animals. The high quality data of mangrove ecosystem is very helpful in protecting, monitoring and restoring of mangrove ecosystems. The identified mangrove tree species can be used as value data for various programs for the community and other stakeholders.

The overall accuracy of the results of the OBIA classification method with the multiresolution segmentation (MRS) algorithm is 82.94% from the groundcheck points of 129 points. This value is more than the minimum threshold of the good accuracy value. The level of measuremnet of the accuracy of the classification used must more than 75.00%.

Very high spatial resolution images can increase overall accuracy in assessing image classification. This is evidenced by a study conducted by [22] which found that the maximum likelihood classification method on remote sensing images with a spatial resolution of 30 m only obtained an overall accuracy of 63.1%. Based on study by [24] found the overall accuaracy of 92 % for mangrove ecosystem using UAV technology mapping The overall accuracy value is 75% or more, show that the classification results are eligible to be used as the final results of the study. According to the National Institute of Aeronautics and Space [19], the level of classification accuracy assessment must be not less than 75%.

4. Conclusion

There are seven types of mangroves identified based on the UAV aerial photography map using the OBIA method, including: *Avicennia alba, Avicennia officinialis, Avicennia Marina, Rhizopora apiculata, Nypahfruticans, Scyphipora hydrophylacea, Bruguiera gymnorriza. Avicennia Marina* dominates the research site with a percentage of 33.86% or an area of 7.80 Ha. The second dominant species is *Avicennia officinalis* with an area of 5.04 ha or 21.88% of the research site. Based on the accuracy was obtained at 82.94%. The accuracy test value can indicate that the use of UAV images is better in terms of accuracy and classification accuracy. We suggest to develop deep learning method for identifying vegetation type in the next study.

Acknowledgment

We would like to thank the local community groups, Kampung Nelayan Seberang, Medan, and the field team who provided support during the data collection. We also thank to Indonesia Mapping Community for supporting of aerial photo processing We also give high appreciation to the editor and reviewers team who provided their corrections and comments

References

[1] Ario R, Subardjo P and Handoyo G 2016 *Jurnal Kelautan Tropis* **18** 64
[2] Martiningsih N G A G E, Suryana I M and Sutiadipraja N 2015 *Agrimeta* **5** 90187
[3] Onrizal, Sulistiyono N and Afifuddin Y 2016 *Talenta* **1** 64
[4] Lee C K, Duncan C, Nicholson E, Fatoyinbo T E, Lagomasino D, Thomas N and Murray N J 2021 *Remote Sensing,* **13** 2047
[5] Purwoko A, Onrizal and Mansor M 2009 *International Conference on Natural and Environmental Sciences* 6
[6] Onrizal 2014 *Wanamina* **3** 6
[7] Azizah P N 2017 *Jurnal Riset Daerah* **16** 2685
[8] Hamid A A, Dubovyk O, Abou El-Magd I and Menz G 2018 *Sustainability* **10** 646
[9] Sulistiyono N, Jaya I N S, Prasetyo L B and Tiryana T 2015 *Int. J. Sci. Basic Appl. Res* **24** 350
[10] Bustomi Y, Ramdhani M A and Cahyana R 2012 *Jurnal Algoritma* **9** 171
[11] Paneque-Gálvez J, McCall M K, Napoletano B M, Wich S A and Koh L P 2014 *Forests* **5** 6 1481
[12] Crutsinger G M, Short J and Sollenberger R 2016 *Journal of Unmanned Vehicle Systems* **4** 161
[13] Khatami R, Mountrakis G and Stehman S V 2016 *Remote Sensing of Environment* **177** 89
[14] Timisela W A, Mardiatmoko G and Puturuhu F 2020 *Jurnal Hutan Pulau-Pulau Kecil* **4** 132
[15] Marwati A, Prasetyo Y and Suprayogi A 2018 *Jurnal Geodesi Undip* **7** 36
[16] Wibowo T S and R Suharyadi R S 2012 *Jurnal Bumi Indonesia* **1** 130
[17] Blaschke T 2010 *ISPRS Journal of Photogrammetry and Remote Sensing* **65** 2
[18] Septiani R, Citra I P A and Nugraha A S A 2019 *Jurnal Geografi: Media Informasi Pengembangan dan Profesi Kegeografian* **16** 90

IOP Conf. Series: Earth and Environmental Science 1115 (2022) 012074 doi:10.1088/1755-1315/1115/1/012074

[19] [LAPAN] Lembaga Antariksa dan Penerbangan Nasional 2015 Pusat Pemanfataan Penginderaan Jauh Lembaga Antariksa dan Penerbangan Nasional LI1020020101
[20] Firmansyah S, Gaol J L and Susilo S B 2019 *Jurnal Pengelolaan Sumberdaya Alam dan Lingkungan (Journal of Natural Resources and Environmental Management)* **9** 746
[21] Floreano D and Robert JW **521** 460
[22] Chen CF, Son N T, Chang N B, Chen C R, Chang L Y, Valdez M and Centeno G 2013 *Remote Sens* **5** 6408
[23] Foody G M 2002 *Remote Sensing of Environment* **80** 185
[24] Thoha A S, Lubis O A, Hulu D N L, Sari T Y and Mardiyadi Z 2022 *IOP Conf. Series: Earth and Environmental Science* **977** 012102

Tensile properties of edible films from various types of starch with the addition of glycerol as plasticizer: a review

M Sartika, F R Rambe, S Z D M Parinduri, H Harahap*, H Nasution, M Lubis, Taslim and R Manurung

Department of Chemical Engineering, Universitas Sumatera Utara, Medan 20155, North Sumatra, Indonesia

*Email: hamidah@usu.ac.id

Abstract. The tensile properties of edible films are very important from their application as packaging for food products. Edible films are made from natural materials so that they are safe for the environment and can be consumed directly with packaged products. Starch used in the preparation of edible films can come from different sources to produce films with different characteristics. It is known that starch-based edible films have several weaknesses, such as being brittle, which causes the film to easily crack. Efforts should be made to improve the mechanical properties, elasticity, and hydrophobicity of edible films by mixing them with other additives such as glycerol as a plasticizer. The use of different starch sources and glycerol concentrations will affect the properties and characteristics of the edible film formed. Different starch sources significantly affect the properties of the thickness and strength of the film. Meanwhile, the addition of glycerol affects the tensile strength and flexibility of the film. Therefore, this review aims to examine the effect of different types of starch with the addition of glycerol on the tensile properties of the resulting film. Studies from various literature were carried out and several research results were compared and analysed regarding the tensile properties of edible films.

1. Introduction

Vegetables, meat, fruits and a variety of processed foods are easily degraded and damaged. Such damage can occur as a result of mechanical damage, so packaging materials that act as effective protectors are required to avoid food product damage [1,2]. Edible film is a type of packaging that consists of a thin layer that coats food and protects it from the environment (water vapor, gases, light, microorganisms) while also going to prevent mass transfer and preserving food quality [1,3]. As a result, the tensile properties of edible films are critical due to their role as a protector in the packaged product. Another benefit of edible film is that it is environmentally friendly because it is composed of natural ingredients including starch, protein, cellulose, or lipid and can be consumed directly with packaged products [4,5].

Starch is a polymeric carbohydrate consisting of 1,4-D-unit glucose bonds and it is the common polysaccharide found in the production of edible films [6,7]. Starch-based films are typically colourless, transparent, tasteless, and odourless. This film plays a role in replacing conventional plastics as food packaging because of its biodegradability, abundance, resistance to degradation, and excellent film-forming ability. In addition, another advantage compared to non-starch films is its edible nature and low permeability to oxygen, as well as being more cost-effective [8,9].

Starch used in the preparation of edible films can come from a variety of sources, including grains, agricultural waste, fruits, and tubers [10], such as corn [1], potato [11], wheat, rice and cassava [12], taro [4], mung beans, oat, soybean, jackfruit seeds [13], durian seeds [14], which gives variations in shape, size, structure, and chemical composition [15,16].

Different properties of starch can cause the formation of edible films with varying properties. For example, cereal starch contains mostly lipids in its granules, whereas starch from tubers contains almost no lipids. Amylose-lipid complexes can form in the presence of lipids, causing changes in starch such as increased swelling capacity (swelling), increased gelatinization temperature, solubility, and disruption of granules [6]. Gelatinization is the process by which starch granules expand due to heat and water, preventing them from returning to their original shape [10].

The primary components of starch granules are two polysaccharides: amylopectin and amylose. Amulopectin and amylose have different molecular properties and film-forming properties due to differences in structure, molecular weight, and composition [7,8,15]. Amylose's ability to form films and coatings is more closely related to its dominant linear nature [17]. Films with a higher amylose content typically have better film-forming properties, such as elongation, mechanical strength, and gas barrier properties [7]. Meanwhile, amylopectin is associated with an increase in crystallinity so that the film is less sensitive to moisture [15].

Starch-based edible films have several flaws, including being brittle, which causes the film to crack easily. As a result, efforts must be made to improve edible film mechanical properties, elasticity, and hydrophobicity by combining them with other additives such as glycerol as a plasticizer [3]. A number of researchers have revealed that adding plasticizers to the film mixture will increase the elasticity, flexibility, flexibility, and toughness of the film, as well as reduce brittleness and prevent cracking of the film [6,18-21]. The plasticizer reduces the stiffness of the edible film by reducing the internal forces between the polymer chains [5]. Figure 1 shows the research trend of providing edible films derived from starch with glycerol added as a plasticizer.

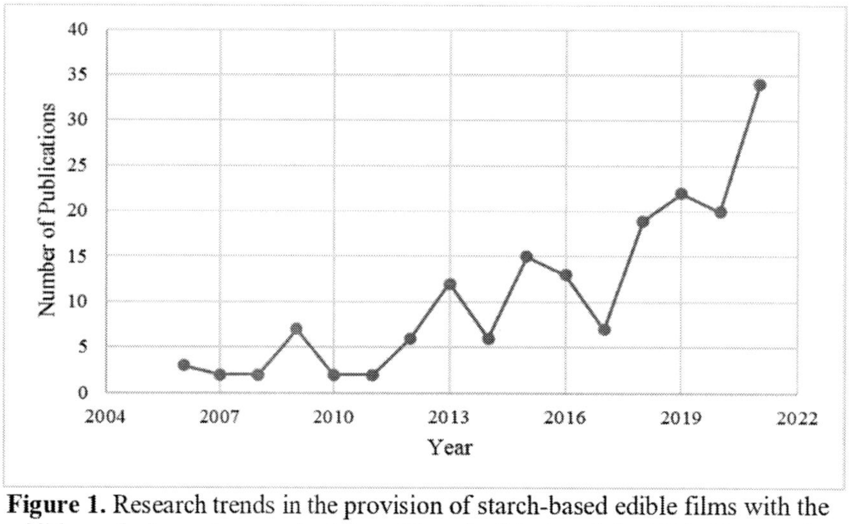

Figure 1. Research trends in the provision of starch-based edible films with the addition of glycerol as a plasticizer (Google Scholar, July 2022, Keywords: edible film, starch, glycerol)

According to the explanation above, the use of various starch sources and concentrations of glycerol will affect the properties of the edible film formed. The characteristics of a good edible film as defined by the Japan International Standard are shown in Table 1 (1975) [22,23].

Table 1. The characteristics of edible films as defined by the Japan International Standard (1975)

Characteristics	Value
Thickness	Max 0.25 mm
Tensile Strength	Min 0.39 MPa
Elongation	< 10% bad
	> 50% good

The lack of information regarding the effect of different sources of starch on the manufacture of edible films makes it necessary to have an article that provides a comprehensive review of this problem. Furthermore, different concentrations of glycerol as a plasticizer made a significant difference in the tensile properties of the film. Therefore, this article aims to summarize the findings of recent research on the use of starch and different concentrations of glycerol in the manufacture of edible films and to examine the effects of different sources of starch and glycerol's concentration on the tensile properties of films.

2. Discussion

In general, the tensile properties of edible films, such as thickness, elongation at break, and tensile strength are the parameters measured to check the film's resistance in protecting the packaged product from mechanical damage. Several souuces of starch have been studied in the production of edible films, including soybean starch [24], jackfruit seeds starch [25], lesser yam starch [26], arrowroot starch [27], taro starch [28], and sago starch [29] with the addition of various concentrations of glycerol. The edible films of various types of starch were studied for their tensile properties, such as: thickness, elongation at break and tensile strength, which are presented in Table 2 below.

Table 2. Effect of starch type and glycerol concentration on the tensile properties of edible films.

Starch Source	Starch Concentratio n (%)	Glycerol Concentration (%)*	Thickness (mm)	Tensile Strength (MPa)	Elongation (%)	References
Soybean	3	50	0.073	-	-	[24]
Jackfruit seeds	3	2[a]	0.11	4.11	101.53	[25]
Lesser yam	3.5	22.5	0.09	0.24	15.55	[26]
Arrowroot	3	50	0.2	-	-	[27]
Taro	3	25	0.17	10.63	60.21	[28]
Sago	3	40	0.185	0.814	38.92	[29]

*glycerol concentration based on starch mass (w/w)
[a]glycerol concentration based on aquadest volume (w/v)

2.1. Effect of starch source and glycerol concentration on edible film thickness

Thickness is a physical property of edible film that affects or determines other properties such elongation, tensile strength, ability to withstand the transmission of water vapor, gases, and other volatile compounds [25]. The type and composition of the material used have a greater impact on the thickness of the edible film [20].

From Table 2 above, it is known that all types of starch used provide edible films with thicknesses that meet Japan International Standards. Arrowroot starch gave edible film with the highest thickness of 0.2 mm, followed by sago and taro starch at 0.185 and 0.17 mm, respectively. The difference in the thickness of the edible film is influenced by several factors, including the physical properties of the solution (density, viscosity, surface tension, drying time), volume solution, mold's area, the total amount of solids in the solution, and the technique of film formation [21,27]. In addition, the increase in thickness may be due to an increase in particle size. The film thickness increases as the total solid content in the film suspension increases [30]. Glycerol addition, which is 40% in sago starch and 50% in arrowroot starch, has the greatest thickness. The main factor is glycerol's ability to bind water. The higher the concentration of glycerol, the slower water evaporates. As a result, some of the water contained in the film solution is bound by glycerol, resulting in a thicker edible film [26]. The high viscosity of the film constituents causes an increase in total solids (density), which affects , which affects the thickness and viscosity of the edible film [28].

The lowest thickness was found in the use of soybean starch, which was 0.073 mm. Although the addition of glycerol to soybean starch is relatively high, the resulting thickness is the lowest. This is due to the very small size of soybean starch granules, which ranges from 0.7-4 μm, resulting in a smaller swelling ability compared to starch from other botanical sources [31]. The thicker the edible film, the stronger the ability to withstand gas and water vapor migration and protect the product from physical damage due to its denser structure.

2.2. The effect of starch source and glycerol concentration on edible film tensile strength

Tensile strength are used to characterize the mechanical properties of the film because they are important indicators of film strength and flexibility. [24]. From Table 2 above, it is known that taro starch has the highest tensile strength value compared to other starch sources, which is 10.63 MPa. This is influenced by the characteristics of the two types of starch which have the highest amylose content compared to the amylose content in other sources. The amylose content of starch was classified as very low with a value of 9%, low with a value of 9-20%, medium with a value of 20-25%, and high with a value of 25-30% [21]. It is known that taro starch has an amylose content of 21.98% [4]. Amylopectin influences the film's stability, whereas amylose influences the material's compactness [10]. Increasing the amylose content in the film suspension increased the number of hydroxyl groups, resulting in a stiffer film. The stronger the edible film, the higher the amylose content and starch concentration [21]. A high amylose content reduces starch's ability to gelatinize and preserves the integrity of swelling starch granules [32].

The use of plasticizers in film forming also has an effect on tensile strength. The addition of plasticizers to edible films can increase their elasticity, flexibility, and brittleness. Plasticizer molecules used in film production can weaken the bonds between protein chains. Plasticizer molecules will enter and interact with monomer-monomer bonds. The film elongates and flexes as a result of this condition, resulting in decreased mechanical resistance, elasticity, and other inhibitory properties [20].

2.3. The Effect of starch source and glycerol concentration on the percent elongation of edible film

The change in the maximum length of the edible film during stretching is referred to as elongation. The flexibility of edible films is critical in the food coating process. Food packaging will improve as edible films with an elastic and flexible texture become easier to shape [33]. The elongation of the film is inversely proportional to its tensile strength. The addition of plasticizer will affect the elongation properties of the film. Films with high plasticizer content will provide high elongation properties as well. Plasticizers work by widening the molecule's vacant space and weakening the hydrogen bonds of polymer chains to reduce intermolecular forces and increase film flexibility [34]. This is in accordance with the results in Table 2 above that the film from lesser yam starch has the lowest elongation value, which is 15.55%, with the addition of glycerol, which is only 22.5%. The highest elongation value was obtained in the film derived from jackfruit seed starch, which was 101.53% with the highest addition of glycerol at 2% (w/v). In addition, the increase in elasticity of the jackfruit seed starch film was also due to the addition of 1.5% extract of rosella flower in the film formulation, which increased the flexibility of the film. Rosella flower extract can replace the bonds between starch polymers during film formation, reducing the stiffness of the edible film [25]. Furthermore, using glycerol as a plasticizer can lower intermolecular forces and increase biopolymer mobility, resulting in a softer and more flexible edible film [35].

3. Conclusion

Tensile properties of edible films are critical because they determine the film's ability to protect food from mechanical damage during storage. Edible films derived from various types of starch with the addition of different concentrations of glycerol are known to have thickness, tensile strength, and elongation properties that meet Japanese International Standards, except for films from lesser yam starch. Films derived from jackfruit seed starch and taro tubers generally provide the most optimum tensile properties when compared to films from other starch sources.

References

[1] Aisyah Y, Irwanda L P, Haryani S and Safriani N 2018 *IOP Conf. Series: Materials Science and Engineering* **352** 1
[2] Espitia P J P and Otoni C G 2018 *Bio-based Materials for Food Packaging* 125
[3] Ningrum R S, Sondari D, Amanda P, Widyaningrum B A, Burhani D, Akbar F and Sampora Y 2020 *Jurnal Sains Materi Indonesia* **21** 164
[4] Wulandari Y, Harini N and Warkoyo 2019 *Food Technology and Halal Science Journal* **1** 22
[5] Juandi N, Rahmi and Helwati H 2016 *Jurnal Natural* **16** 6
[6] Santacruz S, Rivadeneira C and Castro M 2015 *Food Hydrocolloids* 1
[7] Thakur R, Pristijono P, Scarlett C J, Bowyer M, Singh S P and Vuong Q V 2019 *Int. J. OF Bioogical Macromolecules* **132** 1079
[8] Lauer M K and Smith R C 2020 *Comprehensive Reviews in Food Science and Food Safety* 1-53
[9] Liu Z 2005 *Innovations in Food Packaging* 318
[10] Gabriel A A, Solikhah A F and Rahmawati A Y 2021 *Journal of Physics: Conference Series* **1858**
[11] Wibowo C, Erminawati, Wicaksono R and Haryanti P 2020 *Food Research* **4** 1905
[12] Rachtanapun P, Pankan D and Srisawat D 2012 *Journal of Agricultural Science and Technology* **2** 252
[13] Dewi R, Putri A and Fitrianto R 2020 *IJRIE International Journal of Research Innovation and Entrepreneurship* **1**
[14] Ginting M H S, Kristiani M, Amelia Y and Hasibuan R 2016 *International Journal of Engineering Research and Applications* **6** 33
[15] Cazon P, Velaquez G, Ramirez J A and Vazquez M 2017 *Food Hydrocolloids* **68** 136
[16] Basiak E, Lenart A and Debeaufort F 2017 *International Journal of Biological Macromolecules* **98** 348
[17] Kramer M E 2009 *Edible Films and Coatings for Food Applications*
[18] Regubalan B, Pandit P, Maiti S, Nadathur G T and Mallick 2018 *Bio-based Materials for Food Packaging* 105
[19] Walfathiyyah A, Kusuma A P, Cahya F N, Qusyairi N and Wahyuningtyas D 2016 *5th ICRIEMS Proceedings* 1
[20] Maruddin F, Malaka R, Baba S, Amqam H, Taufik M and Sabil S 2020 *IOP Conference Series: Earth and Environmental Science* **492** 1
[21] Hatmi R U, Apriyanti E and Cahyaningrum N 2020 *E3S Web of Conferences* **142** 1
[22] Permata M M 2020 *Tinjauan sistematis: pengaruh jenis pati dan plasticizer terhadap karakteristik* (edible film Institut Pertanian Bogor)
[23] Santoso B, Dwiyanti R, Wijaya A, Priyanto G, Hermanto and Syaiful F 2021 *Current Nutrition & Food Science* **17** 876
[24] Salarbashi D, Tajik S, Ghasemlou M, Shojaee-Aliabadi S, Noghabi M S and Khaksar R 2013 *Carbohydrate Polymers* **98** 1127
[25] Ningsih P, Rahmawati S, Santi N M N, Suherman and Diah A W M 2021 *International Journal of Design and Nature and Ecodynamics* **16** 691
[26] Siskawardani D D, Warkoyo W and Siwi A A P 2020 *Food Technology and Halal Science Journal* **3** 26
[27] Muthi'ah, Handayani C B and Widyastuti R 2021 *Journal of Food and Agricultural Product* **1** 58
[28] Siskawardani D D, Warkoyo W, Hidayat R and Sukardi 2020 *IOP Conference Series: Earth and Environmental Science* **458** 1
[29] Rusmawati D A, Yuliasih I and Sunarti T C 2020 *IOP Conference Series: Earth and Environmental Science* **472** 1
[30] Ploypetchara T and Gohtani S *Journal of Food Science and Technology*
[31] Stevenson D G, Doorenbos R K, Jane J L and Inglett G E 2006 *Starch/Staerke* **58** 509

[32] Phrukwiwattanakul P, Wichienchotand S and Sirivongpaisal P 2014 *International Journal of Food Properties* **17** 1965
[33] Sedyadi E, Aini S K, Anggraini D and Ekawati D P 2016 *Biology, Medicine, & Natural Product Chemistry* **5** 33
[34] Putra A D, Johan V S and Efendi R 2017 *Jom Fakultas Pertanian* **4** 1
[35] Gujar S, Pandel B and Jethoo A S 2014 *International Quaterly Scientific Journal* **13** 425

IOP Conf. Series: Earth and Environmental Science 1115 (2022) 012076 doi:10.1088/1755-1315/1115/1/012076

The role of various plastisizers and fillers additions in improving tensile strength of starch-based bioplastics: A mini review

A Safitri, P S D Sinaga, H Nasution*, H Harahap, Z Masyithah, Iriany and R Hasibuan

Department of Chemical Engineering, Universitas Sumatera Utara, Padang Bulan, Medan 20155, Indonesia

*Email: halimatuddahliana@usu.ac.id

Abstract. The general public uses plastic extensively, particularly in food packaging. Plastic raw materials are generally petroleum derivatives with advantages such as high tensile strength, low cost, lightweight, ease of manufacture, and application. However, there are other disadvantages to plastic, which is not easily biodegradable. As a result, researchers have expressed a strong desire to develop biodegradable plastics in order to reduce the accumulation of non-biodegradable plastics in the environment. One of the candidates for producing bioplastic using natural resources is biodegradable plastics. Bioplastics based on starch are a common material used in the production of bioplastics. However, because bioplastics have lower tensile properties than conventional plastics, fillers are used to increase the strength of bioplastics. Fillers are used to reduce plastic cost, shrinkage during the setting process, and to improve tensile strength and hardness. As a result, the tensile and morphology of several starch-based biodegradable plastics with chitosan, CMC fiber, and clay as fillers will be reported in this review. According to our mini-review, sorbitol and CMC are very good plasticizers for starch-based bioplastics, and CMC has a higher crystallinity form, promoting greater interaction between the cellulose chains and starch-based matrix.

1. Introduction

Plastic production and consumption are increasing due to its use in both domestic and industrial applications, such as packaging systems and the health sector [1]. Plastic has become an indispensable part of daily life, and humans appear to be completely reliant on it. Plastic is a synthetic polymer that can be formed by heating and then molded to meet specific requirements [2]. Plastic is a macromolecular synthetic material derived primarily from petroleum. The most common petroleum-based plastics are polyvinylchloride (PVC), polyethylene terephthalate (PET), polyethylene, polystyrene, polyester, and polypropylene [3]. The reliance on plastic and petroleum products is one of the reasons for researching alternative biomass as a raw material in the manufacture of plastics. Typically, biomass is composed of natural or organic materials such as polysaccharides, proteins, and lipids [4]. Polysaccharides such as starch, chitin, lignin, and cellulose are commonly used in bioplastic synthesis, while natural proteins and lipids such as gelatin, casein, gluten, plant oils, and animal fats can also be used [5].

The primary benefit of starch-based materials is their biodegradability, which results in faster degradation and fewer landfill requirements. However, starch has poor mechanical and thermal properties as well as low water resistance [6-8]. By incorporating a filler material, these properties can be improved without the use of specialized equipment or at a high cost. Carboxymethyl cellulose (CMC), chitosan, and clay are three intriguing fillers. Due to the formation of strong hydrogen bonds between starch and filler, the filler used is expected to increase the tensile strength and viscosity properties of the starch matrix.[9] discovered that incorporating CMC into biocomposite films increased viscosity by up to 50% and tensile strength by up to 37%. The CMC structure contains three major functional groups, which improves molecular interactions in bioplastic, resulting in better film properties. [10] discovered that adding chitosan to corn starch-based bioplastics improves their water absorption, tensile strength, and flexibility. [11] reviewed research on other fillers in the form of clay. Tensile strength increased with increasing bentonite concentration (2.5%), according to the study. However, increasing the amount of clay will reduce the tensile strength of the film due to the formation of clay lumps.

Plasticizers are another alternative material that can be added to increase the shelf life, elasticity, and limitations of the use of bioplastic products [12-15]. The space occupied between the starch polymer chains can be filled in the presence of a plasticizer, reducing the crystallinity of the plastic. When starch is added to plasticizer, it becomes more flexible and can be mixed with a variety of polymeric materials for a variety of applications. The purpose of this article is to discuss the use of various fillers and plasticizers to modify the tensile strength and morphology of starch-based bioplastics.

2. Method

The method used in writing this article is a literature review, which was conducted by collecting various publications on the scope of starch-based bioplastic production published between 2004 and 2022. Secondary data in the form of journals from the internet whose sources of information are indexed by reputable journals were used to collect the data and information for this study. Several sources were used to obtain references, including Google Scholar, SCOPUS, Science Direct, and Springer Link.

3. Discussion

3.1. Tensile strength of bioplastics with the addition plasticizers

The addition of filler and plastisizer during bioplastics production has a significant impact on tensile strength, a critical parameter of mechanical quality. The primary sources of starch used in the production of bioplastics are agricultural and plant sources of starch. The tensile strength of bioplastic with sorbitol, glycerol, and a mixture of both materials is shown in Table 1.

Table 1. The tensile strength of corn starch bioplastic with the addition of sorbitol, glyserol and the mixture of both materials

	Sorbitol	Glycerol	Sorbitol/Glycerol	References
30%	13.62 MPa	2.53 MPa	5.74 MPa	[16]
40%	3.13 MPa	6.49 MPa	-	[17]
50%	3.13 MPa	4.95 MPa	-	[17]
60%	4.48 MPa	1.7 MPa	2.60 Mpa	[16]

The maximum tensile strength of the bioplastic was obtained at a low percentage of the plasticizer, according to Table 1 for the tensile strength analysis. Because of the nature of the properties of glycerol and sorbitol, which are easily bound to molecules, using more than 30% plasticizer reduces tensile strength [18]. The higher the water content in the bioplastic, the greater its tensile strength because it increases the plastic's flexibility. This is because plasticizer molecules disrupt starch cohesiveness, reduce hydrogen interactions, and increase polymer mobility [19].

3.2. Tensile strength of bioplastics with the addition Chitosan, CMC fiber, and Clay

Each filler's tensile strength in Table 2 has a different value. Using CMC as a filler at 45% yielded the highest tensile strength value. The results showed that the addition of cellulose, in this case CMC, increases the attractive forces between molecules, causing a thin layer to form [20]. This condition is caused by hydroxyl groups forming intermolecular and hydrogen bonds, which result in the formation of a thin layer of reinforcing fibers.

Table 2. The tensile strength of corn starch bioplastic with the addition of Chitosan, CMC fiber, and Clay with 1 ml and 2 ml Glycerol

Glycerol	1 ml	2 ml	References
Chitosan: 35%	3 Mpa	3.22 MPa	[22]
Chitosan: 40%	3.14 Mpa	3.47 MPa	[22]
Chitosan: 45%	3.2 Mpa	3.69 MPa	[22]
CMC: 25%	4.07 Mpa	3.62 MPa	[19]
CMC: 20%	8.43 Mpa	3.75 MPa	[19]
CMC: 16,67%	6.34 Mpa	3.3 MPa	[19]
Clay: 25%	6.86 Mpa	2.86 MPa	[19]
Clay: 20%	7.82 Mpa	2.32 MPa	[19]
Clay: 16,67%	3.5 Mpa	1.67 MPa	[19]

The higher tensile strength value resulted from the higher energy required to deform the angles and distances of the bonds between the atoms of the polymer chain. This energy was generated by the positive interaction of starch and CMC elements, specifically starch OH and CMC-COOH [21]. The type of starch source, starch content, and additives such as plasticizers and fillers all have an impact on how well tensile strength in bioplastics demonstrates its properties. Glycerol is the most widely used plasticizer because it has the best interaction ability when combined with different types of starch, with or without fillers [22]. The type and concentration of plasticizer have a significant impact on the bioplastic's tensile strength when stretched. Amylose and amylopectin form a brittle and stiff bioplastic film with an inhomogeneous composition distribution in the absence of a plasticizer [23-24].

3.3. The morphology of bioplastics

Surface microstructure and 3D micrographs of AFM with RMS roughness of starch bioplastics were revealed by SEM analysis. The presence of granules (remaining starch particles) on the surface indicates that the starch was not fully gelatinized during the forming process. This finding supports a previous report [25], which stated that the depicted SEM image contains insoluble remnants of swollen starch granules. The fracture surface has some voids (black dots), which contributes to the low impact and tensile strength.

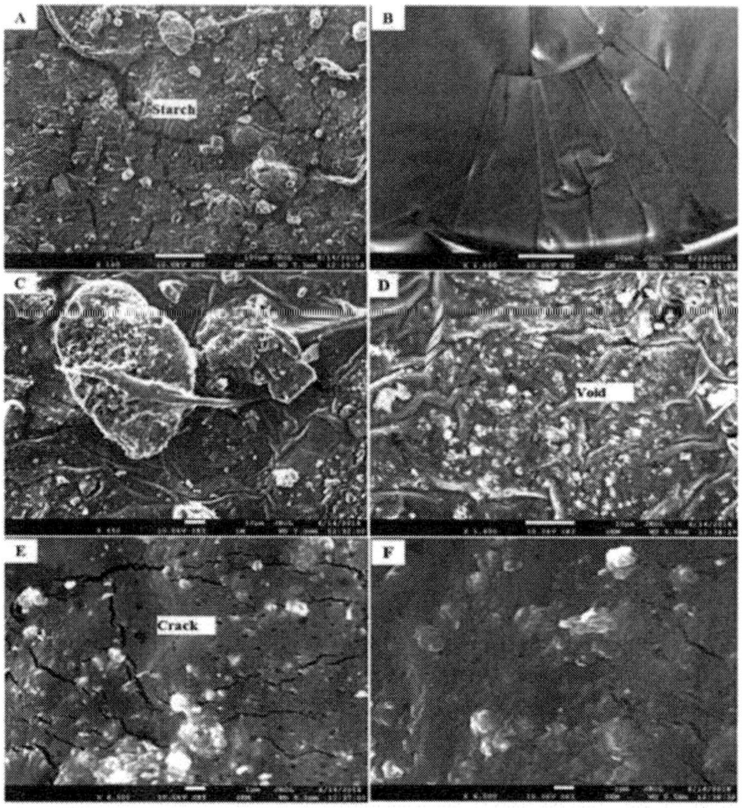

Figure 1. SEM photo of starch bioplastic by [26] (A) Starch bioplastic (mag.160), (B) Starch bioplastic (mag. 1800), (C) Presence of granules (mag. 650), (D) Presence of voids (mag. 1600), (E) – (F) Presence of cracks (mag. 6500)

Ernita et al [22] examined the SEM results for the use of various fillers. Figure 2 shows white granules that are less homogeneous in comparison to the other materials. One of them is partially dissolved starch or chitosan that did not dissolve completely during mixing. Figure 2B depicts a biodegradable plastic with varying chitosan concentrations. Cracks abound on the plastic surface. Water can be absorbed by the cavities formed by the cracks. As a result, the hydrophilicity of the plastic increases.

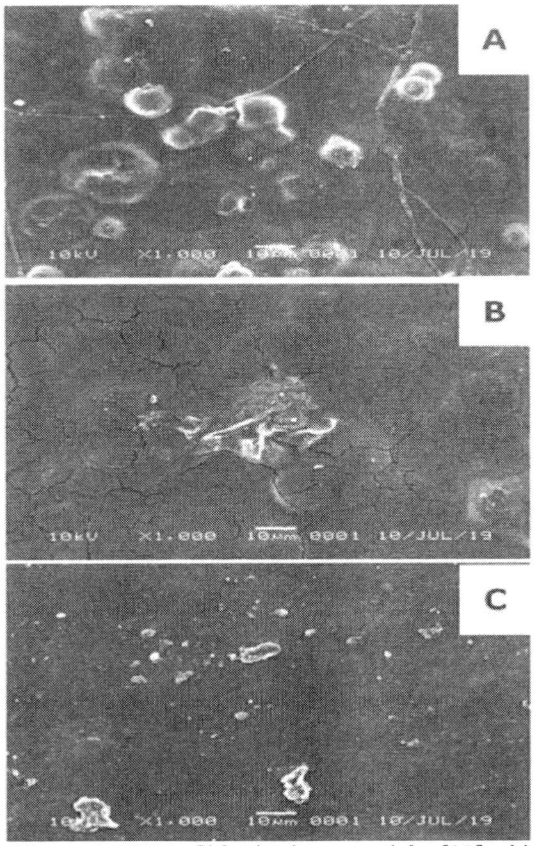

Figure 2. SEM of bioplastic research by [22] with (A) chitosan, (B) chitosan- isopropylacrylamide (C) chitosan-Poly isopropylacrylamide.

The addition of the appropriate amount of plasticizer to the starch-based film aids in particle distribution. Plasticizer can improve the film's structural stability and surface quality. As a result, the highest concentration of plasticizers used in the production of starch films is an incoherent thin film that is difficult to peel off. Surface morphology revealed numerous voids, edges, and holes, as well as poor interfacial adhesion. In composite bioplastics, minor crack propagation indicates poor component bonding. On the fracture surface, some voids (black dots) can be seen, which contributes to the low impact and tensile strength values. This shows that the interaction between the molecules that make up the bioplastic is minimal.

4. Conclusion

Because of its compatibility with amylose, glycerol and sorbitol are frequently used as plasticizers for starch films, promoting tensile properties by interfering with amylose packing and thus decreasing the intermolecular forces between the starch molecules. Tensile strength can be increased by adding 20% CMC to the starch matrix. Formulations containing up to 20% CMC have no effect on the film's flexibility. The structure of film complexes formed during gelatinization was found to have a strong influence on granule swelling and starch polymer dissolution, and paste viscosity was found to be primarily dependent on a disperse phase's particle size.

References

[1] Hidayat Y A, Kiranamahsa S and Zamal M A 2019 *Energy* **3** 350
[2] Paterson H 2019 *Sustainable Earth* **2** 1
[3] Mohanan N, Montazer Z, Sharma P K and Levin D B 2020 *Frontiers in Microbiology* **11**
[4] Wróblewska-Krepsztul J, Rydzkowski T, Borowski G, Szczypiński M, Klepka T and Thakur V K 2018 *International Journal of Polymer Analysis and Characterization* **23** 383
[5] Shafqat A, Tahir A, Mahmood A, Tabinda A B, Yasar A, and Pugazhendhi A 2020 *Biocatalysis and agricultural biotechnology* **27** 101540
[6] Le Corre D, Bras J and Dufresne A 2010 *Biomacromolecules* **11** 1139
[7] Albuquerque P B, and Malafaia C B 2018 *International journal of biological macromolecules* **107** 615
[8] Paixão L C, Lopes I A, Barros A K D and Santana A A 2019 *Journal of Applied Polymer Science* **136** 48263
[9] Santos V P, Marques N S, Maia P C, Lima M A B D, Franco L D O and Campos-Takaki G M D 2020 *International journal of molecular sciences* **21** 4290
[10] Wang W, Xue C and Mao X 2020 *International Journal of Biological Macromolecules* **164** 4532
[11] Méitén N, Konan L K, Tognonvi M T, Doubi B I H G, Gomina M and Oyetola S 2021 *Carbohydrate Polymers* **254** 117322
[12] Mali S, Grossmann M V E, García M A, Martino M N and Zaritzky N E 2004 *Carbohydrate Polymers* **56** 129
[13] Vanin F M, Sobral P J A, Menegalli F C, Carvalho R A and Habitante A M Q B 2005 *Food Hydrocolloids* **19** 899
[14] Ibrahim M I J, Sapuan S M, Zainudin E S and Zuhri M Y M 2019 *International journal of food properties* **22** 925
[15] Ehirej A, Sapuan S M, Jawaid M and Zahari N I 2017 *Starch Staerke* **69** 1
[16] Hazrol M D, Sapuan S M, Zainudin E S, Zuhri M Y M and Abdul Wahab N I 2021 *Polymers* **13** 242
[17] Ball esteros-Mártinez L, Pérez-Cervera C and Andrade-Pizarro R 2020 *NFS journal* **20** 1
[18] Tavares K M, de Campos A, Luchesi B R, Resende A A, de Oliveira J E and Marconcini J M 2020 *Carbohydrate polymers* **246** 116521
[19] Sugiharto A, Syarifa A, Handayani N and Mahendra R 2021 *Jurnal Bahan Alam Terbarukan* **101** 31
[20] Mariana M, Alfatah T, HPS A K, Yahya E B, Olaiya N G, Nuryawan A and Ismail H 2021 *Journal of Materials Research and Technology* **15** 2287
[21] Abdullah A H D, Putri O D, Fikriyyah A K, Nissa R C, Intadiana S 2020 *Polymer-Plastics Technology and Materials* **5912** 1250
[22] Ernita L, Riza M and Syaubari S 2020 *Jurnal Rekayasa Kimia & Lingkungan* **151** 45
[23] Alonso-González M, Felix M and Romero A 2022 *Industrial Crops and Products* **180** 114767
[24] Abdullah A H D, Sri P, Myrtha K, Oceu D P and Rani H F 2018 *Journal of biological Sciences* **191** 57
[25] Ruggero F, Carretti E, Gori R, Lotti T and Lubello C 2020 *Chemosphere* **246** 125770
[26] Amin M R, Chowdhury M A and Kowser M A 2019 *Heliyon* **5** 2009

The effect of distance planting of red mangrove (*Rhizophoraapiculata*) on Pulau Sembilan, PangkalanSusu District, Langkat Regency

Yunasfi[1,2*], K Siregar[1] and A Dalimunthe[1]

[1]Faculty of Forestry, University Sumatra Utara, Medan, North Sumatra, Indonesia
[2]Center of Excellence for Mangrove, Universitas Sumatera Utara, Medan 20155, Indonesia

*Email: yunasfijamhar@yahoo.co.id

Abstract. Mangroves are a very important coastal resource for socio-economic development In addition, mangrove habitats contribute to complex food webs and energy transfer. This research was conducted in the mangrove forest area of Pulau Sembilan Village, Pangkalan Susu. This research activity was carried out for 3 months from October to December 2022. Observations in this study were carried out every two weeks. This study was based on a completely randomized design (CRD) with 5 treatments with 6 replications. The treatment in question is the spacing with P0 = 46 meters, P1 = 47 meters, P2 = 48 meters, P3 = 49 meters and P4 = 50 meters. From the research on the effect of plant spacing on *Rhizophoraapiculata*, the highest seedling height growth was obtained in the P0 treatment (40m spacing) with a value of 213.2 cm or average 35.5 cm. The highest increase in plant diameter was found in the P0 treatment with a value of 7.59mm or average 1.27 mm. The highest added value was found in treatment P2 with a value of 37.0 cm. The increase in leaf length is 55.4 cm which is found in the P4 treatment.

1. Introduction

Indonesia has a mangrove area of 4.255 million ha, including the largest in the world. Red mangrove is one type of mangrove that absorbs water but is able to prevent the entry of salt through the filter found in the roots. Triterpenoid compounds play an important role in protecting red mangroves from the influence of salt. In addition, red mangrove leaves which are the site of photosynthesis also contain many secondary metabolites. A close relative of *Rhizophora*, the white fire mangrove contains steroids in its leaves. The presence of steroid content in this red mangrove is the basis for antioxidant testing [1].

Mangroves are a very important coastal resource for socio-economic development. Many of the ecological services of mangroves lie in protecting the coast from the sun, the greenhouse effect, and the ravages of hurricanes, floods, sea level rise, waves and coastal erosion. Mangroves contribute significantly to the global carbon cycle and produce a large amount of litter in the form of fallen leaves, twigs and other debris. In addition, mangrove habitats contribute to complex food webs and energy transfer. Most communities depend on local resources for their livelihoods. Mangroves are a source of high value commercial products and fishery resources, as well as a growing location for ecotourism development [2]. The loss of biodiversity in the mangrove ecosystem can disrupt the function of the ecosystem and functionally disappear from the earth. Changes in biodiversity and

IOP Conf. Series: Earth and Environmental Science 1115 (2022) 012077 doi:10.1088/1755-1315/1115/1/012077

ecosystem functions in mangroves can directly or indirectly impact the millions of people who depend on mangroves for ecosystem services (eg, coastal protection) and livelihoods. biodiversity and ecosystem functions, mangroves and other aquatic ecosystems. In addition, there is currently substantial evidence from other ecosystems that the loss of biodiversity has a considerable impact on the functioning of one ecosystem and consequently on the survival of humans and other organisms [3]. This study aims to study and determine the effect of spacing on the cardinal directions that are useful for increasing growth for the *R.apiculata* mangrove ecosystem. This test is expected to increase the growth of mangroves and have an impact on the continuity of the mangrove ecosystem.

2. Material and Methods

2.1. Time and place
This research was conducted in the mangrove forest area of Pulau Sembilan Village, Pangkalan Susu. This research activity was carried out for 3 months from October to December 2021. Observations in this study were carried out every two weeks.

2.2. Tools and materials
The tools used in the field research were caliper, rope, measuring tape, compass, tape measure, cloth tape measure, wooden stake made of bamboo, permanent marker, book, pen, oven, knife and polybag. The material used is *R. apiculata* seeds which have relatively the same size as many as 30 seeds and the required data from institutions related to research.

2.3. Research procedure

2.3.1. Measurement of distance from the beach and determination of the cardinal directions. Measurement of the distance from the beach to the nursery as far as 40 to 44 meters as for the treatment: P0 = 46 meters, P1 = 47 meters, P2 = 48 meters, P3 = 49 meters and p4 = 50 meters from the beach with 6 replicates of R. apiculata each.

2.3.2. Selection of seeds and planting of R. apiculata seeds. The seeds of R. apiculata used in this study were 2 month old seedlings. The requirements for seeds are that the seeds have relatively the same size, both in height and number of leaves.The treatments in this study were 6 treatments and 5 replications. The number of seeds required is 30 seeds.

2.3.3. Experimental design. R. apiculata seedlings were planted in the mangrove forest area of Pulau Sembilan, PangkalanSusu District, Langkat Regency. This study used a completely randomized design (CRD). Data collection was carried out after planting the seedlings in the field, during the following time:
a. Days 15 c. Days 45
b. Days 30d. Days 60
e. Days 75 f. Days 90

The linear model in complete random design is:

$$Y_{ij} = \mu + \alpha_i + \sum_{ij}$$

Y_{ij} = Observations on the treatment of i-th and j-th replications
μ = General average value of observations
α_i = The effect of treatment factors to -i; I = 1,2,....... t dan j= 1,2,......r
$\sum_I$ = Effect of error treatment to-i and repeat to-j

2.4. Parameters of R. apiculata seeds measured

2.4.1. Seedling height (cm). Height measurement was carried out by measuring the height of the seedlings from the base of the stem to the highest plant growth point using a meter. Observations were made every two weeks.

2.4.2. Seedling diameter (mm). Measurement of plant stem diameter was carried out using a caliper with units of mm. The diameter measurement was carried out parallel to the markers on each side of the seedling and measured every two weeks.

2.4.3. Number of leaves (strands). The number of leaves was observed by counting the fully grown and fully formed leaves every two weeks. The leaf that counts is the leaf that has fully opened

2.4.4. Leaf Area (cm). Leaf area measurements were carried out at the end of data observation using the millimeter block paper method by placing the leaves on millimeter block paper then the leaves were drawn following the leaf pattern. Then count the number of squares in the leaf area that has been drawn in order to get the result of the leaf area.

2.4.5. Plant wet weight (g). Determination of wet weight was carried out by weighing all R. apiculata seedlings that had been harvested and divided into several parts, namely leaves, stems and roots using analytical scales at the time of observation and harvested.

2.4.6. Plant dry weight (g). After measuring the wet weight of the R. apiculata seedlings, they were put in an oven at 75°C for a period of 48 hours until they were constant and expressed in grams (g). After the oven is done, then the seeds will be weighed using an analytical balance.

3. Results and Discussion

3.1. Seedling height (cm)

Measurement of the growth rate of mangroves is important to know because if these trees grow well and form mangrove forest vegetation, then their function is very large for the lives of living things around them.The following is a table of data on the height growth of *R. apiculata* seedlings, it can be seen the changes and growth. The following results of the average height growth of seedlings can be seen in Table 1.

Table 1. Growth of *R. apiculata* seedlings for 12 weeks

Treatment	Observation to						Total (cm)	Average (cm)
	I	II	III	IV	V	VI		
P0	34.1	34.9	35.3	35,4	36.3	37.2	213.2	35.5
P1	33.8	34.1	34.5	34,5	34.8	34.8	206.4	34.4
P2	31.5	31.8	32.3	32,3	32.6	32.6	193.0	32.2
P3	28.5	29.2	30.2	30,2	30.6	30.6	179.4	29.9
P4	31.8	32.5	33.2	33,9	34.5	35.5	201.3	33.6
Total	159.6	162.5	165.5	166,4	168.6	170.7	993.2	165.6

From the table results it can be seen that the growth of seedling height R. apiculate the highest from other treatments was P0 with a value of 213.2 cm with an average of 35.5 cm and the treatment of P3 plant seeds resulted in the lowest seedling growth of the other treatments with a value of 179.4 cm with an average of 29.9 cm. The analysis of the survival percentage of R. apiculata was 100% where all seedlings were able to survive for 90 days. In general, a good high growth response is obtained at low salinity. This happens because the mangrove plant is not a plant thatrequire salt (salt demand) but plants that are salt tolerant [4].

In the table it can be seen that mangroves grow and increase in height along with the length or age of mangroves, this is in accordance with statement [5] which states that Identification of mangrove age is assumed from tree height. The older the mangrove, the higher its height. Plant growth can also be seen from the increase in height and diameter, but at a certain age the increase in height is almost always followed by an increase in diameter.

Mangrove area is an area with high primary productivity. Primary productivity greatly affects the growth of mangroves. One indicator of mangrove growth is influenced by the sediment in which it lives which contains macro and micronutrients, oxygen, and fresh water to maintain a balance of salt levels in its physical. Limiting factors for mangrove production and growth include temperature and sunlight, salinity, anoxia and tides, bioturbation, and nutrient availability [5].

The increase in the age of the mangrove itself has an effect on the nutrient content in the sediment. As the age increases, the number of leaves also increases. These fallen leaves are then decomposed by biota associated with mangroves and make the area overgrown with mangroves fertile. This is in accordance with the opinion of [6] which states that the sediment around the mangrove then mixes with the fallen litter and is deposited in the sediment. These conditions make mangrove forests as a contributor of nutrients to other ecosystems in the vicinity.

3.2. Seedling diameter (mm)

Growth development of seedling diameter *R. apiculata*measurements were made starting in the second week until the following week (90 days). In each measurement, *R. apiculata* seedlings increased in each measurement made during the study. At the time of the initial measurement, the growth was still not very visible, however, on the third observation and so on, the growth of the seedlings became more clearly visible. The average increase in diameter of *R. apiculata* seedlings can be seen in Table 2.

Table 2. Increase in diameter of *R. apiculata* seedlings for 12 weeks

Treatment	Observation to						Total (mm)	Average (mm)
	I	II	III	IV	V	VI		
P0	1.24	1.25	1.26	1.27	1.28	1.29	7.59	1.27
P1	1.16	1.17	1.19	1.21	1.23	1.24	7.20	1.20
P2	1.14	1.15	1.17	1.2	1.22	1.25	7.13	1.19
P3	1.08	1.09	1	1.13	1.15	1.15	6.71	1.12
P4	1.06	1.08	1.1	1.11	1.13	1.14	6.62	1.10
Total	5.68	5.74	5.83	5.92	6.01	6.07	35.25	5.875

The diameter table shows that the highest increase in plant diameter was found in P0 treatment with a value of 7.59 mm and an average value of 1.27 mm. The lowest value of diameter increase was found in P4 treatment with a value of 6.62 and an average of 1.10mm.The growth in diameter and the incremental growth in diameter are influenced by factors that influence growth, including the substrate, the substrate referred to here is a substrate containing soft mud [7]. According to [8] tree diameter growth is influenced by environmental factors, where environmental factors are influenced by the potential of the soil for plant growth. Land potential seen fromThere are two aspects, namely the physical condition of the soil and the chemical aspect of the soil.

Soil is the main medium for growing plants where plants will grow well in water conditions, and nutrients that depend on the conditions of soil physico-chemical and biological factors.. Mineral and organic matter are essential elements for plant growth. Elements C, H, O, N and S and P are very important components of protoplasm and cell walls [9]. Growth stem diameter is influenced by factors where it grows, especially nutrients that play an important role in cell division activity and the development of plant meristematic tissue which results in stem enlargement. Diameter can be influenced by the intensity of light received, age and number of leaves associated with the photosynthesis process [10].

The factors that can affect the growth of the number of leaves on *R. apiculata* seedlings, namely the temperature is known to the type of *Rhizophora* sp. grows optimally at a temperature of 26^0C-28^0C and production will decrease if the temperature is higher. Fresh leaves on mangrove plants will fall if the temperature is below the optimum and mangrove plants will also delay the production of new leaves if the temperature in the environmental mangrove forest is above the optimum temperature. Leaves are the most important part of plants, this is because leaves are a tool used by plants to obtain food in addition to utilizing their roots to absorb nutrients from the soil. If in a plant there are many leaves, the photosynthesis process that takes place will be more and more and more food will be produced [11].

3.3. *Number of leaves (strands)*

The growth and increase in the number of leaves, *R. apiculata* seedlings at the initial observation (second week) did not experience an increase in the number of leaves. However, at the second observation (fourth week) the leaves began to experience an increase in the number of leaves. The addition of the number of leaves two to four strands. The increase in the number of leaves of *R. apiculata* seedlings can be seen prominently in the fourth week to the following week. And the most increase in the number of leaves can reach four to eight strands.Theincrease in the number of leaves of *R. apiculata* seedlings can be seen in Table 3.

Table 3. Increase in the number of leaves of *R. apiculata* seeds for 12 weeks

Treatment	Observation to						Total (strands)	Average (strands)
	I	II	III	IV	V	VI		
P0	4.2	4.3	4.8	4.7	5.0	5.5	28.5	4.7
P1	5.2	5.5	5.8	6.0	6.3	6.5	35.3	7.1
P2	5.7	5.8	6.0	6.3	6.5	6.7	37.0	7.4
P3	4.0	4.3	4.5	4.7	4.8	5.3	27.7	5.5
P4	4.7	4.8	5.0	5.2	5.3	6.0	31.0	6.2
Total	23.7	24.8	26.2	26.8	28.0	30.0	159.5	30.9

In the Table 3. it can be concluded that the highest added value is found in treatment P2 with a value of 37.0 cm and the average value is 7.4 cm. While the increase in the number of leaves was found in treatment P3 where the added value was 27.7 cm cm with an average value of 5.5 cm. Counting the number of leavesperformed on all seedlings. In addition, the condition of the leaves was also observed. The decrease in the number of leaves from the initial number of leaves was due to the creation of stress conditions in the seedlings by the flooding treatment. Based on observations, the number of leaves on the seedlings decreased due to leaf fall or fall [12]. If the nitrogen supply is sufficient, the plant leaves will grow large and expand the surface available for photosynthesis so that the increased photosynthesis rate will produce large amounts of photosynthate. the nutrient that affects the growth and development of leaves is nitrogen [13].

3.4. *Leaf area index (cm)*

Measurement of leaf area of *R. apiculata* seedlings at the initial observation of seedling leaf area has not clearly seen a change in leaf area at the time of planting. The increase in leaf area began to be seen in the third observation, and at the fourth observation the increase was more striking. The following table shows the results of the study on leaf area growth for 90 days. The increase in leaf area of *R. apiculata* seedlings on average for 12 weeks can be seen in Table 4.

Table 4. Increase in leaf length of *R. apiculata* Seedlings for 12 weeks

Treatment	Observation to						Total (cm)	Average (cm)
	I	II	III	IV	V	VI		
P0	7.7	7.9	8.1	8.3	8.6	8.8	49.5	8.3
P1	7.1	7.4	7.6	7.8	8.0	8.3	46.1	7.7
P2	8.3	8.5	8.8	9.1	9.3	9.5	53.5	8.9
P3	8.0	7.8	8.1	8.3	8.5	8.8	49.5	8.3
P4	8.8	9.0	9.2	9.5	9.3	9.6	55.4	9.2
Total	39.9	40.6	41.8	42.9	43.8	450	253.9	50.6

In the table above, it can be seen that the highest value of leaf area increase was 55.4 cm with an average value of 9.2 cm which was found in treatment P4 while the lowest value was found in treatment P1 and P3 where the value was 49.5 cm and the average of 8.3 cm. One of the most important elements of solar radiation is light intensity. The increase in leaf area is an effort by plants to efficiently capture light energy for photosynthesis normally under conditions of low light intensity. The expansion of leaf blades in plants is the role of nitrogen in organic matter, so that it affects the photosynthesis process of plants. Plants that are adequately supplied with N will form broad leaves with high chlorophyll content.

The difference in leaf area at each research station is influenced by limiting factors. It is suspected that the limiting factor for this growth is light. Dim light will result in a slow rate of photosynthesis, so that it can inhibit the growth process, one of which is the addition of leaf area. Plants need nutrients called plant nutrients. The function of plant nutrients cannot be replaced by other elements and if there is no plant nutrient, then metabolic activities will be disrupted or stop altogether.

3.5. Plant wet weight (g)

Data collection on the wet weight of *R. apiculata* seedlings was carried out after 90 days of observation were completed. Wet weight is the pure weight of *R. apiculata* seedlings. The data taken is the data from weighing the roots, stems, leaves on the seeds of *R. apiculata* before drying (oven) the seeds. The following is a diagram of the wet weight of *R. apiculata* seedlings before oven roasting can be seen in Figure 1.

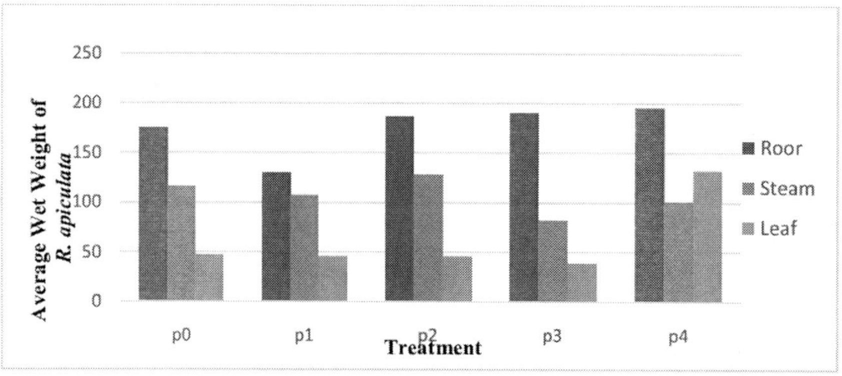

Figure 1. Diagram of the average wet weight of *R. apiculata* seedlings for 12 weeks

From the Figure 1, it can be concluded that the highest wet weight was found in the P0 treatment, namely the wet weight of the roots.The decrease in wet weight and dry weight of plant roots in each shade variation treatment was thought to be due to the transfer of results from photosynthesis. Where the results of photosynthesis are used for metabolic activities that are above the soil surface.

Root volume is closely related to macro and micro nutrients, N absorbed by plants plays an important role in supporting plant vegetative growth such as roots. P element plays a role in the formation of a good root system. Provision of N elements can improve plant vegetative growth and the formation of protein and chlorophyll. Cell division, root development [13].

3.6. *Plant dry weight (g)*

Data collection on the dry weight of *R. apiculata* seedlings was carried out after 90 days of observation were completed. Dry weight is the weight of *R. apiculata* seedlings after being roasted. The data taken is the data from weighing the roots, stems, leaves on the seeds of *R. apiculata* after drying (oven) the seeds. The following is a diagram of the dry weight of the seeds of *R. apiculata* after roasting, which can be seen in Figure 2.

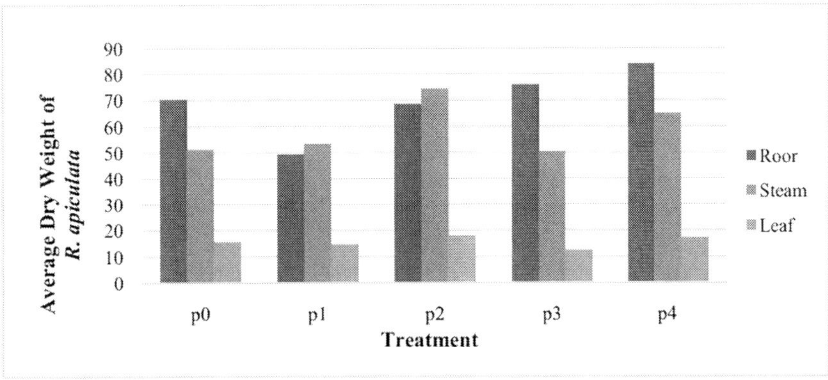

Figure 2. Diagram of the average dry weight of *R. apiculata* seedlings for 12 weeks

Good and normal plant growth is indicated by a balanced shoot-root ratio. This indicates that the shoots and roots of the plant will be sturdy and will not collapse easily due to the plant's root systemable to support shoot growth. The dry weight of the plant is the resultant of three processes, namely the accumulation of assimilate through photosynthesis, the decrease of assimilate due to respiration and accumulation to the food reserves. The increase in dry weight of the plant showed that the growth of the plant was getting better. The dry weight of the plant is strongly influenced by the optimal process of photosynthesis. The dry weight formed reflects the amount of photosynthesis as a result of photosynthesis, because dry matter is highly dependent on the rate of photosynthesis. Larger assimilate allows the formation of larger plant biomass. The dry weight of a plant is influenced by optimal photosynthesis because the dry weight of a plant depends on the amount of carbohydrate accumulation in the plant body [14].

4. Conclusion

From research *R. apiculta* carried out with each plant spacing for 90 days (3 months), the percentage of mortality was found during the study with the percentage of plant growth being 100%. From the results of the ANOVA research on the treatment of plant spacing on plant seeds, *R. apiculata*There was a tendency towards growth in height, diameter, leaf area, number of leaves, wet weight and dry weight of plants during the 90 days research process.

Acknowledgments

According to the guidelines Inquiry regarding University of North Sumatra Talenta Contract Number : 6789/UN5.1.R/PPM/2021, to the ivestigateInstitut the College of North Sumatra, 16 June 2021.

References

[1] Paryanto E and Kwartiningsih WA 2015 *Purification Journal***15**
[2] Kathiresan K2012 *International Journal of Marines***2**
[3] Mukherjee N, SutherlandWJ, Khan M N 2014 *Evology And Evolution*
[4] Hutahaean E, Kusmana C and DewiHR 1999*Journal of Tropical Forest Management***5**1
[5] Chrisyariati I, Hendro B and Suryani 2014 *Diponegoro Journal of Maquares***3** 66
[6] Indrawati A, Hartoko A and Soedarsono P 2013*J Man I res***2**28
[7] Angger RK, Kustani A and Hilmanto R 2016*Sylva Lestari Journal***4**
[8] Budiman M, Hardiansyah G and Darwati H 2015*Sustainable Forest Journal*
[9] Rizki and Novi 2017 *Journal of Biology and Biological Education***3** 48
[10] Darwati H N and Astiani D 2021*Sustainable Forest Journal***9** 690
[11] Yunasfi and SimanungkalitJ M 2022*The Effect of Planting Distance on Seeds Rhizphora Apiculata from The Coast to The Northeast on Pulau Sembilan, Pangkalan Susu District, Langkat* (Medan: Faculty of Forestry, University of North Sumatra)
[12] Permatasari I and Kusmana C 2011 *Journal of Tropical Silviculture***2**(3)
[13] Trisnawati, Wardat and Yulia A E2017*Let's Faperta***4**
[14] Ismoyo U B, Hendrarto and Suryani 2017 *Analysis of Organic Matter and Soil*

Solar tracker design on solar panel for stm32 microcontroller based on battery charging system

H H Rangkuti[1], N P Sinaga[1] and F Ariani[2]

[1]Electrical Engineering, Universitas Sumatera Utara, Medan, Indonesia.
[2]Mechanical Engineering, Universitas Sumatera Utara, Medan, Indonesia

*Email: helmitom@yahoo.co.id

Abstract. This paper describes about solar energy is one of the energy sources that won't terminated. Solar energy can be easily converted into electrical energy by using solar panels. Solar panels that are placed horizontally on the ground, the solar panel cannot absorb the light perfectly. Therefore, solar panels require an automatic *solar tracking system* to increase the efficiency of the solar panels. In this study, a solar tracker has been designed using a light dependent resistor (ldr) sensor based on the stm32 microcontroller. From the results of the study, the increase in power obtained from the use of a *solar tracker* was 27.97%. In *static solar panels*, the time required to fully charge the 12v, 7ah *lead acid* battery is 5 hours 10 minutes, while using a *solar tracker* the *lead acid* battery *charging time* is 3 hours 30 minutes. The use of a *solar tracker* saves battery charging time for 1 hour 40 minutes compared *to static solar panels*.

1. Introduction

Solar energy is one of the energy sources that won't terminated. And with this energy, we can obtain electrical energy without burning fossil fuels such as oil, gasoline and gas. Solar energy can be easily converted into electrical energy by using solar panels. Solar panels that are placed horizontally on the ground, the solar panel cannot absorb the light perfectly, and also the efficiency of using solar panels is only 14-16% depending on the amount of light absorbed by the solar panels. To get the maximum electric current, the solar panel must always be perpendicular to the incoming light [1]. Therefore, solar panels require an automatic *solar tracking system* that is accurate enough to increase the efficiency of the *solar panels*. This solution requires the right control system, sensors, and motors to make the *solar tracker* movement more accurate. The microcontroller must have the required *software* so that when the weather changes, the controller can quickly change the position of the solar panels [2]

In previous studies, *a two-axis solar tracker* system has been designed using *a light dependent resistor* (ldr) *sensor* using an arduino uno controller where the *solar tracker* moves according to the direction of the sun, with an increase in efficiency of 10-40% [3]. However, the arduino controller still has drawbacks, namely that it can only perform simple processing, not yet at the stage of high-power efficiency and performance [4].

This research is devoted to making a *solar tracker* by using the stm32 arm cortex-m microcontroller. From the *datasheet* information obtained, this microcontroller has a *clock* frequency ranging from 32mhz to 400 mhz and only requires a voltage of 2-3,6v so that it can be applied to more complex microcontroller systems, high performance and better power efficiency than the ATMega microcontroller or Arduino board.

2. Solar Energy

The sun is a material composed of very hot gas with a diameter of 1.39 x 109 m, and a distance of 1.5 x 1011 m from the earth. The sun has an effective surface temperature of 5762 K. The temperature in the sun's core area ranges from 8x106 - 40x106 K and its density is estimated to be 100 times greater than water. The sun is essentially a continuous fusion reactor with its constituent gases held in place by the force of gravity. The energy emitted by the sun comes from fusion reactions. Energy is produced in the sun's interior and sent to the surface and then radiated out into space.

The sun is the main source for life on earth, the energy produced by the sun in the form of heat energy and light energy is used by living things to meet their needs. Earth receives 175x105 Watts of solar radiation in the outer atmosphere. Approximately 30% of the total radiation is reflected back into space, where the remaining 70% is absorbed by clouds, oceans, and land [5].

2.1. Solar cell principle

Solar panels have a p-n junction working principle, which is a junction between p-type and n- type semiconductors. This semiconductor consists of atomic bonds in which there are electrons as the basic constituent. The n-type semiconductor has an excess of electrons (negative charge) while the p- type semiconductor has an excess of holes (positive charge) in its atomic structure. The following is the working principle of solar cells on Figure 1.

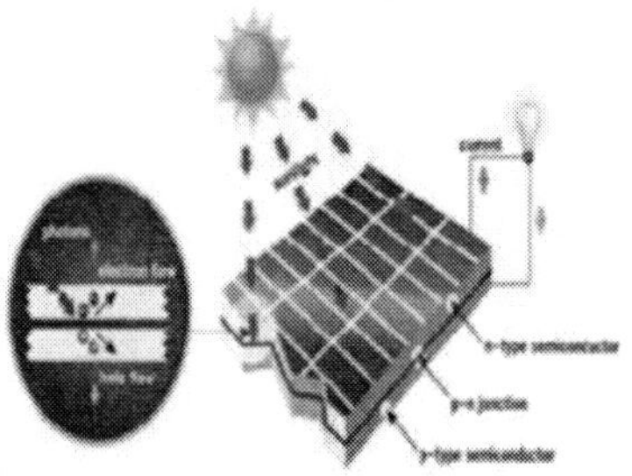

Figure 1. Solar Cell Principle

The condition of excess electrons and holes can occur by doping the material with dopant atoms. When the p-type and n-type semiconductors are in contact, the excess electrons will move from the n-type semiconductor to the p-type so as to form a positive pole on the n-type semiconductor, and vice versa the negative pole on the p-type semiconductor. As a result of the flow of electrons and holes, an electric field is formed which when sunlight hits the p-n junction arrangement it will push electrons to move from the semiconductor to the negative contact, which is then used as electricity, and vice versa the hole moves towards positive contact waiting for the electrons to come.

$$P_{out} = V_{oc} \times I_{sc} \text{ x FF} \quad (1)$$

$$F = \frac{Voc - \ln(Voc + 0{,}72)}{Voc + 1} \quad (2)$$

$$P_{in} = G \times A \quad (3)$$

$$\eta = \frac{\quad}{Pin}\ x\,100\% \quad (4)$$

$$\eta = \frac{Voc\ xIsc}{G\,x\,A}\ x\,100\% \quad (5)$$

Explanation : P_{out} - solar panel output power (Watt)
P_{in} - solar panel input power (Watt)
η - solar panel efficiency (%)
V_{oc} - open network voltage (V)
I_{sc} - short circuit current (A)
G - solar radiation intensity (W/m²)
A - solar panel cross-sectional area (m²)
FF - fill factor based on solar panel material

2.2. Factors that Affecting the Performance of Solar Panels

The main factors that affect the performance of the solar cell panels are [6]:

1. Materials for making solar cell
   Solar panels consist of solar cells which are arranged in series and parallel. The materials for making solar cells also vary, such as [8]:
   a. Crystalline as Cells
   b. Thin film
   c. Multijunction cells
2. Load Resistance
   The battery voltage is the operating voltage of the solar cell panel module, if the battery is directly connected to the solar cell module.
3. Sun light Intensity
   The greater the proportion of sunlight will produce a large current, but moving downwards which indicates a decrease in current and power.

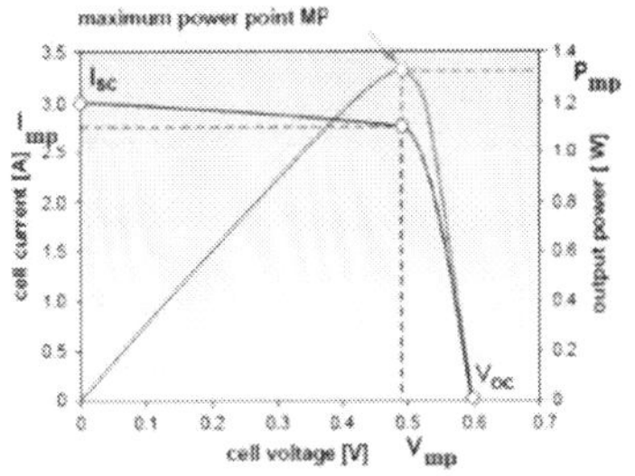

Figure 2. The relationship between solar radiation intensity and output power

4. Orientation and Tilt of panels or PV Arrays The orientation of the PV series (array) towards the sun is optimally important so that the PV panels / series can produce maximum energy.

2.3. Solar tracker

Solar Tracker is a combination of a system that is capable to detect and follow the direction of the sun in order to maximize the reception of solar energy. This solar tracker will be applied to photovoltaic or solar panels. The purpose of providing tracking on photovoltaic is in order to optimize the output of PV. The more perpendicular the PV is to the sun, the greater the output power generated every day. The electrical energy produced will increase when compared to static solar panels. The solar tracker consists of several important components such as sensors, controllers, motors, batteries, and photovoltaics.

2.4. LDR (Light Dependent Resistant)

LDR is an electronic component that has a resistance that can change according to changes in light

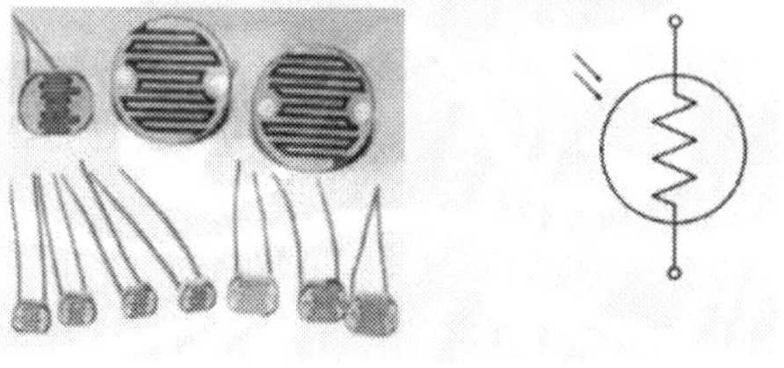

Figure 3. Construction form and schematic drawing of LDR sensor

2.5. Microcontroller STM32 ARM Cortex-M

STM32 ARN Cortex-M is a microcontroller produced by STMicroelectronics with a 32-bit RISC (Reduced Instruction Set Computer) processor. The STM32 microcontroller was developed by the Advanced RISC Machine (ARM) which was formerly known as the Acorn RISC Machine. There are several series of ARM Cortex processors from those that have simple functions to currently widely used in smartphones ranging from the smallest series of ARM Cortex-M, ARM Cortex-R to ARM Cortex-A. There are also ARM secure core and ARM Machine Learning series. The RISC architecture consists of the same large register file, loading and saving architecture, simple addressing modes and the same length instruction fields. Because of these characteristics ARM has high performance, simple code, low power consumption and small size. ARM can execute an instruction set called Thumb which consists of 32-bit to 16-bit instructions. Therefore, 16-bit and 32-bit instructions can be combined without affecting processor performance.

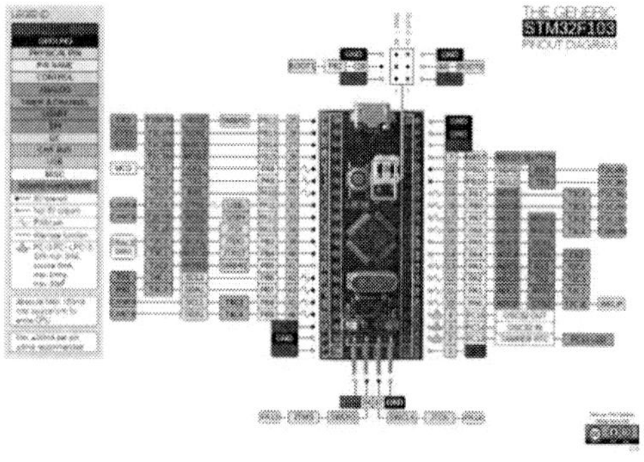

Figure 4. STM32 construction

2.6. *Linear actuator*
By controlling the voltage level, it is also possible to control the position and speed of the linear actuator in Figure 5 [15].

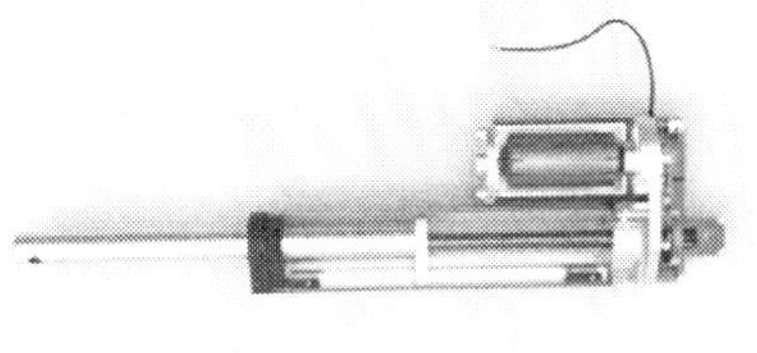

Figure 5. Linear actuator construction

2.7. *Solar charge controller*
Solar Charge Controller is an electronic circuit that regulates the process of charging the battery or battery circuit. The main component in a solar charge controller is the Buck Regulator.

In buck regulators, the average output voltage Va, is less than the input voltage-Vs., so the name "buck" has become a popular regulator. The circuit diagram of a buck regulator using a BJT is shown in Figure 6.

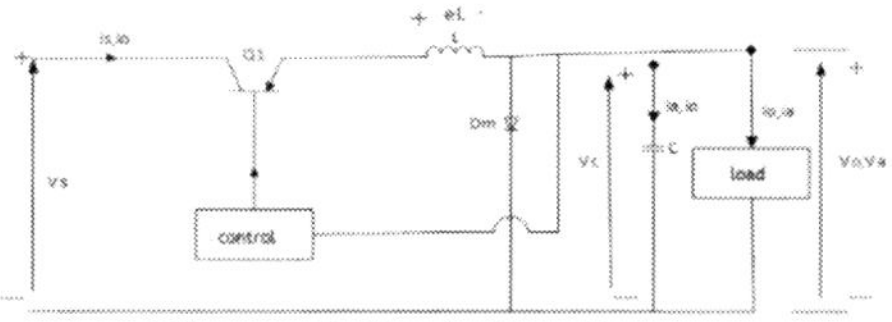

Figure 6. Buck Regulator

2.8. *Batteries*
Batteries are devices that contain electrical cells that can store electrical energy that can store energy that can be converted into power. Batteries generate electricity through chemical processes on Figure 7.

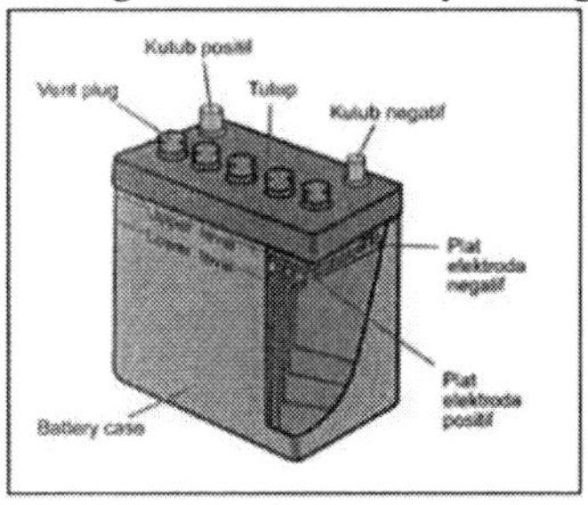

Figure 7. Batteries Construction

3. Research Methods
In this research that aims to produce a battery charging system using Solar Panel Tracking with the STM32 ARM Cortex-M microcontroller, several stages of the research method are used as follows with the research flow shown in Figure 8.

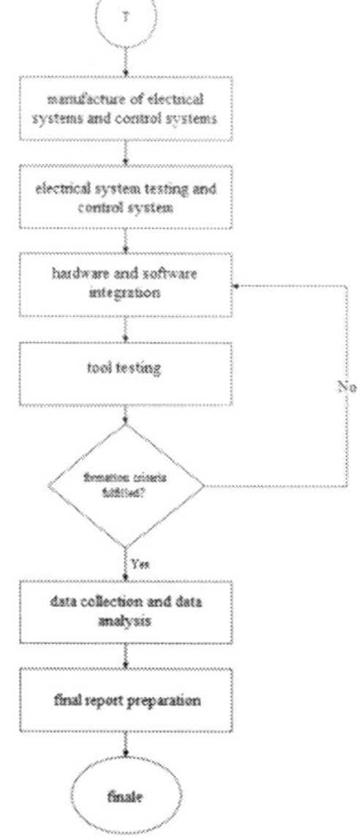

Figure 8. Research Path

3.1. Design of Solar Tracking and Battery Charging System

The equipment used in the design are:

a. The microcontroller circuit used is the STM32 ARM Cortex-M. Microcontroller
b. Using 2 Linear Actuators as the main driver of the solar tracker.
c. The main energy source is the Solar Panel type SP-80-P36 with a maximum power of 80-Watt peak (WP). Watt Peak is the highest unit of power that can be produced by solar panels under standard conditions, namely radiation (E) equal to 1000 W/m2, air mass (AM) equal to 1.5 and temperature 25°C.
d. The study was conducted using lead acid batteries with specifications 12V, 7AH.
e. The Solar Charge Controller used uses the principle of the Buck DC-DC Converter electronic circuit.
f. The research was conducted on the 4th floor of the Department of Electrical Engineering, University of North Sumatra.
g. Solar tracker design is made in accordance with the block diagram that can be seen on Figure 9.

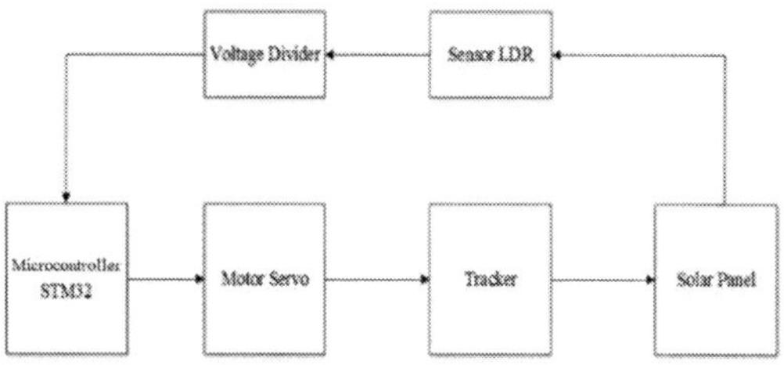

Figure 9. Tool Design Diagram Block

IOP Conf. Series: Earth and Environmental Science 1115 (2022) 012078 doi:10.1088/1755-1315/1115/1/012078

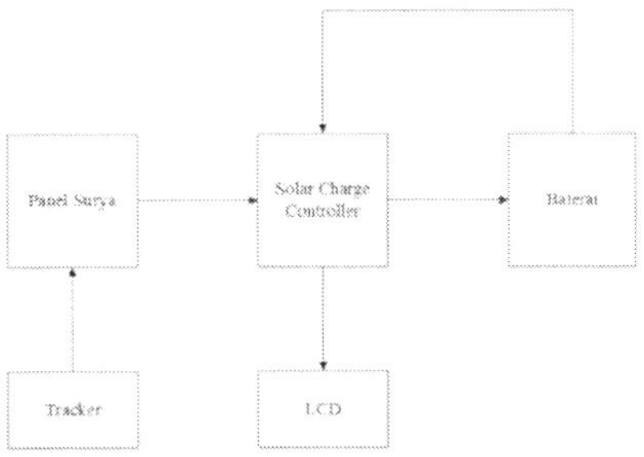

Figure 10. Battery charging system block diagram

Data retrieval in the form of current, voltage, solar panel output power and battery-charged capacity can be done when the tool has been completed.

3.2. Arduino IDE program development

At this stage, a program is made that will be entered on the STM32 microcontroller which will control the entire system of the tool automatically.

The procedure in programming in the IDE is to determine the program algorithm which includes how the tool will work, sketch the program, compile it, choose the communication path, and upload the program. The procedure carried out is as follows on Figure 11 and 12.

1. Writing program instructions
2. Program Saving
3. Installing type board STM32
4. Installing driver ST-Link
5. the type of board to the system used by clicking Tools – Board – "Generic STM32F1 series"
6. Set port on ST-Link option
7. Perform compilation (verify) to check the results of the program
8. Upload the program

Figure 11. Arduino IDE program development

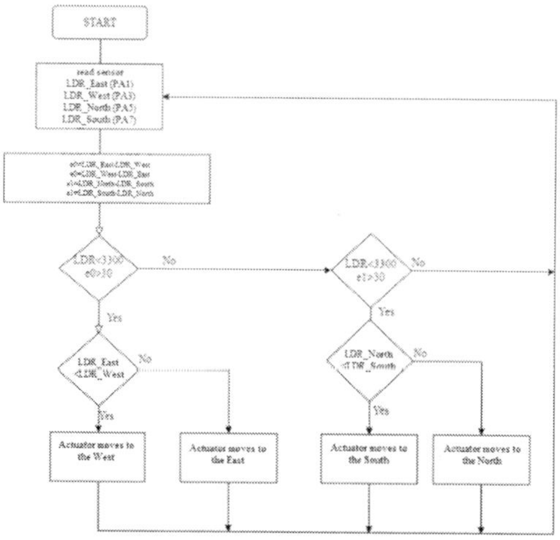

Figure 12. Flowchart solar tracker

3.3. Mechanical system manufacturing

At this stage, the authors utilize the existing solar tracker framework in the Power System Analysis Laboratory and then modify it to fit the expected tracker working principle. The mechanical framework of this system can be seen in Figure 13.

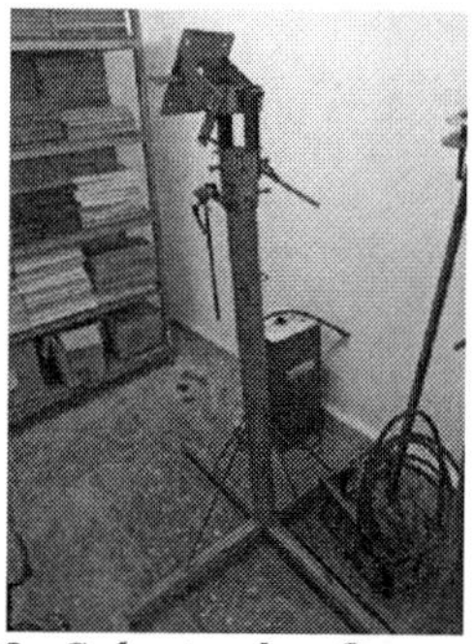

Figure 13. Solar tracker framework

The mechanical frame is then connected to two hydraulic dc motors, each of which moves on the azimuth and zenith axes as shown in Figure 14.

Figure 14. Solar tracker framework after modification

IOP Conf. Series: Earth and Environmental Science 1115 (2022) 012078 doi:10.1088/1755-1315/1115/1/012078

3.4. Electrical system manufacturing

The electrical system is needed to connect the controller and actuator so that the system can work properly. There are several electrical systems used in the manufacture of this solar tracker, including:

1. Power supply
2. LDR Sensor Circuit
3. Signal Conditioning Circuit
4. DC motor drivers

The workflow of the electrical system can be seen in Figure 15.

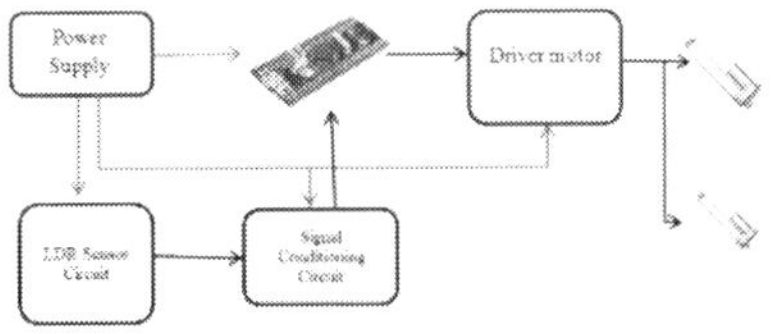

Figure 15. Solar tracker electrical system

The working principle of this system is to compare the readings of the four LDR voltage values. The LDR sensor is a type of resistor whose resistance value depends on the intensity of the light it receives. The resistance value will decrease when the light is bright and the resistance value will be high when it is dark. In other words, the LDR sensor functions to deliver electric current if it receives a certain amount of light intensity and inhibits electric current in dark conditions. The LDR sensor will read the resistance value based on the intensity of the sunlight it receives. Then the microcontroller will read the voltage value on the resistance (RV1.RV2, RV3, RV4) which functions as a voltage divider that is connected in series with the LDR sensor (resistance source). This voltage is used as a microcontroller input. Microcontroller is a semiconductor component in which there is a microprocessor system such as the Arithmetic and Logic Unit (ALU), memory, and input/output devices. To be able to run commands, the microprocessor requires code or machine language code that is used as a medium of communication between the programmer and the machine. After being understood by the machine through the program compile process, the microcontroller will process the input voltage and compare the input voltage value according to the program that has been made. However, the microcontroller cannot directly drive the motor, therefore a DC motor driver is used to take over the task of the microcontroller as a motor drive. The output of the microcontroller which is the input for the driver is in the form of signals 0 and 1. The motor driver is an electronic circuit that functions to supply the current used by an electronic device. The driver circuit used is type L298 with an H-bridge circuit. It is called an H bridge because the principle of disconnecting and connecting an electric motor with four transistors that resembles the letter H. The motor driver works based on the input signal given by the microcontroller, namely 0 and 1. If the motor driver input 1 is 1 and input 2 is 0, then actuator 1 will move vertically up, but if input 1 is 0 and input 2 is 1, the actuator will move vertically down. Likewise for the actuator 2 input settings. Thus, the LDR sensor will continuously track the greatest intensity of sunlight until it reaches a condition where the 4 microcontroller input voltages have the same value. If the values of the four sensors are the same, it can be concluded that the solar panel is in a position perpendicular to the sun and the motor will stop moving. Linear actuators are positioned to regulate 2 different directions. Actuator 1 is to regulate the vertical movement of the solar panels and Actuator 2 is to regulate the horizontal movement of the solar panels.

The illustration of the movement of this system is depicted when the sun is in position, then actuator 1 will move up vertically until it reaches its setpoint. After that, actuator 2 will move to adjust horizontally until it reaches the four LDR voltages of the same value.

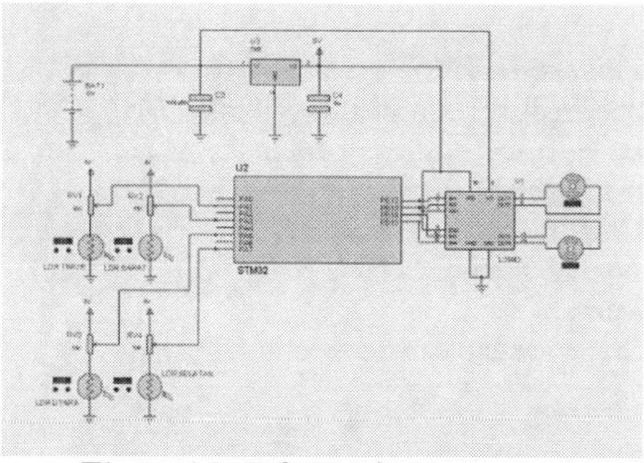

Figure 16. Solar tracker system

4. Results and Discussion

4.1. Static Solar Panel Test

This test is carried out to observe the output of the Solar Panel in the form of DC voltage and DC current, while the power is obtained by multiplying the current and voltage. At the time of testing, measurements were also made of the intensity of solar radiation, ambient temperature, and the intensity of sunlight. The test is carried out in a state of the solar panel parallel to the ground surface. Table 4.1 shows data collection in the form of solar panel output currents and voltages.

Table 1. Result

JAM (WIB)	TEGANGAN (V)	ARUS (A)	DAYA (W)	SUHU (°C)	INTENSITAS RADIASI MATAHARI (W/m2)	INTENSITAS CAHAYA MATAHARI (Lux)
10.30	14.08	2	37	30.42	155.6	42000
10.45	14.10	2.08	29.23	31.74	479.4	44000
11.00	14.01	1.22	17.09	31.64	529.4	35600
11.15	13.76	1.02	45.19	31.92	501.9	43000
11.30	13.92	1.55	32.95	31.66	168.1	37200
11.45	15.83	2.3	19.67	32.38	204.4	15900
12.00	15.08	2.08	39.28	32.92	628.1	43100
12.15	14.49	1.02	22.41	31.89	155.6	14000
12.30	14.47	1.02	22.72	32.61	511.9	22100
12.45	16.27	0.89	22.28	32.10	288.1	26500
13.00	17.37	0.78	26.17	32.43	393.1	34800
13.15	19.64	0.66	22.09	33.44	203.1	14100
13.30	19.20	0.67	14.11	32.67	196.9	15900
13.45	19.36	0.73	21.98	31.64	133.1	8000
14.00	17.75	1.1	14.43	32.20	196.9	11200
14.15	17.29	1.01	12.70	31.79	143.1	9000
14.30	17.66	1.02	13.39	31.79	168.1	12200
14.45	18.05	1.1	10.16	31.26	69.4	5000
15.00	14.27	0.50	7.14	31.23	58.1	7000

Based on Table 1, it is found a graph of changes in the intensity of solar radiation against time at the time of data collection.

Data regarding ambient temperature, voltage and current output of solar panels. The graph of the relationship and influence of temperature on the current and output voltage of the solar panel can be seen in Figure 17 and 18.

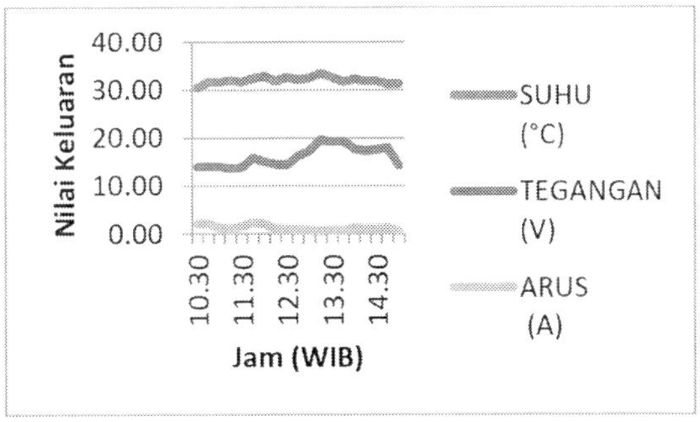

Figure 17. Temperature, voltage and current graph

From the observations, the current and voltage obtained from the solar panel output, which means we can calculate the solar panel output power. The solar panel output power data table is shown in Table 1.

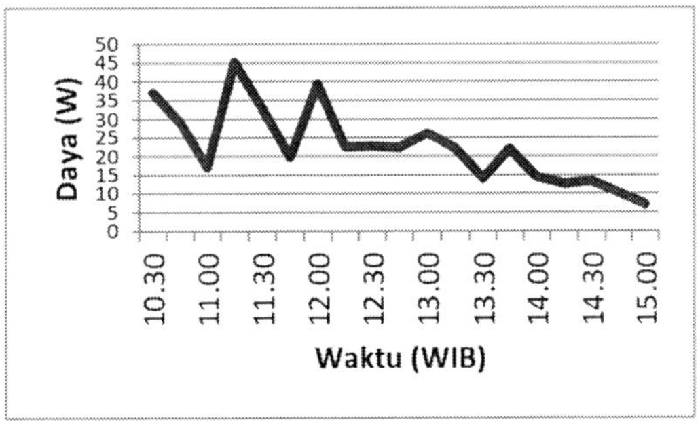

Figure 18, Power-to-time graph

4.2. Testing with solar tracker

The results of the tests and observations can be seen in Table 2.

Table 2. Results of the tests with solar tracker

WAKTU (WIB)	TEGANGAN (V)	ARUS (A)	DAYA (W)	Intensitas Radiasi Matahari (W/m2)	SUHU (°C)	INTENSITAS MATAHARI (Lux)
10.30	17.63	2.61	46.01	546.9	31.74	44600
10.45	18.35	1.88	14.21	238.1	31.84	43400
11.00	14.10	2.07	29.19	621.9	32.02	41800
11.15	14.85	3.13	46.48	369.4	31.33	43900
11.30	15.56	4.15	64.57	616.9	32.59	41200
11.45	13.84	1.02	51.82	145.6	31.20	13200
12.00	16.95	3.11	52.71	143.1	32.00	48600
12.15	14.23	1.07	15.23	564.4	32.67	12200
12.30	14.80	1.88	27.82	159.4	32.90	19500
12.45	15.49	1.74	26.95	235.6	31.89	21300
13.00	16.28	1.56	25.40	168.1	31.59	34200
13.15	15.2	1.47	22.34	203.1	32.85	14100
13.30	17.09	1.42	24.27	249.4	32.05	14400
13.45	15.57	1.01	15.73	199.4	32.07	7200
14.00	17.36	1.04	18.05	153.1	31.82	14000
14.15	17.03	1.02	17.37	148.1	31.89	11000
14.30	16.46	1.02	16.79	173.1	32.33	12300
14.45	18.42	1.55	28.55	188.1	31.82	5000
15.00	14.07	0.49	6.89	121.9	31.26	4000

Table 2 above shows the data taken with the solar panel tracking system. Testing is carried out from 10.30WIB-15.00WIB on the 4th floor of the Electrical Engineering Masters building.

In Table 2, the ambient temperature, voltage and output current of the solar panel have been obtained using a tracking system. Thus, the output power generated by the solar panels can also be calculated. The graph of changes in temperature, current and voltage can be seen in Figure 19.

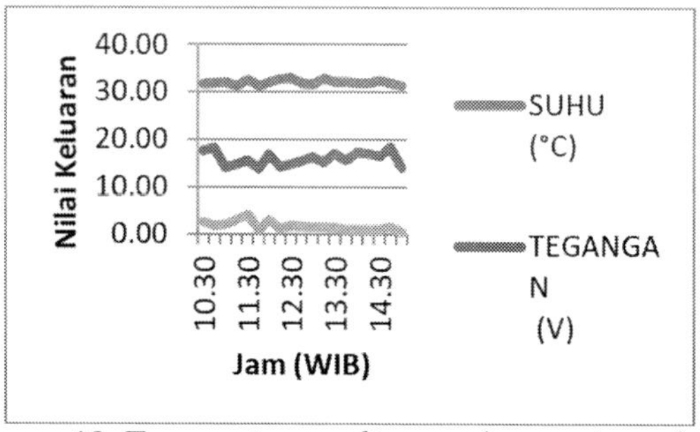

Figure 19. Temperature, voltage and current graph

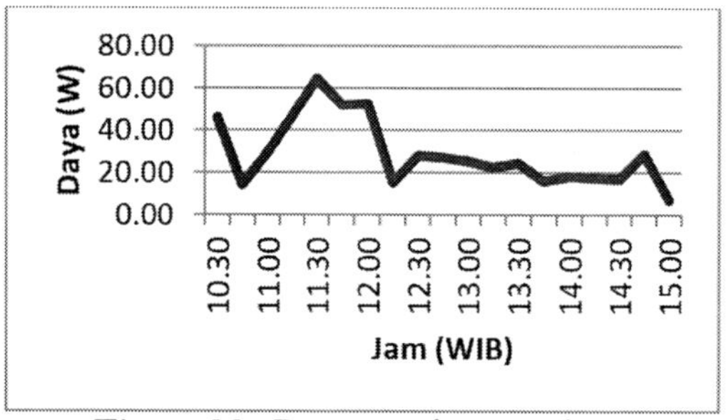

Figure 20. Power-to-time graph

Figure 19 shows that at 10.30 WIB - 10.45 WIB there is a decrease in the output power of the solar panels.

4.3. Power comparison of static tracker and tracking system

After the intensity of sunlight received by the solar panel, the output power of the solar panel must also be compared to determine the efficiency of using a solar tracker based on the output power of the solar panel. Tracker power data can be seen in Table 3.

Table 3. Comparison of static solar panel power and using a solar tracker

WAKTU (WIB)	DAYA STATIC (W)	DAYA TRACKER (W)
10:30	37	46.01
10:45	29.33	14.21
11:00	17.09	29.19
11:15	45.19	46.48
11:30	32.95	64.57
11:45	19.67	51.82
12:00	39.28	52.71
12:15	22.41	15.23
12:30	22.72	27.82
12:45	22.28	26.95
13:00	26.17	25.40
13:15	22.09	22.34
13:30	14.11	24.27
13:45	21.98	15.73
14:00	14.43	18.05
14:15	12.70	17.37
14:30	13.39	16.79
14:45	10.16	28.55
15:00	7.14	6.89
Daya rata-rata	22.63	28.96

4.4. Comparison of battery charging time

From Table 4, it can be seen that using a tracking system can save 1 hour 40 minutes of time to charge a 12 V, 7Ah lead acid battery. The time efficiency can be calculated in the following way.

Table 4. Comparison of Time

WAKTU (WIB)	KAPASITAS TERISI BATERAI (Ah) TANPA TRACKER	KAPASITAS TERISI BATERAI (Ah) MENGGUNAKAN TRACKER
10.30	0.00	0.00
10.45	0.35	0.90
11.00	0.83	1.63
11.15	1.10	1.98
11.30	1.47	2.83
11.45	2.01	3.12
12.00	2.50	3.77
12.15	2.92	4.22
12.30	3.15	4.77
12.45	3.42	5.22
13.00	3.61	5.65
13.15	3.80	6.12
13.30	3.95	6.42
13.45	4.08	6.73
14.00	4.39	7.00
14.15	4.66	7.00
14.30	4.93	7.00
14.45	5.18	7.00
15.00	5.56	7.00
15.40	7.00	7.00

$$time\ efficiency = \left|\frac{charging\ time\ tracker - charging\ time\ static}{charging\ time\ static}\right| x100\% \tag{6}$$

$$time\ efficiency = \left|\frac{210\ minute - 310\ minute}{310\ minute}\right| x100\% \tag{7}$$

$$time\ efficiency = 32{,}25$$

5. Conclusion

Based on the results of the study, it could be concluded that the design of the solar tracker system using the STM32 microcontroller with LDR sensor has been successfully carried out with the average power produced by a static solar panel is 22.63 W, while the average power generated by a solar panel using a solar tracker is 28.96 W and the increase in power generated by solar panels with a solar tracking system is 27.97%. While charging the 12V,7Ah lead acid battery takes 5 hours 10 minutes to fully charge 7Ah using a static solar panel, and using a solar panel with a tracking system the battery can be fully charged for 3 hours 30 minutes. So, the use of solar panels with a tracking system saves 1 hour 40 minutes or

32.25% of the time required for a static solar panel to fully charge a 12V,7Ah lead acid battery. Based on the conclusions, it can be suggested to pay attention to the placement of the solar tracker, because the placement of the solar tracker greatly affects the movement of the tracker and for development, it is very important to use the Internet of things to monitor the movement of the solar tracker.

References

[1] Dannes K, Ferdyan and Muhammad R 2018 Jurnal Teknik ITS **7** 319
[2] Asparuhova, Katya and Ivan 2018 R*esearch and Design of Effective Positioning Algorithm for Solar Tracking System* (IEEE)
[3] Mustafa, Falah I, Al-Ammri A S and Farouk F A 2017 IEEE **17**
[4] Ashari and Hasim S T M 2018 *ARM CORTEX-M Sebagai Media Pembelajaran Mikrokontroler (*Fakultas Teknik Universitas Negeri Yogyakarta: Yogyakarta)
[5] Duffie, John A and Beckman A W 2013 *Solar Engineering Of Thermal Processes Four edition* (New Jersey:John Wiley & Sons,Inc)
[6] Patel and Mukund R 1999 *Wind and Solar Power System US Merchant Marine Academy Kings Point* (CRC Press New York USA)
[7] Gonen and Turan 2014 *Electric Power Distribution Engineering 3th* (Taylor and Francis Group : CRC Press,hal 8202014)
[8] Ross and Carolyn 2009 *Solar Electric System Design, Operational and Installation* (The Energy Trust of OregeonInc: USA)
[9] Green M A 1982 *Solar cells Operating Principles, Technology and System Application* (New Jersey: Prentice-Hall 1982)
[10] Zuhal 1990 *Dasar Teknik Tenaga Listrik Dan Elektronika Daya* (Jakarta: Penerbit PT Gramedia)
[11] Rasyid and Muhammad H 1999 *Elektronika Daya Edisi Bahasa Indonesia Jilid 1* (Jakarta: PT Prenhallindo)
[12] Santoso and Elvia H 2014 *Rancang Bangun Solar tracking system menggunakan kontrol PID pada sumbu Azimuth* (Fakultas Teknologi Industri Institut Teknologi Sepuluh November Surabaya)
[13] Mutaqqin and Rus 2017 *Analisa Performansi dan Monitoring Pembangkit Listrik Tenaga Surya* (Departemen Teknik Fisika FTI-ITS Fakultas Teknologi Industri Institut Teknologi Sepuluh November Surabaya)
[14] Tanaya M H Y and Nityasa 2016 *Rancang Bangun Solar Tracker Dual Axis guna Optimalisasi Panel Surya untuk Penerangan Kapal* (Fakultas Teknologi Kelautan Institut Teknologi Sepuluh November Surabaya)
[15] Ruiz-Rojaz, Edrey D and Varquez Gonzalez *Mathematical Model of a Linear Actuator with Prosthesis Aplications* IEEE200

Growth response of black soybean (*Glycine max* L. Merril) with a multi canopy cropping system on measurement chlorophyll and leaf area index

R Rahmadina[1], I Nurwahyuni[1*], E Elimasni[1] and D S Hanafiah[2]

[1]Department of Biology, Faculty of Mathematics and Natural Sciences, Universitas Sumatera Utara, Medan, North Sumatra 20155, Indonesia
[2]Department of Agriculture, Faculty of Agriculture, Universitas Sumatera Utara, Medan, North Sumatra 20154, Indonesia.

*Email: isnaini@usu.ac.id

Abstract. Black soybean is one of the sources of food that is needed by the people of Indonesia, but its existence is very limited due to the limited area of production that is suitable for its growth so it results in less than optimal production. This study aims to develop new innovations in soybean planting systems by looking at their vegetative growth in terms of the amount of chlorophyll and leaf area index and whether or not it affects vegetative growth. The method in this study used randomized block design with 3 factorials consisting of 4 different cultivars, spacing, and pruning. The parameters measured in this study were chlorophyll a, chlorophyll b, total chlorophyll ratio, and leaf area index. The results obtained on the leaf area index found in the cultivar Detam 4 Prida (K4P2J1) with a total of 121, 33 cm while the amount of chlorophyll in the cultivar Detam 4 Prida (K4P1J1) was 6.37 mol/m2. This study concludes that each treatment is very influential on the growth of soybean plants with a multi-canopy planting system, especially on the cultivar type and spacing.

1. Introduction

Soybean (*Glycine max* (L.) Merril as a food source that has a high protein content as a raw material is classified as the third category after rice and corn as a food source, so that the need is always increasing every year. The existence of this soybean is one of the daily consumption for the people of Indonesia because it has a delicious and savory taste and has considerable benefits for health. However, the need is not balanced with the resulting production because in the last five years there has been a decline in production from 2018 to 2022. This can be seen from the high level of imports made from abroad to reach 47.77% in 2021. Soybean imports are the largest carried out in the United States with a total of 86 [1].

Soybean production in Indonesia is still relatively low when compared to other countries, so efforts are needed to increase production by increasing the production harvest area through an increase in the planting system combined with the verticulture system which is an innovation from the monoculture planting system on several different cultivars in one field. Which is known as a multi-canopy cropping system. The soybeans used in this cropping system are black soybeans Detam 1, Mutiara 2, Detam 3 prida, and Detam 4 prida. Black soybeans have better content than yellow soybeans because they have 6 times higher flavonoid content and 15 times higher antioxidants than yellow soybean [2].

Multi-canopy planting system is a system designed vertically with a mixture of planting on two cultivars that have different strata of crown height and are mutually compatible in the same land plot. The use of different cultivars in the same land must be adjusted to the spacing, fertilization system, good pest control so that the expected results and can inhibit the formation of new race variants in pests and diseases. This system can also increase the growth of the soybean plant which is measured based on the physiological properties of the plant which can be seen in the chlorophyll content and leaf area index formed from the results of its growth [3,4].

The content of chlorophyll contained in a plant is one of the main components in the formation of the rate of photosynthesis. If the chlorophyll content increases, the rate of photosynthesis will also increase. This increase in chlorophyll can be done by giving water and nutrients to plants as needed because if there is a reduction in water use, there will be a decrease in chlorophyll concentration which can inhibit the formation of chlorophyll, nutrients, nitrogen, and magnesium hormones that play a role in chlorophyll synthesis [5].

Chlorophyll content can also increase when the leaf area increases and this occurs if the plant is in shaded conditions. This increase in leaf area can occur to form even more trifoliate leaves, but can reduce the ratio of the leaf area. This happens because plants with multi-canopy systems have crowns that are mutually shaded so that the leaves expand their surface area in absorbing light and light capture efficiency by increasing the area per unit of light captured and reducing the number of leaves in balancing the limited amount of light. the occurrence of a balance that occurs in the photosynthesis process so that it can increase the rate of photosynthesis [6].

Based on this explanation, this study aimed to determine the effect of the multi-canopy planting system on several different cultivars of black soybean on the chlorophyll content and leaf area index.

2. Materials and Method

This research was conducted on Batangkuis Street, Deli Serdang Regency, North Sumatra from April to June 2022. This study used a randomized block design (RAK) with 3 factorials consisting of 4 different cultivars (K1=Detam 1;K2= Mutiara 2; K3=Detam 3 Prida; K4=Detam 4 Prida), spacing (J1= 20X40cm; J2=20X20cm), and pruning (P1=No Pruning, P1=Pruning). The study was conducted using 2 x replications. The parameters measured in this study were chlorophyll a, chlorophyll b, total chlorophyll ratio, and leaf area index. The tools and materials used in this study were black soybean cultivars: Detam 1, Mutiara 2, Detam 3 Prida, Detam 4 Prida, KCL fertilizer, urea, compost, SP-36, @GDM fertilizer, Antracol 70 WP, water, label paper , plastic ropes, hoes, stakes, water pumps. This research was conducted with a quantitative descriptive method that uses direct field research and quantitatively tested the data using the Analysis of variance test and then tested with *Duncan's Multiple Range Test* (DMRT) with a significant 5%.

3. Result and Discussions

3.1. *Research site conditions*

Black soybean (*Glycine max* L.) is one of the plants from the legominosae family that requires suitable environmental conditions for its growth. Black soybeans are quite tolerant of areas with hot temperatures with rainfall of about 100-400 mm3 per month, so it is very good to be planted in dry climates with rainfall between 100-200 mm/month. Soybean growth is very influential on temperature, humidity, and light intensity so that the research location is in the area of Jalan Kayu Besar, Kecamatan Batang Kuis, North Sumatra Province, Indonesia is very suitable for the growth of these plants because it has a rainfall of about 150-250 mm/month, a temperature of about 300-350, with a humidity of about 70%-77%, a soil pH of about 5-6.5, and light intensity under normal conditions and has a shade that is not too high so that it can affect the growth of these plants. Therefore, this location is very suitable to be used as an object of research.

3.2. Chlorophyll content

Leaf chlorophyll content was measured based on the level of greenness of the leaves that had been carried out at 4 WAP, where at this time the soybean leaves of each cultivar had a different greenish level. Each cultivar has a high level of greenness of the leaves so that it is very influential on the growth of the soybean. The high level of greenness of the leaves resulted in the high amount of chlorophyll produced. This is because the green color of the leaves is thought to be influenced by the dry weight of the root nodules and also the N-fixing bacteria that can fix N2 from the air and the very high uptake of N and Mg nutrients in these plants so that it can increase the availability of very important nitrogen nutrients. in the formation of chlorophyll [7].

The high amount of chlorophyll in each cultivar had an insignificant effect on several treatments such as spacing and pruning while for some other cultivars such as Mutiara 2 and Detam 4 Prida had a significant level with a significant difference in these treatments, where in each - each spacing greatly affects the formation of chlorophyll while the pruning treatment has no effect on the formation of the chlorophyll. In the treatment of close plant spacing with a size of 20 x 20 cm, it greatly affects the formation of chlorophyll in Mutiara 2 cultivar with a chlorophyll amount of 3.04 mol/m2, while in Detam 4 Prida it affects the normal spacing of 20 x 40 cm with a total chlorophyll of 6.37. mol/m^2. Observation data. The content of chlorophyll based on the results of research on black soybean plants can be seen in table 1 below:

Table 1. Chlorophyll total

Duncana

Treatment	*N*	Subset for alpha = 0.05 1	 2
K2P1J2	3	3.0367	
K4P2J2	3	3.4033	3.4033
K2P2J1	3	3.5667	3.5667
K1P1J2	3	3.6700	3.6700
K1P2J2	3	3.7967	3.7967
K4P1J2	3	3.9000	3.9000
K2P2J2	3	4.0067	4.0067
K4P2J1	3	4.1433	4.1433
K1P1J1	3	4.1900	4.1900
K2P1J1	3	4.2967	4.2967
K3P1J1	3	4.3967	4.3967
K1P2J1	3	4.9200	4.9200
K3P2J1	3	5.1700	5.1700
K3P1J2	3	5.1967	5.1967
K3P2J2	3	5.5733	5.5733
K4P1J1	3		6.3367
Sig.		.111	.067

Means for groups in homogeneous subsets are displayed.
a. Uses Harmonic Mean Sample Size = 3,000.

Based on the results of the study, it was found that plants in the highest position or the edge in the planting position received more sun exposure so that they could increase the amount of chlorophyll content when compared to the shaded position or the deepest part of a plant. Photoreduction of protochlorophyllide to chlorophyll a and proceeds to esterification of phytol to form a pigment chlorophyll a through the catalysis of the enzyme chlorophyllase. This change must use sunlight so that the formation of chlorophyll can occur in large quantities if the plant is exposed to a high enough amount, but this amount of sunlight if it occurs too high can actually damage the chlorophyll content and enzymes that play a role in it [8,9].

3.3. *Leaf area index*

Based on the results of the research that has been done, which can be seen in Table 2, the results show that the growth of length and width on each leaf with different cultivars gives a very significant leaf area index. It can be seen that there is a significant difference in the length, width, and index of the leaf area measured at 4 WAP. Based on these results, it can be seen in Table 2 that the highest leaf area index was found in the Detam 4 Prida cultivar with a value of 121.33 cm.

Based on these results, it can be seen that a plant will have a different reaction if the light intensity it receives is also different, this is related to the presence of shade between the heigh of the plant with different cultivars because it has a tight spacing. A plant will have two ways of adapting to its environment through the respiration density that needs to be reduced to maintain carbon balance as well as compensation points and increasing leaf area index in reducing the secondary metabolites used [10, 11]. This is related to the weight of the seeds which can be heavier if they have a large enough leaf area, because the leaf surface is wide enough to receive the intensity of sunlight so that more photosynthate is produced so that it can produce more seed weight and number of seeds. Vice versa if the number and size of the leaves are not broad or inappropriate, the CO_2 transport process from the leaves will not work well because the photosynthesis process and the amount of carbohydrates produced as food reserves will decrease so that the results that accumulate to the weight and number of seeds will experience a reduction [12,13].

Table 2. Space leaf

Duncana

		Subset for alpha = 0.05
Treatment	*N*	1
K3P1J2	3	55.3333
K2P1J1	3	62.6667
K1P1J1	3	64.0000
K1P2J2	3	70.6667
K1P2J1	3	71.3333
K1P1J2	3	77.3333
K4P2J2	3	82.6667
K2P2J2	3	83.3333
K2P1J2	3	84.6667
K2P2J1	3	88.6667
K3P1J1	3	89.3333
K4P1J2	3	91.3333
K3P2J1	3	96.6667
K4P1J1	3	99.3333
K3P2J2	3	102.0000
K4P2J1	3	121.3333
Sig.		.292

Means for groups in homogeneous subsets are displayed.
a. Uses Harmonic Mean Sample Size = 3,000.

Based on the results of research from several treatments that have been carried out on several parameters related to soybean plant growth from several cultivars, it was found that there was a significant effect between different cultivar treatments, plant spacing and pruning on plant growth parameters both the amount of chlorophyll and leaf area index. This is because each cultivar has a different height so that it is mutually shaded between plants coupled with a fairly close spacing so that plant growth adapts to each other to get sunlight, one of which is on the leaves by expanding the leaf index so that the leaf surface can make it easier to get sunlight and it facilitates the process of photosynthesis so that it can produce a lot of chlorophyll.

This is also supported by appropriate environmental factors such as rainfall that is not too high, humidity with an amount of 70-77 %, pH 4.5-6.5, and normal light intensity to facilitate plant growth. This is also related to the absorption of nutrients from the soil in the *N* element and also from plants which are very high so that it supports the photosynthesis process. This is also related to the absorption of nutrients from the soil in the *N* element and also from plants which are very high so that it supports the photosynthesis process. This is also related to the absorption of nutrients from the soil in the N element and also from plants which are very high so that it supports the photosynthesis process (Table 3).

Table 3. ANOVA

		Sum of Squares	*df*	Mean Square	F	Sig.
Chlorophyll a	Between Groups	12,436	15	.829	.786	.683
	Within Groups	33,754	32	1.055		
	Total	46,190	47			
Chlorophyll b	Between Groups	7,452	15	.497	1,139	.365
	Within Groups	13,961	32	.436		
	Total	21,413	47			
Chlorophyll total	Between Groups	34,663	15	2,311	.916	.556
	Within Groups	80,726	32	2,523		
	Total	115,389	47			
Space leaf	Between Groups	12836.583	15	855,772	.211	.999
	Within Groups	129613333	32	4050,417		
	Total	142449,917	47			

4. Conclusion

Based on the results of research and discussion that has been carried out, it is found that the treatment of different cultivars, spacing, and pruning affect the number of leaf area indexes, but the amount of chlorophyll only affects the cultivar and spacing. This is due to the existence of a multi-canopy system that has shading between plants of different heights and a fairly tight spacing so that these plants compete with each other to adapt to environmental factors so that these plants can survive.so that the plant increases its leaf area index. This can cause the amount of chlorophyll produced to be large.

Acknowledgment

This study was funded by the Indonesian Ministry of Education, Culture, Research and Technology under the scheme of Penelitian Disertasi Doktor (PDD) fiscal year of 2022.

IOP Conf. Series: Earth and Environmental Science 1115 (2022) 012079 doi:10.1088/1755-1315/1115/1/012079

Reference

[1] BPS 2021 *Tanaman Pangan* 2019-2021 (Sumatera Utara: Badan Pusat Statistik Provinsi Sumatera Utara).
[2] Fauzan M 2021 *Jurnal Ilmiah Mahasiswa Pertanian* **1** 4
[3] Widyastuti Y P L, Suwarno B W, and Aswidinnoor H 2020 *Agronomy Journal* **112** 6
[4] Hahm J, Kim K, Qiu Y, and Chen M 2020 *Nature Communications* **11** 1660
[5] Nurcahyani E, Apriyanti D, Wahyuningsih S, and Mahfut 2020 *Analytical and Environmental Chemistry* **5** 2
[6] Chairudin C, Efendi E, and Sabaruddin S 2015 *Jurnal Floratek* **10** 1
[7] Sutopo A 2019 *Jurnal Citra Widya Edukasi* **11** 1
[8] Lawendatu G P), Pontoh J, and Kamu S V 2019 *Chemistry Progress* **12** 2
[9] Jiao Q, Sun Q, Zhang B, Huang W, Ye H, Zhang Z, Zhang X, and Qian B 2022 *Remote Sensing* **14** 1
[10] Kinasih E M, Zubaidah S, and Kuswantoro H 2017 *Prosiding Seminar Nasional Pendidikan Sains* **2017**
[11] Verma B, Prasad R, Srivastava K P, Yadav A S, Singh P, and Singh K R 2022 *Computers and Electronics in Agriculture* **192** 2022
[12] Sinuraya A M, Barus A, and Hasanah Y 2015 *Jurnal Agroekoteknologi* **4** 1
[13] Widiastuti E and Latifah E *Jurnal Ilmu Pertanian Indonesia* **21** 2

IOP Conf. Series: Earth and Environmental Science 1115 (2022) 012080 doi:10.1088/1755-1315/1115/1/012080

Occurrence of microbial community on plastic wastes in Terjun Landfill, North Sumatra

E Munir[1,2*], D Suryanto[1], Y Pasaribu[1], S Mubtasima[1], A Hartanto[1], A Lutfia[1] and A F Nasution[1]

[1]Department of Biology, Faculty of Mathematics and Natural Sciences, Universitas Sumatera Utara, Medan, North Sumatra 20155, Indonesia
[2]Center of Excellence for Mangrove, Universitas Sumatera Utara, Medan 20155, Indonesia

*Email: erman@usu.ac.id

Abstract. Plastisphere is a term for any microbial community that grow and develop on the surface of plastic. There was still a lot of plastics that was readily disposed in landfills, which would makes it feasible to discover bacteria that could decompose plastic and be used in bioremediation. The present study aims to determine the number of microbial density (colonies) and obtainining a number of microbial (bacteria, fungi) isolates from four different types of plastics namely low-density polyethylene (LDPE), linear low-density polyethylene (LLDPE), high-density polyethylene (HDPE) and polypropylene (PP). The density of bacterial population was the highest in LDPE, followed by LLDPE, PP, and HDPE. Meanwhile, the density and number of fungal population was exceptionally low in this study. A greater portion of gram-negative bacteria was larger than gram-positive bacteria with some filamentous fungi obtained from the isolation. The visual appearance using SEM analysis reveal the presence of mixed community between bacteria, fungi and yeast population.

1. Introduction

Multiple microbial communities that are found attached to the surface of plastic is referred to as a "plastisphere" in practical term. Plastic transforms into a resilient surface for microbial colonization, encouraging the development of microbial biofilms that are advantageous to the bacteria. The properties of the substrate or the kind of plastic have a significant impact on the developing microbial population [1]. The physical properties of a plastic, such as hydrophobicity, particle shape, roughness, crystallinity and surface charge play a very important role in the selection of microbial communities at the beginning of colonization. However, pioneer microbes can also influence the determination of microbial communities in plastics [2].

There are various types of plastics based on the materials used in production, such as PETE (Polyethylene terephthalates), HDPE (High-density polyethylene), PVC (Polyvinyl chloride), LDPE (Low-density polyethylene), PS (Polystyrene) and so on. These plastics are very potential in polluting the environment that has a negative impact on the health of living things, especially on humans. For instance, the PS-type plastic may, over time, be neutrotoxic, release certain carcinogenic substances into humans in small doses, and have cytogenetic, carcinogenic, and hematological consequences [3]. Different types of plastic products play a significant role in determining the variety of pollution or environmental contamination that occurs. Furthermore, the principal reason of plastic pollution is quite complicated, especially given the numerous human activities that contribute to its accumulation through a variety of pathways [4].

Currently, efforts in biodegradation using microorganisms are one of the solutions in decomposing plastic waste in the environment. Biodegradation is defined as the decomposition of materials mainly by fungi or bacteria. Biodegradation techniques have been extended to break down artificial products, especially plastics [5]. Srikanth et al reported some important basidiomycetous fungi with plastic-degrading ability such as *Agaricus bisporus, Pleurotus abalones, Pleurotus ostreatus*, through production of laccase [6]. Members of *Pseudomonas*, *Enterobacteriaceae*, and *Moraxella* were able to degrade plastic polymers from bacteria group. The biodegradation process involved hydroxylation of C-C bonds to release primary or secondary alcohols which are later oxidized to ketones [7].

Terjun landfill is the final disposal site for all parts of the city which is located in Medan City, North Sumatra. The landfill employs a controlled landfill and sanitary landfill system, where the system takes the form of daily soil cover (sanitary landfill) or at least sometimes (controlled landfill), with a pile of plastic waste thickness of 20–30 cm. This study described the occurrence of bacteria and fungi as a community being associated with plastic wastes that may be useful for further investigation as plastic-degrading microbes.

2. Materials and Method

2.1. Sampling site

Sample collection of plastic waste of various types, i.e. low-density polyethylene (LDPE), linear low-density polyethylene (LLDPE), high-density polyethylene (HDPE) and polypropylene (PP) were obtained from Terjun Landfill, Medan Marelan, Medan City, North Sumatra (Figure 1). The samples were stored in sterile containers and transported to the laboratory for experiments.

Figure 1. Field observation of Terjun Landfill designated as a sampling site for plastic-associated microbial community (bacteria, fungi).

2.2. Isolation and enumeration of microbes associated with plastic wastes

The plastic wastes were cut into smaller fragments (±1 g) and put into a 250-mL flask with each containing 50 mL of nutrient broth (NB) for bacterial enrichment and potato dextrose broth (PDB) for fungal enrichment. The flasks were agitated at 120 rpm and incubated at ambient temperature for 7 days. One mL of each enrichment culture was serially diluted until 10^{-4} [8]. Approximately 0.1 mL from each dilution was spread into nutrient agar (NA) for bacterial growth and potato dextrose agar (PDA) for fungal growth supplemented with antibiotics. The culture media were incubated at ambient temperature from 24 hr (bacteria) until 120 hr (fungi). Number of colonies from the highest dilution represented the population density as expressed in colony-forming unit (CFU) per mL. Different bacterial or fungal colonies were transferred into new medium and characterized qualitatively.

2.3. *Surface visualization of microbial community on a plastic waste*

Surface morphology of the representative plastic waste was visualized by using the Hitachi table top scanning electron microscope (SEM–TM3000, Japan) operated at the Integrated Laboratory of Universitas Sumatera Utara, Medan, Indonesia.

3. Results and Discussion

The viable count of microbial community associated with different types of plastic is presented in Figure 1. The highest number of bacterial colonies was observed in LDPE (550 CFU/mL), followed by LLDPE (375 CFU/mL), PP (312 CFU/mL), and HDPE-type plastics (228 CFU/mL). In addition, the highest number for fungal colonies also followed the similar trend as bacterial population however none of them was recovered from PP-type plastics. In terms of culturable isolates or morphotypes, the highest number was obtained from bacterial group in PP (9 isolates), followed by HDPE (8 isolates), and in both LDPE and LLDPE-type plastics (6 isolates) as presented in Figure 2. In fungal group, the highest number of isolates was obtained from LLDPE (4 isolates) followed by both LDPE and HDPE (2 isolates). In total, we successfully recovered 29 bacterial isolates and 9 fungal isolates from plastic wastes in Terjun landfill.

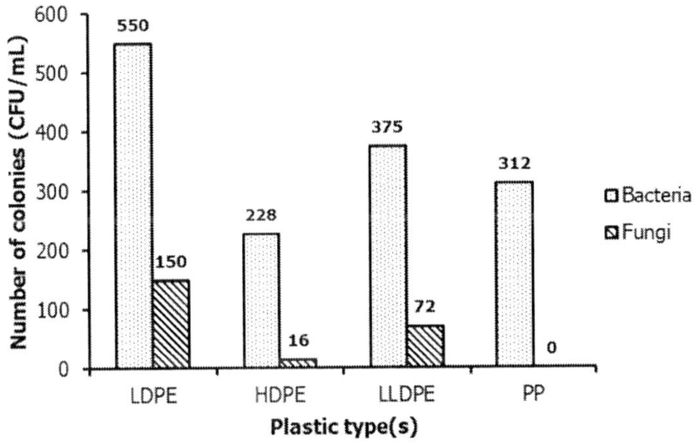

Figure 2. Number of microbial (bacteria, fungi) colonies from aerobic plate count colonizing some types of plastics. LDPE = Low-density polyethylene, HDPE = High-density polyethylene, LLDPE = Linear low-density polyethylene, PP = Polypropylene.

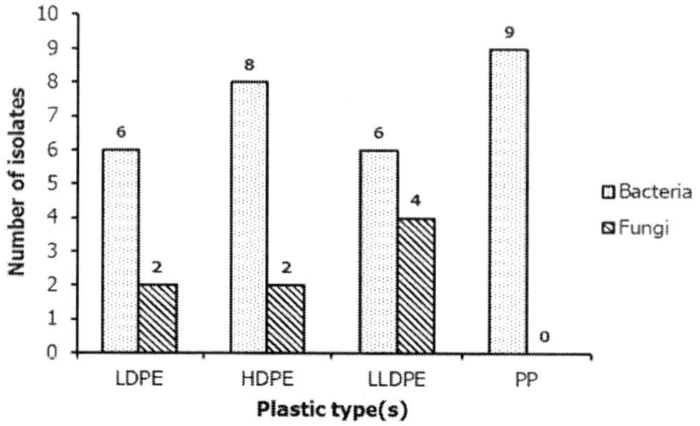

Figure 3. Number of microbial (bacteria, fungi) isolates or morphotypes based on types of plastics. LDPE = Low-density polyethylene, HDPE = High-density polyethylene, LLDPE = Linear low-density polyethylene, PP = Polypropylene.

The occurrence of microbial community on plastic surface may be determined by the types of plastic, the composition of indigenous microbial species, and the selection of isolation media. Most studies recovered a higher number of microbial density and isolates from soil samples within the area of landfill, waste recycling site or natural site rather than from plastic wastes [9-11]. Optimization of isolation medium was also attempted to recover a higher number of isolates from plastic environments. In this study, we used both NA and PDA as general purpose media to support the growth of non-selective group and non-fastidious microorganisms. Ojha et al suggested the use of Czapek-Dox broth supplemented with either HDPE or LDPE as carbon sources in the isolation medium to screen a maximum number of fungal isolates from landfill soils [12]. Other study by Haedar et al employed the isolation medium supplemented with 2% (w/v) of polyethylene glycol (PEG) to screen the indigenous plastic-degrading microbes from soil environments [13].

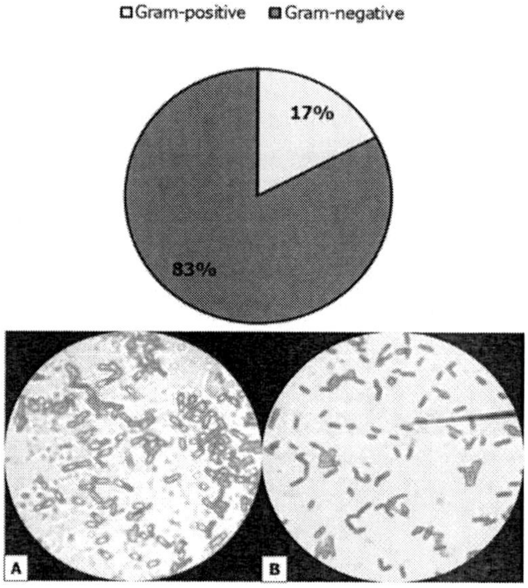

Figure 4. Proportion of gram-positive and gram-negative bacterial isolates from plastic wastes. Microscopical observation of (A) Gram-positive rods and (B) Gram-negative rods. Magnification at 100×.

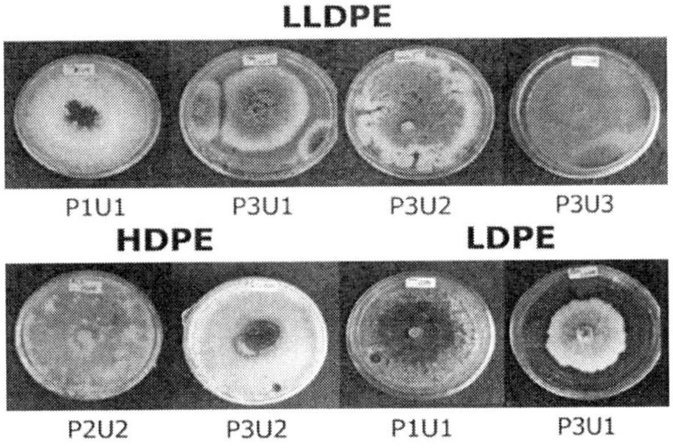

Figure 5. Colony appearance of fungal isolates from plastic wastes.

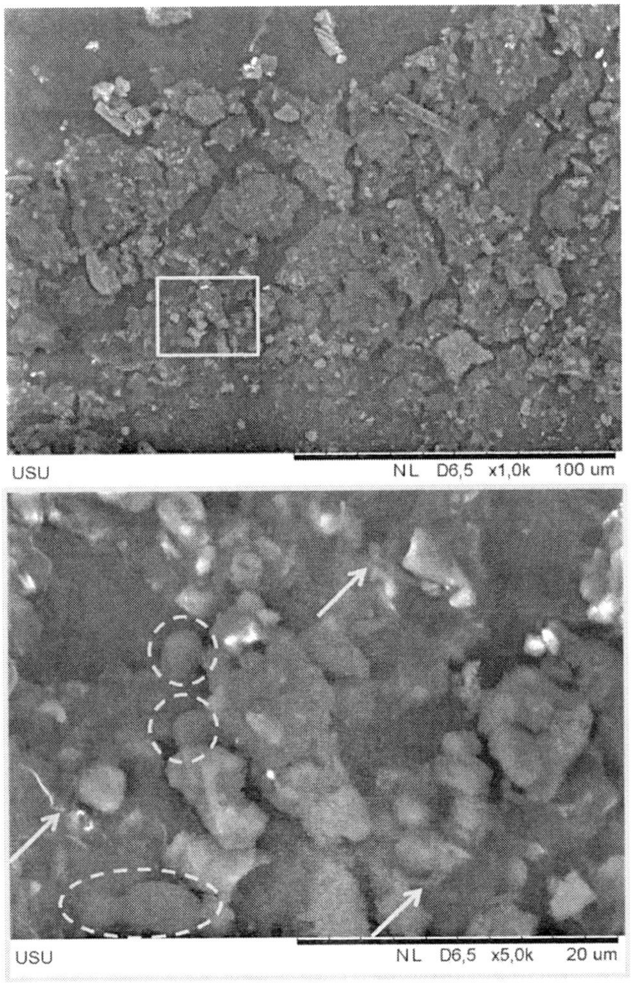

Figure 6. SEM image of plastic waste with indication of bacterial and fungal occurrence. Yellow-dashed circles show the presence either bacteria or yeast population. Arrows show a thread-like objects or mycelium by fungal population.

Gram-negative bacteria represent a far greater proportion of all bacterial isolates than gram-positive group (Figure 4). Members of *Pseudomonas* and *Enterobacter* have been reported to be the most prevalent gram-negative bacteria as opportunistic pathogen in the landfill in the form of bioaerosols or a mixture of waste originating from the hospital and urban environment [14,15]. In addition, the potential of gram-negative bacteria, for example *Pseudomonas* spp. as plastic-degrading bacteria have been noticed especially in the laboratory investigation, i.e. *Pseudomonas aeruginosa* [16], *P. citronellolis* [17], *P. chlororaphis, P. fluorescens, P. protegens, P. stutzeri, P. vesicularis* and so on [18]. Few number of fungal isolates were obtained from the plastic surfaces with a variety of colony appearances (Figure 5). Some notable filamentous fungal species have been reported as plastic-degrading fungi i.e. *Aspergillus nomius* and *Trichoderma viride* [19], *Cephalosporium* sp. and *Mucor* sp. [20], *Penicillium simplicissimum* [21], and etc [22].

The preliminary evidence of microbial community on plastic wastes was analyzed using scanning electron microscopy (SEM). The microscopical image shows the assemblage of mixed community between a network of thread-like objects possibly identified as fungal mycelium with some spherical cells resembling the morphology of bacteria or yeast (Figure 6). The surface of the plastic waste in the figure is typically rough, which may be the result of physical damage or soil particles (dirts) that may explain the dynamical topography. The most common microorganisms in plastic-associated biofilms are bacteria and fungus, although they differ greatly in terms of their physical characteristics, rates of development, sensitivity to environmental factors, and substrate utilization [23]. Several aspects still need to be investigated, including the microbial species that develop on the surface of the plastic waste. Additionally, the use of visualization methods including the electron microscopy is helpful for verifying the physical alterations of the plastic that will be treated with microbial isolates that were recovered from landfill environment.

4. Conclusion

A total of 29 bacterial isolates and 8 fungal isolates were recovered from the surface of plastic wastes in Terjun landfill, Medan City. The portion of gram-negative bacteria was the most prevalent group with some filamentous fungi. The evidence of microbial community was witnessed through SEM analysis showing the presence of bacteria- or yeast-like objects and thread-like objects resembling fungal mycelium.

Acknowledgment

This project is fully funded by the Indonesian Ministry of Education, Culture, Research and Technology under scheme of *Penelitian Dasar Kompetitif Nasional* (PDKN) year 2022.

References

[1] Zhang S J, Zeng Y H, Zhu J M, Cai Z H and Zhou J 2022 *J Hazard Mater* **421** 126780
[2] Du Y, Liu X, Dong X and Yin Z 2022 *Comput Struct Biotechnol J* **20** 975
[3] Alabi O A, Ologbonjaye K, Awosolu and Alalade O E 2019 *J Toxicol Risk Assess* **5** 2572
[4] Schmaltz E, Melvin E C, Diana Z, Gunady E F, Rittschof, Somarelli J A, Virdin J and Daly D M M 2020 *Environ Int* **144** 106607
[5] Dsouza G C, Ryna SS , Varun U , Aniruddh S , Anupama V. Joshi A, Lingayya H, Keshamma E 2021 *Heliyon* **7** E07008
[6] Srikanth M, Sandeep T S R S, Sucharitha K and Godi S 2022 *Bioresour Bioprocess* **9** 42
[7] Rachmawati C A, Anggara M, Djohan, Susanto A B, Bibin B A 2021 *Indonesian Journal of Marine Sciences* **26** 247
[8] Spina F, Tummino M L, Poli A, Prigione V, Ilieva V, Cocconcelli P, Puglisi E, Bracco P, Zanetti M and Varese G C 2021 *Environ Pollut* **274** 116548
[9] Ayeni T O, Arotupin D J and Ayo O E 2022 *Bull Nat Res Cent* **46** 182
[10] Muhonja C N, Makonde H, Magoma G and Imbuga M 2018 *PLoS ONE* **13** e0198446

[11] Sangale M K, Shahnawaz M and Ade A B 2019 *Scientific Rep* **9** 5390
[12] Ojha N, Pradhan N, Singh S, Barla A, Shrivastava A, Khatua P, Rai V and Bose S 2017 *Scientific Rep* **7** 39515
[13] Haedar N, Clara T, Fahruddin, Abdullah A, Fausiah S and Rapak M T 2019 *IOP Conf Ser Earth Environ Sci* **1341** 022023
[14] Nair A T 2021 *Aerobiologia (Bologna)* **37** 185
[15] Flores-Tena F J, Guerrero-Barrera A L, Avelar-Gonzalez F J, Ramirez-Lopez E M and Martinez-Saldana M C 2007 *Rev Latinoam Microbiol* **49** 25
[16] Tamnou E B M, Arfao A T, Nougang M E, Metsopkeng C S, Ewoti O V N, Moungang L M, Nana P A, Takang-Etta L R A, Perriere F, Sime-Ngando T and Nola M 2021 *Environ Challeng* **3** 100056
[17] Bhatia M, Girdhar A, Tiwari A and Nayarisseri A 2014 *Springerplus* **3** 497
[18] Wilkes R A and Aristilde L 2017 *J Appl Microbiol* **123** 582
[19] Munir E, Harefa R S M, Priyani N and Suryanto D 2018 *IOP Conf Ser Earth Environ Sci* **126** 012145
[20] Chaudhary A K and Vijayakumar R P 2019 *Environ Develop Sustain* **22** 4495
[21] Sowmya H V, Ramalingappa, Krishnappa M and Thippeswamy 2014 *Environ Develop Sustain* **17** 731
[22] Venkatesh S, Mahboob S, Govindarajan M, Al-Ghanim K, Ahmed Z, Al-Muhlm N, Gayathri R and Vijayalakshmi S 2021 *J King Saud Univ Sci* **33** 101362
[23] Wang L, Tong J, Li Y, Zhu J, Zhang W, Niu L and Zhang H 2021 *Environ Res* **196** 110371

Preparation and application of heterogeneous catalyst based on KOH impregnated activated carbon from rubber seed kernel for biodiesel synthesis

Iriany, Taslim*, O Bani, A Aldi and S Rahmadani

Department of Chemical Engineering, Faculty of Engineering
Universitas Sumatera Utara, Medan 20155, North Sumatera, Indonesia

*Email: taslim@usu.ac.id

Abstract. Research to explore the potential of rubber seeds kernel as a carbon source for the preparation of heterogeneous catalysts has been carried out. This study aimed to characterize rubber seed kernel carbon, impregnate it with KOH, and apply it as a heterogeneous catalyst in biodiesel synthesis. In this research, rubber seed kernel was powdered then carbonized in a furnace for 4 h at 500°C, followed by activation with HCl for 1 h and dried. The resulting activated carbon was impregnated with KOH solution for 3 h at 80°C. After pre-drying, it was calcined at 500°C for 3 h. The rubber seed kernel powder, activated carbon, and KOH impregnated carbon were characterized using SEM-EDX and FTIR. KOH impregnated carbon was used as a heterogeneous catalyst in coconut oil transesterification for biodiesel synthesis. The parameters studied were catalyst load (2 - 4%) and reaction temperature (55 - 65°C). Molar ratio of oil to methanol and time were set at 1:9 and 2 h, respectively. The highest yield (96.43%) was obtained at 3% catalyst load and reaction temperature of 60°C.

1. Introduction

Indonesia is one of the countries with enormous rubber plantation. Rubber plantation area in Indonesia is reaching 3.683 million hectares, but it mainly produces latex as raw material for rubber industry. Aside of latex, rubber plantations also produce rubber seeds, up to 1 kg/m^2. Rubber seed contains 40 – 50 %wt. oil composed of palmitic acid (13.11%), stearic acid (12.66%), arachidic acid (0.54%), oleic acid (39.45%), linoleic acid (33.12%) and other fatty acids. Meanwhile, rubber seed kernels are currently underutilized and only become plantation waste. Rubber seed kernel ash contains carbon (64.45%), hydrogen (8.23%), nitrogen (3.63%) and sulphur (0.32%) [1]. The high carbon content of rubber seed kernel enables its use as raw material for activated carbon.

Activated carbon has low alkalinity. As such, it cannot be directly employed as catalyst for biodiesel production. For such usage, activated carbon needs to be modified through alkaline impregnation. Alkaline, such as KOH, are easily attached on the surface of activated carbon because there are many micropores. The bond on activated carbon is inert and the surface is easy to modify. Thus, it is suitable as a catalyst support [2].

Preparation of carbon-based catalyst usually goes through several stages, namely carbonization, activation, impregnation, and calcination. Carbonization is usually affected by temperature and depends on the type of raw material used [3]. Activated carbon can be made from lignocellulosic-biomass through thermal decomposition or carbonization. During the process, decomposition of volatile matters in material will trigger pores formation.

Carbon activation is carried out using HCl solution. Metals and other impurities dissolve in HCl, resulting in increased number of pores. In addition, the volume and diameter of the pores also increase. KOH impregnation aims to bind potassium ions to activated carbon surface and can shift the activated carbon surface morphology [4]. The time required for impregnation varies depending on the carbon source and the type of metal ion used. Additionally, it also depends on carbonization temperature and the concentration of KOH used [2]. KOH impregnated carbon can act as heterogeneous catalyst for biodiesel production. Some advantages of such catalyst are it is easy to obtain, cheap, easy to separate from liquid phase products, has high activity and selectivity, and can be reused [5].

Catalyst supports based on biomass for biodiesel production have been published in literatures, such as shea nutshell [2], candlenut shell [4], flamboyant pods [6,] nipa palm shell [7], palm shell [8] and coconut shell [9]. So far, the utilization of carbon from the rubber seed kernel (RSK) as catalyst support for transesterification has not been carried out. Therefore, this study aimed to investigate RSK as carbon source, impregnate it with KOH, and examine it as heterogeneous catalyst in biodiesel transesterification.

2. Materials and Methods

2.1. Materials

Coconut oil was obtained from vegetable oil shop in Medan. Rubber seed kernel (RSK) was obtained from Labuan Batu Utara, Sumatera Utara. KOH and methanol were supplied from local market.

2.2. Catalyst preparation

Rubber seed kernel (RSK) is cleaned of dissolved impurities through washing with water. Then it was dried at 130°C for 2 h in an oven followed by milling and sieved to 100 mesh to produce rubber seed kernel powder (RSKP). A furnace operated for 4 h at 500°C was used to carbonize RSKP [4]. The residual carbon was leached with 0.1 N HCl at a ratio of 1:6 (w/v) and neutralized with deionized water. The activated carbon formed was dried for 2 h at 120°C in an oven. Ten g activated carbon was impregnated with 50 g KOH dissolved in 150 mL water for 24 h at ambient temperature and agitation speed of 250 rpm. KOH impregnated carbon was filtered and dried in an oven and calcined in a furnace at 500°C for 3 h. SEM-EDX and FTIR were used to characterize the resulting carbon which was ready to be used as catalyst in biodiesel synthesis.

2.3. Transesterification

Fifty g oil was added into a flask fitted with a thermometer and condenser. Methanol was mixed with catalysts in a glass beaker for 60 min. Then, this mixture was added into the flask and heated to 55 – 65°C under constant stirring speed of 350 rpm for 120 min [4]. The amount of methanol was adjusted to molar ratio of oil to methanol (MROM) 1:9 and catalyst load was varied from 2 – 4wt.% of the oil. After transesterification, mixture was separated by filtration and purified by washing. The purified final product was examined for methyl ester content, density and viscosity, and values were compared to available standards.

3. Results and Discussion

3.1. SEM-EDX analysis

Figure 1 depicts RSK, RSKP, and carbon from RSKP. RSK and RSKP is light brown, while the carbon was black.

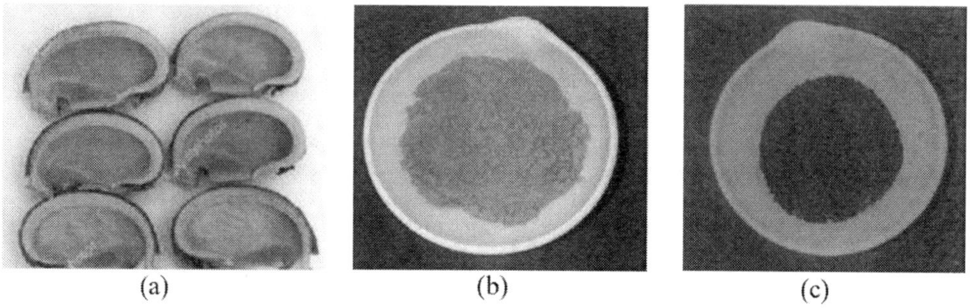

Figure 1. (a) RSK, (b) RSKP and (c) carbon from RSKP

Figure 2 depicts surface morphology of RSKP before and after carbonization using SEM. In Figure 2(a), RSKP surface was irregular and has empty spaces or cavities between solid particles and the pores are not clearly visible. In Figure 2(b), activated carbon from RSKP, carbonized at 500°C for 4 h and activated with HCl solution, formed very clear pores compared to pre-carbonized RSKP. HCl solution can dissolve impurities on the carbon surface, enlarging pore surface which is important as attachment sites for potassium during KOH treatment [4].

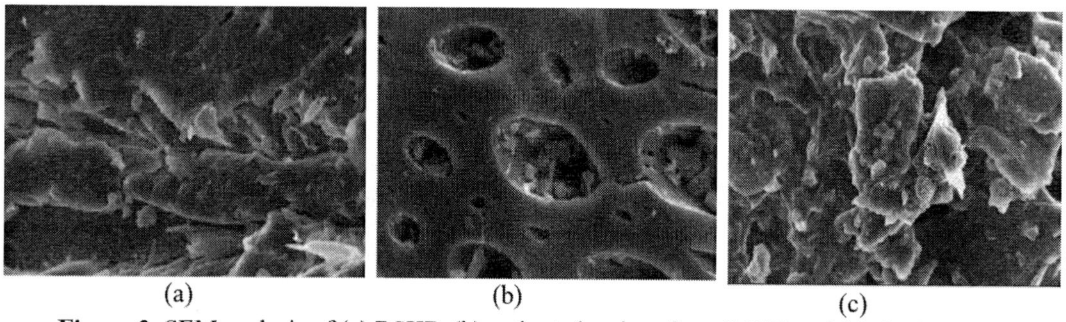

Figure 2. SEM analysis of (a) RSKP, (b) activated carbon from RSKP and (c) the KOH impregnated carbon

In Figure 2(c), KOH impregnated carbon had reduced surface area because the pores are occupied by potassium. During impregnation, potassium ion diffuses into the carbon pores. It is then calcined or thermally activated to remove moisture and aid potassium adherence to catalyst surface.

EDX results of RSKP, activated carbon from RSKP, and the KOH impregnated carbon are in Table 1. In Table 1, the constituent of RSKP and activated carbon from RSKP are mainly C and O, and several other elements in trace amounts such as Cu, Zr and K. Carbon composition increased after carbonation from 59.31 to 88.02%. During carbonization, volatile matters are thermally degraded thus pores are formed and carbon surface area is increased. In addition, the fused carbon will produce more stable bonds while the other components are degraded or vaporized [10]. After impregnation treatment with KOH, carbon composition decreased from 88.02 to 53.20% because it reacted with KOH to form K_2O and K_2CO_3 [11]. Both compounds formed, K_2O and K_2CO_3, are active components that catalyze the transesterification reaction.

$$C + KOH \rightarrow K_2O + K_2CO_3 + 2H_2$$
$$C + K_2O \rightarrow 2K + CO$$
$$2C + K_2CO_3 \rightarrow 2K + 3CO$$

In Table 1, potassium content of the activated carbon increased after impregnation with KOH. This confirms that potassium has been well dispersed on the activated carbon surface. This phenomenon has also been reported in several literatures [8,12]. Potassium composition in RSKP was initially negligible but increased to 1.18% after carbonization and increased to 17.93% after impregnation with KOH.

Table 1. Elemental composition of RSKP, the activated carbon and the impregnated carbon

Element	Composition (%)		
	RSKP	Activated carbon	Impregnated carbon
C	59.31	88.02	53.20
O	39.94	10.04	24.36
Cu	0.31	0.40	0.78
Mg	-	0.06	0.12
Al	-	-	0.15
Si	-	-	2.96
Zn	-	0.31	0.50
K	-	1.18	17.93
Zr	0.43	-	-

3.2. FTIR analysis

Figure 3 shows FTIR results of carbon from RSKP before impregnation in the black spectrum and carbon after KOH impregnation in the red spectrum. Carbon before KOH impregnation had lower peak intensity than the activated carbon after impregnation. The appearance of a peak at wave number 1353 cm^{-1} in the red spectrum indicates the formation of a carbonate group ($-CO_3$). Potassium carbonate, K_2CO_3 is the carbonate compound formed on activated carbon after impregnation with KOH [13]. Modification of activated carbon with KOH, can neutralize the carbon surface [11]. In addition, there is also a K-O bond indicating that there is K_2O at wave number of 649 cm^{-1} [14].

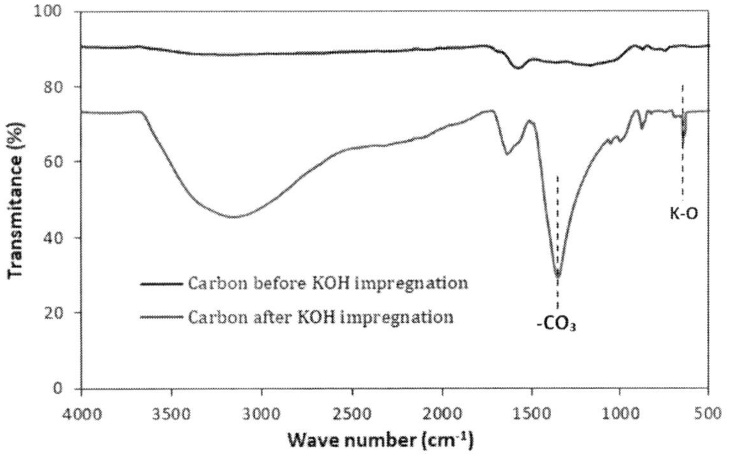

Figure 3. FTIR analysis

3.3. Transesterification

In order to study the effect of catalyst load on biodiesel yield, the catalyst load was varied in the range of 2-4% at MROM 1:9, 120 min and 60°C. In Figure 4, biodiesel yield maximized at 3% catalyst load. More active sites are available as catalyst amount is increased, leading to better contact with the reactant and resulting in higher yield. However, further increase in catalyst amount renders the reaction mixture viscous and difficult to stir, increasing mass transfer resistance which reduce yield [15-17]. The use of

3% catalyst load was the optimum condition with 96.43% yield. This biodiesel yield is close to that using carbon catalyst from candlenut shell which yielded 96.65% [4] but superior to that using carbon catalyst from flamboyant pods at 89.81% [6].

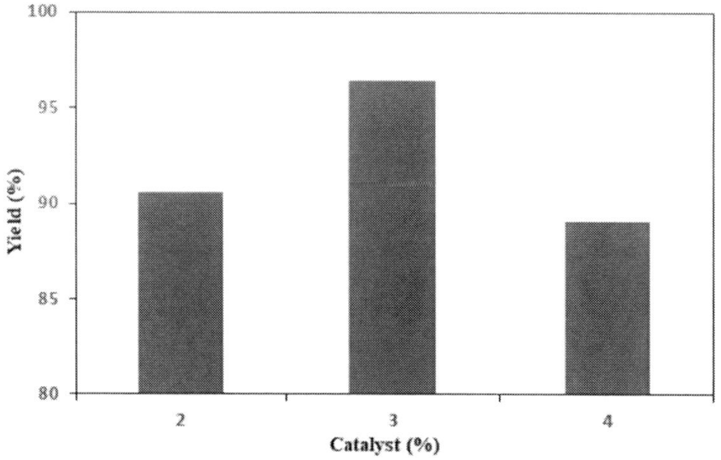

Figure 4. Biodiesel yields at various catalyst loads

Several runs were carried out to examine the effect of temperature on biodiesel yield. Observations were made by adjusting the reaction temperature from 55 – 65°C at 3% catalyst load, reaction time of 120 min, and MROM 1:9. In Figure 5, as reaction temperature increased from 55 to 60°C, the biodiesel yield also increased due to higher collision rate between reactant molecules. In addition, the oil viscosity decreases which lower mass transfer resistance, thus accelerating reaction and increasing yield. However, at 65°C yield decreased due to methanol vaporization. The same incident was also reported by other researchers [9,18]. The highest yielding temperature in this research was 60°C with a yield of 96.43%. Some researchers also proposed that highest yielding temperature was 60°C [9,14].

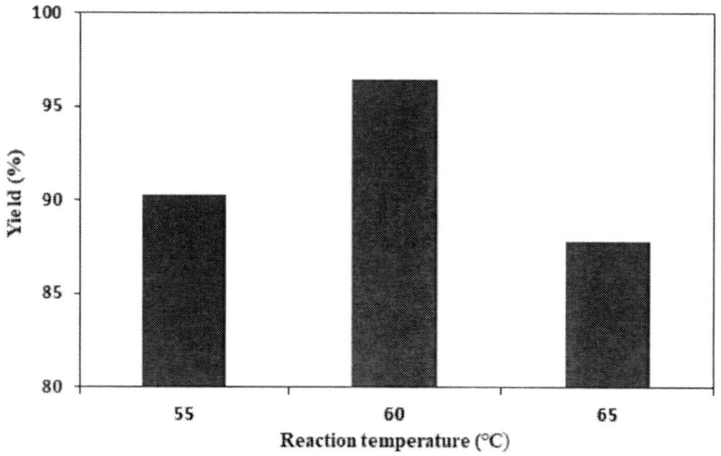

Figure 5. Biodiesel yields at various temperatures

3.4. Biodiesel Properties

Some of the biodiesel properties were analyzed and were found to meet the European standard (EU 14214) as in the following Table.

Table 2. Biodiesel properties

Properties		This study	EN 14214
Methyl ester content	%	99.17	≥ 96.5
Density	kg/m^3	872	860-900
Kinematic viscosity	mm^2/s	4.79	3.5-5.0

4. Conclusion

RSK can be used as a carbon source via carbonization. After calcination and KOH impregnation, the carbon catalyst contains K_2O and K_2CO_3 as active compounds which catalyze biodiesel transesterification. Highest yield of 96.43% was achieved at 3% catalyst load, MROM 1:9, 60°C and 120 min. The biodiesel obtained has also met the European standard.

References

[1] Hassan, S N A M, Ishak M A M, Ismail K, Ali S N and Yusop M F 2014 *Energy Procedia* **52** 610
[2] Dejean A, Ouédraogo I W, Mouras S, Valette J and Blin J 2017 *Energy Sustain Dev* **40** 103
[3] Naji S Z and Tye C T 2022 *Energy Convers Manag X* **13** 100152
[4] Taslim, Bani O, Iriany, Aryani N and Kaban G S 2018 *Key Eng Mater* **777** 262
[5] Zabeti M, Ashri W M, Daud W and Aroua M K 2009 *Fuel Process Technol* **90** 770
[6] Dhawane S H, Kumar T and Halder G 2016 *Renew Energy* 89 506
[7] Taslim, Bani O, Iriany, Audina E and Hidayat R 2021 *IOP Conf Ser Earth Env Sci* **912** 012094
[8] Baroutian S, Aroua M K, Abdul Raman A A and Nik Sulaiman N M 2010 *Fuel Process Technol* **91** 1378
[9] Buasri A, Chaiyuta N, Loryuenyong V, Rodklum C, Chaikwan T, Kumphan N, Jadee K, Klinklom P and Wittayarounayut W 2012 *Sci Asia* **38** 283
[10] Kwiatkowski M, Kalderis D and Diamadopoulos E 2017 *J Phys Chem Solids* **105** 81
[11] Yang L, Xue-Wei C, Li C, Xiang-Li L and Yuan W K 2013 *J Indus Eng Chem* **19** 784
[12] Abu-Jrai A M, Jamil F, Al-Muhtaseb A H, Baawain M, Al-Haj L, Al-Hinai M, Al-Albri M and Rafiq S 2017 *Energy Convers Manag* **135** 236
[13] Gohain M, Devi A and Deka D 2017 *Ind Crop Prod* **109** 8
[14] Taslim, Bani O, Iriany, Lestari S and Ginting L 2019 *AIP Conf Proceed* **2085** 020062-1
[15] Chen G Y, Shan R, Yan B B, Shi J F, Li S Y and Liu C Y 2016 *Fuel Process Technol* **143** 110
[16] Molaei D A and Ghasemi M 2012 *Fuel Process Technol* **97** 45
[17] Noiroj K, Intrapong P, Luengnaruemitchai A and Jai-in S 2009 *Renew Energy* **34** 1145
[18] Tan Y H, Abdullah M O, Nolasco-Hipolito C and Taufiq-Yap Y H 2015 *Appl Energy* **160** 58

IOP Conf. Series: Earth and Environmental Science 1115 (2022) 012082 doi:10.1088/1755-1315/1115/1/012082

Analysis of the effect of seed growth (*Rhizophora apiculata*) on the distance of land from the beach in of Pulau Sembilan Pangkalan Susu District Langkat Regency

Yunasfi[1,3]*, M L Purba[1] and A Fadhilah[2]

[1] Faculty of Forestry, Universitas Sumatera Utara, Medan, Indonesia
[2] Faculty of Agriculture, Universitas Sumatera Utara, Medan, Indonesia
[3]Center of Excellence for Mangrove, Universitas Sumatera Utara, Medan 20155, Indonesia

*Email: yunasfijamhar@yahoo.co.id

Abstract. Mangrove is a renewable resource that provides services related to the environment, such as abrasion protection, seawater control of incursion, reduction of sea wave height and speed, cleaning of water contaminants, among others, important to support living beings. Its objectives include to study and evaluate the impact of planting distance from the beach which is useful for increasing growth for mangrove ecosystem type *Rhizophora apiculata* and this study based on a complete randomized design (RAL) with treatment distance (J) consisting of 5 treatments (a) P0 (planting distance from the beach 41 meters), (b) P1 (planting distance from the beach 42 meters), (c) P2 (planting distance from the beach 43 meters), (d) P3 (distance from the beach 44 meters) (e) P4 (planting distance from the beach 45 meters). The percentage of plant growth is 100%. Observational Data on the measurement of height, diameter, leaf count, leaf size, and leaf area, that the treatment at planting distance can increase the growth of seedlings *R. apiculata*. Based on the results of the study the effect of planting distance on *R. apiculata* obtained in the treatment of P2 (planting distance from the beach 43 m) a seedling height value that was highest growth with a value of 198.0 cm or average 33.0 cm. The highest value of plant diameter obtained from the treatment of P3 (spacing 44 m) with a value of 7.74 mm or average 1.29 mm . The highest added value was found in treatment P1 with a value total number of 39.2 strands or average 6.5 strands. The increase in leaf area is 50.9 cm or average 8.4 cm which is found in the P2 treatment.

1. Introduction

Mangroves are a regenerative resource that provides a variety of products obtained from many types of flora and wildlife as well as environmental services, like abrasion protection, seawater preventing intrusion, reduction of sea wave height and speed, cleaning of water contaminants, among others, important for support living beings. Therefore, mangroves play an important role as bio capital for development, especially to improve the welfare of towns along the coast. Currently, mangrove forests are growing in 124 tropical and subtropical nations covering about 15.2 million ha, or approximately 36% (5.4 million ha) are planted throughout the nations of Southeast Asia. Southeast Asia's mangrove forests mostly grow the present time in Indonesia is around of the mangrove's 3.2 million acres (about 59%) forest area of Southeast Asia, either 21% of the world's mangrove forests). The Southeast Asia the region, particularly in Indonesia, not only features vast mangrove forests, but also has many types of mangroves. Mangrove forests' primary habitat is along the coast [1].

Indonesia is a country with the fourth longest coastline in the world with a coastline of 95,181 km. As a result of the diversity of its ecosystems, Indonesia has the potential to produce remarkable coastal natural resources. The mangrove ecosystem is one of the coastal habitats in Indonesia. Mangrove ecosystems are found around river mouths or in areas where land meets the sea. Mangroves are a collective of plants with a wide range of species from distinct tribes and the similar capacity to adapt to environments impacted by tides [2].

Forests of mangrove in the biodiverse regions of Southeast Asia offer a wide range of ecological functions on this supports millions of people. Fisheries are improved by mangroves, protect coastal areas and keep carbon at the greatest density globally of any ecosystem. Mangroves have been subjected to large-scale deforestation due the world's need for the commodity, and earlier research has detected growth of cultivation as a key driver. Mangroves are lower than previously reported figures, with losses occurring at an average rate of 0.18% year. There were mangroves covering almost 100,000 acres in all logged aquaculture was responsible for 30% of the entire change in the forest during the research period. in Malaysia and Indonesia, the sustainable conversion of mangroves into oil palm fields, was cited as a new, escalating, and less well-known hazard to mangrove ecosystems [3]. As a result, conducting research to learn about running effective red mangrove (*R. apiculata*) nursery to ensure that mangrove restoration activities in the future can achieve success. This study aims to determine the varying approaches to treating red mangrove nursery types of mangroves in their growth and development, based on the how far it is from the coast. Furthermore, this investigation tracked the development of red mangrove saplings planted on the field directly (*direct seeding*). The findings of this study's information will be used to determine the type of place to grow that is suitable for red mangrove plants. This test is expected to increase the growth of mangroves and have an impact on the continuity of the mangrove ecosystem.

2. Materials and Methods

2.1. Time and place
The mangrove forest area of Pulau Sembilan Village, Pangkalan Susu, was the site of this study experiments on seeding *R. apiculata* seedlings that had been planted from the coast as far as 41 meters to 45 meters. From October to December 2022, a three-month period, this research activity was conducted.

2.2. Tools and materials
The caliper, rafiah rope, tape measure, cloth tape measure, bamboo-made bamboo stick, permanent marker, book, pen, oven, knife, and polybag were the tools utilized in the field investigation. It is made of *R. apiculata* seeds, which can have up to 30 seeds and are comparatively the same size and the required data from institutions related to research.

2.3. Research procedure
2.3.1. Distance measurement from the beach. Measurement of the the length of the nursery's driveway from the beach 37 to 41 meters measured using a meter, as for the procedure : P0 = 41 meters, P1 = 42 meters, P2 = 43 meters, P3 = 44 meters and P4 = 45 meters from the beach with each replication 6 plants each *R. apiculata*.

2.3.2. Seedling selection and planting R. apiculata. The selection of *R. apiculata* seedlings was 30 seeds with relatively the same size, with 4 fully opened initial leaves and plant heights ranging from 30 cm to 45 cm. Then the seeds are released within the polybag, do it slowly in order to protect the seeds' roots. Seedlings were planted with a distance of 1 x 1 meter and 6 seedlings were planted for each plot. A stake is also installed on each side of the seedling so that it can be pegged at every measurement from the start to the completion of the research period.

2.3.3. Experimental design. Planting seedlings *R. apiculata* was carried in the Pangkalan Susu District, Langkat regency, which is part of Pulau Sembilan's mangrove forest. An entirely random design was used in this study (RAL). Following the germination of seedlings in the field, data collection takes place during the following period :

a. 15thdays	b. 30thdays	c. 45thdays
e. 75thdays	d. 60thdays	f. 90thdays

The linear Model in complete random design is: $Y_{ij} = \mu + \alpha_1 + \sum_{1,2}$

Y_{ij} = Observations on the treatment of i-th and j-th replications
μ = General average value of observations
α_i = The effect of treatment factors to -i; I = 1,2,....... t dan j= 1,2,......r
$\sum_I$ = Effect of error treatment to-i and repeat to-j

2.3.4. Parameters of seedlings R. apiculata measured. Parameters of seedlings *R. apiculata* observed were stem length, width, and number of leaves, and the maximum leaf width. Stem height measurement is done by using a cloth meter, measuring the diameter of the stem and the width of the widest leaves using a Califer.In this study, observations were made every two weeks over the course of three months. Six times of data collection were done by keeping track of various parameters, namely :

1. Seedling height (cm)
   In the measurement of seedling height is done by measuring the height of the seedlings from the base of the lower stem 3 cm above ground level (given ajir) to the highest growing point by using a meter every two weeks.
2. Seedling diameter (mm)
   In the measurement of the diameter of the seedlings are measured using a Califer with units of mm. Diameter measurements are also carried out at ketingian 3 cm above ground level or pegged to ajir which has been given every two weeks.
3. Number of leaves (strands)
   Counting the number of Leaves is done by counting the leaves that have grown and fully formed every two weeks.
4. Leaf area index (cm)
   Leaf area measurements were carried out at the end of data observation using the millimeter block paper method by placing the leaves on millimeter block paper then the leaves were drawn following the leaf pattern. Then count the number of squares in the leaf area that has been drawn in order to get the result of the leaf area.
5. Plant wet weight (g)
   Done by weighing the entire plant that has been divided into several parts of the leaves, stems, and roots using scales at harvest time before being placed in an oven with a temperature of 75°C for 48 hours until constant and expressed in units of grams (g)/plant.
6. Plant dry weigth (g)
   Done by weighing the entire plant using a scale after being removed from the oven.

2.4. Data collection

Primary data were taken directly from the field, in the form of measurements made on seedlings *R. apiculata*, height data, diameter data, number of leaves and the widest leaf width.

2.5. Harvesting

Harvesting is carried out on seedlings after observation of R. apiculata the 6th time to be observed the development of growth of these seedlings, then removed manually.

2.6. Data Analysis

The Data obtained from the results of the study were analyzed with the fingerprint (Analysis of variance) and if there are different results ($P < 0.05$) between the treatment then done Duncan multiple range test (DMRT) error Level 5%.

3. Results and Discussion

3.1. Seedling height (cm)

Height growth is one of the parameters observed to ascertain the impact of spacing the development of *R. apiculata* seedlings. This growth process can also be measured from the overall weight of the plant. The height growth of the *R. apiculata* type of mangrove can be caused by the addition of a network that causes the stems of the seedlings to become longer. Table 1. Shows the results of the growth of the average height of seedling.

Tabel 1. Seedling height increase *R. apiculata*

Treatment	Observation to						Total (cm)	Average (cm)
	I	II	III	IV	V	IV		
P0	21.2	21.3	21.4	21.5	21.6	22.1	128.9	21.4
P1	28.7	29.0	29.5	33.0	34.8	37.2	192.2	32.0
P2	29.9	30.5	32.5	33.2	35.4	36.4	198.0	33.0
P3	26.0	26.3	28.0	32.2	33.3	35.4	181.2	30.2
P4	22.3	22.6	23.7	26.3	28.8	32.1	155.8	25.9
Total (cm)	128.1	129.6	135.1	146.2	153.9	163.0	856.0	244.6
Average (cm)	25.6	25.9	27. 0	29.2	30.7	32.6	171.2	48.9

Table 1 shows that spacing affects the growth of height *R. apiculata* where observations during 1 time in the first weekly development of the seedling stem height is still low when observed at the time to 3 in the third week, show high accretion. This is believed to be the result of competition to grow and develop on each seedling. In Table 1 can be seen treatment 2 (planting distance from shore 43 M) has the highest value of increase in seedlings *R. apiculata* compared with other treatments with a value of 198.0 cm with average 33.0 cm and on treatment 1 (planting distance from the coast 42 m) has the lowest seed height increase with a total of 128.9 cm and average 21.4 cm. Meanwhile, in the research of [4], that P1 treatment (planting distance from shore 41 M) gave the highest measurement of seedling height growth compared to alternative methods of treatment, namely 233.1 cm with a 38.84 cm average /12 week. P2 treatment (distance 43m) produced the lowest seedling height value growth the seedling height value that was the lowest., namely 220.4 cm, with a 36.73 cm/12 week average. According to [4] this is caused by the growth-related variables and percentage of mangrove life are minimal wave movement, brackish salinity, silt (soil), extensive intertidal (tidal) zones.

The results of the analysis of seed high variance showed that the value of GIS. > 0.05 that is equal to 0.000 where H0 is accepted and H1 is rejected so that it is concluded that the effect on planting distance at a distance of 41 m–45 m has a significant effect on the growth of seedling height *R. apiculata*. This happens because the fight for plant growth space will be exacerbated by close spacing. The rarer the tree, the more competition for nutrients and sunlight. The less competition for nutrients and sunlight, the faster the increase in growth in plants [5]. Setting a planting distance of 1 m x 1 m is usually aimed at protecting the coast from abrasion (*Green belt*). This spacing provides space for seedlings *R. apiculata* to get the sun light [6].

3.2. Seedling diameter (mm)

Increase in diameter in seedlings *R. apiculata* the average on each observation showed the diameter of the stem widened.Table 2 shows that the diameter of the seedling *R. apiculata* has increased diameter. Increase in the highest average diameter of the treatment P3 as big as 7.74 mm and the lowest increase in diameter is in the treatment of P4 of 6.46 mm. Based on the results of measuring the diameter of the stem in this study by proving the table can be seen that the treatment of spacing impact on the development of seedling stem diameter *R. apiculata*. While the research [4] get results which P1 therapy (distance 41m) increased when compared to other treatments, the largest seed diameter is equal to With a 1.32 average, 7.9 is obtained. As well as P3 therapy (distance 44 m) brought about the lowest value of increased seed diameter in comparison to other treatments, namely 6.87 with a mean of 1.15 according to [4] this happens since the seeds were placed come from the same parent, allowing the seed quality is the same. The results of analysis of seed diameter variation (Appendix 2) showed that the value of GIS > 0.05 that is equal to 0.000 where H0 is accepted and H1 is rejected so that it is concluded that the effect on planting distance at a distance of 41 m – 45 m has a significant effect on the growth of seedling diameter *R. apiculata*. This is in accordance with the statement Herpinawati [7] changes in the size of the diameter of the seedlings are influenced by the content of organic matter or nutrients in the substrate. This mangrove ecosystems may have a various species compositions or growth rates Several elements may have an impact on the growth performance because of the growing environment of mangrove seedlings causing inconsistent growth rates such as environmental conditions [8]. The typical diameter of seedling *R. apiculata* for 6 observation (Table 2).

Tabel 2. *R. apiculata* seedlings average diameter for six observations

Treatment	Observations to						Total (mm)	Average (mm)
	I	II	III	IV	V	VI		
P0	1.22	1.23	1.25	1.26	1.28	1.29	7.53	1.25
P1	1.09	1.14	1.15	1.17	1.19	1.27	6.99	1.16
P2	1.20	1.24	1.29	1.31	1.35	1.35	7.74	1.29
P3	0.95	1.1	1.15	1.21	1.23	1.25	6.88	1.14
P4	0.96	1.05	1.06	1.11	1.13	1.15	6.46	1.07
Total (mm)	5.42	5.76	5.89	6.06	6.17	6.31	35.6	5.93
Average(mm)	1.08	1.15	1.17	1.01	1.02	1.26	7.12	1.18

3.3. Number of leaves (strands)

According to the findings of 6 observations, demonstrated that the therapy had an effect on the increase in the number of leaves of planting distance that the difference is not much different, where the treatment 1 (planting distance from the beach 42 meters) has almost the same value in the treatment of P2 (planting distance from the beach 43 meters). The results of the average's growth number of leaves of see the seedlings in Table 3.

Tabel 3. *R. apiculata* seedlings average diameter for six observations

Treatment	Observation to						Total (strands)	Average (strands)
	I	II	III	IV	V	VI		
P0	4	4.3	5.3	6.7	7.5	9.2	37.0	6.1
P1	3.7	4.8	5.7	8	8.3	8.7	39.2	6.5
P2	4	4.3	5.2	6.2	7.3	8	34.5	5.7
P3	3.8	4.8	5.7	6	7.5	8.5	36.3	6.0
P4	3.2	3.7	4.5	4.9	6	7.2	29.4	4.9
Total (strands)	18.2	22.0	26.3	31.7	36.7	41.5	176.4	29.4
Average (strands)	3.6	4.4	5.2	6.3	7.3	8.3	35.2	5.8

Table 3 shows that the number of leaves of seedlings *R. apiculata* experienced accretion. Increase in the average number of leaf blades had height namely in the treatment of P1 39.2 cm and the lowest increase in the number of strands is in the treatment of P4 29.4 cm. In accordance which states that the increase in the number of leaves is one form of seedling growth *R. apiculata* and leaf abortion is a form of adaptation to the environment. After the propagule I planted it take 4-5 month to produce 6 leaves. This addition of leaf blades indicates the growth rate of seedlings. Where this growth rate is directly proportional to the diameter of the xylem vessels. The increase the diameter of the plant also has an impact on how many leaves there are seedling it self, the wider the diameter of the seedling, the wider the diameter of the xylem.

The number of mangrove leaves is a determining factor in the fertility of mangrove trees, because the leaves are a place of photosynthesis that produces nutrients for mangrove trees. If the number of leaves is large, then the fertility of mangrove trees can be guaranteed. The influence of significant ecological factors on the number of Leaves shows that mangrove fertility is sensitive to ecological factors, especially salinity. Regarding the variables that may influence the development of the number of leaves on seedlings of *R. apiculata*, it is known that the type *Rhizophora* sp. grows best at a temperature of 26°C to 28°C and that higher temperatures will result in lower production [9].

3.4. Leaf Area index (cm)

The growth of increasing the length of the leaves is an increase in the tissues on the leaves that are affected by the surrounding environment. Table 4 shows the findings of increase in the average leaf length of seedlings.

Tabel 4. Average leaf area of seedlings *R. apiculata* for 6 observations

Treatment	Observation to						Total (cm)	Average (cm)
	I	II	III	IV	V	VI		
P0	6.9	7.1	7.7	8.5	8.9	9.1	48.0	8.0
P1	7.0	7.4	7.8	8.2	9.7	9.9	50.0	8.3
P2	7.6	7.8	8.3	8.6	9.2	9.4	50.9	8.4
P3	7.2	7.5	7.9	8.6	9.5	9.1	49.7	8.2
P4	6.8	7.1	7.3	7.6	8.7	8.9	46.3	7.7
Total (cm)	35.4	36.9	38.8	41.5	46.0	46.3	244.9	40.8
Rata-rata (cm)	7.0	7.3	7.7	8.3	9.2	9.2	48.9	8.1

Table 4 shows that the increase of leaf length seeds *R. apiculata* experienced accretion. Increase in the average number of leaf blades had height namely in the treatment of P2 of 50.9 cm with an average value of leaf length increase of 8.4 cm, while the lowest value of leaf length increase in P4 of 46.3 cm with an average value of 7.7 cm. Then Each type of plant needs the intensity of light that hits the plant. Intensity has a major role in Plant Physiology processes such as photosynthesis, respiration, growth length of leaves so the availability of certain light will determine the level of crop production, and its development.

3.5. Plant wet weight (g)

The results of the research on wet weight of *R. apiculata* for 12 weeks are presented in Figure 1. Overall, the wet weight of the plant did not show a significant difference.

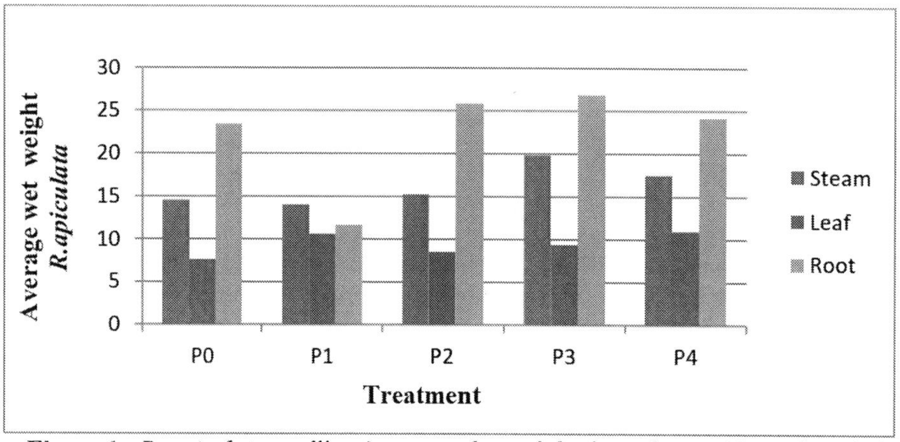

Figure 1. *R. apiculata* seedlings' average dry weight throughout a 12-week period

Figure 1, on the observation of wet weight on seedlings *R.apiculata* obtained the highest value of the stem produced in the treatment of P3 with an average value of 19.80 g while the lowest value of the stem 14.01. In the observation of leaves the highest value is found in P1 with 10.66 g while the lowest value in P2 is equal to 7.6 g. In the observation of the root of the highest value found in P3 is equal to 26.91 g while the lowest value in P1 is equal to 11.68 g. With that leaf area becomes the main parameter because the growth rate of photosynthesis is in many ways strongly influenced by leaf area. If a leaf's surface increases, the assimilation produced will also be greater. A large leaf area will result in an increase in the dry the plant's weight [10].

3.6. Plant dry weight (g)

The results of the research on dry weight of *R. apiculata* for 12 weeks are presented in Figure 2. Overall, the dry weight of the plant did not show a significant difference.

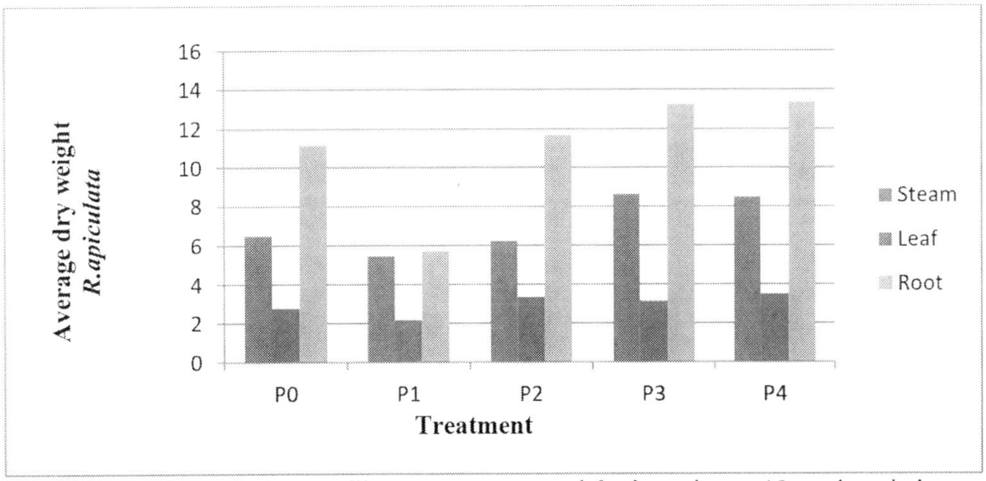

Figure 2. *R. apiculata* seedlings' average wet weight throughout a 12-week period

Figure 2 show that the results of thoughts on the dry weight of seedlings *R.apiculata* obtained the highest value of the stem produced in the treatment of P3 is with an average value of 8.61 g while the lowest value of the stem is at P1 with a value of 5.4 g. In the observation of leaves the highest value is found in P4 with 3.49 g while the lowest value in P1 is equal to 2.16. In the observation of the root of the highest value found in P4 is equal to 13.33 g while the lowest value in P1 is equal to 5.69 g. Plant dry weight gain is produced by nutrients that have been absorbed by the roots, the greater the dry weight of the plant when the physiological processes of plants are more efficient and better [11].

4. Conclusion

Results of research on planting distance on mangrove seedlings *R. apiculata* for 6 observations showed that there is an increase that can affect the growth of seedlings *R. apiculata.* Anova test results show the value of Sig. > 0.05 it was concluded that the treatment of spacing significantly or significantly affect the increase in stem diameter, stem height, number of leaves and leaf width of seedlings *R. apiculata.* Anova test results show the value of Sig. > 0.05 it was concluded that the spacing treatment had no real or significant effect on the Wet Weight of plants and dry weight of plants.

Acknowledgments
According to the guidelines Inquiry regarding University of North Sumatra Talenta Contract Number : 6789/UN5.1.R/PPM/2021, to the investigate Institut the College of North Sumatra, 16 June 2021.

References

[1] Hastuti E D and Hastuti R B 2018 *IOP Conf. Series: Earth and Environmental Science* **130** 012015
[2] Herpinawati D, Zulkifli and Sarno 2010 *Maspari Journal* **1** 59
[3] Juwari D, Ruhiyat M I, Aipassa 2020 *Energy and Environment Research* **10** 1
[4] Yunasfi and Simanungkalit J M 2022 *IOP Conference Series: Earth and Environmental Science* **977** 1 012091
[5] Kusmana C 2018 *IOP Conf. Series: Earth and Environmental Science* **196**
[6] Pranata R A J, Patandean and Ahmad Y 2016 *Makasar.Journal of Physical Science and Education* **12** 1 88 – 95
[7] Halidah 2009. The effect of high inundation and spacing on the growth of *Rhizophora mucronata* (Lamk) saplings on the west coast of south Sulawesi
[8] Richards D R and Friessa D A 2012 Department of Geography, National University of Singapore, Singapore 117570 and Department of Animal and Plant Sciences, The University of Sheffield, Sheffield S10 2AS, United Kingdom. PNAS **113** 344

IOP Conf. Series: Earth and Environmental Science 1115 (2022) 012082 doi:10.1088/1755-1315/1115/1/012082

[9] Noor T, Batool N, Mazhar R and Ilyas N 2015 *National Institute for Genomics and Advanced Biotechnology, (NARC)* **2**
[10] Suryaningrum R, Purwanto E and Sumiyati 2016 *Agrosciences* **18** 2 33
[11] Permatasari I and Kusmana C 2011 *Journal of Tropical Silviculture* **2** 3 181

IOP Conf. Series: Earth and Environmental Science 1115 (2022) 012083 doi:10.1088/1755-1315/1115/1/012083

Utilization of unmanned aerial vehicle (UAV) in identifying the characteristics of the riparian ecosystem of the Percut River, North Sumatra Province

B Slamet[1,2*], D Adinda[1], O P J Nduru[1] and Samsuri[1]

[1]Faculty of Forestry, Universitas Sumatera Utara, Kampus 2 USU Bekala, Simalingkar A, Pancur Batu, Deli Serdang 20353, North Sumatra, Indonesia
[2]Center of Excellence for Mangrove, Universitas Sumatera Utara, Medan 20155, Indonesia

*Email: bejo@usu.ac.id

Abstract. The Percut River border area has changed its function. Changes in land use result in a decrease in the area and types of vegetation that grow on river borders. Considering that the area is quite long and inaccessible in some locations, it can identify riparian vegetation by utilizing remote sensing technology. This study aims to identify the characteristics of the Percut river riparian vegetation using a UAV. Eight vegetation indices were used to analyse land cover types in this riparian ecosystem, namely Green-red ratio (GR), Green-red vegetation index (GRVI), RGB-based vegetation index (RGBVI), Visible atmospherically resistant index (VARI), Simple blue-green ratio (BGI2), Excess green index (ExG), Normalized green-blue difference index (NGBDI) and Modified green-red vegetation index (MGRVI). The analysis of the vegetation index based on RGB images shows that not all indices used are good in separating vegetation from other land covers. Analysis of riparian vegetation characteristics based on the RGB index is recommended using the RGBVI index (RGB-based vegetation index).

1. Introduction

The increasing development and population growth rate in urban areas causes a high intensity of land-use change-one of the areas affected by land cover changes in the riparian ecosystem. If the riparian vegetation is lost, the entire ecological function of the riparian vegetation will be lost. The multiple stressors, such as natural disturbances and human activity accumulatively, could work together to impact riparian communities and affect the condition of riparian vegetation [1].

Pressure on riparian vegetation will increase if there is only a small amount of forest land cover in the watershed. Forest land cover in the Percut watershed is reported to be less than 30 percent of the total watershed area [2]. The built-up area in the Percut watershed is also reported to have increased, which indicates a decrease in the vegetated area [3]. Furthermore, based on the pollution index technique, the water quality status of the Percut River is classified as moderately polluted. However, some of the chemical parameters of the downstream (urban) river water significantly exceed the quality threshold level [4].

Riparian vegetation generally consists of the forest but can also be other types of vegetation [5]. Riparian areas contain vegetation types that represent nearly every terrestrial ecosystem on earth. The vegetation includes grasslands, reeds, shrubs, aquatic vegetation, and flooded lowland tropical forests [6]. Riparian vegetation is an indicator of environmental quality and plays a role in maintaining the integrity of riverbanks, and can also function as a green line [7,8]. Riparian vegetation has an essential role in ecosystems and biota diversity, but this area is vulnerable to exploitation because of its strategic location. Prevention and reduction of the degradation rate needs to be done so as not to disturb riparian vegetation [1].

Some riparian zones are difficult to reach due to road access and many obstacles, so technology is needed to reach these areas. Remote sensing technology enables large-scale research on riparian vegetation. Remote sensing offers a cost-effective approach to monitoring. As riparian buffer zones and wetlands exhibit extreme variations in width, length, spatial complexity, soil, and vegetation cover, mapping their hydrology and land cover requires high spatial and spectral resolution data. The availability of high spatial resolution images and UAVs significantly increases the capacity to map riparian buffers, wetlands, and other ecosystems [9].

UAVs offer a viable alternative to conventional platforms for obtaining high-resolution remote sensing data at lower costs and increased operational flexibility. The UAV has a sophisticated design to carry small payloads and an integrated flight control system [10]. UAV has the potential to classify detailed land cover on a local scale, one of which is the riparian ecosystem [11]. The Percut River urban riparian ecosystem has often been under much stress. Identification of land cover in this riparian ecosystem can be made by utilizing a UAV. Several classification constraints are often encountered because of the visual appearance of the same color, even though the land cover type is different. This study aims to identify the characteristics of the urban riparian ecosystem in the Percut River by utilizing the UAV.

2. Method

This research was carried out from November 2021 to February 2022. The research location was carried out in the Percut River urban riparian ecosystem. Administratively, this location is included in the area of Medan City and Deli Serdang Regency, North Sumatra Province (figure 1).

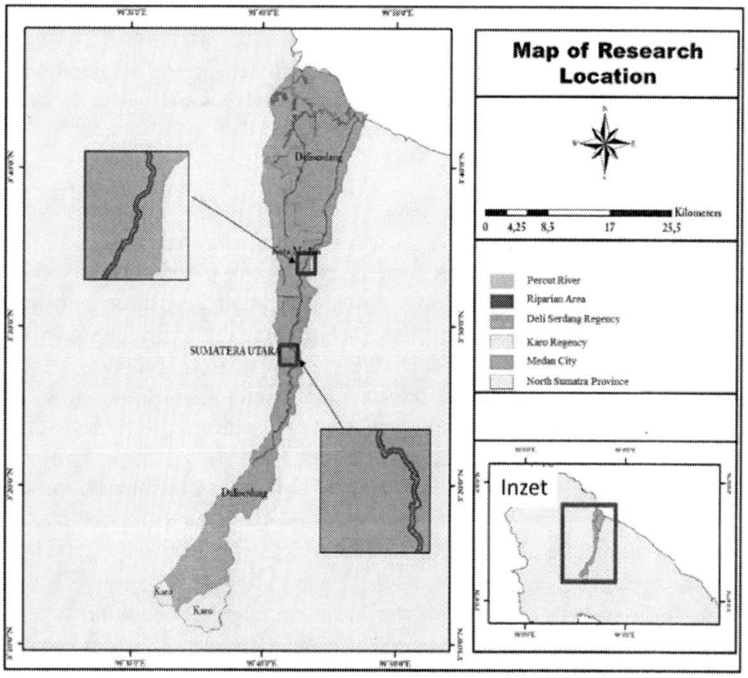

Figure 1. Map of research location

2.1. Research tools and materials

The tools used in this research are camera, GPS, and DJI Phantom 4 Pro drone. We also use DJI Go software, Pix4D Capture, Fast Stone Photo Resizer, and ArcGIS 10.8.

2.2. Research procedure

2.2.1. A field survey was carried out to get the selected location to fly the UAV. This stage is assisted by utilizing visual image information from Google Earth. The selected location is a location with a riparian ecosystem that has vegetation even though its condition has been degraded. The next stage is to plan a flight plan at the selected location. The flight plan was created using the Pix4D Capture software. The flight is carried out from 100 meters above the ground with 80% side overlap and 80% forward overlap. After completing the flight path plan, the next step is to record photos at the selected location using the DJI Phantom 4 Pro Drone.

2.2.2. Aerial photo data processing. The stages in processing aerial photos into orthophoto mosaics are as follows:

a. Align photos

This is the stage of automatically identifying and correcting camera position disturbances in each photo. The align photos process produces an initial 3D model as a point cloud. The ground control point was chosen as a reference for photo pairs to produce a correct camera position with high accuracy in the preselection.

b. Dense point clouds

Dense point clouds are a collection of points in the number of thousands to millions of points obtained from the photogrammetric processing of aerial photographs.

c. Build mesh (3D model) and build texture

Is a photo processing with a three-dimensional model (as it looks natural) that form the basis of making DEM (Digital Elevation Model). This section's resulting point clouds are classified as terrain (soil) and vegetation cover surface based on vertical coordinates. Building texture is forming a 3D physical model from the visual appearance in the photo coverage area. This stage aims to provide texture and color to the 3D model so that it can approach the actual state of the object.

d. Build DEM

DEM is an elevation model that is displayed in digital form as raster data composed of thousands of pixels and even more (picture elements), with each pixel having a height value.

e. orthophoto mosaics.

An orthophoto is an aerial photo that has been rectified/corrected for geometric errors.

2.2.3. Vegetation Indices from UAV. The vegetation indices were made by utilizing the pixel band value. Eight vegetation indices were used to analyze land cover types in this riparian ecosystem (table 1).

Table 1. Vegetation index used in this study

Vegetation Index	Formula	Author
GR (Green-Red ratio)	$GR = \frac{R}{G}$	[12]
GRVI (Green - red vegetation index)	$GRVI = \frac{G - R}{G + R}$	
MGRVI (Modified green - red vegetation index)	$MGRVI = \frac{G^2 - R^2}{G^2 + R^2}$	[13]
RGBVI (RGB-based vegetation index)	$RGBVI = \frac{G^2 - (B \times R)}{G^2 + (B \times R)}$	[13]
VARI (Visible atmospherically resistant index)	$VARI = \frac{G - R}{G + R - B}$	[14]
BGI_2 (Simple blue green ratio)	$BGI2 = \frac{B}{G}$	[15]
ExG (Excess green index)	$ExG = 2G - R - B$	[16]
NGBDI (Normalized green-blue difference index)	$NGBDI = \frac{G - B}{G + B}$	[17]

The calculation results of each index are grouped based on the type of land cover; from each land, cover calculated, the median value and standard deviation. Based on these statistical parameters, the index that best differentiates land cover is then analyzed.

3. Result and Discussion

The analysis of the vegetation index on selected riparian ecosystems in the Percut River shows that not all vegetation indices used to have good visualization for riparian vegetation characteristics (Figure 2). Aerial photos recorded using the DJI Phantom 4 Pro Drone provide excellent spatial resolution, but the number of bands is limited. Although RGB cameras can provide high spatial information about surfaces, their spectral resolution is limited. UAVs with multispectral sensors can be used to collect radiation data from spectral bands with almost no overlap and have near-infrared wavelength data. So that it can produce more accurate and precise information compared to colour index using an RGB digital camera [18]. Visual vegetation identification activities from orthophoto recordings with drones are sometimes faced with almost the same appearance even though the cover type is different. A visual band-based vegetation index was developed to find alternative solutions for vegetation identification.

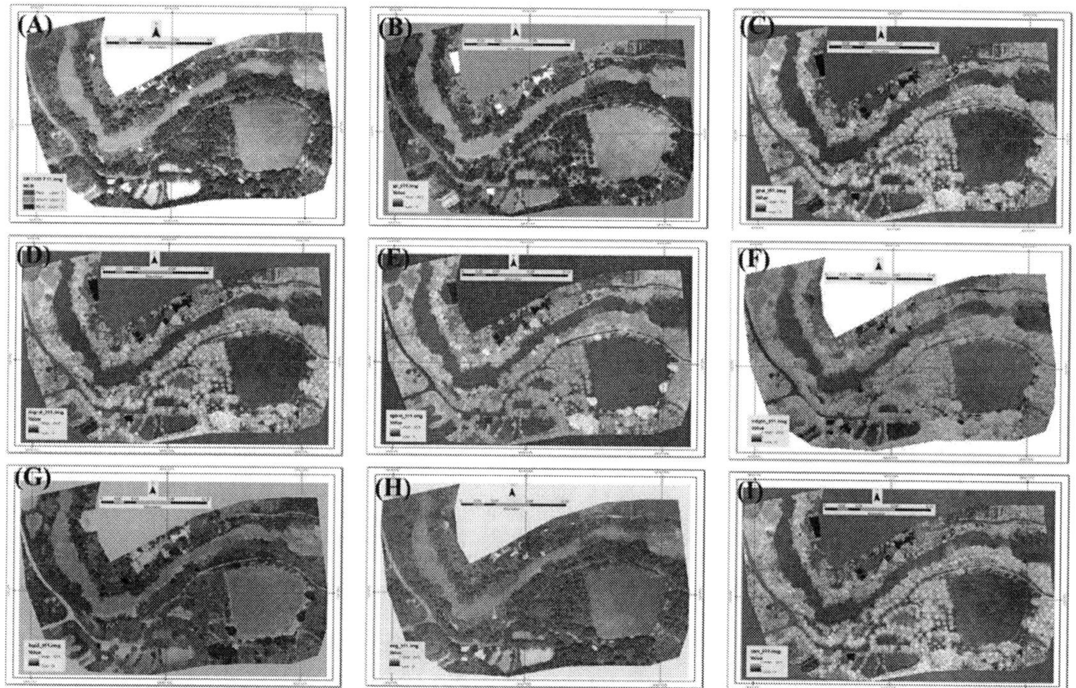

Figure 2. Orthophoto display (A), GR/green-red ratio (B), GRVI/green-red vegetation index (C), MGRVI/modified green-red vegetation index (D), RGBVI/RGB-based vegetation index (E), NGBDI/normalized green-blue different index (F), BGI2/simple blue green ratio (G), ExG/excess green index (H), VARI/ visible atmospherically resistant index (I)

RGBVI can function strongly to differentiate plant growth from NDVI. There is a saturation of the NDVI value at a later stage of growth. At the same time, RGBVI produces a range of reasonable values representing plant growth variability. In the RGBVI index, vegetation generally has a white visual appearance [19]. Vegetation that is quite dense has a lighter white color than the sparse vegetation (Figure 3).

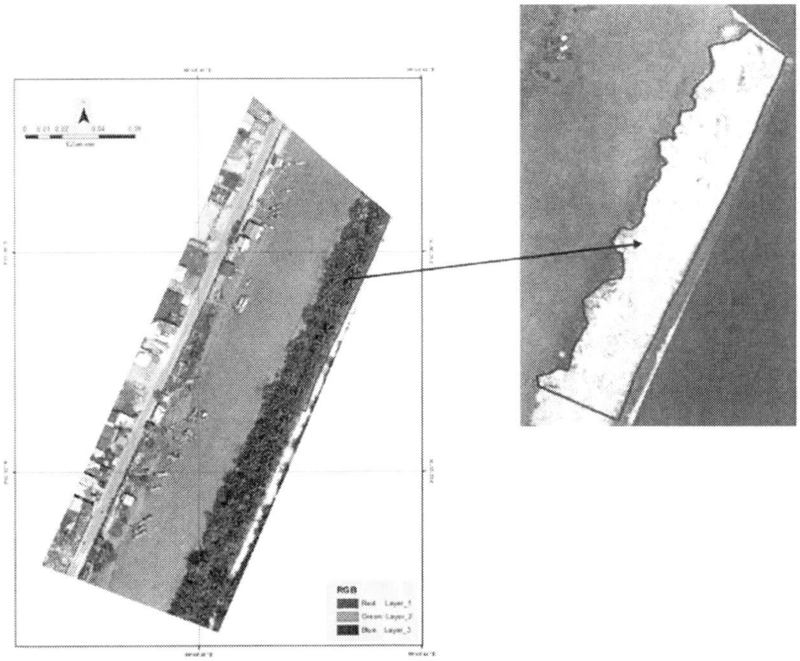

Figure 3. Orthophoto visual display and RGBVI index for vegetation cover types

The orthoimage visual display shows that the water and road cover types have the same brown color appearance (Figure 4). Water and roads on the BGI_2 index have medians and standard deviations that are not significantly different.

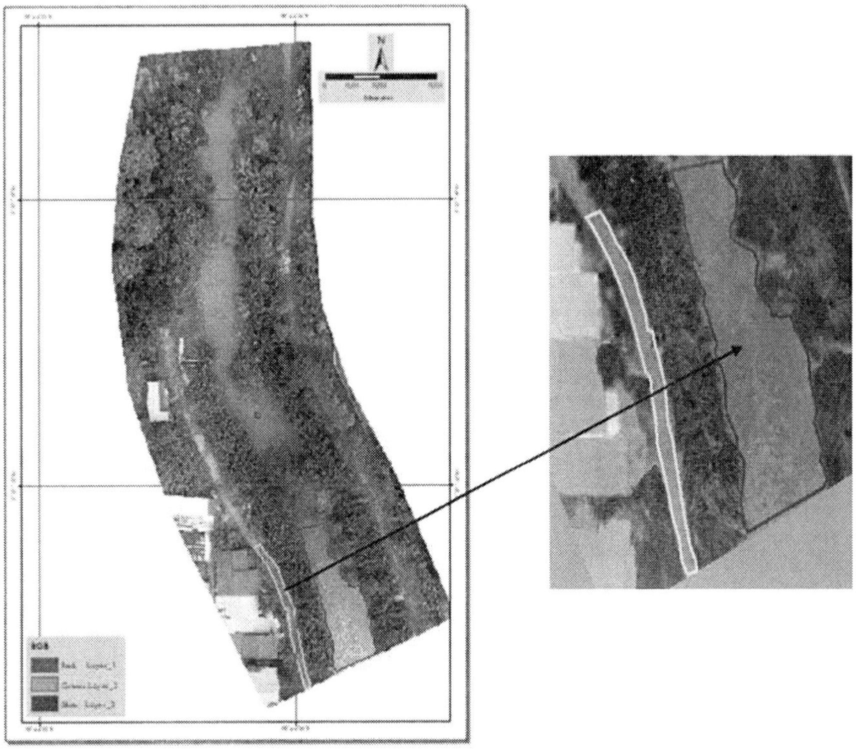

Figure 4. Display of orthoimage and BGI_2 index for water and road cover types

The mean value and standard deviation obtained are used as comparison material to show the difference between the eight vegetation indices used. RGBVI performed well in identifying riparian vegetation cover (Table 2). The GRVI, MGRVI, NGDBI, and VARI indices also provide a reasonably good value for vegetation cover identification but still have values that overlap with other land covers (Figure 5). The RGBVI index for vegetated cover shows a value that is significantly different from other land covers.

Analysis of riparian vegetation characteristics based on the RGB index is recommended using the RGBVI index (RGB-based vegetation index). The other seven vegetation indices used in this study have not properly distinguished the vegetation cover of built-up land, open land, rocks, settlements, water, and roads. The limitations of the UAV camera sensors used caused the limitation of vegetation indices.

Table 2. The mean value and standard deviation of the RGB-based vegetation index on various types of land cover in the riparian ecosystem of the Percut River

Land cover Type	Pixel median							
	GR	GRVI	MGRVI	RGBVI	NGBDI	BGI_2	ExG	VARI
Water Body	114 ± 33,27	92 ± 22,18	99 ± 20,18	82 ± 19,46	3 ± 16,81	50 ± 29,21	33 ± 24,45	62 ± 32,75
Vegetation	84 ± 25,35	123 ± 22,15	35 ± 20,06	123 ± 20,16	0 ± 25,51	11 ± 25,33	95 ± 25,46	100 ± 38,41
Road	103 ± 22,92	98 ± 18,41	08 ± 14,81	79 ± 11,97	7 ± 11,59	49 ± 19,70	53 ± 29,01	81 ± 29,84
Settlement	116 ± 37,56	88 ± 30,11	91 ± 33,83	64 ± 22,19	4 ± 20,91	58 ± 22,82	18 ± 58,96	55 ± 35,79
Open Area	125 ± 41,29	88 ± 27,64	93 ± 25,43	76 ± 20,13	5 ± 12,47	53 ± 24,76	25 ± 24,75	50 ± 29,36
Build up area	99 ± 39,12	104 ± 29,71	09 ± 27,26	81 ± 25,74	3 ± 16,38	59 ± 26,77	57 ± 37,21	29 ± 7,84
Rock	145 ± 23	75 ± 20	81 ± 21	61 ± 10	46 ± 1	166 ± 1	164 ± 2	57 ± 14

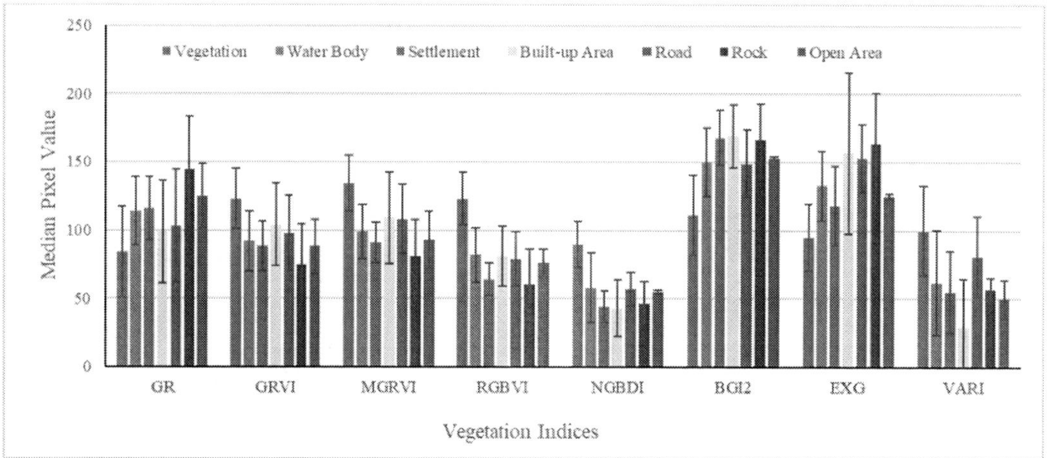

Figure 5. The median value and standard deviation of the RGB-based vegetation index on various types of land cover in the riparian ecosystem of the Percut River

4. Conclusion

Riparian vegetation analysis based on eight RGB-based vegetation indices was used from the UAV. There is one index that is the best, namely RGBVI. RGBVI index for vegetated cover shows a median value significantly different from other land covers. The other seven indices, namely GRVI, MGRVI, NGDBI, VARI, GR, BGI2, and ExG, have not given satisfactory results because the pixel values of vegetation cover still intersect with non-vegetation cover.

Acknowledgments
This research received support from the Universitas Sumatera Utara through the TALENTA scheme based on the Agreement/Contract Number: 369/UN5.2.3.1/PPM/SPP-TALENTA USU/2021 dated June 18, 2021.

References
[1] Sabater S, Elosegi A and Ludwig R 2018 Multiple stressors in river ecosystems (USA: Elsevier)
[2] Slamet B, Sinaga M T and Basyuni M 2018 in ICOSTEERR SCITEPRESS (Science and Technology Publications, Lda) **136**
[3] Slamet B, Syahputra O K H, Kurniawan H, Saraan M and Harahap M M 2021 *IOP Conference Series: Earth and Environmental Science* **912** 012089
[4] Machairiyah, Nasution Z and Slamet B 2020 *Limnotek: Perairan Darat Tropis di Indonesia* **27** 1
[5] Salemi L F, Groppo J D, Trevisan R, Marcos de Moraes J, de Paula Lima W and Martinelli L A 2012 *Journal of Hydrology* **454** 195
[6] Merritt D M 2013 *Treatise on geomorphology* **219**
[7] Gonzalez Del Tanago M, Martinez-Fernandez V, Aguiar F C, Bertoldi W, Dufour S, Garcia de Jalon D, Garofano-Gomez V, Mandzukovski D and Rodriguez-Gonzalez P M 2021 *J Environ Manage* **292** 112730
[8] Macfarlane W W, Gilbert J T, Jensen M L, Gilbert J D, Hough-Snee N, McHugh P A, Wheaton J M and Bennett S N 2017 *J Environ Manage* **202** 447
[9] Klemas V 2014 *Journal of Coastal Research* **30** 5
[10] Klemas V V 2015 *Journal of Coastal Research* **31** 5
[11] Daryaei A, Sohrabi H, Atzberger C and Immitzer M 2020 *Computers and Electronics in Agriculture* **177**
[12] Gamon J A and Surfus J S 1999 *The New Phytologist* **143** 1
[13] Bendig J, Yu K, Aasen H, Bolten A, Bennertz S, Broscheit J, Gnyp M L and Bareth G 2015 *International Journal of Applied Earth Observation and Geoinformation* **39** 79
[14] Gitelson A A, Kaufman Y J, Stark R and Rundquist D 2002 *Remote Sensing of Environment* **80** 76
[15] Zarco-Tejada P J, Berjón A, López-Lozano R, Miller J R, Martín P, Cachorro V, González M R and de Frutos A 2005 *Remote Sensing of Environment* **99** 271
[16] Woebbecke D M, Meyer G E, Von Bargen K and Mortensen D A 1995 S *Transactions of the ASAE* **38** 271
[17] Du M and Noguchi N 2017 *Remote Sensing* **9** 289
[18] Jang G, Kim J, Yu J-K, Kim H-J, Kim Y, Kim D-W, Kim K-H, Lee C W and Chung Y S 2020 *Remote Sensing* **12** 1
[19] Bareth G, Bolten A, Gnyp M, Reusch S and Jasper J 2016 *ISPRS - International Archives of the Photogrammetry, Remote Sensing and Spatial Information Sciences* **XLI-B8** 837

Fiber Quality of Yellow Bamboo (*Bambusa vulgaris* vitata) from Forest Area with Special Purpose Pondok Buluh, Simalungun Regency, North Sumatera Province

R Hartono[1]*, F V A Purba[1], A H Iswanto[1], T Priadi[2], J Sutiawan[1]

[1] Faculty of Forestry, Universitas Sumatera Utara, Kampus USU 2 Bekala, Pancur Batu, Deli Serdang 20353, Sumatera Utara, Indonesia

[2] Forest Product Department, Faculty of Forestry and Environment, IPB University, Jl. Ulin, Kampus IPB Dramaga, Bogor 16680, Indonesia.

Email: rudihartono@usu.ac.id

Abstract. Forest Area with Special Purpose (FASP) Forest Training Pondok Buluh has a forest area of around 1.272 ha, with much plant species diversity, such as Yellow Bamboo. Bamboo can be used as a material that has the potential as a raw material to replace wood in several uses, one of which is pulp and paper. This study aims to analyze the dimension and derivation of fiber dimensions of the yellow bamboo that grows in FASP Pondok Buluh. The method used in this study is to get the maceration process of the bamboo sticks from yellow bamboo, then observe under a microscope to see the dimension of the fiber, then calculate the value of the derivative of its fiber dimensions. The results of this study obtained the average dimension of yellow bamboo fiber is 1945 μm in fiber length, fiber diameter is 12.75 μm, lumen diameter is 4.20 μm, and fiber wall thickness is 4.28 μm. Then calculated the fiber dimension derivative and obtained a Runkle ratio of 2.32, felting power of 152.61, muhlsteph ratio of 88.73%, coefficient of rigidity of 0.34, and flexibility ratio of 0.32. Based on the result of the derivative of the fiber dimensions, yellow bamboo gets a score of 225-275, which is included in quality class II. This means that the yellow bamboo that grows in the Pondok Buluh area can be used as a raw material for pulp and paper.

1. Introduction

Around 1500 species originating from 75 genera have been identified worldwide, and 154 species from 24 genera are found in Indonesia [1]. The amount of bamboo that grows in Indonesia is more than 10% of the total bamboo species in the world [2]. Indonesia is one of the countries with the largest bamboo production. According to data from forestry statistics in 2020, bamboo production in Indonesia reached 11,303,317 stems, with the majority of production coming from the island of Java [3].

Bamboo is a versatile plant and is widely used by people in Indonesia. Bamboo is often used to build houses, bridges, etc. Bamboo is a plant that comes from the Poaceae family. The advantage of this plant is that it can reach its maximum height within a growing period of two to 4 months [4]. With this speed of growth, bamboo is a material that has the potential to replace wood in several uses because it is driven by the increasing demand for wood materials from year to year [5].

For bamboo to be used for its intended use, it is necessary to know the basic properties of bamboo, such as physical, mechanical, and anatomical properties, considering that various types of bamboo have various properties [6, 7, 8]. The age factor of bamboo also influences the moisture content in bamboo. In research [9] suggests that the moisture content in bamboo will continue to decrease with the increasing age of bamboo. Moisture content also decrease with increasing of bamboo height [8]

The anatomical structure of bamboo also influences its physical properties of bamboo. These include the number of vascular bundle proportions and the percentage of parenchyma. Density and moisture content is influenced by the percentage of vascular bundles, where the proportion of vascular bundles from the base to the top of the bamboo stem increases [1, 10], causing the moisture content to decrease. The density increases from the base to the top of the stem [6, 10]. Furthermore, fiber dimensions are fundamental in industrial purposes such as manufacturing composite boards and the pulp and paper industry.

In the pulp and paper industry, fiber dimensions are measured in the form of fiber length, fiber diameter, and wall thickness. The longer the fiber, the pulp produced will have good strength, while the thinner the fiber wall will produce a paper that is more solid and has a high cracking power [11]. This will affect the value of the derivative of the fiber, which will then be based on the scoring results of the value of the derivative of the fiber.

The fiber derivative values include the runkle ratio, felting power, muhlsteph ratio, rigidity coefficient, and flexibility ratio. Each of these criteria will later become a reference for whether or not bamboo is suitable as raw material for pulp and paper. The basic requirement for a material to be used as raw material is to refer to the runkle number produced. If the runkle number produced is more than 1, then the material will be less suitable as raw material for pulp and paper [11, 12].

Yellow bamboo is one type commonly used, given its unique color, so it is often used as handicrafts and furniture and can be used as medicine [13]. This bamboo can also be used for constructional and industrial purposes. The results of previous research stated that yellow bamboo has good MOE, MOR, and density values for furniture and light construction [4, 6, 7]. In research [12], yellow bamboo is also suitable for use as raw material for pulp and paper because it has a Runkle value of less than 1.

Researchers from various regions have carried out various studies on yellow bamboo. The results of these studies indicate diversity in terms of fiber dimensions and physical and mechanical properties. However, there is still little research on the properties of yellow bamboo from FASP Pondok Buluh. The purpose of this study was to analyze the nature and quality of Yellow bamboo fiber from Forest Area with Special Purpose (FASP) Pondok Buluh so that it can add information as the purpose of using this bamboo in the future.

2. Materials and methods

2.1. Material

A total of three yellow bamboo sticks (*Bambusa vulgaris* vitata) from the same clump were taken with a height of up to 15 m and an average diameter of 7.7 cm at FASP Pondok Buluh (02°46'08" North Latitude. 98°59'03" BT) as shown in Figure 1.

2.2. Sample Preparation

The disk-shaped bamboo samples with a thickness of 5 cm were taken at the 1st, 3rd, 5th, 7th, and 9th meters, then chopped into the size of matchsticks, and then the fiber separation process was carried out by maceration using the Schultze method (Silitonga et al. 1973) in [14]. The samples were boiled with a mixture (H_2O_2 + CH_3COOH) solutions with a ratio of v/v at a temperature of ±95 °C. Then free from acidification with distilled water, dehydrated with graded alcohol (20%, 40%, 60%, and 96%) and given safranin.

Figure 1. Yellow bamboo sample from FAST Pondok Buluh

2.3. Determination Dimension of Fiber

Fiber dimensions were observed using a ZEISS Primostar microscope. The fiber dimensions measured include fiber length, fiber diameter, and lumen diameter. 4x magnification is used to measure fiber length and 40x to measure fiber and lumen diameter. The fiber wall thickness is obtained from the difference between the fiber diameter and the lumen diameter and then divided by two. The calculated fiber dimension derivative value are runkell ratio (2x cell wall thickness/lumen diameter), felting power (fiber length/fiber diameter), muhlsteph ratio ((fiber diameter squared-lumen diameter squared)/fiber diameter squared) x 100%, coefficient of rigidity (fiber wall thickness/fiber diameter) and flexibility ratio (lumen diameter/fiber diameter). Each height was measured in 30 replicates.

24. Criteria for Suitability as Raw Material for Pulp and Paper

The dimensions of fiber and its derivatives were calculated at various heights of yellow bamboo. Then the score was calculated and compared to its suitability with the Indonesian Timber Assessment Criteria as raw material for Pulp and Paper (Silitonga et al. 1973) in [14].

3. Results and discussion

3.1. Fiber Dimension

Measuring fiber dimensions includes several parts: fiber length, fiber diameter, lumen diameter, and fiber wall thickness. Measurement of fiber dimensions can be seen in Table 1. The measured fibers are intact, not broken, or moldy.

Table 1. Average of Yellow Bamboo Fiber Dimension

Height (m)	Fiber Length (μm)	Fiber Diameter (μm)	Lumen Diameter (μm)	Wall Thickness (μm)
1	2161 ± 264	13.47 ± 1.97	5.68 ± 1.80	3.89 ± 0.33
3	1951 ± 143	13.56 ± 1.38	5.59 ± 1.74	3.98 ± 0.26
5	1929 ± 187	12.74 ± 1.30	4.27 ± 1.60	4.23 ± 0.25
7	1857 ± 207	12.36 ± 0.85	2.95 ± 0.78	4.73 ± 0.18
9	1827 ± 162	11.61 ± 1.32	2.50 ± 0.78	4.55 ± 0.43
Average	1945 ± 193	12.75 ± 1.36	4.20 ± 1.34	4.28 ± 0.29

Based on Table 1, it can be seen that only yellow bamboo with a height of 1 m has a fiber length of more than 2000 μm, while at other heights, it is below 2000 μm. The average length of yellow bamboo fiber is 1945 μm, which is included in the medium-sized fiber class with quality class II.

Figure 2 shows a decrease in the length of the bamboo fiber from the base to the top. This trend of fiber length was also found in studies [7, 15]. A different trend was found in studies where the length of the bamboo fiber will increase towards the middle of the stem, then decrease towards the top of the stem [16-18].

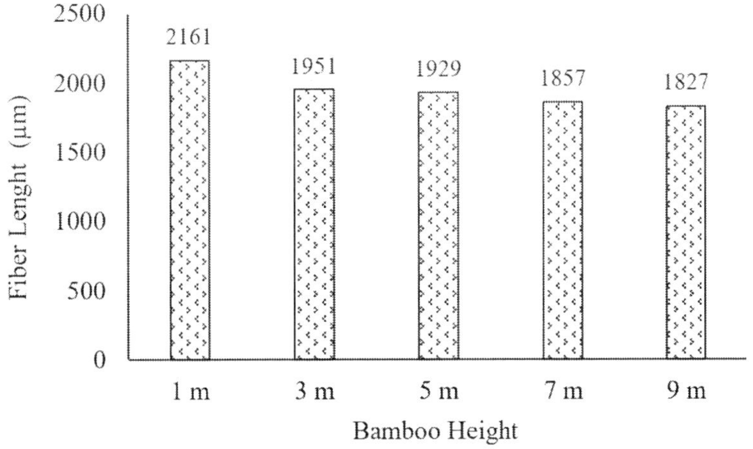

Figure 2. Fiber length at various height

The average fiber diameter obtained is 12.75 µm. The largest diameter is at the 3rd-meter height of 13.56 µm, and the smallest is at the 9th-meter height of 11.61 µm. The average lumen diameter measurement result is 4.20 µm, with the largest lumen diameter found at the 1st meter height of 5.68 µm, and the smallest lumen diameter at the 9th-meter height at 2.50 µm. Mature fibers generally consist of a cell wall and lumen. Fiber walls can contain lignin so that they can be very thick and close to the fiber lumen. In the measurement of fiber wall thickness, the average fiber wall thickness was 4.28 µm, with the highest value found at the 7th-meter height of 4.73 µm. The smallest wall thickness was found at the 1st-meter height of 3.89 µm. The ratio between fiber wall thickness and fiber lumen diameter must be considered in selecting pulp and paper raw materials because it will affect the Runkle ratio value. A good fiber used as raw material for pulp and paper, in general, is a fiber with a lumen diameter larger than the fiber wall thickness [19].

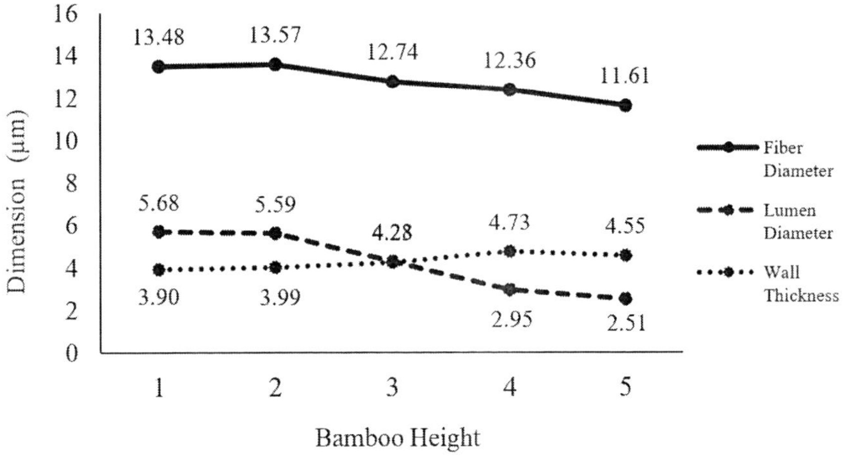

Figure 3. Diameter, lumen, and wall thickness of bamboo fiber at several heights.

A comparison of the fiber dimensions observed with other studies can be seen in Table 2. The results of the study of fiber length obtained an average of 1945 μm. The observed fiber length looks shorter when compared to other studies. In fiber diameter, it can be seen that the observed fibers are larger than the study [16] but smaller than other studies. In the lumen diameter, it can be seen that the observed fibers are larger when compared to the study [15] but smaller when compared to other studies. In the fiber wall thickness, it can be seen that the observed fibers are thicker than [15, 20] but thinner than the study [15]. Loiwatu and Manuhuwa [21] stated that there was a significant effect between the place of growth and the dimensions of the bamboo fiber.

Tabel 2. Comparison of Yellow Bamboo fiber dimension with the various investigation

Citation	Location	Fiber Lenght (μm)	Fiber Diameter (μm)	Lumen Diameter (μm)	Wall Thickness (μm)
Wahab *et al.* *(2010)*	Sabah, Malaysia	3600	16.9	2.5	7.1
Azeez *et al.* *(2016)*	Ado Ekiti, Nigeria	2074	16.9	8.9	4
Yormann *et al.* *(2020)*	Buenos Aires, Argentina	2262.62	14.21	5.77	2.68

3.2. Fiber Derivatives Value

The fiber derivative value is calculated based on the fiber dimensions that have been measured. The results of the calculation of fiber derivative values can be seen in Table 3. After determining the value of fiber derivatives, the quality class of fiber as raw material for pulp and paper is presented in Table 4. Runkle ratio (RR) on Yellow Bamboo ranged from 1.37-3.64, with an average Runkle value of 2.32. For good pulping, the fiber with a Runkle value is smaller than one because it has a thin cell wall and a large lumen diameter so that the fiber can be completely flat in the pulp sheet and the bonds between the fibers are good [22]. In this case, yellow bamboo fiber has an RR of more than 1, which means that if it is used as a raw material for pulp and paper, more chemicals are needed to break down the lignin contained in the fiber wall.

Table 3. Fiber Derivatives Value of Yellow Bamboo

Height (m)	*Runkle Ratio*	*Felting Power*	*Muhlsteph Ratio (%)*	*Coefficient of Ridigity*	*Flexibility Ratio*
1	1.37	160.34	82.24	0.29	0.42
3	1.43	143.83	83.00	0.29	0.41
5	1.98	151.40	88.74	0.33	0.34
7	3.21	150.19	94.31	0.38	0.24
9	3.64	157.32	95.35	0.39	0.22
Rata-rata	2.32	152.61	88.73	0.34	0.32

The felting power (FP) value on Yellow Bamboo, based on Table 3, has an overall average value of 152.61. A high feeling of power indicates that the fibers are arranged more tightly to produce a paper with a relatively high tear strength [23]. Table 4 shows the felting power gets a score of 100 at each height because the value is more than 90.

Table 4. Yellow Bamboo Quality Class based on LPHH criteria, 1976

Height	Fiber Derivatives Value						Total	Class
	Fiber Length	RR	FP	MR	RC	FR		
1	100	-	100	25	25	25	275	II
3	50	-	100	25	25	25	225	II
5	50	-	100	25	25	25	225	II
7	50	-	100	25	25	25	225	II
9	50	-	100	25	25	25	225	II

The muhlsteph ratio (MR) value on Yellow Bamboo from the 1st to the 9th meter ranged from 82.24-95.35%. The value of the muhlsteph ratio could affect the pulp's density. Table 4 shows that each height score is 25 and is in class III because the MR value is more than 60.

The coefficient of rigidity (CR) in Yellow Bamboo fiber ranges from 0.29 to 0.39, with an average value of 0.33. A low CR value will produce a paper that is not stiff and has good fiber bonds [24]. Based on Table 4, the score for CR is 25 and is in class III because the CR value obtained is more than 0.15.

The flexibility ratio (FR) in yellow bamboo ranges from 0.22 to 0.42, with an average of 0.32. The flexibility ratio is one of the important criteria in determining the feasibility of raw materials such as pulp and paper. A high flexibility ratio will have high tensile strength with good bonding between fibers. The score based on the FR value obtained is 25 and is in class III because the FR value obtained is less than 0.5.

Overall, the accumulated score of fiber quality at all heights is in grade II as shown at Table 4. This indicates that Yellow bamboo fiber (*Bambusa vulgaris* vittata) taken from FASP Pondok Buluh is suitable for use as raw material for pulp and paper.

4. Conclusion

Yellow Bamboo (*Bambusa vulgaris* vitata) obtained lengths ranging from 1827-2161 μm with an average of 1945 μm, fiber diameters between 11.61-13.56 μm with an average of 12.75 μm, lumen diameter ranging from 2.50-5.68 μm with an average of 4.20 μm, and fiber wall thickness ranged from 3.89-4.73 μm with an average of 4.28 μm. The value of the derivative of the fiber dimensions on Yellow Bamboo with a height of the 1st meter to the 9th meter obtained an average Runkle ratio value of 2.32, felting power 152.61, muhlstep ratio 88.73%, coefficient of rigidity 0.34 and flexibility ratio 0.32. Yellow Bambu taken from FASP Pondok Buluh is included in quality class II with a total score of 225-275, which means it is still feasible if used as raw material for pulp and paper.

5. Acknowledgements

We would like to express my sincerely thanks to the Directorate of Research and Community Service. Ministry of Research and Technology/National Research and Innovation Agency, for grand of Penelitian Dasar Unggulan Perguruan Tinggi (PDUPT) (No: 059/E5/PG.02.00/PT/2022)

References

[1] Maulana M I, Jeon W S, Purusatama B D, Kim J H, Prasetia D, Yang G D, Savero, M A, Nawawi D S, Nikmatin S, Sari R K, Febrianto F, Lee S H, Kim N H 2022 *BioResources* **17**(1) 1442-53
[2] Mainaki R, Maliki R Z 2020 *Jurnal Kajian Ilmu dan Pendidikan Geografi* **4**(1) 44-54
[3] Badan Pusat Statistik 2020 *Statistik Produksi Kehutanan* (Jakarta: BPS-Statistic Indonesia) p 22-24
[4] Awalluddin D, Ariffin M A M, Osman M H, Hussin M W, Ismail M A, Lee H S, Lim N H A S 2017 *In MATEC Web of Conferences* **138** 01024
[5] Javadian A, Smith I F, Saeidi N, Hebel D E 2019 *Frontiers in Materials* **6**(15) 1-18.

[6] Adam N, Jusoh I 2019 *Transactions on Sci.Techno* **6**(1-2) 95-101.
[7] Siam NA, Uyup MKA, Husain H, Awalludin MF 2019 *BioReseurces* **14**(2) 3925-43.
[8] Hartono R, Iswanto A H, Priadi T, Herawati E, Farizky F, Sutiawan J, Sumardi I 2022 *Polymers* **18** 6848.
[9] Ye F, Fu W 2018 *J. Mater Civil Eng* **30**(2) 05017004.
[10] Villareal J F, Marasigan O S, Mendoza R C, Alipon M A, Abasolo W P 2020 *Philippine J. Sci,* **149** 1005-13.
[11] Sugesty S, Kardiansyah T, Hardiani H 2015 *Procedia Chemistry* **17** 194-199.
[12] Egbewole Z T, Rotowa O J, Omoake P O 2015 *PAT* **11**(2) 188-202.
[13] Nuriyatin N, Sofyan K. 2011 *Jurnal Penelitian Hasil Hutan* **29**(4) 287-300.
[14] Aprianis Y, Rahmayanti S 2009 *Jurnal Penelitian Hasil Hutan* **1** 11-20.
[15] Wahab R, Mustapa M T, Sulaiman O, Mohamed A, Hassan A, Khalid I 2010 *Sains Malaysiana* **39**(4) 571-79.
[16] Yormann G E, Rúgolo Z E, Apóstolo N M 2020 *Flora* **263** 151523.
[17] Wang S, Lin S, Li W, Ding Y 2016 *Forest Products Journal* **66**(5-6) 319-25.
[18] Darwis A, Iswanto A H, Jeon W S, Kim N H, Wirjosentono B, Susilowati A, Hartono R 2020 *Bioresources* **15**(3) 6617-26.
[19] Sadiku N A, Abdukareem K A 2019 *Maderas Ciencia y tecnologia* **21**(2) 239-48.
[20] Azeez M, Andrew J E, Sithole B B 2016 *Maderas. Ciencia y tecnología* **18**(1) 65-78.
[21] Loiwatu M, Manuhuwa E 2008 *Agritech* **28**(2) 76-83.
[22] Rusch F, Ceolin G B, Hillig É 2019 *Pesqui. Agropecu. Trop* **49** 55007
[23] Nasser R A, Hiziroglu S, Abdel-Aal M A, Al-Mefarrej H A, Shetta N D, Aref I M 2015 *Measurement* **62** 179-86
[24] Sadiku N A, Oluyege A O, Ajayi B 2016 *J. Bamboo Ratt* **15** 33-43.

Fiber Quality of Betung Bamboo (*Dendrocalamus asper*) from Forest Area with Special Purpose (FASP) Pondok Buluh, Simalungun, North Sumatera

R Hartono[1*], F Farizky[1], J Sutiawan[1], I Sumardi[2] and E Suhesti[3]

[1]Faculty of Forestry, Universitas Sumatera Utara, Jl. Tri Dharma Ujung No. 1, Kampus USU, Medan 20155, Indonesia
[2]School of Life Sciences and Technology, Institut Teknologi Bandung. Jl. Ganesa No.10, Bandung 40132, Indonesia.
[3]Faculty of Forestry, Universitas Lancang Kuning, Jl. Yos Sudarso KM 8, Umban Sari, Pekan Baru 28266, Indonesia.

[*]Email: rudihartono@usu.ac.id

Abstract. The FASP Pondok Buluh is located in Dolok Panribuan District, Simalungan Regency, Sumatera Utara, which has uniqueness due to the number of bamboo species found in this area, such as Betung Bamboo. This bamboo has a long fiber and is suitable for pulp and paper materials. This research aims to measure the dimension of Betung Bamboo and its fiber derivatives value for the eligibility of pulp and paper material. A total of 3 stems of Betung Bamboo were cut, and samples were taken in the form of a disc with a length of 5 cm at heights of 1, 3, 5, 7, 9, 11, 13, and 15 meters. Then the sample was cut into sticks, and separated fibers were made using the maceration process. Furthermore, the dimensions of the sample were measured using a light microscope, namely fiber length, fiber diameter, and fiber wall thickness, and then the measurement for the fiber derivatives value of the bamboo fiber. The results of this research obtained the average dimension of bamboo fiber is 2402 μm for fiber length, fiber diameter is 18.18 μm, lumen diameter is 8.73 μm, and fiber wall thickness is 4.91 μm. The fiber length graph shows the decreased fiber length along the bottom to the top of the bamboo culm and the wavy trend for the fiber diameter on some bamboo heights. The measurement of the fiber derivatives values obtained the average Runkle ratio 1.15, felting power 133.63, mulhsteph ratio 76.91%, rigidity coefficient 0.27, and flexibility ratio 0.48. Then the quality class of bamboo fiber was determined. A score of 275-325 was obtained with quality class II, meaning that Betung Bamboo is grown in the Pondok Buluh area. It is suitable for use as raw material for pulp and paper because it has long fibers, is easy to flatten when milled, and has good fiber bonds.

1. Introduction

Indonesia has the second largest bamboo production after China, with 154 species from 1250 to 1500 bamboo species found in the world. Data shows that bamboo production reaches 11.3 million stems in Java and becomes the largest bamboo producer, followed by Bali and Nusa Tenggara with a total production of 14 thousand stems, and the Kalimantan region with a total production of 875 stems [1].

Bamboo is a member of the Gramineae family and is also known as gigantic grass. It grows in clumps and is made of several stems that develop progressively over time, beginning with bamboo shoots and immature stems and maturing at 4-5 years of age [2]. The general public is familiar with bamboo due to its beneficial features, which include strong stems, ductile, straight, flat, hard, easy to split, easy to shape, easy to work with, and lightweight is portable. Given the variety of attributes, each bamboo species possesses, the properties of bamboo, such as its physical, mechanical, and anatomical characteristics, must be investigated to determine how bamboo will be used in the future [3].

The usage of bamboo in construction is directly correlated with its physical and mechanical characteristics. According to several studies, bamboo's mechanical characteristics change depending on height. According to several earlier research, bamboo's mechanical strength increases from the bottom to the top of the bamboo stem. This results from the bamboo stems' rising density from the bamboo's base to its apex [4-7]. Adam and Jusoh [6] stated that the bamboo's water content impacts its strength, which decreases from the base to the top of the bamboo. The amount of moisture also correlates with bamboo's physical characteristics, namely the proportion of vascular bundles, which will grow in number from the base to the top of the bamboo [8].

The anatomical characteristics of bamboo also have an impact on a variety of uses. These are used as a source of pulp and paper. Bamboo fibers are easy to bleach because they tend to be long, making them suitable for pulp manufacture [9]. One of the criteria used to evaluate the quality of the paper is the bamboo fiber's size. According to previous research [10], the pulp quality will depend on the dimensions of the fiber, specifically the fiber length, diameter, lumen thickness, and fiber wall. The dimensions of this fiber will affect the bond between the fibers to produce a smoother paper, high tensile strength, and good folding resistance on paper.

The fiber dimension will have an impact on the fiber's derived value. Some criteria to calculate the fiber derivatives include the Runkle ratio, felting power, mulhtstep ratio, rigidity coefficient, and flexibility ratio. The Runkle ratio is one of the important criteria to determine the suitability of material as the raw material for pulp and paper. The Runkle ratio will affect the strength properties of the paper produced. The good material must have a Runkle ratio of less than 1 [11]. A previous study reported that the pulp and paper produced enhanced with a lower Runkle ratio, while the stiffness of the paper produced decreased with a greater pulp rigidity coefficient [12].

One bamboo species widely known and used by the community is Betung bamboo (*Dendrocalamus asper*). The properties of bamboo Betung has been carried out for many purposes, such as for construction, pulp, and paper. Bamboo Betung has a fiber length of 4.693 µm, a fiber diameter of 25 µm, lumen diameter of 7 µm, and fiber wall thickness of 9 µm [13]. A study on the dimensions of bamboo Betung fibers was also carried out by [7] with a fiber length of 2996 µm, the fiber diameter of 23.3 µm, lumen diameter of 6,1 µm, and fiber wall thickness of 8.6 µm. Another study was also conducted by [14] with a fiber length of 2355.65 µm. Thus the difference in the place of growth influences the dimensions of the fiber. This is also explained by [15] that the place of growth significantly affects fiber length, fiber diameter, lumen diameter, and cell wall thickness due to several factors such as nutrient conditions and rainfall at the growing site.

Thus, there have been many studies on the properties of bamboo Betung. However, there is still less research on the properties of bamboo Betung from Pondok Buluh Forest Area with Special Purpose (FASP). This study aims to analyze the quality of Betung bamboo fiber from FASP Pondok Buluh to add information about the bamboo and its intended use in the future.

2. Materials and Methods

2.1. Materials

Three stems of Betung Bamboo (*Dendrocalamus asper*) were taken with an approximate length of 18 m and 17 cm diameter from FASP Pondok Buluh (02° 46' 12" LU, 98° 59' 03" BT). The stems were taken from the same clumps, then cut into several subsections from the height of 1st, 3rd, 5th, 7th, 9th, 11th, 13th, and 15th meters.

2.2. *Sample preparation*
A sample disc (5 cm x 5 cm) was taken from each subsection away from the nodes. The disc then turned into a matchstick-sized, and the separated fiber was made through the maceration process based on [16]. The sample was cooked with a mixed solution (v/v) of hydrogen peroxide (H_2O_2) and acetic acid (CH_3COOH) at a temperature of ±95°C until it softened. Then the sample was freed from acidity with distilled water, dehydrated with alcohol concentrations of 20%, 40%, 60%, and 96%, and treated with drops of safranin.

2.3. *Determination dimension of fiber*
The fiber dimensions were measured using an Apple iPad and a Zeiss Primostar light microscope. The fiber length was measured at a magnification of 4x, while the fiber and lumen diameter was measured at 40x. The fiber wall thickness was obtained with the formula $w = \frac{d-l}{2}$. The observed fiber must be the full-sized fiber with less mold.

2.4. *Calculation of fiber derivatives value*
The following formula can calculate the derived value of fiber dimensions:

$$Runkle\ Ratio = \frac{2w}{l} \quad (1)$$

$$Felting\ Power = \frac{L}{d} \quad (2)$$

$$Mulhsteph\ Ratio\ (\%) = \left(\frac{d^2 - l^2}{d^2}\right) x\ 100 \quad (3)$$

$$Ridigity\ Coefficient = \frac{w}{d} \quad (4)$$

$$Flexibility\ Ratio = \frac{l}{d} \quad (5)$$

The notation (w) indicates the fiber wall thickness, (l) the diameter of the lumen, (L) the length of the fiber, and (d) the diameter of the fiber.

2.5. *Data analysis*
The results of fiber dimensions and their derivative were analyzed in a description by comparing the Indonesian Timber Assessment Criteria as Raw Materials for Pulp and Paper (Silitonga et al, 1972) in [17]. The Indonesian assisment criteria was shown in Table 1.

Table 1. Requirements and grades of wood fiber as raw material for pulp and paper

No.	Description	Class 1		Class 2		Class 3	
		Requirement	Score	Requirement	Score	Requirement	Score
1	Fiber Length	>2000	100	1000 - 2000	50	<1000	25
2	Runkle Ratio	<0,25	100	0,25 - 0,50	50	>0,5 - 1	25
3	Felting Power	>90	100	50 - 90	50	<50	25
4	Mulsteph Ratio	<30	100	30 - 60	50	>60	25
5	Ridigity Coefficient	<0,10	100	0,1 - 0,15	50	>0,15	25
6	Flexibility Ratio	>0,8	100	0,5 - 0,8	50	<0,5	25
		450 - 600		225 - 449		<225	

3. Results and Discussion

3.1. Fiber dimension

Measuring fiber dimensions includes several criteria: fiber length, fiber diameter, lumen diameter, and fiber wall thickness. The average results of the fiber dimension measured are presented in Table 2. The average bamboo fiber length from the first to the fifteenth meter is 2402 m. According to the categorization of wood fiber length, Betung Bamboo is categorized as a long fiber, and according to the evaluation of raw materials for pulp and paper, Betung Bamboo fiber is included in the quality class I. Figure 1 shows a decreasing trend in the length of the bamboo fiber from the base to the top. The trend in fiber length is similar to other earlier research [7,8,18]. However, this study negatively correlates with some investigations. This study reported that the longest fiber length of bamboo was found in the middle section of the bamboo. This has concluded that each bamboo brings its characteristic [16,19]

Table 2. Average of Betung Bamboo fiber dimension

Height (m)	Fiber Length (µm)	Fiber Diameter (µm)	Lumen Diameter (µm)	Wall Thickness (µm)
1	2561 ± 347	16.67 ± 1.69	6.29 ± 2.13	5.19 ± 0.36
3	2522 ± 250	16.28 ± 2.89	7.55 ± 1.85	4.37 ± 0.59
5	2533 ± 253	19.40 ± 1.70	8.88 ± 2.58	5.26± 0.52
7	2514 ± 420	19.41 ± 1.66	9.22 ± 2.53	5.09 ± 0.56
9	2426 ± 336	16.33 ± 2.99	7.79 ± 2.28	4.27 ± 0.45
11	2347 ± 193	20.50 ± 3.36	10.42 ± 3.98	5.04 ± 0.62
13	2275 ± 318	16.70 ± 3.47	8.78 ± 2.11	4.63 ± 0.74
15	2040 ± 175	20.70 ± 2.18	10.92 ± 2.12	5.40 ± 0.37
Average	2402 ± 287	18.18 ± 2.49	8.73 ± 2.45	4.91 ± 0.53

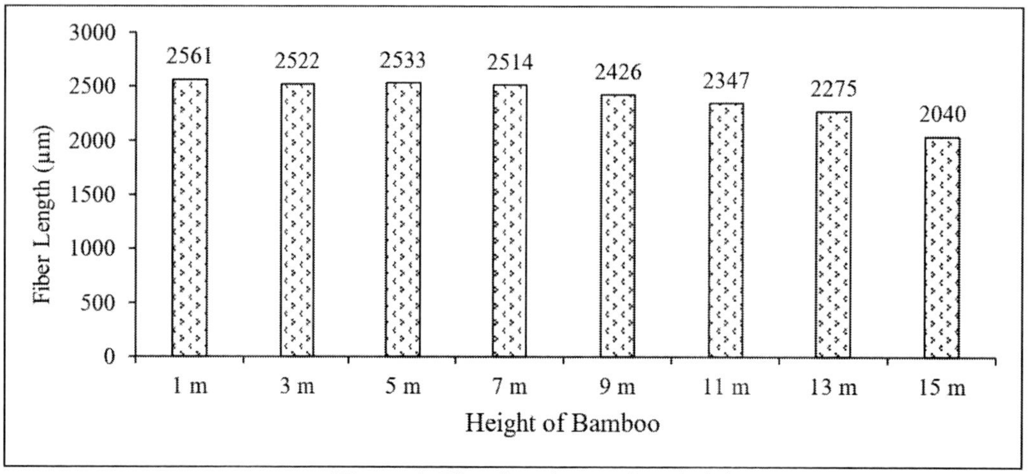

Figure 1. Fiber length at various height

Based on Table 2, the average diameter of the Bamboo Betung fiber is 18,18 µm, so it is classified as a medium diameter fiber. The average fiber wall thickness obtained is 4,91 µm, and the average lumen diameter is 8,73 µm. According to Figure 2, the fiber and lumen diameter is increased from the base to the top of the stem. A previous study stated no significant differences in fiber and lumen diameter along the bamboo culms [16,19].

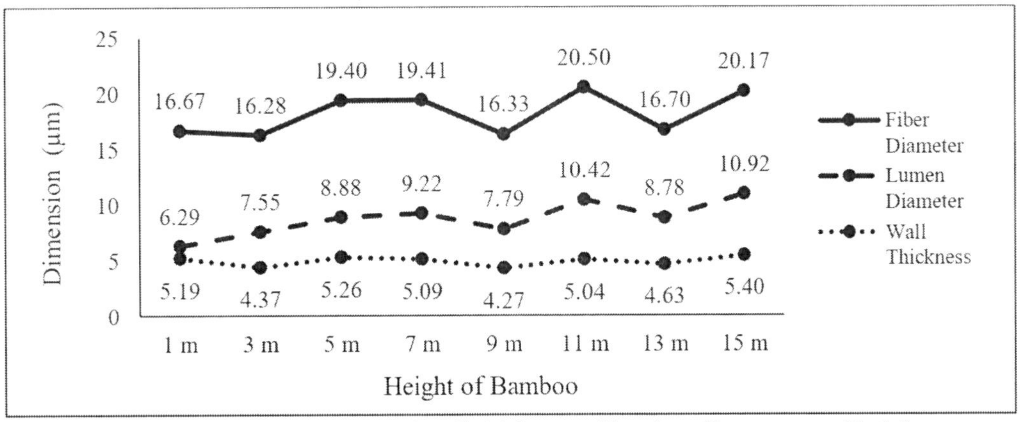

Figure 2. Diameter, lumen, and wall thickness of bamboo fiber at several heights

Table 3 shows a comparison between the Betung Bamboo observed with other investigations. Table 3 shows that, compared to other studies, the fiber length of bamboo Betung observed is shorter. The diameter of the bamboo fiber appears to be greater, but it is smaller compared to previous investigations [20]. Compared to other investigations [20], the fiber lumen diameter observed appears to be greater. The fiber wall thickness observed appears thicker, but compared to other research, it appears thinner. This demonstrates how the size of bamboo fiber will vary depending on where bamboo is grown [15,21].

Tabel 3. Comparison of Betung Bamboo fiber dimension with the various investigation

Citation	Region	Fiber Length (µm)	Fiber Diameter (µm)	Lumen Diameter (µm)	Fiber Wall Thickness (µm)
Fatriasari and Hermiati	Bogor, Indonesia	4693	26	7	9
Loiwatu and Manuhua	Seram, Indonesia	3900	4,91	3,10	0,9
Siam *et al*	Selangor, Malaysia	2998	23.3	6,1	8,6
Jansiri *et al*	Nakhon Ratchasima, Thailand	2890	24,85	6,66	17,62

3.2. Fiber derivatives value

The fiber derivatives value is determined based on the fiber dimension. The result of the fiber derivatives value is presented in Table 4. The obtained Runkle ratio (RR) has a range of 0.97-1.65, with the 11th meter having the smallest value and the 1st meter having the biggest value. To accurately measure the quality of the produced paper, the Runkle ratio is the most crucial parameter that needs to be examined. The flexibility of the paper generated will be affected by the Runkle ratio value, which must be less than 1. If the Runkle ratio value is high, more chemicals are required to break down the lignin during the fiber smelting process [22].

Table 4. Fiber derivatives value of Betung Bamboo (*Dendrocalamus asper*)

Height	Derivatives Value				
	RR	FP	MR(%)	RC	FR
1	1.65	153.61	85.77	0.31	0.38
3	1.16	154.85	78.53	0.27	0.46
5	1.18	130.61	79.04	0.27	0.46
7	1.10	129.53	77.43	0.26	0.48
9	1.10	148.57	77.26	0.26	0.48
11	0.97	114.50	74.17	0.25	0.51
13	1.05	136.17	72.40	0.28	0.53
15	0.99	101.16	70.70	0.27	0.54
Average	1.15	133.63	76.91	0.27	0.48

The felting power (FP) values obtained ranged from 101.16 to 154.85, with the lowest value at the height of 15th meters and the highest value at the height of 3th meters, resulting in a quality class I. A high felting power value will positively impact the paper's flexural strength and folding durability. A strong link will form between the fibers if the value obtained is high, leading to great folding resistance [13].

The measured muhlsteph ratios (MR) were in the range of 70.70% to 85.77% and classified in quality class III. The 15th meter has the lowest MR value, and the 1st has the highest value. The smoothness of the produced paper and the plasticity between the fibers will be affected by the muhlsteph ratio's value. The resulting muhlsteph ratio has a high value, producing a plastic paper that is difficult to rip when folded.

The rigidity coefficient (RC) is obtained between 0.25-0.31 and classified in the quality class III. The smallest RC value was found at the 11th meter, and the largest value was found at the 1st meter. According to Sugesty et al [12], a good rigidity coefficient value is if it has a low value because the lower the rigidity coefficient value, the fiber will be more flexible, and the paper produced will not easily be torn when given a tensile load.

The obtained flexibility ratio value ranges from 0.38 to 0.54, classifying it in quality class III. The 1st meter has the lowest FR value, and the 15th has the highest FR value. The flexibility ratio influences the flexibility and stiffness of the manufactured paper. The paper produced has high tensile strength and a high flexibility ratio [23].

After obtaining the value of the fiber derivative, the quality score of bamboo fiber as the suitable raw material for pulp and paper was determined. The Score bamboo fiber quality score can be seen in Table 5.

Tabel 5. Quality score of Betung Bamboo fiber

Height	Score of Fiber Derivative Value						Total	Class
	Fiber Length	RR	FP	MR	RC	FR		
1	100	-	100	25	25	25	275	II
3	100	-	100	25	25	25	275	II
5	100	-	100	25	25	25	275	II
7	100	-	100	25	25	25	275	II
9	100	-	100	25	25	25	275	II
11	100	25	100	25	25	50	325	II
13	100	-	100	25	25	50	300	II
15	100	25	100	25	25	50	325	II

According to Table 5, the fiber of Betung Bamboo from 1st to 15th meter is classified as class II (with a score of 275 to 325). This concludes that the fiber of Betung Bambo (*Dendrocalamus asper*) from FASP Pondok Buluh is suitable as the raw material for pulp dan paper due to the long fiber and high felting power.

4. Conclusion
The average value of fiber length is 2402 μm, fiber diameter is 18.18 μm, lumen diameter is 8.73 μm, and fiber wall thickness is 4.91 μm. The results of determining the value of the derivative of the fiber based on the dimensions of the measured fiber are that the average value of Runkle ratio is 1.15, felting power is 133.63, mulhstep ratio is 76.91%, rigidity coefficient is 0.27, and flexibility ratio is 0.48. Bamboo fiber Betung (*Dendrocalamus asper*) from FASP Pondok Buluh has a score of 275-325 and belongs to quality class II, which means it is suitable if used as raw material for pulp and paper.

Acknowledgements
We would like to express my sincerely thanks to the Directorate of Research and Community Service, Ministry of Research and Technology/National Research and Innovation Agency, for grand of Penelitian Dasar Unggulan Perguruan Tinggi (PDUPT) (No: 059/E5/PG.02.00/PT/2022)

References
[1] Badan Pusat Statistik 2020 *Statistik Produksi Kehutanan* (Jakarta: BPS-Statistic Indonesia) 22
[2] Ndale F X 2013 *AGRICA* **7** 22
[3] Zakikhani P, Zahari R, Sultan M T B H H and Majid D L A A 2017 *BioResources* **12** 2479
[4] Javadian A, Smith I F, Saeidi N and Hebel D E 2019 *Frontiers in Materials* **6** 15 1
[5] Awalluddin D, Ariffin M A M, Osman M H, Hussin M W, Ismail M A, Lee H S and Lim N H A S 2017 *MATEC Web of Conferences* **138** 01024
[6] Adam N and Jusoh I 2019 *Transactions on Sci.Techno* **6** 95
[7] Siam N A, Uyup M K A, Husain H and Awalludin M F 2019 *BioReseurces* **14** 2 3925
[8] Villareal J F, Marasigan O S, Mendoza R C, Alipon M A and Abasolo W P 2020 *Philippine J. Sci* **149** 1005
[9] Rusch F, Ceolin G B and Hillig É 2019 *Pesqui Agropecu Trop.* **49** 55007
[10] Yusuf S, Syamani F A and Fatriasari W 2018 *IOP Conf. Ser: Earth and Envir. Sci* **141** 012039.
[11] Nasser R A, Hiziroglu S, Abdel-Aal M A, Al-Mefarrej H A, Shetta N D and Aref I M 2015 *Measurement* **62** 179
[12] Sugesty S, Kardiansyah T and Hardiani H 2015 *Procedia Chemistry* **17** 194
[13] Fatriasari W and Hermiati E 2008 *Jurnal Ilmu dan Teknologi Hasil Hutan* **1** 67.
[14] Maulana M I, Jeon W S, Purusatama B D, Kim J H, Prasetia D, Yang G D, Savero, M A, Nawawi D S, Nikmatin S, Sari R K, Febrianto F, Lee S H and Kim N H 2022 *BioResources* **17** 1442
[15] Praptoyo H and Wathoni F 2013 *Prosiding Masyarakat Peneliti Kayu Indonesia* XVI (Balikpapan) 21
[16] Darwis A, Iswanto A H, Jeon W S, Kim N H, Wirjosentono B, Susilowati A and Hartono R 2020 *Bioresources* **15** 6617
[17] Aprianis Y and Rahmayanti S 2009 *Jurnal Penelitian Hasil Hutan* **1** 11
[18] Sharma M, Sharma C L and Laishram D 2017 *J. the Indian Academy of Wood Sci* **14** 79
[19] Kumar Y B, Sharma M and Sharma C L 2015 *Int J. Adv Res* **3** 479
[20] Loiwatu M and Manuhuwa E 2008 *Agritech* **28** 76
[21] Wang S, Lin S, Li W and Ding Y 2016 *Forest Products Journal* **66** 319
[22] Jansiri S, Deenu A, Puangsin B, Sungkaew S and Kamthai S 2021 *Polymer Composites* **42** 5090
[23] Egbewole Z T, Rotowa O J and Omoake P O 2015 *PAT* **11** 188

Determination of potential fishing ground of mackerel fish (*Rastrelliger kanagurta*) using satellite imagery in the Malacca Strait

A Fadhilah*, **D Kurniawan, Z A Harahap and R Leidonald**

Department of Aquatic Resources Management, Faculty of Agriculture, Universitas Sumatera Utara, Medan, North Sumatra 20155, Indonesia

*Email: amanatul.fadhilah@usu.ac.id

Abstract. Mackerel fish (*Rastrelliger kanagurta*) is a fishery commodity in high demand by the community which is caught in the Malacca Strait. The number of mackerel catches tends to fluctuate every month because it is influenced by different mackerel fishing grounds. Determination of fishing ground tends to use traditional methods such as seeing ripples and a flock of birds flying above the water surface. This study aims to determine the effect of sea surface temperature and chlorophyll-a concentration on mackerel catches and determine the value of primary productivity of the VGPM model every month and also to determine the potential fishing ground to catch mackerel fish based on the sea surface temperature (SST) and chlorophyll-a concentration and Aqua MODIS satellite image data (SST and Chlorophyll-a) were downloaded from *oceancolour.gfc.nasa.gov* while the catch data were obtained from the Belawan Ocean Fishery Port, the data was taken from 2016 to 2020. The influence of SST and Chlorophyll-a concentration was analyzed and visualized on a map. It was found that every month the potential fishing ground for catching mackerel fish has a different location due to the influence of sea surface temperature and fluctuating chlorophyll-a caused by seasonal changes that occur in Indonesia, where the average SST ranged from 30.01 to 31.34°C and the concentration of chlorophyll-a ranged from 1.48 – 2.28 mg/m^3 from 2016 – 2020 with a high net primary productivity ranged from 802.12 – 1.341.59 mgC/m^2/day from 2016 to 2020. The potential area for catching mackerel (*Rastrelliger* kanagurta) caught using purse *seines in* the Malacca Strait is in the coordinates range of 980 10"51.6" – 1000 19"4.8" (*longitude*) and 30 28" 48" – 4 0 56"34.8" (*latitude*) and perpendicular horizontally to the Langkat Regency to Asahan Regency shoreline with a distance under 12 miles to 50 miles. Information on potential areas for catching mackerel can help fishermen in fishing activities so that they are more efficient and catch maximum results.

1. Introduction

Mackerel fish are pelagic fish that live in coastal and offshore. These fish live in groups and these fish eat plankton, copepods, and crustaceans. The size of mackerel fish ranges from 184 mm-300 mm (18.4 cm – 30 cm) and has a larger gill filter because the plankton it eats has a large size [1]. Fish life cannot be separated from the influence of various conditions of the aquatic environment. Oceanographic parameters such as sea surface temperature, chlorophyll-a concentration and primary productivity, affect various fish activities such as fish growth, spawning, metabolism, and other activities. The presence of fish and the determination of potential fishing grounds are influenced by the parameters of aquatic oceanography [2]. The way to determine fishing grounds carried out by fishermen in general are still traditional, so they are less effective. Determination of fishing grounds is only based on hereditary experience until now by looking at natural signs, such as the presence or absence of flocks of birds at sea level, froth at sea level and other.

Malacca Strait is a tropical climate area. This condition is related to its position near the equator. Rainfall mainly on the East and North Coast reaches an average of 1000 mm to 2000 mm per year, while in the central, West and South coasts the rainfall is higher, reaching 2000 mm to 3000 mm per year. The average maximum temperature reaches 23 °C to 35 °C, with air humidity reaching 65% to 75%. In general, the Melaka Strait has an average height of 125 m above sea level [3]. This study aims to find out the influence of chlorophyll-a and sea surface temperature on the number of mackerel fish (*Rastrelliger kanagurta*) catches and find out potential fishing grounds based on chlorophyll-a and sea surface temperature values.

2. Materials and Methods

This research was conducted in May - October 2021 and data retrieved at the Belawan Ocean Fisheries Port (PPS Belawan), Medan, North Sumatra.

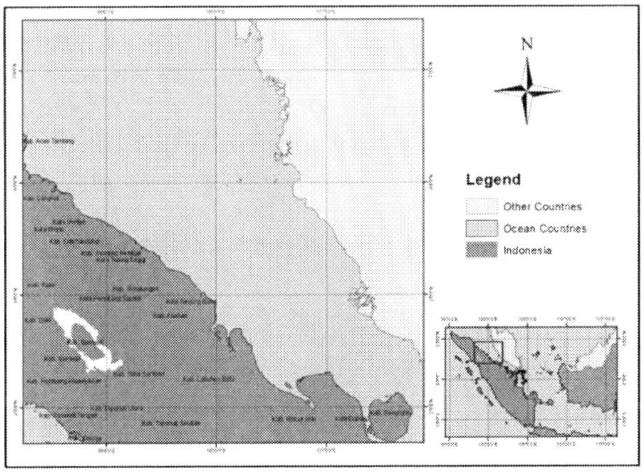

Figure 1. Research Location in the Malacca Strait

2.1. Tools and materials

The tools and materials used in this study were personal AMD A8 laptops (2.20 GHz) 4 GB Ram. Microsoft Office Excel 2013 software, SeaDAS 7.3.2 software, Ocean Data View (ODV) software, SPSS 17 software and ArcMap 10.3. Sea surface temperature data and chlorophyll-a are taken from Aqua MODIS satellite imagery with spatial resolution of 4 km and temporal monthly taken from 2016 - 2020. The fish catch data is taken from data owned by PPS Belawan from 2016 - 2020. Additional data was also taken attenuation coefficient data (Kd), photosynthetically active radiation (PAR) and sea level current data.

2.2. Research procedures

The research method used by the survey method by observing the symptoms that occurred from 2016 to 2020 based on sea surface temperature data and chlorophyll-a obtained from satellite imagery Aqua MODIS downloaded through the http://oceancolorgsfc.nasa.gov site as well as catch data obtained from the Belawan Ocean Fisheries Port (PPS Belawan). Processing of chlorophyll-a distribution data, SST and data through several stages, namely image data collection, cropping, distribution data processing, data control, and data visualization. The data was processed and found on average monthly from 2016 - 2020. Furthermore, the catch data that has been calculated the value of catch per unit effort (CPUE) is also calculated on average per month and compared with existing SST and chlorophyll-a data with graphs and calculate pearson correlation values using SPSS 17.

2.3. Data analysis

Pelagic fish is usually caught with fishing gear that has a different ability to catch a type of fish in a fishing ground [4]. Formula for that CPUE is as follows:

$$CPUE = \frac{Ci}{Fi} \quad \text{...} \quad (1)$$

Note:
CPUE = catch per unit effort
Ci = number of catches of the type of fishing gear i
Fi = number of fishing gear arrest attempts i

Determination of potential fishing ground is based on chlorophyll-a distribution and sea surface temperature anomalies from satellite imagery. The indicators used are the distribution of SST and chlorophyll-a [4].

Table 1. Assessment of determination of potential fishing ground of arrest fish based on sea surface temperature values

No	SST Categories	Criterion	Score/Weight	Category
1	Optimum	24 - 26 °C	6	Potential
2	Moderate	26 - 30 °C	4	Keep
3	Suboptimal	< 24 °C or > 30 °C	2	Less Potential

Table 2. Assessment of determination of potential fishing ground of arrest fish based on chlorophyll-a concentration values

No	Chlorophyll-a	Criterion	Score/Weight	Category
1	High	>0,2 mg/m^3	6	Potential
2	Moderate	0,1 – 0,2 mg/m^3	4	Keep
3	Low	< 0,1 mg/m^3	2	Less Potential

Vertically Generalized Production Model (VGPM) is a model used to obtain the primary productivity value (mgC / m^2 / day) of a water at euphotic depth by looking at the relationship between several parameters such as chlorophyll-a concentration and the depth of these waters [5].

Primary productivity values at the research site obtained through the approach of chlorophyll-a surface concentration obtained from the Aqua-MODIS satellite with the addition of other parameters [5]. The equations of this VGPM model are:

$$PP_I = \text{Chl-a x } P^B_{opt} \text{ (0.66125 x)} \frac{I0}{I0+4.1} \text{x } Z_I \text{ x } Dl \text{ ....} \quad (2)$$

Note: PP_{eu} = the value of primary productivity of the waters; Chl-a = chlorophyll-a concentration value; *I0* = the rate of transmission of light to the sea; P^B_{opt} Z_{eu} = depth of the euphotic zone; *dl* = average length of time of eye irradiation in one day; = optimum level of photosynthesis in a water column

3. Results and Discussion

3.1. Sea surface temperature
The distribution of sea surface temperatures in the waters of the Strait of Malacca based on analysis of Aqua-MODIS satellite imagery averages each month has values that range from 30.01 - 31.34 °C, the lowest average in December and the highest average in April. The distribution of SST using satellite imagery is considered appropriate because it can record SST in large fishing ground at the same time. States that the distribution map of SST in the water has been widely applied in the field of fisheries and the use of marine biological resources [6].

Table 3. Average Value of SST in The Malacca Strait in 2016 – 2020

Month	Sea Surface Temperature (°C)					
	2016	2017	2018	2019	2020	Average
Jan	30.53	29.82	30.12	30.28	29.95	30.14
Feb	30.79	29.91	30.06	30.79	30.18	30.35
Mar	31.03	30.51	30.88	31.25	30.93	30.92
Apr	31.69	30.95	30.90	31.58	31.56	31.34
May	31.70	30.99	30.61	31.54	30.95	31.16
Jun	31.65	31.35	31.41	31.13	31.13	31.33
Jul	31.14	31.02	30.64	30.85	30.51	30.83
Ags	31.27	30.84	30.90	30.44	31.45	30.98
Sep	31.09	30.85	30.88	30.43	30.62	30.78
Oct	30.68	30.81	30.41	30.20	30.88	30.60
Nov	30.22	30.32	30.35	30.55	30.33	30.35
Dec	29.96	30.10	30.29	29.63	30.09	30.01

In April, May and June the average sea surface temperature ranges from 31.16–31.34°C. In that month the waters of the Strait of Malacca experienced summer, resulting in sea surface temperatures that increased from an average of 30 °C to 31 °C. After June the average sea surface temperature returned to 30 °C and at optimum temperatures will have a good impact on small pelagic fish organisms in the waters of the Strait of Malacca. States that high sea surface temperatures in the water can disrupt the balance of ecosystems in sea waters because every marine organism is able to adjust to the temperature at a certain limit [7].

Fluctuations that occur where in December to February the value of sea surface temperatures tends to decrease and the temperature again increases in the middle of the year, namely from April to February August. States that in the western season the monsoon tends to move north to northeast and occasionally changes towards the northwest while in the eastern season it tends to move south to southwest and occasionally changes towards the southeast. Changes in wind direction and seasons are expected to affect the average annual temperature of a place [8].

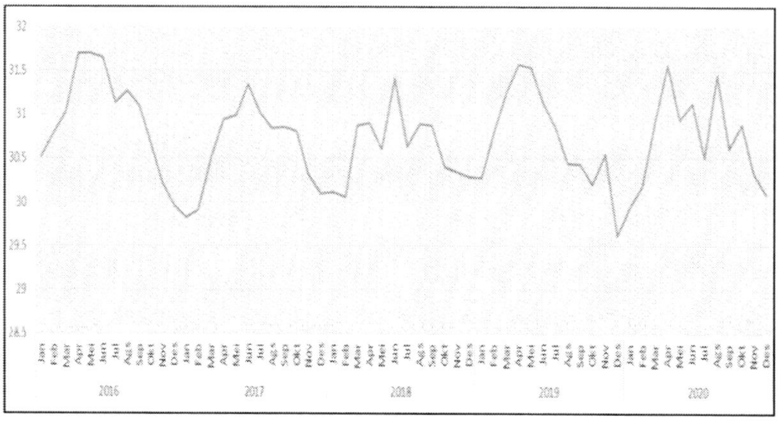

Figure 2. Fluctuations in SST Value in The Waters of the Strait of Malacca 2016–2020

3.2. Chlorophyll-a

Chlorophyll-a concentration values in the waters of the Strait of Malacca based on analysis of Aqua-MODIS satellite imagery each month have values that range from 1.48 – 2.28 mg /m^3 at the lowest average in March and the highest average in November. Chlorophyll-a is a pigment contained in phytoplankton that serves as a binder of solar energy so that the process of photosynthesis occurs. The importance of phytoplankton chlorophyll as an ecological factor lies in its usefulness as a standing-stock measure of phytoplankton and a potential measure of the photosynthesis of a waterway. States that chlorophyll-a is a pigment capable of photosynthesis and is present in all phytoplankton organisms [9].

Table 4. Average Value of Chlorophyll-a Concentration in The Malacca Strait in 2016 – 2020

Month	Chlorophyll-a concentration (mg/m^3)					
	2016	2017	2018	2019	2020	Average
Jan	1.99	1.97	1.42	2.18	2.08	1.93
Feb	2.15	1.79	2.07	2.07	1.47	1.91
Mar	1.44	1.64	1.70	1.22	1.42	1.48
Apr	1.31	1.82	1.48	1.29	1.56	1.49
May	1.42	1.98	1.60	1.17	1.34	1.50
Jun	1.50	2.30	1.78	1.92	1.87	1.87
Jul	1.30	2.08	1.51	1.56	2.11	1.71
Ags	1.16	1.89	1.81	1.30	2.14	1.66
Sep	1.25	2.36	1.76	0.82	2.01	1.64
Oct	1.06	1.99	2.12	1.71	1.66	1.71
Nov	1.48	2.84	2.62	2.22	2.25	2.28
Dec	1.54	2.52	2.33	2.06	2.70	2.23

The high content of chlorophyll-a in the waters of the Strait of Malacca indicates good phytoplankton growth as well. This is because phytoplankton have colored pigments and phytoplankton can carry out photosynthesis and act as a food source for other organisms. In accordance, which states that chlorophyll-a is an indicator of phytoplankton abundance in the waters that play a role in the process of photosynthesis. Phytoplankton contribute greatly to knowing primary productivity in the waters [10].

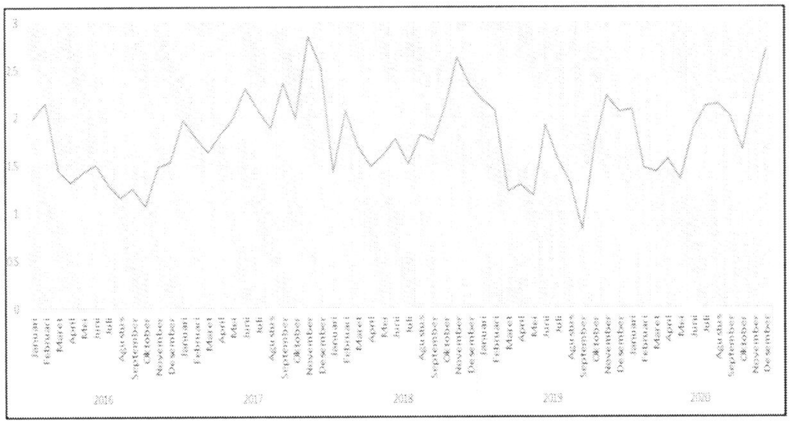

Figure 3. Fluctuations in Chlorophyll-a Values in the Waters of the Strait of Malacca 2016 – 2020

The high content of chlorophyll-a in the waters of the Strait of Malacca indicates good phytoplankton growth as well. This is because phytoplankton have colored pigments and phytoplankton can carry out photosynthesis and act as a food source for other organisms. In accordance, which states that chlorophyll-a is an indicator of phytoplankton abundance in the waters that play a role in the process of photosynthesis [10]. Based on the results of the analysis, it is known that the content of chlorophyll-a concentration is higher on average in the coastal waters of the Strait of Malacca, this is because in coastal waters there is a lot of human activity and the confluence of estuary waters with salty waters. In accordance, which states that the distribution and high concentration of chlorophyll are strongly associated with the oceanographic condition of a water [11].

3.3. *Productivity primer waters VGPM model*

Vertically Generalized Production Model (VGPM) is a model used to obtain the primary productivity value (MgC / m^2 / day) of a water at euphotic depth by looking at the relationship between several parameters such as chlorophyll-a concentration and the depth of these waters [5].

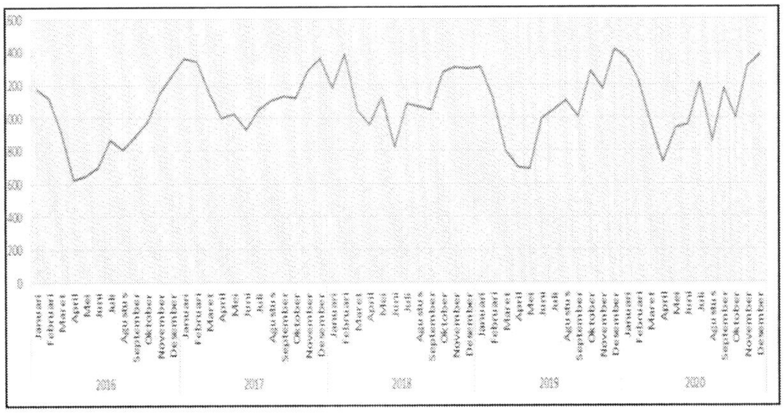

Figure 4. VGPM Model Primary Productivity Value Fluctuation Graph from 2016 – 2020

Based on the analysis conducted the value of the primary productivity concentration of the VGPM model in the waters of the Strait of Malacca every month is estimated to have a value that ranges from 802.12 - 1341.59 mgC / m^2 / day. Good primary productivity will have a good impact on organisms in these waters. Based on the lowest and highest values of the monthly average that has been analyzed the waters of the Strait of Malacca fall into the category of eutrophic. In accordance, which states that the classification of fertility rates is: 0-200 mgC / m^2 / day including oligotrophic, 200-750 mgC / m^2 / day including mesotrophy and more than 750 mg C / m^2 / day including eutrophic [12].

3.4. Catch and CPUE of Mackerel (Rastrelliger kanagurta)

The value of catch per catch effort (*catch per unit effort*) of Mackerel (*Rastrelliger kanagurta*) caught in the waters of the Strait of Malacca and landed in PPS Belawan is relatively volatile every month with values ranging from 584.20-1245.76 kg/Trip. Mackerel live in swarms and enters the waters of the estuary to forage in the form of plankton, copepods, and crustaceans, detritus and *filamenteus algae* [1]. Mackerel which is widely caught in the waters of the Strait of Malacca because the water conditions are suitable for mackerel (*Rastrelliger kanagurta*) live this because the waters of the Strait of Malacca have a high chlorophyll-a content so that the food source of mackerel, namely plankton, is well available in the Malacca Strait.

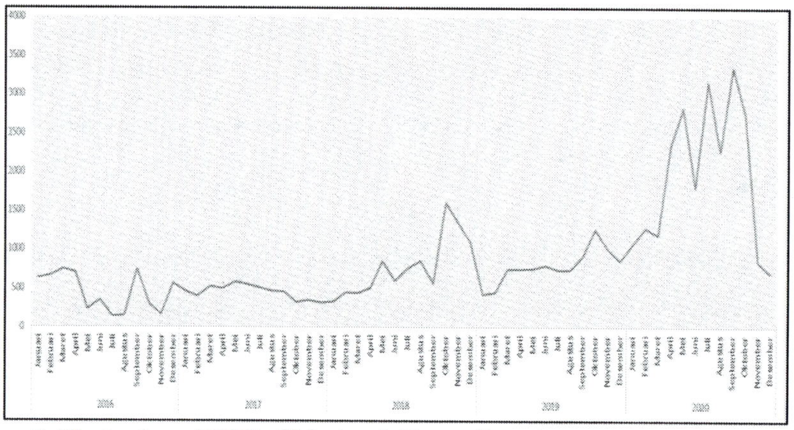

Figure 5. Fluctuations CPUE of Mackerel (*Rastrelliger kanagurta*) from 2016– 2020

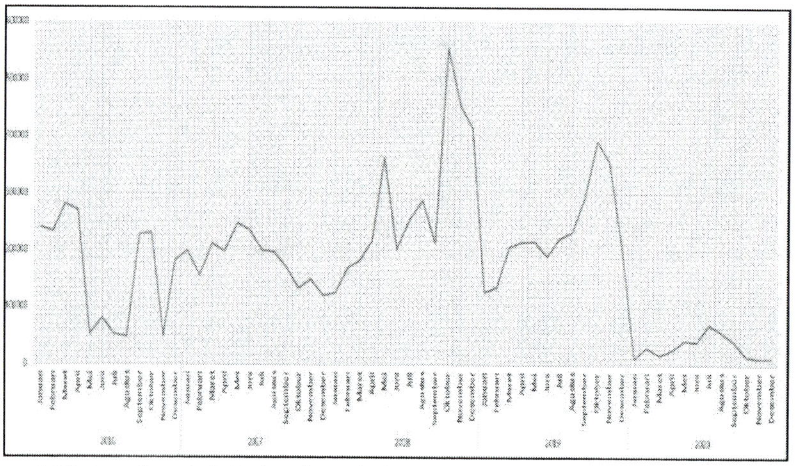

Figure 6. Fluctuations in Mackerel catch (*Rastrelliger kanagurta*) from 2016 – 2020

3.5. *Pearson correlation analysis*
Sea surface temperature has an important role in the survival of organisms in the waters. A good temperature will affect the metabolic system of the organism and its breeding in the waters. In this study, it is known that the pearson correlation value between SST and chlorophyll-a is -0.369 which means that the two have a relationship that is not strong but opposite each other. This means that if the temperature increases then the value of chlorophyll-a will decrease. In accordance, which states that the influence of sea surface temperature on phytoplankton growth will indirectly affect the concentration of chlorophyll-a waters [8].

Table 5. *Pearson* correlation analysis

		SST	Chlorophyll-a	VGPM
SST	*Pearson Corelation*	1	-.396	-.948
	Sig. (2-*tailed*)		.001	.000
	N	64	64	64
Chlorophyll-a	*Pearson Corelation*	-.396	1	.631
	Sig. (2-*tailed*)	.001		.000
	N	64	64	64
VGPM	*Pearson Corelation*	-.948	.631	1
	Sig. (2-*tailed*)	.000	.000	
	N	64	64	64
Mackerel CPUE	*Pearson Corelation*	.050	.073	-.003
	Sig. (2-*tailed*)	.695	.565	.980
	N	64	64	64

Chlorophyll-a is an indicator of phytoplankton abundance in the waters that play a role in the process of photosynthesis [10]. Chlorophyll-a has a unidirectional and not strong (weak) relationship with mackerel CPUE. If the chlorophyll-a value increases in the Malacca Strait then the CPUE value of mackerel will be affected but not significantly. This is because if the value of chlorophyll-a is high enough then phytoplankton in the waters is also quite high.

Primary productivity is the speed at which photosynthesis or carbon binding occurs. Primary productivity as a degree of storage of solar energy in organic form, as a result of photosynthesis and chemosynthesis of primary producers [13]. If primary productivity in the waters increases then the CPUE of the fish will decrease and if the primary conductivity decreases then the CPUE will increase even if it is not significant.

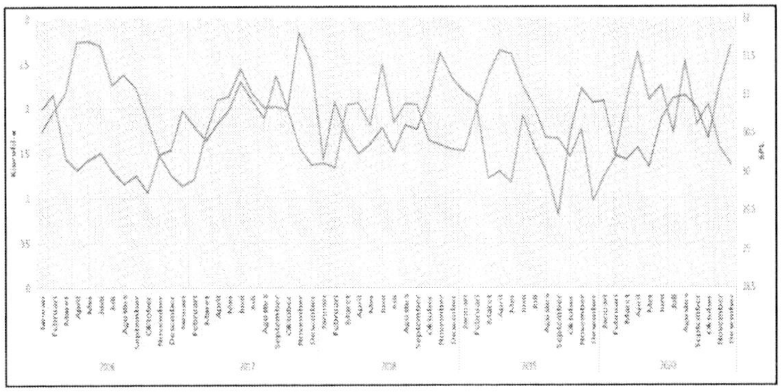

Figure 7. The relationship of sea surface temperature with Chlorophyll-a

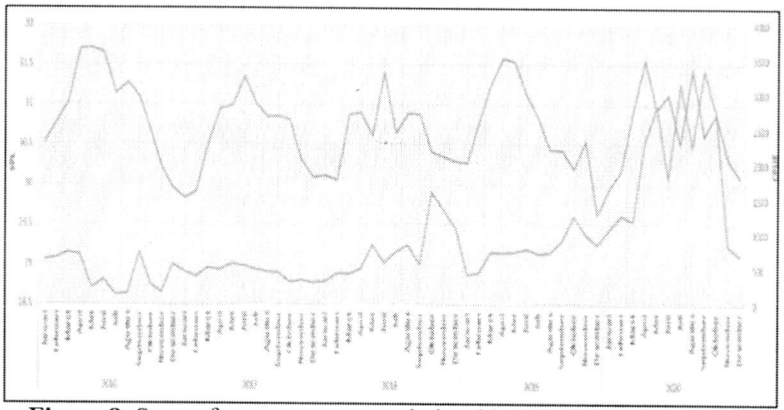

Figure 8. Sea surface temperature relationship with Mackerel CPUE

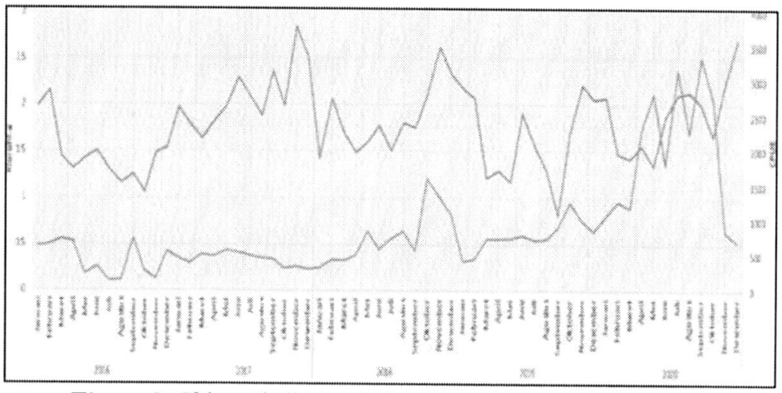

Figure 9. Chlorophyll-a's relationship with Mackerel CPUE

There are aspects that affect sea surface temperatures, namely hydrological cycles, rainfall intensity and consequently have a less favorable impact on ecosystems in the environment [7]. The results of the analysis that has been carried out on the relationship between sea surface temperature with chlorophyll-a concentration value and primary productivity value based on pearson analysis should be on temperature and chlorophyll-a if the temperature rises at the optimum point (Close to the value of 30 °C) then the value of chlorophyll-a concentration will increase and followed by the productivity value primer increased as well, but found in the figure 7 that shows a mismatch, namely in 2017 and 2018, this is due to monsoon winds that bring seasonal changes that make certain months and years. experience high rainfall and in certain months with little rainfall that occurs. There are aspects that affect sea surface temperatures, namely hydrological cycles, rainfall intensity and consequently have a less favorable impact on ecosystems in the environment [7].

3.6. Potential fishing ground of Mackerel Fish (Rastrelliger kanagurta)

Remote sensing technology can be used to determine the position of potential fish fishing ground by identifying several indicators of aquatic fertility rates and feeding abundance for fish. Some of the parameters that are often used as indicators are sea surface temperature [14]. Determination of Mackerel Fish fishing grounds based on the analysis of sea surface temperature and chlorophyll-a concentration value can be seen that the strait of Malacca has varying fluctuations in the value of chlorophyll -a and SST so that every month it is estimated that pelagic fishing ground will change along with changes in SST values and chlorophyll-a concentration values that occur in the Waters of the Strait of Malacca.

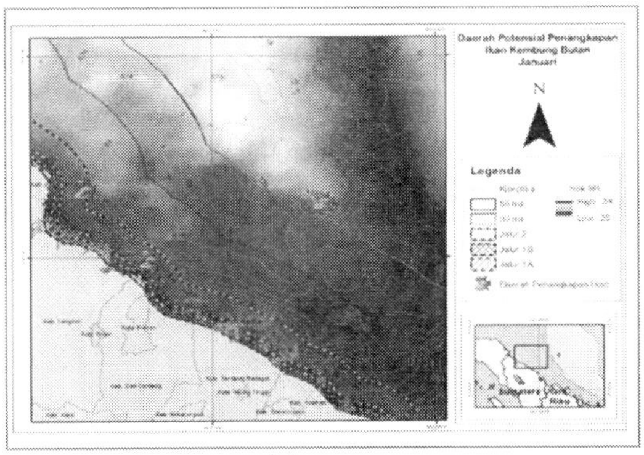
Daerah Potensial Penangkapan Ikan Kembung Bulan Januari
N
Legenda

(a)

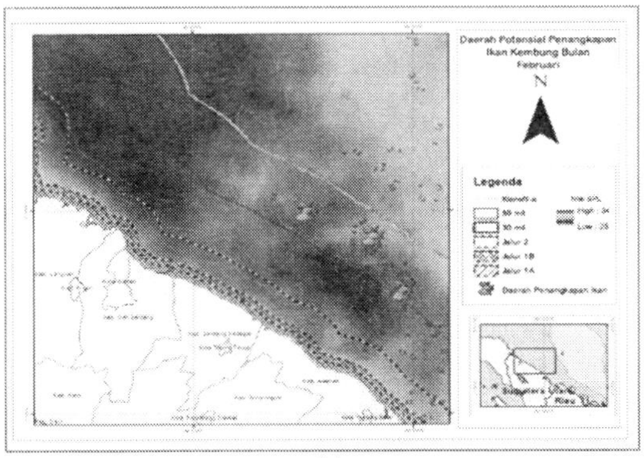
Daerah Potensial Penangkapan Ikan Kembung Bulan Februari
N
Legenda

(b)

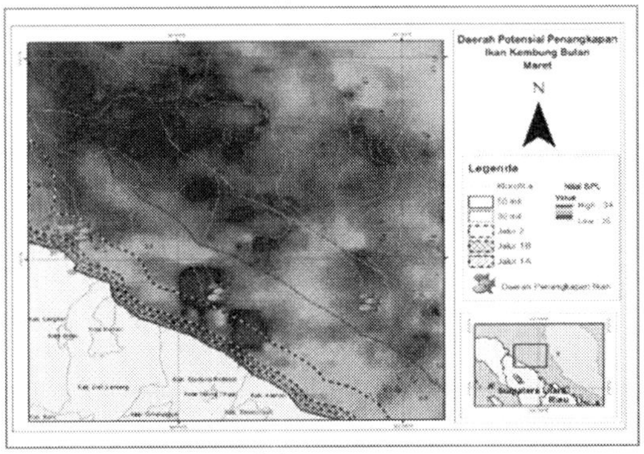
Daerah Potensial Penangkapan Ikan Kembung Bulan Maret
N
Legenda

(c)

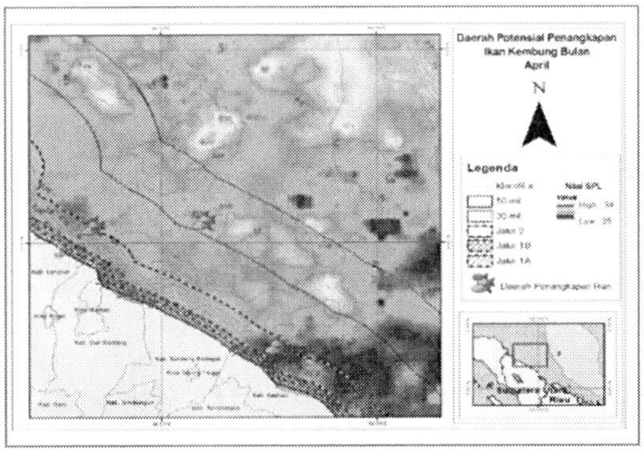
Daerah Potensial Penangkapan Ikan Kembung Bulan April
N
Legenda

(d)

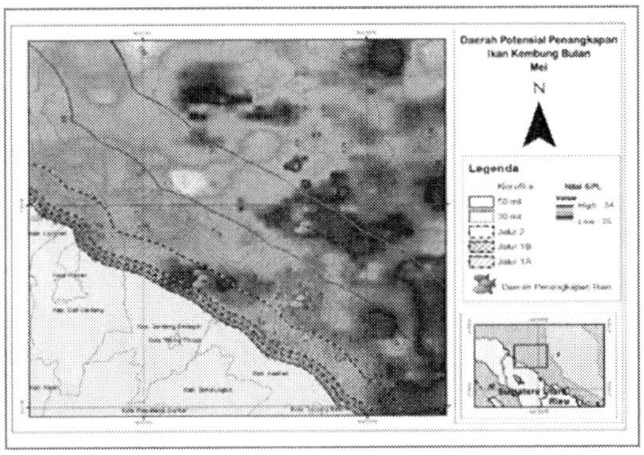
Daerah Potensial Penangkapan Ikan Kembung Bulan Mei
N
Legenda

(e)

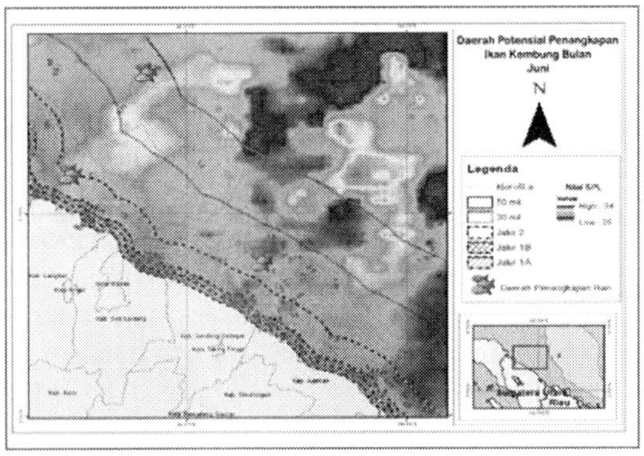
Daerah Potensial Penangkapan Ikan Kembung Bulan Juni
N
Legenda

(f)

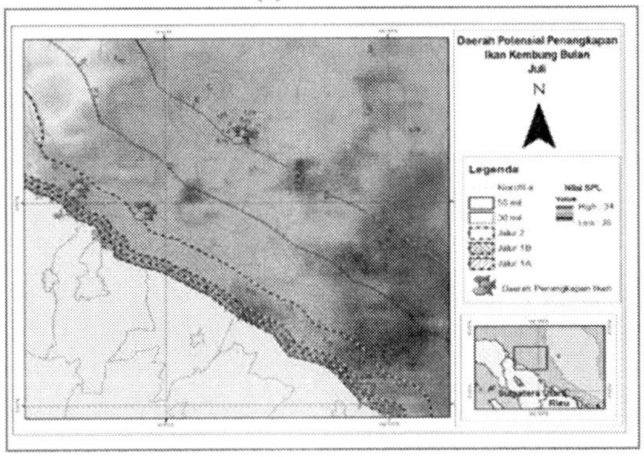
Daerah Potesial Penangkapan Ikan Kembung Bulan Juli
N
Legenda

(g)

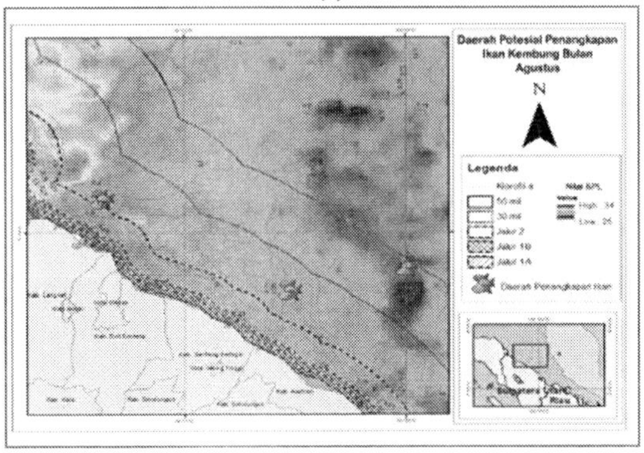
Daerah Potesial Penangkapan Ikan Kembung Bulan Agustus
N
Legenda

(h)

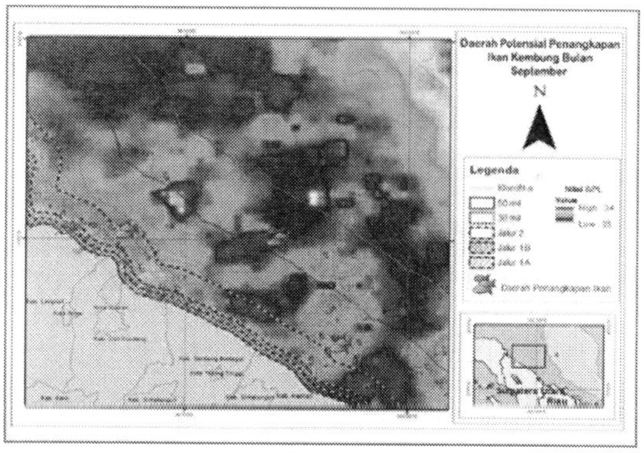

(i)

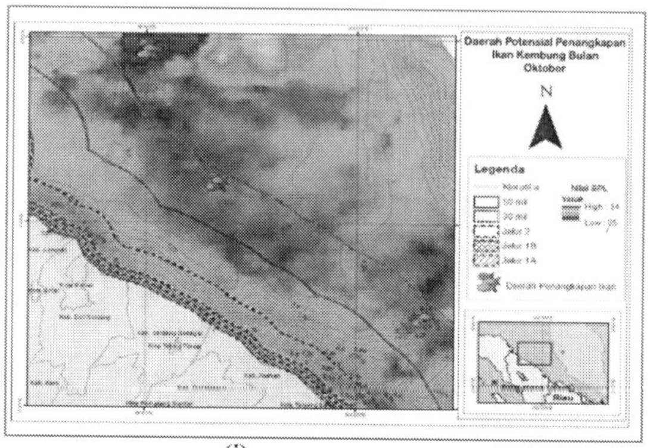

(j)

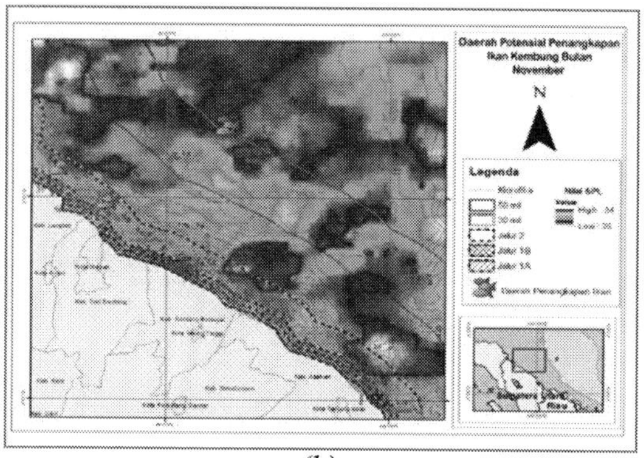

(k)

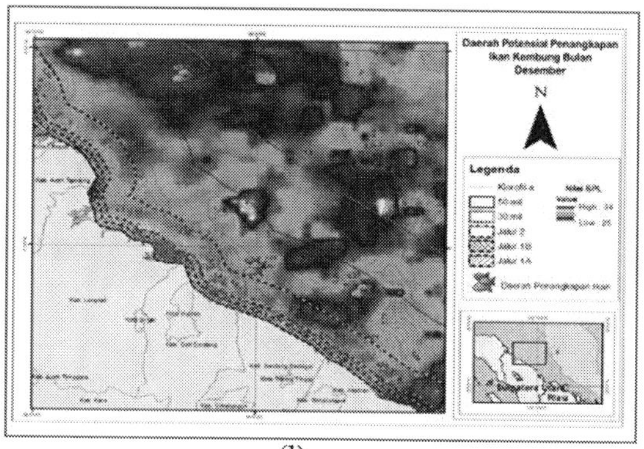

(l)

Figure 10. Potential Fishing ground of Mackerel Fish (*Rastrelliger kanagurta*) (a) January (b) February (c) March (d) April (e) May (f) June (g) July (h) August (i) September (j) October (k) November (l) December

Mackerel fish is a pelagic fish that lives in coastal and offshore waters. These fish live in swarms and enter the waters of the estuary to forage in the form of plankton, copepods, and crustaceans [1]. In determining the coordinates of potential fishing ground of mackerel fish based on chlorophyll-a content and sea surface temperature that is visible and interpreted through the map that has been made. It can be seen that the average sea surface temperature value per month is chosen a location that has a temperature suitable for mackerel to live and a good chlorophyll-a content for small pelagic fish where through the concentration value of chlorophyll-a will be known phytoplankton content and become a food source for mackerel.

The climate in the waters of the Belawan Sea is affected by changes in wind direction and is a water crossed by monsoon currents. In the western season the monsoon tends to move north to northeast and occasionally changes towards the northwest while in the eastern season it tends to move south to southwest and occasionally changes towards the southeast. Changes in wind direction and seasons are expected to affect the average annual temperature of a place [8]. Small pelagic fishing grounds in the Malacca Strait, especially mackerel (*Rastrelliger kanagurta*) change every month. This is due to changes in temperature and chlorophyll-a concentration that occur in the waters of the Strait of Malacca. Fluctuations that occur due to the monsoon wind that occurs in Indonesia, especially in the waters of the Strait of Malacca, the west and east monsoon winds affect the intensity of rainfall.

4. Conclusion

The value of sea surface temperature which on average ranges from 30, 01 – 31.34 °C and the average chlorophyll-a concentration is between 1.48 – 2.28 mg /m^3 from 2016 – 2020. Malacca Strait have a fairly high content of net primary productivity values ranging from 802.12 - 1. 341.59 mgC/m^2/day from 2016 to 2020. Potential fishing ground of mackerel fish (*Rastrelliger kanagurta*) caught using purse seine fishing gear Malacca Strait in the coordinate range of 98°10'51.6" – 100°19'4.8" (*longitude*) and 3°28'48" – 4° 56'34.8" (*latitude*) and perpendicular horizontally from the coastline of Langkat Regency to Asahan Regency with a distance of under 12 to reach 50 miles from the coastline.

Acknowledgments

The author thanked the Belawan Ocean Fishery Port for its assistance in collecting research data as well as conducting interviews with fishermen.

References

[1] Utami N, Redjeki F S, and Supriyantini E 2014 *Journal of Marine Research* **2** 3
[2] Saeri M 2013 *Transnational Journals* **4** 2
[3] Demena E Y, Miswar E and Musman M 2017 *Scientific Journal of Marine and Fisheries Students Unsyiah* **2** 1
[4] Bukhari, Adi W and Kurniawan 2017 *Journal of Capture Fisheries* **1** 3
[5] Wahyudi A 2017 *Utilization of Agro Float and Satellite Data in Biophysical Monitoring Application of The West Indian Ocean Waters of Sumatra Island* (Bogor: Thesis Institut Pertanian Bogor).
[6] Alfajri, Mubarak and Mulyadi A 2017 *Environmental Dynamics of Indonesia* **4** 1
[7] Azizah A and Wibisana H 2020 *Marine Journal* **13** 3
[8] Panjaitan R R, Mulyadi A and Ghalib M 2017 *Analisis Distribusi Suhu Permukaan Laut dan Konsentrasi Klorofil-a di Perairan Belawan Kota Medan Provinsi Sumatera Utara* (Pekanbaru; Skripsi Universitas Riau)
[9] Nufus N, Karina S and Agustina S 2017 *Scientific Journal of Marine and Fisheries Students Unsyiah* **2** 1
[10] Nuzapril M, Susilo S B and Panjaitan J P 2017 *Journal of Remote Sensing* **14** 1
[11] Nuzapril M, Susilo S B and Panjaitan J P 2017 *Journal of Fisheries and Marine Technology*, **8** 1
[12] Tarigan S M and Wiadnyana N N 2013 *National Marine Journal* **8** 2
[13] Sunaryo A 2017 *Journal of Marine and Fisheries Extension* **11** 2
[14] Yulianto D, Muskananfola M R and Purnomo P W 2014 *Journal of Management of Aquatic Resources* **3** 4

Performance testing of the thermal conductivity measuring equipment on particle board isolators

Azhari[1,3*], T I Nasution[1] and Rossy Nurhasanah[2]

[1]Department of Physics, Faculty of Mathematics and Natural Science, Universitas Sumatera Utara, Jalan Bioteknologi No.1 Kampus USU, Medan 20155, Indonesia
[2]Department of Information Engineering, Faculty of Computer Science and Information Engineering, Universitas Summatera Utara, Medan 20155, Indonesia
[3]Integrated Research Laboratory, Universitas Sumatera Utara, Medan 20155, Indonesia

*Email: azhari@usu.ac.id

Abstract. We investigated the value of thermal conductivity of particle board using a tool designed with the principle of converting electrical energy into heat energy by varying temperatures. This study aims to determine the value of thermal conductivity of oil certain particle board and to what extent it can function as a thermal insulation material and to determine the performance of the thermal conductivity test equipment. The size of this tool is 10 mm of thickness, 65 mm of width, 140 mm of length and 40 mm of height, which is equipped with the function of measuring the conductivity value of the material. The average error value of the temperature sensor compared to the standard thermocouple is 0.8%. By entering the value of the area and thickness of the sample, the average conductivity value of the composite material is 0,028 W/m°C. The low material conductivity value will be used as a building construction material. The data show that the sensitivity, accuracy, and precision of this tool are very good for testing the conductivity of a material.

1. Introduction

Each material has different characteristics, both in terms of physical properties, mechanical properties or chemical properties. Physical properties are the properties of a material that we can observe directly, while to determine its mechanical and chemical properties it cannot be seen directly. One of these physical properties is the material's ability to conduct heat. A material has different heat conduction properties such as conduction, convection and radiation. Due to their low cost and ease of processing, polymer-based composites are frequently employed in the micro-electronic packaging industry as underfill materials and thermal interface materials. The ability of electronic devices to manage their internal heat directly depends on the thermal conductivity (TC) of the materials used in microelectronic packaging. The polymer's low thermal conductivity (0.2 $Wm^{-1} K^{-1}$) prevents it from dissipating heat as quickly as necessary for high-power devices. To increase the thermal conductivity of polymer composites, high thermal conductivity nanofillers are utilized, such as boron nitride nanosheets (BNNS), graphene, and carbon nanotubes [1-3].

The development of science and technology, various types of thermal insulation materials have been produced for various purposes, for example in industry and buildings. In determining the value of the thermal conductivity of the material, thermal insulation is very important to overcome the energy losses that are too large due to heat transfer from the system to the environment. The use of thermal insulation is growing not only to avoid heat leakage, but also to prevent fires, control temperatures, increase production and create safer and better working conditions. Because the thermal insulation function is very important in the use of heat energy which must be as efficient as possible, it is necessary to use a thermal insulation material that has a low thermal conductivity and tends to be constant, which is closest to dry air. The most air-containing material provides a structure that resists heat transfer by air so that it has excellent thermal insulation. Based on the literature of previous researchers, the value of thermal conductivity depends on density, porosity, pore diameter, fibre diameter, and fluid in the pores[4-6].

Several studies that have been carried out include the insulation properties of sample such as graphene, carbon nanotube, and BN-Composite [7-9]. The way to find out the value of the thermal conductivity of a material, we can test it by using a material's thermal conductivity tester. In research to examine the conductivity value of an existing material, currently it is hampered by the unavailability of thermal conductivity test equipment for this research. Another obstacle that arises is the amount of costs required to purchase or procure this thermal conductivity test equipment. On that basis, it is necessary to have an engineering thermal conductivity device that is able to assist research at a low cost.

Conductivity is the ability of a material to conduct heat. This is one of the parameters needed in the characteristic properties of a material. In most operations, heat input and output are required, to achieve and maintain the required state during the process [10]. This research aims to determine the thermal conductivity of composite fibre and the extent to which thermal insulation materials can function as insulating materials. To measure the thermal conductivity used equipment designed in such a way by using a heat source from an electric element, the calculation uses the principle of heat transfer [11].

The problem that arises is the number of materials that have not identified the thermal conductivity value of these materials because it is still difficult to make a tool that can measure thermal conductivity with a small bias. Based on these data, it is necessary to carry out a test to determine the value of the thermal conductivity of a material so that later it can become a new alternative material whose function is to replace materials commonly used today [11,12]. Several previous studies have made a tool that is used to measure the value of the conductivity of a material such as determining the thermal conductivity of Metals Using a Microcontroller Based Experiment Kit. However, based on the results of the study, it was found that the problem is in the form of a very large bias due to a good insulator system so that it is not good so that a lot of heat energy is wasted into the environment [13].

2. Research Method

2.1. Research design

The method used in this research is the experimental method. Experiments were carried out to create a design for thermal conductivity devices using the applicable heat-energy equation. The research flowchart can be described as in Figure 1.

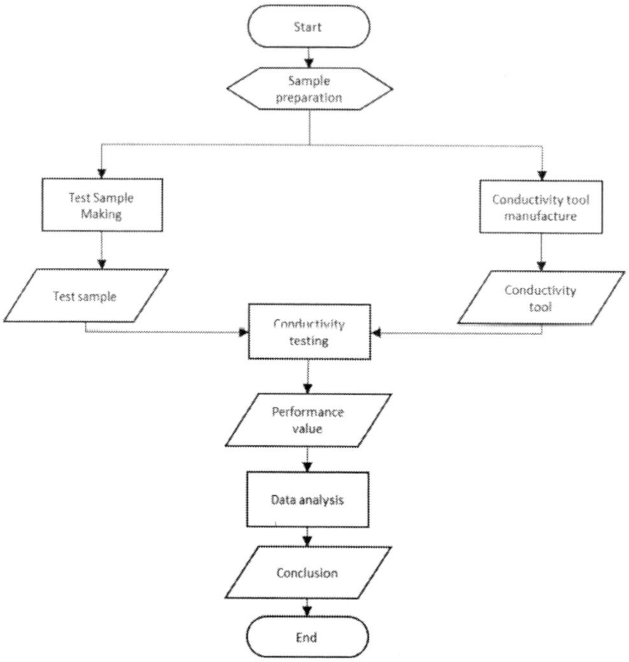

Figure 1. Flowchart

2.2. Manufacturing of conductivity test equipment

The principle of calculating the conversion of electrical energy is used to make a conductivity test instrument. The electrical energy generated (W) by the power supply depends on the value of the current (*i*) Voltage (*V*) and time (*t*).

$$W = Vi \qquad (1)$$

The electrical energy will be turned into heat (Q) by the heater which is affected by the value of mass (m), specific heat (c), and changes in temperature (ΔT):

$$Q = mc\Delta T \qquad (2)$$

Furthermore, the heat received by the heater will be transferred by conduction by the sample/material and will be received by other parts. The heat transfer (*q*) that occurs will satisfy by the temperature of high temperature (T_t), temperature of low temperature (T_r), thickness sample (d), cross-sectional area of the object (A), and thermal conductivity (k).

$$q = \frac{kA(T_t - T_r)}{d} \qquad (3)$$

To find the value of the conductivity of oil palm fiber material, Fourier's law is used, namely:

$$k = \frac{qd}{A(T_t - T_r)} \qquad (4)$$

The temperature value read by the thermocouple sensor will be forwarded and processed by the microprocessor so that the conductivity value appears digitally. The design of the thermal conductivity of the material is designed as shown in Figure 2.

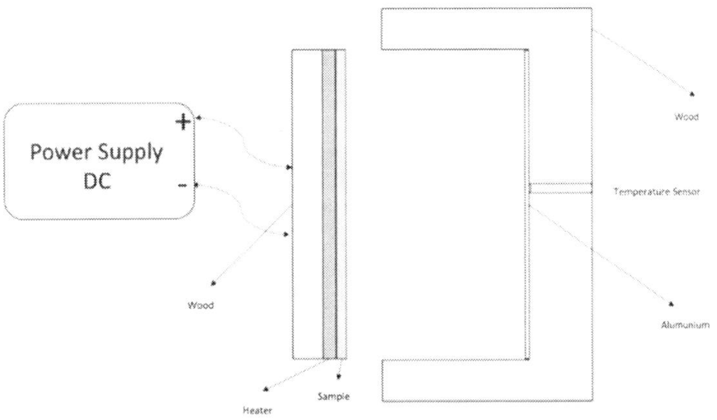

Figure 2. Design of conductivity testing

Two pieces of glass plate with dimension of 65 mm x 140 mm x 15 mm is placed a wire thermocouple is installed to measure the temperature generated by the heater. Another plate is installed at the top as a cover sample for placing wire thermocouple to measure the temperature on the other side. Electrical regulator are used to regulate the electric voltage flowing to the test equipment.

2.3. Overall tool design

In designing a system to measure conductivity of a material based on the ATmega328 microcontroller, the author uses two DS18B20 temperature sensor [14]. To find out how fast and how much temperature changes in an object can conduct heat, how much temperature can change in the material, we must know the thermal conductivity of the material. Electrical carriers (electrons or holes), lattice waves (phonons), electromagnetic waves, spin waves, or other excitations can all carry heat energy through solids [15,16]. The function of the DS18B20 sensor is to detect the initial temperature of the heater (T_1) and the heater temperature after passing through a material in other side (T_2). The difference of those temperature is ΔT. The circuit system involves several circuit components that help design this system, include the ATmega328 Microcontroller, relay and DS18B20 Sensor. Figure 3 shows the design of conductivity test.

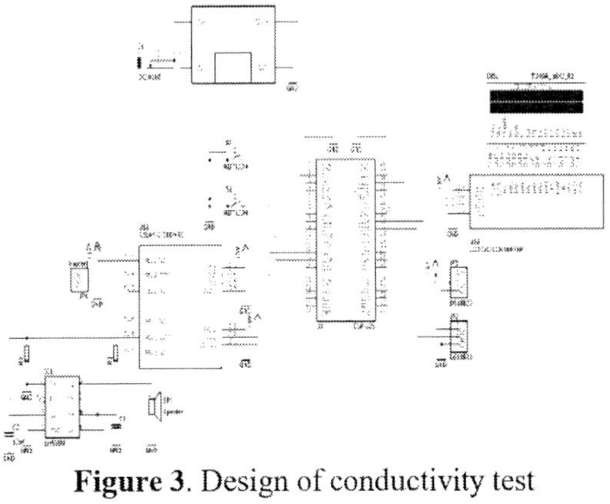

Figure 3. Design of conductivity test

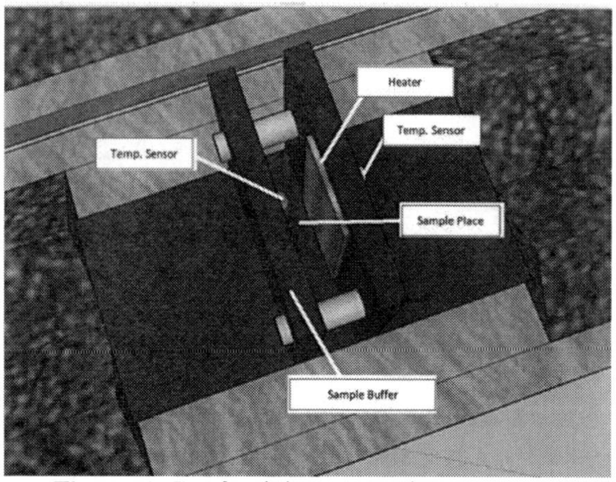

Figure 4. Conductivity test equipment

Figure 4 is the design of the conductivity test equipment. The data obtained from testing the temperature value with standard tools can then be used to determine the accuracy and precision of the DS18B20 sensor. The sample is placed in a layer consisting of two temperature sensors and between heaters. So that when the tool is turned on, the heater will provide heat to the material which will be calculated by the temperature sensor for the temperature difference between the left and right layers. By entering the value of the area and thickness of the sample, the conductivity value of the material can be determined by equation (4).

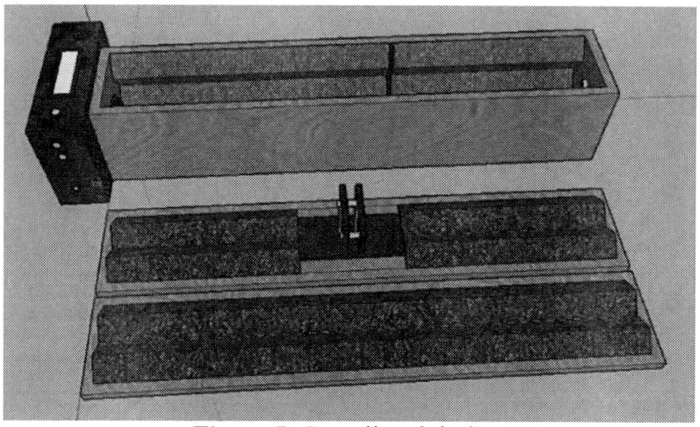

Figure 5. Overall tool design

Figure 5 is the overall design of the conductivity test equipment. The location of the conductivity test equipment is at the top of the case. The control of each tool uses an ATMega328 microcontroller which can control the tool to carry out testing.

3. Result and Discussion

The electrical energy (W) produced is determined based on the value of the power (P) supplied by the power supply. Electrical energy (W) will then be transformed into heat (Q) by the heater, which is influenced by the mass of the heater (m), the specific heat of the heater (c), and the temperature difference between the heater surfaces (ΔT). Furthermore, the heat received by the heater will be transferred to the conduction process by the sample/material to be tested. The heat of propagation (q) will be influenced by the object's conductivity (k), surface area (A), sample length (L), and the difference in temperature at the top and bottom of the sample (ΔT). So to find the value of the conductivity of oil palm fiber material, Fourier's law is used in equation (4).

Figure 6. Conductivity tester

Figure 7. The process of calibrating the temperature

Figure 6 shows an illustration of how the process of reading the temperature difference value (ΔT) of the tested sample using 2 DS18b20 sensors placed on both sides of the sample, namely on the heater and on the sample so that the value of the temperature difference in the sample layer can be determined. The results are then processed using a microprocessor according to equation (3) and (4). The value shown by the microprocessor is the conductivity value of the sample being tested in real time. Prior to testing, the DS18b20 temperature sensor was calibrated using a standard thermometer as shown in Figure 7.

Table 1 shows the results of the comparison of the temperature sensor test with a standard thermometer. From these data, it can be seen that the test between temperature sensors on both sensors 1 and 2 has good accuracy and precision values as evidenced by the bias value of the sensor which ranges from 0.8% to 0.9%. The conductivity test equipment is located in the middle of the test equipment that processes data with the source of the temperature value test.

IOP Conf. Series: Earth and Environmental Science 1115 (2022) 012087 doi:10.1088/1755-1315/1115/1/012087

Table 1. The validation of the temperature sensor

Standard Thermometer (°C)	DS18B20 Sensor(°C)		Deviation of Sensor (%)	
	1	2	1	2
68.4	67.5	68.5	1.3	0.15
66.5	66.0	67.0	0.8	0.75
65.0	64.5	65.5	0.8	0.77
63.4	63.5	64.0	0.2	0.95
62.2	62.0	62.5	0.3	0.48
61.0	61.0	61.5	0.0	0.82
59.5	59.5	60.0	0.0	0.84
58.3	58.5	59.0	0.3	1.20
56.6	57.5	57.5	1.6	1.59
55.4	56.0	56.5	1.1	1.99
54.4	55.0	55.5	1.1	2.02
53.5	54.5	54.5	1.9	1.87
52.9	53.5	53.5	1.1	1.13
52.4	52.5	52.5	0.2	0.19
51.6	51.9	51.5	0.6	0.19
50.9	50.5	51.0	0.8	0.20
50.4	50.0	50.0	0.8	0.79
49.7	49.5	49.5	0.4	0.40
49.1	48.5	49.0	1.2	0.20
48.6	48.0	48.5	1.2	0.21
Average			0.9	0.8

The sample is placed between 2 temperature sensors (ΔT) and a heater. To obtain test data, five intervals will be given with reference in units of time (Table 2). The electrical energy of this system is supplied from heater with specification of masimal power of 50 W and 12 V AC/DC with duration 5 minutes. Calor energy per second of the heater can be calculated with the formulas. When the tool is turned on, the heater will provide heat to the material which will be calculated by the temperature sensor so that the temperature difference between the left and right layers of the sample can be analyzed in minute intervals. The control of each tool uses an ATMega328 microcontroller which can control the tool to carry out testing. The average conductivity value of the composite material is ,028 W/moC, which can be calculated by entering the area and thickness of the sample (Figure 8). This tool that can be used to assess the conductivity value of an insulating material used in building construction is the conductivity test equipment. A material will be more pleasant to utilize as a building construction material the lower its conductivity value.

Table 2. Value of sample conductivity

Name of Sample	Geometry of sample		Temperature Difference (°C)
	Surface Area (m^2)	Thickness (m)	
Composite-1	0.000450	0.00300	30.2
Composite-2	0.000375	0.00250	32.3
Composite-3	0.000300	0.00200	33.4
Composite-4	0.000250	0.00150	35.6
Composite-5	0.000200	0.00100	36.2

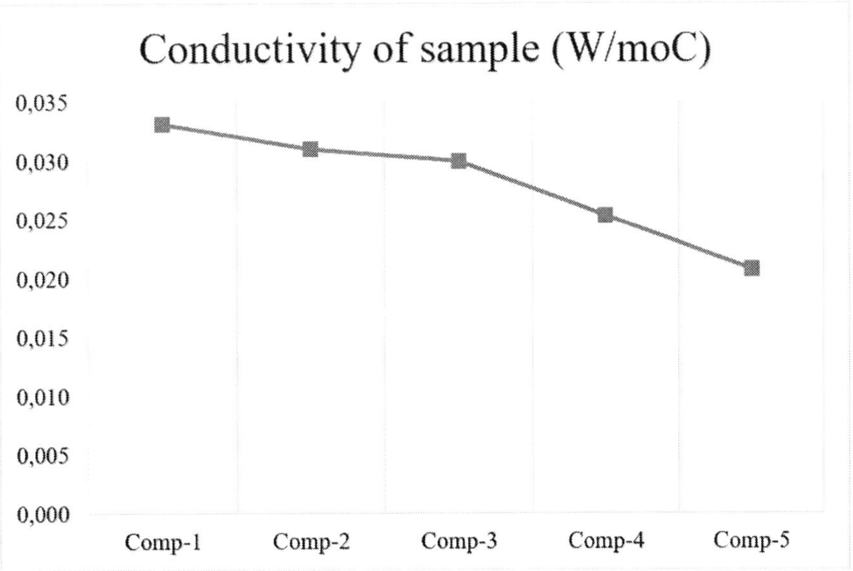

Figure 8. Conductivity value of sample

4. Conclusion

The DSB1860 sensor works both in terms of accuracy and precision with an average error value ranging from 0.8% to 0.9%. The conductivity test equipment that has been made has worked well which is characterized by a low average error value when compared to standard tools. By entering the value of the area and thickness of the sample, the average conductivity value of the composite material is 0,028 W/m°C. The lower the conductivity value of a material, the more comfortable the material will be to be used as a building construction material.

Acknowledgments

The author would like to thank Universitas Sumatera Utara for grant this research in the scheme of Dosen Muda in accordance with the research contract TALENTA Universitas Sumatera Utara funding which number of contract: 6789/UN5.1.R/PPM/2021, dated June 16, 2021.

References

[1] Liu Z, Sun X, Xie J, Zhang X and Li J 2022 *Int. J. Heat Mass Transf* **195** 123031
[2] Wang X, Gu W and Lu H 2022 *Int. Commun. Heat Mass Transf* **136** 106198
[3] Bustomi F and Ghofur A 2021 *Jtam Rotary* **3** 233
[4] Ali M and Kurniawan R 2013 *J. Desiminasi Teknol* **1** 59
[5] Okzama K R and Arwizet 2019 *J. Multidiscip. Res. Dev* **1** 4 906

[6] Ma X, Wu S, Yi Z, Peng D and Zhang J 2019 *Int. J. Heat Mass Transf.* **137** 790
[7] Wang T Y and Tsai J L 2016 *Comput. Mater. Sci.* **122** 272
[8] Zhan H, Zhang G, Zhang Y, Tan V B C, Bell J M and Gu Y 2016 *Carbon N. Y.* **98** 232
[9] Huang L, Zhu P, Li G, Zhou F, Lu D, Sun R and Wong C 2015 *Journal of Materials Science: Materials in Electronics* **26** 6 3564
[10] Sucipto S, Priangkoso T and Darmanto D 2013 *Majalah Ilmiah MOMENTUM* **9** 1
[11] Syukri R and Arwizet K 2019 *Ranah Res. J. Multidiscip. Res. Dev.* **1** 4 921
[12] Dharmajati E S 2017 *Inov. Fis. Indones.* **6** 73
[13] Guerra V, Wan C and McNally T 2019 *Prog. Mater. Sci.*, **100** 170
[14] Azhari, Simanjuntak D, Hakim L and Sabar 2022 *J. Phys. Conf. Ser.* **2193** 1
[15] Tritt T M 2005 *Thermal conductivity: theory, properties, and applications* (New York: Springer Science and Business Media)
[16] Yang Y 2007 *Thermal conductivity, Physical properties of polymers handbook* (New York: Springer)

The effect of fermentation time on fruits as a producer electrical energy

R Sigalingging [1*], V C S Panjaitan[1] and C Sigalingging[2]

[1]Department of Agricultural and Biosystem Engineering, Universitas Sumatera Utara, Prof. A. Sofyan No. 3 Kampus USU, Medan, Indonesia Republic-20155.
[2]Department of Food Science and Technology, Universitas Nahdlatul Ulama Sumatera Utara, Jl. H. A. Manaf Lubis No. 2, Helvetia, Medan, Sumatera Utara.

*Email: riswanti@usu.ac.id

Abstract. The need for electrical energy is growing, and so is the demand for alternative energy. Fruits with high acidity can generate electrical power. The fermentation process can enhance the acidity of the fruit. The lower the pH, the higher the voltage and current generated. This study aims to obtain electrical energy by testing the potential of fruits for induced current, voltage, power, resistance and electrical energy. The experimental parameters are current, voltage, power, resistance and electrical energy. Three fruits (soursop, mango, and tamarind) with three different fermentation times (9, 11, and 13 days) were used in a two-factor, fully randomised design procedure. The results show a tamarind fruit with a fermentation time of 13 days, the highest voltage, current, power and electrical energy values of 5.62 Volts, 1.52 mA, 8.54 mW, 672.30 mWh and a resistance value of 1.800 ohms. Soursop fruits with a fermentation time of 9 days provided minimum voltage, current, power, and electrical energy values of 4.52 volts, 0.93 mA, 4.20 mW, and 265.26 mWh, and a resistance value of 2.118 ohms. The type of fruit and the varying fermentation time had a very significant effect at the 1% level on current, voltage, resistance, power and energy of electricity.

1. Introduction

The increasing electricity demand can be seen in the number of villages that have not received electricity and the quality of the electricity supply that can only last a few hours (there is a power outage). At the end of 2019, the installed national power plant capacity was 69.6 GW or almost close to 70 GW, with the hope that in 2024 it could increase by 35000 MW or 20.62 GW. Therefore, in 2020 the government provided a budget of 133.9 billion rupiahs. In other words, the 2020 budget increased by 34.7% from the 2019 budget [1]. In 2019 it was also noted that the electrification ratio in Indonesia has also increased by 98.89% [2]. The increasing electrification ratio in Indonesia is evidence that the government has made many efforts to provide electricity needs and also continues to develop other alternatives to increasing and stabilising the electrification ratio in Indonesia.

As increasing electrical energy is needed, innovations are needed to find alternative electrical energy, such as fruits that can be used in the broader community. Atina [3] has reported that the nature of the electrolyte could produce electricity if the fruit has acidic properties. The content of acid and water is found in fruits, so when two different metals are put into a fruit solution, there is a potential difference between the metal and water; this will cause an electrode potential to generate an electric current. It can be seen that the principle of the voltaic cell is in line with this concept, that fruits have the potential to be used as an alternative source of electrical energy.

Fruits with a low pH are fruits with a high acidity level. Fruits with a low pH or high acidity can generate electrical energy because the lower the pH, the more acidic the fruit will lead to a higher voltage and electric current [4]. So to increase the acidity of the fruit, a fermentation process is carried out. Fermentation produces acidity in the material because the final product substances include large amounts of lactic acid, acetic acid, and small amounts of organic acids [5].

Soursop fruit has a high level of acidity with a pH of 5.45, which is relatively low [6]. The acid content in soursop fruit is ascorbic acid, malic acid, and citric acid [7]. Mango fruit is classified as having a low pH of 3.9-4.6 [8], and its acid content is ascorbic acid, gallic acid, chlorogenic acid, vanilla acid, and ferulic acid [9]. Tamarind has a high level of acidity because it has a pH below four which is relatively low [10]. Bustan [11] reported that there are organic acids in tamarind, namely citric acid, tartaric acid, and malic acid.

The electrochemical process is a chemical process that converts soursop, mango, and tamarind as alternative electrical energy or changes chemical reactions into electrical energy, with the most important parts being the electrodes and electrolyte solution. According to Marianti et al. [12] the reaction produces an electric voltage in a chemical process called oxidation-reduction (redox). When two electrodes are immersed in a fruit solution containing a high acidity, there will be a potential difference between the electrode and the fruit solution, producing electrical energy. Landis (1990) also reported that when both ends of the conductor there is a continuous redox reaction, there will be an exchange of charge carriers from the electrode to the electrolyte solution and vice versa, namely from the electrolyte solution to the electrode, causing the flow of charge carriers (electric current) in the circuit of the two electrodes. In other words, the electromotive force of the cell is the result of changes in chemical energy through the redox reaction Zn(s) + Cu^{2+}(aq) → Zn^{2+}(aq) + Cu(s)

Chemical reactions that occur between electrolyte solutions such as ascorbic acid ($C_6H_8O_6$) in soursop and mango with Cu and Zn electrodes in producing electrical energy are:

Cathode: $Cu(s) + 4H^+ + C_6H_6O_6^{2-}(aq) + 2e^- \rightarrow CuC_6H_6O_6(s) + 2H_2O(l)$
Anode : $Zn(s) + C_6H_6O_6^{2-}(aq) \rightarrow ZnC_6H_6O_6(s) + 2e^-$
Reaction : $Zn(s) + Cu(s) + 4H^+(aq) + 2\ C_6H_6O_6^{2-}(aq)) \rightarrow ZnC_6H_6O_6(s) + CuC_6H_6O_6(s) + 2H_2O(l)$.

Chemical reactions that occur between electrolyte solutions such as tartaric acid ($C_4H_6O_6$) in tamarind fruit with Cu and Zn electrodes in producing electrical energy are:

Reaction: $Zn(s) + Cu(s) + 4H^+(aq) + 2C_4H_4O_6^{2-}(aq) \rightarrow ZnC_4H_4O_6(s) + CuC_4H_4O_6(s) + 2H_2O(l)$

Cathode: $Cu(s) + 4H^+ + C_4H_4O_6^{2-}(aq) + 2e^- \rightarrow CuC_4H_4O_6(s) + 2H_2O(l)$
Anode : $Zn(s) + C_4H_4O_6^{2-}(aq) \rightarrow ZnC_4H_4O_6(s) + 2e^-$

The low pH of soursop (*Annona muricata* Lin.), mango (*Mangifera indica)*, and tamarind (*Tamarindus indica)* are caused by high acid content. The acid content of each of these fruits (soursop, mango, and tamarind) have the potential to produce electrical energy. Therefore, the study on the electrical potential produced by soursop, mango, and tamarind has been done by varying the fermentation time.

2. Materials and Methods

2.1. Materials

Soursop (F1), mango (F2), and tamarind (F3) fruits, as much as 3 kg each with a maturity level of 50% as an electrolyte, green LED as load and indicator, aquadest, sugar, yeast, work as a mixture for making substrates, tissue as a container cleaner, labels function as markers for each jar were used. An electric blender to grind fruits, a digital balance to measure the mass of fruits, a digital multimeter to measure electric current and voltage, a connecting cable to connect electrodes with electrolytes, and a basin as a fruit container were used. Other tools were a jar as a container for the fruit during fermentation, the duct tape to glue the jar, the knife to cut the fruit, the electrodes (Cu and Zn) as conductors of electric current, the stopwatch to adjust the steaming time, measuring cup to measure aquadest and solutions, pH meter to measure the pH of fruits, the container as a place for fruit solutions when measuring electricity were used.

2.2. Experimental procedures

The electrolyte solution used in this study was a solution of soursop, mango, and tamarind. A fruit that has the potential as a producer of electrical energy is a fruit that has a high level of acid content. According to Atina [3], the electrolyte's nature can produce electricity if the fruit has acidic properties. Soursop, mango and tamarind fruits with a high acid or a low pH, and the acid content found in soursop, mango, and tamarind. Due to the high acid content in soursop fruit, mango, and tamarind, these fruits have the potential to produce electrical energy when two different metals are added to the solution. There is a potential difference between the metal and water; this will cause an electrode potential to generate an electric current.

The electrode is vital in the electrochemical process because it is a conducting medium or transmits electricity into a solution so that a chemical reaction changes into electrical energy. In this study, the electrodes used were Cu electrodes as the cathode (positive charge) and Zn electrodes as anodes (negative charge). The electrolyte solution was a solution of soursop, mango, and tamarind fruit which had been treated with variations in fermentation time (9 (D1), 11 (D2), and 13 (D3) days). The electrical characteristics of the electrolyte on the C-Zn, Cu-Al, and Cu-Zn electrodes are that the Cu-Zn pair produces the most significant voltage compared to other electrode pairs in seawater electrolytes [13]. Hendri et al. [14] also reported that using Cu and Zn electrodes produced higher voltages in electrochemical cells.

Electrical measurements on soursop, mango, and tamarind are arranged in series. Seven Cu and Zn electrodes, each with an electrode area of 9cm×3cm, will be connected to a digital multimeter in the fruit solution. There were seven containers; each container of fruit solution was put (300 ml) with a container volume of 14cm×10cm×5cm. Then the circuit will be connected to the green LED as an actual load with a green LED voltage of 1.9 - 4.0 volts. When the LED can light up means, the fruit has the potential to produce electrical energy.

The following are the stages of experimental procedures carried out, namely:

1. Preparation phase

Firstly, clean the soursop, mango, and tamarind with water, then drain. Separate the skin, seeds and fruit from soursop, mango, and tamarind, then cut into small pieces of 1 × 1 cm. Steam soursop fruit for 15 minutes, mango and tamarind for 30 minutes.

Cool the soursop, mango, and tamarind in the open air until completely cold. The last is puree soursop, mango, and tamarind using a blender separately.

2. Fermentation stage and time continuity

Firstly, make mushroom growing media (Subtract), where the substrate comes from a mixture of sugar, yeast (*Saccharomyces*, sp), and aquadest. Then put 134 grams of sugar, 1300 ml of distilled water, and 74 grams of yeast (saccharomyces, sp) into a jar (for one jar) and leave the substrate for 8 hours. Divide the pulp of soursop, mango, and tamarind into jars with a mass of 1000 grams. Measure the pH of the soursop, mango and tamarind pulp to observe the voltage and current values obtained. Put the substrate into a jar containing soursop fruit pulp, mango and tamarind. Cover the jar with tape and make sure there is no environmental influence. Mark each container with a different length of fermentation time using a label. The last is to observe the solution periodically and continuously, namely 9 days, 11 days, and 13 days.

3. Measurement of electrical properties

The first is to open the jar containing the solution and measure the pH after the completion of the fermentation process using a pH meter. Divide as much as 300 ml of soursop, mango, and tamarind fruit solution into a container of 7 pieces to assemble a series. Insert a pair of electrodes (Cu and Zn) that have been connected to the connecting cable in the seven containers containing a solution of soursop, mango, and tamarind. Connect the digital multimeter to the electrodes (Cu and Zn), which are connected to the connecting cable in the seven containers containing a solution of soursop, mango, and tamarind. Record the results of the current and voltage measurements produced by the electrical circuit that appears on the digital multimeter in the container containing the fruit solution. Connect the green LED with a series circuit as an actual load and note the length of time the green LED is on. Calculate the value of power, solution resistance, and electrical energy produced in a series circuit using a predetermined formula (Eqs. 1, 2 and 5). It was repeating the activity every time the fermentation time was set. The experiment was repeated three times and analysed with statistics (ANOVA).

$$P = V\ I \quad (1)$$

where P is power (watt); V is voltage (volt), and I is current (Ampere)

$$R_{solution} = \frac{1}{G} \quad (2)$$

where $R_{solution}$ is the electrical resistance of solution (Ohm)
According to Dole [15] the value of electrical conductivity (G) can be obtained using Equation

$$G = \frac{1000}{C} \times \frac{L}{R \times A} \quad => \quad G = \frac{1000 \times K}{C} \quad (3)$$

where: G is electrical conductivity ($Ohm^{-1}cm^{-1}$); C is solution concentration (mol/litre).

To obtain the value of the conductivity of the type of solution (K), you can use the formula in Equation

$$K = \frac{L}{R \times A} \quad (4)$$

Where: L is the distance between electrodes (cm); A is area of the container (cm^2) and K is conductivity of the type of solution ($Ohm^{-1}cm^{-1}$)

$$E = P\ t \quad (5)$$

Where: E is electrical energy produced (Wh); P is power (watt); t = LED flash time (h)

3. Results and Discussions

3.1. Temperature and pH during fermentation process

The environmental temperature must be considered during the fermentation process because the temperature factor affects the fermentation process. Kumalasari [16] states that if the ambient temperature is too low, it can slow down the fermentation process, and if the temperature is too high, it can cause the microbes to die. The temperature during the fermentation process was shown in Table 1.

Table 1. The environmental temperature

Day	Temperature (°C)
1	29
2	29
3	24
4	25
5	26
6	27
7	27
8	28
9	29
10	28
11	28
12	28
13	27

Table 2. pH data before and after fermentation

Treatment	Average (pH) Before	Average (pH) After
F1D1	4.20	3.13
F1D2	4.07	3.10
F1D3	3.87	3.03
F2D1	4.13	3.00
F2D2	4.10	2.97
F2D3	4.03	2.90
F3D1	2.90	2.80
F3D2	2.83	2.77
F3D3	2.80	2.73

Information: F1 = soursop fruit, F2 = mango fruit, F3 = tamarind fruit, D1 = 9 days fermentation, D2 = 11 days fermentation, D3 = 13 days fermentation

Table 1 shows that the average ambient temperature during the fermentation process is around 27°C, where this ambient temperature is the temperature for yeast growth. The yeast growth temperature is between 20-30°C [17]. Environmental temperature is also very influential on the fermentation process because the storage temperature, which is the ambient temperature, is one of the essential factors that affect the survival of bacteria. Therefore, ideal storage conditions are needed to maintain the viability of bacteria because the ideal storage temperature so that to benefit from the work of a probiotic (microorganisms or bacteria) [18].

In this study, fermentation variation was 9 (D1) days, 11 (D2) days, and 13 (D3) days for each fruit. The pH value of the fruit solution obtained from the treatment before and after fermentation was displayed in Table 2. A high acidity level in the material will have a low pH, affecting the value of the resulting voltage and current. Low voltage makes the voltage and current also high, so the pH value obtained will be inversely proportional to the current and voltage values. The lowest pH value has a high acidity level, namely in the F3D3 treatment (Javanese tamarind fruit with 13 days fermentation) of 2.73, while the highest pH is in the F1D1 treatment (Soursop fruit with 9 days fermentation) of 3.13.

3.2. *Electrical energy*

Electric current is the amount of electric charge that flows per unit time through an electric conductor in one direction due to the influence of an external force. At the same time, voltage is the energy required to move a unit of electric charge from one place to another. If the voltage increases, the electric potential will increase because of the increase in electrons released by the electrolyte ions. Power is the amount of work done per unit of time. Electrical power is also the amount of energy absorbed in an electrical circuit. The data obtained for the electrical power produced by soursop, mango, and tamarind is the product of the voltage and electric current produced. Electrical energy can occur because there is a high and low potential difference. The occurrence of electricity is due to the potential difference between high potential and low potential and the transfer of charge on a conductor from low potential to high potential. The resistance in an electrolyte solution comes from electrostatic forces between ions or friction and collisions between ions and solvent molecules; in other words, resistance is a force that opposes the flow of electric current.

Table 3. Current, voltage, resistance, power and energy of electric

Treatment	Current (mA)	Voltage (volt)	Resistance (Ohm)	Power (mW)	Time (h)	Energy (mWh)
F1D1	0.93	4.52	2.118	4.20	63	265.26
F1D2	0.96	4.62	1.776	4.46	65	289.01
F1D3	1.13	5.10	1.675	5.75	72	414.22
F2D1	0.97	4.58	1.378	4.43	64	283.29
F2D2	0.98	4.61	1.374	4.51	65	291.49
F2D3	1.21	5.29	1.269	6.42	74	477.50
F3D1	0.99	4.66	2.286	4.60	65	298.99
F3D2	1.04	4.80	2.234	4.99	67	334.54
F3D3	1.52	5.62	1.800	8.54	79	672.30

Information: F1 = soursop fruit, F2 = mango fruit, F3 = tamarind fruit, D1 = 9 days fermentation, D2 = 11 days fermentation, D3 = 13 days fermentation

Table 3 and Figure 1 show that the F3D3 treatment (Javanese tamarind fruit with 13 days of fermentation) was the highest electric current, voltage, power and energy value (1.52 mA, 5.62-volt, 8.54 mW and 672.30 mWh), and the F1D1 treatment (Soursop fruit with 9 days of fermentation) was the lowest value. The value of the electric current obtained is directly proportional to the value of the electric voltage. If the voltage generated is high, the electric current generated will also be high, and vice versa; if the electric current obtained is low, the voltage value generated is also low. This is in line with the sound of Ohm's law. The highest electric current value is also influenced by the length of fermentation time. The 13 days fermentation time is the highest electric current value compared to the electric current value in 11 days and 9 days of fermentation because the longer the fermentation time, the current value will also increase. The highest electric current value is produced by tamarind fruit because the acid content is higher than soursop and mango. A high value of acid content can produce a high value of electric current because if a material has a high acid content, it will have a high concentration of H^+. The high concentration of H^+ ions cause an increase in electrical conductivity in the electrolyte solution because it can increase electrical conductivity. This happens because electrical conductivity has the ability of a liquid to conduct electricity, which is related to the movement of ions in an electrolyte solution. Thus, when the material has a high acid content, it will contain a lot of H^+ ions, which causes the ions to move quickly in solution, which shows a sizeable electrical conductivity which will produce a large amount of electricity as well. The length of fermentation time can increase electrical conductivity [19]. This happens because of the nature of a solution that is more acidic. There is an increasing concentration of H^+ so that the long fermentation time can produce increased electrical conductivity, which causes the value of the electric current also generated to increase.

Table 3 obtained the value of the calculation of the resistance of the solution produced by soursop, mango, and tamarind. The highest solution resistance value for each fruit was in the 9-day fermentation treatment. The highest solution resistance for soursop fruit was 2.118 Ohm, mango fruit was 1.378 Ohms, and tamarind fruit was 2.286 Ohm. If the voltage and current values are high, the resistance value is low. The lowest solution resistance value was obtained in the 13-day fermentation treatment for each fruit, with a resistance value for soursop fruit of 1.675 Ohm, mango fruit at 1.269 Ohms, and tamarind fruit at 1.800 Ohms. The highest inhibition value in each treatment was found in tamarind fruit because tamarind fruit has more acid content than soursop and mango fruits. The difference in the amount of this acid content affects the electrical results obtained, causing tamarind fruit to be superior to soursop and mango. Addition, statistical analysis shows that the type of fruit and the varying fermentation time had a very significant effect at the 1% level on current, voltage, resistance, power and energy of electricity, while the interaction showed an insignificant effect on current, voltage, resistance, power and energy of electricity.

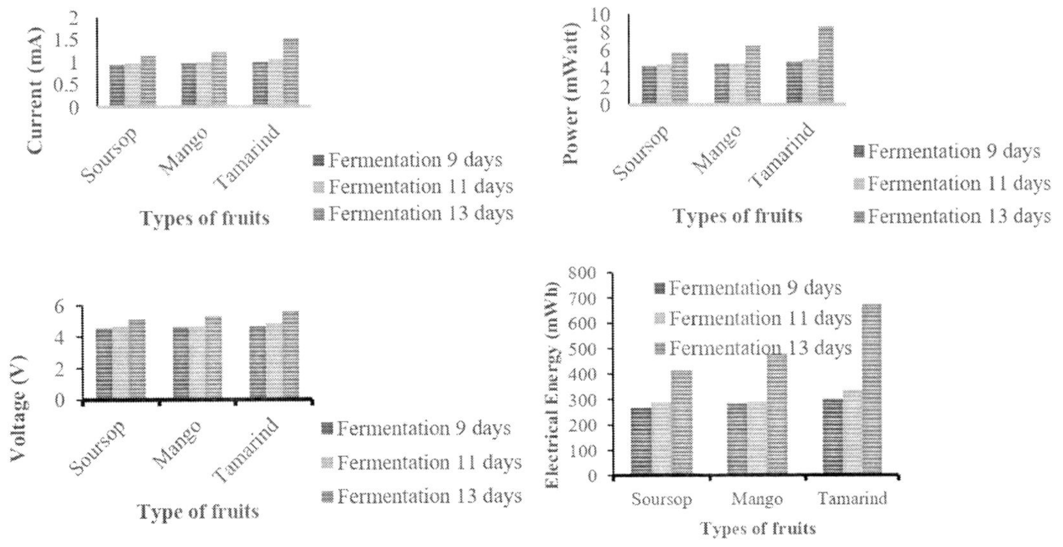

Figure 1. Dependence among current, voltage, power and energy of electricity and types of fruits with varying fermentation time

4. Conclusions

After 13 days of fermentation, tamarind fruit obtained the highest voltage, current, power, and electrical energy of 5.62 volts, 1.52 mA, 8.54 mW, and 672. 30 mWh. Meanwhile, the tamarind fruit's highest resistance value was obtained with nine days of fermentation of 2.286 Ohms. Fermentation time affects the voltage and current values obtained; 13 days of fermentation is the best than 11 days and 9 days of fermentation. Voltage and current increase the longer the fermentation time. Fruits that have a high level of acid content have the potential to produce electrical energy. The electrical energy produced by soursop, mango, and tamarind in the variation of fermentation time has a very significant effect; this is due to the different levels of acid content in each fruit.

References

[1] Rahayu L N and Windarta J 2022 *Jurnal Energi Baru dan Terbarukan* **3** 88
[2] Badan Pusat Statistik 2020 *Rasio Elektrifikasi Jakarta: Badan Pusat Statistik*
[3] Atina A 2015 *Jurnal Ilmiah Matematika dan Ilmu Pengetahuan Alam* 12
[4] Miranda C A and Afrida J 2020 *Jurnal Phi Jurnal Pendidikan Fisika dan Terapan* **1** 18
[5] Sudarmadji S, Suhardi and Haryono B 1989 *Analisa bahan makanan dan pertanian* (Liberty Yogyakarta bekerja sama dengan Pusat Antar Universitas Pangan dan Gizi Universitas Gadjah Mada)
[6] Espinosa I, Ortiz R, Tovar B, Mata M and Montalvo E 2013 *Journal of Food Quality* **36** 10
[7] Okigbo R N and Obire O 2009 *International Journal of Wine Research* **1** 1
[8] Penteado A L 2017 a mini-review *Vigilância Sanitária em Debate* **5** 127
[9] Vithana M D, Singh Z, Johnson S K and Gupta R 2019 *Journal of the Science of Food and Agriculture* **99** 1126
[10] Nassereddin R A and Yamani M I 2005 *Journal of food protection* **68** 773
[11] Bustan M 2008 *Karakterisasi Ekstrak Air Daging Buah Asam Jawa (Tamarindus indica L)* [Skripsi] (Fakultas Farmasi Universitas Sanata Dharma Yogyakarta)
[12] Marianti M, Yanto S and Sukainah A 2021 *Jurnal Pendidikan Teknologi Pertanian* **7** 1
[13] Ayu G A, Rahmayanti D and Nindy E 2015 *Jurnal Ilmiah Penelitian dan Pembelajaran Fisika* 1
[14] Hendri Y N 2015 *Pillar of Physics* 6
[15] Dole M 1935 *Journal of the American Chemical Society* **57** 2731
[16] Kumalasari I J 2011 *Pengaruh variasi suhu inkubasi terhadap kadar etanol hasil fermentasi kulit dan bonggol nanas (Ananas sativus)* [Skripsi] (Universitas Muhammadiyah Semarang Semarang)
[17] Ferdiaz S 1992 *Mikrobiologi Pangan 1*(Jakarta: Gramedia Pustaka Utama)
[18] Lee Y K 2009 Selection and maintenance of probiotic microorganisms *Handbook of Probiotics and Prebiotics; Lee YK, Salminen S Eds* 177
[19] Supratomo S, Laga A, Tahir M, Mochtar A and Salengke S 2019 *Journal of Engineering and Applied Sciences* **14** 1515

IOP Conf. Series: Earth and Environmental Science 1115 (2022) 012089 doi:10.1088/1755-1315/1115/1/012089

Soil amelioration using several types of humic substances extracted from andisol, spodosol, peat and lignite to increase the growth of corn plants (*Zea mays*)

E Rahmi[1], Ernawita[2], H Satriawan[2*] and Suwardi[3]

[1]Forestry Department, Faculty of Agriculture, Universitas Almuslim, Jl. Almuslim, Bireuen 24261, Aceh, Indonesia.
[2]Department of Natural Resources and Environmental Management, Graduate Program, Universitas Almuslim. Jl. Almuslim, Bireuen 24261, Aceh, Indonesia.
[3]Department of Soil Science and Land Resources, Faculty of Agriculture, IPB University. Jl. Meranti IPB Dramaga Campus, Bogor, Indonesia

*Email: satriawan_80@yahoo.co.id

Abstract. Humic substances are products formed during the decomposition process of organic maerials which can be found in aquatic, sedimentary and soil environments. Humic substances can be used as an ameliorant due to its important roles in improving soil fertility and plant growth. The aim of the study was to determine the mechanism of humic substances in increase the growth of maize plants; to obtain the optimum types and doses of humic in increasing plant growth. The research was conducted in a greenhouse of IPB. The study used a completely randomized design (CRD) which consists of two treatments, namely humic types (Andisol, Spodosol, peat and lignite) and humic doses (15 and 30 liters/ha). The results showed that the administration of humic acid from lignite at a dose of 30 liters/ha was the best treatment to increase the growth of corn plants, which was characterized by better stimulation of leaf width, number of leaves, plant biomass weight, and root dry weight. Thus, humic substances are useful in improving the growth and productivity of corn plants.

1. Introduction

Corn is a commodity with high economic value that can be used as food, feed, and raw materials for the processing industry [1]. Nowadays, one of the government focus in agriculture including expanding corn cultivation to meet the national corn demands and to reduce import which continue to increase. In the context of developing agricultural commodities, it is necessary to improve land quality to support plant growth and yields. Soil fertility is an important factor determining productivity and can be enhance by adding fertilizers and ameliorants.

Humic substances are one of ameliorant material which now widely used to increase plant growth. Humic substances in the soil environment can be found in soils that have high organic matter such as Andisols, Spodosols, peat and lignite [2]. One of the factors that influence variations in the content of humic acid is the content of organic matter [3]. In addition, humic substances are also contained in lignite and potentially used as ameliorants [4].

It is important to notes that sources of humic raw materials are abundant and easy to find. So that it can be used as economics potential source that can be extracted for humate. Different sources of materials allow different characteristics of humic compound produced. Improvement of soil quality with humic soil amendments is expected to improve soil properties so as to increase maize production. Considering the many roles of humic materials on plant growth and variations in the content of extracted materials, it is necessary to study the effect of various types of humic materials on plant growth. Hence, the aim of the study was to obtain the best types of humic materials and doses to increase the growth of corn plants.

2. Materials and Methods

This research was conducted in the greenhouse of the Permanent Nursery, Bogor Agricultural University. The material used is Latosol soil as the planting medium. Humic materials used were Andisol, Spodosol, peat, and lignite humic materials. For planting, talenta corn seed, basic fertilizer of Urea, TSP, and KCL were used. The characteristics of the Latosol soil used in this study had an acid pH of 4.27, an Al-dd content of 3.64 me/100g, a very low C-organic of 0.28% and an available P of 8.18 ppm. The characteristics of the humic materials used in this study are presented in the Table 1.

Table 1. Chemical characteristics of humic substances from various soil materials

Parameters	Humic acid			
	Andisol	Spodosol	Peat	Lignite
Ash contents (%)	36.8	34.0	23.3	31.5
Carbon (%)	25.4	35.6	48.8	37.3
Nitrogen (%)	2.3	3.5	1.5	2.4
C/N Ratio	11.0	10.2	32.5	14.5
Carboxyl group (meq/g)	2.6	2.2	3.6	3.0
Total acidity (meq/g)	6.5	5.1	6.4	6.1
Phenol group (meq/g)	3.9	2.9	3.0	3.2

Source: Rahmi et al [9]

The study used a completely randomized design (CRD) which consisted of two treatments. The first treatment was humic acid, namely Andisol, Spodosol, peat, and lignite humic acids. The second treatment was the dose of humic acid with two levels, namely 15 and 30 liters/ha. Each treatment was repeated three times and compared with the control (planting media without humic material treatment) so that in total there were 27 experimental units. The combination of humic acid type and dosage is presented in Table 2.

The planting medium used is Latosol soil material. The collection was carried out in a composite manner at a depth of 0-20 cm. The soil was dried and sieved using a 5 mm sieve and weighed 4 kg of soil. Then the soil material is mixed with basic fertilizer. The basic fertilizers used were Urea (45%N), TSP (36 %P_2O_5), KCl (60% K_2O). The plant used was Talenta corn variety and each pot consisted of three plant seeds. Maintenance activities are carried out in the form of weeding and watering. Parameters observed were vegetative parameters, namely plant height, number of leaves and leaf width which were measured every week for 5 weeks. At harvest time, root weight and plant biomass were measured.

Table 2. Treatment of humic acid types and humic doses tested on soil materials

Treatments	Humic acids	Humic dosage (liter/ha)
H0K0	Control	0
H1K1	Andisol	15
H1K2	Andisol	30
H2K1	Spodosol	15
H2K2	Spodosol	30
H3K1	Peat	15
H3K2	Lignite	30
H4K1	Lignite	15
H4K2	Lignite	30

Note: H = type of humate; and
K = humate dose

Data were calculated and processed with ANOVA using SPSS 2.1 software. Significant result further tested using Duncan Multiple Range Test (DMRT) at a 5% significance level.

3. Results and Discussion

3.1. The effect of humic acid on corn plant growth

Figure 1 shows that the administration of the type of humic acid and the dose of humic acid affected plant height and leaf width at 5 WAP and had no effect on the number of leaves. H1K1 treatment (andisol humic acid and dose of 15 liters/ha) was the best treatment to increase plant height. H4K2 treatment (lignite humic acid and a dose of 30 liters/ha) was the best treatment to increase leaf width. The higher the dose of humic acid given, the more the leaf width of the plant increased. This is because the higher administration of humic acid can affect cell elongation. Humic acid contains compounds that act as plant growth stimulants such as auxin, gibberellins and Indole Acetic Acid (IAA) which plays role in plant physiological processes by increasing cell permeability [5-7]. So that humic acid is able to stimulate the growth of leaf width and has a good influence on the development of leaf width of corn plants.

Meanwhile, the results of statistical analysis showed that the application of humic acid had no significant effect on the number of leaves of plants aged 5 WAP. These results are in accordance with research of [8] which stated that humic acid treatment did not show significant results on the number of leaves on plants aged 2-8 WAP. Although it had no statistical effect, the H4K2 treatment had more leaves than the other treatments. While the H3K2 and H0K0 treatments had the least number of leaves. The more the number of doses given, the more the number of leaves observed on plants aged 5 WAP.

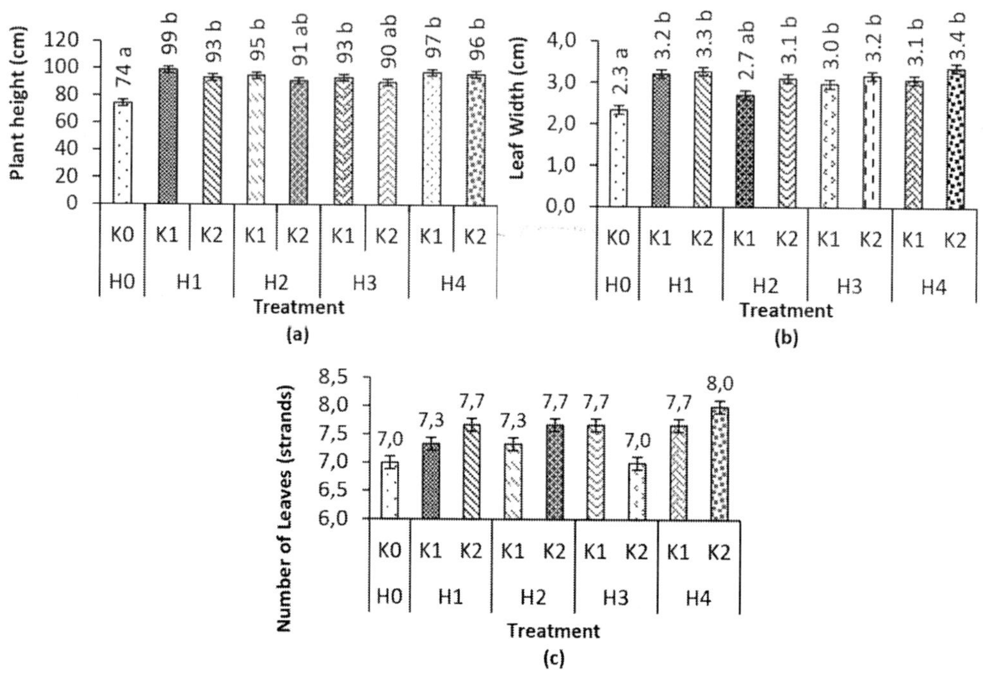

Figure 1. Effect of humic acid on plant height (a), leaf width (b), and number of leaves (c) on corn at 5 weeks after harvest (WAP)

The treatment of lignite humic acid (H4) and andisol humic acid (H1) showed a better effect. It is suspected that there were other dissolved nutrients from lignite and Andisol materials. On the other hand, the humic acid treatment of peat (H3) resulted in the lowest corn plant growth (plant height, leaf width and number of leaves) compared to other humic types. Low plant growth in humic acid treatment from peat due to decomposition of lignin-rich wood from peat under anaerobic conditions, in addition to producing phenolic acids, also produces aliphatic acids, most of which are toxic to plants. Humic acid from peat produced by FTIR has more aliphatic functional groups than humic acid from Andisol soil (Figure 2), namely at wave numbers 2920.23 cm^{-1}, 2848.86 cm^{-1}, and 1458.31 cm^{-1}. The large number of occurrences at these wave numbers causes humic acid from peat to be dominated by aliphatic compounds [9].

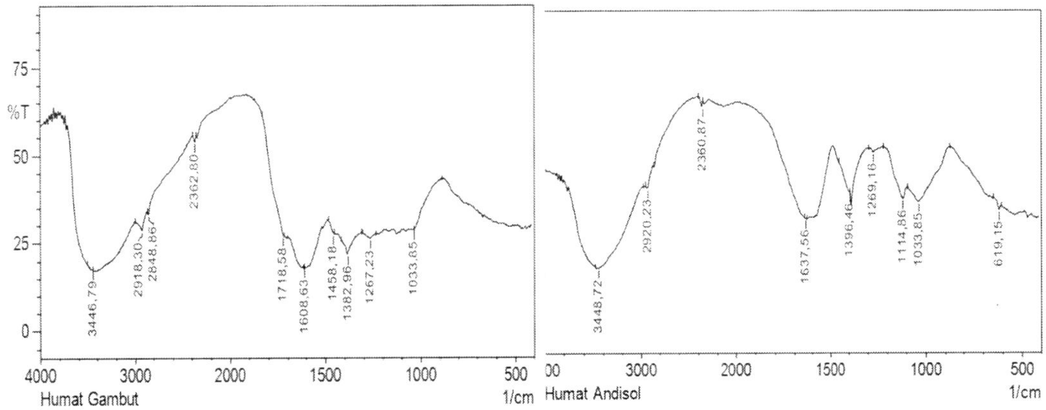

Figure 2. Results of FTIR identification of humic substances from peat and andisol

Several research results show that aliphatic compounds are phytotoxic to plants which cause stunted plant growth; these compounds have an effect on inhibiting the supply of nutrients in the soil [10]. In addition, the C/N ratio in peat is much higher than other humic acids (Table 1). High C/N values can interfere with plant growth, due to the slow decomposition process.

3.2. Effect of humic acid application on dry weight production of plant biomass

The effect of humic acid administration on the dry weight production of corn plant biomass (leaves and stems) is presented in Figure 3. The results of statistical analysis showed that the administration of humic acid and humic dose was significantly different from the control (H0K0) but did not differ between treatments given humic acid. Figure 1 shows that the administration of a dose of 30 L/ha in the treatment of humic acid types H1, H2, and H4 gave the best dry weight. The humic acid treatment of lignite (H4K2) showed the best dry weight. [11-12] stated that the dry weight of plant biomass is the result of accumulation of organic compounds that have been successfully synthesized by plants from inorganic compounds. Nutrients that have been absorbed by the roots, both used in the synthesis of organic compounds and those that remain in ionic form in plant tissues will contribute to the increase in plant dry weight. This is in accordance with the research of [13] which stated that with the addition of 10 L/ha humic acid there was an increase in corn production by 10%, one of which was through an increase in plant biomass weight. The effect of giving humic substances on root biomass production showed the same pattern as the effect on plant dry weight.

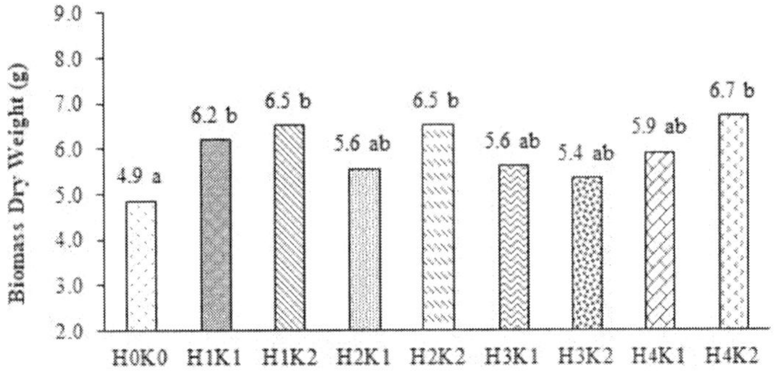

Figure 3. Effect of humic compound application on dry weight of maize leaf and stem biomass based on DMRT test results at 5% significance level

The application of humic acid was not significantly different to the root dry weight parameter between treatments, but showed a significant difference with the control (Figure 3). Although statistically there was no significant difference, the administration of humic acid in the H4K2 treatment tended to produce better root dry weight compared to other treatments. The administration of humic acid in the H4K2 treatment increased root dry weight by 34% compared to the control (Figure 3). This increase in root weight value was due to well-developed roots, one of which was indicated by the increasing number of roots (Figure 4). This is in accordance with research [14,15] that humic substances are able to stimulate root growth. Increasing the number of roots has a good effect on plants, because the absorption and cruising power of roots are more optimal to absorb nutrients in the soil. This causes plant roots to reach previously unreachable nutrients, resulting in higher nutrient absorption. Furthermore [16,17] stated that good root development of corn plants can increase the growth and production of corn plants.

Root dry weight data showed a pattern of increase and yield differences which were in line with the dry weight data of maize plant biomass (Figure 3 and Figure 4). Root weight tends to be positively correlated with leaf width of maize plants. This shows that the increase in root weight will further increase the width of the corn leaves. Root weight can be used as an indicator of nutrient uptake. The higher root weight indicates the ability of the roots to absorb nutrients and supply nutrients to plant leaves better, resulting in more optimal plant growth. Based on this, humic acid is able to have a positive influence on plant growth

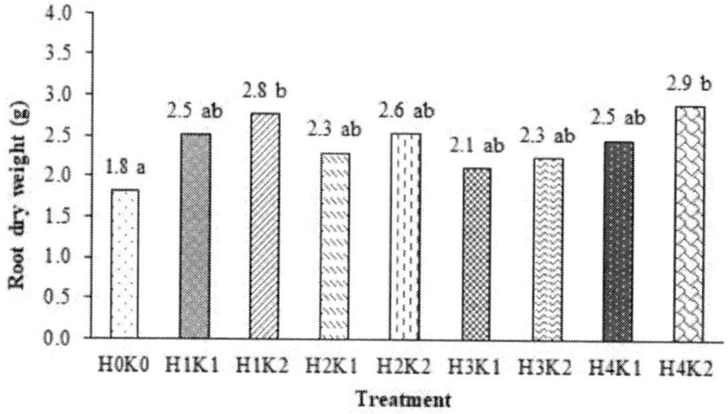

Figure 4. Effect of humic compound application on root dry weight of maize based on DMRT test results at 5% significance level

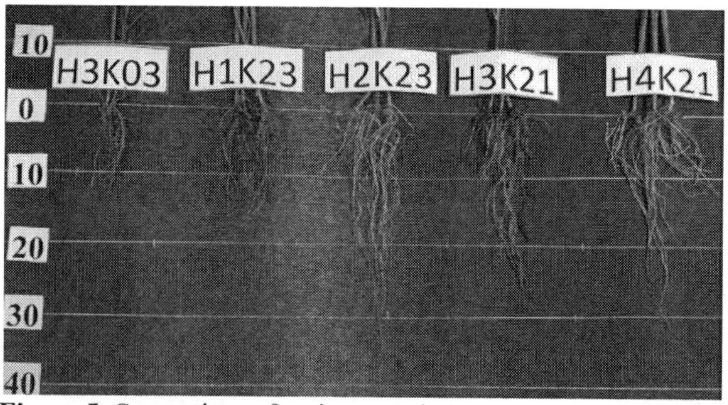

Figure 5. Comparison of maize roots between treatments 5 WAP

Humic acid functional groups containing oxygen atoms such as –OH and –COOH and are the most reactive groups [18]. De-protonation of humic acid functional groups will reduce the ability to form hydrogen bonds, both between molecules and between molecules so that the main functional groups tend to be in the form of –COO which can act as ligands in complex formation [19,20]. With the de-protonation of the functional group, the number of negative charges of the functional group increases so that its ability to bind to metal ions increases. In addition, the phenol-OH functional group and the COOH-containing high oxygen group can form stable complexes with metal ions that can function as ligands in the formation of chelate complexes in humic acid [21]. Complex formation and chelation play an important role in increasing soil fertility. Chelating can increase the mobility of cations resulting in their availability to plants. The release of plant nutrients by weathering of soil minerals is usually a slow process. However, complex formation tends to accelerate the decomposition process of soil minerals and thereby accelerate the release of dissolved nutrients [22]. Based on this, the application of humic acid can increase the ability of the soil to absorb and exchange cations which affect its availability to plants.

The functional group of lignite humic acid is dominated by the phenolic hydroxyl group rather than the carboxyl group. According to [23] the phenol group has properties that are more stable and not easily differentiated than the –COOH group, this causes a more stable total acidity value of humic lignite compounds. The high total acidity is considered to play a very important role in chemical reactivity in the soil. According to [24,25] states that only part of the nutrients such as nitrogen contained in humic substances can be absorbed by plants. Based on this, the direct effect seen with the administration of humic acid is due to the hormone content contained in humic acid. This is in line with what was stated by [26] that humic acid is considered to have activity similar to the hormone auxin. Furthermore [27] stated that the compounds that trigger plant growth in humic substances are numerous, such as vitamins, amino acids, auxins, cytokinins, and Indole Acetic Acid (IAA). Auxin in the roots will increase turgor pressure so that cytokinins in the roots will be transported to the plant crown. The content of auxin in plants will increase cell wall permeability which will increase the absorption of nutrients to form chlorophyll which is needed in the photosynthesis process.

The administration of humic acid from lignite at a dose of 30 l/ha showed a better effect on the development of corn plants, especially on the parameters of biomass and root weight. This is due to the possibility related to the development process of the original material during the weathering of the organic material. In addition, when it is associated with the characteristics of the properties of the material, the characteristics are quite good. One of them is the relatively low C/N ratio (Table 1), because the high C/N ratio of humic acid can interfere with plant growth. The characteristics of humic acids such as functional groups and total acidity also affect the growth of corn plants.

4. Conclusion

It can be concluded that the administration of humic substances from lignite at a dose of 30 liters/ha (H4K2) is the best treatment to increase the growth of corn plants characterized by stimulation of leaf width, plant biomass weight, and better root growth. Mainly through improving root conditions, namely in root biomass so that nutrient uptake is better. The mechanism of humic substances in influencing plant growth is through their activity similar to the hormones auxin and cytokinin.

References

[1] [BKP] Badan Ketahanan Pangan 2018 *Bul. Pasokan dan Harga Pangan* 3
[2] Sanchez-Cortez, Francioso S, Garcia-Ramos O, Marzadori J V and Gessa C C 2000 *Proceedings 10th International Meeting of the International-Humic Substances Society Touluse* France
[3] Soil Survey Staff 2000 *Agency for International Development* (United States Department of Agriculture)
[4] Adiwiganda R, Koendadiri A S and Poeloengan Z 1993 *Buletin PPKS* 1 163
[5] Chen Y, Nobili M D and Aviad T 2004 *Soil Organic Matter in Sustainable Agriculture* (Florida, US: CRC Press)

[6] Zandonadi D B, Santos M P, Busato J G, Peres L E P and Façanha A R 2013 *Theoretical and Experimental Plant Physiology* **25** 13
[7] Trevisan S, Francioso O, Quaggiotti S and Nardi S 2010 *Plant signaling & behavior* **5** 635
[8] Tobing R 2009 Pengaruh Aplikasi Senyawa Humat terhadap Sifat Kimia Tanah Vertisol dan Pertumbuhan Tanaman Jagung (*Zea mays*) (Bogor: IPB)
[9] Rahmi E, Suwardi S and Sumawinata B 2018 *J Soil Sci Agroclimatol* **15** 35
[10] Tsutsuki K and Kondo R 1995 *Soil Sci and Plant Nutr* **41** 515
[11] Heil C A 2005 *Harmful Algae* **4** 603
[12] Ghosh M and Singh S P 2005 *Asian J Energy Environ* **6** 18
[13] Suwardi, Dewi E M and Hermawan B A 2009 *Jurnal Zeolit Indonesia* **8** 44
[14] Atiych R M, Lee S, Edwards C A, Arancon N Q and Metzger, J D 2002 *Bioresource Technol* **84** 7
[15] García A C, van Tol de Castro T A, Santos L A, Tavares O C H, Castro R N, Berbara R L L and García-Mina J M 2019 *Journal of Environmental Quality* **48** 1622-1632
[16] Radite S and Simanjuntak B H 2020 *Agriland* **8** 72
[17] Suwardi and Wijaya H 2013 *Jurnal Ilmu Pertanian Indonesia* **18** 79
[18] Amini S 2015 Carbon dynamics in salt-affected soils (Griffith School of Environment Science, Environment, Engineering and technology Griffith University)
[19] Alimin N, Santosa S J and Noegrohati S 2005 *Jurnal Ilmu Dasar* **6** 1
[20] Murtaza B, Zaman G, Imran M, Shah G M, Amjad M, Ahmad N and Murtaza G 2019 *Communications in soil science and plant analysis* **50** 1013
[21] Spark K M, Wells J D and Johnson B B 1997 *Soil Research* **35** 89
[22] Setyowati D and Ulfin I 2007 *Jurnal Akta* Kimia *Indonesia* **2** 85
[23] Wang M, Zhu Y, Cheng L, Andserson B, Zhao X, Wang D and Ding A 2018 *Journal of Environmental Sciences* **63** 156
[24] Varanini Z and Pinton R 1995 *Progress in botany* 97
[25] Victolika H, Sarno and Ginting Y C 2014 *J Agrotek Trop* **2** 297
[26] Nardi S, Pizzeghello D, Muscolo A and Vianello A 2002 *Soil Biol Biochem* **34** 1527
[27] Conselvan G, Fuentes D, Merchant A, Peggion C, Francioso O and Carletti P 2018 *Plant and Soil* **426** 17

Simulation model in determining forestry and plantation circumstances

N Sembiring*, **H L Napitupulu, M T Sembiring and A Ishak**

Departement of Industrial Engineering, Faculty of Engineering, Universitas Sumatera Utara, Indonesia

*Email: nurhayatipandia68@usu.ac.id

Abstract. Plantation and forestry systems consist of several dynamic subsystems as well as complexes that are integrated. Plantation and forestry productivity is largely determined by the interactions of each of these sub-systems. Simulation models can illustrate interactions between sub-systems in detail to be analyzed and evaluated in predicting any conditions that may occur. The model is indispensable because it leads to significant savings in time and money. This article is compiled by selecting and analyzing the previous research methods selected to draw conclusions on interactions between sub-systems to achieve effective and efficient plantation productivity. The final results show that how sub-systems affect forestry and plantation productivity can be demonstrated by hybrid simulation models. Therefore, hybrid simulation methods can be used to determine the integration process of each party in affecting the productivity of plantations and forests. The hybrid simulation model incorporated in two distinct approaches, namely Agent Based Modeling and System Dynamics. The models of hybrid simulation are useful as an illustration of interactions between each sub-system taking into account influential input variables.

1. Introduction

Manage modern plantation industries well, have very high growth rates, produce better quality wood, short rotational length, and maintain robotization technology for supply chain defense [1]. Plantation and forestry systems consist of integrated dynamic sub-systems and also complex. The productivity of plantations and forestry is largely determined by the interaction of each sub-system. Moreover, forest management is done for various purposes. Basically, forest processing aims to obtain the benefits of products and services sourced from forest resources optimally. Therefore, a tool that can assist forest managers during the decision-making process must be provided [2].

Forest and plantation management has productivity and interactions between species influenced by climate change that affect longstanding forest and plantation productivity over its influence on humidity availability, nutrient cycles and temperature-limited rates of clean photosynthesis. Climate change has the potential to undermine efforts to achieve a sustainable balance and create considerable improbability in forest and plantation management planning and it outcomes [3].

Based under certain conditions to prepare for disruption, some researchers have used dissimilar modelling outfits for simulate that systems. It improve their responsiveness and durability. These tools of modelling that used are fuzzy modelling, hybrid models, linear programming, and discrete event simulations. By focusing on aggregate level analysis and studies isolating certain parts of the supply chain are the objects of that researchers. Blending various tools for purposes will capture operational and strategic attentions in the recommendations and decision-making process [4].

This paper aims to evaluate various approaches in simulations used for forest management ranging from previous researchers who published previous information, methods used, provided brief descriptions of some model examples, and years of research published to be considered in decision making and future improvement.

2. Method

This study uses literature review methods to develop a framework of thought consistent with previous research findings, and theories to complete the formulation of the problem. Figure 1 shows the arrangement of the article review.

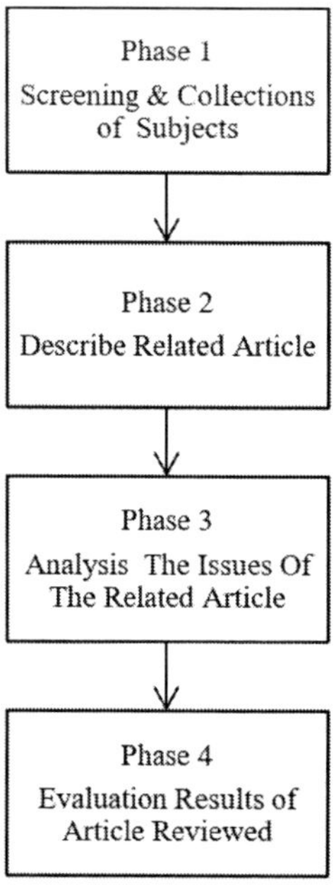

Figure 1. Research methods using for literature review

In this study, the method was to investigate selected prior literature for discussion and evaluation. Results based on the article review method are described in this section. Phase or step one is to collect all reports of earlier research results for investigation. The first step is to collect reports on previous research results associated to the themes deliberated for investigation. This article examines a hybrid simulation model in determining forestry and plantation conditions. A review of various approaches in forest and plantation management simulations. The article selected from various approaches in forest and plantation management simulation aims to determine the appropriate model for eucalyptus growth to determine optimal forestry and plantation conditions.

Phase 2. All reports from earlier studies collected are analyzed. The next phase or step of this research is to analyze the collection of related article from previous studies. Papers that are accumulated from various publishers must have a subject that fits of this research. The reviewed papers related to the hybrid simulation model in determining forestry and plantation conditions were accumulated to be reviewed one by one.

Phase 3. All reports from previous studies selected for the methods used in the studies analyzed. The next step is to take the core contained in the collection of studies. To take the essence of the review, researchers need to analyze each article that has been accumulated in the second phase based on a predetermined theme. The essence of the reviewer who is concerned in this paper is the author, the year of publication, the factors in the study, the purpose of the researcher, the object under study, the method used, and the place of journal of the scientific work. Abstract taking is done to get important evidence about the research discussion. Thus, we can find papers that suit the subject of this research from several publishing countries.

Phase 4. Evaluate the relevance of all previous research finding reports with stated research objectives. The fourth step in this research is to evaluate the results of the selected papers. Each article is selected and analyzed to extract the results that are the essence of the presentations in this paper review. To get a thorough understanding of the contents of the selected article, an analysis is carried out on the background, the research objectives, then the methods, discussions, and finally conclusions.

3. Research Results

After the analysis and evaluation of the research, among the different types of simulation methods available indicate that hybrid simulation methods are capable of controlling factors that affect plantation and forestry productivity. Hybrid simulation is a mixture of dynamic model simulation and discrete simulation. The advantages of this method are able to explain the advantages and disadvantages of each related simulation. In implementing forestry and plantation systems to provide optimal results, hybrid simulation methods can also accept many input establishments to describe simulation models that are very close to their original form. But the weakness of the hybrid simulation model is that in the mathematical algorithms used in the system representation process, it reduces the number of scheming parameters used. Some of literature that reviewed about hybrid simulation model research group in plantation and forestry determination are presented in Table 1.

Table 1. Hybrid simulation model research group in plantation and forestry determination

Author	Method	Outcome	Volume and Year
C. Peng, et al	The TRIPLEX model is used by integrating soil– century 4.0 carbon-nitrogen model, TREENYD3 growth and yield model, and 3-PG forest production model	The TRIPLEX model is proficient of providing a reasonable estimate of the Diameter at Breast Height, height, and density of stems for jack pine stances in northern Ontario. Analysis of the sensitivity suggests that a variety of responses to identical changes in climate drivers can be expected [8].	153 (2002)
M. E. Harmon and B. Marks.	STANDCARB Model	This simulation shows for the ecosystem that provides the lion's share of U.S. timber supply that there are no real problems. But through this simulation if utilization standards are lowered, it is possible to store more in forest products and ecosystems [9].	32 (2002)
B. Richardson, et al	Hybrid and Process-Based Models.	Through a review of several factors that affect a system for weed control treatment on tree growth and predicting the effects of non-plant vegetation and survival, researchers should be able to produce a simulation model that comprises each subsystem, ongoing processes, and interactions between subsystems [10].	79 (2006)
P. Miehle, et al	Four Process-Based Modelling Approaches	From a process-based model, Forest-DNDC achieved the greatest model efficiency in the 2 years after planting and was the most reliable performance model, while CABALA achieved the greatest model competence after 8 years of plant growth. Both 3-PG models performed best for 6-year-old plantations, but during the simulation era the precision of their predictions relied heavily on location fertility estimates [11].	220 (2009)

Author	Method	Outcome	Volume and Year
P. M. Feikema, et al	3PG+ Forest Growth Model	3PG+ integration into CAT, and provides 3PG+ spatial applications in various areas with mixed land use for study of carbon and water movement in forest systems and resulting in improved application accuracy [12].	260 (2010)
P. Lasch, et al	4C Ecophysiological Forest Growth Model	Appropriate contributions to regional CO2 mitigation by SRC aspen plantations and carbon sequestration are below the likelihood of climate change, but in regional water budgets there may be negative impacts [13].	10 (2010)
J. A. Rodríguez-Suárez, et al	3PG Process-Based Model	The 3PG model for plant forest management is very useful and stands out as a tool that simulates plant growth. simplicity and the small number of parameters required as input data [14].	129 (2010)
M. Ueyama, et al	Terrestrial Biosphere Model, Eddy Flux Measurements, and Tree-Ring Analyses.	Model simulations show harvesting is a more significant determinant of current carbon sequestration than inter-yearly climate abnormalities or enlarged CO2 concentrations in the atmosphere and forests now recovering from harvesting [15].	222 (2011)
P. R. Grace and B. Basso.	The 3PG Simulation Model.	Using available growth data, a simple process-based simulation model is ready to be adjusted to develop reliable carbon sequestration estimates through forest management [16].	105 (2012)
S. Sawano,et al	Forest Evapotranspiration Simple Model	On this model requires little meteorological input compared to process-oriented models and has a simple structure, that is why it is considered appropriate for regional scale applications. To be assimilated into other monthly forest ecosystem models this model has great potential [17].	309-310 (2015)
L. Cao, et al	UAV-LiDAR systems and PB Models (3-PG)	The 3-PG model demonstrates good recital in simulating forest structural parameters (R2 > 0.80). observed between UAV-LiDAR technology and 3-PG models have a relatively close linear relationship (R2 > 0.6). The results show to assess the structural parameters of forests under an ongoing climate both of these approaches have potential [18].	12 (2019)
C. Liu, et al	The 3PG model combines the Monte Carlo chain Markov (MCMC) method with Bayesian posterior estimation theory and sensitivity analysis.	This scheme can be used to analyze the uncertainty of the multi-parameter complex stand growth model and to calibrate the parameters. The Monte Carlo chain Markov method (MCMC) optimizes the 63 parameters of the most sensitive 3PG models, the parameters affect the spread of stands, the probability distribution and the default parameter [19].	11 (2020)
C. Alegria, et al	Modeling of species distribution and productivity.	The resulting map allows for the identification of the best locations for regulatory support of timber production, species greening, fire hazard mitigation, and landscape planning taking into	12 (2021)

Author	Method	Outcome	Volume and Year
		account species diversity. Modeling uses environmental covariate to obtain these maps, such as climate features, so that projections in upcoming climate change circumstances can be made [21].	
E. Duchateau, et al	Point Process Modeling Approach to Represent Interaction Between Trees.	The most significant predictor tree in clarifying the observed distance between the found trees is the diameter and species of the two closest neighbors. Since the treatment of singering, in a short period of outcomes showed that the action had little substantial effect on the interaction of trees [20].	12 (2021)
E. B. Searle, et al	ZELIG-CFS	Factors used for simulation with ZELIG-CFS i.e. erect volume, distribution of stem diameter or height, and species composition showed post-harvest treatment achieved volume comparable to natural treatment in a simulated period of 70 years [22].	12 (2021)

The paper collected in the matrix of tables in accordance with the keywords then conducted analysis and evaluation of the 15 journals. It can be determined that most of the research conducted from journals that have been studied are in 2010 and 2021. Publisher of scientific works dominated by Elsevier.

Figure 2 shows the most research related to hybrid simulation model determining forestry and plantation conditions published in 2010 and 2021. In 2010 and 2020, each accounted for 20% of the fifteen papers concise. By an article published in 2021, hybrid simulation methods are able to control factors that affect plantation and forestry productivity. The hybrid method is also able to explain the advantages and disadvantages of each related simulation.

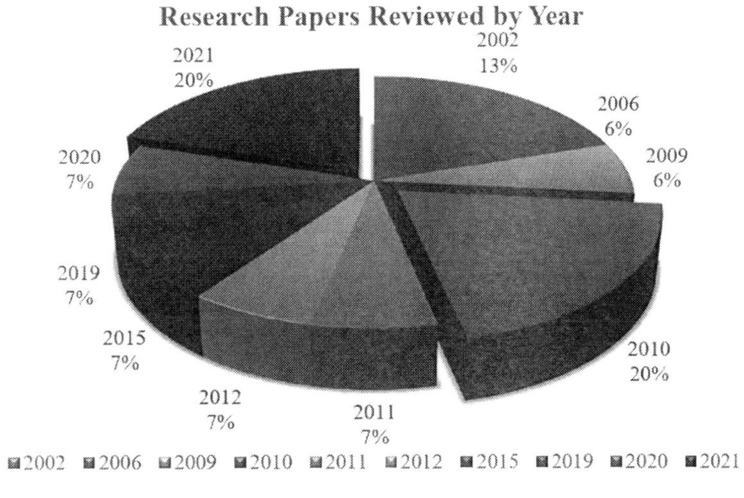

Figure 2. Research papers reviewed by year

3.1. Forestry and plantation conditions

In the 21st century, abiotic environmental conditions in forests and plants changed significantly compared to the last few hundred years. The dynamics of forest conditions are determined by forest abiotic factors ranging from temperature restrictions, water restrictions, wind disturbances and even fires that are factors of climate change causing various effects [5].

3.2. Types of approaches adopted in modelling simulations

To avoid the limitations of the empirical model, it need a mechanistic model that can represent physiological tree processes. Mechanistic models have been widely used, especially for various studies because these models require broad parameters[2]. Knowledge-based systems are also called rule-based systems where components are modelled and interfaces between them do not have to be mathematically represented in specific cases of modelling. Through the use of symbolic representations of information, a systems modeling approach is used to simulate the logical processes of human experts [2]. Empirical model (EM) is a highly recommended model for resolving issues attached to sustainable forest management. This model is based on statistical analysis with variables available on forest inventory [5]. Simulation using a hybrid model that combines the features of two simulation paradigms to take advantage of each approach. Integration of two concepts can be beneficial when allowing for a combination of properties and forces that otherwise belong to a single concept [6].

3.3. Determining factor in forestry and plantation

High level of professional experience, knowledge and assessment is required in building and managing productive crop forests. Therefore, factors that related to the productivity of forests and also crops are; location of affordable access, quality of location (soil depth and drainage; physical and chemical composition of the soil (including pH); number and pattern of availability of annual soil lengas; frequency and nature of wind, storms and fires are common and occasional; and the general climate of the area.), species selection, genetic inputs, nursery practices, distance and age control, planting simulation, post-planting care, holistic management , consideration of biodiversity [7].

3.4. Advantages and disadvantages of dynamic model simulation

Dynamic models are mostly in the form of computer software or mathematical regression models. This model is used to optimize the development of forest and plantation conditions showing predictions of dynamic variables that are used as evaluations for management to prepare themselves and material for analysis in dealing with future conditions in the future. The advantages of this dynamic model can be used as a comparison to get the shape of the model with the smallest error value. However, the weakness of the dynamic model is that in order to get the accuracy value that is closest to the actual condition, it is necessary to re-calibrate the use of this model [7].

4. Conclusion

Through previous research on forest and plantation management can be a source of information for decision-making management and drive the productivity value of existing systems. It is important to make decisions in planning that will be done using the insight to predict the conditions to be experienced with various variable interactions. In plantation and forestry management to predict any possible conditions with various variables that can influence the system using a hybrid simulation model approach is very easy. It was concluded that by paying attention to variable inputs that influence hybrid simulation methods help plantation and forestry management in studying processes and inter-stakeholder relationships that occur in forest and plantation systems in understanding the processes and connections between stakeholders that arise in plantation systems and forest.

References

[1] McEwan A, Marchi E, Spinelli R and Brink M 2020 *J. For. Res* vol **31** 339
[2] Twery M J and Weiskittel A R 2013 *Environ Model Find.* 379
[3] Seely B C, Welham, and K Scoullar 2015 *PLoS One* **10** 1
[4] Childerhouse P and Al Aqqad M 2020 *Network Resilience Modelling : a New Zealand Forestry Supply Chain Casel* **31** 291
[5] Fontes L, Bontemps J D, Bugmann H, Van Oijen M, Gracia C, Kramer K, Lindner M, Rötzer T and Skovsgaard J P 2011 *For. Syst.* **3** 8
[6] Khanzadi M, Nasirzadeh F, Mir M And Nojedehi P 2017 *Prediction and Improvement of Labor*

Productivity Using Hybrid System Dynamics and Agent-Based Modeling Approach
[7] Sembiring N, Napitupulu H L, Sembiring M T, Ishak A and Radot R R 2020 *Proc. Int. Conf. Ind. Eng. Oper. Manag.* **59** 809
[8] Peng C, Liu J, Dang Q and Apps M J 2002 *Modelo Triplex* **153** 109
[9] Harmon M E and Marks B 2002 *Can. J. For. Res.* **32** 868
[10] Richardson B, Watt M S, Mason E G and Kriticos D J 2006 *Forestry*, vol **79** 29
[11] Miehle P 2009 *Ecol. Modell.* 5 **734**
[12] Feikema P M, Morris J D, Beverly C R, Collopy J J, Baker T G and Lane P N J 2010 *For. Ecol. Manage.* **260** 663
[13] P Lasch, Kollas C, Rock J and Suckow F 2010 e *Reg. Environ Chang.* **10** 83
[14] Rodríguez-Suárez J A, Soto B, Iglesias M L and Diaz-Fierros F 2010 *Eur. J. For. Res.* **129** 573
[15] Ueyama M, Kai A, Ichii K, Hamotani K, Kosugi Y and Monji N 2011 *Ecol. Modell.* **222** 3216
[16] Grace P R and Basso B 2012 *Agric. Syst.* **105** 1
[17] Sawano S, Hotta N, Tanaka N, Tsuboyama Y and Suzuki M 2015 *Ecol. Modell.* **309** 93
[18] Cao L, Liu K, Shen X, Wu X and Liu H 2019 *IEEE J.Sel. Top. Appl. Earth Obs. Remote Sens.* **12** 4175
[19] Liu C, Zheng X, and Ren Y 2020 *Forests* **11** 1
[20] Duchateau E, Schneider R, Tremblay S, Dupont-Leduc L, and Pretzsch H 2021 *Forest* **12** 1
[21] Alegria C, Roque N, Albuquerque T, Fernandez P and Ribeiro M M 2021 *Forests* **12** 3
[22] Eric B S, Bell F W, Guy R L, Mathieu F, Jennifer D, Rita S S, Marco M, Holly D D 2021 *Forests* **12** 755

Designing eucalyptus replanting model using hybrid simulation method : a review

N Sembiring*, **M I Mansurin, M D A Banurea, C A Tarigan and T Waruwu**

Departement of Industrial Engineering, Faculty of Engineering, Universitas Sumatera Utara, Indonesia

*Email: nurhayatipandia68@usu.ac.id

Abstract. The main objective of this paper is to find out the concept of designing the eucalyptus plant replanting model using a hybrid simulation method. Logically, single simulation is much easier to apply than hybrid simulation because its procedure is simpler and easier to do. However, it is possible for hybrid simulation to be the best solution in solving problems. Based on previous studies, hybrid simulation which combines agent based model and dynamics system is the right solution in answering this research study object. The eucalyptus plant replanting model is computerized and uses a causal loop diagram. Furthermore, this model will be simulated in two ways, namely (1) a state chart area and (2) stock and flow. This systematic review identified twelve articles and one book related to the topic discussed. In addition, with this hybrid simulation system, it is hoped that the availability of eucalyptus plant as an essential raw material in the pulp and paper mill can be maintained, and it can increase the company's profitability. Here we provide details of Hybrid Simulation using the Agent-Based Modeling and System Dynamics Modeling that underlines our research study.

1. Introduction

At this time, replanting of eucalyptus plant is implemented uncoordinated. This is because pulp and paper mill doesn't have a replanting planning system that can guarantee a steady increase in pulp production in the following year. Whereas, it might help to ensuring the availability of eucalyptus as raw material for pulp, maintaining annual wood production and assisting maintenance of eucalyptus supply chain in eucalyptus forestry. The absence of a replanting planning system causes the production of raw materials from the replanting area does not fulfill the expected results. This is influenced by pests, diseases, and soil types. Accordingly we need a decision support tools for making plan of replanting eucalyptus to increase pulp production steadily every year and making the pulping industry get the higher profitability. The design model of the eucalyptus plant replanting was carried out using the hybrid simulation method. The designed model integrates two methods, known as Agent-Based Model and System Dynamics. Those two models can support stakeholders to making decision in the upstream and downstream rubber supply chain because the model can show qualitative and quantitative information [1-3].

Modeling is a way to solve problems and find the right solution. In this research study, those two models above are chosen as the best solution for its object. Agent-based model or abbreviated as ABM has become widely applied methodology in industrial disciplines because it is more flexible and does not require the concept of mathematical equations. System dynamics is a mathematical modeling concept for determining, comprehending, discussing, and solving the complicated problems [4,7]. A very different approaches from agent-based models and system dynamics (SD) when modeling a process, turns out that they can be combined by seeing the two simulations display mutually supportive points of view.

Finally, the aim of this paper is to know how to design the best model used in replanting eucalyptus plants. Whether through performing hybrid simulations that combine agent-based model and system dynamics, the supply chain of eucalyptus as the essential raw material for pulp and paper mills can be fulfilled or not. So, here we explain details of the combined hybrid simulation that underlines our research.

2. Materials and Methods

This research study uses twelve articles and one book from diverse source. These materials are found by using Publish or Perish (PoP) application and VOSviewer. This application aims to know what is the position of the topic that will be examined, whether it has been widely researched or not. By this application we can get more articles relevant to our research study.

First, open PoP and click Google Scholar as a source of metadata in this methods. Google Scholar is one of the featured applications under the auspices of the largest search engine company, Google. This feature provides scientific literature in the form of journals and scientific articles that have been published by the authors from various disciplines. Publication type of this research is journal with the keyword are replanting of eucalyptus using hybrid simulation, hybrid simulation, agent based models, and system dynamics. This research was conducted from 2000 until 2021, and determined to only fifty journal results. After that, save the results in the Result as RIS / RefManager format, and set your file name yourself.

Second, open VOS viewer then do this following procedures :

1. Click create.
   This step asked you to choose type of data. Because we have collected some metadata from PoP, so we select create a map based on text data. Then click next command.
2. Choose data source.
   In this step select read data from the reference manager file, because our supported file types is RIS manager. Then click next.
3. Select files
   Browse the file that will be analyzed. Then click finish.
4. Choose fields.
   In this research, the data will be analyzed based on the title and abstract fields. Then click next.
5. Choose counting method.
   Select binary counting as counting method in this research. Then click next.
6. Choose threshold.
   Enter number four in case of minimum number of occurrences of a term. Of the fifty terms, thirteen meet the threshold. Then click next.
7. Choose number of terms.
   Keep entering thirteen terms to be selected. Then click next and finish.

The analyzed results are presented on 3 methods, there are network visualization, overlay visualization, and density visualization. To make it easier for us to cite the journal later, then save the journal into Mendeley. By clicking on File on Mendeley then select Import, Research Information System (*.ris). Browse the journals that have been analyzed in VOS viewer before, click open, and then the journals will automatically be saved in Mendeley.

The research study methods are divided into four steps as follows:

2.1. Identification of related subjects

The first step is to identify what theme would be discussed. This research study will discuss replanting of eucalyptus using the hybrid simulation methods. Agent based model and system dynamics are two different simulation which is chosen and combined as the hybrid simulation in this research.

2.2. Searching and collecting the articles

The second step is to search and collect articles from diverse publisher. Articles are searched using PoP and VOS viewer application. This application help us to know what articles would be relevant to our topic. Finally, the result articles collected in one file, and move it into Mendeley as a references.

2.3. Filtering and Analysis of The Selected Articles
The third step is to filtering articles out of the collected articles. After reading a few of the article's abstract and discussion clearly, twelve articles and one book are selected out of the fifty collected articles based on the relevance of the article's title and its abstract with this research objectives. Then, the selected papers are analyzed based on IMRaD format, that is the introduction, methods, results, and discussions.

2.4. Evaluation and discussion
The fourth step in this research is to evaluate and discussed the results of the selected papers from the third stage. The results of the evaluation and discussion will be used as a reference to answering this paper's purpose. The stages of this research can be seen from Figure 1.

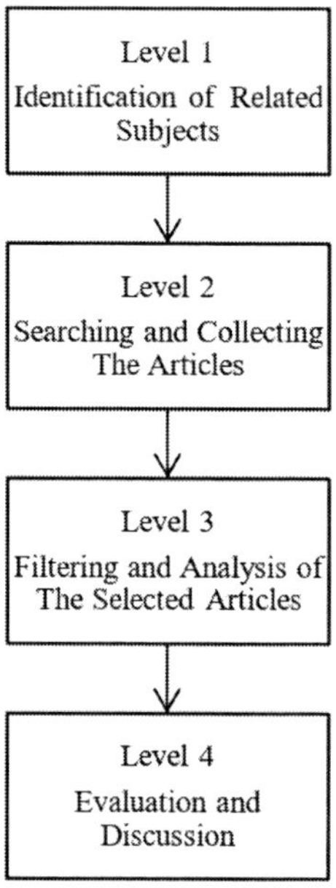

Figure 1. Stages of Research Implementation

3. Results and Discussion
Hybrid simulation is simulation modelling combines two or more different simulations. There are usually three simulations often used : discrete event, agent based model, and system dynamics [2]. In this research study, the planning of eucalyptus replanting was carried out using an agent-based model and system dynamics. Combining different simulation methods can solve this problem.

Agent-based model, also known as multi-agent system in engineering disciplines. Agent-based modeling is more flexible than dynamic systems, because it does not require the concept of mathematical equations. Additionally agent-based models can vary from highly detailed models to very abstract models [7]. System Dynamics is a mathematical modeling concept for determining, comprehending, discussing, and solving the complicated problems. Basically, it uses causal relationships in constructing a model of complex system, as a dynamic basis in recognizing and understanding the behavior of the system. With dynamic system simulation, system behavior and changes in the value of system variables can be known then be used as decision support in designing system improvement policies [10,11]. A very different approaches from agent-based models and system dynamics (SD) when modeling a process, turns out that they can be combined by seeing the two simulations display mutually supportive points of view. For more details about hybrid simulation combining agent based model and system dynamics model can be seen in the Table 1.

IOP Conf. Series: Earth and Environmental Science 1115 (2022) 012091 doi:10.1088/1755-1315/1115/1/012091

Table 1. Research articles on combination of agent based model and system dynamics on Hybrid Simulation

Paper By	Research's scope	Result	Year
Bo Zhang, Donald L. DeAngelis	An overview of agent-based models in plant biology and ecology	Model is a representation of the real system in a simple form by involving the influential components only and used to analyze the existing system. Models are classified into five general groups, starting with the functional structural plant modeling (FSPM) at a low spatial resolution and area scale. Although all population models are based on individual tree characteristics, Agent-Based Modeling of non-spatial gaps, are averaged over plant interactions, has a resolution and area of approximately 0.01 to 0.08. Agent-Based Modeling gaps spatially explicit using a finer and often wider resolution. Regional Agent-Based Modeling extracts information about plant interactions and integrates. And for more information, this modeling system has become an widely applied methodology in many areas such as biology [4].	2020
Brian Dangerfield	Systems Thinking and System Dynamics: a primer	This paper explains about the building blocks or diagram of the mappings which is the basic idea of systems thinking. It known as causal loop diagrams (CLD). This diagrams classified into two groups, they are simple positive feedback loop, and simple negative feedback loop. Some examples of simple positive feedback loop are (1) quality improvement of a product will increase product sales, and (2) the company's cash balance will increase as the company's income increases from the sale of products or services. Whereas, some examples of simple negative feedback loop are (1) reducing the number of employees has an impact on increasing the number of employee recruitment, and (2) company's cash balance will decrease when employee wages increase too [5].	2016
Céline Bérard	Group Model Building Using System Dynamics: An Analysis of Methodological Frameworks	The system is a collection of components that are interrelated and organized each other. The system is often associated with three dimensions, namely (1) the number of components involved, (2) the number of interrelationships between components, (3) and the relationship between system functions and system components. A structured system used to solve complex problems is called dynamic system. This system solves the problem with a qualitative and quantitative approach to conceptualize the underlying feedback [6].	2017
Ilya Grigoryev	AnyLogic 8 in Three Days	Many cases in the real world do not have the right solution. In fact, sometimes the solutions that are decided on actually cause new problems to come and result in changes that are too dangerous and even cause destruction. This is where a model is needed. Modeling is a way to solve problems and find the right solution. Simulation modeling requires a software uses a special programming language. This simulation modeling can help us in determining an analysis with high quality dynamic systems. System dynamics are used in long-term, and strategic models. Example of SD models such as people, products, and events. One of the multi-method models is Agent-based models. Example of ABM model such as units of equipment, products, ideas, investments, pieces of land, people in different roles [7].	2018
James Duggan	System Dynamics Modeling with R	A complex system must have the best foundation and a proofer flow direction, known as stocks and flows diagram. *Stocks and flows diagram* are the building design and the key concepts of dynamic system models [8].	2016

Table 1. Research articles on combination of agent based model and system dynamics on Hybrid Simulation (Continued)

Paper By	Research's scope	Result	Year
John D. Sterman	All models are wrong: reflections on becoming a systems scientist	In this study it was found that, computer simulation model design is more preferred than formal model design. Formal model design takes a very long time, whilst computer simulation is much more effective, has a wide range, is more efficient in terms of time and cost, and can solve problems in a more structured way so that the results obtained look more real. And more importantly system dynamics designed through computer simulations can help build our intuition and mental simulation abilities [9].	2002
M. W. Lake	Trends in Archaeological Simulation	In the last 15-25 years, system dynamics and agent based models have become a modeling concepts that are widely used and prioritized in various fields. System dynamics uses mathematical equation formulas to determine concrete changes over time. This system is more scalable and closely related to mathematization, so it can analyze a change quantitatively. Meanwhile, agent based model is generally referred to multi-agent because it was developed with cellular automata by Complexity Science. Agent based models are more flexible than dynamic systems, because they do not require the concept of mathematical equations [10].	2014
Navid Ghaffarzadegan, et.al.	How Small System Dynamics Models Can Help the Public Policy Process	System dynamics is a unique simulation model, and can solve some important problems. A lot of system dynamics contribution could be shown in decision making because its main characteristic is the acquisition of decisions with a feedback approach, called as causal loop diagram. Despite of these benefits, system dynamics also has limitations in the decisions obtained will be considered in much more depth, its cause-and-effect relationships by customers [11].	2011
Saikou Y, et.al.	Toward a Formalism of Modeling and Simulation Using Model Theory	Formally, to advance modeling and simulation we need to define the boundaries of ambiguity. The concept of formalization allows us to automate a design. Formalization also allows us to create and compose a model and simulation precisely based on predefined references [12].	2013
Xinyou Yin and Paul C. Struik	Modelling the crop: from system dynamics to systems biology	Based on this research study we know that although System Dynamics has been widely used in empirical land modeling, the parameters used still have a little biological side [13].	2010

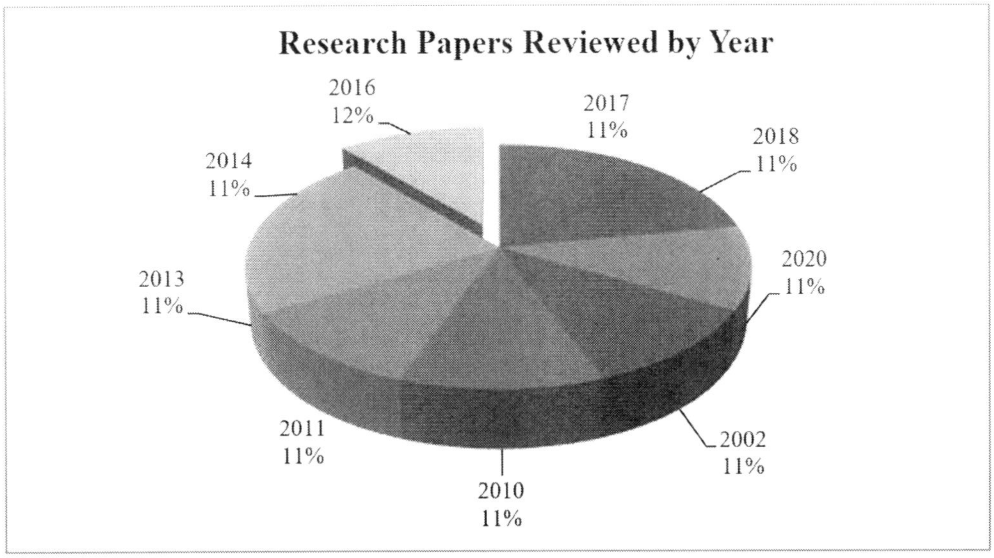

Figure 2. Research papers reviewed by year

Based on the attached graph, it is known that from 2002 to 2020 there were 10 journals that were reviewed. And information was also obtained that the most research on hybrid simulation based on agent-based models and system dynamics was found in 2016 which is about 12% of the 10 journals reviewed. Meanwhile, for 2002, 2010, 2011, 2013, 2014, 2017, 2019, and 2020 have the same percentage rate, which is around 11% of research every year. The two journals analyzed in 2016 explain about the building blocks or diagrams of the mappings which is the basic idea of systems thinking. The diagrams known as causal loop diagrams (CLD). It was also explained that in a complex system must have the best foundation and a proofer flow direction, known as stocks and flows diagrams. Stocks and flows diagrams are the building design and the key concepts of dynamic system models. The making of stocks and flows diagrams in the simulation model will later be concurrent with the creation of a computerized state chart area. [5-9]. Conceptually, the design model for the replanting eucalyptus plant using Agent Based Model and System Dynamics methods is described as follows.

a. Computerized model

System model for eucalyptus replanting is described by the Agent Based Model method and System Dynamics method. The top level agent is the beginning of a simulation model which is created, executed and will be connected to the components in the Agent Area. This model is influenced by the behavior of the agent Area, which is consisted of land area, agent location, and plant harvesting age [9].

b. Making causal loop diagram

Replanting of eucalyptus, eucalyptus land area, wood production, and pulp production is the components of causal loop. The relationship that shows a causal relationship in this model is, if eucalyptus tree replanting increases, so land area, wood production, and pulp production will increase too and vice versa [5].

After getting the best model, then the eucalyptus replanting simulation model was carried out, its model simulation will be explained fully as follow.

a. Making State Chart area

State Chart area is filled with process and auction logic in each state that describes how the land behaves. In the State Chart area, land will be prepared for replanting eucalyptus plantation and determining the harvest time of eucalyptus.

b. Making Stock and Flow

The simulation model is made by describing stock and flow. The influencing variables are plant age, growth rate, land area before replanting, wood potential, wood production, and also pulp production. The model is simulated from year to year according to the age of the plant in each forest area, and dynamically affects the output of wood and pulp. The area of eucalyptus planting area will be influenced by plant age, growth rate, and model time [8].

4. Conclusion

Increasing demand and limited supply chain of eucalyptus plantation caused unfulfilled target of pulping industry. Due to these problems replanting eucalyptus plants using hybrid simulation methods is the best solution. The hybrid simulation is carried out using Agent-Based Model and System Dynamics methods. Agent-Based Model in this model is found in the use of a state chart Area, depicted working upstream eucalyptus replanting system. System dynamics in this model is found in the use of stock and flow diagrams, working to give description of the downstream eucalyptus replanting system.

References

[1] Thompson K 2009 *System Dynamics Review: The Journal of the System Dynamics Society* **24** 433
[2] Magliocca N R 2020 *Land* **9** 1
[3] Nguyen L K N, Howick S and Megiddo I 2021 *Oper. Res. Soc. 10th Simul. Work. SW* 385–394
[4] Zhang B and Deangelis D L 2020 *Ann. Bot.* **126** 539
[5] Dangerfield B 2016 *Discret. Simul. Syst. Dyn. Manag. Decis. Mak.* 9781118349 26
[6] Bérard C 2017 *J. Polym. Res.* **20** 35
[7] Ilya G 2021 *AnyLogic in Three Days.*
[8] Duggan J 2016 *An Introd. to Syst. Dyn. Syst. Dyn. Model.* 1
[9] Sterman J D. 2002 All models are wrong: Reflections on becoming a systems scientist *Syst. Dyn. Rev.* **18** 501
[10] Lake M W 2014 *J. Archaeol. Method Theory* **21** 258
[11] Ghaffarzadegan N, Lyneis J and Richardson G P 2011 *Syst. Dyn. Rev.* **27** 22
[12] Diallo S Y, Padilla J J, Gore R, Herencia-zapana H and Tolk A 2014 *Complexity* **19** 56
[13] Yin X and Struik P C 2010 *J. Exp. Bot.* **61** 2171

Development of cricket flour-enriched cookies

F Tedjakusuma[1], J Linggadiputra[2], A D Cahya[1] and R Surya[1*]

[1]Food Technology Department, Faculty of Engineering, Bina Nusantara University, Jakarta 11480, Indonesia
[2]Chemistry Department, Faculty of Mathematics and Natural Science, Brawijaya University, Malang 65145, Indonesia

[*]Email: reggie.surya@binus.edu

Abstract. The need for high-protein foods strives for the food sector to find an efficiently produced protein sources. One of the most efficient and fast reproducing protein sources are insects containing high levels of protein, nutrients and vitamins. Crickets are one of the most consumed insects, especially in Southeast Asia. Despite the potential of crickets as sources of protein, cricket flour is very little used and studied in Indonesia. This study aimed to develop the best formulation for cricket flour-enriched cookies with a high nutrient and consumer acceptability. As the main ingredient of cookies, wheat was substituted by 5%, 10% and 20% of cricket flour. Among the tested formula, cookies with the 5% and 10% cricket flour showed the highest overall organoleptic acceptability. These cookies are significantly higher in protein and lower in fiber and carbohydrate compared to cookies made with wheat flour.

1. Introduction

Insects are considered as the future foods due to their high nutritional values and rapid reproduction rate. A wide range of edible insect species, with their high contents in proteins, fat, fiber, vitamins and minerals, is believed to play a significant role in addressing food insecurity and environmental destruction linked to classic animal husbandry. Compared to cattle, insects exhibit a significantly higher feed conversion ratio and emit less greenhouse gases. Therefore, the act of eating insects (entomophagy) would not only be beneficial for human health, but also for the environment and economics [1].

House cricket (*Acheta domiscus*) is a type of popular edible insect, especially in the South-East Asian regions. The efficiency of house cricket production is far more efficient than the production of pork and beef. For comparison, the production process to yield 1kg of edible cricket would need 2.1 kg of dried feed while the production of pork and beef would require 9.1 and 25 kg of dried feed respectively. House cricket has a high protein content (20-25 g/100g fresh weight) that is comparable to common sources of protein like beef, egg, or chicken. Besides, house cricket is also rich in B2, B3 and B12 vitamins, as well as other minerals like sodium, phosphorous, calcium, potassium, zinc and iron [2]. One way to implement crickets into food is by processing it into cricket flour. Cricket flour can be used as an alternative for conventional flour due its high protein and nutrient contents [3].

In comparison to other animal proteins (poultry, pig, or beef), insect protein production has lower environmental impacts and a higher economic values. Much less water is required to manufacture protein from insects [4]. Furthermore, substantially smaller quantities of land are required for its production. Another advantage is the ability to use waste products from the agri-food business as feed. It enables the reduction of production-related expenses and the process is independent of the feedstock utilized as human food [5]. Furthermore, such manufacturing produces lesser amounts of hazardous gases to the environment, such as CO_2 or methane. All of these advantages make cricket-based food not only affordable, healthful and useful, but also ecologically benign [6].

Cookies are small, sweet cake typically round and flat made from the mixture of flour, eggs, sugar, fat and other addition that baked until crispy or chewy. Cookies are generally accepted and consumed by a wide range of consumers, particularly those aged 18-24 years old [7]. This study, focusing on the application of cricket flour in the production of cookies for protein fortification, aimed to develop the best formulation for cricket flour-enriched cookies with a higher protein content and consumer acceptability.

2. Methods

2.1. Production of cricket flour

The production of cricket flour was done as previously described [8] with some modifications. The crickets used in this study were wingless juvenile crickets aged 28-30 d old. The crickets were let fast for 9 h prior to freezing at -20°C for 12 h. The frozen crickets were then boiled in demineralized water at 100°C for 5 min prior to oven drying at 90°C for 9 h. The dried crickets were powdered using a blender, sifted to remove unwanted particles (particularly leg fragments) and kept in a sealed airtight glass jar at 4°C prior to utilization in cookie dough.

2.2. Production of cricket flour-enriched cookies

The formulation of cricket flour-enriched cookies was based on a standard recipe book that has been validated by the authors through a series of trial-and-error experiments [9]. Briefly, cricket flour was used to substitute wheat flour (brand "Jawara", Bungasari Flour Mills, Cilegon, Indonesia) to enrich the protein content in cookies. Table 1 represents the formulation of cricket flour-enriched cookies used in this study. All the ingredients were mixed to form cookie dough and tablespoons of the dough were rolled into balls prior to chilling in the fridge (4°C) for 15 min. The balls were then flattened using fingers and placed in the oven at 170°C for 10 min. The cookies were left to cool completely and kept in a sealed airtight glass jar at room temperature prior to the proximate and organoleptic analyses.

Table 1. Formulation of cricket flour (CF)-enriched cookies

Ingredient	Control (0%)	Cookies CF5%	Cookies CF10%	Cookies CF20%
Wheat flour (g)	100	95	90	80
Cricket flour (g)	0	5	10	20
Caster sugar (g)	65	65	65	65
Margarine (g)	60	60	60	60
Chocolate powder (g)	10	10	10	10
Vanilla extract (g)	6	6	6	6
Salt (g)	2	2	2	2
Baking powder (g)	1	1	1	1
Egg	3	3	3	3

2.3. Proximate analysis

The proximate analysis was performed on cricket flour-enriched cookies using standardized methods established by the Association of Official Analytical Chemists (AOAC) and American Oil Chemists Society (AOCS) as previously described [10]. The moisture content was measured by gravimetry following oven-drying at 135°C for 2 h. The ash/mineral content was determined by gravimetry after calcination at 550°C. The fat content was analyzed by gravimetry following sample extraction in Soxhlet apparatus with petroleum ether at 40-60°C. The protein content was analyzed using Kjehldahl method. Total dietary fiber was determined following a series of *in vitro* digestion with human digestive enzymes and ethanol precipitation [11]. The carbohydrate content was obtained by difference.

2.4. Organoleptic analysis

The organoleptic analysis performed in this study consisted of hedonic rating analysis on 5 different organoleptic parameters of cricket flour-enriched cookies: aroma, color, taste, texture and overall, as previously described [12]. The analysis was done by asking 128 untrained panelists aged 18-45 years old to rate the degree at which they like each parameter using a Likert scale of 1 (do not like it at all) to 9 (like it very much). Commercial chocolate cookies without cricket flour of a well-known brand were used as a comparison.

2.5. Statistical analysis

The data obtained from the chemical and organoleptic analyses were analyzed using software SPSS version 26.0. One-way ANOVA was used to compare the means in all groups of cricket flour-enriched cookies. The difference between two groups was analyzed using post hoc Tukey HSD test at the significance level of 0.05.

3. Results and Discussions

The interest of using insect in food fortification is articulated around its high levels of protein [13]. Indeed, the nutritional profile of cricket flour is significantly different from wheat flour. Compared to wheat flour, cricket flour is characterized by a significantly higher content of protein, fat and ash/mineral (Table 2). Protein and fat in cricket flour were three and nine times as high as the ones in wheat flour. In contrast, carbohydrate and fiber were found to be lower in cricket flour than in wheat flour. In general, cricket flour contains higher levels of essential amino acids compared to wheat flour with an approximate ratio of 5:4 (Table 3). Histidine is found to be at similar levels between cricket flour and wheat flour. Interestingly, cricket flour is lower in phenylalanine compared to wheat flour. In both flour, methionine is the limiting amino acid found in the least amount.

Table 2. The nutritional profile of wheat flour and cricket flour (dry basis)

Parameter	Wheat flour [14]	Cricket flour [15]
Protein (%)	15.1	45.8
Fat (%)	2.73	23.7
Ash/mineral (%)	1.56	4.3
Carbohydrate (%)	71.2	19.6
Fiber (%)	10.6	5.7

Table 3. Essential amino acid profile of wheat flour and cricket flour (dry basis) (n=3)

Essential Amino Acid (mg/g protein)	Wheat flour [16]	Cricket flour [17]
Histidine	23.46±0.36	23.37±0.25
Isoleucine	33.99±0.4	40.91±0.33
Leucine	65.76±0.59	72.09±0.73
Lysine	21.35±0.51	55.84±0.55
Methionine	16.75±1.09	19.67±0.29
Phenylalanine	46.84±0.57	34.68±0.35
Threonine	26.13±0.34	36.84±0.11
Valine	37.97±0.34	59.48±0.46
Total	272.25	342.88

Table 4 shows that the addition of cricket flour in cookies formulation increased significantly the protein and mineral content in the cookies, along with the energy content. In contrast, the fiber and carbohydrate content were lower compared to cookies made from wheat flour. The protein content increased by 1.43%, 3.19% and 5.65% for cookies supplemented with 5%, 10% and 20% of cricket flour respectively. The obtained results were consistent with the findings in previous studies demonstrating improvement in protein content in muffins, buns and sourdough bread made with cricket flour [18-24].

Table 4. Nutritional profile of cricket flour (CF)-enriched cookies (dry basis) (n=3)

Parameter	Control CF0%	Cookies CF5%	Cookies CF10%	Cookies CF20%
Energy per 100 g (kcal)	484.04^{a}±4.32	486.18^{b}±3.18	487.41^{c}±3.26	493.64^{d}±4.21
Protein (%)	8.13^{a}±0.62	9.56^{b}±0.45	11.32^{c}±0.68	13.78^{d}±0.94
Fat (%)	20.08^{a}±1.01	20.22^{a}±1.08	20.53^{ab}±1.15	21.44^{b}±1.13
Ash/mineral (%)	0.28^{a}±0.01	0.28^{a}±0.01	0.32^{b}±0.01	0.34^{c}±0.01
Fiber (%)	3.81^{a}±0.14	3.45^{b}±0.18	3.49^{b}±0.16	3.05^{c}±0.11
Carbohydrate (%)	67.70^{a}±0.89	66.49^{b}±0.61	64.34^{c}±0.94	61.39^{d}±0.86

*) Different letters in the same row indicate significant difference among samples ($p<0.05$)

According to our findings, cricket flour was dark brown in color and had an odor resembling fermented shrimp that could potentially lead to negative perception of the consumers towards cookies. Therefore, in our formulation, chocolate powder and vanilla extract were used to mask the flavor of cricket flour. Table 5 shows the organoleptic acceptability of cricket flour-enriched cookies. In general, the supplementation of cricket flour tended to reduce all the parameters of organoleptic acceptability, except for texture. Compared to control (cookies without cricket flour), the cookies supplemented with 5% and 10% of cricket flour were generally accepted by the panel. In contrast, the cookies supplemented with 20% of cricket flour exhibited lower organoleptic acceptability compared to control in all parameters except for texture. Such a phenomenon was due to the strong shrimp-like odor of cricket flour that did not suit the perception of the panel on cookies as a sweet food.

Table 5. Organoleptic acceptability of cricket flour (CF)-enriched cookies

Parameter	Control CF0%	Cookies CF5%	Cookies CF10%	Cookies CF20%	Commercial cookies
Aroma	6.36^{b}±0.74	6.77^{b}±0.82	6.50^{b}±0.68	4.58^{a}±0.47	6.69^{b}±0.72
Color	7.02^{c}±0.81	6.94^{bc}±0.62	6.33^{ab}±0.58	6.07^{a}±0.45	7.38^{c}±0.60
Taste	6.58^{b}±0.52	6.47^{b}±0.49	6.38^{b}±0.47	5.36^{a}±0.55	8.02^{c}±0.65
Texture	5.98^{a}±0.59	6.13^{a}±0.63	6.44^{a}±0.67	6.24^{a}±0.76	7.80^{b}±0.59
Overall	6.69^{b}±0.50	6.61^{b}±0.71	6.55^{b}±0.68	5.52^{a}±0.57	7.77^{c}±0.63

*) Different letters in the same row indicate significant difference among samples ($p<0.05$)

Enriching cookie dough with cricket flour was also shown to result in darker colored cookies. Another studies also demonstrated that the addition of cricket flour into biscuits brought specific characteristics including dark color and fishy and earthy flavor [25,26]. Apart from the natural dark color of cricket flour, it was suggested that the increase of free amino acids from cricket flour in the dough would contribute to the higher intensity of the Maillard non-enzymatic browning reaction during the cooking process [27]. Cookies of darker color also tended to obtain lower organoleptic acceptance due to the impression of overcooking. Compared to commercial cookies, cricket-flour enriched cookies exhibited a lower level of consumer acceptability. To overcome this, it would be necessary to perform further research on cookie formulation in order to improve the organoleptic acceptability of cricket flour-enriched cookies. In addition, adding spices with strong aroma, such as cinnamon or cloves, could also be considered to mask the shrimp-like odor of cricket flour. Furthermore, it would also be interesting to apply cricket flour in the production of salty food products, such as salty crackers or meatballs, since its odor would support the organoleptic profile of such food products.

4. Conclusions

The findings obtained in this study support the possibility of using cricket flour in the production of cookies for protein fortification. Substituting 5% and 10% of wheat flour with cricket flour resulted in cookies that could be accepted by consumers. The organoleptic acceptability of cricket flour-enriched cookies decreased as the amount of cricket flour used in the formulation increased. Further study is needed to improve the cookie formulation for a better consumer acceptability.

References

[1] Lange K and Nakamura Y 2021 *J. Food Bioact.* **14** 4
[2] Bawa M, Songsermpong S, Kaewtapee C and Chanput W 2020 *J. Insect Sci.* **20** 10
[3] Igual M, García-Segovia P and Martínez-Monzó J 2020 *J. Food Eng.* **282** 110032
[4] Van Huis A 2013 *Annu. Rev. Entomol.* **58** 563
[5] Dobermann D, Swift J A and Field L M 2017 *Nutr. Bull.* **42** 293
[6] Duda A, Adamczak J, Chełmińska P, Juszkiewicz J and Kowalczewski P 2019 *Foods* **8** 46
[7] Mudgil D, Barak S and Khatkar B S 2017 *LWT* **80** 537
[8] Fernandez-Cassi X, Supeanu A, Vaga M, Jansson A, Boqvist S and Vagsholm I 2019 *J. Insects Food Feed* **5** 137
[9] McKenney S 2017 *Sally's Baking Addiction: Irresistible Cookies, Cupcakes and Desserts for Your Sweet-Tooth Fix* (New York: WellFleet Press)
[10] Eden W T and Rumambarsari C O 2020 *J. Phys. Conf. Ser.* **1567** 022033
[11] McCleary B V 2003 *Proc. Nutr. Soc.* **62** 3
[12] Villanueva N D, Petenate A J and Da Silva M A 2000 *Food Qual. Pref.* **11** 363
[13] Churchward-Venne T A, Pinckaers P J, van Loon J J and van Loon L J 2017 *Nutr. Rev.* **75** 1035
[14] Godswill A C 2019 *J. Food Sci.* **2** 43
[15] Montowska M, Kowalczewski P Ł, Rybicka I and Fornal E 2019 *Food Chem.* **289** 130
[16] Tkachuk R 1966 *Cereal Chem.* **43** 2017
[17] Araujo R R S, Fagundes M M A, Viana A M F, Paulino A H S, Silva M E and Santos E M 2022 *J. Insects Food Feed* **8** 409
[18] González C M, Garzón R and Rosell C M 2019 *Innov. Food Sci. Emerg. Technol.* **51** 205
[19] Kowalski S, Mikulec A, Mickowska B, Skotnicka M and Mazurek A 2022 *LWT* **159** 113220
[20] Zielińska E, Pankiewicz U and Sujka M 2021 *Antioxidants* **10** 1122
[21] Pambo K O, Okello J J, Mbeche R M, Kinyuru J N and Alemu M H 2018 *Food Res. Int.* **106** 532
[22] Nissen L, Samaei S P, Babini E and Gianotti A 2020 *Food Chem.* **333** 127410
[23] Burt K G, Kotao T, Lopez I, Koeppel J, Goldstein A, Samuel L and Stopler M 2020 *J. Culin. Sci. Technol.* **18** 201
[24] Khatun H, Van der Borght M, Akhtaruzzaman M and Claes J 2021 *Foods* **10** 2750
[25] Biró B, Sipos MA, Kovács A, Badak-Kerti K, Pásztor-Huszár K and Gere A 2020 *Foods* **9** 1561
[26] Tan H S, Verbaan Y T and Stieger M 2017 *Food Res. Int.* **92** 95-105
[27] Grossmann K K, Merz M, Appel D, De Araujo M M and Fischer L 2021 *Food Chem.* **364** 130336

IOP Conf. Series: Earth and Environmental Science 1115 (2022) 012093 doi:10.1088/1755-1315/1115/1/012093

The effect of porang proportion on the characteristics of vegetarian tempeh patties burger

H Herawati*, **E Kamsiati and S Widowati**

Research Center for Agroindustry, National Research and Innovation Agency, Laboratorium Pengembangan Teknologi Industri Agro dan Biomedika (LAPTIAB), Gedung 614, KST BJ. Habibie Serpong, Tangerang Selatan 15314, Indonesia

*Email: herawati_heny@yahoo.com

Abstract. The development of veggie burger patties products is increasing along with changes in people's lifestyles that are more concerned with a healthy product. Tempeh is a source of vegetable protein that has various health benefits. Glucomannan from porang tubers acts as a texture-forming material in food products. This study aimed to determine the effect of the proportion of porang on the characteristics of the veggie burger made from tempeh. This study was arranged in a completely randomized design with treatment proportions of porang:tempeh (75:25); (50:50); (25:75) compared to control (100% tempeh). Parameters observed included colour, cooking loss, cooking yield, hardness and proximate content. The results showed that the treatment had a significant effect on the colour of the burger before and after frying, cooking loss, cooking yield, and moisture content. The use of porang had increased the water content of patties burger. The proportion of porang:tempeh (25:75) produced a patties burger with the lowest cooking loss and the highest cooking yield. The proximate content of burger patties in the proportion of porang:tempeh (25:75) had a value that was not significantly different from the control (100% tempeh).

1. Introduction

Burger patties are burger fillings made from ground beef that is restructured to be round, flat and solid. Currently, the burger patties variants consist of beef, chicken, fish, and vegetable protein sources. Vegetable protein sources are an option in developing burger patties for vegetarian consumers. The trend of developing vegetarian burger patties products is increasing along with changes in people's lifestyles that are more concerned with health. The development of burger patties products from vegetable protein ingredients provides an opportunity to provide burger patties for vegetarian consumers or those who want to reduce the consumption of animal products.

Sources of vegetable protein that are widely used in making burger patties are various beans such as soybeans, kidney beans, chick peas, lentils, and other types of beans. Other materials are also used as texturizers and stabilizers as well as natural coloring agents to produce nutritional and physical characteristics and color appearance as well as burger patties from meat. The formulation of ingredients by mixing several types of raw materials is also carried out to produce the desired burger patties product [1,2].

Porang is a tuber that is rich in carbohydrates, especially soluble fiber in the form of glucomannan. Soluble fiber, glucomannan in porang tubers has a function as fiber for the body. In the food industry, glucomannan is used as a stabilizer, texture-forming material. Porang flour is also used as an additive in making sausages [3]. Tempeh is a superfood. Tempeh is a source of protein that is cheap, easy to obtain and has a complete nutritional content. Tempeh is a product of soybean fermentation by the fungus *Rhizopus* sp. Tempeh protein is easier to digest, compared to other soy products. Tempeh also contains phytochemical compounds that are beneficial for health such as isoflavones and bioactive peptides that are beneficial for health [4]. Previous studies on burger patties from plant-based products have been developed, including the development of non-meat burgers from a mixture of red beans, corn and oyster mushrooms [1]; soybean tempeh, red beans and green beans [5]; and tempeh and tofu [6]. While the use of porang in the manufacture of food products has previously been reported, including the use of gel from porang flour in making sausages [3], the use of porang flour in the manufacture of artificial meat [7]. The formulation will affect the nutritional content of the resulting product [8]. This study aimed determined the effect of proportion of porang with tempeh on the characteristics of burger patties.

2. Materials and Methods

This research was conducted at the Agricultural Postharvest Research Center Laboratory in July-September 2021. The materials used were porang tubers obtained from Klangon Madiun Village, salt, spices, Supertempeh brand tempeh and cooking oil purchased at Superindo Bogor. The tools used were scales, blender, basin, mould, steamer pan, Teflon fryer. The tools used for analysis included oven, furnace, penetrometer, Soxhlet flask, Khjeldahl tube. This study was arranged in a completely randomized design with three replications. The treatments tested were the proportion of porang: tempeh in a ratio (75:25); (50:50); (25:75) compared to control (100% tempeh). The process of making porang-tempeh burgers was as follows: porang was soaked in 8% NaCl solution for 6 hours, then drained, then washed, drained and then mashed with a blender.

Tempeh was steamed for 15 minutes, then mashed, mixed with porang according to the treatment, added with spices and tapioca, then mixed until evenly distributed. The next, dough was molded like burger meat with a diameter of 12.5 cm. It was placed on a baking sheet and steam for 30 minutes. Samples were stored in the freezer for 24 hours and then thawed and analyzed. Cooking burger patties was fried pan for 5 minutes. Parameters observed included the color of burger patties before and after frying, cooking loss, cooking yield, hardness, and proximate content (water, ash, fat, protein, and carbohydrates). Color analysis used a Minolta CR 300 chromameter, with parameters L (Lightness), a (redness), b (yellowness), C (chromaticity index) and H (Hue). Analysis of hardness was run by penetrometer, water content by gravimetric method, ash content by gravimetric method, fat content by was run Soxhlet method, and protein content by Kjeldahl method. The data obtained were analyzed by Analysis of Variety at a 95% confidence interval, if the results were significantly different, it was continued with Duncan's test. Color difference was tested before and after cooking and compared with pairwise comparison test. Data processing was sing SPSS software.

3. Results and Discussion

3.1. Colour

The brightness (L) of the color of the veggie burger before frying ranged from 48.71 to 64.28. The use of porang tended to lower the brightness of the resulting burger. Tempeh is white, so it gives a bright appearance to raw burgers, so the addition of porang tends to lower the brightness of the resulting burger. The level of redness (a) of the raw burger ranged from 17.43-28.02. The use of porang tended to reduce the level of redness of the resulting burger. Porang has a yellow color, while tempeh that has been steamed will produce a reddish-yellow white color so that the use of porang tended to reduce the level of redness of the resulting burger.

Table 1. Colour of the burger patties before and after frying

Treatments	L	a	b	C	H
Before Frying					
Porang:tempeh (75:25)	48.7067a	17.4267a	90.7000d	-.2100a	17.4267a
Porang:tempeh (50:50)	49.6933b	17.7100a	87.5100c	.7733b	17.6933a
Porang:tempeh (25:75)	54.5533c	28.0200b	81.7100a	4.0367d	27.2067b
Tempeh (100)	64.2867d	27.3233b	84.5467b	2.5133c	27.2067b
After Frying					
Porang:tempeh (75:25)	46.0967b	15.8667a	88.6633d	0.3733a	15.8633a
Porang:tempeh (50:50)	45.3433a	17.3633a	85.3867c	1.4033b	17.3067a
Porang:tempeh (25:75)	49.3967c	25.7133b	83.1167b	3.2267c	25.5067b
Tempeh (100)	51.3900d	32.6033c	78.3033a	6.5933d	31.9300c
Pairwise comparison	Bn	Tn	Bn	Tn	Tn
Correlation value	0.939	0.893	0.711	0.023	0.900

The same letter in the same column behind the mean value does not show a significant difference ($P > 0.05$).

In the yellow color index (b), the use of porang tended to increase the yellowness of the resulting burger. As previously explained, the porang tubers are yellow, therefore producing a higher yellow index in the burger. The color of the burger after frying can be seen in Table 1. From Table 1, it can be seen that the brightness value of the veggie burger after frying ranged from 45.34 to 51.39. The brightness of the burgers has decreased compared to its raw form. Likewise, the level of redness and yellowness decreased after frying except for the control burger (100% tempeh).

From the results of the comparison of colors before and after frying with the t test, it can be seen in Table 1. From Table 1, the brightness and yellowness values of the burgers were significantly different between before and after frying. In Table 1 it can also be seen that the use of porang tended to decrease the Chromaticity and Hue indices, the same as the brightness and redness values which decrease with the use of porang. The value of C is the value obtained from the coordinates of the *a* and *b* values. The higher the chroma value caused the stronger intensity of the resulting color.

3.2. *Cooking loss and yield*

The value of cooking loss and cooking yield of vegetarian burger patties can be seen in Table 2, the value of cooking loss ranges from 10.48-23.27. The lowest cooking loss was in the proportion of porang:tempeh (25:75) and the highest was in porang:tempeh (75:25). A high proportion of porang (75:25) resulted in a higher cooking loss because porang contains a lot of water so that during cooking there is a higher water loss compared to the lower proportion of porang.

Table 2. Cooking loss and yield of patty burger analysis

Treatments	Cooking loss	Cooking yield
Porang:tempeh (75:25)	23.2700c	76.7300a
Porang:tempeh (50:50)	16.5300b	83.4700b
Porang:tempeh (25:75)	10.4800a	89.5200c
Tempeh (100)	12.5300ab	87.3450bc

The same letter in the same column behind the mean value does not show a significant difference ($P > 0.05$).

The use of porang 50% still resulted in a higher cooking loss value than the control, while at 25% the cooking loss value was lower than the control. In the manufacture of analog meat analogue of the combination of porang flour with soy protein isolate, a decrease in the proportion of porang flour tended to reduce the water holding capacity of the meat analog produced [9]. The results of previous studies on the interaction of konjac glucomannan with soy protein isolate showed that the interaction between glucomannan molecules and protein isolates through hydrogen bonds increased the gel strength of the matrix formed [10]. Cooking yield of veggie burgers ranged from 76.73-89.52. The lowest value was in the porang:tempeh treatment (75:25). While the highest ratio was porang: tempeh (25:75). This is in accordance with the value of cooking loss Burgers that had high cooking loss have a low cooking yield value.

3.3. Moisture content
The moisture content of veggie burgers ranges from 48.52-63.64%. The results of the analysis of variance showed that the treatment had an effect on the moisture content of the resulting burger. A higher proportion of porang resulted in a higher water content. Porang has a higher water content than tempeh, so the use of porang tends to produce a higher water content. The content of glucomannan contained in porang is able to absorb water up to 200 times its weight and has great swelling properties in water, its swelling power reaches 138-200%.

Table 3. Moisture content and hardness of patties burger

Treatments	Moisture content	Hardness
Porang:tempeh (75:25)	63.6367b	0.3670a
Porang:tempeh (50:50)	61.7533b	0.3830a
Porang:tempeh (25:75)	52.0000a	0.2930a
Tempeh (100)	48.5167a	0.3730a

The same letter in the same column behind the mean value does not show a significant difference (P>0.05).

In the hardness parameter, the proportion of porang:tempeh did not produce a significantly different hardness on the hardness of the resulting veggie burger. The larger the number shown on the penetrometer scale, the softer the fruit. The larger the number shown on the penetrometer scale, the softer the sample.

3.4. Proximate analysis
From the proximate analysis on the best treatment compared to the control, it was found that the use of porang resulted in water, ash, protein, fat and carbohydrate content which were not significantly different from the control.

Table 4. Moisture content and hardness of patties burger

Proximate analysis	Patties burger	Control	Burger non-meat (Red bean, Corn, Mold)
Moisture	52.0000	48.5167	63.18 ± 0.04%;
Ash	0.4933	0.9367	5.83 ± 0.15%;
Protein	17.7530	18.0900	4.50 ± 0.15%;
Fat	4.1750	6.1150	11.01 ± 0,23%
Carbohydrate	25.5787	26.3416	15.48 ± 0.17%.

The same letter in the same column behind the mean value does not show a significant difference (P>0.05).

When compared with previous studies on non-meat burgers from a mixture of red beans, corn and oyster mushrooms, the water content of the porang:tempeh veggie burger (25:75) was lower. The composition of raw materials affects the nutritional value of the resulting burger. Different formulations of raw materials produce burgers with different nutritional values [8]. Furthermore, on the value of protein content, the porang tempeh burger (25:75) resulted in a higher protein content value, which was 17.75% compared to the non-meat burger, which was 11.01 ±0.23 (Table 4). Protein is one of the important nutrients for the body. Burger patties are generally made from animal or vegetable products that are rich in protein, such as meat, fish, chicken, or various nuts. In the manufacture of non-meat burgers, the proportion of red beans is higher than corn and mushroom, resulting in a higher protein content value in the resulting burger patties [1].

4. Conclusion

The results showed that the proportion of porang with tempeh gave a significantly different effect on the color value of the burger before and after cooking. The level of brightness and redness tended to decrease with the increase in the proportion of porang, while the level of yellowness tended to increase. The treatment also had a significantly different effect on the cooking loss and cooking yield of the resulting burger patties. The higher proportion of porang resulted in higher cooking loss and lower cooking yield on the resulting burger patties. However, the proportion of porang tempeh (25:75) resulted in a lower cooking loss value and a higher cooking yield than the control and the values were not significantly different. Burger patties that use the higher percentage of porang had a higher moisture content and the results of the analysis of variance showed significantly different results.

Acknowledgment

We acknowledge to the DIPA of the Agricultural Research and Development Agency for carrying out this research activity.

References

[1] Fadly D and Purwayantie S 2019 *Jurnal Gizi Dan Kesehatan* **3** 19
[2] Kyriakopoulou K, Keppler J K and van der Goot A J 2021 *Foods* **10** 600
[3] Anggraeni D A, Widjanarko S B and Ningtyas D W 2014 *Jurnal Pangan dan Agroindustri* **2** 214
[4] Astawan M, Wresdiyati T, Widowati S, Bintari S H and Ichsani N 2013 *Pangan* **22** 241
[5] Utami N, Jumirah and Siagian A 2014 *Skripsi USU* **4** 48
[6] Herawati H and Kamsiati E 2021 *IOP Conf. Ser.: Earth Environ. Sci.* **653** 012103
[7] Lindriati T, Herlina H and Emania J N 2018 *Jurnal Teknologi Pertanian Andalas* **22** 175
[8] Menezes B, Sawaly M and Al-ansari T 2021 12th International Exergy, Energy and Environmental Symposium (IEEES-12) 20-24 December 2020 Doha Qatar 116
[9] Lindriati T, Nafi A and Sari Z G 2020 *Jurnal Teknologi dan Industri Pertanian Indonesia* **11** 75
[10] Wang M, Yao M N, Jian W J, Sun Y J and Pang J 2010 *Agricultural Sciences in China* **9** 1538

IOP Conf. Series: Earth and Environmental Science 1115 (2022) 012094 doi:10.1088/1755-1315/1115/1/012094

The physicochemical properties of soy sauce made from tempeh

H L Zahidah and D Lo*

Food Technology Department, Faculty of Engineering, Bina Nusantara University, Jakarta, Indonesia 11480

*Email: diana.lo@binus.edu

Abstract. Tempeh is a fermented product that uses Rhizopus sp. to bind raw material into a sliceable cake. Soy sauce is a fermented food product with umami and salty flavor made by *Aspergillus* sp. for koji fermentation and then soaked in salted water. Tempeh has a shelf life of three days, but soy sauce has a shelf life of six months at room temperature. Therefore, we can increase the shelf life of unconsumed tempeh by turning it into soy sauce. This study compared the physicochemical properties of soy sauce made from tempeh. This study has two variables: length of tempeh fermentation times (three days and four days) and salt solution concentration (15%, 20%, 25%). We also use soy sauce made from soybean with *Aspergillus* sp. as a positive control. After the soy sauce was fermented for one month, pH, total dissolved solids, °Brix value, yield, color, acceptance test and just-about-right test were analyzed. Results from the acceptance analysis showed that soy sauce made from 3-days tempeh with 20% of salt has significantly higher acceptance compared to other sample with value of 5.37±1.73 (range 1-9) with represent neutral. This soy sauce has pH of 6.73±0.06, total dissolved solids of 9.73 ± 0.02% with 11.7 ± 0.02°Brix. For color analysis, soy sauce made from 3-days tempeh with 20% of salt showed higher lightness than soy sauce made from koji with 20% of salt. Meanwhile for just-about-right test, very salty taste attribute needs to be reformulated.

1. Introduction

Soy sauce is widely produced in China and Japan and often used as condiment for food dishes in Asian countries and is also gaining popularity in Western countries. Soy sauce as a condiment can serve to get taste desired by consumers and increase their appetite. Soy sauce products are thin textured liquid with a dark brown color and have a strong umami, salty flavor and caramel-like taste that enhances the overall aroma and savory flavor to various foods. Foods that can be added with soy sauce vary widely from cooked to raw foods, including sushi, sashimi, dish, meat and fried noodles [1].

Indonesia is an agricultural country with high soybean production and is becoming the biggest soybean market in Asia. Apart from that, Indonesia is known as the highest tempeh producer in the world. Tempeh serves as a staple food protein source, which is made from fermented soybeans with the help of *Rhizopus* spp. and this product is affordable and nutritious, besides it has attractive flavor, texture and superior digestibility [2]. In Indonesia, 50% of soybean consumption is used for tempeh production, 40% for tofu and 10% for others, including tauco and soy sauce [3]. Although tempeh has a great production potential, soybean tempeh tends to get their qualities reduced before reaching consumers' hands due to several factors such as its limited shelf life, uneven distribution of population, as well as the remote shipping distance to consumers. After 48 hours, the *Rhizopus* fungus found in tempeh perishes and other fungi grows and unpleasant odor may develop due bacterial growth that breaks down the protein in tempeh. These issues make it difficult to ship tempeh to distant areas, hence, other suitable method is needed to extend the shelf life of tempeh. Therefore, soy sauce made from tempeh can be a good alternative product as it can last for about 3 years in glass bottles. Tempeh can be uses as an ingredient for soy sauce production due to its high protein content with a range of 14.77-22.73%. Tempeh was selected as a raw material for making soy sauce in the present study because of its high protein content. This study aims to determine the effect of differences in fermentation time and salt solution concentration on the physicochemical and organoleptic properties of the resulting soy sauce.

2. Materials and Method

2.1. Materials

The materials used for this research including non-GMO soybeans (USA), koji starter, tempeh mold, water, salt, baking soda and 1% lactic acid. The equipment used for making soy sauce were pans, stirrer, jars, digital scales, blenders, cheesecloth, trays, plastic, pH meters, ATAGO Digital Pocket Refractometer PAL-COFFEE (BX/TDS) and measuring cups.

2.2. Soy Sauce Making

2.2.1. Tempe mold and koji fermentation. For making soy sauce two variables were measured: length of fermentation time (three days using tempeh mold, four days using tempeh mold and four days using koji starter) and concentration of salt solution (15%, 20% and 25%). Initially, 500 grams of soybeans were soaked for 24 hours for each treatment. Next, the soybean hulls were peeled and washed with clean water, boiled for 30 minutes in 1L of water. The soybeans treated with tempeh mold was boiled in 1L of water with the addition of 1% lactic acid. Next, soybeans were strained and cooled. For making soy sauce with tempeh mold, soybeans were added with 1 gram of tempeh mold and wrapped with plastic. Meanwhile, for the treatment of soy sauce with koji starter, soybeans were spread on the tray and added with 2 grams of koji starter and stirred well until the koji starter was evenly distributed. Soybeans that were added with tempeh mold were fermented for three and four days, whereas soybeans added with koji starter were fermented for four days. Soybeans that are added with koji starter were fermented in a room with a cold temperature and stirred every 6-8 hours during fermentation process to reduce the temperature of fermented soybeans.

2.2.2. Moromi fermentation and preparation of soy sauce. 500 grams of fermented soybeans were placed into each jar according to their treatment. Then, 1L of brine was prepared following the concentration that has been set for each treatment, i.e. 15% (15 g/ 100 mL), 20% (20 g/ 100 mL) and 25% (25 g/ 100 mL), respectively. Next, brine solutions were poured into jars. Jars that were filled with fermented soybeans and brine solution were closed half-tightly to allow the air to enter the jar. Then, all soy sauce treatment went through the second stage of fermentation process, also known as moromi fermentation, for about a month. After the second stage of fermentation process, all the soy sauce treatments were filtered, cooked at pasteurization temperature for few minutes and filtered using cheesecloth.

2.3. pH and Color Analyses

2.3.1. pH analysis. Two pH analyses were conducted for this study using a pH meter. The pH meter was calibrated using a buffer solution with pH 6.86 [4]. The first pH analysis was performed after the first stage of fermentation process. Three samples were measured (three days tempeh mold, four days tempeh mold, four days koji starter). This process used 100 grams of soybeans for each treatment. Initially, soybeans was blended with 100 mL water to make 1:1 ratio. Next, the pH value of each sample was measured. The pH value for fermented soybeans with tempeh mold treatment was increased by adding baking soda and equalized towards pH value of fermented soybeans with koji treatment. Then, the amount of baking soda, required for increasing and equalizing pH in 100 grams of fermented soybeans with tempeh mold treatment was determined.

The second analysis was performed after the soy sauce was cooked and strained with cheesecloth, the pH of each treatment was measured three times by using pH meter. Then, data was analyzed by using one-way ANOVA and Duncan test.

2.3.2. Color analysis. Color analysis was performed by using Color Meter instrument by the hunter method system also known as the CIELab. Nine different samples were evaluated in this analysis and those samples were designed with two different variables, where the first variable was length of fermentation time (three days tempeh mold, four days tempeh mold, four days koji starter) and second variable was the concentration of salt solution (15%, 20%, 25%). Data were obtained thrice for each sample. The measurement was taken by bringing the sample closer to the camera of the color meter and the results were displayed as L*, a* and b*. L* value indicates lightness, a* value red and green, while b* value represents blue to yellow [5]. Subsequently, data was analysed statistically using one-way ANOVA and Duncan test.

2.4. Total dissolved solids and brix analyses. In these analyses, nine different treatment combinations of soy sauce was measured for total dissolved solids and brix. The treatment combinations were designed by two variables, including length of fermentation time (three days tempeh mold, four days tempeh mold, four days koji starter) and the concentration of salt solution (15%, 20%, 25%). Initially, each sample was diluted with water from 5 mL of soy sauce sample to 10 mL. Each sample was measured three times for both total dissolved solids and brix level by using ATAGO Digital Pocket Refractometer PAL-COFFEE (BX/TDS). Total dissolved solids analysis aimed to determine the amount of solids dissolved in soy sauce samples [6]. Small drops of each sample was dripped on the refractometer prism. After that, the instrument will display the result of total dissolved solids and brix value for each sample. Then, data was collected and analysed by using one-way ANOVA and Duncan test.

2.5. Analysis of yield

Yield analysis is counted from Erviani et al. [7]. The yield of each soy sauce treatment can be calculated by formula (1).

$$\text{Yield (\%)} = \frac{Volume\ filtrate\ (mL)}{Weight\ fermented\ soybean + salt\ (g)} \times 100\% \qquad (1)$$

2.6. Hedonic and just-about-right test

The hedonic and just-about-right test of soy sauce was conducted by using three attributes, including aroma of the sample before tasting, aroma when the sample is tasted and taste. This hedonic test aimed to determine the acceptance level of soy sauce products [8]. Six different combinations with two variables of soy sauce were evaluated, which included the length of fermentation time (three days tempeh mold, four days tempeh mold, 4 days koji starter) and concentration of salt solution (15% and 20%). 30 untrained panellists were used in this study. The panellists were asked to taste the soy sauce sample and give the response by filling the questionnaire. The hedonic response was delivered in numerical scale with the range of 1-9, where 1 represents "dislike extremely" and 9 means "like extremely". Subsequently, data was collected, tabulated, then analyzed statistically with one-way ANOVA and Duncan test. Meanwhile, the response for the just-about-right test is delivered with five assessment criteria. For the flavor attribute, the ratings start from too less salty (score 1) to too much salty (score 5), whereas for aroma attribute the ratings began with too less aroma (score 1) to too much aroma (score 5). Then, it was continued with penalty analysis after the data were obtained.

3. Results and Discussion

In this study, nine different treatment combinations of soy sauce were analyzed with different fermentation times and salt solution concentrations. The different fermentation times consisted of 3 days and 4 days with tempeh mold, as well as 4 days with koji starter. Meanwhile, the salt concentrations used, included 15%, 20% and 25%. The physicochemical analyses that were conducted were pH, color, total dissolved solids, brix and yield, whereas the sensory analyses included hedonic and just-about-right.

Result of pH analysis on Table 1 showed that 3 days fermented tempeh mold with 15% salt concentration showed the highest value, while 3 days fermented tempeh mold with 25% salt concentration showed the lowest. There was a significant decrease in pH value as their fermentation time increased from 3 to 4 days with tempeh mold and 15% salt concentration. Meanwhile, for 3 days and 4 days fermented tempeh mold at the salt concentration of 20% and 25% showed that their pH values were increasing. However, only 3 days and 4 days fermentation with tempeh mold at salt concentration of 20% showed significant increase. For 4-day fermentation koji starter treatment with salt concentration of 15% to 20% showed an increase significantly as their salt concentration increased, while pH value from the salt concentration of 20% to 25% decreased but not significantly.

Table 1. The pH value of soy sauce product from tempeh

Salt Amount	Sample	pH
15%	3 days (Tempeh Mold)	7.63 ± 0.32^c
	4 days (Tempeh Mold)	7.17 ± 0.06^b
	4 days (Koji Starter)	6.53 ± 0.21^a
20%	3 days (Tempeh Mold)	6.73 ± 0.06^a
	4 days (Tempeh Mold)	7.07 ± 0.15^b
	4 days (Koji Starter)	7.27 ± 0.06^b
25%	3 days (Tempeh Mold)	6.50 ± 0.10^a
	4 days (Tempeh Mold)	6.67 ± 0.06^a
	4 days (Koji Starter)	7.20 ± 0.10^b

The result is the mean value ± standard deviation. Different letters show significant difference values ($p < 0.05$).

According to Pangestika et al [9], pH value of soy sauce above 5 can be categorized as the ideal pH after the moromi fermentation process. However, if the pH value turns out to be less than 5, it will negatively affect the sensory characteristics of soy sauce. In addition, this will also cause the activity of lactic acid bacteria to stop and allow mold to grow, for example *Zygosaccharomyces rouxii* that acts to produce alcohol and other various flavor components. Therefore, as the final pH value was above 5 for all soy sauce samples, hence mold growth would not occur, so that the aroma compounds formed were likely to be contributed more by the activity of lactic acid bacteria and maillard reaction during the second stage fermentation process.

For color analysis, the samples were analyzed by applying CIEL*a*b* method and measured by using colorimeter. Among all nine soy sauce samples, 3 days fermented tempeh mold with salt concentration of 25% showed the most brightness, since this treatment had the highest L* value (Table 2). It shows that there were decrease in L* value for 3 days to 4 days fermented tempeh mold with 15% and 25% salt concentration, then this means that the soy sauce samples color gets darker as their fermentation time increases. The decrease was significant for 3 days to 4 days tempeh mold with 25% concentration, whereas the L* value for 3 days to 4 days tempeh mold with 15% concentration did not decrease significantly. However, for 3 days and 4 days fermented tempeh mold at concentration of 20% showed an increase significantly, hence this indicates that the lightness of soy sauce samples for this concentration tends to be brighter. Furthermore, for 4 days fermented koji starter soy sauce treatment showed a significant increase as the salt solution concentration increased. This means that for 4 days koji starter, the soy sauce samples tend to be brighter as the concentration increased.

Table 2. Color analysis on soy sauce products from tempeh

Salt Amount	Sample	L*	a*	b*
15%	3 days (Tempeh Mold)	52.38±0.29[c]	-7.38±1.09[a]	13.86±1.27[b]
	4 days (Tempeh Mold)	52.23±0.29[c]	-7.20±1.45[a]	14.52±0.51[b]
	4 days (Koji Starter)	46.16±0.62[a]	-1.92±1.26[d]	33.50±1.15[d]
20%	3 days (Tempeh Mold)	50.69±1.03[b]	-5.47±0.27[bc]	17.19±1.10[c]
	4 days (Tempeh Mold)	53.06±0.66[c]	-5.44±0.26[bc]	17.72±0.64[c]
	4 days (Koji Starter)	49.90±0.56[b]	-4.35±0.74[bc]	28.21±1.64[e]
25%	3 days (Tempeh Mold)	57.73±1.08[d]	-7.97±0.50[a]	7.67±1.14[a]
	4 days (Tempeh Mold)	52.54±0.47[c]	-5.61±0.03[b]	17.43±1.18[c]
	4 days (Koji Starter)	52.74±1.07[c]	-3.95±0.87[c]	24.15±1.76[d]

The result is the mean value ± standard deviation. Different letters show significant difference values ($p < 0.05$).

Meanwhile, a* and b* value of 3 days to 4 days tempeh mold with concentration of 15%, 20% and 25% showed an increase. An increase in a* value indicates that the samples exhibit more redness, whereas increase in b* value showed that the soy sauce samples tend to be more yellowish. Also, both a* and b* color values for 3 days to 4 days tempeh mold treatment increased significantly at concentration of 25% as their fermentation time increased, whereas for 15% and 20% showed an increase but not significantly. Then, for soy sauce samples of 4 days koji starter treatment showed that they have higher a* and b* color value than 3 days and 4 days tempeh mold treatment, which means the color of soy sauce samples with 4 days koji starter treatment were redder and more yellowish than 3 days and 4 days tempeh mold treatment. The color of soy sauce samples was majorly contributed by high amounts of maillard reaction products, including melanoidins that were formed during thermal treatment and fermentation process [10]. Maillard reaction is a reaction that occurs between an amino group of free amino, peptide chain residue or protein with the carbonyl group of a carbohydrate, this happens if both are heated or stored for a long time, while melanoidins are formed mainly from carbohydrates, phenolic compounds [11].

In total dissolved solids analysis, results showed that 4 days fermented koji starter at 25% salt concentration with value of 26.28±0.02% showed the highest value, whereas 4 days fermented tempeh mold 15% with value of 14.75±0.06% showed the lowest. At the salt concentration of 25%, for samples fermented for 3 days and 4 days tempeh mold, showed an increase in total dissolved solids value from 17.72±0.01% to 20.30±0.05%. An increase in total dissolved solids value might be due to hygroscopic properties from salt solution. This property can attract water and dissolved materials in it, causing the osmotic pressure of the substrate to increase. Hence, this results in osmosis of the material to the salt solution [12]. In addition, an increase in total dissolved solids might be due to the length of fermentation time for 3 days and 4 days tempeh mold 25% treatments and due to dissolution of compounds from the first fermentation stage. According to Iskandar and Darusalam [13], simple compounds are produced during the first fermentation stage and dissolve in the filtrate during the moromi fermentation process. The mold that grows on soybeans will produce enzymes that are able to change protein into amino acids and almost all types of amino acids can be found in soy sauce, including hydrophobic amino acids (valine, leucine and isoleucine) and hydrophilic amino acids (glutamate, glycine and aspartate), as well as aromatic amino acids (phenylalanine). Therefore, total dissolved solids content of soy sauce also comes from the amino acids contained in fermented soybeans [14].

Table 3. Total dissolved solids and °Brix in soy sauce products from tempeh

Salt Amount	Sample	Total Dissolved Solids (%)	°Brix
15%	3 days (Tempeh Mold)	18.29±0.03[d]	22.07±0.03[d]
	4 days (Tempeh Mold)	14.75±0.06[a]	17.95±0.08[a]
	4 days (Koji Starter)	19.91±0.02[f]	23.91±0.03[f]
20%	3 days (Tempeh Mold)	19.46±0.03[e]	23.40±0.04[e]
	4 days (Tempeh Mold)	16.61±0.04[b]	20.11±0.05[b]
	4 days (Koji Starter)	19.87±0.01[f]	23.89±0.01[f]
25%	3 days (Tempeh Mold)	17.73±0.012[c]	21.41±0.01[c]
	4 days (Tempeh Mold)	20.30±0.05[g]	24.37±0.06[g]
	4 days (Koji Starter)	26.28±0.02[h]	31.10±0.02[h]

The result is the mean value ± standard deviation. Different letters show significant difference values ($p < 0.05$).

From data on Table 4 it was shown that there were an increase but not significantly in yield value for 3 days tempeh mold to 4 days tempeh mold treatment. An increase in yield value might be due to the longer fermentation time. The longer the fermentation time, the greater the amount of protein substrate in soybeans that interacts with protease enzymes, where protease enzyme breaks down protein into water-soluble fragments, hence this causes the increase in yield value of soy sauce samples [15]. Therefore, in this study, the best soy sauce yield value was in 4 days tempeh mold with yield value of 118.63±2.97%.

Table 4. Yield value of soy sauce products from tempeh

Sample	Yield (%)
3 days (Tempeh Mold)	114.45±4.52[a]
4 days (Tempeh Mold)	118.63±2.97[a]
4 days (Koji Starter)	116.73±2.30[a]

The result is the mean value ± standard deviation. Different letters show significant difference values ($p < 0.05$).

The result of hedonic test in Table 5 from aroma before tasting sample attribute showed that the soy sauce treatment of 3 days tempeh mold with salt solution of 20% (5.37±1.73) showed the highest panellists' acceptance among other soy sauce treatments and represents neutral, while 4 days koji starter 20% (4.13±1.89) showed the lowest acceptance that indicates slightly dislike according to hedonic scale [8]. The longer the fermentation time showed a significant decrease in aroma in tempeh mold treatment, hence different fermentation time did affect the panellist's acceptance for this attribute. From aroma when tasting sample attribute, the result indicates that 4 days tempeh mold 20% (5.77±1.65) showed higher acceptance than other samples and represents slightly like, yet not significantly different based on statistical analysis by using one-way ANOVA and Duncan test. Meanwhile, for taste attribute, 3 days tempeh mold 15% (5.63±1.63) showed higher liking compared to other treatments, nonetheless the taste attributes for every treatment were not significantly different. Hedonic test for soy sauce treatment with salt solution of 25% was not carried out, since the treatment with salt solution of 20% had better quality.

Table 5. Hedonic test on soy sauce products from tempeh

Salt Amount	Sample	Aroma (Before)	Aroma (When)	Taste
15%	3 days (Tempeh Mold)	5.17 ± 1.90^{ab}	5.67 ± 1.92^{a}	5.63 ± 1.63^{a}
	4 days (Tempeh Mold)	4.60 ± 1.81^{ab}	5.13 ± 1.50^{a}	5.13 ± 1.80^{a}
	4 days (Koji Starter)	4.70 ± 2.09^{ab}	5.27 ± 1.46^{a}	5.30 ± 1.86^{a}
20%	3 days (Tempeh Mold)	5.37 ± 1.73^{b}	5.50 ± 1.64^{a}	5.07 ± 1.98^{a}
	4 days (Tempeh Mold)	4.90 ± 1.49^{ab}	5.77 ± 1.65^{a}	5.53 ± 1.78^{a}
	4 days (Koji Starter)	4.13 ± 1.89^{a}	4.90 ± 1.77^{a}	5.53 ± 2.29^{a}

The result is the mean value ± standard deviation. Different letters show significant difference values ($p < 0.05$).

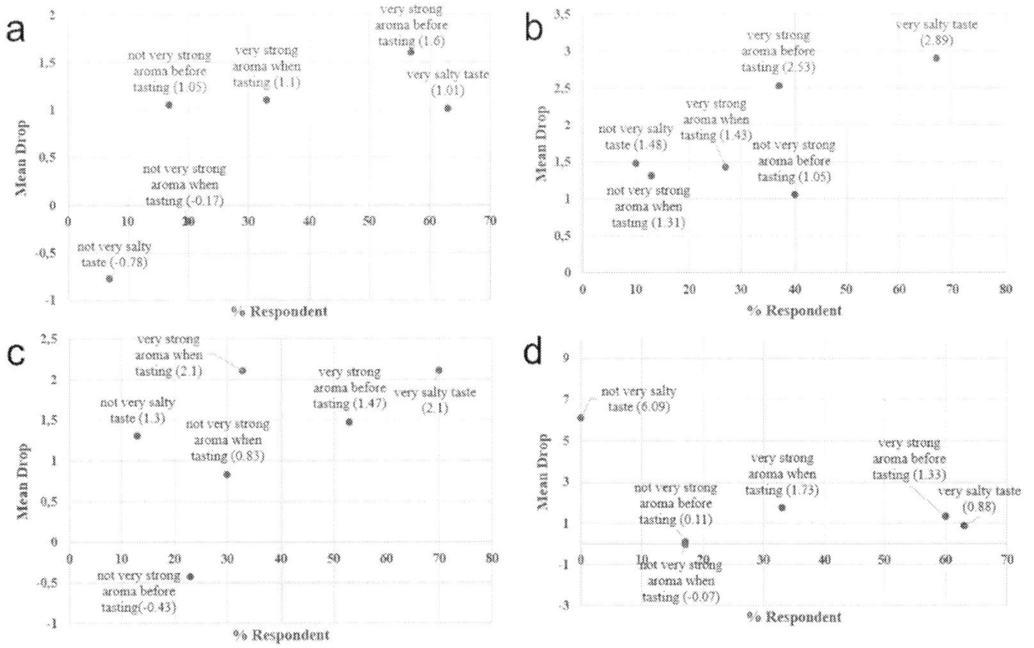

Figure 1. Penalty analysis on soy sauce product from tempeh with 15% salt (left) and 20% salt (right) using tempeh fermented for 3 days (a and b), tempeh fermented for 4 days (c and d)

For the JAR (Just-about-Right) test, the results of the penalty analysis (Figure 1) for soy sauce sample with the treatment of 3 days tempeh mold 15% showed that the attribute that needs for reformulation was "very salty taste", since this attribute on the far right of the penalty analysis graph with the amount of 63% of respondent stated that "very salty taste" with a mean drop value of 1.01. Reformulation for the attribute of "very salty taste" can be done by reducing the salt concentration or by adding seasoning to the soy sauce samples. For soy sauce treatment of 4 days tempeh mold 15%, the result from graph of penalty analysis showed that 70% respondents stated "very salty taste" with mean drop value of 2.1. The "very salty taste" attribute occupies the rightmost part of the graph, so it is necessary to reformulate soy sauce taste. This can be done by reducing the salt concentration. From the results of penalty analysis, the graph shows that for soy sauce 3 days tempeh mold 20%, the attribute that needs to be improved is "very salty taste" with a mean drop value of 2.89 and the number of respondents 67%. For the attribute improvement, it can be done by changing the composition of soy sauce or reducing the concentration of salt solution. For soy sauce treatment of 4 days tempeh mold 20%, the graph shows that "very salty taste" occupies the rightmost part of the graph, this means that "very salty taste" attribute needs to be improved. The number of respondents that chose the "very salty taste" attribute was 63% with a mean drop value of 0.88. The "very salty taste" attribute can be improved by reducing the salt content when making soy sauce samples or by adding other condiments.

4. Conclusion

From this study, it can be concluded that the pH value of soy sauce that showed a significant increase in pH value was shown in the treatment of fermentation for 3 days to 4 days tempeh mold at 20% salt concentration, while in pH value for 3 days to 4 days tempeh mold 15% there was a significant decrease as the fermentation time increased. Then, for color analysis, it was seen that there was a significant increase in the L* value in 3 days to 4 days tempeh mold 20%, whereas for a* and b* value there was an increase for 3 days to 4 days tempeh mold with the concentration of 15%, 20% and 25%, but the treatment that showed significant increase only treatment with concentration of 25%. The values of a* and b* increased significantly due to the high contribution of Maillard reaction products, including melanoidin formed during the heat treatment and fermentation processes. The results of total dissolved solids showed a significant increase for 3 days to 4 days tempeh mold 25%, this is due to the osmosis process in the material to the salt solution and caused by the length of fermentation time or the dissolution of compounds from the first fermentation stage. For the yield value, there was an increase in the treatment of 3 days to 4 days of tempeh mold with concentration of 15% and 20%, this is due the fermentation time, the longer the fermentation time, then the more protein substrates in soybeans that interacted with protease enzymes. Furthermore, from the hedonic test, the soy sauce sample with treatment of 3 days tempeh mold 20% has the highest panelist acceptance (5.37 ± 1.73) compared to other samples and this sample represents neutral. Then, based on the JAR penalty analysis, almost all sample treatments needed improvement on "very salty taste" attribute.

References

[1] Diez-Simon C, Eichelsheim C, Mumm R and Hall R D 2020 A Review *Journal of Agricultural and Food Chemistry* **68** 11612
[2] Romulo A and Surya R 2021 *International Journal of Gastronomy and Food Science* **26** 100413
[3] Razie F and Widawati L 2018 *AGRITEPA* **4** 94
[4] Karangan J, Sugeng B and Sulardi *Jurnal Keilmuan Teknik Sipil* **2** 65
[5] Nuraeni Y, Wijana S and Susilo B 2019 *Jurnal Teknologi Pertanian* **20** 67
[6] Indah A, Isnaini F, Zaida and Nurhadi B 2019 *Jurnal Teknologi Pertanian* **20** 181
[7] Erviani A E and Arif A R 2017 *Int. J. Curr. Res. Aca. Rev.* **5** 25
[8] Meilgaard M C, Civille G V and Carr B T 2016 *Sensory Evaluation Techniques* (Boca Raton, Florida: CRC Press)
[9] Pangestika L M W, Lioe H N and Adawiyah D R *Jurnal Teknologi Pertanian* **22** 1
[10] Liang R, Huang J, Wu X, Fan J, Xu Y, Wu C, Jin Y and Zhou R 2019 *International Journal of Food Properties* **22** 1998
[11] Jiang S, Shi Y, Li M, Xiong L and Sun Q 2019 *Scientific Reports* **9** 11285
[12] Kusumadati W, Satata B and Zubaidah S *Agroscientia* **21** 17-22
[13] Iskandar, A and Darusalam L Y 2020 *Jurnal Teknologi Industri Pertanian* **30** 244
[14] Retnaningsih C, Sumardi, Meiliana and Surya A 2018 *IOP Conference Series Earth and Environmental Science* **102** 12047
[15] Cahyani A, Tari A I N and Asmoro N W 2021 *Jurnal Ilmu dan Teknologi Pangan* **7** 852

IOP Conf. Series: Earth and Environmental Science 1115 (2022) 012095 doi:10.1088/1755-1315/1115/1/012095

The effects of various isoelectric points in tofu processing on physicochemical properties and antioxidant capacities

D Lo, Steviany*, G T Harris, H Febrianta, D J C Hutabarat and C Wijaya

Food Technology Department, Faculty of Engineering, Bina Nusantara University, Jakarta, Indonesia 11480

*Email: stevianyauw2@gmail.com

Abstract. Tofu is a derivative product of soybean that is widely consumed by the society for its high nutritional content and good benefits for health. Research on the physicochemical especially antioxidant content of tofu was still rarely evaluated. The aim of this research is to increase the antioxidant effect of tofu by using Lactobacillus casei FNCC-0090 and Lactobacillus plantarum spp. as coagulants. Tofu was coagulated using Lactobacillus casei FNCC-0900 and Lactobacillus plantarum spp. to reach the isoelectric points of tofu with pH 4.8, 5 and 5.2. The physicochemical of tofu were analysed which consist of total protein content, crude fat content, carbohydrate content, moisture content, ash content, antioxidant capacities, colour, yield and texture of tofu. The results showed that there is a significant difference in moisture content, colour, protein, yield, texture and antioxidant capacities of tofu. While fat, ash, carbohydrate content of tofu did not show a significant difference between sample but showed a significant difference with control. In antioxidant capacity, the antioxidant increased by 11.21% in the radical scavenging capacity, 5.1% in the reduction capacity and 4.9% in the chelating capacity of tofu.

1. Introduction

Tofu is one of the processed foods that are very often consumed in everyday life both as a side dish and as a snack in Asian culture. Tofu is one of the processed products from soybeans that use coagulation process its manufacturing. In making tofu, coagulant is needed to form tofu curds. Acids such as acetic acid can be used as a coagulant to make the protein contained in soy milk reach the isoelectric point to create tofu curds that occurs in protein compounds, namely glycine also β-conglisinin in soybeans [1,2]. Tofu itself known to have a lot of nutritional content. The nutritional content of tofu includes 7.8 grams of protein, 4.6 grams of fat, 1.6 grams of carbohydrates, 124 mg of calcium, 63 mg of phosphorus, 1 mg of zinc, 0.06 mg of vitamin B1 [3].

In this study, tofu was be processed using lactic acid bacteria, namely *Lactobacillus plantarum* spp. and *Lactobacillus casei* FNCC-0900 as coagulant agents to form tofu curds. Lactic acid bacteria were chosen as coagulant agents due to the presence of acid as a coagulant in tofu manufacturing, where acids such as acetic acid and lactic acid can act as coagulants in tofu making [4]. *Lactobacillus plantarum* was chosen as a coagulant agent in this study since *Lactobacillus plantarum* spp. known to be able to metabolize 100% carbohydrates where the more carbohydrates metabolized by bacteria, the more lactic acid was be produced which is able to lower the pH and the faster the coagulation process [5]. In addition, the selection of *Lactobacillus* casei as another coagulant agent in this study was because *L. casei* was able to produce a distinctive aroma to the product so it was expected to improve the sensory quality of the tofu produced [6]. This research is expected to increase the antioxidant capacity contained in tofu because in general the health benefits obtained from tofu such as anti-inflammatory, anticancer, to cardiovascular disease comes from the antioxidants contained in tofu.

2. Materials and Methods

2.1. Materials
The materials used for this research were *Lactobacillus plantarum* spp., isolate *Lactobacillus casei* FNCC-0090 and non-GMO soybean (SBandB, USA). Reagent used were Folin-Ciocalteu 50% (Merck, Germany), Na_2CO_3 (Merck, Germany), gallic acid (Merck, China), $AlCl_3$ (Merck, Germany), 1,1-Diphenyl-2-Picrylhydrazyl (D4313, TCI, Japan), ethylenediaminetetraacetic acid (Merck, Germany), ascorbic acid (Merck, Germany), $FeSO_4$ (Merck, Germany) and Ferrozine (2301-66, HACH, USA). Other chemical used in the analysis were analytical grade.

2.2. Tofu preparation
In tofu making, soybeans are used as the main ingredient. At the beginning of processing, non-GMO soybeans was washed and soaked for 12 hours. After soaking, the soybeans are washed and then peeled. After peeling, the soybean skin is then separated from the soybean seeds. The soybeans are then put into a blender and blended until smooth using water with a ratio of 1:10 of water and soybeans or every 5 liters of water for 500 grams of dry soybeans. After the pure soybeans are formed, the soybean pure are then filtered using tofu cloth. After filtering, the finished soybean extract is then cooked until it boils at approximately 100°C for 3-10 minutes. In control tofu, soybean extract was cooked to approximately 100°C and then vinegar and $CaCl_2$ that have been dilute with soybean extract was added to form tofu curds. For fermented tofu, the soybean extract was cooled down after cooking until it reached room temperature of approximately 25°C before inoculation with *L. plantarum* spp. and *L. casei* FNCC-0900 with a total of 10^6 CFU/ml [7]. After inoculation, the soybean extract was incubated at 37°C until the pH of the soybeans reached 4.8, 5 and 5.2. The clotting process to reach the specified pH varies where at pH 4.8 was incubated for approximately 16 hours, pH 5 was incubated for approximately 14 hours and pH 5.2 was incubated for approximately 12 hours. Tofu curds that have been formed were molded using a tofu mold.

2.3. Antioxidant analysis
The antioxidant activities were carried out in total phenolic, total flavonoid, DPPH radical scavenging activity, reducing power and chelating agent analysis. Total phenolic analysis was carried out according to Murad et al [8], total flavonoid was carried out according to Azizah et al [9]. In antioxidant capacities the DPPH radical scavenging activity was carried out according to Barus et al[10] with modification. Chelating agent analysis was carried out according to Chai et al [11] and reducing power analysis was carried out according to Laetitia et al [12] with modification and using ascorbic acid as standard.

2.4. Physicochemical analysis
For physical analysis, water content was measure according to AOAC [13], texture analysis was carried out using texture analyzer, colour analysis was analysed using a NH310 colorimeter, total yield of tofu was carried out according to Andarwulan et al[14], ash content was carried out according to Lawalata et al[15] Arziyah et al[16], protein content was carried out according to Seftiono [17] Rosaini et al[18] with modification, fat content was carried out according to Pargiyanti et al[19] with modification, carbohydrate content was carried out according to Lawalata et al[15].

3. Results and Discussion
In this study the soybean extract was incubated to lower the pH of soybean to form tofu curds. The pH that used in the tofu making are 4.8, 5 and 5.2. The yield of tofu analysis in Table 1 showed that tofu with pH lower until reaches 4.8 are the one with the highest yield following by tofu with pH 6.03 (control), 5, then 5.2. The yield produced by tofu samples were influenced by several factors which are

the selection of raw materials (soybeans), the coagulation material used, the time and method used during the soaking and milling process, also the sanitation of the sample processing [20]. In a study conducted by Obatolu [21], the results obtained showed that there was a correlation between the yield value and the moisture content value. The moisture content of tofu in this research showed that there is a significant different with moisture content between each tofu. The highest moisture content of tofu is from tofu with pH 4.8 with moisture content 83.67% followed by pH 6.03 (control), pH 5 and last is pH 5.2. According to Anglemier and Montgomery [22], this can occur due to the factor of soaking soybeans. Soaking before processing can affect all analysis results. The soaking process is carried out so that the soybeans can be separated more easily from the skin because the skin that is still attached with the beans will increase the possibility of the tofu rotting faster and produce a rancid smell. However, soaking the beans too long could cause water to enter the cell structure of soybeans and trapped the water inside. This can also cause the soybeans to have a softer texture, thus will help the process of removing the skin from the seeds. Therefore, the result from the yield and moisture content analysis were in line with the yield value of tofu obtained in this research, where tofu pH 4.8 has the highest yield value and the water content value, followed by tofu control, tofu pH 5.2 and tofu pH 5.

Table 1. Physical analysis of tofu at various isoelectric point pH

Analysis	Sample			
	Control pH 6.03	pH 4.8	pH 5.0	pH 5.2
Yield (%)	138.56 ± 10.63^{bc}	148 ± 7.24^{c}	115.64 ± 4.86^{a}	124.66 ± 5.23^{ab}
Moisture content (%)	82.47 ± 0.34^{c}	83.67 ± 0.49^{d}	80.24 ± 0.18^{a}	81.48 ± 0.62^{b}
Ash content (%)	1.089 ± 0.09^{b}	0.316 ± 0.10^{a}	0.346 ± 0.08^{a}	0.404 ± 0.09^{a}
Protein content (%)	9.26 ± 0.73^{c}	1.05 ± 0.32^{a}	2.28 ± 1.17^{b}	2.90 ± 0.54^{b}
Fat content (%)	4.164 ± 0.07^{a}	4.904 ± 1.06^{a}	4.661 ± 0.87^{a}	4.459 ± 1.38^{a}
Carbohydrate content (%)	3.01 ± 0.99^{a}	12.03 ± 0.94^{b}	11.47 ± 1.90^{b}	10.76 ± 1.52^{b}
Texture (N)	1.95 ± 0.02^{c}	0.44 ± 0.16^{a}	1.78 ± 0.06^{c}	1.48 ± 0.01^{b}

$^{a-c}$Different letters indicate a significant difference at $p < 0.05$

Based on the results obtained from the analysis of the ash content, the highest ash content was owned by control tofu with the value of 1.089% while the ash content value was owned by tofu pH 4.8 with the value of 0.316%. This shows that the mineral compounds owned by control tofu is the highest compared to the others. The decrease in ash content in tofu that use *L. plantarum* and *L. casei* as coagulant can also be caused by the presence of organic compounds that are degraded by bacteria during the incubation process. So that the more compounds that are degraded, the less ash content will be produced [23]. This is in line with research conducted by Mulia et al [24] where the results obtained for fermented tofu have a lower value due to the simplification of organic components. This can occur due to the use of H_2O and CO_2 by bacteria during respiration, thereby reducing the carbon content in the sample and causing low ash content. Other than that, in making control tofu, the coagulant used is $CaCl_2$ and vinegar. The use of this coagulant will increase the mineral content inside tofu thus increase the ash content of control tofu.

Based on the results obtained for the protein content in tofu samples, the protein content in control tofu (pH 6.03) has the highest value with the average value of 9.26%. Meanwhile, tofu pH 4.8 has the lowest protein content with the average value of 1.05%. However, tofu at the pH of 5 and pH 5.2 also had low values although not as low as pH 4.8 with the average protein content of 2.89% and 2.28% respectively. Tofu that used *L. plantarum* and *L. casei* as coagulant have a low protein content. This can be caused by several possible reasons. The first is that soybean extract will begin to precipitate and harden when it has reached its isoelectric point which is between the pH of 4.1 to 4.6 [17]. The precipitation of soybean extract could happen because of the presence of a coagulation material which in this study is the mixture of *L. casei* and *L. plantarum* bacteria. These two bacteria will produce lactic acid that will react with the protein contained in soybeans which results in the clumping of tofu. During the incubation process, the acid produced will donate H+ ions so that if it reacts then the negatively charged protein functional groups could change the environment of the solution to be neutral and cause the protein clumping, also reduce the possibility of protein to dissolve in water during the incubation process [25]. According to Poedjiadi and Supriyanti [26], protein that has been denatured and has gone through the clumping process even though it had reached the isoelectric point, it is still possible to dissolve in tofu's whey if the pH of the tofu is outside of the isoelectric point. This concludes that it is possible that the protein contained in tofu is re-dissolved in whey tofu and is not brought during the process of squeezing tofu and molding tofu.

Another factor that could cause the decrease of the tofu protein content is the time duration of soaking soybeans at the beginning of the production process. According to Arziyah et al [16], the longer the soaking process, the lower the protein content of tofu and the higher the water content of tofu. This could occur due to the release of protein bonds contained in soybeans during the soaking process, making it easier for protein components to dissolve in water. Saleh et al [27] also said that the time duration of soaking soybeans influences the protein levels and pH where the longer the soaking of soybeans could cause the protein content to decrease and the water content to increase. Different with the protein content result, the result of fat content analysis showed that the highest fat content is obtained by tofu with the pH of 4.8 with the value of 4.90% and the lowest fat content is obtained by control tofu with the value of 4.16%. There was no significant difference in fat content produced from fermented tofu that probably happened because microorganisms did not use fat as an energy source. Microorganisms tend to eat carbohydrates as their energy source, which is why the value of the fat content in fermented tofu are high can be caused by the impact of the reorganization of the carbohydrate components which is done by microbes [28]. Therefore, the fat content of tofu is higher than the protein content for fermented tofu.

The result from carbohydrate content showed that the highest carbohydrate content is tofu with the pH of 4.8 with the average value of 12.03% and the lowest is obtained by control tofu with a value of 3.01%. There are no significant different on carbohydrate content between tofu that use *L. plantarum* and *L. casei* as their coagulant but there is a significant different with control tofu. The difference produced by the control tofu was caused by the absence of microorganisms in the tofu. Given that in the incubation process, lactic acid bacteria will use carbohydrates as an energy source and decompose carbohydrates into simpler compounds, namely monosaccharides which will later be used to produce lactic acid with the help of the amylase enzyme [29]. The results obtained are also supported by the theory owned by Molin [30] that said *L. plantarum* bacteria can use all carbohydrates contained in an ingredient in the form of glucose or in the form of other simple sugars. The presence of *L. plantarum* and *L. casei* bacteria can increase the use of carbohydrates during the fermentation process. Therefore, this result was in line with the data obtained in this analysis where fermented tofu has an increase in its carbohydrate content.

For the texture and colour analysis of tofu from Table 2, the result showed that the highest texture is obtained by control tofu and tofu with the pH of 5 with the pressure or force of 1.95 N and 1.78 N, respectively. The tofu texture depends on the network formed by soy protein so that it can supply a compact and sturdy texture [25]. According to Meyza et al [31], the water content of tofu affects the texture value produced. This can occur due to too much water trapped in the soy cell structure, making the texture softer and the value lower. This process occurs when the soybeans are soaked before processing. Thus, the longer the soaking process, the softer the resulting texture then the higher the water content [17]. This statement is supported in a study conducted by Obatolu [21] where in this study the results showed a correlation between the water content value and the resulting texture. Tofu with a higher water content has a smoother texture and is easily brittle, while tofu with a lower water content has a coarser and harder texture. All tofu that uses *L. plantarum* and *L. casei* as coagulant was in line with this statement. But for the control tofu, the texture is springy and sturdy even though the water content value is also quite high. The high texture value can also be influenced by the time length of the tofu curd molding process. In a study conducted by Nurhidajah and Suyanto [32], the results obtained showed that the different types of coagulants used had an influence on the texture and color of the tofu. This can be one of the factors that causes control tofu to have a different texture compared to fermented tofu at pH 4.8 and pH 5.2.

Table 2. Color Analysis of Tofu at Various Isoelectric Point pH

Sampel	L*	a*	b*
Control pH 6.03	81,17 ± 0,01^{c}	1,49 ± 0,01^{b}	3,98 ± 0,00^{a}
Tofu pH 4.8	76,44 ± 0,155^{a}	3,18 ± 3,18^{d}	21,42 ± 0,008^{c}
Tofu pH 5.0	77,96 ± 0,005^{b}	2,46 ± 0,04^{c}	18,66 ± 18,66^{b}
Tofu pH 5.2	85,18 ± 0,006^{d}	-1,42 ± 0,15^{a}	24,30 ± 0,003^{d}

$^{a-d}$Different letters indicate a significant difference at $p < 0.05$

For the color analysis, the tool used for the tofu samples is colorimeter with the analysis used were CIE or *Comission Internationale de l'Eclarige*. Based on the results obtained for color analysis, the control tofu showed a different color compared to the fermented tofu. The tofu control has a grayish white color. Meanwhile, tofu with a pH of 5.2 has a bright yellow color, tofu with a pH of 5 has a darker yellow color and tofu with a pH of 4.8 has a dark yellow color (the darkest color). In a study conducted by Dzikunoo [33], tofu that has a higher acid concentration has a more yellow color. So that the more the amount of acid added, the more yellow (dark) the tofu will be produced. So, it can be concluded from the study that the more acidic conditions, it will make the color of the tofu darker. Therefore, this is in accordance with the data obtained where the lower the pH possessed by tofu, the darker the color it has. In the research, Dzikunoo [33] also showed that tofu treated with a regular coagulant ($MgSO_4$) had a pale white color. So, this is in accordance with the data obtained from tofu control in this research.

Based on the results of the total phenolic analysis in Table 3. Showed a significant increase when compared to the control. The lower the pH near the isoelectric point, it showed a significant increase in the total phenolic content of tofu. The results of the analysis obtained are similar with research from Qiamn et al[34] which showed that there was an increase in the total phenolic content of soy milk fermented using *L. casei* 16 at 37°C starting at 8 hours of fermentation which may have occurred due to an increase in the activity of the -glucosidase enzyme. Other review from Romulo and Surya [35] also mention that fermentation process could increase the bioavailability of nutrients especially phenolic compound in soybean such as isoflavone.

Table 3. Total phenolic and flavonoid content of tofu at various isoelectric point pH

pH Sample	Total phenolic (μg GAE/g tahu)	Total flavonoid (μg QE/g tahu)
6.03 (control)	25.51±1.11^{a}	12.20±1.61^{a}
5.2	29.98±2.36^{b}	15.55±1.41^{ab}
5	32.31±1.75^{b}	19.09±0.88^{b}
4.8	37.96±2.92^{c}	26.51±5.74^{c}

$^{a-c}$Different letters indicate a significant difference at $p<0.05$

The results of the total flavonoid analysis showed that there was an increase in total flavonoids from tofu, the lower the pH approached the isoelectric point. Flavonoid compounds can bind to protein compounds [36]. It is known that the bonds between proteins and flavonoid compounds generally occur through hydrogen bonds. The binding of flavonoid compounds and proteins is based on pH where the closer the pH of the product is to the pH of the isoelectric point, the stronger the hydrogen bonds between the protein compounds and the flavonoid compounds formed. The binding of these protein compounds is stronger in casein proteins followed by -lactoglobulin and serum albumin [37].

The protein found in soybeans and their derivatives is classified as a protein globulin. Therefore, the increase in flavonoid compounds contained in tofu through the coagulation process using *L. plantarum* and *L. casei* can occur therefore increase the total flavonoids contained in the product. The coagulation process using *L. plantarum* and *L. casei* which was carried out until it reached pH 4.8 had the highest percentage because pH 4.8 was a pH close to the isoelectric point of soybeans so that at this pH, the flavonoids contained in tofu products were able to bind strongly with protein compounds, leaving only a small amount of flavonoid compounds in tofu whey [37].

For the antioxidant activities, it can be seen, that there are increases in the antioxidant activities of the tofu product in terms of binding free radicals, reducing activity and binding Fe ions (Table 4). An increase in the antioxidant activity of fermented products has long been known. In soybean products and their derivatives, flavonoid and phenolic are compounds that give antioxidant effect. So that the presence of fermentation in tofu products which increases phenolic and flavonoid compounds in tofu will certainly increase antioxidant activity in tofu [35,38,39]. These results are also supported by the results of research on fermented soy milk where there is an increase in the reduction activity also the antioxidant activity of soy milk after fermentation [40].

Table 4. Antioxidant capacity of tofu at various isoelectric point pH

pH sample	DPPH radical scavenging Activity (%)	Reduction power (%)	Ferrous ion chelating activity (%)
6.03 (control)	10.80±2.58[a]	19.51±0.64[a]	4.25 ± 0.52[a]
5.2	16.03±0.62[b]	21.68±0.63[b]	6.19 ± 0.48[b]
5	17.56±1.74[b]	23.34±0.51[c]	7.51 ± 0.16[c]
4.8	22.01±1.27[c]	24.61±0.35[d]	9.15 ± 0.50[d]

[a-d]Different letters indicate a significant difference at $p<0.05$

4. Conclusions

Taken together, our result showed that isoelectric point in fermentation of tofu using *Lactobacillus casei* FNCC-0090 and *Lactobacillus plantarum* spp. could give a significant difference in physicochemical and antioxidant activity of tofu. The tofu that has the higher antioxidant was tofu that fermented until 4.8 of isoelectric point was reached. This result also in line with the carbohydrates content in tofu. Other than that, the ash content of tofu showed that tofu with isoelectric point of 4.8, has a lower ash content than all tofu. Tofu made from fermentation is a promising product, as it has higher antioxidant and can replace common tofu containing calcium oxalate which is harmful for patients with kidney stone.

References

[1] Diniyani N 2013 *Food Science and Culinary Education Journal* **2** 24
[2] Sitanggang A, Alexander R and Budijanto S 2020 **57** 4660
[3] Azhar M 2018 *Pola konsumsi tahu dan tempe pada keluarga prasejahtera (Kasus di kelurahan Way Lunik, Kecamatan Panjang, Bandar Lampung)* (Lampung: Universitas Lampung)
[4] Jianming W, Qiuqian L, Yiyun W and Xi C 2013 *Springer Plus* **2** 1
[5] Bachruddin Z, Agustiani E and Yusiati L 2017 *Buletin Peternakan* **41** 279
[6] Prastyaharasti L and Zubaidah E 2014 *Jurnal Pangan dan Agroindustri* **2** 285
[7] Serrazanetti D, Ndagijimana M, Miserocchi C and Perillo L 2013 *Food Control* **34** 336
[8] Murad M, Abdulah A and Mustapha W 2013 *Am. J. Appl. Sci.* **10** 1315
[9] Azizah D and Kumolowati E F 2014 *Kartika Jurnal Ilmiah Farmasi* **2** 45
[10] Barus T, Titarsole N, Mulyono N and Prasasty V 2019 *Journal of Food Engineering and Technology* **8** 75
[11] Chai T, Mohan M, Ong H and Wong F 2014 *Trop. J. Pharm. Res.* **13** 67
[12] Laetitia L, Canabady-Rochelle H-S, Kesler V, Aymes A, Fournier, F and Girardet J 2015 *Food Chem.* **183** 129

[13] AOAC 2005 *Official Methods of Analysis* (Maryland: Association of Official Analytical Chemists)
[14] Andarwulan N, Nuraida L, Adawiyah DR, Triana RN, Agustin D and Gitapratiwi D 2018 **5** 66
[15] Lawalata V N, Budiastra I W and Haryanto B 2004 *Agritech* **24** 9
[16] Arziyah D, Yusmita L and Ariyetti 2019 *Jurnal Teknologi Pertanian Andalas* **23** 143
[17] Seftiono H 2017 *Jurnal Kesejahteraan Sosial* **3** 85
[18] Rosaini H, Rasyid R and Hagramida V 2015 *Jurnal Farmasi Higea* **7** 120
[19] Pargiyanti 2019 *Indonesian Journal of Laboratory* **1** 29
[20] Koswara S 1992 *Teknologi Pengolahan Kedelai* (Jakarta: Pustaka Sinar Harapan)
[21] Obatolu V A 2008 *Eur. Food Res. Technol.* **226** 467
[22] Anglemier A E and Montgomery M W 1976 *Amino Acids Peptides and Protein* (New York: Mercil Decker Inc.)
[23] Styawati N E, Muhtarudin and Liman 2014 *Smooth cayenne Jurnal Ilmiah Peternakan Terpadu* **2** 19
[24] Mulia D S, Yulyanti E, Maryanto H and Purbomartono C 2015 *Rhizopus oligosporus Sainteks* **12** 10
[25] Aryanti N, Kurniawati D, Maharani A and Wardhani D H 2016 *Jurnal Ilmiah Teknosains* **2** 73
[26] Poedjiadi A and Supriyanti F T 2006 *Dasar-Dasar Biokimia* Edisi Revisi (Jakarta: UI-Press)
[27] Saleh E, Alwi L O and Herdhiansyah D 2020 **1** 185
[28] Kasmidjo 1990 Tempe, *Mikrobiologi dan Biokimia Pengolahan serta Pemanfaatannya* (Semarang: Soegija Pranata Press)
[29] Ferdaus F, Wiajayanti M O, Retnoningtyas E S and Irawati 2008 *Widya Teknik* **7** 1
[30] Molin G 2008 *Lactobacillus plantarum*: The role in food and in human health *Handbook of Fermented Functional Food* ed ER Farnworth (Boca Raton: CRC Press Taylor and Francis Group)
[31] Meyza M I, Nawansih O and Nurainy F 2013 *Jurnal Teknologi and Industri Hasil Pertanian* **18** 62
[32] Nurhidajah and Suyanto A 2012 *Jurnal Pangan dan Gizi* **3** 39
[33] Dzikunoo J, Ayernor G S and Saalia F K 2015 *American Journal of Food Science and Nutrition Research* **2** 138
[34] Qian F, Zhang J, Hou K, Zhang Y, Wang Z, Luo P and Tuo Y 2019 *Food Sci. Nutr.* **8** 48
[35] Romulo A and Surya R 2021 *Int. J. Gastron Food Sci.* **26** 100413
[36] Terahara N 2015 *Nat. Prod. Commun.* **10** 521
[37] Yildirim-Eikoglu S and Erdem Y 2017 *Food Rev. Int.* **34** 665
[38] Chen Y, Xiao H, Zheng J and Liang G 2015 *PLOS ONE* **10** e0121276
[39] Zhu C, Zhang W, Kang N, Zhou G and Xu X *Meat Sci.* **96** 783
[40] Marazza J, Nazareno M, Savoy de Gion G and Garro M 2012 *J. Funct. Foods* **4** 594

IOP Conf. Series: Earth and Environmental Science 1115 (2022) 012096 doi:10.1088/1755-1315/1115/1/012096

Effect of ingredient composition on characteristics of artificial rice made from banana flour

E Kamsiati[1*], H Herawati[1], Setyadjit[1], Sunarmani[1] and I. Mulyawanti[2]

[1]Research Centre of Agroindustry, National Research and Innovation Agency, Indonesia
[2]Indonesian Centre for Agricultural Postharvest Research and Development, Ministry of Agriculture, Indonesia

[*]Email: elmikamsiati@gmail.com

Abstract. Artificial rice is a grain that similar with rice made from flour with the main content of carbohydrates. In the production of artificial rice, the composition of the ingredients will determine the characteristics of the product. This study aimed to determine the effect of raw material composition on the characteristics of artificial rice made from banana flour. This study was arranged in a completely randomized design with treatments including the use of 100% banana flour; 100% banana flour + emulsifier and 90% banana flour + 10% + emulsifier. Parameters observed were colour, expansion ratio, and sensory characteristics. Then, the best product was analysed its nutritional content. The results showed that the formulation had an effect on the Lightness, redness and Chromaticity index of banana grain. The use of emulsifier produced artificial rice which has a higher bulk density. Artificial rice from a formula of 90% banana flour combined with 10% tapioca and emulsifier had the highest average panelist preference. The artificial rice has the characteristics of lightness 36.89, redness 0.03, yellowness 6.03, Chroma 6.06 and Hue 94.71, bulk density 0.61g/mL. The product had 7.65% moisture content, 0.84% fat; 3.07% protein; 86.67% carbohydrates; and 1.77% ash content.

1. Introduction

Banana is a tropical fruit that has a fairly complete nutritional content. Bananas contain lots of carbohydrates, vitamins, and minerals that are beneficial for health. The important mineral content in bananas is potassium. In addition, bananas also contain fiber and resistant starch that beneficial for health [1].

Bananas, especially plantain types, are utilized by processing them into various food products, boiling, steaming, frying, and processing them into chips and mixtures in producing cakes and other types of snacks. Processing of plantain bananas into intermediate products in the form of flour is also carried out

to increase shelf life and convenience of use. The consumption of plantain bananas is mostly used as a snack product. The use of plantain bananas in the production of food as the main carbohydrate source has not been widely reported.

Technology to produce staple foods from local carbohydrate sources as food diversification has been widely practiced in Indonesia. One of the efforts made is the development of artificial rice. Artificial rice is a grain resembling rice made from high carbohydrate local flour which can be processed through formulation and mixing followed by an extrusion process. The development of artificial rice from local flour has previously been reported including the manufacture of corn-based analog rice [2]; the production of instant artificial rice [3], production of analog rice from banana flour using granulation methods [4]; analog rice from non-rice flour [5] and rice analog from arrowroot [6]. The formulation determined the characteristics of the artificial rice produced. This study aimed to determine the effect of the formulation of ingredients on the characteristics of banana artificial rice.

2. Material and Methods

This research was conducted in July-December 2021 at the Research and Development Laboratory of the Indonesian Center of Agricultural Postharvest Research and Development. The materials used sere banana flour obtained from Keltan Setia Bogor, Glycerol Mono Stearate (GMS) and tapioca (Pak Tani). Meanwhile, the equipment used was scale, basin, steamer, filter cloth, extruder, measuring cup and drying oven. The analysis equipment include chromameter, oven, Kjeldahl flask, and Soxhlet. This research consisted of two stage of research. Stage one research was conducted to determine the prospective treatment to get rice with a firm texture. The second stage of treatment was carried out to optimize the results of the first stage of treatment. The first stage of treatment was as follows:

Table 1. Treatment of formula and condition process for producing banana artificial rice

Treatment	Banana flour (%)	Tapioca (%)	Hidrocolloid/HK (%)	Water (%)	Pregelatinisation time (minute)
A	85	15	-	25	25
F	85	15	1	25	25
B	85	15	-	25	20
E	85	15	1	25	15
C	85	15	-	30	10
D	85	15	-	20	15
H	100	-	-	25	15
I	100	-	-	30	15

In the second stage of research, optimization was carried out by adding an emulsifier to strengthen the resulting artificial rice structure. In the second stage of the study, the research was arranged in a completely randomized design with three replications. The process of making banana rice was carried out by the following process, banana flour is mixed with other ingredients according to the treatment, added with water, then pregelatinized, then moulded using a molding extruder with rice die then dried with a dryer to obtain artificial banana rice.

The products that have been obtained were analyzed for its colour, bulk density, and sensory analysis. Next, the product with the best characteristics is selected and analysis for proximate analysis. Colour measurement using a Chromameter instrument (CR 300 Chroma-metter, Minolta, Japan) to obtain the values of L, a, b C, and H. The value of L describes the level of Lightness, a describes the redness and b describes the yellowness. Bulk density was measured by put the sampel into a 25 mL measuring cup. The measuring cup is tapped against the table for 1 minute and the sample is continuously added until the volume reaches 25 mL. The weight of the sample was then weighed and the bulk density was calculated as a comparison between the weight (g) and the sample volume (mL).

Moisture content is obtained by gravimetry methods. Ash content was determined using the total ash method, fat content was obtained using the direct extraction method using Soxhlet. Protein content using semi-micro Kjeldahl method. The data obtained were analyzed using the Analysis of Variety test, if the results were significantly different, then the DMRT test was carried out. Data processing used SPSS software.

3. Result and Discussion

3.1. Preliminary research

The treatment combines the formula with the time of the steaming process. Based on the literature obtained information on the development of banana grain by granulation method using 85% banana flour and 15% tapioca. So that in the initial formulation the formula was used. The results of the first stage of the research descriptively can be seen in Table 2.

From Table 2, can be seen that the use of 15% tapioca and steaming process for 20-25 minutes, produced sticky grain. The use of 100% banana flour combined with the addition of 30% water can produce a grain but not a firm product. So it can be possible to optimize the length of steaming time.

Table 2. Description of the results of the first stage of the research

Treatment	Banana flour (%)	Tapioca (%)	Hidrocolloid (%)	Water (%)	Pregelatinisation time (minutes)	Result
A	85	15	-	25	25	Sticky grain (++)
B	85	15	1	25	25	More sticky grain (+++)
C	85	15	-	25	20	Sticky (+)
D	85	15	1	25	15	Little Sticky Grain (+)
F	85	15	-	30	10	Little sticky grain (+)
G	85	15	-	20	15	Little sticky grain
H	100	-	-	25	15	Grain formed, not firm
I	100	-	-	30	15	Grain formed, but not firm.

3.2. Main research

From the preliminary research, the treatment was selected 100% banana flour, 25% water, and increased pre-gelatinization time to 20 minutes combined with the addition of an emulsifier. In addition, the use of tapioca was also tested in combination with an emulsifier. There were the colour, bulk density organoleptic, and proximate characteristics of the product.

3.3. Colour

The lightness level of banana artificial rice ranged from 34.76-37.67. The results of the analysis of variance showed that the treatments had a significant effect. The use of additives results in an increase in the lightness of the product. Tapioca has a white colour so it produced a brighter product. The means of banana artificial rice colour can be seen in Table 3.

Table 3. The means of banana artificial rice colour

Treatments	L	a	b	C	H
100% banana flour	34.76±1.71a	-2.19±-1.95a	4.89±1.33a	5.50±0.82a	115.82±15.36b
100% banana flour + GMS 0,5%	37.67±1.21b	0.14±0.02b	7.30±1.18b	7.35±1.13b	89.33±8.23a
90% banana flour, 10% tapioca + GMS 0,5%	36.89±0.48b	0.03±0.34b	6.03±0.89ab	6.06±0.89a	94.71±5.06a

Note: the same letter in the same column behind the means showed no significant difference (P>0.05)

Furthermore, the redness level ranges from -2.19-0.14. The results of the analysis of variance showed that the formulation had a significant effect on the redness of the product. The use of food additives tended to increase the redness of the product.

The level of yellowness ranges from 4.89-7.30. The results of the analysis of variance showed that the treatment had a significant effect on the yellowness of the product. The level of yellowness of banana artificial rice increased with the use of food additives, both emulsifiers, and tapioca. Then, on the Chroma index, it ranges from 5.50-7.35. The results of the analysis of variance showed that the treatment had a significantly different effect on the Chroma value. Hue values ranged from 89.33-94.71. Hue values tended to decrease with the use of food additives. The artificial rice produced from this research had brownish gray colour similar with previous studies in producing rice analog from banana flour using granulation method [4].

3.3.1. *Bulk density.*

Banana grains had bulk density, ranging from 0.51 to 0.51 g/ml. The results of the analysis showed that the use of GMS and tapioca increased the bulk density of the product. Bulk density shows the weight of grain in the volume of the container used. Bulk density and expansion ratio of banana artificial rice can be seen in Table 4.

Table 4. Bulk density and expansion ratio of banana artificial rice

Treatments	Bulk density (g/mL)
100% banana flour	0.5375±0.01a
100% banana flour + GMS 0,5%	0.6000±0.01b
90% banana flour, 10% tapioca + GMS 0,5%	0.6100±0.02b

Note: the same letter in the same column behind the means showed no significant difference (P > 0.05)

Banana grain had a lower bulk density than analog rice from a mixture of sorghum, sago, corn and mocaf (modified cassava flours) that 0.67-0.70 g/ml [5]. Likewise, compared to the bulk density of the Sintanur variety rice had a bulk density of 0.86 g/ml [7]. Banana grain was made from banana flour with the addition of water and a pre-gelatinization process. The presence of a pre-gelatinization process causes the starch to undergo a gelatinization process and undergo swelling. That process produces a porous structure in the grain resulting in a lower density. The use of GMS increased the bulk density due to the bond between starch and fat so as to produce a compact structure. The increasing density of grains affected an increase in the weight of the grains per volume unit.

3.3.2. *Sensory characteristics*

Sensory analysis is widely used to test the quality of a food product. Testing by involving panelists to assess the sensory characteristics of a product. One of the consumers' acceptance of a product is influenced by the results of sensory evaluation [8].

The results of the test sensory evaluation with artificial rice banana could be seen in Table 5. The means of panelist assessment of the banana artificial rice colour ranged from 3.40-3.63 (neutral-like). The results of the analysis of variance showed that the treatment did not give a significant effect on the colour of the product. The colour of banana artificial rice was brownish gray, different from original rice in general that have white colour. However, the panelists gave a neutral-like response to this nuanced colour.

Furthermore, on the flavor parameter, the panelist's assessments ranged from 3.13-3.40 (neutral). The results of the analysis of variance showed that the treatment did not had a significant effect. The flavor of banana artificial rice occurred from the aroma of banana flour which is typical of bananas. In the shape parameter, the panelists' ratings ranged from 3.70-3.87 (like). The results of the analysis of variance showed that the treatment did not has a significant effect. The highest mean on formulas that use tapioca and emulsifiers. The formula of 90% banana flour, 10% tapioca with the addition of an emulsifier produces a firmer texture than the other two formulas.

Then, on the overall parameters, the assessment of the panelists ranged from 3.57-3.70. The results of the analysis of variance showed that the treatment did not has a significant effect. Panelists' assessment of the overall parameters of banana artificial rice using 100% banana flour was the same as using banana flour, tapioca and emulsifiers.

Table 5. The result of sensory evaluation of banana artificial rice

Treatment	Colour	Flavour	Shape	Overall
100% banana flour	3.63a	3.13a	3.70a	3.70a
100% banana flour + GMS 0,5%	3.40a	3.40a	3.70a	3.57a
90% banana flour, 10% tapioca + GMS 0,5%	3.40a	3.20a	3.87a	3.70a

Note: the same letter in the same column behind the means showed no significant difference ($P > 0.05$)

Furthermore, banana artificial rice was cooked to determine the panelist evaluation on cooked banana artificial rice. The result could be seen in Table 6. There was a decrease in panelists' assessment of sensory parameters compared to uncooked banana artificial rice.

Table 6. The result of sensory evaluation of cooked banana artificial rice

Treatments	Colour	Flavour	Taste	Texture	Overall
100% banana flour	3.03a	2.87a	2.87a	2.97a	3.00a
100% banana flour + GMS 0,5%	3.23a	3.07a	2.97a	3.10a	3.03a
90% banana flour, 10% tapioca + GMS 0,5%	3.23a	3.23a	3.10a	3.30a	3.33a

Note: the same letter in the same coloumn behind the means showed no significant difference ($P > 0.05$)

On the colour characteristics, the panelists' assessments ranged from 3.03-3.23 (neutral). The cooked banana artificial rice was dark brown in colour, different from white rice which is widely consumed in Indonesia, but the panelists still gave a neutral assessment.

Furthermore, on the characteristics of the aroma, the panelists' assessments ranged from 2.87-3.23 (neutral). The banana rice smelled a bit like steamed bananas, and it did not have a strong aroma like the smell of ripe bananas. As in previous studies on artificial rice from mocaf with the addition of porang, the cooked rice that produced was typical of mocaf aroma which was the raw material [9].

Then for taste, the average panelist assessment ranged from 2.87-3.10 (neutral). For rice made from artificial rice, a combination of corn and cassava, the panelists' assessment of the taste ranged from neutral to a bit like. The taste of artificial rice are from the raw materials used [5].

On the rice texture parameter, the panelists gave an average rating of 2.97-3.30 (neutral). The highest rating on the use of banana flour combined with tapioca and emulsifier. The combination of this formula produces a softer texture of cooked banana artificial rice compared to the other two formulas. Panelists' overall assessment of cooked banana artificial rice ranged from 3.00-3.33 (neutral). The highest rating was on the formula 90% banana flour, 10% tapioca, and emulsifier.

3.3.3. Nutrient content of banana artificial rice.

From the result of the sensory analysis, formula 90% banana flour with 10% tapioca and emulsifier had the highest panelist rating. Furthermore, the product then analyzed for its nutritional content. The nutritional content of banana artificial rice can be seen in Table 7.

Table 7. Nutrition content of banana artificial rice

Nutrition content	Banana artificial rice*	Mocaf-banana-sagoo artificial rice [10]*	Corn artificial rice [11]**	Sintanur rice [12]**
Moisture (%)	7.65±0.02	8.59	9.32	10.61
Fat (%)	0.84±0.00	1.86	1.22	0.28
Protein (%)	3.07±0.02	3.49	6.86	8.07
Carbohydrate (%)	86.67±0.06	83.44	91.54	87.69
Ash (%)	1.77±0.06	2.62	0.38	3.15

Note: * in dry base; ** in dry base

The moisture content of banana artificial rice was quite low, namely 7.65%, lower than the Mocaf-banana-sagoo flour articial rice, corn artificial and rice varieties of Sintanur [10-12]. The maximum moisture content for rice storage is 14%. The low moisture content had potential benefits for a longer shelf life. Moisture content also related to the ability to absorb water.

Furthermore, the fat content of banana artificial rice was 0.84% lower than that of the mocaf-banana-sagoo artificial rice. The fat content of banana flour lower than compared to mocaf flour [13] so it was produced the lower fat content of artificial rice. However, when compared to the Sintanur variety rice, this fat content has a higher value, but the value is still below 1%. Low-fat content, has benefits related to storage, because the fat will undergo oxidation during storage that results in changing of aroma. In addition, the current consumer trend is to prefer low-fat products.

Then on protein content, banana artificial rice had a protein content of 3.07% which was lower than mocaf-banana-sagoo artificial rice, corn artificial rice and Sintanur rice varieties. Banana flour has a lower protein content than corn flour. The nutrient contents of raw materials will affect the nutritional content of the product. Therefore, it is recommended to consume artificial banana rice combined with protein sources to produce sufficient protein intake.

The further characteristic was carbohydrates. The carbohydrate content of banana artificial rice was 86.67%, that it was higher than mocaf-banana-sagoo artificial rice, corn artificial rice and rice varieties of Sintanur. High carbohydrate content was obtained from banana flour combined with tapioca. Carbohydrates in banana flour include starch and resistant starch [10]. Resistant starch has benefits for healthy because its function as fiber.

4. Conclusion

From this research, it can be concluded that the ingredient formula affects the colour characteristics of banana artificial rice before and after cooking and the bulk density of the product. Sensory evaluation showed that the treatment did not give a significant effect. Banana artificial rice that has the highest panelist assessment was in the formula 90% that combined with 10% tapioca and emulsifier. The artificial rice has the characteristics of lightness 36.89, redness 0.03, yellowness 6.03, Chroma 6.06 and Hue 94.71, bulk density 0.61g/mL. The product had 7.65% moisture content, 0.84% fat; 3.07% protein; 86.67% carbohydrates; and 1.77% ash content. The use of banana flour as a raw material in the manufacture of artificial rice provides a choice of carbohydrate sources other than cereals. Artificial rice from banana flour also provides more diverse food choices for consumers. With more diverse food sources, the availability of food sources is more sustainable.

Acknowledgement

We express our gratitude to the Indonesian Center for Postharvest Research and Development, Ministry of Agriculture for the facilities provided in the implementation of this research.

IOP Conf. Series: Earth and Environmental Science 1115 (2022) 012096 doi:10.1088/1755-1315/1115/1/012096

References

[1] Thakorlal J, Perera C O, Smith B, Englberger Land Lorens 2010 *Pac Health Dialog* **16** 49
[2] Hidayat B, Akmal S, Muslihudin M and Suhada B 2017 *Food Sci Qual Manag* 2017 **61** 19
[3] Herawati H, Kusnandar F, Adawiyah D R and Budijanto 2013 S *Pangan* **22** 317
[4] Yudanti R Y and Waluyo S 2015 *Jurnal Teknik Pertanian Lampung* **4** 117
[5] Noviasari S, Kusnandar F and Setiyono A 2017 *Pangan* **26** 1
[6] Damat D, Setyobudi R H, Utomo J S, Vincēviča-Gaile Z, Tain A and Siskawardani D D 2021 *Jordan J Biol Sci.* **14** 389
[7] Sasmitaloka K S, Widowati S and Sukasih E 2020 *J Pascapanen* **17** 1
[8] Destarianto P, Agustianto K, Riskiawan H Y and Kautsar S 2017 *International Conference on Sustainable Information Engineering and Technology (SIET)* 430
[9] Yuwono S S, Febrianto K and Dewi N 2013 *Jurnal Teknologi Pertanian* **14** 175
[10] Mamuaja C F and Lamaega J C E 2015 *J. Ilmu dan Teknologi Pangan* **3** 14
[11] Noviasari S, Kusnandar F and Budijanto S 2013 *J. Teknol. dan Industri Pangan* **24** 194
[12] Widowati S and Sasmitaloka K S 2019 *PANGAN* **29** 87
[13] Rahmiati Y A, Purwanto Y A, Budijanto S and Khumaida N 2016 AGRITECH **36** 459

Antioxidant activity of Salak Sidempuan (*Salacca zalacca*) fruit with different solvents using the DPPH method

N S Siregar[1], E Julianti[2*], J Silalahi[3] and H Sinaga[2]

[1]Department of Agricultural Science, Faculty of Agriculture, Universitas Sumatera Utara, Medan 20155, Indonesia
[2]Department of Food Science, Faculty of Agriculture, Universitas Sumatera Utara, Medan 20155, Indonesia
[3]Faculty of Pharmacy, Universitas Sumatera Utara, Medan 20155, Indonesia

[*]Email: elisa1@usu.ac.id

Abstract. Free radicals in the body are a by-product of the oxidation and burning of cells that take place in the body. If this reaction occurs continuously in the body and does not stop, it will cause oxidative stress, impacting health problems. The presence of antioxidants can reduce oxidative stress that arises. Antioxidants are compounds that can capture free radicals because they can donate one electron. Antioxidants can obtain from fruits, one of which is Sidempuan salak fruit. Based on this, this study aims to determine the antioxidant activity of Sidempuan salak fruit using the DPPH method. Salak sidempuan was obtained from three different villages: Sitinjak Village, Sibangkua Village, and Hutalambung Village. In the DPPH test, three other solvents were used: ethanol, methanol, and n-hexane. The results obtained are that methanol solvents increase the antioxidant ability of salak sitinjak to reduce the oxidation process, where the IC50 value of 31771.67 mg/L is smaller than the use of ethanol as a solvent. Using ethanol as a solvent can increase the ability of Sitinjak bark to reduce the oxidation process by 90.31% inhibition. The use of methanol as a solvent increased the antioxidant capacity of salak sitinjak 82.04 mg vitamin C/100g compared to salak Sibangkua and Hutalambung. These results could conclude that using different DPPH solvents in Sidimpuan salak can produce antioxidant activity despite its weak ability. Among the three Sidimpuan salaks, Sitinjak salak has the best antioxidant activity when using methanol as a solvent.

1. Introduction

Free radicals are relatively unstable molecules with one unpaired electron in their outer orbit. Free radicals are so reactive that they always try to reach a stable state by attracting electrons from other molecules or cells in the body. Free radical molecules can oxidise other substances and cause oxidative damage in the body. The damage caused by ROS, if not stopped, will cause oxidative stress [1,2].

Free radicals affect human health, so the body needs antioxidants to capture and neutralise these free radicals. Further reactions that cause oxidative stress can stop, cell damage can avoid, or the disease induction can stop [3]. Antioxidants have a protective role against oxidative damage in the body due to pathological processes mediated by oxidative stress. Antioxidants play an essential role in inhibiting oxidation processes and contribute to health promotion and are therefore widely used in manufacturing dietary supplements, nutraceuticals and functional food ingredients [4,5].

Antioxidants are substances which, at low concentrations, can delay or prevent the oxidation of the substrate. Antioxidant compounds work through several chemical mechanisms, namely hydrogen atom transfer (HAT), single electron transfer (SET), and the ability to chelate metal transitions [6]. These antioxidants can inhibit oxidation reactions from the normal metabolic processes of living things that cause free radical compounds [7]. Antioxidants can inhibit oxidation reactions and act as hydrogen donors or free radical acceptors, resulting in more stable radicals [8].

Antioxidants are compounds that can scavenge free radicals because they can donate one electron. Various research results show that fruits have antioxidant activity, including dragon fruit, soursop, red guava, wuluh star fruit, strawberries, *mahkota dewa, salak,* rambutan, avocado, *jamblang, kemloko,* apple, banana, mangosteen, green pepper, kiwi, areca nut yaki, and tomatoes [9]. Based on the explanation above, this study aims to analyse the antioxidant activity of Salak Sidempuan fruit using the 1,1-diphenyl-2-picrylhidrazyl (DPPH) method.

2. Method

2.1. *Materials and instruments*
The main ingredient was fresh salak sidempuan fruit taken from the villages of Sitinjak, Sibangkua, and Hutalambung, West Angkola District. The chemicals used were ethanol, methanol, n-hexane, ascorbic acid and DPPH reagent. The instruments used were a 50 mL volumetric flask, pipette, 15 mL tube and a spectrophotometer.

2.2. *Sample preparation*
Separate the flesh of the salak fruit with the seeds. Salak peeled, cleaned, and washed with running water. The pulp of the salak fruit was blended, homogenised and divided into several sub-samples, and each weighed as much as 5 g. This mashed fresh salak sidempuan fruit is used for further testing for its antioxidant capacity.

2.3. *Antioxidant activity test with DPPH*
Ascorbic acid was used to compare the antioxidant activity of sidempuan salak fruit. The antioxidant activity test was determined by DPPH radical immersion using the IKP/K-34 test procedure at the Bogor Mbrio Laboratory. The antioxidant activity test was carried out using a spectrophotometer at 517 nm. The homogenized sample was weighed, put into a 50 mL volumetric flask squeezed with methanol-water to the mark, and then filtered (sample concentration of 1%, to calculate IC50, the sample was made in 3 concentrations of 1%, 0.5% and 0.25%, depending on antioxidant levels in the example). Then, the DPPH solution was made 0.004% in ethanol (made fresh in a dark room). Next, a standard solution of 50 ppm ascorbic acid was made in water (made fresh in a dark room). The solution is pipetted 1 ml of the sample/standard/blank (solvent) into a 15 ml tube. To the solution, 5 mL of 0.004% DPPH solution was added, vortexed for 30 seconds and incubated for 20 minutes in the dark. The absorbance of the sample was measured using a spectrophotometer at 517 nm.

2.4. *Data analysis*
The collected data were analyzed descriptively with the help of the SPPS program. The analysis results are intended to determine the mean and standard deviation of the test results.

3. Results and Discussion
Salak Sidempuan is widely grown in Angkola Barat District, South Tapanuli. Salak fruit in this study was obtained from salak farmers from the villages of Sitinjak, Sibangkua and Hutalambung, located in West Angkola District, which was harvested in the morning and an average of 5 months old. These three salaks were tested for their antioxidant activity using the DPPH method.

The DPPH method is the most common method used to determine antioxidant activity because it is a relatively stable and sensitive free radical in determining antioxidant activity. The DPPH method aims to measure the ability of antioxidants to reduce 1,1-diphenyl-2-picrylhidrazyl (DPPH) and other radicals that are not commonly found in biological systems using a spectrophotometer [8,10]. The DPPH method is based on extracts of antioxidant compounds to absorb DPPH free radicals shown by the naked eye. With a fainter DPPH stain. The fainter the colour of the DPPH solution, the more DPPH is suppressed by the antioxidant compounds of the extract [11].

In the assay, the purple chromogen radical DPPH is reduced by antioxidant or reducing compounds to the corresponding pale yellow hydrazine. The activity is expressed as the inhibitory concentration IC50 (inhibitory concentration), the number of antioxidants required to reduce 50% of the initial DPPH concentration. The lower the IC50, the higher the antiradical efficiency [12].

Table 1. Antioxidant activity of Salak Sidimpuan from Sibakua, Sitinjak and Hutalambung villages with an ethanol solvent

Parameter	Method	Unit	Salak Sidempuan		
			Sibangkua	Sitinjak	Hutalambung
DPPH	Spectrophotometric	% inhibition	63.64±0	90.31±0.07	61.68±0.37
IC50	Spectrophotometric	mg/L	75452.27±49.83	47224.12±337.94	86786.34±725.72
AEAC	Spectrophotometric	mg vitamin.C/100 g	42.24±0	62.335±0,05	43.07±0.25

The results of antioxidant activity with IC50 show that salak Sitinjak can reduce DPPH free radicals better than Sibangkua and Hutalambung salak, although its ability is weak. The calculation results obtained data on the free radical scavenging activity of DPPH from the highest extract in salak Sitinjak. The highest antioxidant capacity in salak Sitinjak when using ethanol as solvent.

Table 2. Antioxidant activity of Salak Sidimpuan from Sibakua, Sitinjak and Hutalambung villages with methanol solvent

Parameter	Method	Unit	Salak Sidempuan		
			Sibangkua	Sitinjak	Hutalambung
DPPH	Spectrophotometric	% inhibition	35,23 ± 0,12	60,39±0	39,20±0,08
IC50	Spectrophotometric	mg/L	57902,68±116,92	31771,67±24,38	52333,26±81,26
AEAC	Spectrophotometric	mg vitamin.C/100 g	46,80±0,15	82,04±0	52,30±0,11

The calculation results obtained data on the free radical scavenging activity of DPPH from the highest extract in salak Sitinjak. The results of antioxidant activity IC50 show that salak Sitinjak can reduce DPPH free radicals better than Sibangkua and Hutalambung salak, although its ability is weak. The antioxidant capacity of salak Sitinjak is higher than the other two salaks using methanol as a solvent.

Table 3. Antioxidant activity of Salak Sidimpuan from Sibakua, Sitinjak and Hutalambung villages with n-hexane solvent

Parameter	Method	Unit	Salak Sidempuan		
			Sibangkua	Sitinjak	Hutalambung
DPPH	Spectrophotometric	% inhibition	35,23 ± 0,12	60,39±0	39,20±0,08
IC50	Spectrophotometric	mg/L	57902,68±116,92	31771,67±24,38	52333,26±81,26
AEAC	Spectrophotometric	mg vitamin.C/100 g	46,80±0,15	82,04±0	52,30±0,11

The calculation results obtained data on the free radical scavenging activity of DPPH from the highest extract in salak Sibangkua. The antioxidant activity with IC50 on the three salaks could not be read using n-Hexan as a solvent. Using n-hexane as a solvent increased the antioxidant capacity of salak Sibangkua compared to the other two salaks.

Different DPPH solvents resulted in various antioxidant activities of salak Sidempuan. These results indicate that polar solvents are suitable for counteracting free radicals in salak Sidempuan fruit. Using methanol as a polar solvent can potentially increase antioxidant activity because it can mediate mast cell stabilisation [13]. Methanol solvent increases the ability of antioxidants to reduce the oxidation process where the IC50 value is more petite in using methanol as a solvent than ethanol solvent. Using a non-polar solvent (n-hexane) causes the IC50 value to be unreadable.

The DPPH value when using ethanol solvent was higher than methanol and n-hexane solvents. DPPH describes the sample's ability to reduce the oxidation process in units of % at a specific concentration. The results of this study illustrate that the use of ethanol as a solvent can increase the ability of salak to reduce the oxidation process. Antioxidant compounds can respond to free radicals or oxidants in different ways. On the other hand, each method categorizes or sequences the study species differently depending on their antioxidant activity [14].

The antioxidant capacity of the three types of salak in this study also resulted in varying antioxidant capacities. Different kinds of DPPH solvents produce various antioxidant capacities. When viewed from the type of solvent used, namely ethanol, methanol, and n-hexane, it has the highest antioxidant capacity in solvents using methanol. The different antioxidant capacities between these three types of solvents can be caused by other reaction mechanisms [15].

The DPPH radical scavenging test is considered to be the simplest colorimetric method. This assay is based on the principle that a radical scavenger or antioxidant will serve as a hydrogen donor for DPPH that is reduced to DPPH. This reaction will change colour intensity [16,17]. These acceptors and donors are required to form intermolecular hydrogen bonds with phenolic compounds and to donate hydrogen atoms to free radicals. Solutions had high polarity, and the ability to contribute hydrogen bonds than antioxidants increased the susceptibility to transfer hydrogen atoms to DPPH radicals [18,17].

The DPPH test with ethanol as a solvent increased the ability of salak sitinjak to reduce DPPH free radicals better than that of salak Sibangkua and Hutalambung, although its ability was weak. The antioxidant capacity of salak Sitinjak compared to ascorbic acid equivalence increased with using methanol as a solvent. The DPPH value when using ethanol solvent was higher than methanol and n-hexane solvents. DPPH describes the ability of salak to reduce the oxidation process in units of % at a particular concentration. Using ethanol as a solvent can increase salak Sitinjak's ability to reduce oxidation.

4. Conclusion

The antioxidant activity of Salak Sitinjak fruit using ethanol as a solvent resulted in better antioxidant activity than Salak Sibangkua and Hutalambung. The antioxidant activity of Salak Sitinjak fruit using methanol as a solvent resulted in better antioxidant activity than Salak Sibangkua and Hutalambung. The antioxidant activity of Salak Sitinjak fruit using n-hexane as a solvent resulted in better antioxidant activity than Salak Sibangkua and Hutalambung.

Acknowledgements

This research was funded by the Directorate General of Research Strengthening and Development, Ministry of Research, Technology and the Higher Education Republic Indonesia through the "*Penelitian Disertasi Doktor*" 2022 scheme.

References

[1] Michael S and Navdeep S C 2014 *Stress Current Biology* **24** 453
[2] Sukweenadhi J, Yunita O, Setiawan F, Kartini, Siagian M T, Danduru A P and Avanti C 2020 *Biodiversitas* **21** 2062
[3] Kikuzaki H, Hisamoto M, Hirose K, Akiyama K and Taniguchi H 2002 *Journal Agriculture Food Chemistry* **50** 2161
[4] Gulcin I 2020 *Springer Nature* **94** 651

[5] Shahidi F and Zhong Y 2015 *Journal of functional foods* **18** 757
[6] Sanchez N F S, Coronado R S, Canongo C V and Carlos B H 2019 *Antioxidant Compounds and Their Antioxidant Mechanism in* (London : Intech)
[7] Nugrahani R, Andayani Y and Hakim A 2020 *Journal of Research in Science Education* 6 194
[8] Moharram H A and Youssef M M 2014 *Journal of Food Science and Technology* **11** 31
[9] Rahmi H 2017 *Jurnal Agrotek Indonesia* 2 34
[10] Munteanu I G and Apetrei C 2021 *Journal of Molecular Sciences* **22** 3380
[11] Fitriansyah S N, Aulifa D L, Febriani Y and Sapitri E 2018 *Activities Pharmacognosy Journal* **10** 447
[12] Boligon A A, Machado M M and Athayde M L 2014 *Medicinal chemistry* **4** 517
[13] Jadid N, Hidayati D, Hartanti S R, Arraniry B A, Rachman R Y and Wikanta W 2017 *AIP Conference* Proceedings **1854** 020019
[14] Chaves N, Santiago A and Alías J C 2020 *Antioxidants* **9** 76
[15] Zhang H, Yang Y F and Zhou Z Q 2018 *Journal of Integrative Agriculture* **17** 256
[16] Brand-Williams W, Cuvelier M E and Berset C 1995 *Food Sci Technol* **28** 25
[17] Thavamoney N, Sivanadian L, Tee L H, Khoo H E, Prasad K N and Kong K W 2018 *J. Food Sci Technol.* **55** 2523
[18] Jabbari M and Gharib F 2012 *J Mol Liq* **168** 36

IOP Conf. Series: Earth and Environmental Science 1115 (2022) 012098 doi:10.1088/1755-1315/1115/1/012098

Sensory profiling of beverages from Tulsi leaves (*Ocimum tenuiflorum Linne*) using Quantitative Descriptive Analysis (QDA)

J Marvell and E Ervina*

Food Technology Department, Faculty of Engineering, Bina Nusantara University, Jakarta 11480, Indonesia

*Email: ervina002@binus.ac.id

Abstract. Tulsi, also known as holly basil, are well known for their medicinal, pharmacological, and health benefits. However, the sensory profiles of food or beverages made from this plant have never been reported before. This study aimed to define the sensory characteristics of beverages made from tulsi leaves using the Quantitative Descriptive Analysis (QDA) method. Fourteen trained panelists were selected and trained, three samples of tulsi beverages were evaluated and the tests were replicated twice. In total, 27 sensory attributes were generated. The dominant attributes include yellow, brown and transparent colors for the visual aspect, representing tea-like beverage. Moreover, both the odor and flavor of basil, mint and grassy were also perceived as key attributes for the tulsi beverages. In addition, the tulsi beverages were also characterized by strong astringent taste and aftertaste. In general, the test was reliable, however, additional training is needed to train the panelists intensively for the selected attributes. The results obtained from this research can be used as a preliminary study to develop new functional food products made from tulsi leaves by considering their sensory profiles.

1. Introduction

The Covid-19 pandemic led to social restrictions and as a consequence, resulted in a significant change in food consumption [1]. Consumers' attitudes toward food intake and behavior were reported to change in a response to the closure of physical workplaces, canteens, restaurants, and cafes. These conditions forced people to consume homemade meals and where they placed more concerns on the nutritional and healthy aspects of what they were consuming [2,3]. Based on a cross-sectional study conducted in different countries, people show significant changes in their dietary consumption where they chose food with more functional benefits for nutrition, health and convenience [1,3,4].

Tulsi (*Ocimum tenuiflorum Linne*), also known as holy basil, is an aromatic shrub belonging to the basil family of Lamiaceae. This plant can be found in India, and in tropical countries such as Indonesia [5]. Tulsi has been recognized as a medicinal herb for curing various diseases such as diarrhea, dysentery, skin irritation and bronchitis [6]. The leave extract of this plant has been reported to possess antidiabetic, anticancer, antimicrobial, and cardioprotective activities, and as such, is of used as traditional medicine [6,7]. Daily consumption of tulsi-leaves extract has been reported able to prevent diseases, and promote general health and wellbeing [5].

The use of tulsi, is still limited in its use as food and/or beverages despite the high potential health benefit from this plant. The leaves have been mainly used in tea by simply boiling the leaves [5] or added into the meals as powdered leaves extract [8]. Tulsi is used in herbal medicine mostly because of the high index of pharmacological and health promoting properties, with a high potential to be developed as functional foods [9]. However, tulsi has quite a strong bitter taste and astringent. Bitter taste has been widely known to be strongly correlated with food rejection and decrease in liking [10, 11] while astringent could also be contributed to dislike. These two sensations become a concern in food or beverages products since they can significantly influence food acceptance [12,13].

The objective of this study is to determine the sensory characteristics of functional beverages made from tulsi leaves using the sensory evaluation method of Quantitative Descriptive Analysis (QDA). The characterization of food and/or beverages made from tulsi leaves have never previously been reported despite the positive health aspects the plant possesses. A selection of trained panelists was used in this study to profile the main sensory attributes. The performance of the panelist in conducting the test and the reliability of the test was also discussed. In addition, a comparison with the existing competitors of tulsi beverages were also evaluated.

2. Material and Methods

2.1. Sample preparation: tulsi beverage

There were three samples evaluated in this study, namely Tulsi-healthy one, Tulsi-raw-herbal and Tulsi-India organic. All samples were tea-beverage products. Tulsi-healthy one was prepared by mixing tulsi leaves as the main ingredient with additional ingredients such as ginger, lemon, honey and palm sugar. The leaves were simmered until its boiling while ginger was extracted by a cold press juicer. Further, lemon juice, honey and palm sugar were added to the formula and diluted with water. The mixture was then pasteurized and stored in bottles at a cold temperature (4°C). The complete formulation of Tulsi-healthy one cannot be disclosed here as the formula was patented by PT. Pharos Indonesia. For comparison, two tulsi tea beverages that already exist in the market were involved in the evaluation. Tulsi-raw herbal (ingredients: tulsi leaves, ginger, made in Indonesia) and Tulsi-India organic (ingredients: tulsi leaves, ginger, lemongrass, lemon, made in the USA). The existing products were prepared by placing the sample (in a tea bag form) in 250 ml of hot water, the samples were then cooled, placed in a bottle and stored at cold temperature (4°C).

2.2. Quantitative Descriptive Analysis (QDA)

QDA was used to profile the sensory characteristic of tulsi beverages. A categorical scale of 1-9 anchored with "weak" and 'strong" was used in this study [14]. The scale was adjusted for each attribute (for example in the case of viscosity, the scale was anchored with "thin/watery" and "thick/viscous"). The panelists were trained with the use of the scale and the test was conducted with two replications at different times.

2.2.1. Panelist selection and training. A total of 14 panelists were selected and trained. The selection and training of the panelists followed ISO standard 8586:2012 for guidelines of selection, training, and monitoring of trained and expert sensory assessors [15]. Basic taste identification and intensity ranking tests were conducted to measure the panelists' sensitivity to different basic tastes of sweet, sour, salty, bitter and umami. The panelist was considered to pass the selection test if they scored a minimum of 80% for both identification and ranking intensity tests. The panelist with a score less than 80% will be further trained and asked to repeat the test until they are able to obtain 80%. Those who did not pass the second attempt will be excluded from the entire test. In our study, all panelists have passed the selection criteria.

To train the panelists with the selected sensory attributes, different references that closely represent the attribute of the samples were prepared and presented (i.e. the reference of aroma and flavor of the ginger using fresh cut ginger as the reference. Grassy attributes for both aroma and flavor were presented using 3-hexenal and the smell of fresh-cut grass). The references were selected by the panel leader and discussed with the panelist. A total duration of 20 hours was divided within 6 weeks (2-4 hours per week) was implemented to conduct the selections and training sessions for the trained panel.

2.2.2. Focus Group Discussion (FGD). The Focus Group Discussion (FGD) was conducted as one of the training sessions. The FGD aims to generalize terms of the sensory profiles of the tulsi beverage. The panelists were asked to describe all the sensory properties of color, aroma, texture, taste and aftertaste of the tulsi beverage that they perceived. During the FGD, the panelists were encouraged to discuss the detailed sensory properties among them in order to standardize the vocabulary of each attribute. All the sensorial properties generated in this session will be further categorized and similar terms were grouped together (i.e. viscous and thick will be grouped under viscosity). Furthermore, the selected terms were further discussed, defined and agreed among the panelists. The agreed terms were then further quantified for their intensity rating using the QDA method.

2.3. Data analysis

The perceived intensity of each attribute based on QDA was analyzed using Analysis of Variance (ANOVA) to evaluate the differences of each attribute across the samples. A pairwise comparison test was conducted using Tukey test if the ANOVA analysis was significant. The panelists' agreement and spider plot of sensory attributes were assessed using Tucker 1 plots and two-way ANOVA generated by PanelCheck ver. 1.4.2 (2016) respectively [16]. Tucker -1 plots is a Principal Component Analysis (PCA) based to unfolded data matrix consisting of individual responses per each attributes [16]. The panelists and replication effect were involved in the models to evaluate panelist performance or disagreement (reproducibility) and test precision (reliability) respectively.

3. Results and Discussion

3.1. Term generation

A total of 27 sensory attributes were selected based on the FGD session. The attributes consist of color (3 attributes), texture (1 attribute), aroma (10 attributes), taste (11 attributes) and aftertaste (2 attributes). The complete attributes are presented on Table 1. All the attributes generated by the trained panelist during the FGD were listed and similar attributes were grouped. The attributes that were cited by more than 10% of the panelists (i.e. cited minimum by two panelists) will be selected to be further discussed and defined for the complete terms. Note that lemongrass aroma, coconut, spices and basil flavor were assessed in the QDA session despite that these attributes were cited by less than 10% of the panelists. These attributes were agreed to be included in the QDA evaluation and perceived by the panelists.

The attributes cited as the highest for color were brown and yellow as these colors were dominant in the tulsi beverages and represented tea-like color. Watery-like texture were cited by all panelists for the texture attribute while ginger, citrus, basil and mint were the most cited aroma 100%, 71.4%, 50%, and 50% respectively. Citrus, ginger and sweet flavor were cited as the most dominant by 92.9%, 71.4% and 64.3% respectively while aftertaste of hot sensation from ginger and astringent were the two most attributes cited as aftertaste. All attributes listed in Table 1 were then defined together by all panelists. A reference representing each attribute was presented in the training sessions. The panelists tasted the samples of tulsi beverages to get them used to with all the assessed attributes in the pretesting session. All the agreed attributes then further quantified in the QDA session.

Table 1. Term generation of tulsi beverage

Sensory Attributes	Profiles	% Cited	Notes, references
Color (4)	Yellow	50.0%	Royal Horticultural Society (RHS) color chart containing 920 different colors [17]
	Brown	85.7%	RHS color chart [17]
	Transparent	14.3%	RHS color chart, see through, not cloudy [17]

Sensory Attributes	Profiles	% Cited	Notes, references
Texture (1)	Viscosity	100%	Watery-like, in contrast of thick
Aroma (10)	Ginger	100%	Smell of fresh ginger
	Citrus	71.4%	Smell of fresh lemon and/or citrus
	Lemongrass	7.1%	Smell of fresh lemongrass
	Grassy	42.9%	Fresh of cut-grass, 3-hexenal [18]
	Honey	21.4%	Smell of honey, sweet-like
	Spices	14,3%	Smell of fresh cloves and/or cinnamon
	Basil	50.0%	Smell of fresh basil
	Mint	50.0%	Smell of mint
	Sweet	21.4%	Smell of sugar, 20% sucrose concentration [19]
	Bitter	14.3%	Smell of caffeine, 0.5% caffeine [19]
Flavor (11)	Ginger	71.4%	Flavor of fresh ginger
	Citrus	92.9%	Flavor of fresh lemon and/or citrus
	Lemongrass	14.3%	Flavor of fresh lemongrass
	Coconut	7.1%	Flavor of coconut, shredded coconut
	Grassy	14.3%	Flavor of fresh cut-grass, 3-hexenal [18]
	Sweet	64.3%	Flavor of sweetness, 20% sucrose [19]
	Honey	28.6%	Flavor of honey, sweet-like
	Spices	7.0%	Flavor of cloves and/or cinnamon
	Basil	7.0%	Flavor of fresh basil
	Mint	14.0%	Flavor of mint-like
	Astringent	21.4%	Taste of tea, tannin, dry-mouthfeel sensation [20]
Aftertaste (2)	Hot sensation	71.4%	Hot sensation of ginger-like, long sensation
	Astringent	42.9%	Aftertaste of tea, tannin, long dry-mouthfeel sensation after sample is consumed [20]

3.2. Sensory profiling

The complete sensory profile of tulsi beverage is presented in Figure 1. Based on two-way ANOVA, Tulsi-healthy one is the most different sample compared to the other two samples. This can be explained due to the additional ingredients used in each sample. Unlike Tulsi healthy-one that was added with fresh ginger, lemon, honey and palm sugar in their formula, Tulsi-raw herbal and Tulsi-India organic did not use fresh ingredients. They are tea-based products, and the ingredients were dried and placed in a teabag together with dried tulsi leaves. This explains why the attributes for both aroma and flavor of ginger, citrus, honey and sweet were significantly higher in Tulsi-healthy one compared to the other samples. In addition, both sweet aroma and sweet taste were significantly different among the samples, showing that Tulsi-healty one as the sweetest with intense sweet aroma since this sample has sugar added in the formulation. In contrast, both Tulsi-raw herbal and Tulsi-India organic did not use any sugar addition in their ingredients resulting a lower sweetness profiles of these two products.

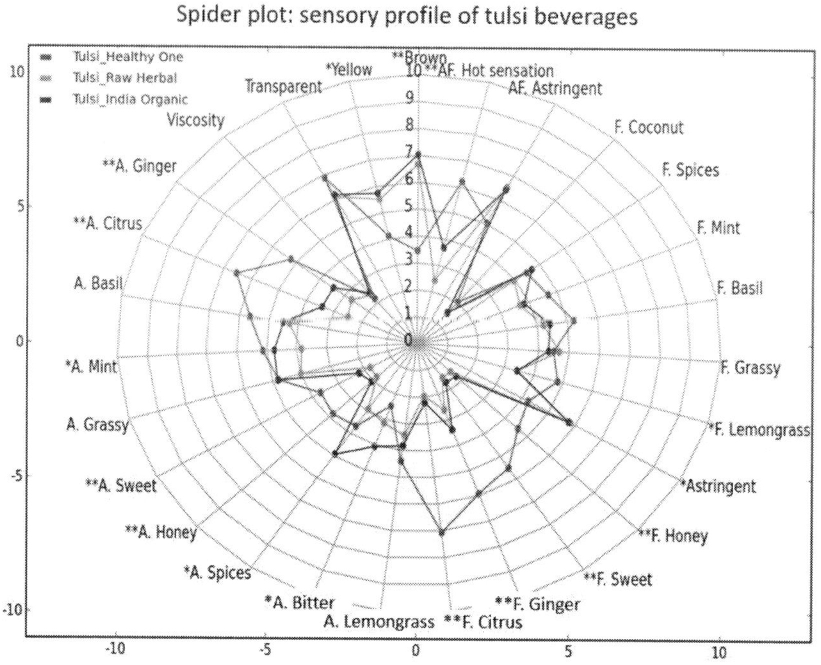

Figure 1. Spider plot of sensory attribute of tulsi beverage (A=Aroma, F=Flavor, AF=Aftertaste, *significant different at $p < 0.05$, **significant different at $p < 0.01$)

There were significant differences in mint aroma, which shows that Tulsi-raw herbal was perceived as the least intense compared to Tulsi-healthy one and Tulsi-India organic. However, there were no significant differences for basil aroma, basil flavor and mint flavor for all the samples. Interestingly, astringent is perceived as more intense in Tulsi-raw herbal and Tulsi-India organic, but the astringent aftertaste did not show any significant differences among the samples. Both Tulsi-raw herbal and Tulsi-India organic are tea-based products, this could contribute to the high tannin and provide intense astringent flavor [20]. The hot sensation aftertaste of ginger was perceived strongest in Tulsi-healthy one compared to the other two samples. This because of fresh ginger extract obtained by cold press was added in Tulsi-healthy one while Tulsi-raw herbal and Tulsi-India organic using dried ginger. The fresh extract ginger obtained by cold press can retain more spicy and pungent odorants compared to dried ginger [21] resulting in a strong aftertaste of ginger (hot, spicy and pungent).

Further, the intensity of basil taste and odor was perceived as quite intense in all the samples. The same goes for grassy attributes. All the samples were made from tulsi leaves, which came from the plant origin. Tulsi variant *Ocimum tenuiflorum L.* is commonly substituted by *Ocimum basilicum L* due to similar basil aroma and flavor from the plants [22]. Additionally, since both species were plant origin, they naturally contain 3-hexenal which contributes to grassy odor and flavor [18].

Based on sensory profiles generated by QDA, all the tulsi beverage samples have yellow, brown and transparent colors for the visual attributes that represent tea-like beverages. The samples were also characterized by the odor and flavor of basil, mint and grassy. The taste of astringent and astringent aftertaste were also contributed significantly to the profiles of the tulsi samples. These attributes may become the main sensory profiles of tulsi beverages and could be further investigated as key attributes in developing new functional food product made from tulsi leaves.

3.3. Panelists agreement and test reliability

The panelist agreement was evaluated using a two-way ANOVA with assessor effect involved in the model. In addition, the test reliability was also assessed by including the replication test in the statistical model. Figure 2 presents the panelists' performance and agreement of the QDA test. Overall, the panelists' agreement shown to be in consensus for 14 attributes of yellow, transparent, citrus aroma, basil aroma, spices aroma, bitter aroma, citrus flavor, ginger flavor, sweet flavor, astringent, spices flavor, aftertaste astringent and aftertaste of hot sensation of ginger). The attributes that were not fully agreed were grassy aroma, honey aroma, mint flavor and coconut flavor. This was showed by significant assessors' effects (Figure 2) which indicates that panelists score these attributes with disagreement. Longer and more intense training is needed to eliminate this non-agreement issue for certain attributes.

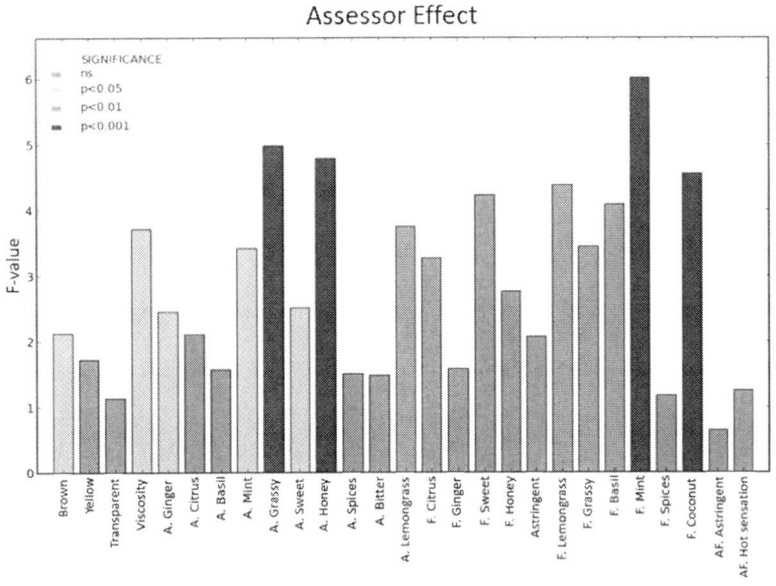

Figure 2. Assessor effect (A=Aroma, F=Flavor, AF=Aftertaste)

However, when the replicate effect was included in the model as seen in Figure 3, there were no significant differences found between replications for all attributes except for the astringent and astringent aftertaste. This shows that the test was reliable as the replication did not show any significant differences for most of the attributes between the first and the second test.

The taste of astringency and astringent aftertaste showed to be perceived significantly different in the first and the second test. Astringency is a sensory sensation caused by consumption of various food and beverages including unripe fruit, tea or red wine [20,23]. The major cause of astringent sensation is tannin, the natural polyphenol found in plant leaves, seeds and fruit skins [23,24]. Astringency is a complex sensation and is still being debated whether this taste is included as a taste sensation or as a trigeminal sensation, or if this sensation triggers mechanosensation, chemosensation or a combination of both [23]. Astringency is even more complex because the sensation involves salivary protein contained in the saliva, thus salivary flow also becomes the main factor in determining intensity perception of this sensation [25]. This attribute may be perceived differently among the panelists due to a complex sensation of this pungent-like and/or mouth-puckering sensation. A repeated training session focusing on this sensation is needed for the panelists to reach a good agreement.

IOP Conf. Series: Earth and Environmental Science 1115 (2022) 012098 doi:10.1088/1755-1315/1115/1/012098

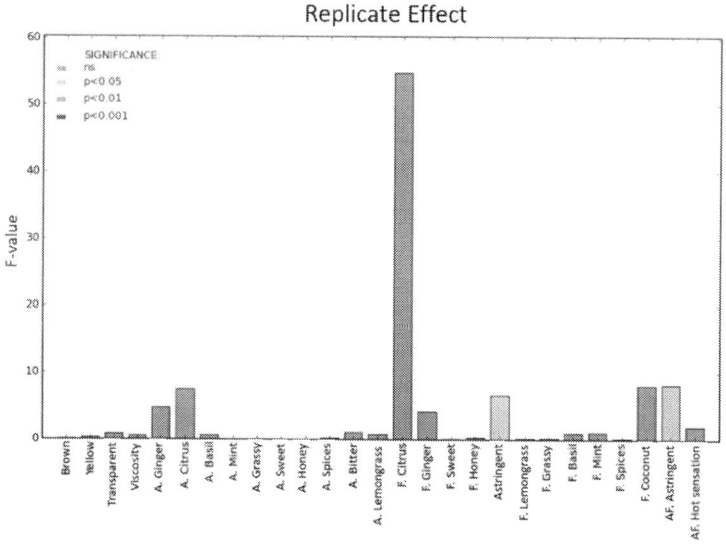

Figure 3. Reliability of QDA based on panelist replication, replicate effect (A = Aroma, F = Flavor, AF = Aftertaste)

4. Conclusion

This study aims to generate complete profiling of tulsi beverages. Based on QDA, a total of 27 sensory attributes were generated. The main characteristic of tulsi can relate to a tea-like visual appearance of yellow, brown and transparent with watery-like texture, while the aroma is characterised by a grassy, mint and basil odor. The flavor is represented by a basil, mint and grassy taste in addition to an astringent and astringent aftertaste. There were some attributes that required additional training since the panelists were not in full consensus for some attributes. However, the reliability of the test is considered to be good based on test repeatability. QDA can be selected as alternative method to define the profile of tulsi beverages. This is the first study investigating sensory profiles of tulsi-beverages and the key profiles generated in this study can be further evaluated especially in developing new food products made from tulsi.

References

[1] Hassen T B, Bilali H E, Allahyari S H, Samman H A and Marzban S 2022 *Frontiers in Public Health* **9**

[2] Sato K, Kobayashi S, Yamaguchi M, Sakata R, Sasaki Y, Murayama C and Kondo N 2021 *Appetite* **165** 105323

[3] Janssen M, Chang B P I, Hristov H Pravst I, Proveta A and Millard J 2021 *Frontiers in Nutrition* **8**

[4] Monroy C G, Gómez I G, Sánchez C M O and Motrico E 2021 *International journal of environmental research and public health* **18** 11130

[5] Cohen M M 2014 *Journal of Ayurveda and integrative medicine* **5** 251

[6] Pattanayak P, Behera P, Das D and Panda S K 2010 *Pharmacognosy reviews* **4** 95

[7] Borah R and Biswas S P 2018 *International Journal of Environment Agriculture and Biotechnology* **3**

[8] Jamshidi N and Cohen M M 2017 *Evidence-based complementary and alternative medicine* **28400848** PMC5376420

[9] Mahajan, N, Rawal S, Poddar M and Verma M 2013 *Biomedicine & Preventive Nutrition* **3** 185

[10] Mennella J A and Bobowski N K 2015 *Physiol Behav* **152** 502

[11] Hayes J E and Johnson S L 2017 *Nutrition Today* **52** S41

[12] Dinnella C, Recchia A, Tuorila H and Monteleone 2011 *Appetite* **56** 633
[13] Pagliarini E, Spinelli S, Lavelli V, Proserpio C 2021 *Food Quality and Preference* **93** 104250.
[14] Puri R, Khamrui K, Khetra Y, Malhotra R and Devraja H C 2016 *Journal of food science and technology* **53** 1238
[15] ISO, ISO 8586. 2012. Guideline for selection, training and monitoring of selected assessors and expert sensory assessors in General guidelines for the selection, training and monitoring of selected assessors and expert sensory assessors. iso.org.
[16] Tomic O, Luciano G, Nilsen A and Hyldig G 2009 *European Food Research and Technology* **230** 497
[17] Ayustaningwarno F, Vogliano V, Verkerk R and Dekker M 2021 *Food Research International* **143** 110230.
[18] Nieuwenburg D V, Groot D J H B and Smeets M A M 2019 *Frontiers in Psychology* **10**
[19] ISO, ISO 3972. 2011. Methodology of sensory analysis, in method of investigating sensitivity of taste. The International Organization for Standardization: Geneva, Switzerland
[20] Pires M A, Pastrana L M, Fuciños P, Abreu C S and Oliveira S M 2020 *Foods* **9** 1124
[21] Yu D X, Guo S, Wang J M, Yan H, Zhang Z Y, Yang J and Duan J A 2022 *Foods* **11**(11).
[22] Jür Jürges G, Beyerle K, Tossenberger M, Häser A and Nick P 2009 *European Food Research and Technology* **229** 99
[23] Jiang Y, Gong N N, and Matsunami H 2014 *Chemical senses* **39** 467
[24] Bajec M R and G J Pickering 2008 *Physiol Behav* **95** 581
[25] Lee C A, B. Ismail, and Z M Vickers 2012 *Journal of Food Science* **77** C381

IOP Conf. Series: Earth and Environmental Science 1115 (2022) 012099 doi:10.1088/1755-1315/1115/1/012099

How agroecology practices in lobu rappa village can improve peasant community welfare

F A Nasution[1*], A Anshari[1], H Thamrin[1], M H Thamrin[1], Y I Indainanto[3] and Y Aqmarina[2]

[1]Universitas Sumatera Utara, Medan, Indonesia
[2]Yayasan Sinar Tani Indonesia, Medan, Indonesia
[3]Universitas Semarang, Semarang, Indonesia

[*]Email: faiz@usu.ac.id

Abstract. There are various strategies to improve the welfare of farming communities. Building an agroecological region is one of them. Agroecology encompasses more than just managing agricultural land; it also includes many other elements and facets that make up the agricultural ecosystem. In Lobu Rappa Village, Aek Songsongan District, Asahan Regency, one of the agroecological regions is situated. The practice of agroecology was initiated by the Sinar Tani Indonesia Foundation. The descriptive qualitative approach is used in this research design. The research method used by the author is direct observation and interviews with agroecological practitioners in Lobu Rappa Village. The three operational activity pillars of organizing, interpreting, and implementing are used to carry out the program. The results showed that the agroecology practice had improved the welfare of the peasants.

1. Introduction

The term "agroecology" has several different meanings, including science, practice, and movement. Agroecology, in general, is the study of how agriculture functions within an interdisciplinary framework for investigating agricultural operations [1]. The practice of agroecological agriculture is one of the agricultural programs that continues to be carried out as a sustainable solution to climate change. Therefore, farming communities as direct actors in agricultural activities must continue to be encouraged and motivated to have the will and determination to do agroecological agriculture [2]. Through various initiatives, such as ongoing education and training activities carried out by various societal components, including the government, universities, and the private sector, this encouragement, and motivation are intended to influence changes in agricultural patterns that lead to healthy agricultural production and is oriented towards natural sustainability [3].

Agroecology is more than just a method of farming, like ecological or organic production. Agroecology is a social movement that promotes equity, relationships, access, resilience, resistance, and sustainability. It has a solid ecological foundation [4]. The growth of the agroecological area can be facilitated by a better understanding of how these numerous variables interact, support, and generate positive feedback from one another [5]. At least three agroecological areas must be taken into consideration for the transition to sustainable agriculture and food systems, with an adaptation of agricultural techniques serving as the key domain; preservation of natural resources and biodiversity [6]. This is a point that requires consideration.

IOP Conf. Series: Earth and Environmental Science 1115 (2022) 012099 doi:10.1088/1755-1315/1115/1/012099

Lobu Rappa Village, Aek Songsongan District, Asahan Regency is one location with the ability to undertake agroecological agriculture. This community is situated in a region of oil palm plantations, which was formerly a section of farmland. This region became one of the areas for building oil palm plantations, both held by the community and companies, as the industry and development of oil palm plantations developed. There are still agricultural land sites in Lobu Rappa Village that are overseen by the Sinar Tani Indonesia Foundation in the middle of the oil palm plantations and booming business. The Sintesa Foundation (Yayasan Sintesa) has been managing the agricultural land since 1990 and has been renamed the Alam Tani Education and Training Center. The location with an area of ± 14 Ha is developing as an agroecological area that can provide benefits to the community [7].

In 2050, 9.1 billion people, according to Wezel, will require more food to be produced. New agricultural techniques that can increase food production without sacrificing sustainable agriculture are therefore the subject of active discussion. It's crucial to assess the potential, effectiveness, replacement, and redesign of agricultural techniques [8]. Agroecological practices and agriculture itself may face obstacles because of intrinsic coupling, which also presents opportunities for some of the most cutting-edge research and sustainable agricultural practices. To reduce the negative effects of agriculture on the environment and to improve peasants' welfare, it is crucial to raise the significance of agroecology [9]. The advantages of agroecological agriculture through education and support for farming communities that want to manage their land in an agroecological way should also be considered. Given that agroecological approaches were able to optimize the productivity of agricultural land, this condition proved to be able to aid improve the welfare of the community.

2. Methodology

The descriptive qualitative approach is used in this research design. The location of this research is in Lobu Rappa Village, Aek Songsongan District, Asahan Regency. Researchers conducted observations and interviews with agroecological practitioners to obtain research results. The concept of Charles O. Jones, who separates program operations into three pillars of activity organizing, interpreting, and implementing is the foundation for the execution of community welfare enhancement initiatives using agroecological techniques in Lobu Rappa Village [10]. The goal of organizing is to create program operations with qualified and skilled human resources. Interpretation is done to ensure that the software follows technical guidelines and that the desired outcomes are attained. Implementation is required in the interim so that the program follows a set schedule.

3. Result and Discussion

The process of applying agroecology has various significant developments from the level of organizational management, participation, and increasing economic benefits. The implementation of agroecology creates awareness of Lobu Rappa Village farmers about the importance of strengthening the organization to create a common understanding in building systems, increasing capabilities and utilization outside of agricultural products that increase economic independence. However, the learning process in presenting a modern agricultural system requires a process and effort, both in terms of material and time. Consistency and willingness to change are the keys to positive impact agroecology. There are three main points that this program can improve the welfare of farmers.

IOP Conf. Series: Earth and Environmental Science 1115 (2022) 012099 doi:10.1088/1755-1315/1115/1/012099

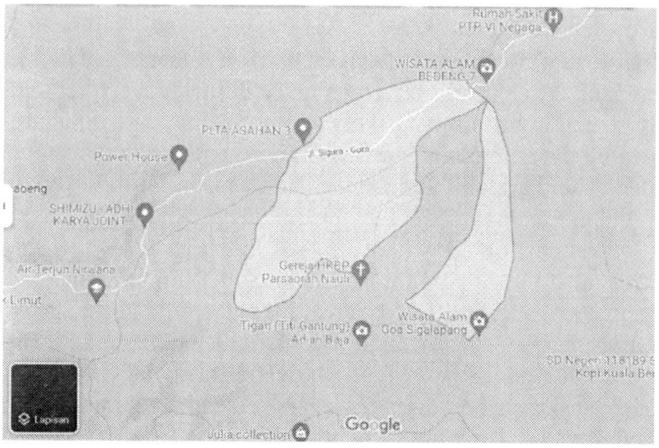

Figure 1. Research Location

3.1. Organizing

The Sintesa Foundation's initiative has been put into action continuously since 1992. The first step at the time was to construct a Micro Hydro Power Plant (PLTMH) in Lobu Rappa Village as a source of electrical energy. The village community actively participates in the creation of this MHP, which serves as the cornerstone for establishing the organization of the Lobu Rappa Village community. The Sintesa Foundation then joined the community in Lobu Rappa Village and neighboring communities to promote agrarian issues. The Sintesa Foundation then urged the community to become economically independent, stressing that the land held should be utilized to the greatest extent feasible for the community's good. This series of events served as the planning phase for a campaign to foster agroecology in the neighborhood.

3.2. Interpreting

The Sintesa Foundation creates an agroecological region and takes control of it on its own during the first stage of interpretation. This is a portfolio that may be used in numerous settings, both big and little. By combining ecological, social, and economic aspects as intensification, Sintesa Foundation creates agroecological regions. According to Jonathan Mockshell and Josey Kamanda, who outlines the roadmap for agroecological intensification, the notion of agroecology has advanced significantly in recent years [11]:

Table 1. Agroecological Intensification Roadmap

	Intensification Roadmap
Concept	• By employing ecological processes that conserve natural resources,peasants can increase agricultural production • Approaching rural development holistically, taking into account all aspects of the environment and people • Adapting nature's practices • Consider animal husbandry to be a field of study in ecology.
Vision	• Food sovereignty, increasing welfare, preserving the environment • Creation of viable alternatives to the dominance of traditional agricultureand the agroindustry

Actor	• Non-governmental organizations • Civil society • Researcher • Academics
Practice	• Mulch, tumpeng sari, crop rotation • Integrated soil and nutrient management • Soil and water conservation • Biological control methods, integrated pest management techniques, andresponsible pesticide usage • Utilization of organic inputs and balanced, more effective fertilizers
EcologicalDimension	• Land use: land sharing (less land set aside for conservation and use of lessintensive production methods to preserve biodiversity) • Spatial planning: mixed farming and multi-purpose crops (such as agroforestry and cover crops) • Landscapes: building Ecosystem services that support resilient agroecosystems • Agricultural practices: biological connections to boost production in various farming systems
Economic Dimension	• Efficiency: efficiency as a ratio; yield potential gaps; and land usage efficiency • Seed system: seed • Input usage: low external input usage
Social Dimension	• Knowledge generation: generating local knowledge with participation • Peasants: small-scale peasants • Livelihood support: assistance for small-scale rural household livelihood

3.3. Implementing

The Sintesa Foundation actively invites and instructs the community to adopt agroecology during the implementation stage. Because it can maximize the yield of agricultural products, this is portrayed as an endeavor to increase welfare. In addition, the Sintesa Foundation regularly arranges community-wide agroecological activities and training so that people can take control of their agroecology. If actively engaged in developing a sustainable and wholesome agroecological area while transforming local agriculture into an asset to supply food and promote the area. Reviving small farms as part of agroecology can, at the absolute least, point the way to new agriculture [12].

From the observations made, the researchers found the fact that the Sinar Tani Indonesia Foundation had been practicing agroecology in Lobu Rappa Village in the decade of the 90s. At that time the practice of agroecology became a rarity because of the widespread conversion of agricultural land into plantation land. This resulted in the location of the agroecology practice being a training ground for many groups who wanted to practice the same thing.

Peasants' decisions to embrace agroecological techniques are influenced by two factors: copying and learning. Adopt agroecology, it is required to manage various risks and align economic interests [13]. It is crucial to implement specific methods to increase movements' symbolic and material capacities and their potential for long-term development [14]. Following that, the community's enthusiasm appears to be positive because many of them are cultivating their property in an agroecological manner. The welfare of the locals has been impacted, and this can yet be improved such that Lobu Rappa is recognized as one of the top agroecological travel destinations. The Sintesa Foundation is still working to spread the use of agroecology, and the number of people who do so is increasing.

4. Conclusion

If applied correctly and responsibly, agroecological techniques in peasants' communities will have beneficial effects. The agroecological approach that integrates all these aspects has become a symbol of resistance to conventional farming systems in addition to its potential to improve human welfare. This will ultimately bring up the socio-economic strength of the peasants' community. The Sintesa Foundation with peasants in Lobu Rappa Village has been practicing agroecology and this has boosted the welfare of peasants. It is important to develop the modernization of agroecological farming systems in Lobu Rappa Village in this research to see the future.

Acknowledgments

Thank you to Pusat Kajian Hak Asasi Petani (PUSKAHAP) Faculty of Social and Political Sciences, Universitas Sumatera Utara, for allowing the writing of this essay.

References

[1] Uphoff N 2013 *Agroecological innovations: increasing food production with participatory development* (Routledge)
[2] Kerr R B, Kangmennaang J, Dakishoni L, Nyantakyi-Frimpong H, Lupafya E, Shumba L, Msachi R, Boateng G O, Snapp S S and Chitaya A 2019 *Agric Ecosyst Environ* **279** 109
[3] Wee C S, Ariff M S B M, Zakuan N, Tajudin M N M, Ismail K and Ishak N 2014 R*ev Integr Bus Econ Res* **3** 378
[4] Gliessman S 2013 *Sustain Food Syst* **37** 19
[5] Mier y Terán Giménez Cacho M, Giraldo O F, Aldasoro M, Morales H, Ferguson B G, Rosset P, Khadse A and Campos C 2018 *Agroecol Sustain Food Syst* **42** 637
[6] Wezel A, Brives H, Casagrande M, Clément C, Dufour A and Vandenbroucke P 2016 *Agroecol Sustain Food Syst* **40** 132
[7] Sintesa 2022 *Pusdiklat Alam Tani* SINTESA
[8] Wezel A, Casagrande M, Celette F, Vian J F, Ferrer A and Peigné J 2014 *A review Agron Sustain Dev* **34** 1
[9] Subercaseaux D, Moreno-Calles A I, Astier M and de Jesús Hernández L J 2021 *Sustainability Science and Transdisciplinarity Sustainability* **13** 3257
[10] Jones C O 1994 *Pengantar Kebijakan Publik (Public Policy)* (Jakarta: PT Raja Graf Persada)
[11] Mockshell J and Kamanda J 2018 Int J Agric Sustain **16** 127
[12] Altieri M A and Nicholls C I 2020 *Agriculture J Peasant Stud* **47** 881
[13] De La Cruz S M and Dessein J 2021 *Ecol Econ* **187** 107091
[14] Van den Berg L, Behagel J H, Verschoor G, Petersen P and Gomes da Silva M 2022 *J Rural Stud* **89** 140

Substitution of wheat flour with almond drink dregs-based flour and its effect on cracker quality

J A Santoso and F H Pamungkaningtyas*

Food Technology Department, Faculty of Engineering, Bina Nusantara University, Jakarta, 11480 Indonesia

*Email: fathyah.hanum@binus.ac.id

Abstract. Cracker is one of the popular snacks in Indonesia. Not only are they easy to carry and ready to eat, but they are also inexpensive, have a long shelf life, and are consumed by all age groups. Typical snacks on the market are poor in nutritional value as they are majorly high in carbohydrates and fat, for example, fish crackers, bread, and sweet confectionaries. On the other hand, Indonesians' mean fibre intake is inadequate compared to the fibre Daily Allowance Intake in Indonesia. The increasing trend of vegetable-based drinks such as almond milk resulted in the rise of almond drink dregs – a by-product of almond milk resulting from the filtration step of almond milk processing. Incorporating almond drink dregs-based flour improved the nutritional profile of crackers, particularly the protein and fibre content, and did not affect the moisture content. The wheat and almond drink dregs-based flour ratio of 1:4 resulted in the highest protein and fibre content and the least force to deform the cracker. There was no significant impact of flour substitution on the organoleptic properties of crackers.

1. Introduction

The global consumption of snacks has evolved over the last three years. Among many kinds of snacks, savoury biscuit is a popular snack that 69% of people consume biweekly or more [1]. In Indonesia, the volume per capita of cookies and crackers increased from 2020 (3.7 kg) to 2021 (3.9 kg), and it is projected to reach 5.4 kg in 2026 [2]. Not only are they easy to carry and ready to eat, but they are also inexpensive, have a long shelf life, and are consumed by all age groups [3].

A report by Mondelez International explained that the percentage of consumers who considered "healthy" attribute to snacks increased by 4% in 2021 compared to the previous year [1]. In Indonesia, typical snacks on the market are poor in nutritional value as they are majorly high in carbohydrates and fat, for example, fish crackers, bread, and sweet confectionaries. On the other hand, Indonesians' mean fibre intake is inadequate compared to the fibre Daily Allowance Intake in Indonesia.

The increasing trend of vegetable-based drinks such as almond milk resulted in the rise of almond drink dregs–a by-product of almond milk resulting from the filtration step of almond milk processing. Almond is an excellent source of protein and fibre, approximately 25% and 11%, respectively [4]. Solid waste of almond drinks is believed to contain a valuable amount of such nutrition. However, besides having low economic value, fresh dregs have a short shelf life. The almond drink dregs can be processed into flour to prolong its shelf life and be used as an alternative to substitute wheat flour as an ingredient for crackers [3].

To date, scientific reports on the valorisation of almond dregs into crackers have not been available. The utilization of almond dregs is a way to increase their value and alter the cracker's nutritional content.

IOP Conf. Series: Earth and Environmental Science 1115 (2022) 012100 doi:10.1088/1755-1315/1115/1/012100

However, the substitution of wheat flour for cracker processing must meet national regulation. According to SNI 2973:2011 about the quality standard of crackers, a cracker should have a maximum of 5% moisture content and a minimum of 4.5% protein [5]. This research investigated the effect of the wheat and almond dregs flour ratio on the protein, fibre, and moisture content. In addition, the evaluation of the hardness and sensory of the crackers was included.

2. Methodology

2.1. Production of almond flour
The almond flour was made from the solid waste (dregs) of almond drink production, particularly the cake obtained in filtration step. The residue was dried using an oven at 110°C for 30 minutes and ground. The powder was sieved and stored in a container.

2.2. Production of cracker
Crackers were made by mixing the dry ingredients such as wheat flour, almond flour, salt, sesame seeds, and baking powder. Water and oil were mixed with dry ingredients to form a dough which was then rolled to obtain a 0.5 cm-thick layer. The cracker formulation is presented in Table 1. The dough was moulded and then baked in an oven for 20 minutes at 170°C. The crackers were then cooled to room temperature and stored in a closed container.

Table 1. Cracker formulation.

Material	Control	A1	A2	A3	A4
Wheat flour (g)	50	40	30	20	10
Almond flour (g)	0	10	20	30	40
Black sesame seeds (g)	5	5	5	5	5
Salt (g)	1	1	1	1	1
Baking powder (g)	2	2	2	2	2
Oil (g)	5	5	5	5	5
Water (g)	25	25	25	25	25

2.3. Analysis

2.3.1. Moisture content. The moisture content of cracker was measured using gravimetric method [6]. The moisture content was determined based on the amount of moisture evaporated during the oven drying. The measurement was done in triplicate.

2.3.2. Protein content. The protein content of almond flour (AF) and crackers was analysed using Kjeldahl method [7]. The analysis was done in duplicate.

2.3.3. Crude fibre content. The crude fibre of almond flour and five types of crackers was analysed according to Fajri [2] and done in duplicate. The sample (1 g) was mixed with 50 mL sulfuric acid 0.3 N in an Erlenmeyer flask and then heated for 30 minutes at 70°C. The solution was filtered and the residue was washed with warm aquadest, 50 mL sulfuric acid 0.3 N and 25 mL ethanol 96%. The residue was placed into a dry porcelain tin and the weight was measured. The samples were dried in a furnace for two hours at 600°C. The crude fibre was calculated based on the weight difference of the residue.

2.3.4. Hardness analysis. The hardness of cracker was measured using a texture analyser (Shimadzu, Japan) with three point bent fixture. The sample was placed on a holder and the hardness was measured by the required force (in Newton) loaded on the cracker by the plunger. The analysis was done in triplicate [8].

2.3.5. Sensory evaluation. The sensory of cracker was evaluated using a hedonic test involving 35 untrained panellists. Seven-points Likert scale was used to assess the sensory attributes which were colour, texture, aroma, taste, and overall organoleptic characteristic. The points were as following: (1) very bad, (2) bad, (3) fairly bad, (4) neutral, (5) fairly good, (6) good, (7) very good.

2.3.6. Statistical analysis. One-way Analysis of Variance (ANOVA) was utilized to investigate whether there was a quality difference between the sample groups and Post-hoc Tukey test was performed to determine where the difference came from. The analysis was conducted using SPSS software.

3. Result and Discussion

3.1. Moisture, protein, and crude fibre content

Almond drink by-products are normally revalorized for livestock feeding, renewable energy production such as biofuel, and carbon active. There has been a growing interest in exploring their bioactive compounds and suitability as functional ingredients for food applications [9]. During the production of almond milk, almonds were soaked in water for hours and then dehulled. The dehulled almond kernels were mixed with water and ground using a blender for some minutes resulting in a slurry mixture. The slurry was filtered in which layers of muslin were traditionally used to separate the dregs from the almond milk [10]. The food waste of such products includes the hulls and the dregs. The dregs consisted of grounded almond kernels along with almond skin. The latter was reported to contain several bioactive compounds and explored the most [9].

The substitution of almond drink dregs-based flour significantly improved the protein and fibre content of crackers. The higher concentration of almond flour added, the higher the protein and fibre content. The moisture content, however, was comparable among different sample groups. Figure 1 illustrates the effect of varying wheat and almond flour ratio on the moisture, protein, and fibre content.

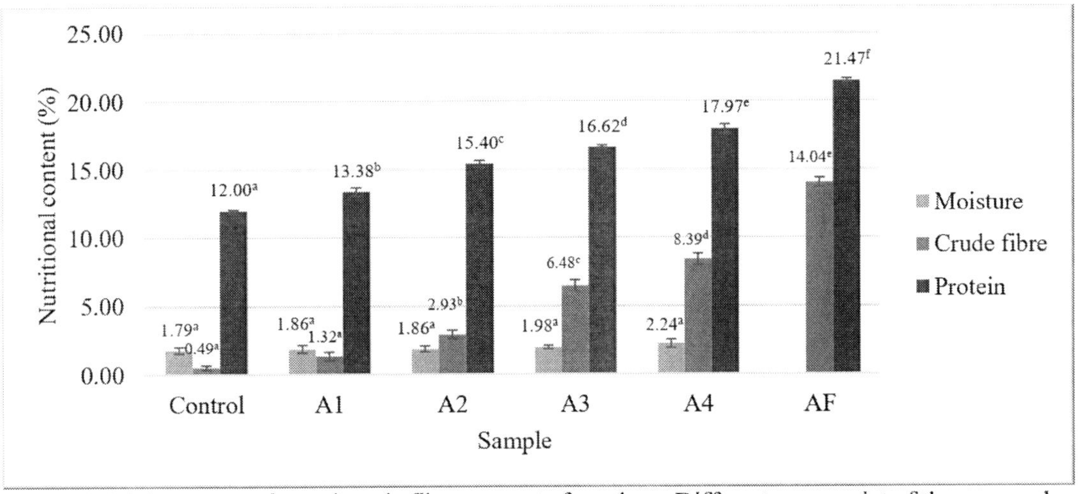

Figure 1. Moisture, protein, and crude fibre content of crackers. Different superscript of the same color bars showed significant difference according to Tukey HSD.

No significant difference was observed in the moisture content of all groups, although the control had the lowest moisture content. The substitution of 40 g wheat flour with almond drink dregs-based flour resulted in an insignificant increase in moisture content.

The upward trend in the protein content of substituted crackers was in line with the high protein content of almond drink dregs-based flour. The protein content of all samples (12.00-17.97%) met the national protein standard of cracker [5]. The protein content of almond drink dregs-based flour (AF) was approximately 21.47%, while the wheat flour contained 16% protein. As mentioned, almond kernels, one of the constituents of the almond drink dregs, have exceptional nutrients representing a dense protein source [9–11]. Previous in vivo and in vitro studies reported that almond protein has high digestibility, although the protein quality is lower than FAO/WHO recommendation [12,13].

A similar trend was also observed in the fibre content. Compared to control samples, the cracker made of wheat flour and almond flour ratio of 4:1 was not significantly different. However, the fibre content analysis of the rest of the crackers with a higher concentration of almond flour showed a significant rise. The almond drink dregs-based flour contained 14.62% crude fibre. Therefore, incorporating almond drink-based flour in crackers resulted in a promising result to improve the fibre profile of the product. Another study reported that the total, soluble, and insoluble dietary fibre content of biscuits increased when alkali extract rich in dietary fibre from almond shells [9,14].

3.2. Hardness

The hardness of crackers is presented in Figure 2. As can be seen in the chart, the wheat and almond ratio of 4:1 resulted in a proportional hardness between the modified cracker and the control. However, the addition of higher almond flour concentration noticeably reduced the hardness of cracker. Among the crackers, sample A4, with the highest almond flour concentration, required the minimum amount of force to deform the cracker.

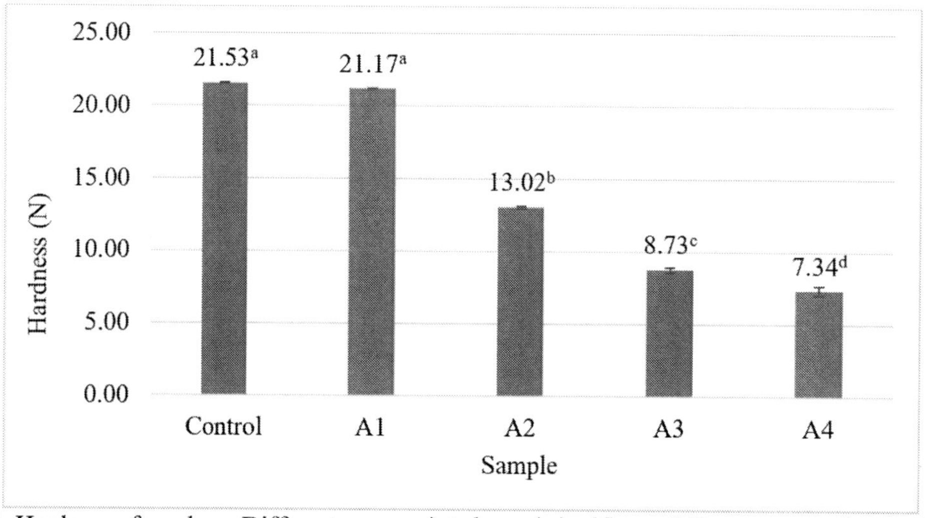

Figure 2. Hardness of crackers. Different superscript showed significant difference according to Tukey HSD.

This finding agreed with previous study that explains the hardness of biscuit or cracker may be linked with the fibre content. The fibre in almond by-products, such as skin, influenced the dough's water absorption. The higher amount of almond by-product was added, the water absorption of dough increased [15].

3.3. Sensory evaluation

The sensory evaluation was done through a hedonic test of five different crackers, and 35 untrained panellists were involved. The result of the sensory evaluation is presented in Table 1. Statistical analysis showed that no significant difference among samples was observed in all tested attributes. As can be seen in Table 1, all sensory qualities of each piece were given approximately a score of 5, which indicates that all crackers were good. A similar finding was reported. Compared to the control, the fortification of biscuits using almond by-products showed no significant effect on the flavour, texture, and overall acceptance [15].

Table 2. Sensory evaluation of crackers[a].

Sample[b]	Colour	Aroma	Taste	Texture	Overall
Control	5.43 ± 0.86	4.66 ± 1.22	4.11 ± 1.65	4.74 ± 1.46	4.97 ± 1.25
A1	5.06 ± 1.17	5.23 ± 1.33	4.74 ± 1.59	4.97 ± 1.32	5.09 ± 1.23
A2	4.97 ± 1.36	5.21 ± 1.49	4.97 ± 1.68	5.40 ± 1.10	5.37 ± 1.20
A3	4.89 ± 1.70	4.80 ± 1.67	4.69 ± 1.69	5.09 ± 1.44	5.09 ± 1.50
A4	4.66 ± 1.49	4.77 ± 1.62	4.20 ± 1.77	4.71 ± 1.61	4.66 ± 1.53

[a] The test used 7-points likert scale in representing (1) very bad, (2) bad, (3) fairly bad, (4) neutral, (5) fairly good, (6) good, (7) very good.
[b] Crackers with wheat and almond flour ratio (A1) 4:1, (A2) 3:2, (A3) 2:3, (A4) 1:4.

4. Conclusion

In conclusion, incorporating almond drink dregs-based flour improved the nutritional profile of cracker, particularly the protein and fibre content, and did not affect the moisture content. The wheat and almond drink dregs-based flour ratio of 1:4 resulted in the highest protein and fibre content and the least force to deform the cracker. There was no significant impact of flour substitution on the organoleptic properties of crackers. The substitution of wheat flour with almond drink dregs-based flour would reduce the environmental damage due to a large amount of organic waste while valorising the food waste and improving the nutritional profile of snacks.

Acknowledgement
This work is supported by Research and Technology Transfer Office, Bina Nusantara University.

References

[1] Mondelez International 2021 *The Third Annual State of Snacking*
[2] Statista 2022 *Snack Food*
[3] Özer E A 2022 *Journal of Food Processing and Preservation* **46**
[4] Barreca D, Nabavi S M, Sureda A, Rasekhian M, Raciti R, Silva A S, Annunziata G, Arnone A, Tenore G C, Süntar İ and Mandalari G 2020 *Nutrients* **12**
[5] National Standardization Agency of Indonesia 2011 *Syarat Mutu Kreker*
[6] Hamzah Y, Jumat N A and Sembok W Z W 2013 *13th ASEAN Food Conference* 1
[7] Yus Nasution A, Novita E, Nadela O and Putri Arsila S 2020 *Journal Of Pharmacy and Science* **4** 6
[8] Shimadzu 2013 Texture Analyzer EZ Test Creating Culinary Science
[9] Garcia-Perez P, Xiao J, Munekata P E S, Lorenzo J M, Barba F J, Rajoka M S R, Barros L, Mascoloti S R, Amaral J S, Prieto M A and Simal-Gandara J 2021 *Foods* **10** 1
[10] Kundu P, Dhankhar J and Sharma A 2018 *Current Research in Nutrition and Food Science Journal* **6** 203
[11] Esfahlan A J, Jamei R and Esfahlan R J 2010 *Food Chemistry* **120** 349
[12] Ahrens S, Venkatachalam M, Mistry A M, Lapsley K and Sathe S K 2005 *Plant Foods for Human Nutrition* **60** 123
[13] Martínez-Padilla E, Li K, Blok F H, Skejovic J M, Vargas-Bello-Pérez E and Lykke P I 2020 *Foods* **9** 1
[14] Kacem I, Martinez-Saez N, Kallel F, ben Jeddou K, Boisset H C, Ellouze C S and del Castillo M D 2017 *European Food Research and Technology* **243** 2115
[15] Pasqualone A, Laddomada B, Spina A, Todaro A, Guzmàn C, Summo C, Mita G and Giannone V 2018 *LWT* **89** 299

IOP Conf. Series: Earth and Environmental Science 1115 (2022) 012101 doi:10.1088/1755-1315/1115/1/012101

The challenges of implementing agroecology as a social movement by members of The *Serikat Petani* Indonesia

R P K Sinaga*, M A Lubis, F A Nasution and E P H Sembiring

Universitas Sumatera Utara, Medan 20155, North Sumatra, Indonesia

*Email: randasinaga@usu.ac.id

Abstract. The disintegration of peasants' social, economic, and cultural systems, as well as environmental harm, have been facilitated by conventional agricultural models and the existence of the green revolution. The Serikat Petani Indonesia (SPI, Indonesian Peasants Union) has made agroecology a social movement issue in an effort to resist the capitalistic agricultural system and has become one of the milestones in the realization of food sovereignty. SPI's initiatives to persuade its members to adopt an agro-ecological model of agriculture rather than a conventional one, however, cannot be implemented effectively. There are still issues that need to be resolved before SPI member may adopt agroecology, despite the fact that instruction is being implemented on both a conceptual and practical level. As an illustration, the SPI base at Pamah Village, Serdang Bedagai Regency, North Sumatra is one instance where the program for agroecological application has not been successful. As a reflection on the problems that SPI member peasants encounter in implementing agroecological agriculture, the author is interested in conducting research to see and analyze the types of challenges peasants in Pamah Village face while adopting agroecology. This study adopts a qualitative methodology to accomplish its goals, and its data-gathering methods include participatory observations, in-depth interviews, literature and documentary studies, participatory observations, and a limited group discussion. The data is then examined using the Narrative Research Analysis approach after the field data has been collected. Based on the study's findings, it is known that the failure to adopt agroecological agricultural models as an alternative is a strategic issue brought on by low-intensity organizations in the grassroots sector, as well as the quality of efforts to stimulate the involvement of the peasants in analyzing the problems that are entrenched and to choose alternative goals that are also not fully achieved.

1. Introduction

In today's social movements, political institutions, and intellectual circles, the agro-ecological farming system has become a populist narrative[1]. This is inextricably linked to the two paradigm drivers that drive the need for agricultural system alternatives [2]. The first paradigm emerges from international social movements seeking a sovereign food system, as well as from other paradigms emerging as a type of industrial compromise to perfect agricultural systems being reformed in the middle of a civilization crisis [3]. Aside from these two opposing viewpoints, what matters most is how peasants, as the primary actors in agriculture, may live a decent existence, as well as the construction of a decent food system for all human beings on the planet.

In Indonesia, agricultural problems cannot be separated from the shift in agricultural programs from the Old Order government, which focused on radical structural improvements in ownership and land rights, to the New Order government's green revolution program, which focused on agricultural production [4]. Whereas the green revolution's success has improved output, it has also resulted in social stratification in rural areas and inequalities between rich and poor peasants. This has become a structural influence as a result of the agricultural industrialization process, which has systematically removed small peasants. Millions of peasants in Indonesia have quit their jobs in the last ten years as a result of the long-term impact (from 2003 to 2013). According to the Agricultural Census Results Report by Badan Pusat Statistik Republik Indonesia, the number of agricultural business households in Indonesia was 26.14 million in 2013, a 16.32 percent reduction from 31.23 million in 2003. In addition to a decrease in the number of rural families, the agricultural industry saw a decrease in employment the following year. According to data from the Central Statistics Agency's Monthly Report on Socioeconomic Data (2016; 2018), the number of individuals aged 15 and over working in agriculture declined by 3.29 million (8.39 percent), from 39.22 million in August 2013 to 35.93 million in August 2017 [5].

The problem's challenge is to develop agriculture that can adapt to and respond to climate change while also increasing food diversity and lowering greenhouse gas emissions [6]. The agro-ecological paradigm was promoted by experts and practitioners of the peasant social movement and the environmental movement in response to the challenge. According to Wezel [7], the term agroecology can be used in several ways, including as a science, a movement, and a practice. Model agroecology offers an integrative framework for studying agricultural activities. Agriculture is an aspect of the ecological context in this framework.

This agroecological issue has also begun to become a populist topic in Indonesia. The Serikat Petani Inonesia (SPI, Indonesian Peasants Union) is one of the groups that has made agroecology a social movement issue in order to reject the capitalistic agriculture system and become one of the milestones in the fulfillment of food sovereignty. According to the SPI, this agroecological concept is defined as a farming method that integrates environmental and socioeconomic components of agricultural communities. A farming mechanism capable of meeting the following criteria: (1) economic profit; (2) social benefits for farming families and communities; and (3) long-term environmental conservation. The goal was to break peasants' reliance on external inputs and market lords who controlled rural resources. The implementation of agroecological agriculture stems from a family farming heritage that appreciates, guarantees, and preserves nature's sustainability in order to re-realize agricultural culture as life.

SPI's attempt to encourage members to convert from a conventional agricultural model to an agroecological model cannot be implemented optimally. Although instruction is being implemented on both an ideological and practical level, there are still obstacles in the way of SPI peasants practicing agroecology. The SPI base at Pamah Village, Serdang Bedagai Regency, North Sumatra, is one example. Peasants in Pamah Village who are members of the SPI have been programmed to practice agroecology, although this has yet to be achieved. As a result, the author wishes to do study on the difficulties that peasants in Pamah Village experience when employing agro-ecology.

2. Methodology

As a reflection on the problems that SPI member peasants encounter in implementing agroecological agriculture, this paper is based on the results of research that uses qualitative approach with narrative research method. This research method focusing on studying one or two individuals, gathering data through the collection of their stories, reporting individual experiences, and chronologically ordering the meaning of those experiences (or using life course stages) [8].

Using field research techniques[9], the research was conducted at SPI base locations, namely Pamah Village, Silinda District, Serdang Bedagai Regency, North Sumatra Province in July 2021. As for deepening information, researchers also dug up information on the Regional Council (DPW) of North Sumatra SPI based in Medan City.

The study using data collection techniques in the form of literature studies, in-depth interviews, observation of participation as observers. As for the determination of interview informants, this study

IOP Conf. Series: Earth and Environmental Science 1115 (2022) 012101 doi:10.1088/1755-1315/1115/1/012101

uses purposive sampling techniques with informant based on the required information. In addition, the data analysis techniques in this study use the Narrative Research Analysis approach.

3. Results and Discussion

Based on the study's findings, it is known that peasants in Pamah Village only consideragroecology as one of the SPI movement's issues and that it is "agricultural model of the times it used to be" that has been adopted by earlier generations. Due to the fervent demands of the SPI to implement agriculture that they get as members, peasants in Pamah Village do not ideologically oppose agroecology. Agroecology as a step toward achieving food sovereignty, a strategy of protecting the environment, and a response to capitalism's agricultural paradigm [10]. However, the general public has not yet made the decision to move to an agroecological agriculture paradigm.

Peasants in Pamah Village typically run agricultural businesses with a focus on food crops (rice, corn, and yams) and forest crops. Pamah Village is located in the middle of an oil palm plantation area (owned by a private company, which is also an adversary of SPI peasants in Pamah Village in agrarian conflict) (teak trees). Pamah Village occupies an area of approximately 10.64 km^2 and has 914 inhabitants, or 18.76 percent of the total population of Silinda District, the majority of whom are peasants. Most recently, SPI peasants in Pamah Village held a member meeting in July 2021, coinciding with the time the research was taking place. As seen in the figure below.

Figure 1. SPI member meeting in Pamah Village

According to the study's findings, the peasants in Pamah Village emphasized certain plant varieties as a sort of shared social attitude (even before becoming members of SPI), in order to preserve the ecosystem around communities (at least not participating in environmental destruction). Where the peasants in Pamah Village believe that oil palm plantations and other types of deforestation have contributed to the environmental harm around their settlements, such as the decline or extinction of many sorts of fish in the river flow in the village).

On the other hand, peasants in Pamah Village continue to run agriculture according to a conventional model, which includes continuing to use fertilizers and chemical drugs, which indirectly contribute to the destruction of soil quality on agricultural land and the existence of health risks for peasants. Although peasants are aware of the negative effects, switching to a more ecologically friendly agricultural model is difficult for them due to their lack of technical knowledge and expertise [11].

As a matter of fact, SPI has carried out an intensively executed agroecology teaching program in Bogor Regency (SPI Education and Training Center), involving representatives from all members at the provincial level. Numerous SPI base areas have adopted agroecological farming as a result of these trainings, but overall results have fallen short of expectations. This serves as an example of how SPI's efforts to ensure that graduates of its agroecology education can develop agroecological practices and sciences on any basis across its membership were not successful in ensuring that the output of agroecology education can be developed as a follow-up. Including SPI participants in Pamah Village who haven't had visits from participants in agroecology education.

Based on the situation, it can be concluded that the agroecology education program has been repeated, with no study of the situation to evaluate the results of educational achievements prior to agroecology, before conducting the next education. Whereas the outcomes of the scenario analysis can be regarded as developing education with a learning strategy that is based on the realities of the region and the challenges of each institution. As a reflection of the reality of formal agroecological education in Brazil, that it is critical to develop learning approaches based on the realities of their respective regions, institutional challenges, and dialogue between educators and students [12].

Essentially, the use of agroecology is not confined to resistance to the agricultural system dominated by capitalism. Not to that extent, but rather as an endeavor to restructure the food agricultural system ecologically in order to give birth to the structure of a new way of life [13]. As a result, agroecology education is an important component of efforts to organize society through an educational process that would give birth to the development of food sovereignty.

Peasants in Pamah Village continue to have a condition of low agricultural productivity while transitioning to an agricultural model that does not utilize fertilizers and chemical medications, in addition to the problem of understanding and skills in a more environmentally friendly agricultural model. Although peasants recognize that organic agricultural products have a higher selling value than conventional agricultural products, they also understand the incapacity to access markets. This is basically a contraction of the existing scenario in Pamah Village, where SPI peasants have organized cooperatives to ease production and access to peasant marketing. However, so far, the cooperative institution has only been able to enable the production and trading of one commodity, palm sugar (both in the form of powder, as well as liquid). Researchers see this as a weak encouragement on the part of SPI organizers in encouraging the birth of SPI members in Pamah Village in developing businesses that are not only trapped in the discourse market, but rather how to run a cooperative as a forum to facilitate the needs of its members, including when it has switched to using agroecological models. Because, in principle, the output coming from the agroecological model does not have to be fixed on a market orientation, but rather on how the chain model that is managed can prioritize the requirements of peasants in the locality network as a food sovereignty concept. It's as understood, that agroecology as a social movement can emerge in the peasant household of La Via Campesina with the existence of a 'peasant-to-peasant movement' as an advantage of the development of a social Movement that takes the scale of agroecology [14]. As a result of this example, SPI peasants in Pamah Village should be able to overcome production issues and gain access to marketing.

In looking specifically at the problems of peasants for the application of agroecology by SPI, the author focuses his views on the problem of strategic movement of agroecological application in each of the SPI member bases, which have the character of regional realities and their respective institutional challenges. It's as understood, that strategic issues have resulted in a sluggish pace of progress in adopting alternative agricultural production models [15]. Wherever a movement is formed, it must be able to arrange things that can propagate the best practices and identities in order to stimulate the use of agroecology. It is important to underline the importance of educational programs in educating members and generating forms of identity. Regular evaluation is required in the implementation of more ingrained instruction, much as there were issues with agroecology in Pamah Village. Because the movement's application to the resistance related to agrarian disputes differs, it will have a different impact on the peasants' drive to develop direct action and instrumental techniques that fueled the growth of resistance. So, as a movement in agricultural output, the method employed must have a complementary aspect.

4. Conclusion

The adoption of agroecological farming models by SPI members in Pamah Village as a substitute is a strategic issue, according to the study's conclusions. Due to the low level of grassroots organizing and the effectiveness of efforts to engage the peasant masses in reading the ingrained problem and choosing alternative goals that are also not optimally achieved, the movement has had limited success in implementing alternative agricultural production models. This is demonstrated by the fact that the movement (cooperative) has a low level of institutionalization due to its strategic strength.

Acknowledgements
I would like to thank the Serikat Petani Indonesia, PUSKAHAP (Center for Agrarian Studies and peasants' Rights) USU, and also the reviewers for the suggestions that greatly improved this contribution.

References

[1] Van den Berg L, Behagel J H, Verschoor G, Petersen P and Gomes da Silva M 2022 *J. Rural Stud.* **89** 140
[2] Mockshell J and Kamanda J 2018 *Int. J. Agric. Sustain.* **16** 127
[3] Giraldo O F and Rosset P M 2018 Agroecology *J. Peasant Stud.* **45** 545
[4] Al A and Awan P 2002 Perebutan Kuasa Tanah *Yogyakarta: Lappera Pustaka Utama*
[5] Sinaga R P K 2022 *J. Peasant. Rights* **1** 20
[6] [6] Mier y Terán Giménez Cacho M, Giraldo O F, Aldasoro M, Morales H, Ferguson B G, Rosset P, Khadse A and Campos C 2018 *Agroecol. Sustain. Food Syst.* **42** 637
[7] Wezel A, Bellon S, Doré T, Francis C, Vallod D and David C 2009 Agroecology as a science, a movement and a practice. A review *Agron. Sustain. Dev.* **29** 503
[8] Creswell J W and Poth C N 2016 *Qualitative inquiry and research design: Choosing among five approaches* (Sage publications)
[9] W Lawrence N 2014 *Social Research Methods: Qualitative and Quantitative Approaches*
[10] Bloomfield M J 2014 *Env. Polit.* **23** 263
[11] Anibaldi R, Rundle-Thiele S, David P and Roemer C 2021 *Land* **10** 386
[12] Rivera-Ferre M G, Gallar D, Calle-Collado Á and Pimentel V 2021 *J. Rural Stud.* **88** 138
[13] Gliessman S 2013 *Agroecol. Sustain. Food Syst.* **37** 19
[14] Rosset P M and Martínez-Torres M E 2012 *Ecol. Soc.* **17**
[15] Pahnke A 2015 *J. Peasant Stud.* **42** 1087

Antioxidant activity and microbial contamination of jeruju leaf tea with ginger and lemongrass variant packaging in storage period

R Batubara[1,2,4,6*], K Aulia[1], T Ismanelly Hanum[3], O Affandi[1,4] and M Ulfa[1,5]

[1]Faculty of Forestry, Universitas Sumatera Utara, Jln. Tri Dharma Ujung No. 1 Kampus USU, Medan 20155, Indonesia
[2]Center of Excellence for Mangrove, Universitas Sumatera Utara, Medan 20155, Indonesia
[3]Faculty of Pharmacy, Universitas Sumatera Utara, Jl. Tri Dharma No. 4 USU, Medan 20155, Indonesia
[4]Center for Social Forestry Studies, Universitas Sumatera Utara, Medan 20155, Indonesia
[5]JATI-Sumatran Forestry Analysis Study Center, Universitas Sumatera Utara, Medan 20155, Indonesia
[6]Center of Excellence for Mangrove, Universitas Sumatera Utara, Medan 20155, Indonesia

*Email: ridwantibb@yahoo.com

Abstract. The jeruju plant *(Acanthus ilicifolius)* is a plant that is affected by the tides and usually gets a higher supply of fresh water. Jeruju plants can be processed into processed products in the form of mangrove tea. This study aims to determine the antioxidant activity and analysis of microbial contamination of ready-to-drink jeruju leaf tea based on storage time. Determination of antioxidant activity was carried out by the 2,2-diphenylpicrylhdrazyl (DPPH) method using UV-Visible Spectrophotometry. The microbiological analysis test was carried out based on the Coliform Most Probable Number (MPN). The results of the antioxidant activity test have different antioxidant activity values at each storage time. The results of microbiological analysis testing with a storage period of 3 months contain coliform bacteria, so they are not suitable for consumption.

1. Introduction

The use of mangroves by the surrounding community is used as clothing, food and shelter [1]. Jeruju (*Acanthus ilicifolius*) is a type of mangrove which is classified as a herbaceous plant [2]. Jeruju is a type of mangrove plant that has many benefits, one of which is as a medicinal plant. The jeruju plant lives in the middle to late zone of the mangrove ecosystem, which is an area that is affected by tides and usually gets a higher supply of fresh water. Jeruju leaves (*A. ilicifolius*) can be processed into mangrove tea products. This mangrove tea processing has been developed in several villages in the mangrove area in Indonesia. This is very much in accordance with the habits of the surrounding community who always consume tea or coffee before carrying out their daily activities [3].

Tea is a natural ingredient that can be used as a drink for everyday. The content of substances owned by tea in the form of flavonoid bioactive substances makes the taste of tea more distinctive. The leaf part of a plant is the part that contains many chemical compounds that are useful for providing a fresh taste, can restore body health and have been proven to have no negative impact [4]. Chemical compounds contained in green tea are tannins, catechins, galocatechins, flanols and simple polyphenols. Polyphenols are active compounds in tea that act as antioxidants. Polyphenol compounds that act as antioxidants are found in many plants [5].

Several previous studies have reported the traditional benefits of jeruju leaves and their pharmacological activities [6,7]. *Achanthus ilicifolius L.* is also used to treat allergic diseases and helminthiasis [8]. Ethanol extract, aqueous extract and n-hexane extract of *A. ilicifolius* leaf have anthelmintic activity against *Ascaridia galli* and *Pheretima posthuma* worms [9]. Hot water extract polysaccharide *A. ilicifolius* has antioxidant activity [10].

The innovation of developing tea made from jeruju leaves is a way to get the benefits of these plants easily and practically. For practical purposes, tea is packaged in the form of ready-to-drink tea. This innovation requires a series of studies. The study is related to antioxidants and their stability while stored in the packaging and their relation to microbial contamination in the product. This information is important in the context of the safety of consumption of the product made.

Functional food is not only function as food or drink but has other health effects. These functional foods or drinks are usually made from plants that contain substances or compounds that are clinically proven to be beneficial to health. Functional additives that will be added to jeruju tea in this study are ginger and lemongrass. This addition is also to improve the organoleptic side of the product to be made. The aim of the research was to study jeruju leaf tea in ready-to-drink packaging and its storage stability.

2. Materials and Method

2.1. Study period and place
This research was conducted from June to October 2021. The Sampling of jeruju leaves was carried out in Paya Pasir Village, Medan Marelan, Medan, North Sumatra. The extraction is carried out at the Forest Products Technology Laboratory, Faculty of Forestry, Universitas Sumatera Utara. Antioxidant activity testing was carried out at the Research Laboratory, Microbiological testing was carried out at the Microbiology Laboratory, Faculty of Pharmacy, Universitas Sumatera Utara.

2.2. Sample preparation
The Sampling of jeruju leaves was carried out in Paya Pasir Village, Medan Marelan District, Medan City, North Sumatra. Jeruju leaves were taken using simple equipment such as knives, scissors, burlap and carried out randomly (random sampling). The simplicia was made on each sample, namely jeruju, jasmine, ginger and lemongrass leaves. The sample is cleaned of adhering dirt using water [1]. Spread on the cloth until the water is absorbed [2]. Samples were dried in a drying cabinet until dry and brittle [3]. The dried samples were made into powder using a blender [4]. The simplicia that has become powder is put into a container that is protected from the sun before further processing [5].

2.3. Extract making
The making of jeruju leaf tea extract and flavor variant extracts in the form of ginger and lemongrass was done by brewing 12 grams of simplicia powder with 50 ml of hot water.

2.4. Making of supplementary solution
2.4.1. Preparation of sugar solution. The sugar solution was made by dissolving 1 kg of granulated sugar into 500 ml of water then heated and stirred continuously until the sugar was completely dissolved and a TSS of about 69-72 obrix was obtained. Then packaged in sterile bottles and pasteurized at 70°C for 30 minutes.

2.4.2. Preparation of Na-Benzoate solution. 5000 ppm Na-Benzoate solution was prepared by dissolving 0.5 g of Na-Benzoate into 100 ml of water, then stirring continuously. After that, it is packed in a sterile bottle.

2.5. Jeruju leaf tea storage

Storage is a preparation that can last for different periods of time before starting to decrease/lose its nutritional content [11]. The storage process is carried out on jeruju leaf tea that has been packaged and stored at 10^{0}C. During the storage time of jeruju leaf tea, antioxidant activity tests and microbiological tests were carried out. The process of storing jeruju leaf tea was carried out for 3 months with tests carried out once a month.

2.6. Antioxidant test

Antioxidant testing of jeruju leaf tea (*A. ilicifolius*) was carried out using the DPPH free radical scavenging method [12]. Tests are carried out at the beginning, after storage of the product and after the storage period.

2.7. Microbial contamination test

Microbiological analysis of jeruju leaf tea aims to determine whether the jeruju leaf tea is still proper for consumption or not. The microbiological analysis of jeruju leaf tea was carried out based on the research of [13] as follows:

2.7.1. Presumption test. The presumption test was carried out by making a series of test tubes each containing Lactose Broth (LB) media and Durham tubes with dilutions of 10-1, 10-2, 10-3. Add 1 mL of the sample using a sterile pipette to each test tube containing the media. Beat until the mixture is homogeneous. Incubate in an incubator at 37°C for 24 hours. Observe each tube, if the results show the formation of gas can be seen by the presence of air bubbles in the Durham tube then it shows a positive result. Next, do the affirmation test.

2.7.2. Confirmation test. Confirmation test is carried out by taking one drop of water from the tube and the test that shows a positive presumptive test into a tube containing BGLB (Brilliant Green Lactose Broth) media. All tubes were incubated at 37°C for 24 hours. Observe each tube, a positive result in the confirmation test is indicated by the formation of gas. If the result is positive, the confirmation test is continued with the complete test.

2.7.3. Complete test. The complete test was carried out by taking one ose from a positive BGLB (Brilliant Green Lactose Broth) tube, using a streak method on EMB (Eosin Metyhlen Blue) media. Incubate the EMB (Eosin Metyhlen Blue) plate at 37oC for 24 hours. If the result shows the formation of colonies on EMB (Eosin Metyhlen Blue), a positive result is indicated by the formation of metallic green colonies. The results can be ascertained that the tested samples contain Escherichia coli.

2.7.4. Calculation of coliform MPN value. Determination of the coliform APM value and the fecal coliform APM value in packaged tea beverage samples was directly indicated by the presence of gas bubbles in the Durham tube, namely by counting the number of tubes that produced gas in the presumptive test for three dilution series.

2.8. Data analysis

The research data are displayed in the form of tables and figures. Antioxidant data were compared before and after storage. Microbial contamination data determines whether the tea is suitable for consumption again, shown in the form of tables and figures as well.

IOP Conf. Series: Earth and Environmental Science 1115 (2022) 012102 doi:10.1088/1755-1315/1115/1/012102

3. Results and Discussion

3.1. Microbiological analysis (microbial contamination) after storage
The testing of microbiological analysis on jeruju leaf tea (*A. ilicifolius*) in ready-to-drink packaging using the Most Probable Number (MPN) method to get the number of coliform bacteria by going through the stages of estimating, confirming and reinforcing tests used to identify pathogenic bacteria (*E. coli*). The results of this test are important to see whether it is still suitable for consumption or not.

3.1.1. Presumption test results. The MPN testing was carried out to determine the presence of coliform bacteria in jeruju leaf tea (*A. ilicifolius*) in ready-to-drink packages based on storage time for three months. Tests were carried out in the first, second and third months after storage. The presumptive test was carried out to determine the presence or absence of coliform bacteria (Gram-negative bacteria) in the form of acid and gas caused by lactose fermentation by coliform bacteria.

The results of the presumptive test were declared positive if there was turbidity or the presence of gas bubbles in the Durham tube which was placed upside down after being incubated for 24 hours and 48 hours. The results of the observational test on jeruju leaf tea (*A. ilicifolius*) in three flavors with a storage period of three months indicating the presence of coliform bacteria can be seen in Table 1.

Table 1. Presumption test results of ready-to-drink Jeruju Leaf Tea (*A. ilicifolius*) in three flavor variants using LB media with three months of storage

Sample	Storage Period (Month)			
	0	1	2	3
Jeruju	-	-	-	+
Jeruju+Lemongrass	-	-	-	+
Jeruju+Ginger	-	-	-	+

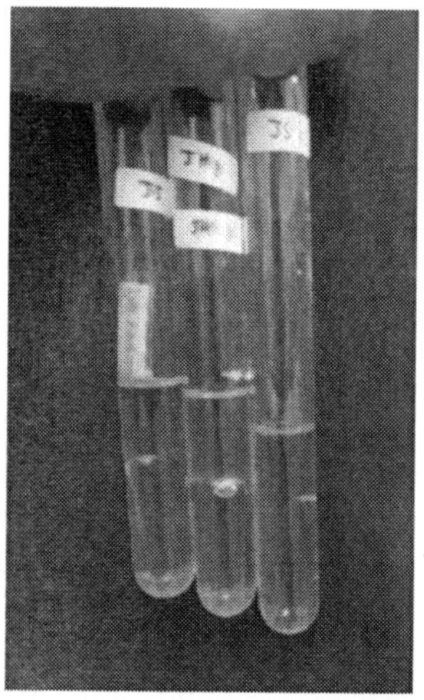

Figure 1. Observation of positive tubes presumption test results using lb media on three flavour variants of Jeruju Leaf Tea (*A. ilicifolius*) ready to drink packages with three months of storage

Based on Table 1. the results of the observational test of jeruju leaf tea (*A. ilicifolius*) in three flavor variants with three months of storage were declared positive. This is due to the turbidity and gas bubbles in the Durham tube which has been incubated for 24 hours and 48 hours. Positive test results were obtained from the first dilution with the first replication. Because the results of the presumptive test showed positive results, the test continued to the confirmation test. The results of the presumption test can be seen in Figure 1.

3.1.2. Confirmation tests results. The follow-up test of this presumptive test is the confirmation test. It is carried out in the same way as the conjecture test. The confirmation test uses to detect coliform (gram-negative) bacteria in water, food and other products. The media used was able to inhibit the growth of Gram-negative bacteria and increase the growth of coliform bacteria, to determine the presence or absence of coliform bacteria (Gram-negative bacteria) in the form of acid and gas caused by lactose fermentation by coliform bacteria. This test was carried out to confirm that the gas formed was caused by coliform bacteria and not caused by coliform bacteria and not due to the cooperation of several other species to produce gas. The sample used in the confirmation test is the result of a positive predictive test, while the results of the confirmation test with BGBB media can be seen in Table 2.

Table 2. Jeruju leaf tea confirmation test results (*A. ilicifoliu*) on three variants of ready-to-drink packaging flavors using BGLB media with 3 months of storage

Sampel	1/10			1/100			1/1000			MPN/100ml	MPN	MPN
	U1	U2	U3	U1	U2	U3	U1	U2	U3	Index	Value	Limit
Jeruju	1	0	0	0	0	0	0	0	0	4	4000	<2/100 ml
Jeruju+Lemongrass	1	0	0	1	0	0	0	0	0	7	7000	<2/100 ml
Jeruju+Ginger	1	0	0	1	0	0	1	0	0	11	11000	<2/100 ml

Based on Table 2, the results of the confirmation test on jeruju leaf tea (*A. ilicifolius*) in three flavor variants using BGLB media with a storage time of three months that had been incubated for 24 hours and 48 hours at 37°C showed positive results which were indicated by the presence of turbidity and gas formation in the inverted Durham tube. The MPN value obtained from the results of the confirmation test shows a different amount in each sample.

In the test results confirming the sample of jeruju leaf tea without a taste variant (original) in the first dilution and the first dilution, there was a color turbidity in the test tube. Meanwhile, in the second and third dilutions with three replications, no gas or cloudy color was formed in the incubated media. The MPN value obtained from the confirmation test of the jeruju leaf tea sample showed the number of coliform bacteria of 3600 per 100 ml.

In the test results confirming the sample of jeruju leaf tea + lemongrass in the first and second dilutions with each repetition to one, there was gas formed in an inverted Durham tube and color turbidity in the test tube. Meanwhile, in the third dilution with three replications, no gas or color turbidity was formed. The MPN value obtained from the confirmation test of the jeruju leaf tea + lemongrass sample showed a number of coliform bacteria of 7400 per 100 ml.

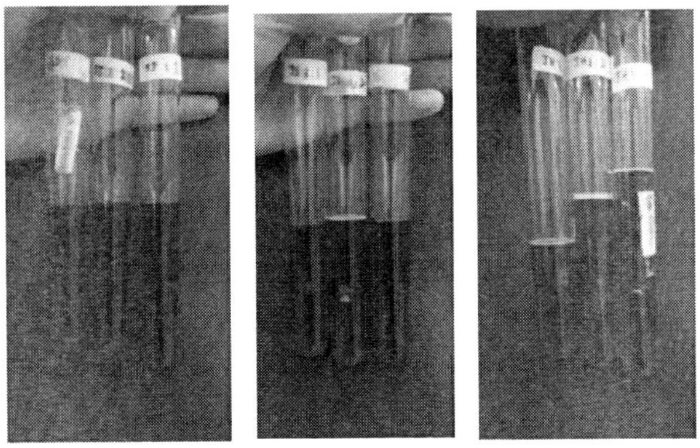

Figure 2. Observation of positive tubes of confirmation test results using bgbb media on three flavor variants of Jeruju Leaf Tea (*A. ilicifolius*) ready-to-drink packages with three months of storage

In the test results confirming the sample of jeruju + ginger leaf tea in the first, second and third dilutions with each repetition to one, there was gas formed in the inverted Durham tube and color turbidity in the test tube. The MPN value obtained from the confirmation test of the jeruju + ginger leaf tea sample showed the number of coliform bacteria at 11000 per 100 ml. From each sample that was tested for confirmation and showed a positive result, a reinforcing test was then carried out.

3.1.3. Complete test result. Further testing of the confirmation test or reinforcement test to determine the type of fecal coliform bacteria present in all positive confirmation test samples. The results were seen when there were fecal coliform bacteria in the media and indicated by the growth of colonies with metallic green color and pink colors. These results indicate that the positive sample contained *E. coli* so that all samples, namely jeruju leaf tea, jeruju leaf tea + lemongrass and jeruju leaf tea + ginger in ready-to-drink packages with a storage time of three months were not suitable for consumption. The growth of fecal coliform bacterial colonies can be seen in Figure 3.

The MPN value of jeruju leaf tea (*A. ilicifolius*) in three flavor variants with a storage period of three months did not meet the requirements for consumption. This is because the MPN values in the three flavors of jeruju leaf tea with a storage period of three months exceeded the maximum limit of microbial contamination. The MPN value is different for each flavor variant. Therefore, jeruju leaf tea (*A. ilicifolius*) in three flavor variants with a storage period of three months is not suitable for consumption because it contains coliform bacteria as evidenced by a reinforcement test.

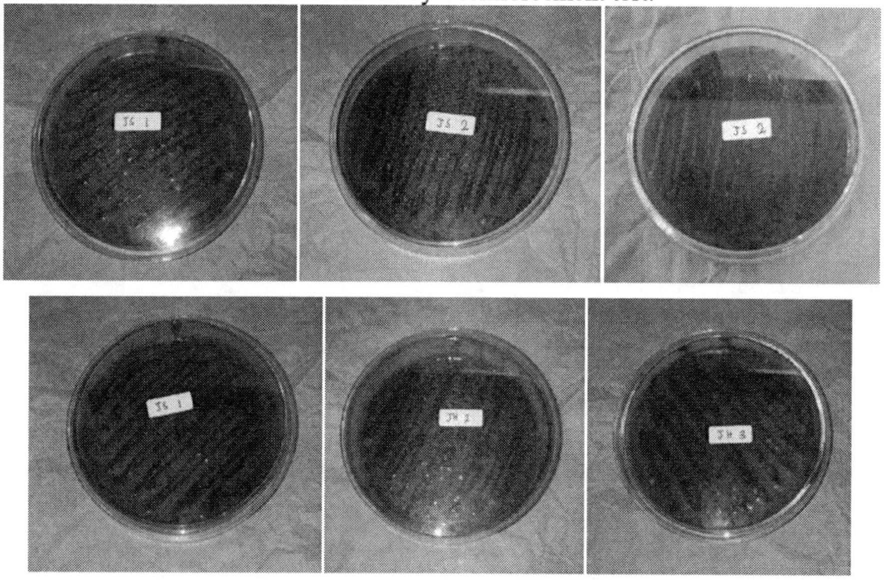

Figure 3. Observation of reinforcing test results using emb media on three flavor variants of jeruju leaf Tea *(A. ilicifolius)* ready to drink packages with three months of storage

Escherichia coli bacteria are the cause of diarrheal disease suffered by all ages. Bacteria *E. coli* produce toxins that can adhere to and damage the mucosal cells of the small intestine. The most common clinical symptoms in cases of this infection include abdominal cramps, watery diarrhea, low-grade fever, nausea and feeling unwell. These bacteria are classified by their characteristic virulence properties and each group causes disease through different mechanisms [14].

In this research, several methods have been carried out to avoid the emergence of microbes such as adding Na-Benzoate to tea solution, pasteurization, sterilization of equipment, as well as packaging and storage at low temperatures (10°C). Sodium benzoate has the characteristics of being stable, odorless, completely white, soluble in water and ethanol. Sodium benzoate is more widely used because it is more soluble and works optimally at a pH of 2.5-4.0 so it is often used to preserve various foods and beverages that have a low pH. Sodium benzoate does not cause negative effects when used in small amounts. The level of use of sodium benzoate ranges from 0.05-0.1% [15].

According to [16], pasteurization is a thermal process with moderate temperatures (Mild Heat Treatment) given to food products. The purpose of pasteurization is to kill certain vegetative microbes, namely pathogens and inactivating enzymes because the pasteurization process does not kill all vegetative microorganisms and spore-forming microorganisms so the pasteurized products must be packaged or stored at low temperatures with the addition of preservatives, modified atmosphere packaging, pH adjustment, or regulation of water activation to control microbial growth.

The packaging process carried out in this research is using polyethylene terephthalate (PET) bottles, where this packaging serves to extend shelf life and can protect products from external causes of damage such as microbes or insects as well as to maintain the quality of food or beverages. This is in accordance with [17] statement that PET is clear, strong, chemical and heat resistant and has good electrical properties. PET is used as packaging for carbonated drinks, juice bottles and bedding.

In this study, a sterilization process was also carried out, both before the beverage manufacturing process, during the manufacturing process and during testing. However, there are several factors that cause the emergence of microbes in drinks, namely the contamination of ingredients during the process of making packaged tea and the packaging process that is less sterile. Based on [18] factors that can cause food damage include water content in food, temperature (heating or cooling), time and air. In addition, the condition of food materials stored at room temperature (±28°C) allows and even supports microbial growth, resulting in microbial metabolism such as removing toxins or poisons that cause food spoilage. In this study, the storage process of jeruju leaf tea (*A. ilicifolius*) in three flavors based on storage time was carried out at a low temperature (10°C).

3.2. Antioxidant test

In addition to the microbial contamination test to see the stability in the storage period, antioxidant tests were also carried out on ready-to-drink tea. The test results can be seen in Table 3. The test was carried out until it was still suitable for consumption, in this case, it was only for 2 months of storage, while for the 3rd month it was no longer tested.

Antioxidant stability during shelf life cannot be maintained. The longer the shelf life, the lower the antioxidant activity. The antioxidant activity of jeruju leaf tea at the beginning of the test was in the moderate category referring to [12], in accordance with the results of research by [19].

Table 3. Test results for antioxidant content of jeruju leaf tea bags

Tea Ingredients Formula	Month - (µg/mL) 0	1	2
Jeruju	152.67	192.50	226.02
Jeruju + Ginger	131.22	133.96	161.23
Jeruju + Lemongrass	124.47	199.89	230.04

4. Conclusion

The storage period of up to 2 months of ready-to-drink jeruju tea is still suitable for consumption, but after 3 months of storage, it has been contaminated with microbes so that it is no longer suitable for consumption. The antioxidant content of tea during storage is not stable, the longer it is stored the lower the content.

Acknowledgment

This research was funded by TALENTA USU 2021 on the scheme of *Penelitian Dasar*.

References

[1] Rosyada A, Sofwan M and Muflihati 2018 *Jurnal Hutan Lestari* **6** 62
[2] Wahwakhi S, Kusmana C and Iswantini D 2017 *Aquaculture, Aquarium, Conservation & Legislation* **10** 1186
[3] Mardhia D, Rahman F, Aan S, Fitri A and Djodi A 2019 **6** 348
[4] Ramadhani F, Umi B and Joko S 2020 *Jurnal Ilmu-ilmu Pertanian* **4** 21
[5] Analuddin A, Septiana A and Harlis W O 2018 *Jurnal Ilmu Teknologi Peternakan Tropis* **5** 60
[6] Singh A, Duggal S and Suttee A 2009 *International Journal of Phytomedicine* **1** 1
[7] Saranya S, Ramanathan T, Kesavanarayanan K S and Adam A 2015 *American-Eurasian J. Agric. & Environ. Sci.* **15** 243
[8] Sardar P K, Dev S, Bari M A Al, Paul S, Yeasmin M S, Das A K and Biswas N N 2018 *Clinical Phytosci.* **4** 34
[9] Husori D I, Sumardi, Heron T, Selviani G and Sri R N 2018 *Journal of Applied Pharmaceutical Science* **8** 164
[10] Zhang T, Tian Y, Jiang B, Miao M and Mu W 2014 *LWT- Food Sci. Technol.* **56** 9
[11] Badan Pengawas Obat dan Makanan 2010 **5**
[12] Molyneux P 2004 *Songklanakarin J Sci Technol* **26** 212
[13] Yuliana A and A Saeful 2016 *Jurnal Kesehatan Bakti Tunas Husada* **15** 1
[14] Tuntun M 2016 *Jurnal Kesehatan* **3** 497
[15] Andriani D, Revianti S and Prananingrum W 2020 *Biodiversitas Journal of Biological Diversity* **21** 2521
[16] Khurniyati M I dan E Teti 2015 *Jurnal Pangan dan Agroindustri* **3** 523
[17] Suminto S 2017 *Jurnal Desain Produk* **3** 26
[18] Arini L D D 2017 *JITIPARI (Jurnal Ilmiah Teknologi dan Industri Pangan UNSRI)* **2** 15
[19] Wang Y, Zhu H and Tam N F Y 2014 *Plant Soil* **374** 549

Shaping ethnobotanical tourism on the coastal landscape through Halobanese oral traditions at Banyak Island

L Andriany[1], T Lubis[2*], Amalia[2], A F Abus[3] and Delima[4]

[1]Universitas Islam Sumatera Utara, Medan, Indonesia
[2]Universitas Sumatera Utara, Medan, Indonesia
[3]Aceh Landscape Development Centre, Langsa, Indonesia
[4]Universitas Negeri Medan, Medan, Indonesia

[*]Email: tasnimlubis@usu.ac.id

Abstract. This paper focused on exposing the unique and typical plants at Pulau Banyak (Banyak Island), which strongly relates to Halobanese oral traditions in shaping ethnob0otanical tourism. The benefit of knowing the typical plants in Banyak Island will inform the natural resources as both natural wealth and cultural heritage owned by the Halobanese. Further, this information also benefits the Halobanese in their confidence and motivation to move forward with ethnobotanical tourism. This is a qualitative study and applied ethnography method. The data were the documentation of typical plants in Pulau Banyak, which have a strong relationship with food security and daily needs for Halobanese, such as mangrove, coconut, sagoo, mangrove and pandan leaves. The results were evaluated by defining the conceptual connection between shaping ethnobotanical tourism and the Halobanese oral heritage, which informed the local flora. The result of the study showed that the typical plants in the Banyak Islands area consist of naturally grown plants and can be grown well. The natural growing plant is mangrove, coconut, sagoo, rumbia and pandan leaves. The plant that can be grown well is taro and yam. Consequently, ethnobotanical tourism is an ideal program to build Halobanese human resources and revitalize their typical plant.

1. Introduction

The archipelago in Indonesia has a lot of flora and fauna that benefit food sources, medicine and tourism value. Tourism is an activity that can provide physical and spiritual satisfaction to tourists. The selection of tourist areas want to visit is expected to provide information that provides not only from a physical point of view but also from a mental point of view. The concept of tourism generally includes beautiful scenery, typical culinary delights and comfort, for example, on the island of Bali, which is already very famous for its natural beauty, cultural traditions, culinary delights and the friendliness of its people [1]. Several other places in Indonesia also have beautiful natural scenery and unique culinary culture. However, the attitude and behaviour shown by the Balinese people are still very far away. A friendly attitude, serving and being aware of tourism is still a must because it dramatically affects the comfort of tourists who want to visit tourist areas [2].

Plants are biological resources humans have used worldwide for a long time. Human interaction with plants is so important that interest in studying plants has arisen throughout human history. This plant is often referred to as Botany, with a broad coverage ranging from molecular and cellular structure, origin, diversity and its classification system up to the function of plants in nature and their role in human life. This knowledge need is increasing along with increasing human dependence on plants [3].

The influence of tribes with diverse experience, point of view and perception of nature affects and produces various shapes and characters landscape. In many traditional societies with the practices of local wisdom and indigenous knowledge, which is still awake in everyday life community, the existence of the landscape is very stable and capable of continuously providing support for life people living in the landscape. Landscapes from community activities and perceptions influenced by local culture are cultural landscapes [4].

The objective of this study was to describe the unique and typical plants at Pulau Banyak (Banyak Island), which have a strong relationship with Halobanese oral traditions in shaping ethnobotanical tourism. The vegetation plants in Banyak island significantly contribute to the Halobanese needs, especially for foodstuff and daily equipment such as mat, sweeping, etc. The Halobanese use of characteristic plants is also informed by oral traditions that are passed down from generation to generation. On formal occasions, the pantoon, songs and even dance include information about the vegetation plants in Banyak island.

2. Methods

The ethnography method was used in this study [5,6]. The method can be seen as shown in Figure 1. The steps consist of 12 steps.

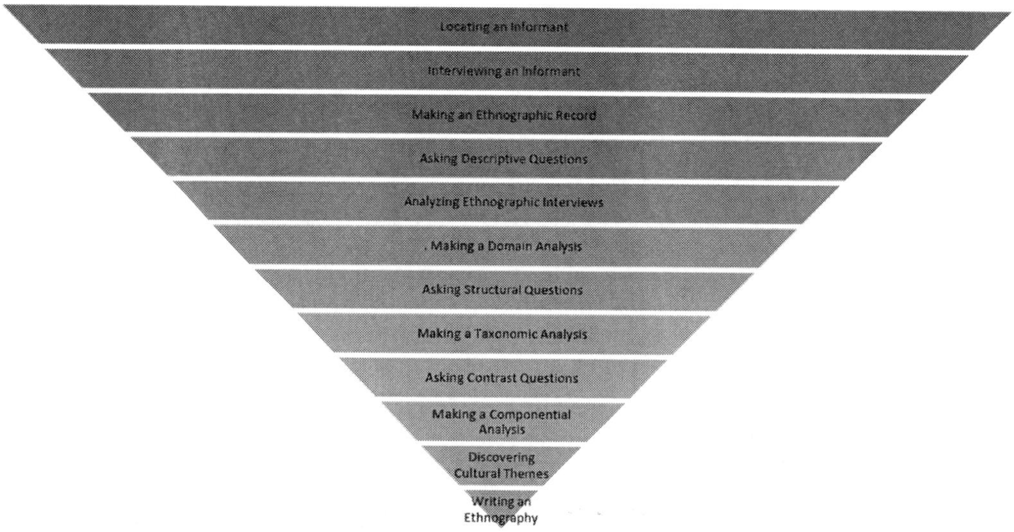

Figure 1. Ethnography method

The data was gathered through documenting the native plant species on Banyak Island. In-depth interviews and participant observation were undertaken to get additional knowledge about the typical plants. Domain, taxonomy and componential analyses were utilized to examine the data. In domain analysis, the semantic relationship between ethnobotany tourism and Halobanese oral traditions is part of [7]. The scheme can be seen as shown in Figure 2.

Figure 2. Semantic relationship in domain analysis

IOP Conf. Series: Earth and Environmental Science 1115 (2022) 012103 doi:10.1088/1755-1315/1115/1/012103

In taxonomy analysis, ethnobotany is classified into natural and grown plants. In componential analysis, the description of natural and grow plants informed the Halobanese using for their daily life and also become an asset to attract the visitors to know more about their unique and typical food security as their local wisdom [8,9].

The research location is precisely in the villages of Asantola and Haloban. The Haloban language is endangered and is only spoken in two villages in Pourku sub-district. On Banyak Island, there are three tribes: the Aneuk Jamee Tribe, the Haloban Tribe and the Nias Tribe. The Aneuk Jamee tribe inhabits three villages in Banyak Island sub-district, namely Pulau Balai, Pulau Baguk and Teluk Nibung. The Haloban tribe inhabits two villages in the West Banyak Island sub-district: Haloban and Asantola. Meanwhile, the Nias tribe inhabits two villages in West Banyak Island, namely the villages of Ujung Sialit and Suka Makmu. The research location can be seen as shown in Figure 3.

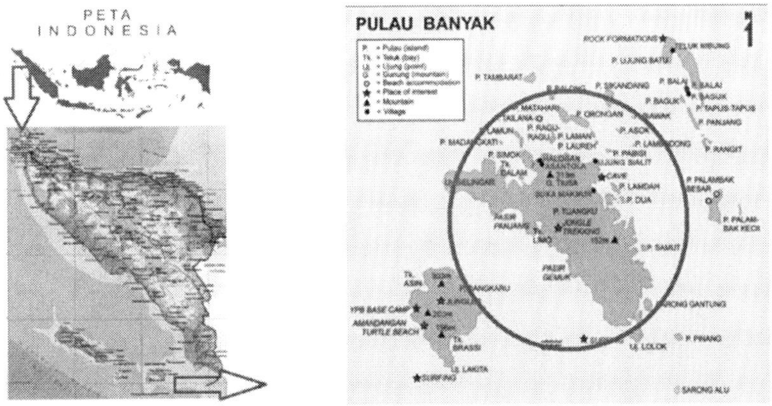

Figure 3. Location of study

3. Result and Discussion

The typical plants in Banyak Island are mangrove, coconuts, sago, durian, nangka fruit, pinang/betel nut, Nipah palm fruit, bamboo, rattan, timber from forest trees, corn/Ubi (edible roots, sweet potato), kaladi, glaga (wild sugar cane or grasses. The functions and benefits of mangroves, which generally have been widely known to make mangrove resources, are threatened. This is due to easy access and high biodiversity value [10,11]. Though Mangrove rehabilitation is also carried out, the cause of the damage to mangrove forests continues to occur and increase [12,13]. That matter causes the potential for mangrove forest loss to be faster than the loss of terrestrial tropical forests and coral reefs [14]. This is the cause of the decline in the productivity of mangrove and coastal ecosystems, which is one of the centres of environmental problems in the world [15].

The oral tradition of Halobanese Pantun, which informed the natural resources and what they do with them, are pendendang (pantoon singers) dendang sikambang, talibun, buai and langser. The Urang Pulo (called for Halobanese) has musical arts, which are under the term sikambang, comprised of five categories: They are (i) the vocal music/dendang, comprising the buai (lullaby), talibun (bardic intoning) and dikie perempuan (women's Muslim dikie singing), (ii) the vocal and instrumental music comprising sikambang singing with biola and gandang playing, dikie laki (men's zikir singing with gandang playing), nasyid/main rebana (men's singing of mainly religious texts with frame drum playing) and popular music (with the pop band playing) (iii) the vocal music with dancing, including the randai dances in baralek processions with gandang playing) (iv) the vocal and instrumental music with dance, including the sacred. Tari Adok and the couples dances: Bungkus, Pondok Batu, Anak Pandan, Sampayan, Mak Inang, Perak-Perak, Lelehusen, Pulau Pinang, Siantung, Sikambang, Yadoi-Yadoi and Bacarai Kasih (B.M.: Bercerai Kasih) and (v) instrumental music and dancing with a caller,including the Gelombang Duabaleh ("twelve waves"), Padang ("sword") and Langser dances with biola and gandang accompaniment.

Adok dance is a dance in pairs accompanied by oral literature in the form of singing rhymes accompanied by beats and dances. The origin of Adok dance originates from West Sumatra. This dance was first developed during the reign of the first king of Haloban, who is believed to have been sent to Haloban by Queen Pagaruyung, Bundo Kanduang, to find my Pour Kingdom at the request of the Four Datuks Haloban (tribal chief) in the 16th century. Adok is defined as a musical instrument or, For example, a drum-like Malay drum that only exists in West Sumatra (Below the lyric of Haloban lullaby) as shown in Table 1.

Table 1. Lyric of Haloban lullaby

Haloban language	English
Bue buyung bue o bue	Sway, little boy, sway
Anak mudo dibuekan	Young child is being swayed
Tanjak-tanjak Palembang dari jao	Palembang is seen protruding from a far
Ayun anak buekan bue	Sway, child sway
Anak mudo dibuekan salamo ado	Young child in the sling, some day
Banyak-banyak batanam ubi tanah taleh keladi	You will plant a great deal of cassava and taro
Ambik pandan anyam katika	Take the pandan leaves and weaving mat
kambangkanlalu dialaman	When finished) lay it on the ground of the yard
Tasuo isuk kato batingka	If we have a misunderstanding in the near future
Baok bagurau kahalaman	Please don't have hard feelings while we sit on the mat
Pulau asok pulau kiramat	Pulau Asok is a sacred island
singgahan urang sipangai	Where wise people stop by
Pintakkan doa baksalamat	Pray for blessings
nantingga jangan batangisi	who stay please don't cry

In the lyrics, the lexicon taro and cassava are mentioned to inform one of Halobanese food stuff and be one of the leading food too. The pandan leaves that are used to make the mat are easily obtainable in the vicinity of their dwelling region. Not only is it useful to them in their day-to-day lives, but it also has value.

Responsible for nature tourism with how to conserve the environment and improve the welfare of local communities [14,15]. Ecotourism activities can be carried out in the area with the potential of natural resources. The wrong one is a rural area that is still naturally owned by village resources so that it can support ecotourism activities. The managers and tourists expect the end of the ecotourism trip can manage natural tourism resources in an environmentally friendly manner [14].

A village will have potential different village resources. Various village resources can be optimized, for example, local culture, local plants and home gardens. Indonesia has the potential for ecotourism area is quite large. The area is spread over land (in conservation areas) and at sea (in the form of the marine national park). Based on Indonesia's diverse potential, it can attract domestic tourists abroad. Potential development needs to be done in the implementation of ecotourism to increase the diversity of owned tourist attractions.

According to Yayasan Pulau Banyak, the region supports many habitats, such as coral reefs and mangroves. It is a marine protected area with the designations of Recreation Park and Multiple Use Management Area (Taman Wisata Alam). It established an environmental program in 1997 to investigate, conserve and raise awareness of the anthropological and ecological richness of the region.

Ecotourism activities in rural areas usually provide attractions tourism in the form of local culture related to the use of plants local. So, tourists can see the attractions tourism in the form of the use of local plants growing in rural areas. A house yard is a vital aspect of the agricultural system, where plants are picked based on the type of crops grown and managed to serve several roles, including the requirement for medicinal, ornamental and plants that provide products for populations with varying income levels. The community in the yard house usually plants local plants as a side income. The results of the house's yard can be used for food needs and education conservation for tourists in tourist villages.

Human resources have become an integral part of the development and management of ethnobotanical tourism. In response to the growing interest of tourists in cultural heritage, destination management organizations, museums and other cultural institutions are attempting to unlock and share the intangible cultural legacy of locals before it is lost. This includes specialized knowledge regarding the uses and practices of vanishing urban space, flora and foodstuffs, apparel and labor patterns that were more compatible with local, sustainable production [16,17].

4. Conclusion

Ethnobotanical tourism is an ideal program to build Halobanese human resources and revitalize their typical plant. The oral tradition has a strong relationship in informing the typical plant in this area and the natural wealth owned by the community, including the function and sometimes how to get them. Revitalizing the oral tradition is a straight line with revitalizing a community's natural resources. The preserving effort of vegetation and typical plants is an asset to shaping ethnobotanical tourism on the coastal landscape at Banyak Island.

References

[1] Setiadi I G W and Siringoringo R L 2017 E *JUMPA* **3** 387
[2] Abus A F, Lubis T and Abus N A A 2021 *IOP Conf. Ser. Earth Environ. Sci.* **922** 12035
[3] Lubis T, Perangin-angin A B, Dardanila, Nasution T, Hasrul S, Ramlan and Abus A F 2021 *IOP Conf. Ser. Earth Environ. Sci.* **922** 12039
[4] Lubis T, Dardanila, Nasution T, Zulkarnain, Hasrul S, Ramlan and Abus A F 2021 T *IOP Conf. Ser. Earth Environ. Sci.* **926** 12029
[5] Spradley J 1979 *The Ethnographic Interview* (Orlando: Harcourt Brace Jovanovich College Publishers)
[6] Spradley J 1980 *Participant Observation* (USA: Rinehart and Winston)
[7] Shaumiwaty S, Lubis M A, Lubis T, Dardanila, Purba A, Nasution T, Ramlan and Hasrul S 2020 *J. Phys. Conf. Ser.* **1663** 12066
[8] Zulkarnain, Lubis T, Ramlan, Dardanila, Hasrul S, Shaumiwaty and Saputra N 2021 *IOP Conf. Ser. Earth Environ. Sci.* **807** 22007
[9] Lubis T, Ramlan, Dardanila, Hasrul S, Nasution T and Abus A F 2022 *IOP Conf. Ser. Earth Environ. Sci.* **977** 12099
[10] Valiela I, Bowen J L and York J K 2001 *Biosciences* **51** 807
[11] Onrizal 2005 *War. Konserv. Lahan Basah* **13** 5
[12] Primavera J H 2005 *Sciences (New. York).* **310** 57
[13] Dave R 2006 *Trop. Resour. Bull.* **25** 7
[14] Gao B 2018 *IOP Conf. Ser. Mater. Sci. Eng.* **371**
[15] Abus A F, Lubis T and Abus N A A 2022 *IOP Conf. Ser. Earth Environ. Sci.* **977** 12100
[16] Mansfield C and Topler J 2021 *Ann. Ser. hist. sociol* **31** 197
[17] Lubis T, Abus A F, Ramlan, Nasution T and Saputra N 2022 *IOP Conf. Ser. Earth Environ. Sci.* **1047** 12035

IOP Conf. Series: Earth and Environmental Science 1115 (2022) 012104 doi:10.1088/1755-1315/1115/1/012104

Local wisdom in natural resources management of freshwater fish In Lau Teba

H Thamrin*, **H Ridho, R P K Sinaga, F U Ritonga, F A Nasution and C P Purwanto**

Universitas Sumatera Utara, Medan, North Sumatra 20155, Indonesia

*Email: pungkut@usu.ac.id

Abstract. The management of freshwater resources of jurung fish with local wisdom in the Lau Teba River makes one of the watersheds like a pond but without using cages or other barriers. Jungle fish remain in their natural habitat in the river without being blocked by cages or other barriers. The purpose of this study was to explore and analyze the management of the freshwater fish of Jurung with the Local Wisdom approach, as well as the potential of natural resources as a tourist attraction. The form of research used in this study used a descriptive method with a qualitative approach. Data was collected using interviews and observations at the research site. The results show that the concept of natural resource management with local wisdom makes nature choose its own resources. Jurung fish managed with the concept of local wisdom will be selected through natural processes. Included in the process of taking fish using a rod, we do not choose the fish we take, but fish that are taken based on the natural selection process which is a tourist attraction.

1. Introduction

Nature is created with various things available and can be managed by humans. Everythingthat comes from nature and can be used by humans is called natural resources. Natural resources canbe used for various purposes and meet the needs of human life. Natural resources are foundeverywhere, such as in the soil, water, land surface, air, and so on. Natural resources based on theirtype can be divided into two, namely biological or biotic natural resources and non-biological/abioticnatural resources. Biological natural resources are natural resources that come from living things.Biological natural resources can come from animals or plants. Examples of biological natural resources are chickens, cows, vegetables, rice, corn, cotton, wood, tea, coffee, and fish. Non-biologicalnatural resources are natural resources that do not come from living things. Examples of non-biological natural resources are water, sunlight, air, land, mining materials, petroleum, and natural gas.

Nature has provided many natural resources that can be utilized, but that does not mean that humans can just enjoy them. In the utilization of natural resources, there are several things that must be considered, one of which is maintaining the preservation of natural resources. Natural resource management is very influential with the availability of production of these natural resources. Conservation of natural resources is very important for sustainable natural resource management. Utilization of natural resources that is not accompanied by conservation can result in the exhaustion of production on these natural resources. Indonesia is known throughout the world as a country that has rich and abundant natural resources. Natural resources are renewable (renewable) and non-renewable (non-renewable), as well as in the form of natural capital (natural resource stock), such as watersheds, lakes, protected areas, coasts, swamp and peat areas, and others, as well as Natural resources in the form of commodities (natural resource commodities) such as wood, rattan, mining minerals, oil and gas, fish, and others, are evenly distributed throughout the sovereign territory of the Unitary State of the Republic of Indonesia (NKRI).

Natural resource management in Indonesia has several traditional concepts, namely natural resource management with local wisdom. Local wisdom is a form of environmental wisdom that existsin social life in a place or area [1]. So, it refers to a particular locality and community. Local wisdom is a value system or behavior of local people in interacting with the environment in which they live wisely [2]. The concept of local wisdom makes wild animals into livestock but still in their natural habitat, such as doing fish farming in the river. The concept of managing natural resources with local wisdom, one of which is applied in the Lau Teba River in Kutagajah Village, Kutambaru District, Langkat Regency, North Sumatra Province, namely in the form of managing freshwater resources of Jurung fish with the concept of local wisdom.Jurung fish is one of the freshwater fish that can be found in river areas. Water resources have a role and function to meet the needs of human life and support their activities [3]. Therefore, the use of water resources, especially freshwater resources, is important to meet the needs of the surrounding community.

Water resources are part of the watershed system (DAS), which includes sub-systems of land resources, forest resources, social and cultural resources, and water resources themselves. Water resource management is inseparable from watershed management; thus a good watershed management strategy will produce good water resources as well [4]. One of the good managements of water resources is to apply the concept of local wisdom. The management of freshwater resources of jurung fish with local wisdom in the Lau Teba River makes one of the watershed areas like a pond but withoutusing cages or other barriers. The jurung fish remain in their natural habitat in the river without being obstructed by cages or other barriers.

The jurung fish are free to move anywhere in the Lau Teba River, but the jurung fish will flockto the management area with local wisdom when given food. While jurung fish gather in the management area with local wisdom, there is an opportunity to pick them up using a fishing rod. However, the process of taking jung fish using a fishing rod should not be too much because it can damage the concept of local wisdom. The concept of resource management with local wisdom can produce food sources that do not consume production. Integrated water resources management requires the involvement of various parties [5]. The management of freshwater resources of jurungfish with local wisdom requires involvement, especially to make the area a tourist spot.

Apart from being a source of food, the concept of managing freshwater resources of jung fish with local wisdom can also become a tourist attraction. The view of a group of jung fish in the management area is an attraction for tourists to come to Lau Teba. In addition, close to the management area there is a river battle which is the confluence of the Wampu River and the Lau Teba River. Fresh water resource management with local wisdom creates sustainable natural resources. The concept of managing the freshwater resources of jung fish with local wisdom also has its own charm for tourists to see firsthand how the concept of local wisdom is made. The purpose of this research is to explore and analyze the management of the freshwater fish of Jurung with the Local Wisdom approach, as well as the potential of natural resources as a tourist attraction.

2. Method

The form of research used in this study used a descriptive method with a qualitative approach. According to Burhan, descriptive qualitative research aims to describe, summarize various conditions, situations, or phenomena of social reality that exist in the community that is the object of research and seek to draw that reality to the surface of various characteristics, characters, traits, models, signs or pictures about certain conditions, situations, or phenomena [6]. This research was conducted in Kuta Gajah Village, Kutambaru District, Langkat Regency, North Sumatra Province. There is the Lau Teba River which is used as a location for managing freshwater resources for Jurung fish with local wisdom. The Jurung fish management area with local wisdom is located near the confluence of the Wampu and Lau Teba rivers. Data collection techniques used are in-depth interviews, observation, and documentation.

3. Results and Discussion

The Lau Teba River is in the Gunung Leuser National Park (TNGL) area. Gunung Leuser National Park is a nature conservation area in Indonesia with an area of 1,094,692 hectares. Administratively, Gunung Leuser National Park is in the provinces of Aceh and North Sumatra. The province of Aceh which was delineated by the Gunung Leuser National Park included the areas of Subulussalam, South Aceh, Aceh Singkil, Central Aceh, Gayo Lues, Bener Meriah, Aceh Tamiyang, while the North Sumatra Province which was delineated included the districts of Dairi, Karo and Langkat.

In accordance with the Minister of Forestry Regulation Number: P.03/Menhut-II/2007, currently the manager of GLNP is the Technical Implementation Unit (UPT) of the Directorate General of Forest Protection and Nature Conservation (Ditjen PHKA) of the Ministry of Forestry, namely the Gunung Leuser National Park Center (BBTNGL) led by the Head of Balai Besar (echelon II level). Gunung Leuser National Park has 2 global-scale statuses, namely as a Biosphere Reserve in 1981 and as a World Heritage in 2004. Both statuses were determined by UNESCO and the World Heritage Committee (WHC) at the suggestion of the Government of Indonesia after going through a series of rigorous selection processes.

On the other hand, this national park has also received attention because of the rampant casesof illegal logging in several locations that violate environmental reservations. Most of the TNGL area has a steep topography and a soil structure and texture that is prone to landslides. This was proven when flash floods destroyed the natural tourist area of Bukit Lawang several years ago. To further protect GLNP from further damage, an area called the Leuser Ecosystem Area was formed. This area of 2.6 million hectares covers a flatter area around TNGL and functions as a buffer.

The national park which takes its name from Mount Leuser with an altitude of 3,404 meters above sea level. This national park includes native ecosystems from the coast to high mountains covered by dense forests typical of tropical rain. Gunung Leuser National Park is also one of the areas with high biodiversity on the island of Sumatra. In addition, in Gunung Leuser National Park there are also many objects and natural tourist attractions, including the Sumatran Orangutan Observation Center, Bukit Lawang in the Bukit Lawang Bahorok Nature Tourism Area, Langkat Regency, North Sumatra.

The existence of the management of freshwater resources of jung fish with local wisdom in the Lau Teba River adds to the list of objects and natural tourist attractions in Gunung Leuser National Park. To reach the management location area, you must walk through the forest of Gunung Leuser National Park for two hours from the last hamlet, namely Kinagkong Hamlet. Traveling through the forest is quite exhausting because the tracking path is quite extreme, especially if you pass it in the rainy season, the tracking path becomes slippery and muddy. The atmosphere of the forest that is still beautiful and natural will be felt when walking through the Gunung Leuser National Park to the management location.

There are two cottages by the river where the management site is located. The hut is a simplehut made of wood and thatched roof. The two cottages are used as a shelter for managers and can also be used as cottages for tourists who come. The location of the Lau Teba River, which is the location forthe management of Jurung fish with local wisdom, is about twenty meters below the hut. The location with quite difficult terrain makes the Lau Teba River still rarely visited by tourists. The management is still planning to make a path that is easy for visitors to pass.

3.1. Management of jurung fish freshwater resources with the concept of local wisdom

Jurung fish that are managed with the concept of local wisdom become like domesticated livestock but live in the open. The basic concept of local wisdom in managing freshwater resources is the use of the instincts and instincts of animals to find food. Animals in the wild look for their own food, such as the jurung fish in the Lau Teba River looking for food. Taking advantage of this, the manager provides food in the location area that will be used as a management area. In choosing a location for natural resource management with local wisdom, it can also not be arbitrary. The location that is used as a management area with local wisdom, especially for fish management with local wisdom, must first be ascertained by the fish population in that location. In this case, the location of the management of Jurung fish with local wisdom, namely the Wampu River and the Lau Teba River, is quite well known for its large population of river fish. In addition, the location used as the management area must be in accordance with the natural habitat of the animal. This is to avoid animals in reading the concept of management with local wisdom that will be carried out on them.

The concept of natural resource management with local wisdom makes nature select its own resources. Jungle fish that are managed using the concept of local wisdom will be selected by natural processes. Included in the process of taking fish using a fishing rod, we do not select the fish we take, but fish that are taken based on the natural selection process.

The location of management with local wisdom must be in a deep river flow. A river that is deep enough can eliminate the fear of the jurung fish to go out looking for food because if the river flow applied to the concept of local wisdom is too shallow, the Jurung fish tend to be afraid to goout looking for food and prefer to hide. Jungle fish is one of the fish that is sensitive to movement. So the location that is set to apply the concept of local wisdom must be able to eliminate the sensitivity of the Jurung fish to movement. Thus, the jurung fish feel comfortable and have no problem showing themselves even though there are many movements carried out on the banks of the river, including the movement of feeding the jung fish.

The management of jung fish with local wisdom in the Lau Teba River is managed by Mr. Pardamean Sitepu and his who are local natives. Both are residents of Kinangkong Hamlet, Kutagajah Village, Kutambaru District, Langkat Regency, North Sumatra Province. They manage jurung fish with local wisdom in the Lau Teba River because they see the natural potential that supports them. The feeding of the jurung fish is carried out continuously at the same location for approximately one year. The food that is given in the process of managing Jurung fish with local wisdom is leftover food such as rice and leftover side dishes mixed in a bucket and then thrown into the river which is the location of the management.

At the beginning of the process of giving food to apply the concept of local wisdom, of course, there have not been any jurung fish who have come to take the food. After two months of continuous feeding at the same location point, some jurung fish began to dare to come to take food. Feeding is done at least three times a week.

After five months of continuous feeding, there were more and more jurung fish coming to collect food at the management location. Jungle fish are like tame animals and understand that Pak Pardamean and his wife will provide food. The routine feeding began to create a relationship between Pak Pardamean and his wife and the Jurung fish which they applied the concept of local wisdom to. Even when Pak Pardamean and his wife arrived at the riverbank with a bucket used for feeding, the jung fish began to gather and appear as if they didn't want to miss their food. After the food is thrown into the river, the jung fish scramble to eat it. The sight of the attraction of the jungly fish fighting overthe thrown food ensued.

In the sixth month, Pak Pardamean conducted an experiment in taking jurung fish while being fed using a fishing line with the same bait as the food given. Pak Pardamean managed to get twenty jurung fish from the experiment. However, after the retrieval experiment, there were not as many jurung fish that usually gathered during the feeding process as usual. Based on that experience, Mr. Pardamean concluded that when you want to take fish in the concept of local wisdom, you must pay attention to the number of fish taken. Because it affects the instincts of fish, which seem to have been able to read the concept of management with local wisdom.

Feeding is still being carried out after the retrieval experiment. Because of the consistent feeding, the jurung fish still gather to take food as usual. This shows that there is a time lapse in the retrieval process with the re-gathering of the jurung fish in the management location area. In addition, the process of managing jurung fish with local wisdom is also influenced by the rainfall that falls. If the rainfall in the management area is high enough, the volume of river water will also increase. This causes many jungling fish to be invisible. However, after the rain stopped and the water volumereturned to normal, the jurung fish began to appear again.

3.2. Management with local wisdom as an alternative for management of natural resources and tourism objects

Jurung fish management with local wisdom can be an alternative in traditional natural resource management. In this management concept with local wisdom, the capital costs incurred are very minimal considering that the feeding of jurung fish only uses food scraps. The key to success in the concept of local wisdom is the selection of the right management location according to the animal'shabitat and the consistency of feeding at the same location point.

Management of jurung fish with local wisdom is one of the sources of food that can be utilized without destroying and depleting production. It is quite difficult to get jungling fish if fishing in a river that is not a management location. Compared to catching fish using stun or fish bombs, cultivation with the concept of managing Jurung fish with local wisdom is much better in maintaining the availability of production, namely the Jurung fish itself. If we use a stun or fish bomb, we cannot choose the fish we will take. All fish within range of the stun or fish bomb will die and be caught.

Sustainable natural resource management can be applied with the concept of local wisdom. This concept keeps the animals in their natural habitat and when you want to take the animal for consumption, you must pay more attention to the number of animals taken so as not to damage the concept of managing the jungle fish with the concept of local wisdom.

In addition to being used as a source of food, the view of jung fish can also attract the attention of tourists to witness these attractions. This makes the concept of managing jung fish with local wisdom in the Lau Teba River a special attraction for tourists. This kind of management concept is rarely found in other places combined with the Gunung Leuser National Park area which has beautiful and beautiful natural scenery. The management location which is between two rivers, namelythe Wampu River and the Lau Teba River, is also an added value for a tourist attraction.

The Bahorok River, which is one of the most popular tourist attractions in Gunung Leuser National Park, is one of the Wampu tributaries. However, to make the location of the management of Jurung fish with local wisdom as a tourist attraction, there are still many things that need to be considered. One of them is road access to locations that are still extreme. If the rainfall is heavy enough, then the road to the Gunung Leuser National Park to the location will be very slippery. This also causes many nuisance animals to come out such as pacets and other animals. A long and extreme path must be taken to get to the management site. That makes tourists who come to the management location on the Lau Teba River only tourists who can go on the trip. Not surprisingly, the management location is mostly visited by young people who still have the physical endurance to get through it.

In addition to the extreme path, the facilities and infrastructure at the management location are also inadequate. In the area of the location there are only two cottages as the main stopover for tourists. Electricity in the management area is also quite difficult, Mr. Pardamean and his wife use batteries and solar panels to store electricity. Likewise, electricity can only be used for lights in the cottage as a source of lighting there. This condition is classified as still minimal because electricity cannot be used for tourists, for example in charging cellphones and others. The facilities and infrastructure for the management of jung fish with the concept of local wisdom in the Lau Teba River are still far from being compared to Bukit Lawang and Tangkahan which also presents the natural beauty of Gunung Leuser National Park and includes the flow of the river there. The management of jurung fish with local wisdom remains an added value for the attraction of the tourist attraction on the Lau Tebah River.

By making the management of jurung fish with local wisdom on the Lau Teba River a tourist attraction, it can generate income for the surrounding community, especially those in Kinangkong Hamlet, Kutagajah Village, Kutambaru District, Langkat Regency, North Sumatra Province. The arrival of tourists encourages the economy of the surrounding community and can also introduce the concept of managing resources with the concept of local wisdom to tourists and the wider community.

The management of freshwater resources of jurung fish with local wisdom in the Lau Teba River can be a role model for the management of other natural resources in other areas. The concept ofmanagement with local wisdom can produce sustainable natural resources. With the application of the concept of local wisdom, we can better preserve nature so that natural resources can be maintained andenjoyed by our children and grandchildren in the future.

4. Conclusion

From the description above there are several conclusions that can be drawn, namely: First, Nature provides natural resources to be managed for the benefit of human life. In the management of natural resources there is a concept of natural resource management with local wisdom that manages natural resources traditionally and prioritizes natural processes. The concept of resource management with local wisdom is one of the managements of natural resources that does not damage nature and does not deplete the production of these natural resources. Second, the concept of natural resource management with local wisdom makes nature choose its own resources. Forest fish managed with the concept of local wisdom will be selected through natural processes. Included in the process of taking fish using a rod, we do not choose the fish we take, but fish that are taken based on the natural selection process which is a tourist attraction.

Acknowledgment

Thank you to Pusat Kajian Hak Asasi Petani (Puskahap) Fakultas Ilmu Sosial dan Ilmu Politik Universitas Sumatera Utara espe PUSKAHAP for facilitating the writing of this article

References

[1] Tamalene M N, Al Muhdhar M H I, Suarsini E and Rochman F 2014 Indonesia *Int J Plant Res* **4** 1
[2] Sidiq R, Sofro S, Sulistyani A and Achmad W 2022 *Webology* **19**
[3] Chaiphar W, Sakolnakorn T P N and Naipinit A 2013 *Thailand J Sustain Dev* **6** 16
[4] Wassie S B 2020 *Environ Syst Res* **9** 1
[5] As'ari R, Rohmat D, Maryani E and Ningrum E 2019 *IOP Conference Series: Earth and Environmental Science* **243** 12002
[6] Leavy P 2017 *Research design: Quantitative, qualitative, mixed methods, arts-based, and community-based participatory research approaches* (Guilford Publications)

AUTHOR INDEX